2024 RELEASE

NUTRITION

for Healthy Living

McGraw Hill

Traci L. Kec

NUTRITION FOR HEALTHY LIVING

Published by McGraw Hill LLC, 1325 Avenue of the Americas, New York, NY 10019. Printed in the United Kingdom.

Some ancillaries, including electronic and print components, may not be available to customers outside the United States.

This book is printed on acid–free paper.

1 2 3 4 5 6 7 8 9 CPI 29 28 27 26 25 24

ISBN 978–1–266–80731–2
MHID 1–266–80731–4

Cover Image: *Olga Miltsova/123RF*

The Internet addresses listed in the text were accurate at the time of publication. The inclusion of a website does not indicate an endorsement by the authors or McGraw Hill LLC, and McGraw Hill LLC does not guarantee the accuracy of the information presented at these sites.

mheducation.com/highered

Brief Contents

Alexandra Rowley/Photodisc/Getty Images

Meet the Author

Derek Keck

Traci L. Keck, MS, RDN, LD received her BS in food and nutrition with a specialization in dietetics and MS in food and nutrition from Texas Tech University. She is currently Assistant Professor of Practice for the Department of Nutrition and Dietetics, as well as the Coordinated Program in Dietetics, at The University of Texas at San Antonio (UTSA). Traci has taught a variety of food and nutrition courses including introductory nutrition for majors and nonmajors. She has diverse professional experience as a registered dietitian nutritionist in both acute and long-term health care settings. Additionally, Traci has worked as a research dietitian at UTSA for studies related to childhood obesity prevention and as a consulting dietitian in the San Antonio, Texas, community. Outside the classroom, Traci enjoys gardening, hiking, volunteering on her church's media team, and spending time with her family.

To Wendy. Thank you for your guidance and inspiration.

Welcome to the 2024 Release of *Nutrition for Healthy Living*

We think of ourselves as consumers when we purchase homes, cars, computers, and foods. We are also consumers of nutrition-related information. Nearly every day, we are literally bombarded with messages in media and from acquaintances concerning nutrition, foods, and health. Much of this information is unreliable, and often it is intended to promote sales of products or services. Unfortunately, many consumers lack the knowledge and skills needed to analyze such information critically and decide whether or not to apply it to their decision-making process.

Helping students become better-informed consumers, particularly as this relates to food and nutrition, is the foundation of ***Nutrition for Healthy Living****. This major theme flows throughout the textbook by providing students with practical information, critical thinking skills, and the scientific foundation needed to make better-informed choices about their diet and health. By reading* ***Nutrition for Healthy Living****, not only will students learn basic principles of nutrition, they will also be able to evaluate various sources of nutrition information critically and to apply sound nutrition practices to improve their lives.*

Traci L. Keck

bdspn/iStockphoto.com

lynx/Iconotec/Glow Images

McGraw Hill connect®

A complete course platform

Connect enables you to build deeper connections with your students through cohesive digital content and tools, creating engaging learning experiences. We are committed to providing you with the right resources and tools to support all your students along their personal learning journeys.

65%
Less Time Grading

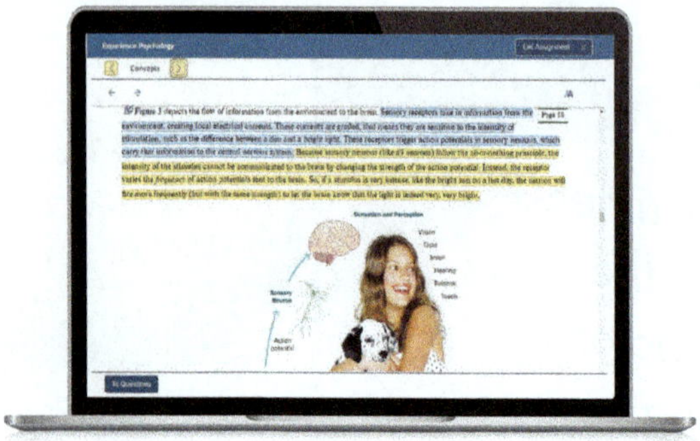

Laptop: Getty Images; Woman/dog: George Doyle/Getty Images

Every learner is unique

In Connect, instructors can assign an adaptive reading experience with SmartBook®. Rooted in advanced learning science principles, SmartBook delivers each student a personalized experience, focusing students on their learning gaps, ensuring that the time they spend studying is time well spent. **mheducation.com/highered/connect/smartbook**

Study anytime, anywhere

Encourage your students to download the free ReadAnywhere® app so they can access their online eBook, SmartBook, or Adaptive Learning Assignments when it's convenient, even when they're offline. And since the app automatically syncs with their Connect account, all of their work is available every time they open it. Find out more at **mheducation.com/readanywhere**

"I really liked this app—it made it easy to study when you don't have your textbook in front of you."

Jordan Cunningham, a student at *Eastern Washington University*

Effective tools for efficient studying

Connect is designed to help students be more productive with simple, flexible, intuitive tools that maximize study time and meet students' individual learning needs. Get learning that works for everyone with Connect.

SUPPORT AT every step

Education for all

McGraw Hill works directly with Accessibility Services departments and faculty to meet the learning needs of all students. Please contact your Accessibility Services Office, and ask them to email **accessibility@mheducation.com**, or visit **mheducation.com/about/accessibility** for more information.

Affordable solutions, added value

Make technology work for you with LMS integration for single sign-on access, mobile access to the digital textbook, and reports to quickly show you how each of your students is doing. And with our Inclusive Access program, you can provide all these tools at the lowest available market price to your students. Ask your McGraw Hill representative for more information.

Solutions for your challenges

A product isn't a solution. Real solutions are affordable, reliable, and come with training and ongoing support when you need it and how you want it. Visit **supportateverystep.com** for videos and resources both you and your students can use throughout the term.

Updated and relevant content

Our new Evergreen delivery model provides the most current and relevant content for your course, hassle-free. Content, tools, and technology updates are delivered directly to your existing McGraw Hill Connect® course. Engage students and freshen up assignments with up-to-date coverage of select topics and assessments, all without having to switch editions or build a new course.

Personalized Teaching and Learning Environment

Saves students and instructors time while improving performance.

Victoria Shibut/vicushka/123RF

NutritionCalc Plus

NutritionCalc Plus is a **powerful dietary analysis tool** featuring more than 106,000 foods from the reliable and accurate ESHA Research nutrient database, which is comprised of data from the latest USDA Standard Reference database, manufacturers' data, restaurant data, and data from literature sources. NutritionCalc Plus allows users to track food and activities, and then analyze their choices with a robust selection of intuitive reports. The interface was updated to accommodate ADA requirements and modern mobile experience native to today's students.

Virtual Labs

While the sciences are hands–on disciplines, instructors are now often being asked to deliver some of their lab components online, as full online replacements, supplements to prepare for in–person labs, or make–up labs.

These simulations help each student learn the practical and conceptual skills needed, then check for understanding and provide feedback. With adaptive pre–lab and post–lab assessment available, instructors can customize each assignment.

From the instructor's perspective, these simulations may be used in the lecture environment to help students visualize processes, such as digestion of starch and emulsification of lipids.

Dietary Analysis Case Studies in Connect®

Pkchai/Shutterstock

One of the challenges instructors face with teaching nutrition classes is having time to grade individual dietary analysis projects. To help overcome this challenge, assign auto–graded dietary analysis case studies. These tools require students to use NutritionCalc Plus to analyze dietary data, generate reports, and answer questions to apply their nutrition knowledge to real–world situations. These assignments were developed and reviewed by faculty who use such assignments in their own teaching. They are designed to be relevant, current, and interesting!

Tegrity: Lectures 24/7

Tegrity in Connect is a tool that makes class time available 24/7 by automatically capturing every lecture. With a simple one–click start–and–stop process, you capture all computer screens and corresponding audio in a format that is easy to search, frame by frame. Students can replay any part of any class with easy–to–use, browser–based viewing on a PC, Mac, or other mobile device.

Educators know that the more students can see, hear, and experience class resources, the better they learn. In fact, studies prove it. Tegrity's unique search feature helps students efficiently find what they need, when they need it, across an entire semester of class recordings. Help turn your students' study time into learning moments immediately supported by your lecture. With Tegrity, you also increase intent listening and class participation by easing students' concerns about note–taking. Using Tegrity in Connect will make it more likely you will see students' faces, not the tops of their heads.

Assess My Diet

Auto–graded personalized dietary analysis
Students are using NutritionCalc Plus to analyze their own dietary patterns. But how can instructors integrate that information into a meaningful learning experience? With Assess My Diet, instructors can now assign autograded, personalized dietary analysis questions within Connect. These questions refresh their memory on the functions and food sources of each nutrient and prompt the students to evaluate their own eating behaviors. Students can compare their own nutrient intakes to current Dietary Reference Intakes and demonstrate their ability to perform calculations on their own data, such as percentage of calories from saturated fat. They can compare the nutrient density of their own food selections to see which of their food choices provides the most fiber or iron. A benefit of the Assess My Diet question bank is that it offers assignable content that is personalized to the students' data, yet it is still auto–graded. It saves time and keeps all assignments in one place.

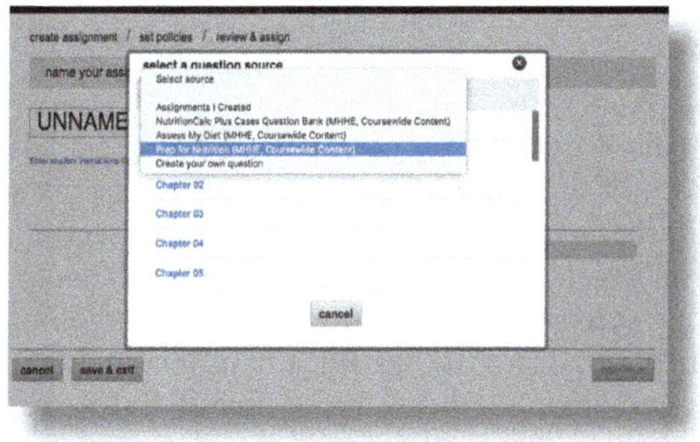

Prep for Nutrition

To help you level–set your classroom, we've created Prep for Nutrition. This question bank highlights a series of questions, including Basic Chemistry, Biology, Dietary Analysis, Mathematics, and Student Success, to give students a refresher on the skills needed to enter and be successful in their course. By having these foundational skills, you will feel more confident your students can begin class, ready to understand more complex concepts and topics. Prep for Nutrition is course–wide for ALL nutrition titles and can be found in the Question Bank dropdown within Connect.

ReadAnywhere®

Read or study when it's convenient for you with McGraw Hill's free ReadAnywhere app. Available for iOS or Android smartphones or tablets, ReadAnywhere gives users access to McGraw Hill tools including the eBook and SmartBook or Adaptive Learning Assignments in Connect. Take notes, highlight, and complete assignments offline—all of your work will sync when you open the app with WiFi access. Log in with your McGraw Hill Connect username and password to start learning—anytime, anywhere!

Writing Assignment

Available within Connect and Connect Master, the Writing Assignment tool delivers a learning experience to help students improve their written communication skills and conceptual understanding. As an instructor you can assign, monitor, grade, and provide feedback on writing more efficiently and effectively.

Remote Proctoring & Browser-Locking Capabilities

connect + proctorio

Remote proctoring and browser–locking capabilities, hosted by Proctorio within Connect, provide control of the assessment envi–ronment by enabling security options and verifying the identity of the student.

Seamlessly integrated within Connect, these services allow instruc–tors to control the assessment experience by verifying identification, restricting browser activity, and monitoring student actions.

Instant and detailed reporting gives instructors an at–a–glance view of potential academic integrity concerns, thereby avoiding personal bias and supporting evidence–based claims.

Create

Your Book, Your Way

McGraw Hill's Content Collections Powered by Create® is a self–service website that enables instructors to create custom course materials—print and eBooks—by drawing upon McGraw Hill's comprehensive, cross–disciplinary content. Choose what you want from our high–quality textbooks, articles, and cases. Combine it with your own content quickly and easily, and tap into other rights–secured, third–party content such as readings, cases, and articles. Content can be arranged in a way that makes the most sense for your course, and you can include the course name and information as well. Choose the best format for your course: color print, black–and–white print, or eBook. The eBook can be included in your Connect course and is available on the free ReadAnywhere® app for smartphone or tablet access as well. When you are finished customizing, you will receive a free digital copy to review in just minutes! Visit McGraw Hill Create—www.mcgrawhillcreate.com—today and begin building!

McGraw Hill GO

McGraw Hill GO is an easy–to–use, quick–to–set–up eBook+ that lives within the LMS. GO equips instructors and students with essential course materials: trusted, assignable content and auto–graded chapter questions that sync directly into the LMS. GO makes it easier to keep up with progress and direct attention where it matters.

OLC-Aligned Courses

Implementing High–Quality Online Instruction and Assessment through Preconfigured Courseware

In consultation with the Online Learning Consortium (OLC) and our certified Faculty Consultants, McGraw Hill has created pre–configured courseware using OLC's quality scorecard to align with best practices in online course delivery. This turnkey courseware contains a combination of formative assessments, summative assessments, homework, and application activities, and can easily be customized to meet an individual's needs and course outcomes. For more information, visit https://www.mheducation.com/highered/olc.

Polling

Every learner has unique needs. Uncover where and when you're needed with the new Polling tool in McGraw Hill Connect. Polling allows you to discover where students are in real time. Engage students and help them create connections with your course content while gaining valuable insight during lectures. Leverage polling data to deliver personalized instruction when and where it is needed most.

NewsFlash Articles

NewsFlash brings the real world into your classroom! These activities in Connect tie current news stories to key concepts. After interacting with a contemporary news story, students are assessed on their ability to make the connections between real–life events and course content.

Test Builder in Connect

Available within Connect, Test Builder is a cloud–based tool that enables instructors to format tests that can be printed, administered within a Learning Management System, or exported as a Word document. Test Builder offers a modern, streamlined interface for easy content configuration that matches course needs, without requiring a download.

Test Builder allows you to:

- access all test bank content from a particular title.
- easily pinpoint the most relevant content through robust filtering options.
- manipulate the order of questions or scramble questions and/or answers.
- pin questions to a specific location within a test.
- determine your preferred treatment of algorithmic questions.
- choose the layout and spacing.
- add instructions and configure default settings.

Test Builder provides a secure interface for better protection of content and allows for just–in–time updates to flow directly into assessments.

Ingram Publishing/SuperStock

Nutrition for Healthy Living, 2024 Release

Writing an introductory college–level nutrition textbook and revising the text on a regular basis are not easy tasks. Wendy Schiff authored the first five editions of *Nutrition for Healthy Living.* In anticipation of her retirement, Wendy invited me, Traci Keck, to work with her to revise the 6th edition of this textbook. For the 2024 Release of *Nutrition for Healthy Living,* I relied on my experience working with Wendy and my skills as a registered dietitian and university educator to present basic and relevant information about nutrition. My experi–ences teaching non–nutrition majors and dietetic students also provide valuable insights into the diversity, as well as the needs, interests, and capabilities of today's students. In addition, reviews by numerous nutrition faculty helped shape and refine the contents of this textbook.

Nutrition Information for Today's Students

Nutrition for Healthy Living is intended for students who are interested in learning about nutrition for personal reasons, as well as students considering majoring in nutrition, nursing, or other health– and science–related fields. *Nutrition for Healthy Living* is an introductory textbook that appeals to students who represent a broad range of academic backgrounds—English majors, as well as biology and nursing majors. Using this textbook to accompany an introductory nutrition course is intended to spark students' interest in adopting healthier dietary practices and possibly even inspire some students to consider nutrition as their major.

Nutrition for Healthy Living is not only scientifically up–to–date, but also includes consumer– and clinically–oriented content and features. The text is visually appealing and fun to read, engages students' interest, is well organized, and has features that contribute to the pedagogy without being distracting.

Diversity, Equity, and Inclusiveness

Previous editions of *Nutrition for Health Living* consciously aimed for cultural, ethnic, and color balance in art and language. However, the 2024 Release has received even closer attention to ethnic and cultural diversity, color, and the gender spectrum. It addresses the inadequately diverse representation seen in many textbooks in hopes that all readers will feel a sense of inclusion and familiarity. In the language of ethnicity, race, and color, this edition follows the recent guidelines of the American Medical Association (*JAMA,* August 21, 2021). Evidence–based standards discussed throughout the text and used by professionals for nutrition and health assessment, including Dietary Reference Intakes for energy needs and all macronutrients and micronutrients, waist circumference, and others, are based on sex–specific reference values. Therefore, in the 2024 Release, gender–specific pronouns and nouns are minimized and balanced, where some remain and others were deleted where they were adding nothing to the meaning of the writing.

Hero Images/Getty Images

Science That Is Accessible for Students

Nutrition is an "offspring" science that requires a basic under–standing of certain chemical and physiological concepts and terms. However, students from a wide variety of academic backgrounds often enroll in introductory nutrition courses, and in many instances, they have not taken college–level science courses prior to this nutrition course.

Ignorance about chemistry and physiology contributes to food faddism and health quackery. *Nutrition for Healthy Living* introduces basic scientific principles that relate to the study of nutrition in a manner that every college student can understand. The textbook, for example, introduces and defines terms that relate to nutrition, foods, and dietetics, including medical terminology. Additionally, the text includes numerous illustrations that are designed to help students understand basic, as well as complex, physiological concepts that relate to nutrition and digestive system anatomy and physiology. By providing a solid scientific foundation, nutrition educators can more easily dispel commonly held but inaccurate beliefs, such as "When you're physically inactive, muscle turns into fat," and "Cellulite is a special type of body fat."

Students who use this textbook will be prepared to make intelligent, practical choices that may result in improved nutrition and health. Furthermore, the students will be able to take more advanced nutrition courses, should they decide to major in the field of nutrition and dietetics.

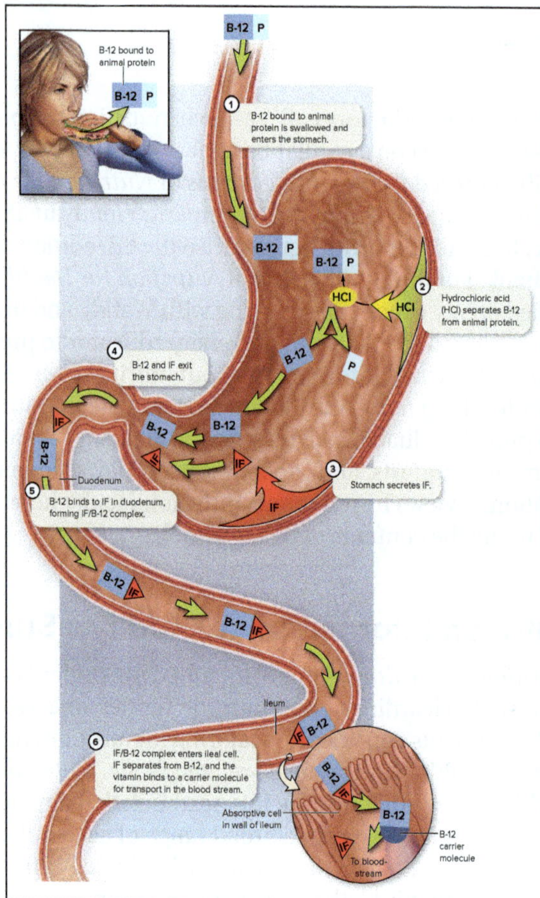

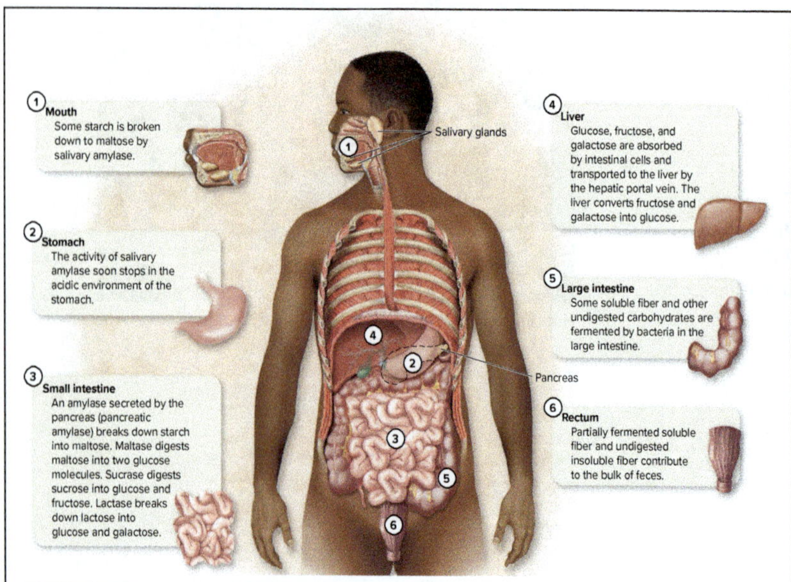

Brief Organization and Traditional Approach

Burke/Triolo Productions/Artville/Getty Images

Nutrition for Healthy Living follows a more traditional approach to the study of nutrition in that the textbook's organization focuses on providing basic information about each class of nutrients rather than certain functions, tissues, or diseases. Additionally, the textbook integrates health information, particularly information about diet-related chronic diseases, within each chapter where it is appropriate, rather than relegate it to chapters near the end of the textbook. For example, the chapters that discuss nutrients provide basic information first and then present applications, including the nutrient-related health effects of certain lifestyle practices, particularly dietary choices.

Instructors often do not have the time to cover all the material in their textbooks. *Nutrition for Healthy Living* organizes core content into 13 chapters, which helps make teaching introductory nutrition more manageable and fits the time frame of most courses better than textbooks with 15 or more chapters. Chapter 2 is unique among nutrition textbooks in that the chapter focuses on understanding basic scientific research and evaluating the reliability of nutrition- and health-related sources and messages, especially those promoting dietary supplements. Chapters 5 through 9 provide fundamental information about the nutrients, and integrate health information, particularly diet-related chronic diseases, within each chapter where it is appropriate, rather than relegate it to a single chapter near the end of the textbook. Chapter 6, for example, discusses fats and other lipids before focusing on atherosclerosis, including risk factors for the disease. Although some nutrition-related topics were important to cover, they did not warrant using a full chapter. Thus, topics such as global nutrition concerns, alcohol and alcohol abuse, and eating disorders are presented in "Nutrition Matters," the last section of each chapter.

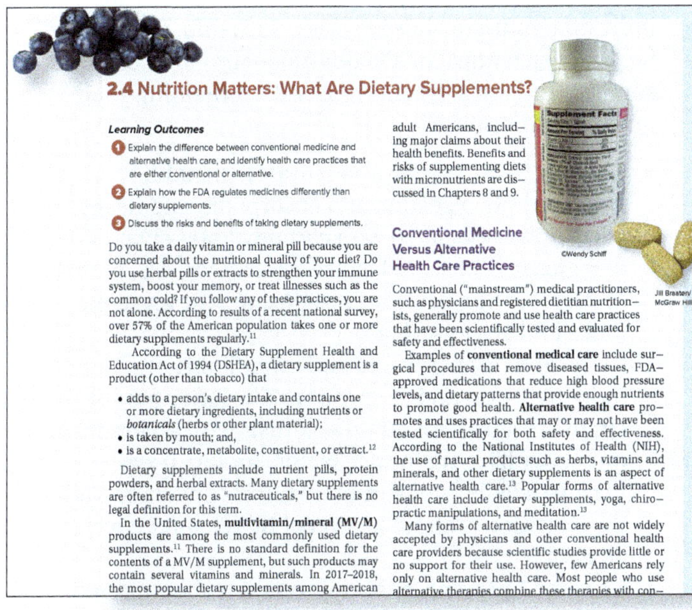

2.4 Nutrition Matters: What Are Dietary Supplements?

Learning Outcomes

1. Explain the difference between conventional medicine and alternative health care, and identify health care practices that are either conventional or alternative.
2. Explain how the FDA regulates medicines differently than dietary supplements.
3. Discuss the risks and benefits of taking dietary supplements.

Do you take a daily vitamin or mineral pill because you are concerned about the nutritional quality of your diet? Do you use herbal pills or extracts to strengthen your immune system, boost your memory, or treat illnesses such as the common cold? If you follow any of these practices, you are not alone. According to results of a recent national survey, over 57% of the American population takes one or more dietary supplements regularly.[11]

According to the Dietary Supplement Health and Education Act of 1994 (DSHEA), a dietary supplement is a product (other than tobacco) that

- adds to a person's dietary intake and contains one or more dietary ingredients, including nutrients or *botanicals* (herbs or other plant material);
- is taken by mouth; and,
- is a concentrate, metabolite, constituent, or extract.[12]

Dietary supplements include nutrient pills, protein powders, and herbal extracts. Many dietary supplements are often referred to as "nutraceuticals," but there is no legal definition for this term.

In the United States, **multivitamin/mineral (MV/M)** products are among the most commonly used dietary supplements.[11] There is no standard definition for the contents of a MV/M supplement, but such products may contain several vitamins and minerals. In 2017–2018, the most popular dietary supplements among American adult Americans, including major claims about their health benefits. Benefits and risks of supplementing diets with micronutrients are discussed in Chapters 8 and 9.

©Wendy Schiff

Conventional Medicine Versus Alternative Health Care Practices

Conventional ("mainstream") medical practitioners, such as physicians and registered dietitian nutritionists, generally promote and use health care practices that have been scientifically tested and evaluated for safety and effectiveness.

Jill Braaten/McGraw Hill

Examples of **conventional medical care** include surgical procedures that remove diseased tissues, FDA-approved medications that reduce high blood pressure levels, and dietary patterns that provide enough nutrients to promote good health. **Alternative health care** promotes and uses practices that may or may not have been tested scientifically for both safety and effectiveness. According to the National Institutes of Health (NIH), the use of natural products such as herbs, vitamins and minerals, and other dietary supplements is an aspect of alternative health care.[13] Popular forms of alternative health care include dietary supplements, yoga, chiropractic manipulations, and meditation.[13]

Many forms of alternative health care are not widely accepted by physicians and other conventional health care providers because scientific studies provide little or no support for their use. However, few Americans rely only on alternative health care. Most people who use alternative therapies combine these therapies with con-

Chapters: Basic Descriptions

Nutrition for Healthy Living covers the basic material instructors need in a format that is logical and practical for nearly all introductory nutrition courses:

- **Chapter 1** introduces students to nutrition and nutrients, and presents 10 key nutrition concepts, such as "most naturally occurring foods are mixtures of nutrients" and "variety can help ensure the nutritional adequacy of a diet." The Nutrition Matters section of Chapter 1 discusses global nutrition concerns, including malnutrition and sustainable agriculture.
- **Chapter 2** presents basic information about scientific methodology as it relates to nutrition research and provides tips for becoming a more wary consumer of nutrition- and health-related information. The Nutrition Matters section of Chapter 2 focuses on dietary supplements; however, information concerning specific dietary supplements is incorporated where it is relevant throughout the book.
- **Chapter 3** discusses dietary standards and guidelines, food groups and guides, and how to use information provided on food packages. This chapter's Nutrition Matters section presents information about traditional dietary practices of several cultural and ethnic groups that comprise much of the U.S. population and the health benefits of following traditional diets.
- **Chapter 4** presents basic principles of chemistry and human physiology as they apply to the study of nutrition but at a level that students can easily understand. The text includes numerous illustrations to help students understand basic physiological concepts that relate to nutrition and digestive system anatomy and physiology. Because students and courses vary in the depth of chemistry prerequisites, this chapter features some flexibility, so professors can choose to skip the chemistry section if they prefer. This chapter also provides information about intestinal microbiota, and common gastrointestinal disorders and diseases, primarily in the Nutrition Matters section.

- **Chapters 5, 6, 7, 8, and 9** present basic and practical information about each major class of nutrients and individual nutrients in each class. Such information includes each nutrient's major functions in the body, food sources, and roles in achieving and maintaining good health. Furthermore:
  - **Chapter 5** discusses carbohydrates, including carbohydrate digestion and absorption, and health conditions associated with carbohydrates such as diabetes mellitus and lactose intolerance. The Nutrition Matters section provides insights into the glycemic index and glycemic load.
  - **Chapter 6** discusses fats and other lipids, including lipid digestion and absorption. The latter part of the chapter focuses on atherosclerosis with an emphasis on recognizing major risk factors for the disease. Because alcohol consumption affects lipid metabolism, the Nutrition Matters section of this chapter covers alcohol.
  - **Chapter 7** presents basic information about proteins, including protein digestion and amino acid absorption, and protein metabolism. The Nutrition Matters section features information about protein undernutrition.
  - **Chapter 8** covers fat– and water–soluble vitamins, including deficiency and toxicity disorders. The Nutrition Matters section of the chapter examines the role of certain vitamins in chronic disease prevention.
  - **Chapter 9** features water, as well as essential minerals, including minerals that may be dietary essentials. Bottled water is discussed in the Nutrition Matters section of the chapter.
- **Chapters 10, 11, 12, and 13** focus on applying basic nutrition information for important concerns or different age groups:
  - **Chapter 10** covers body composition and weight management. The Nutrition Matters section of the chapter provides basic information about the three major eating disorders.
  - **Chapter 11** presents information relating to physical fitness; developing a practical program for lifelong physical fitness is the focus of this chapter's Nutrition Matters section.
  - **Chapter 12** features information about food–borne illness and food safety concerns. This chapter's Nutrition Matters section provides food safety information for global travelers.
  - **Chapter 13** discusses nutrition during pregnancy, lactation, infancy, childhood, and the older adult years. The Nutrition Matters section of this chapter examines the role of diet in promoting longevity.

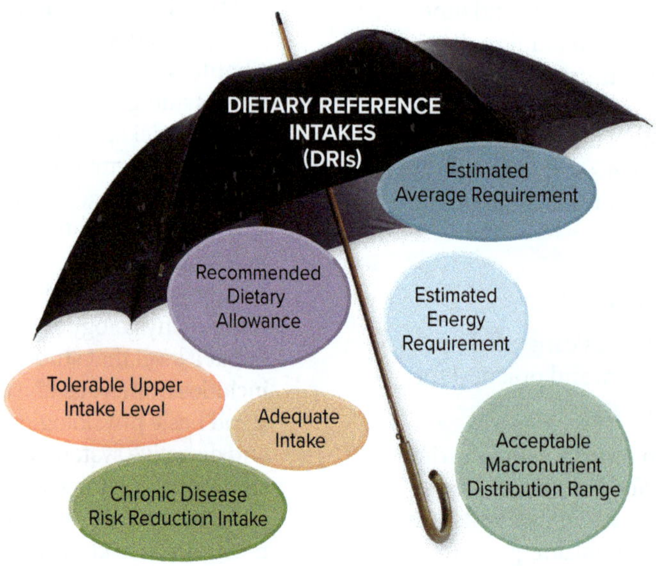

Ryan McVay/Photodisc/Getty Images

Consumer/Student Focus

Regardless of their backgrounds, students are consumers of nutrition information from a wide variety of sources, including popular magazines, friends, diet books, infomercials, social media, and the Internet. Often times these students arrive in class with many misconceptions about their diet and health. As nutrition educators, we seek to identify these beliefs and to impart sound, reliable nutrition and health information. We also strive to equip our students with the tools they need to make intelligent, informed food– and nutrition–related decisions beyond the classroom. In addition to devoting an entire chapter to evaluating nutrition–related information (Chapter 2), the consumer emphasis is integrated throughout the book.

Food & Nutrition Tips present practical suggestions that apply material discussed in a section. These tips provide information students can use every day, while they are in college— and for the rest of their lives.

Food & Nutrition tips

Ways to preserve the vitamin content of food include:

- Eat fresh fruits and vegetables along with their vitamin-rich edible peels or skins whenever possible.
- Cook fresh or frozen vegetables by microwaving, steaming, or stir-frying to minimize vitamin losses from cooking. Vegetables generally have high water content; therefore, add no water or just a small amount when microwaving vegetables.
- Do not overcook vegetables and minimize reheating because prolonged heating reduces vitamin content.
- Do not add margarine or butter to vegetables during cooking because fat-soluble vitamins and phytochemicals may enter the fat and be discarded when the fat is drained before serving. Fat in foods can enhance the body's absorption of fat-soluble vitamins; therefore, you can add some fat, such as olive oil or soft margarine, to vegetables after they are cooked.
- Store canned foods in a cool place. Canned foods can vary in the amount of nutrients they contain, largely because of differences in storage times and temperatures. If the can has been on the shelf for an extended period of time, the food's vitamin content may have deteriorated, and its taste and texture may be less than desirable.

Did YOU Know?

Impossible Foods manufactures a meat substitute made from soy protein called Impossible Beef. The meat substitute is used to replace ground beef in recipes and burger patties. Impossible Beef products have a texture and color similar to raw beef. In fact, these meat substitutes are a bright red color; "bleed" red, blood-like juices when raw; and taste like beef when cooked! The unique red coloring and umami, "beefy" flavor come from soy *leghemoglobin*, which is a heme molecule found in the roots of soybeans.[23] Impossible Foods uses yeast genetically engineered with the gene for leghemoglobin to create large amounts of the plant heme to use in their products. A 4-oz. Impossible Beef burger pattie provides 19 grams of high-quality soy protein.

Sundry Photography/Shutterstock

Did You Know? This margin feature notes interesting nutrition–related tidbits that relate to information presented in that section of the chapter. Some of these features dispel commonly held beliefs about food and nutrition that are inaccurate.

My Diverse Plate. This feature introduces students to foods and dietary practices that are ethnically and culturally diverse.

MY DIVERSE PLATE

Although the kiwano (*Cucumis metuliferusis*) is native to southern Africa, this odd-looking relative of the cucumber is grown in California. The green pulpy seeds can be scooped out of the rind and eaten as a snack or added to salads. A single kiwano is a good source of protein (almost 4 g), iron (2.3 mg), and vitamin C (11 mg). The pulp, however, has a slimy texture and not much flavor. In Africa, kiwano is generally considered to be an edible weed that is not eaten unless other foods are unavailable.

©Wendy Schiff

Recipe for Healthy Living is a practical application of nutrition and food information that will appeal to most college students. Each chapter features an easy–to–make, kitchen–tested recipe that helps bring the chapter's content to life (e.g., "Apricot Quinoa"). In addition to the pie graph for macronutrient content and a bar chart to illustrate % Daily Value for energy and key nutrients in a serving of the food, this feature also indicates which MyPlate food groups the major ingredients in the dish represent. By trying the recipes, students can develop basic food preparation skills and may be inspired to cook more foods from scratch.

Recipe for Healthy Living

Apricot Quinoa

The ancient Incas of South America cultivated quinoa and relied on the protein-rich vegetable as a dietary staple. Today, cooked quinoa is often added to vegetable salads or served as an accompaniment to chicken or pork. This quinoa recipe makes about two ½-cup servings. Each serving provides about 240 kcal, 7 g protein, 3 g fat, 4.5 g fiber, 3 mg iron, 2 mg zinc, and 80 mcg folate.

©Wendy Schiff

INGREDIENTS:

1⅓ cups water
½ cup quinoa
⅓ cup dried apricots, chopped
¼ tsp ground cinnamon
1 Tbsp maple syrup

PREPARATION STEPS:

1. Add water to a small saucepan and bring to a boil. Add quinoa to boiling water.
2. Cover saucepan and cook quinoa over medium heat until all the water is absorbed (about 12–15 minutes).
3. Remove from heat, fluff quinoa with fork, and stir chopped apricots into the quinoa.
4. Place quinoa into a bowl, drizzle with maple syrup, and sprinkle cinnamon on the quinoa.
5. If desired, garnish with fresh fruit. Serve warm

Approximate Percent of Calories from Macronutrients

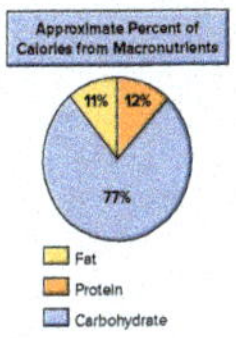

U.S. Department of Agriculture (USDA)

Percent of Daily Value/Serving

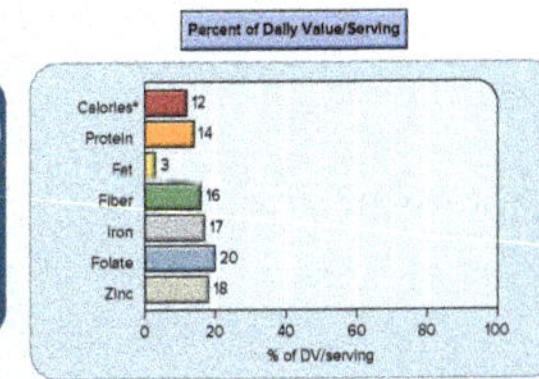

*2000 daily total kcal

Real People, Real Stories feature information about "real" people, many of whom have actually recovered from or are currently living with nutrition–related conditions such as celiac disease, type 1 diabetes, and hypertension. This feature is designed to help students recognize the daily challenges people with such conditions face and the role diet and regular physical activity play in achieving and maintaining good health.

REAL PEOPLE | REAL STORIES

Traci Keck

Annie Pietsch

Annie Pietsch is a student at the University of Texas at San Antonio (UTSA) who is majoring in nutrition and dietetics. When Annie is not focused on her studies, she enjoys running for exercise, spending time with her family, and shopping and eating ice cream with her friends.

While she was in high school, Annie decided she wanted to become a registered dietitian nutritionist because she wanted to help her grandmother, who had type 2 diabetes. Before she started college, Annie had no idea that her lifestyle was about to change dramatically.

Annie's health began to decline during the spring semester of her first year at UTSA. She became unusually tired when exercising as well as walking to class. According to Annie, "Walking up a flight of stairs felt like running 5 miles." She also developed an unusual craving for sugar-sweetened "soda," ice cream, and fast-food chicken. "I knew something was wrong when I was losing weight instead of gaining weight after eating this way."

At the beginning of spring break, Annie went home to visit her family. Her parents were immediately concerned with Annie's weight loss and overall physical

Pedagogy

Key Terms and **Definitions** Key terms are noted in boldface text within the narrative. Many unfamiliar terms have italicized pronunciations in parentheses appearing next to them. Definitions of key terms are provided in a box, generally in the margin of the page where the terms first appear in the chapter. A glossary of these key terms is at the end of the textbook.

Essential Fatty Acids

The body cannot make two PUFAs, **alpha–linolenic** (*al'–fah lin'–o–len'–ik*) **acid** and **linoleic** (*lin'–o–lay'–ik*) **acid.** These lipids are **essential fatty acids** because they must be supplied by the diet. Alpha–linolenic acid is an **omega–3 fatty acid.** The "3" refers to the position of the first double bond that appears in the fatty acid's carbon chain, when you start counting carbons at the omega end of the molecule (see Figure 6.1c). Cells use alpha–linolenic acid to synthesize two other omega–3 fatty acids, eicosapentaenoic (*eye'–koss–uh–pen'–tah–ee–no'–ik*) acid (EPA) and docosahexaenoic (*doe'–koss–uh–hex'–uh–ee–no'–ik*) acid (DHA). However, the body's ability to make EPA and DHA from alpha–linolenic acid is limited.

alpha-linolenic acid an essential fatty acid

linoleic acid an essential fatty acid

essential fatty acids lipids that must be supplied by the diet

References *Nutrition for Healthy Living* includes in–text citations and extensive lists of references in Appendix G. References provide readers with access to sources of information for more in–depth understanding or for topics that spark students' interest.

Chapter References

See Appendix G.

End–of–Chapter Summary This feature provides a brief review of the main points of each major section of the chapter.

Summary

6.1 Introducing Lipids

- Lipids are needed for energy, proper growth and development, nerve functioning, and the production of bile and several hormones. Major lipids are triglycerides, phospholipids, and sterols. In addition to being a source of fuel, insulates the body against cold temperatures, and protects against damaging blows.

Glossary

A

absorption process by which substances are taken up from the GI tract and enter the bloodstream or the lymph
Acceptable Macronutrient Distribution Ranges (AMDRs) macronutrient intake ranges that are nutritionally adequate and may reduce the risk of diet-related chronic diseases
acid-base balance maintaining the proper pH of body fluids
acid group acid portion of a compound
acids substances that donate hydrogen ions
added sugars sugars added to foods during processing or preparation
adenosine diphosphate (ADP) high-energy compound; by-product of ATP use
adenosine triphosphate (ATP) molecule that stores energy
Adequate Intakes (AIs) dietary recommendations for nutrients that scientists do not have enough information to set RDAs
adipose cells fat cells; specialized cells that store fat
adolescence life stage in which a child matures physically into an adult
aerobic conditions that require free oxygen
aerobic exercise physical activities that
anencephaly type of neural tube defect in which the brain does not form properly or is missing
anorexia nervosa (AN) severe psychological disturbance characterized by self-imposed starvation
antibodies infection-fighting proteins
antidiuretic hormone (ADH) hormone that participates in water conservation
antioxidant substance that gives up electrons to radicals to protect cells
appetite desire to eat appealing food
arterial plaque lipid-filled deposits that form under the lining of certain arteries
arteries vessels that carry blood away from the heart
ascorbic acid vitamin C
atherosclerosis long-term disease process in which plaques build up inside arterial walls
athlete triad condition that is characterized by low energy availability, abnormal reproductive function, and reduced bone mineral density in athletes

B

bacteria simple single-celled microorganisms
bioelectrical impedance technique of estimating body composition in which a device measures the conduction of a weak electrical current through the body
bioengineered food (BE food) food that contains genetic material that has been modified by certain scientific methods
bioengineering scientific techniques that alter an organism's hereditary material
biological activity describes vitamin's degree of potency or effects in the body
biotechnology using living things to manufacture new products
blood alcohol concentration (BAC) percentage of alcohol in the bloodstream
body mass index (BMI) numerical value that reflects body weight adjusted for height
buffer substance that can help maintain the normal pH of a solution
bulimia nervosa (BN) severe psychological condition characterized by repeated episodes of binge eating followed by unhealthy behaviors to prevent weight gain

C

caffeine naturally occurring stimulant drug
capillaries smallest blood vessels

Assessment of Learning

Pixtal/age fotostock

One of our primary goals as nutrition educators is to ensure that our students leave the introductory nutrition course with a better understanding of the nutrition principles and concepts needed to improve their diet and health. To assess how well faculty are achieving that goal, many colleges and universities have implemented Coursewide Student Learning Outcomes as a way to measure what students have learned upon completing an introductory nutrition course. *Nutrition for Healthy Living* has been developed around the following student learning outcomes for nutrition courses.

Coursewide Student Learning Outcomes

1. Identify functions and sources of nutrients.
2. Demonstrate basic knowledge of digestion, absorption, and metabolism.
3. Apply current dietary guidelines and nutrition recommendations.
4. Analyze and evaluate nutrition information scientifically.
5. Relate roles of nutrients in good health, optimal fitness, and chronic diseases.
6. Summarize basic concepts of nutrition throughout the life span.
7. Evaluate a personal diet record using a computer database.

Chapter Learning Outcomes

Each major section of a chapter opens with a list of section–specific learning outcomes that build upon the broader coursewide outcomes. The Learning Outcomes help students prepare for reading the section and clarify major concepts they are expected to learn. These measurable outcomes are further supported by assessment methods and study aids found within the chapters and in Connect. Learning Outcomes can be used to help instructors identify specific content that requires more refined or focused teaching methods.

4.3 The Digestive System

Learning Outcomes

1. Identify major organs of the digestive system, and describe primary functions of each organ.
2. Identify the accessory organs of the digestive system and the roles these organs play in digestion.
3. Discuss the overall processes of digestion, absorption, and transport; and waste elimination.

Quiz Yourself

This pretest, composed of five true–or–false questions, appears at the beginning of each chapter; answers to the quiz are provided at the end of the chapter. The purpose of Quiz Yourself is to stimulate interest in reading the chapter. By taking the quiz, students may be surprised to learn how little or how much they know about the chapter's contents.

Quiz Yourself

How much protein is recommended for optimal health? Can people obtain enough protein by eating only plant foods? What happens if you eat more protein than your body needs? Before reading Chapter 7, take the following quiz to test your knowledge of protein. The answers are at the end of the chapter.

1. Animal foods such as meat and eggs are almost 100% protein. _____ T _____ F
2. Foods made from soybeans can be sources of high-quality protein. _____ T _____ F
3. Americans typically consume more protein from animal sources than from plant foods. _____ T _____ F
4. Registered dietitian nutritionists generally recommend that people with liver or kidney disease take amino acid supplements to increase their protein intake. _____ T _____ F
5. People can nourish their hair by using shampoo that contains protein. _____ T _____ F

Concept Checkpoint

The Concept Checkpoint feature includes review questions, many of which involve critical thinking skills, posed at the end of major section headings. Such questions enable students to test their acquisition of information presented in the section. Answers to the questions in each Concept Checkpoint are located in Appendix F.

***Concept* CHECKPOINT**

20. What is hunger? What is satiety?
21. Discuss the roles of leptin and ghrelin in regulating hunger.
22. Under what conditions would having "thrifty genes" benefit a person?
23. What is the set-point theory?
24. What is the difference between hunger and appetite?
25. Describe how the environment influences a person's food intake and physical activity level.
26. Provide at least three examples of ways that physiological, psychological, and environmental factors can influence eating behavior.

See Appendix F for responses.

Critical Thinking

The Critical Thinking feature involves higher–level cognitive skills, including applying, analyzing, synthesizing, and evaluating information. This assessment features a series of thought–provoking questions or activities at the end of the chapter. The questions and activities can help students develop higher–level cognitive skills using nutrition–related content. Acquiring and/or sharpening these skills can help students become better consumers of nutrition–related information.

Critical Thinking

1. Analyze your weekly physical activity habits. Does your participation in various physical activities meet the minimum recommendations? If not, which physical activities are you willing to include in your weekly routine to improve your fitness level?
2. Calculate your target heart rate zone for moderate–intensity as well as vigorous–intensity activities.
3. Why do human cells rely far more on glucose and fat for energy than on protein?
4. What advice would you give a runner concerning fluid intake before and during a marathon?
5. One of your friends is a student athlete who uses ginseng, nitrate, and caffeine to enhance their physical performance. What would you tell your friend concerning the pros and cons of taking these substances?
6. Sunny is an athlete who weighs 143 pounds. According to the information in Section 11.4, calculate the recommended amount of protein (g) she should consume daily.

Personal Dietary Analysis

Many chapters include an end–of–chapter activity for analyzing personal eating habits. Most of these activities require the use of a dietary analysis software program, such as NutritionCalc Plus. Students can gain insight into their eating behaviors by completing this activity.

Personal Dietary Analysis

Using the DRIs

1. Refer to your 1- or 3-day food log from the "Personal Dietary Analysis" feature in Chapter 3.
   a. Find the RDA values for vitamins under your life stage/sex group category in the DRI tables (see Appendix H). Write those values under the "My RDA" column in the table of this analysis activity.
   b. Review your personal dietary assessment. Find your 3-day average intakes of vitamins A, E, C, D, folate, B-12, thiamin, riboflavin, and niacin. Write those values under the "My Average Intake" column of the table.
   c. Calculate the percentage of the RDA you consumed for each vitamin by dividing your intake by the RDA amount and multiplying the figure you obtain by 100. For example, if your average intake of vitamin C were 100 mg/day, and your RDA for the vitamin were 75 mg/day, you would divide 100 mg by 75 mg to obtain 1.25. To multiply this figure by 100, simply move the decimal point two places to the right, and replace the decimal point with a percentage sign (125%). Thus, your average daily intake of vitamin C was 125% of the RDA. Place the percentages for each vitamin under the "% of My RDA" column.
   d. Under the ">, <, or =" column, indicate whether your average daily intake was greater than (>), less than (<), or equal to (=) the RDA.
2. Use the information you calculated in the first part of this activity to answer the following questions:
   a. Which of your average vitamin intakes equaled or exceeded the RDA value?
   b. Which of your average vitamin intakes was below the RDA value?
   c. What foods would you eat to increase your intake of the vitamins that were less than the RDA levels? (Review sources of certain vitamins in Chapter 8.)
   d. Turn in your completed table and answers to your instructor.

Practice Test

Each chapter ends with a series of 10 or more multiple–choice questions that test students' comprehension and recall of information presented in the chapter. Answers to the test questions are in Appendix F. The multiple–choice questions prepare students for classroom exams because they are similar in type and format to those in the test bank. In many instances, the test questions are correlated to the Coursewide Student Learning Outcomes and Chapter Learning Outcomes.

Practice Test (Section numbers are listed below in red.)

Select the best answer.

1. (6.1) Fats in foods
   a. add taste and contribute to "mouth feel."
   b. are digested and absorbed in the stomach.
   c. carry water–soluble nutrients.
   d. need to be eliminated to have a healthful diet.
2. (6.2) Solid fats generally have a high proportion of _____ fatty acids.
   a. unsaturated
   b. saturated
   c. polyunsaturated
   d. monounsaturated
3. (6.2) A saturated fatty acid has
   a. one double bond within the hydrocarbon chain.
   b. two double bonds within the hydrocarbon chain.
   c. no double bonds within the hydrocarbon chain.
   d. three or more double bonds within the hydrocarbon chain.
5. (6.2) Trans fatty acids are
   a. naturally in many foods.
   b. a by–product of the hydrogenation process.
   c. essential to good health.
   d. found primarily in fatty fish.
6. (6.2) Phospholipids
   a. do not have fatty acids in their chemical structures.
   b. lack glycerol in their chemical structures.
   c. do not occur naturally.
   d. are partially water–soluble.
7. (6.2) Cholesterol is
   a. metabolized for energy.
   b. found only in animal foods.
   c. not made by the human body.
   d. harmful to health.
8. (6.3) The primary site of triglyceride digestion and absorption is the
   a. stomach.

What's New in this Release

Jozsef Szasz-Fabian/123RF

The 2024 Release of *Nutrition for Healthy Living* continues the textbook's beautiful, pedagogically based illustrations and creative page layouts that are designed to facilitate learning. Furthermore, each chapter has several new photos and many of the diagrams and illustrations have been modified to make them easier to follow and understand. Such images help draw students' attention to the narrative, stimulate their interest in topics, and relate content to the "real world." It is important to note that the use of brand–name products in photos is for example representation only and does not constitute an endorsement.

Most chapters have new critical thinking questions and all chapters have several new practice test questions at the end of chapters. The practice test questions are now organized and identified by chapter sections to help students quickly locate the correct answers within the chapter as they study.

Today's students also need to be aware of the factors contributing to nutrition and health inequities in America. Thus, *Nutrition for Healthy Living* 2024 Release has five Nutrition and Health Equity features that discuss these issues and how social determinants of health influence a person's health and nutrition status.

The following information highlights several of the updates that are in each chapter.

Nutrition & Health *Equity*

Equity and equality are concepts often misunderstood, especially when we consider the health and nutrition status of all Americans. *Health equality* provides the same resources and opportunities to all people across a population, such as the option to purchase health insurance. However, people experience social determinants of health (SDOH), conditions in which we are born, grow, live, learn, work, worship, and age, that influence health. The differences in SDOH contribute to some minority groups *not* having safe and affordable housing, and access to quality education and health care. Thus, higher rates of chronic disease are observed in these populations.[1] It is important to understand that not all minority populations live in unsafe housing, and many *do* have access to quality education and health care!

Health equity addresses the barriers from SDOH and provides the resources needed to create an opportunity for all people to obtain equity—for example, access to affordable health insurance. Therefore, health equity allows for everyone to be as healthy as possible, regardless of race, education, gender identity, sexual orientation, employment, neighborhood, and whether they have a disability.

The Nutrition and Health Equity feature in this textbook discusses some of the inequities present in America. Understanding these inequities and their complex contributing factors is essential for identifying long-term solutions.

Chapter 1: The Basics of Nutrition

- Two Nutrition and Health Equity features provide an introduction to equity versus equality, social determinants of health, and current health insurance inequities in America.
- Updated Figure 1.2, which is a pie chart that has the 10 leading causes of death in the United States.
- Figures 1.4 (American's changing eating habits) and 1.6 (Energy and nutrient comparison) have been updated and revised to make the information easier for students to grasp.
- The chapter Nutrition Matters section Undernutrition—A Worldwide Concern now includes a discussion of the double burden of malnutrition that occurs in low–income developing nations.

Chapter 2: Evaluating Nutrition Information

- The information on human intervention studies and epidemiological studies has been revised for clarity to make the information easier for students to grasp.
- This chapter has been revised to provide a greater emphasis on social media and "influencers" as a source of misinformation.
- There are two new Did You Know? features, including one that discusses a fraudulent "cure" for COVID–19 and one identifying accurate diet tracking applications for smartphones.

Chapter 3: Planning Nutritious Diets

- Discussion of the nutritional qualities of plant milk alternative beverages and the FDA's labeling guidance for these beverages. Guidelines for FDA "healthy" product claims are also discussed.
- Figure 3.9 MyPlate Recom–mendations; the graph data bar colors now align with the corresponding colors represented by the various MyPlate food groups.
- New Did You Know? feature on healthy, quick, and easy snacks for college students on the go.
- The Nutrition Matters section has been revised to discuss the health benefits of traditional diets and foods of the cultural and ethnic groups that comprise much of the U.S. population.

Chapter 4: Body Basics

- Figure 4.1 (Atom) is revised to include neutrons along with protons and electrons.
- New Food and Nutrition Tips feature that discusses alliinase, an enzyme in garlic that provides garlic's unique flavor and smell.
- Content on the microbiome and diet has been revised and updated.

Chapter 5: Carbohydrates

- Content on added sugars revised to include 1 teaspoon of sugar equals 4 grams added sugar.
- Table 5.3 (How Much Added Sugar Is in That Food?) now includes grams of added sugar in addition to teaspoons of added sugar.
- Figure 5.8 (Regulating blood glucose); step 7 has been revised to make it easier for students to follow.
- New content discussing how to use the percent daily values on a Nutrition Facts Panel to evaluate carbohy–drates in a food product.

- New content discussing the role of diet quality, refined carbohydrates, and ultraprocessed foods in weight gain.
- Table 5.12 (Glycemic Index and Load) revised with 2021 data from the international tables of glycemic index and glycemic load values. This table was also revised to include a typical serving size of a food and grams of carbohydrates to make the information easier for students to grasp.

Chapter 6: Fats and Other Lipids

- Essential fatty acid deficiency has its own subsection for emphasis for students.
- Table 6.2 (Comparing Fats and Oils) has been revised to include avocado oil.
- New Table 6.3 (Food Sources Rich in Saturated, Unsaturated, and Trans Fats) lists foods rich in these fats.
- New Did You Know? feature discusses physical and mental health benefits of cooking meals at home and resources for learning basic cooking skills.
- New My Diverse Plate discusses trying a variety of ethnic cuisines and provides examples of lower-fat ethnic dishes to choose.

Chapter 7: Proteins

- Table 7.3 (Approximate Protein Content of Some Commonly Eaten Foods/Portion) has been revised and updated.
- New content discusses the pseudocereals amaranth and buckwheat as sources of high-quality protein.
- Two new Did You Know? features that discuss vegetarian meat substitutes rich in soy leghemoglobin and ways to meet protein needs during illness and recovery from significant illnesses.

Chapter 8: Vitamins

- New content discusses mushrooms exposed to UV light as a rich source of vitamin D in the diet, and serum folate levels and depression risk.
- High-dose vitamin C for the prevention of the cold virus in high-intensity athletes that are at higher risk for virus exposure has been added.

Chapter 9: Water and Minerals

- New content discusses the calcium content of plant milk alternative beverages.
- New Did You Know? feature discusses the FDA's voluntary guidance for the food industry to reduce the amount of salt added to processed foods.
- New Nutrition and Health Equity feature that discusses hypertension in Black Americans and community-based education programs that have proven success in lowering blood pressure levels in this population.
- New Did You Know? feature discusses magnesium therapy and migraine headaches.

Chapter 10: Energy Balance and Weight Control

- New content discussing the stigma of obesity and weight bias.
- New Nutrition and Health Equity feature that discusses body composition of different racial and ethnic populations, and using body mass index and waist circumference in these populations.
- Two new Did You Know? features discussing how many Americans do not seek weight management treatments from their health care providers, and emotional eating and the benefits of practicing mindfulness while eating.
- New content discussing setting goals as part of a weight management plan and the benefits of intuitive eating.

Chapter 11: Nutrition for Physically Active Lifestyles

- New content discussing energy bar ingredients.
- Table 11.6 (Popular Energy Bars and Gels) was revised to include added sugar and fiber content.
- Nutrition Matters introduces the term cross-training.

Chapter 12: Food Safety Concerns

- New content discusses the similarities and differences between the signs and symptoms of illness from foodborne illness, influenza, and COVID-19.
- New Did You Know? feature discusses norovirus and alcohol-based hand sanitizers.
- Table 12.10 (Foods to Store for Emergency Situations) has been revised and updated.

Chapter 13: Nutrition for a Lifetime

- New Did You Know? feature discusses morning sickness and treatments for the condition.
- Table 13.2 (Sample Menu for a 25-Year-Old Pregnant Female) was revised and updated.
- New content discusses why not to give plant milk alternative beverages to infants in the first year of life.
- Figure 13.11 (Prevalence of obesity among American children aged 2 to 19 years); Table 13.10 (Environmental Factors That Contribute to Childhood Obesity); and Table 13.12 (Percentages of American Adults Ages 71 Years and Older Who Follow USDA Dietary Recommendations) are revised and updated.
- New content discussing the treatment of childhood obesity that includes the entire family adopting a healthy lifestyle.

Pixtal/age fotostock

Acknowledgments

The development of an accurate and current manuscript for *Nutrition for Healthy Living,* 2024 Release, was facilitated by the input of the board of advisors and numerous college instructors. The board of advisors were an invaluable resource providing manuscript reviews and wisdom and guidance for the new Nutrition and Health Equity feature. We offer our sincere thanks to the following colleagues who provided feedback that we used to help us revise the content.

Board of Advisors

Carmen Blakely
Associate Professor for Nutrition and Dietetics, Kent State University

Amy Jamieson, Department Chair and Faculty
Exercise and Sports Studies at the University of California Santa Barbara

Reviewers

Angela Harter Alger
California State University, Chico

Nancy J. Allen
Danville Area Community College (DACC)

Karen Breshears
University of Central Missouri

Ann E. Embry
Western Kentucky University

Michelle Higgins
Erie Community College

Michelle L. Hoffer
Community College of Denver

Wimberley Kay Krueger
Blinn College District

Jennie Porter
CSU Fresno

Carole A. Sloan
Henry Ford College and *Wayne State University*

Kathryn Wagner
Maryville University

Christine N. Walters
Oklahoma State University–Oklahoma City

Shuo Zhang
Houston Community College

Savannah E. Zinebi
Houston Community College

My very special thanks is necessary to the individuals who helped make *Nutrition for Healthy Living* more real and interesting by contributing their stories and photographs of themselves to Real People, Real Stories features.

Many McGraw Hill staff members invested a great deal of time and effort into the development and production of this new release of *Nutrition for Healthy Living.* My gratitude is extended to all the members of the McGraw Hill editorial, design, production, and marketing team who provided their enthusiastic support and encouragement to me during this revision.

A few members of the team deserve special recognition. Lauren Vondra served as my Portfolio Manager. Under her very capable direction, I was able to focus my attention on the preparation of this release. Thanks as well to Tami Hodge, Executive Marketing Manager; David Hash, Lead Designer; Rachael Hillebrand, Assessment Content Project Manager; Laura Fuller, Buyer; and Lorraine Buczek, Content Licensing Specialist. Paula Patel, Content Project Manager, had the difficult task of managing the chapters as they progressed through production, and she helped convey my wishes concerning the art and layout to the compositor. Paula's efforts were instrumental in the production of a beautiful textbook.

Thank you Rachel Williams for your keen eye and review of the manuscript. Last, but not least, my Product Developer, Darlene Schueller, deserves my heartfelt thanks for the hard work, long hours, and extraordinary dedication she invested in the production of *Nutrition for Healthy Living.*

Traci L. Keck

D. Hurst/Alamy Stock Photo

Contents

©Wendy Schiff

1 The Basics of Nutrition 1

Isabelle Rozenbaum/PhotoAlto/Getty Images

2 Evaluating Nutrition Information 31

C Squared Studios/Photodisc/ Getty Images

Ingram Publishing/Alamy Stock Photo

Thesomegirl/iStock/Getty Images

7 Proteins 222

8 Vitamins 265

9 Water and Minerals 322

C Squared Studios/Photodisc/Getty Images

10 Energy Balance and Weight Control 377

lynx/iconotec.com/Glow Images

12 Food Safety Concerns 463

13 Nutrition for a Lifetime 497

Fruit bowl: Verdina Anna/Flickr/Getty Images

NUTRITION

for Healthy Living

CHAPTER 1

The Basics of Nutrition

Hill Street Studios/Blend Images/Getty Images

Quiz Yourself

Take the following quiz to test your basic nutrition knowledge; the answers are at the end of the chapter.

1. There are four classes of nutrients: proteins, lipids, sugars, and vitamins. ______ T ______ F
2. Proteins are the most essential class of nutrients. ______ T ______ F
3. All nutrients must be supplied by the diet because they cannot be made by the body. ______ T ______ F
4. Vitamins are a source of energy. ______ T ______ F
5. Milk, avocados, and blueberries are examples of "perfect" foods that contain all nutrients. ______ T ______ F

1.1 Nutrition: The Basics

Learning Outcomes

1. Explain why it is important to learn about foods and nutrition.
2. Identify factors that influence personal food choices.
3. Identify the leading causes of death in the United States and lifestyle factors that contribute to them.
4. List the six classes of nutrients, and identify a major role of each class of nutrient in the body.
5. Explain how to determine whether a substance is a nutrient, an essential nutrient, a nonnutrient, or a phytochemical.

When you were an infant and a young child, your parents or other adult caregivers were the "gatekeepers" of your food; they chose what you ate and prepared it, and you probably ate most of it. If you balked at eating steamed broccoli or baked salmon, they may have told you, "Eat your vegetables if you expect to get dessert" or "Make it a happy (clean) plate; finish that fish!" As you grew older, your **diet,** your usual pattern of food choices, came increasingly under your control. Today, your diet is more likely to be composed of foods that you enjoy, as well as can afford, and probably those you can prepare easily or obtain quickly. Your family's ethnic and cultural background may also play a role in determining what you eat regularly. For example, do you eat tamales, tripe, goat, tofu, or kimchi because you ate these foods as a child? Numerous other factors influence your food choices, including friends and food advertising, as well as your beliefs and moods (Fig. 1.1).

diet usual pattern of food choices

nutrients chemicals necessary for proper body functioning

Food is a basic human need for survival. You become hungry and search for something to eat when your body needs **nutrients,** the life–sustaining substances in food. Nutrients are necessary for the growth, maintenance, and repair of your body's cells. Eating food supplies your body with nutrients and satisfies your hunger, but it also can be comforting, especially if consuming certain foods soothes your anxiety or ends your boredom. Furthermore, what you choose to eat may have social, economic, ceremonial, or religious significance. For example, lobster, caviar, prime rib steak, wagyu beef, and angulas (baby eels) are often associated with special occasions and people who have high incomes. Other foods, such as a frosted layer cake topped with lit candles, stuffed roasted turkey served with candied sweet potatoes, and communion wafers, are associated with birthday celebrations, seasonal holidays, and religious rites. Thus, food is more than just something to eat.

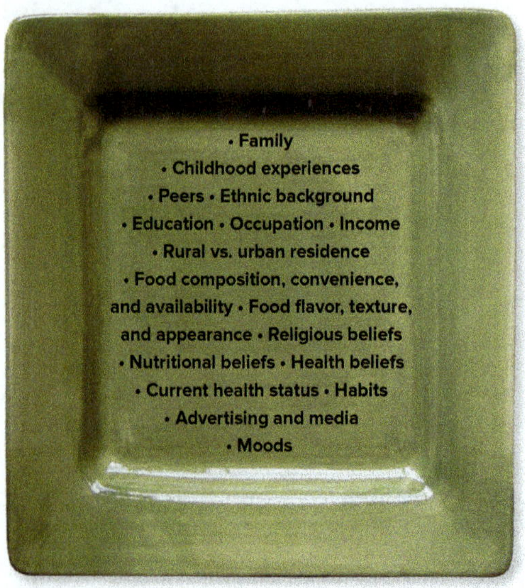

Figure 1.1 What influences your eating practices? Numerous factors influence food choices, including food advertising, peers, income, moods, and personal beliefs. ©Wendy Schiff

You were not born with the ability to select the appropriate mix of nutrients your body requires for proper functioning. To eat well, you need to learn about the nutritional value of foods and the effects that your diet can have on your health.

Simply having information about nutrients and foods and their effects on health may not be enough for people to change ingrained food–related behaviors; a person must be motivated to make such changes. Some people become motivated to improve their diets because they want to lose or gain weight. Others are so concerned about their health that they are motivated to change their eating habits in specific ways, such as by eating fewer salty or fatty foods. Many people, however, do not care if the food they eat is beneficial or harmful to their health.

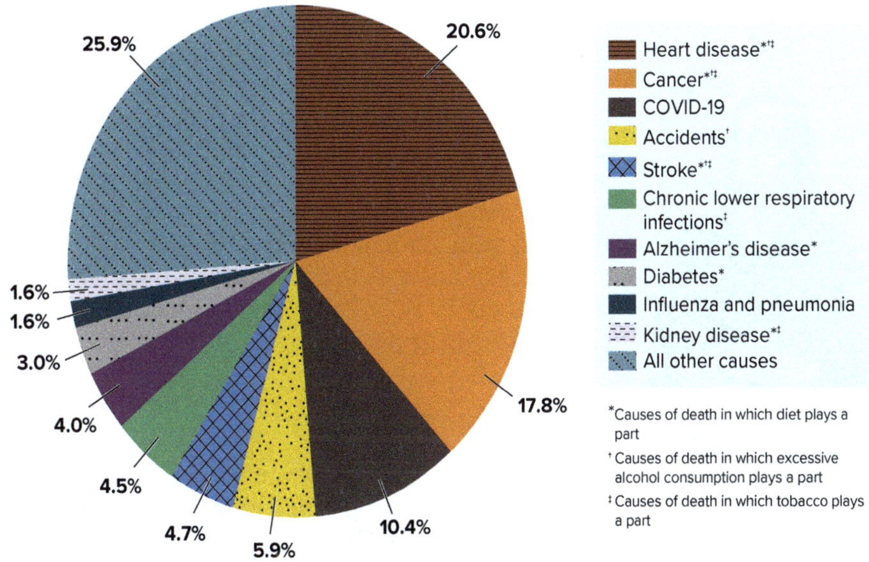

Figure 1.2 Approximate percentages of leading causes of death (U.S.), 2020. Dietary factors contribute to many of the 10 leading causes of death in the United States.
Source: National Center for Health Statistics, National Vital Statistics System: Data Brief 427. Mortality in the United States, 2020. 2021. www.cdc.gov/nchs/data/databriefs/db427-tables.pdf#4 Accessed:June 3, 2022.

Why Learn About Nutrition

Why should you care about your diet? The foods and beverages you usually select for meals and snacks contribute to your health now and in the future. In the United States, poor eating habits contribute to several leading causes of death, including heart disease, some types of cancer, stroke, and type 2 diabetes (Fig. 1.2). Such diseases and many other medical conditions are **chronic,** which means they are long term and, often, difficult to treat. Therefore, making efforts to prevent chronic health problems, especially when one is young, is crucial to a person's future physical health and emotional well-being.

People who consume a healthy diet are more likely to have proper immune responses to agents that cause infectious diseases, which include the viruses that cause seasonal flu and COVID-19. Recently, a team of scientists analyzed dietary information that was provided by almost 75,000 adults for 12 years. According to the results of the analysis, the people who had taken steps to consume healthier diets were less likely to die prematurely (too early) from all causes, and, especially, heart disease, than the people who did not improve their diets.[2] "Healthier" diets generally supplied more whole grains and vegetables and fewer salty foods. Chapter 3 provides more information about the composition of healthy diets.

Are you concerned about the nutritional quality of your diet? The fact that you are taking this course indicates you have a strong interest in nutrition and a desire to learn more about the topic. A major objective of this textbook is to provide you with the basic information you need to better understand how your diet can influence your health. Managing your diet is your responsibility. We will not tell you what to eat to guarantee optimal health: No one can make that promise. After reading this textbook and learning about foods and the nutrients they contain, you can use the information to make informed decisions concerning the foods you eat. Furthermore, you will be able to evaluate your diet and decide if it needs to be changed.

Each chapter of this textbook begins with "Quiz Yourself," a brief true-or-false quiz to test your knowledge of the material covered in the chapter.

Nutrition & Health Equity

Equity and equality are concepts often misunderstood, especially when we consider the health and nutrition status of all Americans. *Health equality* provides the same resources and opportunities to all people across a population, such as the option to purchase health insurance. However, people experience social determinants of health (SDOH), conditions in which we are born, grow, live, learn, work, worship, and age, that influence health. The differences in SDOH contribute to some minority groups *not* having safe and affordable housing, and access to quality education and health care. Thus, higher rates of chronic disease are observed in these populations.[1] It is important to understand that not all minority populations live in unsafe housing, and many *do* have access to quality education and health care!

Health equity addresses the barriers from SDOH and provides the resources needed to create an opportunity for all people to obtain equity—for example, access to affordable health insurance. Therefore, health equity allows for everyone to be as healthy as possible, regardless of race, education, gender identity, sexual orientation, employment, neighborhood, and whether they have a disability.

The Nutrition and Health Equity feature in this textbook discusses some of the inequities present in America. Understanding these inequities and their complex contributing factors is essential for identifying long-term solutions.

chronic refers to diseases and many other medical conditions that are long term and, often difficult to treat

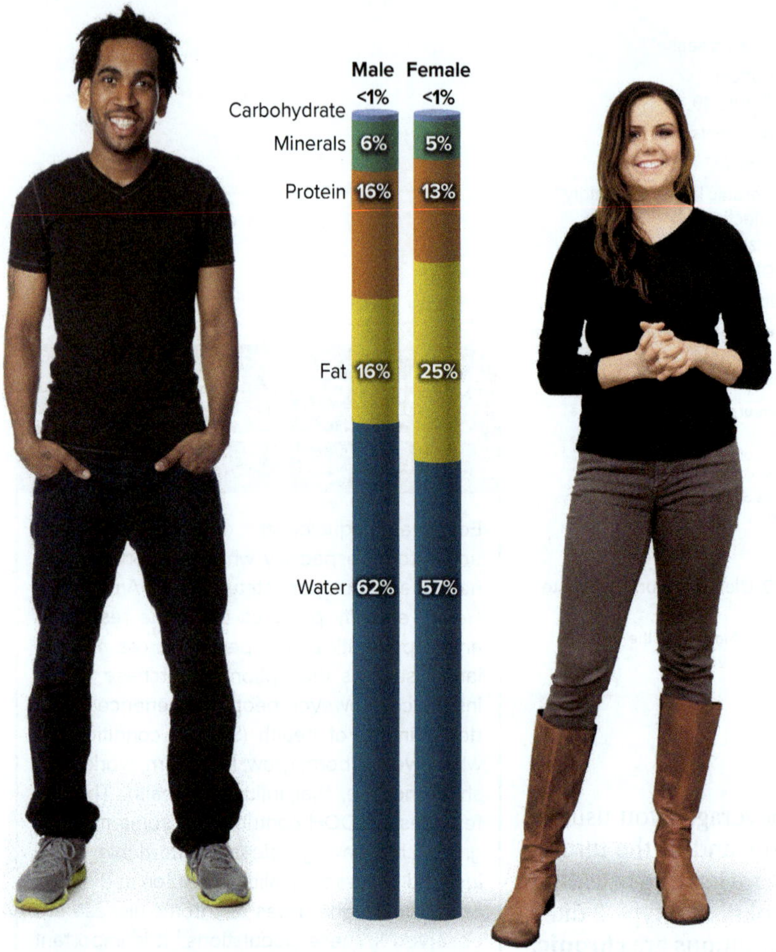

Figure 1.3 Comparing composition. These illustrations present the approximate percentages of nutrients that comprise the bodies of a healthy young male and female. Note that the amount of vitamins in the human body is so small, it is not shown.
Male: Blend/Image Source; Female: Aaron Roeth Photography/McGraw Hill

Each major section of a chapter ends with "Concept Checkpoints," a series of questions that can help you determine whether you understood the major concepts in the section. Answers to the "Concept Checkpoints" are given in Appendix F. At the end of each chapter, you will find the answers to the opener quiz, as well as a group of multiple–choice questions that test your understanding of the material in the chapter. The answers to those questions are provided in Appendix F. References for information cited in chapters are in Appendix G.

Introduction to Nutrition

Nutrition is the scientific study of nutrients, chemicals necessary for proper body functioning, and how the body uses them. Understanding nutrition requires learning about chemistry. **Chemistry** is the study of the composition and characteristics of matter, and changes that can occur to it. Matter is anything that takes up space and has mass or weight (on Earth). The air you breathe, this textbook, and even your body consist of chemicals and are forms of matter. "There are chemicals in our food!" This statement may sound frightening, but it is true. Food is matter; therefore, it contains chemicals, some of which are nutrients.

There are six classes of nutrients: carbohydrates, fats and other lipids, proteins, vitamins, minerals, and water. Your body is composed of these nutrients (Fig. 1.3). Although an average healthy young male and female have similar amounts of vitamins, minerals, and carbohydrates in their bodies, the young female has less water and protein, and considerably more fat than the male.

Table 1.1 presents major roles of nutrients in your body. In general, your body uses certain nutrients for energy, growth and development, and regulation of processes, including the repair and maintenance of cells. A **cell** is the smallest living functional unit in an organism, such as a human being. There are hundreds of different types of cells in your body. Cells do not need food to survive, but they need the nutrients in food to carry out their metabolic activities. **Metabolism** is the total of all chemical processes that occur in living cells, including chemical reactions (changes) involved in generating energy, making proteins, and eliminating waste products. You will learn more about metabolism in Chapter 4 (Body Basics).

Understanding nutrition also involves learning about human *physiology,* the study of how the body functions. Chapter 4 prepares you for the study of nutrition by presenting basic information about chemistry and human physiology. Chapters 5 (Carbohydrates), 6 (Fats and Other Lipids), 7 (Proteins), 8 (Vitamins), and 9 (Water and Minerals) provide information about the functions of nutrients in the body.

What Is an Essential Nutrient?

The body can synthesize (make) many nutrients, such as the lipids cholesterol and fat, but about 50 nutrients are essential in the diet. An **essential nutrient** must be supplied by food because the body does not synthesize the nutrient or make enough to meet its needs. Water is the most essential nutrient.

nutrition scientific study of nutrients and how the body uses these substances

chemistry study of the composition and characteristics of matter and changes that can occur to it

cell smallest living functional unit in an organism

metabolism total of all chemical processes that take place in living cells

essential nutrient nutrient that must be supplied by food

TABLE 1.1 Major Functions of Nutrients in the Body

Nutrient	Major Functions
Carbohydrates	Source of energy (most forms of carbohydrate)
Lipids	Source of energy (fat)
	Cellular development, physical growth and development
	Regulation of body processes (certain chemical messengers, for example)
	Absorption of certain vitamins
Proteins	Production of structural components, such as cell membranes, and functional components, such as enzymes
	Cellular development, growth, and maintenance
	Regulation of body processes (certain chemical messengers, for example)
	Transportation of many nutrients
	Immune function and fluid balance
	Source of energy
Vitamins	Regulation of body processes, including cell metabolism
	Maintenance of immune function, production and maintenance of tissues, and protection against agents that can damage cellular components
Minerals	Regulation of body processes, including fluid balance and metabolism; formation of certain chemical messengers; structural and functional components of various substances and tissues; physical growth, maintenance, and development
Water	Maintenance of fluid balance, regulation of body temperature, elimination of wastes, and transportation of substances
	Participant in many chemical reactions

There are three key features that help identify an essential nutrient:

- If the nutrient is missing from the diet, a **deficiency disease** occurs as a result. The deficiency disease is a state of health characterized by certain abnormal physiological changes. *(Physiological* refers to the functioning of the body.) Visible or measurable changes are referred to as *signs* of disease. Disease signs include rashes, failure to grow properly, and elevated blood pressure. *Symptoms* are subjective complaints of ill health that are difficult to observe and measure, such as dizziness, fatigue, and headache.
- When the missing nutrient is added to the diet, the abnormal physiological changes are corrected. As a result, signs and symptoms of the deficiency disorder resolve as normal functioning is restored, and the condition is cured.
- After scientists identify the nutrient's specific roles in the body, they can explain why the abnormalities occurred when the substance was missing from the diet.

deficiency disease state of health that occurs when a nutrient is missing from the diet

If you wanted to test your body's need for vitamin C, for example, you could avoid consuming foods or vitamin supplements that contain the vitamin. When the amount of vitamin C in your cells became too low for them to function normally, you would develop physical signs of *scurvy,* the vitamin C deficiency disease. Early in the course of the deficiency, tiny red spots that are actually signs of bleeding under the skin (tiny bruises) would appear where the elastic bands of your clothing applied pressure. When you brushed your teeth, your gums would bleed from the pressure of the toothbrush. If you cut yourself, the wound would heal slowly or not

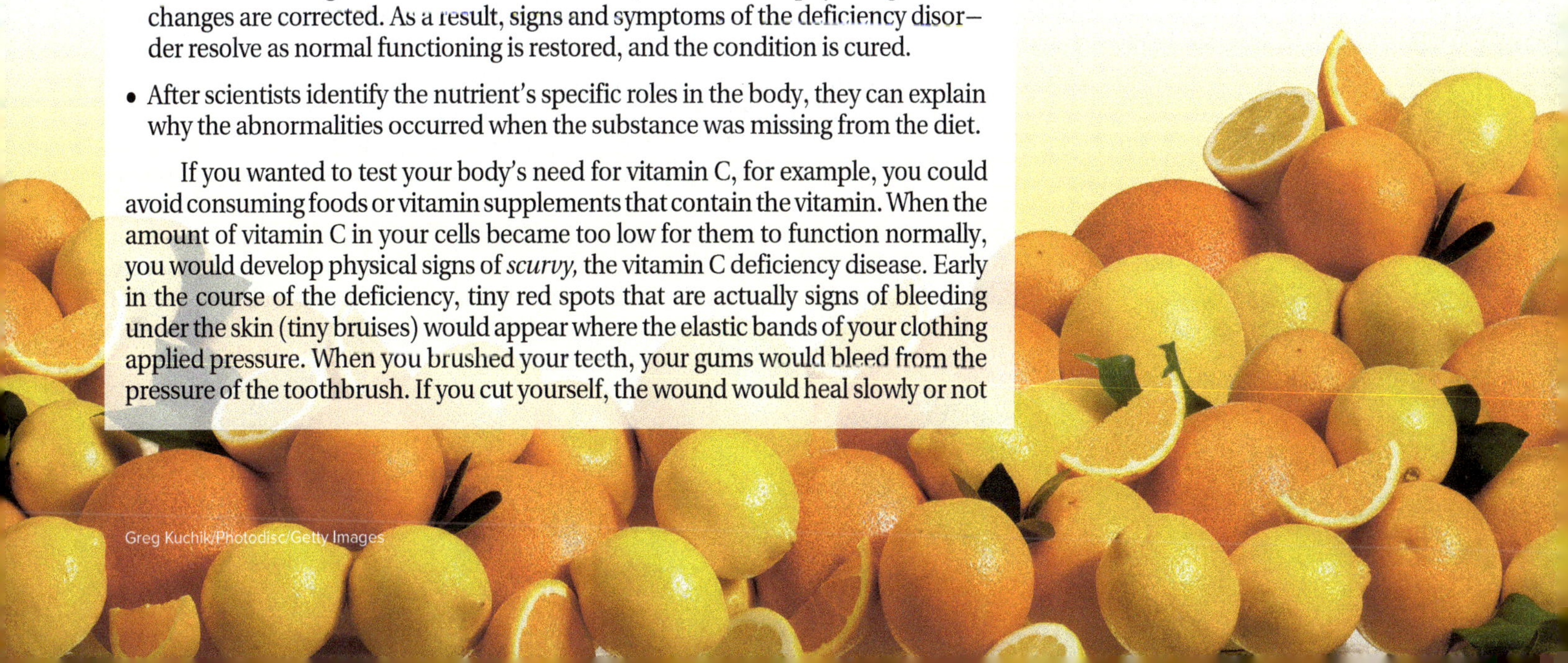

Greg Kuchik/Photodisc/Getty Images

TABLE 1.2 Essential Nutrients for Humans

Water	**Glucose†**	**Fats that contain linoleic and alpha-linolenic acids**
Vitamins:	**Minerals:**	
A	Calcium	
B vitamins	Chloride	
Thiamin	Chromium	**Components of proteins (amino acids):**
Riboflavin	Copper	
Niacin	Iodine	
Pantothenic acid	Iron	
Biotin	Magnesium	Histidine
Folate	Manganese	Isoleucine
B-6	Molybdenum	Leucine
B-12	Phosphorus	Lysine
C	Potassium	Methionine
Choline*	Selenium	Phenylalanine
D**	Sodium	Threonine
E	Sulfur	Tryptophan
K	Zinc	Valine

*The body makes choline but may not make enough to meet needs. Often classified as a *vitamin-like* compound.

**The body makes vitamin D after exposure to sunlight, but a dietary source of the nutrient is often necessary.

†A source of glucose is needed to supply the nervous system with energy and spare protein from being used for energy.

at all. If you started consuming vitamin C–containing foods again, the deficiency signs and symptoms would disappear within a few days as your body recovered. By reading about vitamin C in Chapter 8, you will learn that one of the physiological roles of vitamin C is maintaining a substance in your body that literally holds cells together. This substance is also needed to produce scar tissue for wound healing. Thus, vitamin C meets all the required features of an essential nutrient.

Table 1.2 lists nutrients that are generally considered to be essential. Fortunately, the human body is designed to obtain these substances from a wide variety of foods. Chapter 3 provides information about ways to plan nutritious diets.

What Are Nonnutrients? Some foods contain nonnutrients—substances that are not nutrients, yet they may have healthful benefits. Plants make hundreds of nonnutrients called **phytochemicals** (*phyto* = plant), which give plants, including fruits and vegetables, their specific color, taste, and aroma. The phytochemical anthocyanin gives strawberries their red color, and the phytochemical allicin contributes to the distinct taste and odor of garlic. Many phytochemicals are antioxidants that may reduce risks of heart disease and certain cancers. An **antioxidant** protects cells and their components from being damaged or destroyed by exposure to certain harmful environmental and internal factors. Some vitamins also function as antioxidants (see Chapter 8).

Not all phytochemicals have beneficial effects on the body. Some phytochemicals, such as nicotine in tobacco leaves, ricin in castor beans, and oxalic acid in rhubarb leaves, are toxic or can interfere with the absorption of nutrients. Scientific research that explores the effects of phytochemicals on the body is ongoing. Table 1.3 lists several phytochemicals that are currently under scientific investigation, identifies rich food sources of these compounds, and indicates their biological effects on the body, including possible health benefits. Note that few animal foods are sources of phytochemicals.

phytochemicals compounds made by plants that are not nutrients

antioxidant substance that gives up electrons to radicals to protect cells

dietary supplements nutrient preparations, certain hormones, and herbal products that are taken to increase total dietary intake

Dietary Supplements

Many Americans take dietary supplements such as vitamin pills and herbal extracts to improve their health. The Dietary Supplement Health and Education Act of 1994 (DSHEA) allows manufacturers to classify nutrient supplements and certain herbal products as foods.[3] The DSHEA defines a **dietary supplement** as a product (excluding tobacco) that contains a vitamin, a mineral, an herb or other plant product, an amino acid, or a dietary substance that supplements the diet by increasing total intake. According to scientific evidence, some dietary supplements, such as vitamins and certain herbs, can have beneficial effects on health. However, results of scientific testing also indicate that many popular dietary supplements are not helpful and may even be harmful. The "Nutrition Matters" section in Chapter 2 ("What Are Dietary Supplements?") discusses dietary supplements. Information about specific dietary supplements, including those that contain phytochemicals, is also woven into chapters where it is appropriate.

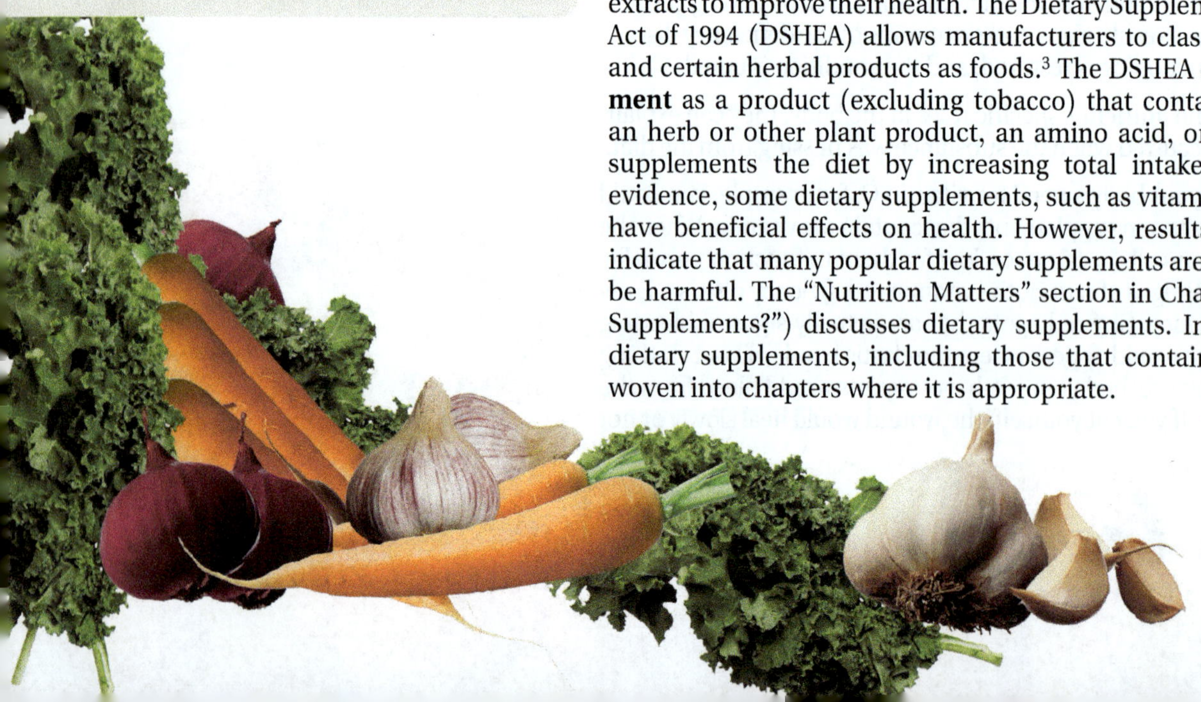

Carrots: Isabelle Rozenbaum/Photo Alto/SuperStock; Kale, purple striped garlic bulbs: Stockdisc/Getty Images; Garlic cloves, garlic bulb, red onion: C Squared Studios/Photodisc/Getty Images

TABLE 1.3 Phytochemicals of Scientific Interest

Classification and Examples	Rich Food Sources	Biological Effects/Possible Health Benefits
Carotenoids		Antioxidant activity; may reduce risk of macular degeneration (a major cause of blindness)
Alpha-carotene, beta-carotene, lutein, lycopene, zeaxanthin	Orange, red, yellow fruits and vegetables; egg yolks; liver	
Phenolics		Antioxidant activity; may inhibit cancer growth, may reduce risk of heart disease
Quercetin	Apples, tea, red wine, onions, olives, raspberries, cocoa	
Catechins	Tea, chocolate, plums, apples, berries, pecans	
Anthocyanins	Red, blue, or purple fruits and vegetables	
Resveratrol	Red wine, purple grapes and grape juice, dark chocolate	
Isoflavonoids	Soybeans and other legumes	
Lignans	Flaxseed, berries, whole grains, bran, nuts	
Ellagic acid	Raspberries, strawberries, cranberries, walnuts, pomegranates	
Organosulfides		Antioxidant effects; may improve immune system functioning and reduce the risk of heart disease
Isothiocyanates, indoles, allylic sulfur compounds, including allicin	Garlic, onions, cruciferous vegetables (broccoli, cauliflower, cabbage, kale, bok choy, collard and mustard greens)	
Alkaloids		Stimulant effects
Caffeine	Coffee, tea, kola nuts, cocoa	
Capsaicinoids		May provide some pain relief
Capsaicin	Chili peppers	
Fructooligosaccharides	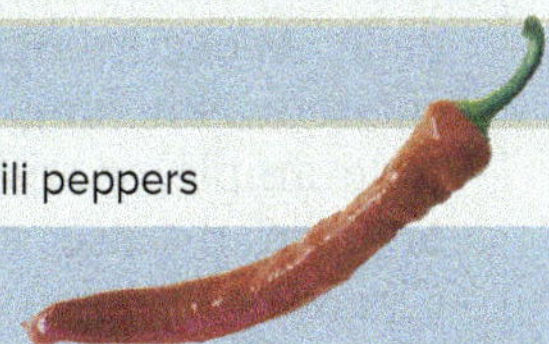	May stimulate the growth of beneficial bacteria in the human intestinal tract
	Onions, bananas, asparagus, wheat	

Apple, strawberries: Burke Triolo Productions/Artville/Getty Images; Raspberries: Burke/Triolo Productions/Getty Images; Plums: Scott Bauer/ARS/U.S. Department of Agriculture (USDA); Red grapes: Jules Frazier/Photodisc/Getty Images; Almonds: C Squared Studios/Photodisc/Getty Images; Red chili pepper: Heide Benser/Getty Images

Concept CHECKPOINT

1. Identify at least two of the ten leading causes of death that are diet related.
2. Identify at least four factors that influence your eating habits.
3. List the six major classes of nutrients.
4. What are three key factors that determine whether a substance is an essential nutrient?
5. Define *phytochemical* and *dietary supplement.*

See Appendix F for responses.

Did *YOU* Know?

Your genetic makeup influences the effects of diet on your health, as well as susceptibility to some diseases including obesity, heart disease, diabetes, cancer, and specific food allergies. *Nutritional genomics* is a relatively new area of nutrition research that explores complex interactions among gene functioning, diet and other lifestyle choices, and the environment. Section 7.9 provides more information about nutritional genomics.

risk factor personal characteristic that increases a person's chances of developing a disease

lifestyle usual way of living, including dietary practices and physical activity habits

1.2 Factors That Influence Americans' Health

Learning Outcomes

1. Explain why people should be concerned about their lifestyle and risk factors for chronic diseases.
2. Compare Americans' current typical diet to the population's typical eating patterns in 1970.
3. Identify the main nutrition-related goal of *Healthy People 2030.*

As mentioned in the beginning of this chapter, poor eating habits contribute to several of the leading causes of death in the United States. As we showed in Figure 1.2, heart disease is the leading cause of death for all Americans, and cancer is the second leading cause of death. In 2021, these two diseases accounted for almost 38% of all deaths.[4]

Conditions such as heart disease and cancer are *chronic* diseases. Chronic diseases usually take many years to develop and have complex causes. A **risk factor** is a personal characteristic that increases your chances of developing a chronic disease. For example, genetic background or family history is an important risk factor for heart disease. If your father's father had a heart attack before he was 55 years old and your mother is being treated for having a high blood cholesterol level (a risk factor for heart disease), your family history indicates you have a higher-than-average risk of having a heart attack. For many people, however, having a family history of a chronic disease does not mean that they definitely will develop the condition. Other risk factors that contribute to health are age, environmental conditions, psychological factors, access to health care, and lifestyle practices.

Lifestyle is a person's usual way of living that includes dietary practices, physical activity habits, use of drugs such as tobacco and alcohol, and other typical patterns of behavior. Your lifestyle may increase or reduce your chances of developing a chronic disease or delay its occurrence for years, even decades. Poor diet, cigarette smoking, and excess alcohol consumption, for example, are risk factors that increase the likelihood of heart disease, stroke, and many forms of cancer. Cigarette smoking is the primary cause of preventable cancer deaths, but dietary habits and physical activity patterns also contribute to the development of certain cancers.[5,6] Additionally, poor diet and lack of physical activity can result in obesity, a condition characterized by the accumulation of too much body fat. Obesity is a risk factor for numerous health problems, including heart disease, certain cancers, type 2 diabetes, and *hypertension* (chronic high blood pressure).

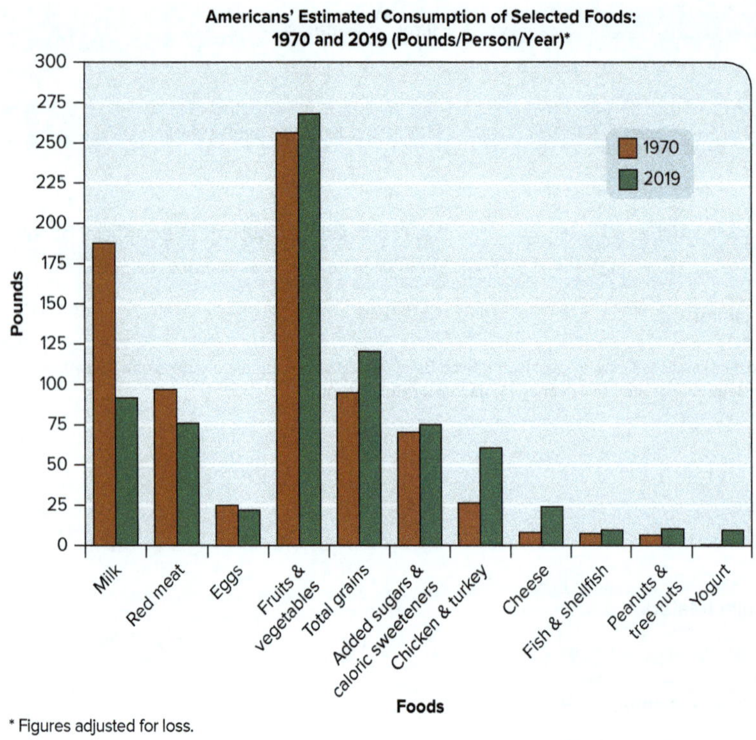

Figure 1.4 Americans' changing eating habits, comparing 1970 to 2019.
Source: U.S. Department of Agriculture, Economic Research Service: *Food availability (per capita) data system.* 2019. www.ers.usda.gov/data-products/food-availability-per-capita-data-system/food-availability-per-capita-data-system/#Loss-Adjusted%20Food%20Availability. Accessed: March 25, 2023.

Our Changing Eating Habits

Americans' diets have changed over the past 50 years (Fig. 1.4). In general, the population now eats more peanuts and tree nuts, poultry, fish and shellfish, yogurt, and cheese than amounts that they ate in 1970. Nevertheless, Americans have reduced their intakes of other protein-rich foods, including fluid milk, red meat (beef, veal, pork, and lamb), and eggs, since 1970. Although today's Americans consume more fruits and vegetables, they do not eat the recommended amounts.[7] Compared to amounts eaten in the past, Americans now consume more foods that contain too much added sugars. The population's current diet also supplies more food energy from total grains, including flour and cereal products, than in 1970, but foods made from *refined* grains, such as white bread and pizza dough, make up the majority of these products.

Raw foods often undergo some form of processing, such as refining, canning, freezing, or cooking, before they are eaten. Processing can make a food more nutritious, safer to eat, and less likely to spoil. However, some forms of processing remove nutrients and phytochemicals that were naturally in the food. Other forms of processing add unhealthy amounts of sodium, sugar, and certain fats to foods.

In 1987–1988, adult Americans consumed an average of 1785 Calories/day (excluding calories from alcohol). By 2017–March 2020 (prepandemic), adult Americans consumed an average of 2144 Calories/day (excluding calories from alcohol). Thus, adults' average food energy intake increased by almost 17% between the two time periods.[8,9] If a person's energy intake is more than needed, especially for physical activity, their body fat increases. Nationwide surveys indicate that Americans have more body fat than in previous decades. Dietary practices, however, should not receive all the blame for this unhealthy finding; during the same period, we have become increasingly dependent on various labor–saving gadgets and machines that make our lives easier but also reduce the amount of energy we need to expend to avoid unwanted weight gain. Chapter 10 examines weight management in detail.

Healthy People

Since the late 1970s, health promotion and disease prevention have been the general focus of public health efforts in the United States. A primary component of such efforts is developing educational programs that can help people prevent chronic and infectious diseases, birth defects, and other serious health problems. In many instances, it is more practical and less expensive to prevent a serious health condition than to treat it.

In 2020, the U.S. Department of Health and Human Services (HHS) issued *Healthy People 2030,* a report that includes national health promotion and disease prevention objectives to be met by 2030. *Healthy People 2030* overarching goals encourage Americans to:

- attain healthy, thriving lives and well–being that are free of preventable disease, disability, injury, and premature death;
- eliminate health disparities, achieve health equity, and attain health literacy to improve the health and well–being of all;
- create social, physical, and economic environments that promote attaining the full potential for health and well–being for all;
- promote healthy development, healthy behaviors, and well–being across all life stages; and
- engage leadership, key constituents, and the public across multiple sectors to take action and design policies that improve the health and well–being of all.[12]

The main nutrition–related goal of *Healthy People 2030* is to improve health by promoting healthy eating and making nutritious foods available. To help meet this goal, *Healthy People 2030* has several nutrition–related objectives, some of which are listed in Table 1.4. More information about these and other nutrition objectives are at the government's website (https://health.gov/healthypeople/objectives–and–data/browse–objectives/nutrition–and–healthy–eating).

Nutrition & Health *Equity*

People who have regular professional medical care may reduce their chances of developing serious chronic diseases that lead to disability, lower quality of life, and premature (early) death. However, people who lack health insurance are more likely to skip routine medical care because they cannot afford its high cost. As a result, these persons have a higher risk of becoming ill with chronic and debilitating diseases that may have been preventable. It is important to address health care inequities by increasing access to both health insurance and regular, high-quality medical care to improve and maintain good health for all Americans.

The Affordable Care Act of 2010 increased health insurance coverage for the uninsured, especially people who identify themselves as Black, Hispanic, Asian, American Indian, and Alaska Natives. In 2020, most Americans (91%) who were younger than 65 years of age had health insurance (public or private plans).[10] Individuals with Asian ancestry were more likely to have health insurance (92%). However, some minority populations still had lower rates of health insurance coverage than the majority of the population, including those who identified themselves as Black (88%), Hispanic or Latino (78%), and American Indian or Alaska Native (72%).[11] These lower rates of health insurance are due to the cost of insurance still being too high for some people, others not attempting to obtain health insurance because they think they cannot afford it, and some states not expanding Medicaid insurance coverage to include low-income people.[11] Medicaid coverage was previously only for older adults and people with disabilities.

People with low incomes that need regular health care often face the question of what is more important—buying nutritious food or paying for physical exams and prescription medications. The "Nutrition Matters" section of this chapter discusses some major food programs that the U.S. government sponsors for low–income populations.

Donenko Oleksii/Shutterstock

Image Source/Getty Images

TABLE 1.4 Some *Healthy People 2030* Objectives: Nutrition and Healthy Eating

Number	Objective
03	Reduce the proportion of adults who are obese
04	Reduce the proportion of obese children and adolescents
06	Increase fruit intake of people who are 2 years of age and older
07	Increase the intake of vegetables by people who are 2 years of age and older
09	Increase whole grain intake by people who are 2 years of age and older
10	Reduce intake of added sugars by people who are 2 years of age and older
12	Reduce sodium intake by people who are 2 years of age and older
13	Increase calcium intake by people who are 2 years of age and older

Source: https://health.gov/healthypeople/objectives-and-data/browse-objectives/nutrition-and-healthy-eating

Concept CHECKPOINT

6. What is a risk factor?
7. Explain how your lifestyle can affect your health.
8. Discuss how Americans' food choices have changed since 1970.
9. Identify the main nutrition-related goal of *Healthy People 2030*.

See Appendix F for responses.

1.3 Metrics for Nutrition

Learning Outcomes

1. Identify basic units of the metric system often used in nutrition.
2. Use the caloric values of macronutrients and alcohol to estimate the amount of energy (kcal) in a serving of food.
3. Estimate the percentage of calories that each macronutrient contributes in a serving of food.

Scientists classify specific nutrients according to their chemical composition and major functions in the body. Nutrients can also be classified based on how much of them are in food. Americans usually refer to length in terms of inches and feet, weight in pounds, and amounts of food in familiar household measures (e.g., teaspoons, tablespoons, cups). However, scientists and health care professionals generally use metric values to report length (***meter***), weight (***gram***), and volume (***liter***). The following section provides a basic review of the metric system. Appendix A provides common English–to–metric and metric–to–household unit conversions.

TABLE 1.5 Common Metric Prefixes in Nutrition

kilo- (k) = one thousand (1000)
deci- (d) = one-tenth (0.1)
centi- (c) = one-hundredth (0.01)
milli- (m) = one-thousandth (0.001)
micro- (μ or mc) = one-millionth (0.000001)

Metric Basics

The metric prefixes ***micro–***, ***milli–***, ***centi–***, ***deci–***, and ***kilo–*** indicate whether a measurement is a fraction or a multiple of a meter (m), gram (g), or liter (l or L) (Table 1.5).

There are approximately 2.54 centimeters (cm) per inch. To obtain your approximate height in centimeters, multiply your height in inches by 2.54. For example, a person who is 5′5″ in height (65″) measures about 165 cm (65 × 2.54) in length. There are approximately 28 g in an ounce and 454 g in a pound. A kilogram (***kilo*** = 1000) equals 1000 g or about 2.2 pounds. To determine your weight in kilograms (kg), divide your weight in pounds by 2.2. A person who weighs 130 pounds, for example, weighs about 59 kg.

Assume that a small raisin weighs 1 gram. If you cut this raisin into 1000 equal pieces, then each piece weighs 1 milligram (***milli*** = 1000). Thus, 1000 milligrams (mg) equal 1 gram (g). Imagine cutting a small raisin into 1 million equal pieces. Each piece of raisin would weigh 1-millionth of a gram, or a microgram (µg or mcg). Amounts of nutrients in blood are often reported as the number of milligrams or micrograms of the substance per deciliter of blood. For example, a normal blood glucose level for a healthy fasting person is 90 milligrams/deciliter (90 mg/dl).

A kilogram weighs about 2.2 pounds.
Comstock Images/Alamy Stock Photo

What Is a Calorie?

Running, sitting, studying—your body uses energy even while sleeping. Every cell in your body needs energy to carry out its various activities. As long as you are alive, you are constantly using energy. A calorie is the heat energy necessary to raise the temperature of 1 g (1 ml) of water 1° Celsius (C). A calorie is such a small unit of measurement, the amount of energy in food is reported in 1000-calorie units called kilocalories or Calories. Thus, a **kilocalorie (kcal)** or **Calorie** is the heat energy needed to raise the temperature of 1000 g (a liter) of water 1° Celsius (C). A small apple, for example, supplies 40,000 calories or 40 kcal or 40 Calories. If no number of kilocalories is specified, it is appropriate to use "calories." In this textbook, the term *calories* is interchangeable with *food energy* or simply *energy*.

A gram of carbohydrate (sugars and starches) and a gram of protein each supply about 4 kcal; a gram of fat provides about 9 kcal (Fig. 1.5). Although alcohol is not a nutrient, it does provide energy; a gram of pure alcohol furnishes 7 kcal. If you know how many grams of carbohydrate, protein, fat, and/or alcohol are in a food or a meal, you can estimate the number of kilocalories it provides. For example, if a serving of food contains 15 g of carbohydrate, 3 g of protein, and 7 g of fat, multiply 15 g by 4 (the number of kcal each gram of carbohydrate supplies). Next, multiply 3 g by 4 (the number of kcal each gram of protein provides). Then multiply 7 g by 9 (the number of kcal each gram of fat supplies). By adding the three caloric values (60 kcal from carbohydrate, 12 kcal from protein, and 63 kcal from fat), you will determine that this food provides 135 kcal/serving.

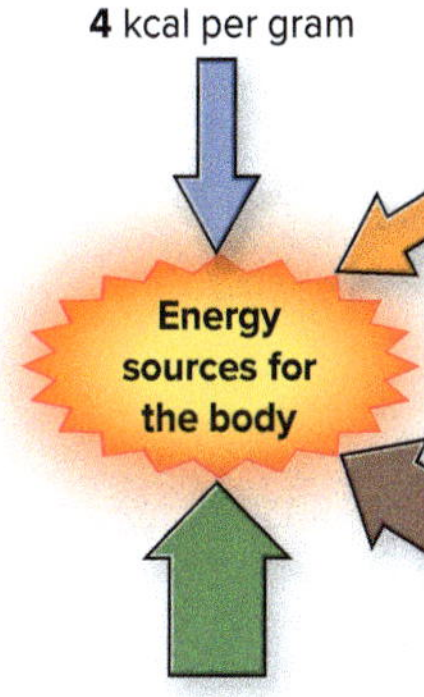

Figure 1.5 Energy sources for the body. Most forms of carbohydrate supply 4 kcal/g; protein also provides 4 kcal/g. Fat supplies 9 kcal/g, and alcohol (a nonnutrient) provides 7 kcal/g.

Macronutrients and Micronutrients

Carbohydrates, fats, and proteins are referred to as **macronutrients** because the body needs relatively large amounts (grams) of these nutrients daily. Vitamins and minerals are **micronutrients** because the body needs very small amounts (milligrams or micrograms) of them to function properly. In general, a serving of food supplies grams of carbohydrate, fat, and protein, and milligram or microgram quantities of vitamins and minerals. It is important to understand that macronutrients supply energy for cells, whereas micronutrients do not. Although the body requires large amounts of water, this nutrient provides no energy and is not usually classified as a macronutrient.

Amounts of nutrients present in different foods vary widely, and even the same food from the same source can contain different amounts of nutrients. Therefore, food composition tables and nutrient analysis software generally indicate average amounts of nutrients in foods. By using these tools, however, you can obtain approximate values for each nutrient measured and estimate your nutrient intake.

kilocalorie (kcal) or **Calorie** heat energy needed to raise the temperature of 1 liter of water 1° Celsius; measure of food energy

macronutrients nutrients needed in gram amounts daily and that provide energy; carbohydrates, proteins, and fats

micronutrients vitamins and minerals

Slice of sunflower seed bread
©Wendy Schiff

Concept CHECKPOINT

10. Health professionals generally use which metric values to report volume, weight, and length?

11. A person weighs 198 pounds. How many kilograms does this person weigh?

12. A slice of sunflower seed bread supplies approximately 10 g of wheat starch, 1 g of sugar, 3 g of fat, 3 g of protein, and 10 g of water. Based on this information, estimate the number of kilocalories this food provides.

13. Which nutrients are classified as macronutrients? Which are classified as micronutrients?

See Appendix F for responses.

1.4 Key Nutrition Concepts

Learning Outcomes

1. Give examples of foods that contain a lot of empty calories and foods that are energy-dense and/or nutrient-dense.
2. Discuss key basic nutrition concepts, such as the importance of eating a variety of foods and why food is the best source of nutrients.

Before learning about the nutrients and their roles in health, it is important to grasp some key basic nutrition concepts (Table 1.6). The content in the chapters that follow will build upon these key concepts and can help you make more informed choices concerning your dietary practices.

TABLE 1.6 Key Basic Nutrition Concepts

• Most naturally occurring foods are mixtures of nutrients.
• Variety can help ensure the nutritional adequacy of a diet.
• There are no "good" or "bad" foods.
• Enjoy eating all foods in moderation.
• For each nutrient, there is a range of safe intake.
• Food is the best source of nutrients and phytochemicals.
• There is no "one size fits all" approach to planning a nutritionally adequate diet.
• Foods and the nutrients they contain are not cure-alls.
• Malnutrition includes *under*nutrition as well as *over*nutrition.
• Nutrition is a dynamic science.

Concept 1: Most Naturally Occurring Foods Are Mixtures of Nutrients

Which foods do you think of when you hear the words *protein or carbohydrate*? You probably identify meat, milk, and eggs as sources of protein; and potatoes, corn, and rice as sources of carbohydrate. Most naturally occurring foods, however, are mixtures of nutrients. In many instances, water is the major nutrient in foods. For example, 8 fluid ounces of fat–free milk is 99% water by weight, but it is an excellent source of protein and supplies carbohydrate and several vitamins and minerals. A 6–ounce raw white potato is 79% water and only about 18% carbohydrate by weight. The potato also supplies iron and potassium (minerals) and vitamins C and niacin.

Most commercially processed foods are also mixtures of nutrients. Sweet snacks, for example, are sources of nutrients other than sugar, a carbohydrate. Although sugar comprises about 44% of the weight of a chocolate with almonds candy bar, over one–third of the sweet snack's energy is from fat. The candy bar also contains small amounts of protein, iron, calcium, vitamin A, and the B–vitamin riboflavin. Figure 1.6 compares the energy, water, protein, carbohydrate, fat, and calcium contents of raw broccoli (1 cup), rice with vegetables (3/4 cup), 8 fluid ounces of fat–free milk, and a chocolate–frosted, cake–type doughnut (3.5–inch diameter). You can find information about the energy and nutrient contents of foods by searching the "What's in the Foods You Eat *Search Tool*" that is available at the U.S. Department of Agriculture's website: www.usda.gov.

Concept 2: Variety Can Help Ensure the Nutritional Adequacy of a Diet

No natural food is "perfect" in that it contains all nutrients in amounts that are needed by the body. To help ensure the nutritional adequacy of your diet,

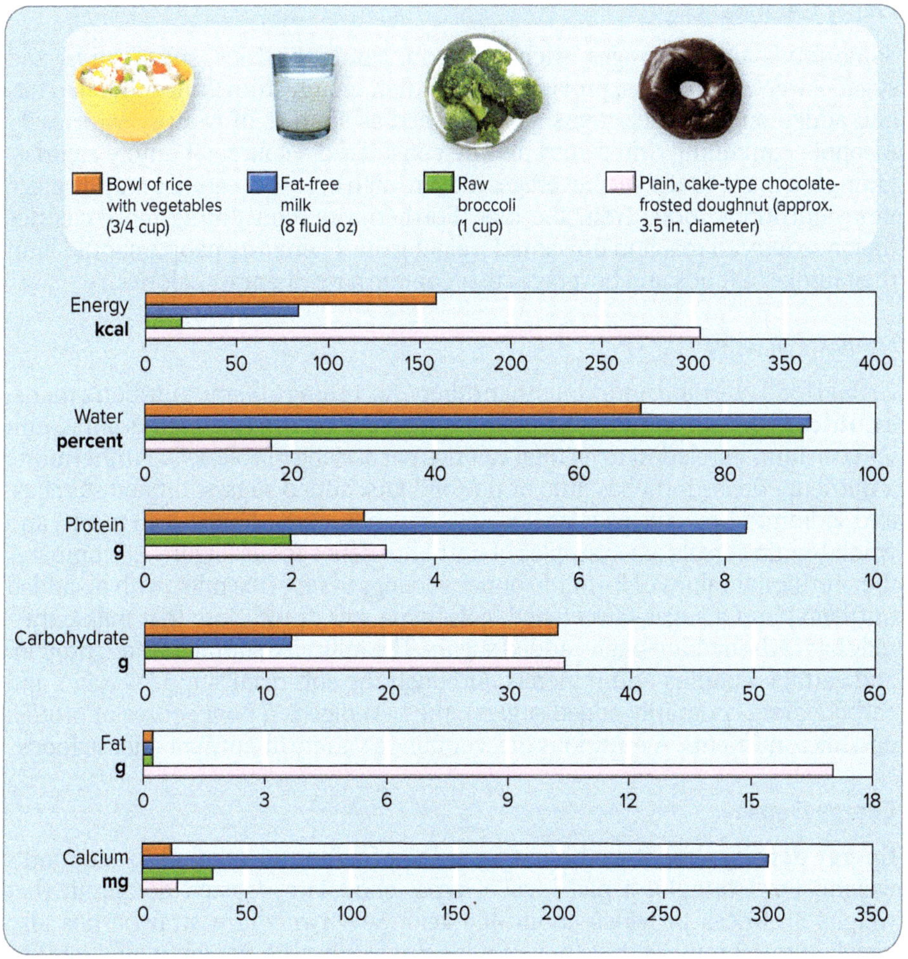

Figure 1.6 Energy and nutrient comparison. Most foods are mixtures of nutrients. A serving of rice with vegetables, fat-free milk, raw broccoli, and a plain, cake-type, chocolate-frosted doughnut contribute very different amounts of energy, water, protein, carbohydrate, fat, and calcium to diets.
Rice and vegetables: spaxiax/Shutterstock; Milk: Christopher Kerrigan/McGraw Hill; Broccoli: spafra/iStock/Getty Images; Doughnut: ©Wendy Schiff

choose a diet that contains a variety of foods from each food group that is shown in MyPlate (Fig. 1.7). Chapter 3 (Planning Nutritious Diets) provides information about the food groups and practical menu planning tools, such as MyPlate.

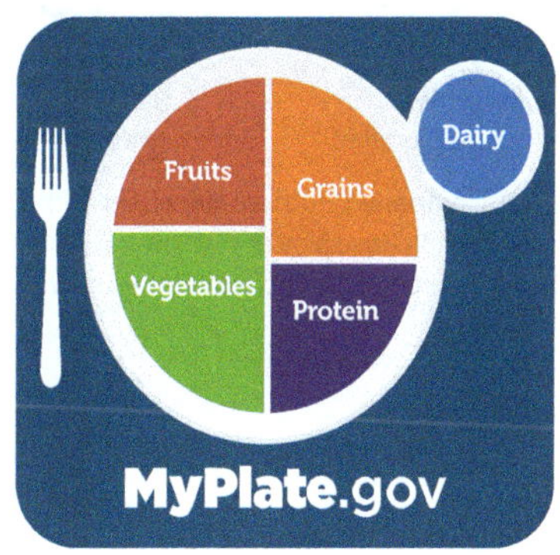

Figure 1.7 MyPlate.
U.S. Department of Agriculture

Concept 3: There Are No "Good" or "Bad" Foods

Are some foods "good" and others "bad" for your body? If you think there are such foods, which ones are good and which are bad? Do you sometimes feel guilty about eating "junk foods"? What is a junk food? Should pizza, chips, candy, doughnuts, and ice cream be classified as junk food?

All foods have nutritional value. For example, many people think pumpkin pie is a junk food. Pumpkin pie, however, is a good source of protein, the mineral iron, and the phytochemical beta–carotene that the body can convert to vitamin A. Although pies, doughnuts, and ice cream contain a lot of fat and added sugar, these foods may also supply small amounts of protein, vitamins, and minerals to diets. Healthy diets, however, generally contain limited amounts of such foods.

A food *is* bad for you if it contains toxic substances or is contaminated with bacteria, viruses, or microscopic animals that cause food–borne illness. Chapter 12 (Food Safety Concerns) focuses on food safety concerns, including major types of food–borne illnesses and ways to prevent them.

Because pumpkin pie is made with eggs and milk, it is a good source of protein.
©Wendy Schiff

TABLE 1.7 Nutrient Density: Key Beneficial Nutrients

Protein
Fiber
Vitamins: A, C, and E
Minerals: Calcium, iron, potassium, and magnesium

Source: Drewnowski A, Fulgoni VL (III): Nutrient density: Principles and evaluation tools. *American Journal of Clinical Nutrition* 99(suppl):1223S, 2014.

8 fluid ounces
Fat-free milk, not fortified

8 fluid ounces
Sugar-sweetened soft drink

Figure 1.8 Comparing nutrient densities. Although the milk and soft drink have similar water contents, milk is a much better source of protein and certain micronutrients.
©Wendy Schiff

Empty Calories

Some foods and beverages, such as bacon, candy, pastries, snack chips, and sugar–sweetened drinks, provide much of their energy from unhealthy solid fats and added sugars. Such items are described as sources of **"empty calories."**[13] Alcohol–containing drinks may also be considered a source of empty calories. Eating too many foods and beverages that are high in empty calories may displace more nutritious foods from the diet. Furthermore, consuming more calories than needed can result in unwanted weight gain. Therefore, people should limit their intake of foods and beverages that contain a lot of empty calories.

Nutrient Density

Certain foods are more nutritious than others. According to some nutrition experts, a **nutrient–dense** food supplies more key beneficial nutrients, such as potassium and calcium, in relation to its total calories per serving (Table 1.7). Furthermore, a nutrient–dense food has little or no solid fats, added sugars, refined starches, and sodium.[14] Broccoli, leafy greens, fat–free milk, orange juice, lean meats, and whole–grain cereals are examples of nutrient–dense foods. Figure 1.8 compares the nutritional values of 8–fluid–ounce servings of fat–free milk (with no added nutrients) and a sugar–sweetened, cola–type soft drink. Note that milk naturally has carbohydrate but no added sugars. The milk also supplies water, protein, and certain vitamins and minerals. Although the soft drink supplies water and carbohydrate (primarily added sugars), the beverage is a poor source of protein and micronutrients. A nutritious diet contains a variety of nutrient–dense foods.

Energy Density

Energy density describes the energy value of a food in relation to the food's weight. For example, a plain, cake–type, chocolate–frosted doughnut that weighs 3 ounces provides about 400 kcal. Seven medium strawberries also weigh about 3 ounces, but they provide only 28 kcal. A person would have to eat 100 of the strawberries to obtain the same amount of food energy that is in the doughnut (Fig. 1.9). Therefore, the doughnut is an energy–dense food in comparison to the berries. In general, high–fat foods such as doughnuts are energy dense because they are concentrated sources of energy. Most fruits are not energy dense because they contain far more water than fat.

Not all energy–dense foods are high in empty calories. Nuts, for example, are high in healthy kinds of fat and, therefore, energy dense. However, nuts are

3-ounce chocolate cake-type doughnut with chocolate frosting

100 medium strawberries

Figure 1.9 Energy density. A person would need to eat 100 medium strawberries to obtain the same amount of energy as in this chocolate doughnut.
Doughnut, strawberries: ©Wendy Schiff

empty calories energy supplied by unhealthy solid fats, added sugars, and/or alcohol

nutrient dense describes a food or beverage that contains more key beneficial nutrients in relation to its total calories

energy density energy value of a food in relation to the food's weight

also nutrient dense because they contribute protein, vitamins, minerals, and fiber to diets. Most forms of fiber are classified as carbohydrates.

A nutritionally adequate diet can include foods such as brownies when such foods are consumed in moderation. Michael Lamotte/Cole Group/Photodisc/Getty Images

Concept 4: Enjoy Eating All Foods in Moderation

Dietary **moderation** involves obtaining enough nutrients from food to meet one's needs and balancing calorie intake with calorie expenditure, primarily by physical activity. This can be accomplished by choosing nutrient-dense foods, limiting serving sizes of energy-dense foods, and incorporating moderate- to vigorous-intensity physical activities into your daily routine. Although moderation requires planning meals and setting aside time for physical activity daily, it can help you achieve your health and fitness goals. If, for example, you overeat during a meal or snack, you can regain dietary moderation and balance by eating less energy-dense food and exercising more intensely during the next 24 hours.

Eliminating all sources of empty calories from your diet is not generally recommended or necessary. If your diet is comprised primarily of nutrient-dense foods and meets your nutritional needs, including some items that supply empty calories adds enjoyment to living when they are consumed in moderation. Physically active individuals, such as athletes in training programs, often find it difficult to consume enough energy from nutrient-dense foods to sustain healthy body weights, unless they include some foods that are sources of empty calories in their diets.

Concept 5: For Each Nutrient, There Is a Range of Safe Intake

By eating a variety of nutrient-dense foods, you are likely to obtain adequate and safe amounts of each nutrient. The **physiological dose** of a nutrient is the amount that is within the range of safe intake and enables the body to function optimally. Consuming less than the physiological dose can result in marginal nutritional status. In other words, the person's body has just enough of the nutrient to function adequately, but that amount is not sufficient to overcome the added stress of infection or injury. If a person's nutrient intake falls below the marginal level, the individual is at risk of developing the nutrient's deficiency disease. For example, recommended daily intakes of the B-vitamin niacin are 16 mg for males and 14 mg for females. People whose diets contain little or no niacin are at risk of developing *pellagra,* the vitamin's deficiency disease.

Most people require physiological amounts of micronutrients. A **megadose** is an amount of a vitamin or mineral that greatly exceeds the recommended amount of the nutrient. When taken in high amounts, many vitamins act like drugs and can produce unpleasant and even toxic side effects. For example, physicians sometimes use megadoses of the B-vitamin niacin to treat high blood cholesterol levels, but such amounts may cause painful facial flushing and liver damage. Minerals have very narrow ranges of safe intakes.

Many consumers take megadoses of vitamin and/or mineral supplements without consulting physicians because they think the micronutrients will prevent or treat ailments such as the common cold or heart disease. For most people, consuming amounts of nutrients that exceed what is necessary for good health is economically wasteful and could be harmful to the body. "More is not always better," when it relates to optimal nutrition.

In their natural states, most commonly eaten foods do not contain toxic levels of vitamins and minerals. You probably do not need to worry about consuming toxic levels of micronutrients, unless you are regularly eating large amounts of foods that are fortified with these nutrients. The diagram shown in Figure 1.10 illustrates the general concept of deficient, safe, and toxic intake ranges for nutrients such as vitamins and minerals. Chapters 8 and 9 provide more information about micronutrients, including deficiencies and toxicities.

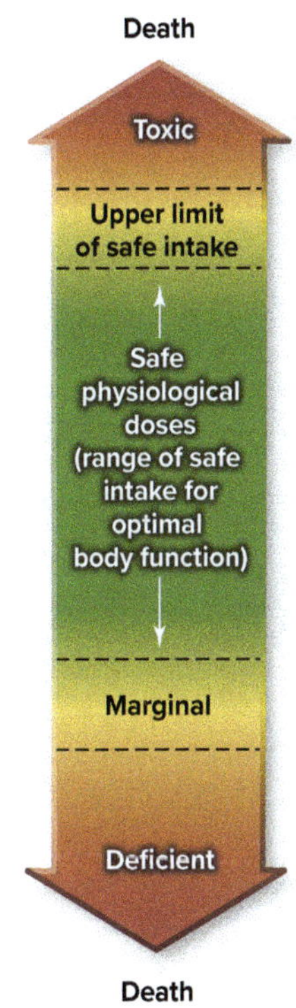

Figure 1.10 Intake continuum. For each nutrient, there is a range of safe intake.

moderation obtaining adequate amounts of nutrients while balancing calorie intake with calorie expenditure

physiological dose amount of a nutrient that is within the range of safe intake and enables the body to function optimally

megadose amount of a vitamin or mineral that greatly exceeds the recommended amount

MY DIVERSE PLATE

Traditional Mexican meals typically include some form of corn, such as tortillas made from corn meal; beans; chili peppers; and rice. Meals may also feature eggs, pork, chicken, beef, and fish, as well as a wide variety of fruits and vegetables. Main dishes, dips, sauces, and snacks often incorporate fresh avocados, tomatillos, papayas, pineapples, limes, mangoes, chayote, cherimoya, jicama, tomatoes, garlic, onions, plantains, and the fruit (*tuna*) and young stems (*nopalitos*) of the prickly pear cactus that have been prepared for consumption. Cilantro; oregano; cumin; parsley; and various dried peppers, such as chile de árbol, and chipotle, are used to add distinctive flavors to Mexican cuisine. Because the traditional Mexican diet includes a variety of fruits and vegetables, it is high in fiber, vitamins C and A, potassium, magnesium, and beneficial phytochemicals.

Prickly pear cactus with its spiny green stems (*nopales*) and dark purple fruit (*tuna*). *Nopalitos* are nopales that are prepared for cooking or eating raw
Cactus, nopalitos: ©Wendy Schiff

Concept 6: Food Is the Best Source of Nutrients and Phytochemicals

The most natural, reliable, and economical way to obtain nutrients and beneficial phytochemicals is to base your diet on a variety of "whole" and *minimally processed* foods. Plant foods naturally contain a variety of nutrients and phytochemicals, but processing the foods often removes some of the most healthful parts. For example, a wheat kernel is stripped of its germ and outer layer (bran) during refinement into white flour (Fig. 1.11). Wheat germ is a good source of thiamin, vitamin E, iron, phosphorus, magnesium, zinc, fiber, and healthy fats. Wheat bran contains fiber and certain phytochemicals, and it is a good source of niacin, iron, copper, zinc, and selenium. The endosperm that remains is primarily starch (a form of carbohydrate) with some protein and very small amounts of micronutrients and fiber. By replacing refined grain products, such as white bread, with whole-grain products, you can increase the likelihood of obtaining a wide variety of nutrients and phytochemicals.

In addition to eating food, many people take nutrient supplements in the form of pills, powders, bars, wafers, or beverages. The human body, however, is designed to obtain nutrients from foods. In some instances, nutrients from food are more available, that is, more easily digested and absorbed, than those in supplements.

Figure 1.11 What is white flour? During refinement, a wheat kernel is stripped of its nutrient-rich germ and bran. The endosperm (white flour) that remains is mostly starch.
Wheat: Photodisc/Getty Images; Flour: Michael Scott/McGraw Hill

It is important to understand that nutrient supplements do not contain everything one needs for optimal nutrition. For example, they do not contain the wide variety of phytochemicals found in plant foods. Although dietary supplements that contain phytochemicals are available, they may not provide the same healthful benefits as consuming the plants that contain these compounds. Why? Nutrients and phytochemicals may need to be consumed together to provide the desirable effects in the body. Food naturally contains combinations of these chemicals in very small amounts and certain proportions. There is nothing "natural" about gulping down handfuls of supplements.

Some individuals have increased needs for certain nutrients, particularly micronutrients. People who have chronic illnesses, digestive disorders that interfere with nutrient absorption, and certain inherited disorders may require supplemental nutrients. Additionally, many older adults may need higher amounts of vitamins and other nutrients than those found in food. Because it is often difficult to plan and eat nutritious menus each day, taking a dietary supplement that contains a variety of vitamins may be advisable, especially for older adults. In general, there appears to be little danger in taking a supplement that provides 100% of recommended amounts of the micronutrients daily.[15] However, healthy adults should consider taking such supplements as a dietary "insurance policy" and not a substitute for eating a variety of nutrient-dense foods.

Concept 7: There Is No "One Size Fits All" Approach to Planning a Nutritionally Adequate Diet

By using food guides presented in Chapter 3, you can individualize your diet so that it is nutritionally adequate and suits your food likes and dislikes, budget, and lifestyle. Individualizing a diet does not mean only eating foods that "match" your blood type, hair color, personality, or shoe size. If someone promotes a diet based on such personal traits, steer clear of the diet and the promoter. Consider this: Human beings would not have survived as a species for thousands of years if their diets had to be matched to physical characteristics or personalities.

Physicians often prescribe special diets, sometimes referred to as *medical nutrition therapies,* for people with chronic health conditions such as diabetes. Even the nutritional needs of healthy people vary during different stages of their lives. Chapter 13 (Nutrition for a Lifetime) provides information about the importance of diet during pregnancy, childhood, and other stages of the life cycle.

Concept 8: Foods and the Nutrients They Contain Are Not Cure-Alls

Although specific nutrient deficiency diseases, such as scurvy, can be cured by eating foods that contain the nutrient that is missing or in short supply, nutrients do not "cure" other ailments. Diet is only one aspect of a person that influences their health. By making certain dietary changes, however, a person may be able to prevent or forestall the development of certain diseases, or possibly lessen their severity if they occur.

Although there is no legal definition for "functional foods," such products may have health benefits beyond simply helping to meet basic nutritional needs.[17] Functional foods are often manufactured to boost nutrient intakes, reduce the risk of disease, or help manage specific health problems. For example, consumers who want to increase their calcium intake can purchase orange juice that has the mineral added to it. Beverages that are formulated to contain a variety of nutrients may be a convenient way to supplement a person's diet. Certain peanut butter substitutes contain oils that may lower the risk of heart disease. Although some functional foods can help Americans improve their health, such products should not replace whole and minimally processed foods and beverages from healthy people's diets.

Food & Nutrition tips

Advertisers may use the term "superfood" to describe a dietary supplement or particular food for marketing purposes. However, there is no scientific definition and the Food and Drug Administration (FDA) does not have a legal definition for this term. No dietary supplement contains every substance needed for good health. Furthermore, no naturally occurring food is perfect (a so-called superfood) in that it contains all of the nutrients required for human survival. Certain foods, however, can add more key beneficial nutrients and other healthful substances, including phytochemicals, to diets than others. Such foods include:

- berries;
- seafood;
- legumes (soybeans and split peas, for example);
- yogurt;
- whole-grains; and
- tree nuts.[16]

To improve your diet, consider eating these foods in meals or as snacks regularly.

©Wendy Schiff

This product is a functional food because it was manufactured with oils that may reduce the risk of heart disease.
©Wendy Schiff

Concept 9: Malnutrition Includes *Under*nutrition as Well as *Over*nutrition

malnutrition state of health that occurs when the body is improperly nourished

Malnutrition is a state of health that occurs when the body is improperly nourished. Everyone must consume food and water to stay alive, yet despite the abundance and variety of nutritious foods, many Americans consume nutritionally poor diets and suffer from malnutrition as a result. Some people select nutritionally inadequate diets because they lack knowledge about nutritious foods or the importance of nutrition to health. Low-income people, however, are at risk for malnutrition because they have limited financial resources for making wise food purchases. Other people who are at risk of malnutrition include those who are addicted to drugs such as alcohol, are hospitalized or in long-term care facilities, or have chronic medical conditions that reduce the body's ability to absorb or use nutrients (Fig. 1.12). This chapter's "Nutrition Matters" section discusses the international problem of *under*nutrition.

Figure 1.12 Malnutrition and long-term care. People who are in long-term care facilities are at risk of undernutrition.
Stockbyte/Getty Images

Although many people associate malnutrition with undernutrition and starvation, *over*nutrition, the long-term excess of energy or nutrient intake, is also a form of malnutrition. Overnutrition is often characterized by obesity (having an excessive and unhealthy amount of body fat). People who are obese may also be undernourished if they consume diets that lack adequate amounts of essential nutrients, especially micronutrients. Taking megadoses of micronutrients can also cause overnutrition, especially in people who have no physiological need for the extra nutrients.

You may be surprised to learn that overnutrition is associated with more deaths throughout the world than undernutrition.[18] Obesity is widespread in countries where most people have the financial means to buy plenty of food, have an ample supply of energy-dense foods, and obtain little exercise. Chapter 10 provides information about obesity.

Concept 10: Nutrition Is a Dynamic Science

As researchers continue to explore the complex relationships between diets and health, nutrition information constantly evolves. As a result, dietary practices and recommendations undergo revision as new scientific evidence becomes available and is reviewed and accepted by nutrition experts. Unfortunately such changes can be confusing to the general public, who expect medical researchers to provide definite answers to their nutrition-related questions and rigid advice concerning optimal dietary practices.

Even nutrition educators find it difficult to keep up with the vast number of research articles published in scientific journals. Chapter 2 explains how nutrition research is conducted using scientific methods. Furthermore, Chapter 2 provides information to help you become a better consumer of nutrition and health information that appears in popular sources such as the Internet and social media, infomercials, and magazines.

Concept CHECKPOINT

14. Identify at least five of the key nutrition-related concepts presented in this section.
15. What is the difference between a food that supplies a lot of empty calories and a nutrient-dense food?
16. What is the difference between a physiological dose and a megadose of a nutrient?
17. Explain why a person who has too much body fat could be undernourished.

See Appendix F for responses.

1.5 Nutrition Matters: Undernutrition—A Worldwide Concern

Learning Outcomes

1. Discuss factors that contribute to undernutrition, and identify regions of the world where people are most likely to be undernourished.
2. Describe how undernutrition affects overall health, growth, and development during pregnancy, infancy, and childhood.
3. Define the double burden of malnutrition (DBM) and the type of countries that may experience DBM.
4. Discuss undernutrition in the United States.
5. Identify major federal food assistance programs in the United States and the populations served by each program.
6. Define biotechnology as it relates to food production.
7. Discuss how sustainable agriculture can improve the environment.

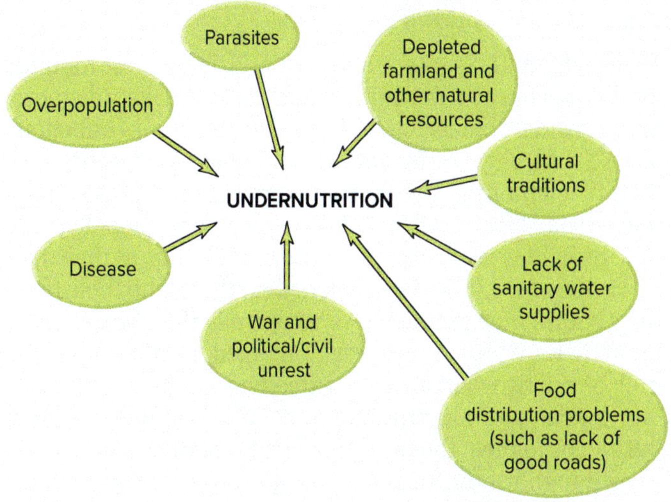

Figure 1.13 Factors that contribute to chronic undernutrition. Many factors, including war, disease, and overpopulation, contribute to undernutrition in developing countries.

Eight billion people inhabit the Earth,[19] and an estimated 800 million of these people are chronically undernourished.[20] Chronic undernutrition is a condition that occurs when a person's long-term energy and nutrient intakes are insufficient to meet their needs. Chronically undernourished people are underweight and often hungry.

Many factors contribute to chronic undernutrition (Fig. 1.13). Poverty, civil unrest, wars, and lack of access to nutritious food are major contributors, particularly in Africa and certain regions of South America and Asia. In Africa, the AIDS epidemic has left millions of people impoverished and living in uncertainty. Furthermore, unfavorable weather conditions and crop failures can cause regional food shortages. Consequences of the COVID-19 pandemic, including unemployment with less income to buy food and food supply interruptions, increased the number of people living in poverty globally.[21]

According to the World Health Organization (WHO), at least 2 billion people use drinking water that may be contaminated with feces.[22] Such water is unsafe to drink because it can contain bacteria and viruses that cause diarrhea and serious infectious diseases that include cholera, typhoid, dysentery, and polio. Every year, an estimated 485,000 people die as a result of developing diarrhea after consuming unclean water (Fig. 1.14).[22]

Figure 1.14 Unsafe water source. In developing countries, lack of clean cooking and drinking water contributes to the spread of potentially deadly infectious diseases.
Stockbyte/Getty Images

Undernutrition During Periods of Growth

When undernutrition occurs during periods of rapid growth, such as pregnancy, infancy, and childhood, the long-term effects can be devastating. In 2020, over 149 million children who were under 5 years of age were *stunted*.[23] The height of a child with stunting is much lower than normal when compared to children who are the same age but adequately nourished. Additionally, an estimated 49 million children in this age group were severely underweight (*wasted*). Throughout the world, undernutrition contributes to almost 50% of deaths of children under 5 years of age.[24]

Undernutrition During Pregnancy

Females who are undernourished during pregnancy (*maternal undernutrition*) are more likely to experience difficult labor, have excessive bleeding after delivering their infants, and die as a result of pregnancy.[25] Maternal undernutrition can also result in the birth of premature babies (infants born too early in pregnancy) and low-birth-weight (LBW) newborns. These babies often suffer from serious health problems that increase their risk of dying during their first year of life. The vast majority of LBW babies are born in developing countries. Chapter 13 provides more information about the importance of adequate nutrition during pregnancy.

Undernutrition During Infancy

As explained in Chapter 13, breast milk is the best food for infants because it is sanitary and nutritionally adequate, and provides babies with immunity to some infectious diseases. Ideally, babies should be exclusively breastfed for their first 6 months and then consume breast milk in addition to solid foods well into their second year. Globally, only 44% of babies consume only breast milk during their first 6 months of life.[26] Undernourished mothers who breastfeed their babies may produce less breast milk than breastfeeding females who are properly nourished. Furthermore, the milk of poorly nourished breastfeeding mothers may contain inadequate amounts of some vitamins and essential fatty acids to meet the nutritional needs of their infants.[25]

Although infant formulas are nutritious substitutes for breast milk, they do not provide immunity to diseases, and they are generally expensive. To increase the volume of infant formula that they can afford to give their children, poor parents in developing countries often add excessive amounts of water. This practice dilutes the nutritional value of the formula and increases the likelihood of contaminating it with disease-causing microbes. Infants who drink formula that has been mixed with unsanitary water can develop diarrhea that causes loss of body water (*dehydration*) and death.

Undernutrition During the Preschool Years

In undernourished preschool children, nutrient deficiencies are responsible for delayed physical growth and development, blindness, impaired intellectual development, wasting, and premature death. Many chronically undernourished young children are stunted and may not achieve their full potential height by the time they reach maturity (Fig. 1.15).

The body needs the micronutrient vitamin A for proper vision, growth, and immune system function, especially protection against infectious diseases. In developing countries, vitamin A deficiency causes as many as 500,000 cases of blindness among young children annually.[27]

In the United States and other developed countries, children are usually well nourished and vaccinated against common childhood diseases, such as measles. In poorer nations, however, many children are malnourished and not protected against infectious agents, including the virus that causes measles. Measles can be a life-threatening illness in undernourished children because their immune systems do not function normally.

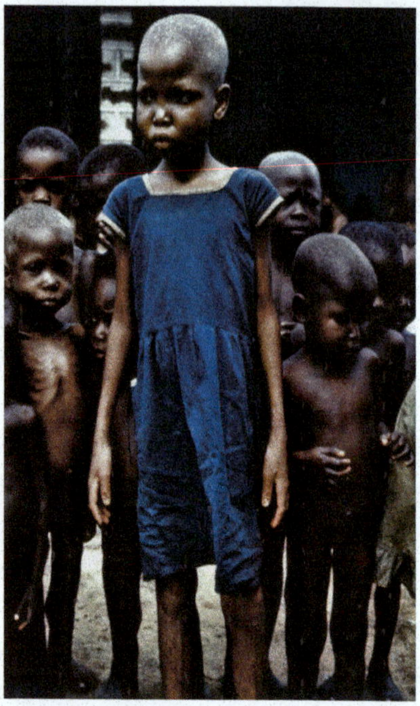

Figure 1.15 Chronic undernutrition. This photograph shows a group of young children who have signs of stunted growth and wasting. The children are outside a Nigerian orphanage during the late 1960s.
Dr. Lyle Conrad/Centers for Disease Control and Prevention

Undernutrition in the United States

Undernutrition also occurs in wealthy, developed nations such as the United States. In some instances, undernutrition is not due to poverty in these countries. For example, people suffering from anorexia nervosa (an eating disorder) and chronic alcoholism are undernourished despite often having enough money to purchase food. Nevertheless, Americans with low incomes have a higher risk of malnutrition than members of the population who are in higher income categories. In 2020, 11.4% of the population were living at or below the U.S. Department of Health and Human Services' *poverty guideline*.[28] In 2022, the poverty guideline was $27,750 for a family of four.[29]

Most American households are *food secure,* which means the people in those households have access to and can purchase sufficient food to lead healthy, active lives. In 2020, **food insecurity** was reported in an estimated 10.5% of all households in the United States.[30] Food insecurity describes individuals or families who are concerned about running out of food or not having enough money to buy more food. People who are unemployed, work in low-paying jobs, have excessive medical and housing expenses, or are homeless often experience food insecurity. Food insecurity may also affect older adult Americans who live on fixed incomes, especially if they are forced to choose between purchasing nutritious food and buying life-extending medications.

Charities and churches in many cities operate food pantries and "soup kitchens" to feed food-insecure people.

food insecurity situation in which individuals or families are concerned about running out of food or not having enough money to buy more food

In addition to the help provided by the private sector, low-income individuals can obtain food aid from federal food assistance programs. Table 1.8 summarizes information about some of the major federally subsidized food programs in the United States. Although not every eligible food-insecure adult has access to or takes advantage of the aid, federal food assistance programs can protect most American children from hunger and undernutrition. For more information about the federal government's various food assistance programs, visit www.nutrition.gov/topics/food-security-and-access/food-assistance-programs.

World Food Crisis: Finding Solutions

Reducing hunger and eliminating all forms of malnutrition is a major goal of the United Nations. The World Food Program (WFP) and United Nations Children's Fund (UNICEF) are agencies within the United Nations that provide nutrient-dense food for undernourished populations. In 2020, UNICEF distributed over 47,000 tons of ready-to-use therapeutic food (RUTF) to treat severe undernutrition in young children in developing countries.[31] For example, the RUTF Plumpy'Nut is an energy- and nutrient-dense paste made from a mixture of peanuts, powdered milk, oil, sugar, vitamins, and minerals.

The Double Burden of Malnutrition

Over the past 30 years, the impoverished regions of South and East Asia and sub-Saharan Africa experienced economic growth and development. As a result, these populations now consume more highly processed and energy-dense foods and sugar-sweetened beverages because they are easy to obtain and low in cost.[32] Improvements in transportation and technology also allow people to perform less manual labor and physical activity resulting in lower overall calorie expenditure. Scientists attribute the increase in the number of people who are overweight and obese living in these regions to all these lifestyle changes.[32]

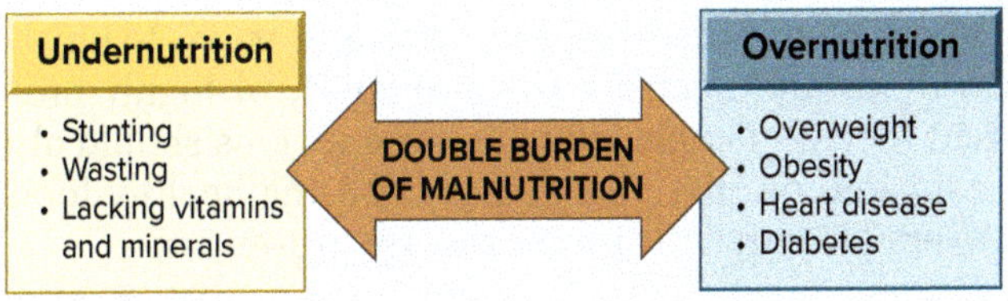

Figure 1.16 **Double burden of malnutrition.** Undernutrition and overnutrition contribute to the double burden of malnutrition experienced by some developing countries.

Therefore, these developing areas of Asia and Africa have an increasing number of people with *overnutrition* and chronic diseases, including obesity, heart disease, and diabetes, in addition to the chronic problem of undernutrition. The health systems of these regions struggle to manage the burden of undernutrition and overnutrition at the same time, called the *double burden of malnutrition* (DBM) (Fig. 1.16).[33]

The WHO recommends that support agencies help address DBM in affected regions by encouraging food

TABLE 1.8 Some Major Federally Subsidized Food Programs in the United States

Program	General Eligibility Requirements	Description
Supplemental Nutrition Assistance Program (SNAP)	Low-income individuals and families	Participants use an electronic benefit transfer (debit) card to purchase allowable food items.
Commodity Distribution Program	Certain low-income groups, including pregnant women, preschool-age children, and the elderly	In some states, state agencies distribute U.S. Department of Agriculture (USDA) surplus foods to eligible people.
Women, Infants, and Children (WIC)	Low-income pregnant or breastfeeding women, new mothers who are not breastfeeding, infants, and children under 5 years of age who are at nutritional risk	Participants receive checks or vouchers to purchase a variety of nutritious foods, such as milk, cheese, fruits and vegetables, certain cereals, and infant formula, at grocery stores. Nutrition education and support for breastfeeding mothers are also provided.
Child Nutrition Programs School Breakfast Program, National School Lunch Program, Summer Food Service Program, Child and Adult Care Food Program, Special Milk Program	Low-income children of school age Low-income adults (Child and Adult Care Food Program)	Certain schools and child-care facilities receive subsidies from the government to provide free or reduced-price nutritionally balanced meals and snacks.
Older American's Act Nutrition Program	Age 60 or older (no income guidelines)	Provides grants for sites to provide nutritious congregate and home-delivered meals
Food Distribution Program on Indian Reservations	Low-income American Indian or non-Indian households on reservations; members of federally recognized Native American tribes	Distribution of monthly food packages. This program is an alternative to SNAP and provides grants for a nutrition education component.

supplies that increase the production and access to nutritious foods and decrease the availability of highly processed foods.[34] People living in these regions should also be educated on diet quality and choosing healthy foods for overall health and chronic disease prevention.

The Promise of Biotechnology

Biotechnology involves the use of living things—plants, animals, microbes—to manufacture new products. Biotechnology in agriculture has led to the development of crops that have more beneficial characteristics than conventional crops. Biotechnology offers a variety of ways to help alleviate the world food crisis.

Bioengineering techniques involve scientific methods that change an animal or plant's hereditary material (*genes* or DNA). For example, genes that produce a desirable trait, such as disease resistance, are transferred from one organism into the DNA of a second organism, favorably altering its genes.

Cotton, corn, and soybeans are the most common bioengineered (BE) crops grown in the United States.[35] These crops may have been genetically altered to improve their flavor, resist pests, increase yields, and improve their immunity to plant diseases. Farmers primarily use BE crops to decrease pesticide use and increase crop yields. Bioengineering of food crops can also enhance the plants' nutritional qualities, such as increased vitamin A and protein. In July 2021, the Philippines was the first country to approve the commercial production of Golden Rice.[36] This bioengineered rice is enhanced with beta-carotene, which the body converts to vitamin A. In developing countries, vitamin A deficiency is the leading cause of blindness. Chapter 8 discusses vitamin A in more detail.

Bioengineering has also been used to improve animals that are eaten by humans. In 2019, the U.S. Food and Drug Administration (FDA) approved the sale of foods made with AquAdvantage Salmon, a bioengineered form of Atlantic salmon. Additionally, the eggs of the salmon can be imported into the United States and raised in FDA-approved facilities. AquAdvantage Salmon grow faster than their non-bioengineered counterparts (Atlantic salmon).[37]

Compared to traditional white rice, bioengineered Golden Rice is rich in beta-carotene, which provides it a yellow-orange color.
JIANG HONGYAN/Shutterstock

Figure 1.17 Approved label icon for BE foods.
U.S. Department of Agriculture (USDA)

In the United States, the National Bioengineered Food Disclosure Law requires a national standard for identifying foods that are bioengineered. The U.S. Department of Agriculture (USDA) released the standard in 2018.[38] According to the standard, a **bioengineered food** (BE food) contains genetic material that has been modified by certain scientific methods. Furthermore, a BE food is not found in nature or is not the result of conventional breeding techniques. A BE food may be commonly referred to as a *genetically modified organism* (*GMO*) or having ingredients from GMOs. The USDA, however, does not recognize the term "genetically modified organism" for labeling purposes.

As of January 2022, food manufacturers must use an approved method to identify a BE food or a food that contains BE ingredients. One such method is having the package logo shown in Figure 1.17. For a list of BE foods, visit the USDA's Agricultural Marketing Service website (www.ams.usda.gov) and search for "list of bioengineered foods."

The FDA regulates the safety of bioengineered foods. According to a report published by the National Academy of Sciences in 2016, there is no scientific evidence that indicates genetically modified crops are unsafe for humans to consume or harmful to the environment.[39]

Feeding the World, Protecting Natural Resources

Our current system of food production relies primarily on conventional agricultural methods. In general, conventional

biotechnology using living things to manufacture new products

bioengineering scientific techniques that alter an organism's hereditary material

bioengineered food (BE food) food that contains genetic material that has been modified by certain scientific methods

farming requires considerable amounts of water and pesticides that can harm the environment. Irrigation systems often remove fresh water from rivers and other natural sources at a faster rate than it is restored. This activity reduces water flow to many communities. The water that runs off conventional farms can carry with it precious topsoil and potentially harmful pesticides that pollute waterways. Such farming methods also release greenhouse gases, especially carbon dioxide and methane, which contribute to global warming. Furthermore, the need for new farmland often requires cutting down trees so that forests can be converted to croplands. The loss of forests eliminates native animal and plant habitats. About 50% of the Earth's land (excluding Greenland and Antarctica) is used for food production; very little suitable land remains to be farmed.[40] What can be done to feed the world's population without destroying the Earth's natural resources?

This row of young citrus trees is obtaining water via a drip irrigation system. Drip irrigation helps conserve water because it delivers small amounts of water close to the trees' roots over a long period and reduces the amount of irrigation water that is lost by evaporation.
Tim McCabe/Natural Resources Conservation Service/USDA

What Is Sustainable Agriculture? **Sustainable agriculture** involves farming methods that meet the demand for more food without depleting natural resources and harming the environment. The challenge is finding ways for farmers and ranchers to make the conversion from primarily conventional farming techniques to sustainable agriculture. Farming needs to be profitable for farmers and ranchers, so any switch from conventional to sustainable agricultural methods must not reduce their profit margins.

To solve the problems created by conventional agricultural methods, an international team developed the following points for establishing a universal policy:[40]

- Stop expanding agricultural activity, especially into tropical forests and grasslands.
- Find ways to improve crop yields on existing farms. Biotechnology in agriculture has led to the development of crops that supply higher yields, resist pests, or are tolerant of drought conditions. By increasing food production or modifying the nutritional content of foods, biotechnological advances can help reduce the number of undernourished people in the world.
- Rely more on nonchemical methods of pest management (see Chapter 12).
- Eat less meat. Sixty percent of the world's crops (primarily grains) are grown for human consumption. Most of the remaining crops are used to feed cattle and other farm animals. It takes about 30 pounds of grains to produce 1 pound of hamburger. By reducing the consumption of meat, especially beef, more grains and other plant foods could be produced to feed people. Raising grass-fed beef also spares grains for human consumption because grass is not eaten by people.
- Increase fish supply, including finfish and shellfish. Fish provide 17% of animal protein consumed globally and are also an important source of protein for developing countries.
- Reduce food waste—about 33% of food is wasted. In many instances, the food spoils before it is eaten or it is thrown out as garbage. Smaller portion sizes and better menu planning can reduce the amount of food that people waste each day.
- Find ways to use natural resources and pesticides more efficiently. Use irrigation systems, such as drip irrigation, that apply water directly to a plant's base instead of spraying it into the air, where much of the water evaporates.

Taking Action

Poverty and hunger have always plagued humankind; the causes of poverty and hunger are complex and, therefore, difficult to eliminate. Nevertheless, certain social, political, economic, and agricultural changes can reduce the number of people who are chronically hungry. In the short run, wealthy countries can provide food aid to keep impoverished people from starving to death. Families and small farmers in underdeveloped nations need to learn new and more efficient methods of growing, processing, preserving, and distributing nutritious regionally produced foods. Additionally, governments can support

sustainable agriculture farming methods that do not deplete natural resources or harm the environment while meeting the demand for food

programs that encourage breastfeeding and fortify locally grown or commonly consumed foods with vitamins and minerals that are often deficient in local diets.

In the long run, population control is critical for preserving the Earth's resources for future generations. Impoverished parents in poor countries often have many children because they expect only a few to survive to adulthood. When people are financially secure, adequately nourished, and well educated, they tend to have fewer, healthier children. Thus, long–term ways to slow population growth include providing well–paying jobs, improving public education, and increasing access to health care services.

Concept CHECKPOINT

18. How do unfavorable environmental and political factors in low income countries affect the health status of people living in those nations?

19. What effects can undernutrition have on the health of pregnant females and young children?

20. What is the "WIC program"?

21. Explain the difference between conventional and sustainable agricultural methods.

22. What is a bioengineered food?

See Appendix F for responses.

Chapter References

See Appendix G.

Summary

1.1 Nutrition: The Basics

- Lifestyle choices, including poor eating habits and lack of physical activity, contribute to the development of many of the leading causes of premature deaths for American adults, including heart disease, cancer, stroke, and diabetes. However, you may be able to extend your years of healthy life and improve your quality of life by applying what you learn about nutrition and the role that a nutritious diet plays in achieving and maintaining good health.
- There are six classes of nutrients: carbohydrates, lipids, proteins, vitamins, minerals, and water. The body needs certain nutrients for energy, growth and development, and regulation of processes, including the repair and maintenance of cells. The human body can synthesize many nutrients, but about 50 nutrients are dietary essentials that must be supplied by food.
- Nonnutrients, which include phytochemicals, are substances in food that may have healthful benefits. Many phytochemicals are antioxidants that protect cells from being damaged or destroyed by exposure to certain environmental factors. However, some phytochemicals are toxic. The Dietary Supplement Health and Education Act of 1994 (DSHEA) allows manufacturers to classify herbal products and nutrient supplements as dietary supplements.

Burke/Triolo Productions/Artville/Getty Images

1.2 Factors That Influence Americans' Health

- Heart disease is the leading cause of death for all Americans. Chronic diseases, such as heart disease, are complex conditions that have multiple risk factors. A risk factor is a personal characteristic that increases a person's chances of developing diseases. In many instances, people can live longer and healthier by modifying their diets, increasing their physical activity, and altering other aspects of their lifestyles. Objectives to improve the status of Americans' health, including nutrition-related programs, are an important focus of *Healthy People.*

1.3 Metrics for Nutrition

- Scientists and health care professionals generally use metric values when measuring volume, weight, and length. The metric prefixes *micro-, milli-, deci-, centi-,* and *kilo-* indicate whether a measurement is a fraction or a multiple of a meter, gram, or liter. Approximately 28 grams are in an ounce and 454 grams are in a pound; a kilogram equals 1000 grams or about 2.2 pounds. Each gram equals 1000 milligrams or 1 million micrograms.
- Every cell needs energy. A Calorie is the heat energy needed to raise the temperature of 1 liter of water 1° Celsius (C). Calories or kilocalories (kcal) are used to indicate the energy value in food. If no number of kilocalories is specified, it is appropriate to use "calories." A gram of carbohydrate and a gram of protein each supply about 4 kcal; a gram of fat provides about 9 kcal. Although alcohol is not a nutrient, a gram of pure alcohol furnishes 7 kcal.
- Carbohydrates, fats, and proteins are referred to as macronutrients because the body needs relatively large amounts of these nutrients daily. Vitamins and minerals are micronutrients because the body needs very small amounts. Although the body requires large amounts of water, this nutrient provides no energy and is not usually classified as a macronutrient.

1.4 Key Nutrition Concepts

- There are several key points to understanding nutrition. Most naturally occurring foods are mixtures of nutrients, but no food contains all the nutrients needed for optimal health. Thus, nutritionally adequate diets include a variety of foods from all food groups. Instead of classifying foods as "good" or "bad," people can focus on eating all foods in moderation and limiting empty calories. For each nutrient, there is a range of safe intake. Healthy people should rely on eating a variety of whole and minimally processed foods to meet their nutrient needs instead of taking dietary supplements. Foods and the nutrients they contain are not cure-alls. There is no "one size fits all" approach to planning a nutritionally adequate diet. Malnutrition includes overnutrition, as well as undernutrition. Finally, nutrition is a dynamic science; new scientific information about nutrients and their roles in health is constantly emerging. Therefore, ways the science of nutrition is applied, such as dietary recommendations, also change.

1.5 Nutrition Matters: Undernutrition—A Worldwide Concern

- Poverty and undernutrition are commonplace in many developing countries. Low income people must often cope with infectious diseases and polluted water supplies. In developing countries, poor sanitation practices and lack of clean cooking and drinking

Burke/Triolo Productions/Artville/Getty Images

water contribute to diseases and deaths. When undernutrition occurs during periods of rapid growth, such as pregnancy, infancy, and childhood, the long-term effects can be devastating. In undernourished children, nutrient deficiencies are responsible for stunted physical growth, delayed physical development, blindness, impaired intellectual development, and premature death. Chronic undernutrition depresses the body's immune functioning, increasing the risk of death from infectious diseases, such as measles, especially in childhood. Low income areas of Asia and Africa may experience an increasing number of people with *overnutrition* and chronic diseases, including obesity, heart disease, and diabetes, in addition to the chronic problem of undernutrition. The health systems of these regions struggle to manage the burden of undernutrition and overnutrition at the same time, which is called the double burden of malnutrition (DBM).

- Reducing hunger and all forms of malnutrition is a major goal of the United Nations. Biotechnological advances in agriculture have led to the development of improved food crops and animals. Conventional farming methods can cause soil loss and add pesticides to water supplies. Sustainable agriculture refers to farming methods that do not deplete natural resources or harm the environment while meeting the demand for food.

Recipe for Healthy Living

Food Preparation Basics (Yes, You *Can* Cook!)

By learning how to prepare dishes, experiment with recipe ingredients, and use a variety of spices and herbs as seasonings, you can make home-cooked meals that are more tasty, more appealing, and lower in unhealthy types of fat, sugar, salt, and calories than the usual choices at fast-food restaurants. Additionally, you can save money by making your meals and snacks instead of purchasing them from restaurants and convenience stores.

At the end of each chapter, you will find the "Recipe for Healthy Living," a nutritious, easy-to-prepare recipe. Each recipe includes a list of ingredients, instructions, and some information concerning the energy and selected nutrient contents in a serving of the product. The "MyPlate" icon indicates which of the USDA's food groups are primarily represented in the recipe (see Figure 1.7). The circle graph shows approximate percentages of total calories for each macronutrient in the food. The bar graph indicates approximate percentages of Daily Values (DVs) for energy and some key nutrients. Daily Values are a set of nutrient standards used for food labeling purposes (see Chapter 3).

Even if you've had little or no cooking experience, you can learn the basics of preparing foods. Some cooks don't measure ingredients; they know from experience how to estimate amounts of foods and seasonings to add when preparing dishes. Until you feel confident with your food preparation skills, it is best to follow recipes and measure ingredients carefully.

You'll need some basic food preparation equipment to get started. You don't have to spend a lot of money, but buy well-made stainless steel (rustproof) cooking utensils and mixing bowls that will last for decades. Baking pans should also be stainless steel. A square or rectangular tempered-glass baking dish can be used for a variety of cooking needs, including heating foods in a toaster oven or microwave oven.

Understanding how to use household measurements is a good place to begin when learning how to cook. Purchase a set of stainless steel measuring spoons that includes ⅛ teaspoon (tsp), ¼ tsp, ½ tsp, 1 tsp, and 1 tablespoon (Tbsp) measures. You'll also need a set of measuring cups that includes the following measures: ¼ cup, ⅓ cup, ½ cup, and 1 cup. These cups are used to measure dry ingredients such as flour or sugar. Finally, purchase a two-cup glass or clear plastic measuring pitcher that is marked to indicate fluid ounces. This pitcher is used for measuring liquid ingredients such as water, milk, and oil.

To measure dry ingredients, fill the appropriate measuring cup or spoon to the top, and skim off the excess with the straight edge of a knife. To measure liquid ingredients, use a liquid measuring pitcher. Fill the pitcher to the desired amount, and place it on a level surface. Crouch down so you are at eye level with the fluid's level, and then carefully add more or remove some fluid as needed.

HOUSEHOLD UNITS

Common household units often used for measuring food ingredients and their commonly used abbreviations are listed below. Ounces (oz) are a measure of weight; *fluid* ounces are a measure of volume. Appendix A provides information about English-to-metric and metric-to-household unit conversions.

COMMON HOUSEHOLD UNITS FOR MEASURING FOOD INGREDIENTS

3 tsp = 1 Tbsp	1 cup = 8 fluid ounces (oz)
4 Tbsp = ¼ cup	1 cup = ½ pint
5 Tbsp + 1 tsp = ⅓ cup	2 cups = 1 pint
8 Tbsp = ½ cup	4 cups or 2 pints = 1 quart
16 Tbsp = 1 cup	4 quarts = 1 gallon

Applesauce Oatmeal Muffins

©Wendy Schiff

The following muffin recipe is simple and will give you an opportunity to practice measuring dry and liquid ingredients. This recipe makes 12 small muffins, so you'll need a muffin tin with 12 muffin cups. Each muffin supplies approximately 168 kcal, 4.7 g fat, 27.0 g digestible carbohydrates, 3.6 g protein, 1.5 g fiber, and 5.8 mg iron.

INGREDIENTS:

- 1½ cups quick oats, enriched, uncooked
- 1¼ cups all-purpose enriched flour
- 1 tsp baking powder
- ¾ tsp baking soda
- 1 tsp ground cinnamon
- ¼ tsp ground nutmeg
- 1 cup unsweetened applesauce
- ½ cup fat-free milk
- ½ cup brown sugar
- 3 Tbsp canola oil
- 1 medium egg
- vegetable oil cooking spray

Approximate Percent of Calories from Macronutrients

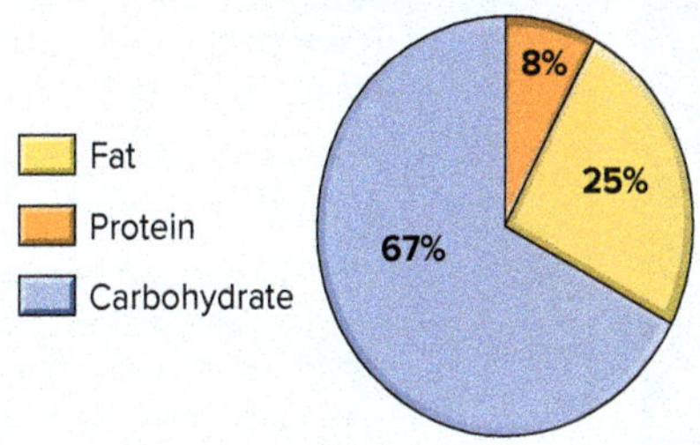

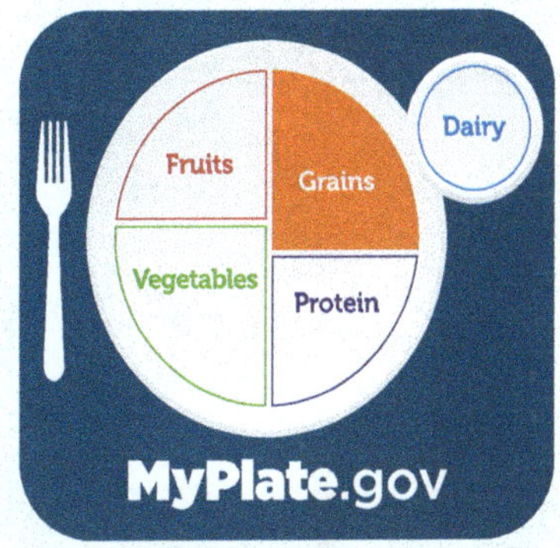

US Department of Agriculture or USDA

Percent of Daily Value/Serving

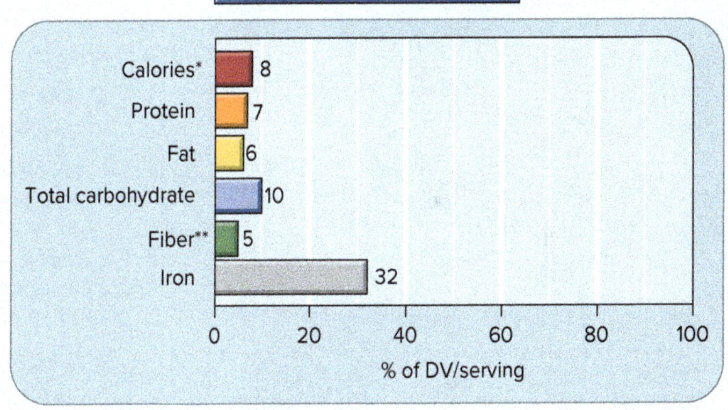

*2000 daily total kcal
**Component of total carbohydrate

PREPARATION STEPS:

1. Preheat oven to 400° F. Apply vegetable oil cooking spray lightly on the bottom of each cup of the muffin tin.
2. Combine the first six ingredients in a large bowl. Stir until well mixed. Use a large spoon to form a depressed area (a "well") in the center of the dry ingredients.
3. In another bowl, break the egg by firmly tapping the center of the shell on the edge of the bowl. Using two hands, hold the cracked egg over the bowl and spread the two sides of the shell apart to let the egg's contents fall into the bowl. The goal is to keep any bits of egg shell from dropping into the food. Use a fork to gently beat the egg to blend the white and yolk into a yellow mixture.
4. Add applesauce, milk, brown sugar, and oil to the beaten egg, and stir with a large spoon until ingredients are well mixed.
5. Pour the applesauce-containing mixture into the "well" of dry ingredients. Stir just until the dry ingredients have been moistened. Do not be concerned if small lumps of dry ingredients are in the batter. Do not beat or overmix the ingredients; otherwise, the muffins will form air tunnels and have peaked rather than rounded tops.
6. Spoon batter into muffin cups until each cup is half full of batter.
7. Bake 20 minutes or until muffins are golden brown. Use pot holders to protect your hands when taking the muffin tin out of the oven. Cool muffins for about 5 minutes before removing them from the tin. Muffins can be stored in a closed container in the refrigerator for up to a week. Before eating, warm each muffin in a microwave oven for about 15 seconds.

Tip: For easier cleanup, soak the bowl that contained the batter in cold, soapy water.

Personal Dietary Analysis

1. For 1 week, keep grocery store receipts, including receipts for foods bought at warehouse clubs and supercenters.

  a. How much money did you spend on foods purchased at these places?

  b. Which foods were the most expensive items purchased?

  c. How much money did you spend on foods and beverages that contain high amounts of empty calories such as salty snacks, cookies, soft drinks, and candy? ______

  d. What percentage of your food dollars were spent on foods that were rich sources of empty calories? ______ (Divide the amount of money spent on such foods by the total cost of food for the week. Move the decimal point over two places to the right and place a percent sign after the number.)

  e. How much money did you spend on nutrient-dense foods such as whole-grain products, nuts, fruits, and vegetables? ______

  f. What percentage of your food dollars were spent on nutrient-dense foods? ______ (Divide the amount of money spent on nutrient-dense foods by the total cost of food for the week. Move the decimal point over two places to the right and place a percent sign after the number.)

2. For 1 week, keep a detailed log of your convenience store food purchases, including the item(s) purchased and amount of money spent for each purchase.

  a. What types of foods and beverages did you buy? ______________

  b. How many soft drinks did you consume each day? ______

  c. How much money did you spend on foods and beverages that were purchased at convenience stores? ______

  d. Based on this week's food and beverage expenditures at convenience stores, estimate how much money you spend on such purchases in a year. ______

3. For 1 week, keep a detailed log of your usual fast-food restaurant consumption practices, including fast-food purchases at your campus's restaurants. List the types of food and beverages you purchased and amount of money you spent.

  a. According to your weekly record, how often do you buy food from fast-food places? ______

  b. What types of foods did you usually buy? ______________

  c. How much money did you spend on fast foods? ______

  d. Based on this week's expenditures, estimate how much money you spend on fast-food purchases in a year. ______

Complete the Personal Dietary Analysis activity in Connect.

Grocery store receipt
Comstock/PunchStock

Critical Thinking

1. Identify at least six factors that influence your food and beverage selections. Which of these factors are the three most important to you? Explain why.
2. Consider your current eating habits. Explain why you think your diet is or is not nutritionally balanced.
3. "Everything in moderation." Explain what this statement means in terms of your diet.
4. If you were at risk of developing a chronic health condition that could be prevented by changing your diet, would you make the necessary changes? Explain why or why not.
5. One of your friends takes megadoses of various vitamin and mineral supplements. Explain why this person has a high risk of being malnourished.
6. List five packaged foods you would donate to a local food pantry. Explain why you chose these particular foods to donate.
7. Explain why you would or would not eat a bioengineered food.

Practice Test (Section numbers are listed below in red.)

Select the best answer.

1. (1.1) Diet is a
   a. practice of restricting calorie intake.
   b. technique for limiting empty calorie intake.
   c. method of reducing portion sizes.
   d. usual pattern of food choices.
2. (1.1) Which of the following conditions is the second leading cause of death in the United States?
   a. Cancer
   b. Stroke
   c. Heart disease
   d. Diabetes
3. (1.1) The nutrients that provide energy are
   a. vitamins, minerals, and proteins.
   b. lipids, proteins, and minerals.
   c. fats, proteins, and carbohydrates.
   d. proteins, fats, and water.
4. (1.1) _____ refers to all chemical processes that occur in living cells.
   a. Metabolism
   b. Catabolism
   c. Anatomy
   d. Physiology
5. (1.2) In the United States, the primary cause of preventable cancer deaths is
   a. physical inactivity.
   b. tobacco use.
   c. high-fat diet.
   d. excessive alcohol intake.
6. (1.2) Which of the following is not a goal or objective of *Healthy People 2030*?
   a. Eliminate health disparities and achieve health equity for all people.
   b. Increase fruit and vegetable intake of people who are 2 years of age or older.
   c. Promote healthy development and behaviors across all life stages.
   d. Increase sodium intake of people who are 2 years of age or older.
7. (1.3) Lena weighs 80 kg. What is her weight in pounds?
   a. 75 pounds
   b. 220 pounds
   c. 132 pounds
   d. 176 pounds
8. (1.3) A serving of food contains 10 g carbohydrate, 2 g protein, and 11 g fat. Based on this information, a serving of this food supplies _____ kcal.
   a. 281
   b. 225
   c. 147
   d. 165
9. (1.3) Yosan's meal supplies 50 g carbohydrate, 16 g protein, 30 g fat, and 100 g water. Which of the following statements is true about Yosan's meal?
   a. Fat provides the most food energy.
   b. Carbohydrate provides the most food energy.
   c. Water provides the most food energy.
   d. Protein provides almost 50% of total calories.

10. (1.4) Which of the following foods is energy and nutrient dense?

a. Fat–free yogurt
b. Sugar–sweetened soft drink
c. Peanut butter
d. Iceberg lettuce

11. (1.4) Which of the following foods is not a source of phytochemicals?

a. Hamburger
b. Raspberries
c. Chocolate
d. Broccoli

12. (1.4) Trevor consumes a diet that contains different foods from each food group. He is practicing the nutrition concept of

a. moderation
b. malnutrition
c. variety
d. empty calories

13. (1.4) Which of the following foods is the most nutrient dense?

a. Potato chips
b. Celery
c. Butter
d. Fat–free milk

14. (1.4) Which of the following statements is false?

a. A megadose is generally less than a physiological dose of a nutrient.
b. Megadoses of nutrients may behave like drugs in the body.
c. In general, physiological doses of nutrients are safe to consume daily.
d. People should avoid taking megadose amounts of vitamin and mineral supplements.

15. (1.5) The _____ Program enables eligible participants to use a special debit card to purchase food at authorized stores.

a. Nutritious Food Purchase
b. Supplemental Nutrition Assistance
c. Healthy Diets for All
d. Eat Better for Less

16. (1.5) Which of the following actions has been proposed to help solve environmental problems created by conventional agricultural methods?

a. Having populations shift to diets based primarily on meat, fish, and poultry
b. Converting tropical jungles into much–needed farmland
c. Developing ways to use irrigation systems more effectively
d. Resisting efforts to increase food production through biotechnology

Answers to Quiz Yourself

1. There are four classes of nutrients: proteins, lipids, sugars, and vitamins. **False** (Section 1.1)
2. Proteins are the most essential class of nutrients. **False** (Section 1.1)
3. All nutrients must be supplied by the diet because they cannot be made by the body. **False** (Section 1.1)
4. Vitamins are a source of energy. **False** (Section 1.3)
5. Milk, avocados, and blueberries are examples of "perfect" foods that contain all nutrients. **False** (Section 1.4)

Design Elements: (Stacked books): Siede Preis/Photodisc/Getty Images; (Nuts): C Squared Studios/Photodisc/Getty Images; (Recipe box, Raspberries, Blueberries): ©Wendy Schiff; (Pineapple): Stockbyte/Getty Images; (Radishes): karandaev/iStock/360/Getty Images; (MyPlate icon): US Department of Agriculture or USDA

CHAPTER 2

Evaluating Nutrition Information

Aaron Roeth Photography/McGraw Hill

Quiz Yourself

Before reading Chapter 2, test your knowledge of scientific methods and reliable sources of nutrition information by taking the following quiz. The answers are at the end of the chapter.

1. Scientists use anecdotes as scientific evidence to support their findings. ______ T ______ F
2. Popular health-related magazines typically publish articles that have been peer reviewed. ______ T ______ F
3. By conducting observational epidemiological studies, medical researchers can determine risk factors that may influence health. ______ T ______ F
4. Dietary supplements include vitamin pills, as well as products that contain echinacea, ginseng, and garlic. ______ T ______ F
5. In general, registered dietitian nutritionists are reliable sources of food and nutrition information. ______ T ______ F

McGraw Hill connect®

A wealth of proven resources are available on Connect® including SmartBook®, NutritionCalc Plus, and many other dynamic learning tools. Ask your instructor about Connect!

2.1 Nutrition: Science for Consumers

Learning Outcomes

1. Explain the basic steps of the scientific method.
2. Explain why nutrition information derived from anecdotes is not evidence based.
3. Discuss the differences between experimental and epidemiological research studies.
4. Recognize the importance of having controls when performing experiments.
5. Explain the difference between a direct (positive) relationship and an inverse (negative) relationship between two variables.
6. Discuss why similar scientific studies often have different results.

We consume food, and we consume nutrition information. There seems to be no way to avoid nutrition information: If you browse shelves of books and magazines at a bookstore; surf the Internet, including social media; watch television; listen to the radio or a podcast series; or shop for groceries, you will encounter information about the subject. Even family members and friends contribute to the flood of nutrition information. Much of the information from such popular sources is not **evidence based,** that is, it is not supported by scientific evidence. The challenge for you and other consumers is to understand how scientists collect evidence about nutrition and health and how to analyze this information to determine whether it is factual and based on high-quality scientific evidence (reliable) or misinformation that is unsupported by the facts. The information in this chapter will help you become a more critical and careful consumer of nutrition- and health-related information.

evidence based information that is based on results of high-quality scientific studies

anecdotes reports of personal experiences

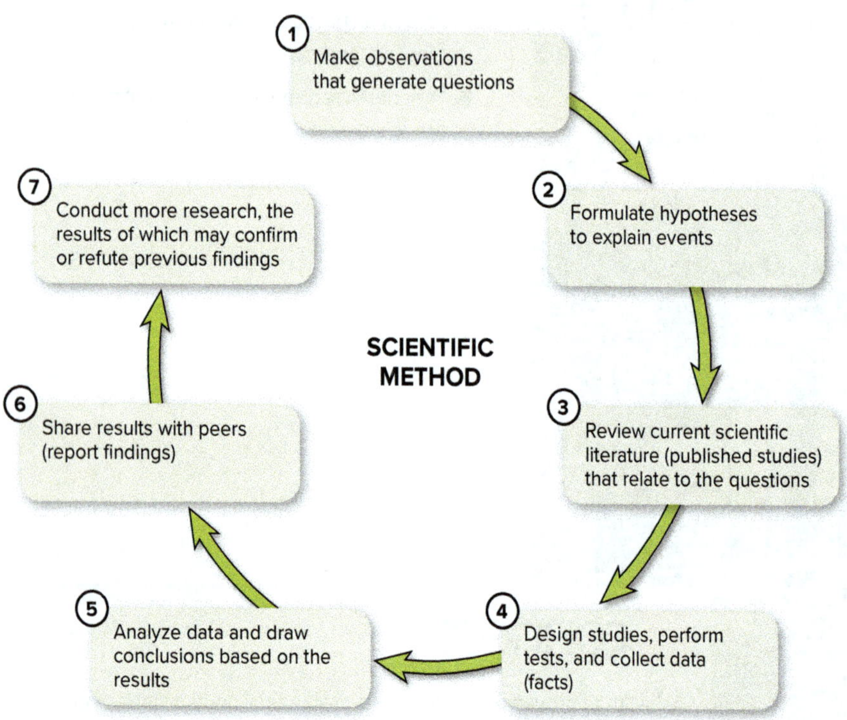

Figure 2.1 Scientific method. Nutrition scientists generally follow these steps when conducting research.

The Scientific Method

In the past, nutrition facts and dietary practices were often based on intuition, common sense, "conventional wisdom" (tradition), or **anecdotes** (reports of personal experiences). Today, *registered dietitian nutritionists (RDNs)* and other nutrition experts discard conventional beliefs, explanations, and practices when the results of current scientific research no longer support them.

Scientists ask questions about the natural world, such as: "How do cells make proteins?" and "What factors increase the risk of stomach ulcers?" To obtain answers for their questions, researchers design studies that follow generally accepted methods. Figure 2.1 presents the scientific method that is often followed by researchers who conduct experiments and other kinds of research.

Laboratory Experiments

An experiment is a systematic way of testing a hypothesis (question). Because of safety, as well as ethical concerns, nutrition scientists often conduct laboratory experiments on small mammals before performing similar research on humans. An experiment that uses whole living organisms, such as mice, is called an *in vivo* experiment. Nutrition researchers also perform controlled laboratory experiments on cells or other components derived from living organisms. These studies are *in vitro* or "test tube" experiments. Researchers are cautious when drawing conclusions from results of *in vitro* experiments because cells removed from a living thing may not function the same way they do when they are in the entire life form.

Using the Scientific Method: Animal Experimentation Experiments generally involve the basic steps shown in Figure 2.1. A team of nutrition scientists, for example, makes observations and generates questions that result in the development of a hypothesis. According to their hypothesis, consuming "chemical X" in charcoal–grilled meat is harmful. Before testing this hypothesis, the scientists review published scientific studies that provide information about the role of chemical X in the development of disease.

After conducting the literature review, the researchers design an *in vivo* experimental study to provide answers relating to their hypothesis. The scientists divide 100 genetically similar, healthy, 3–week–old mice into two groups of 50 mice. One group **(treatment group)** is fed a certain amount of chemical X daily for 52 weeks; the second group **(control group)** does not receive the treatment during the period (Fig. 2.2).

Why is a control group necessary? Having a control group enables scientists to compare results between the two study groups to determine whether the treatment had any effect. A **variable** is a characteristic or other factor that can change and influence an outcome. Variables include physical activity, air temperature, blood pressure, and sodium intake. Many variables can influence the outcome of an experimental study. Thus, scientists who want to determine the effect or effects of a single variable, such as chemical X intake, need to control the influence of other variables. Therefore, all other conditions, including timing of feedings and amount of handling by caregivers, must be the same for both groups of mice. If researchers design an experiment in which they fail to control variables that are not being tested, their findings are likely to be unclear or inaccurate.

treatment group group being studied that receives a treatment

control group group being studied that does not receive a treatment

variable personal characteristic or other factor that changes and can influence an outcome

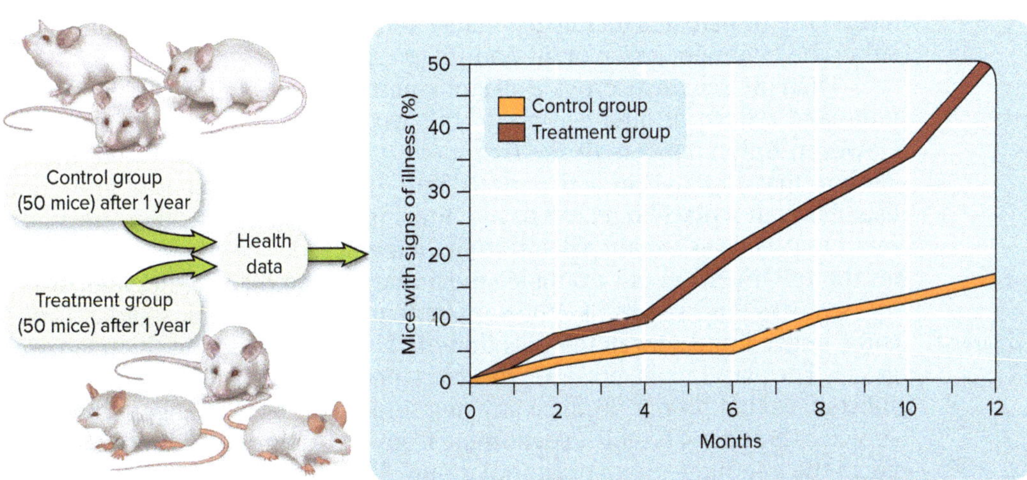

Figure 2.2 An *in vivo* experiment. Based on the findings shown in the graph, is it safe for mice to consume chemical X daily for a year?

For the duration of this study, the scientists examine the mice regularly for signs of health problems and record data (facts). If the data show that the mice in the treatment group are as healthy as the mice in the control group at the end of this experiment, the researchers may conclude that mice can safely consume the amount of chemical X used in the study on a daily basis for a year.

Medical researchers must be careful when applying the results of *in vivo* animal studies to people because of the physiological differences between humans and laboratory animals. Nevertheless, scientists are often able to determine the safety and effectiveness of treatments by conducting research on laboratory animals before engaging in similar testing on humans. Two types of research that involve human subjects are *experimental* (intervention) studies and *observational* studies. Such investigations can provide clues about the causes, progression, and prevention of diet-related diseases.

Human Research: Experimental (Intervention) Studies

Nutrition scientists conduct experimental studies to test an *intervention,* a type of change or treatment, including diet changes, addition of a dietary supplement, or other lifestyle changes. The scientist then observes the outcomes of these changes. When conducting an experimental study involving human subjects, researchers usually randomly divide a large group of people into treatment and control groups. This kind of experiment is called **randomized controlled trial (RCT).** Random assignment helps ensure that the members of the treatment and control groups have similar variables, such as age, weight, and other characteristics. All subjects will be instructed to maintain their usual lifestyle during the duration of the study, except for the activities required by their participation in the research.

In an RCT, a large group of human subjects is often needed because unlike laboratory mice and rats that are bred specifically for research purposes, individual humans vary greatly within a population. Furthermore, researchers can expect that some of their subjects will not provide all of the necessary physical data, such as blood or urine samples, and others will drop out of the study. When a large number of subjects fully participate in the study, scientists are more likely to have findings that result from the treatment and not simply by chance.

A group of nutrition scientists, for example, want to test the hypothesis that taking a dietary supplement that contains garlic reduces blood cholesterol levels. The researchers recruit 200 adults and randomly assign each person to either the treatment group or the control group.

Then the scientists provide all study participants with the same instructions and a form of intervention, in this example, a bottle containing pills. However, only members of the treatment group actually receive the intervention that contains an active ingredient. Subjects in the control group are usually given a **placebo.** Placebos are not simply "sugar pills"; they are a fake treatment, such as a sham pill, injection, or medical device. The placebo mimics the treatment. In our example study, the placebo dietary supplement pill looks, tastes, and smells like the supplement pill that contains garlic. Garlic is the active ingredient in the pills that are given to subjects in the treatment group. The placebo pill, however, has *inert* ingredients; that is, the pill contains substances that do not produce any measurable physical changes. Thus, our subjects should not be able to determine from examining the pills whether they are in the treatment group or control group. Providing placebos to members of the control group enables scientists to compare the extent of the treatment's response with that of the placebo.

randomized controlled trial (RCT) experimental study involving humans that usually randomly divides a group of people into treatment and control groups

placebo fake treatment, such as a sham pill, injection, or medical procedure

Did *YOU* Know?

Male with signs of pellagra
Centers for Disease Control and Prevention

Today, the idea that something missing in diets can cause a nutrient deficiency disease is widely accepted. A hundred years ago, however, it was a novel idea that most medical experts dismissed because they thought only "germs" caused disease. In the early 1900s, the disease pellagra was widespread in the United States, especially in southern states. Each year, thousands of Americans died from this dreaded illness. Most medical experts thought pellagra was an infectious disease because it often occurred where people lived in close quarters, such as prisons, orphanages, and mental health institutions.

Dr. Joseph Goldberger, a physician who worked in a federal government laboratory, observed that not everyone who was exposed to people suffering from pellagra developed the condition. Dr. Goldberger also noted that prisoners ate a diet that was typically eaten by other people with pellagra. The diet emphasized corn bread, hominy grits (a corn product), molasses, potatoes, cabbage, and rice. At the time, this monotonous, low-protein diet was associated with poverty throughout the southern United States. He also observed that people who did not develop pellagra had higher incomes and ate more meat, milk, and fresh vegetables. Based on his observations, Goldberger rejected the medical establishment's notion that pellagra was an infectious disease. He proposed the idea that pellagra resulted from the lack of something in poor people's diet and the missing dietary factor was in meat, milk, and other foods eaten regularly by people with high incomes. To test his dietary hypothesis, Goldberger gave these foods to children in two Mississippi orphanages and patients in a Georgia mental institution who were suffering from pellagra, and they were cured of the disease. Many members of the medical establishment, however, rejected his finding that a poor diet was the cause of pellagra.

To satisfy his critics, Dr. Goldberger enrolled a group of healthy Mississippi prison inmates in an experiment that involved consuming the corn- and molasses-based diet commonly eaten in the southern states at the time. After a few months, more than half of the inmates developed pellagra. Once again, however, many of Goldberger's critics rejected his finding that poor diet was the cause of pellagra.

In 1916, Dr. Goldberger experimented on himself and some volunteers during what they called a "filth party." The group applied secretions taken from inside the nose and throat of a patient with pellagra into their noses and throats; they also swallowed pills made with flakes of skin scraped from the rashes of people with the disease. Additionally, Goldberger and one of his colleagues gave each other an injection of blood from a person who had pellagra. If pellagra were infectious, filth-party participants should have contracted the disease—but none of them did. Despite the results of Dr. Goldberger's extraordinary experiment, a few physicians still resisted the idea that pellagra was associated with diet.[1]

Dr. Goldberger died in 1929—8 years before Dr. Conrad Elvehjem and his team of scientists at the University of Wisconsin isolated a form of the vitamin niacin from liver extracts. Elvehjem and his colleagues discovered niacin cured "black tongue," a condition affecting dogs that was similar to pellagra.[2] Soon after Elvehjem's discovery, other scientists determined that niacin was effective in treating pellagra, and the medical establishment finally accepted the fact that the disease was the result of a dietary deficiency. Chapter 8 provides more information about niacin and pellagra.

Dr. Joseph Goldberger
Centers for Disease Control and Prevention

Burke Triolo Productions/Artville/Getty Images

Double-Blind Studies

Human experimental studies are usually **double-blind studies;** that is, neither the investigators nor the subjects are aware of the subjects' group assignments. Codes are used to identify a subject's group membership, and this information is not revealed until the end of the study. Maintaining such secrecy is important during the course of a human study involving placebos because researchers and subjects may try to predict group assignments based on their expectations. If the investigators who interview the participants are aware of their individual group assignments during the study (a single-blind study), they may unwittingly convey clues to each subject, perhaps in the form of body language, that could influence the subject's belief about being in the experimental or control group. Subjects who suspect they are in the control group and taking a placebo may report no changes in their condition because they expect a placebo should have no effect on them. On the other hand, subjects who think they are in the treatment group could insist that they feel better or have more energy as a result of the treatment, even though the treatment may not have produced any measurable changes in their bodies. Ideally, subjects should not be able to figure out their group assignment while researchers are collecting information from them.

Did *YOU* Know?

Prevalence refers to the number of people in a particular population who have a disease. In the United States, for example, more people have diabetes than tuberculosis, so diabetes is a prevalent disease in this country.

What Are Placebo and Nocebo Effects? People may report positive or negative reactions to a real or sham treatment because of their expectations and not as a result of the treatment. If, for example, a patient believes a form of treatment will improve their health, the patient is more likely to report positive results for the therapy. Such wishful thinking is called the **placebo effect.** On the other hand, a patient who expects a medical treatment will be harmful may report unwanted effects of the therapy, even though the treatment was harmless. In this case, the reported side effects indicate a *nocebo* (*no-see'-bow*) effect.

People who take certain herbal products or use other unconventional medical therapies to prevent or treat diseases are often convinced the products and treatments are effective, despite the general lack of scientific evidence to support their beliefs. Such personal findings may be examples of the placebo effect. However, placebos can produce beneficial physiological and psychological changes, particularly in conditions that involve pain.[3] Because subjects in the control group believe they are receiving a real treatment, their faith in the "treatment" can stimulate the release of chemicals in the brain that alter pain perception, reducing their discomfort. Therefore, when people report that a treatment was beneficial, they may not have been imagining the positive response, even when they were taking a placebo.

Human Research: Epidemiological Studies

For decades, medical researchers have noted differences in rates of chronic diseases and causes of death among various populations. The most common type of diabetes, for example, occurs more frequently among American Indian/Alaskan Native, Hispanic, and non-Hispanic African-American adults than among non-Hispanic, white American adults.[4] To understand why this difference exists, medical researchers rely on the findings of epidemiological studies.

Epidemiology (*eh-peh-dee-mee-ah'-luh-jee*) is the study of the occurrence, distribution, and causes of health problems in populations. Epidemiologists often use data collected from physical examinations and blood samples of a specific population to determine factors that affect the group's health. In 2017, for example, U.S. public health officials discovered 30 new cases of lead poisoning among a state's 50,000 preschool-aged children (*occurrence*). Epidemiologists determined that 95% of the affected children

double-blind study experimental design in which neither the participants nor the researchers are aware of each participant's group assignment

placebo effect positive response to a placebo

epidemiology study of the occurrence, distribution, and causes of health problems in populations

lived in a major city of the state; the remaining children lived in smaller cities or towns (*distribution*). To determine why children who were living in the major city were more likely to have lead poisoning than other children in the state, the epidemiologists compared the living conditions of the children with lead poisoning to those who were healthy. As a result of their investigation, the scientists determined that lead was in the drinking water. This water was from a source that had not been properly treated to limit erosion of the lead pipes that carried the water into homes. As a result, the water damaged the pipes, enabling the poisonous mineral to enter the water (*cause*).

Epidemiologists often use data collected from physical examinations and laboratory tests (for example, analysis of blood components) of people to determine factors that affect the population's health. Additionally, they may obtain other kinds of information by conducting surveys. Such surveys typically include questions about subjects' personal and family medical histories, environmental exposures, and health practices. Since 1999, National Health and Nutrition Examination Survey (NHANES) researchers have used interviews, physical examinations, and blood tests to assess the health and nutritional status of a nationally representative sample of Americans. Data collected from NHANES help epidemiologists determine the prevalence of major diseases, risk factors for such diseases, and national standards for measurements that are associated with health status, including height, weight, and blood pressure.

In cases involving chronic diseases such as heart disease and cancer, it is often difficult to determine a single variable that is responsible for the development of the condition. Multiple factors, including a person's *genetic susceptibility* (inherited proneness), exposure to certain environmental conditions, and diet, as well as other lifestyle practices, can influence their likelihood of developing a disease.

Nutritional epidemiologists study how long-term dietary choices influence people's health. By conducting studies that explore differences in dietary practices and disease occurrences among populations, nutritional epidemiologists may learn much about the influence of diet on health. If one group of people is more likely to develop a certain health disorder than another group and the two populations consume very different diets, the scientists can suggest the role that diet plays in this difference.

During a dietary survey, trained interviewers often use a variety of serving-size aids to help study participants recall portion sizes of the foods and beverages they consumed.
Faye McDonald Smith/NHANES/CDC

Observational Epidemiological Studies

Most epidemiological research is observational; scientists collect diet and health information and look for associations between these factors and health outcomes. In a **case–control study,** individuals with a health condition (cases) such as heart disease or breast cancer are matched to persons with similar characteristics who do not have the condition (controls). Information such as personal and family medical histories, eating habits, and other lifestyle behaviors are collected from each participant in the study. By analyzing the results of case–control studies, researchers identify factors that may have been responsible for the illness. Scientists, for example, may be able to identify dietary practices that differ between the two groups, such as long-term fruit and vegetable intakes (Fig. 2.3).

In a **cohort study,** epidemiologists collect and analyze various kinds of information (variables) about a large group of people (a cohort) over time. The scientists are generally interested in making associations between exposure to a specific factor and the subsequent development of health conditions. Cohort studies can be *retrospective* or *prospective.*

Retrospective and Prospective Cohort Studies Retrospective means "to look back" and prospective means "to look forward" in time (Fig. 2.4). In a retrospective cohort study, researchers collect information about a group's past exposures

case-control study study that compares individuals who have a health condition with persons who have similar characteristics but are healthy

cohort study study that measures variables of a group of people over time

Figure 2.3 Case-control study. People with a health condition, such as females with breast cancer (cases), are matched with females who have similar characteristics but are healthy (controls). By conducting case-control studies, scientists may be able to identify long-term dietary practices that differ between the two groups, which may be associated with the development of the health problem.

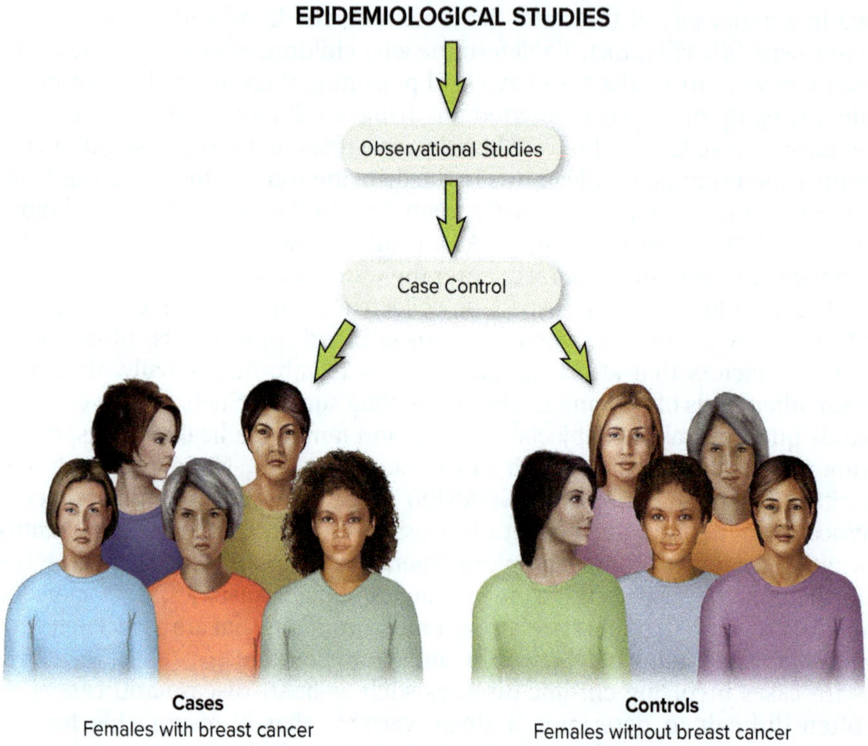

Figure 2.4 Cohort studies. Epidemiologists collect and analyze various kinds of information about a large group of people, a cohort, over time. The scientists make associations between exposure to a specific factor and the subsequent development of health conditions. Cohort studies can be retrospective or prospective.

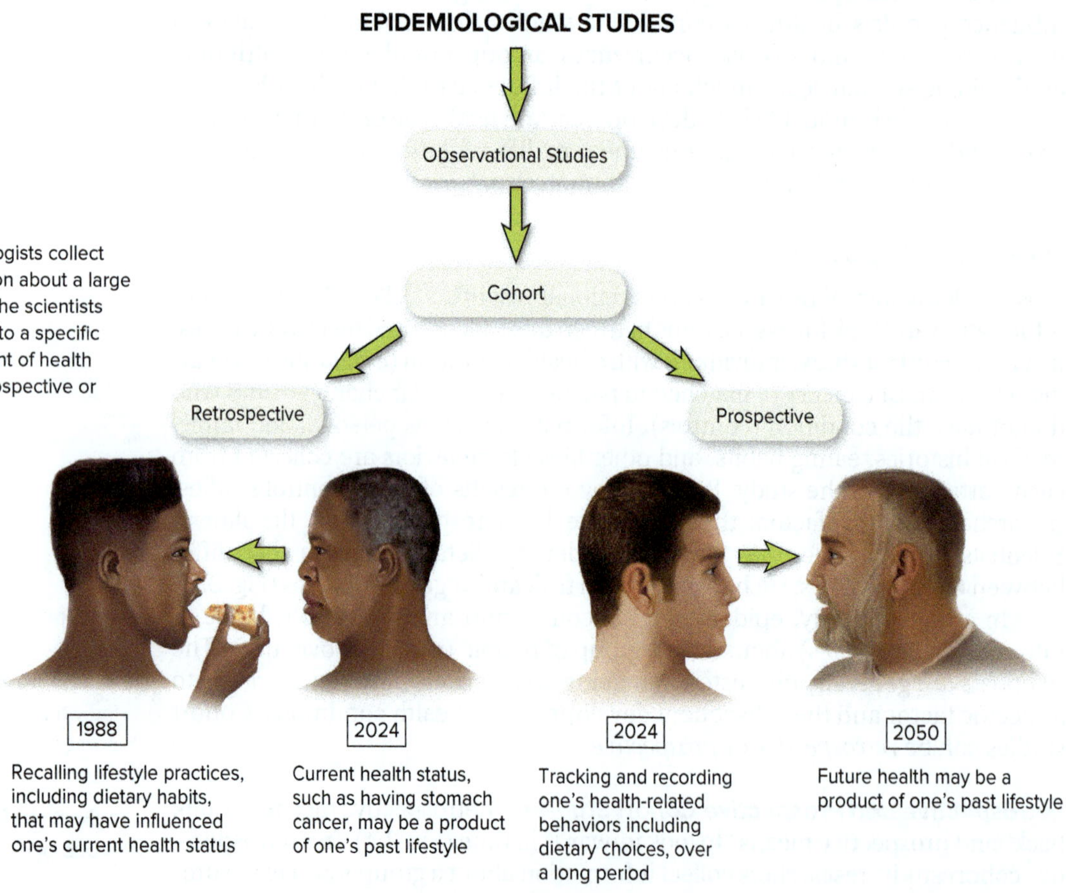

and identify current health outcomes. For example, nutritional epidemiologists might examine whether a group of people who have stomach cancer consumed more charcoal–broiled meat (the exposure) in the past than a group of people with similar characteristics who do not have stomach cancer. Retrospective dietary studies typically collect data about peoples' past food–related behaviors. Researchers, for example, may ask adult subjects to remember which foods were eaten, how much of the items was consumed, and how the foods were prepared when they were teenagers. Because individuals' long–term memories are often faulty, people who participate in retrospective cohort studies may provide inaccurate dietary information to the investigators. Thus, the findings of such retrospective studies can show relationships between past dietary practices and current health conditions, but they do not determine that the dietary habits caused the health problems.

In a prospective cohort study, a group of healthy people are interviewed, tested, and then followed over a time period. Any diseases that subjects eventually develop are recorded. At the end of the study, the scientists analyze the data they obtained and try to identify links between exposures and diseases that occurred between the beginning and end of the study period.

The Framingham Heart Study that began in 1949 in Framingham, Massachusetts, is one of the most well–known prospective studies. At the beginning of the study, over 5200 healthy participants (males and females) underwent extensive physical examinations and questioning about their family and personal medical histories, as well as their lifestyle practices. Over the following years, a group of medical researchers periodically collected data concerning each participant's health and, if the person died, cause of death. The scientists analyzed this information and found relationships between a variety of personal characteristics and health outcomes. Findings from the Framingham Heart Study identified numerous risk factors for heart disease, including elevated blood cholesterol levels, cigarette smoking, and hyper–tension. Today, medical researchers are still collecting information from the original Framingham Heart Study participants, as well as their descendants.

Limitations of Epidemiological Studies

Epidemiological studies cannot establish *causation,* that is, whether one factor causes another. Instead, these studies determine if simultaneously occurring events within a population may be correlated. A **correlation** is a relationship between variables. A correlation occurs when two variables change over the same period; for example, when a population's intake of sugar–sweetened soft drinks increases, the percentage of overweight people in the population also increases (Fig. 2.5a).[5] In this case, the correlation is *direct* or *positive* because the two variables—body weight and regular soft drink consumption—are changing in the same direction: They are both increasing. An *inverse* or *negative* correlation occurs when one variable increases and the other one decreases (the variables change in opposite directions). An example of an inverse correlation is the relationship between fruit intake and hypertension; as a population's fruit consumption increases, the percentage of people with hypertension in that population decreases (Fig. 2.5b).[6]

What appears to be a correlation between a behavior and an outcome could be a coincidence, which is a chance happening rather than a true relationship between variables. For example, in a survey of lemonade consumption in Colorado over a 10–year period, we might observe that fewer people drank lemonade during the winter than during the summer months. In a survey of snow skiing accidents in Colorado during the same 10–year period, we might also find that snow skiing accidents were more likely to occur during the winter than in the summer. Thus, as lemonade consumption declined, snow skiing accidents

correlation relationship between two variables

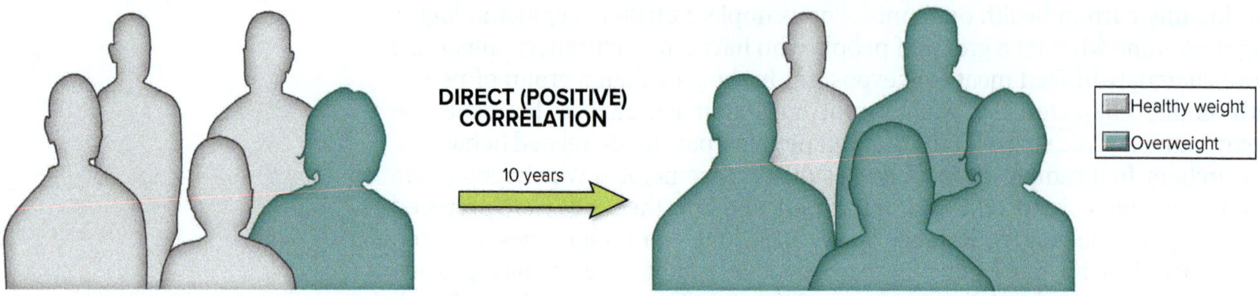

(a) As a population's intake of sugar-sweetened soft drinks increases (↑), the percentage of overweight people in that population increases (↑).

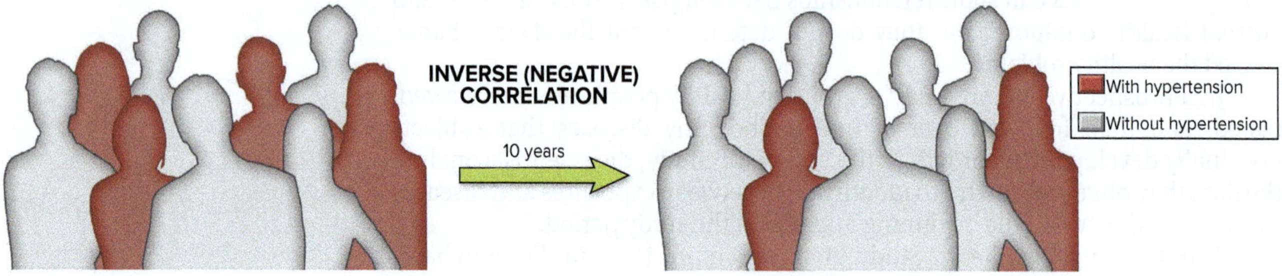

(b) As a population's fruit intake increases (↑), the percentage of people in that population with hypertension decreases (↓).

| **Figure 2.5** Examples of (a) direct (positive) and (b) inverse (negative) correlations.

increased. Does this mean lemonade consumption is inversely correlated to skiing accidents and people who do not drink lemonade have a greater risk of having a skiing accident at this time of year? It is more likely that the relationship between snow skiing and lemonade drinking is coincidental because both activities are associated with seasonal weather conditions. Although this example is obviously far-fetched, it illustrates the problems scientists can have when analyzing results of epidemiological studies.

Analyzing Data, Drawing Conclusions, and Reporting Findings

C Squared Studios/Photodisc/Getty Images

Nutrition researchers use a variety of statistical methods to analyze data collected from observations and experiments. These methods may enable the researchers to find relationships between the variables and health outcomes that were studied. As a result, scientists can determine whether their hypotheses are supported by the data.

When an experiment or study is completed and the results analyzed, researchers summarize the findings and seek to publish articles with information about their investigation in scientific journals. Before articles are accepted for publication, they undergo **peer review,** a critical analysis conducted by a group of "peers." Peers are investigators who were not part of the study but are experts involved in related research. If peers agree that a study was high-quality and well conducted, its results are fairly represented, and the research is of interest to the journal's readers, these scientists are likely to recommend that the journal's editors publish the article. Examples of peer-reviewed medical and nutrition journals include the *Journal of the Academy of Nutrition and Dietetics, The New England Journal of Medicine, Journal of the American Medical Association,* and *Journal of the American College of Nutrition.*

peer review expert critical analysis of a research article before it is published

Research Bias

Scientists expect other researchers to avoid relying on their personal attitudes and biases ("points of view") when collecting and analyzing data and to evaluate and report their results objectively and honestly. This process is important

Did *YOU* Know?

By the middle of the twentieth century, the leading causes of death in the United States shifted from infectious diseases to chronic diseases, particularly heart disease. At that time, medical researchers began to suspect that certain dietary components played a major role in the development of this major killer of Americans. Two scientists, John Yudkin and Ancel Keys, supported different hypotheses concerning dietary "causes" of heart disease. Yudkin thought the consumption of added sugars caused heart disease, while Keys thought the disease was caused by diets that contained too much fat, saturated fat, and cholesterol. By the 1980s, Yudkin's sugar hypothesis had been largely dismissed by medical and nutrition scientists, and many Americans heeded nationally promoted recommendations to reduce cholesterol, total fat, and saturated fat intakes.

In September 2016, a paper published in *JAMA Internal Medicine* exposed the likelihood that in the mid-1960s, people in the sugar industry used their influence to divert attention away from sugar's role in the development of heart disease.[7] This paper revealed correspondence between individuals who worked for the Sugar Research Foundation (SRF) and a small group of well-known and respected nutrition scientists. SRF was closely associated with the sugar industry. In 1965–1966, SRF paid the nutrition researchers to prepare a paper that reviewed scientific literature concerning the role of sugar and lipids in the development of heart disease. In 1967, *The New England Journal of Medicine* published the two-part paper, which challenged the strength and credibility of the scientific evidence that identified sugar consumption as a contributor to heart disease. The paper also blamed dietary cholesterol and saturated fat as the cause of the disease. Although the 1967 article listed the authors' affiliations and sources of support for their research, it did not mention that the authors had received money from the SRF for the paper's preparation. Today, reliable scientific journals require authors to disclose their sources of funding, including organizations and for-profit companies established by the food and wellness industries to promote specific products.

In the mid-1960s, the sugar industry used its influence to divert attention away from sugar's role in the development of heart disease.
Iconotec/Glowimages

because much of the scientific research that is conducted in the United States is supported financially by the federal government, nonprofit foundations, and drug companies and other private industries. Some funding sources can have certain expectations or biases about research outcomes, and as a result, they are likely to finance studies of scientists whose research efforts support their interests. The beef industry, for example, might not fund scientific investigations to find connections between high intakes of beef and the risk of certain cancers. On the other hand, the beef industry might be interested in supporting a team of scientists whose research indicates that a high-protein diet that contains plenty of beef is useful for people who are trying to lose weight.

Scientists whose research efforts depend on food, dietary supplement, or pharmacological industry support may emphasize positive research findings and minimize negative findings. In these instances, the scientists may have "conflicts of interest" because they have to choose between reporting the facts and keeping their financial support.

Peer-reviewed nutrition journals usually require authors of articles to disclose their affiliations and sources of financial support. Such disclosures may appear on the first page or at the end of the article. By having this information, readers can decide on the reliability of the findings. Although peer review helps ensure that the scientists are as ethical and objective as possible, it is impossible to eliminate all research bias.

Spreading the News After the results of a study are published in a nutrition-related journal or reported to health professionals attending a meeting of a nutrition or medical society, the media (e.g., Internet and television news sources, newspapers, and magazines) may receive notice of the findings. If the

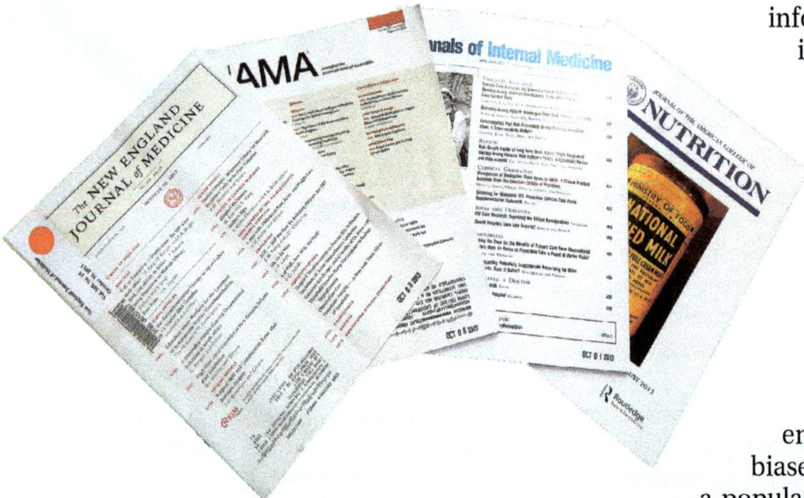

Examples of peer-reviewed medical and nutrition journals. Many journals have online versions and contain articles that can be downloaded for free or available for purchase.
©Wendy Schiff

information is simplistic and sensational, such as a finding that drinking green tea can result in weight loss, it is more likely to be reported in the popular press. In many instances, you learn about the study's results when they are reported in a television or radio news broadcast as a 15- or 30-second "sound bite." Such sources generally provide very little information concerning the way the study was conducted or how the data were collected and analyzed.

Popular sources of nutrition information, such as magazines and the Internet, generally do not subject articles or blogs to peer review or other scientific scrutiny, and as a result, they may feature faulty, biased information. For example, a health news column in a popular magazine may report findings from a few nutrition journal articles that support the use of garlic supplements for reducing blood cholesterol levels. However, you may conclude that the column is biased if it excludes results of other studies that do not indicate such benefits. You can often distinguish a peer-reviewed scientific journal from a popular magazine simply by looking at their covers and skimming their pages. Compared to scientific journals, magazines typically have more colorful, attractive covers and photographs, and their articles are shorter and easier for the average person to read (Fig. 2.6).

It is important to keep in mind that sensational media coverage of a medical "breakthrough" is not necessarily an indication of the value or quality of research that resulted in the news story or magazine article. The results of one study are rarely enough to gain widespread acceptance for new or unusual findings or to provide a basis for nutritional recommendations. Thus, the findings obtained by one research team must be supported by those generated in other studies. If the results of several scientific investigations conducted under similar conditions confirm the original researchers' conclusions, then these findings are more likely to be accepted by other nutrition scientists.

Figure 2.6 Judging by the cover. Consumers can be trained to look for features, such as "busy," brightly colored covers, that distinguish popular sources of nutrition information from peer-reviewed scientific journals. Also, popular sources of nutrition information, such as these magazines, are readily available at supermarkets.
©Wendy Schiff

Confusion and Conflict

One day, the news highlights dramatic health benefits from eating garlic, dark chocolate, brown rice, or cherries. A few weeks later, the news includes reports of more recent scientific investigations that do not support the earlier findings. When consumers become aware of conflicting results generated by nutrition studies, they often become confused and disappointed. As a result, some people may mistrust the scientific community and think nutrition scientists do not know what they are doing.

Consumers need to recognize that conflicting findings often result from differences in the ways various studies are designed. Even when investigating the same question, different groups of scientists often conduct their studies and analyze the results differently. For example, the numbers, ages, and physical conditions of subjects; the type and length of the study; the amount of the treatment provided; and the statistical tests used to analyze results typically vary among studies. Additionally, individual genetic differences often contribute to a person's response to a treatment. Not only are people genetically different, but they also have different lifestyles, and they typically recall dietary information and follow instructions concerning health care practices differently. These and other factors can influence the results of nutrition research involving human subjects.

The science of nutrition is constantly evolving; old beliefs and practices are discarded when they are not supported by more recent scientific evidence, and new principles and practices emerge from the new findings. By now you should understand that science involves asking questions, developing and testing hypotheses, gathering and analyzing data, drawing conclusions from data, and, sometimes, accepting change.

Concept CHECKPOINT

1. What is the next step in the scientific method after conducting a research study and collecting the data?
2. Why is it important to have a control group when conducting experimental research?
3. What is the major difference between a prospective study and a retrospective study?
4. What is a placebo? Why are placebos often used in studies involving human subjects?
5. What is the difference between a direct correlation and an inverse correlation between two variables?
6. What is a peer-reviewed article?
7. Explain why results of similar studies may provide different findings.

See Appendix F for responses.

2.2 Nutrition Information: Fact or Fiction

Learning Outcomes

1. Explain why there is so much nutrition misinformation.
2. Discuss how people can become more critical and careful consumers of nutrition information, including information on the Internet.
3. Identify common red flags that are signs of nutrition misinformation.

While surfing social media one afternoon, you stop and watch a video by an *influencer* (a person who has a reputation for knowledge on a specific topic) who promotes FatMegaMelter, a dietary supplement for losing weight. According to the influencer, the supplement contains a chemical derived from an herb that grows naturally in South Africa. This amazing chemical reduces the appetite for fattening foods, enabling an overweight person taking FatMegaMelter to lose up to 30 pounds in 30 days, without the need to exercise more or eat less. The influencer also claims to have lost a lot of weight after they started taking FatMegaMelter pills. A few days later, a friend mentions losing 3 pounds since they began taking this product a week ago. You would like to lose a few pounds without resorting to restricting your food intake or exercising. Should you take FatMegaMelter? The supplement helped the influencer and your friend; will it help you?

Although the influencer's health history appears to be compelling evidence that the weight–loss supplement is effective, her information is a **testimonial,** a personal endorsement of a product. People are usually paid to provide their testimonials for advertisements; therefore, their remarks may be biased in favor of the product. Your friend's experience with taking the same weight–loss product is intriguing, but it is an anecdote and not *proof* that FatMegaMelter promotes weight loss. When your source of nutrition information is a testimonial, anecdote, or advertisement, you cannot be sure that the information is based on scientific facts and, therefore, reliable.

MY DIVERSE PLATE

A taboo (*tah-boo'*) is a practice that is forbidden by a religion or culture. Taboos often exclude the consumption of certain foods and beverages, such as prohibitions against eating pork or drinking alcohol. Other food taboos are practiced by people with specific roles within a community, for example, tribal chiefs or warriors. In some cultures, pregnant females avoid specific foods because the population generally believes that eating such items may prolong labor or cause birth defects. Furthermore, females who are breastfeeding may refuse to eat a particular food because they accept a traditional belief that consuming it will reduce their milk production.

In many instances, food taboos have no basis in scientific fact, that is, there are no nutrition-related or scientific reasons why a particular food should be prohibited by one group of people and allowed by another group.[8] Nevertheless, food taboos tend to be cultural traditions that are passed down from generation to generation.

Religions often incorporate food prohibitions into their laws of personal conduct. For example, Muslims and Jews who follow their religions' dietary laws do not eat pork. Pigs can harbor tiny parasitic worms that can cause illness (*trichinosis*) in people who eat raw or improperly cooked pork (see Chapter 12). In addition to health concerns, economic and environmental factors also may have contributed to the dietary pork taboo among followers of Islam and Judaism. Many centuries ago, these two religions originated among nomadic populations living in the Middle East, where it was not practical to raise pigs for food. The animals eat a wide variety of foods, and compared to humans, pigs have a much higher requirement for water. As result, they can compete with people for whatever food and water are available. Thus, it is likely multiple factors contributed to the development of religious laws that prohibited the consumption of pork.

Anatolii Tsekhmister/123RF

testimonial personal endorsement of a product

Be wary of social media influencers using only testimonials and anecdotes to promote a product.

Become a Critical Consumer of Nutrition Information

People may think they have learned facts about nutrition by reading popular magazine articles or best–selling books; visiting Internet websites; or watching social media videos, television news, infomercials, or home shopping programs. In many instances, however, they have been misinformed. To be a careful consumer, do not assume that all nutrition information presented in the popular media is reliable. The First Amendment to the U.S. Constitution guarantees freedom of the press and freedom of speech, so people can provide nutrition information that is not true. Thus, the First Amendment does not protect consumers with freedom from nutrition misinformation or false nutrition claims.

The U.S. Food and Drug Administration (FDA) can regulate nutrition– and health–related claims on product labels, but the agency cannot prevent the spread of health and nutrition misinformation published on the Internet, on social media, or in books or pamphlets, or presented in television or radio programs. As a consumer, you are responsible for questioning and researching the accuracy of nutrition information, as well as the credentials of the people making nutrition–related claims.

Promoters of worthless nutrition products and services often use sophisticated marketing methods to lure consumers. For example, some promoters of dietary supplements claim their products are "scientifically tested," or they include citations to what appear to be scientific journal articles in their ads or articles. Consumers, however, cannot be certain if the information is true. In some instances, these products have been scientifically tested, but much of the research that has been conducted using reliable methods has shown that most dietary supplements, other than vitamins and minerals, provide little or no measurable health benefits. Nevertheless, promoters of nonnutrient dietary supplements continue to sell their goods to an unsuspecting, trusting public. The "Nutrition Matters" section at the end of this chapter focuses on dietary supplements.

Consumers also need to be alert for promoters' use of **pseudoscience,** the presentation of information masquerading as factual and obtained by scientific methods. In many instances, pseudoscientific nutrition or physiology information is presented with complex scientific–sounding terms, such as "enzymatic therapy" or "colloidal extract." Such terms are designed to convince people without science backgrounds that the nutrition–related information is true. Often, promoters of nutrition misinformation try to confuse people by weaving false information with facts into their claims, making the untrue material seem credible too.

Although people's lives have improved as a result of scientific advancements in medicine, the general public tends to mistrust scientists, medical professionals, and the pharmaceutical industry. Promoters of nutrition misinformation exploit this mistrust to sell their products and services. For example, they may tell consumers that scientists are hiding the cure for chronic diseases such as cancer, and physicians rely on costly diagnostic methods and treatments for serious diseases because they are more interested in making money than doing what is best for their patients, such as recommending a dietary supplement. Are some scientists and physicians driven by the desire to make money from people's illnesses, and do they hide information about natural cures from them?

Over the past 100 years, people's lives have been greatly improved by contributions of medical researchers. As a group, scientists and physicians are dedicated to improving people's health and saving lives. Physicians strive to diagnose and treat diseases using scientifically tested and approved

pseudoscience presentation of information masquerading as factual and obtained by scientific methods

techniques. Moreover, physicians may face a malpractice lawsuit if they fail to diagnose and treat a condition effectively. Additionally, physicians have much to gain from treating their patients kindly and effectively. Consider this: If you follow a physician's advice and have positive results, are you likely to be that doctor's patient for a long time and recommend the practitioner to others?

If your car is not functioning properly, you probably would want people who have the best training, tools, and equipment to determine the problem and repair it. If you think something is wrong with your body, it is prudent to seek information and opinions from medical professionals who have the best training and experience to diagnose and treat health disorders.

What motivates promoters of nutrition-related products and information?

Ask Questions

If you are like most people, you do not want to waste your money on things you do not need or that are useless or potentially harmful. How can you become a more careful, critical consumer of nutrition-related information or products? The following questions should help you evaluate various sources of nutrition information:

- *What motivates the authors, promoters, or sponsors to provide the information? Do you think they are more interested in your health and well-being or in selling their products?* Salespeople often have favorable biases toward the things they sell, and, therefore, they may not be reliable sources of information about these products. A clerk in a dietary supplements outlet store, for example, may wear a white lab coat to look as though they have a science or medical education background, but you should keep in mind that the clerk was hired to sell dietary supplements and may have little or no scientific training. Furthermore, salespeople who work in such outlets may be unwilling to inform customers about the potential health hazards of taking certain products, particularly when they earn a commission from each sale.
- *Is the source scientific, such as an article from a peer-reviewed nutrition journal?* In general, popular sources of nutrition information, such as the Internet, best-selling books, and articles in magazines, are not peer-reviewed by scientists. Additionally, radio or TV programs that promote nutrition information may actually be sophisticated advertisements for nutrition-related products.
- *If a study is mentioned (cited), how was the research conducted? Did the study involve humans or animals? If people participated in the study, how many subjects were involved in the research? Who sponsored the study?* As mentioned earlier in this chapter, epidemiological studies are not useful for finding cause-and-effect relationships. Additionally, the results of studies involving large numbers of human subjects are more reliable than studies of animals. Sponsors may influence the outcomes of the studies they fund.
- *To provide scientific support for claims, does the source refer to articles in respected nutrition or medical journals or mention reliable experts?* Be careful if you see citations in popular nutrition or health books and magazines. Promoters of nutrition misinformation may refer to scientific-appearing citations from phony medical journals to convince people that their information is reliable. Furthermore, be wary of nutrition experts introduced or

If this person appeared on a television show or in a magazine advertisement, would you consider him to be a reliable source of nutrition information? Explain why you would or would not trust his nutrition-related advice.
Stockbyte/Stockdisc/Getty Images

identified as "Doctor" because they may not be physicians or scientists. Furthermore, a so–called nutrition expert who is referred to as "Doctor" may not have a doctorate degree (Ph.D.) in human nutrition from an accredited university. Such experts may have obtained their degrees simply by purchasing them on the Internet or through a mail–order outlet, without having taken appropriate coursework or graduated from an accredited university or college.

Practicing medicine without the proper training and licensing *is* illegal. However, providing nutrition information and advice without the proper training and licensing *is* legal. **Quackery** involves promoting useless medical treatments, such as copper bracelets to treat arthritis.

To obtain information about a nutrition expert's credentials, enter the person's name at an Internet search engine and evaluate the results. For example, is the person associated with an accredited school of higher education or a government agency such as the Centers for Disease Control and Prevention or U.S. Department of Agriculture? You can also visit www.quackwatch.org and submit an "Ask a Question" e–mail requesting information about a person's credentials from the site's sponsors.

Did *YOU* Know?

In May 2020, a federal court in eastern Oklahoma issued a temporary restraining order that stopped the owners of a business from marketing and distributing a silver-containing product as a cure or treatment for COVID-19. According to the federal court's allegations, the product contained "colloidal silver," which is generally not recognized as safe and effective for preventing or treating COVID-19 (or any other medical condition).[9] Furthermore, the Food and Drug Administration considered the Oklahoma company's colloidal silver preparation to be an unapproved new drug. In response to the court's swift action, the business owners removed their questionable product's website from the Internet and provided online instructions for customers to obtain refunds.

Look for Red Flags

To become a critical consumer of nutrition information, you need to be aware of red flags, clues that indicate a source of information is unreliable. Common red flags include the following:

1. **Promises of quick and easy remedies for complex health–related problems:** "Our product helps you lose weight *without* exercising or dieting," or "Garlic cures heart disease."
2. **Claims that sound too good to be true:** "Our all–natural product blocks fat and calories from being absorbed, so you can eat everything you like and still lose weight." Such claims are rarely true. Remember, if the claim sounds too good to be true, it probably is not true.
3. **Scare tactics that include sensational, frightening, false, or misleading statements about a food, dietary practice, or health condition:** "Dairy prod–ucts cause cancer," or "Eating sugar causes hyperactivity."
4. **Personal attacks on the motives and ethical standards of registered dietitian nutritionists or conventional scientists:** "Dietitians and physicians don't want you to know the facts about natural cures for cancer, diabetes, and heart disease because they'll lose money if you get well." Such statements indicate unsubstantiated biases against bona fide nutrition experts and the scientific community.
5. **Statements about the superiority of certain dietary supplements or uncon–ventional medical practices:** "Russian scientists have discovered the count–less health benefits of taking Siberian ginseng," or "Colon cleansing with herbs is the only cure for intestinal cancer."
6. **Testimonials and anecdotes as evidence of effectiveness:** "I lost 50 pounds in 30 days using this product," or "I rubbed this vitamin formula on my scar

quackery promotion of useless medical treatments

and it disappeared in days." Reliable nutrition information is based on scientific evidence, not testimonials and anecdotes.

7. **Information that promotes a product's benefits while overlooking its risks:** "Our all–natural supplement boosts your metabolism naturally so it won't harm your system." Anything you consume, even water, can be toxic in high doses. Beware of any source of information that fails to mention the possible side effects of using a dietary supplement or nutrition–related treatment.
8. **Vague, meaningless, or scientific–sounding terms to impress or confuse consumers:** "Our *all–natural, clinically tested, patented, chelated* dietary supplement works best."
9. **Sensational statements with incomplete references of sources:** "Clinical research performed at a major university and published in a distinguished medical journal indicates food manufacturers add ingredients to their products that make you hungry and fat," or "Millions of Americans suffer from various nutritional deficiencies." Which "major university" and "distinguished medical journal"? What study reported that "millions of Americans" are deficient in nutrients?
10. **Recommendations based on a single study:** "Research conducted at our private health facility proves coffee enemas can cure cancer."
11. **Information concerning nutrients or human physiology that is not supported by reliable scientific evidence:** "Now, you can combine certain foods based on your blood type," or "People with alkaline bodies don't develop cancer."
12. **Results disclaimers, usually in small or difficult–to–read print:** "Results may vary," or "Results not typical" (Fig. 2.7). Disclaimers are clues that the product may not live up to your expectations or the manufacturer's claims.

Figure 2.7 Results disclaimer.

Using the Internet Wisely

You can find abundant sources of information about nutrition and the benefits of dietary supplements on the Internet. However, you must be careful and consider the sources. Who or what organization sponsors the site? Is the information intended to promote sales? Be wary if the site discusses benefits of dietary supplements and enables you to purchase these products online. Furthermore, a site is likely to be unreliable if it includes comprehensive disclaimers such as "The manufacturer is not responsible or obligated to verify statements." Also avoid sites that publish disclaimers such as "The nutrition and health information at this site is provided for educational purposes only and not as a substitute for the advice of a physician or registered dietitian nutritionist. The author and owner of this site are not liable for personal actions taken as a result of the site's contents."

Be wary of websites that are authored or sponsored by one person, or sites that promote or sell products for profit (*.com) because such sources of information may be biased. In general, websites sponsored by nationally recognized health associations such as the Academy of Nutrition and Dietetics (www.eatright.org) and nonprofit organizations such as the American Cancer Society (www.cancer.org) are reliable sources of nutrition information (*.org). Government agencies (*.gov) such as the Centers for Disease Control and Prevention (www.cdc.gov) and nationally accredited colleges and universities

Did *YOU* Know?

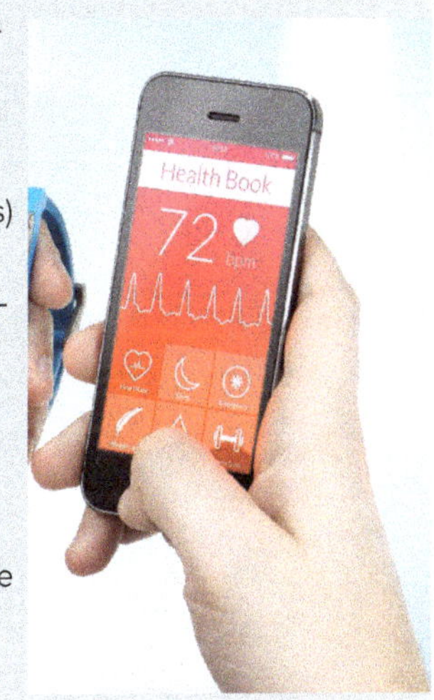

Alexey Boldin/Shutterstock

In 2019, researchers reviewed the top 7 smartphone diet tracking applications (apps) consumers used. The scientist compared a nutrient analysis of foods from each app to the USDA nutrient database, the standard database for professional nutrient analysis programs. The number of calories and carbohydrates in the foods reported by the smartphone applications was primarily accurate; however, the amount of fat and protein reported in foods was underestimated.[10] Nonetheless, the applications were determined to be easy to use and helped the user track their food and beverage intake and monitor progress towards achieving their diet and health-related goals. Choosing the most reliable food, nutrition, and fitness-related apps to use can be difficult. You can make an informed decision by searching online for reliable reviews and ratings of these apps before using them.

TABLE 2.1 Tips for Searching Nutrition Information on the Internet

To Be a Careful Consumer of Internet Sources of Information:
1. Use multiple sites, especially government sites, as well as the sites of nationally recognized nutrition- or health-related associations such as the American Diabetes Association (www.diabetes.org) and the American Heart Association (www.heart.org). Use PubMed (www.ncbi.nlm.nih.gov/pubmed/) as a resource when searching for science-based published articles.
2. Rely primarily on sites that are managed or reviewed by qualified health professionals. Blogs might be fun and interesting to read, but they are not necessarily reliable.
3. Look for the HONcode symbol at the bottom of the main page of the website. The Health On the Net Foundation is a nonprofit, international organization that promotes the HONcode, a set of principles for standardizing the reliability of health information on the Internet. For more information, visit the organization's website (www.hon.ch/).
4. Do not trust information at a site that does not indicate valid sources, such as well-respected peer-reviewed scientific journals or nationally recognized medical centers. Contributing authors and their credentials should be identified; when they are, perform an online search of the scientific journals, as well as the authors' names and credentials, to determine their validity.
5. Be wary of sites that have surveys for you to complete, advertisements for diet-related products, and promotions in pop-up windows and that provide online diagnoses and treatments.
6. Do not trust a site that includes attacks on the trustworthiness of the medical or scientific establishment.
7. Be wary of commercial sites (*.com) with links to government sites or the sites of well-known medical, nutrition, or scientific associations. An unreliable *.com site can be linked to reliable sites without having received their endorsements.

Health On the Net Foundation. www.hon.ch/en

Source: Adapted from: National Institutes of Health: *How to evaluate health information on the Internet: Questions and answers.* 2022. https://ods.od.nih.gov/Health_Information/How_To_Evaluate_Health_Information_on_the_Internet_Questions_and_Answers.aspx. Accessed: April 7, 2023.

(*.edu) that have nutrition and dietetics programs are also excellent sources of credible nutrition information. Table 2.1 presents some tips for using the Internet to obtain reliable nutrition information.

The Federal Trade Commission (FTC) enforces consumer protection laws and investigates complaints about false or misleading health claims that appear on the Internet. For information to help you evaluate nutrition and health–related claims, visit the agency's website (www.consumer.ftc.gov/articles/0167–miracle–health–claims). To complain about a product, you can complete and submit the FTC's complaint form at the website or call the agency's toll–free line (1–877–382–4357).

Food & Nutrition tips

If you are looking for a recipe for guacamole, cranberry chutney, or sweet and sour cabbage, simply use the Internet—it is like having a cookbook at your fingertips. Recipes for nearly every food and from various countries and ethnic groups are available on the Internet, and you can find cooking tips from this vast resource.

When searching the Internet for recipes, you need to recognize that many food manufacturers use the sites to promote their products in recipes. If the brand name of a product is mentioned in the recipe, you can usually substitute another company's product. Additionally, recipes may not provide accurate information about the nutrients and calories in a serving of food, and you cannot be certain that the recipes have been tested for quality. Therefore, it is a good idea to check more than one site for a recipe and compare the information.

Concept CHECKPOINT

8. How do testimonials and anecdotes contribute to misinformation?

9. List at least three red flags that may indicate a questionable source of nutrition information.

10. List at least three tips for using the Internet as a reliable source of nutrition information.

See Appendix F for responses.

2.3 Reliable Nutrition Experts

Learning Outcome

1. Explain how to identify reliable nutrition experts.

If you have questions about food or nutrition, where do you find factual answers? Although many states regulate and license people who call themselves nutritionists, you cannot always rely on someone who refers to themselves as a "nutritionist" or "nutritionalist" for reliable nutrition information because there are no standard legal definitions for these descriptors. Should you ask a physician for nutrition advice? Physicians are not necessarily the best sources of nutrition information because most doctors do not have extensive college coursework in the subject.

If your university or college has a nutrition or dietetics department, you are likely to find nutrition experts, including professors, registered dietitians, and registered dietitian nutritionists, who are faculty members. A **registered dietitian (RD)** or **registered dietitian nutritionist (RDN)** is a college-trained health care professional who has extensive knowledge of foods, nutrition, and *dietetics,* the application of nutrition and food information to help treat many health-related conditions. The titles "registered dietitian (RD)" and "registered dietitian nutritionist (RDN)" are legally protected. This means people cannot legally refer to themselves as RDNs unless they have been certified by the appropriate accrediting agency.

You can also locate RDNs by consulting online directories, contacting your local dietetic association or dietary department of a local hospital, or visiting the Academy of Nutrition and Dietetics website (www.eatright.org) or the Dietitians of Canada website (www.dietitians.ca).

©Wendy Schiff

Becoming a Registered Dietitian Nutritionist

There are three major professional divisions for RDNs: clinical dietetics, community nutrition, and food systems management. Clinical dietitians can work as members of medical teams in hospitals or clinics. Community dietitians typically work in public health settings or as dietary counselors in private practice or with wellness programs. Food systems management dietitians direct food service systems in hospitals, schools, and settings where healthy foods are prepared and served to employees or residents of long-term care facilities. Although most RDNs work in health care settings, some are educators or researchers.

registered dietitian (RD) or **registered dietitian nutritionist (RDN)** college-trained health care professional who has extensive knowledge of foods, nutrition, and dietetics

As of January 2024, all registered dietitian nutritionists must have completed a master's degree program approved by the Accreditation Council for Education in Nutrition and Dietetics (ACEND) of the Academy of Nutrition and Dietetics, the largest organization of dietitians in the United States. Dietetics majors are required to take a wide variety of college–level courses, including food and nutrition sciences, organic chemistry, biochemistry, anatomy, physiology, microbiology, statistics, management, and communications classes. If you are attending a major university, you can check the course or program catalog at your school to determine whether an accredited dietetics program is offered.

In addition to taking the required college coursework, the student dietitian nutritionist must obtain at least 1000 hours of supervised practice (training), which is often in a health care facility, such as a university hospital. Such training (a *supervised practice program* or *SPP*) prepares the student dietitian nutritionist for an entry–level position upon completion of a graduate degree in dietetics. Student dietitian nutritionists who attend colleges or universities that have accredited dietetics programs but do not offer SPPs can apply for admittance into a *dietetic internship (DI)*. Dietetic internships are accredited training programs that begin after the student graduates. For more information about internships, visit www.eatrightpro.org/acend.

After completing an SPP or DI, the student dietitian nutritionist has to pass the national registration examination to become certified as an RDN. After becoming certified, RDNs are also required to fulfill continuing education requirements to maintain their certification. The Academy of Nutrition and Dietetics provides more information about becoming an RDN and roles these professionals play in health care, community, and other settings.

Did *YOU* Know?

After becoming a credentialed registered dietitian nutritionist, a person can follow a wide variety of career paths. RDNs work primarily in hospitals or other medical settings, but they may also find jobs in:

Prostock-studio/Shutterstock

- Day-care centers, school districts, prisons, long-term care facilities, and universities as managers of food service operations;
- Corporations as developers and managers of corporate wellness programs;
- Pharmaceutical industrial settings as consumer affairs specialists and marketing representatives;
- Sports nutrition as nutrition "coaches" involved in the training of personal fitness enthusiasts, professional sports teams, and members of the military;
- Food and agriculture industry as recipe developers, supermarket dietitians, culinary medicine experts, and advocates for farmers and sustainable farming practices;
- Universities as educators, research scientists, and wellness specialists for campus health centers;
- Governmental agencies and national health organizations as administrators and professional staff of food- and/or nutrition-related programs and advocates for public policy development;
- Media as writers, advisors, and experts who prepare and convey nutrition-related content and programming to the public; and
- Private practice as trained experts who provide dietary and nutrition counseling.

Concept CHECKPOINT

11. What is the difference between a nutritionist and a registered dietitian nutritionist?
12. List three ways of locating reliable nutrition experts.

See Appendix F for responses.

2.4 Nutrition Matters: What Are Dietary Supplements?

©Wendy Schiff

Jill Braaten/ McGraw Hill

Learning Outcomes

1. Explain the difference between conventional medicine and alternative health care, and identify health care practices that are either conventional or alternative.
2. Explain how the FDA regulates medicines differently than dietary supplements.
3. Discuss the risks and benefits of taking dietary supplements.

Do you take a daily vitamin or mineral pill because you are concerned about the nutritional quality of your diet? Do you use herbal pills or extracts to strengthen your immune system, boost your memory, or treat illnesses such as the common cold? If you follow any of these practices, you are not alone. According to results of a recent national survey, over 57% of the American population takes one or more dietary supplements regularly.[11]

According to the Dietary Supplement Health and Education Act of 1994 (DSHEA), a dietary supplement is a product (other than tobacco) that

- adds to a person's dietary intake and contains one or more dietary ingredients, including nutrients or *botanicals* (herbs or other plant material);
- is taken by mouth; and,
- is a concentrate, metabolite, constituent, or extract.[12]

Dietary supplements include nutrient pills, protein powders, and herbal extracts. Many dietary supplements are often referred to as "nutraceuticals," but there is no legal definition for this term.

In the United States, **multivitamin/mineral (MV/M)** products are among the most commonly used dietary supplements.[11] There is no standard definition for the contents of a MV/M supplement, but such products may contain several vitamins and minerals. In 2017–2018, the most popular dietary supplements among American adults were vitamin D, fish oil or "omega 3," vitamin C, and calcium.[11]

Some people use dietary supplements, particularly those containing nutrients, because the products are recommended by their physicians or registered dietitian nutritionists. Physicians, for example, may prescribe a prenatal MV/M supplement for their pregnant patients. Most dietary supplements are available without the need for prescriptions, so people can take them without their doctors' knowledge and approval.

Table 2.2 provides information about some non-micronutrient dietary supplements that are popular among adult Americans, including major claims about their health benefits. Benefits and risks of supplementing diets with micronutrients are discussed in Chapters 8 and 9.

Conventional Medicine versus Alternative Health Care Practices

Conventional ("mainstream") medical practitioners, such as physicians and registered dietitian nutritionists, generally promote and use health care practices that have been scientifically tested and evaluated for safety and effectiveness.

Examples of **conventional medical care** include surgical procedures that remove diseased tissues, FDA-approved medications that reduce high blood pressure levels, and dietary patterns that provide enough nutrients to promote good health. **Alternative health care** promotes and uses practices that may or may not have been tested scientifically for both safety and effectiveness. According to the National Institutes of Health (NIH), the use of natural products such as herbs, vitamins and minerals, and other dietary supplements is an aspect of alternative health care.[13] Popular forms of alternative health care include dietary supplements, yoga, chiropractic manipulations, and meditation.[13]

Many forms of alternative health care are not widely accepted by physicians and other conventional health care providers because scientific studies provide little or no support for their use. However, few Americans rely only on alternative health care. Most people who use alternative therapies combine these therapies with conventional medical care, called *complementary health care*.[13] For example, a female who has trouble falling

multivitamin/mineral (MV/M) describes a dietary supplement that contains vitamins and minerals

conventional medical care health care practices that are widely accepted and used by mainstream medical practitioners

alternative health care health care practices that are not widely accepted and used by conventional medical practitioners

TABLE 2.2 Some Popular Non-micronutrient Dietary Supplements

		Some Known Health Effects	
Dietary Supplement	**Major Claims**	**Benefits**	**Risks**
Alfalfa	Treats kidney problems and improves urine flow Relieves arthritis symptoms	May be effective in lowering total cholesterol and "bad" cholesterol levels, but little scientific evidence that it has other benefits	Consuming alfalfa seeds may be hazardous to health; may increase the likelihood of blood clots; and may lower blood sugar levels, which could be dangerous for people with diabetes
Bitter orange	Causes weight loss and relieves heartburn, upset stomach, and constipation	Applying the oil to skin may help certain skin infections, but there is little scientific evidence to support other health benefit claims.	May increase heart rate and blood pressure, which could result in heart attack and stroke
Chondroitin sulfate	Relieves joint damage associated with osteoarthritis	Studies show "mixed" results concerning effectiveness for treating osteoarthritis, which means results of some studies support the benefits, but results of other studies show no benefits. Often combined with glucosamine	May interfere with blood clotting; should be avoided by males who have a history of prostate cancer
Coenzyme Q-10	Increases exercise tolerance, prevents heart disease and cancer, and reverses signs of aging	More research is needed to determine health benefits.	Not FDA approved for treating cancer; low toxicity but may interfere with certain blood thinners
Echinacea  Ingram Publishing/SuperStock	Boosts immune system Prevents the common cold and reduces cold symptoms	Studies show "mixed" results, regarding the usefulness for preventing colds.	Generally safe for short-term use but may provoke allergic response or intestinal upset
Evening primrose oil	Treats eczema (a common skin condition) Relieves arthritis and menopausal symptoms	Little scientific evidence to support health benefit claims	Long-term safety has not been established.
Feverfew	Prevents migraine headaches	Studies generally indicate mixed results, but feverfew extract may be useful in preventing migraines.	No serious side effects, but may cause gastrointestinal upsets. Pregnant females should avoid feverfew.
Flaxseed and flaxseed oil	Acts as a laxative (flaxseed) Reduces blood cholesterol levels	Nonanimal source of an omega-3 fatty acid Lack of scientific evidence to support flaxseed as a treatment for constipation May benefit people with heart disease, but more research is needed.	Generally safe but may cause diarrhea; seeds should be taken with plenty of water to avoid constipation. Pregnant females should avoid using flaxseed and flaxseed oil.
Garcinia cambogia	Suppresses appetite, promotes weight loss Relieves joint pain and symptoms of intestinal upset	Studies show no significant results regarding the usefulness for weight loss.	Can cause liver damage, headache, and gastrointestinal upset.
Garlic	Lowers blood cholesterol levels	Evidence that garlic reduces blood cholesterol and blood pressure levels is weak.	Raw garlic may cause allergic reaction and unpleasant body odor. May "thin" the blood (make blood clots less likely to form)
Ginger  Stockbyte/Getty Images	Treats "morning sickness" and other forms of nausea Relieves stomach upsets Relieves joint and muscle pain	Can treat morning sickness and may reduce other forms of nausea More studies are needed to support other health benefit claims.	In general, few side effects when taken in small amounts; powdered ginger may cause intestinal upset. Pregnant females should consult with their physician before taking ginger.

TABLE 2.2 Some Popular Non-micronutrient Dietary Supplements *(continued)*

		Some Known Health Effects	
Dietary Supplement	**Major Claims**	**Benefits**	**Risks**
Ginkgo biloba	Improves memory Reduces the risk of Alzheimer disease and other forms of dementia	Lack of scientific evidence to support health benefit claims	May increase the risk of bleeding and cause allergic reactions, headaches, intestinal upsets, and nausea Ginkgo seeds are toxic.
Ginseng (Asian)	Boosts overall health Treats erectile dysfunction (male impotence)	Lack of scientific evidence to support most health benefit claims; may lower high blood sugar levels in some people with diabetes	May cause diarrhea, headaches, sleep disturbances, itching, and nervousness Long-term use may increase risk of toxicity. Pregnant females should avoid ginseng.
Glucosamine sulfate	Relieves joint damage associated with osteoarthritis	Some scientific evidence supports health benefit claims for osteoarthritis, but more research is needed.	Some products are not labeled accurately. Appears to be safe when taken by mouth but may cause mild gastrointestinal side effects, including nausea and heartburn
Green tea  ©Wendy Schiff	Prevents or treats cancer Promotes weight loss Reduces blood cholesterol levels	Green tea extracts may be useful in treating genital warts. Lack of scientific evidence that green tea protects against cancer or reduces cancer cell growth Lack of reliable evidence to support other health benefit claims, including promoting weight loss	In a few cases, green tea extracts were associated with liver problems. Caffeine content may cause sleep disturbances, irritability, and digestive upset.
Kava	Reduces stress	Potential harm of using kava outweighs any benefit.	Toxic—can damage the liver May cause heart problems and yellowed skin
Melatonin	Treats some sleep disorders Prevents jet lag Treats cancer, headaches, and irritable bowel syndrome Delays the aging process	May be effective for treating certain sleep disorders and preventing jet lag Lack of evidence that the supplement is useful for other conditions	Probably safe for short-term use, but long-term effects of its use are unknown. Interacts with several prescription medications and other herbs that produce sedation (calming effects) Common side effects include drowsiness, dizziness, headache, and nausea
St. John's wort  Jennifer Anderson/USDA-NRCS Plants Database	Reduces depression	At this time, most studies do not support using the supplement to treat depression.	May interact with other herbal products and many medications, including prescription antidepressants and oral contraceptives

Source: National Institutes of Health, National Center for Complementary and Integrative Health: *Health topics A-Z*. www.nccih.nih.gov/health; National Institutes of Health: *Drugs, herbs and supplements*. https://medlineplus.gov/druginfo/herb_All.html; National Institutes of Health, National Center for Complementary and Integrative Health: *Herbs at a glance*. www.nccih.nih.gov/health/herbsataglance. Accessed: April 4, 2023.

Scientific studies have shown that meditation can help reduce elevated blood pressure levels.
Witthaya Phonsawat/123RF

asleep decides to take a dietary supplement that contains melatonin in addition to a prescribed anti–anxiety medication, without getting a doctor's approval.

Some forms of alternative health care, such as meditation, mindfulness, yoga, and tai chi, are gaining acceptance among conventional medical practitioners, primarily because there is growing scientific evidence for the practices.[14,15] Health care professionals who practice **integrative medical care** develop treatments that coordinate the use of both conventional medical therapies and alternative health care approaches. For example, a registered dietitian nutritionist may provide conventional dietary counseling to a young person who has high blood pressure and needs to lose 25 pounds. Typically, such counseling includes tips for modifying food preparation practices and increasing physical activity. If the patient tells the dietitian that they overeat when anxious, the dietitian may refer the patient to another health care professional who teaches meditation or provides stress management training. Thus, this dietitian is integrating conventional medical care with alternative health care practices.

How Are Dietary Supplements Regulated?

The FDA is the federal agency that is responsible for ensuring the safety and effectiveness of medications and health–related devices, including dietary supplements.[16]

integrative medical care medical care provided by health care professionals that coordinates conventional medical therapies with alternative health care approaches

The FDA strictly regulates the development, production, and marketing of new medications (drugs). A drug is a substance, natural or human–made, that alters body functions. As a result, drugs can produce both beneficial and harmful effects on the body. Before marketing a new drug, the manufacturer must submit evidence to the FDA indicating that the product has been tested extensively and is safe and effective. If FDA experts have serious concerns about a medication's side effects or if they question its usefulness, the agency may reject the manufacturer's petition to sell the product.

When consumed, many dietary supplements act as drugs in the body. However, the FDA does not regulate dietary supplements as medications. The agency classifies dietary supplements as a nontraditional category of foods and uses different sets of rules to regulate dietary supplements, conventional foods, and medications.[16] As a result, dietary supplement manufacturers can bypass most of the strict regulations that the FDA applies to the introduction of new medications into the marketplace. Supplement manufacturers, for example, generally do not need FDA approval before manufacturing or marketing their products. Additionally, the manufacturers are not required to provide the FDA with scientific evidence indicating their products provide measurable health benefits. Manufacturers, however, must notify the FDA and provide the agency with information about the safety of any supplement that contains dietary ingredients that were not marketed in dietary supplements prior to 1994, unless the substance had been used in foods.

According to the FDA, dietary supplements are not intended to prevent, diagnose, treat, or cure diseases.[17] Thus, dietary supplement manufacturers are not allowed to make claims that are appropriate for medications, such as "reduces pain," "prevents polio," or "treats diabetes." The agency regulates the labeling of dietary supplements, including health–related claims that are permitted on labels. The "Food and Dietary Supplement Labels" section of Chapter 3 provides information about supplement labeling.

The FDA has set standards designed to help ensure the identity, purity, strength, and composition of dietary supplements. The agency, however, does not test dietary supplements to determine whether they meet such quality standards. Manufacturers of dietary supplements can have their products tested for quality by certain nongovernmental agencies, including U.S. Pharmacopeia and ConsumerLab.com. Products that pass this testing may include "seals of approval" on labels, but these seals are no guarantee that the dietary supplements are safe or effective.

The FDA's role in reviewing the safety of dietary supplements begins after products enter the marketplace.[17] Supplement manufacturers are required to keep records concerning reports they receive about serious adverse (negative) health effects that may have been caused by their products. Furthermore, the manufacturers must

The USP (U.S. Pharmacopeia) seal of approval does not guarantee that dietary supplements are safe or effective.
©Wendy Schiff

also inform the FDA about such reports. Consumers and health care professionals can also report health problems that are possibly associated with supplement use directly to the FDA.

When the FDA determines that a particular supplement presents a significant or unreasonable risk of harm, the agency alerts consumers about the risk and seeks to recall the product, that is, initiates efforts to have it removed from the market.[17] In most instances, the manufacturer voluntarily recalls the product after determining there is a problem with it or being notified by the FDA about the problem. In some cases, however, the FDA requests a recall. If the FDA determines that the manufacturer's response to the recall is inadequate, the agency can take enforcement steps, such as initiating legal action against the company to seize products or stop producing the items.

Using Dietary Supplements Wisely

When used properly, many dietary supplements, particularly micronutrient products, are generally safe. Herbal supplements, however, are made from plants that may have toxic parts. Comfrey, pennyroyal, sassafras, kava, lobelia, and ma huang are among the plants known to be highly toxic or cancer-causing. Products containing material from these plants should be avoided. The use of botanical products can also evoke allergic or inflammatory responses that often result in signs and symptoms of skin, sinus, or respiratory illnesses. Herbal teas may contain pollens and other parts of plants that can cause allergies, particularly in people who are sensitive to the herbs or their related species. Echinacea (purple cone flower) is related to ragweed, a plant that is often associated with seasonal respiratory allergies. When people who are allergic to ragweed pollen take supplements that contain echinacea, they may develop allergic responses that mimic symptoms of the common cold (watery eyes, runny nose, and sneezing). In some instances, inflammatory responses, such as asthma attacks, can occur after exposure to echinacea or other plants. Asthma can be a life-threatening condition. Therefore, people who have asthma or allergies should be very careful when using botanical supplements.

Consumers also need to be aware that medicinal herbs may contain substances that affect the usefulness or safety of prescription or over-the-counter medications as well as other herbs. Such interactions can produce unwanted and even dangerous side effects (see Table 2.2). Ginkgo biloba, for example, can interact with aspirin, increasing the risk of bleeding. Garlic, ginseng, and chondroitin sulfate supplements can also increase bleeding. Kava acts as a sedative (calming agent) and can amplify the effects of anesthetics and other medications used during surgery. Therefore, provide your physicians and other health care professionals with a list of the supplements you are taking.

If you use or are thinking about using one or more dietary supplements:

- Determine whether the supplement is necessary. Some people have medical reasons for taking dietary supplements that contain one or more micronutrients.
- Discuss your need for the supplement with your physician or a registered dietitian nutritionist before you purchase or use the product. This action is particularly important if you are pregnant, are breastfeeding a baby, or have a chronic medical condition such as diabetes or heart disease.
- Consult a physician before giving such products to children. Treat dietary supplements as drugs: Store them away from children.
- Consult a physician as soon as you develop signs and symptoms of a serious illness. Using supplements to treat serious diseases instead of seeking conventional medical care that has proven effectiveness is a risky practice. In these instances, delaying or forgoing useful medical treatment may result in the worsening of the condition or even be life threatening.
- Be wary of claims made about a supplement's benefits and investigate the claims used to promote the product. The following government websites provide reliable information about dietary supplements:

  U.S. National Library of Medicine https://medlineplus.gov/druginfo/herb_All.html

  Food and Drug Adminstration www.fda.gov/Food/DietarySupplements/default.htm

National Institutes of Health, National Center for Complementary and Integrative Health https://nccih.nih.gov/health/herbsataglance.htm

- Determine hazards associated with taking the supplement. Information about the risks and benefits of various dietary supplements can be found at the **Office of Dietary Supplements'** website: https://ods.od.nih.gov/.
- Avoid using dietary supplements as substitutes for nutritious foods. Plant foods provide a wide array of phytochemicals, many of which may have health benefits when taken in their natural forms—foods. When these substances are isolated from plants and manufactured into supplements, they may lose their beneficial properties.

If you experience negative side effects after using a particular dietary supplement, it is a good idea to be examined by a physician immediately. Furthermore, you as well as your physician should report the problem to the FDA's MedWatch program by calling (800) FDA-1088 or visiting the agency's website: www.fda.gov/safety/medwatch-fda-safety-information-and-adverse-event-reporting-program.

It is important to recognize that the manufacturing of dietary supplements is a profitable industry in the United States. According to estimates provided by the dietary supplements' industry, Americans spent over $48 billion on such products in 2022.[18] Results of one major study indicate that healthy people do not benefit from taking dietary supplements, and in some cases, using these products may be harmful to health.[19] Thus, people who are in good health are wasting their money by purchasing dietary supplements—money that could be better spent on natural sources of nutrients and phytochemicals, particularly fruits, vegetables, and whole-grain cereals.

Concept CHECKPOINT

13. List at least three examples of alternative health care practices.

14. Explain why a prescription medication that is used to treat the common cold is not a dietary supplement.

15. List at least two examples of non-micronutrient dietary supplements and the health risks and benefits of taking such supplements.

See Appendix F for responses.

Photodisc/Getty Images

Chapter References

See Appendix G.

Summary

2.1 Nutrition: Science for Consumers

- Scientists ask questions about the natural world and follow generally accepted methods to obtain answers to these questions. Nutrition research relies on scientific methods that may involve making observations, asking questions and developing possible explanations, performing tests, collecting and analyzing data, drawing conclusions from data, and reporting on the findings. Other scientists can test the findings to confirm or reject them.
- Epidemiology is the study of the occurrence, distribution, and causes of health problems in populations. By studying differences in dietary practices and disease occurrences among populations, epidemiologists can suggest nutrition-related hypotheses for the prevalence of certain diseases. Epidemiological studies can establish a correlation between two variables. However, a limitation of these studies is the what appears to be a correlation between two variables may be a coincidence rather than a true relationship.

- Researchers summarize the findings and seek to publish articles with information about their investigations in scientific journals. Before articles are accepted for publication, they undergo peer review. Scientists generally do not accept a hypothesis or the results of a study until they are supported by considerable research evidence. Media coverage of a medical breakthrough is not necessarily an indication of the value or quality of research that resulted in the news story. More research is often necessary for scientists to determine whether the results are valid and can be generalized.
- Consumers may think scientists do not know what they are doing when conflicting research findings are reported in the media. Conflicting findings often result because different teams of researchers use different study designs when investigating the same hypothesis. Furthermore, each team of scientists may analyze the results differently. Other factors, such as genetic and lifestyle differences, can also influence the results of nutrition research involving human subjects. The science of nutrition is constantly evolving.

2.2 Nutrition Information: Fact or Fiction

- Although testimonials and anecdotes are often used to promote nutrition-related products and services, consumers cannot be sure that this information is reliable or based on scientific facts. Personal observations are not evidence of a cause-and-effect relationship because many factors, such as lifestyle and environment, can influence outcomes.
- Popular magazine articles, best-selling books, Internet websites, social media, television news reports, and other forms of media are often unreliable sources of nutrition information. Consumers need to be careful and question the reliability of such sources because the First Amendment to the U.S. Constitution guarantees freedom of the press and freedom of speech.
- Consumers need to ask questions to determine the author's reasons for promoting the information. Consumers should also look for red flags, such as scare tactics and claims that sound too good to be true.
- Much of the popular nutrition information that is on the Internet is unreliable and intended to promote sales. Websites sponsored by nonprofit organizations, nationally recognized health associations, government agencies, and nationally accredited colleges and universities are generally reliable sources of information.

2.3 Reliable Nutrition Experts

- Although some states regulate and license nutritionists, there is no standard legal definition for "nutritionist" in the United States. For reliable food, nutrition, and dietary information, consumers can consult persons with degrees in human nutrition from accredited institutions of higher learning, such as nutrition instructors and registered dietitian nutritionists.

Purple coneflower is also known as *Echinacea*.
©Wendy Schiff

2.4 Nutrition Matters: What Are Dietary Supplements?

- A dietary supplement is a product (other than tobacco) that adds to a person's dietary intake, contains one or more dietary ingredients, is taken by mouth, and may be a concentrate, metabolite, constituent, or extract.
- Dietary supplements include nutrient pills, protein powders, and herbal extracts. In many instances, people do not need dietary supplements. Healthy people should focus on obtaining nutrients and phytochemicals from foods, particularly fruits, vegetables, and whole–grain cereals. Before taking a dietary supplement, discuss the matter with your physician or a registered dietitian nutritionist.

Recipe for Healthy Living

Blueberry Chicken Salad

©Wendy Schiff

You probably have enjoyed eating fresh ripe blueberries that have been sprinkled on cereal or baked into pies. Blueberries are rich in the antioxidant, phytochemical anthocyanin, which gives the berries their blue color and contributes to their health benefits. Epidemiological research studies show that eating 1/3 cup of blueberries a day may reduce your risk of developing heart disease, diabetes, and dementia (decline in memory and thinking abilities).[20] Blueberries may also help people improve their thinking skills and manage their body weight.[21]

The following blueberry salad recipe provides a single serving and supplies approximately 615 kcal, 38 g of protein, 42 g of fat, 6 g of fiber, 23 mg of vitamin C, and 600 RAE of vitamin A.

INGREDIENTS:

3 Tbsp oil and vinegar dressing, no salt
1/4 cup chopped red onion
2 cups romaine lettuce, lightly shredded
1 cup raw spinach leaves, lightly packed
3/4 cup cooked, cubed, chilled chicken or turkey meat
1/2 cup fresh blueberries
1 Tbsp unsalted, roasted sunflower seeds
2 Tbsp crumbled blue cheese
Pinch of ground black pepper

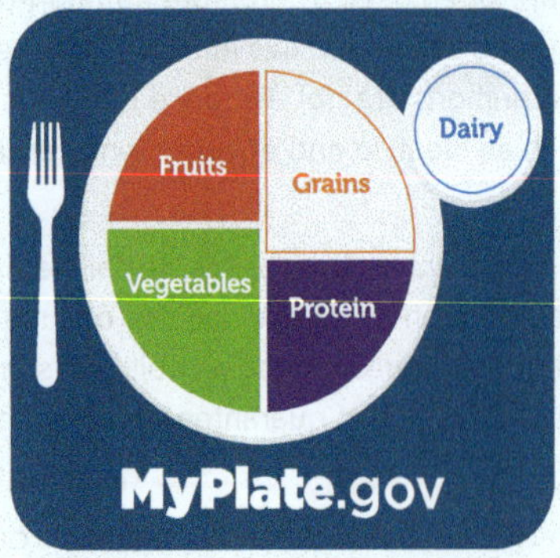

US Department of Agriculture or USDA

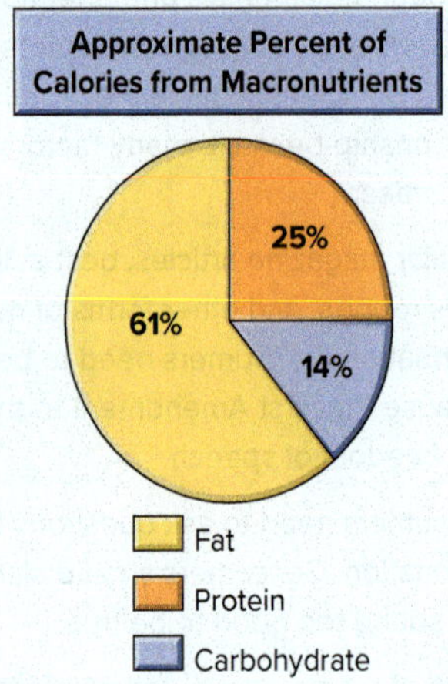

PREPARATION STEPS:

1. Place chopped red onion in a small bowl and pour the oil and vinegar dressing over the onions. Set aside.
2. Wash greens in cold water and pat dry. Lightly shred lettuce and tear spinach into small pieces, toss greens together, and arrange them on a dinner plate to form a "bed."
3. Place chopped, cooked poultry on top of the greens. Add crumbled blue cheese and blueberries.
4. Pour red onion and salad dressing mixture over the salad. Sprinkle with sunflower seeds and add dash of black pepper. Serve immediately.

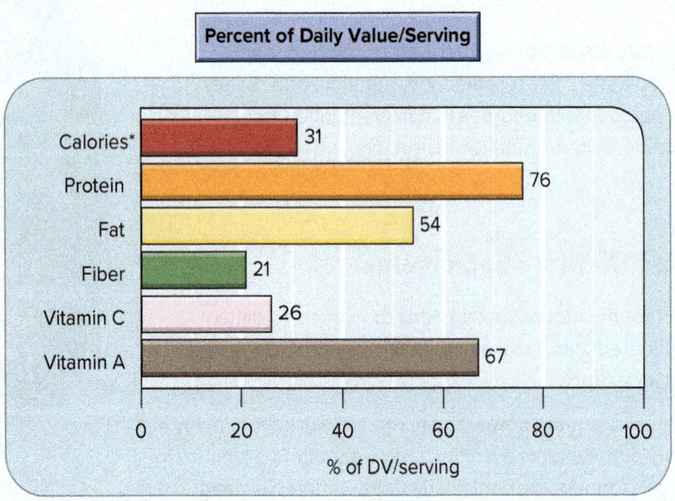

*2000 daily total kcal

Critical Thinking

1. An online health news source reports the results of a study in which people who took fish oil and vitamin E supplements daily did not reduce their risk of heart attack. Moreover, the researchers stopped the study when they determined the supplements increased the subjects' risk of stroke! Explain how you would determine whether this information is reliable.
2. Evaluate the facts of a trendy diet or weight-loss program. What are the author's or creator's credentials and motivations? Are testimonials or anecdotes used to promote the diet or program? Are research citations from a peer-reviewed journal? Does the diet or program have any red flag claims?
3. A group of scientists conduct a study to determine risk factors for breast cancer. According to their results, females who consume soy foods have a lower risk of breast cancer than females who do not eat soy foods. Think of two questions this finding is likely to generate that could be answered by further scientific investigation.
4. Search the Internet for sites that promote or sell one of the dietary supplements listed in Table 2.2. Analyze the website's content for signs the nutrition- or health-related information is unreliable. Is the website a reliable source of information? Explain why it is or why it is not.
5. Do you personally use or know someone who uses dietary supplements, such as vitamins, to treat or prevent illnesses? If you have, describe the situations and discuss which supplements were used. How will you use the information you learned in this chapter to change your supplement use?

Practice Test (Section numbers are listed below in red.)

Select the best answer.

1. (2.1) The first step of the scientific method usually involves
   a. gathering data.
   b. developing a hypothesis.
   c. identifying relationships between variables.
   d. making observations.
2. (2.1) In 2022, a team of scientists begins an epidemiological study of 1000 people who are between 18 and 30 years of age to determine whether their consumption of seafood will influence their state of health during the next 25 years. This study is an example of a(an)
   a. case-control study.
   b. prospective study.
   c. retrospective study.
   d. experimental study.

G.K. & Vikki Hart/Photodisc/Getty Images

3. (2.1) An aspect of _____ involves studying causes of health problems in a population.
   a. epidemiology
   b. technobiology
   c. diseasiology
   d. censusology
4. (2.1) Comparing the diets of individuals with liver cancer to diets of individuals who have very similar characteristics but are healthy would be an example of a
   a. prospective study.
   b. placebo study.
   c. retrospective study.
   d. case-control study.
5. (2.1) Generally, epidemiological studies
   a. establish causation.
   b. prove correlations.
   c. cannot determine if one factor causes another.
   d. are science-based research efforts that examine two variables.
6. (2.2) Which of the following journals does not have peer-reviewed articles?
   a. *Journal of the American Medical Association*
   b. *The New England Journal of Medicine*
   c. *Journal of the American College of Nutrition*
   d. *Men's Journal*

7. **(2.2)** The government agency that enforces consumer protection laws by investigating false or misleading health-related claims is the
   a. Federal Trade Commission (FTC).
   b. Environmental Protection Agency (EPA).
   c. Agricultural Research Service (ARS).
   d. Centers for Disease Control and Prevention (CDC).

8. **(2.2)** A friend tells you that her fingernails used to be brittle and weak. However, she states that her nails became strong and beautiful after she began taking a daily supplement of the B vitamin biotin. This kind of nutrition-related information is
   a. an unbiased report about biotin's usefulness.
   b. a scientifically valid health claim.
   c. a personal endorsement of a product.
   d. an anecdote.

9. **(2.2)** Which of the following websites is most likely to provide biased and unreliable nutrition information?
   a. The site of a nationally recognized health association (*.org)
   b. A site that promotes or sells dietary supplements (*.com)
   c. The site of a U.S. government agency (*.gov)
   d. An accredited college or university's site (*.edu)

10. **(2.3)** Lingyun is concerned his diet may be inadequate to meet his nutritional needs. He should consult a ______ to evaluate his diet.
    a. doctor
    b. registered dietitian nutritionist
    c. trainer at the gym
    d. nutritionist

11. **(2.4)** Which of the following substances would be classified as a dietary supplement according to the Dietary Supplement Health and Education Act of 1994?
    a. Tobacco
    b. Melatonin
    c. Chicken liver
    d. Plain yogurt

12. **(2.4)** The _____ is the federal agency that tries to ensure the safety and effectiveness of medications.
    a. FDA
    b. FTC
    c. EPA
    d. USGS

13. **(2.4)** Which of the following activities is an alternative health care practice?
    a. Drinking green tea to lose weight
    b. Taking ibuprofen to treat a headache
    c. Consuming more calcium-rich foods to help increase bone mass
    d. Eating fewer salty snacks to reduce sodium intake

Answers to Quiz Yourself

1. Scientists use anecdotes as scientific evidence to support their findings. **False.** (Section 2.1)
2. Popular health-related magazines typically publish articles that have been peer-reviewed. **False.** (Section 2.2)
3. By conducting observational epidemiological studies, medical researchers can determine risk factors that may influence health. **True.** (Section 2.1)
4. Dietary supplements include vitamin pills, as well as products that contain echinacea, ginseng, and garlic. **True.** (Section 2.4)
5. In general, registered dietitian nutritionists are reliable sources of food and nutrition information. **True.** (Section 2.3)

CHAPTER 3

Planning Nutritious Diets

Quiz Yourself

Before reading the rest of Chapter 3, test your knowledge of nutrient standards, recommendations, and guides, as well as food labels, by taking the following quiz. The answers are at the end of the chapter.

1. According to the latest U.S. Department of Agriculture food guide, fruits and vegetables are combined into one food group. ______ T ______ F
2. According to the recommendations of the *Dietary Guidelines for Americans*, it is acceptable for certain adults to consume alcoholic beverages in moderation. ______ T ______ F
3. Last week, Jacara didn't consume the recommended amount of vitamin C for a couple of days. Nevertheless, he is unlikely to develop scurvy, the vitamin C deficiency disease. ______ T ______ F
4. Canned carrots are an example of ultra-processed food. ______ T ______ F
5. The Nutrition Facts panel on a food label provides information concerning amounts of energy, fiber, and sodium that are in a serving of the food. ______ T ______ F

Monkey Business Images/Shutterstock

3.1 From Requirements to Standards

Learning Outcomes

1. Explain the difference between a nutrient requirement and nutrient standard.
2. Identify the various Dietary Reference Intakes and explain how they can be used.

When you shop for groceries, do you sometimes feel overwhelmed by the vast array of foods that are available? If your answer is yes, your response is not surprising, considering the typical supermarket offered an average of 35,000 items in 2021.[1] Every time you enter a supermarket, you are likely to find hundreds of food and beverage items that were not on the shelves during your last visit to the store. Food manufacturers typically focus on developing new products that support good health and sustainability and are convenient to prepare and reasonably priced.[2]

Chapter 1 introduced some key nutrition concepts, including the need for dietary adequacy, moderation, balance, and a variety of foods. Chapter 2 described how you can become a more careful consumer of nutrition information. However, you are also a consumer of food. With so many grocery items from which to choose, what are the primary factors that influence your food purchases? Do you select foods simply because they taste good, are reasonably priced, or are easy to prepare? Do you ever consider the effects certain foods may have on your health before you purchase them?

Your lifestyle reflects your health-related behaviors, including your dietary practices and physical activity habits. Americans of all ages may reduce their risk of chronic disease by adopting nutritious diets and engaging in regular physical activity. However, consumers need practical advice to help them make decisions that can promote more healthy lifestyles.

Chapter 3 discusses nutrient standards, including how the standards are established and used. The information in this chapter also presents practical ways to plan a nutritionally adequate, well-balanced diet using tools such as the Dietary Guidelines and MyPlate. Furthermore, Section 3.5 explains how to interpret and use nutrition-related information that appears on food and dietary supplement labels.

requirement smallest amount of a nutrient that maintains a defined level of nutritional health

What Is a Nutrient Requirement?

By using research methods discussed in Chapter 2, scientists have been able to estimate the amount of many nutrients required by the body. A **requirement** can be defined as the smallest amount of a nutrient that maintains a defined level of nutritional health.[3] In general, this amount, when consumed daily, prevents the nutrient's deficiency disease. The requirement for a particular nutrient varies to some degree from person to person. A person's age, sex, general health status, physical activity level, and use of medications and drugs are among the factors that influence their nutrient requirements.

Nutrient requirements vary and change throughout the lifespan.
Lane Oatey/Blue Jean Images/Getty Images

Many nutrients are stored in the body, including vitamin D and most minerals. Major storage sites include the liver, body fat, and bones. Other nutrients, such as vitamin C and most B vitamins, are not stored by the body. For optimal nutrition, you need to consume enough of those nutrients to maintain storage levels. Your body uses its nutrient stores much like you can use a savings account to help manage your money. When you have some extra cash, it is wise to place the money in a savings account so you can withdraw some of the reserves to meet future needs without going into debt. When your consumption of certain nutrients is more than enough to meet your needs, the body stores the excess. When your intake of a stored nutrient is low or needs for this nutrient become increased, such as during recovery from illness, your body withdraws some from storage. As a result of having optimal levels of stored nutrients, you may recover more quickly and avoid or delay developing deficiencies of those nutrients.

Dietary Reference Intakes

Dietary Reference Intakes (DRIs) encompass a variety of daily energy and nutrient intake standards that nutrition experts in the United States use as references when making dietary recommendations. DRIs are intended to help people reduce their risk of nutrient deficiencies and excesses, prevent disease, and achieve optimal health.[4] The seven standards (Fig. 3.1) are the Estimated Average Requirement (EAR), Recommended Dietary Allowance (RDA), Adequate Intake (AI), Tolerable Upper Intake Level (UL), Chronic Disease Risk Reduction Intake (CDRR), Estimated Energy Requirement (EER), and Acceptable Macronutrient Distribution Range (AMDR).[5]

A group of nutrition scientists, the **Food and Nutrition Board (FNB)** of the Institute of Medicine, develops DRIs. Periodically, members of the board adjust DRIs as new information concerning human nutritional needs and dietary adequacy becomes available. You can find tables for the latest DRIs in Appendix H of this book. The following sections provide basic information about the various DRI standards. It is important to become familiar with these terms because we refer to them in this and other chapters.

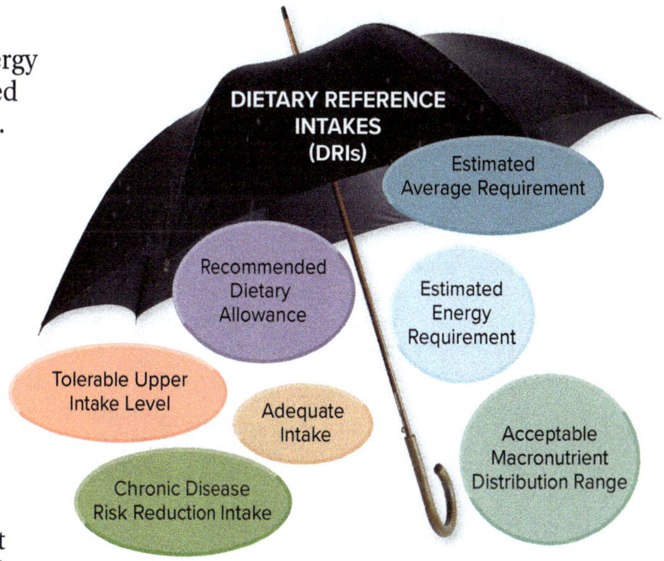

Figure 3.1 Dietary Reference Intakes. The Dietary Reference Intakes (DRIs) encompass a variety of terms that represent standards for energy and nutrient recommendations.
Ryan McVay/Photodisc/Getty Images

Estimated Average Requirement

An **Estimated Average Requirement (EAR)** is the daily amount of the nutrient that meets the needs of 50% of healthy people who are in a particular *life stage/sex group.* Life stage/sex groups classify people according to age, sex, and whether females are pregnant or breastfeeding. A typical 20–year–old female college student, for example, would be classified as a female, between 19 and 30 years old, and not pregnant or breastfeeding. Because the EAR meets the needs of 50% of healthy people, it is not a nutrient goal for individuals.

To establish an EAR for a nutrient, the FNB identifies a physiological marker, a substance in the body that reflects proper functioning and can be measured. This marker indicates whether the level of a nutrient in the body is adequate. A marker for vitamin C, for example, is the amount of the vitamin in certain blood cells. When these cells contain nearly all the vitamin C they can hold, the body has an optimal supply of the vitamin. Thus, a physician can diagnose whether a patient is vitamin C deficient by taking a blood sample from the person and measuring the vitamin C content of certain blood cells.

Dietary Reference Intakes (DRIs) various energy and nutrient intake standards for Americans

Food and Nutrition Board (FNB) group of nutrition scientists who develop DRIs

Estimated Average Requirement (EAR) amount of a nutrient that meets the needs of 50% of healthy people in a life stage/sex group

Recommended Dietary Allowances (RDAs) daily nutrient recommendations that meet the needs of 97 to 98% of healthy people in a life stage/sex group

Recommended Dietary Allowances

The **Recommended Dietary Allowances (RDAs)** are standards for recommending daily intakes of several nutrients. RDAs meet the nutrient needs

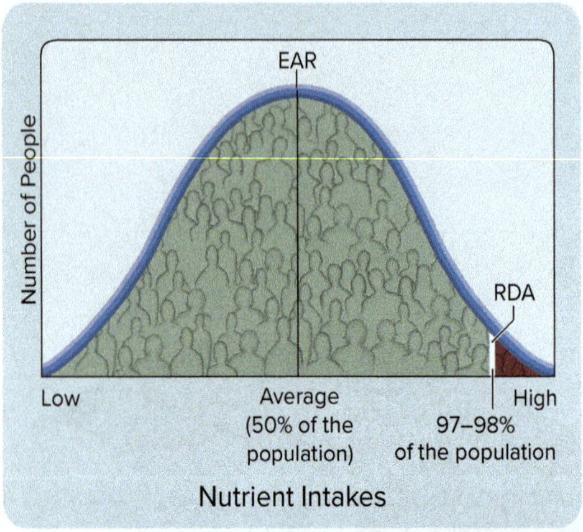

Figure 3.2 Establishing RDAs. A nutrient's RDA is high enough to meet or exceed the requirements of 97 to 98% of the population for the nutrient. In other words, about 98% of the population will have their needs for the nutrient met by just consuming the RDA amount.

of nearly all healthy individuals (97 to 98%) in a particular life stage/sex group. To establish an RDA for a nutrient, nutrition scientists first determine its EAR. Then scientists add a "margin of safety" amount to the EAR that allows for individual variations in nutrient needs and helps maintain tissue stores (Fig. 3.2). For example, the adult EAR for vitamin C is 60 mg for females who are not pregnant or breastfeeding and 75 mg for males.[3] However, the adult RDA for vitamin C is 15 mg higher than the EAR: 75 mg for females who are not pregnant or breastfeeding and 90 mg for males. Thus, the margin of safety for vitamin C is 15 mg. Because smoking cigarettes increases the need for vitamin C, smokers should add 35 mg to their RDA for the nutrient.

Adequate Intakes

In some instances, nutrition scientists are unable to develop RDAs for nutrients because there is not enough information to determine how much is required by people. Until such information becomes available, scientists set **Adequate Intakes (AIs)** for these nutrients. To establish an AI, scientists record eating patterns of a group of healthy people and estimate the group's average daily intake of the nutrient. If the people who are being observed show no signs of the nutrient's deficiency disorder, the researchers conclude that the average level of intake must be adequate and use that value as the AI (Fig. 3.3). Vitamin K is one of the nutrients that has an AI instead of an RDA.

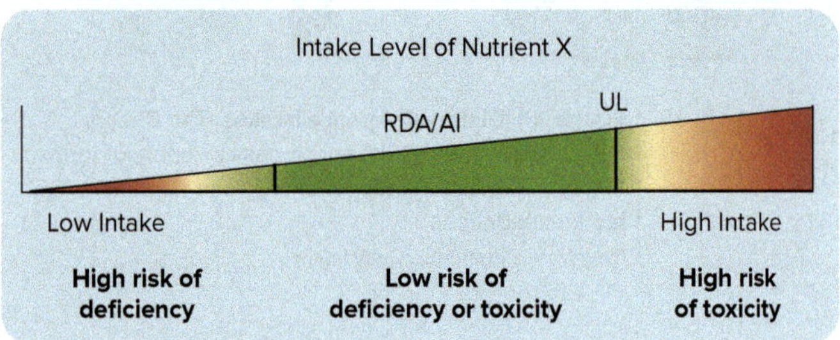

Figure 3.3 Adequate Intakes and Upper Levels. Consuming too much or not enough of a nutrient can cause health problems.

Tolerable Upper Intake Level

Nutrition scientists also establish a **Tolerable Upper Intake Level (Upper Level or UL)** for many vitamins and minerals. The UL is the highest average amount of a nutrient that is unlikely to harm most people when they consume that amount daily (see Figure 3.3).[4] The risk of a toxicity disorder increases when a person regularly consumes amounts of a nutrient that is more than its UL. The UL for vitamin C, for example, is 2000 mg/day for adults.

Estimated Energy Requirement

The **Estimated Energy Requirement (EER)** is the average daily energy intake that meets the needs of a healthy person who is maintaining their weight. Registered dietitian nutritionists (RDNs) can use EERs to evaluate an individual's energy intake. The EER takes into account the person's physical activity level, height, and weight, as well as sex and life stage. Because the EER is an average figure, some people have energy needs that are higher or lower. Chapter 10 provides formulas for calculating your EER.

Adequate Intakes (AIs) dietary recommendations for nutrients that scientists do not have enough information to set RDAs

Tolerable Upper Intake Level (Upper Level or UL) highest average amount of a nutrient that is unlikely to be harmful when consumed daily

Estimated Energy Requirement (EER) average daily energy intake that meets the needs of a healthy person maintaining their weight

Acceptable Macronutrient Distribution Ranges (AMDRs) macronutrient intake ranges that are nutritionally adequate and may reduce the risk of diet-related chronic diseases

Acceptable Macronutrient Distribution Ranges

The results of scientific research suggest that food energy sources (macronutrients) are associated with risk of developing certain diet—related chronic diseases, such as heart disease. **Acceptable Macronutrient Distribution Ranges (AMDRs)** indicate ranges of carbohydrate, fat, and protein intakes that provide adequate amounts of vitamins and minerals and may reduce the risk of diet—related chronic diseases.[4] The AMDR for carbohydrates, for example, is 45 to 65% of total energy intake. If a person's total energy intake is 2000 kcal/day, then their recommended carbohydrate intake is 900 to 1300 kcal/day. To obtain the range of energy from carbohydrate, multiply total kcal/day (in this case, 2000 kcal) by 0.45 (45%) and then multiply 2000 kcal by 0.65 (65%). Table 3.1 lists adult AMDRs for a person who consumes 2000 kcal/day.

Chronic Disease Risk Reduction Intakes

The **Chronic Disease Risk Reduction Intakes (CDRRs)** is the newest DRI category. CDRRs are nutrient intake levels that reduce chronic disease risk within a healthy population. For example, excess dietary sodium intake increases the risk of high blood pressure and heart disease. According to the CDRR for sodium, adults should limit their sodium intake to less than 2300 mg/day to reduce the risk of chronic disease.[5]

TABLE 3.1 Acceptable Macronutrient Distribution Ranges: Adults (2000 kcal/day)

Macronutrient	AMDR (% of Total Energy Intake)	Calories
Carbohydrate	45–65	900–1300
Protein	10–35	200–700
Fat*	20–35	400–700

*Fat intake should include essential fatty acids (see Chapter 6).

Applying Nutrient Standards

Table 3.2 summarizes information about the DRIs. RDNs refer to DRIs for planning nutritious diets for groups of people and evaluating the nutritional adequacy of a population's diet. Nevertheless, RDAs and AIs are often used to evaluate an individual's dietary practices.[4] Your diet is likely to be nutritionally adequate if your average daily intake for each nutrient meets the nutrient's RDA or AI value. If your diet consistently supplies less than the EAR for a nutrient, you may be at risk of eventually developing the nutrient's deficiency disorder (see Figure 3.3). On the other hand, if your intake of a nutrient is consistently above its UL, you are at risk of developing that nutrient's toxicity disorder. Nutrient toxicity disorders are more likely to occur when people take high doses of individual nutrient supplements, particularly vitamins and minerals. If you do not take large doses of nutrient supplements and you

Chronic Disease Risk Reduction Intakes (CDRRs) nutrient intake levels that reduce chronic disease risk within a healthy population

TABLE 3.2 Summary of Dietary Reference Intakes

Standard	Definition	Example
Estimated Average Requirement (EAR)	Amount of a nutrient that meets the needs of 50% of healthy people who are in a particular life stage/sex group	**Vitamin C** 75 mg/day for males and 60 mg/day for females ages 19 through 50 years
Recommended Dietary Allowance (RDA)	Amount of a nutrient that meets the needs of nearly all healthy individuals in a particular life stage/sex group	**Vitamin C** 90 mg/day for males (nonsmokers) and 75 mg/day for females ages 19 through 50 years (nonsmokers)
Adequate Intake (AI)	Amount of a nutrient that is considered to be adequate based on the population's typical intakes, but there is not enough scientific information available to determine an RDA for the nutrient at this time	**Vitamin C** 40 mg/day for infants from birth through 6 months of age
Tolerable Upper Intake Level (UL)	Highest average amount of a nutrient that is unlikely to harm most people when the amount is consumed daily	**Vitamin C** 2000 mg/day for adults
Estimated Energy Requirement (EER)	Average daily energy needs of healthy people maintaining their weight; considers physical activity level, height, weight, and life stage/sex group	**Estimated Energy Intake** 3000 kcal/day for active males and 2400 kcal/day for active females ages 21 through 24 years
Acceptable Macronutrient Distribution Range (AMDR)	Range of carbohydrate, fat, and protein intakes that provide adequate amounts of micronutrients and may reduce the risk of developing certain diet-related chronic diseases	See Table 3.1
Chronic Disease Risk Reduction Intake (CDRR)	Nutrient intake levels that reduce chronic disease risk within a healthy population	**Sodium** 2300 mg/day limit for adults

eat reasonable amounts of food, your risk of developing a nutrient toxicity disorder is low.

Nutritional standards have a variety of commercial applications. Pharmaceutical companies refer to DRIs when developing formulas that replace breast milk for infants and special formulas for people who cannot consume regular foods. As a result, babies can be nourished by consuming commercially prepared formulas, and adults who are unable to swallow can survive for years on formula feedings administered through tubes inserted into their bodies.

For nutrition labeling purposes, the Food and Drug Administration (FDA) uses RDAs to develop a set of standards called *Daily Values (DVs)*. Consumers may find DVs useful for comparing the nutritional contents of similar foods. The "Food and Dietary Supplement Labels" section of this chapter provides more information about nutritional labeling, including DVs.

Concept CHECKPOINT

1. Describe how scientists establish the RDA for a nutrient.
2. What is the difference between an RDA and an AI?
3. Explain how an EER differs from an AMDR or CDRR.
4. Discuss how registered dietitian nutritionists, pharmaceutical companies, and the FDA use nutrient standards.

See Appendix F for responses.

3.2 Major Food Groups

Learning Outcomes

1. List major food groups, and identify foods that are typically classified in each group.
2. Distinguish foods that are ultra-processed from those that are minimally processed, and explain why healthy diets contain mostly raw and minimally processed foods.

Grains include products made from wheat, rice, corn, barley, and oats.
©Wendy Schiff

For most people, the DRIs are not practical for planning menus, so nutrition experts develop more consumer-friendly food (dietary) guides. In general, such guides classify foods into major food groups according to their natural origins and key nutrients. Major food groups are usually grains, dairy products, fruits, vegetables, and protein-rich foods. In most instances, dietary guides also provide recommendations concerning amounts of foods from each group that should be eaten daily. The following points identify major food groups, summarize key features of each group, and provide serving sizes according to the U.S. Department of Agriculture (USDA).

- ***Grains*** include products made from wheat, rice, and oats. Pasta, noodles, and flour tortillas are members of this group because wheat flour is their main ingredient. A serving of a grain food is equivalent to 1 slice of bread, 1 cup of ready-to-eat cereal, or ½ cup of cooked rice, pasta, or cereal such as oatmeal. Although corn is a type of grain, it is often used as a vegetable in meals. Cornmeal and popcorn, however, are usually grouped with grain products.

  Carbohydrate (starch) and protein are the primary macronutrients in grains. Dietary guides generally recommend choosing foods made with whole grains instead of refined grains as they naturally contain more fiber and other various nutrients. According to the FDA, whole grains are the intact, ground, cracked, or flaked seeds of cereal grains, such as wheat, buckwheat, oats, corn, rice, wild rice, rye, and barley.[6]

In the United States, refined grain products are commonly consumed and can be good sources of several vitamins and minerals when they have undergone enrichment or fortification. In general, **enrichment** replaces some of the nutrients that food loses during processing. **Fortification** is the addition of any nutrient to food to boost its nutrient content, such as adding calcium to orange juice, vitamins A and D to milk, and numerous vitamins and minerals to ready-to-eat cereals, and snack and energy bars. When grains undergo processing, they can lose nutrients, particularly vitamins, minerals, and fiber. Enriched grains have specific amounts of iron and four B vitamins added to them. Although these nutrients are replaced, other nutrients, such as fiber, that were lost during refinement are not replaced.

Dairy products, especially yogurt and hard cheeses, are excellent sources of calcium, protein, phosphorus, and riboflavin. Additionally, milk is often fortified with vitamins A and D.
D. Hurst/Alamy Stock Photo

- ***Dairy products*** include milk and products made from milk that retain (keep) their calcium content after processing, such as yogurt and hard cheeses. Such dairy foods are also excellent sources of protein, phosphorus (a mineral), and riboflavin (a B vitamin). Most of the milk sold in the United States is fortified with vitamins A and D, so it is a good source of these vitamins.

  Foods and beverages made from soybeans (soy "milk" and soy "cheese") can substitute for cow's milk if they are fortified with calcium and other micronutrients. Other beverages often called "milks" made from nuts, cereals, and other plants such as almond, cashew, oat, rice, and coconut lack some macronutrients and micronutrients cow's milk provides. However, as of February 2023, the FDA is drafting label guidance to determine if manufacturers can label these beverages as "milks."[7] Milk-based desserts such as ice cream, pudding, and frozen yogurt are often grouped with dairy products, even though they often have high added sugar and fat contents. Although cream cheese, cream, and butter are made from cow's milk, they are not included in this group because they have little or no calcium and are high in fat.

  Most dietary guides recommend choosing dairy products that have most of the fat removed because milk supplies *saturated* fat that is associated with an increased risk of cardiovascular disease (CVD). Thus, people should consume milk and dairy products made with fat-free or low-fat milk. (Fat-free milk may also be referred to as nonfat or skim milk.) Compared to whole milk, which is about 3.25% fat by weight, low-fat milk contains only 1% fat by weight and is often called "1% milk."

  One cup of fat-free milk generally substitutes for 1 cup of plain, low-fat yogurt or frozen yogurt; 2 cups of low-fat cottage cheese; 1½ ounces of natural cheese such as Swiss or cheddar; or 1 ounce of processed cheese, such as American cheese. To obtain about the same amount of calcium as in 1 cup of fat-free milk, you would have to eat almost 1⅔ cups of regular vanilla ice cream. This amount of ice cream provides 490 kcal and about 26 g of fat, whereas 1 cup of fat-free milk supplies about 85 kcal and less than 1 g of fat.

Dry beans and peas, eggs, tofu, nuts, and seeds are protein-rich foods that can substitute for meat.
©Wendy Schiff

- ***Protein-rich foods*** include beef, pork, lamb, fish, shellfish, and poultry. Beans, lentils, eggs, nuts, and seeds are included with this group because these protein-rich foods can substitute for meats. A serving of food from this group generally equals 1 ounce of meat, poultry, or fish; ¼ cup cooked dry beans, dry peas or, lentils; 1 egg; 1 tablespoon of peanut butter; or ½ ounce of nuts or seeds. One-fourth cup of tofu, a food made from soybeans, can substitute for 1 ounce of meat, fish, or poultry. Most dairy products are also protein rich.

  Foods in the protein group are rich sources of micronutrients, especially iron, zinc, and B vitamins. In general, the body absorbs minerals, such as iron and zinc, more easily from animal foods than from plants.

- **Fruits** include fresh, dried, frozen, sauced, and canned fruit, as well as 100% fruit juice. One serving of food from this group generally equals 1 cup of fruit or fruit juice, or ½ cup of dried fruit, such as raisins or dried apricots.[8] Most

enrichment replacement of some nutrients that were removed during processing

fortification addition of any nutrient to food to boost its nutritional content

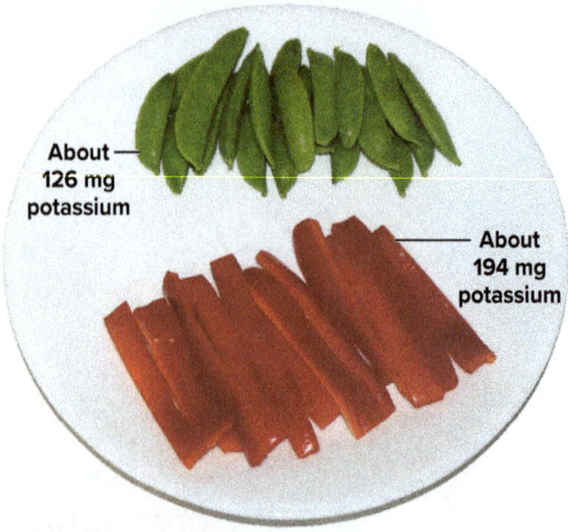

Figure 3.4 Comparing sugar snap peas to red bell peppers. The nutritional content of foods within each group often varies.
Traci Keck

fruits are low in fat and good sources of phytochemicals and micronutrients, especially the mineral potassium and vitamins C and folate. Additionally, whole or cut–up fruit is a good source of fiber. Although 100% juice is a source of phytochemicals and can count toward one's fruit intake, the majority of choices from this group should be whole or cut–up fruits.[8] Whole or cut–up fruits are healthier options than juices because they contain more dietary fiber.

- ***Vegetables*** include fresh, cooked, canned, frozen, and dried/dehydrated vegetables, and 100% vegetable juice. Vegetables may be further grouped into dark green, orange and red, starchy, and "other" categories. Beans, peas, and lentils are also in the vegetable group as well as in the protein–rich foods group. One serving of food from this group generally equals 1 cup of raw or cooked vegetables, 1 cup vegetable juice, or 2 cups of uncooked leafy greens, such as salad greens. This means if you eat 1 cup of raw carrots and drink 1 cup of tomato juice, you have consumed 2 servings from the vegetables group. Many vegetables are good sources of micronutrients, fiber, and phytochemicals. Furthermore, many vegetables are naturally low in fat and energy.

It is important to note that the nutritional content of foods within each group often varies. For example, 1 cup raw sugar snap peas and 1 cup raw red bell pepper slices each supply about 25 kcal. However, the peas contribute about 126 mg of potassium, whereas bell peppers supply about 194 mg of the mineral to diets (Fig. 3.4). Therefore, dietary guides generally recommend that people choose a variety of foods from each food group when planning daily meals and snacks.

Other Foods

The seeds and almonds in this trail mix make it a good source of "healthy" fats.
Thomas Northcut/Photodisc/Getty Images

Oils include canola, corn, and olive oils, as well as other fats that are liquid at room temperature. Certain spreadable foods made from vegetable oils, such as mayonnaise, soft or "tub" margarine, and salad dressing, are also classified as oils. Peanuts and peanut butter, walnuts, sunflower seeds, olives, avocados, and some types of fish have high fat contents, but these kinds of fat are "healthy" fats that do not contribute to heart disease (see Chapter 6). Oils are often good sources of fat soluble vitamins.

Solid fats, such as beef fat, butter, stick margarine, and shortening, are saturated fats that are fairly hard at room temperature. Although cream and coconut oil are liquid or soft at room temperature, these foods are classified as solid fats because they are rich sources of saturated fat. It is recommended to limit consumption of foods rich in solid fats. Chapter 6 discusses how dietary fats can affect health.

Foods that contain high amounts of sugar added during processing or preparation include candy, regular soft drinks, jelly, and other foods. It is recommended to limit consumption of these foods and beverages because they typically supply energy but few or no micronutrients. The alcohol in alcohol–containing beverages is another source of empty calories.

Levels of Food Processing

Raw foods often undergo different levels of processing, including refinement, before they are eaten (see Chapter 1). Raw carrots, for example, usually are trimmed, washed, and chilled for sale as fresh produce. When refrigerated, the raw carrots can keep their desirable taste and texture for a few weeks. The produce, however, eventually dries out, becomes rubbery, and spoils. To extend their "shelf life," raw carrots can be processed by being frozen or canned. Minimally processed foods, such as canned carrots, typically contain few refined ingredients (Fig. 3.5a).

An **ultra–processed food** contains numerous *additives* which are ingredients that were added to a food during production to not only extend its shelf

solid fats fats that are fairly hard at room temperature

ultra-processed food food that contains numerous ingredients that were added to improve desirable characteristics

life but also enhance other characteristics, including flavor, texture, color, nutritive value, and ease of preparation (convenience).[9] The commercially made, frozen carrot cake shown in Figure 3.5b is an example of an ultra–processed food that requires no preparation other than taking it out of the freezer and thawing. Although the cake contains carrots, the ingredient's list displays numerous other additives that help maintain its sensory appeal while frozen (Fig. 3.5b). Note that sugar is the cake's first ingredient.

Nutrition experts are concerned about the long–term negative health effects of eating diets that supply high amounts of ultra–processed foods. The majority of adult Americans have unhealthy amounts of body fat; the population's excessive consumption of ultra–processed foods may be associated with weight gain and obesity (see Chapter 10).

As of April 2023, the FDA had no standard definitions for categorizing processed foods, according to their level of processing. To evaluate whether a food is minimally processed or ultra–processed, consumers need to read the ingredients' list on the package label. In general, an ultra–processed food contains several additives that are not readily available when people prepare foods at home (see the ingredients list displayed in Figure 3.5b). For more information about food additives, see Chapter 12.

(a) **INGREDIENTS:** CARROTS, WATER, SEA SALT, CALCIUM CHLORIDE.

(b) **INGREDIENTS:** SUGAR, CREAM CHEESE (PASTEURIZED MILK AND CREAM, CHEESE CULTURE, SALT, CAROB BEAN GUM), CARROTS, SOYBEAN OIL, BLEACHED WHEAT FLOUR, EGGS, MOLASSES, BUTTER, WATER, INTERESTERIFIED SOYBEAN OIL, WALNUTS, APPLES, RAISINS, WHEAT STARCH, NONFAT DRY MILK, GROUND CINNAMON, ORANGE PEEL, FOOD STARCH MODIFIED, SALT, PRUNE JUICE, SODIUM BICARBONATE, EGG WHITES, MONO AND DIGLYCERIDES, HYDROGENATED COTTONSEED OIL, NATURAL AND ARTIFICIAL FLAVORS, DISTILLED MONOGLYCERIDES, GUAR GUM, SODIUM CASEINATE (MILK) GINGER, SODIUM ACID PYROPHOSPHATE, RED 40, BLUE 1, BLUE 2, RED 3, YELLOW 6, YELLOW 3S, PROPYLENE GLYCOL MONOESTER, LECITHIN, PHOSPHORIC ACID, GROUND NUTMEG, MONOCALCIUM PHOSPHATE, POLYGLYCEROL ESTER OF FATTY ACIDS.

Figure 3.5 Processed food. (a) These canned carrots are an example of a minimally processed food. (b) This commercially prepared carrot cake is an example of an ultra-processed food.
Carrots, cake, ingredients label: ©Wendy Schiff

Concept CHECKPOINT

5. List at least three foods that are generally classified as grain products.
6. What is the difference between nutrient fortification and nutrient enrichment?
7. List at least four foods that are generally classified as dairy products.
8. Why can beans, eggs, and nuts be a substitution for meats?
9. According to the information in this section of this chapter, how many cups of dried apricots are nutritionally equivalent to 2 cups of fresh apricots?
10. Why is it important to consume a wide variety of foods from all the different food groups?
11. Identify at least two foods that are classified as solid fat.
12. What is the difference between a minimally processed food and an ultra-processed food?

See Appendix F for responses.

3.3 Dietary Guidelines

Learning Outcomes

1. List at least three overarching guidelines of the *2020–2025 Dietary Guidelines for Americans,* and provide recommendations of each one.
2. Identify features of a healthy eating pattern.
3. Apply the Dietary Guidelines to improve the nutritional quality of diets.

Heart disease, cancer, hypertension (chronically elevated blood pressure), and diabetes mellitus (commonly referred to as diabetes) are among the leading causes of disability and death among Americans. According to a considerable amount of scientific evidence, risk of these diseases is strongly linked with certain lifestyles, particularly poor dietary choices and lack of regular physical activity. As required by law, the U.S. Department of Health and Human Services (HHS) and the USDA publish the *Dietary Guidelines for Americans* (Dietary Guidelines), a set of general nutrition–related lifestyle recommendations that are intended for healthy children and adults.[10] The Dietary Guidelines are designed to promote good health and reduce the risk of major nutrition– related chronic health conditions, such as obesity and CVD. These guidelines are updated every 5 years.

TABLE 3.3 *2020–2025 Dietary Guidelines for Americans*: Overarching Guidelines

• Follow a healthy dietary pattern at every life stage.
• Customize and enjoy nutrient-dense food and beverage choices to reflect personal preferences, cultural traditions, and budgetary considerations.
• Focus on meeting food group needs with nutrient-dense foods and beverages, and stay within calorie limits.
• Limit foods and beverages higher in added sugars, saturated fat, and sodium, and limit alcoholic beverages.

TABLE 3.4 A Healthy Eating Pattern[10]

Includes:
• A variety of vegetables;
• Fruits, especially whole fruits;
• Grains, especially whole grains;
• Fat-free or low-fat dairy products;
• A variety of protein foods; and
• Oils.
A healthy eating pattern limits:
• Saturated fats, trans fats, added sugars, sodium, and alcohol

A healthy dietary pattern includes a variety of vegetables and whole fruits.
Roman Iegoshyn/romastudio/123RF

The Dietary Guidelines form the basis of the federal government's nutrition policy, which serves as a framework for national, state, and local health promotion as well as food and nutrition programs. Dietary Guidelines are also used for menu planning by individuals and health care professionals.[10] Furthermore, food and beverage manufacturers can apply the guidelines to develop healthier products that appeal to consumers.

To revise the previous edition of the dietary guidelines, officials with the HHS and the USDA appointed a Dietary Guidelines Advisory Committee (DGAC) comprised of nationally recognized medical and nutrition experts. DGAC members reviewed the scientific evidence concerning the role of foods in health before preparing the *Scientific Report of the 2020 Dietary Guidelines Advisory Committee.*[11] This publication contained recommendations for the development of the *2020–2025 Dietary Guidelines for Americans.* The federal government incorporated some but not all of the DGAC's recommendations into the *2020–2025 Dietary Guidelines for Americans.*

According to the *Dietary Guidelines for Americans*, a healthy diet contains fruits, vegetables, whole grains, low–fat or nonfat dairy products, seafood, legumes, and nuts.[10] Additionally, a healthy diet contains limited amounts of refined grains, red and processed meats, sodium, and sugar–sweetened foods and beverages. Most Americans can obtain health benefits by choosing foods and beverages that follow healthy dietary patterns.[10]Such eating patterns should be flexible so food choices can be varied and individualized to meet a person's preferences, medical needs, and economic situation

Table 3.3 presents the four overarching guidelines of the 2020–2025 Dietary Guidelines. Table 3.4 lists key features of a healthy eating pattern, according to the Dietary Guidelines. The following information includes several key recommendations for following a healthy diet. You can learn more about the Dietary Guidelines by visiting http://health.gov and searching for "Dietary Guidelines."

Follow a Healthy Dietary Pattern at Every Life Stage

Across the lifespan, people should choose a healthy dietary pattern that has an appropriate number of calories to achieve and maintain a healthy body weight, is nutritionally adequate, and reduces the risk of diet–related chronic diseases. A healthy dietary pattern contains:

- Vegetables of all types, including red, orange, and dark green; beans, peas, and lentils; and starchy vegetables.
- Fruits, especially whole or cut–up fruit.
- Grains, at least half of which are whole–grain products.
- Dairy, especially fat–free or low–fat milk, yogurt, and cheese, and/or fortified soy beverages and yogurt as alternatives.
- Protein foods, such as lean meats, poultry, and eggs; seafood; dried beans, peas, and lentils; nuts; seeds; and soy products.
- Oils, especially vegetable oils and oil–rich protein foods, such as seafood and nuts.
- Less than 10% of total calories from saturated (solid) fat, and less than 10% of total calories from added sugars.
- Less than 2300 mg of sodium per day.

Recommendations for Specific Population Groups

The 2020–2025 Dietary Guidelines focus on dietary recommendations for people who are in specific life stage groups. The following information presents some of those recommendations.

Infants from birth to 6 months of age should:

- Consume only breast milk or iron–fortified infant formula.

Infants 6 months to 12 months of age should:

- Begin appropriate low–sodium, nutrient–dense foods, in addition to breast milk or infant formula.
- Consume potentially allergenic foods, including peanuts, if approved by a qualified health care provider.
- Consume foods rich in iron and zinc to support proper growth and development.
- Not consume foods and beverages that contain added sugars.

Infants 12 months of age to toddlers 23 months of age should:

- Drink cow's milk or fortified soy beverages.
- Eat a variety of low–sodium, nutrient–dense foods from all food groups.
- Not consume foods and beverages that contain added sugars.

Africa Studio/Shutterstock

Children 2 to 18 years of age should:

- Limit consumption of foods and beverages that contain added sugars.

Females who can become pregnant, are pregnant, or are breastfeeding should:

- Consume 8 to 12 ounces of seafood per week from a variety of seafood types.
- Not eat certain large fish, including shark and swordfish, because they may contain high amounts of methylmercury.
- Consume iron–rich foods or take an iron supplement, if recommended by a qualified health care provider.
- Not consume alcohol, and limit caffeine intake.

Females who can become pregnant should obtain 400 mcg of folic acid each day.

BananaStock/Alamy Stock Photo

Adults 60 years and older should:

- Consume adequate amounts of protein–rich food.

Chapter 13 provides more information about nutrient needs and dietary recommendations during pregnancy, infancy, childhood, and the older adult years.

Applying the Dietary Guidelines

The Dietary Guidelines include several food and nutrition–related messages for consumers, such as "Consume more nutrient–dense fruits and vegetables"

(Table 3.5). Table 3.6 suggests practical ways people can apply the Dietary Guidelines' recommendations to their usual food choices. However, making recommended dietary and other lifestyle changes does not always reduce risk factors for disease. For example, some people with hypertension may find their blood pressure remains dangerously elevated after several months of exercising, limiting their salt intake, and maintaining a healthy weight for their height. In this case, genetic factors may be influencing people's health more than their lifestyle, and medication may be necessary to reduce their blood pressure.

TABLE 3.5 Selected Messages from *2020–2025 Dietary Guidelines for Americans*

- Increase the variety of protein foods consumed; choose more plant sources of protein including beans, peas, lentils; and incorporate about 8 ounces per week of various seafood into meals.
- Consume more nutrient-dense fruits and vegetables.
- Choose 100% whole grain and enriched grain products. Make at least half your grains whole grains.
- Choose lower-fat versions of milk, yogurt, and cheese, or fortified soy beverages.
- Season foods with herbs and spices, and add little to no salt; compare sodium in foods and choose the foods with the lowest sodium content.
- Drink water, fat-free and low-fat milk, and 100% juice instead of drinks with added sugars.
- Choose nutrient-dense foods that provide potassium, dietary fiber, calcium, and vitamin D, which are "nutrients of public health concern," because Americans tend to consume them in limited amounts.
- Consume as little cholesterol and trans fats as possible while following a healthy diet.
- If one consumes alcohol, the beverage should be consumed in moderation and only by adults of legal drinking age (see Chapter 6). Certain individuals should not consume alcohol, especially pregnant females.

TABLE 3.6 Applying the Dietary Guidelines to One's Usual Food Choices

If One Usually Eats:	Consider Replacing with:
White bread and rolls	Whole-wheat bread and rolls
Ready-to-eat cereals with added sugar	High-fiber cereal with no more than 5 g added sugar per serving; sweetened with berries, bananas, peaches, or other fruit
Cheeseburger, French fries, and a regular (sugar-sweetened) soft drink	Roasted chicken or turkey sandwich, beans, fat-free or low-fat milk, or soy milk
Potato salad or cole slaw	Leafy greens, marinated vegetable salad, or three-bean salad
Regular soft drinks and other sugar-sweetened beverages	Water, fat-free or low-fat milk, or 100% fruit juice
Breaded and fried meat, fish, or poultry	Broiled or roasted meat, fish, or poultry
Fatty meats such as barbecued ribs, sausage, and hot dogs	Chicken, turkey, or fish; lean meats such as ground round
Whole or 2% milk, cottage cheese with 4% fat, or yogurt made from whole milk	1% or fat-free milk, low-fat cottage cheese (1% fat), or low-fat yogurt
Ice cream	Frozen yogurt or "lite" ice cream
Creamy salad dressings or dips made with mayonnaise or sour cream	Oil and vinegar dressing, reduced-fat salad dressings, or dips made from low-fat sour cream or plain yogurt
Salt added to season foods	Herbs, spices, or lemon juice
Doughnuts, pastries, chips, or salty snack foods	Unsalted nuts, dried fruit, or a small bran muffin or whole-grain bagel topped with peanut butter or other nut butter

Concept CHECKPOINT

Respond to the following points according to recommendations of the Dietary Guidelines (2020–2025 version).

13. What are two overarching guidelines of the Dietary Guidelines?

14. A healthy person who is 23 years of age should limit his sodium intake to less than _____ per day.

15. List at least three of the key recommendations of the Dietary Guidelines.

16. Replace solid fats with _____.

17. Consume less than _____% of total calories from saturated fat.

18. Consume less than _____ % of total calories from added sugars.

19. What percentage of your intake of grain foods should be whole grains?

20. How much alcohol should pregnant females consume?

21. Which nutrients are "of public health concern" in the American diet? Explain why these particular nutrients are in this category.

See Appendix F for responses.

3.4 Food Guides

Learning Outcomes

1. Use MyPlate to develop nutritionally adequate daily menus.
2. Compare MyPlate with Food Lists for menu planning.

For over 100 years, the USDA has issued specific dietary recommendations for Americans. In 1943, the USDA published the first food guide based on RDAs for the general public to use. In the mid–1950s, the USDA simplified the original food guide to create the "Basic Four" food groups: milk, meats, fruits and vegetables, and breads and cereals. The recommendations of the Basic Four provided the foundation for an adequate diet. Although the food guide supplied only about 1200 to 1400 kcal/day, people could add extra servings of food to the basic diet plan if they had higher energy needs.

The USDA released several updated food guides after the "Basic Four." In the late 1980s, the USDA determined that a new food guide design was needed to promote diet–related messages of variety, proportion, and moderation. The agency chose a pyramid design to represent these messages. The USDA's last pyramid–shaped food guide, the *MyPyramid Plan*, was released in 2005 (Fig. 3.6).

The MyPyramid Plan was a *food guidance system,* which was based on the *Dietary Guidelines for Americans, 2005.* In addition to providing food and nutrition information, the MyPyramid.gov website emphasized the importance of physical activity and enabled consumers to monitor their activity levels. In 2011, the USDA replaced the MyPyramid Plan with *MyPlate,* another interactive dietary and menu planning guide.

Figure 3.6 MyPyramid Plan.

MyPlate

MyPlate USDA's interactive Internet dietary and menu planning guide

MyPlate (www.myplate.gov) includes a variety of food, nutrition, and physi– cal activity resources for consumers that are based on the recommendations of

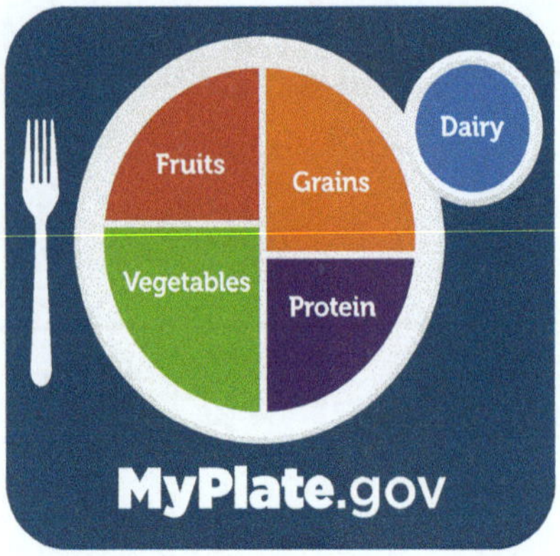

Figure 3.7 **MyPlate.**
U.S. Department of Agriculture

the Dietary Guidelines. MyPlate differs from the MyPyramid Plan in that it no longer has six food groups shown in boxes or stripes within a pyramid. MyPlate focuses on five different food groups: fruits, vegetables, protein foods, grains, and dairy (Fig. 3.7). It is important to note that MyPlate does not have a portion of the plate for "oils." According to the USDA, "oils" is not a food group.[12] The government agency, however, notes the need for fat in the diet because some kinds of fat are dietary essentials.

To learn more about MyPlate's five food groups, visit www.myplate.gov and click on "Eat Healthy" in the menu bar to obtain a list of food groups. Click on each food group to find practical information about foods in the group, such as how much food should be eaten, scientifically supported health benefits of foods, and helpful food-related tips. The www.myplate.gov website has a variety of other helpful diet-related tools, including games, recipes, quizzes, video clips, and "MyPlate Plan," an interactive tool for developing nutritionally adequate daily food plans.

MyPlate Plan

MyPlate Plan has 12 different nutritionally adequate daily food patterns that supply from 1000 to 3200 kcal/day for people who are 2 years of age and older. Each food pattern is individualized to meet a person's age, sex, and physical activity level. Other personal information, such as food likes and dislikes, cultural background, and food budget, can also be considered when developing an individual's dietary plan.

To use MyPlate Plan as your personalized menu planning guide, visit https://www.myplate.gov/myplate-plan. The plan requests information about your age, sex, weight, height, and estimated level of physical activity. MyPlate provides two estimates of your daily energy needs, including "maintain your current weight" or "achieve a healthy weight" (gain or lose weight). Your personal MyPlate Plan also indicates how much food you should eat from each of the food groups daily to meet the recommended energy level. Table 3.7 indicates MyPlate's food intake recommendations for average healthy young adults who consume 1800 to 3200 kcal/day.

TABLE 3.7 MyPlate: Recommendations for Average, Healthy 20-Year-Old Adults

Nelea33/Shutterstock

MyPlate Guidelines (Daily)	Females	Males
Energy	1800–2400 kcal	2600–3200 kcal
Fruit	1.5–2 cups	2–2.5 cups
Vegetables	2.5–3 cups	3.5–4.0 cups
Grains	6–8 oz	9–10 oz
Protein foods	5.0–6.5 oz	6.5–7 oz
Dairy	3 cups	3 cups
Oils*	5–6 tsp	6–7 tsp

*Although "Oils" is not a MyPlate food group, the USDA recommends daily amounts of oils.

MyPlate dietary patterns emphasize nutrient-dense foods and beverages that contain little or no empty calories. After a person consumes recommended amounts of nutritious foods from each food group (and oils), fewer than 15% of calories remain.[10] According to information about healthy eating patterns (2020–2025 version of the Dietary Guidelines), these extra calories may be consumed as limited amounts of saturated fat and added sugars. The 2000 kcal dietary pattern, for example, has fewer than 300 kilocalories remaining, which is almost the same amount of energy in two 12-ounce sugar-sweetened, cola-flavored, carbonated soft drinks. Instead of choosing to eat more sources of empty calories, people can use up these remaining calories by choosing more servings of healthy foods such as dried beans and peas, lentils, seeds, nuts, fresh fruits, and vegetables.

All MyPlate plans limit intakes of saturated fat, sodium, and added sugars.
https://www.myplate.gov/sites/default/files/2020-12/CircleButton_Limit.jpg

Overall, MyPlate can be helpful for planning menus because it promotes food variety, nutritional adequacy, and moderation. You can also use dietary analysis software programs, including NutritionCalc Plus, to evaluate your diet. Section 3.7 provides more information about such software.

A computer or smartphone and Internet access are necessary to use the program. Some people, particularly older adults, may be unfamiliar with smartphones and personal computers and may find the interactive www.myplate.gov website challenging and frustrating to use.

It can be difficult using MyPlate to classify menu items that combine a variety of foods from more than one group, such as pizza, sandwiches, and casseroles. A slice of pizza, for example, has thin crust made with wheat flour (grains), tomato sauce (vegetable), and cheese (dairy). The first step is to determine the ingredients and classify each into an appropriate food group. Estimate the number of cups or ounces of each ingredient and record the amounts contributed from a particular food group. The slice of pizza may provide ¼ cup of a vegetable, 2 ounces of grains, and ¼ cup of dairy. How many servings of foods from dairy and fruit groups are supplied by the mango lassi recipe in the "Recipe for Healthy Living" feature at the end of this chapter?

How do you judge portion sizes when using MyPlate without keeping handy a collection of measuring cups and a scale for weighing foods? Figure 3.8 provides convenient ways to estimate typical portions using familiar objects, including a tennis ball and bar of soap.

Classifying foods that combine ingredients that represent portions of different food groups is challenging. This slice of pizza, for example, has crust (grains), tomato sauce and tomatoes (vegetables), and cheese (dairy).
©Wendy Schiff

MyPlate for Weight Management

MyPlate includes a "Strive for a healthy weight" recommendation for adults. Following a healthy diet, however, is just one aspect of achieving a healthy weight. You should also spend some time each week engaging in physical activities, especially those that are moderately intense. Such activities can include brisk walking, dancing, and swimming. Chapter 11 provides more information about physical activity and the importance of a physically active lifestyle.

Other Food Guides

The USDA's original Food Guide Pyramid inspired the development of other food pyramids for people who follow cultural and ethnic food traditions that differ from the mainstream American ("Western") diet. The "Nutrition Matters" section at the end of Chapter 3 examines the traditional dietary practices of certain cultures and ethnic groups, and how they influence American

MY DIVERSE PLATE

Tomatillos

For centuries, the people of Central America have grown and eaten tomatillos ("husk-tomatoes"). Purchase tomatillos that are dry and firm and have a tight-fitting husk. After the papery husk is removed, the fruit can be roasted, chopped, and mixed with ground chili peppers to form sauces, such as salsa verde (green sauce). Tomatillos can also be added to soups, salads, or omelettes. Three medium tomatillos provide about 33 kcal, 2 g fiber, and 12 mg vitamin C.

©Wendy Schiff

dietary practices. The "Nutrition Matters" section also includes illustrations of the traditional Latin American Diet (see Figure 3.18), Mediterranean Diet (see Figure 3.19), and the Asian Diet Pyramids (see Figure 3.20). Health Canada, the federal agency responsible for helping Canadians achieve better health, also has a food guide, "Canada's Food Guide." Go to this website: https://food-guide.canada.ca/en/ to access this interactive guide.

Do Americans Follow Dietary Recommendations?

Analysis of government food consumption data indicates that the overall diet quality of Americans is improving, but many Americans do not follow the Dietary Guidelines.[13] In 2017, the typical diet of Americans who were 2 years of age and older did not provide recommended amounts of fruit, vegetables, and dairy foods (Fig. 3.9).[14] Furthermore, the diet generally contained too much added sugars, fats, and sodium.[13] It is apparent that the public needs to learn more about the importance of choosing a variety of foods and applying MyPlate to everyday menu planning.

Food Lists and Carbohydrate Counting

People with diabetes or obesity who are trying to manage their disease must control how many calories they consume, and often, modify the macronutrient contents of their diet. The **Food Lists for Diabetes and Weight Management** (formerly called the Exchange System) can be used to estimate the energy, protein, carbohydrate, and fat content of foods. This information is helpful for menu planning. The Food Lists method for menu planning categorizes foods into three groups: carbohydrates, proteins, and fats.[15] The foods within each group have similar macronutrient composition, regardless of whether the food is from a plant or an animal.

Food Lists For Diabetes and Weight Management method of classifying foods into lists based on macronutrient composition

Within each macronutrient group, the Food Lists suggest choices for each type of food. The specified amount of a food listed for a choice provides about the same amount of macronutrients and calories as each of the other

Figure 3.8 Estimating portion sizes. You can use familiar items such as these to estimate portion sizes. Dice, soap: Christopher Kerrigan/McGraw Hill; Mouse: Amos Morgan/Photodisc/Getty Images; Tennis ball: ©Wendy Schiff; Baseball: Ryan McVay/Photodisc/Getty Images; Yo yo: Stockdisc/Getty Images

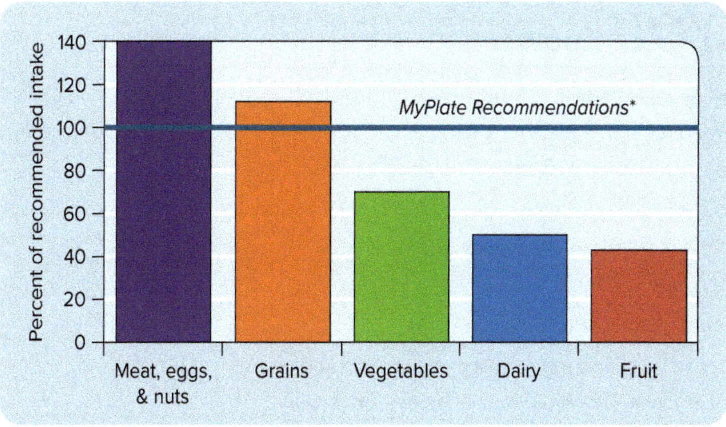

*Data based on a 2000-kcal diet.

Figure 3.9 MyPlate Recommendations. The typical American diet does not provide recommended amounts of fruit, vegetables, and dairy products.

Source: U.S. Department of Agriculture, Economic Research Service: *Food availability (per capita) data system: Consumption compared to dietary recommendations.* Last updated January 2020. https://www.ers.usda.gov/data-products/food-availability-per-capita -data-system/interactive-charts-and-highlights/#calorie. Accessed: June 24, 2022.

specified amounts of foods in that list. For example, according to the fruit list (carbohydrate group), an orange is equivalent to a small apple, a kiwifruit, one–half of a fresh pear, or one–half of a large grapefruit. Within the protein group, equivalent food choices for the very lean protein list include 1 ounce skinless chicken or turkey breast, 2 egg whites, 1 ounce fat–free cheese, or ¾ cup nonfat cottage cheese. This equality allows people to plan a wide variety of nutritious menus by exchanging one food choice for another within each list.

Using Food Lists for Diabetes and Weight Management for menu plan–ning has lost much of its popularity. Now, many people with diabetes and obesity are using other tools, especially *carbohydrate counting* ("counting carbs"). Registered dietitian nutritionists are trained to create individu–alized, nutritionally adequate meal plans using the Food Lists method or

Did *YOU* Know?

Many foods make healthy, quick, and easy snacks that can help you meet your MyPlate food group recommendations. Use small reusable containers with lids to transport snacks in backpacks, purses, and briefcases.

- 1 cup raw veggies such as mini carrots, cherry tomatoes, broccoli and cauliflower pieces, bell pepper slices, or sugar snap peas.
- 1 ounce (about a handful) of unsalted nuts.
- 1 to 2 cups lightly salted, plain popcorn.
- A piece of fresh fruit, including a medium-sized apple, pear, tangerine, 2 kiwifruit, 1 cup of grapes, or 1 banana.

Ken Karp/McGraw Hill

carbohydrate counting for clients seeking to lose weight or manage their diabetes diets. The American Diabetes Association (www.diabetes.org/) offers information about counting carbohydrates. You will learn more about diabetes in Chapter 5.

Concept CHECKPOINT

22. Americans generally do not eat enough ____ to meet MyPlate recommendations.
23. Explain how to use MyPlate to evaluate the nutritional adequacy of an individual's daily food choices.
24. Americans generally consume more _____ than MyPlate recommendations.
25. Describe how the Food List method differs from MyPlate for menu planning.

See Appendix F for responses.

3.5 Food and Dietary Supplement Labels

Learning Outcomes

1. Use the Nutrition Facts panel to make more nutritious food choices.
2. Identify nutrition-related claims the FDA allows on food and dietary supplement labels.

Figure 3.10 What's in a food? Food labels provide specific information, including the food's identity and the container's net weight or volume.

Consumers can use information on food labels to determine the ingredients in a packaged food or drink and to compare the number of calories (energy) and nutrient contents. In the United States, the FDA regulates and monitors information that can be placed on food labels, including claims about the health benefits of ingredients. Today, nearly all foods and beverages sold in grocery stores must have labels that provide the product's name, manufacturer's name and address, and amount of product in the package (Fig. 3.10). Producers and sellers of fresh and frozen fruits and vegetables; fresh poultry, fish, and shellfish; and a few other food items must declare the product's *country of origin* either on the packaging or where the product is located in stores. Furthermore, the ingredients list is a very important component of a label. Products that have more than one ingredient must display a list of the ingredients in descending order according to weight.

Nutrition Facts Panel

Consumers can find specific nutrition-related information about the contents of many packaged foods simply by reading the products' labels. The FDA requires food manufacturers to use a special format, the Nutrition Facts panel, to display information about the energy and nutrient contents of products (Fig. 3.11).

The Nutrition Facts panel shows the amount of a serving size, in household units, such as ounces or cups, as well as grams, and the number of servings in the entire container. The serving size is a reasonable amount of the food

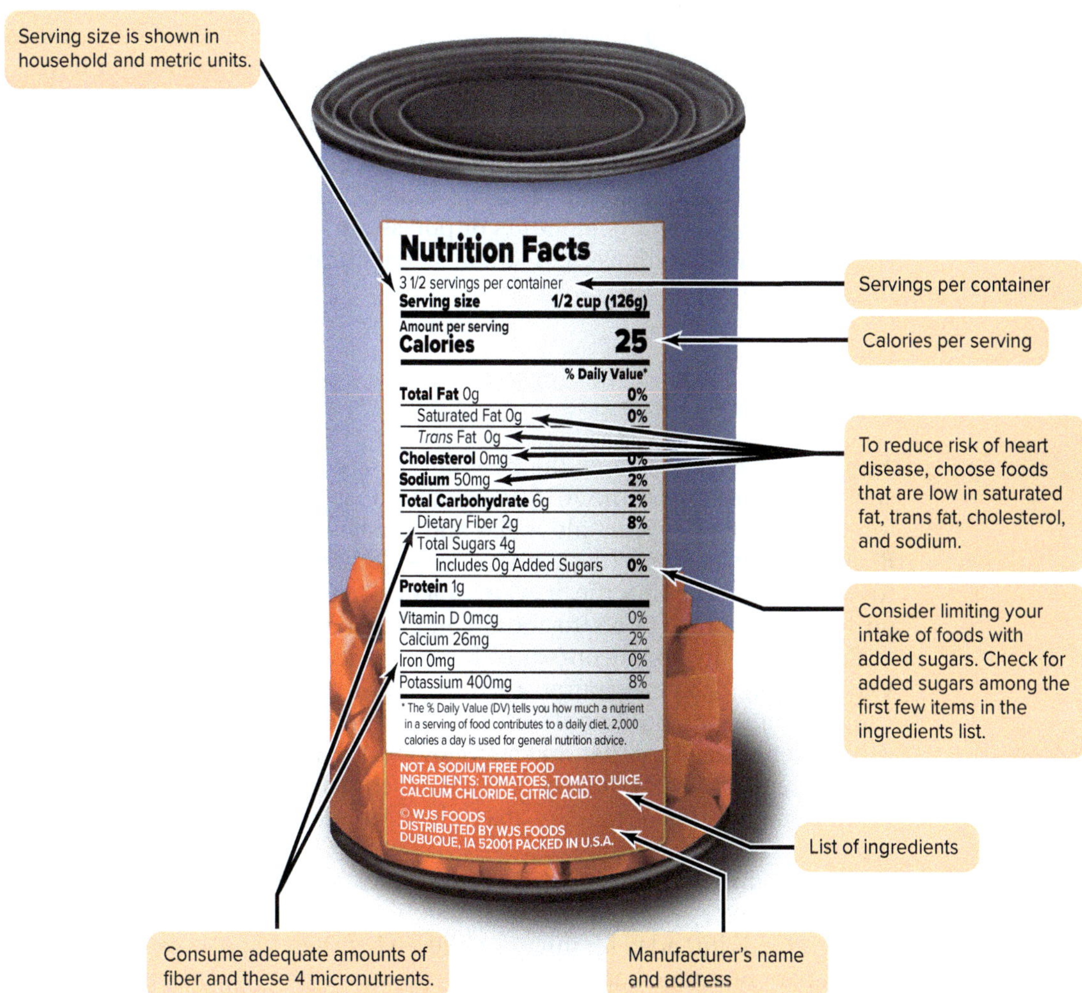

Figure 3.11 The Nutrition Facts panel. The Nutrition Facts panel displays information about the serving size, servings per container, and energy and certain nutrient contents of a packaged food product.

Did *YOU* Know?

According to The Patient Protection and Affordable Care Act of 2010, restaurants and similar retail food establishments with 20 or more national locations must display calorie content information for standard retail food items as labels on menus or menu signs. Furthermore, the restaurants are required to provide written information about their products' total calories, fat, saturated fat, cholesterol, sodium, total carbohydrates, sugars, fiber, and protein contents when their customers request it. This nutrition-related information is typically provided at the restaurant's website.

Results of one study indicated that full-service, as well as fast-food, restaurant patrons who reported using menus' calorie information to make menu choices had lower daily caloric intakes than customers who read the information but chose not to use it when purchasing restaurant foods and beverages.[16]

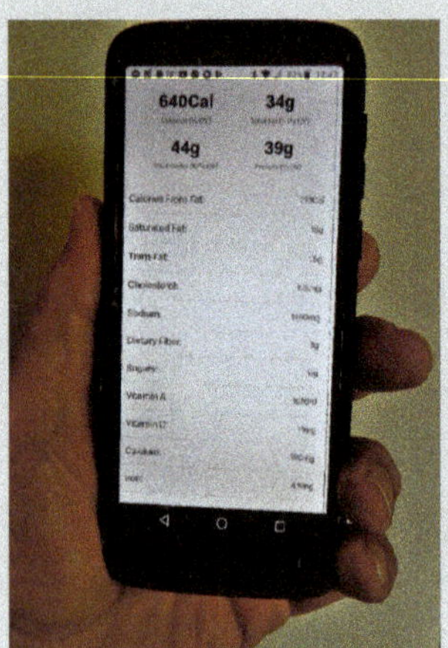

©Wendy Schiff

or beverage that a person typically consumes at one time. For example, a serving size for a 12-ounce can of soft drink is 12 ounces (the entire can).

The Nutrition Facts panel requires most food manufacturers to provide information about the food's total fat, saturated fat, trans fat, cholesterol, sodium, total carbohydrate, fiber, total sugars, added sugars, protein, vitamin D, potassium, calcium, and iron contents. Additionally, the panel must display the total amount of energy, indicated as the number of calories, in a serving.

The panel uses grams (g) and milligrams (mg) to indicate amounts of nutrients in a serving of food. Food manufacturers can also include amounts of polyunsaturated and monounsaturated fats, as well as micronutrients that are not required to be listed on the label. If the manufacturer has fortified the food with the nutrients or made claims about the product's nutrient contents, listing these particular food components is required.

Fresh foods such as chopped meat and poultry and their major cuts are required to have a Nutrition Facts label. Fresh fruits and vegetables, fish, and shellfish are not required to have Nutrition Facts labels. However, many food suppliers and supermarket chains provide consumers with information about their products' nutritional content on posters or shelf tags displayed near the foods.

The Nutrition Facts panel displays the number of calories per serving in very bold print, so it is highly visible. Under "Total Sugars," ". . . Added Sugars" differentiates between ingredients that are added to foods or beverages to sweeten them, such as high-fructose corn syrup, and sugars that are naturally in a food or beverage, such as those in fruit and cow's milk. Such information is helpful for people who want to monitor their intake of empty calories and compare foods for their added sugar contents. The Nutrition Facts panel also displays amounts of potassium, calcium, iron, and vitamin D. Many Americans have less-than-adequate intakes of these particular micronutrients.

Daily Values

Nutrient standards such as the RDA and AI are sex-, age-, and life stage-specific. For example, the RDA for vitamin C is 75 mg/day and 65 mg/day for nonsmoking 18-year-old males and females, respectively. The vitamin's RDA increases to 80 mg/day for 18-year-old pregnant females. Because the RDAs and AIs are so specific, it is not practical to provide nutrient information on food labels that refers to these complex standards. To help consumers evaluate the nutritional content of food products, FDA developed the **Daily Values (DVs)** for labeling purposes. Compared to the RDAs, the DVs are a more simplified and practical set of nutrient standards. The adult DV for a nutrient is based on a standard diet that supplies 2000 kcal/day. Most nutrients have DVs, including total fat, total carbohydrate, added sugars, fiber, vitamins, and several minerals. As of January 2021, there were no DVs for total sugars and trans fat.

Appendix B lists the DVs for people over 4 years of age, which is used for foods and beverages that adults consume. Three other sets of DVs are used on labels of foods intended for infants, children between 1 and 4 years of age, and pregnant or breastfeeding females.

Daily Values (DVs) set of nutrient intake standards developed for labeling purposes

The DVs are generally based on the highest RDA or AI for a particular nutrient, but in some instances, they are based on recommendations of public health experts. For example, the RDA for carbohydrate is 130 g/day for people over 1 year of age. The DV for carbohydrate, however, is 275 g/day.[17] This amount reflects general dietary recommendations that carbohydrates (other than fiber and added sugars) can contribute 55% of a person's total energy intake, or 1100 kcal (275 g × 4 kcal/g of carbohydrate) of a 2000 kcal/day diet. For people older than 1 year of age, no RDA or AI has been set for daily fat intake. However, DV for fat is 78 g/day. This amount meets the general recommendation that fat intake can be as much as 35% of a person's total energy intake for a 2000 kcal/day diet.

When evaluating or planning nutritious menus, your goal is to obtain at least 100% of the DVs for fiber, vitamins, and most minerals each day. On the other hand, you may need to limit your intake of foods that have high %DVs of total fat, saturated fat, added sugars, and sodium. High intakes of these nutrients may have negative effects on your health. Thus, your goal is to consume less than 100% of the DV for total fat, saturated fat, added sugars, and sodium each day. The general rule of thumb: A food that supplies 5%DV or less of a nutrient is a low source of the nutrient; a food that provides 20%DV or more is a high source of the nutrient.[18]

Percents of DVs are designed to help consumers compare nutrient contents of packaged foods to make more healthful choices. However, most people do not eat just packaged foods. Fresh fruits and vegetables, as well as many full-service-restaurant meals, do not have labels or menus with information about %DVs per serving. Therefore, many consumers will underestimate their nutrient intakes if they do not consider the contribution that unlabeled foods make to their diets.

It is important to note the description of a serving size and the number of servings per container when using nutritional labeling information to estimate your intakes of energy, fiber, and nutrients in the food. A common mistake people make when using a Nutrition Facts panel is assuming that the information applies to the entire package, which is not always true. For example, the Nutrition Facts panel on a package of food indicates there are four servings in the container. If you eat all the container's contents, you must multiply the information concerning calories, fat, and other food components by four. Why? Because you ate four servings and the nutritional information on the Nutrition Facts panel applies to only *one* serving. For some packaged foods, the FDA has developed a Nutrition Facts label format that displays calorie and nutrient information for part, as well as all, of the package's contents (Fig. 3.12).

Nutrition Facts

2 servings per container
Serving size **1 cup (255g)**

	Per serving		Per container	
Calories	**220**		**440**	
		% DV*		**% DV***
Total Fat	5g	**6%**	10g	**13%**
Saturated Fat	2g	**10%**	4g	**20%**
Trans Fat	0g		0g	
Cholesterol	15mg	**5%**	30mg	**10%**
Sodium	240mg	**10%**	480mg	**21%**
Total Carb.	35g	**13%**	70g	**25%**
Dietary Fiber	6g	**21%**	12g	**43%**
Total Sugars	7g		14g	
Incl. Added Sugars	4g	**8%**	8g	**16%**
Protein	9g		18g	
Vitamin D	5mcg	25%	10mcg	50%
Calcium	200mg	15%	400mg	30%
Iron	1mg	6%	2mg	10%
Potassium	470mg	10%	940mg	20%

* The % Daily Value (DV) tells you how much a nutrient in a serving of food contributes to a daily diet. 2,000 calories a day is used for general nutrition advice.

Figure 3.12 Other acceptable label formats. A Nutrition Facts label may display calorie and nutrient information for part as well as all of the package's contents.
Source: Guidance for Industry: Food Labeling Guide. www.fda.gov/regulatory-information/search-fda-guidance-documents/guidance-industry-food-labeling-guide

Health- and Nutrition-Related Claims

To make consumers more likely to buy their foods, manufacturers often promote products as having certain health benefits or high amounts of nutrients. A health claim describes how a food, food ingredient, or dietary supplement may reduce the risk of a nutrition-related condition. The FDA permits food manufacturers to include certain health claims on food labels (Fig. 3.13). For example, an allowable health claim may state, "Diets low in saturated fat may reduce the risk of heart disease."

The FDA only allows claims on food labels that:

- indicate the product has health benefits when it is consumed with other foods that make up a daily diet;
- are complete, easy to understand, honest, and not misleading;
- refer to a product that has 10% or more of the DVs for fiber, protein, calcium, iron, potassium, or vitamin D *before* being fortified with nutrients. A food manufacturer, for example, wants to fortify its orange juice with calcium to boost the juice's calcium content from 2%DV to 35%DV. The FDA

Figure 3.13 Label claims. The FDA permits food manufacturers to include certain health claims on food labels.
©Wendy Schiff

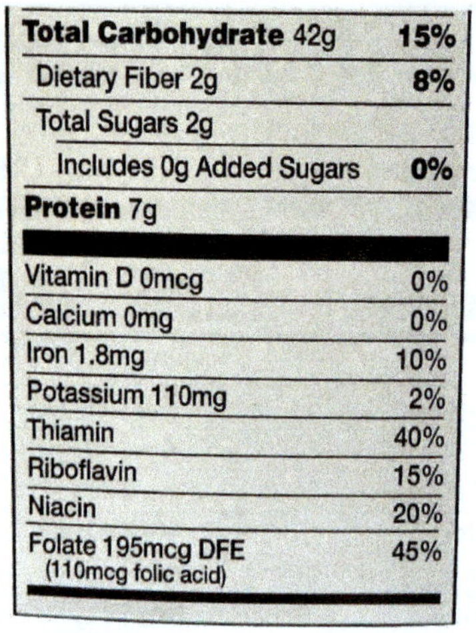

Total Carbohydrate 42g	**15%**
Dietary Fiber 2g	**8%**
Total Sugars 2g	
Includes 0g Added Sugars	**0%**
Protein 7g	
Vitamin D 0mcg	0%
Calcium 0mg	0%
Iron 1.8mg	10%
Potassium 110mg	2%
Thiamin	40%
Riboflavin	15%
Niacin	20%
Folate 195mcg DFE (110mcg folic acid)	45%

According to the Nutrition Facts panel, a serving of this food is a "high source" of which nutrients?
©Wendy Schiff

would not allow the manufacturer to place a health claim about the juice's calcium content on the label because the original juice had less than 10%DV for the vitamin;

- are for a product intended for people who are 2 years of age or older;
- use *may* or *might* to describe the relationship between the product and disease. For example, "Diets containing foods that are good sources of potassium and that are low in sodium may reduce the risk of high blood pressure and stroke" is an allowable claim. However, the claim "Reduces the risk of stroke" would not be permitted on a label;
- do not quantify any degree of risk reduction. For example, a claim that states, "Reduces risk of cancer by 41%" would not be allowed because it specifies the degree of risk reduction; and
- indicate that many factors influence disease.

The FDA requires specific wording for certain health claims that are allowed on labels. Table 3.8 lists some permissible health claims that can be used for labeling purposes. For more information, visit FDA's website at www.fda.gov and search for "qualified health claims."

Bangkok Click Studio/Shutterstock

TABLE 3.8 Examples of Permissible Health Claims for Food Labels

Dietary Factor/Health Condition	Example of Permissible Health Claim
Certain lipids and heart disease	"While many factors affect heart disease, diets low in saturated fat and cholesterol may reduce the risk of this disease."
Diet and heart disease	"Diets low in saturated fat and cholesterol and rich in fruits, vegetables, and grain products that contain some types of dietary fiber, particularly soluble fiber, may reduce the risk of heart disease, a disease associated with many factors."
Calcium, exercise, and osteoporosis (a disease that weakens bones)	"Regular exercise and a healthy diet with enough calcium help teen and young adult white and Asian females maintain good bone health and may reduce their high risk of osteoporosis."
Sodium (a mineral) and high blood pressure	"Diets low in sodium may reduce the risk of high blood pressure, a disease associated with many factors."
Folate (a B vitamin) and neural tube defects (conditions in which the skull and spine do not form properly before birth)	"Healthful diets with adequate folate may reduce a female's risk of having a child with a brain or spinal cord defect."
Fruits and vegetables and risk of cancer	"Foods that are low in fat and contain dietary fiber, vitamin A, or vitamin C may reduce the risk of some types of cancer, a disease associated with many factors. Broccoli is high in vitamins A and C, and it is a good source of dietary fiber."

Source: U.S. Food and Drug Administration: *Food labeling guide.* 2013. https://www.fda.gov/regulatory-information/search-fda-guidance-documents/guidance-industry-food-labeling-guide. Accessed: June 27, 2022.

Structure/Function Claims

A structure/function claim describes the role a nutrient or dietary supplement plays in maintaining a structure, such as bone, or promoting a normal function, such as digestion. The FDA allows structure/function claims such as "calcium builds strong bones" or "fiber maintains bowel regularity" (Fig. 3.14). Structure/function statements cannot claim that a nutrient, food, or dietary supplement can be used to prevent or treat a serious health condition. For example, the FDA would not permit a claim that a product "promotes low blood pressure" because that claim implies the product has drug-like effects and can treat high blood pressure.

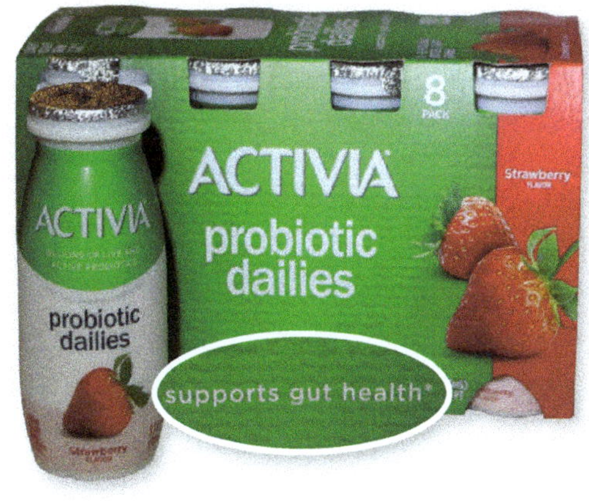

Figure 3.14 Structure/function claim. Structure/function statements cannot claim that a nutrient, food, or dietary supplement can be used to prevent or treat a serious health condition.
Traci Keck

Nutrient Content Claims

The FDA permits manufacturers to include claims on labels that describe levels of nutrients in packaged foods. Such nutrient content claims can use the terms *free, high,* or *low* to describe how much of a nutrient is in the product. Additionally, nutrient content claims can use terms such as *more* or *reduced* to compare amounts of nutrients in a product to those in a similar product. This claim is often used for an item that substitutes for a *reference food,* which is a similar and more familiar food. For example, "reduced-fat" popcorn has considerably less fat than its reference food, regular popcorn.

Table 3.9 lists some legal definitions for common nutrient content claims that are allowed on labels. Note that a product may contain a small amount of a nutrient such as fat or sugar, yet the Nutrition Facts panel can indicate the amount as "0 g." For example, the Nutrition Facts panel may indicate that a serving of food supplies "0 g" of trans fat, even though the food actually supplies less than 0.5 g of trans fat. As a result, it is possible to consume some trans fats from foods even though labels indicate a serving of each food does not

TABLE 3.9 Legal Definitions for Common Nutrient Content Claims (April 2023)

Sugar	• **Sugar free:** The product provides less than 0.5 g of sugar per serving. • **Reduced sugar:** The food contains at least 25% less sugar per serving than the reference food. • **No added sugar:** The food contains no sugar or sugar-containing ingredients such as jam, jelly, or concentrated fruit juice added during processing.
Calories	• **Calorie free:** The food provides fewer than 5 kcal per serving. • **Low calorie:** The food supplies 40 kcal or less per serving. • **Reduced or fewer calories:** The food contains at least 25% fewer kcal per serving than the reference food.
Fat	• **Fat free:** The food provides less than 0.5 g of fat per serving. • **Low fat:** The food contains 3 g or less fat per serving. Two-percent milk is not "low fat" because it has more than 3 g of fat per serving. The term *reduced fat* can be used to describe 2% milk. • **Reduced or less fat:** The food supplies at least 25% less fat per serving than the reference food.
Cholesterol	• **Cholesterol free:** The food contains less than 2 mg of cholesterol and 2 g or less of saturated fat per serving.
Fiber	• **High fiber:** The food contains 5 g or more fiber per serving. Foods that include high-fiber claims on the label must also meet the definition for low fat. • **Good source of fiber:** The food supplies 2.5 to 4.9 g of fiber per serving.
Meat and poultry products regulated by USDA	• **Extra lean:** The food provides less than 5 g of fat, 2 g of saturated fat, and 95 mg of cholesterol per serving. • **Lean:** The food contains less than 10 g of fat, 4.5 g of saturated fat, and 95 mg of cholesterol per serving.

contain this type of fat. To learn more about the FDA's regulations concerning nutrient claims, visit the agency's website (www.fda.gov) and search for "food labeling guide."

Healthy Product Claims Manufacturers can also use the nutrient content claim "healthy" on food products based on individual nutrients supplied by the product, including total fat, saturated fat, sodium, cholesterol, and some vitamins and minerals. However, foods rich in refined grains and added sugars, including white bread, sugar-sweetened cereals, and yogurts, currently qualify for the "healthy" claim. Olive oil, which is rich in healthy unsaturated fats, does *not* qualify for the claim because of the oil's total fat content.

In September 2022, FDA officials proposed new guidelines to update the "healthy" label claim to reflect current scientific evidence and dietary guidelines.[19] The updated claim would consider the nutrient density and variety of nutrients, including the type of fat, not just total fat, supplied by the food. Naturally, nutrient-dense foods—including raw fruits and vegetables, whole grains, low-fat dairy products, lean meats, and foods rich in healthy monounsaturated and polyunsaturated fats (see Chapter 6), such as salmon, nuts, and olive oil—would qualify for the "healthy" claim. However, these foods must not exceed saturated fat, sodium, and added sugar daily value (DV) guidelines for their specific food group. For example, a 3/4-cup serving of yogurt that supplies 5%DV (2.5 grams) or less of added sugar and 10%DV (2 grams) or less of saturated fat would qualify as a "healthy" food.

Food & Nutrition tips

Creamy salad dressings, such as ranch or Caesar, are higher calorie condiments because of the vegetable oils they contain. "Light" versions of these dressings often have more water and sugar added to replace some fat in the dressing. However, you can make your own calorie-reduced salad dressing by adding about ¼ cup water to a jar of regular creamy salad dressing, then stir or shake the mixture.

Olga Nayashkova/Shutterstock

Other Descriptive Labeling Terms

A food can be labeled *light* or *lite* if it has at least one-third fewer kilocalories or half the fat of the reference food. For example, a tablespoon of lite pancake syrup has one-third fewer kcal than a tablespoon of regular pancake syrup. A tablespoon of light mayonnaise has less than half the fat of regular mayonnaise. The term *light* may also describe food characteristics such as texture and color; for example, "light brown sugar." As of June 2022, the FDA was reviewing the use of "natural" on food labels.

Dietary Supplement Labels

According to federal law, every dietary supplement container must be properly labeled (Fig. 3.15). The label must include the term "dietary supplement" or a similar term that describes the product's particular ingredient, such as "herbal supplement" or "vitamin C supplement." Dietary supplement labels are also required to display the list of ingredients, manufacturer's address, and suggested dosage. Furthermore, the label must include facts about the product's contents in a special format—the "Supplement Facts" panel (see Figure 3.15). The panel provides information about the serving size, amount per serving, and percent Daily Value (%DV) for ingredients, if one has been established. DVs are standard desirable or maximum intakes for many nutrients, but DVs have not been established for nonnutrient products, including herbal supplements.

According to the FDA, dietary supplements are not intended to treat, diagnose, cure, or alleviate the effects of diseases. Therefore, the agency does not permit manufacturers to market a dietary supplement product as a treatment or cure for a disease, or to relieve signs or symptoms of a disease.

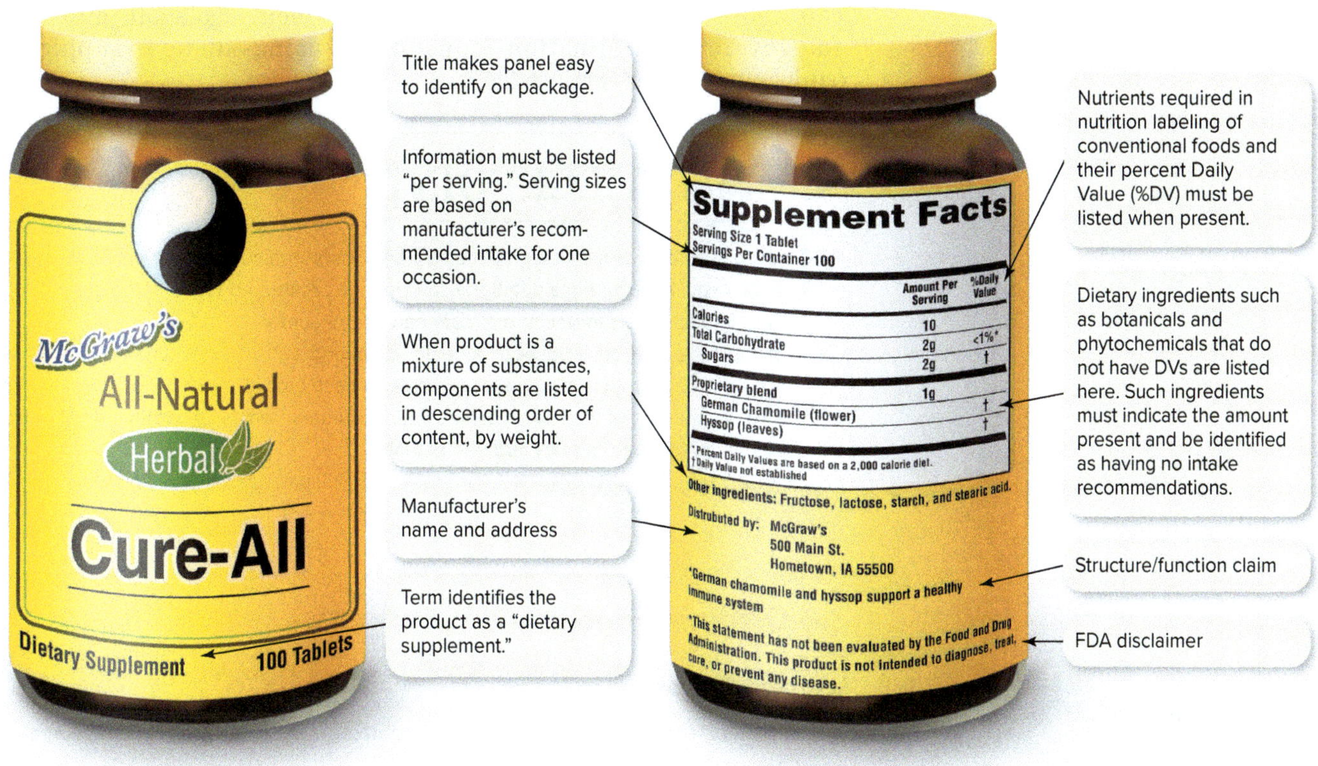

Figure 3.15 Supplement Facts label. A nutrient supplement label must list the product's ingredient(s), serving size, amount(s) per serving, suggested use, manufacturer and the company's address, and %DV, if one has been established. If a health claim appears on the supplement's label, the claim must be followed by the FDA disclaimer.

Although such products generally cannot prevent diseases, some can improve health or reduce the risk of certain diseases or conditions. Thus, the FDA allows supplement manufacturers to display structure/function claims on labels. Manufacturers of iron supplements, for example, may have a claim on the label that states: "Iron is necessary for healthy red blood cell formation." If the FDA has not reviewed a claim, the label must include the FDA's disclaimer indicating that the claim has not been evaluated by the agency (Fig. 3.16).

The FDA does not require dietary supplement manufacturers or sellers to provide evidence that labeling claims are accurate or truthful before they appear on product containers. However, manufacturers that include structure/function claims on labels must notify the FDA about the claims within 30 days after introducing the products into the marketplace. If FDA officials question the safety of a dietary supplement or the truthfulness of claims that appear on supplement labels, manufacturers are responsible for providing the agency with evidence that their products are safe and the claims on labels are honest and not misleading.

The FDA requires dietary supplement manufacturers to evaluate the purity, quality, strength, and composition of their products before marketing them. The regulations are designed to result in the production of supplements that contain the ingredients listed on the label, are wholesome, contain standard amounts of ingredients per dose, and are properly packaged and accurately labeled. Visit the FDA's *Supplement Your Knowledge* initiative to

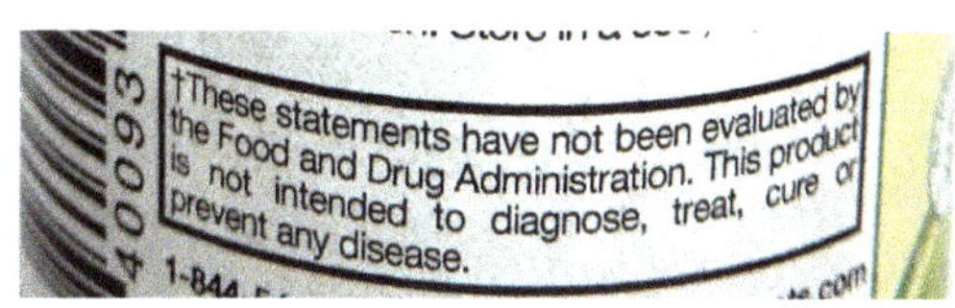

Figure 3.16 Label disclaimer. The FDA permits dietary supplement manufacturers to include certain health-related claims on their product labels. However, the label of products bearing such claims also must display this disclaimer.
©Wendy Schiff

learn more about dietary supplements and their regulation, https://www.fda.gov/food/information-consumers-using-dietary-supplements/supplement-your-knowledge.

Concept CHECKPOINT

26. Identify at least one limitation of using %DVs to determine your nutrient intakes.
27. Explain how you can use nutritional information provided on food and dietary supplement labels to become a more careful consumer.
28. What is the difference between a health claim and a structure/function claim? What is a nutrient content claim? Give an example of each type of claim.
29. Discuss the role of the FDA in protecting consumers from false nutrition and health claims on food and dietary supplement labels.

See Appendix F for responses.

3.6 Organic Food

Learning Outcomes

1. Explain differences between the production of organic foods and the production of conventional foods.
2. Discuss the USDA's three labeling categories for organic foods.

By the late twentieth century, emphasis on increasing agricultural production resulted in an inexpensive and abundant food supply in the United States. However, the rise of agribusiness also resulted in social, economic, and environmental costs. Rural agricultural communities experienced a dramatic decline in the number of small farms, as the farms' owners could not compete with the production capabilities and financial resources of large, commercially run farms (agribusiness).

Instead of producing a variety of crops, big farms often focus on growing corn, soybeans, or wheat. These crops require conventional farming methods that include heavy use of fertilizers and products to control pests (*pesticides*). In some parts of the country, large farms also need considerable amounts of water for irrigating crops. As a result, underground water supplies are being depleted in these regions.

The rise in agribusiness helped fuel interest in sustainable agriculture. As mentioned in Chapter 1, sustainable agriculture focuses on producing adequate amounts of food without reducing natural resources, such as the water supply, and harming people as well as the natural environment.[20] Such agricultural methods promote crop variety, soil and water conservation, and recycling of plant nutrients. Additionally, sustainable agriculture can support small farms, particularly organic farms.

Technically, organic substances have the element carbon bonded to hydrogen (another element) in their chemical structures. Therefore, all foods are organic because they contain substances composed of carbon bonded with hydrogen. The term *organic,* however, also refers to certain agricultural methods that can promote sustainability. Organic farming and the production of **organic foods** do not rely on the use of antibiotics, hormones, synthetic fertilizers and most synthetic pesticides, genetic improvements, or ionizing radiation.[23] Table 3.10 compares organic and conventional agricultural systems.

Food & Nutrition tips

You can reduce your exposure to pesticides in food by:

- washing and scrubbing all fresh fruits and vegetables under running water, avoiding special produce washes and detergents to remove residues, which are not more effective than using water;
- peeling and trimming fruits and vegetables before eating them;
- trimming the fat from meat and skin of poultry and fish because some pesticide residues accumulate in fat; and
- eating a variety of foods; this reduces the likelihood of excessive exposure to a single pesticide.[21, 22]

organic foods foods produced without the use of antibiotics, hormones, synthetic fertilizers and most synthetic pesticides, genetic improvements, or spoilage-killing radiation

TABLE 3.10 Comparing Organic and Conventional Farming Systems

Organic	Conventional
Synthetic fertilizers are not allowed.	Limited restrictions on fertilizers
Sewage sludge products are not allowed.	Sludge products may be used on some fields.
Restrictions on use of raw manure on fields used for food crops	Few restrictions on raw manure use for edible crop fields
Most synthetic pesticides are not allowed; natural pest management practices are encouraged.	Any government-approved pesticide may be used according to label instructions. Natural pest management practices may also be used.
Genetically modified organisms (GMOs) are not allowed.	Government-approved GMOs are permitted.
Feeding livestock mammal and poultry by-products and manure is not allowed.	Certain mammal and poultry by-products are allowed in livestock feed.
Use of growth hormones and antibiotics in livestock production is not allowed.	Government-approved hormone and antibiotic treatments are permitted.
Food irradiation (a food safety method) is not allowed.	Food irradiation may be used.
Detailed record keeping and site inspections by regulators are required.	Some records are required, but no on-site checks by regulators are necessary.

Over the past 40 years, the popularity of organic foods has increased in the United States as many Americans have become concerned about the environment and the safety and nutritional value of the food supply. Sales of organic foods have increased steadily since the 1990s, even though these products are usually more expensive than the same foods produced by conventional farming methods. According to the Organic Trade Association, Americans spent over an estimated $63 billion on organic foods and beverages from 2020 to 2021.[24]

People who purchase organically grown foods often think the products are better for their health and more nutritious than conventionally produced foods. A few well-designed studies have compared nutrient and phytochemical contents of organically grown foods to their conventionally grown counterparts. The researchers noted small to moderate increases in some nutrients in organic foods, including phytochemical polyphenols in fruits and vegetables and omega-3 fatty acids in dairy products (see Chapter 6).[25] In general, organic food crops are not more nutritious than conventionally grown food crops.[25] Organic crops, however, may contain fewer pesticides than conventionally grown crops.[25] Nevertheless, more research is needed to determine whether there are health advantages to eating organic foods.

Labeling Organic Foods

To protect consumers, the USDA developed and implemented rules for the organic food industry. A food product cannot be labeled "organic" unless its production meets strict national standards. For labeling purposes, organic food manufacturers can use the circular "USDA Organic" symbol on the package (Fig. 3.17). This symbol indicates that the products meet USDA's standards for organic food. According to the USDA, there are three organic labeling categories (Table 3.11). Note that certain foods can have the organic symbol on the package, yet they may contain small amounts of ingredients that are not considered organic by the USDA. For more information about the government's organic labeling standards, visit the USDA's Agricultural Marketing Service's website (www.ams.usda.gov) and search for "organic labeling standards."

Figure 3.17 Organic food logo. Foods that have been certified "organic" may use the USDA's symbol.
©Wendy Schiff

TABLE 3.11 Organic Labeling Categories

"100% Organic" (may use USDA organic seal)	100% certified organic ingredients Must identify organic ingredients
"Organic" (may use USDA organic seal)	Contains at least 95% certifed organic ingredients Remaining 5% of ingredients are on USDA's list of allowed ingredients Must identify organic ingredients
"Made with organic ______" (may not use USDA organic seal)	Contains 70 to 95% certified organic ingredients Must identify organic ingredients

Source: U.S. Department of Agriculture, Agricultural Marketing Service: *Labeling Organic Products*. ND. www.ams.usda.gov/rules-regulations/organic/labeling. Accessed: April 11, 2023.

Concept CHECKPOINT

30. Jeremy prefers to drink organic milk instead of milk that is not organic. What is the difference between the two types of milk?

31. Jeremy only purchases cereals that have the organic seal on the package. Explain why such cereals may not be 100% organic.

See Appendix F for responses.

3.7 Using Dietary Analysis Software

Learning Outcomes

1. Explain why different dietary analysis software programs may provide different values for energy and nutrient contents for the same food item.
2. Use the U.S. Department of Agriculture's websites to estimate the nutritional value of various foods.

How much selenium, magnesium, and niacin are in an ounce of Swiss cheese? Have you ever wanted information about nutrients in a food that are not listed on the Nutrition Facts panel? In the past, people relied on food composition tables, lists of commonly eaten foods that provide amounts of energy, fiber, macronutrients, and several micronutrients. Today, people can determine the energy and nutrient contents of their food choices by using a dietary analysis software program. Furthermore, people with Internet access can obtain the information from certain websites.

Dietary analysis software and websites can be quick and easy tools for determining nutrient and energy contents of a specific food. However, the values provided by these resources are not necessarily exact amounts. The same type of plant food may vary in nutrient content depending on hereditary factors, age, growing conditions, and production methods. Therefore, scientists generally analyze several samples of a particular food to determine their nutrient contents, and then the researchers average the results. For example, if the amount of energy in three Valencia oranges that each weighs about 4 ounces (120 g) were 55, 60, and 62 kcal, respectively, the value listed in the food composition table for a Valencia orange weighing 4 ounces would be 59 kcal, the average of the three. In many instances, values for certain nutrients are missing. This occurs when accurate data concerning the complete nutrient analysis of the food are unavailable.

Valencia oranges and other produce may vary in nutrient content depending on numerous factors, including growing conditions.
Savany/E+/Getty Images

The following section discusses some government–sponsored websites that provide practical tools for evaluating food intakes and physical activity habits. The Personal Dietary Analysis feature at the end of this chapter provides an opportunity for you to practice using dietary analysis software.

Government-Sponsored Dietary Analysis Websites

The USDA sponsors some websites to help you assess the energy and nutrient contents of your food intake. The "What's in the Food You Eat *Search Tool*" at www.usda.gov is one such site. Another USDA–sponsored site that provides extensive information regarding the energy and nutrient content of food is FoodData Central. You can access this nutrient database by visiting https://fdc.nal.usda.gov/. To keep current, USDA–sponsored websites are updated regularly to provide nutritional information about new products.

Concept CHECKPOINT

32. Identify at least two reliable sources of information about the energy and nutrient contents of foods and beverages.

See Appendix F for responses.

3.8 Nutrition Matters: Traditional Diets

Learning Outcomes

1. Discuss the traditional diet of certain cultures and ethnic groups living in the United States.
2. Identify religion-based dietary restrictions.

Wherever you live or travel in the United States, you're likely to find restaurants that serve a wide variety of ethnic fare, such as Italian, Thai, Ethiopian, or Middle Eastern dishes. Although your primary food selection and cooking habits probably reflect your cultural/ethnic heritage, you likely enjoy foods from other cultures and ethnic groups. However, restaurants in the United States may serve ethnic cuisine that is not *traditionally* prepared. Instead, these meals may contain larger portions of food, including more meat; refined grains; and very little fruits or vegetables. These dishes may also include added fat, sugar, and salt, which is not part of traditional cuisine preparation.

Traditional ethnic diets are often based on dishes containing small amounts of animal foods and larger amounts of plant foods, including locally grown fruits, vegetables, and whole grains. Dietary pyramids or plans developed by governmental agencies or private organizations have plant foods as the core of a healthy diet. Scientific research shows that eating patterns rich in plant foods, including vegetables, fruits, whole grains, legumes, nuts, and seeds, are associated with a lower incidence of chronic disease.[11] However, traditional foods are typically the first to be abandoned as immigrants *assimilate,* that is, blend into the general population over time. After an immigrant population has assimilated fully, the prevalence of chronic diseases such as CVD, type 2 diabetes, and high blood pressure often increases among them, partly as a result of adopting less healthy eating practices. This Nutrition Matters section examines the traditional dietary practices of certain cultures and ethnic groups, how they influence the American diet, and the benefits of following traditional diets. Learning about traditional diets may pique your interest to try new foods!

Northwestern European Diet

Immigrants from northwestern European regions or countries such as the United Kingdom, Scandinavia, and Germany traditionally consumed meals that featured a large portion of beef or pork served with a smaller portion of potatoes that were either boiled or mashed. This "meat-and-potato" diet became the mainstream American diet, often referred to as a "Western" diet, which provides large amounts of animal protein and fat, and lacks fruits, whole grains, and a variety of green vegetables. Such diets are associated with high rates of serious chronic diseases, particularly CVD and type 2 diabetes, which are discussed in later chapters of this textbook.

Latin American Diet

The Hispanic or Latino (people with Spanish ancestry) population is now the largest ethnic minority group in the United States. The traditional Mexican diet includes corn, beans, chili peppers, avocados, papayas, and pineapples. Many supermarkets in the United States sell other plant foods that are often incorporated into Mexican meals, such as fresh chayote, cherimoya, jicama, plantains, and cactus leaves and fruit. Such fruits and vegetables add fiber and a variety of nutrients, phytochemicals, and vivid colors to Mexican dishes.

Authentic Mexican meals also include whole-grain corn tortillas, quinoa, rice, fish, shellfish, poultry, and meats. Dishes also have unique flavors from the spices used in cooking, including chili peppers, garlic, cinnamon, and the herbs cilantro, oregano, and parsley.

Scientific research shows that Hispanics who follow a traditional Latin American diet rich in fruits, vegetables, and seafood (Fig 3.18) have lower blood pressure levels than Hispanics who follow a Western diet.[26]

Figure 3.18 **Latin American Diet Pyramid.**
©2009 Oldways Preservation and Exchange Trust. Used with permission. www.oldwayspt.org

Italian and Mediterranean Diets

The traditional Italian diet of pasta and other grain products, olive oil, fish, nuts, fruits, and vegetables is healthier than the Western diet. Pasta, a product made from wheat flour and water, is the core of the traditional Italian diet. To many Americans, *pasta* is spaghetti topped with tomato sauce, meatballs, and grated Parmesan cheese. However, Italians eat a variety of different forms of pasta, such as penne, linguini, acini de pepe, and rotini, along with sauces that are often meatless. Pizza, a dish from southern Italy, is one of the most frequently consumed foods in the United States. Unlike traditional Italian pizza that has a thin crust and is lightly covered with tomatoes, basil (a leafy herb), and mozzarella cheese, many Americans choose thick-crust pizza topped with tomato sauce, shredded mozzarella cheese, and dotted with pork sausage or pepperoni.

The Mediterranean Diet Pyramid shown in Figure 3.19 is based on traditional dietary practices of Mediterranean countries, such as Greece, southern Italy, and the island of Crete. Whole grains, a wide variety of fruits and vegetables, beans, and potatoes form the foundation of this diet. Red meat and sugary foods, which are at the top of the pyramid, are rarely eaten. Main dishes often include seafood and poultry, and wine may be included with meals. Although the Mediterranean Diet Pyramid allows as much as 35% of total calories as fat in the diet, much of the fat is from olive oil. Olive oil is a rich source of a type of fat that reduces rather than increases the risk of CVD. Chapter 6 provides more information about oils and fats and their roles in health.

Figure 3.19 **Mediterranean Diet Pyramid.** The Mediterranean Diet Pyramid is based on traditional dietary practices of Greece, southern Italy, and the island of Crete.

African Heritage Diet

The traditional African diet is rich in nutrient–dense leafy green vegetables, such as kale, collards, mustard, chard, and dandelion. In addition, beans of all kinds, black–eyed peas, peanuts, flat breads, whole–grain millet and sorghum, kiwano "horned fruit," plantains, okra, and tuber vegetables, including yams and sweet potatoes, provide a variety of vitamins, minerals, and dietary fiber (Fig. 3.20). Hearty soups and stews are made from these foods and seafood, including fin fish and shellfish, seasoned with curry spices, peppers, garlic, onion, and fresh herbs. This traditional diet promotes overall health because it is rich in plant foods.

In 2022, researchers studied 586 people that enrolled in a six–week cooking course to learn about the traditional African diet rich in plant foods and how to prepare these traditional foods. Participants were also encouraged to be physically active every day. Scientists noted upon completion of the cooking course, participants had lower blood pressure levels and had decreased their waist circumference and total body weight. The course participants also increased their leafy green vegetable, fruit, and overall vegetable intake, as well as their physical activity compared to their habits prior to taking the cooking course.[27] For more information about African heritage diets and foods, visit https://oldwayspt.org/traditional–diets/african–heritage–diet.

Figure 3.20 **Traditional African foods.**
James Gathany/CDC

Asian Diet

Traditional Asian foods, such as Chinese, Japanese, Vietnamese, Thai, and Korean cuisines, are similar and generally feature large amounts of vegetables, rice, or noodles combined with small amounts of meat, fish, or shellfish. The variety of vegetables and soy foods used in Asian dishes adds color, flavor, texture, phytochemicals, and nutrients to meals. Additionally, Asian dishes often include flavorful sauces and seasonings made from plants, such as soy sauce, rice wine, ginger root, garlic, scallions, peppers, and sesame seeds. The Asian Diet Pyramid, shown in Figure 3.21, illustrates the traditional Asian dietary pattern, which generally provides inadequate amounts of calcium from milk and milk products. However, using calcium–rich or calcium–fortified foods can add the mineral to diets.

Traditional Chinese food preparation methods, particularly steaming and stir–frying, tend to preserve the vitamins and minerals in fresh vegetables. Stir–frying involves cooking foods in a lightly oiled, very hot pan for a short period of time. Unlike Western methods of deep–fat frying or boiling vegetables, stir–frying vegetables keeps them crisp and colorful.

Figure 3.21 Asian Diet Pyramid. The Asian Diet Pyramid dietary pattern generally provides inadequate amounts of calcium from milk and milk products. ©2018 Oldways Preservation and Exchange Trust. Used with permission. www.oldwayspt.org

Rice is the staple food in the traditional Japanese diet. Additionally, fish, poultry, pork, and foods made from soybeans provide protein in this diet. The Japanese people eat sushi, small pieces of raw fish or shellfish that are usually served rolled in or pressed into rice and served with vegetables and seaweed.

Some of the longest-lived people in the world reside on Okinawa, a tiny island south of the main Japanese islands. The traditional diet of fresh vegetables, minimal amounts of salt and animal protein (mainly from pork and fish), and moderate amounts of fat may protect the island's population from premature heart disease and stroke. Younger Okinawans, however, have adopted more Western dietary practices, such as eating fast food, and as a result, their life expectancy is not as long as that of their grandparents.[28]

Traditional Asian food patterns may contain too much sodium. High sodium intakes increase the risk of hypertension. In many parts of Japan, the general population consumes too much sodium. Thus, hypertension is a serious public health problem in these areas.[29]

Native American Diet

In the past, some Native Americans were hunter-gatherers, traveling to where food was available and ripe following the seasonal weather patterns. They depended on wild vegetation, berries, fish, and game for food. Other Native Americans grew vegetable crops, including tomatoes, corn, green leafy vegetables, beans, and squash. In general, the traditional Native American diet was low in sodium and fat and high in fiber. During the last half of the twentieth century, many Native Americans abandoned their traditional diets and adopted the typical Western diet. The negative health effects of this lifestyle change have been significant. Today, chronic diseases, including obesity and type 2 diabetes, are extremely prevalent among Native Americans, whereas in the past, these conditions rarely affected this population.

The traditional native Alaskan diet was composed of fatty fish and sea mammals, game animals, and a few plants. Alaskan natives who still follow traditional dietary practices have CVD rates that are lower than those in the general North American population.

The USDA Indigenous Food Sovereignty Initiative supports the redevelopment of traditional food production for Native Americans and native Alaskans to improve the population's overall health and decrease the incidence of chronic disease. You can learn more about this initiative at www.usda.gov/tribalrelations/usda-programs-and-services/usda-indigenous-food-sovereignty-initiative.

Religious Diets

Many religions require members to follow strict food handling and dietary practices that often include the prohibition of certain foods and beverages (see Table 3.12). According to Jewish dietary laws, for example, meat and poultry products must be kept separate from milk products. Milk products are neither used to prepare foods that contain meat or poultry, nor are they served with them. A cheeseburger, for example, is not kosher. *Kosher* refers to a specific procedure concerning killing, butchering, and preparation activities that makes food acceptable for the religion's followers to eat. Fruits, grains, vegetables, eggs, and fish with fins and scales are "neutral" foods that can be eaten with meals that contain either

Symbol for a kosher food. ©Wendy Schiff

TABLE 3.12 Religious Influences on Dietary Practices

Religion	Dietary Practices*
Buddhist	Meat is avoided; vegetarianism is encouraged.
Eastern Orthodox	Meat and fish restrictions; fasting and specific food abstinence during certain holidays
Hindu	Beef is forbidden, but dairy products are "pure" for consumption. Pork may be restricted. Alcohol is avoided. Fasting is often encouraged.
Islam	Pork; birds of prey; reptiles; insects, except locusts; most gelatins; and alcohol-containing beverages are prohibited (*ha-raam*). Ritual killing of animals that are permitted as food (*ha-lal*) Fasting from all food and drink (daytime) during month of Ramadan and certain other religious holidays
Jewish	Only kosher foods are acceptable. Tref (*trayf*) refers to prohibited foods. Pork and shellfish are prohibited. Eating meat with dairy is prohibited. Consuming blood is forbidden. Raw meat is soaked in cold water to remove blood, salted for 1 hour, and then rinsed. Eggs, fruits, and vegetables can be eaten with either meat or dairy foods. Eggs, however, are inspected to make sure they do not contain blood specks. Only fish with fins and scales can be eaten. Only land animals that have split hooves and chew their cud can be eaten, and only the front half of the cud-chewing animal is used. Ritual killing of certain animals is required. Fasting and specific food restrictions for certain holidays
Mormon	Alcohol, coffee, and tea are avoided. Fasting is practiced regularly.
Roman Catholic	Fasting before communion; fasting and specific food abstinence during certain holidays
Seventh Day Adventist	Animal product consumption generally limited to milk, milk products, and eggs (lacto-ovo vegetarianism) Alcohol and beverages containing stimulants are prohibited.

* Some religions have extensive rules governing food-related practices, but many people do not follow their religion's dietary guidelines fully or at all.

meat or dairy products. However, vegetables cooked with meat become a "meat" food and cannot be served with milk; peaches served with cottage cheese become a "milk" food and cannot be eaten with meat or poultry. Today many American Jews do not follow their religion's com–plex dietary laws as closely as their ancestors did.

Bagels with smoked salmon (*lox*), pickled herring, cream cheese, dill pickles, corned beef, and pastrami are popular among the Ashkenazi, the predominant group of Jews in America. Although many non–Jews enjoy eating these traditional Ashkenazic foods, such items should be consumed in moderation due to their sodium content.

Concept CHECKPOINT

33. Discuss the health benefits from following the traditional Latin American, Mediterranean, African Heritage, and Asian diets.

34. Discuss at least two religion-related food restrictions.

See Appendix F for responses.

Chapter References

See Appendix G.

Summary

3.1 From Requirements to Standards

- A requirement is the smallest amount of a nutrient that maintains a defined level of health. Numerous factors influence nutrient requirements. Scientists use information about nutrient requirements and storage capabilities to establish specific dietary recommendations. The Dietary Reference Intakes (DRIs) are various energy and nutrient intake standards for Americans. An Estimated Average Requirement (EAR) is the amount of the nutrient that meets the needs of 50% of healthy people in a particular life stage/sex group. The Recommended Dietary Allowances (RDAs) meet the needs of nearly all healthy individuals (97 to 98%) in a particular life stage/sex group. When nutrition scientists are unable to determine an RDA for a nutrient, they establish an Adequate Intake (AI) value. The Tolerable Upper Intake Level (UL) is the highest average amount of a nutrient that is unlikely to harm most people when the amount is consumed daily. The Estimated Energy Requirement (EER) is used to evaluate a person's energy intake, and the Acceptable Macronutrient Distribution Ranges (AMDRs) are ranges of carbohydrate, fat, and protein that are nutritionally adequate and may reduce the risk of diet-related chronic diseases. The Chronic Disease Risk Reduction Intakes (CDRRs) are nutrient intake levels that reduce chronic disease risk within a healthy population.
- DRIs can be used for planning nutritious diets for groups of people and evaluating the nutritional adequacy of a population's diet. RDAs and AIs are often used to evaluate an individual's dietary practices. For nutrition labeling purposes, the FDA uses RDAs to develop Daily Values (DVs).

3.2 Major Food Groups

- Food guides generally classify foods into groups according to their natural origins and key nutrients. Such guides usually feature major food groups.

3.3 Dietary Guidelines

- The Dietary Guidelines is a set of general nutrition-related lifestyle recommendations designed to promote adequate nutritional status and good health, and to reduce the risk of major chronic nutrition-related health conditions.

3.4 Food Guides

- Myplate.gov is an online, interactive menu planning tool that is based on the U.S. Dietary Guidelines. Most Americans do not follow the government's dietary recommendations.
- Food Lists and carbohydrate counting are tools that people can use for planning menus.

alinamd/123RF

3.5 Food and Dietary Supplement Labels

- Consumers can use information on food labels to determine ingredients and compare nutrient contents of packaged foods and beverages. The FDA regulates and monitors information that can be placed on food labels, including claims about the product's health benefits. Nearly all foods and beverages sold in supermarkets must be labeled with the product's name, manufacturer's name and address, amount of product in the package, and ingredients listed in descending order by weight. Furthermore, food labels must use a special format for listing specific information on the Nutrition Facts panel.
- The Daily Values (DVs) are a practical set of nutrient standards for labeling purposes. The nutrient content in a serving of food is listed on the label as a percentage of the DV (%DV). Most nutrients have DVs. A dietary goal is to obtain at least 100% of the DVs for fiber, vitamins, and minerals (except sodium) each day.
- The FDA permits food manufacturers to include certain health claims on food labels. However, the agency requires that health claims meet certain guidelines and, in some instances, use specific wording. A structure/function claim describes the role a nutrient plays in the body. Structure/function statements cannot claim that a nutrient or food can be used to prevent or treat a serious health condition.

©Wendy Schiff

3.6 Organic Food

- Organic farming and the production of organic foods do not rely on the use of antibiotics, hormones, synthetic fertilizers and most synthetic pesticides, genetic improvements, or ionizing radiation. The USDA developed and implemented rules for the organic food industry. A food cannot be labeled "organic" unless its production meets strict national standards.

3.7 Using Dietary Analysis Software

- Dietary analysis software and websites can be quick and easy tools for determining nutrient and energy contents of a specific food.

3.8 Nutrition Matters: Traditional Diets

- Traditional ethnic diets are often based on dishes containing small amounts of animal foods and larger amounts of plant foods, including locally grown fruits, vegetables, and whole grains. These diets are associated with lower risk of developing chronic disease. However, traditional foods are typically abandoned as people immigrate to other countries and assimilate into the general population. The prevalence of chronic diseases such as cardiovascular disease, type 2 diabetes, and high blood pressure often increases among immigrants, partly as a result of adopting unhealthy eating practices. Many religions require members to follow strict food handling and dietary practices that often include the prohibition of certain foods and beverages.

Recipe for Healthy Living

©Wendy Schiff

Mango Lassi

Lassi (*luh-see*) is a simple yogurt-based beverage that originated in India. Lassi is usually made and served before a meal, but the drink can also be a refreshing, nutritious snack. To test a mango for ripeness, gently squeeze the mango in your hand. A ripe mango will give slightly to pressure.

This recipe makes about four ½-cup servings. Each serving supplies approximately 97 kcal, 4 g protein, 19 g carbohydrate (excluding forms of fiber), 0.4 g fat, 1 g fiber, 122 mg calcium, and 30 mg vitamin C.

INGREDIENTS:

1 ripe mango
1 cup plain, fat-free yogurt
1 Tbsp sugar
6 ice cubes

Ian 2010/Shutterstock

PREPARATION STEPS:

1. Wash and peel mango. Remove fruit pulp from mango, and discard large seed and peel.
2. Dice mango pulp and place in blender.
3. Add yogurt, sugar, and ice cubes to blender.
4. Blend ingredients until smooth.
5. Serve immediately or refrigerate for up to 24 hours.

Approximate Percent of Calories from Macronutrients

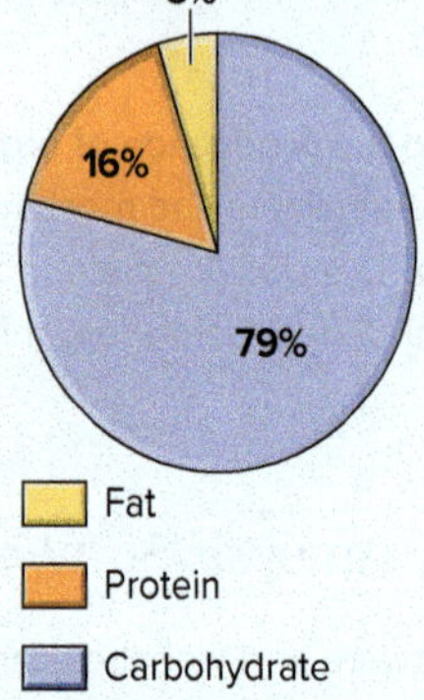

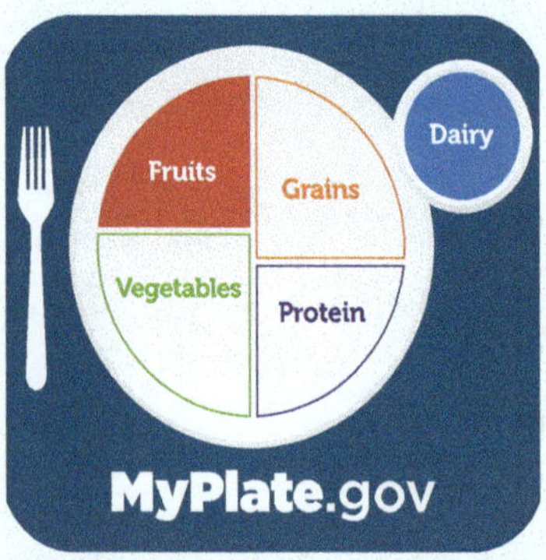

US Department of Agriculture or USDA

Percent of Daily Value/Serving

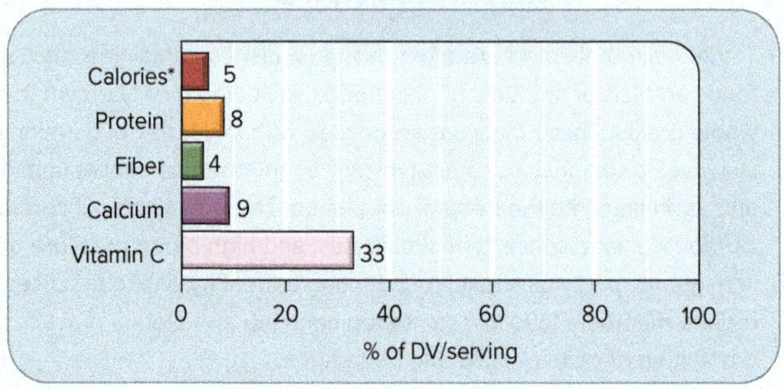

*2000 daily total kcal

Personal Dietary Analysis

I. Record Keeping

A. 24-Hour Dietary Recall

1. Recall every food and beverage that you have eaten over the past 24 hours. Recall how much you consumed and how it was prepared.

   a. How easy or difficult was it to recall your food intake?

B. Three-Day Diet Record

1. Without changing your usual diet, keep a detailed log of your food and beverage intake for 3 days; one of the days should be Friday or Saturday. Use a separate log for each day.

II. Analysis

Using nutritional analysis software, analyze your daily food intakes and answer questions in Part III of this activity. Keep the record on file for future applications.

A. Computer-Generated Dietary Analysis

1. Load the software into the computer, or log on to software website.
2. Choose the DRIs or related nutrient standard from Appendix H of this book, based on your life stage, sex, height, and weight.
3. Enter the information from the 3-day food intake record. Be sure to enter each food and drink and the specific amounts.
4. The software program will give you the following results:

   a. The appropriate RDA (or related standard) for each nutrient

   b. The total amount of each nutrient and the kilocalories consumed for each day

   c. The percentage intake compared with the standard amount for each nutrient that you consumed each day

III. Evaluation of Nutrient Intakes

Remember it is not necessary to consume the maximum of your nutrient recommendations every day. A general standard is meeting at least 70% of the standards averaged over several days. It is best not to exceed the Upper Level (if set) over the long term to avoid potential toxic effects of some nutrients.

A. For which nutrients did your average intake fall below the recommended amounts, that is, to less than 70% of the RDA/AI?

B. For which nutrients did your average intake exceed the Upper Level (if a UL has been set)?

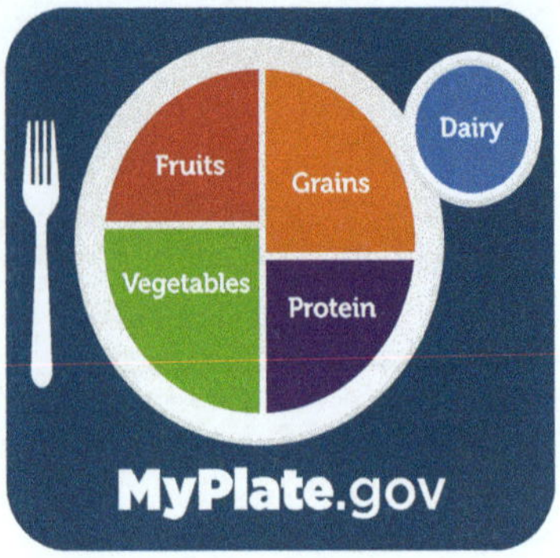

U.S. Department of Agriculture

IV. MyPlate

A. Refer to your 3-day diet record. Classify each food item in the appropriate food group of MyPlate. For each food group, indicate whether you ate the recommended amount daily for your sex, age, height, weight, and physical activity level. Note that some of your food choices—pizza, for example—may contribute to more than one food group. Enter a minus sign (–) if your total falls below the MyPlate recommendation or a plus sign (+) if it equals or exceeds the daily recommendation for each food group.

Mc Graw Hill connect® Complete the Personal Dietary Analysis activity in Connect.

Critical Thinking

1. Your friend takes several dietary supplements daily, and as a result, their vitamin B-6 intake is 50 times higher than the RDA for the vitamin. You would like to convince them to stop taking the supplements. To support your advice, which nutrient standards would you show them? Explain why.
2. Why should consumers use MyPlate to plan menus instead of the DRIs?
3. How do your added sugars, sodium, and saturated fat intakes compare to the recommendations of the latest Dietary Guidelines?
4. Examine Table 3.6. Which foods in the left-hand column do you eat regularly? Why are those foods listed in that column?
5. The ingredient list on a package of crackers includes vegetable oil. What can you do to learn which type of vegetable oil is in the product?
6. According to an article in a food science journal, an 8-oz serving of fat-free milk contains 15 mcg of folate (a B vitamin). An article in a different food science journal reports that an 8-oz serving of fat-free milk contains 12 mcg of folate. Explain why both sources of information can be correct.
7. Visit the USDA's website (https://fdc.nal.usda.gov/index.html) to access FoodData Central, which consists of several databases for searching the nutritional content of foods. To practice, use the "SR Legacy" database to find the number of kilocalories and the amounts of fiber, vitamin C, iron, and caffeine in 1 cup of raw jicama, 1 cup of reduced-fat milk with added vitamins A and D, and ¼ cup of dry roasted, salt-added sunflower seed kernels. Tip: Search for the name of the food and do not include serving size; for example, "raw jicama." You select the serving size after you choose a food from the database.
8. Consider the ingredients that are used to make a typical cheese pizza. What changes can you make to the pizza recipe to make the food more healthy?
9. Evaluate your dietary grain intake and identify three specific changes you can make to increase your whole-grain intake.

Practice Test (Section numbers are listed below in red.)

Select the best answer.

1. (3.1) The amount of a nutrient that should meet the needs of almost all healthy people in a particular group is the
   a. Estimated Average Requirement (EAR).
   b. Recommended Dietary Allowance (RDA).
   c. Nutrient Requirement (NR).
   d. Tolerable Upper Intake Level (UL).
2. (3.1) Which of the following statements is true?
   a. AMDRs have been developed for all nutrients.
   b. ULs meet the nutrient needs of nearly all healthy people.
   c. RDAs contain a margin of safety.
   d. AIs are requirements for certain nutrients.
3. (3.1) The Estimated Energy Requirement (EER)
   a. has a margin of safety.
   b. does not account for a person's height, weight, or physical activity level.
   c. is based on the average daily energy needs of a healthy person.
   d. reflects a person's actual daily energy needs.
4. (3.1) A diet is likely to be safe and nutritionally adequate if
   a. average daily intakes for nutrients meet RDA or AI values.
   b. intakes of various nutrients are consistently less than EAR amounts.
   c. nutrient intakes are consistently above ULs.
   d. vitamin supplements are included.
5. (3.1) Nutritional standards, such as the RDAs, are *not*
   a. used to develop formula food products.
   b. the basis for establishing DVs.
   c. used to evaluate the nutritional adequacy of diets.
   d. the basis for developing AIs.

Pixtal/age fotostock

6. (3.2) Which of the following foods is a good source of protein and contains a lot of saturated fat?
   a. Avocado
   b. Carrots
   c. Cheese
   d. Peanuts
7. (3.2) Fruit is generally a good source of all of the following substances, except
   a. fiber.
   b. vitamin C.
   c. phytochemicals.
   d. protein.
8. (3.2) Which of the following foods is typically grouped with dairy products?
   a. Cheese
   b. Egg
   c. Butter
   d. Cream
9. (3.2) Chidubem wants to improve his diet by eating fewer foods that are rich sources of added sugars. To help him achieve his dietary goal, Chidubem should reduce his intake of
   a. whole milk.
   b. grape jelly.
   c. whole-grain bread.
   d. corn oil.
10. (3.3) The *Dietary Guidelines for Americans* is
    a. revised every year.
    b. a set of general nutrition-related recommendations.
    c. published by the Centers for Disease Control and Prevention.
    d. used to develop DVs.
11. (3.4) Which of the following statements is true?
    a. According to MyPlate, vegetable oils are grouped into the "Fats and Oils" food group.
    b. The MyPlate menu planning guide cannot be individualized to meet a person's food preferences.
    c. A person can use MyPlate to evaluate their diet's nutritional adequacy.
    d. MyPlate was the first food guide developed for Americans.
12. (3.4) The Food Lists method of menu planning
    a. classifies foods in the same groups as MyPlate.
    b. has food choices based on macronutrient contents.
    c. is useful only for people who have diabetes.
    d. incorporates high-protein foods with high-carbohydrate foods.

13. (3.5) Which of the following information is *not* provided by the Nutrition Facts panel?

a. Percentage of calories from alcohol
b. Amount of total carbohydrate per serving
c. Serving size
d. Amount of trans fat per serving

14. (3.5) Daily Values are

a. for people who consume 1200 to 1500 kcal diets.
b. based on the lowest UL for each nutrient.
c. dietary standards developed for food-labeling purposes.
d. used to evaluate the nutritional adequacy of a population's diet.

15. (3.5) Robert is evaluating dietary supplement label claims. Which of the following claims is *not* a structure/function claim the FDA allows on food packages?

a. Maintains a healthy heart
b. Builds strong bones
c. Supports the immune system
d. Lowers blood pressure

16. (3.6) Lina would like to eat more organically grown foods. To be certain that the foods she purchases are at least 95% organic, Lina should

a. look for the organic logo.
b. check the produce's protective packaging for current "picked on" dates.
c. read the ingredients list on the label.
d. examine the produce for signs of mold and insect damage.

17. (3.7) Which of the following can alter the nutrient content of a fruit or vegetable?

a. FDA regulation
b. Scrubbing under running water
c. Growing conditions
d. Product labeling

18. (3.8) People who follow the traditional Latin American diet may experience

a. increased risk for CVD.
b. increased risk for cancer.
c. lower blood pressure levels.
d. lower vitamin and mineral intake.

19. (3.8) Badih is a devout follower of Islam. Therefore, he will *not* consume

a. red meat.
b. beer.
c. enriched grains.
d. coffee.

Answers to Quiz Yourself

1. According to the latest U.S. Department of Agriculture food guide, fruits and vegetables are combined into one food group. **False.** (Section 3.4)
2. According to the recommendations of the *Dietary Guidelines for Americans*, it is acceptable for certain adults to consume alcoholic beverages in moderation. **True.** (Section 3.3)
3. Last week, Jacara didn't consume the recommended amount of vitamin C for a couple of days. Nevertheless, he is unlikely to develop scurvy, the vitamin C deficiency disease. **True.** (Section 3.1)
4. Canned carrots are an example of ultra-processed food. **False.** (Section 3.2)
5. The Nutrition Facts panel on a food label provides information concerning amounts of energy, fiber, and sodium that are in a serving of the food. **True.** (Section 3.5)

Design Elements: (Stacked books): Siede Preis/Photodisc/Getty Images; (Nuts): C Squared Studios/Photodisc/Getty Images; (Recipe box, Raspberries, Blueberries): ©Wendy Schiff; (Pineapple): Stockbyte/Getty Images; (Radishes): karandaev/iStock/360/Getty Images; (MyPlate icon): US Department of Agriculture or USDA

CHAPTER 4

Body Basics

DragonImages/iStock/Getty Images

Quiz Yourself

To test your knowledge of the material covered in Chapter 4, take the following quiz; the answers are at the end of the chapter.

1. The atom is the smallest living unit in the body. ______ T ______ F
2. The stomach produces gastric juice that contains hydrochloric acid. ______ T ______ F
3. Digestion begins in the stomach. ______ T ______ F
4. The gut microbiota are the various microbes, primarily the bacteria, that live in the gastrointestinal tract. ______ T ______ F
5. Symptoms of Irritable Bowel Syndrome (IBS) include constipation and diarrhea. ______ T ______ F

4.1 Nutrition: Chemistry Foundations

Learning Outcomes

1. Define key basic chemistry terms, including *element, ion, chemical bond,* and *enzyme.*
2. Explain the difference between an acid and a base.
3. Explain the role of enzymes in many chemical reactions.

In Chapter 1, we defined chemistry as the study of the composition and characteristics of matter, and the changes that it can undergo. We also defined human physiology as the study of how the human body functions. Principles of chemistry and human physiology form the foundation for the scientific study of nutrition. The foods you eat and the air you breathe provide nutrients and oxygen, the raw materials (matter) that your cells need to survive and function. By reading Section 4.1, you will learn some basic chemistry concepts to help you understand how the matter in food becomes the raw materials for building, fueling, and sustaining healthy bodies. In Section 4.2, you will learn about body structures and functions so you can understand the roles of nutrients in the body and why these particular chemicals are so important. Section 4.3 focuses on the digestive system and how it functions. Finally, Section 4.4 covers some common health problems that affect the digestive system, including heartburn and irritable bowel syndrome.

Parts of this chapter may seem to be filled with unfamiliar terms and their definitions. However, learning the meaning of these terms can help you understand the nutrition information that is in the following chapters.

Watermelons contain 1600 chemicals that function as carbohydrates, proteins, lipids, and antioxidants in the melon.
Superheang168/Shutterstock

Basic Chemistry Concepts

Do you or does someone you know only eat organic foods because they do not contain chemicals such as additives or pesticides? It is true that chemicals are in your food, but they are not different from those made by a laboratory scientist. For example, watermelon, a type of fruit, contains over 1600 chemicals, including tropinone, coumarin, sucrose, curcubitacin, citrullin, oleic acid, phloretin, and ginkgolic acid. These chemicals are not harmful and function as carbohydrates, proteins, lipids, and antioxidants in the melon.[1] Chemicals also make up every other aspect of your environment; air, water, rocks, and other forms of matter contain chemicals. In fact, you are a complex collection of chemicals, much of which is organized into cells. The following sections provide basic information about some chemistry concepts that apply to the study of nutrition.

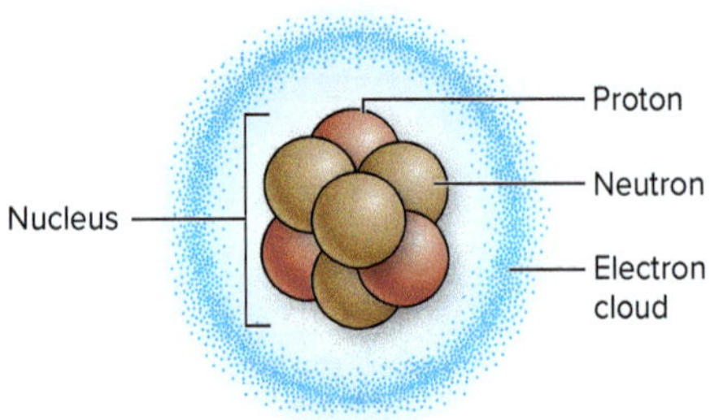

Figure 4.1 An atom. Matter is composed of atoms that contain particles, including protons, neutrons, and electrons.

From Atoms to Compounds

Matter is composed of atoms that contain certain particles, including protons, neutrons, and electrons (Fig. 4.1). The central region of the atom, called the *nucleus,* is made of **protons,** positively charged particles, and **neutrons,** particles that carry no charge. **Electrons** are small negatively charged particles that form a cloud surrounding the nucleus. The number of electrons surrounding the nucleus equals the number of protons within the nucleus. Thus, the negative and positive charges cancel out each other, making an atom neutral, which means it has no electrical charge.

More than 100 different types of atoms exist, and each type is an **element,** a substance that cannot be separated into simpler substances by ordinary chemical or physical means. Elements are the "building blocks" of matter. Table 4.1 lists several elements, most of which are essential for human nutrition.

protons positively charged particles in the nucleus of an atom

neutrons particles in the nucleus of an atom that carry no charge

electrons small, negatively charged particles that surround the nucleus of an atom

element each type of atom; substance that cannot be separated into simpler substances by ordinary chemical or physical means

An element is essential if the body cannot function normally without it and the element must be supplied by the diet. Note that chemists use letters as symbols to represent elements. For example, the symbols for carbon, nitrogen, and sodium are C, N, and Na, respectively.

Minerals are elements, such as calcium, iron, and potassium, that are found in the Earth's crust. Many minerals are essential nutrients. However, not every mineral is in living things or is necessary for life. Your external environment has natural and human–made forms of matter that may contain elements such as mercury (Hg), aluminum (Al), and cadmium (Cd). The human body does not need these minerals to function properly, and they can be quite toxic. Chapter 9 discusses the importance of various minerals to health.

Molecules When atoms interact, they may share electrons and rearrange themselves, forming a chemical bond. A **chemical bond** is an attraction that holds atoms together and forms a **molecule.** When illustrating the structure of molecules, chemists often use straight lines to show the bonds (Fig. 4.2a).

Molecules can contain the same element or different elements. For example, an oxygen molecule forms when two oxygen atoms bind together, whereas a water molecule forms when two hydrogen atoms bond to an oxygen atom. To identify a particular molecule without using lines to draw its chemical structure, chemists use a chemical formula. The chemical formula for an oxygen molecule is O_2; the subscript "2" indicates the presence of 2 oxygen atoms. The chemical formula for a water molecule is H_2O. The chemical formula for the simple sugar *glucose* is $C_6H_{12}O_6$. Judging from its formula, how many carbon, hydrogen, and oxygen atoms are in a glucose molecule?

Some atoms form single bonds, but a few atoms can form multiple bonds. Each carbon atom (C) has four bonding sites (see Figure 4.2a), and as a result, carbon atoms can bond to each other by single, double (Fig. 4.2b), and even triple bonds (Fig. 4.2c). Carbon's ability to bond to other carbon atoms as well as form multiple bonds with a neighboring carbon atom contributes to the formation of a vast array of organic (carbon–containing) compounds. **Compounds** are molecules that contain two or more different elements in specific proportions.

Solutions A **solution** is an evenly distributed mixture of two or more compounds. In living things, water is often the **solvent,** the primary component of solutions. Your body, for example, is about 60% water. A substance that dissolves in the solvent is a **solute.** Many beverages and foods are solutions that have water as the solvent (Fig. 4.3a). A sports drink, for example, is a solution that is mostly water, the solvent. The drink has relatively small amounts of sugar, minerals, colorings, and flavorings (solutes) dissolved in the water.

The **solubility** of a substance describes its ability to dissolve, that is, form a solution, in a solvent. Many naturally occurring substances, including simple carbohydrates such as sugar and all mineral elements, dissolve in water. Other substances, such as fat, are insoluble and will not dissolve in water (Fig. 4.3b). Your blood has high water content, a characteristic that makes it

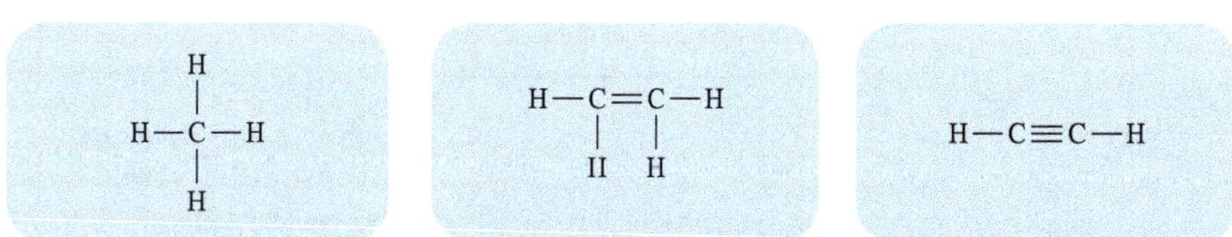

(a) (b) (c)

Figure 4.2 Chemical bonds. Some atoms form single bonds (*a*), but a few atoms can form multiple bonds. Carbon atoms can bond to each other by single, double (*b*), and even triple bonds (*c*).

TABLE 4.1 Nutrition-Related Elements in the Body

Element	Symbol
Hydrogen	H
Oxygen	O
Carbon	C
Nitrogen	N
Calcium	Ca
Phosphorus	P
Potassium	K
Sulfur	S
Sodium	Na
Chloride	Cl
Magnesium	Mg
Iron	Fe
Iodine	I
Copper	Cu
Zinc	Zn
Manganese	Mn
Cobalt	Co
Chromium	Cr
Selenium	Se
Molybdenum	Mo
Fluoride*	F

*Although fluoride is not essential, the mineral helps strengthen teeth and bones.

minerals elements that are found in the Earth's crust

chemical bond attraction that holds atoms together

molecule matter that forms when two or more atoms interact and are held together by a chemical bond

compounds molecules that contain two or more different elements in specific proportions

solution evenly distributed mixture of two or more compounds

solvent primary component of a solution

solute lesser component of a solution that dissolves in solvent

solubility describes a substance's ability to dissolve and form a solution

(a) (b)

Figure 4.3 Solubility. The solubility of a compound describes how easily it forms a solution. Many substances, including table sugar, dissolve in water. (*a*) Sports drinks are beverages in which sugar remains dissolved in water. Other substances, such as fat, are insoluble in water and will not dissolve in it. (*b*) Vinaigrette is an example of a food in which oil and water (balsamic vinegar) do not form a solution. In this particular product, note how the oil separates from the vinegar and forms a layer on top of the darker balsamic vinegar layer.

easier for the body to transport and eliminate water–soluble substances than water–insoluble materials.

Ions When an atom (or group of atoms) gains or loses one or more electrons, it has an electrical charge and is called an **ion** (Fig. 4.4). If an atom gains one electron, it becomes an ion with a negative charge because electrons are negatively charged. If an atom loses an electron, it becomes an ion with a positive charge because it has an extra proton and protons are positively charged. A negative charge is indicated with a minus sign ($^-$), and a positive charge is indicated with a plus sign ($^+$), after the chemical symbol or formula. For example, a molecule of ammonia is neutral, that is, has no electrical charge, because it has the same number of protons and electrons. If the molecule loses an electron, it becomes the positively charged ion ammonium. The formula for the positively charged *hydrogen ion* is simply **H^+.**

When most mineral elements, including sodium and potassium, dissolve in water, they form solutions containing ions that can conduct electricity. Thus, sodium and potassium are called **electrolytes.** Electrolytes have many important functions in the body, including helping to maintain proper fluid balance.

What Are Acids and Bases? **Acids** are substances that lose H^+ when dissolved in water; **bases** are substances that remove and accept H^+ when dissolved in water. Many of the chemicals you encounter daily are either acidic or basic (*alkaline*). As you can tell by their names, phosphoric acid in soft drinks and

ion atom or group of atoms that has a positive or negative charge

H^+ chemical symbol for hydrogen ion

electrolytes ions of minerals that conduct electricity when they are dissolved in water

acids substances that donate hydrogen ions

bases substances that accept hydrogen ions

(reaction can go in either direction)

$$NH_3 + H_2O \longleftrightarrow NH_4^+ + OH^-$$

Ammonia **Water** **Ammonium ion** **Hydroxide ion**

Figure 4.4 What is an ion? An ion is an atom or group of atoms that loses or gains one or more electrons and, as a result, has an electrical charge. In this reaction, ammonia and water react to form the positively charged ammonium ion and the negatively charged hydroxide ion.

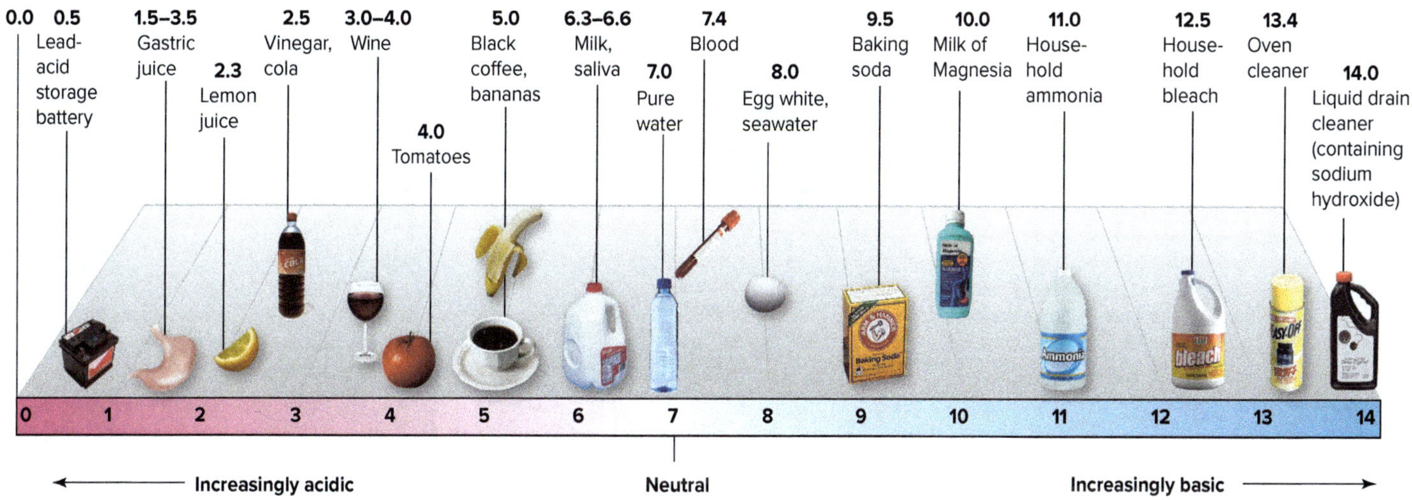

Figure 4.5 **pH scale.** Chemists use the pH scale to measure the concentration of hydrogen ions in a solution. Adapted from: Australian Academy of Science. (2017). www.science.org.au/curious/everything-else/what-gives-lemon-its-sour-taste.

Battery: Ingram Publishing/Alamy Stock Photo; Lemon: Burke/Triolo Productions/Artville/Getty Images; Cola: Oleksiy Mark/Scanrail/iStock/Getty Images; Wine: Burke/Triolo Productions/Brand X Pictures/Getty Images; Tomato: Stockdisc/PunchStock/Getty Images; Banana: Stockdisc/Getty Images; Coffee: Fuse/Digital Stock/Getty Images; Milk: Bob Coyle/McGraw Hill; Bottle: Gilmanshin/Shutterstock; Blood sample: SPL/Science Source; Egg: Siede Preis/Photodisc/Getty Images; Baking soda: Stephen Frisch/McGraw Hill; Milk of Magnesia bottle: ©Wendy Schiff; Ammonia: Jacques Cornell/McGraw Hill; Bottle of bleach: Richard Hutchings/Digital Light/McGraw Hill; Oven cleaner: Ken Karp/McGraw Hill; Bottle of drain cleaner: Matt Meadows/McGraw Hill

ascorbic acid (better known as vitamin C) are acids. Baking soda, household ammonia, and sodium hydroxide, an ingredient of oven and drain cleaners, are bases.

pH measure of the acidity or alkalinity of a solution

Chemists measure the concentration of hydrogen ions **(pH)** in a watery solution by using the pH scale. The scale ranges from 0 to 14. Substances with low pH values contain more H^+ and are more acidic than substances with high pH values. A solution that has a pH of 7 is neutral because it has the same number of H^+ as OH^-. As the pH of a solution increases above 7, the solution becomes more basic (alkaline). Thus, a drain cleaner that contains sodium hydroxide with a pH of about 14 is more alkaline than egg white, which has a pH of about 8.

Examine the pH scale shown in Figure 4.5. Note that black coffee has a pH of about 5.0 and tomatoes have a pH of about 4.0. Thus, tomatoes are more acidic than black coffee. Although it may seem confusing, a solution with a pH of 2.0 has a *higher* H^+ concentration and is *more* acidic than a solution with a pH of 12.0.

Pure water has equal concentrations of H^+ and OH^-, so it is neither acidic nor basic. Thus, pure water has a pH of 7.0 and is *neutral*. By adding an acid to an alkaline solution, the pH of the solution can become neutralized, which means its pH is 7.0 or almost 7.0.

Your body must maintain its *acid–base balance* to function properly. Under normal conditions, the pH of your blood ranges from 7.35 to 7.45, which is slightly alkaline. To control its pH within normal limits, blood contains buffers—ions or molecules that accept excess OH^- or H^+ when necessary. Acids form naturally as by-products of cellular activity, but if the pH of blood begins to fall, the excess H^+ must be eliminated to prevent *acidosis*, a condition that can be deadly. The lungs remove excess H^+ from blood by means of chemical reactions that result in the release of carbon dioxide (CO_2) and H_2O in exhaled air. Kidneys also participate in the buffering system by removing or neutralizing H^+ from the blood when forming urine. Therefore, healthy lungs and kidneys help maintain the normal pH of blood and other body fluids.

Did *YOU* Know?

Blueberries, raspberries, red cabbage, and strawberries contain the pigment *anthocyanin*. This pigment is used as a natural dye for coloring yarns and can also be used as a crude pH meter. Anthocyanin is red when it is in solutions that have a pH of less than 4. The pigment, however, loses its red color at higher pH values and appears blue-green or green. Aside from being a colorful phytochemical, anthocyanin has antioxidant activity and may provide health benefits as a result (see Table 1.3).

Red cabbage
Pixtal/age fotostock

Figure 4.6 Chemical reaction. A chemical reaction occurs when vinegar (an acid) combines with baking soda (a base). This reaction forms sodium acetate (a salt), water, and carbon dioxide.
©Wendy Schiff

What Is a Chemical Reaction?

Most molecules can undergo **chemical reactions,** processes that change the arrangement of atoms in the molecules. The elements that comprise the molecules that react are never destroyed, but they combine with other elements to form new molecules or compounds. When elements or compounds combine to form new substances, a synthesis reaction has occurred. The new substances often have physical and chemical characteristics (properties) that are quite different from those of the reactants. Decomposition reactions involve the breaking down of molecules. **Digestion,** the process by which large molecules in food are broken down into smaller ones, requires decomposition reactions.

You can observe a simple chemical reaction by combining the reactants vinegar (an acid) and baking soda (a base). As soon as the vinegar makes contact with the soda, the powdery baking soda disappears and a fizzy liquid forms (Fig. 4.6). This particular reaction produces carbon dioxide gas, which forms bubbles in the fizzy liquid, and sodium acetate. Thus, carbon dioxide and sodium acetate are two of the products that result when vinegar and baking soda react. Sodium acetate is a **salt,** a substance that forms when an acid reacts with a base. Table salt (sodium chloride) is actually one type of salt. Sodium chloride forms when an acid, such as hydrochloric acid (HCl), reacts with a base, such as sodium hydroxide (NaOH) (Fig. 4.7).

$$HCl + NaOH \longrightarrow NaCl + H_2O$$

HCl	NaOH	NaCl	H_2O
Hydrochloric acid	Sodium hydroxide (base)	Sodium chloride (table salt)	Water

Figure 4.7 What is a salt? A salt forms when an acid reacts with a base.

In recipes for baked goods that require baking soda, you will find an acid ingredient such as lemon juice, buttermilk, or cream of tartar (tartaric acid) included to react with the soda. The carbon dioxide gas that forms "raises" the mixture, giving it a light, airy structure after baking.

Metabolism Metabolism refers to the sum of all chemical reactions that occur in living cells. *Catabolic reactions* involve breaking down molecules. Catabolism, for example, occurs during digestion. *Anabolic reactions* involve synthesizing new compounds. Repairing damaged muscle tissue after injury is an example of anabolism.

Enzymes Living things contain thousands of chemicals, and life depends upon chemical reactions. However, many of these reactions occur slowly or do not occur spontaneously. An **enzyme** is a molecule (usually a protein) that *catalyzes* (speeds up) a particular chemical reaction. Enzymes are recyclable; they do not become part of the products of a reaction, and as a result, one enzyme molecule can catalyze many reactions. Figure 4.8 illustrates an enzyme's action.

chemical reactions processes that change the atomic arrangements of molecules

digestion process by which large food molecules are mechanically and chemically broken down

salt substance that forms when an acid and a base combine

enzyme molecule (usually a protein) that speeds up a particular chemical reaction

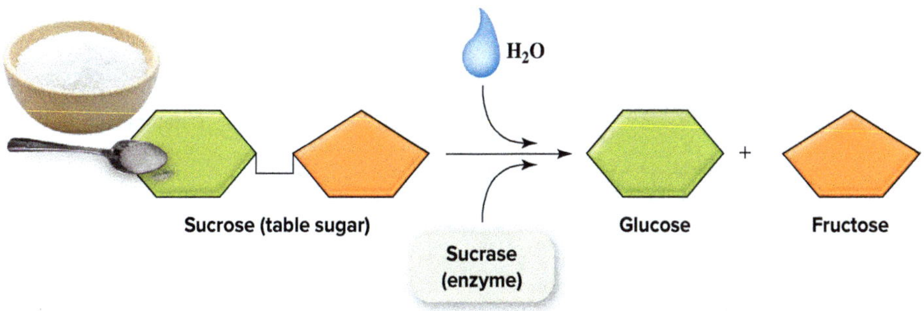

Figure 4.8 Enzyme action. The enzyme sucrase catalyzes the breakdown of sucrose into glucose and fructose.
koosen/iStock/Getty Images

In general, the names of most enzymes end with *–ase.* Sucrase, for example, is the enzyme that catalyzes the reaction that breaks down the carbohydrate sucrose (table sugar) to its component simple sugars, glucose and fructose (see Figure 4.8). Additionally, each enzyme usually has a specific action. For example, sucrase breaks down sucrose, but the enzyme does not affect lactose, the type of sugar in milk.

Enzymes are sensitive to environmental conditions, including pH, temperature, and the presence of certain vitamins and minerals. If the pH or temperature is too high or too low, the enzyme will not function. Raw food contains enzymes, but cooking food usually destroys them.

Concept CHECKPOINT

1. Define the following terms: *electron, proton, element, chemical bond, molecule, compound, solution, solvent,* and *solute.*
2. What is an ion?
3. Explain the difference between an acid and a base.
4. What is pH?
5. What is a chemical reaction?
6. What is an enzyme?
7. What factors can alter an enzyme's activity?

See Appendix F for responses.

4.2 Basic Physiology Concepts

Learning Outcomes

1. Define *cell, tissue, organ,* and *organ systems.*
2. List the organ systems, identify major organs or tissues in each system, and describe primary functions of each system.

The human body is often compared to a complex machine, such as a car. Like a car, the body has numerous interrelated working parts and requires a source of fuel to operate. Additionally, the body has to be able to cool itself, eliminate waste products, and rely on lubrication to keep operating smoothly. Unlike most machines, however, the human body can make many of its spare parts, enabling the body to repair and maintain itself for long periods. When the body functions properly, this wondrous machine may be taken for granted and expected to perform optimally. Nevertheless, the quality of the fuel that powers the human body can affect performance, much like the quality of gasoline that runs a car.

Food & Nutrition tips

A chemical reaction occurs when you cut, crush, mash, or chew whole raw garlic. The enzyme alliinase in the garlic combines with large molecules of alliin, which has a mild flavor. Alliinase breaks down large alliin molecules into small allicin molecules, resulting in garlic's intense odors and tastes.

How you process garlic for a recipe affects the type of garlic flavor the finished food will have. For example, roasting whole garlic before using it in recipes destroys the alliinase in the garlic, resulting in mild-flavored garlic. However, a more robust garlic flavor is created in vinegar-based salad dressings and meat marinades by crushing the garlic cloves and waiting before adding the vinegar. This allows allicin and the desired garlic flavor to form in the crushed garlic before the vinegar destroys alliinase, stopping the chemical reaction.

Maks Narodenko/Shutterstock

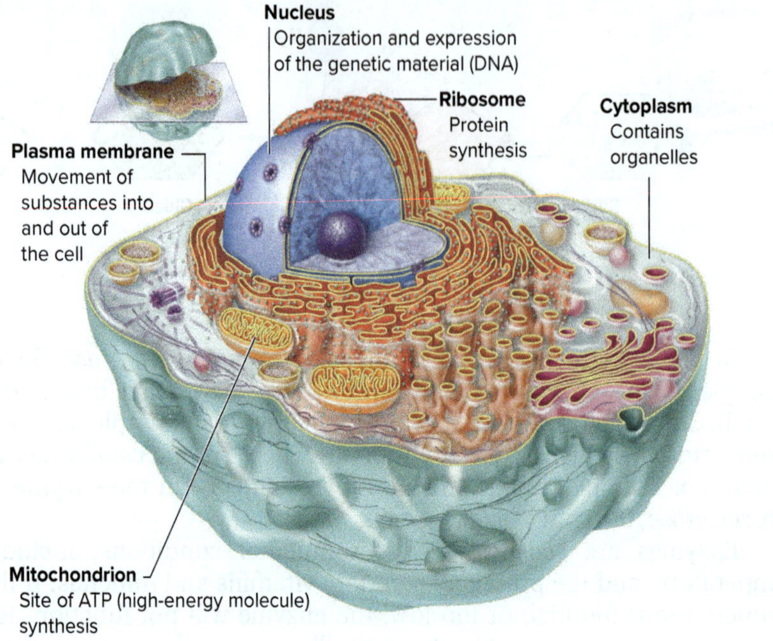

Figure 4.9 **Typical human cell.** A typical human cell contains certain structures and various organelles, including mitochondria, ribosomes, and a nucleus.

Anatomy is the scientific study of cells and other body structures; **physiology** is the scientific study of how cells and body structures function. This section provides some basic information about human anatomy and physiology, including the organization of the body into systems. By learning about human anatomy and physiology, you may appreciate the complexity of your body and be amazed at the variety of metabolic activities that occur within you to keep you alive, physically active, mentally alert, and healthy.

The Cell

A cell is the smallest living functional unit in an organism. Your body has about 100 trillion cells that can be classified into numerous cell types. Each type of cell has a specific function. For example, muscle cells are necessary for movement, red blood cells transport oxygen and carbon dioxide, and certain white blood cells protect the body from disease–causing bacteria and viruses.

Most human cells contain several different types of **organelles,** structures that have specific functions (Fig. 4.9). The nucleus, for example, is an organelle that contains **DNA,** the molecule that provides coded instructions for making proteins. DNA enables the nucleus to control various cellular activities, including cell division and enzyme production. Mitochondria are organelles that play a major role in the generation of energy. Ribosomes are structures involved in the assembly of proteins. The nucleus, mitochondria, and other organelles are surrounded by a jelly–like fluid (a gel) called cytoplasm. Many chemical reactions take place in the cytoplasm, including some reactions necessary to make proteins. Each human cell has a plasma membrane that defines the boundaries of the cell and holds the cytoplasm in place. The plasma membrane also controls the passage of materials into and out of the cell.

From Cells to Systems

Cells that have similar characteristics and functions are usually joined together into larger masses called **tissues.** There are four basic types of tissues: epithelial

anatomy scientific study of cells and other body structures

physiology scientific study of the functioning of cells and other body structures

organelles structures in cells that have specific functions

DNA molecule that provides coded instructions for making proteins

tissues masses of cells that have similar characteristics and functions

organ collection of tissues that function in a related fashion

organ system group of organs that work together for a similar purpose

homeostasis maintenance of an internal chemical and physical environment that is critical for good health and survival

(*ep′–ih–thee′–lee–al*), connective, muscle, and nervous. The cells that line every body surface, including skin and the inside of blood vessels, are epithelial tissues. Fat, bone, and blood are types of connective tissue.

An **organ** is composed of various tissues that function in a related fashion. The brain, for example, is an organ because it contains different forms of nervous tissue that work together to interpret information from the environment, find meaning from this information, and signal responses, such as muscle movements. An **organ system** is a group of organs that work together for a similar purpose. The urinary system, for example, includes two organs, the kidneys and the bladder. The major functions of the urinary system are filtering blood and excreting wastes in urine. A living person is a complete individual life form composed of organ systems that function together. Figure 4.10 illustrates how cells in the body are organized into tissues, organs, and systems. Table 4.2 lists each system's major organs or tissues and summarizes their primary functions. The following sections provide a brief description of each organ system.

All systems in your body must work together in a coordinated manner to maintain good health. When one system fails to function correctly, the functioning of the other systems is soon affected. The body's ability to maintain **homeostasis,** an internal chemical and physical environment that supports life and good health, is critical. Internal conditions such as body temperature and blood pressure normally fluctuate throughout the day, but the body strives to maintain such factors within fairly specific limits. Changes in the cell's internal and external environment can disrupt homeostasis, causing sickness and even death if the abnormality persists. A healthy body, however, uses various mechanisms to regain its normal internal status. As you read about the nutrients, you will recognize how many of them play crucial roles in homeostasis, including maintaining proper body temperature, acid–base balance, and tissue fluid levels.

MY DIVERSE PLATE

Throughout the world, humans use nearly every part of an animal for food. "Flesh" is the skeletal muscle tissue of animals, which contains not only protein-rich muscle fibers but also nerves, fat, blood, and blood vessels. In addition to flesh (sometimes referred to as "meat"), people often incorporate organs, structures, and blood of butchered animals, especially cattle, pigs, and sheep, into main dishes. Edible organs may be called "variety meats" or *offal* (*awh′-ful*). Offal includes the heart, brains, lungs, liver, tongue, kidneys, small intestine (*chitterlings*), pancreas and thymus (*sweetbreads*), stomach (*tripe*), and testicles (*prairie* or *Rocky Mountain oysters, lamb* or *calf fries,* or *huevos del toro*). Structures from butchered livestock, including oxtails and the ears, rectums, and snouts of hogs, are also used for food. Certain kinds of offal are highly nutritious, especially the liver, heart, kidneys, pancreas, thymus, and lungs.

Raw beef tongue
©Wendy Schiff

Cardiovascular System

The major components of the cardiovascular ("circulatory") system are the heart, blood, and blood vessels (Fig. 4.11). The main function of the cardiovascular system is circulating blood throughout the body. The human heart is a four–chambered muscular pump that keeps blood moving through

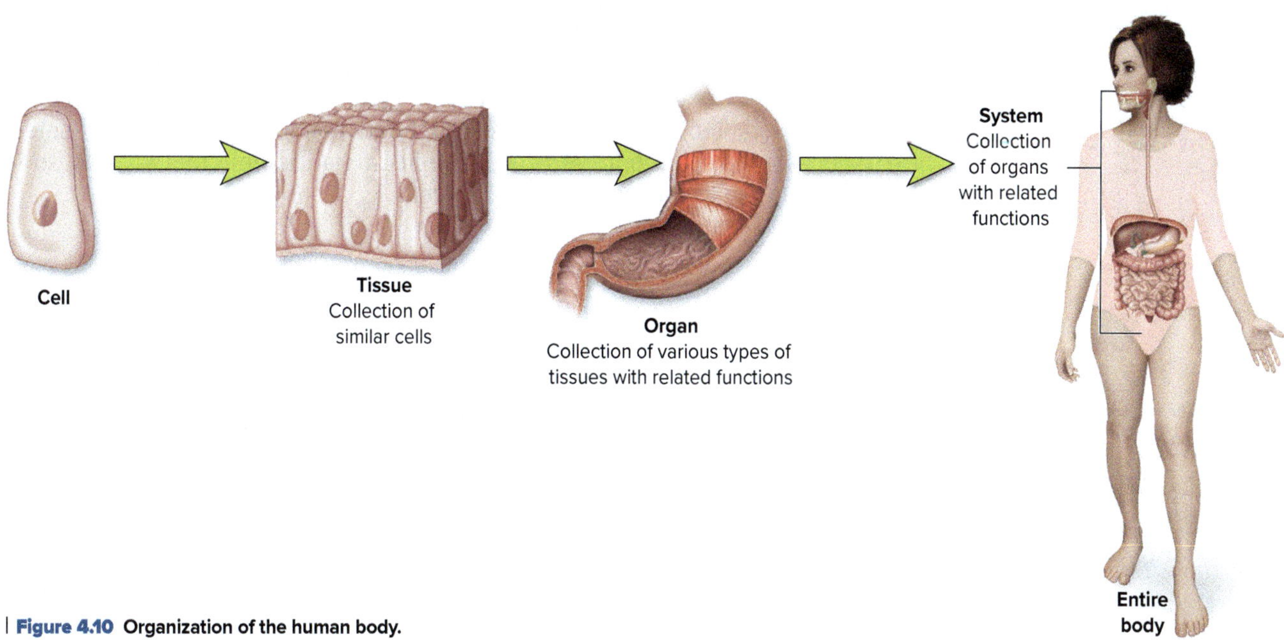

Figure 4.10 Organization of the human body.

TABLE 4.2 The Organ Systems of the Human Body

System	Major Organs or Tissues	Primary Functions
Digestive	Mouth, salivary glands, esophagus, stomach, intestines, pancreas, liver, gallbladder	Digestion and absorption of nutrients; elimination of wastes
Cardiovascular	Heart, blood vessels, blood	Circulation of blood throughout the body
Respiratory	Nose, pharynx, larynx, trachea, bronchi, lungs	Exchange of oxygen and carbon dioxide
Lymphatic/ Immune	Lymphatic fluid, white blood cells, lymph vessels and nodes, spleen, thymus, fat cells	Defense and immunity against infectious agents, fluid balance, white blood cell production, absorption of fat-soluble nutrients from intestinal tract
Urinary	Kidneys, bladder	Elimination of salts, water, and wastes; maintenance of fluid balance and pH balance
Muscular	Muscles	Movement and stability of the body
Skeletal	Bones, tendons, ligaments	Support, movement, protection, and production of blood cells
Nervous	Brain, spinal cord, nerves, sensory receptors	Thought processes, regulation and coordination of various body activities, detection of changes in external and internal environments
Endocrine	Glands or tissues that secrete hormones (chemical messengers)	Regulation and coordination of many body activities, including growth, nutrient balance, and reproduction
Integumentary	Skin, hair, nails	Protection and immunity, regulation of body temperature, vitamin D synthesis
Reproductive	Gonads and genitals	Procreation (having children)

blood vessels. Blood contains red and white blood cells, nutrients, other substances, and *plasma,* the watery portion of the blood. *Serum* is the yellowish liquid that remains after plasma has formed blood clots. Blood vessels form a network of tubes that help circulate blood throughout the body. **Arteries** carry blood away from the heart. Arteries branch into smaller and smaller vessels until they form **capillaries,** a network of tiny blood vessels with walls that are only one cell thick. The thin capillary walls enable nutrients and oxygen to move out of the blood and into cells, and carbon dioxide and other waste products to pass from cells and into the blood. After this exchange occurs, the deoxygenated (oxygen–poor) blood enters **veins** for the return trip to the heart.

After entering the right side of the heart, deoxygenated blood is pumped to the lungs, via the pulmonary arteries. In the lungs, red blood cells release carbon dioxide, a cellular waste product, and pick up oxygen from inhaled air. Cells need oxygen to obtain energy. Hemoglobin, an iron–containing protein in red blood cells, carries most of the oxygen in blood. The oxygenated (oxygen–rich) blood then returns to the heart via the pulmonary veins, so it can be pumped to the rest of the body's cells.

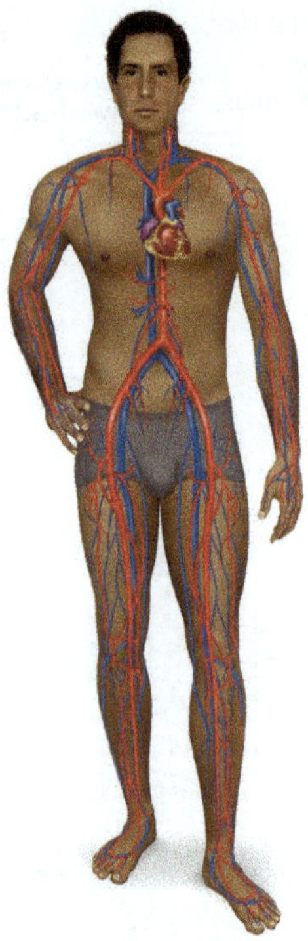

Figure 4.11 Cardiovascular system.

arteries vessels that carry blood away from the heart

capillaries smallest blood vessels

veins vessels that return blood to the heart

Respiratory System

Lungs, the primary structures of the respiratory system, enable the body to exchange gases, particularly oxygen and carbon dioxide (Fig. 4.12). As mentioned in the previous paragraph, blood circulates through the lungs, picks up oxygen from inhaled air, and releases carbon dioxide, a waste product that forms when cells obtain energy. By exhaling, you eliminate carbon dioxide from your body.

Lymphatic/Immune System

This system includes a network of lymphatic vessels and lymph nodes (Fig. 4.13). As blood circulates in the body, some watery fluid (plasma) leaks out

foods is more likely to contribute to a sense of "fullness" than consuming carbohydrate–rich foods.

Small Intestine

The small intestine is a tightly–coiled hollow tube that extends from the stomach to the large intestine. On average, the organ measures about 11 feet (3.3 meters) long because of the effects of muscle tone in living persons.[2] The small intestine is "small" because the tube's diameter is only about 1 inch (2.54 cm), about half the width of the large intestine.

The small intestine has three sections (Fig. 4.25). The first, the **duodenum** (*do–wah–dee'–num*), is only about 10 inches (25.4 cm) long. Within the duodenum, the acidic stomach contents mix with alkaline fluids secreted by the pancreas, gallbladder, and the duodenum. This process neutralizes the acidity of chyme and enables enzymes that function in more alkaline conditions to work. The middle segment of the small intestine is the **jejunum** (*jeh–ju'–num*). The last portion of the small intestine is the **ileum** (*il'–lee–um*). Most digestion and nutrient absorption occurs in the duodenum and jejunum.[8]

A **lumen** is a hollow space in an organ or structure that is surrounded by walls, such as the lumen of the small and large intestines. Each day, the small intestine secretes approximately 1½ quarts (1500 ml) of watery fluids into the lumen. This fluid lubricates the intestinal walls, facilitating the passage of chyme. The cells lining the small intestine also produce a layer of mucus that protects the tissue from being damaged by chyme as it moves through the tract.

In addition to involving enzymes, many of the major chemical reactions that occur during digestion need water. Such reactions are *hydrolytic* (*hydro* = water; *lytic* = breakdown). Water in intestinal fluid contributes H^+ and OH^- ions. These ions react with certain nutrients and become part of the products. Sucrose, for example, undergoes hydrolysis in the small intestine (see Figure 4.8).

The small intestine relies on peristalsis and *segmentation* to help digestion. Segmentation involves regular contractions of ringlike intestinal muscles followed by muscular relaxations to mix chyme within a short portion of the small intestine (Fig. 4.26). The alternating pressure forces chyme to move back and forth within the segment.

Absorbing and Transporting Nutrients

The lining of the small intestine is not smooth. It has circular folds that look like rings, and the surface is covered by tiny, fingerlike projections called **villi** (singular, villus). Each villus has an outer layer of epithelial cells called *absorptive cells* (Fig. 4.27). Furthermore, each absorptive cell has tiny, hair–like microvilli at the end of the cell that face the lumen of the small intestine and are exposed to chyme (see Figure 4.27). The microvilli may be referred to as the intestinal "brush border." The circular folds, villi, and microvilli increase the surface area of the lumen, which helps increase nutrient absorption.

As chyme passes through the small intestine, the pancreas secretes enzymes into the lumen that break down complex forms of macronutrients. Additionally, microvilli have enzymes that complete the breakdown of carbohydrates and proteins into smaller fragments.

The process of absorption occurs in a variety of ways. Some nutrients require the help of transport proteins or pumping mechanisms within the absorptive cell's membrane to enter the cell. Other kinds of nutrients can simply *diffuse* (pass into) into these cells. Such diffusion happens when

Did *YOU* Know?

When you eat foods or drink beverages, you swallow some air. Burping expels most of this air before it enters the stomach. If some air manages to enter your stomach and small intestine, it mixes with chyme and bubbles through it, often producing rather loud, gurgling sounds that can be embarrassing.

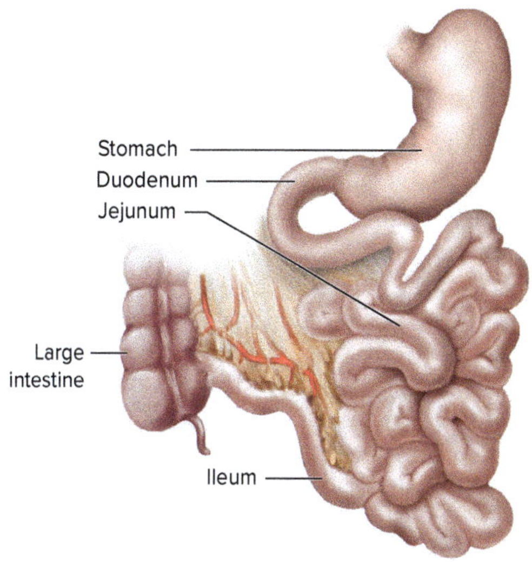

Figure 4.25 **Small intestine.**

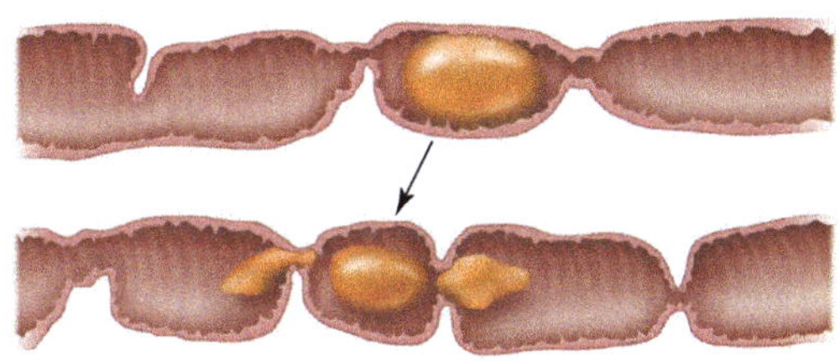

Figure 4.26 **Segmentation.** Segmentation helps mix chyme in the small intestine.

duodenum first segment of the small intestine

jejunum middle segment of the small intestine

ileum last segment of the small intestine

lumen open space within a structure such as the small intestine

villi (singular, villus) tiny, fingerlike projections of the small intestinal lining that participate in digesting and absorbing food

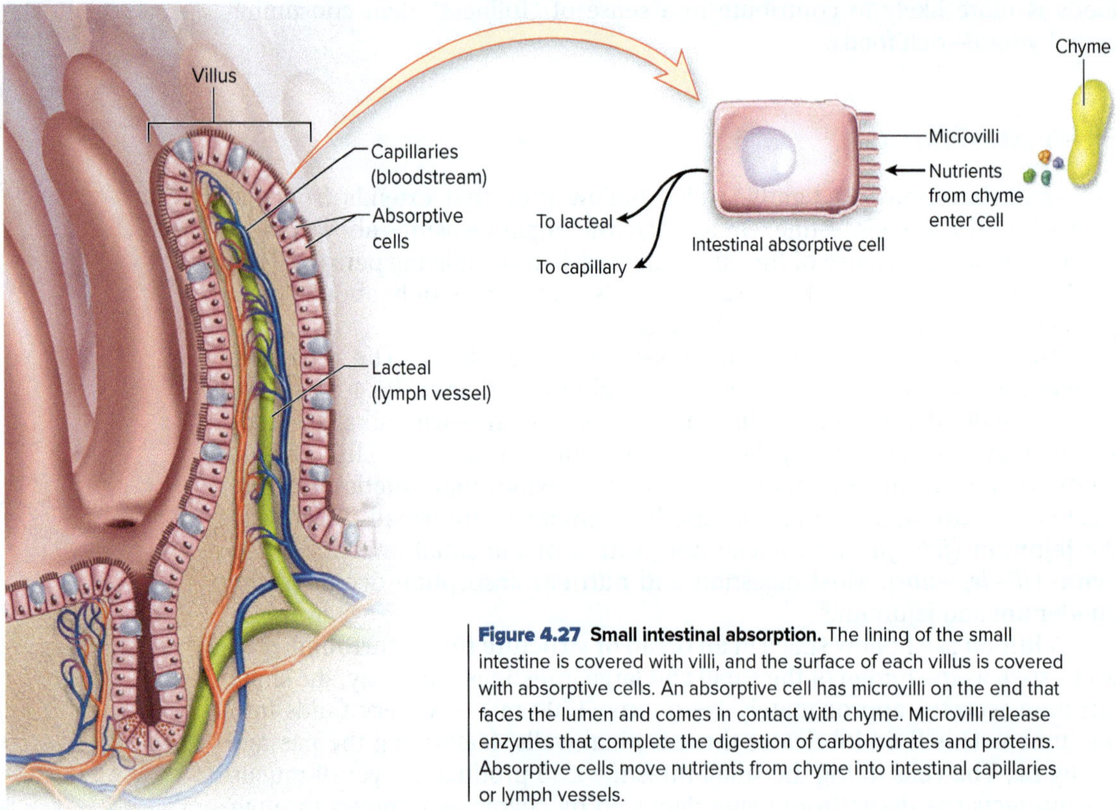

Figure 4.27 Small intestinal absorption. The lining of the small intestine is covered with villi, and the surface of each villus is covered with absorptive cells. An absorptive cell has microvilli on the end that faces the lumen and comes in contact with chyme. Microvilli release enzymes that complete the digestion of carbohydrates and proteins. Absorptive cells move nutrients from chyme into intestinal capillaries or lymph vessels.

the concentration of a particular nutrient is higher in chyme than in the absorptive cells.

After being absorbed, water–soluble nutrients, such as sugars and minerals, enter the villus's capillary, which is a structure of the *portal venous system,* and eventually, the *hepatic portal vein* (Fig. 4.28). The portal venous system includes veins that drain blood from the intestines. The hepatic portal vein delivers the nutrient–rich blood directly to the liver, where many of the nutrients undergo processing before they enter the general circulation.

In general, lipids are not water soluble. After entering absorptive cells, cholesterol and most of the other lipids are coated with a layer that contains protein, forming a **chylomicron** (*ky′–lo–my′–cron*). Chylomicrons move into a **lacteal,** a lymphatic system structure in each villus (Fig. 4.29). Lymphatic fluid ("lymph") transports chylomicrons, and eventually the lipid–filled structures enter the bloodstream via a vein near the heart. It takes about 3 to 5 hours for chyme to move from the beginning of the duodenum to the end of the ileum.[8] By the time chyme reaches the middle part of the ileum, most of its nutrient contents have been digested and absorbed. The Carbohydrates, Fats and Other Lipids, and Proteins chapters of this textbook provide more specific information about the digestion and absorption of macronutrients.

About 7.6 to 8.5 quarts (8 to 9 liters) of water from ingested foods and beverages and the secretions of intestinal cells enter into the GI tract daily. Most of the water is absorbed along with other nutrients in the small intestine. Any remaining water and the undigested material that reaches the end of the small intestine must pass through another sphincter before entering the large intestine. This sphincter prevents the contents of the large intestine from reentering the small intestine.

chylomicron particle formed by small intestinal cells that transports lipids in the bloodstream

lacteal lymph vessel in villus

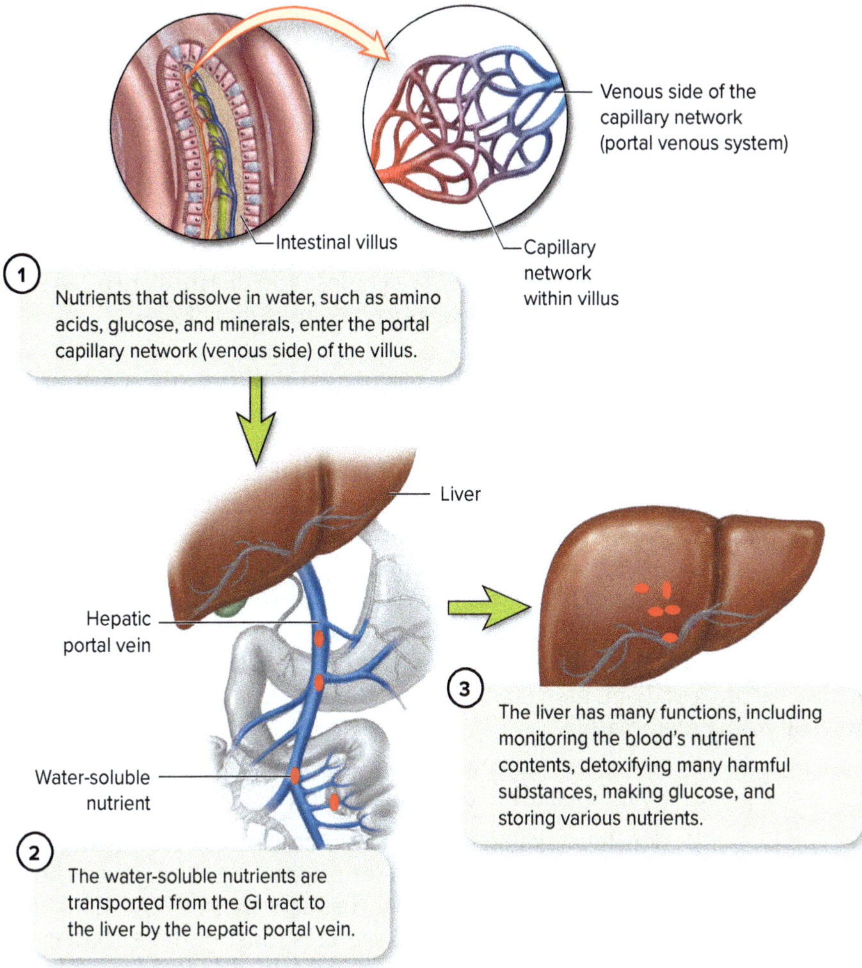

Figure 4.28 Water-soluble nutrient transport from the small intestine to the liver.

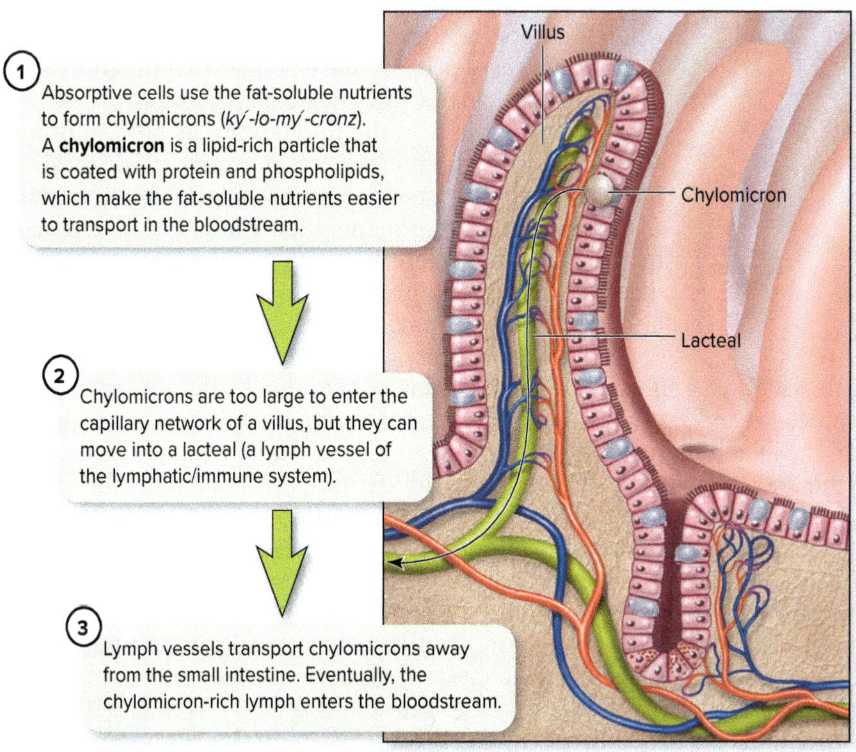

Figure 4.29 Fat-soluble nutrient transport from the small intestine to the rest of the body.

What Is Malabsorption? Absorptive cells only live for 3 to 5 days. While digestion is occurring, older absorptive cells are shed into the lumen and added to the contents of chyme. These cells are digested along with chyme. Newly formed absorptive cells constantly replace those that have been shed. The new cells can absorb nutrients from the dead cells and recycle some of their contents. However, the small intestine's high cell turnover rate leads to relatively high nutrient needs for these tissues. If the nutrients needed for cell division are lacking, fewer dead absorptive cells can be replaced, and nutrient *malabsorption* (*mal* = *poor*) can occur as a result. Thus, weight loss and diarrhea are often signs of chronic nutrient deficiency disorders, as well as intestinal diseases.

Intestinal Tract: Immune Function Agents of infection, such as harmful microbes and viruses, can be in food and water. As mentioned earlier, special cells in the intestinal tract produce mucus. In the intestines, mucus forms a barrier that reduces the likelihood that the agents of infection are absorbed. The ileum also has small areas of lymphatic system cells (lymphoid tissue) called *Peyer's (pie'–erz) patches.* Peyer's patches have immune system cells that can produce antibodies when they are exposed to potentially infectious agents, especially bacteria and viruses. Antibodies are proteins that recognize and bind to the harmful agents, which help destroy the agents before they can enter the bloodstream. **Inflammation** is a sign of the immune system's response to an agent of infection or a disease process. Pain, tissue redness and swelling, loss of function, and fever can result from inflammation.

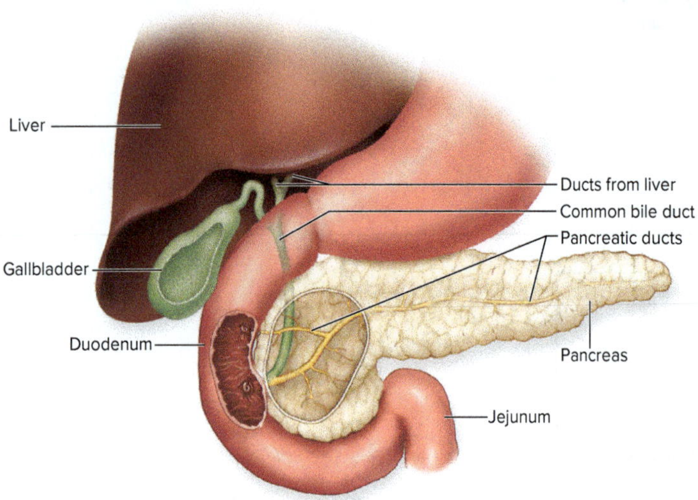

Figure 4.30 Pancreas, liver, and gallbladder. The pancreas, liver, and gallbladder play important roles in digestion.

Pancreas, Liver, and Gallbladder

The liver, gallbladder, and pancreas are accessory organs of the GI tract that play major roles in digestion, even though chyme does not move through them (Fig. 4.30). The pancreas produces and secretes most of the enzymes that break down carbohydrates, protein, and fat in the GI tract. In addition to digestive enzymes, the pancreas secretes *bicarbonate ions* (HCO_3^-) that neutralize HCl in chyme when it enters the duodenum. This is a critical step in the digestion process because the enzymes that function in the small intestine do not work in acidic conditions.

When chyme is in the duodenum, its fat and protein content triggers the release of a hormone *cholecystokinin* (*co'–lee–sis'–toe–ky'–nin*) from small intestinal cells. Cholecystokinin enters the bloodstream and circulates to the pancreas, where it stimulates the organ to secrete digestive enzymes into the duodenum. Cholecystokinin also helps delay stomach emptying.

The liver processes and stores many nutrients. This organ also makes cholesterol and uses this lipid to make **bile,** a substance that prepares fat and fat–soluble vitamins for digestion and absorption. Bile flows from the liver into the gallbladder, where it is stored until needed. When acidic chyme and, particularly, fat are in the duodenum, the small intestine sends a hormonal signal to the gallbladder, and as a result, the gallbladder contracts, releasing bile into the *common bile duct,* the tube that connects the gallbladder with the duodenum (see Figure 4.30). Chapter 6 provides more information concerning bile's role in digestion.

In the United States, many adults develop gallstones within their gallbladders. Gallstones usually consist of cholesterol; they can be small and grainy, like particles of sand, or as large as a coin (Fig. 4.31). When a gallbladder that contains stones contracts or a gallstone lodges in the common bile duct

inflammation sign of an immune system response to an agent of infection or a disease process

bile substance that is produced by the liver to prepare fat and fat-soluble vitamins for digestion and absorption

or one of the ducts that carries bile from the liver to the gallbladder, it causes considerable pain in the right upper part of the abdomen. If the stone moves out of the duct, the discomfort ends, but in some cases, the duct remains blocked and bile backs up into the liver or pancreas. When this occurs, surgery to remove the diseased gallbladder is necessary to prevent damage to the liver or pancreas. After surgery, bile drips from the liver directly into the small intestine. Having excess body fat increases the risk of gallstones, so keeping your weight at a healthy level can reduce your chances of developing the condition.

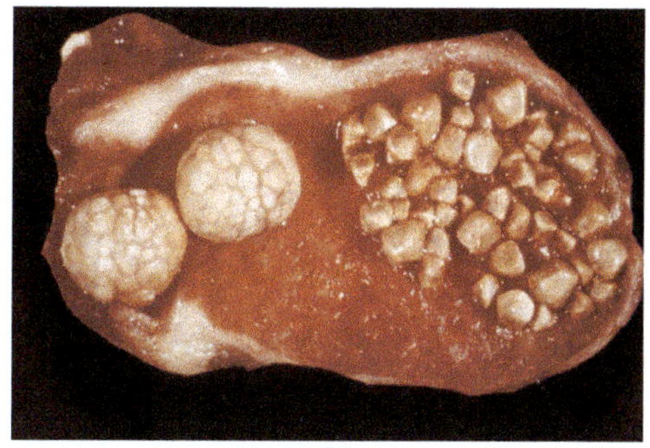

Figure 4.31 Gallstones. This diseased gallbladder has been cut open to show gallstones, which can vary in size. The stones usually consist of cholesterol.
Clinical Photography, Central Manchester University Hospitals NHS Foundation Trust, UK/Science Source

Cystic Fibrosis Cystic fibrosis (CF) is an inherited incurable disease that is usually diagnosed in early childhood. In CF, certain cells produce thick, sticky mucus that blocks passageways, particularly in the respiratory and digestive systems. People with CF often suffer from serious breathing problems and respiratory infections. The pancreatic ducts of an affected person may also become blocked by thick mucus, which interferes with the organ's ability to deliver digestive enzymes to the small intestine (see Figure 4.30). As a result, the digestion of nutrients, especially fat, is impaired. To overcome the malabsorption problem, patients with CF can take capsules that contain pancreatic enzymes with their meals. The capsules protect the enzymes from being destroyed by stomach acid.

Large Intestine

The large intestine is shorter than the small intestine (about 40 inches long or 101 centimeters),[9] but as you can see in Figure 4.25, its diameter is wider. The large intestine's major sections are the *colon* and the *rectum* (Fig. 4.32). Cells that line the large intestine produce a couple of layers of mucus. The mucus helps protect and lubricate the walls of the colon and rectum.

Under normal circumstances, very little carbohydrate, protein, and fat escape digestion and absorption in the small intestine and enter the large intestine. However, the large intestine has no villi, so little additional absorption other than some water and minerals occurs in this structure.

In a healthy person, it can take 30 to 40 hours for any remaining undigested material to move through the large intestine and become semisolid *feces* (stools).[11] An estimated 46% of feces (its wet weight) consists of bacteria that normally live in the large intestine.[12] Feces also contain undigested fiber from plant foods; a small amount of water, protein, and fat; and some mucus and cells shed from the walls of the intestinal tract.

Did *YOU* Know?

If you have a house cat, you probably are aware that the pet occasionally vomits a "furball," a tightly tangled bit of fur shaped like a caterpillar. The medical term for a furball is trichobezoar (*trik-oh-bee'-zore*). *Tricho-* refers to hair; a bezoar is a dense mass of undigested material that becomes trapped in the digestive tract.

Healthy people rarely form bezoars; however, people who have difficulty chewing food, consume large amounts of plant foods and insufficient water, or have disorders that delay stomach emptying may form bezoars. Additionally, people who engage in *pica* (the consumption of nonfood items, including stones and clay) or have *trichotillomania* (compulsive hair plucking) and then swallow the hair (*trichophagia*) are also at risk of developing bezoars.[10]

Bezoars can form anywhere in the human gastrointestinal tract, but they are most likely to develop in the stomach. Small bezoars can pass through the GI tract and are eliminated in the feces. Some bezoars can become large enough to damage the walls of the stomach or block the intestines.

By using an endoscope to view the inside of a patient's stomach, a physician can see a gastric bezoar and remove the mass with special instruments. Other effective treatments include the ingestion of enzymes and carbonated beverages that soften or dissolve the bezoar, enabling it or its remnants to be "flushed" out of the digestive tract naturally. Surgery, however, may be necessary to remove large bezoars.

Elimination

Feces remain in the **rectum,** the lower section of the large intestine, until muscular contractions move the material into the anal canal and then out of the body through the anus (defecation). The *external anal sphincter* that allows feces to be expelled is under voluntary control, so a healthy person can determine when to relax the sphincter and have a bowel movement. Young children must reach the stage of physical and emotional maturity in which they

rectum lower section of the large intestine

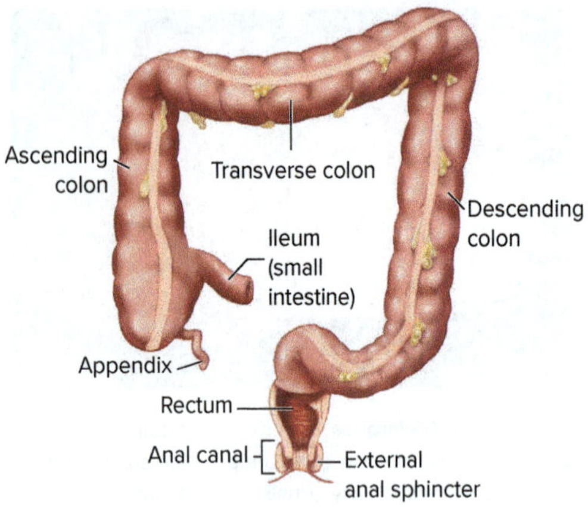

Figure 4.32 Large intestine.

are able to relax or tighten their external anal sphincter voluntarily. The timing of this stage varies but generally occurs in healthy children by 4 years of age.

Figure 4.33 shows the key structures of the digestion system. This figure also summarizes the processes of digestion, absorption, and elimination.

Microbes in Your Digestive Tract

The duodenum and jejunum of a healthy person's small intestine usually have few microorganisms residing in them. On the other hand, the large intestine is home to viruses, fungi, and vast numbers of various *species* (types) of bacteria, which are collectively called **gut microbiota.** Interestingly, the composition of the bacterial species (diversity) varies from person to person, due to individual differences, including diet and other lifestyle factors, age, health status, emotional stress, and antibiotic use.[13] Scientists are studying the roles that gut microbiota play in the prevention and treatment of certain infectious and inflammatory diseases.

Under normal conditions, gut micro–biota maintain a balance with each other that benefits their human host. Intestinal bacteria can make vitamin K, thiamin, and biotin, which their human host may be able to absorb, and break down (ferment) undigested food. As a result of their metabolic activity, intestinal bacteria also produce substances that colon cells can use for energy, and gases that may be expelled through the anus. In healthy persons, gut microbiota play an important role in maintaining good health by limiting the growth of harmful species of intestinal bacteria and yeast (a type of fungi). However, not all of the bacteria's metabolic activities benefit the host. The bacteria can produce potentially harmful products, particularly when they ferment dietary proteins and the components of these proteins (amino acids).

Researchers discovered that some people who consume the mainstream American ("Western") diet high in animal proteins, saturated fat, and ultra–processedfoodshighinsugar,salt,andother additives could develop a less diverse gut microbiota, including fewer beneficial intestinal bacteria.[13] Scientific evidence shows the Western diet may also contribute to **intestinal dysbiosis,** a condition characterized by the imbalance of the normal diversity of microbial populations in the GI tract.[13,16] Starvation, antibiotic use, and excessive emotional stress are other factors that can cause this condition. A person's risk of developing heart disease, diabetes, liver diseases, and obesity is linked to the diversity of their intestinal bacteria.[13,17] Nonetheless, consuming a Mediterranean–type diet (see Chapter 3), which is rich in healthy fats (see Chapter 6) and antioxidant–rich, high–fiber foods, is associated with a lower risk of developing chronic disease. High–fiber diets, such as the Mediterranean diet, seem to reduce the level of protein fermentation, thus the harmful products of protein fermentation. People who follow the Mediterranean eating pattern also have higher levels of beneficial gut bacteria and a more diverse microbiota population.[13] Chapter 5 discusses the benefits of fiber and high–fiber diets in more detail.

Did *YOU* Know?

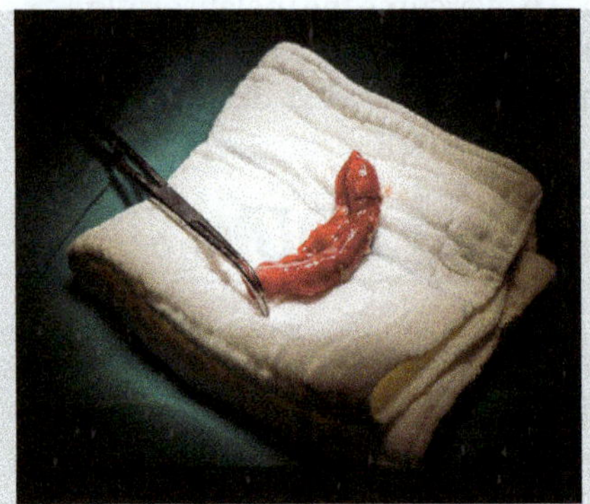

Sitthisombat/123RF

About 5% of Americans experience an inflamed appendix (*appendicitis*) at some point in their lives.[14] The inflammation may be the result of an infection, a blockage within the appendix, or a physical injury to the abdomen. If a diseased appendix is not properly diagnosed and treated, the patient can experience life-threatening complications and even death.

For years, medical experts thought the appendix was useless—it was simply a worm-shaped remnant of the human intestinal tract that formed during embryonic development (see Figure 4.32). In the past, surgeons often removed a healthy appendix while they performed abdominal operations. Recently, however, medical researchers have discovered the appendix has immune system function because it has an abundance of lymphoid tissue that produces an antibody called immunoglobulin A (IgA).[15] This antibody helps prevent bacteria and viruses from sticking to the lining of the intestinal tract and gaining entry into the bloodstream. IgA also helps regulate the balance of gut microbiota.[15] Additionally, the appendix serves as a "storage facility" for various kinds of bacteria that can be added to the colon's microbiota, when necessary. Thus, the appendix appears to help maintain good health of the large intestine and, therefore, the entire person.

gut microbiota various microbes, mostly bacteria, that reside in the GI tract

intestinal dysbiosis condition characterized by the imbalance of the normal diversity of microbial populations in the GI tract

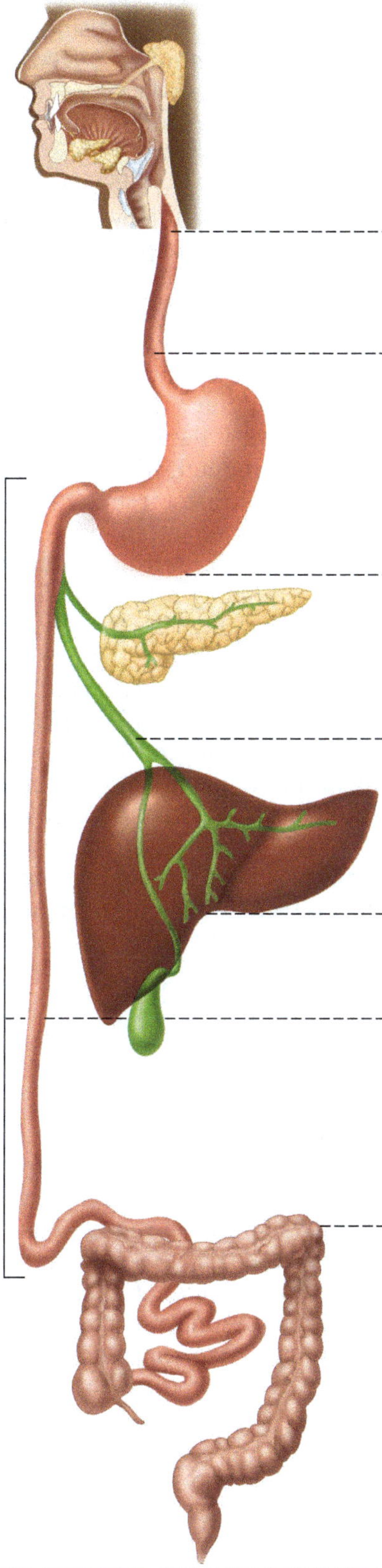

Organ	Secretions	Digestive Functions
Mouth and salivary glands	Saliva	Moistens food
	Mucus	Lubrication; helps bolus stick together
	Salivary amylase	Begins starch digestion
	Lingual lipase	Begins some lipid digestion in stomach
Teeth		Chew food
Tongue		Has taste buds; moves food within mouth
Esophagus		Moves food to stomach by peristaltic waves
Stomach		Stores, mixes, dissolves, and continues digestion; regulates emptying of chyme into small intestine
	HCl	Kills microbes; activates pepsinogen to pepsin
	Pepsin	Digests proteins
	Mucus	Lubricates and protects stomach wall
Pancreas		Secretes enzymes and bicarbonate
	Enzymes	Digest carbohydrates, fats, and proteins
	Bicarbonate	Neutralizes HCl entering small intestine from stomach
Liver		Receives water-soluble products of digestion from small intestine
		Synthesizes and secretes bile
	Bile salts	Aid fat digestion
Gallbladder		Stores and concentrates bile between meals
		Releases bile during meals
Small intestine		Digests and absorbs most substances; mixes and propels contents
	Enzymes	Food digestion
	Water	Maintains fluidity of intestinal contents
	Mucus	Lubricates inner walls; provides a barrier against infectious agents (for example, bacteria and viruses)
Large intestine		Stores and concentrates undigested matter; absorbs some minerals and water; eliminates wastes
		Provides environment for gut microbiota
	Mucus	Lubricates inner walls; aids defecation

Figure 4.33 Summary of digestive system organs and their major functions.

Did *YOU* Know?

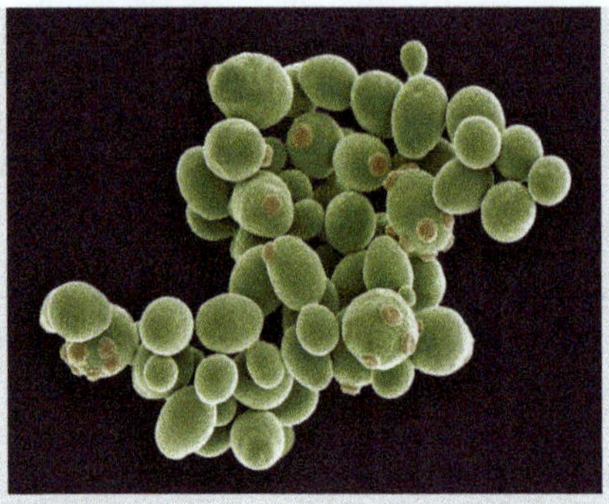

Brewer's yeast
Steve Gschmeissner/Science Photo Library/Getty Images

The production of beer, wine, and liquor relies on yeast—single-celled organisms that are classified as fungi. Brewers, for example, use "brewer's yeast" (*Saccharomyces cerevisiae*) to ferment sugars, forming ethanol ("alcohol"). In rare instances, however, a person's intestinal tract becomes a living brewery.

In addition to the many different bacterial species that normally live in the colon, relatively small numbers of fungi, including *S. cerevisiae and Candida albicans,* may reside in the large intestine. In a healthy person, the various bacterial species limit the growth of the fungi. Taking antibiotics to treat bacterial infections, however, can destroy both undesirable and desirable species of bacteria, causing intestinal dysbiosis. In such cases, yeast multiply and overpopulate the colon. The yeast ferment sugars, producing alcohol as a by-product of their metabolism. If the alcohol is absorbed and enters the bloodstream, the affected person develops signs and symptoms of alcohol intoxication, such as impaired physical and intellectual functioning and mood changes, without having consumed any alcohol. This condition is called gut fermentation syndrome or "auto-brewery syndrome." Treatment generally includes taking antifungal medications to kill the yeast. The Nutrition Matters section of Chapter 6 has more information about alcohol consumption and its effects on health.

Currently, researchers are testing the safety and effectiveness of using **gut microbiota transplantation (GMT),** commonly referred to as "fecal transplant," to treat certain intestinal disorders, such as *Clostridioides difficile* ("C–diff") infection. C–diff bacteria are a usual but potentially harmful resident of the colon. The population of C–diff is normally kept in check by the beneficial populations of gut bacteria. When antibiotic use kills most of the bacteria in the colon, C–diff is often able to survive and multiply. This bacterium secretes toxins that inflame and destroy tissues of the large intestine, causing chronic diarrhea that resists many conventional forms of treatment. GMT involves taking some feces from a healthy person and introducing it into the large intestine of the patient. The transfer can be accomplished by giving the patient an enema or capsule that contains the donor's processed feces and is placed in the patient's colon for a period.[18] According to the results of many studies, GMT is a highly effective treatment for cases of C–diff infection that are not cured by conventional medical therapies.[18]

The use of fecal transplants may also benefit people with ulcerative colitis or Crohn's disease, which are discussed later in this section. As is often the case, more research is needed to determine the long–term safety and effectiveness of GMT.

These products are probiotics.
©Wendy Schiff

Probiotics and Prebiotics

Probiotics are live, beneficial intestinal microbes (primarily bacteria or yeast) that have been grown (*cultured*) under laboratory conditions. Products that contain these microbes are also referred to as probiotics. Many brands of yogurt, cultured buttermilk, and other fermented foods, for example, are probiotics because they contain "live and active" cultures of the beneficial bacteria (see the Recipe for Healthy Living feature at the end of this chapter). Probiotics are also available in pills (dietary supplements). *Lactobacillus* and *Bifidobacterium* are the groups of bacteria that are most often used in probiotics.[19] When consumed, adequate amounts of the bacteria must survive the acidic conditions of the stomach and digestive enzymes in the upper small intestine before reaching the large intestine, where they can grow and multiply.

According to several studies, probiotics may help prevent the diarrhea that often occurs when antibiotics are taken, and probiotics may be useful in treating people with IBS (see Section 4.4).[19] However, scientific information about the long–term safety and effectiveness of using probiotics is lacking. Thus, the FDA has not approved any health claims for probiotics.

As mentioned earlier, a person's diet influences the kinds of bacteria that make up their gut microbiota. **Prebiotics** are forms of dietary fiber

gut microbiota transplantation (GMT) fecal transplants

probiotics products that contain healthful microbes; beneficial gut microbes

prebiotics forms of dietary fiber that are poorly digested by humans but support the growth of beneficial gut microbes

that are poorly digested by humans but support and promote the growth of probiotics in the colon. Prebiotics are in a wide variety of foods, particularly fruits, vegetables, and whole grains. Soybeans, berries, garlic, barley, kale, and legumes are rich sources of prebiotics. Chapter 5 discusses fiber, including prebiotics.

Some species of the gut microbiota can be harmful if they enter other parts of the body or contaminate food. People should wash their hands after having bowel movements to reduce the likelihood of spreading dangerous microbes from their intestinal tract to others. Chapter 12 discusses microorganisms that cause common food–borne infections and ways to limit exposure to them.

Did *YOU* Know?

A human being is an **omnivore,** an organism that can digest and absorb nutrients from plants, animals, fungi, and even bacteria. Additionally, the healthy human GI tract responds to dietary changes, such as alterations in the nutritional composition or amounts of food consumed, by increasing the production of various digestive enzymes. Consider this: How would people be able to survive in extreme environments, including deserts and permanently frozen regions, if they were unable to digest foods from a wide variety of sources?

What Is Inflammatory Bowel Disease?

Inflammatory bowel disease (IBD) is the general name for a group of chronic diseases that cause inflammation and swelling of the intestines. The inflammation disrupts digestion and nutrient absorption, and damages the intestines. An estimated 3 million adult Americans have either of the two most common forms of IBD, which are **ulcerative colitis (UC)** and **Crohn's disease (CD)**.[20] The majority of persons who are diagnosed with IBD are between 15 and 35 years of age.[21]

Medical experts do not know what causes IBD, but people with the disease may have immune systems that react abnormally to bacteria or viruses in their digestive tract. Genetic factors, cigarette smoking, and poor diet are also thought to play a role in the development of IBD. The typical Western diet, which is high in animal protein and ultra–processed foods and low in fiber (plant foods), is a risk factor for IBD.[22] This particular diet negatively alters the kinds of bacteria that make up the gut's microbiota. Some scientists think this change (dysbiosis) triggers an abnormal response by the intestinal tract's immune cells, resulting in inflammation that leads to IBD.[22]

Ulcerative Colitis

Ulcerative colitis is the most common type of IBD. The condition is characterized by the formation of *ulcers* (sores) in the inner lining of the large intestine. As a result of the chronic inflammation, the colon's lining erodes away, leading to painful abdominal cramps, rectal bleeding, diarrhea, weight loss, and anemia.

Although emotional stress does not cause UC, people with the disease report that stressful situations seem to worsen their symptoms. Furthermore, food allergies do not cause UC. However, some people with the disease learn to avoid certain foods and beverages because consuming them seems to trigger *flares* (bouts of signs and symptoms) of the illness. Dietary recommendations generally include eating smaller meals more often, drinking more liquids but avoiding carbonated drinks, and eating high–fiber foods. When people with UC experience flares, however, they usually are advised to follow a low–fiber diet.

There is no cure for UC. Treatment usually includes medications to reduce inflammation and avoiding foods and beverages that seem to trigger flares. People with UC must take the medication for the rest of their lives, unless their large intestine is so severely diseased that surgical removal of the organ is necessary.

Patients with UC have an increased risk of colon cancer, especially when their entire colon is affected and they have had the disease for 8 or more years.[23] Patients can reduce their risk of colon cancer by managing the inflammation with medication and having regular screening (testing) to detect colon cancer.

omnivore organism that can digest and absorb nutrients from plants, animals, fungi, and bacteria

ulcerative colitis (UC) common form of inflammatory bowel disease

Crohn's disease (CD) common form of inflammatory bowel disease

Did *YOU* Know?

Using "high colonics" and other types of enemas (coffee enemas, for example) to "cleanse" your colon is not necessary because the large intestine does not need to be cleansed. Furthermore, frequent enemas may deplete the body of vital minerals, including sodium and potassium. Check with your physician before trying enema treatments.

Crohn's Disease

According to the National Institutes of Health (NIH), more than 500,000 Americans have Crohn's disease (CD), another form of IBD.[24] Unlike ulcerative colitis (UC), Crohn's disease can affect the entire gastrointestinal tract, including the mouth. In most cases, however, the last section of the small intestine (the ileum) and beginning of the colon are inflamed. CD damages deeper layers of the intestines than does UC. Diarrhea, painful abdominal cramps, loss of appetite, and weight loss are typical signs and symptoms of the chronic disease.

Treatment for CD is similar to that of UC: taking medications to reduce the inflammation and promote healing as well as avoiding foods and beverages that cause "flares" (bouts of signs and symptoms). There is no cure for the disease. In severe cases, surgery to remove badly damaged portions of the intestinal tract is necessary. People with Crohn's disease have an increased risk of colon cancer.[24] The "Real People, Real Stories" feature in this section introduces Alicia Anne, a young woman who was diagnosed with CD when she was 25 years of age.

©Wendy Schiff

REAL PEOPLE | REAL STORIES

Alicia Anne

Alicia Anne loves her work; the young woman is a hair stylist/makeup artist in an upscale beauty salon. According to Alicia, "I've always been heavy and struggled with my weight." Soon after she became a beautician, Alicia noticed that she was losing weight rapidly—she had lost about 50 pounds in 3 months. She attributed the weight loss to her busy lifestyle and demanding occupation that required her to be standing for 8 hours per day. Although Alicia was thrilled with her new appearance, her "guts" hurt; she had diarrhea almost constantly; and she had very little energy. Furthermore, she was not hungry and consumed only fluids.

When Alicia discovered blood in her stools, she realized something was seriously wrong with her body, so she went to a hospital's emergency room (ER). After undergoing a computed tomography (CT) scan, Alicia was diagnosed with an inflamed ileum. The ER physician gave her some pain medication and told her to be examined by a *gastroenterologist,* a type of physician who is specially trained to diagnose and treat diseases of the gastrointestinal tract and liver. The gastroenterologist determined that Alicia had ulcers throughout her large intestine and chronic inflammation had severely damaged her ileum. She suffered from Crohn's disease (CD). Her initial treatment included prednisone, a powerful anti-inflammatory medication. Within a few months of the treatment, Alicia began feeling better and strong enough to resume her usual work schedule.

Today, Alicia takes more than one medication to treat CD, and she monitors what she eats and drinks, so she can identify foods that cause "flares" (bouts of painful intestinal cramps followed by diarrhea). "I used to be able to eat whatever I wanted," Alicia says. "Now, I have to constantly assess how my food choices affect my body. It's been difficult to live with Crohn's, but I'm determined to do whatever I can to feel better."

Concept CHECKPOINT

11. Describe what happens to the water- and fat-soluble nutrients in a cheese sandwich after it is eaten and as it moves through the digestive tract.

12. Provide an example of mechanical digestion and chemical digestion.

13. Eliot choked on a piece of hot dog. Explain why choking can occur while eating.

14. What keeps stomach contents from reentering the esophagus?

15. List the sections of the small intestine. Where does most digestion and absorption occur?

16. How would removal of the pancreas affect digestion?

17. Explain why it is important to have proper concentrations of microbes residing in a person's large intestine.

18. What are ulcerative colitis and Crohn's disease?

See Appendix F for responses.

4.4 Nutrition Matters: Gut Reaction

Learning Outcomes

1. Identify some common gastrointestinal disorders, including colorectal cancer, and discuss these conditions, including possible causes, preventive measures, and useful treatments.

Even though you may be healthy, you have probably experienced occasional bouts of constipation, vomiting, diarrhea, or heartburn. This section provides general information about these and some other common intestinal problems, including colorectal cancer.

Constipation

Many Americans, especially older adults, think they are constipated if they do not have a bowel movement at least once a day. According to the American Gastroenterological Association, it is not necessary to have a daily bowel movement. Although the normal frequency of bowel movements varies individually, a healthy person should have at least three bowel movements per week.[25] When bowel movements occur less frequently and/or are difficult to eliminate, or the feces are dry or hard, the condition is called constipation.

Constipation often affects older adults and pregnant females. Other factors that influence the frequency of bowel movements include lack of dietary fiber; low water intake; certain medications and health conditions (for example, diabetes and Parkinson's disease); and changes in your typical routine, such as taking a long trip or having major surgery. Furthermore, constipation can result when people regularly ignore their normal bowel urges and avoid making a trip to the bathroom to defecate when it's not convenient.

Although occasional constipation is a common health problem, chronic constipation can cause discomfort and may contribute to the development of inflamed hemorrhoids and diverticula. Chapter 5 provides more information on these common intestinal conditions.

If you feel uncomfortable because your bowel habits have changed or you have hard, dry stools that are difficult to eliminate, discuss the matter with your physician. In many instances, adding more fiber–rich foods to your diet and drinking more watery fluids are all that is needed to become more "regular." Chapter 5 provides information about dietary fiber, including rich food sources.

Diarrhea

Diarrhea is a condition characterized by frequent bowel movements with loose stools. Diarrhea occurs when more water than normal is secreted into the GI tract, and the extra water softens and dilutes the stools. In most cases, diarrhea results from bacterial or viral infections of the intestinal tract. The infectious bacteria or viruses produce irritating or toxic substances that increase the movements (motility) of the GI tract. As a result, the GI tract propels chyme more rapidly through it, absorbing less water than normal in the process. Increased GI motility also enables the large intestine to rapidly eliminate the watery feces and the toxic material it contains.

Cases of severe diarrhea require medical attention because frequent watery stools can deplete the body's water volume, causing dehydration and excessive losses of electrolytes, such as sodium and potassium. Treatment generally includes providing fluids that contain sodium, potassium, and simple sugars such as glucose. Prompt treatment of severe diarrhea—within 24 to 48 hours—is especially crucial for infants and older adults because they can become dehydrated quickly by the loss of body water. In adults, diarrhea that is accompanied by bloody stools or lasts more than 7 days may be a sign of a serious intestinal disease, and a physician should be consulted.

Vomiting

Not long after eating something toxic or drinking too much alcohol, you begin to feel queasy or "sick to your stomach." You soon become well aware that your body has an effective way of removing the harmful food or beverage—vomiting. Although vomiting is an unpleasant experience, it often prevents toxic substances from entering your small intestine, where they can do more harm or be absorbed. Vomiting can also be a response to intense pain, food allergies, rotating movements of the head (motion sickness), hormonal changes in pregnancy (morning sickness), and migraine headaches.[26] Vomiting may also be a sign of very serious conditions, including intestinal blockages, appendicitis, and cancer.

Vomiting occurs when the vomiting center in the brain interprets information from various nervous system receptors concerning the physical and chemical conditions of the stomach, small intestine, and bloodstream. When a toxic chemical is detected, the center initiates vomiting by contracting the abdominal muscles, expelling the contents of the stomach and duodenum forcefully out of the body via the mouth.

Vomiting generally does not last more than 24 hours. Repeated vomiting, however, can result in dehydration and the loss of electrolytes (such as sodium and potassium), especially if it is accompanied by diarrhea. Treatment includes avoiding solid food until the condition resolves. Additionally, sipping small amounts of water or clear liquids, including noncarbonated beverages such as sports drinks, can help prevent dehydration. If the affected person is able to retain small amounts of watery fluids, then they can try to drink increasing amounts of the liquids until the vomiting subsides completely.

Adults should contact a physician if their vomiting lasts for more than a day or dehydration develops. Signs of dehydration include decreased urination, sunken eyes, and dry mouth.[26] Dehydration is likely to occur when an adult has been unable to "keep fluids down" for more than 12 hours. Contact a physician immediately if the vomit is bloody (looks like coffee grounds) or vomiting is accompanied by other signs and symptoms, such as confusion, fever, or severe abdominal pain. Children who suffer from vomiting and/or diarrhea are likely to develop dehydration more rapidly than adults who are suffering from these conditions. Signs of dehydration in young children include sunken eyes, dry mouth, decreased urination, and dark yellow urine. Contact a physician if a child has vomiting and diarrhea that persists for more than a few hours and shows signs of dehydration.

Gastroesophageal Reflux Disease (GERD)

Many American adults complain of having occasional heartburn ("acid indigestion"), a gnawing pain or burning sensation generally felt in the upper chest, under the breastbone. This discomfort is not the result of a heart problem but is caused by the passage of acidic contents from the stomach into the esophagus ("acid reflux"). Because the esophageal lining does not produce much protective mucus, the acid quickly destroys the tissue, causing pain and, sometimes, bleeding. Many factors can contribute to heartburn or worsen the condition, including being pregnant, smoking cigarettes, having excess body fat, drinking alcohol, and consuming certain foods.

Although many people think heartburn is a trivial health problem, frequent chronic heartburn can be a symptom of **gastroesophageal reflux disease (GERD).** If not treated properly, acid reflux damages the lining of the lower esophagus and contributes to the development of esophageal ulcers (sores). Such ulcers can damage blood vessels in the wall of the esophagus, causing bleeding. Signs of bleeding from the esophagus and stomach include black, tarry stools. In severe cases, the loss of blood from a bleeding ulcer can be deadly.

People who suffer from GERD have a higher risk of esophageal cancer than people who do not have a history of this condition. If you or someone you know has GERD, typical dietary advice for treating the condition includes consuming smaller, more frequent meals that are low in fat; not overeating at mealtimes; and limiting intake of foods that relax the lower esophageal sphincter, such as spicy or greasy foods, tomatoes and tomato products, peppermint, caffeine, alcohol, and chocolate.[27] Table 4.3 lists other recommendations for reducing the risk of heartburn and managing GERD. Taking over–the–counter antacids can

gastroesophageal reflux disease (GERD) condition that can result when acid reflux damages the wall of the lower esophagus and causes ulcers

TABLE 4.3 Recommendations to Reduce the Risk of Heartburn

- If you have too much body fat, lose the excess weight.
- Do not lie down within 3 hours after eating a meal.
- Do not overeat at mealtimes.
- Avoid smoking cigarettes.
- Elevate the head of your bed 6 to 8 inches higher than the foot of the bed.
- Do not wear tight belts or clothes with tight waistbands.
- Learn to recognize foods that cause heartburn.

neutralize excess stomach acid and relieve the discomfort of heartburn within minutes, but these products do not prevent heartburn. People who have GERD should see a physician for treatment, which can include medications that inhibit stomach acid production, preventing heartburn.

Peptic Ulcer

A peptic ulcer is a sore that occurs primarily in the lining of the stomach (gastric ulcer) or the upper small intestine (Fig. 4.34). The typical symptoms of a peptic ulcer are dull or burning pain in the upper abdominal area that occurs when the stomach is empty, such as between meals or during the night. The pain is relieved briefly by taking antacids or eating. The pain results when HCl acid in the stomach comes in contact with and digests the lining of the organ, forming one or more sores. It is not unusual for the sores to damage the wall of a blood vessel, causing bleeding. Furthermore, an ulcer may erode through the stomach or intestinal wall and allow GI contents to leak into the abdominal cavity, resulting in a potentially life–threatening infection. Therefore, it is important to recognize ulcer symptoms and obtain treatment early.

Physicians detect peptic ulcers by performing a clinical examination called upper endoscopy. The first step of this procedure involves administering a medication that relaxes the patient. Then the physician inserts a special flexible scope into the mouth, down the esophagus, and into the stomach and upper small intestine. The scope is equipped with a video camera that transmits images of the lining of the esophagus, stomach, and upper small intestine to a screen that the physician views for the presence of ulcers, eroded areas, or cancerous tumors. The scope also enables the physician to use tools to treat areas of bleeding or remove pieces of tissue (biopsy) for microscopic examination.

Helicobacter pylori (*H. pylori*) is a type of bacteria that can live in parts of the stomach and is largely responsible for the development of stomach ulcers. *H. pylori* damages the thick protective mucous coating that covers the inner lining of the stomach. When this occurs, stomach acid comes in contact with the stomach lining and destroys it. In addition to infection with *H. pylori,* other factors are associated with the development of peptic ulcers, particularly smoking cigarettes, heavy consumption of alcohol, and use of NSAIDs (nonsteroidal anti–inflammatory drugs) such as aspirin, ibuprofen, and naproxen.[28]

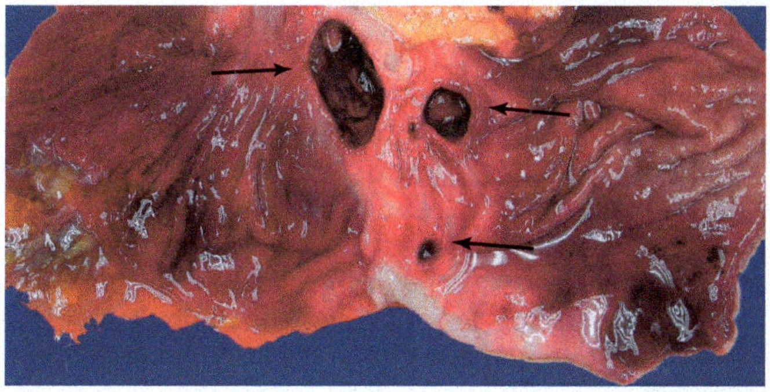

Figure 4.34 Gastric ulcers. Gastric ulcers are peptic ulcers that occur in the stomach.

Dietary approaches do not seem to help in the prevention or treatment of peptic ulcers.[28] Instead, people may be able to prevent peptic ulcers by avoiding the regular use of NSAIDs. People infected with *H. pylori* can take antibiotics, as well as medications that reduce stomach acid production. This treatment is highly effective for combating *H. pylori* infections and healing peptic ulcers.

Irritable Bowel Syndrome (IBS)

About 10 to 15% of Americans suffer from irritable bowel syndrome (IBS).[29] IBS is characterized by intestinal cramps, bloating, gas, and abnormal bowel function, particularly diarrhea, constipation, or alternating episodes of both. After bowel movements, the affected person may feel as though elimination of stools was incomplete. It is important to note that IBS is not a form of inflammatory bowel disease (IBD). Unlike IBD, IBS does not permanently damage the large intestine or lead to colon cancer.[30]

The cause of IBS is unknown, but gastrointestinal infections, and emotional stress may trigger the disorder.[30] Furthermore, the colons of people with IBS may be more sensitive to nervous stimulation and contract more than the colons of people who do not have this condition.

Therapy is individualized and may include elimination diets that focus on determining which foods are most likely to contribute to IBS symptoms. A low–FODMAP diet may help some people with IBS.[31] FODMAPs are poorly digested forms of carbohydrate, including fructose in fruit and high–fructose corn syrup; lactose in milk; and oligosaccharides in wheat, onions, and legumes. Chapter 5 provides more information about FODMAPs. Treatment may also include learning stress management strategies, obtaining psychological counseling, and taking antidepressants and other medications.

Colorectal Cancer

Cancer that develops in either the colon or the rectum is often called **colorectal cancer.** In the United States, colorectal cancer is the third–leading cause of cancer deaths among adults.[32] (Lung cancer is the leading cause of cancer deaths among adult Americans.) Table 4.4 lists major signs and symptoms of colorectal cancer.[33] Table 4.5 lists major risk factors for colorectal cancer.[34]

According to the American Cancer Society, adults who have an average risk of colorectal cancer should undergo a regular screening test for the disease, beginning at age 45.[35] People who have an average risk of this form of cancer do not have ulcerative colitis, Crohn's disease, or a family or personal history of colorectal cancer. People who have one or more of the major risk factors listed in Table 4.5 should check with their physicians for advice concerning when to begin regular colorectal screenings to detect the disease.

TABLE 4.4 Major Signs and Symptoms of Colorectal Cancer

- Change in usual bowel movements (persistent diarrhea, constipation, or pencil-thin stools)
- Sensation that bowel movements are incomplete
- Bloody bowel movements or dark stools
- Abdominal cramps or gut pain
- Unusual tiredness
- Involuntary weight loss
- Anemia

TABLE 4.5 Major Risk Factors for Colorectal Cancer

- Being over 50 years of age
- Having colorectal *polyps* (clumps of tissue)
- Having a family history of colorectal cancer
- Having ulcerative colitis or Crohn's disease
- Having type 2 diabetes
- Eating a lot of red meat (beef, pork, and lamb) and processed meat (e.g., hot dogs and "deli" meats)
- Having excess body fat and/or being physically inactive
- Smoking cigarettes and/or drinking alcohol
- Having African or Ashkenazi Jewish ancestry

Treatment for colorectal cancer usually involves surgery, chemotherapy, biological therapy, or radiation therapy. Surgery removes the diseased areas of the large intestine. Chemotherapy uses drugs that target and kill cancer cells. Biological therapy involves using antibodies, which are special proteins that detect and bind to colorectal cancer cells. This treatment interferes with the cancer cells' ability to grow and spread. Radiation therapy uses high–energy rays to kill cancerous cells. Although surgery is the most common treatment for colorectal cancer, people with the disease often undergo a combination of treatments.

Figure 4.35 shows the digestive system organs that are affected by inflammatory bowel disease, colorectal cancer, and the other digestive system problems that are discussed in this chapter. Food allergies and celiac disease also affect the digestive system; those conditions are discussed in Chapter 7. The "Nutrition Matters" section of Chapter 8 provides more general information about cancer and the role of diet in cancer development.

colorectal cancer cancer that develops in the colon or the rectum

Concept CHECKPOINT

19. List at least three possible causes of constipation.
20. Explain why excessive diarrhea and vomiting can be dangerous.
21. Describe at least three steps people can take to reduce their risk of heartburn and GERD.
22. Discuss at least three steps people can take to reduce their risk of peptic ulcers.
23. Describe common signs and symptoms of irritable bowel syndrome.
24. List at least three signs and symptoms of colorectal cancer.

See Appendix F for responses.

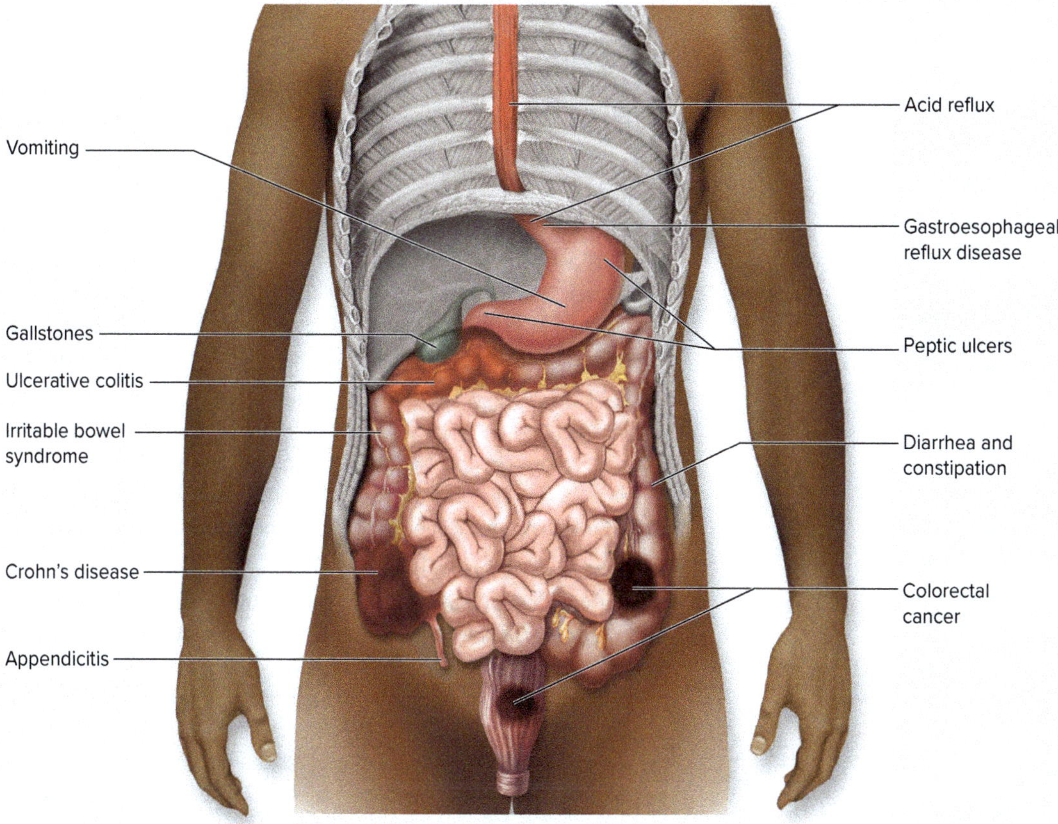

Figure 4.35 Organs affected by major inflammatory bowel diseases and common digestive system disorders. (Some of the liver is not shown.)

Chapter References

See Appendix G.

Summary

4.1 Nutrition: Chemistry Foundations

- Matter, including food, is composed of chemicals. Over 100 different types of atoms exist, and each type is an element. Atoms of elements may react with each other to form chemical bonds that hold the atoms together in a new arrangement called a molecule. Molecules that contain two or more different elements are called compounds. A solution is an evenly distributed mixture of two compounds. In living things, water is the solvent, the primary compound of a solution.
- Ions, acids, and bases play important physiological roles in the body. An ion is an atom or group of atoms that has a positive or negative electrical charge. Electrolytes are ions of minerals that can conduct electricity. An acid is a molecule that donates hydrogen ions (H^+); a base is a molecule that accepts H^+. Chemists use the pH scale to describe the H^+ concentration of a watery solution.
- Most molecules undergo chemical reactions that change their arrangement of atoms, forming new molecules or compounds. When elements or compounds combine to form new substances, a synthetic reaction has occurred. The new substances often have physical

and chemical properties that are quite different from those of the reactants. Decomposition reactions, such as those occurring during digestion, involve the breaking down of molecules. Enzymes are molecules (usually proteins) that speed up rates of chemical reactions. Enzymes do not become part of the products of a reaction and, as a result, can catalyze many reactions.

4.2 Basic Physiology Concepts

- Anatomy is the scientific study of cells and other body structures; physiology is the scientific study of how cells and body structures function. A cell is the smallest functioning unit in a living organism. Cells that have similar characteristics and functions are usually joined together into larger masses called tissues. An organ is composed of various tissues that function in a related fashion. An organ system is a group of organs that work together for a similar purpose.
- The cardiovascular system involves the pumping action of the heart to circulate blood in blood vessels throughout the body. The respiratory system enables the body to obtain oxygen and eliminate carbon dioxide. The lymphatic/immune system helps maintain fluid balance, absorb certain nutrients, and defend the body against infectious disease. The urinary system filters and excretes unneeded substances from blood and maintains proper fluid balance.
- The muscular and skeletal systems enable the body to move within its environment and provide support and protection for the body. The nervous system produces intellectual and emotional responses, and controls and regulates many body functions. The endocrine system produces hormones that convey information to target cells; hormones regulate a variety of physiological activities. Hair, nails, and skin are structures of the integumentary system. Skin protects against minor injuries and infectious disease–causing agents, and helps maintain body temperature. The main function of the reproductive organs is to produce children. The primary roles of the digestive system are the breakdown of large food molecules into smaller components (nutrients) and the absorption of nutrients into the bloodstream.

4.3 The Digestive System

- Digestion is a mechanical and chemical process that breaks down large food components into nutrients that can be absorbed. The mechanical aspects of digestion include the chewing action of teeth as well as involuntary muscular activity, including peristalsis. Chemical digestion includes the actions of enzymes and substances such as hydrochloric acid and bile.
- Digestion begins in the mouth with the mechanical action of teeth and some minor chemical action of salivary amylase. The esophagus conveys food from the mouth to the stomach, where it mixes with gastric juice and is referred to as chyme. The chyme moves from the stomach to the duodenum of the small intestine. In the small intestine, enzymes complete the process of digestion. Cells that line the villi absorb the end products of digestion and transfer them to blood or lymph. Any remaining undigested material, some water, and intestinal bacteria are eliminated from the body as feces.
- The human intestinal tract contains vast numbers and diverse kinds of microbes, particularly bacteria that reside in the large intestine. Gut microbiota play important roles in health, and imbalances in the diversity and numbers of the bacteria can cause infection, inflammation, and, possibly, certain chronic diseases.
- Millions of Americans have inflammatory bowel disease (IBD). The two most common forms of IBD are ulcerative colitis and Crohn's disease. Currently, there is no cure for IBD.

4.4 Nutrition Matters: Gut Reaction

- Common gastrointestinal disorders include constipation, vomiting, diarrhea, heartburn, peptic ulcer, and irritable bowel syndrome. Chronic constipation can cause discomfort and may contribute to the development of diverticula. Frequent chronic heartburn can be a symptom of GERD. People with GERD are at risk of esophageal cancer. Peptic ulcers can cause life-threatening bleeding. Although irritable bowel syndrome is common, the condition does not result in inflammation of the large intestine. Colorectal cancer is the third-leading cause of cancer deaths among adults in the United States.

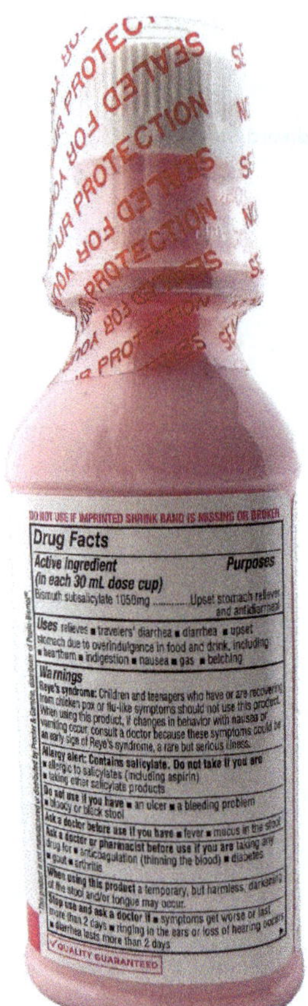

The active ingredient in this product is bismuth subsalicylate (*biz'-muth, sub'-sa-lis'-eh-late*), which is taken by mouth to treat occasional gastroesophageal reflux for people who are 12 years of age and older.
©Wendy Schiff

Recipe for Healthy Living

Easy Orange-Strawberry Smoothie

When made with fruit and low-fat Greek yogurt that contains "live and active" cultures, smoothies provide beneficial probiotics and are a low-fat, protein-, calcium-, and potassium-rich snack. Experimenting with different fruits and flavored yogurts makes it easy to individualize smoothies. You'll need a sturdy blender to combine the ingredients.

The following recipe makes one serving that supplies approximately 200 kcal, 18 g protein, 4 g fat, 306 mg calcium, 605 mg potassium, 1.3 mg zinc, and 47 mg magnesium.

Baiba Opule/123RF

INGREDIENTS:

¼ cup orange juice, calcium fortified
6 oz container low-fat, plain Greek yogurt that has live and active cultures
1 cup fresh, sliced strawberries, washed, with leafy "caps" removed (can substitute frozen strawberries)
2 ice cubes

PREPARATION STEPS:

1. Place all ingredients, except ice cubes, in blender.
2. Blend until smooth. Add ice cubes and blend again, until ice is crushed. If smoothie is too thick, add 1 to 2 tablespoons water to reach desired consistency. Serve.

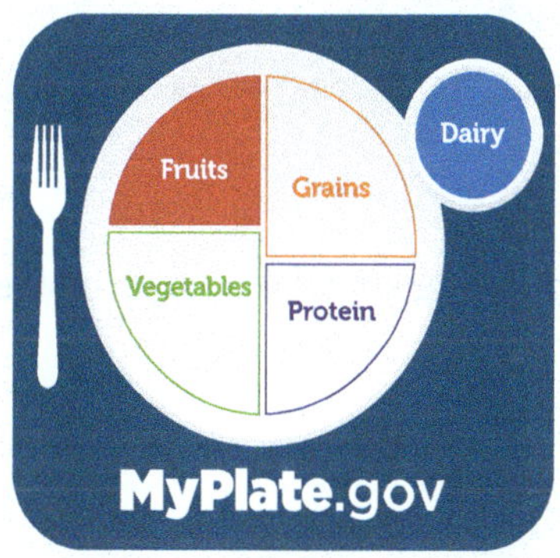

US Department of Agriculture or USDA

Approximate Percent of Calories from Macronutrients

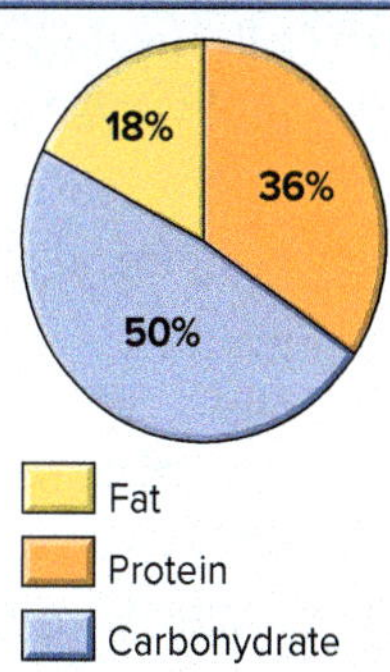

Percent of Daily Value/Serving

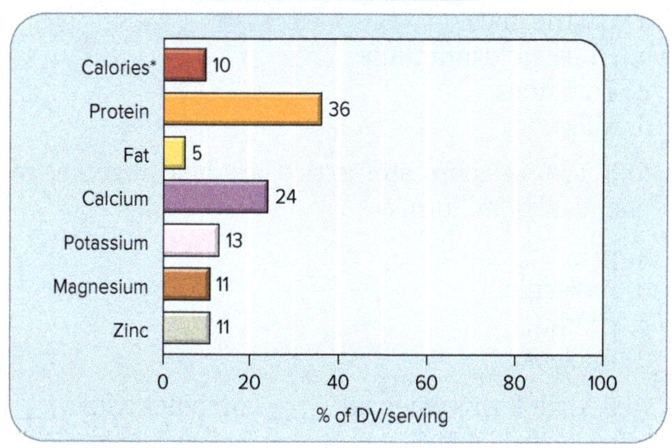

*2000 daily total kcal

Critical Thinking

1. Marcella adds a teaspoon of baking soda to a cup of water to relieve her "acid indigestion." Explain why this remedy works, even though physicians do not recommend it (baking soda contains unhealthy amounts of sodium).
2. Why would a person's stomach feel uncomfortably full several hours after eating a bacon-topped double cheeseburger, a large serving of French fries, and a milkshake?
3. Explain why the label for a multivitamin supplement recommends taking the pill with meals, particularly meals that contain some fat.

©Wendy Schiff

4. Miguel had a serious condition that required removal of the upper third of his small intestine. Based on this information, is Miguel at risk for developing multiple nutrient deficiencies? Explain why or why not.
5. Your mother has persistent heartburn, but she does not think her discomfort is serious enough to be investigated by her physician. Do you agree or disagree with her attitude about heartburn? Explain your position.

Practice Test (Section numbers are listed below in red.)

Select the best answer.

1. (4.1) Which of the following substances is insoluble in water?
   a. Iron
   b. Sugar
   c. Fat
   d. Sodium
2. (4.1) Which of the following substances has the lowest pH?
   a. Gastric juice
   b. Household ammonia
   c. Tomatoes
   d. Milk
3. (4.1) Two or more atoms that are held together by a chemical bond form a
   a. pH.
   b. molecule.
   c. proton.
   d. nucleus.
4. (4.1) A salt forms when an _____ combines with a _____.
   a. electrolyte; proton
   b. acid; base
   c. element; pH
   d. enzyme; mineral
5. (4.2) A _____ is a group of similar cells that perform similar functions.
   a. ribosome
   b. tissue
   c. cytoplasm
   d. system
6. (4.2) Which of the following statements is true?
   a. Arteries carry blood away from the heart.
   b. Hemoglobin carries most of the enzymes in the blood.
   c. Cells need carbon monoxide to obtain energy.
   d. The liver is the primary organ of the cardiovascular system.
7. (4.2) Chemical messengers in the body are
   a. enzymes.
   b. cells.
   c. capillaries.
   d. hormones.
8. (4.3) Which of the following organs is an accessory organ of the digestive system?
   a. Heart
   b. Pancreas
   c. Bladder
   d. Spleen

9. (4.3) Chemical digestion begins in the
   a. liver.
   b. stomach.
   c. mouth.
   d. pancreas.
10. (4.3) Tiny, fingerlike projections of the small intestine that absorb nutrients are called
    a. salivary glands.
    b. villi.
    c. lymph nodes.
    d. organelles.
11. (4.3) A lacteal is a
    a. lymph vessel within each villus.
    b. form of carbohydrate in milk.
    c. muscular structure that regulates digestion in the large intestine.
    d. specialized cell.
12. (4.3) Which of the following habits will help Dario increase the diversity of his microbiome?
    a. Consume easy–to–digest ultra–processed foods.
    b. Perform colonic enemas to remove harmful bacteria from the colon.
    c. Consume a high protein, mainstream American ("Western") diet.
    d. Consume a high–fiber Mediterranean–type diet.
13. (4.3) The stomach secretes
    a. salivary amylase.
    b. hydrochloric acid.
    c. bile.
    d. cholecystokinin.
14. (4.3) Peristalsis
    a. is a common intestinal infection.
    b. interferes with lipid absorption in the small intestine.
    c. stimulates red blood cell formation in bone marrow.
    d. helps move food/chyme through the digestive tract.
15. (4.3) Which of the following conditions primarily affects the large intestine?
    a. Ulcerative colitis
    b. GERD
    c. Gallstones
    d. Peptic ulcers
16. (4.4) Which of the following practices increases the risk of peptic ulcer?
    a. Chewing gum
    b. Smoking cigarettes
    c. Drinking orange juice
    d. Eating a low–fiber diet
17. (4.4) Which of the following habits will most likely improve Salmai's chronic GERD?
    a. Limit consumption of alcohol, greasy foods, and tomato products
    b. Lie down after eating meals
    c. Consume spicy foods
    d. Drink caffeinated beverages, such as tea

Answers to Quiz Yourself

1. The atom is the smallest living unit in the body. **False.** (Section 4.2)
2. The stomach produces gastric juice that contains hydrochloric acid. **True.** (Section 4.3)
3. Digestion begins in the stomach. **False.** (Section 4.3)
4. The gut microbiota are the various microbes, primarily the bacteria, that live in the gastrointestinal tract. **True.** (Section 4.3)
5. Symptoms of Irritable Bowel Syndrome (IBS) include constipation and/or diarrhea. **True.** (Section 4.4)

Design Elements: (Stacked books): Siede Preis/Photodisc/Getty Images; (Nuts): C Squared Studios/Photodisc/Getty Images; (Recipe box, Raspberries, Blueberries): ©Wendy Schiff; (Pineapple): Stockbyte/Getty Images; (Radishes): karandaev/iStock/360/Getty Images; (MyPlate icon): US Department of Agriculture or USDA

CHAPTER 5

Carbohydrates

Quiz Yourself

Do "carbs" cause diabetes or unwanted weight gain? Would sweetening your cereal with honey be a healthier choice than using table sugar? Does eating sugary foods make children hyperactive? Check your knowledge of carbohydrates by taking the following quiz. The answers are at the end of the chapter.

1. Compared to table sugar, honey is a natural and far more nutritious sweetener. ______ T ______ F
2. Ounce per ounce, sugar provides more energy than starch. ______ T ______ F
3. Eating a high-fiber diet may reduce your blood cholesterol level. ______ T ______ F
4. The average American consumes 80% of their energy intake as added sugars. ______ T ______ F
5. The results of clinical studies indicate that eating too much sugar makes children hyperactive. ______ T ______ F

Ryan McVay/Photodisc/Getty Images

5.1 Introducing Carbohydrates

Learning Outcomes

1. Identify the two major kinds of carbohydrates in human diets.
2. Explain how plants make carbohydrates.
3. Identify roles of carbohydrates in the body.

When you are bored, excited, or in a good or bad mood, do you reach for something sweet to eat? Can you imagine celebrating birthdays, weddings, or holidays without cakes, candies, or cookies? If you are like many Americans, you enjoy eating sweets and may even describe yourself as having a "sweet tooth."

Why do humans, even newborn infants, prefer foods that taste sweet? The pleasant and sometimes irresistible taste of sugar is a clue that the food contains **carbohydrates.** Carbohydrates are a class of nutrients that includes substances cells can use for energy. Without a steady supply of energy, cells cannot function and they die.

Carbohydrates are generally grouped into simple or complex forms. **Simple carbohydrates** ("sugars" or "simple sugars") include *glucose.* Glucose is a very important simple carbohydrate because it is a primary energy source for human cells. Starches and most kinds of fiber (dietary fiber) are *complex carbohydrates* because they contain many simple carbohydrate molecules. Plants make sugars, starches, and fiber and store these carbohydrates in their stems, leaves, roots, flowers, fruits, and seeds. Plants use the sun's energy to combine the carbon, oxygen, and hydrogen atoms from carbon dioxide (*carbo—*) and water (*—hydrate*) (Fig. 5.1). Energy from the sun is stored in the bonds that hold the carbon and hydrogen atoms together. Human cells can break down some of

carbohydrates class of nutrients that includes glucose, a major source of energy for the body

simple carbohydrates group of carbohydrates that includes sugars

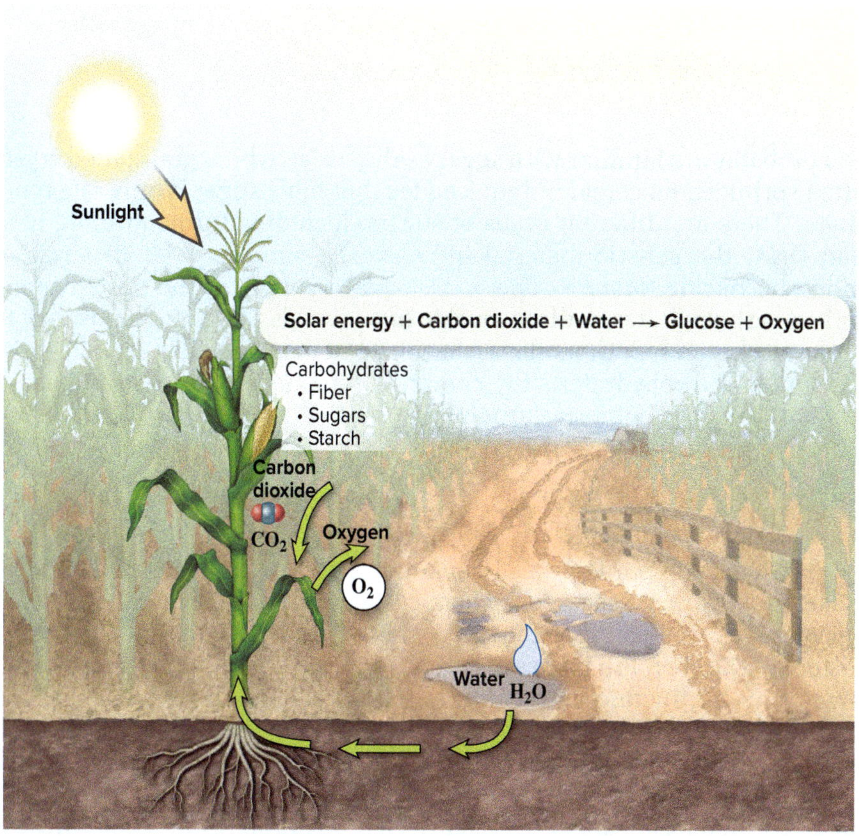

Figure 5.1 Carbohydrates. Plants use the sun's energy to combine carbon, oxygen, and hydrogen atoms from carbon dioxide and water to make glucose. As a result of this process, oxygen gas is released. Plants can use glucose to make fiber, starch, and other sugars.

high-fructose corn syrup (HFCS) caloric sweetener that is often added to food

monosaccharide simple sugar that is the basic chemical unit of carbohydrates

disaccharide simple sugar composed of two monosaccharides

glucose monosaccharide that is a primary fuel for muscles and other cells; "dextrose" or "blood sugar"

those bonds, releasing energy that powers various forms of cellular work, including contracting muscles, making vital compounds, and building bone tissue.

On average, each American consumed about 72 pounds of *added* caloric sweeteners, in 2021.[1] Such sweeteners include honey, maple syrup, and refined sugars, such as **high–fructose corn syrup (HFCS)** and table sugar. Added caloric sweeteners are sources of empty calories that offer little nutritional value other than simple carbohydrates to diets. Although people need some glucose for energy, consuming too many calories from carbohydrates, as well as the other macronutrients, can lead to extra body fat.

This chapter provides information about the major roles of carbohydrates in the body and rich food sources of carbohydrates, including simple sugars and complex carbohydrates.

Concept CHECKPOINT

1. What substances do plants need to make carbohydrates such as the starch in a potato or sugar in a peach?
2. Explain why a plant will die if it is not exposed to light.
3. Why does a human cell need glucose?

See Appendix F for responses.

5.2 Simple Carbohydrates: Sugars

Learning Outcomes

1. List the three most important dietary monosaccharides for humans.
2. Identify the basic monosaccharide components of sucrose, maltose, and lactose.
3. Explain the difference between a nutritive sweetener and a nonnutritive sweetener.
4. List three nutritive and three nonnutritive sweeteners.

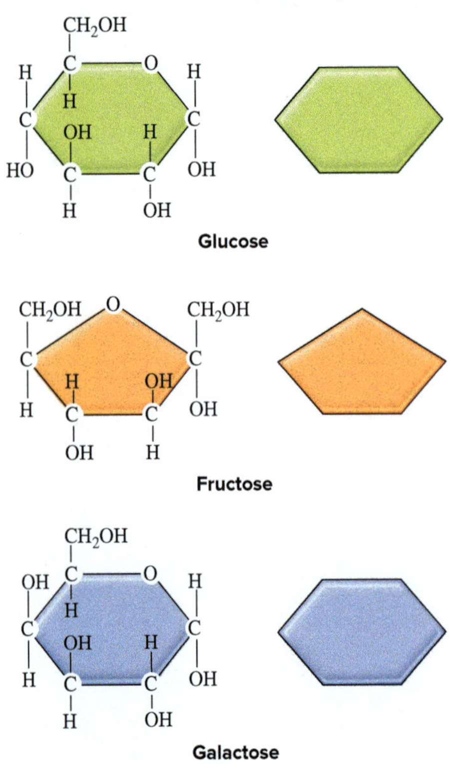

Figure 5.2 Monosaccharides. These chemical symbols for glucose, fructose, and galactose indicate the number and arrangement of carbon, hydrogen, and oxygen atoms. For simplicity, we will omit the atoms and just show the color symbol for each monosaccharide.

You probably are familiar with sugar as the sweet, white, granulated crystals often sprinkled on cereal or into iced tea, but table sugar is only one type of sugar. There are different types of sugars, including forms in milk, blood, and DNA, the genetic material in cells. The simplest type of sugar, the **monosaccharide** (*mono* = one; *saccharide* = sugar), is the basic chemical unit of carbohydrates. A **disaccharide** (*di* = two) is a sugar composed of two monosaccharides. Both monosaccharides and disaccharides are categorized as simple carbohydrates. By combining monosaccharides, plants and animals can form complex carbohydrates.

Monosaccharides

The three most important dietary monosaccharides are glucose, fructose, and galactose. Figure 5.2 shows the chemical structures of glucose, fructose, and galactose, as well as the geometric symbols used to represent them in this textbook. The chemical names of carbohydrates, particularly sugars, often end in *ose.* Gluc*ose,* fruct*ose,* and sucr*ose* are sugars commonly found in foods.

Fruits and vegetables, especially berries, grapes, corn, and carrots, are good food sources of **glucose.** Glucose is the most important monosaccharide in the human body because it is a primary fuel for muscle and other cells. In fact, red blood and nervous system cells, including brain cells, must use glucose for energy under normal conditions. Thus, a healthy body maintains its *blood glucose* levels carefully. Glucose is also called dextrose and may be referred to as blood sugar.

Fructose (fruit sugar or levulose) is naturally found in fruit, honey, and a few vegetables, particularly cabbage, green beans, and asparagus. Because fructose tastes much sweeter than glucose and is easily made from corn, food manufacturers use large amounts of high–fructose corn syrup (HFCS) as a food additive to satisfy Americans' demand for "regular" soft drinks, candies, and baked goods. The body has little need for fructose, but it can convert the monosaccharide to fat or glucose.

Unlike glucose and fructose, **galactose** is not commonly found in foods. Galactose is a component of lactose, the form of carbohydrate in milk. After a female gives birth, special glands in her breasts convert glucose into galactose, which is necessary for production of lactose, the form of carbohydrate in breast milk.

fructose monosaccharide in fruits, honey, and certain vegetables; "levulose" or "fruit sugar"

galactose monosaccharide that is a component of lactose

maltose disaccharide composed of two glucose molecules; "malt sugar"

sucrose disaccharide composed of a glucose and a fructose molecule; "table sugar"

lactose disaccharide composed of a glucose and a galactose molecule; "milk sugar"

Disaccharides

Disaccharides include maltose, sucrose, and lactose. **Maltose** (malt sugar) has two glucose molecules bonded together (Fig. 5.3a). Few foods naturally contain maltose. **Sucrose** (table sugar) consists of a molecule of glucose and one of fructose (Fig. 5.3b). **Lactose** (milk sugar) forms when a galactose molecule bonds to a glucose molecule (Fig. 5.3c). Although most animal foods are not sources of carbohydrate, milk and some milk products, such as yogurt and ice cream, contain lactose.

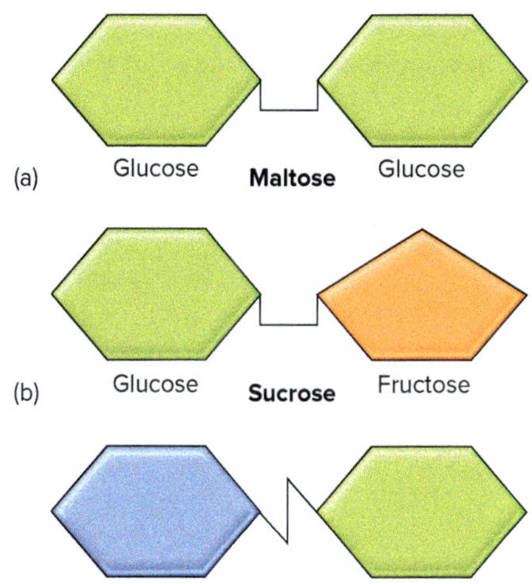

Figure 5.3 Disaccharides.

Sucrose

Although sucrose occurs naturally in honey, maple syrup, carrots, and pineapples, much of the sucrose in the American diet is refined from sugar cane and sugar beets. The refining process strips away the small amounts of vitamins and minerals in sugar cane and sugar beets. "Raw sugar," turbinado sugar, and some forms of brown sugar are not as fully processed from sugar cane as white sugar. These sweeteners contain a small amount of molasses, which contributes to their flavor, color, and nutritional value. Because refined sucrose has the reputation of being a "junk food," some manufacturers have used creative names, such as "evaporated cane juice," to disguise the presence of table sugar in their product's ingredients list. The Food and Drug Administration (FDA) advises food manufacturers to identify the sweetener as "cane sugar" or "sugar" instead of "evaporated cane sugar."[2]

Some people claim that refined white sugar is poisonous and other sweeteners, including honey and maple syrup, are nutritionally superior to table sugar. However, these claims are not true. Table sugar does not contain toxic substances; in fact, it is almost 100% carbohydrate. Table 5.1 compares the nutritional value of honey and maple syrup with certain forms of sucrose. Note that none of the sweeteners is a good source of protein, vitamins, or minerals. The simple sugars in honey and maple syrup are not superior to those that comprise sucrose, and your body does not distinguish whether glucose or fructose came from honey, maple syrup, or table sugar.

A tablespoon of white table sugar and maple syrup is almost 100% sucrose; a tablespoon of honey has glucose, fructose, water, and a small amount of sucrose. A tablespoon of honey contains more protein and micronutrients than a tablespoon of white sugar and maple syrup, but the amounts are insignificant. For example, you would have to eat a cup of honey to obtain 1.0 g of protein, 1.7 mg of vitamin C, and 1.4 mg of iron. That amount of honey supplies over 1000 kcal! Although honey contains phytochemicals, substances in plant foods that may

Did *YOU* Know?

Bees make honey by consuming the sucrose-rich nectar from flowers and digesting most of it into glucose and fructose. The bees regurgitate (vomit) this material within the beehive, and eventually, it is collected by beekeepers for human processing and packaging.

DieterMeyrl/iStock/Getty Images

TABLE 5.1 Nutritional Comparison of Selected Sweeteners

Sugar/Syrup Tablespoon 1	Water %	Kcal	Protein g	Carb g	Vit. C mg	Calcium mg	Niacin mg	Potassium mg	Iron mg	Zinc mg
Honey	17	64	0.06	17.3	0.1	1	0.025	11	0.09	0.05
White granulated sugar	0	49	0	12.6	0	0	0	0	0.01	0
Maple syrup (100% maple)	32	52	0.01	13.4	0	20	0.016	42	0.02	0.29
High-fructose corn syrup (HFCS)	24	53	0	14.4	0	0	0	0	0.01	0
Brown sugar	1	52	0.02	13.5	0	11	0.015	18	0.10	0
Molasses	22	58	0	15.0	0	41	0.186	293	0.94	0.06

Source: Data from USDA, Agricultural Research Service, FoodData Central, *FoodData Central* 2019.

nutritive sweeteners substances that sweeten and contribute energy to foods

added sugars sugars added to foods during processing or preparation

provide health benefits, the amounts are too small to make this sticky sweetener a valuable source of these compounds. Table sugar, maple syrup, and honey are sources of empty calories; thus, their consumption should be limited.

Nutritive and Nonnutritive Sweeteners

Sugars are **nutritive sweeteners** because they are carbohydrates that contribute energy to foods. Each gram of a mono– or disaccharide supplies 4 kcal. **Added sugars** ("added sweeteners"), such as sucrose and HFCS, which is chemically similar to sucrose, are often added to foods during processing or preparation. In baked cereal products, added sugars contribute to the browning and tenderness of the food. Sugar also serves as a preservative by inhibiting the growth of molds and bacteria that would otherwise cause food spoilage. If one of the nutritive sweeteners listed in Table 5.2 is the first or second ingredient listed on

TABLE 5.2 Sugars and Other Nutritive Sweeteners

Agave syrup	Polydextrose
Brown sugar	Raw sugar
Coconut sugar	Rice Syrup
Confectioner's or powdered sugar	Sorbitol*
Corn sweeteners, corn syrup, high-fructose corn syrup (HFCS), cultured corn syrup	Table sugar (sucrose)
Date sugar	Maltodextrin
Dextrose	Turbinado sugar
Fructose (levulose)	Maltose, high-maltose corn syrup
Fruit juice concentrate or concentrated fruit juice sweetener	Xylitol*
Glucose	Mannitol*
Honey	Maple syrup
Invert sugar	Molasses
Lactose	

©Wendy Schiff

*Alcohol forms of sugars

TABLE 5.3 How Much Added Sugar Is in That Food?

Food	Serving Size	Kcal	Approximate Teaspoons Added Sugars*	Added Sugars, g
Energy drink	16 fl oz	212	12.5	50
Chocolate milkshake	11.5 fl oz	410	11	44
Cola, sugar-sweetened, canned	12 fl oz	150	9	36
Sweet tea	12 fl oz	120	8	32
Lemonade	8 fl oz	120	7	28
Apple pie, double crust	⅛ of a 9″ diameter pie	600	6	24
Chocolate candy bar with almonds	1.45 oz	210	4	16
Yogurt, 1%, vanilla low-fat	6 oz	150	4	16
Chocolate chip cookies, commercial brand	1 package (39 g)	200	3.5	14
Sugar-frosted cornflakes	1 cup	150	3.5	14
Chocolate milk (soy beverage)	1 cup	150	3	12
Ice cream, vanilla	½ cup	111	2	8

*One teaspoon added sugar equals 4 grams of sugar.
Source: Data from USDA Agriculture Research Service, *FoodData Central*, 2019.

a product's label, the food probably contains a high amount of added sugar. The amount of added sugar in a food can be reported in teaspoons or grams. Each teaspoon of sugar supplies 4 grams of sugar. For example a chocolate milkshake contains 11 teaspoons of sugar, which equals 44 grams of sugar (11 teaspoons multiplied by 4 grams of sugar per teaspoon equals 44 grams of sugar). Table 5.3 indicates the amounts of added sugars that are in typical servings of commonly consumed foods and beverages. The following section provides information about other kinds of sweeteners.

Alternative nutritive sweeteners include sugar alcohols: *sorbitol, xylitol (zigh′–lih–tol),* and *mannitol.* Unlike sugars, sugar alcohols do not promote dental decay. Thus, these compounds are used to replace sucrose in products such as sugar–free chewing gums, breath mints, and "diabetic" candies. Sugar alcohols are not fully absorbed by the intestinal tract, and as a result, they supply an average of 2 kcal/g. However, sugar alcohols may cause diarrhea when consumed.

Food & Nutrition tips

Honey can contain spores, the inactive life stage, of the deadly bacterium *Clostridium botulinum* that resist being destroyed by food preservation methods. These spores can become active bacteria within an infant's intestinal tract and produce a poison that is extremely toxic to nerves. According to experts at the Centers for Disease Control and Prevention (CDC), honey should not be fed to children younger than 12 months of age or used to sweeten infant foods because it may cause botulism poisoning.[3] Older children and adults can eat honey without being concerned about botulism because the mature stomach produces enough acid to destroy the bacterial spores.

Honey and table sugar have similar nutritional value.
Sugar bowl: MaraZe/Shutterstock; Honey: Burke/Triolo Productions/Brand X Pictures/Getty Images

Nonnutritive Sweeteners

Some people try to control their caloric intake by reducing their consumption of foods and beverages sweetened with nutritive sweeteners such as sugar. **Nonnutritive sweeteners** (also referred to as sugar replacers, sugar substitutes, or "artificial" sweeteners) are substances added to food that sweeten the item while providing few or no kilocalories. **High–intensity sweeteners** are a group of nonnutritive sweeteners that have an extremely sweet taste when compared to the same amount of table sugar. Thus, a very small amount of a high–intensity sweetener is needed to sweeten a food, and such sweeteners supply little or no energy per serving.

Nonnutritive sweeteners may help people control their energy intake and manage their body weight.[4] Consumers, however, need to recognize that most

nonnutritive sweeteners substances that sweeten foods while providing few or no kilocalories

high-intensity sweeteners group of nonnutritive sweeteners that are extremely sweet tasting compared to the same amount of sugar

TABLE 5.4 Comparing Nonnutritive Sweeteners

Sweetener	Comparison to Sugar/ Serving*	Brand Name	Kilocalories/ serving
Aspartame	200 times sweeter	NutraSweet®, Equal®	Nearly 0
Saccharin	200–700 times sweeter	Sweet'N Low®, Sugar Twin®, Necta Sweet®	0
Acesulfame-K	200 times sweeter	Sunett®, Sweet One®	0
Neotame	7000–13,000 times sweeter	Newtame®	0
Sucralose	600 times sweeter	Splenda®	0
Highly purified stevia leaf extracts (rebaudioside A)	200–400 times sweeter	Truvia®, SweetLeaf®, Rebiana	0
Advantame	20,000 times sweeter	Advantame	0
Monk fruit	100–250 times sweeter	Monk Fruit in Raw®, PureLo®	0
Thaumatin	2000–3000 times sweeter	Talin	0

Traci Keck

*Source: U.S. Food and Drug Administration: *Aspartame and other sweeteners in food. 2023.* www.fda.gov/food/food-additives-petitions/aspartame-and-other-sweeteners-food. Accessed: September 2, 2023.

"sugar–free" or "diabetic" foods are not calorie free. Furthermore, results of some studies suggest that artificially sweetened foods and beverages can promote excess calorie consumption. Why? More research is needed, but the taste of products that contain artificial sweeteners may interfere with a person's ability to regulate their intake of sugary foods and beverages.[4]

The FDA allows the use of nine high–intensity sweeteners: saccharin, aspartame, acesulfame–K, sucralose, Neotame, Advantame, and commercially purified extracts (steviol glycosides) from leaves of the stevia plant, extracts from *Siraitia grosvenorii* Swingle fruit (luo han guo or monk fruit), and thaumatin proteins from *Thaumatococcus danielli Katemfe fruit.*[5] Table 5.4 provides information about these high–intensity sweeteners.

In the United States, the safety of nonnutritive sweeteners has been under public and scientific scrutiny for decades. In 1970, the FDA banned the nonnutritive sweetener *cyclamate* after research indicated that the substance caused bladder cancer in mice. In the 1980s, panels of experts at the FDA and the National Academy of Sciences reviewed the scientific evidence and determined that cyclamate did not increase the risk of cancer in humans. Nevertheless, the ban on this food additive continues in the United States. Although *saccharin* has been in use for over 100 years, its safety has also been questioned. Despite the concern, most of the scientific evidence indicates that saccharin is safe when consumed in typical amounts.

Figure 5.4 Warning label for people with PKU. People with PKU need to be concerned about their phenylalanine intake. Artificially sweetened foods that contain aspartame carry this warning on the label because aspartame contains phenylalanine. ©Wendy Schiff

Aspartame, better known by its trade names "NutraSweet®" or "Equal®," consists of phenylalanine and aspartic acid, two amino acids, and the molecules that comprise proteins. Some people must avoid aspartame and certain protein–rich foods because they have phenylketonuria (PKU) (*fen'–nul–keet'–en–yur'–e–ah*), a rare inherited disorder. People with PKU cannot metabolize phenylalanine properly. If an infant with PKU is not treated with a special diet, phenylalanine and its metabolic by–products accumulate in the child's bloodstream and cause severe brain damage.[6] To alert people with PKU about the presence of aspartame in foods, the FDA requires manufacturers of products containing the nonnutritive sweetener to include a warning on the label (Fig. 5.4). The "Real People, Real Stories" feature in Chapter 7 is about a college student who has PKU.

Aspartame has been blamed for causing a variety of health problems, including cancer, certain immune system diseases, and chronic headaches.

TABLE 5.5 Acceptable Daily Intakes for Nonnutritive Sweeteners

Nonnutritive Sweetener	Amount*
Aspartame	75 packets
Saccharin	45 packets
Advantame	4920 packets
Acesulfame-K	23 packets
Sucralose	23 packets
Stevia**	27 packets
Monk fruit***	Not determined
Thaumatin****	Not determined

*Number of table-top packets for a person who weighs 132 pounds
**Steviol glycosides
***Siraitia grosvenorii Swingle
****Thaumatococcus danielli Katemfe fruit

Source: U.S. Food and Drug Administration: *Additional information about high intensity sweeteners permitted for use in food in the United States.* 2018. www.fda.gov/food/food-additives-petitions/additional-information-about-highintensity-sweeteners-permitted-use-food-united-states. Accessed: August 3, 2022.

However, no scientifically reliable studies have linked typical aspartame consumption to any health disorder.[7]

Sucralose, sold under the brand name "Splenda®," is made from a molecule of sucrose that has been chemically modified to escape digestion and absorption. As a result, sucralose sweetens foods and beverages without increasing their caloric value. Because the sweetener is not digested or absorbed by the intestinal tract, it is excreted in feces unchanged. During normal cooking conditions, sucralose is not destroyed by heat, a feature that makes it better for sweetening baked products than aspartame.

Do nonnutritive sweeteners pose a danger to health? A group of international health and safety organizations, including the FDA, has established *Acceptable Daily Intakes* (*ADI*) for certain nonnutritive sweeteners (Table 5.5). In July 2023, the experts from the International Agency for Research on Cancer (IARC), the Joint Food and Agriculture Organization (FAO), and World Health Organization (WHO) expert committee on food additives issued a statement that classified aspartame as "possibly carcinogenic to humans" based on limited, but not convincing evidence. However, the committee did not raise safety concerns for the current use of aspartame in the food supply and did not change the ADI for aspartame. According to the Academy of Nutrition and Dietetics, nonnutritive sweeteners are safe when consumed "within acceptable daily intakes, even during pregnancy."[8] Nevertheless, researchers continue to investigate the safety and usefulness of artificial sweeteners.

Concept CHECKPOINT

4. Identify the three most important dietary monosaccharides.
5. What are the chemical names for blood sugar, table sugar, milk sugar, and malt sugar? Which monosaccharides form each molecule of maltose, lactose, and sucrose?
6. What is the difference between a nutritive sweetener and a nonnutritive sweetener?
7. Parents of a child with PKU can give their child either a beverage sweetened with sucralose or one containing aspartame. Which drink should they choose? Explain your answer.

See Appendix F for responses.

Did *YOU* Know?

Wheat, jackfruit, figs, and raisins naturally contain small amounts of a rare monosaccharide called allulose. The chemical structure of allulose is similar to fructose. Because extracting large amounts of this sugar from foods is difficult, food scientists use enzymes to convert fructose into allulose. Food manufacturers add allulose to sweeten baked, frozen, and liquid foods and beverages. When compared to sucrose, allulose has a similar texture, but it provides only 70% of the sweetness.

Jackfruit naturally contains allulose.
Dinodia/Pixtal/age fotostock

Although the human intestinal tract absorbs allulose, the body metabolizes very small amounts of the sugar. Therefore, the monosaccharide does not increase blood glucose levels and supplies little energy (0.4 kcal/g).[9] After being absorbed, allulose circulates in the bloodstream, and the kidneys eliminate the simple sugar in urine.

In October 2020, the FDA determined that allulose did not have to be declared as an added sugar on the Nutrition Facts panel when used as a product ingredient.[10] Nevertheless, the amount of allulose per serving must be included in the declaration of total grams of carbohydrate line of the nutrition label.

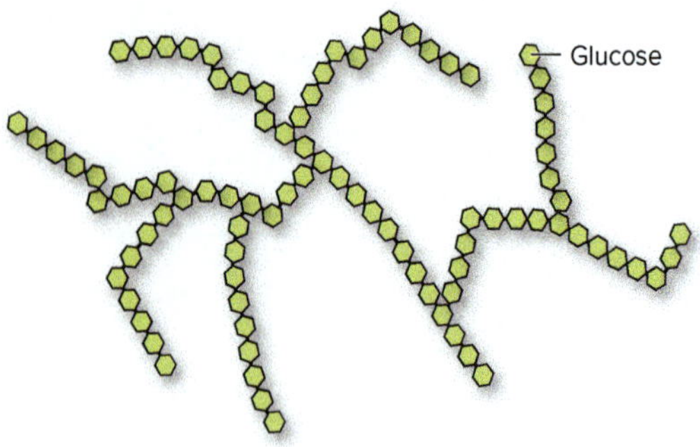

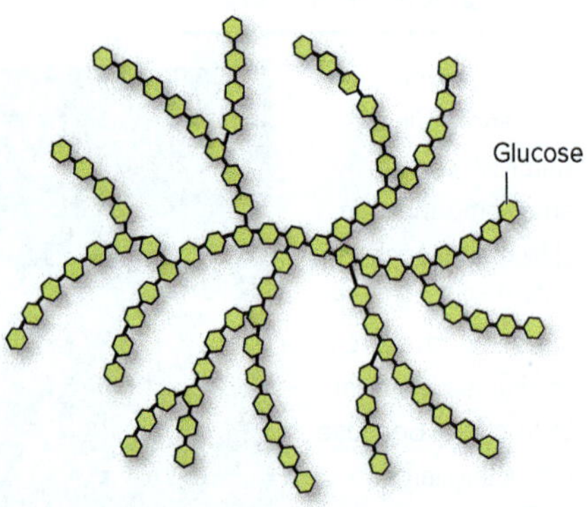

Figure 5.5 Starch and glycogen. (a) *Amylopectin* is the major form of starch made by plants. (b) The chains of glycogen are more highly branched than those of amylopectin.

5.3 Complex Carbohydrates

Learning Outcomes

1. Explain the difference between a simple sugar and a polysaccharide.
2. Explain the difference between starch and glycogen.
3. Identify rich food sources of starch, soluble fiber, and insoluble fiber.

Complex carbohydrates (polysaccharides) are composed of monosaccharides bonded together, forming large chainlike structures. Plants and animals use complex carbohydrates to store energy or make certain structural components such as stems and leaves. The most common dietary polysaccharides consist of hundreds of glucose molecules and include digestible and nondigestible forms.

Starch and Glycogen

Starch and **glycogen** are polysaccharides that contain hundreds of glucose molecules bound together (Fig. 5.5). Plants store glucose as starch, primarily in the form of amylopectin, in their seeds, roots, and fleshy underground stems called tubers (Fig. 5.5a). Rich food sources of starch include bread and cereal products made from wheat, rice, barley, and oats; vegetables such as corn, squash, beans, and peas; and tubers such as potatoes, yams, taro, cassava, and jicama. Sports drinks and sports or energy bars often include *modified starches* such as maltodextrin and dextrin. Regardless of its source, each gram of starch supplies 4 kcal.

The human body stores limited amounts of glucose as glycogen (see Figure 5.5b). Muscles and the liver are the major sites for glycogen formation and storage. Although muscles contain glycogen, most animal foods (for example, meat or the flesh of fish and poultry) are not sources of this complex carbohydrate because muscle glycogen breaks down soon after an animal dies.

Fiber

In addition to storing energy as starch, plants use complex carbohydrates to make supportive structures and protective seed coats that contribute to the fiber content of your diet. Most forms of **dietary fiber (fiber)** are complex carbohydrates composed of monosaccharides. The monosaccharides are connected by bonds that humans cannot digest (see Figure 5.6). Cellulose, hemicellulose, pectin, gums, beta–glucans, and mucilages are polysaccharide forms of fiber. Lignin is the only type of fiber that is not carbohydrate. Although fiber is not digested, it may be metabolized (fermented) by gut microbiota (see Chapter 4). Any fiber that is not used by the microbes contributes to stools (fecal residue) that are eventually eliminated during bowel movements.

There are two types of dietary fiber, **soluble fiber** and **insoluble fiber.** Soluble types of fiber, such as pectins, gums, and beta–glucans, dissolve or swell in water. Insoluble forms of fiber, such as cellulose and lignin, generally do not dissolve in water. Oat bran and oatmeal, beans, apples, carrots, oranges and other citrus fruits, and psyllium (*sill'–e–um*) seeds are rich sources of soluble fiber; whole–grain products, including brown rice, contain high amounts of insoluble fiber. Table 5.6 provides information about the

complex carbohydrates (polysaccharides) compounds composed of monosaccharides bonded together, forming large chainlike structures

starch storage polysaccharide in plants

glycogen storage polysaccharide in animals

dietary fiber (fiber) nondigestible plant material; most types are polysaccharides

soluble fiber forms of dietary fiber that dissolve or swell in water

insoluble fiber forms of dietary fiber that generally do not dissolve in water

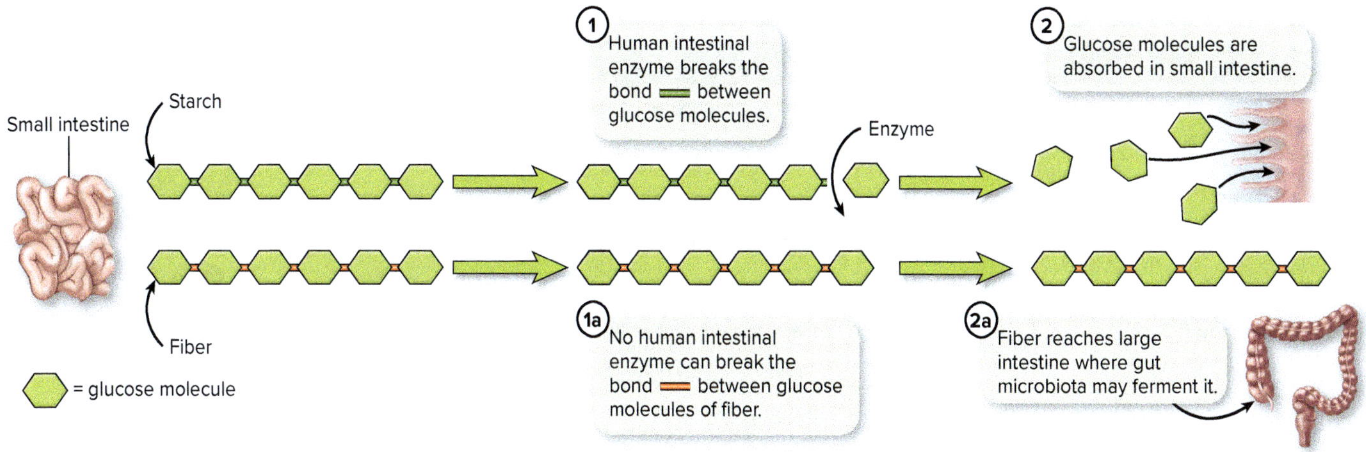

Figure 5.6 Starch digestion and fiber fermentation. Humans do not make intestinal enzymes that digest fiber such as cellulose.

solubility of various types of fiber, effects of fiber in the body, and major food sources of soluble and insoluble fiber. Although the foods listed in Table 5.6 are rich sources of either soluble or insoluble fiber, plant foods usually contain both forms.

Sources of starch, including grains, lose much of their fiber contents when they undergo refinement. Such grains may include wheat, buckwheat, oats, corn, rice, wild rice, rye, barley, bulgur, millet, and sorghum. If a whole-grain product is made from ground, cracked, or flaked cereal grains, the food must contain the starchy endosperm, oily germ, and fiber-rich bran seed components in the same relative proportions as they exist in the intact grain (see Figure 1.11).[11]

Although fiber is not digested by humans, soluble and insoluble fiber provide important health benefits. Soluble fiber can help reduce blood cholesterol levels, and insoluble fiber may ease bowel movements. The "Carbohydrates and Health" section of this chapter provides information about the benefits of adding more fiber to your diet and practical ways to increase your fiber intake.

Table 5.7 lists common foods that are sources of dietary fiber. Note that only plant foods provide fiber; animal flesh contains muscle fibers, which are digestible proteins.

Did *YOU* Know?

The fiber content of different forms of a food can vary widely. For example, an unpeeled raw apple that weighs 6 ounces (about 3 inches in diameter) has 4.4 g of fiber. However, 6 ounces of applesauce contains 2.0 g of fiber, and a 6-fluid ounce serving of apple juice provides only 0.4 g of fiber.

Tadeusz Wejkszo/iStock/360/Getty Images Movingmoment/123RF

TABLE 5.6 Classifying Fiber

Type	Component(s)	Physiological Effects	Food Sources
Insoluble	Cellulose, hemicelluloses	Increases fecal bulk and speeds fecal passage through GI tract	All plants Wheat, rye, brown rice, vegetables
	Lignin	Increases fecal bulk, may ease bowel movements	Whole grains, wheat bran
Soluble	Pectins, gums, beta-glucans, mucilages, some hemicelluloses	Delays stomach emptying; slows glucose absorption; can lower blood cholesterol	Apples, bananas, citrus fruits, carrots, oats, barley, psyllium seeds, beans, and thickeners added to foods

TABLE 5.7 Dietary Fiber Content of Common Foods

Food	Fiber (g)	Food	Fiber (g)
Split peas, cooked (1 cup)	16.3	Beans, green snap, cooked (1 cup)	4.0
Black beans, cooked (1 cup)	15.0	Banana, sliced (1 cup)	3.9
Oat bran, raw (1 cup)	14.5	Prunes, dried uncooked (5 prunes)	3.5
Kidney beans, canned (1 cup)	11.0	Orange, raw (1 orange)	3.4
Chickpeas, canned (1 cup)	10.6	Strawberries, raw, sliced (1 cup)	3.3
Baked beans, canned (1 cup)	10.4	Carrots, raw, grated (½ cup)	3.1
Kellogg's All-Bran® cereal (½ cup)	9.9	Barley, cooked (½ cup)	3.0
Raspberries, raw (1 cup)	8.0	Peanuts (1 ounce)	2.7
Blackberries (1 cup)	7.6	Broccoli, chopped, cooked (½ cup)	2.6
Green peas, frozen, cooked (½ cup)	7.2	Whole-grain bread, toasted (1 slice)	1.9
Kellogg's Raisin Bran® (1 cup)	6.7	Romaine lettuce, shredded (1 cup)	1.0
Apple, raw with skin 3″ diameter	4.4	Iceberg lettuce, shredded (1 cup)	0.9
Baked potato, with skin (approx. 6½ oz)	4.0	White bread (1 slice)	0.8

Source: Data from USDA Agricultural Research Service, *FoodData Central*, 2019.

Carrots: C Squared Studios/Photodisc/Getty Images; Lettuce: Burke Triolo Productions/Artville/Getty Images; Raspberries: Burke/Triolo Productions/Getty Images; Potato: Christopher Kerrigan/McGraw Hill; Banana: ©Wendy Schiff; Strawberries: Burke/Triolo Productions/Artville/Getty Images

Concept CHECKPOINT

8. What is starch? What is glycogen?
9. What is dietary fiber? Identify at least two rich food sources of soluble fiber and two rich food sources of insoluble fiber.

See Appendix F for responses.

5.4 What Happens to Carbohydrates in the Body?

Learning Outcomes

1. Describe the major steps involved in digesting and absorbing starches and sugars.
2. Explain how the body regulates blood glucose.
3. Explain what happens when cells do not have glucose to use for energy.

If you eat cooked oatmeal made with milk and sweetened with a little brown sugar for breakfast, what happens to the carbohydrates in these foods? The carbohydrates in oats are primarily starch and fiber; mixing milk and brown sugar with the cereal adds lactose and sucrose. Figure 5.7 summarizes what happens to carbohydrates in the digestive tract.

The small intestine is the main site for carbohydrate digestion and absorption, but a minor amount of starch digestion begins in the mouth, as **salivary amylase** converts some of the oat starch molecules into maltose. Salivary amylase does not function when it is in an acid environment. Thus, starch digestion stops soon after the food enters the stomach.

salivary amylase enzyme secreted by salivary glands that begins starch digestion

pancreatic amylase enzyme secreted by the pancreas that breaks down starch into maltose molecules

Starch digestion resumes in the small intestine. In the small intestine, an amylase secreted by the pancreas **(pancreatic amylase)** breaks down the remaining polysaccharides in oat starch into maltose molecules. The enzyme **maltase** digests maltose into glucose molecules. The final products of starch digestion, glucose molecules, are absorbed into intestinal blood vessels and transported to the liver via the hepatic portal vein. Under normal conditions, the process is very efficient and nearly all the starch is digested. The complex carbohydrates that remain are primarily forms of fiber.

The molecules of sucrose in brown sugar and lactose in milk are too large to enter the bloodstream directly from the intestinal tract. The small intestinal enzyme **sucrase** splits each sucrose molecule, forming one glucose and one fructose molecule in the process (see Figure 5.7). Additionally, the enzyme **lactase** breaks down the lactose from milk into glucose and galactose molecules. Intestinal cells absorb galactose, fructose, and glucose, and the hepatic portal vein transports them to the liver. The liver converts fructose and galactose into glucose. The liver can use glucose to make glycogen or fat, but if the body needs energy, the organ releases glucose into the bloodstream.

The small intestine of humans does not make digestive enzymes that digest fiber, including the fiber in oats. Dietary fiber eventually enters the large intestine and "friendly" intestinal bacteria that reside in the large intestine can break down (ferment) the soluble fiber and metabolize the fermentation products for energy. Soluble fiber is sometimes referred to as *viscous fiber* because it usually forms a gooey, semisolid mass in the intestinal tract that is rapidly fermented by bacterial action. On the other hand, insoluble or

maltase enzyme that splits maltose molecule

sucrase enzyme that splits sucrose molecule

lactase enzyme that splits lactose molecule

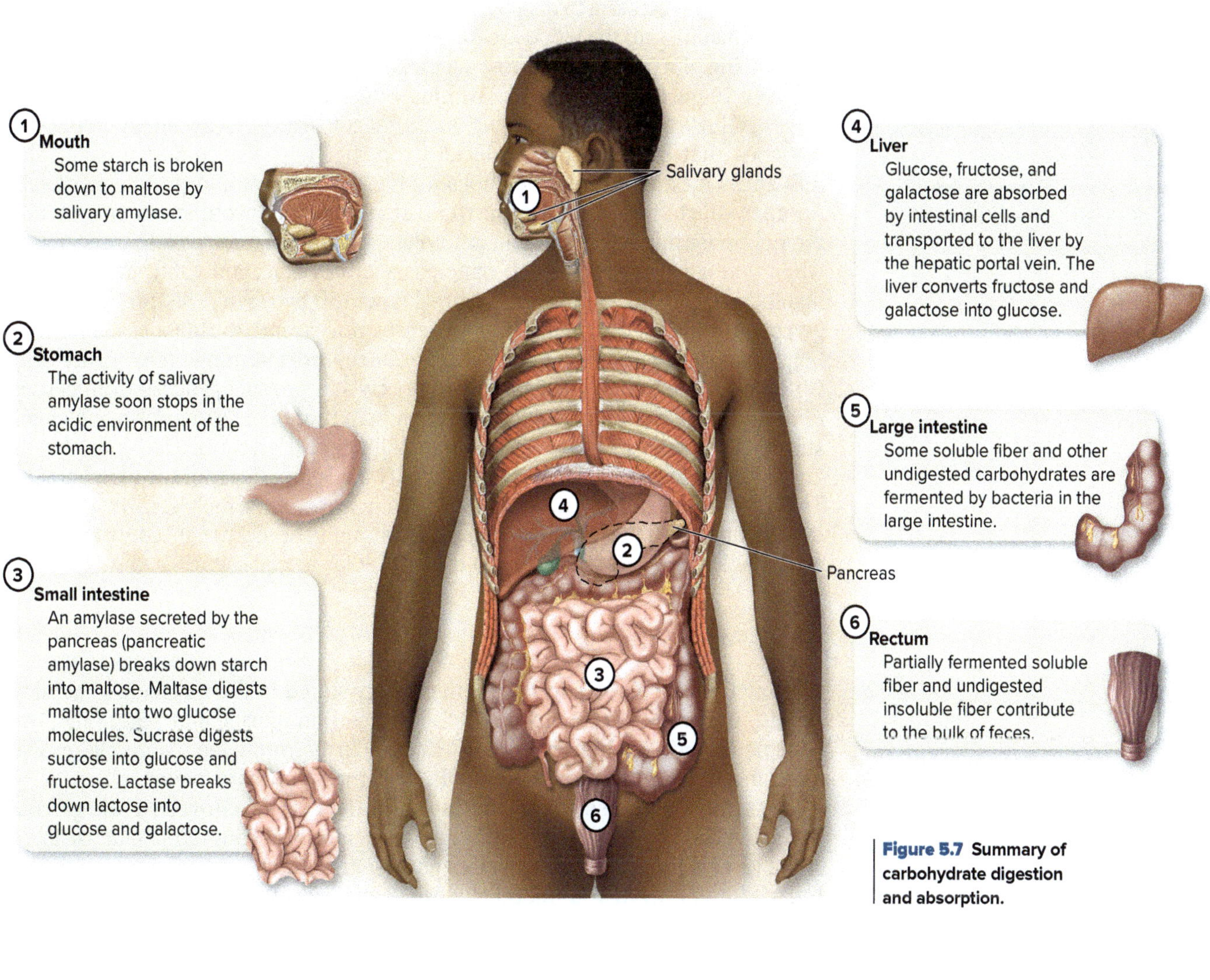

Figure 5.7 Summary of carbohydrate digestion and absorption.

fermentation–resistant fiber does not break down completely and, as a result, contributes to softer and easier–to–eliminate bowel movements.

At one time, scientists thought fiber was a nonnutrient because it had no nutritional value. Recent scientific evidence indicates that the body, particularly the cells that line the large intestine, can use by–products produced by the bacterial metabolism of fiber for energy. According to estimates, a gram of fiber adds 2 kcal to human diets.[12] The average American who is 2 years of age and older consumes only 16.4 g of fiber daily; therefore, fiber contributes relatively little to a typical person's energy intake.[13]

Did *YOU* Know?

As mentioned in Chapter 4, prebiotics are forms of carbohydrate in plants that are poorly digested by humans but can be metabolized by beneficial gut microbes (probiotics). The nondigestible carbohydrates are short-chain polysaccharides composed of several monosaccharides bound together in chains (*oligosaccharides*). Fructans are oligosaccharides or other polysaccharides that contain mostly fructose molecules. Fructans include inulin and fructooligosaccharides (FOS). Galactooligosaccharides (GOS) are oligosaccharides that have mostly galactose molecules in their short chains.

When the FOS and GOS that have escaped digestion in the small intestine reach the colon, they support and promote the growth of the probiotics, especially lactobacilli *(lak'-toe-bah-sill'-ee)* and bifidobacteria *(biff'-ih-doe-bak-tear'-ee-ah)*.[14] Certain lactobacilli secrete chemicals called postbiotics that prevent the growth of harmful bacteria, viruses, and fungi in the colon. Benefits of FOS and GOS include increased calcium absorption, reduced constipation, and protection against infections.[15] Scientists are studying other potential health benefits of adding oligosaccharides as well as combinations of oligosaccharides and probiotics to a wide variety of foods. The FDA advises manufacturers to declare certain nondigestible carbohydrates, including inulin, as dietary fiber on Nutrition Facts panels.

Maintaining Blood Glucose Levels

Glucose is such an important cellular fuel that its blood level is carefully maintained by hormones. Hormones are chemicals that convey messages concerning specific responses to target cells. The pancreas, a digestive system organ shown in Figure 4.30, contains beta cells, clusters of special cells that produce **insulin,** and groups of alpha cells that produce **glucagon.** These two hormones play key roles in regulating blood glucose levels. Figure 5.8 illustrates the effects of insulin and glucagon on blood glucose levels in a healthy person, after a meal that is followed by fasting.

Although individual blood glucose levels vary, normal blood glucose levels after sleeping overnight or not eating for several hours (fasting) are between 70 mg/dl and 99 mg/dl. After you eat digestible carbohydrates as part of a meal or snack, your intestinal tract absorbs the glucose from the food and it eventually enters the bloodstream. In response to the rising blood glucose level, your pancreas releases insulin into the bloodstream. Insulin helps lower your blood glucose level by allowing glucose to enter most cells. A healthy person's blood glucose level is between 70 mg/dl and 140 mg/dl within 2 hours after eating (see Figure 5.8).

Insulin stimulates your liver to store glucose as glycogen by activating enzymes that bind glucose molecules together (see Figure 5.8). Insulin also influences fat and protein metabolism by promoting fat and protein synthesis. Another effect of insulin's action: You feel satisfied with your snack or meal and are no longer hungry. When you ignore your hunger signals and do not eat, and when you sleep overnight, the pancreas secretes glucagon. Glucagon opposes insulin's effects by promoting the breakdown of glycogen. This process, called **glycogenolysis** (*lysis* = break down), releases glucose into the bloodstream, which helps maintain your blood glucose level in the normal range (see Figure 5.8). Glucagon also stimulates liver and kidney cells to produce glucose from certain *amino acids,* the basic molecules that make up proteins. Furthermore, glucagon stimulates **lipolysis** (*lipo* = fat), the breakdown of triglyceride (fat) into *glycerol* and *fatty acids.* As a result, glycerol and fatty acids rapidly enter the bloodstream. The liver uses glycerol to produce glucose, and most cells, including muscle cells, can metabolize fatty acids for energy. Although the body can convert certain amino acids into glucose, it cannot use fatty acids to make glucose.

Glucose for Energy

Cells metabolize glucose to release the energy stored in the molecule's chemical bonds. As a result of this process, cells form carbon dioxide and water (Fig. 5.9). Glucose is a primary fuel for the body's cells. Furthermore, red blood cells, as well as brain and other nervous system cells, burn mostly glucose for energy.

As previously discussed, fructose and galactose are converted into glucose in the liver. What happens to liver glucose depends on your blood glucose level. If the level is low, your liver releases glucose into the bloodstream, so cells can burn it for energy. If your blood glucose level is normal and you are resting, the

insulin hormone that helps regulate blood glucose levels

glucagon hormone that helps regulate blood glucose levels

glycogenolysis glycogen breakdown

lipolysis fat breakdown

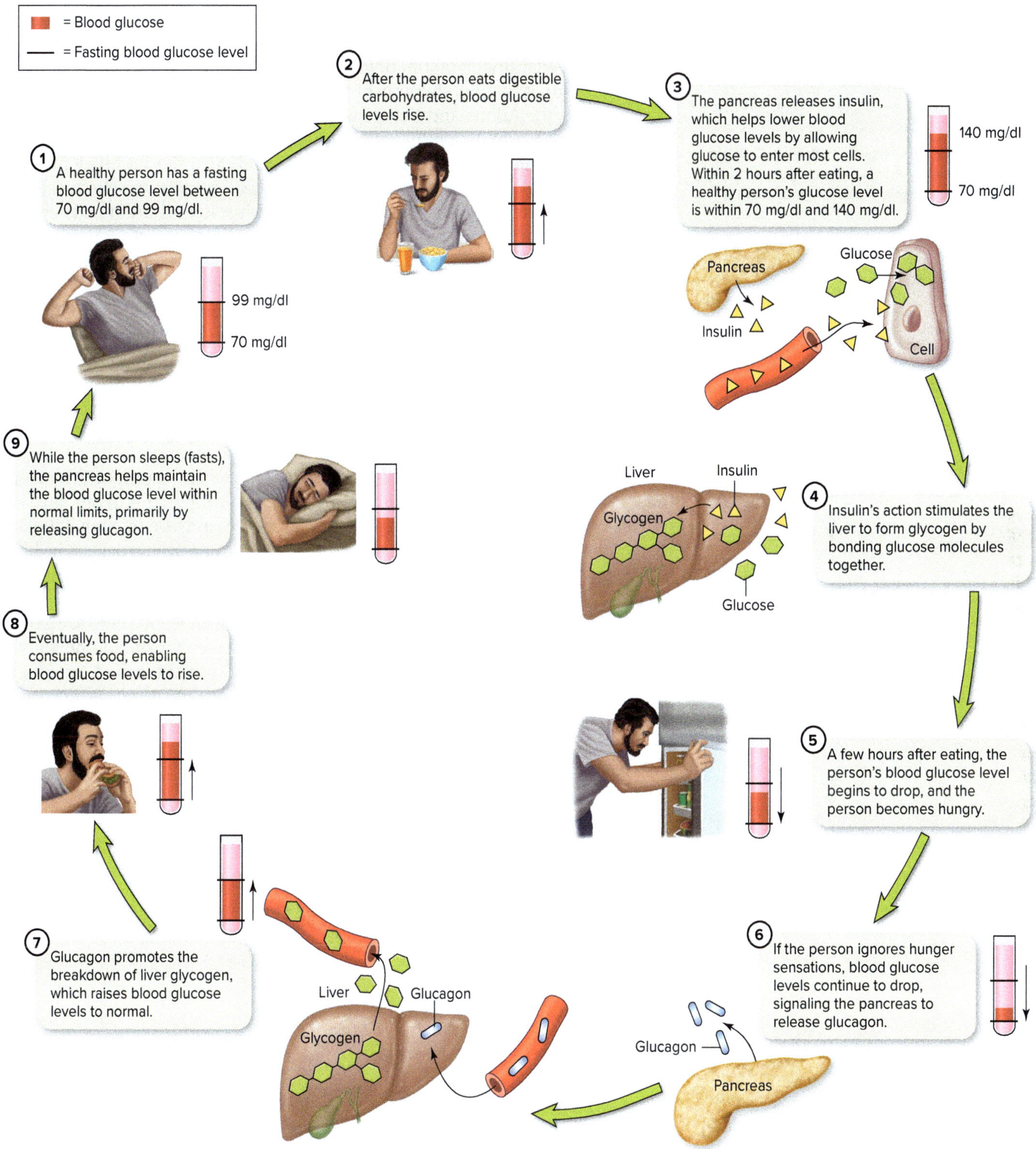

Figure 5.8 Regulating blood glucose. Insulin and glucagon are key hormones in maintaining normal blood glucose levels.

excess glucose is stored as glycogen (see Figure 5.5b). When glycogen storage is at maximum capacity, your liver converts the extra glucose into fat. The fat can be released into the bloodstream; adipose (*ad'–eh–pose*) cells (fat cells) remove and store the fat. In people who are obese and/or have insulin resistance, the fat that forms in the liver can remain and accumulate in the organ.

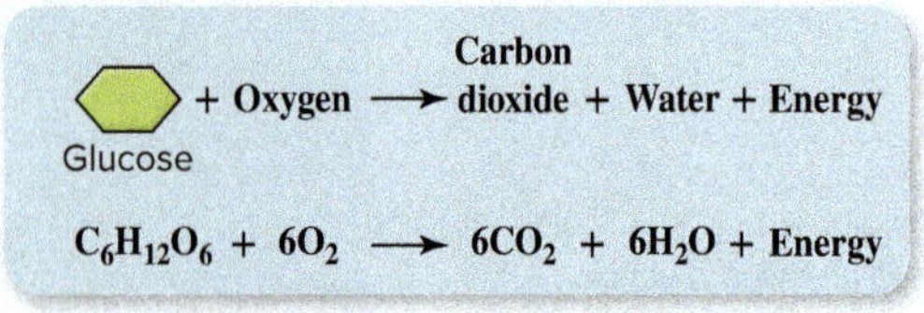

Figure 5.9 Releasing energy from glucose. Cells use oxygen to release the energy stored in glucose. As a result of this process, cells produce carbon dioxide and water.

This abnormal condition is called *nonalcoholic fatty liver disease,* which is discussed in a later section of this chapter.

Cells need a small amount of glucose to metabolize fat for energy properly. When a person is fasting, starving, or following a very low–carbohydrate/high–protein diet (the Atkins and Ketogenic diets, for example), their cells must use greater–than–normal amounts of fat for energy. Under these conditions, there is not enough glucose available for cells to metabolize the fat efficiently, and cells form more *ketone bodies* than usual. **Ketone bodies** are chemicals that result from the incomplete breakdown of fat. Muscle and brain cells can use ketone bodies for energy, but a condition commonly called *ketosis* occurs when these compounds accumulate in the blood. In cases of poorly controlled type 1 diabetes, the production of ketone bodies can be so excessive that the blood becomes acidic, and **ketoacidosis** results. If not treated, ketoacidosis can cause loss of consciousness and even death. The "Type 1 Diabetes" section of this chapter provides more information about ketoacidosis.

The Recommended Dietary Allowance (RDA) for carbohydrate is 130 g/day.[16] This amount of carbohydrate is enough to prevent ketosis. (The RDAs and other DRIs have been discussed in Chapter 3.) To estimate your daily carbohydrate intake, complete the Personal Dietary Analysis at the end of this chapter.

Under normal conditions, human cells obtain a small proportion of their energy needs by converting certain amino acids, the components of proteins, into glucose. Starvation, however, dramatically alters the body's energy metabolism. Starvation diets lack sources of energy such as glucose and amino acids. The body, however, needs glucose to fuel vital activities such as breathing, transmitting nerve impulses, and pumping blood. To meet the body's energy needs, the starving person's skeletal muscles sacrifice amino acids from their proteins for glucose production. Using muscle proteins as a source of energy extends the starving person's survival time but results in muscle wasting, weakness, and, eventually, death. Chapter 7 provides information about proteins.

Concept CHECKPOINT

10. Ishaan ate some whole-wheat roti bread with fruit jam spread for a snack. As this snack passed through his digestive tract, discuss what happened to the starch, sucrose, and fiber in the food.

11. What effects do insulin and glucagon have on blood glucose levels?

12. What is a ketone body? Under what conditions does the body form excessive ketone bodies?

See Appendix F for responses.

5.5 Carbohydrate Consumption Patterns

Learning Outcomes

1. Identify the major sources of added sugars in American diets.
2. List three practical ways to reduce a person's added sugar consumption.

In developing nations, millions of people rely on diets that supply the majority of their energy from relatively unprocessed carbohydrates, especially complex carbohydrates from whole grains, beans, potatoes, corn, and other starchy vegetables. In industrialized nations, people tend to eat more highly refined starches and added sugars. According to the *Dietary Guidelines for Americans*, nutritionally adequate diets should provide amounts of fat that are within

ketone bodies chemicals that result from incomplete fat breakdown

ketoacidosis condition that occurs when the body forms excessive ketone bodies

the Acceptable Macronutrient Distribution Range (AMDR: 45 to 65% of total calories).[16] The typical diet of adult Americans supplies 48% of calories from carbohydrates, much of which is sugars (nutritive sweeteners) added to foods during processing (Fig. 5.10).

As mentioned in Chapter 1, added sugars are sources of empty calories. In 2021, the average American consumed about 21 teaspoons (88 g) of added sugars per day, which is almost 18% of the energy in a diet that supplies 2000 kcal/day.[1] People should limit their intake of added sugars to no more than 10% of their total calorie intake. For example, a diet that supplies about 2000 kcal/day, this is 12 teaspoons (48 g) added sugar per day (about 200 kcal).[17]

A 12–fluid ounce can of a cola–flavored, sugar–sweetened soft drink contains about 37 g of added sugar.[18] Recall, each gram of carbohydrate, and thus sugar, supplies 4 kcal, so this soft drink contributes about 150 kcal of added sugars to the diet. By drinking only one can of the cola, a person who needs 2000 kcal daily will meet three–quarters of their daily limit for the intake of added sugars.

A 12–fluid ounce serving of 100% orange juice supplies about the same amount of carbohydrate as 12 fluid ounces of a sugar–sweetened cola. Because they both contain simple sugars, should you drink regular soft drinks instead of fruit juices? Unlike colas and other soft drinks, 100% fruit juices, such as orange, grapefruit, and cranberry juice, contribute water–soluble vitamins and antioxidant phytochemicals to your diet. Thus, the sugars that are naturally in fruit are not considered added sugars (empty calories). However, 100% fruit juices are rich in calories and a can easily be over consumed and contribute to excess kcals and weight gain.

The main sources of added sugars in Americans' diets are sugar–sweetened beverages, snacks, and "sweets."[17] Sugar–sweetened beverages include carbonated "regular" soft drinks, fruit "ades," flavored vitamin–water products, and sports and energy drinks. Although energy drink consumption has increased, carbonated soft drink consumption has declined since 1999–2000. How many fluid ounces of energy, sports, and/or sugar–sweetened soft drinks do you consume each day?

Average Percentage of Calories per Person (Adult Americans, One Day)

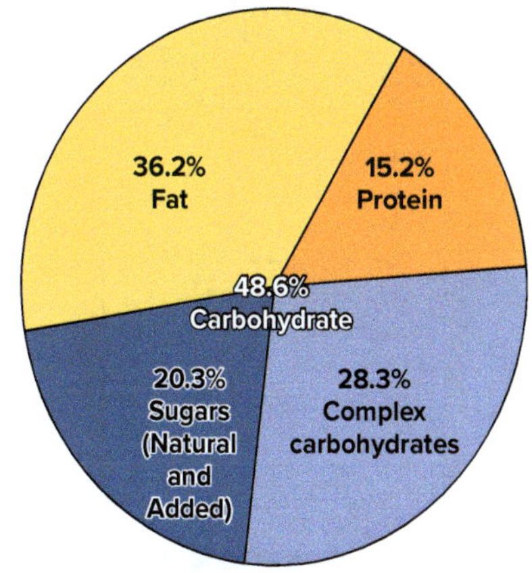

Figure 5.10 Average macronutrient intakes. This graph shows the adult American's average intake of macronutrients as approximate percentages of total Calories (2080) from macronutrients on one day in 2017 to March 2020. (Percentage of total energy from alcohol is not included.)

Source: U.S. Department of Agriculture, Agricultural Research Service: Nutrient intakes from food and beverages: Mean amounts consumed by individuals: By gender and age. *What We Eat in America, NHANES 2017–March 2020.* 2023. www.ars.usda.gov/ARSUserFiles/80400530/pdf/usual/Usual_Intake_gender_WWEIA_2017_March%202020.pdf September 4, 2023.

Increasing Your Intake of Healthy Carbohydrates

Sugar–sweetened beverages, cookies, chips, and many other types of processed snack foods contain large amounts of refined starches and added sugars. Such foods may satisfy your hunger and thirst, but they may be crowding out more nutritious items from your diet. If you frequently purchase foods and beverages from vending machines, convenience stores, or fast–food restaurants, you are probably eating high amounts of unhealthy carbohydrates. Some fast–food restaurants and college cafeterias sell plain yogurt, fresh fruit, fat–free or low–fat milk, and 100% fruit juices. With a little planning, you can prepare your own portable, tasty, and nutritious snacks. For example, place whole fresh fruit or small plastic containers filled with chunks of fresh fruit and pieces of vegetables into your purse or backpack to eat during the day. At home, keep a bowl of fresh grapes, pears, bananas, or other easy–to–eat fruit available for handy snacks, and eat fresh fruit for dessert. Most fruits contain a variety of antioxidants, and they have less fat and more fiber, vitamins, and minerals than pastries or chips.

Food & Nutrition tips

- Replace soft drinks with plain water.
- Make plain water more interesting to drink by adding to it a slice of lemon or lime, or a few fresh or frozen berries.
- Add 1 part club soda to 1 part orange or other 100% fruit juice to make a refreshing carbonated drink.
- Read the label for information about juice content when selecting a fruit juice product. Fruit "drinks," "punches," "blends," "cocktails," or "ades" often contain added sugars and only 10% fruit juice.

Concept CHECKPOINT

13. What are the main sources of added sugars in American diets?

14. Instead of drinking orange juice, should you choose a beverage called "Orange-ade"? Explain why or why not.

See Appendix F for responses.

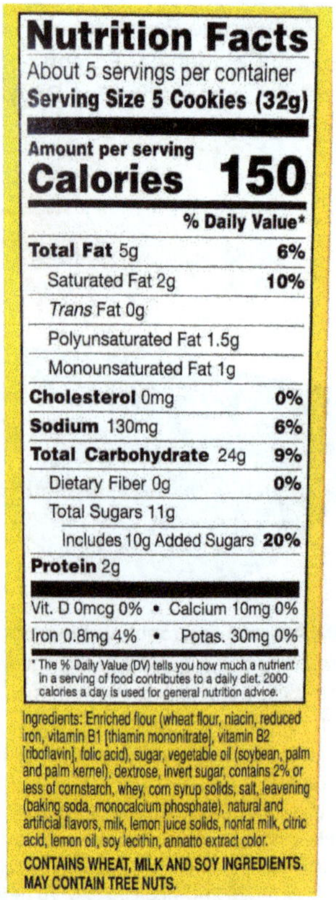

Figure 5.11 **Nutrition Facts panel.**
©Wendy Schiff

INGREDIENTS: Whole-wheat flour, Water, Brown sugar, Wheat gluten, Cracked wheat, Wheat bran, Yeast, Salt, Molasses, Soybean oil, Calcium propionate (preservative), Mono–and diglycerides, Lecithin, Reduced-fat milk

Figure 5.12 **Ingredients list.** Note the different sources of carbohydrate that are in these crackers, compared to the label in Figure 5.11.

5.6 Understanding Nutrient Labeling: Carbohydrates and Fiber

Learning Outcome

1. Compare carbohydrate contents of packaged foods by using the Nutrition Facts panel.

Consumers can learn how much total carbohydrate, sugars, and dietary fiber are in packaged foods and beverages by reading the Nutrition Facts panel of a food label. As indicated in Figure 5.11, grams of total carbohydrate in a serving (5 cookies) are listed, and under it, grams of fiber, total sugar, and added sugars.

According to the label shown in Figure 5.11, 24 g of total carbohydrate are in one serving; of this amount, there are 11 g of total sugars and 10 g of added sugars. Each serving supplies no grams of fiber. How can you estimate the grams of starch in the serving of the food? In this example, add the number of grams of total sugars (11 g) with that of fiber (0 g) and subtract this amount from grams of total carbohydrate (24 g), and you will find that starch comprises 13 g of the carbohydrate in a serving of the food.

Recall from Chapter 3 that the Daily Values on a Nutrition Facts panel can be used to compare the nutrient content of packaged foods. A serving of food that supplies 5%DV or less is a low source of a nutrient, and a serving of food that supplies 20%DV or more is a high source of a nutrient. What is your assessment of the added sugars in the product shown in figure 5.11?

If you are interested in the kinds of added sugars used to make a product, read the ingredients list at the bottom of the panel. Consider the ingredients used to make crackers (Fig. 5.12). Can you identify the sources of fiber and added sugars in the product?

Concept CHECKPOINT

15. According to the Nutrition Facts panel, a serving of ready-to-eat cereal contains 44 g of total carbohydrate, 5 g of dietary fiber, 15 g of total sugars, and 3 g of added sugars. Estimate the grams of starch in the serving of the cereal.

16. You compare the Nutrition Facts panel for two packaged foods. One provides 25%DV for added sugars, and the other provides 8%DV for added sugars. Which product is the better choice? Why?

See Appendix F for responses.

5.7 Carbohydrates and Health

Learning Outcomes

1. Discuss the effects of excess carbohydrate consumption on health.
2. Discuss differences between type 1 and type 2 diabetes, including treatments.
3. List at least three signs or symptoms of diabetes.
4. List at least four risk factors for developing type 2 diabetes and ways to reduce the likelihood of developing the disease.
5. List at least four serious health problems associated with poorly controlled diabetes, including poorly controlled diabetes during pregnancy.

6 Compare hyperglycemia with hypoglycemia.

7 Identify the primary signs of metabolic syndrome, the impact the condition has on health, and how one can reduce the risk of developing the syndrome.

8 Explain the difference between lactose intolerance and milk allergy.

9 Discuss the health benefits of including fiber-rich foods in one's diet.

Carbohydrates seem to get a lot of "bad press." Promoters of low–carbohydrate/high–protein diets often blame sugars and starches for causing unwanted weight gain and diabetes. Many Americans think consuming sugary foods causes depression and hyperactive behavior, and is addictive. On the other hand, "carbs" are welcomed by athletes as an inexpensive and efficient source of energy. What have scientists learned about the roles of carbohydrates in health?

Did *YOU* Know?

Tooth decay is clearly associated with consuming certain carbohydrates, particularly simple sugars that stick to teeth. If a person does not follow good dental hygiene practices, the debris becomes food for bacteria that live on teeth. As the bacteria metabolize carbohydrate for their energy needs, the acid they produce damages tooth enamel and results in decay.

Do Carbohydrates Cause Weight Gain?

Americans weigh more now than they did 30 years ago. To lose the excess weight, many people follow low–carbohydrate diets, such as the Atkins diet or the Paleo diet. Are carbohydrates responsible for the epidemic of excess body fat and weight gain in the United States? Are low–carbohydrate diets useful for weight loss?

"Calories do count" because people gain body fat when their intake of food energy from macronutrients and the nonnutrient alcohol exceeds their output of energy for various physiological needs.[19] Researchers reported in 2022 that the macronutrient content of people's diet, such as a low–carbohydrate, high–fat, or high–protein diet, may not increase or decrease body weight much. However, the quality of foods consumed, including the amount of ultra–processed foods, does influence body weight.[20] Diets rich in ultra–processed foods and refined carbohydrates are associated with weight gain and obesity, a chronic health condition characterized by unhealthy amounts of body fat.

Foods that contain large amounts of refined carbohydrates do not satisfy hunger as well as those that contain more protein or fat. As a result, a person may become hungrier sooner after eating a meal or snack that contains a lot of added sugars and refined starches than if they ate a high–protein, high–fat meal or snack. Thus, people following a carbohydrate restricted, high–protein, high–fat diet can lose weight in the short term because the diet keeps their appetite under control by reducing hunger. However, scientists report that many people do not follow these diets for more than 6 months because it is difficult.[21]

Metabolism also plays a major role in the development of obesity, a chronic health condition characterized by unhealthy amounts of body fat. When a person consumes excess carbohydrate, their body converts some of the glucose into fat, but much of the excess is "burned" as a biological fuel. As a result, dietary fat is spared from being used as a fuel and stored in fat cells.[22] Thus, eating too much carbohydrate indirectly contributes to excess body fat.

Combining Carbohydrate with Fat

Foods that contain a lot of added sugars and solid fats tend to be energy dense. Thus, people can reduce their energy intake by eating fewer energy–dense foods. Fats supply 9 kcal/g compared to only 4 kcal/g of carbohydrates, so adding even small amounts of fat to food can dramatically increase its energy content. People, however, tend to blame carbohydrates for their unwanted

Many people find it difficult to resist foods that contain a lot of sugar and fat, such as these pastries.
Africa Studio/Shutterstock

weight gain because starches and sugars are often combined with hidden fats, such as butter or shortening in processed foods. Fats make foods taste rich, creamy, and difficult to resist. Moreover, the sweet taste of sugar masks the bland taste of fat. Although you would not consider eating spoonfuls of plain sugar, flour, shortening, or butter, your mouth waters at the sight of candy bars, fruit pies, doughnuts, and other baked or fried foods. If you are like many people, you probably find it is difficult to resist eating these foods.

Although consumption of added sugars has increased over the past several years, other dietary changes have occurred as well. Americans consume an excess of calories from all foods, and the higher intake contributes to the increased prevalence of obesity in the United States. In 2017–2018, adult males consumed an average of 2485 kcal/day, and adult females consumed an average of 1849 kcal/day.[13] These amounts were about 300 kcal higher than the average daily calorie intakes of adult males and females about 30 years earlier.[23] Therefore, overall dietary habits may be partly to blame for rising rates of overweight and obesity among the U.S. population.

Did *YOU* Know?

Many people think sucrose is addictive. Addiction is characterized by an uncontrolled need (compulsion) to take a substance and the development of withdrawal signs and symptoms when the substance is not taken.

There is no scientific evidence that people can become addicted to foods, particularly those containing sucrose, as cigarette smokers become addicted to nicotine. Certain nervous system cells make a chemical messenger called *dopamine* that is associated with pleasurable sensations. Eating a tasty food causes the release of dopamine in the reward areas of the brain, similar to the response that occurs in the brain of a person using an addictive drug, such as nicotine. The release of dopamine, however, is not enough of a response to cause addiction.[28]

Factors that regulate eating behavior are very complex and include a person's hormones, environment, past experiences, senses, and mood.[28] What appears to be an "addiction" to a particular food or type of food may be a learned response to situations that the person thinks are difficult and stressful, such as worrying about a test. Nevertheless, researchers are continuing to study the possible effects of eating certain foods, particularly chocolate and "sweets," on mental health.

Sugar-Sweetened Beverages

Over the past 35 years, the percentage of obese Americans rose dramatically. Some nutrition scientists think Americans' intake of added sugars is largely responsible for the population's rising rates of obesity as well as type 2 diabetes. Sugar–sweetened beverages are one of the major sources of added sugars in the diets of Americans who are 2 years of age and older.[24]

According to results of large epidemiological studies, risks of obesity, type 2 diabetes, heart disease, and stroke are associated with the highest intakes of sugar–sweetened beverages.[25] People who drink sugar–sweetened beverages do not reduce their consumption of solid foods enough to avoid gaining weight.[26] Excess body fat is a risk factor for type 2 diabetes, heart disease, and stroke. Furthermore, consuming high amounts of beverages sweetened with sucrose and/or fructose may cause weight gain and abnormal blood lipid levels and may damage the liver, resulting in a condition commonly called "fatty liver."[27] However, consuming added sugars, such as sucrose and fructose, in amounts that support the recommendations of the Dietary Guidelines (less than 10% of total calories) is not likely to contribute to the risk of chronic diseases, particularly obesity or fatty liver.

Nonalcoholic Fatty Liver Disease

Under normal conditions, the liver does not store fat, but the organ can accumulate the lipid in response to alcohol consumption and develop fatty liver. Fatty liver also occurs in people who drink little or no alcohol (**nonalcoholic fatty liver disease [NAFLD]**). Risk factors for developing NAFLD include obesity, insulin resistance, elevated blood lipids, and excessive carbohydrate and fat intake.[29] Although there is no specific treatment for NAFLD, losing excess body fat, avoiding alcohol, and eating a healthy diet may allow the liver to heal and avoid inflammation. According to experts at the National Institutes of Health, an estimated 24% of adult Americans have NAFLD.[30]

If people with NAFLD continue to have fat accumulate in their liver, the fat damages liver cells, causing inflammation of the organ (*nonalcoholic steatohepatitis* [*stee'–a–toe–hep'–ah–tie'–tis*]). The inflammation can result in cirrhosis (*seh–roe'–sis*)) of the liver, a condition in which liver cells die and are replaced with scar tissue that has no metabolic function. Eventually, the scar tissue hardens, and the affected person experiences liver failure.

nonalcoholic fatty liver disease (NAFLD) abnormal accumulation of fat in the liver that is not related to alcohol intake

People with liver failure require a liver transplant to survive. An estimated 1 to 6% of Americans have nonalcoholic steatohepatitis.[30]

What Is Diabetes?

Diabetes mellitus (diabetes) is a group of serious chronic diseases characterized by high levels of blood glucose. There are two major types of diabetes mellitus: type 1 and type 2 diabetes. About 5% of people with diabetes have type 1; in the past, this form of diabetes was called "juvenile diabetes" because it was diagnosed more often in children and young adults. Although type 1 diabetes is more likely to occur in children and young adults, the disease can strike at any age.[31] The majority of people with diabetes have type 2, which used to be called "adult–onset diabetes." Gestational Diabetes is a type of diabetes that can occur during pregnancy.

The primary sign of diabetes is **hyperglycemia** (*hyper* = excess; *glycemia* = blood glucose), abnormally elevated blood glucose levels. A person's blood glucose levels are usually measured after they have not eaten (have fasted) for about 12 hours. Normal fasting blood glucose levels are 70 to 99 mg/dl (Table 5.8). People with **pre–diabetes** ("impaired glucose tolerance") have fasting blood glucose levels that are 100 to 125 mg/dl. Although a person with pre–diabetes does not have diabetes, they have a high risk of cardiovascular disease and eventually developing type 2 diabetes. Individuals who have fasting blood glucose levels of 126 mg/dl or more have diabetes.

Some people with diabetes experience hyperglycemia because their beta cells do not produce any insulin or do not produce enough to meet their needs. In other cases, the affected person produces some insulin, but their body does not respond properly to the hormone (*insulin resistance*), and hyperglycemia results. Major signs and symptoms of hyperglycemia include excessive thirst, frequent urination, blurred vision, and poor wound healing (Table 5.9). Over time, hyperglycemia damages nerves, organs, and blood vessels. In fact, poorly controlled diabetes is a major cause of heart disease, kidney failure, blindness, and lower limb amputations. In 2020, diabetes was the eighth leading cause of death in the United States.[32]

In the United States, the prevalence of diabetes is increasing at an alarming rate (Fig. 5.13). In 2019, an estimated 37.3 million Americans had diabetes;

TABLE 5.8 Classifying Diabetes Mellitus

Blood Glucose Level (Fasting)	Classification
70 to 99 mg/dl	Normal
100 to 125 mg/dl	Pre-diabetes
126 mg/dl or more	Diabetes

TABLE 5.9 Signs and Symptoms of Diabetes Mellitus

- Elevated blood glucose levels
- Excessive thirst
- Frequent urination
- Blurry vision
- Vaginal yeast infections (females)
- Foot pain, abdominal pain
- Numbness, especially in the feet
- Impotence (males)
- Sores that do not heal
- Increased appetite with weight loss*
- Breath that smells like fruit*
- Fatigues easily*
- Confusion*

*Typical symptoms of poorly controlled type 1 rather than type 2 diabetes.

diabetes mellitus (diabetes) group of serious chronic diseases characterized by high levels of blood glucose

hyperglycemia abnormally high blood glucose level

pre-diabetes condition characterized by fasting blood glucose levels that are 100 to 125 mg/dl

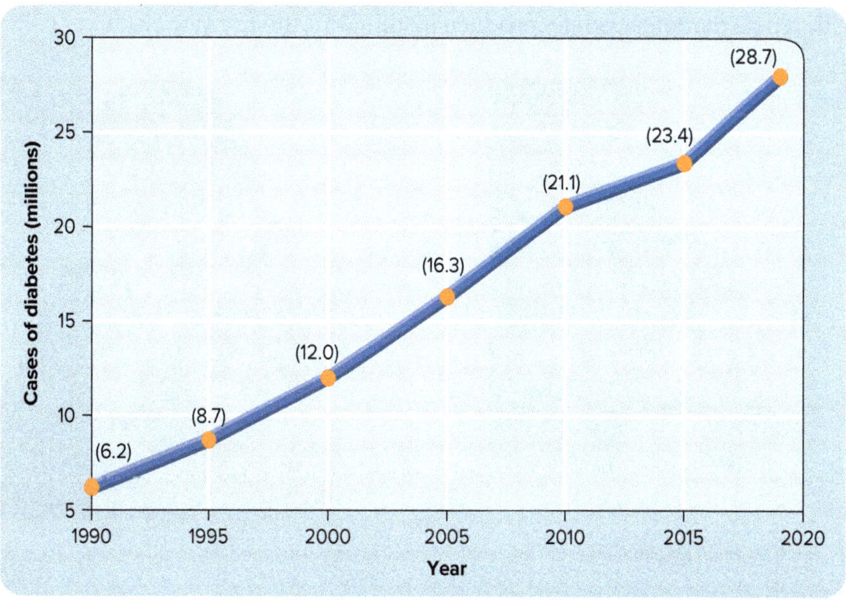

Figure 5.13 Estimated number of Americans diagnosed with diabetes (millions). Between 1990 and 2019, the prevalence of diabetes increased dramatically in the United States.

only 28.7 million of these persons were diagnosed with diabetes, and the remainder did not know they had the disease.[33] An estimated 96 million adult Americans had pre–diabetes in 2019. The prevalence of diabetes increases with advancing age; an estimated 29% of Americans who are 65 years of age or older have diabetes.[33] Public health officials are very concerned about the increasing number of children and adolescents who have diabetes, particularly type 2 diabetes.

Type 1 Diabetes

Type 1 diabetes is an *autoimmune* disease that occurs when certain immune system cells malfunction and do not recognize the body's own beta cells. As a result, the immune system cells attack and destroy many of the beta cells, and the affected person must obtain insulin, usually by injecting the hormone into their body, regularly (Fig. 5.14). It is not clear why the immune cells of some individuals malfunction, but genetic susceptibility and environmental factors, particularly exposure to certain viral intestinal infections, are associated with the development of type 1 diabetes. Furthermore, infants who drink cow's milk or cow's milk–based formulas have a greater risk of developing type 1 diabetes than breastfed babies.[34] The role of dietary factors in the development of type 1 diabetes continues to undergo scientific study.

People with undiagnosed or poorly managed type 1 diabetes usually develop ketoacidosis. Signs and symptoms of ketoacidosis include excessive thirst, frequent urination, and a blood glucose level of more than 250 mg/dl. Individuals with the condition may also have nausea, vomiting, fatigue, and confusion. The breath of a person with ketoacidosis typically has a "fruity" or acetone odor; acetone is a chemical that is often in fingernail polish remover. Ketoacidosis is a potentially life–threatening condition that requires immediate medical care. If untreated, ketoacidosis can lead to coma (loss of consciousness) and death. Proper management of diabetes, including regular testing of blood glucose levels, is crucial for preventing ketoacidosis.

At present, there is no cure for type 1 diabetes, but the disease can be managed. You can learn about Annie Pietsch and how she manages her type 1 diabetes by reading the "Real People, Real Stories" feature in this chapter.

Type 2 Diabetes

The most common form of diabetes is type 2 diabetes. Beta cells of people with type 2 diabetes usually produce insulin, but the hormone's target cells are

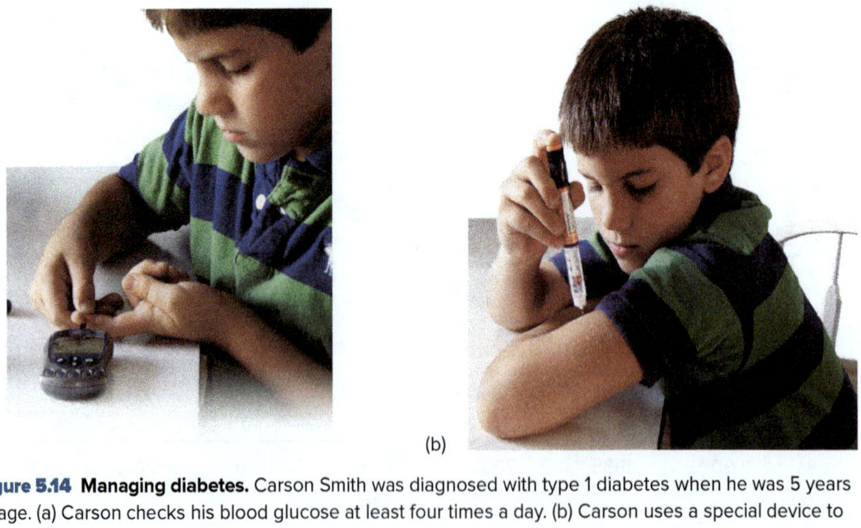

(a) (b)

Figure 5.14 Managing diabetes. Carson Smith was diagnosed with type 1 diabetes when he was 5 years of age. (a) Carson checks his blood glucose at least four times a day. (b) Carson uses a special device to inject the insulin into his body.

gestational diabetes form of diabetes that pregnant females can develop

insulin–resistant, which means they do not respond properly to the hormone and do not allow glucose to enter them. As a result, the level of glucose in the bloodstream becomes abnormally elevated and the typical signs and symptoms of diabetes occur (see Table 5.9).

Over the past 30 years, the number of adults and children with type 2 diabetes has reached epidemic proportions in the United States. Certain people have greater risk of type 2 diabetes than others. Individuals who are physically inactive (sedentary), have excess body fat, and are genetically related to a close family member with type 2 diabetes are more likely to develop the disease than persons who do not have these characteristics. Additionally, Americans who have Hispanic, Native American, Asian, African, or Pacific Islander ancestry are more likely to develop type 2 diabetes than Americans who are not members of these racial/ethnic groups.[33] The American Diabetes Association has an online questionnaire that you can take to assess your risk of type 2 diabetes: www.diabetes.org/risk–test.

Diabetes During Pregnancy During pregnancy, a female with poorly controlled diabetes and her developing offspring (*embryo* or *fetus*) can develop serious health problems. The first 8 weeks of pregnancy are a critical period in a human embryo's development because its organs are forming. If a female with diabetes lacks good control of her blood glucose level, her embryo can develop birth defects as a result. By managing her blood glucose properly, however, the pregnant diabetic female can reduce her chances of giving birth to a baby with birth defects. Additionally, she will be less likely to experience serious health problems, such as high blood pressure, during pregnancy.

About 9% of pregnant American females develop a form of diabetes **(gestational diabetes),** usually after the fifth month of pregnancy.[39] Pregnant females who have a family history of type 2 diabetes, are overweight, or have high blood pressure are more likely to experience gestational diabetes than pregnant females who do not have these characteristics. The fetus of a female with gestational diabetes receives too much glucose from its hyperglycemic mother. As a result of obtaining the excess glucose, the fetus gains weight rapidly and can be abnormally heavy at birth, weighing about 9 pounds or more. Giving birth to such a large infant is risky for the female, as well as the infant, because it may prolong the birth process and cause the baby to be injured during delivery.

After giving birth, most females who had gestational diabetes recover and have normal blood glucose levels. Although the new mothers are healthy, they have an increased risk of developing type 2 diabetes later.[40]

A healthy pregnancy typically lasts about 40 weeks. Pregnant females with poorly controlled diabetes or gestational diabetes are more likely to have miscarriages (death of an embryo/fetus before the 20th week of pregnancy), stillbirths (delivery of a dead baby), and premature deliveries (born before the 37th week of pregnancy) than females with diabetes who manage their blood glucose properly. Furthermore, females with poorly controlled diabetes or gestational diabetes are more likely to give birth to babies who are too large and have difficulty controlling their own blood glucose levels than mothers who do not have diabetes or develop the condition during pregnancy. Thus, pregnant females who have diabetes are encouraged to monitor blood glucose levels carefully to minimize risks to both themselves and their fetuses.

Females who have adequate *prenatal* (before birth) medical care during pregnancy usually undergo screening to detect gestational diabetes. Treatment for females who are diagnosed with gestational diabetes generally includes a special diet to help manage blood glucose levels and regular physical activity. Some mothers–to–be will also need to monitor their blood glucose levels and give insulin injections to themselves regularly.

Nutrition & Health *Equity*

Only 12% of American adults have *health literacy,* the ability to find, understand, and use health information to manage their health and make appropriate health decisions.[35] Health literacy also includes *numeracy,* the ability to understand and work with numbers. It is important for people with diabetes to have diabetes health literacy. However, minority populations, including people of African ancestry, Hispanic ancestry, and Asian and Pacific Islander ancestry, often have lower diabetes health literacy. Lack of health literacy in these populations is associated with uncontrolled blood glucose levels, poor diabetes management, and development of chronic disease complications such as heart disease, nerve damage, kidney damage, and blindness.[36]

Recent scientific research reports people with diabetes may not attend their diabetes-related medical appointments,[37] and often are not referred to a registered dietitian for nutrition education.[38] Both of these factors may contribute to lower diabetes literacy and poor disease management. Therefore, people with low diabetes literacy may not understand the complex routines needed to manage diabetes, such as checking blood glucose levels multiple times per day, taking multiple diabetes medications as prescribed, and using numeracy to adjust medication dosages, read food labels, and calculate grams of carbohydrate in foods.[36]

Addressing the inequity of low diabetes health literacy in affected populations requires further research to find ways to increase attendance of diabetes medical appointments and increase referrals to registered dietitians. Medical professionals and diabetes education programs should have clear communication and education strategies, including:[35,36]

- Use plain language education materials and Internet-based education tools.
- Provide spoken and written education in the client's preferred language.
- Avoid using medical terms and jargon when discussing diabetes.
- Write out acronyms such as "Body Mass Index (BMI)."
- Use pictures as part of diabetes education.
- Assess diabetes knowledge and understanding by asking clients to explain in their own words what they know or learned.
- Involve clients in their personal health care decisions.

TABLE 5.10 Classifying Diabetes According to A1c Values

Diagnosis*	A1c Level
Normal	4.5–<5.7%
Pre-diabetes	5.7–6.4%
Diabetes	6.5% or above

*Any test for diagnosing diabetes requires a second test to confirm the diagnosis, unless the patient has signs and symptoms of diabetes.

A hamburger with fries and a sugar-sweetened beverage is an example of a typical meal for a "Western" diet. Baked salmon served with pasta and a variety of fruits and vegetables is an example of a healthier meal.
Hamburger plate and drink: serezniy/123RF; Lean meal: Oleksandra Naumenko/123RF

hypoglycemia condition that occurs when the blood glucose level is abnormally low

Controlling Diabetes

To avoid or delay serious health complications, people who have diabetes need to achieve and maintain normal or near–normal blood glucose levels. Many people with diabetes rely on daily blood testing to monitor their blood glucose levels (see Figure 5.14a). Physicians can measure *glycated hemoglobin,* also called glycosylated hemoglobin or *hemoglobin A1c* ("HbA1c"), to determine their patients' average blood glucose levels over longer periods. Results of this blood test can provide information about a patient's long–term management of the condition.

Hemoglobin is the compound in red blood cells that carries oxygen and carbon dioxide. Glucose in the bloodstream can bind to hemoglobin, especially when blood glucose levels are elevated. About 4.5 to <5.7% of a healthy person's hemoglobin is bound to glucose (glycated hemoglobin). A person with poorly controlled diabetes or undiagnosed diabetes often has blood glucose levels that are much higher than normal. As a result, this individual will have a higher percentage of glycated hemoglobin, and thus an elevated HbA1c (Table 5.10). Most people with diabetes should strive to maintain their HbA1c level below 7%.[41]

Proper blood glucose management involves monitoring blood glucose levels regularly and carefully following a special diet that usually includes counting grams of carbohydrates. Including physical activity in one's daily routine is also recommended. People with pre–diabetes and type 2 diabetes who are overweight can often reduce their insulin resistance by losing small amounts of excess body fat. Additionally, exercise increases glucose uptake by muscles, which reduces blood glucose levels and improves the body's insulin response. In some instances, however, people with type 2 diabetes need oral medication to lower their blood glucose levels. Commonly prescribed medications decrease the amount of glucose the liver releases into the bloodstream and reduce insulin resistance, allowing increased glucose uptake by cells. Some people with type 2 diabetes do not produce enough insulin and may require insulin injections.

Promoters of certain weight reduction diets claim that people can lose weight or control diabetes by following *low glycemic index* diets. However, some medical experts question the value of the glycemic index for predicting the body's blood glucose level after eating. To learn more about the glycemic index and glycemic load, read the "Nutrition Matters" section of this chapter.

Can Diabetes Be Prevented?

There is no way to prevent type 1 diabetes. By losing some excess body fat and increasing physical activity levels, people with pre–diabetes can prevent or delay the onset of type 2 diabetes and lower their risk of heart disease and stroke.[42] Medical experts refer to these actions as *therapeutic lifestyle changes* (*TLC*).

Certain eating habits may help prevent type 2 diabetes. The typical American diet ("Western diet") contains high amounts of saturated fat and refined sugars and starches. Such diets may increase the risk for type 2 diabetes and other serious chronic diseases. On the other hand, diets that contain fewer sugar–sweetened drinks, less saturated fat, and more fiber–rich whole grains, fruits, and vegetables are healthier than the Western diet.[43] The "Mediterranean diet" is a healthy eating pattern that may help prevent type 2 diabetes or delay its onset.[44] The Mediterranean diet pyramid is illustrated in Figure 3.19.

What Is Hypoglycemia?

Hypoglycemia (*hypo* = low) is a condition that occurs when the blood glucose level is too low to provide enough energy for cells. Hypoglycemia may be diagnosed when the blood glucose level is less than 70 mg/dl.[45]

REAL PEOPLE | REAL STORIES

Traci Keck

Annie Pietsch

Annie Pietsch is a student at the University of Texas at San Antonio (UTSA) who is majoring in nutrition and dietetics. When Annie is not focused on her studies, she enjoys running for exercise, spending time with her family, and shopping and eating ice cream with her friends.

While she was in high school, Annie decided she wanted to become a registered dietitian nutritionist because she wanted to help her grandmother, who had type 2 diabetes. Before she started college, Annie had no idea that her lifestyle was about to change dramatically.

Annie's health began to decline during the spring semester of her first year at UTSA. She became unusually tired when exercising as well as walking to class. According to Annie, "Walking up a flight of stairs felt like running 5 miles." She also developed an unusual craving for sugar-sweetened "soda," ice cream, and fast-food chicken. "I knew something was wrong when I was losing weight instead of gaining weight after eating this way."

At the beginning of spring break, Annie went home to visit her family. Her parents were immediately concerned with Annie's weight loss and overall physical condition. "I could only walk a few steps and had to rest." Within a few days, she could no longer speak normally to her mother—she just mumbled. Annie's mother became so alarmed, she rushed her daughter to a hospital. While in the emergency room, Annie was diagnosed with type 1 diabetes and diabetic ketoacidosis. "When I came into the E.R., my blood glucose was 1700 mg/dl." The young woman was given an intravenous insulin drip to lower her blood glucose level.

Today, Annie is feeling fine. She manages her blood glucose levels with insulin injections. She still runs, and she enjoys all the activities she did prior to her diagnosis and an occasional treat of ice cream. "I test my blood glucose level before all meals and snacks. I have learned to 'count carbs.' For every 10 grams of carbs, I inject myself with 1 unit of insulin. It's easier to count carbs using a Nutrition Facts panel." For many people, counting carbohydrates is made easier by using the information on a Nutrition Facts label and eating "whole foods" (unrefined or lightly processed foods). Examples of whole foods that contain 10 to 15 grams of carbohydrates include ½ of a small banana, 8 fl oz of milk, and a slice of bread.

After she becomes a registered dietitian nutritionist (RDN), Annie wants to obtain certification as a diabetes educator so she can teach others who have diabetes. "People often think I have diabetes because I ate the wrong foods or ate too much sugar. I tell them that type 1 diabetes is an autoimmune disease; you can't control your risk of developing type 1 diabetes. Every person with diabetes has unique challenges and critical choices to make when managing their health," Annie says.

The Diabetes Link (TDL) has chapters throughout the United States that provide support groups for college students with type 1 diabetes. Members of TDL are able to connect with their peers, support each other, and have access to diabetes experts through TDL's national organization. To learn more about TDL, visit https://thediabeteslink.org.

Did *YOU* Know?

A diabetic alert dog (DAD) is a service dog that can save the life of an individual with type 1 diabetes by alerting other people when the diabetic person's blood sugar level is abnormal.[46] People with type 1 diabetes can develop hypoglycemia ("low blood sugar") quickly, and the effects of hypoglycemia can be dangerous, even deadly. To learn how to detect hypoglycemia, dogs being trained to become DADs are exposed to samples of perspiration and exhaled breath that are provided by people with diabetes who are experiencing hypoglycemia.[47] In a small study of eight females with type 1 diabetes who were hypoglycemic, scientists found high levels of a particular chemical (*isoprene*) in samples of their exhaled air.[45] Additional research is needed to determine whether isoprene can be used to train DADs effectively.

Realistic Reflections

Did *YOU* Know?

For over 50 years, medical researchers have been testing the effectiveness of transplanting a pancreas or pancreatic islet (*i'-let*) cells, which contain insulin-producing beta cells, into the bodies of people with type 1 diabetes. The expectation was to help patients achieve good control over their blood glucose levels while reducing their need to have daily insulin injections. A major limitation of such procedures has been the need to place the donor's islet cells into the patient's liver, which does not have the same blood supply as the pancreas.

In 2017, scientists at the University of Miami's medical school published a report in *The New England Journal of Medicine* that described a 43-year-old woman who had been using insulin for over 2 decades to treat type 1 diabetes.[49] The woman had bouts of severe hypoglycemia, and she was often not aware that her blood glucose level was too low. In an experimental treatment, islet cells were taken from a person who had died and transplanted into the *omentum* of the woman with diabetes. The omentum is a fatty structure that drapes over the stomach and small intestine (see Figure 10.3). Unlike the liver, the omentum has the same blood supply as the pancreas.

After the procedure, the patient was given medication to suppress rejection of the transplanted cells. She is likely to need anti-rejection medications for the rest of her life. Within 3 weeks after the surgery, the patient was able to stop giving herself insulin injections. After 1 year, she still had good control over her blood glucose levels and had not experienced hypoglycemic episodes. The scientists involved in this experiment will continue to follow their patient's progress to determine whether this method of transplanting islet cells is both safe and effective as a long-term treatment for type 1 diabetes.

In response to rapidly declining blood glucose levels, the adrenal glands secrete epinephrine (see Figure 4.18 [endocrine system]). You may be more familiar with epinephrine's common name, *adrenaline.* Like glucagon, epinephrine increases the supply of glucose and fatty acids in the bloodstream, but the hormone can also make a person with hypoglycemia feel irritable, restless, shaky, hungry, and sweaty. If the blood glucose level drops too low, the affected person can develop seizures, and they may lose consciousness and die.

Hypoglycemia is a major concern for people who give themselves insulin injections or take medication to raise the amount of insulin made by their bodies. Having too much insulin in the bloodstream contributes to hypoglycemia.

Treating hypoglycemia generally includes taking pills or a gel that contains glucose or consuming a serving of a sugary food or beverage that supplies 15 grams of sugar and little fat or protein.[45] Eating a tablespoon of sugar or honey, for example, can help raise the blood glucose level to normal. To prevent hypoglycemia, people with diabetes should eat regular meals and snacks that contain appropriate amounts of carbohydrate and monitor their blood glucose levels as often as recommended by their health care professionals.[48]

Metabolic Syndrome

People who are overweight often have excess abdominal fat, which can be dangerous, especially when it is accompanied by insulin resistance, hypertension, and elevated blood lipid (triglyceride and cholesterol) levels.

An estimated one in three Americans has **metabolic syndrome,** a condition characterized by three or more of the signs listed in Table 5.11.[50] A **syndrome** is a group of signs and symptoms that occur together and indicate a specific health

TABLE 5.11 Signs of Metabolic Syndrome

Sign	Defining Value
Large waist circumference	≥ 40 inches (males) ≥ 35 inches (females)
Chronically elevated blood pressure (hypertension)	≥ 130 mm Hg systolic (upper value) or ≥ 85 mm Hg diastolic (lower value) or Drug treatment for hypertension
Chronically elevated fasting blood fats (triglycerides)	≥ 150 mg/dl or Drug treatment for elevated triglycerides
Low fasting high-density lipoprotein cholesterol (HDL cholesterol)	< 40 mg/dl (males) < 50 mg/dl (females) or Drug treatment for reduced HDL
High fasting blood glucose	≥ 100 mg/dl or Drug treatment for elevated glucose

Source: National Heart Lung and Blood Institute: *Health topics: Metabolic syndrome.* Updated January 2019. www.nhlbi.nih.gov/health-topics/metabolic-syndrome. Accessed August 6, 2022.

metabolic syndrome condition that increases risk of type 2 diabetes and CVD

syndrome group of signs and symptoms that occur together and indicate a specific health problem

problem. Compared to people who do not have metabolic syndrome, individuals with this condition have five times the risk of type 2 diabetes and almost twice the risk of heart and blood vessel (cardiovascular) disease.[50]

Although genetic factors play a major role in the development of metabolic syndrome, excess abdominal fat, lack of physical activity, and insulin resistance are the primary risk factors for the condition.[51] People may lower their likelihood of developing metabolic syndrome by avoiding obesity, being physically active, and increasing their intake of fruits, vegetables, and other fiber–rich foods, such as whole–grain cereals.

Individuals who already have metabolic syndrome may reduce their risk of cardiovascular disease (CVD) by lowering their elevated blood pressure, glucose, insulin, and triglyceride levels. Lifestyle changes that can help manage these levels include losing excess weight, exercising regularly, and reducing intakes of salt, saturated fat, and simple sugars. If such changes do not alleviate the condition, medication may be necessary to manage blood pressure and blood lipid levels. Chapter 6 provides more information about cardiovascular disease and "heart healthy" eating patterns.

Did *YOU* Know?

FODMAPs (fermentable oligo-, di-, and monosaccharides and polyols) are forms of carbohydrate that the GI tract may not fully digest or absorb, especially when a person eats too much of them. When FODMAPs are in the lumen of the colon, they attract water, and the bacteria that reside in the large intestine ferment the substances for their energy needs. Side effects of the bacteria's activity often include abdominal cramps, diarrhea, bloating, and gas.

FODMAPs include fructose, lactose, and sugar alcohols such as xylitol and sorbitol. Wheat products, rye, various kinds of fruits and vegetables, beans, and high-lactose dairy foods are often excluded in low-FODMAP diets because they are rich sources of FODMAPs. Registered dietitians can help plan a nutritionally adequate low-FODMAP diet, which is followed for 4 to 8 weeks, after which FODMAPs are reintroduced in the diet to establish the amount of FODMAPS a person can tolerate.[52] People with irritable bowel syndrome and, possibly, inflammatory bowel disease may benefit from such diets.[52]

Carbohydrate Intolerance

In the United States and throughout the world, many people are unable to digest lactose, fructose, and/or other kinds of carbohydrates completely. This condition is called carbohydrate intolerance (also referred to as carbohydrate maldigestion or malabsorption). People who suffer from **lactose intolerance** do not produce enough lactase, the enzyme that breaks lactose into glucose and galactose. When a lactose–intolerant person consumes lactose, the disaccharide is not completely digested and absorbed by the time it enters the large intestine. Bacteria that reside in the large intestine break down lactose and produce irritating gases and acids as metabolic by–products. As a result, a lactose–intolerant person usually experiences intestinal cramps, bloating, gas, and diarrhea within 30 minutes to 2 hours after consuming milk or other lactose–containing products.

Normally, infants produce lactase, but by the time children are 2 years old, their small intestine begins to produce less of the enzyme. Many older children and adults, particularly those with African, Asian, and Eastern European ancestry, are lactose intolerant and experience some degree of abdominal discomfort after drinking milk.

Lactose-intolerant people may be able to eat yogurt without experiencing digestive tract discomfort. ©Wendy Schiff

Lactose intolerance is not the same as milk allergy, which is an immune system response to cow's milk proteins. Milk allergy is most likely to occur during infancy; lactose intolerance is more likely to occur during adolescence (the teen years) or in adulthood.[53]

Milk and milk products are often excellent sources of protein, many vitamins, and the minerals calcium and phosphorus. What can people with lactose intolerance do to achieve a nutritionally adequate diet without drinking milk? Lactose–intolerant people are often able to eat hard cheeses and yogurt without experiencing any digestive tract discomfort. Milk loses most of its lactose content when it is processed to make aged cheeses, such as cheddar and Swiss. The bacteria used to make yogurt convert much of the lactose in milk to lactic acid, and the microbes assist with the digestion of the remaining lactose even after the yogurt is eaten.[54]

Many people with lactose intolerance can consume small amounts of milk without experiencing intestinal discomfort. People who cannot tolerate even limited amounts of fresh fluid milk can probably drink milk that has been pre–treated with lactase to reduce its lactose content. Most large supermarkets sell fresh lactase–treated dairy products in the dairy food section (Fig. 5.15).

lactose intolerance inability to digest lactose properly

Figure 5.15 Lactase-treated milk. Fresh lactase-treated milk is often available in the dairy section of supermarkets.
©Wendy Schiff

Also, lactase–containing solutions and pills are available without prescription. A lactose–intolerant person simply adds a small amount of the solution to milk 24 hours before drinking the beverage or takes one of the pills with lactose–containing food. People who cannot tolerate lactose can substitute soy milk for cows' milk because soy milk does not contain lactose.

Does Sugar Cause Hyperactivity?

If you have ever been in charge of a 7–year–old child's birthday party or observed third graders preparing for their Halloween celebration, you can understand why people often blame sugary foods for causing unruly and "hyperactive" behavior. Attention deficit hyperactivity disorder (ADHD) is characterized by impulsivity, hyperactivity, and difficulty paying attention.[55] According to estimates provided by the CDC, about 9% of American children have ADHD.[56] Many children do not outgrow the condition by the time they reach adulthood.

Although scientists do not know what causes ADHD, it probably involves genetic and environmental factors. The results of scientific studies do not indicate that eating sugary foods or drinks increases children's physical activity levels, causes ADHD, or has any other negative effects on their behavior.[55] Birthday and school parties are exciting and happy occasions that typically involve a radical change from a child's usual routine. In these situations, a youngster's excitement and more active behavior are more likely to be the result of the occasion rather than a particular food.

Fiber and Health

Technically, fiber is not an essential nutrient because the human body can live without it. You can live *better,* however, by adding fiber–rich foods to your diet. Eating high–fiber foods may reduce your risk of obesity, diabetes, certain intestinal tract disorders, and cardiovascular disease, which includes heart disease and stroke.

A person's diet, particularly its fiber content, affects their bowel habits. Some kinds of fiber in food attract water and swell in the digestive tract, forming a large, soft mass that applies pressure to the inner muscular walls of the large intestine, stimulating the muscles to push the residue quickly through the tract. People who often eat foods that contain adequate amounts of fiber have softer stools and more regular bowel movements, and they are less likely to strain while having bowel movements than people whose diets lack fiber. Thus, people generally do not need to rely on over–the–counter laxatives to treat constipation. Eating more fiber–rich foods is the natural way to become "regular."

Did *YOU* Know?

Dried plums (*Prunus domestica* L.) are commonly called prunes. The fruit contains fiber, the sugar alcohol sorbitol, and other substances that are mild natural laxatives. Prune juice provides the laxative effect of prunes but lacks the whole fruit's fiber content.

Scott Bauer/ARS/ U.S. Department of Agriculture (USDA)

Fiber and Heart Health

Diets rich in fiber, particularly soluble types of fiber, can reduce the risk of cardiovascular disease by reducing blood cholesterol levels. High blood levels of certain forms of blood cholesterol are associated with increased risks of cardiovascular disease. The liver uses cholesterol to make bile, a substance that helps digest fats. The gallbladder stores bile and releases the substance into the small intestine during meals. Instead of eliminating bile along with fecal matter in bowel movements, the intestinal tract breaks it down and absorbs its components, which eventually enter the liver. The liver recycles the "used" bile components to make new bile. When you eat oat cereal, the soluble fiber in oats interferes with this recycling process because it binds to the used bile components in the intestinal tract and prevents them from being absorbed (see Figure 6.12). Thus, the used bile components are eliminated during bowel movements. As a result, blood cholesterol levels drop as the liver removes cholesterol from the blood to make new bile. The "Recipe for Healthy Living" feature of this chapter has a hot oatmeal recipe.

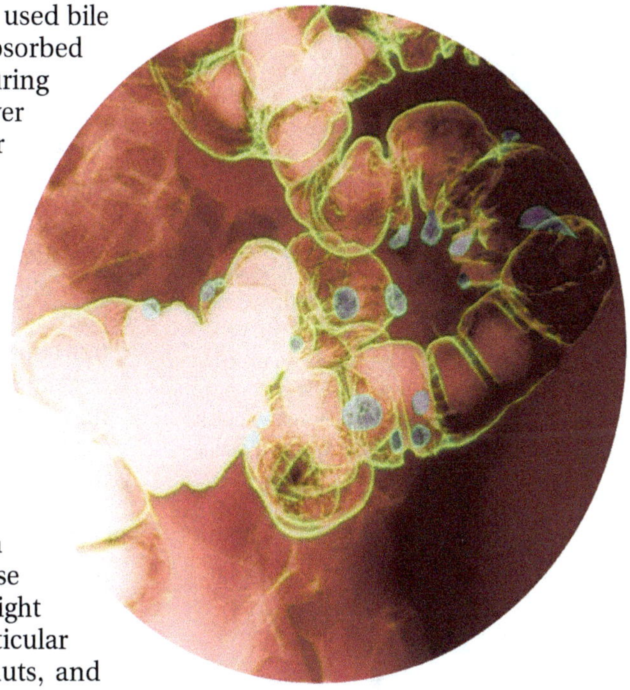

Figure 5.16 **Diverticula.** In this color-enhanced x-ray, the blue areas are diverticula.
Du Cane Medical Imaging Ltd/Science Source

Diverticula and Swollen Hemorrhoids

Diverticula are marble–size sacs that form in the lining of the colon (Fig. 5.16). In most people, diverticula do not produce symptoms, but they can bleed or become painfully inflamed when bacteria or feces are trapped within them. When this condition (*diverticulitis*) occurs, the affected person may need surgery to remove the damaged section of large intestine. The causes of diverticula are unknown, but age is a factor because the sacs are more likely to form in older adults.[57] Population studies suggest genetics is also a factor of diverticular disease. People from Western countries tend to have diverticula in their left colon, whereas those from Asian countries are more likely to have diverticula in their right colon.[57] At one time, health care providers advised people with diverticular disease to not eat certain high–fiber foods, including popcorn, nuts, and sunflower seeds. Now, no special diet is recommended, but individuals with the disease should learn to avoid foods that seem to aggravate the condition. High fiber diets can soften stools and help prevent diverticulitis flare–ups.

Hemorrhoids are clusters of small rectal veins that can become swollen and inflamed, making them likely to bleed and cause discomfort and itching (Fig. 5.17). This condition commonly affects adults and may be more likely to occur when a person sits for long periods or has constipation and strains during bowel movements. Although swollen hemorrhoids are generally not a sign of a serious health problem, you should consult a physician if you experience any bleeding during bowel movements because it may also be a sign of colorectal cancer, a major cancer site for Americans.

diverticula abnormal, small sacs that form in wall of colon

Fiber and Colorectal Cancer

In the United States and throughout the world, colon or *colorectal* cancer (cancer of the two lower portions of the large intestine) is one of the most common cancers.[58] In the early 1970s, a group of scientists noted that rural African populations who typically ate high–fiber diets rarely developed colorectal cancer. When these populations moved to urban areas and adopted relatively low–fiber Western diets, their risk of colorectal cancer increased. As a result of these observations, the scientists suggested that high–fiber diets were protective against colorectal cancer. Today, scientists think a healthy diet that supplies high–fiber foods, such as whole grains, fruits, and vegetables, may reduce the risk of colorectal cancer.[59]

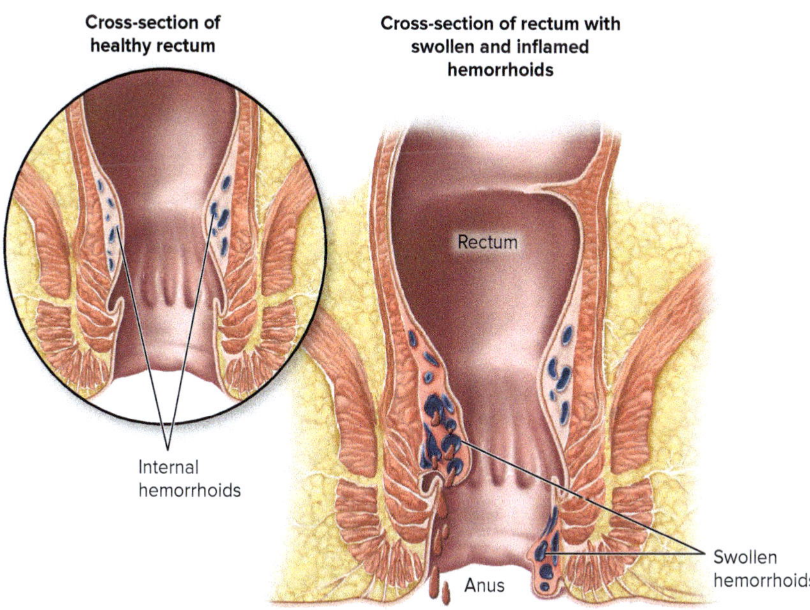

Figure 5.17 **Swollen and inflamed hemorrhoids.** Swollen hemorrhoids can bleed, cause pain, and become itchy.

Fiber and Weight Control

If you are trying to lose excess body fat, you may find it helpful to add more fiber–rich foods to your diet. A serving of a high–fiber food

MY DIVERSE PLATE

Red Lentils

For thousands of years, people living in Mediterranean countries, the Middle East, India, and other countries of southeastern Asia have eaten lentils, a type of *legume*. Peas and beans are also legumes. Lentils, particularly green varieties, can be used in salads and are an excellent source of fiber. Lentils also provide protein, iron, potassium, and folate. Red lentils soften quickly with cooking and can be used in stews, sauces, or dips.

©Wendy Schiff

generally has lower energy content than the same volume of a low–fiber food. As a result, energy intake usually decreases when people switch their consumption from low–fiber to high–fiber diets.

Increasing Your Fiber Intake

The recommended Adequate Intakes (AIs) for fiber are 38 and 25 g/day for young males and females, respectively, but the typical diet of adult Americans supplies only about 17 g of dietary fiber/day.[16] According to the Dietary Guidelines, fiber is a "nutrient of public health concern." You can estimate your daily fiber intake by completing the Personal Dietary Analysis at the end of this chapter. The "Food & Nutrition Tips" feature in this section of the chapter provides tips for increasing the fiber content of diets. Eating too much fiber may interfere with the small intestine's ability to absorb certain minerals. In rare instances, consuming too much dietary fiber results in intestinal blockage, especially if fluid intake is low.

Intestinal bacteria produce gases when they metabolize fiber; therefore, registered dietitian nutritionists recommend that people adjust by gradually increasing their fiber intake. Practices that result in swallowing air, such as eating quickly; drinking carbonated beverages, especially with a straw; and chewing gum, also contribute to intestinal gas.

Food & Nutrition tips

Brown rice.
Yellow Cat/Shutterstock

- Read the ingredients list on the label to find out if a bread or cereal product is whole grain; bran or whole grain should be the first ingredient.
- Don't rely on the product's name or appearance. Check the ingredients. Terms such as *100% wheat, multigrain,* or *stone-ground wheat* are misleading because the product may contain little or no whole grain.
- Brown rice has more fiber and flavor than white rice.
- Substitute whole-wheat pasta for regular pasta, or use half whole-wheat and half regular pasta in pasta dishes.
- Snack on pieces of fresh, frozen, or dried fruit.
- Include more nuts, beans, and seeds in your diet.
- Spread peanut, almond, or soy butter on whole-grain crackers for a fiber-filled snack.
- Add frozen, dried, or fresh fruit such as berries, raisins, or bananas instead of sugar or honey to sweeten cereal or plain yogurt.
- Add a small amount of uncooked oatmeal and wheat germ to raw ground meats when making hamburgers or meatloaf.
- Adding bran, wheat germ, and uncooked oatmeal to pancake or waffle batter also enhances the batter's fiber content.
- Good dietary sources of fiber contain at least 2.5 g of fiber per serving.

Peter Cade/Photodisc/Getty Images

Concept CHECKPOINT

17. What are the signs and symptoms of type 1 and type 2 diabetes?
18. Lata wants to prevent the development of type 2 diabetes. What health-related lifestyle practices can she follow to reduce her risk of this serious metabolic disease?
19. Identify at least three signs of metabolic syndrome.
20. What is lactose intolerance?
21. List at least three ways to increase dietary fiber intake.
22. Discuss the health benefits of including dietary fiber in meals and snacks.

See Appendix F for responses.

5.8 Nutrition Matters: Glycemic Index and Glycemic Load

Learning Outcomes

1. Explain the difference between the glycemic index and glycemic load.
2. Identify foods that have low or high glycemic indexes or glycemic loads.

In the early 1980s, researchers noted that the body digested carbohydrate–rich foods at different rates, and its insulin response to each food varied. Foods that contained large amounts of refined carbohydrates were digested rapidly, and the rapid flow of glucose into the bloodstream raised blood glucose and insulin levels sharply. Other carbohydrate–rich foods that had high fiber contents were digested slowly and did not cause such a dramatic increase in blood glucose and insulin levels. This observation led to the development of the dietary concepts *glycemic index* and *glycemic load.* Currently, the notion that a food's glycemic index could influence health has spurred the publication of popular diet books recommending avoiding foods with high glycemic indices, such as potatoes and bread. What is the glycemic index? Should you avoid foods that have high glycemic indices?

Comstock/PunchStock/Getty Images

The glycemic index (GI) is a way of classifying foods by comparing the rise in blood glucose that occurs after eating a portion of a food that supplies 50 g of digestible carbohydrate to the rise that occurs after eating a standard source of carbohydrate, such as glucose or white bread. A related value, the glycemic load (GL), is the grams of carbohydrate in a serving of food multiplied by the food's GI; this figure is then divided by 100. Compared to the GI, the GL may be a more realistic way of rating foods because the value indicates the relative rise in blood glucose levels after eating a *typical* serving of a carbohydrate—containing food.

Table 5.12 lists average GIs and GLs of several commonly eaten foods. Note that sucrose and other sugary foods, as well as highly refined starchy foods such as cornflakes, baked potatoes, and white rice, have high GIs (70 or more). Raw apples, carrots, spaghetti, fat–free milk, and mixed nuts have low to moderate GIs (less than 70). It is important to note that the GL of a food is usually lower than its GI. For example, the GI for cornflakes is 80, but the GL value for 1/2 cup (30 g) of cornflakes is only 19. Foods with low GLs have values below 10; high GL foods have values of more than 20. It may be helpful to consider the glycemic index and glycemic load of foods. For example, the GI of watermelon is 50, which is moderate; however, the melon's GL is low at 5 because watermelon is 90% water, and a typical serving of the fruit supplies 11 grams, a small amount of carbohydrates.

Critics think the GI and GL standards have limited usefulness as a menu–planning tool because the values can vary too much. Therefore, the glycemic index and glycemic load values shown in Table 5.12 may vary significantly, depending on where the food was grown, its degree of ripeness, or the extent of its processing. Additionally, individuals often experience different blood glucose levels after they eat the same carbohydrate–rich food. Furthermore, the values reflect a single food's effect on blood glucose levels. The effect may be reduced when the food is eaten as part of a meal that contains a mixture of macronutrients and fiber.

TABLE 5.12 Glycemic Index and Load: Average Values of Selected Foods

Food	Glycemic Index* Value	Glycemic Load**		
		Typical Serving Size	Carbs per Serving	Value
Cornflakes, cereal	80	1 cup	24 g	19
Potato, russet, microwaved	77	1 medium	36 g	27
Jelly beans, candy	76	10 pieces	27 g	20
Cola, sugar-sweetened drink	73	12 fl oz can	40 g	29
French fries, fast food	64	1 medium order	48 g	30
Wheat chapati bread	63	1 piece (68 g)	31 g	19
Popcorn, microwave, plain	62	2 cups	12 g	7
White rice, basmati, boiled	60	1 cup	44 g	26
Buckwheat (soba) noodles, cooked	59	1 cup	24 g	14
Watermelon	50	1 cup, cubed	11 g	5
Orange juice	48	8 fl oz	28 g	13
Banana	47	1 medium	27 g	12
Spaghetti, cooked	47	1 cup	38 g	17
Peaches, canned in natural juice	46	1 cup	29 g	13
Fat-free milk	46	1 cup	12 g	5
Milk chocolate candy	45	1 bar (46 g)	26 g	11
Apple, raw	44	1 medium (3 inches)	25 g	11
Carrots, boiled	32	1 cup	12 g	3
Mixed nuts	24	1 oz	7 g	1

*Compared to white bread (GI = 100); **Calculated per serving

Source: Data from Atkinson FS and others: International table of glycemic index and glycemic load value, Supplemental Table 1. *American Journal of Clinical Nutrition* 114(5): 1625–1632, 2021.

MaraZe/Shutterstock

Despite the criticism directed at the use of these indices, epidemiological studies suggest an association between high GI/GL diets and serious chronic diseases.[60] As people in developing countries abandon traditional diets and eat more high GI/GL foods, they become more likely to develop obesity and type 2 diabetes, as well as cardiovascular disease and certain cancers. Low GI diets may improve blood lipid levels and reduce the risk of cardiovascular disease and improve HbA1c levels.[60] People with diabetes can follow a low GI or low GL diet while monitoring their total carbohydrate intake to control their blood glucose levels.[60]

Concept CHECKPOINT

23. Which of these foods has the lowest glycemic index: plain, white bagel; raw apple; orange juice; or microwave popcorn?

See Appendix F for responses.

Chapter References

See Appendix G.

Summary

5.1 Introducing Carbohydrates

- Carbohydrates are an important source of energy for the body. Plants use energy from the sun to make carbohydrates from water and carbon dioxide. Some of the energy is stored in the bonds that hold the carbon and hydrogen atoms together. Cells break down those bonds, releasing the energy that powers various forms of cellular work.

5.2 Simple Carbohydrates: Sugars

- The three most important dietary monosaccharides are glucose, fructose, and galactose. Glucose is a primary fuel for muscles and other cells; nervous system and red blood cells rely on glucose for energy under normal conditions. Lactose and sucrose are major dietary disaccharides.

©Wendy Schiff

5.3 Complex Carbohydrates

- Starch, glycogen, and most forms of dietary fiber are polysaccharides. Although fiber is not digested by humans, soluble and insoluble fiber provide important health benefits.

5.4 What Happens to Carbohydrates in the Body?

- Glucose is the primary end product of carbohydrate digestion. Hormones, particularly insulin and glucagon, maintain normal blood glucose levels. Insulin enables glucose to enter cells, where the sugar is metabolized for energy. Glucagon stimulates the liver to break down glycogen into glucose molecules and release them into the bloodstream.

5.5 Carbohydrate Consumption Patterns

- People in industrialized nations tend to eat less complex, unprocessed carbohydrates and more highly refined carbohydrates than people living in less-developed countries. Healthy Americans should consume diets that furnish 45 to 65% of energy from carbohydrates, primarily complex carbohydrates. Refined carbohydrates are often blamed for causing obesity, diabetes, and hyperactivity, but tooth decay is the only health problem that is clearly associated with eating carbohydrates. Many adults are lactose intolerant because they do not produce enough lactase, the intestinal enzyme needed to digest the disaccharide.

©Wendy Schiff

5.6 Understanding Nutrient Labeling: Carbohydrates and Fiber

- The Nutrition Facts panel provides information about total carbohydrate, total sugars, added sugars, and fiber contents of packaged foods and beverages.

5.7 Carbohydrates and Health

- Foods that contain a lot of added sugars and fats tend to be energy dense. Some nutrition scientists think high intakes of added sugars, especially in regular soft drinks, are largely responsible for Americans' rising rate of obesity.
- Diabetes mellitus is characterized by elevated blood glucose levels. Diabetes can result in cardiovascular disease, kidney failure, blindness, and lower limb amputations. There are two major types of diabetes mellitus, type 1 and type 2 diabetes. Type 2 diabetes is the more common form of the disease. People who are sedentary, have excess body fat, eat Western diets, and have a close relative with type 2 diabetes are at risk of developing this form of the disease.
- Eating fiber-rich foods may reduce your risk of obesity, type 2 diabetes, cardiovascular disease, and certain intestinal tract disorders. High-fiber diets, especially those with whole grains, fruits, and vegetables, are associated with lower risk of colon cancer compared to diets that contain little fiber. Additionally, foods that contain soluble fiber may improve cardiovascular health by reducing cholesterol absorption.

ddsignstock/123RF

5.8 Nutrition Matters: Glycemic Index and Glycemic Load

- The glycemic index (GI) and glycemic load (GL) are ways of classifying certain foods by their effect on blood glucose levels. Results of some epidemiological studies suggest an association between high GI/GL diets and serious chronic diseases.

Recipe for Healthy Living

Berry-Good Hot Oatmeal Cereal

This single-serving hot oatmeal with blueberries recipe is a very good source of complex carbohydrates, fiber, calcium, and iron. For variety, substitute raisins, frozen or fresh strawberries, or raspberries for the blueberries. A serving of the cooked cereal supplies about 270 kcal, 12 g protein, 3 g fat, 52 g carbohydrate, 6 g fiber, 230 mg calcium, and 2.3 mg iron.

Lucy Stein/Image Source/Glow Images

INGREDIENTS:

½ cup frozen or fresh unsweetened blueberries
½ cup quick-cooking oats
¾ cup fat-free milk
1 tsp sugar
pinch of ground cinnamon (optional)

PREPARATION STEPS:

1. Place ingredients in a microwaveable bowl and stir.
2. Microwave on "high" for about 2 ½ minutes.
3. Stir and top with blueberries before eating.

Cleanup tip: After eating, soak cereal bowl in cold water before washing it.

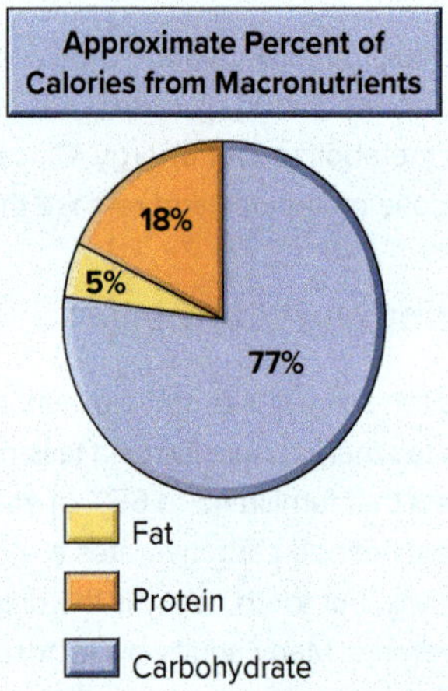

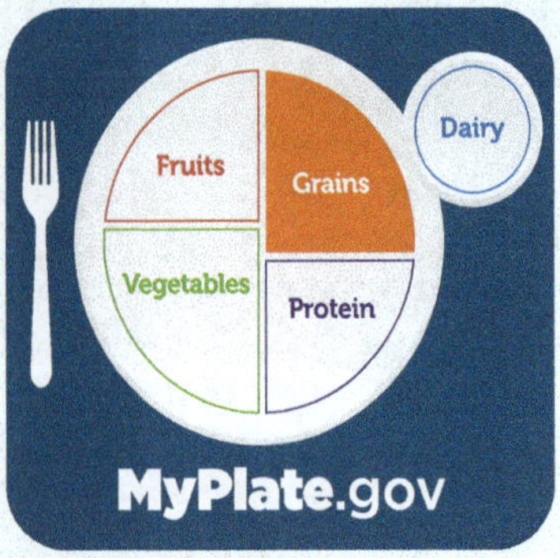

US Department of Agriculture or USDA

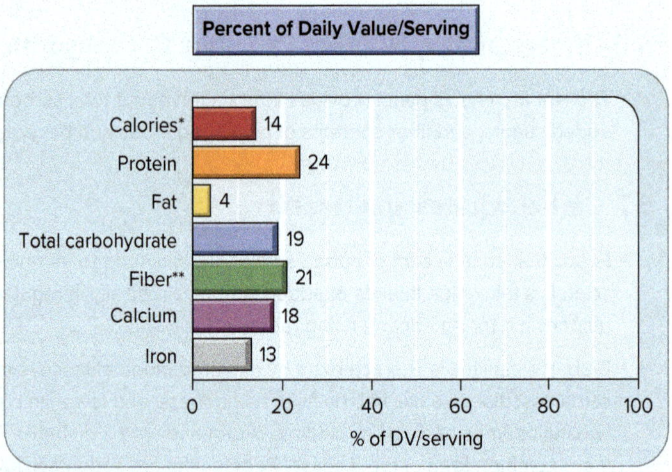

*2000 daily total kcal
**Component of Total carbohydrate

Personal Dietary Analysis

1. Refer to the 3-day food log from the "Personal Dietary Analysis" feature in Chapter 3. List the total number of kilocalories you consumed for each day of record keeping. Add the figures to obtain a total, divide the total by 3, then round the figure to the nearest whole number to obtain your 3-day average for energy intake.

Sample Calculation:

Day 1 2500 kcal

Day 2 3200 kcal

Day 3 2750 kcal

Total kcal **8450 ÷ 3 days = 2817 kcal/day**
(average kilocalorie intake, rounded to the nearest whole number)

Your Calculation:

Day 1 _____ kcal

Day 2 _____ kcal

Day 3 _____ kcal

Total kcal _____ **÷ 3 days =** _____ **kcal/day**
(average kilocalorie intake, rounded to the nearest whole number)

2. Add the number of grams of total carbohydrate eaten each day of the period. Divide the total by 3 and round to the nearest whole number to calculate the average number of grams of total carbohydrate consumed daily.

Your Calculation:

Day 1 _____ g

Day 2 _____ g

Day 3 _____ g

Total _____ g

Total g _____ **÷ 3 days =** _____ **g of total carbohydrate/day**
(average, rounded to the nearest whole number)

3. Each gram of carbohydrate provides about 4 kcal; therefore, you must multiply the average number of grams of total carbohydrate obtained in step 2 by 4 to obtain the number of kcal from total carbohydrate.

Your Calculation:

_____ **g/day × 4 kcal/g =** _____ **kcal from total carbohydrate**

4. To calculate the average daily percentage of kilocalories that carbohydrates contributed to your diet, divide the average kilocalories from total carbohydrate obtained in step 3 by the average total daily kilocalorie intake obtained in step 1; round the figure to the nearest one-hundredth. Multiply the value by 100, drop the decimal point, and add the percent symbol.

C Squared Studios/Photodisc/Getty Images

Sample Calculation:

1692 kcal ÷ 2817 kcal = 0.60

0.60 × 100 = 60%

Your Calculation:

____ kcal ÷ ____ kcal = ____

____ × 100 = ____ %

5. On average, did you consume *at least* the RDA of 130 g of carbohydrate?
Yes _____ No _____

6. Did your average carbohydrate intake meet the recommended 45 to 65% of total energy? Yes _____ No _____

   a. If your average carbohydrate intake was less than 130 g or below 45% of total calories, list five nutrient-dense, carbohydrate-rich foods you could eat that would boost your intake of carbohydrates.

   Foods: ____________________

   b. If your average carbohydrate intake was greater than 130 g or more than 65% of total energy, list the top five foods and beverages that contributed to your carbohydrate intake. ____________________

7. Calculate your average daily intake of fiber by adding the grams of fiber consumed over the 3-day period and dividing the total by 3.

Day 1 ____ g

Day 2 ____ g

Day 3 ____ g

Total ____ g

Total grams ____ ÷ 3 days = ____ g of fiber daily

   a. What was your average daily fiber intake? _____ g

   b. Did your average daily fiber intake meet the recommended Adequate Intakes of 38 and 25 g/day for young males and females, respectively?
   Yes _____ No _____

   c. If your response is yes, list foods that contributed to your fiber intake.

   d. If you did not meet the recommended level of fiber intake, list at least five foods that you would eat to increase your fiber intake to the recommended level.

Mc Graw Hill connect® Complete the Personal Dietary Analysis activity in Connect.

Critical Thinking

1. One of your friends thinks honey is more nutritious and safer to eat than table sugar. They want you to avoid table sugar and use only honey as a sweetener. What would you tell this person about the nutritive value and safety of honey compared to sugar?
2. Maria has diabetes, and she just found out that she is in the sixth week of pregnancy. She wants to know what she can do to increase her chances of having a problem–free pregnancy and healthy baby. What would you tell her about the importance of properly managing her blood glucose level during pregnancy?
3. How did you feel about drinking sugar—sweetened soft drinks, sports drinks, and energy drinks before reading this chapter? Has your opinion changed? If so, explain how.
4. Wyatt is a 5–year–old whose mother thinks sugar causes hyperactivity, so she forbids him from eating sweets. When Wyatt is invited to his friends' birthday parties, his mother gives him a small bag of apple slices to eat instead of cake and ice cream. What would you tell Wyatt's mother concerning sugar intake as a cause of hyperactivity?

Dima Fadeev/Shutterstock

5. Consider the fiber content of your diet. Do you consume enough fiber each day? If your fiber intake is adequate, what foods do you eat regularly that contribute soluble and insoluble fiber to your diet? If your fiber intake is low, list foods you would consume to increase your intake of both types of fiber.
6. Kyran experiences intestinal cramps and gas after drinking a glass of milk. He suspects that he has lactose intolerance. What steps can he take to obtain adequate nutrition without eliminating dairy foods from his diet?

Practice Test (Section numbers are listed below in red.)

Select the best answer.

1. (5.1) _____ is a primary fuel for muscles and other cells.
   a. Protein
   b. Cholesterol
   c. Glucose
   d. HFCS
2. (5.2) Which of the following substances is a disaccharide?
   a. Maltose
   b. Starch
   c. Galactose
   d. Glycogen
3. (5.3) Which of the following substances is a polysaccharide?
   a. Glucose
   b. Glycogen
   c. Glucagon
   d. Galactose
4. (5.3) Dietary fiber
   a. supplies more energy, gram per gram, than fat.
   b. is not digested by the human intestinal tract.
   c. promotes tooth decay.
   d. is only in animal sources of food.
5. (5.3) Insoluble fiber
   a. is in beef and pork.
   b. dissolves or swells in water.
   c. is in whole–grain products, including brown rice.
   d. increases the risk of heart disease.
6. (5.3) Which of the following foods is the best source of soluble fiber?
   a. Cows milk
   b. Whole–grain oat cereal
   c. Sports drinks
   d. Cooked meat

7. (5.4) After he eats a meal, Kyle's pancreas secretes ____, which lowers his blood glucose level.
   a. glucagon
   b. insulin
   c. HFCS
   d. maltase
8. (5.5) _____ are a primary source of added sugars in American diets.
   a. Cured deli meats
   b. Sugar–sweetened soft drinks
   c. Refined cereals
   d. Canned fruits
9. (5.6) A 1–cup serving of Zenong's favorite cereal supplies 12 grams of added sugar (24%DV) and 3 grams fiber (10%DV). Which statement is true for a 1–cup serving of cereal?
   a. The cereal supplies very little added sugar.
   b. The cereal supplies a high amount of added sugar.
   c. The cereal supplies very little fiber.
   d. The cereal supplies a high amount of fiber.
10. (5.7) Which of the following conditions is a common sign or symptom of untreated type 1 diabetes?
   a. Migraine headaches
   b. Obesity
   c. Diverticula
   d. Hyperglycemia
11. (5.7) Type 2 diabetes is
   a. a disease that primarily affects young children.
   b. characterized by severe hypoglycemia.
   c. often associated with excess body weight.
   d. caused by eating refined sugars.
12. (5.7) Which of the following signs is associated with metabolic syndrome?
   a. Low blood pressure
   b. High fasting blood glucose
   c. Low hemoglobin
   d. High fasting HDL cholesterol
13. (5.7) Which of the following conditions is clearly associated with eating certain dietary carbohydrates, especially sticky sugars?
   a. Tooth decay
   b. Type 1 diabetes
   c. Attention deficit hyperactivity disorder
   d. Hypertension
14. (5.7) Ben often experiences abdominal discomfort, bloating, and diarrhea after consuming milk. Which intestinal enzyme could Ben be deficient in?
   a. Galactose
   b. Salivary amylase
   c. Lactase
   d. Lactic acid
15. (5.8) Which of the following foods has a high glycemic index?
   a. Nonfat milk
   b. Cornflakes cereal
   c. Mixed nuts
   d. Raw carrots

Answers to Quiz Yourself

1. Compared to table sugar, honey is a natural and far more nutritious sweetener. **False.** (Section 5.2)
2. Ounce per ounce, sugar provides more energy than starch. **False.** (Sections 5.2 and 5.3)
3. Eating a high-fiber diet may reduce your blood cholesterol level. **True.** (Section 5.7)
4. The average American consumes 80% of their energy intake as added sugars. **False.** (Section 5.5)
5. The results of clinical studies indicate that eating too much sugar makes children hyperactive. **False.** (Section 5.7)

Design Elements: (Stacked books): Siede Preis/Photodisc/Getty Images; (Nuts): C Squared Studios/Photodisc/Getty Images; (Recipe box, Raspberries, Blueberries): ©Wendy Schiff; (Pineapple): Stockbyte/Getty Images; (Radishes): karandaev/iStock/360/Getty Images; (MyPlate icon): US Department of Agriculture or USDA

CHAPTER 6

Fats and Other Lipids

Quiz Yourself

What are trans and omega-3 fats, and which foods contain these fats? How much dietary fat is recommended? Should you avoid eating eggs because they contain cholesterol? Test your knowledge of fat and other lipids by taking the following quiz. The answers are at the end of the chapter.

1. To lose weight, use regular, stick margarine instead of butter because it has 25% fewer calories per teaspoon. ______ T ______ F
2. Egg yolks are a rich source of cholesterol. ______ T ______ F
3. Taking too many fish oil supplements may be harmful to health. ______ T ______ F
4. On average, Americans consume 60% of their calories from fat. ______ T ______ F
5. Increasing your intake of saturated fat will reduce your risk of heart disease. ______ T ______ F

Brent Hofacker/123RF

The oil and the vinegar (which is 95% water) in this salad dressing do not mix.
vincent noel/Shutterstock

6.1 Introducing Lipids

Learning Outcomes

1. Identify the three major kinds of lipids.
2. List at least six functions of lipids in the body.
3. List at least three important functions of lipids in foods.

What do you think when you hear the word *fat* or *cholesterol?* Does "bad," "heart attack," or "deadly" enter your mind? If your answer is yes, are you concerned about the amounts and types of fat in your diet? Do you have higher–than–normal levels of triglycerides and cholesterol in your bloodstream? Do you avoid eating eggs because of their cholesterol content? Does your concern have anything to do with having a family history of heart disease?

Lipids are a class of macronutrients that includes fatty acids, triglycerides (*try–glis'–eh–rides*), phospholipids (*fos–foe–lih'–pids*), and cholesterol. In general, lipids are insoluble in water, but they are soluble in organic solvents such as alcohol and acetone. Consider what happens when you mix vinegar and canola oil to make a vinaigrette salad dressing. Vinegar is 95% water; oil is 100% triglyceride (fat). Therefore, the oil does not dissolve in water to make a solution. Additionally, oil is less dense than water, so it rises to the top of the vinegar in small globules when added to vinegar. The globules join others to form an oily layer that floats on the vinegar until you shake the mixture. Shaking the ingredients mixes them temporarily. When left undisturbed, the oil and vinegar soon separate; hence the saying, "Oil and water don't mix."

lipids class of nutrients that generally do not dissolve in water

You probably know fat (triglyceride) is a major source of energy and eating too many calories from fat can result in excess weight gain, which is unhealthy. Additionally, you may know that cholesterol is associated with heart attacks. However, you may not be aware of the many important roles that fat, cholesterol, and other lipids play in the body (Table 6.1). Lipids are crucial components of the plasma membrane that surround each human cell. In fact, a person cannot claim to be "fat free" because every cell in the body contains fat, as well as other lipids. The layer of fat under your skin (subcutaneous fat) stores energy, insulates you against cold temperatures, and protects you against minor bruising. In addition to storing energy and fat–soluble vitamins, the fat deposits in your body and abdominal region cushion your vital organs from jarring movements and damaging blows.

In food, lipids enhance intestinal absorption of fat–soluble vitamins and phytochemicals. Dietary lipids also provide nonnutritional benefits by contributing to the rich flavor, smooth texture, and appetizing aroma of foods. Whether fat is naturally in food or added to it, the nutrient often makes foods taste more appetizing. For example, if you are used to consuming whole milk that is about 3.25% fat by volume, you will recognize the difference fat makes to "mouth feel" when you drink fat–free milk that contains less than 0.5% fat.

It is not surprising that many Americans are confused about the roles of fat and cholesterol in health. The results from numerous studies indicate that consuming high amounts of certain lipids may increase the risk of serious health conditions, particularly cardiovascular disease (CVD), which includes common forms of heart disease.[1] On the other hand, some dietary fat is essential to good health. By reading this chapter, you will learn about the roles of lipids in your foods and body, as well as their major food sources. Additionally, you will learn how certain lipids may influence your health.

TABLE 6.1 Major Functions of Lipids in the Body

Fats and certain other lipids are important for:

- providing and storing energy (fat);
- maintaining cell membranes;
- producing certain hormones;
- insulating the body against cold temperatures;
- cushioning the body against bumps and blows;
- absorbing and storing fat-soluble vitamins and phytochemicals.

Concept CHECKPOINT

1. Eighty-eight-year-old Rose has more body fat than her 88-year-old cousin, Lily. Explain why having excess body fat can be beneficial for older adults.
2. The recipe for a cake includes ½ cup of water. What would happen to the sensory quality of the baked cake if you replace the water with the same amount of whole milk?

See Appendix F for responses.

6.2 Fatty Acids, Triglycerides, Phospholipids, and Sterols

Learning Outcomes

1. Explain the differences in the chemical structures of saturated, monounsaturated, polyunsaturated, and trans fatty acids; triglycerides; phospholipids; and cholesterol.
2. Identify major food sources of each kind of lipid, including the essential fatty acids, phospholipids, and cholesterol.
3. Discuss the roles of essential fatty acids, phospholipids, and cholesterol in the body.
4. Discuss signs of an essential fatty acid deficiency.

Fatty Acids

Most lipids, particularly triglycerides and phospholipids, have fatty acids in their chemical structures. Fatty acids provide energy for muscles and most other types of cells. As Figure 6.1 illustrates, a fatty acid is composed of a **hydrocarbon chain,** a chain of carbon atoms bonded to each other and to hydrogen atoms. The first carbon in the molecule has three hydrogen atoms attached to it. Chemists call this part of the molecule the **omega** (or methyl) **end.** The last carbon in the fatty acid molecule forms an acid group.

In nature, common fatty acids have even numbers of carbon atoms. *Short–chain* fatty acids have 2 to 4 carbons; *medium–chain* fatty acids have 6 to 12 carbons; and *long–chain* fatty acids have 14 to 24 carbons. The molecules shown in Figure 6.1 contain 18 carbons; thus, they are long–chain fatty acids. Chemists identify a fatty acid by its number of carbon atoms and type of bond between carbon atoms in the hydrocarbon chain. Additionally, these factors influence how various fatty acids can affect a person's health.

Saturation

Fatty acids can be saturated or unsaturated. The carbons in the fatty acid chain, shown in Figure 6.1a, have single bonds between them. Note that each carbon in the chain has two hydrogen atoms attached to it. This is a **saturated fatty acid** because each carbon within the chain is saturated, that is, completely filled with hydrogen atoms.

An **unsaturated fatty acid** has two neighboring carbons within the chain that are missing two hydrogen atoms, and a double bond holds those particular carbons together (Fig. 6.1b). Unsaturated fatty acids can have one or more double bonds within the chain of carbons. The fatty acid illustrated in Figure 6.1b has only one double bond linking two carbon atoms; therefore, it is referred to as a **monounsaturated fatty acid (MUFA).** The fatty acid shown in Figure 6.1c is also unsaturated, but it has three double bonds within its hydrocarbon chain. A fatty acid that has two or more double bonds between carbons is a **polyunsaturated fatty acid (PUFA).**

hydrocarbon chain chain of carbon atoms bonded to each other and to hydrogen atoms

omega end first carbon of a fatty acid chain that has three hydrogen atoms attached to it

saturated fatty acid fatty acid that has each carbon atom within the chain filled with hydrogen atoms

unsaturated fatty acid fatty acid that is missing hydrogen atoms and has one or more double bonds within the carbon chain

monounsaturated fatty acid (MUFA) fatty acid that has one double bond within the carbon chain

polyunsaturated fatty acid (PUFA) fatty acid that has two or more double bonds within the carbon chain

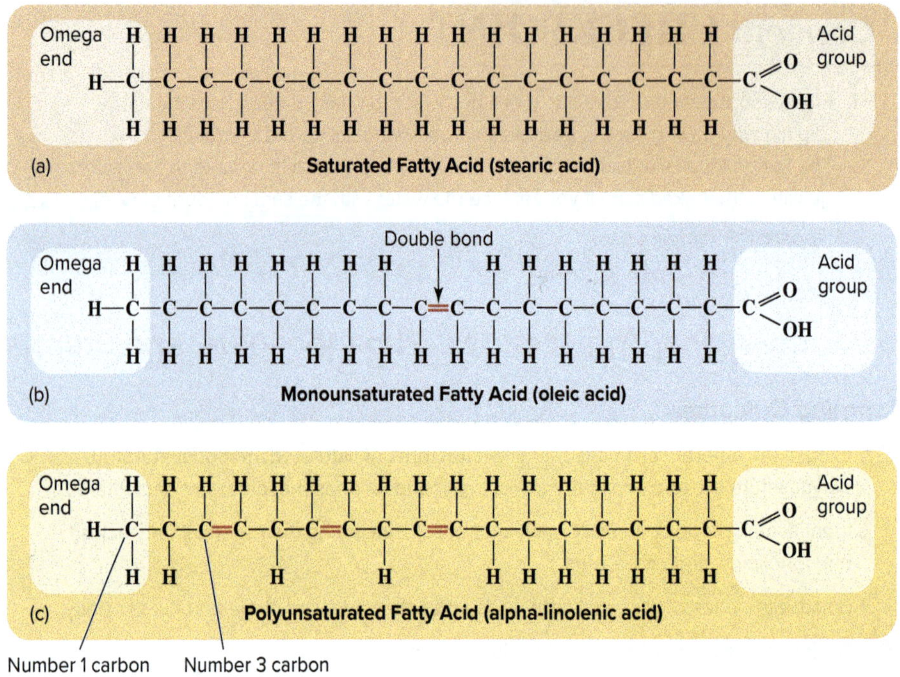

Figure 6.1 Fatty acids. A fatty acid is composed of a hydrocarbon chain. The omega end contains the first carbon in the molecule; the last carbon in the molecule forms an acid group.

What is the difference between fats and oils? Although both substances are triglycerides that contain fatty acids, fats are solid and most oils are liquid at room temperature. Compared to foods that contain high amounts of unsaturated fatty acids, foods that are rich sources of long–chain saturated fatty acids, such as the fatty acid shown in Figure 6.1a (stearic acid), tend to be more solid at room temperature. A pat of butter, for example, contains more long–chain saturated fatty acids than a pat of soft margarine (80% fat) that is rich in unsaturated fatty acids. Thus, butter, which is a "solid fat," keeps its shape better than margarine when it is not refrigerated. The "Lipids and Health: Cardiovascular Disease" section of this chapter discusses the health effects of eating diets that are rich in saturated or unsaturated fats.

Essential Fatty Acids

The body cannot make two PUFAs, **alpha–linolenic** (*al'–fah lin'–o–len'–ik*) **acid** and **linoleic** (*lin'–o–lay'–ik*) **acid.** These lipids are **essential fatty acids** because they must be supplied by the diet. Alpha–linolenic acid is an **omega–3 fatty acid.** The "3" refers to the position of the first double bond that appears in the fatty acid's carbon chain, when you start counting carbons at the omega end of the molecule (see Figure 6.1c). Cells use alpha–linolenic acid to synthesize two other omega–3 fatty acids, eicosapentaenoic (*eye'–koss–uh–pen'–tah–ee–no'–ik*) acid (EPA) and docosahexaenoic (*doe'–koss–uh–hex'–uh–ee–no'–ik*) acid (DHA). However, the body's ability to make EPA and DHA from alpha–linolenic acid is limited.

Linoleic acid is an *omega–6 fatty acid.* Cells can convert linoleic acid to arachidonic (*air'–eh–keh–dah'–nik*) acid (AA). The body uses EPA, DHA, and AA to make several compounds that have hormonelike functions, including *prostaglandins.* Prostaglandins produce a variety of important effects on the body, such as stimulating uterine contractions, regulating blood pressure, and promoting the immune system's inflammatory response. Figure 6.2 shows relationships among the essential fatty acids.

People need small amounts of essential fatty acids for good health. The Adequate Intake (AI) for alpha–linolenic acid is 1.6 g/day for males and 1.1 g/day for females. The AI for linoleic acid is 17 g/day for males and 12 g/day

alpha-linolenic acid an essential fatty acid

linoleic acid an essential fatty acid

essential fatty acids lipids that must be supplied by the diet

omega-3 fatty acid type of polyunsaturated fatty acid

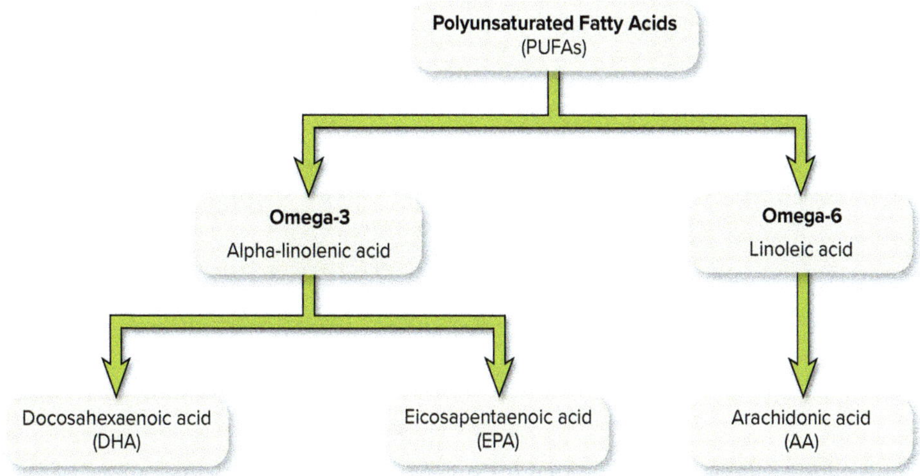

Figure 6.2 Essential fatty acids. Alpha-linolenic acid and linoleic acid are not synthesized by the body; they must be supplied in the diet. The body uses these essential fatty acids to make DHA, EPA, and AA.

for females who are 19 through 50 years of age.[4] These amounts can be met by eating 2 to 3 tablespoons of vegetable fat daily, especially products made with canola and soybean oils, and meals that contain fatty fish, such as salmon and tuna, at least twice a week. The typical American diet provides far more omega–6 foods than foods that contain omega–3 fatty acids.[1]

Essential Fatty Acid Deficiency In the United States, essential fatty acid deficiency is uncommon because most Americans consume plenty of fat in their diets. However, signs of essential fatty acid deficiency include scaly skin, hair loss, and poor wound healing. Infants have poor nervous system development and babies do not grow properly when their diets lack essential fatty acids.

Linoleic acid is the main polyunsaturated fatty acid in the American diet. This omega–6 fatty acid is in vegetable oils, often used for frying foods and making margarines and salad dressings. Additionally, whole–grain products contain linoleic acid. Certain fish are rich sources of omega–3 fatty acids ("omega–3s"), including DHA and EPA (see Table 6.7). Many Americans eat relatively small amounts of omega–3 fatty acids. The "Lipids and Health: Cardiovascular Disease" section of this chapter presents more information about essential fatty acids and their health effects.

Trans Fats

Unsaturated fatty acids usually have the two hydrogen atoms of the double–bonded carbons on the same side of the molecule. This type of structure is called a cis fatty acid. Cis fatty acids, such as oleic acid, have a "kink" or bend where the double bonds are in the carbon chain (Fig. 6.3a). A **trans fatty acid,** such as elaidic acid, has the two hydrogen atoms of the double–bonded carbons on the opposite sides of the molecule (see Figure 6.3b).

Whole milk and whole–milk products, butter, and meat naturally contain small amounts of trans fats. Although the body can use trans fatty acids for energy, these lipids are not essential, and medical researchers have not discovered any positive health effects from consuming them.[5] In the body, trans fats raise blood levels of an unhealthy form of cholesterol, which increases the risk of heart disease.[1]

Until recently, most of the trans fat in processed food (artificial trans fat) resulted from *partial hydrogenation.* **Partial hydrogenation** was a food manufacturing process that added hydrogen atoms to some unsaturated fatty acids

Did *YOU* Know?

The human brain is composed of 60% fat, which makes it one of the most lipid-dense organs in the body.[2] DHA is the main omega-3 fatty acid in the brain, while AA is the organ's main omega-6 fatty acid. Large amounts of essential fatty acids accumulate in the brain and the retina (light-sensitive tissue) of the eye prior to birth and during the first 2 years of life. Essential fatty acids are transferred from a pregnant female's bloodstream to the brain of her fetus (developing human being). In fact, a fetus acquires 80% of its total brain DHA content during the last 3 months of pregnancy.[3]

It is critical that babies consume adequate intakes of essential fatty acids during their first 2 years of life for good vision and normal brain and intellectual development. After birth, babies receive essential fatty acids from the milk of breastfeeding mothers who consume a diet rich in omega-3 and omega-6 fatty acids. Babies can also obtain a supply of essential fatty acids if they consume infant formula that has been fortified with DHA and AA. Eventually, solid foods provide essential fatty acids for infants. Chapter 13 provides more information about nutrition during pregnancy and infancy.

lostinbids/E+/Getty Images

trans fatty acid unsaturated fatty acid that has a trans double bond

partial hydrogenation food manufacturing process that added hydrogen atoms to some unsaturated fatty acids, forming trans fats

Figure 6.3 Cis and trans fatty acids. (a) Oleic acid is a cis fatty acid. (b) Elaidic acid is a trans fatty acid. For simplicity, most of the hydrogen atoms in these molecules are not shown.

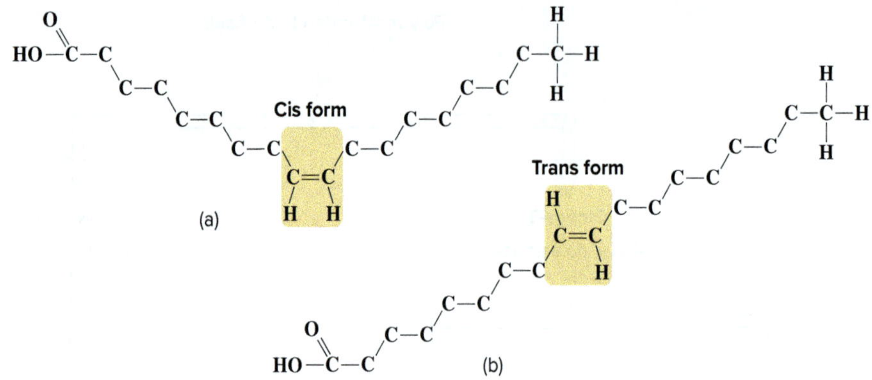

Did *YOU* Know?

Before the ban on partially hydrogenated oils, fast-food outlets relied on the trans fat for deep-fat frying menu items, such as French fries and breaded chicken. Because PHOs can no longer be used in processed or commercially prepared foods, the food industry is now relying on other oils, including high oleic oils, which are rich sources of monounsaturated fatty acids.

Mixa/Alamy Stock Photo

triglyceride lipid that has three fatty acids attached to a glycerol molecule

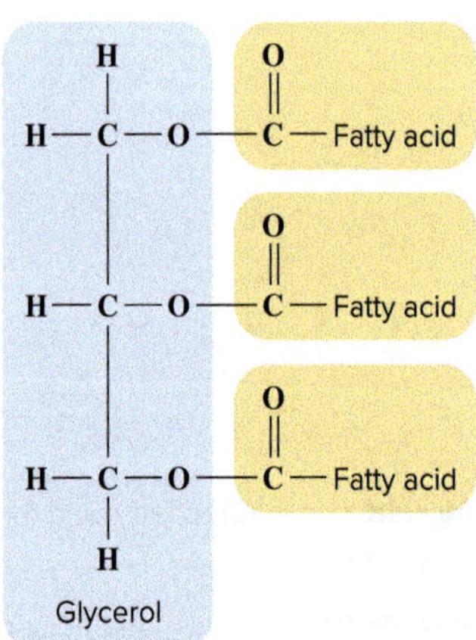

Figure 6.4 Triglyceride.

in liquid vegetable oil. The partial hydrogenation process converted many of the oil's naturally occurring cis fatty acids into trans fatty acids. Oils that contained these artificial fats were called partially hydrogenated oils (PHOs). Structurally, trans fatty acids resemble saturated fatty acids and provide properties of long–chain saturated fatty acids to foods that contain them. Thus, fats that contain a high proportion of PHOs are more solid at room temperature than those with a high proportion of cis fatty acids. As a result of the partial hydrogenation process, vegetable oil could be made into vegetable shortening (a solid fat) or shaped into sticks of margarine.

In 2015, the Food and Drug Administration (FDA) determined that PHOs posed a health risk, and in 2021 the agency banned their use in foods.[6] Why did food manufacturers make products that contained PHOs? Foods made with PHOs can be stored for longer periods than foods that contain cis fatty acids. Trans fatty acids are less likely to undergo oxidation, a chemical process that alters the compound's structure. When oxidized, the fat in food becomes rancid and develops an unappetizing odor and taste. Unsaturated fatty acids that have the cis double–bond arrangement, especially polyunsaturated fatty acids, are very susceptible to oxidation.

Instead of relying on PHOs to extend the shelf life of products, manufacturers can preserve fat and other ingredients in foods by adding antioxidants to them. Certain food additives, vitamins, and plant pigments function as antioxidants (see Chapter 8). Some manufacturers replace artificial trans fats in processed foods with *fully hydrogenated oils,* which are sources of saturated fatty acids, but not trans fats. The use of an artificial lipid, *interesterified* (*in'–ter–eh–stear'–ih–fide'*) oil, to replace PHOs in processed foods may also have undesirable health effects.[7] More research is needed to determine the long–term safety of consuming foods that contain interesterified oil. Processed foods that include "fully hydrogenated" or "interesterified" oil in their ingredient lists are sources of interesterified fat.

Despite the ban on PHOs, trans fat is not likely to disappear completely from Americans' diets because some animal products, such as beef and dairy foods, naturally contain small amounts of this type of fat. People can learn how much trans fat is in a serving of most packaged foods by reading the ingredient list and Nutrition Facts panel on the product's label. According to the Dietary Guidelines, Americans should keep their trans fat intake as low as possible, while eating a healthy diet.[8]

Triglycerides

A **triglyceride** has three fatty acids attached to glycerol (*glis'–er–ol*), a three–carbon compound that is often referred to as the "backbone" of the triglyceride (Fig. 6.4). Triglycerides comprise about 95% of lipids in your body and food.

Triglycerides are often referred to as fats and oils. The body stores energy as triglycerides (fat).

Most triglycerides contain mixtures of unsaturated and saturated fatty acids. In a particular food, such as olives or cheese, the unsaturated and saturated fats occur in different proportions, but one type of fatty acid (saturated, monounsaturated, or polyunsaturated) often predominates. Table 6.2 compares the percentages of saturated, monounsaturated, and polyunsaturated fatty acids in commonly eaten fats. Note that the fat in beef and dairy products contains more saturated than unsaturated fatty acids; olive oil and avocado oil are rich sources of monounsaturated fatty acids; and liquid corn oil and walnut oil contain a greater proportion of unsaturated than saturated fatty acids. Certain animal foods, especially beef and dairy foods such as cheese, cream, and butter, contain higher percentages of saturated fatty acids than most plant fats. Exceptions are fats and oils from tropical plants such as cocoa butter, coconut oil, and palm oil. Tropical fats and oils contain more saturated than

Did *YOU* Know?

Lard is pork fat. In some parts of the United States, lard is often used to make biscuits, pie dough, and refried beans. Lard is high in saturated fat (39%), but it is not as highly saturated as butter (50%).

DNy59/E+/Getty Images

TABLE 6.2 Comparing Fats and Oils: Saturated, Monounsaturated, and Polyunsaturated Fatty Acids

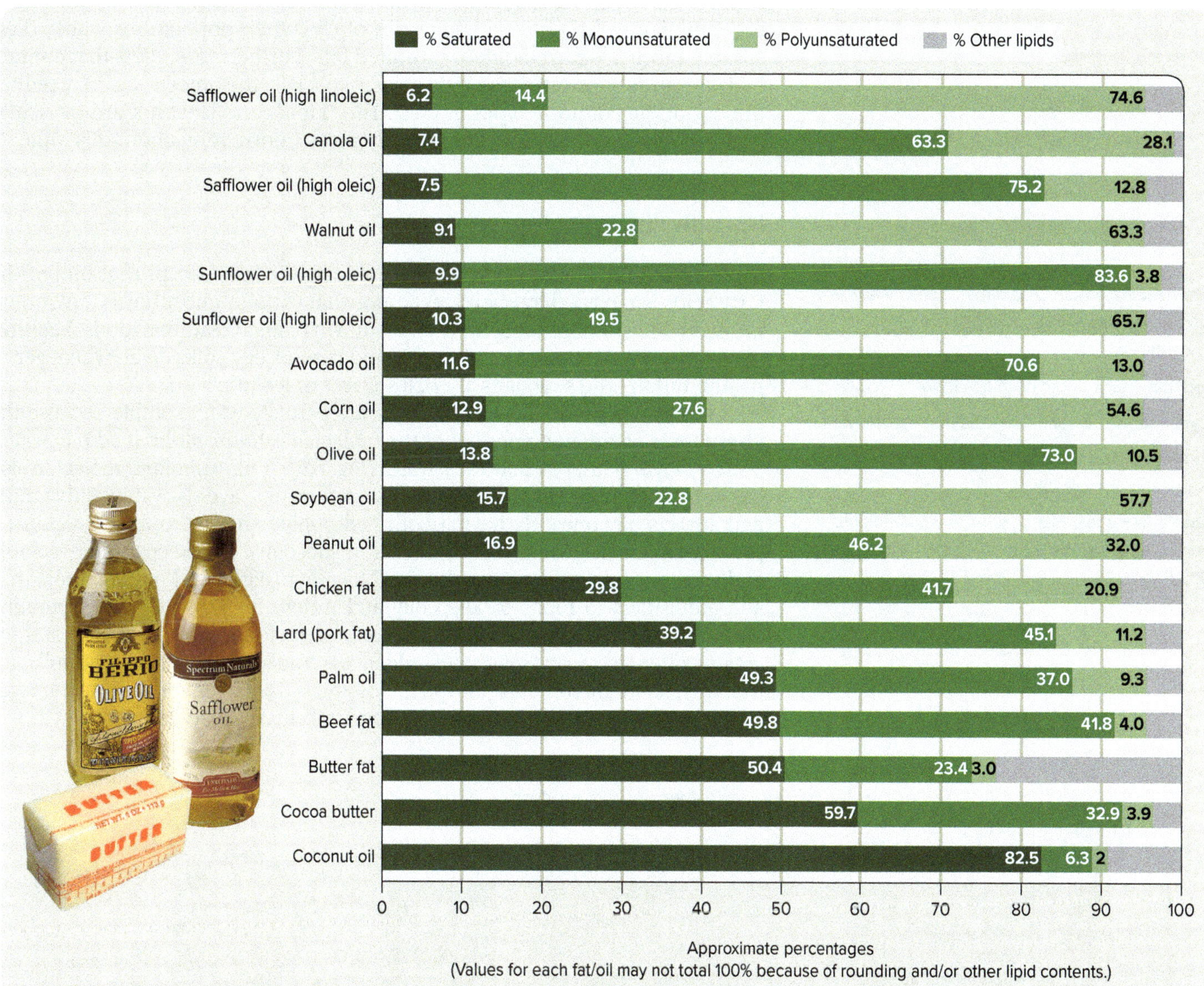

Source: Data from USDA Agricultural Research Service, *FoodData Central*, 2019.

Butter: D. Hurst/Alamy Stock Photo; Cooking oils: Elite Images/McGraw Hill

TABLE 6.3 Food Sources Rich in Saturated, Unsaturated, and Trans Fats

Type of Fat	Food Sources
Saturated	Beef, butter, cream, whole and reduced-fat dairy products, coconut
Monounsaturated	Avocados, olives, peanuts, pumpkin and sesame seeds, almonds, hazelnuts, pecans, fish
Polyunsaturated	Walnuts, flaxseed, fish
Trans	Beef, milk and other dairy products

unsaturated fatty acids. Fats and oils that contain high amounts of saturated or unsaturated fatty acids are commonly called saturated fats or unsaturated fats. Table 6.3 lists foods rich in saturated fats, unsaturated fats, and trans fats.

Why is it important to understand the differences between saturated, unsaturated, and trans fats, and identify foods that contain high amounts of these fats? Populations that consume diets rich in saturated fat and trans fat ("unhealthy fat") have a higher risk of CVD than populations whose diets contain more unsaturated than saturated fat.[1] Furthermore, diets that contain high amounts of unsaturated fatty acids, especially the omega–3s, may reduce the risk of the common types of CVD. The "Lipids and Health: Cardiovascular Disease" section of this chapter provides more information about this topic.

Phospholipids

phospholipid type of lipid that is chemically similar to a triglyceride, except it contains phosphorus

lecithin major phospholipid in food

hydrophilic part of a molecule that attracts water

hydrophobic part of a molecule that avoids water and attracts lipids

emulsifier substance that helps water-soluble and water-insoluble compounds mix with each other

A **phospholipid** is chemically similar to a triglyceride, except that one of the fatty acids is replaced by chemical groups that contain phosphorus and, often, nitrogen. Phospholipids are naturally found in plant and animal foods. **Lecithin** (*less'–uh–thin*) is the major phospholipid in food; egg yolks, liver, wheat germ, peanut butter, and soybeans are rich sources of lecithin.

Unlike triglycerides, phospholipids are partially water soluble because the phosphorus–containing portion of the molecule is **hydrophilic** (*hydro* = water; *philic* = loving); that is, it attracts water (Fig. 6.5). A phospholipid molecule also has a **hydrophobic** (*phobic* = fearing) portion that avoids watery substances and attracts oily ones. By having both hydrophilic and hydrophobic regions, a phospholipid can serve as an **emulsifier,** a substance that keeps water–soluble and water–insoluble compounds mixed together (Fig. 6.6). Manufacturers may add emulsifiers to foods to keep oily and watery ingredients from separating during storage. Processed foods, such as cheese food, salad dressing, and ice cream, often have phospholipids added as emulsifying agents. Egg yolk, for example, naturally contains phospholipids and is used to emulsify oil and vinegar when making mayonnaise or mixing oil and milk in cake batters.

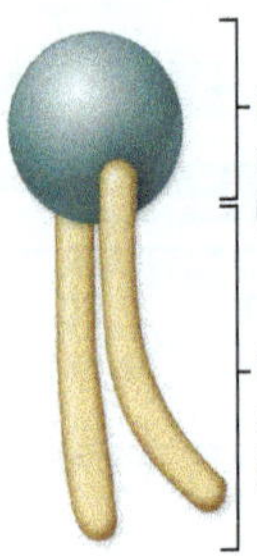

Figure 6.5 Phospholipids have a hydrophilic and a hydrophobic region.

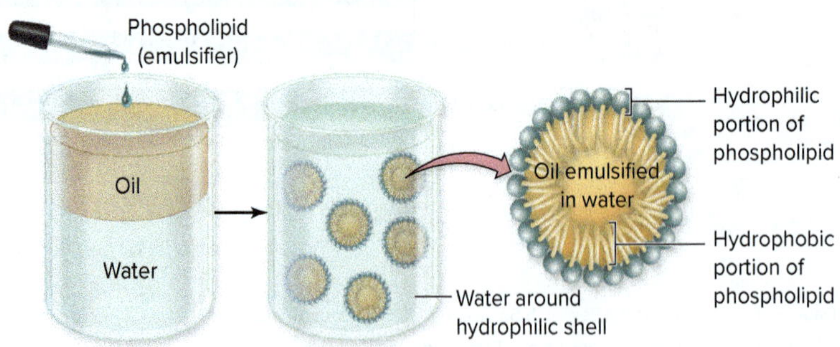

Figure 6.6 Emulsification. Phospholipids act as emulsifiers.

In the body, phospholipids are major structural components of cell membranes. Cell membranes are composed of a double layer that is mostly phospholipids (Fig. 6.7). The chemical structure of the phospholipids enables the membrane to be flexible and function properly. Phospholipids are also needed for normal functioning of nerve cells, including those in the brain. Phospholipid deficiencies among adults are uncommon because the lipids are in a variety of foods and bodies of healthy adults synthesize these compounds.

Lecithin contains **choline** (*co′–leen*), a water–soluble compound that nerves use to produce the neurotransmitter *acetylcholine* (*ah–see′–till–co′–leen*). A neurotransmitter is a chemical that transmits messages between the nerve cells. The human body makes small amounts of choline, but the phospholipid is classified as a vitamin–like nutrient because deficiency symptoms can occur under certain conditions.[9] Chapter 8 provides more information about choline.

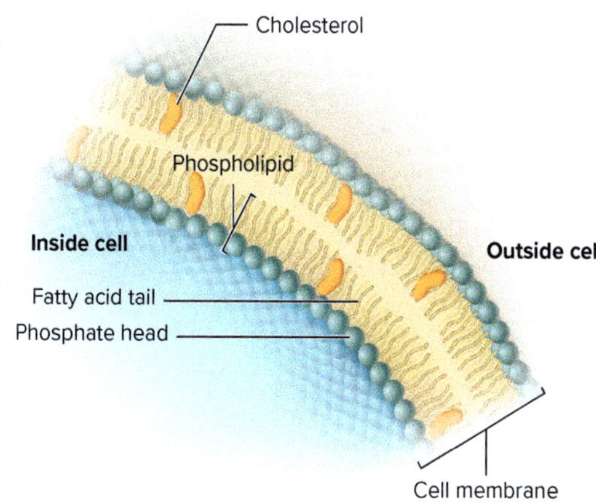

Figure 6.7 Phospholipids in cell membranes. Human cell membranes are composed of a double layer of phospholipids. Cholesterol is embedded in the membrane.

Sterols

Sterols have carbons arranged in rings, which makes them a more chemically complex type of lipid than a triglyceride or a phospholipid (Fig. 6.8). **Cholesterol** is the most well–known sterol. Many people think that cholesterol is unhealthy and foods that contain the lipid should be avoided, but it is a very important nutrient. Cholesterol is a component of every cell membrane in your body (see Figure 6.7). Although cholesterol is not metabolized for energy, cells use the lipid to synthesize a variety of substances, including vitamin D, and steroid hormones such as estrogen and testosterone. The liver uses cholesterol to make bile, an emulsifier that facilitates lipid digestion.

choline water-soluble compound in lecithin

cholesterol sterol in animal foods and precursor for steroid hormones, bile, and vitamin D

Although triglycerides are widespread in foods, cholesterol is found only in animal foods. Egg yolk, liver, meat, poultry, whole milk, cheese, and ice cream are rich sources of cholesterol (Table 6.4). Even if you do not eat animal foods, your body produces cholesterol, primarily in the liver. If your body makes too much cholesterol, the excess can increase your risk of CVD.

Plant Sterols and Stanols

Plants make small amounts of substances that have chemical structures that are similar to cholesterol. These substances are *phytosterols* and *phytostanols*. Although plant sterols and stanols are not well absorbed by the human intestinal tract, these substances compete with cholesterol for absorption by the digestive tract. This competition reduces the amount of cholesterol that is absorbed. As a result, phytosterols and phytostanols are added to certain foods, beverages, and dietary supplements because they may lower elevated

Examples of spreads that contain plant sterols and stanols.
©Wendy Schiff

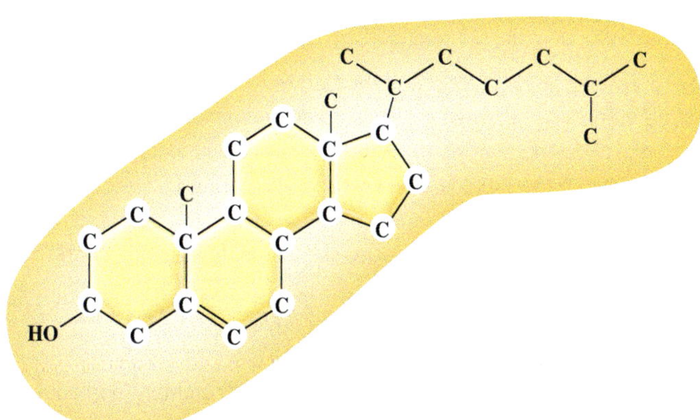

Figure 6.8 Cholesterol. Compared to the triglyceride molecule shown in Figure 6.4, cholesterol has a more complex chemical structure. For simplicity, this illustration shows only the molecule's OH group and carbon atoms.

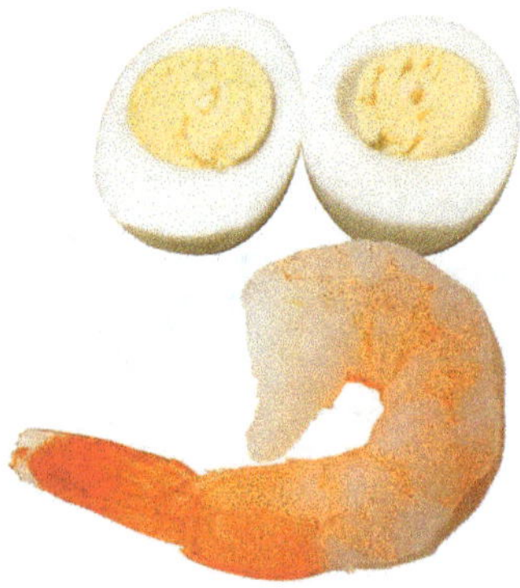

Eggs: Brian Yarvin/Alamy Stock Photo; Hamburger patty, Shrimp: ©Wendy Schiff

TABLE 6.4 Approximate Cholesterol Content of Some Foods/Serving

Food	Serving Size	Cholesterol (mg)
Liver, beef, braised	3 oz	337
Egg, hard cooked	1 large	186
Egg yolk, hard cooked	1	186
Sardines, Atlantic, canned in oil	3 oz	161
Shrimp, cooked	3 oz	161
Turkey, cooked, dark meat, no skin	3 oz	109
Turkey patty, ground, lean, broiled	3 oz	89
Danish fruit-filled pastry	2½ oz	81
Ice cream, soft-serve	½ cup	78
Chicken breast, meat only, cooked	3 oz	77
Beef patty, ground, lean broiled	3 oz	75
Salmon, Atlantic, cooked	3 oz	54
Ham, smoked, cooked	3 oz	36
Egg noodles, cooked	1 cup	46
Cottage cheese	1 cup	38
Hot dog	1	28
Cheddar cheese	1 oz	28
Chocolate milkshake	1 cup	25
Whole milk	1 cup	24

Source: Data from USDA Agricultural Research Service, *FoodData Central*, 2019.

blood cholesterol levels, a risk factor for heart disease. Cholesterol–lowering margarine–like spreads such as Benecol® and Promise Activ® contain plant sterols and stanols. These are functional foods because they have been developed specifically to provide beneficial health effects (see Chapter 1).

Concept CHECKPOINT

3. What are the major lipids in food and the body?
4. What is the difference between a saturated and an unsaturated fatty acid? What is the difference between a monounsaturated and a polyunsaturated fatty acid?
5. Identify at least one food that is a rich source of saturated fat, monounsaturated fat, and polyunsaturated fat.
6. Identify the two essential fatty acids, and list at least three signs of an essential fatty acid deficiency.
7. How does a phospholipid differ from a triglyceride?
8. What is an omega-3 fatty acid?
9. A recipe mixes ¼ cup of oil with ¾ cup of milk. What common food could you add to keep the oil and milk emulsified?
10. Which foods contain cholesterol? List at least three functions of cholesterol in the body.

See Appendix F for responses.

©Wendy Schiff

6.3 What Happens to Lipids in the Body?

Learning Outcomes

1. Discuss the digestion, absorption, and transport of lipids in the body.
2. Describe effects of enterohepatic circulation on cholesterol metabolism.
3. Explain the primary role of adipose cells.

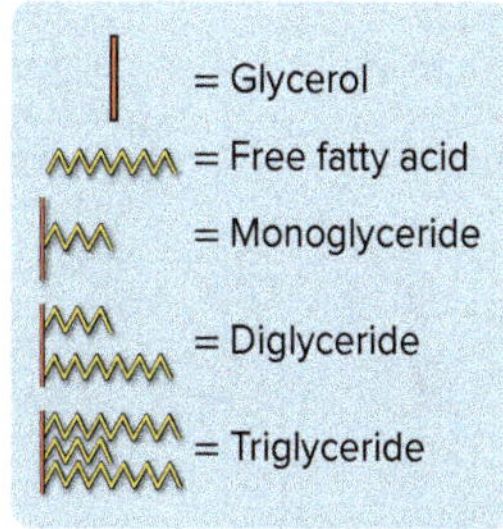

These simple diagrams show structural differences among glycerol and certain lipids.

Lipid Digestion

Compared to sugars and starches, triglyceride digestion and absorption is a more complicated process. Triglycerides and phospholipids need to be broken down by special enzymes called **lipases** before they can be absorbed. When you eat a cheeseburger and French fries, an inactive lipase in saliva mixes with the food. As the food enters the stomach, the organ's acid environment activates salivary lipase, enabling some lipid breakdown to occur. The stomach secretes gastric lipase, which breaks down some fat before it enters the duodenum. The small intestine, however, is the primary site of lipid digestion.

CCK and Bile Salts

As the fatty chyme leaves the stomach and enters your small intestine, it stimulates certain intestinal cells to release the hormone **cholecystokinin** (*ko–leh–sis′–toe–kye′–nin*) **(CCK).** CCK signals the gallbladder to contract and release bile. CCK also signals the pancreas to secrete digestive enzymes, including *pancreatic lipase*, into the duodenum. Bile contains **bile salts**, compounds that enhance digestion and absorption by keeping lipids dispersed in the watery environment of the small intestine (emulsification).

Bile salts have hydrophilic "heads" and hydrophobic "tails" (see Figure 6.5). The tails are oriented inward, so they surround the lipid particle, while the heads face outward and toward the watery components of chyme. Bile salts surround the lipids, forming tiny lipid–rich particles called **micelles** (Fig. 6.9). Such action reduces the size of the large fat droplets, but in doing so, it increases the surface area of the lipids. As a result, pancreatic lipase gains greater access to the lipid molecules and digests them more readily. If bile is not secreted into the duodenum, the lipids in chyme clump together in large fatty globules, making lipid digestion less efficient.

Intestinal Enzyme Action **Pancreatic lipase** digests triglycerides by removing two fatty acids from each triglyceride molecule. This action converts most triglycerides into **monoglycerides,** a single fatty acid attached to the glycerol backbone of the molecule. Some triglycerides are completely broken down into glycerol and *free fatty acid molecules* (fatty acids not attached to glycerol). The process of digesting phospholipids is similar to that of digesting triglycerides. The enzyme phospholipase removes two fatty acids from a phospholipid molecule. The remaining structure ("fragment") contains the lipid's phosphate region. Cholesterol does not undergo digestion; small intestinal cells can absorb cholesterol directly from food. Glycerol, free fatty acids, monoglycerides, and phospholipid fragments are the end products of lipid digestion. Figure 6.10 illustrates the basic steps involved in fat digestion.

Lipid Absorption

The process of absorbing lipids also involves micelles, which transport the fat–soluble end products of digestion to villi in the small intestine. When micelles are close to the tip of a villus, the free fatty acids, monoglycerides, and cholesterol leave them and move into the villus (Fig. 6.11). Under normal conditions,

Did *YOU* Know?

Monoglycerides and diglycerides may be listed as ingredients on food labels. These lipids result from the breakdown of triglycerides, and they may be added to foods during processing to improve the product's texture.

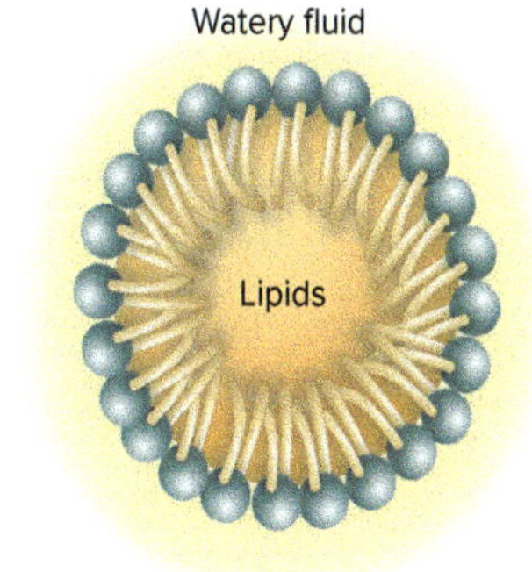

Figure 6.9 Micelle.

lipases enzymes that break down lipids

cholecystokinin (CCK) hormone that stimulates the gallbladder to release bile and pancreas to secrete digestive enzymes

bile salts components of bile that enhance lipid digestion and absorption

micelle lipid-rich particle that is surrounded by bile salts; transports lipids to absorptive cells

pancreatic lipase digestive enzyme that removes two fatty acids from a triglyceride

monoglyceride single fatty acid attached to a glycerol backbone

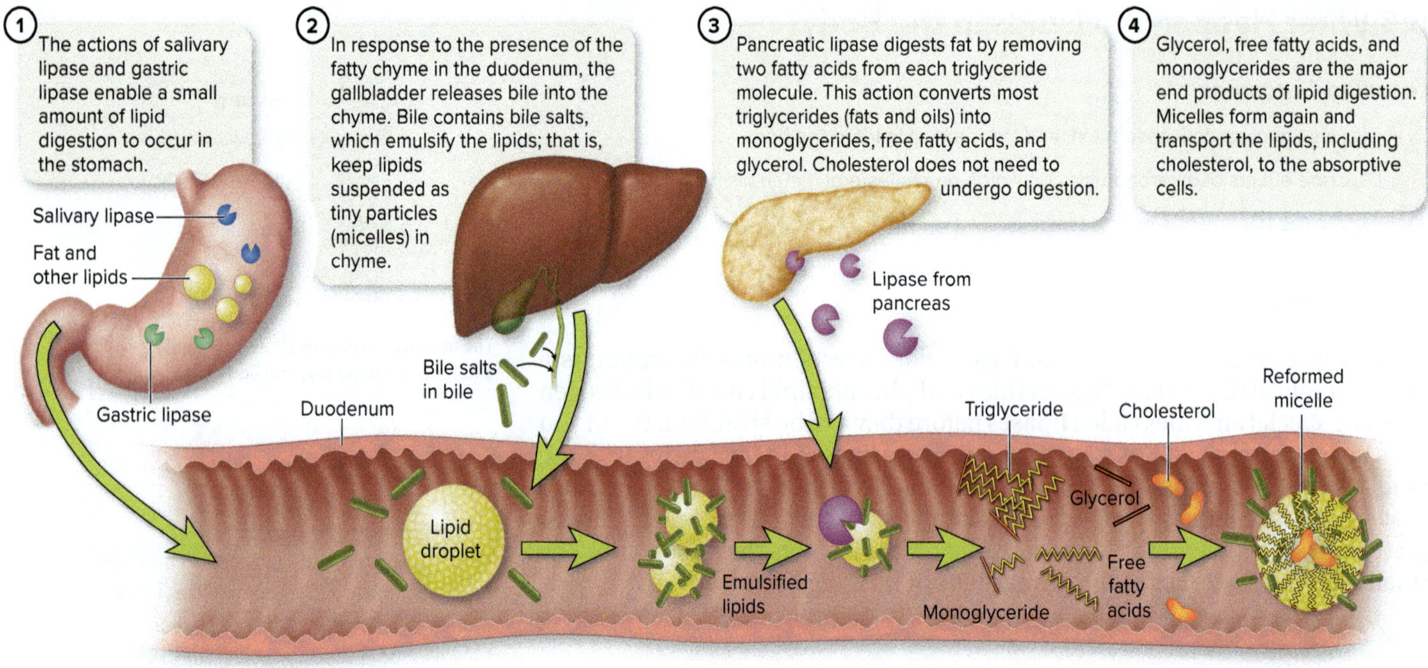

Figure 6.10 Lipid digestion in the duodenum.

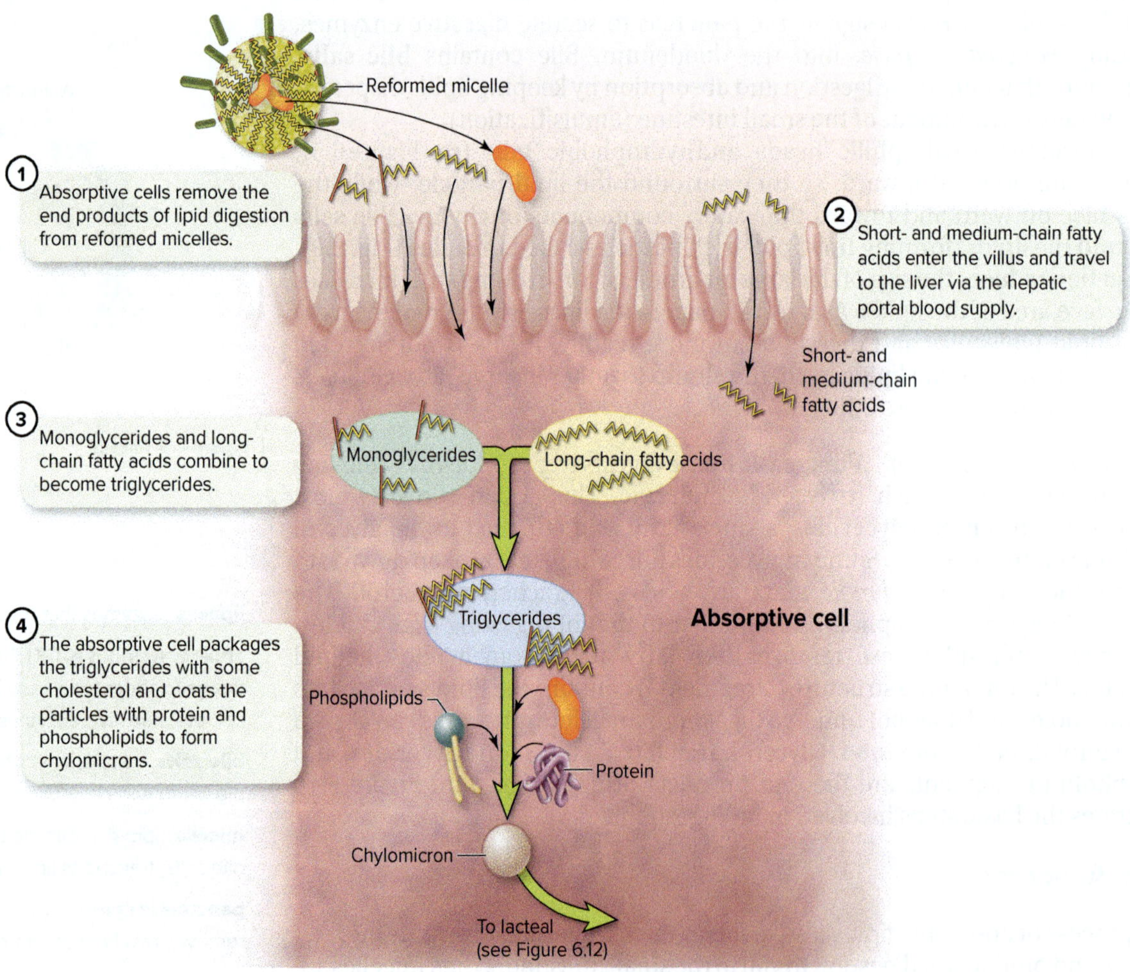

Figure 6.11 Lipid absorption.

the small intestine digests and absorbs nearly all of the triglycerides and phospholipids in food, but on average, only a little more than half of the dietary cholesterol is absorbed.[10]

After being absorbed, short– and medium–chain fatty acids can enter the portal bloodstream of villi directly. Most long–chain fatty acids, glycerol, monoglycerides, and phospholipid fragments are reassembled into triglycerides and phospholipids within the absorptive cells of the small intestine. Cholesterol and the reassembled triglycerides are coated with a thin layer of protein, phospholipids, and cholesterol to form chylomicrons. A *chylomicron* (*ky′–low–my′–kron*) is a type of *lipoprotein.* **Lipoproteins** are water–soluble structures that transport lipids through the bloodstream. Chylomicrons are too large to be absorbed directly into the bloodstream. These lipoproteins must pass through the larger openings of *lacteals (lak′–te–als),* lymphatic system vessels in each villus. Figure 6.11 illustrates key steps in lipid absorption.

The lymphatic system transports chylomicrons to the thoracic duct, where they enter the bloodstream through the left subclavian vein in the chest (Fig. 6.12). As chylomicrons circulate through the body, **lipoprotein lipase,** an enzyme in the walls of capillaries, breaks down chylomicrons' load of triglycerides into fatty acids and glycerol. Nearby cells can then pick up the fatty acids and glycerol molecules to use for energy. Ten to 12 hours after a meal, most chylomicrons have been reduced to small cholesterol–rich remnants. The liver clears these remnants from the bloodstream and uses their contents to synthesize new lipids and other lipoproteins that are released into the general circulation. Figure 6.13 provides a summary of lipid digestion and absorption.

Did *YOU* Know?

Many commonly eaten foods are rich sources of *long-chain triglycerides* (LCTs). *Medium-chain triglycerides* (MCTs) contain more medium-chain fatty acids than long-chain fatty acids. Furthermore, MCTs are made up of saturated fatty acids, whereas LCTs have unsaturated as well as saturated fatty acids in their chemical structures. Although few foods contain considerable amounts of MCTs, coconut and palm kernel oils are naturally rich sources of MCTs.

The human small intestine absorbs MCTs more easily than LCTs. Thus, manufactured products that contain 100% MCTs are often used along with other therapies to treat patients with medical conditions that interfere with fat digestion and absorption. Such conditions include cystic fibrosis and *pancreatitis* (inflammation of the pancreas).

©Wendy Schiff

Recycling Bile Salts

Most bile salts are absorbed in the ileum, where the compounds enter the bloodstream and travel to the liver. The liver uses the bile salts to make new bile. The process of recycling bile from the intestinal tract is called **enterohepatic** (*en′–teh–roe–hih–pah′–tik*) **circulation** (Fig. 6.14). Interfering with enterohepatic circulation can reduce blood cholesterol levels because the liver must use cholesterol to make new bile salts. Plants contain substances, such as soluble fiber, that interfere with cholesterol and bile absorption (see Table 5.6 for common food sources of soluble fiber).

lipoprotein water-soluble structure that transports lipids through the bloodstream

lipoprotein lipase enzyme in capillary walls that breaks down triglycerides

enterohepatic circulation process that recycles cholesterol in the body

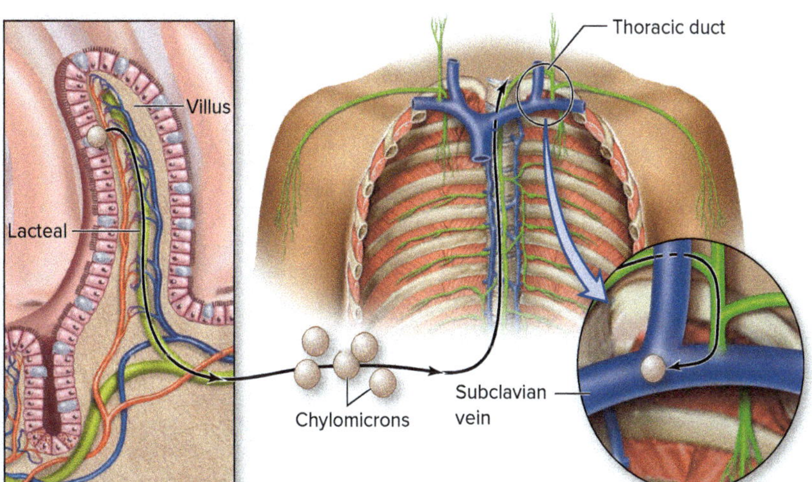

Figure 6.12 Lipid transport. The lymphatic/immune system transports chylomicrons from the small intestine to the thoracic duct, where they enter the bloodstream through the left subclavian vein in the chest.

Figure 6.13 Summary of lipid digestion and absorption. This illustration summarizes the basic steps of lipid digestion, absorption, transport, and elimination.

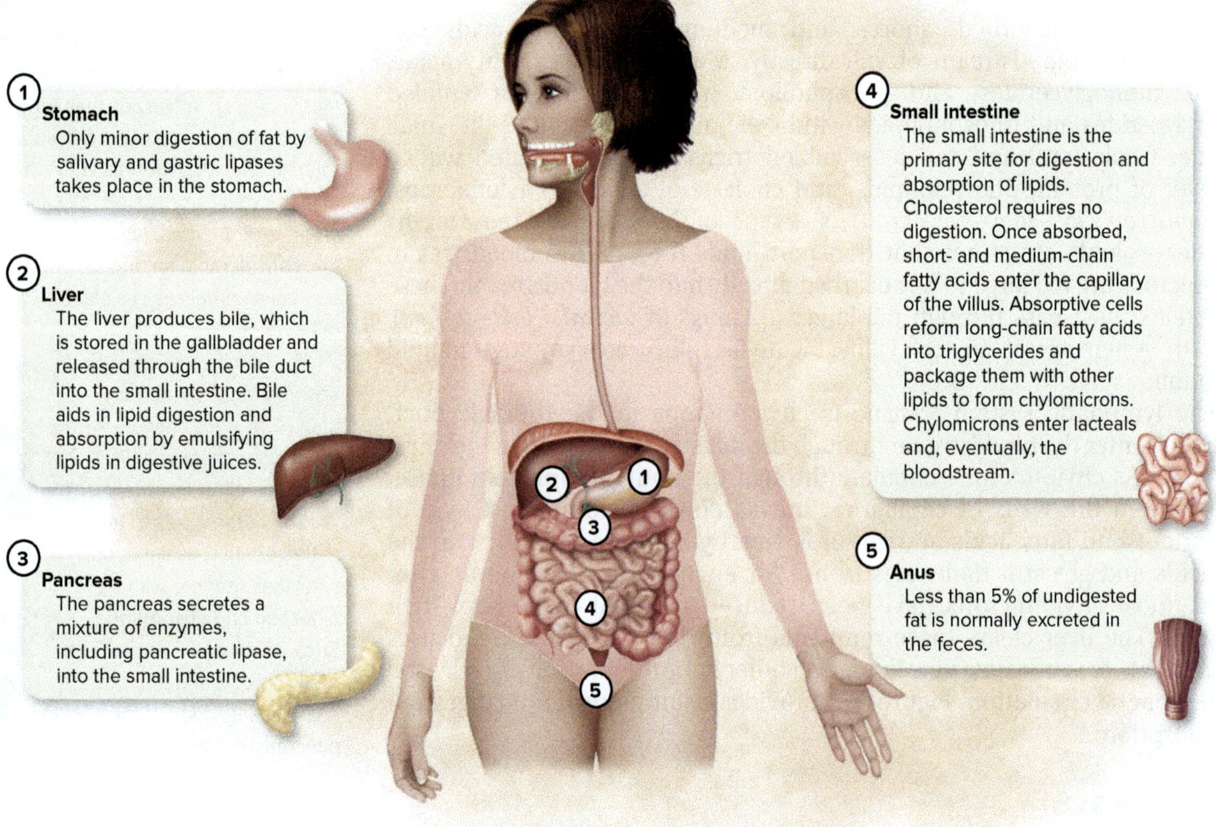

Figure 6.14 Enterohepatic circulation. Enterohepatic circulation is the process of recycling bile salts.

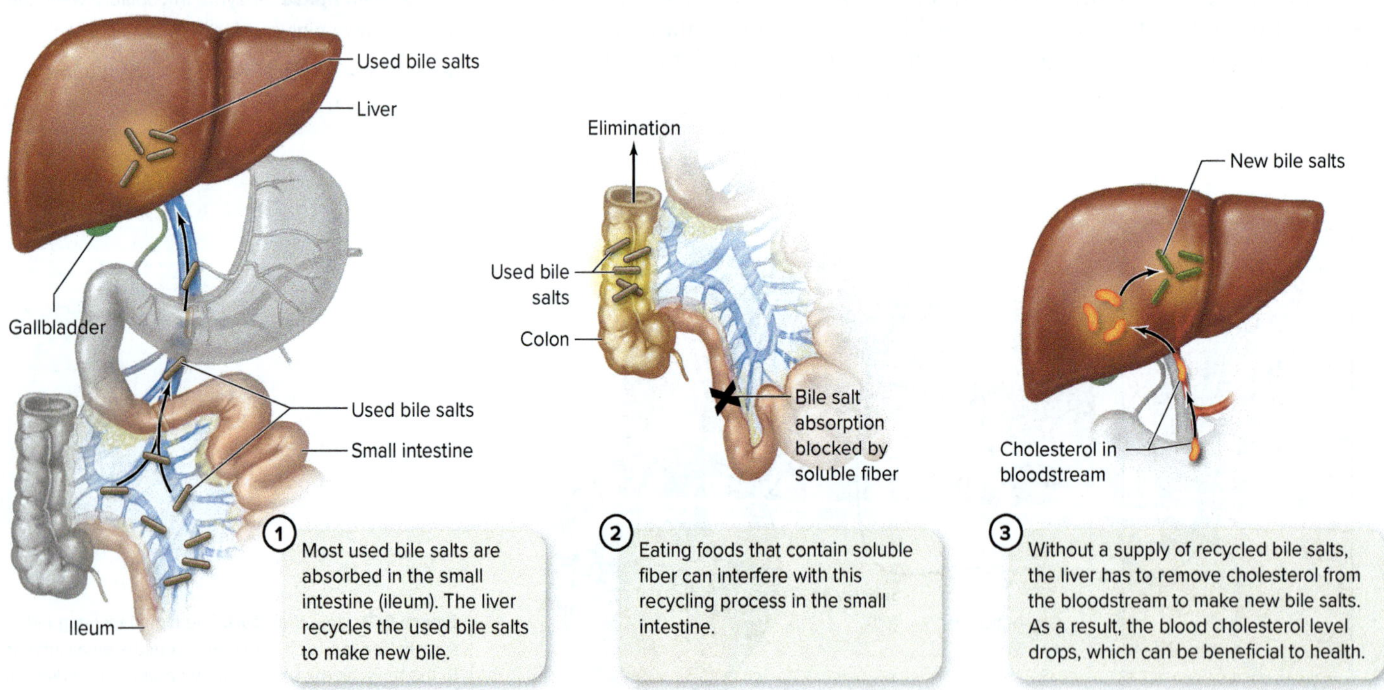

Using Triglycerides for Energy

Most cells can metabolize fatty acids for energy. Fat is more energy dense than carbohydrate or protein: A gram of fat supplies 9 kcal, whereas a gram of carbohydrate or protein provides only 4 kcal. Thus, a high–fat food is a more concentrated source of energy than a high–carbohydrate food. For example, you would need to eat 2½ tablespoons of sugar to obtain the amount of energy in 1 tablespoon of oil.

In many instances, your body does not need the energy from the fat in the food you have just eaten. **Adipose** (*ad′–eh–pose*) **cells,** commonly called fat cells, remove fatty acids and glycerol from circulation and reassemble them into triglycerides for storage. Most cells in your body contain triglycerides, but adipose cells are designed to store large amounts of fat (triglycerides). As illustrated in Figure 6.15, a fat droplet comprises most of an adipose cell's volume. When your body needs energy, adipose cells break down some stored triglycerides into fatty acid and glycerol molecules, and release these substances into your bloodstream. Muscle and other cells then remove the fatty acids from circulation and metabolize them. The liver clears the glycerol molecules from the bloodstream and converts them into glucose molecules that cells can also use for energy.

Eating too many kilocalories from fat contributes to unwanted weight gain, but consuming too much energy from protein and carbohydrates also increases body fat. Why? The body can convert excess glucose and certain amino acids into fatty acids that are used to make triglycerides. Additionally, the nonnutrient alcohol stimulates triglyceride synthesis, and excess body fat can accumulate as a result of alcohol consumption. To learn more about alcohol, read "Nutrition Matters: Drink to Your Health?" (Section 6.7). For more information about energy metabolism, see Chapter 11 and Appendix C.

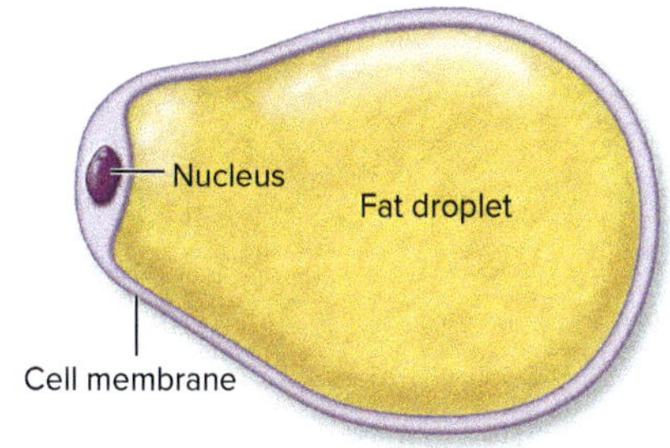

Figure 6.15 **Adipose cells.** A primary function of an adipose cell is storing a fat droplet.

adipose cells fat cells; specialized cells that store fat

Concept CHECKPOINT

11. Describe what happens to the fat in a piece of fried chicken as it undergoes digestion and absorption in a healthy person's intestinal tract. In your description, include the roles of bile, CCK, pancreatic lipase, villi, and chylomicrons.

12. Explain why high-fiber diets can reduce the concentration of cholesterol in a person's blood.

13. Why can a person with a larger amount body fat survive starvation for a longer period than a person with less body fat stores?

See Appendix F for responses.

6.4 Lipid Consumption Patterns

Learning Outcomes

1. Recall the percentage of fat in the typical American diet and the AMDR for fat.
2. Discuss Dietary Guidelines recommendations for fat and cholesterol intakes.

According to the Dietary Guidelines, people should consume diets that contain amounts of fat that are within the Acceptable Macronutrient Distribution Range (AMDR: 20 to 35% of total calories).[8] Furthermore, healthy eating patterns include rich sources of monounsaturated and polyunsaturated fatty acids, such as fish, nuts, and vegetable oils (see Table 6.3).[8] Adults should consume less than 10% of their total calories from saturated fatty acids by replacing

Adults should consume less than 10% of their total calories from saturated fats by replacing foods rich in saturated fat with foods rich in unsaturated fat, including fatty fish, avocados, olive oil, and nuts.
tinalarsson/iStock/Getty Images

foods that are rich sources of saturated fat with foods rich in unsaturated fat. Cholesterol and trans fat intake should be limited to as little as possible, while following a healthy eating pattern.

In 2017–March 2020, average total fat intake among adult Americans was 36% of total calories, and average saturated fat intake among adults was about 12% of total calories.[11] Thus, American adults consume slightly more saturated fat than the recommendation (10% of total calories). Some of the main sources of saturated fat in the American diet are hamburgers, tacos, burritos, pizza, ice cream, cookies, brownies, and donuts.[8] To estimate your average daily intake of total fat, as well as your average daily intake of saturated fat for a 3–day period, complete the "Personal Dietary Analysis" activity at the end of this chapter.

Concept CHECKPOINT

14. On average, fat contributes about ___% of the energy in an adult American's diet.

15. What is the AMDR for fat?

16. According to the Dietary Guidelines, what is the recommendation for saturated fat intake (percentage of total calories)?

See Appendix F for responses.

6.5 Understanding Nutritional Labeling: Lipids

Learning Outcomes

1. Use the Nutrition Facts panel to determine amounts of total fat and various types of fatty acids in a serving of a packaged food.
2. Compare total fat and saturated fat contents of packaged foods.

Figure 6.16 **Information about lipid contents on a Nutrition Facts panel.**
©Wendy Schiff

Consumers can determine how much total fat, saturated fat, trans fat, and cholesterol are in most packaged food products by reading the Nutrition Facts panel. As you can see in Figure 6.16, grams of total fat in one serving of the food (3 flatbreads) are shown, and under it, the number of grams of saturated fat and trans fat. The panel indicates that there are 3 g of total fat in each serving (5% Daily Value [DV]), and of that amount, 1 g is saturated fat (5%DV). Recall, a food that supplies 5%DV or less is a low source of the nutrient (see Chapter 3). Therefore, this food product is low in total fat and saturated fat. These flatbreads also supply 0 g of trans fat in a serving. If a serving of food has less than 0.5 g of a specific type of fat, the amount can be reported as 0 g. By reading the list of ingredients, consumers can learn which fats and oils provided the fat in this product.

Food manufacturers are not required to indicate amounts of monounsaturated and polyunsaturated fat in products. However, the amount of cholesterol in a serving of the food must be listed. In this example, there is no cholesterol in the serving.

What are the sources of fat in this product?
©Wendy Schiff

Concept CHECKPOINT

17. According to the Nutrition Facts panel, a serving of potato chips supplies 150 kcal, and there are 9 g of total fat in the serving of chips. Calculate the number of kcal from fat in a serving of the chips.

See Appendix F for responses.

6.6 Lipids and Health: Cardiovascular Disease

Learning Outcomes

1. Define cardiovascular disease, and explain the process of atherosclerosis.
2. List at least three modifiable risk factors and three nonmodifiable risk factors for atherosclerosis.
3. Distinguish HDL cholesterol from LDL cholesterol.
4. Discuss dietary and other lifestyle actions that can reduce the risk of atherosclerosis.

Cardiovascular disease (CVD) is a group of medical conditions that affect the functioning of the heart and blood vessels. Some forms of CVD are the result of birth defects, inherited conditions, and infections. In this chapter, we focus on some of the more common diseases that CVD encompasses (Fig. 6.17). **Coronary artery disease (CAD)** (also called heart disease) affects the arteries that supply the heart and often leads to heart attack. Carotid artery disease involves abnormal functioning of the carotid arteries in the neck that provide blood to the brain (see Figure 6.21). Peripheral artery disease (PAD) affects the arteries that supply the arms and legs (limbs) and pelvic region. Chronic high blood pressure (*hypertension*) damages arteries that supply many organs, especially the heart, kidneys, and brain.

Heart disease and stroke, the most common forms of CVD, are among the top five leading causes of death in the United States (see Figure 1.2). In 2020, heart disease and strokes were responsible for about 20% of deaths in the United States.[12] Because CAD is the most common cause of death in the United States, we will use the term *heart* disease when referring to CAD.

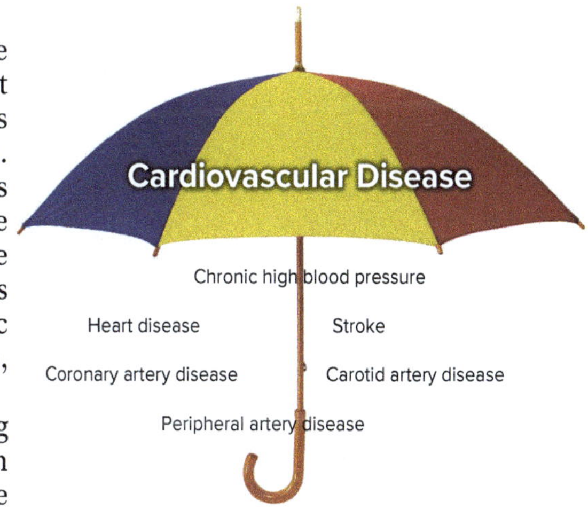

Figure 6.17 Cardiovascular disease.
D. Hurst/Alamy Stock Photo

From Atherosclerosis to Cardiovascular Disease

Most cases of CVD result from **atherosclerosis** (*athero* = lipid containing; *sclerosis* [*skleh–ro′–sis*] = hardening), a chronic process that negatively affects the functioning of arteries. Healthy arteries are flexible and have a smooth lining (Fig. 6.18.1). When something in the bloodstream, such as excess cholesterol or glucose, compounds from cigarette smoke, or certain bacteria, irritates the lining of an artery, a cascade of events begins that results in atherosclerosis. The body's immune system responds to the irritation by producing inflammation within the artery. Inflammation can stimulate healing, but the process can also trigger certain cells within the arterial wall to deposit cholesterol under the artery's lining. As a result, **arterial plaque** forms (Fig. 6.18.2). In addition to cholesterol, arterial plaques contain fat, components of dead cells, and calcium. Plaque interferes with circulation in the affected area of the artery because it narrows and may even block the opening through which blood flows (*lumen*). Furthermore, plaque roughens the normally smooth surface that lines the artery. The rough lining slows blood flow in the area and makes clots more likely to form (Fig. 6.18.3). If a plaque ruptures (tears open), repairing the damage also involves clot formation, and such blood clots can be life threatening (Fig. 6.18.4). Figure 6.19 shows cross sections of a healthy artery and one that is almost completely blocked by plaque.

Blood must be able to clot, especially when blood vessels have been injured; otherwise, a person could bleed to death from a minor bruise. In some instances, however, clots form too readily at the injury site. A **thrombus** is a fixed bunch of clots that remains in place and disrupts blood flow. If a thrombus partially closes off the lumen of an artery that nourishes the heart, the affected section

cardiovascular disease (CVD) group of diseases that affect the heart and blood vessels

coronary artery disease (CAD) a major form of CVD; heart disease

atherosclerosis long-term disease process in which plaques build up inside arterial walls

arterial plaque lipid-filled deposits that form under the lining of certain arteries

thrombus fixed bunch of clots that remains in place

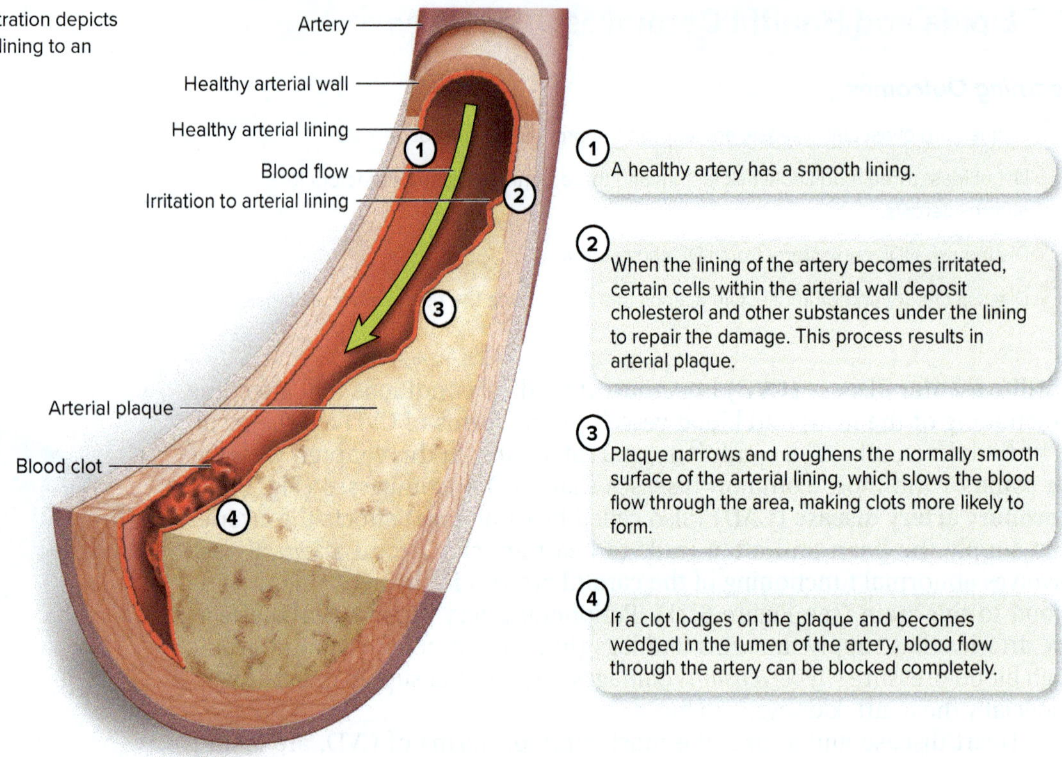

Figure 6.18 Atherosclerosis. This illustration depicts the progression from a normal arterial lining to an atherosclerotic one.

myocardial infarction heart attack

embolus thrombus or part of a plaque that breaks free and travels through the bloodstream

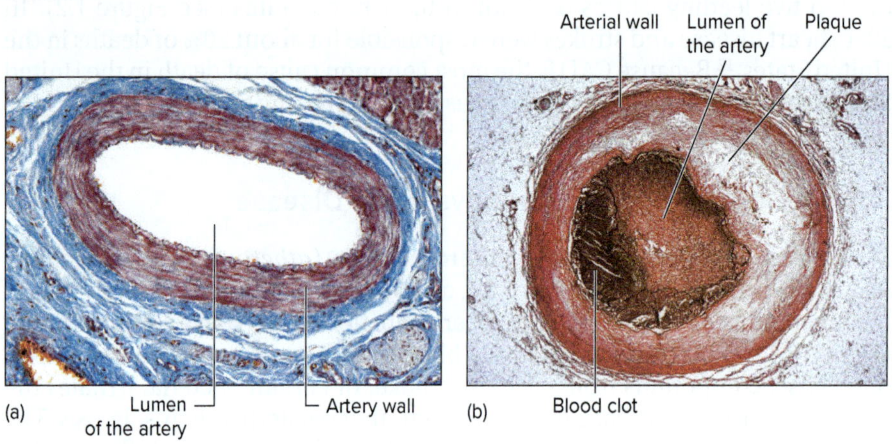

Figure 6.19 Healthy and atherosclerotic arteries. Note the differences between the cross section of a healthy artery (a) and that of an artery partially blocked as a result of atherosclerosis (b).
Healthy artery: Al Telser/McGraw Hill; Clogged artery: Biophoto Associates/Science Source

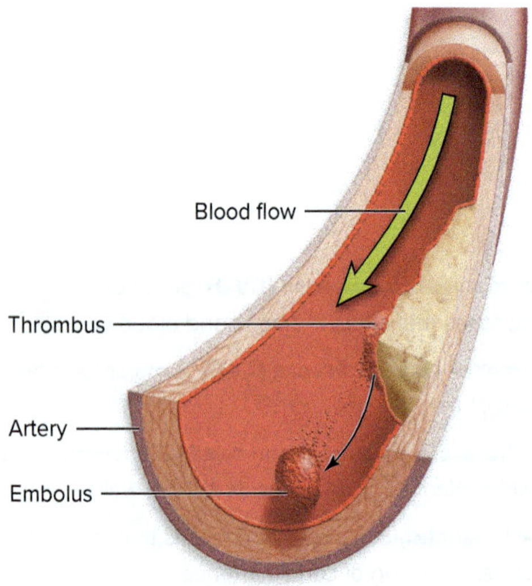

Figure 6.20 Embolus formation. A thrombus or part of a plaque that breaks free from where it formed and travels through the bloodstream is an embolus.

of the heart muscle is unable to receive enough oxygen and nutrients to function properly (see Figure 6.19b). As a result, the affected person typically experiences bouts of chest pain, especially when their heart beats faster, such as during intense emotional states or physical activities. If a thrombus completely blocks blood flow to a section of the heart muscle, the muscle dies and a **myocardial infarction** (*my′–oh–car′–de–al in–farc′–shun*) (heart attack) occurs. Sudden death can result from a severe myocardial infarction. A stroke can happen when a clot blocks an artery in the brain and brain cells that are nourished by the vessel die. When an artery to a limb is blocked, the tissue in the extremity dies, causing gangrene to occur. If the affected area is large, amputation of the gangrenous limb is often necessary to prevent life–threatening infection. A thrombus or part of a plaque that breaks free from where it formed and travels through the bloodstream is an **embolus** (Fig. 6.20). An embolus that lodges in an artery can create the same serious consequences as a stationary thrombus.

Certain arteries are more likely to be damaged by atherosclerosis; in addition to the blood vessels of the heart and brain, the blood supply in the kidneys, eyes, and legs is vulnerable. When atherosclerosis occurs in the carotid arteries in the neck, blood flow to the brain can be decreased and clots can form that travel to the brain, causing a stroke ("brain attack") (Fig. 6.21). Although atherosclerosis can begin during adolescence and young adulthood, the disease usually does not produce signs or symptoms of CVD until decades later.

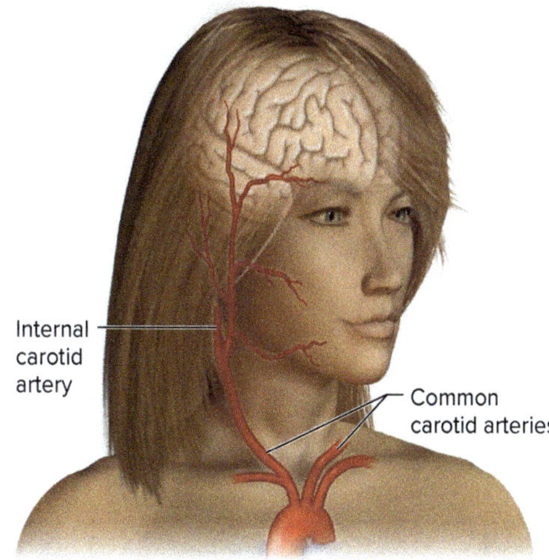

Figure 6.21 Carotid arteries.

Atherosclerosis and Hypertension

In addition to interfering with blood flow, atherosclerosis causes the affected arteries to thicken and lose their flexibility (become stiff). This condition is commonly called "hardening of the arteries." As a result, atherosclerosis contributes to the development of **hypertension,** a chronic condition characterized by abnormally high blood pressure levels that persist even when the person is relaxed. Hypertension is a major risk factor for heart disease and stroke. The heart of a person with hypertension must work harder to circulate blood through abnormally stiff arteries. Furthermore, elevated blood pressure can cause hardened arteries in the brain to tear or burst. Hypertension also damages arteries in other parts of the body, which increases risk of internal bleeding problems and even sudden death, depending on the artery's size and location.

Lipoproteins and Atherosclerosis

Lipoproteins transport cholesterol and triglycerides in the bloodstream, so these structures play major roles in the development of atherosclerosis. In addition to chylomicrons, the body makes three major types of lipoprotein (Fig. 6.22). Each type carries different proportions of protein, cholesterol, triglycerides, and phospholipids.

The protein content of a lipoprotein contributes to its density. A chylomicron, for example, is the largest and least dense kind of lipoprotein. Compared to the other types of lipoproteins, chylomicrons carry much more lipids and very little protein (see Figure 6.22a). **High–density lipoprotein (HDL)** is the smallest and densest kind of lipoprotein because it transports more protein and less lipids than the other lipoproteins (see Figure 6.22b). **Low–density lipoprotein (LDL)** transports more cholesterol than HDL in the bloodstream (see Figure 6.22c).

hypertension abnormally high blood pressure levels that persist

high-density lipoprotein (HDL) lipoprotein that transports cholesterol away from tissues and to the liver, where it can be eliminated

low-density lipoprotein (LDL) lipoprotein that carries cholesterol into tissues

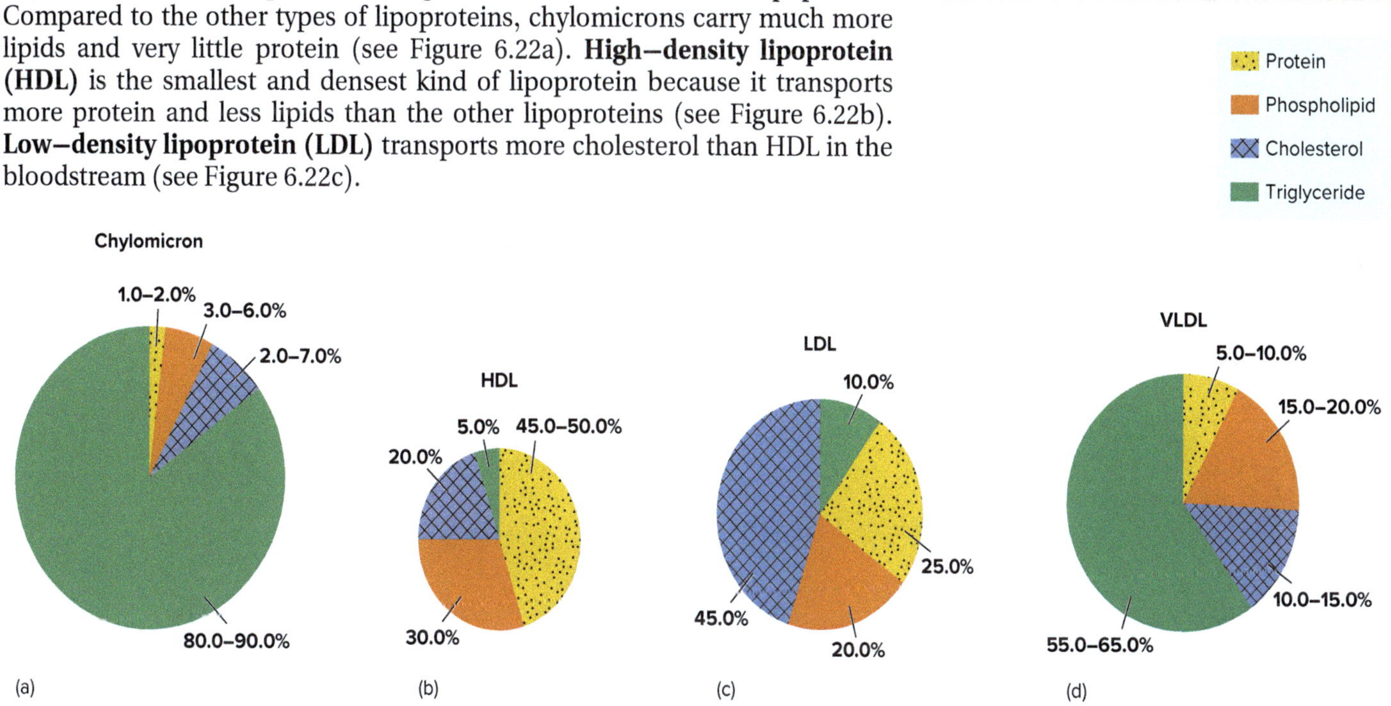

Figure 6.22 **Major lipoproteins.** The composition of a lipoprotein contributes to its size and density. (a) Chylomicron, (b) HDL, (c) LDL, (d) VLDL.

HDL normally conveys lipids away from tissues and to the liver, where they can be processed and eliminated (Fig. 6.23a). Thus, the cholesterol carried by HDL (HDL cholesterol) is often called "good" cholesterol because it does not contribute to plaque formation. LDL transports lipids to tissues, where the nutrients are used to make cell structures and vital compounds (see Figure 6.23b). The cholesterol carried by LDL (LDL cholesterol) is often referred to as "bad" cholesterol because LDL conveys the lipids to the cells in arterial walls that make atherosclerotic plaques. This simple "good" or "bad" classification, however, is no longer helpful in predicting a person's risk of atherosclerosis. Why?

very-low-density lipoprotein (VLDL) lipoprotein that carries much of the triglycerides in the bloodstream

Although LDL transports LDL to cells, there are different types of LDL, and some forms are unhealthy. People with high levels of small, dense LDLs are more likely to develop atherosclerosis than are people with low levels of these particular LDLs.[13] Additionally, chemically unstable substances ("free radicals") can damage cholesterol, especially the cholesterol in small, dense LDLs. The damage results in oxidized LDL cholesterol, which is taken up by the plaque–forming arterial cells (see Figure 6.23c). Cigarette smoking increases the production of oxidized LDL cholesterol.[14]

Medical researchers have discovered that some forms of HDL do not function normally.[13] The malfunctioning HDLs may transport oxidized cholesterol. Like oxidized LDL cholesterol, oxidized HDL cholesterol is also associated with increased risk of atherosclerosis. Therefore, having a high level of HDL cholesterol in the bloodstream is not necessarily a sign of good cardiovascular health, especially if the HDL is transporting oxidized cholesterol.

A third major class of lipoproteins, **very–low–density lipoprotein (VLDL),** may also contribute to atherosclerosis. VLDL carries a larger amount of triglycerides than cholesterol (see Figure 6.22d). A high level of triglycerides in the bloodstream may stimulate the production of small, dense LDL–cholesterol particles that are prone to becoming oxidized. When the oxidized particles bind to the walls of arteries, they promote inflammation and atherosclerosis.[15] Excessive alcohol and refined carbohydrate intakes stimulate VLDL production in the liver.

Risk Factors for Atherosclerosis

The National Heart, Lung, and Blood Institute has identified risk factors for atherosclerosis.[16] The more risk factors a person has, the

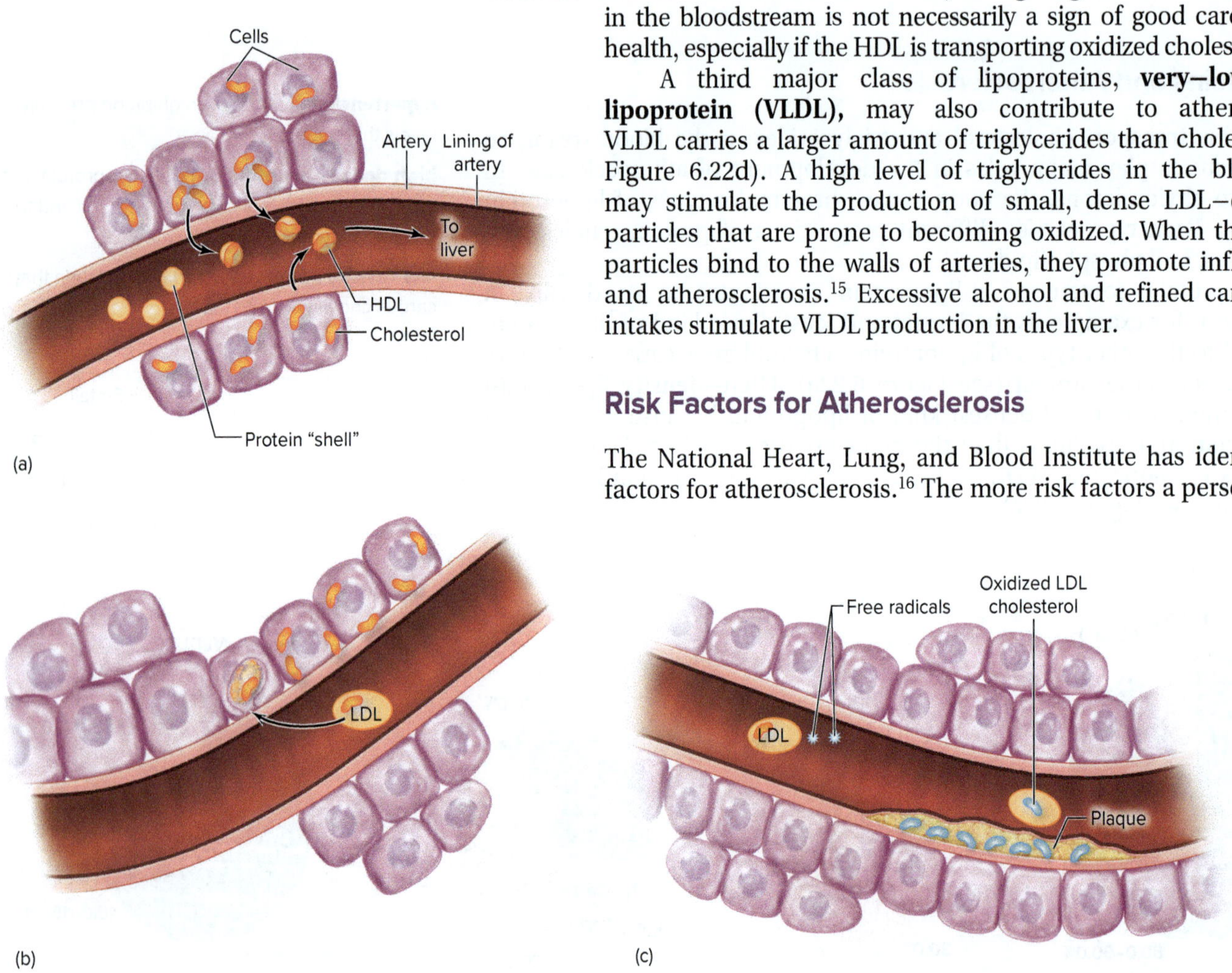

Figure 6.23 HDL, LDL, and oxidized LDL. Lipoproteins transport lipids in the bloodstream. (a) The liver releases protein "shells" into the bloodstream that pick up cholesterol and other lipids from cells. When filled with lipids, the shells are called high-density lipoprotein (HDL). (b) LDL transports cholesterol and other lipids in to tissues. (c) Oxidized LDL cholesterol is harmful because the oxidized lipids it carries are taken up by arterial cells that form plaque.

TABLE 6.5 Atherosclerosis: Major Risk Factors

Nonmodifiable Risk Factors
Family history of CVD
Increasing age
Modifiable Risk Factors
Unhealthy diet
Hypertension
Insulin resistance and diabetes
Elevated blood cholesterol (especially LDL cholesterol)
Excess body fat
Physical inactivity
Tobacco use or exposure to tobacco smoke

greater their likelihood of developing CVD. Table 6.5 lists major risk factors and indicates which ones are nonmodifiable or modifiable. A nonmodifiable risk factor cannot be changed.

Major risk factors for CVD include advanced age, tobacco use, excess body fat, and physical inactivity. ©Wendy Schiff

Nonmodifiable Risk Factors

Advancing age is a major risk factor for CVD. The risk of atherosclerosis increases for males after age 45 and for females after age 55. Having a family history of CVD also influences one's risk of the disease. A person's risk increases when they have a father or brother who was diagnosed with CVD before 55 years of age. Furthermore, one's risk increases when they have a mother or sister who was diagnosed with CVD before 65 years of age.

Genetics and CVD Genetics (family history) is a major risk factor for atherosclerosis and CVD. A person's genes, for example, may code for various physical conditions that increase risk of heart disease, such as hypertension and diabetes. Additionally, genes may influence the way in which the circulatory and immune systems respond to diet. Thus, some people may be protected against the development of atherosclerosis, whereas other people with similar diets develop serious arterial plaques early in life and die prematurely of CVD as a result.

Until recently, much of the research examining the role of diet in the development of atherosclerosis focused on the association between lipids and CVD. Nevertheless, some people with no apparent CVD risk factors still had heart attacks, strokes, and related blood vessel diseases. In many cases, affected individuals were under 40 years of age. This puzzling observation led to the discovery that the amino acid **homocysteine** (*ho'–mo–sis'–teen*) may be associated with CVD. Amino acids are the chemical units that comprise proteins. Although homocysteine is an amino acid, the substance is not found in human proteins, and higher–than–normal blood levels of homocysteine can be toxic. High blood levels of homocysteine may injure arterial walls and contribute to atherosclerosis. A few people have a genetic abnormality that causes the amino acid to accumulate in their bloodstream. These individuals have a higher risk of premature CVD than persons who do not have the genetic defect. Deficiencies of certain B vitamins can also cause homocysteine levels to become elevated. Cells use three B vitamins, B–6, B–12, and folate, to convert homocysteine into safer compounds. Some medical experts think an elevated blood homocysteine level may be a "marker" in blood that indicates a person has CVD. Nevertheless, more studies are needed to determine whether elevated blood homocysteine plays a role in the development of CVD among the general population.

homocysteine amino acid that may play a role in the development of atherosclerosis

Did *YOU* Know?

Electronic cigarettes (e-cigarettes) are handheld, battery-operated devices that smokers use as alternatives to smoking tobacco-containing cigarettes. The liquid inside the e-cigarette contains nicotine, solvents, flavors, and other chemicals, including cannabinoid (CBD) oils and tetrahydrocannabinol (THC). CBD is the chemical found in marijuana plants, and THC is the mind-altering compound of marijuana. The device heats this liquid to create a vapor that can be inhaled and exhaled ("vaping"). Vaping emits various toxic and harmful chemicals.[18] Using e-cigarettes may also increase the person's heart rate and blood pressure and promote thrombus formation.[18] Thus, these devices may contribute to atherosclerosis and poor cardiovascular health.

In 2019, some people that used e-cigarettes and vaping products developed severe lung injuries. As of February 2020, 68 people have died, and over 2600 people were hospitalized for vaping lung injuries. The FDA determined that the e-cigarette additives vitamin E acetate and THC caused the lung injuries. As a result, many manufacturers removed vitamin E acetate from their e-cigarette formulations, but THC is still used. Thus, the overall safety and long-term health effects of e-cigarettes are not fully known. More studies are needed to understand how e-cigarettes affect cardiovascular and overall health. For more information about e-cigarettes, visit www.cdc.gov/tobacco/basic_information/e-cigarettes/about-e-cigarettes.html.

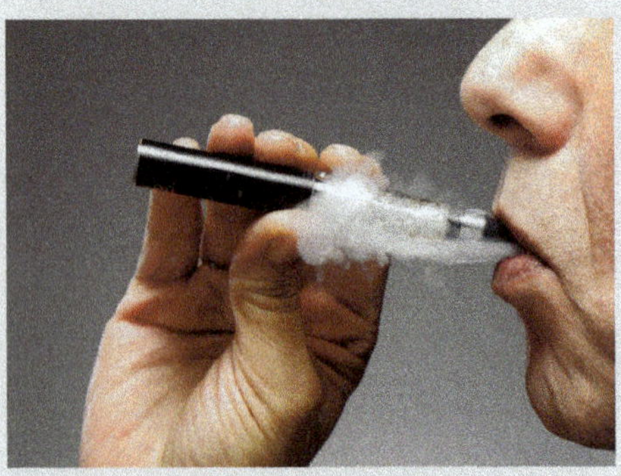

VICTOR DE SCHWANBERG/Science Photo Library/Alamy Stock Photo

Modifiable Risk Factors

Although a person cannot change their age and family history of CVD, several major risk factors for atherosclerosis involve lifestyle choices that can be modified. Hypertension, diabetes, excess body fat, and elevated blood cholesterol are modifiable risk factors that can be influenced by diet and exercise. Many people, for example, can reduce their chances of developing atherosclerosis by avoiding tobacco use, limiting their intake of saturated fat, exercising regularly, and maintaining a healthy body weight.

Nearly one out of two adult Americans has hypertension, chronically elevated blood pressure.[17] Hypertension is often referred to as a "silent disease" because people with the condition frequently feel healthy and do not have obvious symptoms that indicate trouble within their circulatory system. The condition, however, is quite serious because it damages arterial walls and increases the risk of stroke, heart failure, and kidney disease. High intakes of sodium are associated with increased risk of hypertension. Chapter 9 presents information about hypertension, including healthy blood pressure values (see Table 9.10). Chapter 9 also discusses ways to reduce the likelihood of developing the condition.

Diabetes is another modifiable risk factor for atherosclerosis. Adults who have diabetes are two times more likely to develop heart disease or have a stroke than adults without diabetes.[19] Chapter 5 discusses diabetes in detail. Excess body fat, especially in the abdominal region, increases the risk of type 2 diabetes and hypertension; Chapter 10 focuses on sensible ways to lose body fat. Physical inactivity also contributes to excess body fat. Chapter 11 provides suggestions for becoming more physically active. Elevated blood lipids, particularly certain lipoproteins, are a risk factor for atherosclerosis. People who have metabolic syndrome have a higher risk of CVD than people who do not have this condition. Chapter 5 includes information about metabolic syndrome.

Tobacco use is another major risk factor for atherosclerosis that is modifiable. Compared to nonsmokers, smokers have two to four times the likelihood of developing heart disease and having a stroke.[20] Smokers expose other people in their environment to secondhand smoke, which is a combination of the smoke from the burning end of a cigarette and the exhaled air of the person who is smoking. Exposure to secondhand smoke is a risk factor for heart disease.[21] If you smoke, simply improving your diet is unlikely to reduce your risk of atherosclerosis; therefore, if you smoke, make every effort to quit using tobacco products. If you do not smoke, avoid breathing secondhand smoke.

Emotional stress, particularly anger, also plays a role in the development of atherosclerosis.[22] Although individuals respond to stress differently, chronic stress generally causes physical changes in the body that can damage arteries and contribute to atherosclerosis. Furthermore, people who are "stressed out" may make unhealthy food choices, drink too much alcohol, and get inadequate exercise.

It is important to understand that a risk factor is not the same as a *cause* of disease. AIDS, for example, is caused by human immunodeficiency virus (HIV); a person cannot develop AIDS without being infected with HIV. Atherosclerosis, however, is an extremely complex disease process. In most cases, no single cause for the condition can be identified. Instead, having one or more risk factors increases a person's chances of developing the condition.

Assessing the Risk of Atherosclerosis

You may not be able to prevent having a heart attack or stroke some day, but there is plenty of scientific evidence that suggests you can forestall CVD and live a longer, more satisfying life by reducing or eliminating modifiable risk factors for atherosclerosis. Diet, for example, influences the likelihood of atherosclerosis and is highly modifiable.

Medical Testing

To determine your risk of atherosclerosis, it is a good idea to have regular medical checkups in which a physician checks your blood pressure and listens to blood flow in your carotid arteries to assess whether the arteries are becoming blocked (see Figure 6.21). The physician may request a lipoprotein profile to assess your total serum cholesterol level, as well as serum HDL cholesterol, LDL cholesterol, and triglyceride levels. Although this textbook generally refers to "blood cholesterol" or "blood lipids," the amount of lipids in serum or plasma rather than whole blood is usually measured. Serum is the liquid portion of blood; plasma is similar to serum except that it contains clotting factors.

Table 6.6 presents classifications for healthy and unhealthy fasting blood lipid levels. The "desirable" (low risk) range for total cholesterol is less than 200 mg/dl. In 2015–2018, American adults had an average total blood cholesterol level of 191 mg/dl.[23] About 12% of adults have blood cholesterol levels that were 240 mg/dl or higher, which placed them at high risk of cardiovascular disease.[23]

Do you know what your fasting blood cholesterol level is? Even if it is below 200 mg/dl, you may still have a high risk of atherosclerosis. Why? Although knowing the concentration of cholesterol carried by all lipoproteins is important, the amounts of certain lipoproteins in your blood, particularly LDL and HDL, are more critical risk factors. As mentioned earlier, LDL carries cholesterol to cells and HDL transports cholesterol away from cells.

Did *YOU* Know?

Blood lipoprotein levels increase after consuming foods and most beverages, and the levels can remain elevated for a few hours. To obtain accurate information about a person's usual blood lipoprotein level, the individual should be in a fasting state. Thus, a person should fast (avoid eating or drinking anything except plain water) for 12 hours before having blood drawn for a lipoprotein profile.

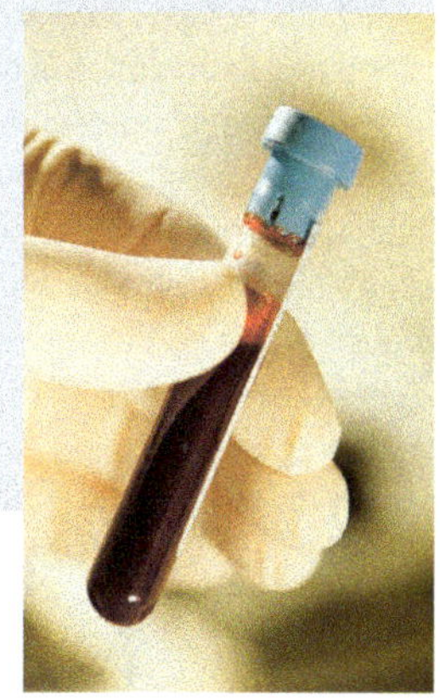

Comstock Images/PunchStock/Getty Images

TABLE 6.6 Classification of Fasting Blood Lipid Levels (According to Risk of CVD)

Total Cholesterol (mg/dl)	Classification
<200	Low risk
200–239	Borderline high risk
≥240	High risk
LDL Cholesterol (mg/dl)	**Classification**
<100	Very low risk
100–129	Low risk
130–159	Borderline high risk
160–189	High risk
≥190	Very high risk
HDL Cholesterol (mg/dl)	**Classification**
<40 (for males); <50 (for females)	High risk
≥60	Low risk*
Triglycerides (mg/dl)	**Classification**
<150	Low risk
150–199	Borderline high risk
≥200	High risk

*Having a very high HDL cholesterol level may increase risk of CVD.

Source of data: National Institutes of Health, National Heart, Lung, and Blood Institute: *ATP III guidelines at-a-glance quick desk reference.* 2001. www.nhlbi.nih.gov/files/docs/guidelines/atglance.pdf. Accessed: April 27, 2023.

Did *YOU* Know?

Sleep apnea, a common breathing disorder among overweight and obese people, may be a risk factor for atherosclerosis.[1] While asleep, a person with sleep apnea stops breathing intermittently or does not breathe deeply. As a result, the person does not feel refreshed after sleeping. Sleep apnea is usually treated with a special breathing device; untreated sleep apnea can contribute to hypertension, heart attack, and stroke. For more information about sleep apnea, visit www.heart.org/en/health-topics/sleep-disorders/sleep-apnea-and-heart-disease-stroke.

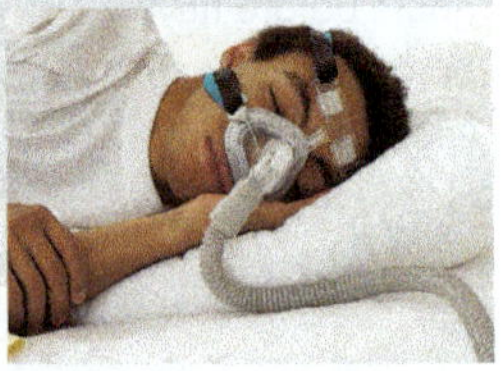

Jean-Paul Chassenet/123RF

high-sensitivity C-reactive protein (hs-CRP) protein produced primarily by the liver in response to inflammation

People with a high ratio of total cholesterol to HDL cholesterol usually have too much LDL cholesterol in their blood and an increased risk of heart disease and stroke. A healthy ratio of total cholesterol to HDL cholesterol is less than 3.5:1; the lower the ratio, the lower the risk of CVD.[24] To calculate your risk of having a heart attack during the next 10 years, use the risk assessment tool at http://static.heart.org/riskcalc/app/index.html#!/baseline–risk.

Someday, scientists may be able to locate genes that are involved in the process of atherosclerosis and develop a blood test that identifies biological markers produced by the abnormal genes. Thus, young people could undergo testing to determine their risk of atherosclerosis well before the signs and symptoms of the condition appeared. Until then, it is wise to have regular health checkups that include blood pressure and lipid measurements (*lipoprotein profile*). The American Heart Association recommends everyone 20 years and older to have their risk factors and lipoprotein profile checked every 4 to 6 years.[25] Ask your physician for a copy of the laboratory results, and keep them along with others in your personal "medical file" for future reference.

C-reactive Protein Chronic inflammation is involved in the development of CVD and other chronic diseases. The liver responds to certain kinds of inflammation by producing and releasing **high–sensitivity C–reactive protein (hs–CRP)** into the bloodstream. People with high levels of hs–CRP are more likely to develop CVD.[26] Thus, elevated hs–CRP may be a marker for atherosclerosis, like homocysteine. If you have a family history of premature CVD, consider having your homocysteine and hs–CRP levels measured.

Coronary Calcium Cigarette smoking is significantly associated with the formation of *coronary calcium,* calcium deposits in arteries of the heart.[27] High amounts of coronary calcium are associated with increased risk of atherosclerosis. Computed tomography (CT) scans, special computerized images that show the body's internal structures, can detect coronary calcium deposits. A CT scan, however, involves exposure to ionizing radiation, so physicians generally reserve its use for other diagnostic purposes.

MY DIVERSE PLATE

MyPlate.gov

Ghee (Indian Clarified Butter)

Ghee is a solid fat that is used for cooking, primarily in India. Traditionally, ghee was made from the milk of a type of buffalo that was native to India and Pakistan, but milk from other animals can be used. To make ghee, butter is heated until its water content evaporates completely and the solids (primarily lactose and some proteins) brown and sink to the bottom. After heating, the fat is strained to remove the solids. Researchers who fed ghee to laboratory rats found improvement in the animals' blood lipid profiles.[29] Clinical studies are needed to determine the effects of ghee on human health.

©Wendy Schiff

Reducing the Risk of Atherosclerosis: Dietary Changes

Specific recommendations for consumption of various types of fat are primarily based on the results of epidemiological studies that examined the effects of dietary lipids on blood lipids and risk of CVD. Populations that consume diets rich in saturated fats generally have higher rates of heart disease than populations that eat less saturated fat. Saturated fat alters the structure of liver cell membranes so they no longer function properly. As a result, the liver removes less cholesterol from the bloodstream.[28] Most saturated fatty acids increase blood cholesterol levels by raising concentrations of both LDL and HDL cholesterol. Trans fats also raise blood cholesterol levels. However, trans fats raise LDL cholesterol while reducing beneficial HDL cholesterol levels.[28]

Monounsaturated fatty acids generally lower LDL cholesterol without reducing HDL cholesterol levels. Foods rich in monounsaturated fat include peanuts and peanut oil, canola oil, olives and olive oil, almonds, and avocados.

Diets containing high amounts of polyunsaturated fatty acids, particularly omega–3 fats, may reduce blood levels of total cholesterol and LDL cholesterol. In some individuals, however, polyunsaturated fat also reduces beneficial HDL cholesterol. Nevertheless, polyunsaturated fatty acids tend to be beneficial because they do not promote atherosclerosis. Foods rich in polyunsaturated fat include corn, soybean, and cottonseed oils, as well as some types of sunflower and safflower seed oils (see Table 6.2).

As previously mentioned, the Dietary Guidelines recommend limiting saturated fat intake to less than 10% of total energy intake by replacing foods that are rich sources of saturated fat with foods that contain high amounts of unsaturated fat. People should limit their trans fat and cholesterol intake as much as possible while eating a healthy diet.

What About Omega-3 and Omega-6 Fats?

Recall, the typical American diet provides far more omega–6 foods than foods that contain omega–3 fatty acids.[1] The primary omega–6 fatty acid, linoleic acid, increases blood clotting and is used to create markers of inflammation in the body. Some inflammation is necessary because it attracts immune system cells to disease–causing microorganisms that have entered the body. Inflammation, however, can also damage the inside of arteries. Clotting is also an important function of blood, but excess blood clotting can increase the risk of strokes, heart attacks, and other serious blood vessel disorders. The role of diets rich in omega–6 fats in preventing or contributing to the development of excessive amounts of inflammation in the body or heart disease is controversial among medical experts.[30, 31] Thus, more research is needed to clarify the pros and cons of following diets that contain high amounts of these particular fatty acids.

Eating foods that supply omega–3 fatty acids may reduce the risk of heart disease and stroke.[1] The body incorporates omega–3 fatty acids into cell membranes and uses these lipids to make compounds that can reduce inflammation, serum triglycerides, and blood clotting. Table 6.7 lists foods that are sources of omega–3 fatty acids, including several species of fish. Certain fatty fish are rich sources of DHA and EPA, long–chain omega–3 fatty acids.

Alpha–linolenic acid is an omega–3 fatty acid that the body can convert to DHA and EPA (see Figure 6.2). Flaxseeds, chia seeds, soybeans, and walnuts are good sources of alpha–linolenic acid. As mentioned earlier, the body can only make small amounts of DHA and EPA from alpha–linolenic acid. Therefore, it is important to include other sources of these long–chain fatty acids in your diet.

To obtain beneficial long–chain omega–3 fatty acids such as DHA, the Dietary Guidelines recommend that Americans eat at least 8 ounces of various seafood a week.[8] Because large species of fish can contain high amounts of the toxic compound *methylmercury,* young children, pregnant and breastfeeding females, and females who are likely to become pregnant should not eat certain kinds of fish, including shark, king mackerel, marlin, and swordfish.[8] Shrimp, canned light tuna, salmon, pollock, and catfish are low in methylmercury. According to the Dietary Guidelines, females who are pregnant or

Salmon is a rich source of omega-3 fatty acids, including DHA and EPA.
Michael Lamotte/Cole Group/Photodisc/Getty Images

TABLE 6.7 Food Sources of Omega-3 Fats

Fish/Shellfish
Herring, salmon, sablefish, anchovies, tuna, bluefish, sardines, catfish, striped bass, mackerel, trout, halibut, pollock, flounder, shrimp, mussels, crab
Oils
Flaxseed, walnut, canola, soybean
Nuts and Seeds
Walnuts, flaxseeds, chia seeds
Other
Algae

Chia seeds: everydayplus/Shutterstock; Walnuts: Author's Image/Glow Images

Did *YOU* Know?

Chicken eggs normally contain very low amounts of omega-3 fatty acids. By feeding hens ground flaxseed, farmers can increase the amount of omega-3 fatty acids in eggs. In supermarkets, such omega-3 enriched eggs are often sold beside cartons of regular eggs. Expect to pay more for the enriched eggs.

©Wendy Schiff

breastfeeding can safely consume up to 12 ounces of low–mercury fish and shellfish weekly (see Chapter 3). The "Recipe for Healthy Living" feature of this chapter has an easy recipe made with salmon.

What if you do not like to eat fish? Taking supplements that contain omega–3s may reduce inflammation and the risk of deadly forms of heart disease.[1] Omega–3 supplements seem to be safe, but these fatty acids may interfere with normal blood clotting. The FDA recommends taking no more than 5 grams per day of omega–3 supplements. Consuming more than these recommendations may increase your risk of stroke.[32] It is important to note that fish liver oil supplements are not the same as fish oil or omega–3 supplements. Fish liver oils, such as cod liver oil, are rich sources of vitamins A and D. These vitamins are toxic when taken in large amounts. Check with your physician before you embark on a campaign to increase omega–3 fatty acids in your diet. The first "Food & Nutrition Tips" in this section provides some ideas for increasing your intake of healthy fats from dietary sources.

Food & Nutrition tips

To increase your intake of "healthy fat," consider the following tips.

- Eat seafood, especially fatty fish, two times a week.
- Bake, grill, or broil fish. Before cooking, marinate fresh fish in olive or canola oil that has been seasoned with a small amount of garlic, pepper, and lemon juice. The light coating of oil on fish can help keep the food from drying out during cooking.
- Add tuna to salads, or mix tuna with a little olive oil and spread on toast.
- If you don't want the flavor of olive oil in a food, use canola oil, soybean oil, or liquid margarines made from these oils for frying or sautéing.
- Sprinkle chopped walnuts, chia seeds, or flaxseeds on salads, yogurt, or cereal, or simply eat the nuts as a snack.
- Spread mashed avocado on bread or toast, instead of butter.
- Add sliced avocados to salads, wraps, and sandwiches instead of cheese.

Egg substitutes that are made with egg whites contain no fat or cholesterol.
©Wendy Schiff

Should People Avoid Eggs?

Eggs are a relatively economical source of protein and many micronutrients. However, egg yolks are the most concentrated source of cholesterol in the typical American diet. The yolk of a large egg contains 5 grams of fat and about 190 mg of cholesterol. To reduce their cholesterol consumption, many Americans eat fewer fresh eggs than in the past. Whole eggs, however, are often hidden in commonly eaten foods, such as salad dressings, noodles, frozen custards, sauces, and baked goods.

Does eating egg yolks increase the risk of atherosclerosis? In general, the cholesterol in food does not have as much effect on blood cholesterol levels of healthy people as the saturated fat does. Why? In a healthy person, the liver produces less cholesterol when large amounts of cholesterol are eaten. On the other hand, eating large amounts of saturated fat increases the liver's cholesterol production. Foods that are high in cholesterol are often also rich sources of saturated fat. Egg yolks, however, have almost twice as much unsaturated fat as saturated fat. Scientists do not agree on whether eggs increase a person's risk of CVD. The American Heart Association recommends healthy people practice "moderation" when eating all foods, including eggs.[33]

Did *YOU* Know?

Preparing home-cooked meals can boost your mood and improve your physical health. A recent study showed college students who prepare meals for themselves and others report an improvement in overall mood and an increased sense of accomplishment from the creative process of cooking. People who prepare and consume home-cooked meals experience health benefits, including higher diet quality, such as increased fruit and vegetable intake; maintaining a healthy body weight; and living a long life.[34]

Do you want to learn how to prepare healthy meals for yourself and others but need to know where to start? Look for meal preparation courses offered at your college or in your local community, and consider taking a course with a friend! When you are ready to start cooking, first choose a food item you enjoy eating and then find a recipe to provide preparation instructions. The internet is a great recipe resource; you can also watch videos showing step-by-step recipe preparation. Next, use the recipe substitutions listed in the Food and Nutrition Tips box to reduce the amount of solid fat, added sugars, and salt in the foods you prepare.[35] Visit the USDA's cooking website for cooking videos, recipes, and other meal preparation resources, https://snaped.fns.usda.gov/nutrition-education/nutrition-education-materials/cooking.

Jack Frog/Shutterstock

Food & Nutrition tips

Recipe Includes	Substitution	Product Changes
Butter or margarine (baked products)	Replace fat with the same amount of applesauce Replace each 1 cup of fat with ½ cup mashed cooked carrots, sweet potato, or squash Replace each cup of fat with ¾ cup vegetable oil (e.g., canola, peanut, and safflower oils)	Reduces calories and unhealthy fats and adds some fiber, phytochemicals, and vitamins Reduces the amount of fat in the recipe by about 20%
Cream	Replace with equal amount of whole milk or half-and-half	Reduces calories and unhealthy fat
Cream cheese (regular)	Replace with equal amount of fat-free or reduced-fat ("Neufchatel") cream cheese	Reduces calories and unhealthy fat
Cheese (full fat)	Replace with equal amount of low-fat or nonfat cheese *Note: Do not replace with nonfat cheese in cooked dishes because the cheese does not melt properly.*	Reduces calories and unhealthy fat
Whole or 2% milk	Replace with equal amount of low-fat milk, fat-free milk, or soy milk. *Note: Other milk alternatives including almond, rice, and coconut contain less protein and may contain less sugar, which will affect the final product's texture and browning in baked goods.*	Reduces calories and unhealthy fat
Evaporated milk	Use fat-free evaporated milk	Reduces calories and unhealthy fat
Condensed milk	Use fat-free evaporated milk *Note: Product will not be as sweet.*	Reduces calories, unhealthy fat, and added sugars
Sour cream	Replace with fat-free or low-fat sour cream Replace with plain, fat-free yogurt	Reduces calories and unhealthy fat
Egg	Replace each whole egg with 2 egg whites, ¼ cup egg substitute, or "flax egg." Create a flax egg by mixing 2 tablespoons raw ground flaxseed meal with 1.5 tablespoons water. Let sit for 5 minutes.	Reduces cholesterol
Vegetable oil	Reduce amount by 25% Replace each 1 cup of oil with 3/4 cup mashed cooked carrots, sweet potato, pumpkin, or squash in baked products	Reduces calories and fat and adds fiber, vitamins, and phytochemicals
Sugar (white, brown, and syrups)	Reduce the amount of sugar by 25–50% Use a sugar substitute that is appropriate for baked products (check labels)	Reduces calories and added sugars
Salt	Reduce the amount by 50–75% Season with fresh herbs or ground herb products that have no added salt	Reduces sodium

Egg whites have no fat or cholesterol. For those who must limit their cholesterol intake, egg whites can often be used in recipes that call for whole eggs. Products that substitute for whole eggs are available in supermarkets. The second Food & Nutrition Tips feature in this section provides practical steps, including recipe substitutions, to reduce the amount of solid fat, added sugars, and salt in foods.

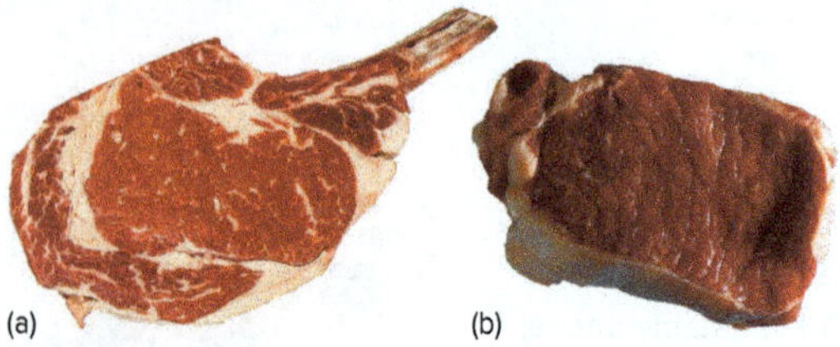

(a) (b)

Figure 6.24 Choosing leaner meats. "Prime" rib steak (a) compared to a lesser grade of steak, such as a "choice" rib eye steak (b). Prime cuts of meat usually have more solid fat that lesser grades of the same cut. Prime steak: Aoldman/iStock/Getty Images; Choice steak: Ingram Publishing/SuperStock

Food Selection and Preparation

You can change your food selection and preparation practices to reduce the amount of unhealthy fat in your diet. Fatty meats such as prime rib steaks are often more tender and expensive than lower–grade leaner cuts (Fig. 6.24). A "prime" rib steak, for example, usually contains large amounts of visible solid fat (white streaks) that surround the muscle and are embedded in its tissue (see Figure 6.24a). Such fat can be difficult to remove from the meat before and after cooking. Cuts of meat that are not graded as "prime" (for example, "choice") are usually leaner, but they often taste drier than higher–graded cuts (see Figure 6.24b).

Certain cooking methods can increase the tenderness of lean cuts of meat, such as pork loins, shoulder "steaks," and roasts. Moist cooking methods, such as pot roasting or tightly covering the baking dish with foil, help tenderize meats without adding fat. In addition to using a moist cooking method, reduce the oven temperature from 350°F to less than 325°F. The meat will take longer to cook, but it is less likely to toughen and dry out. After cooking, avoid eating the visible fat that remains. For example, trim away much of the fat from the meat and do not use pan drippings to make sauces or gravies. Steaming meats and vegetables is a cooking method that does not require adding fat during preparation. Stir–frying pieces of raw vegetables, meat, fish, shellfish, and poultry in small amounts of hot vegetable oil cooks them quickly and preserves micronutrients. Additionally, when you brown ground beef in a pan, drain much of the fat before you add other ingredients to the meat. Purchase skinless chicken pieces or remove the skin and fatty pieces before baking or broiling. Dipping raw foods in batter and deep–fat frying them adds considerable amounts of fat to your diet because breading serves as a sponge that soaks up oil. If you prepare breaded fried foods, place the items on paper towels after cooking them to soak up as much excess fat as possible. Some people think the breading is the "best part" of a fried food, but removing some or all of the breading before eating the item can reduce your overall fat intake.

Although it is easy to peel greasy breading from fried fish, much of the fat that you eat is hidden in foods and beverages. For example, do you drink 2% milk? Fat comprises only 2% of the milk's volume, but the lipid contributes 37% of the beverage's calories. Fat–free milk actually contains less than 0.5% fat by volume, and fat contributes essentially no energy to the beverage.

What about the fat content of the cream cheese, margarine, or butter that you spread on a piece of toast? About 90% of the calories in cream cheese and about 100% of the calories in butter are from fat. Vegetable oils are almost 100% fat, so fat contributes all the energy in most salad dressings. Fried foods, chips, and salad dressings are high in fat; processed meat (such as bacon, sausage, hot dogs, and luncheon meats) and hard cheeses are also fatty foods. Nuts, including peanuts and almonds, have high fat contents, but they generally contain high amounts of healthy monounsaturated fats.

Instead of striving to eliminate all fatty foods from your diet, try reducing your intake of them, especially rich sources of saturated fats. For example, if you drink 2% or whole milk, switch to 1% or fat–free milk. Nearly all the lipids in fat–free milk have been removed; that is why it tastes watery to people who

MY DIVERSE PLATE

Stimulate your overall senses and taste buds with new foods and flavors by trying a new cuisine! Choose dishes lower in fat including:

- Chinese steamed dumplings filled with lean meats and vegetables or fresh spring rolls (rice wrappers filled with raw veggies)
- Japanese miso soup (fermented soybean paste soup) or sukiyaki pot (tofu, beef, vegetables, and noodles)
- Italian minestrone soup (white bean and vegetable soup)
- African braised collard greens, stewed black-eyed peas, and beef oxtail
- Mexican tortilla soup (vegetables, peppers, and tortilla strips in chicken broth) or bean burritos (mashed pinto beans in a flour tortilla)
- Indian aloo gobi (potatoes, cauliflower, and sometimes broccoli cooked in tomatoes, onion, and spices) or chana masala (chickpeas in a spicy tomato-based sauce)

Indian aloo gobi Pikoso.kz/Shutterstock

are not accustomed to drinking it. Except for energy and lipid content, the nutritional value of fat–free milk is basically the same as that of whole and 2% milks. The "Food & Nutrition Tips" features in this section provide more practical suggestions for reducing your saturated fat intake.

Fried foods are often high in fat.
Ingram Publishing/Alamy Stock Photo

Other Dietary and Lifestyle Modifications

In addition to modifying your fat intake, you can reduce your risk of CVD by making other dietary changes. Increasing consumption of foods that are rich sources of fiber, including fruits, vegetables and whole grains. Foods rich in soluble fiber (see Chapter 5), can reduce LDL cholesterol levels without lowering beneficial HDL cholesterol levels.

Consuming high amounts of sucrose or fructose can result in elevated blood triglyceride levels. The body can convert excess fructose into triglycerides.

Food & Nutrition tips

To reduce your intake of unhealthy fats, consider the following tips.

- Purchase lean meats that have "loin" or "round" in their name (pork tenderloin, for example).
- Try replacing some fatty foods with reduced-fat or fat-free alternatives. For example, place a spoonful of yogurt, instead of butter or sour cream, on a baked potato.
- Because most nuts are rich sources of healthy unsaturated fats, replace foods that contain saturated fat with nuts. For example, use peanut or soy nut butters instead of cheese or luncheon meat in sandwiches. Some nut butters contain added sugars, so read the Nutrition Facts panel and ingredients list before selecting these products.
- Pretzels, air-popped popcorn, and most fruits and vegetables are low fat and generally more nutrient-dense than chips, cookies, pastries, and candy bars.
- Purchase brands of microwave popcorn that have little added fat. Buy plain popcorn, and to pop the kernels, use a small amount of hot oil in a covered saucepan or use a hot-air popping machine.
- Use tzatziki (*zäd–zē'–kē*) a Greek condiment made from low-fat yogurt, cucumber, garlic, and dill as a substitute for full-fat sour cream dips such as ranch dip.
- Commercial frostings are high in solid fat and added sugars. Remove most of the frosting from a serving of cake before eating it. You can make your own frosting by mixing soft margarine or smooth peanut butter with a little low-fat milk, powdered sugar, and vanilla flavoring.
- Patronize fast-food restaurants that offer low-saturated-fat menu items, such as salads without added cheese and meat, baked or broiled chicken and fish, low-fat yogurt, and fresh fruit.
- Use less salad dressing on salads. When in restaurants, order salad dressings "on the side" so you can control the amount that is added.
- When folllowing recipes, replace butter with vegetable oils (canola, peanut, or corn oil, for example). Reduce the amount of solid fat in the recipe by about 20% (butter is 80% fat; oil is 100% fat).

Did *YOU* Know?

Garlic, onions, and chives are sources of sulfur-containing compounds, such as allicin, that have antioxidant properties. Fresh garlic and garlic supplements are promoted to lower blood cholesterol levels. However, there is a lack of strong scientific support for consuming fresh garlic or garlic supplements to reduce elevated blood cholesterol.[36] Furthermore, taking dietary supplements that contain garlic may interfere with blood's ability to form clots, which can increase one's risk of bleeding.

C Squared Studios/
Photodisc/Getty Images

Tzatziki is a Greek condiment made from low-fat yogurt, cucumber, garlic, and dill that can be used as a substitute for full-fat sour cream dips, such as ranch dip.
Traci Keck

This product's label displays the American Heart Association's heart-healthy symbol next to a diet-related health claim.
©Wendy Schiff

Did *YOU* Know?

Eggs with brown shells are not more nutritious than eggs with white shells; the color of an eggshell is determined by the breed of hen that laid it. Also, grading does not reflect the nutritional content of an egg. When cracked open and placed on a plate, the membrane of a "AA" grade egg tends to hold the yolk and white together better than membranes of lower-grade eggs.

Pixtal/age fotostock

If your triglyceride level is too high (> 150 mg/dl), consider cutting back on your intake of refined carbohydrates by eating less candy and pastries and drinking fewer sugar–sweetened soft drinks.

Drinking small amounts of alcohol (1 to 2 drinks/day) can raise beneficial HDL cholesterol levels. However, recent scientific research reports that even moderate amounts (1 drink per day) of alcohol intake may increase the risk of some types of CVD and cancer.[8] Therefore, the *Dietary Guidelines for Americans* do not recommend people start drinking alcohol for any reason. People who consume alcohol should discuss their habits with their physician because the health benefits of drinking alcohol may not outweigh the risks. Section 6.7 discusses alcohol metabolism and the drug's effects on health.

Registered dietitian nutritionists often promote the traditional Mediterranean diet for healthy eating, as well as reducing elevated blood lipid levels and the risk of CVD. The Mediterranean diet (see Figure 3.19) recommends eating more fish and seafood, and smaller amounts of red meat; using heart–healthy unsaturated fatty acids; engaging in regular physical activity; and drinking a glass of wine daily. This diet also emphasizes antioxidant– and phytochemical–rich fruits, vegetables, beans, and nuts.

Hypertension is a major risk factor for CVD. Dietary Approaches to Stop Hypertension ("the DASH diet") is a healthy dietary pattern that promotes fruits and vegetables, low–fat dairy products, whole–grain foods, tree nuts, seeds, fish, poultry, and legumes to reduce blood pressure. (*Legumes* include peanuts, split peas, and soybeans.) Unlike the typical American diet ("Western diet"), the DASH diet is high in fiber and low in saturated fat, cholesterol, and refined carbohydrates, such as added sugars. Additionally, the standard DASH diet recommends limiting sodium intake to less than 2300 mg/day. Table 6.8 summarizes how some dietary manipulations and other therapeutic lifestyle changes may alter a person's blood lipid levels.

Excess body fat, especially around the midsection of the body, is associated with unhealthy LDL cholesterol and triglyceride levels. Physical inactivity and excess energy consumption contribute to unwanted weight gain. Performing moderate–intensity physical activity nearly every day and balancing energy intake with energy expenditure each day can help people achieve and maintain healthy body weights. Taking such steps can also reduce elevated LDL cholesterol and triglyceride levels.

TABLE 6.8 Ways to Lower your Risk of CVD

Action	Potential Benefits
Increase physical activity level	Raises HDL levels
Lose excess body fat	Lowers elevated triglycerides, blood pressure, and risk of type 2 diabetes; may increase HDL
Quit smoking	May raise HDL and lower LDL levels
Make specific dietary changes:	
Reduce intake of saturated fats and avoid trans fats	Raises HDL and lowers LDL
Replace saturated fats with polyunsaturated and monounsaturated fats	Lowers LDL

TABLE 6.8 Ways to Lower your Risk of CVD *(continued)*

Action	Potential Benefits
Include some omega-3 and omega-6 fats	Reduces triglycerides; may raise HDL and reduce blood pressure
Replace butter with olive oil or spread made from plant sterols/stanols	Reduces LDL
Eat more whole-grain, fiber-rich foods, especially those containing soluble fiber	Reduces total cholesterol and LDL without altering HDL
Consume foods that are rich sources of antioxidants and phytochemicals, including vegetables, fruit, beans, and nuts.	Reduces risk of heart disease
Reduce added sugar and excessive alcohol intake	Lowers triglyceride levels
Limit sodium intake; adopt DASH diet	Lowers elevated blood pressure

What If Lifestyle Changes Do Not Work?

Some people are unable to lower their risk of CVD significantly by making dietary changes, exercising regularly, and losing excess body fat. If your blood lipids are too high and the levels have remained elevated even after you have made these lifestyle modifications, it is important to discuss additional treatment options with your physician. Millions of Americans take a class of prescription drugs called *statins* to reduce their elevated blood lipid levels. Statins interfere with the liver's metabolism of cholesterol, effectively reducing LDL cholesterol and triglyceride levels as a result. Statins are relatively safe when taken as directed.[37]

Concept CHECKPOINT

18. Define *atherosclerosis, arterial plaque, thrombus,* and *embolus.* Discuss the series of physiological changes that occur in arteries and contribute to the development of CVD.

19. List three kinds of fish that often contain high amounts of methylmercury.

20. List at least two major risk factors for atherosclerosis that are nonmodifiable and at least five that are modifiable.

21. Bernard's HDL cholesterol level is 62 mg/dl. Based on this information, does Bernard have a high risk or low risk of CVD? Explain your answer.

22. What is "hs-CRP"? What can you learn about your risk of heart disease and stroke from having a lipoprotein profile performed on your blood?

23. Suggest at least four ways people can reduce their intakes of saturated and trans fats and increase their intakes of unsaturated fats.

24. What is a statin?

See Appendix F for responses.

6.7 Nutrition Matters: Drink to Your Health?

Learning Outcomes

1. Identify amounts of beer, wine, and distilled spirits in a standard drink.
2. Classify people who drink alcohol according to their usual pattern of consumption.
3. Identify factors that affect alcohol metabolism.
4. Discuss alcohol's effects on health.

What do beer, wine, vodka, whiskey, sake (*sak′–e*), koumiss (*koo′–mis*), and kefir (*keh–feer′*) have in common? These beverages contain *ethanol,* a two–carbon compound that chemists classify as an alcohol (Fig. 6.25). This textbook refers to ethanol simply as "alcohol."

Offering wine in religious ceremonies, toasting the bride and groom with champagne at a wedding, or barhopping with friends on their 21st birthdays—for many Americans, alcohol consumption is a part of celebrating religious rites and life's milestones. When consumed in moderation, alcoholic beverages can make social situations more enjoyable. Many people, however, experience serious problems as a result of their drinking habits. This section focuses on alcohol metabolism, as well as the chemical's effects on the body.

For many Americans, alcohol consumption is a part of celebrating religious rites and life's milestones.
VStock/Alamy Stock Photo

Alcohol Production

Throughout the world, people have been producing and drinking alcoholic beverages for thousands of years. The chemical process that results in alcohol is not complicated and occurs naturally. The process requires certain microbes, warm conditions, and a source of simple sugars. Although some types of bacteria produce alcohol, commercial alcoholic beverage production relies on *yeast,* one–celled fungi (*fun′–ji*) that break down (*ferment*) simple sugars in the absence of oxygen to obtain energy and the metabolic waste product, alcohol. Grains, fruit, and potatoes—just about anything that contains simple sugars—will ferment under the proper conditions. Yeast will even ferment lactose in milk.

$CH_3—CH_2—OH$

Ethanol

Figure 6.25 Ethanol. Ethanol is a simple two-carbon compound.

Alcohol is soluble in water, and alcoholic beverages generally contain a considerable amount of water. Beers are typically 3 to 6% alcohol, wines contain about 8 to 14% alcohol, and malt liquors are about 7% alcohol by volume. Alcohol is poisonous (*toxic*), and yeast die when the concentration of alcohol in the fermenting solution reaches 14 to 16%. The distilling process increases the alcohol concentration of an alcoholic beverage. Distilled spirits (hard liquors) such as whiskey, bourbon, and vodka are generally 40 to 50% alcohol. You can determine the percentage of alcohol in hard liquor by dividing the "proof" declaration on the label by two. Tequila, for example, is "80 proof," or 40% alcohol.

Although each gram of alcohol provides 7 kcal, alcohol is not a nutrient; it is a mind–altering drug that is often classified as a food. Beer and wine contain simple carbohydrates and small amounts of certain minerals and B vitamins. Distilled spirits have essentially no nutritional value other than water. Mixing distilled spirits with juices, cocktail mixes, or other flavorings increases the alcoholic beverage's energy content and may add some nutrients, particularly simple sugars, depending on the ingredients in the mixer. Table 6.9 indicates the grams of alcohol and carbohydrates in certain alcoholic drinks, as well as their caloric content. A standard drink (approximately 12 fluid ounces of beer, 5 fluid ounces of wine, or 1½ fluid ounces of liquor) contains 14 grams of alcohol (Fig. 6.26).

TABLE 6.9 Approximate Alcohol, Carbohydrate, and Energy Contents of Alcoholic Beverages*

Beverage	Amount (fl oz)	Energy (kcal)	Alcohol (g)	Carbohydrates (g)
Beer				
Regular	12.0	153	14	12
Lite	12.0	103	11	5
Table Wines	5.0	123	14	4
Distilled Spirits				
Gin, rum, vodka, whiskey	1.5	97	14	0

* Protein and fat contribute little or nothing to the caloric content.

Source: U.S. Department of Agriculture.

Figure 6.26 What's a standard drink? A standard drink is approximately 12 fluid ounces of beer, 1½ fluid ounces of liquor, or 5 fluid ounces of wine. ©Wendy Schiff

How the Body Processes Alcohol

Alcohol requires no digestion and readily passes through the tissues lining the inside of the mouth, esophagus, stomach, and small intestine. When alcohol is consumed with meals, food delays its absorption from the stomach and slows the rate at which the drug enters the bloodstream. To reduce alcohol's harmful effects, the body detoxifies the simple chemical by converting it into less damaging compounds. Detoxification begins in the stomach, where the enzyme gastric alcohol dehydrogenase metabolizes up to 20% of the alcohol. Most of the remaining alcohol passes through the small intestinal wall and circulates to the liver, the primary site for metabolizing alcohol. Because alcohol is a poison, the liver shifts its metabolic focus from macronutrient metabolism to alcohol detoxification when the compound enters its tissues.

The liver relies on two biochemical pathways to metabolize alcohol. When relatively low doses are consumed, the enzyme alcohol dehydrogenase converts most of the alcohol to acetaldehyde, a substance that is more toxic than alcohol. Another enzyme, aldehyde dehydrogenase, reacts with acetaldehyde to form acetate, which can also be harmful. Acetate, however, can be metabolized to carbon dioxide (CO_2) and water (H_2O), or the molecule can be used to synthesize fatty acids and cholesterol (Fig. 6.27).

If a person consumes excessive amounts of alcohol such as during a drinking binge, the alcohol overwhelms the liver's ability to metabolize the drug using the dehydrogenase pathway. When this occurs, the second method of processing alcohol, the microsomal ethanol oxidizing system (MEOS), takes over. Unlike the alcohol dehydrogenase pathway, MEOS wastes energy in the form of body heat that dissipates into the environment. Thus, alcoholics typically gain little weight from their energy intake when alcohol supplies most of their energy.

Factors That Influence Alcohol Metabolism

You may have noticed that a few of your friends who drink can "hold their liquor" better than others. Why are some people able to drink more alcohol at one time than others? Several factors account for the variability. In addition to the amount and timing of alcohol consumption, personal characteristics such as sex, body size and composition, age, and prior drinking history affect the body's detoxification rate. For example, a healthy person who weighs 154 pounds (70 kg) metabolizes about 1 alcoholic drink per hour. Drinking caffeinated beverages, exercising, or taking vitamins does not increase this rate. To sober up, the drinker must stop consuming alcohol and give their liver time to metabolize the drug.

Alcohol is not stored in the body, and it dissolves easily in the watery tissues of the body. Until the liver can detoxify the alcohol, the toxic chemical circulates

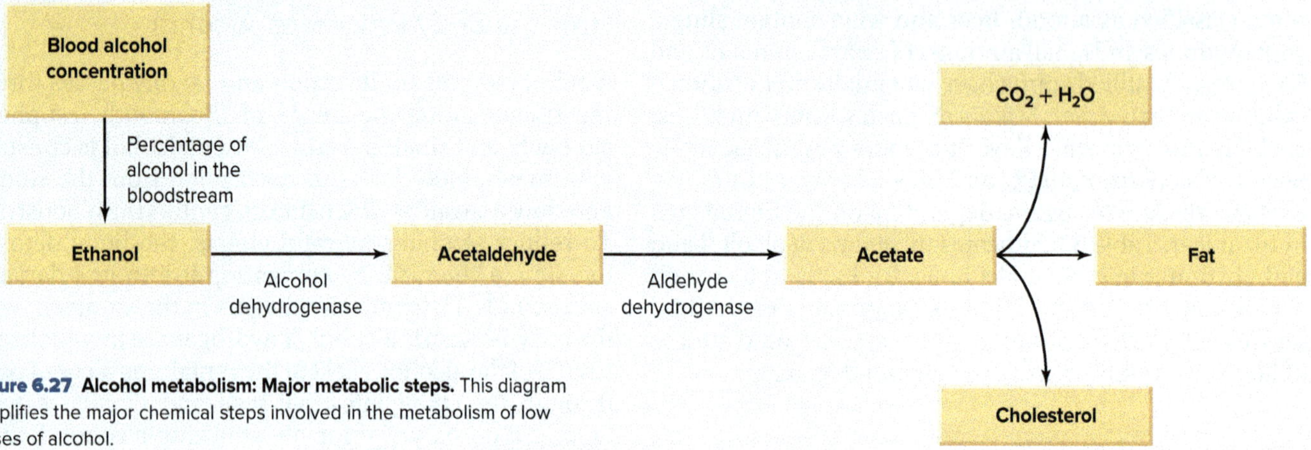

Figure 6.27 Alcohol metabolism: Major metabolic steps. This diagram simplifies the major chemical steps involved in the metabolism of low doses of alcohol.

in the bloodstream and diffuses into the watery fluids within and surrounding cells. The lungs and perspiration eliminate some of the alcohol; that's why you can smell alcohol when you're around someone who's been drinking. The kidneys also filter some of the drug from the bloodstream and eliminate it in urine. To determine whether someone is legally intoxicated as a result of drinking alcohol, law enforcement officials use special devices to analyze the alcohol in blood, urine, or expired air to estimate the person's **blood alcohol concentration (BAC).** BAC is reported as a percentage that indicates the amount of alcohol in the blood. In the United States, a BAC of 0.08% is the legal limit for intoxication for automobile operators who are 21 years of age or older.

blood alcohol concentration (BAC) percentage of alcohol in the bloodstream

Males and females have different physical responses to alcohol. Females tend to become more impaired than males do after drinking the same amount of alcohol, even when differences in body weight are taken into account (Fig. 6.28). The reasons for these sexual differences are unclear, but they probably involve physiological factors including body size and composition. The average male is larger than the average female, and larger people can often drink more alcohol without showing ill effects than smaller individuals because they have bigger livers that detoxify more alcohol at a time. Additionally, a healthy

Figure 6.28 Alcohol consumption and approximate BAC. The accompanying table shows the relationship between the number of alcoholic drinks consumed on an empty stomach per hour and BACs for healthy males and females. For example, a 170-pound male will have his BAC reach 0.05% after he has three drinks, whereas a 140-pound female will have her BAC reach 0.05% after drinking only two drinks. However, alcohol's effects on individuals can vary. In most of the United States and Canada, a person is legally intoxicated when their BAC is 0.08% or higher (as of January 1, 2023).

150–pound male typically has less body fat and more body water than a healthy 150–pound female. After the male drinks a beer, the alcohol diffuses out of his bloodstream and into the water compartments of his body. The female's body has less water, so more alcohol remains in her bloodstream after she drinks a beer. As a result, her BAC rises faster, and she becomes more intoxicated after drinking the same amount of alcohol per hour as her male counterpart. Compared to males who drink heavily, females have a higher risk of serious health problems, especially damage to their liver, brain, and heart, when they abuse the same amounts of alcohol.[38]

Prior alcohol exposure also influences the rate of alcohol metabolism. People who drink regularly develop tolerance. Tolerance occurs as the levels of liver enzymes needed to metabolize alcohol increase, and as a result, the rate of alcohol metabolism increases. Consequently, the regular drinker needs to consume more alcohol at a time to achieve the same mind–altering effects as a person who drinks infrequently. Tolerance can lead to alcohol dependence (alcoholism).

Classifying Drinkers

In 2020, 50% of Americans who were 12 years of age or older reported being current alcohol drinkers during the 30 days prior to the survey.[39] Almost one–quarter of the current drinkers engaged in binge drinking at least once, and 6.4% of current drinkers drank heavily during this period.

What is "moderate drinking" or "binge drinking"? Although there is no universally accepted definition for "light," "moderate," or "heavy" drinking, medical practitioners and alcohol researchers often follow the classification guidelines listed in Table 6.10. Binge drinking is a form of heavy drinking, but the definition of binge drinking often differs according to sex. According to the Centers for Disease Control and Prevention, binge drinking is defined as having four or more drinks during an occasion (about 2 hours) for females and five or more drinks during an occasion for males.[40] Regardless of their backgrounds, binge drinking is a common practice among Americans.

TABLE 6.10 Classifying Drinkers

Classification of Drinker	Amount of Alcohol Consumed (Standard Drinks)	
	Males	**Females**
Moderate	Up to 2 drinks/day	Up to 1 drink/day
Heavy	15 or more drinks/week	8 or more drinks/week
Binge	5 or more drinks/ occasion (about 2 hours)	4 or more drinks/ occasion (about 2 hours)

Source: Centers for Disease Control and Prevention, Alcohol & Public Health: "Frequently asked questions." Updated April 9, 2022. www.cdc.gov/alcohol/faqs.htm. Accessed: April 23, 2023.

The three teenagers who died in this automobile had consumed alcohol before the accident.
Gwinnet County Police Department/Centers for Disease Control and Prevention

Alcohol Use Disorder

A "problem drinker" experiences problems at home, work, and school that are associated with their drinking habits. According to experts with the National Institutes of Health, a person with a severe drinking problem has an alcohol use disorder (AUD).[41] In 2019, about 6% of Americans who were 18 years of age or older had an AUD. Both problem drinkers and people with AUD engage in behaviors that place themselves and others in danger, such as drinking and driving.

A person who is *dependent* on alcohol has an uncontrollable need to drink; is unable to limit their alcohol consumption; suffers withdrawal symptoms, such as shakiness and anxiety, when alcohol is unavailable after a period of heavy drinking; and experiences *tolerance* to the drug. Tolerance develops over time as the drinker's body becomes better able to metabolize the alcohol. As a result, the person with AUD needs to consume more of the drug to obtain the same effects that they experienced prior to becoming dependent. Table 6.11 is a self assessment that lists possible signs of AUD.

Alcohol and Health

Alcohol is a central nervous system depressant. Mild alcohol intoxication often produces pleasant sensations and relaxed inhibitions. Consuming large amounts, however, depresses normal motor functioning, including breathing, and death can result. Alcohol use is often

TABLE 6.11 Signs of Alcohol Abuse

Check all of the boxes that apply.

☐ Had times when you drank more than you intended?
☐ Wanted to reduce or stop drinking, or you tried to, but you couldn't?
☐ Had a strong need or urge to drink?
☐ Felt that drinking or its aftereffects interfered with your ability to take care of your personal responsibilities, do your job, or keep up with school work?
☐ Continued to drink even though it was causing trouble with your family or friends?
☐ Gotten into situations while or after drinking that increased your chances of being harmed, such as driving, swimming, using machinery, walking in a dangerous area, or having unsafe sex?
☐ Continued to drink even though it was making you feel depressed or anxious, or your drinking was causing other health problems?
☐ Continued to drink even after having a memory blackout?
☐ Found that when the effects of alcohol were wearing off, you had withdrawal symptoms, such as trouble sleeping, shakiness, irritability, anxiety, depression, restlessness, nausea, or sweating?

If you checked one or more of these boxes, your alcohol use may already be a concern for you and others. Consider seeking professional help.

Adapted from National Institutes of Health, National Institute of Alcohol Abuse and Alcoholism: *Alcohol use disorder.* ND. www.niaaa.nih.gov/publications/brochures-and-fact-sheets/understanding-alcohol-use-disorder. Accessed: April 27, 2023.

involved in motor vehicle accidents, falls, and drownings, as well as acts of violence and abuse. An estimated 140,000 Americans die each year as a result of excessive alcohol use.[42] Thousands of other people suffer physical injuries and have damaged interpersonal relationships related to excess alcohol consumption.

The potentially harmful physiological effects of abusing the drug vary from person to person, primarily because of differences in overall health, drinking habits, and genetic background. Alcohol affects every cell in the body, and when consumed in excess, the drug damages every system in the body, particularly the gastrointestinal, nervous, and cardiovascular systems. Figure 6.29 summarizes major damaging physiological effects of alcohol.

Alcohol and the Gastrointestinal Tract

If you have consumed distilled spirits without adding mixers, you probably felt a burning sensation as the alcohol entered your throat and stomach. This sensation is an indication

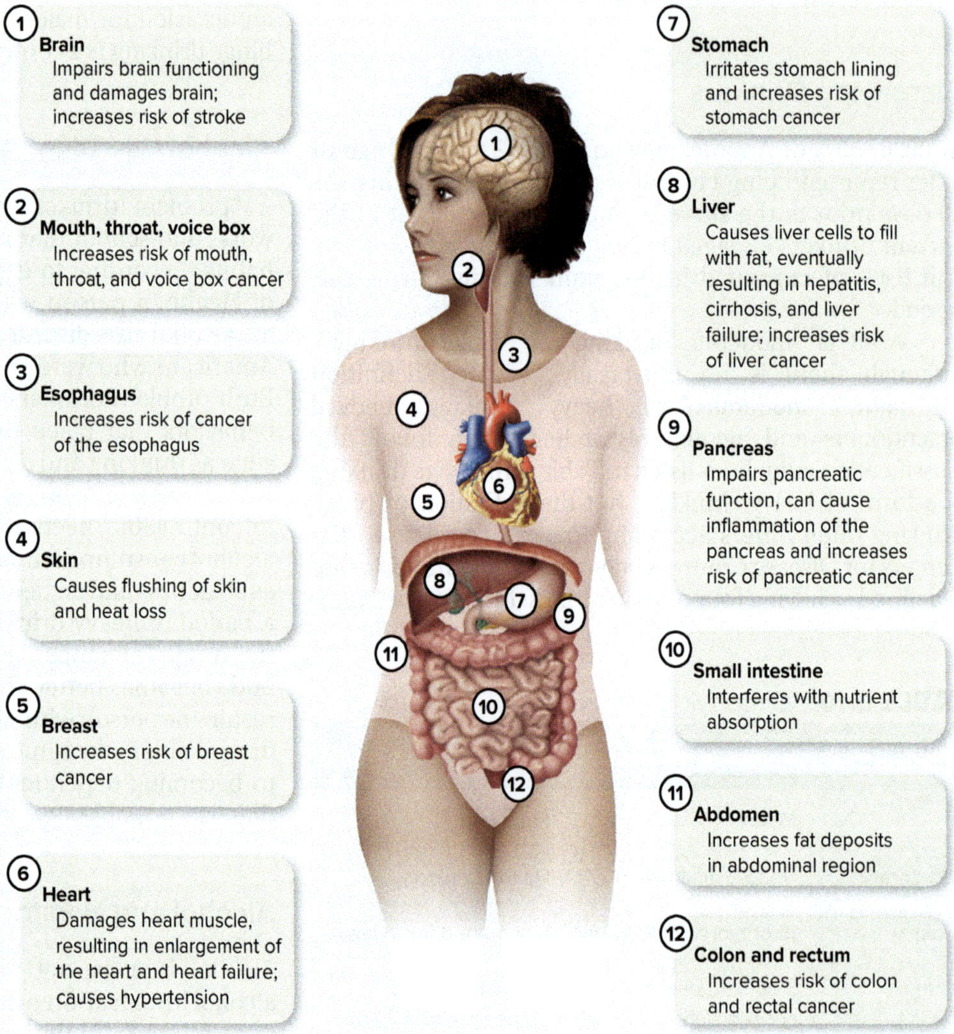

Figure 6.29 Some of alcohol's effects on the body. Chronic alcohol abuse seriously damages several organs and increases the risk of various cancers.

of alcohol's irritating effects on the lining of your gastrointestinal tract. Not surprisingly, chronic drinking contributes to intestinal ulcer formation, particularly in the esophagus and stomach. An ulcer is a sore. Intestinal ulcers can cause chronic bleeding and may penetrate through the intestinal wall. When this occurs, intestinal contents leak into the abdominal cavity, causing serious and often deadly infections. Although the reasons are unclear, chronic alcohol consumption increases the risk of *alcoholic pancreatitis,* a painful and sometimes fatal condition characterized by inflammation and destruction of the pancreas.

People who consume alcohol have an increased risk of cancers of the digestive system. Such cancers include mouth, throat, esophagus, liver, colon, and rectal cancers.[43]

Alcohol and the Brain

Alcohol's effects on the central nervous system, especially the brain, appear within a few minutes of having a drink. Alcohol acts as a depressant, slowing the transmission of messages between nerve cells (neurons). At low BACs (< 0.06%), the drinker is relaxed and less inhibited socially as regions of the brain that control decision making and reasoning ability are depressed. If the person continues to drink and their BAC increases to between 0.08 and 0.15%, the person loses control over voluntary muscles, particularly muscles that move the lips, eyes, and limbs. As a result, the drinker's speech is slurred, their eyes have difficulty focusing on objects, and the person's ability to drive or operate any heavy equipment is seriously compromised. When the drinker's BAC reaches 0.25% or higher, their brain is unable to process information. Higher BACs usually result in loss of consciousness ("passing out"). Moreover, coma and even death can occur as the brain loses control over lung and heart functioning. Table 6.12 lists BAC levels and typical nervous system effects that can occur at each level.

Alcohol can disrupt the ability of neurons in the brain to transmit messages. Long–term consumption of excessive amounts of alcohol can cause the neurons to shrink, and the brain develops other structural abnormalities. Confusion and memory loss are common signs of the extensive brain damage that occurs in a chronic heavy drinker.

Alcohol Poisoning As a college student, you may have observed binge drinking or engaged in the practice. In 2020, Americans who were 18 to 25 years of age were more likely to binge drink than Americans in other age groups.[39] Youthful binge drinking is a serious public health concern because the behavior is often associated with driving while drunk. Furthermore, the practice may increase a person's later risk of alcoholism and can result in death.

Binge drinking has become an expected "rite of passage" for American youth celebrating their 21st birthday. The practice of trying to consume 21 drinks quickly while celebrating "power hour" ("21 for 21") can have deadly consequences. On March 15, 2004, Jason Reinhardt was celebrating his 21st birthday with friends in a bar. Jason rapidly drank 16 shots of alcohol. Although he managed to return to his friends' house and go to bed, his lifeless body was discovered a few hours later. The young man's blood alcohol concentration was 0.361, well within the deadly range (see Table 6.12). What makes binge drinking so dangerous?

TABLE 6.12 Typical Effects of Alcohol at Various BAC Levels (Adults)

Blood Alcohol Concentration	Physiological and Psychological Effects
0.02	Some loss of good judgment, altered mood, relaxation
0.05	Reduced inhibitions, resulting in exaggerated emotional and behavioral responses to situations Impaired judgment, good mood
0.08	Loss of balance, slower-than-normal reaction time, impaired reasoning ability and memory Reduced ability to control one's behavior
0.10	Major impairment of hearing, vision, and muscular coordination Slurred speech and obvious delayed reaction time
0.15	Poor muscular control, vomiting, loss of balance
0.20	Cannot walk without help, mental confusion May pass out
0.25 or above	Loss of consciousness, coma, possible death from respiratory arrest (breathing stops)

Source: Adapted from Centers for Disease Control and Prevention, Injury Prevention and Control, Motor Vehicle Safety: *Impaired driving: Get the facts: BAC Effects.* Page last updated: August 24, 2020. www.cdc.gov/motorvehiclesafety/impaired_driving/impaired-drv_factsheet.html; University of Notre Dame, McDonald Center for Student Well-Being: *Blood alcohol concentration.* ND. https://mcwell.nd.edu/your-well-being/physical-well-being/alcohol/blood-alcohol-concentration/. Accessed: April 23, 2023.

Binge drinking increases a person's BAC rapidly and to a point at which signs of alcohol poisoning occur. An individual suffering from alcohol poisoning is confused, "passes out" and cannot be aroused (comatose), breathes slowly and irregularly, and has pale or bluish skin. Alcohol poisoning can cause the heartbeat to slow down and the lungs to stop functioning, resulting in death. Additionally, if a comatose person vomits, their stomach contents can enter the lungs, causing the person to choke to death. Jason died in his sleep from alcohol poisoning. His breathing rate slowed, his heartbeat became irregular, and his organs gradually shut down while he slept. Thus, it is important to recognize that alcohol poisoning is a life–threatening condition. If you suspect someone has consumed a deadly amount of alcohol, do not waste time trying to estimate how many drinks that person has drunk; call 911 immediately.

Alcohol and the Liver

In addition to harming the brain, alcohol can damage the liver. Some acetaldehyde that forms when alcohol is metabolized enters the complex series of biochemical pathways that eventually produce carbon dioxide, water, and ATP, the primary energy storage molecule for cells. The liver, however, also uses acetaldehyde to make fatty acids for synthesizing triglycerides (see Figure 6.27). Even after a single bout of heavy drinking, fat accumulates in liver cells and causes a condition called "fatty liver." Fatty liver may be reversible; if the affected person avoids alcohol for an extended period, the liver metabolizes the fat, and the organ eventually heals itself. If the person continues to drink, the buildup of fat destroys their liver cells, and tough scar tissue replaces them. This irreversible condition is called *liver cirrhosis* or hardening of the liver (Fig. 6.30). People with AUD are prone to develop *hepatitis,* inflammation of the liver. Hepatitis increases the risk of liver cancer.

The scarred regions of the liver of a person with AUD have no function other than holding the organ together. Under normal conditions, a healthy liver can regenerate sections of itself, but when destruction of liver cells is extensive, the organ begins to fail. In this situation, the affected person will die unless they undergo liver transplantation. Chronic alcohol abuse is a major cause of liver failure among adult Americans.

(a)

(b)

Figure 6.30 Liver cirrhosis. The photograph on the left (*a*) shows a normal liver; the one on the right (*b*) shows a liver damaged by alcoholic cirrhosis.
Arthur Glauberman/Science Source

Alcohol and the Cardiovascular System

Consuming light to moderate amounts of alcohol raises HDL cholesterol levels and decreases platelet stickiness.[44] Platelets are cell fragments involved in the blood–clotting process. Reducing the likelihood of blood clot formation lowers the risk of heart attack and certain types of strokes.

Some medical researchers think drinking beer and red wine is healthier than consuming white wines or spirits. Although the alcohol in beer is the same as that in wine and distilled spirits, red wine and beer have higher levels of certain antioxidants and B vitamins than other alcoholic beverages, which may provide health benefits. The role of moderate alcohol consumption in coronary artery disease prevention is controversial. Drinking small amounts of alcohol seems to reduce the risk of heart disease.[44] However, recent scientific research reports that moderate amounts (1 standard drink per day) of alcohol intake may increase the risk of some types of CVD.[8] Therefore, the *Dietary Guidelines for Americans* do not recommend people start drinking alcohol for any reason. Consuming moderate to excessive amounts is associated with increased risks of hypertension, stroke, heart failure, and weight gain.[44] Therefore, people who consume alcohol should discuss their habits with their physician because the cardiovascular and overall health benefits of drinking alcohol may not outweigh the risks.

Alcohol and Cancer

As mentioned earlier, alcohol increases the risk of oral cavity, esophageal, stomach, liver, pancreatic, and colorectal cancer. Recent scientific research reports moderate amounts (1 standard drink per day) of alcohol intake may increase the risk of cancer.[8] Heavy drinkers who smoke tobacco products have a much greater risk of developing cancers of the oral cavity and esophagus than people who drink less and do not smoke. Furthermore, females who consume one or more drinks daily have a higher risk of breast cancer than females who avoid alcohol or drink less than one standard drink per day.

Alcohol and Interactions with Other Drugs

People who drink alcohol while taking other drugs, including prescription medications and over–the–counter remedies, need to recognize that alcohol's harmful effects may be amplified by the medications. Additionally, alcohol may interact with other drugs, causing serious side effects that do not occur when a drug is consumed alone. For example, combining alcohol with products that contain the pain–reliever acetaminophen can cause severe liver damage and even death.

Effects of Alcohol on Nutritional Status

When consumed in moderation, alcohol stimulates the appetite. Alcohol, however, lowers blood glucose levels and raises blood triglycerides. Chronic, excessive alcohol intake can have adverse effects on the drinker's nutritional intake and status. Many alcoholics consume a considerable portion of their energy as alcohol, which often displaces nutrient–dense foods from their diets and increases their risk of malnutrition. The excess alcohol damages the liver and interferes with the absorption, metabolism, and storage of various micronutrients, including phosphorus, magnesium, potassium, sodium, selenium, chromium, and zinc.[45]

Poor diets contribute to deficiencies of vitamin A, vitamin C, and the B vitamins thiamin and folate among alcoholics. It is not unusual for chronic alcoholics to become thiamin deficient and develop *Wernicke–Korsakoff syndrome,* a brain disorder characterized by mental confusion, memory loss, and uncoordinated muscular movements. The person with this condition typically staggers when trying to walk. Taking thiamin supplements can resolve some signs of the syndrome, but the person must avoid drinking alcohol while being treated.

Although chronic alcohol abuse is associated with an increased risk of bone loss and fractures, light to moderate alcohol drinkers tend to have stronger bones and lower risk of fractures than nondrinkers.[46] More research, however, is needed to determine the effects of alcohol on bone health.

Alcohol and Body Water

Billboard and commercial advertisements for alcoholic beverages often show sweaty, physically active young adults preparing to gulp down beer or liquor to relieve their thirst. These ads are misleading. Alcohol is not a good thirst quencher because it is a diuretic that suppresses the production of antidiuretic hormone (ADH) by the pituitary gland. Without ADH's action, the kidneys produce more urine, and the body loses water and certain vitamins and minerals along with it. If a dehydrated drinker consumes even more alcohol to relieve thirst, this response only increases their water losses. Drinking water and other nonalcoholic drinks is the best way to keep the body well hydrated.

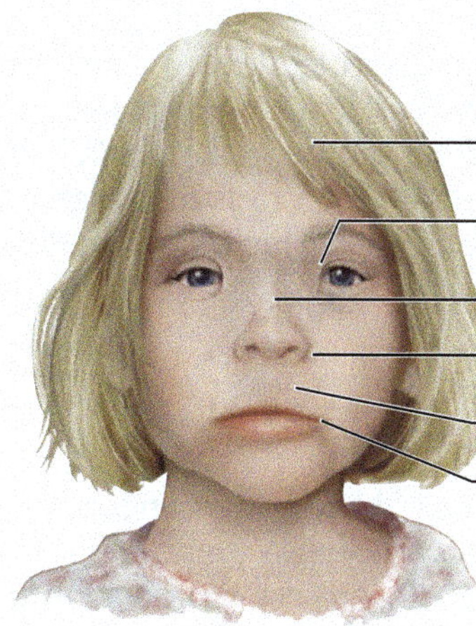

Figure 6.31 Fetal alcohol syndrome. Physical deformities and developmental delays are characteristics of fetal alcohol syndrome, a type of fetal alcohol spectrum disorder.

Fetal Alcohol Spectrum Disorders

When a pregnant female drinks alcohol, her embryo/fetus also "drinks" alcohol because the drug passes freely from the mother's bloodstream into the embryo/fetus's bloodstream. Alcohol is most devastating when it affects an embryo because organs develop during the first 2 months after conception. Unfortunately, many females are not aware that they are pregnant during this early stage of their child's prenatal (before birth) development, and they may drink socially or binge drink. Alcohol is toxic to cells, including rapidly dividing embryonic cells. An infant born with a *fetal alcohol spectrum disorder* (*FASD*) has some degree of developmental abnormalities as a result of its mother's consumption of alcohol during pregnancy. *Fetal alcohol syndrome* (*FAS*) is a devastating form of FASD. A child with FAS is born with certain facial and heart defects, as well as extensive, irreversible damage to its nervous system that causes intellectual disability (Fig. 6.31). Children with FAS also experience delayed and abnormal physical development.

TABLE 6.13 Who Should Avoid Alcohol?

- Females who suspect they are pregnant, know they are pregnant, or are trying to become pregnant
- People who plan to drive or use heavy machinery
- People taking certain over-the-counter or prescription medications
- People with medical conditions that alcohol can aggravate
- People recovering from alcohol use disorder
- People younger than 21 years of age

Source: Adapted from National Institute on Alcohol Abuse and Alcoholism: *Drinking levels defined.* ND. www.niaaa.nih.gov/alcohol-health/overview-alcohol-consumption/moderate-binge-drinking. Accessed: April 23, 2023.

The amount of alcohol that can be safely consumed by a pregnant female has not been determined. Therefore, if you or someone you know is trying to conceive or is pregnant, you or that person should "play it safe" and avoid alcohol. The risks are too high to justify even one drink. Table 6.13 identifies people, including pregnant females, who should not drink alcohol.

Alcohol and Physical Performance

Although some athletes may think consuming a small amount of alcohol before a competitive event might help relieve anxiety, alcohol reduces hand–eye coordination and slows reaction times even when BACs are relatively low (0.02 to 0.05%). Studies indicate conflicting findings concerning the effects of low to moderate amounts of alcohol on strength and endurance. In some instances, low to moderate intakes of alcohol reduce endurance and have negative effects on strength, but in other cases, this level of alcohol consumption produces no detrimental effects on strength and endurance. Alcohol can contribute to dehydration, which impairs muscular performance and causes heat injuries such as heat exhaustion and heat stroke. Chronic alcohol abuse causes muscular wasting that affects skeletal as well as heart muscle. Obviously, such effects will have negative effects on muscular mass, strength, and endurance.

Athletes should learn about alcohol's effects on health and avoid consuming excess alcohol, especially during the 48 hours before an event. After exercise, and until their body recovers its normal fluid status, the athlete should focus on consuming nonalcoholic beverages.

Where to Get Help for Problem Drinking or an Alcohol Use Disorder

If you think you are abusing alcohol or are dependent on the drug, seek help from your personal physician. For information about alcohol abuse, you can contact the National Drug and Alcohol Treatment Referral Routing Service at 1–800–662–HELP. You can also visit the websites of Alcoholics Anonymous (www.aa.org), the Substance Abuse and Mental Health Services Administration (www.samhsa.gov), or Al–Anon (https://al–anon.org/).

Concept CHECKPOINT

25. Summarize the effects of alcohol on the body, including the liver.

26. What is BAC? What is the legal limit for BAC in the United States?

27. Dontel drinks three 12-ounce cans of beer daily. Based on this information, is Dontel a moderate, heavy, or binge drinker?

See Appendix F for responses.

Chapter References

See Appendix G.

Summary

6.1 Introducing Lipids

- Lipids are needed for energy, proper growth and development, nerve functioning, and the production of bile and several hormones. Major lipids are triglycerides, phospholipids, and sterols. In addition to being a source of fuel, insulates the body against cold temperatures, and protects against damaging blows.

6.2 Fatty Acids, Triglycerides, Phospholipids, and Sterols

- Triglycerides, an important fuel for the body, comprise most of the lipid content of your food and body. Lipids can be sources of the essential fatty acids. Furthermore, the fat in food enhances absorption of fat-soluble vitamins and phytochemicals. Dietary lipids also contribute to the appealing flavor, texture, and aroma of foods. Although consuming some lipids is essential for health, high amounts may increase your risk of serious health conditions, including obesity, certain cancers, and CVD.

- Most lipids have fatty acids in their chemical structures. Fatty acids can be saturated or unsaturated, and unsaturated fatty acids can be either monounsaturated or polyunsaturated. The body cannot synthesize the omega-6 fatty acid linoleic acid or the omega-3 fatty acid alpha-linolenic acid; these essential fatty acids must be supplied by the diet. The typical American diet provides more omega-6 fat than omega-3 fat. Fatty fish, canola and soybean oils, walnuts, and flaxseed are sources of the omega-3 fats.
- Trans fatty acid molecules have a different configuration than cis fatty acid molecules. This difference enables fats that contain a high proportion of trans fatty acids to be more solid at room temperature than fats with a high proportion of cis fatty acids. Whole milk and whole-milk products, butter, and meat naturally contain small amounts of trans fats. Diets that contain high amounts of trans fats are associated with an increased risk of heart disease and stroke. Partially hydrogenated oils are banned as food additives.
- A triglyceride has three fatty acids attached to glycerol. Triglycerides comprise about 95% of lipids in the body and in food. Triglycerides usually contain mixtures of unsaturated and saturated fatty acids, but one type of fatty acid (saturated, monounsaturated, or polyunsaturated) tends to predominate. In general, animal fats contain higher percentages of saturated fatty acids than plant fats. Important exceptions are highly saturated cocoa butter, coconut oil, and palm oil.
- Phospholipids have both hydrophilic and hydrophobic regions, and as a result, they are partially soluble in water and can serve as emulsifiers. Phospholipids are the major structural component of cell membranes and are needed for proper functioning of nerve cells, including those in the brain. Lecithin is the major phospholipid in food; egg yolks, liver, wheat germ, peanut butter, and soybeans are rich sources of this compound.
- The sterol cholesterol is a component of every cell membrane. Cells use cholesterol to make a variety of substances, including vitamin D, bile, and steroid hormones such as estrogen and testosterone. Cholesterol is found only in animal foods. Plants synthesize sterols and stanols that are not well absorbed by humans. Plant sterols and stanols, however, may be beneficial to health because they interfere with cholesterol absorption.

Irina Marwan/Moment/Getty Images

6.3 What Happens to Lipids in the Body?

- The triglycerides and phospholipids in food undergo digestion primarily in the upper part of the small intestine. Cholesterol is not broken down and is absorbed into villi. Before leaving the small intestine, triglycerides, cholesterol, and other lipids are coated with a layer that contains protein to form chylomicrons. Chylomicrons enter the lymphatic system of the small intestine and eventually reach the bloodstream. The liver uses lipids to make various lipoproteins, substances that transport lipids in the bloodstream. Enterohepatic circulation enables the liver to recycle bile salts to make new bile.
- Triglycerides and carbohydrates are major sources of cellular energy. If energy is not needed, adipose cells remove fatty acids and glycerol from circulation and use them to synthesize triglycerides for storage. When energy is needed, adipose cells break down some stored triglycerides and release glycerol and fatty acids into the bloodstream.

C Squared Studios/Photodisc/Getty Images

6.4 Lipid Consumption Patterns

- Fat contributes almost 36% of Americans' average daily energy intake. According to the Dietary Guidelines, fat should comprise 20 to 35% of total energy intake; the AMDR for fat is 20 to 35% of total calories.

Ingram Publishing/SuperStock

6.5 Understanding Nutritional Labeling: Lipids

- Consumers can use nutrient labels to determine how much total fat, saturated fat, and trans fat are in packaged food.

6.6 Lipids and Health: Cardiovascular Disease

- CVD affects the heart and blood vessels. In the United States, heart disease (coronary artery disease) is the leading cause of death. Atherosclerosis is a long-term process that can result in heart disease and other common forms of CVD. Numerous risk factors are associated with atherosclerosis; some risk factors are nonmodifiable, but many are related to lifestyle practices that can be altered. Smoking cigarettes, eating certain fats, and being physically inactive are lifestyle practices that increase a person's risk of atherosclerosis and CVD.
- Blood lipid levels can have a major influence on the risk of atherosclerosis. Having high blood levels of HDL cholesterol is usually healthier than having high LDL cholesterol levels because elevated LDL cholesterol contributes to atherosclerosis, whereas elevated HDL cholesterol generally reduces the risk of this condition.
- Exercising and replacing saturated and trans fats with unsaturated fats may reduce LDL levels and increase HDL levels. On the other hand, physical inactivity and eating high amounts of saturated fat may raise LDL and blood triglyceride levels, increasing one's risk of atherosclerosis. Healthy people, however, generally do not experience an increase in their blood cholesterol levels when they eat cholesterol in foods.
- Although oils, fatty spreads, and salad dressings are obvious sources of dietary fat, much of the fat we eat is not visible. Saturated fat contributes much of the calories in fatty meats, processed meat, hard cheeses, and whole milk. If people make dietary modifications, lose excess weight, and exercise regularly, and their blood lipid levels still remain elevated, they should discuss additional treatment options with their physicians. Millions of Americans take prescription statin medications to reduce their elevated blood lipid levels.

ZoonarShullye Serhi/age fotostock

6.7 Nutrition Matters: Drink to Your Health?

- Alcohol (ethanol) is a water-soluble, two-carbon compound that is toxic to cells. Beers are typically 3 to 6% alcohol, and wines contain about 8 to 14% alcohol by volume. Each gram of alcohol provides 7 kcal, but alcohol is not a nutrient; it is a mind-altering drug. A standard drink (approximately 12 fluid ounces of beer, 5 fluid ounces of wine, or 1½ fluid ounces of liquor) contains 14 grams of alcohol.
- Alcohol requires no digestion and readily passes through the tissues lining the inside of the mouth, esophagus, stomach, and small intestine. The body detoxifies alcohol by ultimately converting the chemical into less-damaging compounds. Until the liver has had enough time to metabolize all the alcohol that has been consumed, the amount of the drug that remains circulates in the bloodstream. In the United States, a blood alcohol concentration of 0.08% is the legal limit for intoxication for automobile operators who are 21 years of age or older (as of January 2023). Binge drinking is a major public health problem in the United States.
- Although drinking some alcohol raises HDL, recent scientific research reports that even moderate amounts (1 standard drink per day) of alcohol intake may increase the risk of some types of CVD and cancer.[8] Excessive amounts of the drug damage every system in the body, particularly the gastrointestinal, nervous, and cardiovascular systems. Alcohol has especially devastating effects on an embryo. Infants born with FAS have certain facial and heart defects, as well as extensive, irreversible damage to their nervous systems that causes intellectual disability. Therefore, pregnant females should not drink alcohol.

Sebastian Vera/E+/Getty Images

Recipe for Healthy Living

©Wendy Schiff

Salmon Salad Sandwiches

Are you tired of eating the same foods and looking for a new meal idea? Try something different that's easy to prepare and a rich source of omega-3 fatty acids: salmon salad sandwiches. This recipe calls for canned salmon, but you can use leftover baked salmon to make the salad. This recipe makes enough salad for two sandwiches. Each sandwich provides approximately 406 kcal, 12 g fat, 31 g protein, 4.5 g fiber, and 352 mg calcium.

INGREDIENTS:

- 8 oz canned, pink salmon
- 2 Tbsp pickle relish
- 2 Tbsp minced (finely chopped) sweet onion
- 2 Tbsp lite mayonnaise
- 4 slices whole-wheat bread
- 1 slice fresh tomato (optional)
- 1 piece leaf lettuce (optional)

PREPARATION STEPS:

1. Drain fluid from salmon. In a bowl, break up salmon with a fork, including the small bones. Add other ingredients to the salmon and mix until well blended.
2. Toast slices of whole-wheat bread.
3. Spread ½ cup of salad on one slice of the toasted bread. Add a fresh tomato slice and/or piece of leaf lettuce, if desired.

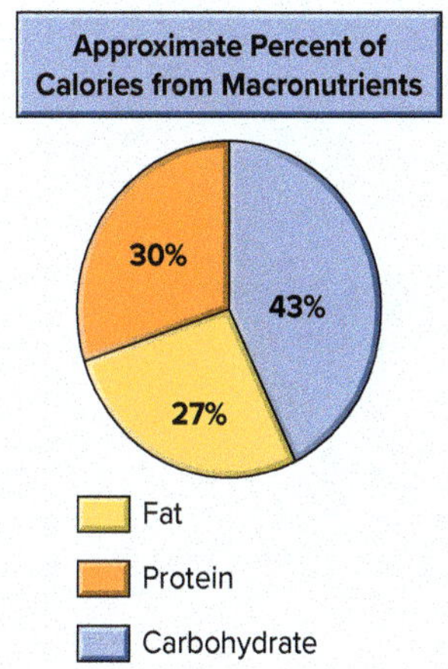

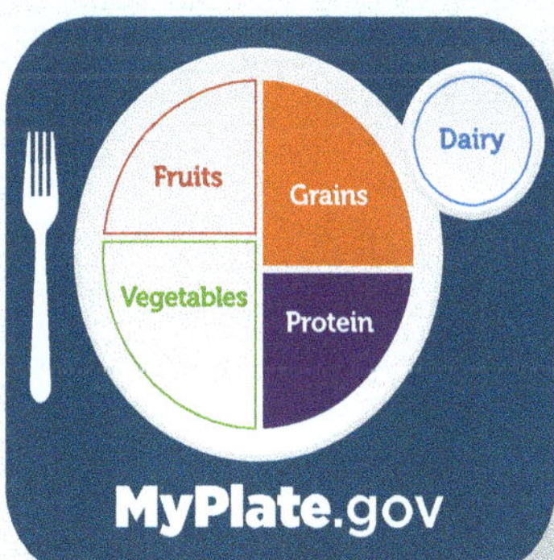

US Department of Agriculture or USDA

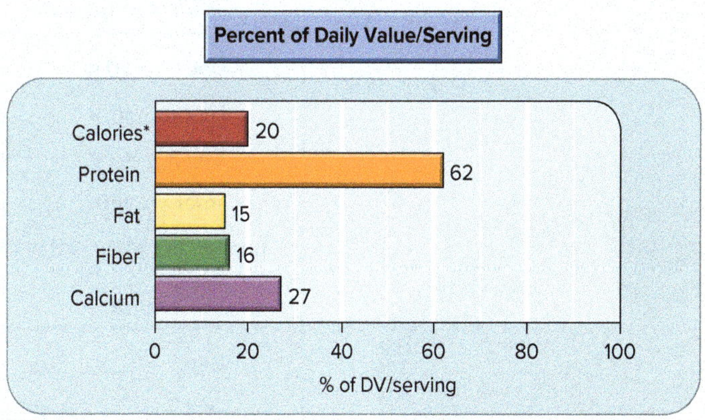

*2000 daily total kcal

©Wendy Schiff

Personal Dietary Analysis

1. Refer to the 3-day food log from the "Personal Dietary Analysis" feature in Chapter 3. List the total number of kilocalories you consumed for each day of record keeping. Add the figures to obtain a total, divide the total by 3, then round the figure to the nearest whole number to obtain your average daily energy intake for the 3-day period.

Sample Calculation:

Day 1 **2000 kcal**

Day 2 **1700 kcal**

Day 3 **2350 kcal**

Total kcal **6050 ÷ 3 days = 2017 kcal/day**

(average kilocalorie intake, rounded to the nearest whole number)

Your Calculation:

Day 1 ______ **kcal**

Day 2 ______ **kcal**

Day 3 ______ **kcal**

Total kcal ______ **÷ 3 days =** ______ **kcal/day**

(average kilocalorie intake, rounded to the nearest whole number)

2. Add the number of grams of fat eaten each day of the period. Divide the total by 3 and round to the nearest whole number to calculate the average number of grams of fat consumed daily.

Sample Calculation:

Day 1 **50 g**

Day 2 **57 g**

Day 3 **42 g**

Total = **149 g**

Total grams 149 ÷ 3 days = 50 g/day (average)

Your Calculation:

Day 1 ______ **g**

Day 2 ______ **g**

Day 3 ______ **g**

Total = ______ **g**

Total grams ______ **÷ 3 days =** ______ **g of fat/day**

(average, rounded to the nearest whole number)

3. Each gram of fat provides about 9 kcal; therefore, you must multiply the average number of grams of fat that you ate daily (step 2) by 9 to obtain the average number of kilocalories from fat.

Sample Calculation:

50 g/day × 9 kcal/g = 450 kcal from fat

Your Calculation:

______ g/day × 9 kcal/g = ______ kcal from fat

4. To calculate the average percentage of calories that fat contributed to your diet, divide the average number of calories from fat obtained in step 3 by the average total daily calorie intake obtained in step 1 and round to the nearest one-hundredth. Multiply the value by 100, drop the decimal point, and add the percent symbol.

Sample Calculation:

450 kcal ÷ 2017 kcal = 0.22

0.22 × 100 = 22%

Your Calculation:

______ kcal ÷ ______ kcal = ______

______ × 100 = ______ %

5. Did your average daily fat intake meet the U.S. Dietary Guidelines recommended 20 to 35% of total energy?

   Yes ______ No ______

6. If your average fat intake was more than 35% of your total energy intake, which foods contributed to your intake of fats?

   Foods: ____

7. Review the log of your 3-day food intake. Calculate your average daily intake of saturated fat by adding the grams of saturated fat consumed over the 3-day period and dividing the total by 3.

Your Calculation:

Day 1 ______ **g**

Day 2 ______ **g**

Day 3 ______ **g**

Total = ______ **g**

Total g ______ **÷ 3 days =** ______ **g of saturated fat daily**

   a. What was your average daily saturated fat intake? ______ g

   b. Was your average daily intake less than the Dietary Guidelines recommended limit for an adult (less than 10% of total calories)? ______ Yes ______ No

   c. If your average saturated fat intake was greater than the recommended limit, list foods that contributed to your intake. ____________________________

Elite Images/McGraw Hill

Elena Shashkina/shashkina/123RF

Using Nutrient Labels: Fats

1. Remove the labels from two different packaged foods that contain fat. Using information from each of the product's Nutrition Facts panel, answer the following questions.

   a. Name of product #1 ________________

   Name of product #2 ________________

   b. How many kcal are in one serving of each food?

   Product 1 ______ kcal
   Product 2 ______ kcal

   c. How many grams of fat, saturated fat, and trans fat are in a serving?

   Product 1

   ______ grams of total fat
   ______ grams of saturated fat
   ______ grams of trans fat

   Product 2

   ______ grams of total fat
   ______ grams of saturated fat
   ______ grams of trans fat

   d. Calculate the percentage of total kcal/serving that are from saturated fat. First, multiply the number of grams of saturated fat by 9 (number of kcal per gram of fat). Then, divide the number of kcal contributed by saturated fat by the total number of kcal per serving. (The Nutrition Label uses the term "Calories" for kcal.) To obtain the percentage, move the decimal point two places to the right, remove the decimal point, and insert % symbol.

   Product 1 ______ grams of saturated fat × 9 kcal = ______ kcal of saturated fat/serving

   ______ % total kcal/serving contributed by saturated fat

   Product 2 ______ grams of saturated fat × 9 kcal = ______ kcal of saturated fat/serving

   ______ % total kcal/serving contributed by saturated fat

   e. Read the list of ingredients for each product. If the food contains saturated fat and trans fat, identify ingredients that contributed saturated fat or trans fat to the product.

   Saturated fat ingredients in Product 1 ________________________________

   ____

   Trans fat ingredients in Product 1 ________________________________

   ____

Saturated fat ingredients in Product 2 ______________________________

____

Trans fat ingredients in Product 2 ______________________________

____

Complete the Personal Dietary Analysis activity in Connect.

Assessment: Evaluating your Saturated Fat Intake

Do you Consume:	Rarely or Never	1 to 2 Times/ Week	3 to 5 Times/ Week	Daily
1. Bacon, hot dogs, sausage, salami, bologna, or other fatty luncheon meat?	0	1	2	3
2. Whole milk?	0	1	2	3
3. 2% milk?	0	1	2	3
4. Ice cream or milkshakes?	0	1	2	3
5. Sour cream or cream cheese?	0	1	2	3
6. Fatty cuts of pork or beef?	0	1	2	3
7. Hard cheeses, such as cheddar or Swiss?	0	1	2	3
8. Butter?	0	1	2	3
9. Stick margarine?	0	1	2	3
10. Gravy, cheese sauce, or cream-based sauce?	0	1	2	3
11. Buttered popcorn or "rinds" ?	0	1	2	3
12. Biscuits, croissants, doughnuts, Danish pastries, pies, cakes, or cookies?	0	1	2	3
13. Products made with lard?	0	1	2	3
14. Cream in your coffee or tea?	0	1	2	3
15. Pizza?	0	1	2	3
16. Creamed soups such as cream of potato soup or New England clam chowder?	0	1	2	3
TOTAL	______	______	______	______

Scoring: Add points in each column and then add those figures together. (The higher the total points, the higher your saturated fat intake.)

My "Saturated Fat Score" is ______

If your score is 30 or more: Your saturated fat intake is probably too high. Note which of these fatty foods you eat more than three times per week. Consider reducing your intake of these items and replacing them with foods that contain healthy unsaturated fats.

Brand X Pictures/Getty Images

Critical Thinking

1. Calculate your risk of having a heart attack by completing the American Heart Association's Heart Attack Risk Calculator at http://static.heart.org/riskcalc/app/index.html#!/baseline–risk. Based on your results, should you be concerned about your risk of CVD? If you are concerned, discuss steps you can take to reduce your risk of atherosclerosis and CVD.
2. Complete the assessment of your saturated fat intake on the previous page. If your score is near 30 or more, identify four ways you can reduce your intake of these items and replace them with foods that provide healthy fats.
3. Plan a meal that supplies 20 to 35% of energy from fat. The meal should include foods from the major food groups and provide 700 to 900 kcal.
4. Recently, your friend Paul's 50–year–old grandfather had a mild heart attack. Paul is very concerned about his risk for having a heart attack. What would you tell him?
5. Develop an infographic or video for college students that describes atherosclerosis and the role that personal choices (lifestyles) play in the development of the disease.
6. If you drink alcohol, consider filling out the checklist shown in Table 6.11. According to this assessment, do you have a problem with your alcohol consumption habits? If your answer is yes, what can you do to obtain help?

Practice Test (Section numbers are listed below in red.)

Select the best answer.

1. **(6.1)** Fats in foods
   a. add taste and contribute to "mouth feel."
   b. are digested and absorbed in the stomach.
   c. carry water–soluble nutrients.
   d. need to be eliminated to have a healthful diet.
2. **(6.2)** Solid fats generally have a high proportion of_____ fatty acids.
   a. unsaturated
   b. saturated
   c. polyunsaturated
   d. monounsaturated
3. **(6.2)** A saturated fatty acid has
   a. one double bond within the hydrocarbon chain.
   b. two double bonds within the hydrocarbon chain.
   c. no double bonds within the hydrocarbon chain.
   d. three or more double bonds within the hydrocarbon chain.
4. **(6.2)** Which of the following statements is true?
   a. Certain fish are rich sources of omega–3 fatty acids.
   b. Omega–3 fatty acids increase the risk of cardiovascular disease.
   c. Trans fats are rich sources of omega–3 fatty acids.
   d. The human body converts dietary fiber into omega–3 fatty acids.
5. **(6.2)** Trans fatty acids are
   a. naturally in many foods.
   b. a by–product of the hydrogenation process.
   c. essential to good health.
   d. found primarily in fatty fish.
6. **(6.2)** Phospholipids
   a. do not have fatty acids in their chemical structures.
   b. lack glycerol in their chemical structures.
   c. do not occur naturally.
   d. are partially water–soluble.
7. **(6.2)** Cholesterol is
   a. metabolized for energy.
   b. found only in animal foods.
   c. not made by the human body.
   d. harmful to health.
8. **(6.3)** The primary site of triglyceride digestion and absorption is the
   a. stomach.
   b. liver.
   c. small intestine.
   d. gallbladder.

9. **(6.4)** According to the *Dietary Guidelines for Americans,* healthy adults should consume less than _____ of their total calories from saturated fatty acids.
   a. 5%
   b. 10%
   c. 20%
   d. 35%

10. **(6.5)** According to the Nutrition Facts panel, a serving of mixed nuts supplies 168 kcal, and there are 14 g of total fat. How many kcal from fat are in a serving of the nuts?
    a. 56
    b. 98
    c. 126
    d. 154

11. **(6.6)** Lipoproteins
    a. are water–insoluble.
    b. transport lipids in the bloodstream.
    c. contain glucose.
    d. are toxic to cells.

12. **(6.6)** Modifiable risk factors for atherosclerosis include
    a. family history.
    b. age.
    c. tobacco use.
    d. racial/ethnic background.

13. **(6.6)** Homocysteine is a(n)
    a. form of folate.
    b. lipid that lowers blood pressure.
    c. possible marker for cardiovascular disease.
    d. essential amino acid.

14. **(6.6)** A young adult has ____________, which makes them at high risk for cardiovascular disease.
    a. low saturated fat intake
    b. low LDL cholesterol
    c. a family history of headaches
    d. high oxidized LDL cholesterol

15. **(6.6)** Jose would like to reduce his saturated fat intake and increase his polyunsaturated and monounsaturated fat intake. Which habit should Jose adopt?
    a. Choose chicken with the skin on instead of rib bone steaks
    b. Cook with coconut oil instead of canola oil
    c. Switch from 2% milk to whole milk
    d. Spread mashed avocado on toast instead of butter

16. **(6.7)** Alcohol metabolism is not influenced by a person's
    a. sex.
    b. level of caffeine consumption.
    c. prior history of alcohol use.
    d. body size and composition.

17. **(6.7)** After attending a college graduation party, a young adult's BAC was 0.16%. This level is _____ times the legal limit.
    a. 2
    b. 4
    c. 3
    d. 5

18. **(6.7)** Brita was a binge drinker during the first 3 months of her pregnancy. Her baby was born with FAS. Which of the following conditions is a sign of FAS?
    a. Chronic vomiting
    b. Iron deficiency anemia
    c. Bleeding gums
    d. Thin upper lip

Burke/Triolo Productions/Brand X Pictures/Getty Images

Answers to Quiz Yourself

1. To lose weight, use regular, stick margarine instead of butter because it has 25% fewer calories per teaspoon. **False.** (Section 6.6)
2. Egg yolks are a rich source of cholesterol. **True.** (Section 6.6)
3. Taking too many fish oil supplements may be harmful to health. **True.** (Section 6.6)
4. On average, Americans consume 60% of their energy from fat. **False.** (Section 6.4)
5. Increasing your intake of saturated fat will reduce your risk of heart disease. **False.** (Section 6.6)

CHAPTER 7

Proteins

Pixtal/age fotostock

Quiz Yourself

How much protein is recommended for optimal health? Can people obtain enough protein by eating only plant foods? What happens if you eat more protein than your body needs? Before reading Chapter 7, take the following quiz to test your knowledge of protein. The answers are at the end of the chapter.

1. Animal foods such as meat and eggs are almost 100% protein. ______ T ______ F
2. Foods made from soybeans can be sources of high-quality protein. ______ T ______ F
3. Americans typically consume more protein from animal sources than from plant foods. ______ T ______ F
4. Registered dietitian nutritionists generally recommend that people with liver or kidney disease take amino acid supplements to increase their protein intake. ______ T ______ F
5. People can nourish their hair by using shampoo that contains protein. ______ T ______ F

7.1 What Are Proteins?

Learning Outcomes

1. List the primary function of proteins in the body.
2. Identify the basic structural unit of proteins and its components.
3. Classify an amino acid as essential or nonessential.

Chickpeas, soybeans, and the grain quinoa in this salad are sources of protein.
Ekaterina Kondratova/nblxer/123RF

Since ancient times, many people have believed that eating animal foods, particularly meat, was necessary for good health and optimal physical performance. Milo of Croton, an ancient Greek Olympian wrestler with extraordinary strength, reportedly consumed about 20 pounds of meat daily. Although accounts of Milo's superhuman capacity for eating meat are unreliable, modern athletes often make protein–rich foods and protein supplements the foundation of their diets. Furthermore, it is not unusual for nonathletes to associate meat with protein and a lack of protein with poor muscular development and weak muscular strength.

Many Americans think a meal is not adequate unless it contains large portions of meat. Although it is true that meat is a rich source of protein, other foods, including eggs, milk, grains, dry beans, peas, and lentils, are often overlooked as sources of protein. For example, the beef patty of a cheeseburger provides protein, as does the melted cheese that tops the burger and the bun that makes the sandwich convenient to eat.

proteins large, complex organic molecules made up of amino acids

Proteins are complex organic molecules that are chemically similar to lipids and carbohydrates because they contain carbon, hydrogen, and oxygen atoms. Proteins, however, are the only macronutrient that contains nitrogen, one of the elements cells need to make a wide array of important biological compounds. Plants, animals, bacteria, and even viruses contain hundreds of proteins.

Protein is an important class of nutrients, but it is not more valuable to your health than other nutrients. Nutrients work together in your body like members of a well–trained basketball team on the playing court. Making one player the star while neglecting to develop the other athletes' skills can have disastrous effects on the team's success. Similarly, overemphasizing one class of nutrients in your diet, such as protein, while ignoring other nutrients can lead to nutritional imbalances that result in serious health problems. By reading this chapter, you will learn about the roles of proteins in your body and their major food sources. You will also learn how proteins influence your health.

Proteins in the Body

Proteins are necessary for muscle development and maintenance, but more than 200,000 different proteins in your body have a wide variety of functions, including those shown in Table 7.1.[1] Skin, blood, nerve, bone—all cells in your body—contain proteins. Structural proteins such as *collagen* are in your cartilage, ligament, and bone tissue. *Keratin* is another structural protein; it is in your hair, nails, and skin. Contractile proteins in your muscles enable you to move, and transport proteins carry many substances in the bloodstream. Proteins are also necessary for your blood to clot properly.

Certain hormones, such as insulin and glucagon, are proteins. Hormones and neurotransmitters are chemical messengers that regulate body processes and responses, such as growth, metabolism, and hunger. *Neurotransmitters* send signals from one nerve cell to another. Some neurotransmitters, such as glutamate, the most abundant neurotransmitter in the brain, are proteins. Nearly

TABLE 7.1 Proteins in the Body

The body uses proteins to make or function as:
• New cells and many components of cells
• Structures such as hair and nails
• Enzymes
• Lubricants
• Clotting compounds
• Antibodies
• Compounds that help maintain fluid and pH balance
• Certain hormones and neurotransmitters
• Energy source (minor, under usual conditions)

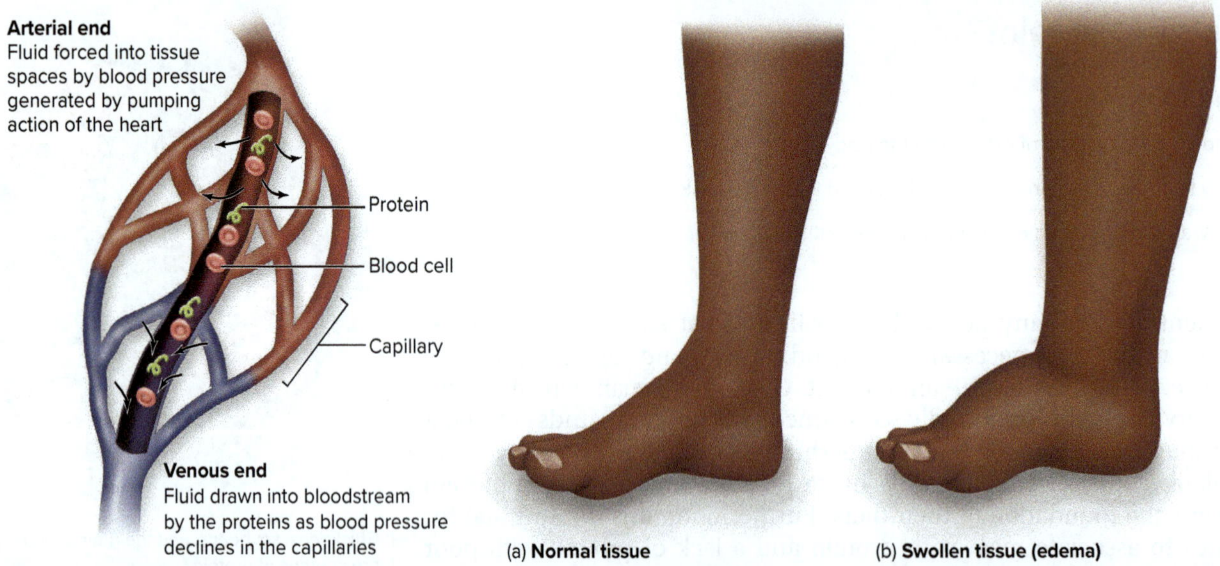

Figure 7.1 Normal fluid balance and edema. Under normal conditions, blood pressure is high enough at the arterial end of a capillary bed to force fluid (mostly water) into the tissue space that surrounds capillaries. At the venous end of the capillary bed, blood pressure is lower, and proteins in the bloodstream help draw fluid back into the capillaries. (a) As a result of the balance between blood pressure and proteins in the bloodstream, the tissues contain normal amounts of fluid. (b) When blood pressure is greater than the counteracting force of the proteins in the bloodstream, fluid remains in the tissues, causing edema.

all enzymes are proteins. Enzymes speed up the rate of (*catalyze*) chemical reactions without becoming a part of the products (see Fig. 4.8). Additionally, infection–fighting **antibodies** are proteins. Although cells can use proteins for energy, normally they metabolize very little for energy, conserving the nutrient for other important functions that carbohydrates and lipids are unable to perform.

In the bloodstream, proteins transport nutrients and oxygen. Proteins in blood, such as *albumin,* also help maintain the proper distribution of fluids in blood and body tissues (Fig. 7.1). The force of blood pressure moves watery fluid out of the bloodstream and into tissues. Blood proteins help counteract the effects of blood pressure by attracting the fluid, returning it to the bloodstream. During starvation and some chronic illnesses, the level of protein in blood decreases, and as a result, some water leaks out of the bloodstream and enters spaces between cells. The resulting accumulation of fluid in tissues is called **edema** (*eh–dee′–mah*).

Proteins also help maintain **acid–base balance,** the proper pH of body fluids. To function properly, blood and tissue fluids need to maintain a pH of 7.35 to 7.45, which is slightly basic.[1] (To review the concept of pH, see Chapter 4.) Metabolic processes can produce acidic or basic by–products. If a particular body fluid becomes too acidic or too basic, cells can have difficulty functioning and may die. A **buffer** can help maintain the normal pH of a solution. Proteins can act as buffers because they have acidic and basic components. For example, if cells form an excess of hydrogen ions (H^+), the pH of tissues decreases. To help restore the pH level to within the normal range, the basic portions of protein molecules bind to the excess H^+, neutralizing the excess ions and raising the pH.

antibodies infection-fighting proteins

edema accumulation of fluid in tissues

acid-base balance maintaining the proper pH of body fluids

buffer substance that can help maintain the normal pH of a solution

amino acids nitrogen-containing chemical units that comprise proteins

Amino Acids

Proteins are composed of smaller chemical units called **amino acids.** The human body contains proteins made from 20 different amino acids (see Appendix D). To understand how the body uses amino acids, it is necessary to learn some basic chemistry that relates to these compounds.

Each amino acid has a carbon atom that anchors a hydrogen atom and three different groups of atoms: the **amino** or **nitrogen–containing group,** the **R group,** and the **acid group.** The chemical structure of the amino acid alanine, shown in Figure 7.2, indicates these three groups. Note that the nitrogen atom is in the amino group. The R group varies with each type of amino acid, so it identifies the molecule as a particular amino acid, such as *serine* or *lysine.* When the nitrogen–containing group is removed, the R group, acid group, and anchoring carbon atom form the "carbon skeleton" of an amino acid (see Fig. 7.2). The carbon skeleton is an important component of an amino acid because the body can convert the carbon skeletons of certain amino acids to glucose and use the simple sugar for energy.

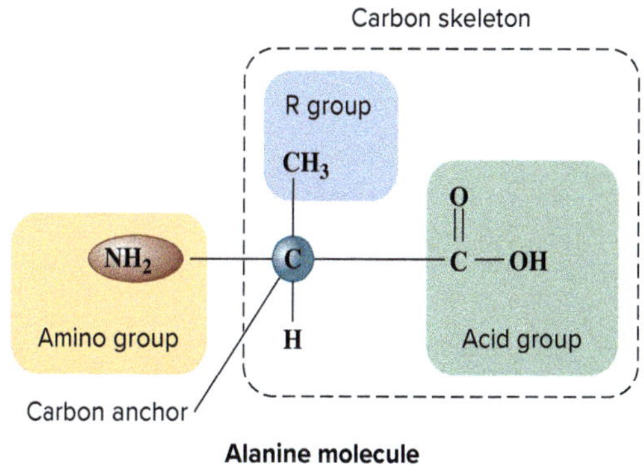

Figure 7.2 Amino acid: Basic chemical structure.

Classifying Amino Acids

Traditionally, nutritionists classify amino acids as either nonessential or essential according to the body's ability to make them. A healthy human body can make 11 of the 20 amino acids that are in its proteins. These compounds are the **nonessential amino acids.** The remaining nine amino acids are **essential amino acids** that must be supplied by foods because the body cannot synthesize them or make enough to meet its needs. Sometimes, nonessential and essential amino acids are referred to as "dispensable" and "indispensable" amino acids, respectively. Table 7.2 lists amino acids according to their classification as essential and nonessential.

Several nonessential amino acids are "conditionally essential," which means they become essential in certain situations. For example, cells can make cysteine from methionine and serine. If a person's methionine and serine intake is inadequate, their body cannot make enough cysteine to meet its needs, and dietary sources of the amino acid are necessary.

Did *YOU* Know?

Taurine is a nonessential, sulfur-containing amino acid that human cells do not use to make proteins. The amino acid, however, has many functions in the body. Humans need taurine to form components of bile, and the amino acid acts as an antioxidant, lowers blood pressure, and reduces inflammation. The human liver can make small amounts of taurine, but the amino acid is a dietary essential for cats and foxes.[2]

Taurine may be added to energy drinks as an "energy-booster," and athletes often consume taurine-caffeine–containing energy drinks in an effort to boost their endurance and performance. There is, however, a lack of scientific evidence supporting the notion that the taurine in these products provides additional performance-related benefits beyond those of the caffeine ingredient. At present, taurine in doses of up to 10 g/day appears to have low toxicity.[2]

©Wendy Schiff

TABLE 7.2 Amino Acids in Human Proteins

Essential		Nonessential	
Histidine	Threonine	Alanine	Cysteine*
Isoleucine	Tryptophan	Aspartic acid	Glutamine*
Leucine	Valine	Asparagine	Glycine*
Lysine		Glutamic acid	Proline*
Methionine		Serine	Tyrosine*
Phenylalanine		Arginine*	

*Under certain conditions, this amino acid can become essential.

Concept CHECKPOINT

1. What is the chemical unit that makes up a protein?
2. List at least four different functions of proteins in the body.
3. Identify the three groups of atoms that make up a typical amino acid.
4. What is the "carbon skeleton" of an amino acid?
5. How many different kinds of amino acids are needed to make human proteins? How many of these amino acids are essential?

See Appendix F for responses.

amino or **nitrogen-containing group** portion of an amino acid that contains nitrogen

R group part of amino acid that identifies the molecule as a particular amino acid

acid group acid portion of a compound

nonessential amino acids group of amino acids that the body can make

essential amino acids amino acids the body cannot make or make enough of to meet its needs

7.2 Proteins in Foods

Learning Outcomes

1. Explain the difference between a high-quality protein and a low-quality protein.
2. Identify foods that are rich sources of high-quality proteins.
3. List at least three amino acids that are often "limiting amino acids" in plant foods.

People often associate animal foods with protein, but beans, legumes, nuts, seeds, grains, and certain vegetables are good sources of protein, too. In fact, nearly all foods contain protein, but no naturally occurring food is 100% protein. Protein comprises only about 20 to 30% of the weight of a piece of beef; 25% of the weight of drained, water–packed tuna fish; and only 12% of an egg's weight. Nevertheless, animal foods generally provide higher amounts of protein than similar quantities of plant foods. A 3–ounce serving of broiled lean ground beef supplies 20 g of protein; a 3–ounce serving of steamed broccoli or cooked carrots provides only about 1 g of protein. In general, most plant foods provide less than 3 g of protein per ounce. Table 7.3 lists some commonly eaten foods and their approximate protein content per serving.

TABLE 7.3 Approximate Protein Content of Some Commonly Eaten Foods/Portion

Food	Serving Size	Protein g/serving
Chicken, breast, roasted, meat only	4 oz	32
Hamburger, 80% lean, broiled	4 oz	29
Pepperoni pizza, regular crust, 14″ pie	2 slices (246 g)	29
Ham, lean, cooked	4 oz	28
Salmon, Atlantic, wild, cooked	4 oz	28
Chicken taco, flour tortilla	1 taco	18
Cottage cheese, 2% low-fat	4 oz	12
Bagel, plain	1 (3½″ diam)	11
Tofu, regular	½ cup	10
Milk, fat-free	1 cup	8
Peanut butter	2 Tbsp	8
Red kidney beans, cooked	½ cup	7
Egg, hard cooked	1	6
American processed cheese	1 oz	5
Mixed nuts	1 oz	5
White rice	1 cup	4
Peas, green	½ cup	4
Banana, small, 6″ long	1	1

Pizza, Plate of tofu: C Squared Studios/Photodisc/Getty Images; Kidney beans: Moving Moment/Shutterstock; Pea pods: Mitch Hrdlicka/Photodisc/Getty Images; Salmon steak: Purestock/SuperStock; Bagels: Oksana Struk/E+/Getty Images; Eggs: Brian Yarvin/Alamy Stock Photo

Certain parts of plants contain more protein than other parts. Seeds, tree nuts, and legumes supply more protein per serving than servings of fruit or the edible leaves, roots, flowers, and stems of vegetables. Tree nuts include walnuts, cashews, and almonds; **legumes** are plants that produce pods that have a single row of seeds, such as soybeans, peas, peanuts, lentils, and beans (Fig. 7.3). A 3–ounce serving of almonds, dry–roasted peanuts, or sunflower seed kernels supplies about 20 g of protein. Many seeds and nuts, however, pack a lot of calories from fat. Snack on just 3 ounces of almonds, dry–roasted peanuts, or sunflower seed kernels, and you will add almost 500 kcal to your diet!

Mature peas, lentils, and most kinds of beans contain more protein and complex carbohydrate than fat. Eating a 3–ounce serving of vegetarian baked beans, for example, adds about 4 g of protein, 14 g of carbohydrate, and less than 1 g of fat to your diet. Although soybeans contain more fat than carbohydrate, soy fat is high in unsaturated fatty acids. The health benefits of unsaturated fatty acids are discussed in Chapter 6.

Figure 7.3 Legumes. Legumes, such as these soybeans, are plants that produce pods that have a single row of seeds.
C Squared Studios/Photodisc/Getty Images

Protein Quality

Foods differ not only in the amount of protein they contain, but also in their protein quality. A **high–quality (complete) protein** contains all essential amino acids in amounts that support protein deposition in muscles and other tissues, as well as a young child's growth.[3] High–quality proteins are well digested and absorbed by the body. Meat, fish, poultry, eggs, and milk and other dairy products contain high–quality proteins. Egg protein generally rates very high for protein quality because it is easy to digest and has a pattern of essential amino acids that closely resembles that needed by humans.

A **low–quality (incomplete) protein** lacks one or more of the essential amino acids or contains inadequate amounts of these nutrients. Furthermore, the human digestive tract does not digest low–quality protein sources as efficiently as foods containing high–quality protein. The essential amino acids that are in relatively low amounts are referred to as *limiting* amino acids because they reduce the protein's ability to support growth, repair, and maintenance of tissues. In most instances, lysine, tryptophan, threonine, and the sulfur–containing amino acids methionine and cysteine are the limiting amino acids in foods.[4]

legumes plants that produce pods with a single row of seeds

high-quality (complete) protein protein that contains all essential amino acids in amounts that support the deposition of protein in tissues and the growth of a young person

low-quality (incomplete) protein protein that lacks or has inadequate amounts of one or more of the essential amino acids

Most plant foods, including vegetables, grains, nuts, seeds, and legumes, are not sources of high–quality proteins. *Quinoa* (*keen'–wa*), amaranth, buckwheat, and soy protein are exceptions. The *pseudocereals* amaranth, quinoa, and buckwheat are prepared and consumed similar to cereal grains but are actually seeds.[5] However, the quality and amount of protein in quinoa seeds are superior to those of many cereal grains.[6] Cooked quinoa is often used as a cereal (see "Recipe for Healthy Living" at the end of this chapter). The quality of soy protein is comparable to that of most animal proteins.[7] Processed soybeans are used to make a variety of nutritious foods, including soy milk, infant formula, and meat substitutes. Furthermore, eating foods made from soybeans may reduce the risk of chronic diseases, such as certain cancers.[7] More research, however, is needed to determine the long–term health benefits of eating diets that contain soy products.

Understanding the concept of protein quality is important. Regardless of how much protein is eaten, a child will fail to grow properly if their diet lacks essential amino acids. The "Vegetarianism" section of this chapter explains how you can obtain these and other essential nutrients by eating only plant foods.

©Wendy Schiff

Concept CHECKPOINT

6. Explain the difference between a high-quality protein and a low-quality protein.
7. Identify at least three dietary sources of high-quality protein and three dietary sources of low-quality protein.
8. List at least three essential amino acids that are most likely to be limiting amino acids.

See Appendix F for responses.

7.3 What Happens to Proteins in the Body?

Learning Outcomes

1. Explain how cells make proteins.
2. Describe what happens to excess amino acids in the body.
3. Explain what happens to proteins as they undergo digestion and absorption in the human digestive tract.
4. Explain the concept of nitrogen balance, and identify conditions in which the body is in a state of positive or negative nitrogen balance.
5. Calculate a person's RDA for protein based on their body weight.

In a television crime series, police in a major city are investigating what could be a homicide. A male has been reported missing by his parents, who suspect foul play and their daughter–in–law's involvement in their son's disappearance. While knocking on the door of the missing male's house, police notice some dried blood on the front porch. The male's wife, who lives in the house, tells police that the blood is from her injured dog. How can police know she is telling the truth? The blood holds important clues. Every organism synthesizes proteins—including those in blood—that are unique. Samples of the blood can be analyzed to determine whether it contains proteins from a dog or another animal, such as a human. We will leave it to your imagination to finish this story, but we will examine a real–life story: how human cells make proteins.

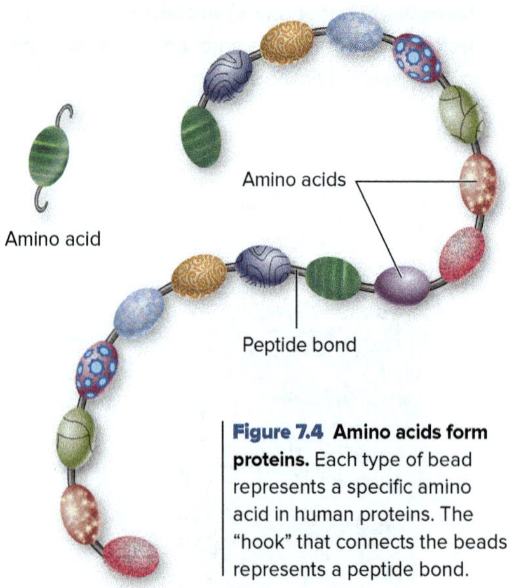

Figure 7.4 Amino acids form proteins. Each type of bead represents a specific amino acid in human proteins. The "hook" that connects the beads represents a peptide bond.

How the Body Uses Amino Acids to Synthesize Proteins

Your body makes proteins by following information coded in your DNA, or deoxyribonucleic (*de–ox′–e–rye′–bow–new–klay′–ik*) acid, the hereditary material in a cell's nucleus. To make proteins, cells assemble the 20 amino acids in specific sequences according to the information provided by DNA. To understand this process, imagine proteins as various chains made from 20 different amino acid "beads." Figure 7.4 illustrates some of these beads and how they can be assembled into chains. Note that each bead has two metal wires that are used to link it with another bead. To make a copy of a particular beaded chain, you would follow directions for connecting the beads in a specific order and length by hooking the metal wires of each bead together. Consider the vast variety of beaded chains composed of different bead sequences and chain lengths that you could make from just 20 different beads.

In living things, the beaded chains are proteins that contain amino acids. An amino acid can connect to another amino acid by a **peptide bond,** a chemical attraction between the acid group of one amino acid and the amino group of another amino acid (Fig. 7.5). A **polypeptide** forms when two or more amino acids join to form a chain (see Fig. 7.4). **Peptides** are small chains of amino acids

peptide bond chemical attraction that connects two amino acids together

polypeptide protein composed of two or more amino acids

peptides small chains of amino acids

that usually contain fewer than 15 amino acids. Dipeptides and tripeptides are compounds that consist of two and three amino acids, respectively. Most naturally occurring proteins are polypeptides composed of 50 or more amino acids.

Figure 7.6 summarizes the basic steps of protein synthesis. Protein synthesis begins with DNA in the cell's nucleus. DNA is a twisted, two–stranded molecule referred to as a double helix. To begin the process, a section of the DNA double helix unwinds, exposing a gene. A **gene** is a portion of DNA that contains information concerning the order of amino acids that comprise a specific protein. *Messenger ribonucleic acid* (*mRNA*), a compound that is chemically similar to DNA, is formed by *transcription* (see Fig. 7.6). The actual production of a protein occurs in the cytoplasm, so mRNA leaves the nucleus and moves to ribosomes—the protein–manufacturing sites in the cytoplasm. Ribosomes translate the gene's coded instructions for adding amino acids to the polypeptide chain. During this translation process, *transfer ribonucleic acid* (*tRNA*) conveys specific amino acids, one at a time, to the ribosomes. At the ribosomes, the amino acid from tRNA is added to the last amino acid, causing the peptide chain to grow longer. After the mRNA is read completely, the ribosome releases the polypeptide, and then the new protein generally undergoes further processing at other sites within the cytoplasm.

Diets that contain low–quality protein can result in poor growth, slowed recovery from illness, and even death. These situations occur because protein synthesis in cells cannot proceed when the supply of amino acids does not have one or more of the essential amino acids needed for constructing the polypeptide chain. When this happens, production of the protein stops. The partially made polypeptide chain is dismantled, and its amino acids are returned to the amino acid "pool," a readily available supply of amino acids that cells can use for future protein synthesis.

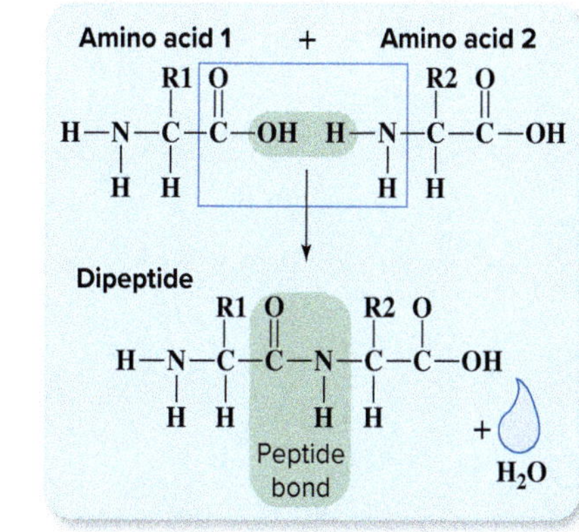

Figure 7.5 Peptide bond. A dipeptide forms when two amino acids bond and a molecule of water is released in the process.

gene portion of DNA

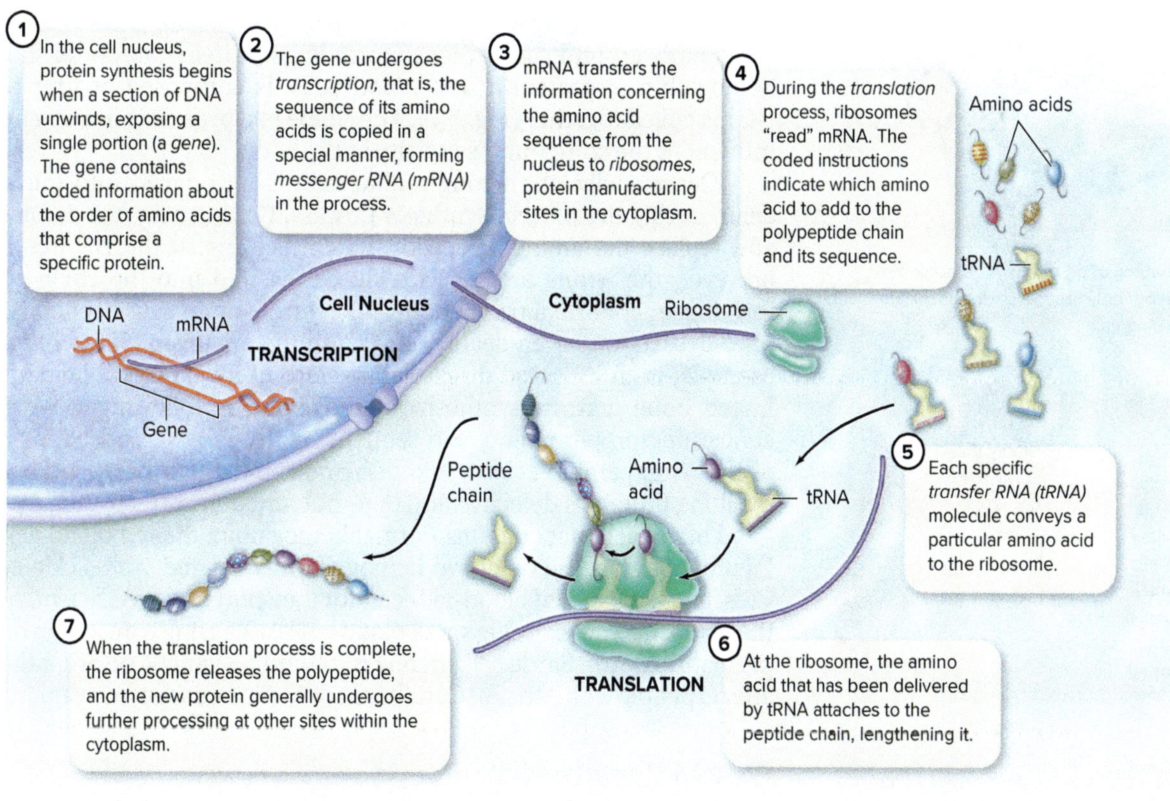

Figure 7.6 Protein synthesis. This illustration summarizes the basic steps of protein synthesis.

Did *YOU* Know?

Collagen is a major structural protein of the bones, ligaments, and skin of mammals that consists primarily of the nonessential amino acids glycine, proline, and hydroxyproline (a form of proline). Dietary sources of collagen include bone broth, nonorgan meats, fish, poultry, and egg whites.

Manufacturers use collagen to make dietary supplements that contain small peptides called *collagen hydrolysates* (*CHs*). The small intestine digests CHs into individual amino acids that can be absorbed along with AAs that were obtained from proteins in food. Therefore, consumers cannot be certain that their bodies will use the amino acids derived from CH supplements to produce collagen or make other proteins in the body. Nevertheless, adequate dietary protein intake from animal and plant sources of food should provide enough amino acids for the body to produce enough collagen without needing CH supplements.

Results of some scientific studies involving human subjects indicated that taking oral collagen supplements increased skin elasticity, may help wound healing, and may be effective for relieving osteoarthritis pain but not the discomfort associated with rheumatoid arthritis.[8,9] As is often the case, more research is needed to determine the usefulness of taking dietary supplements that contain CHs.

©Wendy Schiff

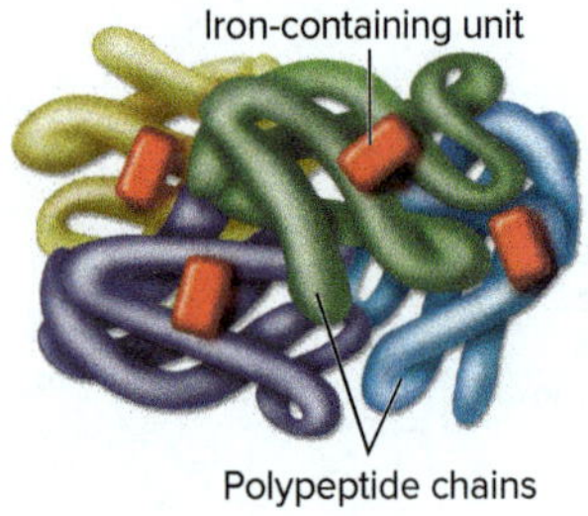

Figure 7.7 Shaping of a protein. A hemoglobin molecule has four polypeptide chains associated with an iron-containing unit.

When assembly of the new protein has been completed, the polypeptide chain coils and folds into a three–dimensional shape that is characteristic of that particular protein. In some instances, more than one polypeptide chain curls around each other to form large protein complexes. For example, hemoglobin, a protein in red blood cells, is composed of four polypeptide chains coiled together (Fig. 7.7). The shape of a protein is important because it influences the compound's activity in the body.

Occasionally, the wrong amino acid is introduced into the amino acid chain during the protein synthesis process. Cells usually check for such errors and replace the amino acid with the correct one. If the DNA code is faulty, however, the wrong amino acid will be inserted into the chain consistently, forming an abnormal polypeptide. Such errors often cause genetic defects that have devastating, even deadly, effects on the organism. *Sickle cell anemia,* for example, is an inherited disease characterized by abnormal hemoglobin. Cells in red bone marrow synthesize hemoglobin by following DNA instructions concerning proper amino acid sequencing. If the DNA codes for the insertion of the wrong amino acid in two of hemoglobin's four polypeptide chains, the resulting protein is defective and does not function correctly. Figure 7.8a shows a red blood cell that contains normal hemoglobin; the red blood cell, shown in Figure 7.8b, has the defective hemoglobin associated with sickle cell anemia. Crescent–shaped red blood cells cannot transport oxygen efficiently. As a result, the abnormal cells can clog small blood vessels, causing pain, organ damage, and premature death. Sickle cell anemia is a common genetic disorder that generally affects people with African, Caribbean, or Mediterranean ancestry.

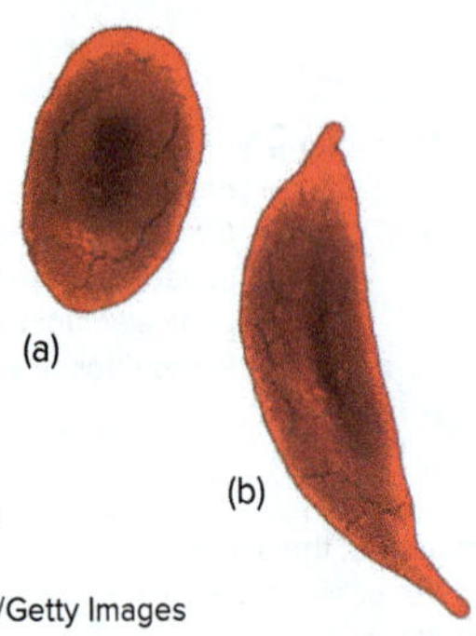

Figure 7.8 Sickle cell anemia. This microscopic view of red blood cells shows (a) normal disc-shaped cell and (b) sickle cell that contains abnormal hemoglobin.
Callista Images/Image Source/Getty Images

denaturation altering a protein's natural shape and function by exposing it to conditions such as heat, acids, and physical agitation

Protein Denaturation

A protein undergoes **denaturation** when it is exposed to various conditions that alter the macronutrient's natural folded and coiled shape (Fig. 7.9).

We often cook protein–rich foods to make them more digestible and safe to eat, but heat also causes the proteins in foods to unfold. The protein in raw egg white, for example, is almost clear and has a jellylike consistency. When you cook egg white, it becomes white and firm as its proteins become denatured. Other treatments often used during food preparation also denature proteins, including whipping or exposing them to alcohol or acid. Wine, for example, is often used in marinades because the alcohol it contains denatures proteins in meat, helping tenderize it. Adding acidic lemon juice to milk denatures ("curdles") the proteins in milk. In your stomach, hydrochloric acid denatures food proteins, making them easier to digest. Denaturation does not "kill" a protein (because proteins are not living), but the process usually permanently alters the protein's shape and functions. Once an egg white has been cooked or milk has curdled, the food cannot return to its original state.

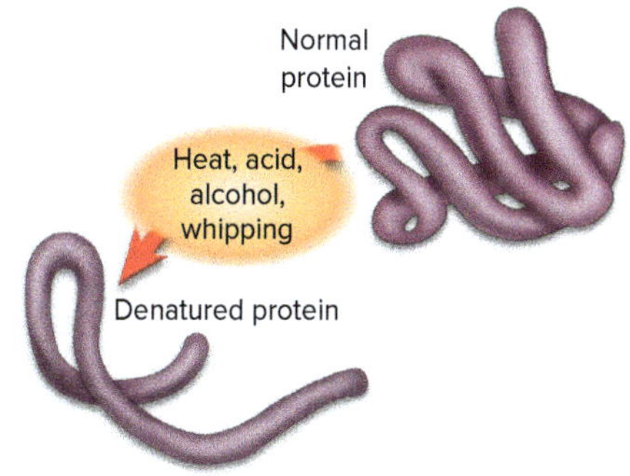

Figure 7.9 **Denaturation.**

Protein Digestion and Absorption

When you eat oatmeal mixed with milk for breakfast, the large proteins in these foods must be digested before undergoing absorption. The chemical digestion of protein begins in the stomach, where hydrochloric acid denatures food proteins and **pepsin,** an enzyme, digests proteins into smaller polypeptides. Soon after the polypeptides enter the small intestine, the pancreas secretes protein–splitting enzymes, including trypsin (*trip'–sin*) and chymotrypsin (*ki'–mo–trip'–sin*). *Trypsin* and *chymotrypsin* break down polypeptides into shorter peptides and amino acids. Enzymes released by the absorptive cells of the small intestine break down most of the shortened peptides into dipeptides, tripeptides, and individual amino acids that are absorbed. Within the absorptive cells, di– and tripeptides are broken down into amino acids. Thus, amino acids are the end products of protein digestion. After being absorbed, the amino acids enter the hepatic portal vein and travel to the liver, where they may enter the general circulation.

By the time cells obtain amino acids from blood, they cannot distinguish the ones that were originally in oat proteins from those that were in milk proteins. The cells, however, now have all the amino acids they need to make human proteins. Protein digestion and absorption are very efficient: Very little dietary protein escapes digestion and is eliminated in feces. Figure 7.10 summarizes protein digestion and absorption.

Did *YOU* Know?

Despite information provided in commercials or advertisements, you cannot "feed" your hair, nails, or skin by using shampoos, conditioners, or lotions containing proteins or other nutrients. Hair, nails, and the outermost layer of skin are not living. By eating a nutritious diet, you will provide your body with the nutrients it needs to make healthy hair, nails, and skin.

Armina-Udovenko/iStock/Getty Images

Protein Turnover

Not all protein must be supplied by the diet. **Protein turnover,** the process of breaking down old or unneeded proteins into their component amino acids and recycling them to make new proteins, occurs constantly within cells. Amino acids that are not incorporated into proteins become part of the amino acid pool that cells can use for future protein synthesis. The amino acid pool is an *endogenous,* or internal, source of nitrogen. The body obtains about two–thirds of its amino acid supply from endogenous sources and the remainder from *exogenous* (dietary) sources.

Transamination and Deamination

A healthy human body can make 11 of the 20 amino acids. The liver is the main site of nonessential amino acid production. Chemical reactions called deamination and transamination are involved in the synthesis of amino acids. **Deamination** is the process of removing the nitrogen–containing group

pepsin gastric enzyme that breaks down proteins into smaller polypeptides

protein turnover cellular process of breaking down proteins and recycling their amino acids

deamination removal of the nitrogen-containing group from an amino acid

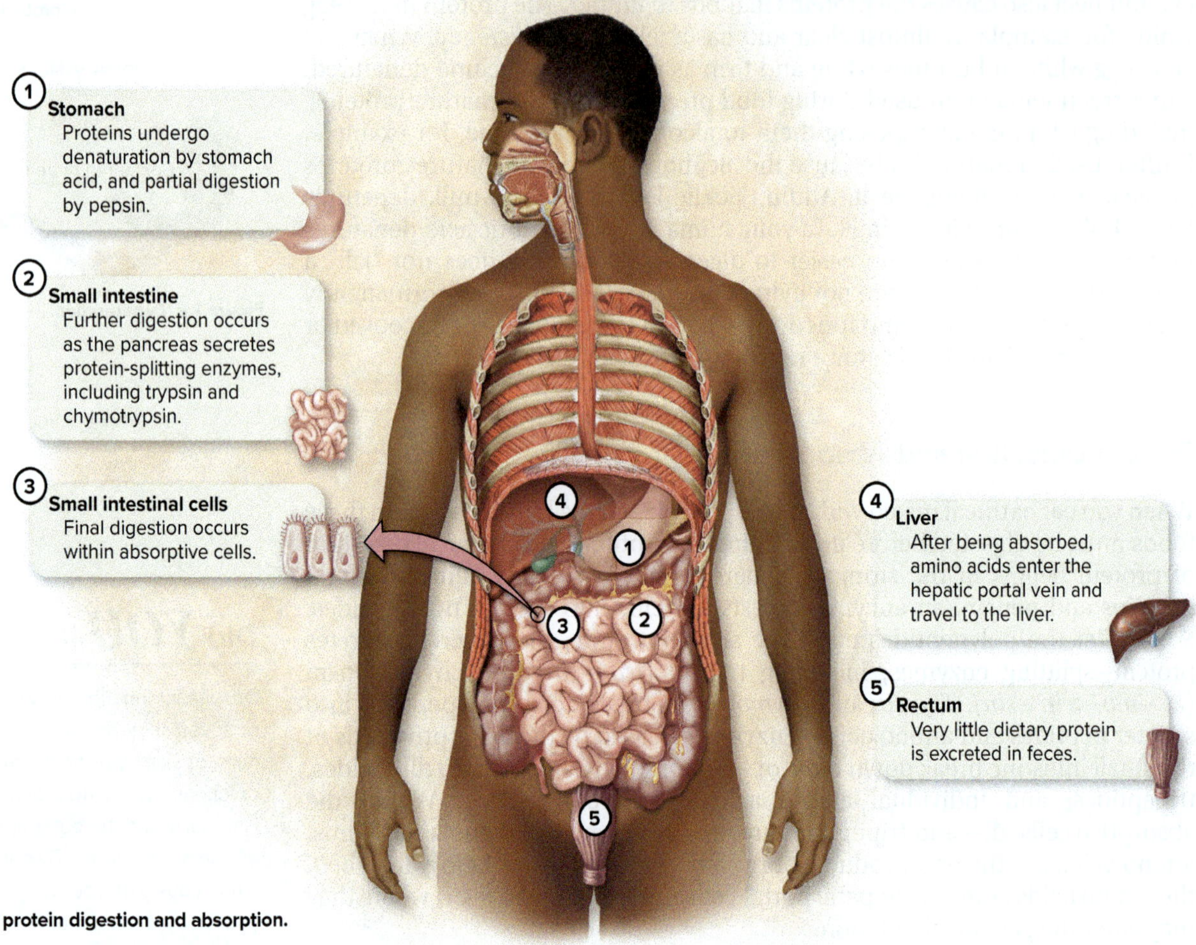

Figure 7.10 **Summary of protein digestion and absorption.**

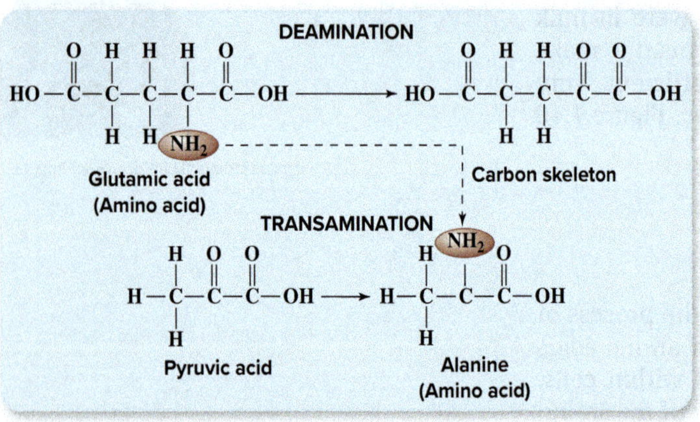

Figure 7.11 **Deamination and transamination.** Deamination is the process of removing the nitrogen-containing group from an unneeded amino acid. Transamination occurs when the nitrogen-containing group is transferred to another substance to make an amino acid.

transamination transfer of the nitrogen-containing group from an unneeded amino acid to a carbon skeleton to form an amino acid

urea waste product of amino acid metabolism

(usually NH_2) from an unneeded amino acid. As a result of deamination, the amino acid that gives up its amino group becomes a carbon skeleton (Fig. 7.11).

Deamination occurs primarily in the liver. After an amino acid undergoes deamination, the carbon skeleton that remains can be used for energy or converted to other compounds, such as glucose. **Transamination** occurs when the nitrogen–containing group is transferred to another substance to make an amino acid. To make the amino acid alanine, for example, liver cells remove the amino group (NH_2) from glutamic acid and transfer it to pyruvic acid (Fig. 7.11). Transamination reactions are reversible.

Liver cells remove excess NH_2 from the bloodstream, forming ammonia (NH_3), a highly poisonous waste product (Fig. 7.12). The liver can use the ammonia to make **urea,** a metabolic waste product that is released into the bloodstream. The kidneys filter urea, small amounts of ammonia, and *creatinine* (a nitrogen–containing waste produced by muscles) from blood and eliminate the compounds in urine.

If you consume more protein than you need, what happens to the extra amino acids? The body does not store excess amino acids in muscle or other tissues. The unnecessary amino acids undergo deamination, and cells convert the carbon skeletons into glucose or fat, or metabolize them for energy (see Fig.7.12).

Nitrogen Balance

Although your body conserves nitrogen by recycling amino acids, each day you lose some protein and nitrogen from your body. Urinary elimination of urea and creatinine accounts for most of the lost nitrogen. Daily nitrogen losses also occur as your nails and hair grow, and when you shed the outermost layer of your skin and cells from your intestinal tract. Your body uses amino acids from foods to replace the lost nitrogen.

Normally, an adult's body maintains its protein content by maintaining **nitrogen balance (equilibrium),** that is, balancing nitrogen intake and protein turnover with losses. During certain stages of life or physical conditions, however, nitrogen intake and retention do not equal nitrogen losses. When the body is in a state of **positive nitrogen balance,** it retains more nitrogen than it loses as proteins are added to various tissues. In this case, a person must eat more protein to satisfy the increased need for the nutrient. Positive balance occurs during periods of rapid growth such as pregnancy, infancy, and puberty, and when people are recovering from illness or injury. Hormones such as insulin, growth hormone, and testosterone stimulate positive nitrogen balance. Performing weight (resistance) training also leads to nitrogen retention. When the body is in a state of **negative nitrogen balance,** the body loses more nitrogen than it retains and protein intake is less than what the body needs. Negative balance occurs during starvation, serious illnesses, and severe injuries. Recovery from the illness or injury results in positive nitrogen balance until nitrogen equilibrium is restored. Figure 7.13 illustrates the concept of nitrogen balance and lists conditions that result in positive and negative nitrogen balance.

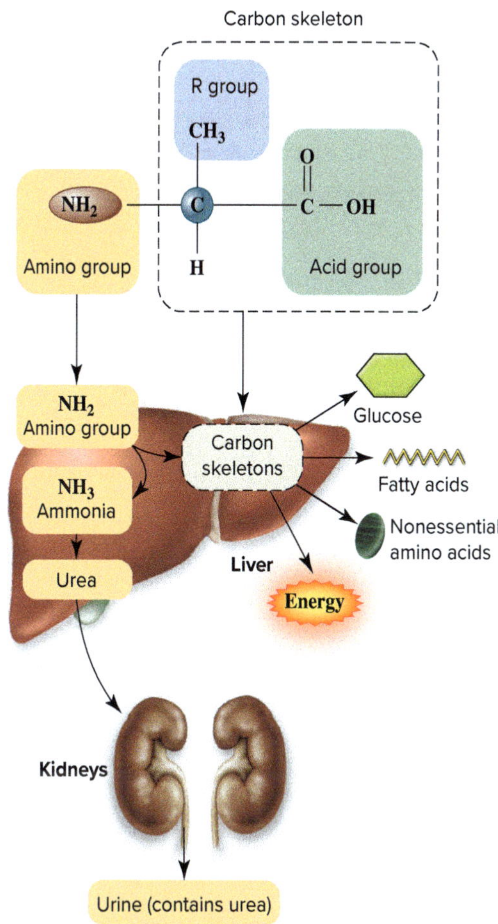

Figure 7.12 Processing unneeded amino acids. In the liver, unnecessary amino acids are deaminated, and the remaining carbon skeletons can be converted to glucose or fat, used to make nonessential amino acids, or metabolized for energy.

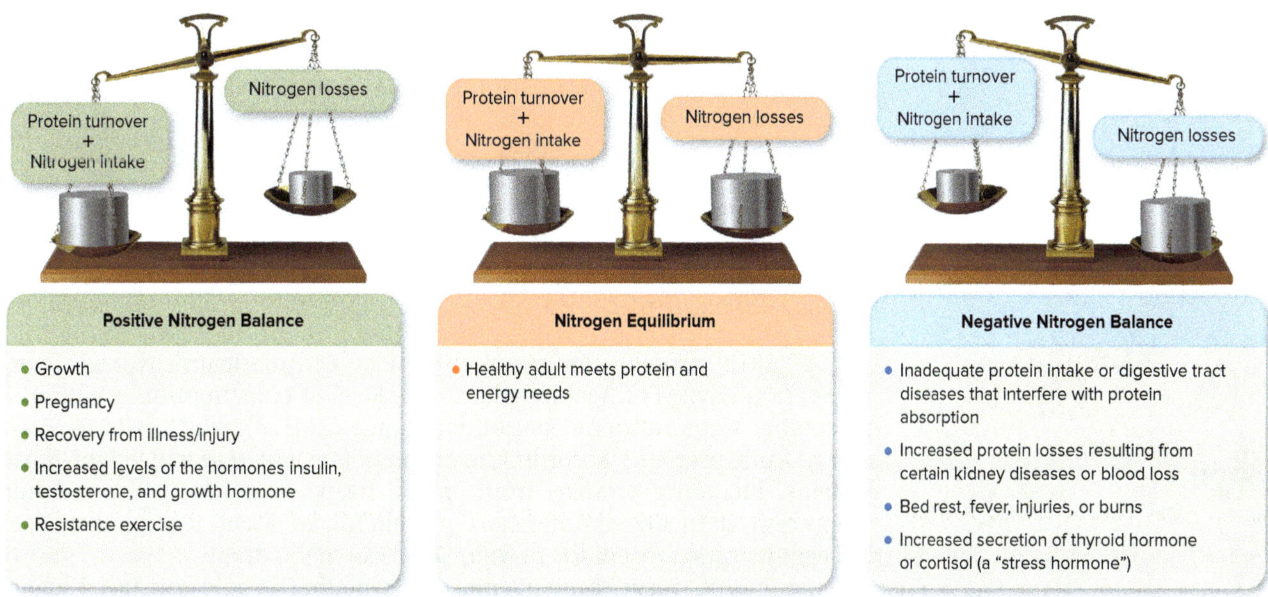

Figure 7.13 Nitrogen balance. This diagram illustrates the concept of nitrogen balance and lists conditions that result in positive and negative nitrogen balance.

nitrogen balance (equilibrium) balancing nitrogen intake with nitrogen losses

positive nitrogen balance state in which the body retains more nitrogen than it loses

negative nitrogen balance state in which the body loses more nitrogen than it retains

How Much Protein Do You Need?

The Estimated Average Requirement (EAR) for protein is 0.66 g of protein/kg of body weight.[4] The EAR for protein increases during pregnancy, breastfeeding, periods of rapid growth, and recovery from serious illnesses, blood losses, and burns. Recall from Chapter 3 that scientists use EARs to establish Recommended Dietary Allowances (RDAs). A healthy adult's RDA for protein is 0.8 g/kg of body weight. By reviewing DRI tables (Appendix H), you will note that the RDAs for protein vary during certain ages and conditions.

To determine your RDA for protein, multiply your weight in kilograms by 0.8 gram. If you are underweight or overweight, use a healthy weight for your height when making this calculation (see the body mass index chart in Chapter 10). For example, a healthy male who is 5-feet, 10-inches tall and weighs 75 kilograms (his weight in pounds divided by 2.2) should consume 60 grams of protein daily (75 kg × 0.8 g) to meet his RDA for the nutrient. The "Personal Dietary Analysis" at the end of this chapter can help you estimate your daily protein intake.

Concept CHECKPOINT

9. Explain the basic steps involved in protein synthesis.
10. Define *denaturation, deamination,* and *transamination.*
11. Describe conditions that can cause the body to be in negative nitrogen balance. Describe conditions in which the body is in positive nitrogen balance.
12. A healthy young female weighs 143 pounds. Calculate her RDA for protein.
13. Explain what happens to proteins in beans as they undergo digestion and absorption in the human digestive tract.

See Appendix F for responses.

7.4 Protein Consumption and Health

Learning Outcomes

1. Compare the percentage of calories that protein contributes in the typical American diet to the AMDR for protein.
2. Identify food groups that contribute most of the protein in the typical American diet.
3. Identify chronic diseases associated with processed meat consumption.

Protein Consumption Patterns

Meat, fish, poultry, dairy, and eggs provide 70–85% of the protein in the American diet.
Oleksandra Naumenko/123RF

In 2017–2018, protein comprised 16% of adult Americans' average energy intake for a day.[10] For healthy adults, this level of consumption is within the Acceptable Macronutrient Distribution Range (AMDR), which is 10 to 35% of energy from protein.[4] According to results of recent research, plant-based proteins, including proteins from grains, beans, legumes, peas, and nuts, supply only 20 to 30% of Americans' protein intake. Meat, fish, poultry, dairy, and eggs provide most of the protein (70–85%) in the diet.[11]

Sources of "meat" protein (muscle tissue that people consume) include unprocessed "raw" meat, fish, and poultry that are usually cooked before eating. In 2015–2016, unprocessed poultry (especially chicken and turkey)

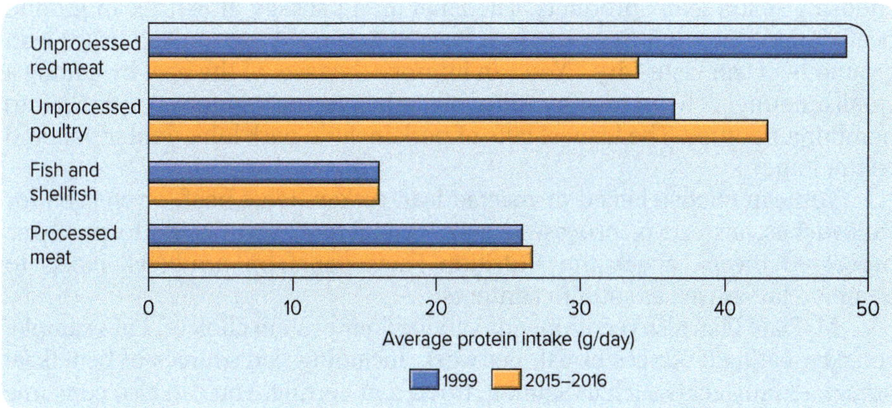

Figure 7.14 Meat protein consumption patterns. From 1999 to 2016, Americans' consumption of unprocessed poultry increased and unprocessed red meat consumption decreased. Fish, shellfish, and processed meat intake remained unchanged.

contributed the largest share of adult Americans' total protein intake from meat (Fig. 7.14).[11] This amount represents an increase of about 14% compared to the population's intake of unprocessed poultry in 1999. On the other hand, unprocessed red meat consumption decreased by 13% between 1999 and 2015–2016. Red meats include beef, pork, and lamb. Adults' intake of fish and shellfish remained consistent from 1999 to 2016 (Fig. 7.14).

Meat Consumption and Health

Heart disease and cancer are the leading causes of death in developed countries, including the United States. In these nations, the typical "Western" diet that contains high amounts of red or *processed* meat is associated with increased risk of certain chronic diseases, particularly cardiovascular disease (CVD)[12] and some forms of cancer.[13] Processed meat is often made from red meat (primarily beef and pork), poultry, and liver that have been salted, cured, fermented, or smoked.[13] Thus, processed meat includes ham, bacon, luncheon or "deli" meat, sausage, and hot dogs.

Processed meat, including ham, bacon, sausage, frankfurters, and deli meats, generally contain high amounts of saturated fat and sodium.
Pixel-Shot/Shutterstock

CVD, which includes heart disease and hypertension, is associated with red meat and processed meat consumption. Recall from Chapter 6 that these meats generally contain a lot of saturated fat and high amounts of added sodium. Excessive intake of saturated fat and sodium increases the risk of CVD.

There is strong scientific evidence linking consumption of red meat and processed meat with increased risk of colorectal cancer.[13] Eating these kinds of meat may also increase the risk of pancreatic cancer and stomach cancer. In 2015–2016, nearly 25% of the meat and poultry consumed by Americans was in the form of processed meat.[14] Therefore, Americans should consider whether decreasing their intake of red meat, as well as processed meat, is necessary.

MyPlate: Recommendations for Protein Intake

Animal sources of protein are often rich sources of saturated fat. According to the Dietary Guidelines and the MyPlate Plan, Americans should choose lean or low–fat protein foods, such as eggs, fish, poultry, and lean meat.[15] The leanest cuts of beef include round and top round, loin, and top sirloin steaks, as well as chuck and arm roasts.

Before cooking a piece of beef, you can reduce its fat content by trimming the visible fat away from the meat. When buying ground beef, consider

MyPlate recommends varying your protein foods by including fish, lean meats and poultry, beans, peas, nuts, and seeds.
Jill Braaten

(a)

Nutrition Facts	
1 serving per container	
Serving size 1 Meal (255g)	
Amount per serving	
Calories	**250**
	% Daily Value*
Total Fat 5g	**6%**
Saturated Fat 0.5g	**3%**
Trans Fat 0g	
Polyunsaturated Fat 1g	
Monounsaturated Fat 3g	
Cholesterol 0mg	**0%**
Sodium 600mg	**26%**
Total Carbohydrate 39g	**14%**
Dietary Fiber 10g	**36%**
Total Sugars 4g	
Includes 2g Added Sugars	**4%**
Protein 13g	

(b)

INGREDIENTS: CHICKPEAS, RED BELL PEPPERS, SPINACH, CARROTS, QUINOA, CHICKEN BREAST, OLIVE OIL, SALT, LEMON JUICE, OREGANO, BASIL.

Figure 7.15 Product Nutrition Facts panel and ingredient list. (a) The Nutrition Facts panel provides information about a product's protein content. (b) A product's protein quality can be judged from the ingredient list.
Nutrition Facts panel Traci Keck

choosing "extra lean" products. The label on a package of extra lean ground beef should state that the meat is at least 95% lean. When cooked, extra lean ground beef can taste "dry." You can improve the taste of the beef by adding a small amount of "heart healthy" olive oil to the raw meat before shaping it into hamburger patties. The leanest cuts of pork include pork loin, tenderloin, and center loin.

You can choose baked or roasted lean turkey, roast beef, or chicken for sandwiches, instead of processed meat products. If you decide to purchase processed meats, check the Nutrition Facts panel on products' labels to compare fat, as well as sodium contents.

MyPlate Plan also recommends varying your protein choices. For example, consider eating 8 ounces of fish per week, including rich sources of beneficial omega–3 fatty acids such as salmon, trout, and herring. You can also consume dishes made with dry beans, split peas, or foods made from soybeans because these foods are rich sources of fiber, vitamins, and minerals. Additionally, consider snacking on unsalted nuts and seeds, such as peanuts, almonds, cashews, walnuts, pecans, and sunflower seeds.

Concept CHECKPOINT

14. What is the AMDR for adult protein intake?

15. Describe how adult Americans' consumption of meat proteins has changed since 1999.

16. Which two categories of meat are associated with increased risk of certain chronic diseases? Identify at least one chronic disease that may develop as a result of eating these kinds of meat.

17. Consider your usual food choices. Using the recommendations of the MyPlate food guide, discuss ways you can reduce your intake of protein from animal foods.

See Appendix F for responses.

7.5 Understanding Nutritional Labeling: Protein

Learning Outcomes

1. Explain how to use the Nutrition Facts panel to determine the grams of protein in a serving of a packaged food.
2. Explain how you can use the ingredient list on a food label to determine the quality of the protein in the product.

You can determine how much protein is in a packaged food product by reading the Nutrition Facts panel on the label. As you can see in Figure 7.15a, one ready–to–eat meal contains 13 g of protein. The Food and Drug Administration (FDA) does not require food manufacturers to include information about the % Daily Value for protein when the food is intended for adult consumption. However, the manufacturer has to display the %DV for protein when the label has a claim about the product's protein contents. Most Americans consume plenty of protein, so protein intake is not a public health concern. The Daily Value for protein (adults) is 50 g. The panel does not provide information about a product's protein quality, but you can judge from the list of ingredients. What are the sources of high–quality and low–quality protein in Figure 7.15b?

Concept CHECKPOINT

18. According to the Nutrition Facts panel on a food package, a serving of the food supplies 5 g of protein. Using the information provided in this section, calculate the %DV of protein for the product.

19. Review the ingredients list in Figure 7.15(b). What are the sources of high-quality and low-quality protein in the product?

See Appendix F for responses.

7.6 Proteins: Economical Considerations

Learning Outcomes

1. Describe ways people can reduce the amount of meat in their diet without sacrificing protein quality.
2. Explain how using mixtures of foods with complementary proteins can help reduce animal protein intake.

Animal foods are among the most expensive items on our grocery lists.
Paul Johnson/E+/Getty Images

If you are like most Americans, your daily protein intake is higher than the RDA. Furthermore, animal foods contribute the largest share of the protein in your diet. Animal foods, including eggs and milk products, supply over 85% of the protein in the American diet.[16] Some of these foods, however, are among the most expensive items on our grocery lists, and you may be able to reduce your food costs if you eat less of them. Furthermore, you may reap substantial health benefits by reducing your animal protein intake and increasing your consumption of plant foods.

Animal foods are among the best dietary sources of essential amino acids, so is it safe to eat less animal protein? Yes! One way you can lower your intake is to include only one animal source of protein in a meal and reduce its serving size. For example, if your breakfast is a 6–ounce slice of ham with two large fried eggs, you are obtaining over 400 kcal and 42 g of high–quality protein. That is enough protein in one meal to meet the RDA for a person who weighs 116 pounds. Instead of eating such a large serving of ham with the fried eggs, have 3 ounces of ham without the eggs, or skip the ham and eat just the eggs. Two fried eggs supply over 12 g of high–quality protein and only 190 kcal.

An easy way to reduce your meat consumption is to replace meat with other high–quality protein sources. Eggs, milk, cheese, and yogurt are animal sources of high–quality protein that you can substitute for meat, fish, or poultry items in your diet. For example, simply have a cheese sandwich instead of eating a submarine sandwich made with various luncheon meats and cheeses. If you are interested in eating less fat, a serving of low–fat cottage cheese or low–fat yogurt makes a protein–rich substitute for the "sub" or cheese sandwich.

Another way to reduce the amount of animal food in your diet and your food costs is to make meals that contain less animal protein and more plant protein. Many commonly eaten menu items provide the proper amounts and mixtures of essential amino acids without relying heavily on animal products. Throughout the world, people with limited access to meat and other animal foods rely heavily on recipes that combine small amounts of animal protein with larger portions of certain plant proteins. Proteins in animal foods contain enough essential amino acids to extend or "beef up" the lower–quality plant proteins in dried peas, mature beans, cereals, and other grain products. Pasta made from white flour, for example, contains cereal (wheat) proteins that have limiting amounts of lysine. By mixing large amounts of cooked pasta with smaller

MY DIVERSE PLATE

Miso

Miso, a seasoning paste made from fermented soybeans, is a staple of traditional Japanese diets. Miso is added to soup (misoshiru) and vegetables. Two tablespoons of miso supply about 70 kcal, 4 g protein, 2 g fat, and 1280 mg sodium.

Although miso adds flavor to foods, the paste is very salty; therefore, it should be used in small amounts.

©Wendy Schiff

Food & Nutrition tips

Commercially canned beans often have considerable amounts of salt added to them. Consider purchasing canned or frozen beans that have little or no added salt. Dried beans do not have salt added to them, but they take a long time to cook, unless you soak them in water for several hours before cooking. The soaking process softens the beans, reducing cooking time and making them more digestible and less likely to contribute to intestinal gas.

anmbph/123RF

amounts of lysine–rich meat, fish, chicken, or cheese, the proteins in the animal foods enhance the quality of the wheat proteins in pasta. As a result, the body can use the amino acids in pasta for growth, repair, and maintenance of tissues.

Pancakes, waffles, crepes, and cornflakes with milk are examples of breakfast foods that extend egg and milk proteins with large amounts of cereal proteins. Many popular Asian dishes mix small amounts of chicken, beef, or shellfish with large portions of rice; Italian dishes often combine pasta with small amounts of cheese or meat sauce. Serving meals that extend the high–quality protein in animal foods is an economical way to feed large numbers of people. For example, you can use 1 pound of ground meat to make a single hamburger for each of your four friends. However, you will have enough chili con carne to feed six or more friends if you combine that pound of ground meat with three cans of kidney beans and a couple of large cans of tomatoes. The ground meat has plenty of cysteine and methionine, the essential amino acids that are low in kidney beans. By mixing plant and animal sources of protein together in chili con carne, the beef protein extends the quality of the protein in the kidney beans. Although the chili recipe calls for adding tomatoes to the meat and beans, tomatoes are botanically classified as fruit, so they add very little protein to the dish. (Imagine eating chili con carne without tomatoes!) If you want to extend the chili even more, add cooked macaroni to the mixture. Macaroni is made from wheat, so it contains cereal proteins that are enhanced by the proteins in the meat and beans. For more practical ways to "stretch" your food dollars without sacrificing the nutritional quality of your diet, read the "Nutrition Matters" section at the end of this chapter.

Combining Complementary Proteins

complementary combinations mixing certain plant foods to provide all essential amino acids without adding animal protein

Although it is not necessary to consume all essential amino acids during a meal for the body to utilize them for growth, certain plant–based recipes ensure that these compounds are consumed during a day. **Complementary combinations** are mixtures of certain plant foods that provide all essential amino acids without adding animal proteins. However, to make dishes that contain complementary amino acid combinations, you must know which plant foods are good protein sources and which essential amino acids are limiting or low in those plant foods. Most plant foods are poor sources of one or more essential amino acids, particularly lysine, tryptophan, cysteine, and methionine. Split peas, for example, are good sources of lysine, but they contain low amounts of cysteine and methionine. Cereal grains such as wheat, rice, and corn are good sources of cysteine and methionine, but they tend to be low in lysine. Wheat germ, however, is a rich source of lysine. Corn is low in tryptophan. Seeds such as sesame and sunflower seeds are generally low in lysine. Walnuts, cashews, almonds, and other tree nuts also contain low amounts of lysine. Although most fruits and some kinds of vegetables are poor sources of protein, they add appealing colors and textures, as well as vitamins, minerals, and phytochemicals, to plant–based meals.

Traditional Asian dishes, such as beef stir-fry, combine small amounts of animal protein with larger amounts of grain and vegetable proteins.
Cobraphotography/Shutterstock

Many cultures have traditional foods that combine complementary plant proteins. For example, a peanut butter sandwich combines two foods that supply complementary plant proteins. Peanuts are a fair source of lysine. Bread contains some methionine, but the grain product is very low in lysine. Serving the two foods together as a peanut butter sandwich provides adequate amounts of these essential amino acids. Table 7.4 lists some other foods that are examples of complementary protein combinations. When menu planning, you can combine a variety of legumes, grains, tree nuts, and seeds with vegetables to prepare dishes that provide adequate mixtures of the essential amino acids. Figure 7.16 shows three categories of plant proteins (legumes, grains, and tree

TABLE 7.4 Complementary Protein Dishes

• Red beans and rice
• Peanut or soy nut butter on bagel, sprinkled with wheat germ
• Hummus (mashed chickpeas/garbanzo beans) with sesame seeds or tahini (ground sesame seed paste)
• Hummus on whole-grain pita bread
• Black beans and cornmeal tortilla
• Split pea soup with toasted whole-wheat bread
• Meatless kidney bean chili with macaroni
• Cornmeal tortilla with black bean salsa
• Peanut butter on whole-grain crackers, sprinkled with wheat germ
• Tofu with brown rice and cashews

Peanut butter on bread provides complementary plant proteins.
David Buffington/Blend Images/Getty Images

Figure 7.16 Complementary combinations. To ensure an adequate mix of amino acids, combine legumes with grains, seeds, and/or tree nuts.
Pistachios: Mark Steinmetz/McGraw Hill; Peanuts: C Squared Studios/Photodisc/Getty Images; Corn, Plate of nuts, legumes, and vegetables: ©Wendy Schiff

Food & Nutrition tips

The following information describes popular soybean foods and tips for how to use them in menu planning.

- Plain tofu has little flavor, so it can be added to a variety of foods, including stir-fried vegetables and scrambled eggs.
- Tempeh is a fermented soybean and grain mixture that can substitute for meat in sandwiches and casseroles.
- Soy nuts are roasted soybeans that are often eaten as a snack. Ground soy nuts form a spread that is used like peanut butter.
- Soy milk is made from crushed soybeans. Soy milk is usually fortified with calcium and vitamins A, D, B-12, and riboflavin. Regular soy milk can substitute for cow's milk as a beverage or in recipes. Soy milk cheeses and yogurt are also available.
- Texturized soy protein (TSP) is made from soybean flour. A TSP product that resembles ground beef can be used to replace half or all of the ground beef in meatloaf, meatball, chili, taco, or meat sauce recipes.
- Soy protein concentrate is a high-protein, high-fiber refined soybean product that is used to boost the protein content of foods.

Mitch Hrdlicka/Photodisc/Getty Images

vegetarians people who eat plant-based diets

semivegetarian (or **flexitarian)** vegetarian who eats animal products, except red meat

lactovegetarian vegetarian who consumes milk and milk products for animal protein

ovovegetarian vegetarian who eats eggs for animal protein

lactoovovegetarian vegetarian who consumes milk products and eggs for animal protein

nuts and seeds) that make complementary combinations when legumes are mixed with grains, seeds, and/or tree nuts.

Not every mixture of plant foods creates a complementary combination. For example, making a fruit salad by combining apples, grapes, and oranges will not provide a complementary mixture of essential amino acids. Fruits are nutritious foods, but they are generally poor sources of protein. Combining Boston, iceberg, and romaine varieties of lettuce with carrots and onions makes a tasty salad, but simply mixing leafy greens with other vegetables does not make a complementary combination because vegetables have small amounts of protein that tend to contain low amounts of essential amino acids. However, adding sunflower seed kernels, kidney or black beans, cashews, and bread cubes to the salad boosts the amount of protein and provides a complete mix of amino acids. To increase the essential amino acid content of the salad even further, you can add a small amount of hard–cooked egg, shredded cheese, or bits of *tofu,* a soybean product, to it. If you are interested in trying foods made from soybeans, the "Food & Nutrition Tips" feature on this page provides information about some of the more popular foods made from soybeans.

Concept CHECKPOINT

20. Give examples of common foods that are high-quality substitutes for meat and foods that extend a source of high-quality protein.

21. Does a recipe that combines apples and oranges with peanuts provide a complementary mixture of proteins? Explain why or why not.

22. A recipe mixes cereals made from wheat, rice, and corn. What plant foods could you add to this combination of cereals to make the recipe a source of high-quality protein?

See Appendix F for responses.

7.7 Vegetarianism

Learning Outcomes

1. Describe different forms of vegetarianism.
2. Discuss pros and cons of vegetarian diets.

Are you or is anyone you know vegetarian? If you are vegetarian, do you eat any animal foods? **Vegetarians** rely heavily on plant foods and may or may not include some animal foods in their diets. Five percent of American adults refer to themselves as vegetarians; 3% of these persons avoid eating any foods from animal sources.[17]

There are many different types of vegetarian diets, including some that contain animal foods. A **semivegetarian** (or **flexitarian**), for example, avoids red meat but consumes other animal foods, including fish, poultry, eggs, and dairy products. Other vegetarians have more restrictive diets, particularly when choosing whether to eat animal foods. A **lactovegetarian** (*lacto* = milk) consumes milk and milk products, including yogurt, cheese, and ice cream, to obtain animal protein. An **ovovegetarian** (*ovo* = egg) eats eggs, and a **lactoovovegetarian** consumes milk products and eggs. **Pescatarians** (*pesce* = fish) eat fish and shellfish but no other sources of animal protein. However, some pescatarians may choose to consume milk and milk products and eggs; thus,

TABLE 7.5 Vegetarian Diets

Type of Vegetarian Diet	Animal Foods Included
• **Semivegetarian (flexitarian)**	All except red meats
• **Lactovegetarian**	Milk and milk products No animal flesh or eggs
• **Ovovegetarian**	Eggs but no other animal foods
• **Lactoovovegetarian**	Milk and milk products and eggs but no other animal foods
• **Pescatarian**	Fish and shellfish but no other animal foods
• **Vegan**	No animal foods
Fruitarian	No animal foods (nuts and seeds only)
Macrobiotic	No animal foods (will eat organically grown whole grains, fruits and vegetables, and soups made with vegetables, seaweed, grains, beans, and miso)

A vegan, or vegetarian, might enjoy this dish: couscous with pomegranate seeds and fresh herbs. Couscous is a grain product.
martinturzak/iStock/Getty Images

they are lactoovo–pescatarians. A **vegan,** or total vegetarian, eats only plant foods. Table 7.5 lists types of vegetarian diets.

Vegetarians have various reasons for eating few or no animal products. Many vegetarians have religious, ethical, and other philosophical beliefs that do not support the practice of killing and eating animals. For others, vegetarianism is a matter of concern for the environment and economics; plant foods are generally less expensive than animal foods. Some vegetarians believe that humans are not physically able to digest animal foods. This is not true. The omnivore's intestinal tract is able to obtain nutrients from both plants and animals, and humans are omnivores. Nevertheless, eating more plant than animal sources of protein may provide important health benefits.

pescatarian vegetarian who eats seafood but no other animal proteins

vegan vegetarian who eats only plant foods

Is Vegetarianism a Healthy Lifestyle?

Vegetarian diets are often lower in saturated fat and energy than "Western" diets that contain animal foods, particularly plenty of red meat. Compared to people who eat meat, vegetarians tend to have a lower risk of obesity, type 2 diabetes, hypertension, and certain cancers.[18] Furthermore, vegans tend to be leaner than nonvegans. It is difficult, however, to pinpoint diet as responsible for vegetarians' health status. Why? Vegetarians often adopt other healthy lifestyle practices, such as exercising regularly; practicing relaxation activities, such as meditation; and avoiding tobacco products and excess alcohol.

Compared to the typical American diet, vegetarian diets provide more fiber, soy protein, folate (a B vitamin), beta–carotene (a phytochemical that the body can convert to vitamin A), vitamins E and C, and the minerals potassium and magnesium.[18,19] Furthermore, vegetarian diets often supply less saturated fat and cholesterol than diets that include animal foods.

Poorly planned, plant–based diets may not contain enough vitamins B–12 and D, minerals iron and calcium, and omega–3 fatty acids to meet a person's nutritional needs.[19] Plant foods contain little or no vitamin B–12, and there are few dietary sources of vitamin D other than fortified cow's milk. Furthermore, mineral nutrients such as calcium and iron are more available from animal

Food & Nutrition tips

To make a vegetarian stir-fry:

- Heat a small amount of peanut or canola oil in a frying pan;
- Add cooked rice and pieces of raw vegetables;
- While the mixture is heating, add a beaten egg to it and stir so the egg cooks thoroughly; and
- Before serving the dish, sprinkle cashews or sunflower seed kernels over the hot rice and vegetable mixture.

TABLE 7.6 Vegetarian Diets: Nutritional Aspects

Advantages	Possible Disadvantages
High in:	**Low in:**
Vitamins C, E, and folate	Vitamins B-12 and D
Beta-carotene	Iron, zinc, and calcium
Fiber	Omega-3 fatty acids
Magnesium and potassium	Certain essential amino acids (children)
Low in:	Energy (children)
Fat (saturated)	
Cholesterol	

than from plant foods. Plants often contain phytochemicals that interfere with the body's absorption of minerals, particularly iron, zinc, and calcium. Recall from Chapter 6 that the body is unable to make much DHA and EPA from alpha–linolenic acid, an essential omega–3 fatty acid.

Vegetarians who do not consume fish may need to obtain DHA and EPA by eating certain algae or taking dietary supplements that contain the algae. Vegans can obtain vitamin B–12, vitamin D, iron, zinc, and several other micronutrients by consuming fortified foods, such as many brands of ready–to–eat cereal, soy and rice beverages, and fortified vegetarian nutritional yeast. Vegetarians can also take a daily multiple vitamin/mineral supplement to compensate for missing nutrients from their diet.

Humans digest animal proteins to a greater extent than plant proteins.[20] Therefore, vegetarians may need to increase their protein intakes. Total vegetarians, including vegan athletes, can obtain adequate amounts of the essential amino acids by eating quinoa, processed soybean products, and foods that combine complementary plant proteins.

In general, plant foods have low energy density: They add bulk to the diet without adding a lot of calories. Children have higher protein and energy needs per pound of body weight than adults. Because plant foods add bulk to the diet, vegan children are likely to eat far less food than adult vegans because they become full sooner during meals. Thus, very young vegans may be unable to eat enough plant foods to meet their protein, micronutrient, and energy needs. Therefore, it is very important for parents or other caretakers to plan nutritionally adequate diets for vegetarian children and monitor the youngsters' growth rates.

Vegan females who breastfeed their infants may produce milk that is deficient in vitamin B–12, particularly if the mothers' diets lack the vitamin. These infants of vegan mothers have a high risk of developing severe developmental delays associated with neurological damage, especially when breast milk is their only source of vitamin B–12.[21] Pregnant vegan females should consult with their physicians about the need to take a vitamin B–12 supplement to reduce the likelihood of having a baby who is deficient in this nutrient. Additionally, vegan mothers who breastfeed their infants may need to provide the babies with a source of vitamin B–12 as well. Table 7.6 summarizes the nutritional advantages and possible nutritional disadvantages of vegetarian diets.

Switching from the typical Western diet to vegetarianism can be a healthy practice for teens; vegetarian youth often eat more fruits and vegetables and less fat than their nonvegetarian peers. On the other hand, female vegetarian teenagers may have a higher risk of unhealthy weight control practices and eating disorders, such as *anorexia nervosa,* than young people who eat meat.[22] Anorexia nervosa ("anorexia") is a serious psychological disorder that can result in starvation and suicide. The Chapter 10 "Nutrition Matters" section provides more information about anorexia nervosa and other eating disorders.

Did *YOU* Know?

There is nothing inherently "bad" about eating small portions of red meat. Red meat is a good source of zinc and iron, minerals that are often less bioavailable from plant foods.

Commercially prepared vegetarian foods, such as these products, are often available in the frozen food section of supermarkets.
©Wendy Schiff

Meatless Menu Planning

Many common menu items can be converted into vegetarian foods by removing the meat, fish, or poultry. For example, pizza and lasagna can be prepared without meat and still provide plenty of protein from the cheese, as well as the crust or pasta. Stir–fried foods can also be a reliable source of protein without adding meat, fish, or poultry. Table 7.7 presents more meatless menu suggestions.

TABLE 7.7 Meatless Menu Ideas

• Cooked pasta with marinara sauce and grated cheese
• Vegetable lasagna with layers of cheese, thinly sliced zucchini, mushrooms, and bell peppers
• Vegetable stir-fry with bits of tofu and cheese
• Grilled vegetable kabobs with cooked rice and black beans
• Black or red bean burritos
• Red bean and corn tacos

John E. Kelly/Photodisc/Getty Images

Commercially prepared vegetarian foods that substitute for meat, fish, and poultry items are often available in the frozen food section of supermarkets. These vegetarian products can look and taste like their nonvegetarian counterparts, but they generally do not contain cholesterol and may be lower in saturated fat. Such foods include soy–based sausage patties or links, soy hot dogs, "veggie" burgers, and soy "crumbles" that look like bits of cooked ground beef. Asian restaurants usually offer vegetarian dishes. Some "Western–style" restaurants offer vegetarian menu items, or their cooks can modify menu items by substituting meatless sauces, omitting meat from stir–fries, and adding vegetables or pasta in place of meat.

With careful planning, vegetarians can overcome the nutritional limitations of limiting some or all animal foods and consume adequate diets. If you are interested in learning more specific details about vegetarian cookery and menu planning, contact a registered dietitian nutritionist (RDN). The MyPlate website also offers suggestions for planning nutritionally adequate meatless meals. For more information, visit www.myplate.gov.

Did *YOU* Know?

Impossible Foods manufactures a meat substitute made from soy protein called Impossible Beef. The meat substitute is used to replace ground beef in recipes and burger patties. Impossible Beef products have a texture and color similar to raw beef. In fact, these meat substitutes are a bright red color; "bleed" red, blood-like juices when raw; and taste like beef when cooked! The unique red coloring and umami, "beefy" flavor come from soy *leghemoglobin,* which is a heme molecule found in the roots of soybeans.[23] Impossible Foods uses yeast genetically engineered with the gene for leghemoglobin to create large amounts of the plant heme to use in their products. A 4-oz. Impossible Beef burger pattie provides 19 grams of high-quality soy protein.

Sundry Photography/Shutterstock

Concept CHECKPOINT

23. Describe how the diets of semivegetarians differ from other vegetarian diets.
24. Identify nutrients that are most likely lacking in a vegan's diet.
25. Explain why vegans must be careful when planning vegan meals for children.

See Appendix F for responses.

7.8 Protein Adequacy

Learning Outcomes

1. Discuss potential health problems that may occur with excessive protein intake.
2. Explain why protein-energy malnutrition is a serious nutritional state, especially for young children.

If some protein is necessary for proper growth and good health, can eating extra amounts of the nutrient make you *extra* healthy or physically fit? Intuitively, the idea of eating more protein to improve your health seems logical, but protein is no different from the other nutrients. If your diet contains adequate amounts of protein, then eating "more is not better."

MY DIVERSE PLATE

Locusts

In parts of Africa, Cambodia, and the Philippines, many people eat locusts, especially when the insects are plentiful. Popular ways to cook locusts include removing the wings and legs before frying with seasonings or placing locusts on skewers and roasting them over hot embers. The legs and wings are removed from the grilled insects before they are cooked or eaten. Over 60% of the dry weight of a locust is protein, so the insect is an excellent source of the macronutrient.

Source: Food and Agriculture Organization, www.fao.org/ag/locusts/en/info/info/faq/.

Simon Murrell/IT Stock Free/Alamy Stock Photo

purines nitrogen-containing chemicals in genetic material

Some weight-loss diets promote large servings of protein-rich foods and few items from other food groups.
from my point of view/Shutterstock

In many parts of the world, the lack of foods containing high–quality proteins is a serious problem, particularly for young children. Protein deficiency interferes with a child's normal growth and development, and contributes to many childhood deaths. The following sections examine protein malnutrition.

Excessive Protein Intake

High–protein diets are generally not recommended for healthy people. Excess intakes of dietary protein may lead to higher–than–normal urinary losses of calcium. However, scientists do not agree on whether the loss of calcium contributes to poor bone health.[24,25]

Protein–rich foods from animals, especially liver, eggs, meat, and certain fish, are rich sources of **purines,** nitrogen–containing chemicals that comprise genetic material, including DNA. The liver breaks down purines to form uric acid. Healthy kidneys remove uric acid from the bloodstream and excrete the waste product in urine. A diet that is high in animal proteins may increase the level of uric acid in the bloodstream and urine. *Gout* is a type of arthritis that can develop in people who have elevated blood levels of uric acid. The condition is characterized by the buildup of uric acid crystals in joints, especially in the foot and big toe. Medication is used to treat gout. Having an excess of uric acid in urine can contribute to the formation of kidney stones. Thus, people who have a history of kidney stones, have kidney disease, or are at risk of kidney disease should check with their physician before eating a high–protein diet.[26]

Excess amino acid or protein intake can lead to dehydration because kidneys need more water to dilute and eliminate the toxic waste products of amino acid metabolism in urine. Dehydration is a potentially life–threatening condition in which the body's water level is too low. People with liver or kidney diseases may need to avoid protein–rich diets and amino acid supplements because metabolizing the excess amino acids is a burden to their bodies.

What About High-Protein Weight-Loss Diets?

Certain popular weight–loss diets, such as the Atkins, Protein Power, and Paleo diets, promote high intakes of protein. While following high–protein diets to lose weight, people often report decreased feelings of hunger and increased sense of fullness (satiety) after meals. High–protein diets that limit total calorie intake appear to promote the loss of excess body fat and have other beneficial effects on health. Such diets are often difficult for people to maintain for long periods.[27] However, after losing weight, high plant protein intake from nuts, legumes, starchy and nonstarchy vegetables, fruits, and nondairy beverages is associated with weight maintenance.[28] Chapter 10 provides information about the safety and effectiveness of various popular weight–loss diets.

What About Protein or Amino Acid Supplements?

For thousands of years, humans have obtained amino acids by eating plants and animals. The use of amino acid and protein supplements as sources of the nutrient is a relatively recent development, and little is known about the long–term safety of using these products. Chapter 11 provides more information about protein needs of physically active people and examines the evidence concerning the value and safety of using certain foods and dietary supplements to enhance physical performance.

Protein Deficiency

Although food insecurity exists in the United States, protein deficiency is uncommon. (See the Chapter 1 "Nutrition Matters" section for information about food insecurity.) People with low incomes, especially older adults, are at risk of protein deficiency. Many Americans who are over 65 years of age have limited incomes and must make difficult choices concerning their expenses. If you were an older adult, chronically ill, and on a low fixed income, what would you think was more important: purchasing nutritious foods or costly prescription medications that control your serious health problems?

Protein Undernutrition: Chronic and Acute Diseases

Protein undernutrition can develop in adults who have chronic diseases, including cancer, heart disease, lung disease, and kidney disease.[29] People with one or more chronic diseases also tend to have mild to moderate amounts of chronic inflammation in their body. Chronic inflammation can change how cells normally use proteins. For example, an affected person's cells may use more proteins for managing the inflammatory process than for more typical needs, such as replacing red blood cells or maintaining healthy bone tissue.[30] Thus, a person with chronic inflammation often has a protein requirement that is higher than normal.

People who have chronic diseases may experience several factors that contribute to lower–than–normal protein consumption. It is not uncommon for such patients to experience a decrease in appetite and the ability to chew and/or swallow properly. These factors reduce the amounts of nutrients, including protein, that affected persons consume. Furthermore, the gastrointestinal tract's ability to digest and absorb proteins can be lower than normal in cases of chronic disease.

Persons with serious injuries or acute diseases, such as pneumonia, COVID–19, broken bones, and body burns, often experience the same risk factors for low protein consumption as people with chronic diseases. Serious health consequences, even death, can occur if cases of chronic and acute protein undernutrition are not treated with appropriate nutritional and medical interventions.[29] Regardless of the cause, registered dietitian nutritionists are trained to help diagnose and treat protein malnutrition.[32]

Did *YOU* Know?

People who experience acute illness or injuries such as the flu, COVID-19, pneumonia, broken bones, and body burns experience generalized inflammation that increases the infected or injured person's energy and protein needs. However, very sick or injured people may have a poor appetite that can reduce the amount of calorie- and protein-rich foods consumed, resulting in weight loss and protein undernutrition.

While recovering from infections or injuries, people should consume a high-calorie, high-protein diet that provides more than the DRIs for calories and protein for their sex and age.[31] Small frequent nutrient- and calorie-dense meals and snacks may prevent weight loss and muscle mass loss, thus reducing the infected or injured person's risk of developing protein undernutrition. To increase calorie intake, people recovering from infections and injuries should add extra servings of fat to meals and snacks, such as olive oil, avocados, salad dressings, butter, and cheese sauce. Dietary protein needs can be met by consuming eggs, milk and other dairy products, meat, fish, poultry, nuts and nut butters, and beans. Furthermore, oral nutritional supplements ("meal replacements") can be a source of additional calories and protein during recovery.

Protein–Energy Malnutrition

As mentioned in Chapter 1, the lack of food is often widespread in poor nations in which populations endure frequent famine resulting from crop failures, political unrest, or civil wars. In these countries, **protein–energy malnutrition (PEM)** affects people whose diets lack sufficient protein, as well as energy. The failure to consume nourishing food also results in vitamin and mineral deficiencies.

When food is limited, it is often more difficult for children to obtain nutritionally adequate diets than for adults to do so. Why? Adults may be able to consume enough plant proteins to meet their protein and energy needs, but children have smaller stomachs and higher energy and protein needs per pound of body weight than adults. They are unable to eat enough plant foods to meet their relatively high protein and other nutrient requirements.

According to the World Health Organization (WHO), PEM is responsible for 45% of deaths that occur in children under 5 years of age.[33] Children whose diets lack sufficient protein and energy do not grow and are very weak, irritable, and vulnerable to dehydration and infections, such as measles, that can kill

protein-energy malnutrition (PEM) condition that results from diets that provide inadequate amounts of protein and energy

them. If these children survive, their growth may be permanently stunted and their intelligence may be lower than normal because undernutrition during early childhood can cause permanent brain damage.

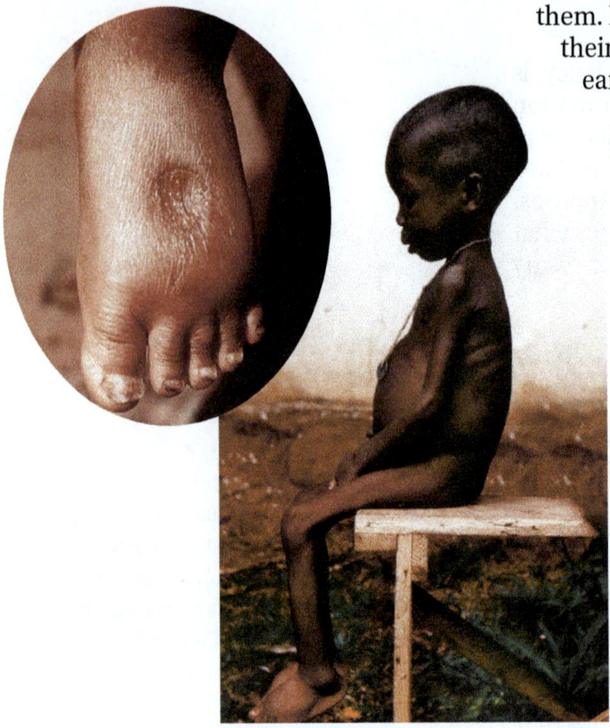

Figure 7.17 Marasmic kwashiorkor. This Nigerian child is suffering from marasmic kwashiorkor.
Dr. Lyle Conrad/Centers for Disease Control and Prevention

Forms of PEM At one time, nutrition experts thought there were only two types of PEM, **kwashiorkor** (*qwash'–e–or'–kor*) and **marasmus** (*mah–raz'–mus*). The distinctions between these conditions, however, are often blurred because protein deficiency is unlikely when a person's energy intake is adequate. Nevertheless, registered dietitian nutritionists generally consider *kwashiorkor, marasmic kwashiorkor,* and *marasmus* as forms of PEM.

Kwashiorkor primarily occurs in low–income countries where mothers commonly breastfeed their infants until they give birth to another child. The older youngster, who is usually a toddler, is fairly healthy until abruptly weaned from their mother's milk to make way for the younger sibling. Although the toddler may obtain adequate energy by consuming a traditional diet of cereal grains, the diet lacks enough complete protein to meet the youngster's high needs, and they soon develop signs of protein deficiency. Children affected by kwashiorkor have stunted growth (see Figure 1.15); unnaturally blond, sparse, and brittle hair; and patches of skin that have lost their normal coloration. Children with kwashiorkor have some subcutaneous (under the skin) fat and swollen cheeks, arms, legs, and bellies that make them look well fed, but their appearance is misleading. An important function of certain proteins in blood is to maintain proper fluid balance within cells and blood vessels, as well as between cells. During starvation, levels of these proteins decline, resulting in edema, which can make the protein–deficient child look overweight instead of thin and undernourished. In many cases, the child suffering from kwashiorkor does not obtain enough energy and eventually develops marasmic kwashiorkor, a condition characterized by edema and wasting (Fig. 7.17). *Wasting* is the loss of organ and muscle proteins as the body tears down these tissues to obtain amino acids for energy metabolism.

Severe PEM causes extreme weight loss and a condition called marasmus, which is commonly referred to as starvation (Fig. 7.18). Obvious signs of marasmus are weakness and wasting. The marasmic person is so thin that their ribs, hips, and spine are visible through the skin. People suffering from marasmus avoid physical activity to conserve energy, and they are often irritable. If marasmus is not treated with a nutritious diet, death results.

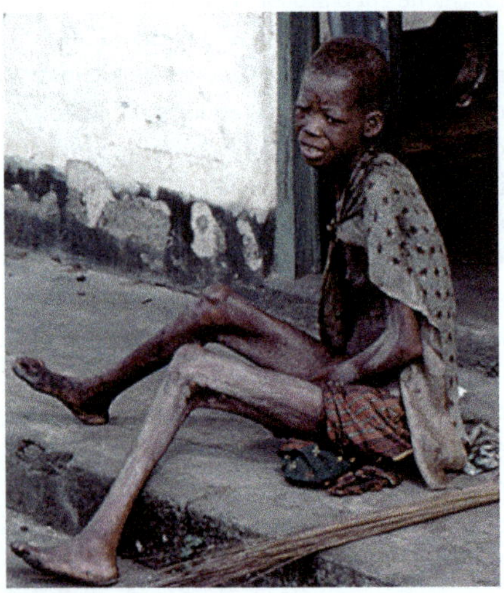

Figure 7.18 Severe protein-energy malnutrition.
Dr. Lyle Conrad/Centers for Disease Control and Prevention

Concept CHECKPOINT

26. In the United States, which groups of people are most likely to suffer from protein deficiency?

27. Why is protein-energy malnutrition a devastating condition for young children?

28. Police bring a 2-year-old child into a hospital; the child is so underweight, his bones are visible through the skin. The police report indicates the child was severely neglected by his parents. According to this information, is this child suffering from kwashiorkor, severe protein-energy malnutrition, or sickle cell anemia? Choose one of these conditions and explain why you selected it.

See Appendix F for responses.

kwashiorkor form of undernutrition that results from consuming adequate energy and insufficient high-quality protein

marasmus starvation

7.9 Protein and Health: Environmental and Genetic Factors

Learning Outcomes

1. Explain the cause of a food allergy, identify foods that are most likely to cause food allergies in vulnerable people, and list three common signs or symptoms of a food allergy.
2. Explain what causes celiac disease, identify foods that a person with celiac disease must avoid, and list three common signs or symptoms of the disease.
3. Explain the cause of phenylketonuria, and discuss how the condition is treated.
4. Distinguish between nutrigenetics and nutrigenomics.

There is no question that cells need amino acids to function properly. In some instances, however, certain amino acids or proteins cause havoc in the body, resulting in serious health problems and even death. The following sections take a closer look at three protein–related conditions that affect the lives of millions of Americans. By following special diets, people with these conditions can live normal and productive lives. This section also includes information about *nutritional genomics,* the science that explores ways genes interact with dietary choices and the environment.

What Is a Food Allergy?

Have you ever experienced an allergic reaction after eating certain foods or drinks? A food allergy is an inflammatory response that results when the body's immune system reacts inappropriately to one or more harmless substances

Did *YOU* Know?

In a healthy person, certain immune system cells make proteins called antibodies when the cells identify foreign (non-self) substances that are potentially troublesome. Such substances are called *antigens.* The antibodies bind to the antigens, destroying the antigens' ability to cause health problems. One particular class of antibodies (*IgE*) triggers an allergic response, which is the immune system's reaction to a harmless substance in the environment. IgE is present in the bloodstream of people who have allergies, including common food allergies. Although various proteins in foods can be antigens, a form of galactose, a monosaccharide (see Chapter 5) simply referred to as *α-Gal* (alpha-Gal), also has the potential to cause allergic responses. Most mammals, including cattle, pork, and sheep, make α-Gal, so it is present in their tissues. Products made from mammals, such as gelatin and dairy products, can also contain α-Gal. Humans do not make α-Gal, so the carbohydrate is normally not present in the human body.

Recently, medical researchers discovered a link between having a history of tick bites and developing α-Gal allergy.[34] The scientists determined that certain kinds of ticks, including the Lone Star tick and black-legged "deer" tick, can have α-Gal in their saliva. When one of these ticks bites a person, α-Gal in the tick's saliva enters the bloodstream, where it triggers the individual's immune system to make anti-α-Gal antibodies. ("Anti" means against, so these antibodies specifically target α-Gal.) As a result, the person becomes *α-Gal IgE-positive.* Whenever the IgE-positive individual eats red meat, a natural source of α-Gal, their immune system cells can quickly form anti-α-Gal antibodies that bind α-Gal, causing an allergic response. Therefore, it is possible to develop an allergy to red meat after being bitten by a tick!

"Deer" tick.
Jim Gathany/CDC

Did *YOU* Know?

According to conventional wisdom, babies who have siblings or parents with peanut allergy should not eat products made with peanuts because the peanuts will trigger allergic responses in the infants. In 2019, a clinical report published by the American Academy of Pediatrics stated that current scientific evidence supports early introduction of infant-safe forms of peanuts to reduce the risk of peanut allergies later in childhood.[35] Thus, caregivers may contribute to the development of peanut allergy by not introducing peanuts into the diets of infants. The results of this study are changing how the medical profession views food allergy prevention, not just for peanuts, but also for tree nuts, eggs, fish, wheat, and other foods that are considered highly allergenic.

©Wendy Schiff

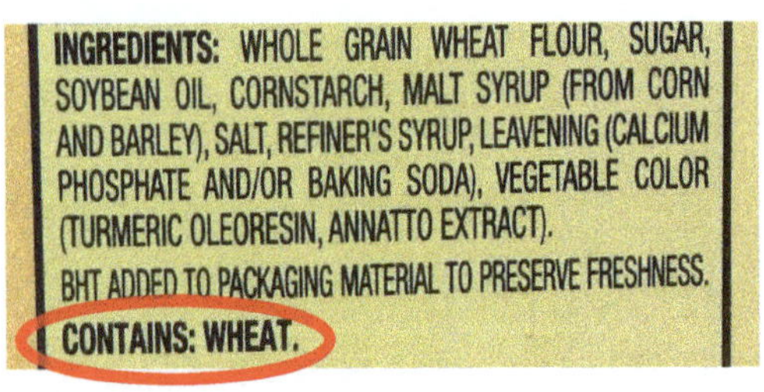

Figure 7.19 **Allergen labeling.**
©Wendy Schiff

(*allergens*) in the food. In most instances, the allergen is a protein. People with food allergies may experience swelling of their lips and tongue as they eat the food or soon after they swallow it. If the allergen in the food that is eaten does not undergo denaturation by stomach acid, the molecule enters the small intestine. Immune system cells in the small intestine may recognize the food protein as a foreign substance and try to protect the body by mounting a defensive response. In other instances, the cells of the small intestine absorb the food allergen, and the body's immune system reacts to it. As a result of the immune response, the person who is allergic to that food experiences typical signs and symptoms.

Common signs and symptoms of food allergies include *hives,* red raised bumps that usually appear on the skin; swollen or itchy lips; skin flushing; a scaly skin rash (eczema); difficulty swallowing; wheezing and difficulty breathing; and abdominal pain, vomiting, and diarrhea. Allergic reactions generally occur within 30 minutes to a couple of hours after eating the offending food. In severe cases, sensitive people who are exposed to food allergens can develop *anaphylactic shock,* a serious drop in blood pressure that affects the whole body. Anaphylaxis (*an–a–pha–lax′–is*) can be fatal, unless emergency treatment is provided. Genetics play a major role in the risk of food allergies; people who have family histories of allergies to foods are more likely to develop these conditions.

Although any food protein has the potential to cause an allergic reaction in a susceptible person, the most allergenic proteins are in cow's milk, eggs, peanuts and other nuts, wheat, soybeans, fish, and shellfish. The FDA declared sesame to be the ninth major allergen in January 2023.[36] Manufacturers are required to identify sesame in the ingredient list of food products. Allergic responses to nonprotein food dyes or other food additives such as *sulfites* can also occur. Sulfites are a group of sulfur–containing compounds that result from the metabolism of certain amino acids. Sulfites can be found naturally in foods, but the compounds are often added to wines, potatoes, and shrimp as a preservative. People who suffer from asthma often develop breathing difficulties after consuming food treated with sulfites. Other sulfite–sensitive people report skin flushing (redness and warmth), hives, difficulty swallowing, vomiting, diarrhea, and dizziness after consuming foods that contain the compounds.

In the United States, approximately 5% of children and 4% of adults suffer from food allergies.[37] Over the past 25 years, the prevalence of food allergies among children under 18 years of age has been increasing in the United States. Children with food allergies are more likely than other children to experience other allergies and asthma. As they mature, children often "outgrow" allergies to wheat, milk, soy, and eggs.[37] On the other hand, allergies to peanuts, tree nuts, fish, and shellfish tend to persist into adulthood.

Accurate diagnosis of a food allergy should be undertaken by an immunologist, a physician who specializes in the diagnosis and treatment of allergies. Skin testing is a reliable way to identify allergens. Although hair analysis, cytotoxic or electrodermal testing, and kinesiology are promoted by alternative medical practitioners to diagnose allergies, these are unproven diagnostic methods.

Treatment of food allergies involves strict avoidance of the offending foods. Parents or caregivers of young children with food allergies should read food labels carefully to check for allergens listed among ingredients. Additionally, they should educate teachers and other adults who associate with the allergic child about the importance of not exposing the youngster to specific foods. In 2004, the U.S. Congress passed the Food Allergen Labeling and Consumer Protection Act (FALCPA), which required food manufacturers to identify potentially allergenic ingredients, such as soy, milk, and peanuts, on product labels (Fig. 7.19).

In early 2020, the FDA approved Palforzia to help reduce the risk of severe allergic reactions to peanut proteins in certain children who have peanut allergy and are 4 through 17 years of age. Palforzia is not a cure for peanut allergy; the medication must be given under tightly controlled medical supervision and with avoidance of dietary peanuts. For more information about this medication, visit www.fda.gov/vaccines–blood–biologics/allergenics/oral–immunotherapy.

What Is Celiac Disease?

Celiac (see′–lee–ak) disease is a common inherited condition that results in poor absorption of nutrients (malabsorption) from the small intestine. People with the disease cannot tolerate foods that contain gluten, a group of related proteins in wheat, barley, and rye. Gluten provides the chewy texture and stiff structure of breads and other baked products made from wheat, barley, and rye. After a person with celiac disease eats foods or is exposed to substances that contain gluten, a component of the protein stimulates the body to mount an immune response in the small intestine that inflames or destroys villi. (Villi are the tiny, fingerlike projections of the small intestine that absorb nutrients [see Figure 4.27].) Even though the affected person's food intake may be nutritionally adequate, malnutrition results because their intestinal tract lacks healthy villi.

The signs and symptoms of celiac disease vary from person to person but usually include abdominal bloating, chronic diarrhea, and weight loss. Children with this condition also experience poor growth due to nutrient malabsorption and protein malnutrition. Some people have no obvious signs or symptoms of the disease, despite the damage occurring to their small intestines. Serious health problems such as anemia (a blood disorder), osteoporosis (weak bones), infertility, liver disease, and intestinal cancer can result from untreated celiac disease.

In the United States, as many as 1 in 141 people has celiac disease, but most of them are not aware that they have the disease.[38] Although the tendency to develop the disease is probably inherited, environmental factors often play a role by triggering the condition. A serious viral infection or severe emotional stress, for example, may activate the disorder in a genetically susceptible person.

No single medical test is effective in determining whether a person has celiac disease. However, blood testing and intestinal biopsies are often used to diagnose the condition. The biopsy involves removing tiny pieces of tissue from the small intestine. The tissue samples undergo microscopic examination to evaluate the condition of the villi. The presence of damaged villi can help confirm the diagnosis of celiac disease.

There is no cure for celiac disease, but persons with the condition can achieve and maintain good health by following a special gluten–free diet very carefully. Many supermarkets have a special section with a variety of gluten–free foods. Table 7.8 lists foods that must be avoided by people with celiac disease as well as those that are safe for them to consume. Although people with the condition can eat corn, rice, and soy products, these foods may be contaminated with gluten if they are processed in factories that also manufacture wheat products. Because gluten may be used to make medications, dietary supplements, and lipsticks, people with celiac disease should read ingredient lists to determine whether such "hidden" sources of gluten are present. The FDA has rules for the use of the claim "gluten–free" on product labels (see www.fda.gov/consumers/consumer–updates/gluten–free–means–what–it–says).

Katie Adams was diagnosed with celiac disease when she was in college. The first "Real People, Real Stories" feature in this section introduces Katie and describes how she manages her diet to avoid signs, symptoms, and complications of celiac disease.

Did *YOU* Know?

Celiac disease is not the same as a condition called *gluten sensitivity*. People who have gluten sensitivity also develop health problems when they eat foods that contain gluten, but their small intestine is not damaged. Some symptoms of gluten sensitivity, particularly tiredness and abdominal pain, are similar to those of celiac disease. Therefore, it is important for people who think they cannot eat foods that contain gluten to undergo medical testing to determine whether they actually have celiac disease. Regardless of whether a person has gluten sensitivity or celiac disease, the treatment is the same—avoiding gluten.

TABLE 7.8 Gluten-Free Diet for Celiac Disease

Avoid foods that contain:
wheat, including enriched, durum, graham, and semolina flours; farina; wheat bran; wheat germ; cracked wheat; and wheat protein
barley
rye
triticale
Unless contaminated with any of the above foods, products that contain the following ingredients are generally safer to eat:
arrowroot, buckwheat, cassava, corn, flax, oats (small amounts), millet, nuts, quinoa, rice, sorghum, soy, and tapioca

REAL PEOPLE | REAL STORIES

©Wendy Schiff

Katie Adams

When Katie Adams was an American college student, she had the opportunity to attend England's renowned Cambridge University during her junior year. Within a few months after Katie began her studies in England, she began to experience frequent bouts of abdominal pain not long after eating. Making matters even worse, she suffered from nearly constant diarrhea. Although her weight was at a healthy level when she first went to England, she began to lose pounds despite eating adequate amounts of food.

Eventually, Katie reached the point at which she could no longer ignore her health problems. While at Cambridge, she was examined by a physician who thought her signs and symptoms were the result of celiac disease. Although Katie was unfamiliar with the disease, she decided to forgo being tested for the condition until she returned to the United States.

After coming home from England, Katie underwent specialized testing by a physician who diagnoses and treats conditions affecting the intestinal tract. This testing detected the presence of antibodies to proteins in gluten. She also underwent an endoscopic examination of her upper gastrointestinal tract. This exam enabled the gastroenterologist to see the condition of her small intestine and remove a small part of the damaged tissue to view under a microscope. As a result of these definitive tests, Katie was diagnosed with celiac disease. The only treatment for celiac disease is complete avoidance of gluten-containing foods.

Eating out is especially difficult for people with celiac disease because of restaurants' widespread inclusion of breads, rolls, pasta, and desserts made from wheat flour in menus. Katie learned how to locate the few restaurants in her community that cater to people with celiac disease by having kitchens that are dedicated to gluten-free food preparation.

Today, Katie maintains her good health by continuing to avoid gluten. Although her diet is very restrictive, she knows from experience that if she "cheats" by eating even a small amount of food that contains gluten, she'll get sick again. After such occasions, she returns to following a gluten-free diet, and within several days, her intestinal tract heals itself.

According to Katie, people who are unfamiliar with celiac disease should take the condition seriously and accommodate the special dietary needs of people who have the disease. For example, if you invite someone with celiac disease to share a meal or snack with you, wash your hands after you prepare foods that contain gluten and don't offer those foods to the person. Additionally, prepare some gluten-free foods for your guest who has celiac disease to eat safely. She says, "It's discouraging when people think I'm simply a 'picky eater' and just making life hard for them. The more people who understand celiac disease, the easier my life becomes!"

Gluten-free foods.
©Wendy Schiff

What Is PKU?

Phenylketonuria (*fen′–il–keet′–n–your′–ee–ah*), or PKU, is a rare genetic metabolic disorder that affects about 1 in 10,000 to 1 in 15,000 infants in the United States.[39] PKU and other *inborn errors of metabolism* occur when cells are unable to produce an enzyme necessary for a metabolic pathway. Children with PKU do not produce the enzyme that converts the essential amino acid phenylalanine (*fen–il–al′–ah–neen*) to other compounds. As a result, phenylalanine or its toxic by–products build up in tissues and damage cells, including nerve cells in the brain. If PKU is not diagnosed and treated within a few weeks of birth, the affected infant can develop intellectual disability within the first year of life.

In the United States, all states require newborns to undergo blood testing for PKU and several other treatable inherited diseases.[40] Infants who have PKU are generally given special formulas that lack phenylalanine. Phenylalanine is essential for growth and development; therefore, young children can consume a small amount of the amino acid. As the children grow and mature, fruits, vegetables, and special low–protein foods can be added to their formula diet. However, children and adults with the disorder need to avoid foods that are rich sources of phenylalanine, such as nuts, milk products, eggs, meats, and other animal foods. Additionally, people with PKU should not consume diet soft drinks and other foods and beverages containing the alternative sweetener aspartame because the sweetener is a source of phenylalanine (see Chapter 5 for information about aspartame).

In the United States, all newborns undergo a simple blood test to detect PKU and several other inborn errors of metabolism. In most instances, the blood sample is taken from the infant's heel ("heel prick").
Noor Haswan Noor Azman/Shutterstock

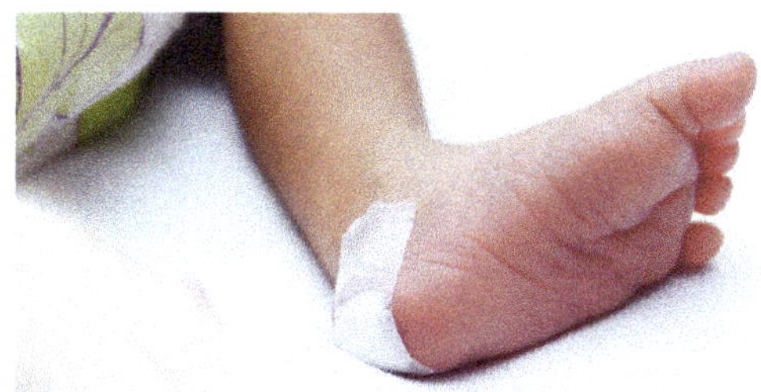

Throughout their lives, individuals with PKU may need to follow the restrictive diet, especially if they cannot take medication that helps their body metabolize phenylalanine. Furthermore, people with the disorder need to undergo frequent blood tests to make sure they are maintaining a healthy concentration of phenylalanine in their bodies. Monitoring blood phenylalanine is especially important during pregnancy. An expectant female who has PKU must carefully control her phenylalanine level to avoid exposing her embryo/fetus to excessive amounts of the amino acid. If she fails to control the concentration of phenylalanine in her blood, she can give birth to a baby with severe birth defects, including a smaller–than–normal brain. The second "Real People, Real Stories" in this section features Dallas Clasen, a young male who has PKU.

What Is Nutritional Genomics?

Human DNA has approximately 23,000 genes that code for protein synthesis ("the human genome").[41] **Nutritional genomics** is a relatively new science that investigates the complex interactions among gene functioning, dietary choices, and the environment. Such interactions influence a person's health status. Nutritional genomics includes **nutrigenetics,** the study of how a person's genetic makeup affects the way their body responds to food. For example, people who are genetically vulnerable to develop celiac disease are healthy until they consume foods that contain gluten. Nutritional genomics also involves **nutrigenomics,** the study of how nutrients and other food components can affect a person's *genetic expression*. Gene expression results in protein synthesis. As mentioned earlier, newborns with a particular genetic defect develop PKU when they consume amounts of phenylalanine that are harmless to babies who do not have the defect. By providing a diet that is low in phenylalanine to newborns with PKU, the infants develop normally.

nutritional genomics science that investigates the complex interactions among gene functioning, dietary choices, and the environment

nutrigenetics study of how a person's genetic makeup affects the way their body responds to food

nutrigenomics study of how nutrients and other food components can affect one's genetic expression

REAL PEOPLE | REAL STORIES

Mark A. Dierker/McGraw Hill

Dallas Clasen

Dallas Clasen recently graduated from Loras College with a degree in management information systems. While he was in college, Dallas ran cross country and participated in triathlons. Today, Dallas is very busy, but unlike most young adults, he has to pay very close attention to his diet. Dallas was born with a rare inherited metabolic disorder, an inborn-error of metabolism called phenylketonuria (PKU).

A few days after birth, Dallas underwent standard newborn blood testing. The results of the test indicated that the level of phenylalanine in his blood was about 40 times higher than the normal amount, a sign of PKU. To avoid developing severe brain damage and other physiological effects of PKU, Dallas was cared for by a physician who specializes in treating children with the disorder. The primary treatment for PKU is a low-phenylalanine diet.

Most foods that are rich sources of protein, especially high-quality animal proteins, contain more phenylalanine than people with PKU can tolerate. Thus, from the time Dallas was a week old, he has consumed a formula that does not contain the amino acid. In addition to the formula, Dallas can eat limited amounts of grain products and most fruits and vegetables. He also consumes special low-phenylalanine foods that resemble "regular" foods but are not available in supermarkets.

Foods that are eaten away from home can present problems for people with PKU. When at a restaurant, Dallas usually orders French fries or "sub" sandwiches that are made with vegetables but do not contain any cheese or meat. Instead of drinking milk, Dallas drinks his special formula or sports drinks that do not contain the nonnutritive sweetener aspartame (see Chapter 5). He is so accustomed to his special diet, he thinks meat looks "gross." In addition to following the special diet, Dallas takes Kuvan®, a prescription medication that stimulates the enzyme that converts phenylalanine to tyrosine. He takes 16 capsules after breakfast with apple juice. The medication does not help everyone with PKU, but it allows Dallas to consume more foods that contain phenylalanine. "I can eat double the amount of the amino acid," Dallas says. To determine whether his special diet and medication are working, each week Dallas pricks his finger and allows a few drops of blood to soak into a special paper. Then he sends the paper to a laboratory that checks the level of phenylalanine in his blood.

In the past, children with PKU were often allowed to eat regular foods after they were about 6 years of age. However, the importance of continuing the low-phenylalanine diet became evident when many of the children experienced learning and behavioral problems as they matured. Dallas is aware of the consequences that can occur if he does not limit his phenylalanine intake, and he accepts the need to follow the special diet for the rest of his life. "Being on a strict diet has not only made me disciplined, it has taught me to do whatever is needed to always take good care of myself," says Dallas. "I have learned that we are all different, anyway. So, accept who you are!"

People with PKU must restrict their phenylalanine consumption by consuming a diet rich in fruits and vegetables, some grains, and very little animal protein.
Evgeniia Trushkova/Shutterstock

Nutritional genomics may explain why special diets or dietary supplements can have different effects on the health of different individuals (Fig. 7.20). A young male, for example, may develop kidney stones while taking a daily dietary supplement that contains 1000 mg vitamin C. However, another male who is the same age and takes the same amount of vitamin C each day does not form kidney stones. Thus, the effectiveness of a specific diet, dietary supplement, medication, or lifestyle intervention may largely depend on one's genes. In the future, physicians may be able to consider such individual differences to provide more personalized treatments and preventive measures.

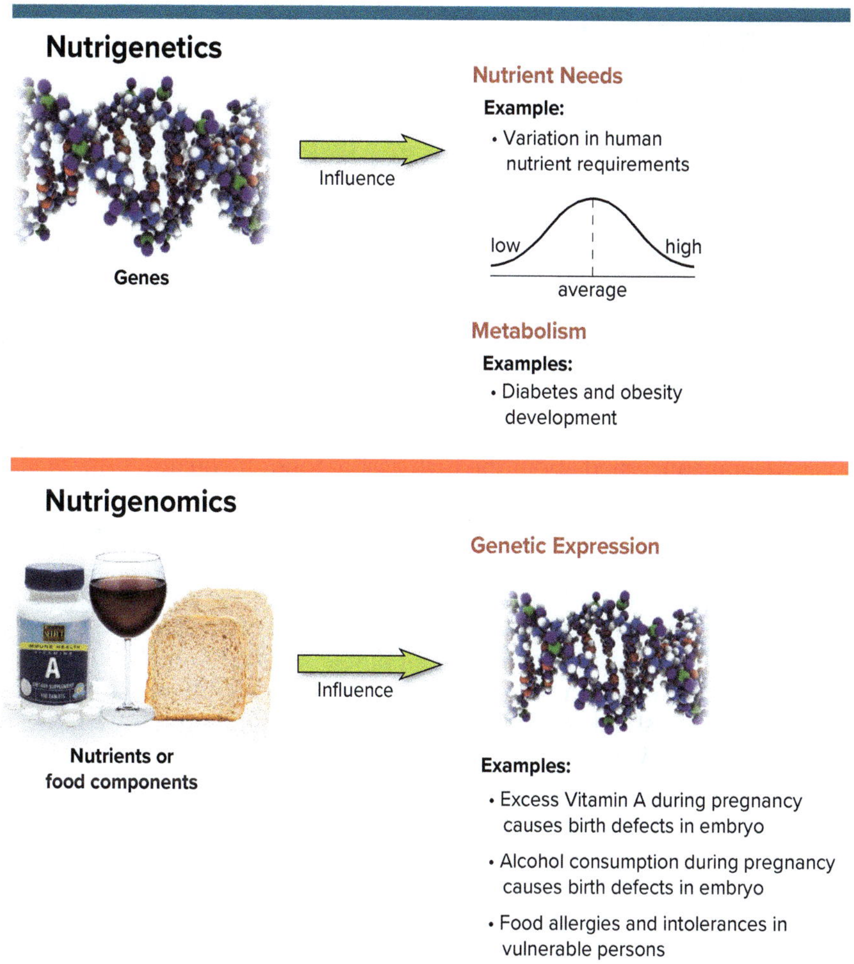

Figure 7.20 Nutritional genomics.
Vitamin A: John Flournoy/McGraw Hill; Glass of wine: Burke/Triolo Productions/Brand X Pictures/Getty Images; Bread: Ingram Publishing/Getty Images; DNA and DNA: Digital Vision/Getty Images

Genetic Testing Kits

Several biotechnology companies sell at–home genetic test kits for individuals to collect a small amount of saliva and mail the saliva sample to the company for genetic analysis. As a result of the saliva analysis, the company develops an "optimal health" plan based on the person's genetic makeup. Consumers, however, are encouraged to use caution when considering purchasing at–home genetic testing kits because the industry of nutrigenetic testing is not well regulated.[42] People who are interested in having genetic testing should consult their physician.

Concept CHECKPOINT

29. List three common signs or symptoms of food allergy and celiac disease.
30. Discuss what parents of infants with PKU can do to help their children grow and develop normally.
31. Explain how nutritional genomics may be used to improve a person's health.

See Appendix F for responses.

7.10 Nutrition Matters: Stretching Your Food Dollars

Learning Outcomes

1. Discuss practical ways to save money when purchasing food.
2. Compare similar foods to determine the most economical product.

Are you concerned about the high cost of food? If you are, how can you lower your food costs without sacrificing proper nutrition? You can trim food costs, and preserve and even improve the nutritional quality of your diet, by analyzing your food buying practices and making a few changes. This section focuses on ways you can eat well for less. Please note that the pricing of products reflects retail food prices during the summer of 2023 in a major Southern city, but you should find them useful for comparing current prices where you live.

Where and What You Buy

Where you buy foods and beverages and what you decide to purchase are major factors in determining your overall food expenditures. If you frequently purchase meals and snacks from vending machines, convenience stores, fast–food outlets, and other restaurants, you may be spending too much on such "convenience" foods. In general, the less time you spend preparing food, the more money you will spend by having someone at a commercial outlet prepare the food for you. Even if a fast–food restaurant hamburger seems like a bargain because it is "only $1.00," you are likely paying a premium for the French fries and soft drink that you purchase to accompany the sandwich. From a nutrition standpoint, fast foods are often energy dense—high in saturated fat and added sugars—and low in micronutrients (except sodium), fiber, and phytochemicals.

You probably already know that supermarkets have the same snack foods and staple items (bread and milk) as convenience stores and usually offer these foods at lower prices. Furthermore, supermarkets have a much larger selection of foods, particularly more nutrient–dense raw or less processed foods, than convenience stores. It may be easier to stop at a convenience store and load up on foods while you load up on gas than to shop in a supermarket. However, you pay extra for the convenience, and your diet may suffer as a result of poor food choices.

A good way to save money is to buy foods in large quantities at "bulk" food stores or at smaller, "no frills," cash–only grocery stores. Although shopping at these places can reduce your food costs, food choices are usually more limited than the selection in supermarkets. It is important to recognize that buying food in bulk is a bargain only if you can use the item before it spoils.

Stockbyte/Getty Images

How to Shop Wisely in Supermarkets

To lower your food costs, plan your meals and snacks before you shop for food. Prepare a shopping list to help avoid needless or impulse purchases and having to return to the store to buy forgotten items. Keep the grocery list in a convenient place, such as on the refrigerator, and jot down items as they become depleted.

When shopping for groceries, you will need to compare *unit costs* of similar food products. A "unit" of a packaged food item is usually indicated in ounces, pounds, cups, or other serving sizes. Food packages and containers are available in different sizes and hold different amounts of food, which can make comparing costs of similar foods difficult. Using a small calculator to compare prices can be helpful. For example, a box of one brand of toasted oat cereal contains 15.0 ounces of cereal and costs $3.85; a box of a different brand of toasted oats that contains 27.0 ounces of cereal costs $6.00. Which box contains cereal that is lower in cost? To answer this question, determine the unit price of the cereal in each box by dividing the cost of the entire package by the number of units of food it contains.

Box #1

$3.85 entire box ÷ 15.0 ounces = approximately $.25/ounce

Box #2

$6.00 entire box ÷ 27.0 ounces = approximately $.22/ounce

In this situation, the larger box of cereal is a better buy, even though it costs more than the smaller box. Some supermarkets indicate unit prices on shelf tags, but the tags may be difficult to read or confusing.

Supermarkets offer a wide variety of prepared foods, including deli items, frozen entrées, fresh cut–up fruit packaged in single–serving plastic bags or cups, and rotisserie chicken. Although these "fast foods" are convenient, food processing and packaging add to the cost of food. Such foods may contain unhealthy amounts of salt, solid fat, or sugar. To save money and improve your diet, rely less on prepared foods and more on "slow foods"—foods that require some of your time and effort to prepare. Figure 7.21 uses various forms of chicken to illustrate this point. A 2.5–lb rotisserie–cooked chicken costs $3.50/lb, but you can save money by purchasing a whole chicken ($1.50/lb) and cooking it yourself. Instead of buying raw, skinless, boneless chicken breasts ($3.99/lb), buy raw split chicken breasts that have the bones and skin ($1.99/lb). Remove the skin and bones after the chicken breasts are cooked. You can use the cooked chicken immediately, freeze the meat in containers for a future meal, or chill the meat before chopping it to make chicken salad.

Eating more fruits and vegetables is the natural way to add micronutrients, fiber, and phytochemicals to your diet. Many supermarkets offer fresh fruit and vegetables that have been trimmed, cut up, and packed into containers. Before you reach for one of these items, compare its price with that of its raw counterpart. For example, a 2–oz, single–serving package of sliced apples sells for $1.00, and a pound of raw apples is priced at $1.99. In this case, buying the raw apples and slicing them yourself is a money–saving practice. To make this determination, you will need to calculate the unit price of each product (Fig. 7.22). An ounce of the packaged apples costs about 45 cents ($1.00 divided by 2.2 oz/package). A pound of raw apples equals 16 ounces, so an ounce of these apples costs about 12 cents ($1.99 divided by 16). Thus, the prepackaged sliced apples cost almost four times more than the bulk raw apples.

Figure 7.21 Convenience—Is it worth it? You may save money by purchasing a raw whole chicken and cutting it into parts or a whole chicken breast and removing the skin and bones.

Roast chicken, Whole raw chicken, Raw skinless chicken, and Raw chicken breast: ©Wendy Schiff

Figure 7.22 Comparing apples to apples. The cost of a 2.2-oz package of apple slices (a) is almost five times the cost of the same amount of fresh, bulk, sliced raw apples (b).

Apple Bites, Sliced apple: ©Wendy Schiff

Produce that is grown in the United States is usually less expensive when it is in season and more plentiful. For example, berries are plentiful in the spring and early summer, tomatoes and melons in late summer, and apples in the fall. "Fresh" produce that is shipped long distances to supermarket suppliers is often picked before its peak ripeness. As a result, the fruits and vegetables may not taste as good or be as nutritious as locally grown produce. Produce, for example, loses some of its vitamin C content when it is bruised or becomes wilted. Therefore, consider buying fresh produce, for example, directly from local farmers, if possible. Although you still need to compare prices of farmers' market items with those of fresh produce sold in supermarkets, you may be able to find reasonably priced produce that is grown within 100 miles of your home at farmers' markets. Buying produce from local farmers helps support your community's economy, too.

An alternative to buying produce directly from growers is to purchase frozen or canned fruits and vegetables. These products are often less expensive than their fresh counterparts. Furthermore, produce that is destined to be frozen is generally harvested when ripe and frozen immediately, maximizing much of its vitamin content and fresh flavor.

The following practices can help you save money while shopping for food.

- Use unit pricing to compare costs of similar packaged products.
- Be wary of appealing offers that are not bargains. For example, items marked "3 for $1.00" or "10 for $10.00" can be confusing to consumers. It is important to understand that you do not need to buy the entire quantity. For example, if a store offers 3 kiwifruits for a dollar, and you cannot use 3 of them, buy one for 34 cents. That price is still a bargain compared to buying kiwifruits that are "2 for $1.00." Also, beware of offers such as "10 oranges for $10." Paying $1 for an orange is not a bargain! The same store may be selling a bag of 10 oranges for $5. At that price, each orange costs only 50 cents.
- Package sizes can be misleading, so ignore the size of a package and compare net weights of package contents to determine the best value.
- Compare the unit price of organic foods and similar conventionally produced items. If an organic food item is more expensive, consider purchasing the conventional product instead.
- Compare unit prices of preportioned snack foods such as chips or cookies in 100–Calorie packages with prices of the typical, larger package–size products. Such preportioned snacks may be convenient and helpful if you are trying to control your calorie intake, but they are usually more costly per ounce than the same foods purchased in larger packages. Furthermore, people may eat more of the preportioned snack foods because they think they are lower in calories than the same items packed in larger containers. Make your own preproportioned snacks. For example, buy a large package of oatmeal cookies and place two–cookie portions into reusable storage bags. When you want a snack, eat the two cookies.
- Use coupons when purchasing brand–name items. However, be aware that many stores offer similar products, often referred to as "store brands," that are less expensive than the brand–name items. Although store brand items do not have name recognition, these products are usually just as nutritious as their corresponding brand–name items.
- Limit your carbonated soft drink, commercially prepared coffee, energy drink, and bottled water purchases. When you are thirsty, locate the nearest water fountain and have a drink for free!
- Check "sell by," "best if used by," or "use by" dates on packages of perishable foods such as eggs and dairy products. Compared to canned and frozen foods, perishable foods lose their appealing sensory qualities, such as fresh taste and appearance, more rapidly during storage. The "sell by" date informs store employees about the length of time the product should be displayed for sale. Obviously, it is best to buy foods before the sell–by date. The "best if used by" or "use by" dates indicate the last day to use the product before it loses its best flavor and other sensory qualities. With the exception of infant formula, the date does not indicate the last day to purchase the food because of safety concerns.[43] Also, read food product labels for storage information. Many products must be refrigerated after opening to delay spoilage. Chapter 12 provides more information about proper food storage practices.
- Have at least one vegetarian meal a week.
- Red meats are among the most expensive protein sources, so have more meals in which chicken, turkey, or fish is the main source of animal protein.
- Canned fish can be a less expensive choice than fresh fish because there is less waste (skin and large bones).
- Remember that any food product is not a good buy if the item spoils before you can use it.

For more ways to save money on food, read the "Food & Nutrition Tips" at the end of this section.

Canned fish, such as tuna, salmon, and sardines, can be a less expensive choice than meat or fresh fish.
©Wendy Schiff

Food & Nutrition tips

- Foods that contain a lot of empty calories often look more appealing when your stomach is empty. To reduce the likelihood of buying such foods on impulse, avoid shopping for groceries when you are hungry.
- Be wary of buying food items that are displayed near or along checkout aisles. They are usually energy-dense foods that are not nutrient dense, such as candy bars, energy drinks, and baked goods.
- Many perishable foods, such as shredded cheese and fluid fat-free milk, can be frozen.
- If you drink coffee, buy a small coffeemaker, a can of ground coffee, and coffee filters to make the beverage. Consider carrying a small thermos of coffee in your backpack. It is much less expensive to make a cup of your own coffee than to buy one from a coffeehouse.
- Canned beans are often sold in large cans that may be cheaper per ounce than beans in smaller cans. If the larger-sized product is less expensive, buy it. After opening the large can of beans, remove the amount needed for the meal, and freeze the remainder in an air-tight container for a future meal.
- For fresh seasonings, grow your own herbs. Basil, oregano, parsley, cilantro, and thyme grow well in flower pots placed in a sunny area. Snip leaves as needed for recipes, while keeping the rest of the plants intact. Snipping off some leaves stimulates the plant to grow more leaves.
- Share potluck dinners with friends. Plan the menu and ask your friends to bring certain items, such as vegetable or fruit salads and whole-grain rolls or breads. Inexpensive entrees (the main food items on the menu) include spaghetti and meat sauce, chicken burritos, vegetarian chili, tuna and noodle casserole, and tacos made with canned tuna fish.

Prepare a shopping list to help avoid needless or impluse purchases.
FG Trade/E+/Getty Images

For fresh seasonings, grow your own herbs such as basil, oregano, parsley, and thyme.
©Wendy Schiff

Concept CHECKPOINT

32. List at least five ways to save money while shopping for food.
33. You can purchase a 3.5-pound bag of frozen sliced peaches for $7.99 or 3.5 pounds of fresh peaches for $2.49 per pound. Considering your situation, which is the more economical way to buy peaches? Explain why.

See Appendix F for responses.

Chapter References

See Appendix G.

Summary

7.1 What Are Proteins?

- Proteins are organic compounds that contain nitrogen, the element cells need to make a wide array of important biological compounds with structural or metabolic functions in the body. Additionally, a relatively small amount of protein contributes to the body's energy needs.
- Proteins are composed of amino acids. The typical amino acid has nitrogen-containing or amino, acid, and R groups. The diet must supply nine of the amino acids (essential amino acids) because the body cannot make them or make enough of them to meet its needs. Under normal conditions, cells can synthesize the remaining amino acids if the raw materials are available.

7.2 Proteins in Foods

- A high-quality or complete protein contains all essential amino acids in amounts that support growth in children. Most animal proteins are high quality. A low-quality or incomplete protein lacks or contains inadequate amounts of one or more of the essential amino acids.
- Tryptophan, lysine, threonine, methionine, and cysteine are often the limiting amino acids in foods.

Jack Hollingsworth/Photodisc/Getty Images

7.3 What Happens to Proteins in the Body?

- Human proteins are composed of 20 different amino acids arranged in various combinations. Cells produce proteins by linking amino acids together in specific sequences that are dictated by instructions coded in DNA. Faulty DNA results in the wrong amino acids being inserted into peptide chains, causing genetic defects. If an essential amino acid is not available when protein synthesis occurs, proteins in muscles and organs can provide the essential amino acids. Otherwise, protein synthesis halts, and the amino acids in the unfinished peptide are removed and returned to the amino acid pool. Excess amino acids are metabolized for energy or converted into body fat.
- Protein turnover is the process of breaking down old or unneeded proteins into their component amino acids and recycling them to make new proteins. The body conserves nitrogen by recycling amino acids, but each day, it loses some protein and nitrogen. In positive nitrogen balance, the body retains more nitrogen than it loses; in negative nitrogen balance, the body loses more nitrogen than it retains.
- The protein requirement increases during pregnancy, breastfeeding, periods of growth, and recovery from serious illnesses, blood losses, and burns. The adult RDA for protein is 0.8 g/kg of body weight.
- Protein digestion begins in the stomach, where hydrochloric acid denatures food proteins and pepsin breaks proteins into polypeptides. In the small intestine, enzymes secreted by the pancreas and absorptive cells digest polypeptides primarily into di- and tripeptides. The absorptive cells convert the peptides into amino acids. The end products of protein digestion, amino acids, travel to the liver. The liver uses the amino acids or releases them into the general circulation.

©Wendy Schiff

7.4 Protein Consumption and Health

- The AMDR for adults is 10 to 35% of energy intake from protein. Total protein consumption as a percentage of daily calories has remained about the same since the early twentieth century, but Americans now eat more animal protein. People can reduce their intake of animal protein without sacrificing the protein quality of their diets.

7.5 Understanding Nutritional Labeling: Protein

- The Nutrition Facts panel on a packaged food product label provides information concerning the amount of protein in a serving of the food.

7.6 Proteins: Economical Considerations

- The average American consumes more protein than the RDA for the nutrient.
- In general, meat, fish, poultry, eggs, milk, and milk products contain high-quality proteins. When compared to animal foods, most plant foods provide low-quality protein. Quinoa and foods made from processed soybeans are good plant sources of essential amino acids.

7.7 Vegetarianism

- Vegetarian diets are based on plant foods and limit animal foods to some extent. Although vegetarians are generally healthier than people who eat Western diets, it is difficult to pinpoint diet as responsible for vegetarians' better health.
- If not properly planned, plant-based diets may not contain enough energy, high-quality protein, omega-3 fatty acids, vitamins B-12 and D, and iron and calcium to meet a person's nutritional needs, especially children's needs.

7.8 Protein Adequacy

- High-protein diets are generally not recommended for healthy individuals.
- PEM affects people whose diets lack sufficient protein as well as energy; children are more likely to be affected by PEM than adults. In impoverished low-income countries, PEM is a major cause of childhood deaths.

7.9 Protein and Health: Environmental and Genetic Factors

- A food allergy is an inflammatory response that occurs when the body's immune system reacts inappropriately to an allergen. Accurate diagnosis of a food allergy should be undertaken by an immunologist. Treatment involves avoidance of foods that contain the allergens.
- Celiac disease is an inherited condition that causes malabsorption of nutrients. People with the disease cannot eat foods that contain gluten because the body's response to the protein damages villi in the small intestine. There is no cure for celiac disease, but people with the condition can be healthy as long as they avoid gluten-containing foods.
- Phenylketonuria (PKU) is a rare genetic metabolic disorder that occurs when cells cannot produce an enzyme that converts phenylalanine to other compounds. If PKU is not detected in early infancy, phenylalanine and its toxic by-products build up and damage tissues, including nerve cells in the brain. Treatment involves a special diet that contains limited amounts of phenylalanine and, in some cases, medication.
- Nutritional genomics is a science that investigates the complex interactions among gene functioning, dietary choices, and the environment. Nutritional genomics includes nutrigenetics, the study of how a person's genetic makeup affects the way their body responds to food. Nutritional genomics also involves nutrigenomics, the study of how nutrients and other food components can affect a person's genetic expression.

7.10 Nutrition Matters: Stretching Your Food Dollars

- In general, the less time a person spends preparing food, the more money they will spend by having someone else prepare the food. Convenience foods are often energy dense—high in solid fat and added sugar. A good way to save money is to buy foods in large quantities at "bulk" food stores, but buying food in bulk is a bargain only if the items can be used before they spoil.

Recipe for Healthy Living

©Wendy Schiff

Apricot Quinoa

The ancient Incas of South America cultivated quinoa and relied on the protein-rich vegetable as a dietary staple. Today, cooked quinoa is often added to vegetable salads or served as an accompaniment to chicken or pork. This quinoa recipe makes about two ½-cup servings. Each serving provides about 240 kcal, 7 g protein, 3 g fat, 4.5 g fiber, 3 mg iron, 2 mg zinc, and 80 mcg folate.

INGREDIENTS:

1⅓ cups water
½ cup quinoa
⅓ cup dried apricots, chopped
¼ tsp ground cinnamon
1 Tbsp maple syrup

PREPARATION STEPS:

1. Add water to a small saucepan and bring to a boil. Add quinoa to boiling water.
2. Cover saucepan and cook quinoa over medium heat until all the water is absorbed (about 12–15 minutes).
3. Remove from heat, fluff quinoa with fork, and stir chopped apricots into the quinoa.
4. Place quinoa into a bowl, drizzle with maple syrup, and sprinkle cinnamon on the quinoa.
5. If desired, garnish with fresh fruit. Serve warm.

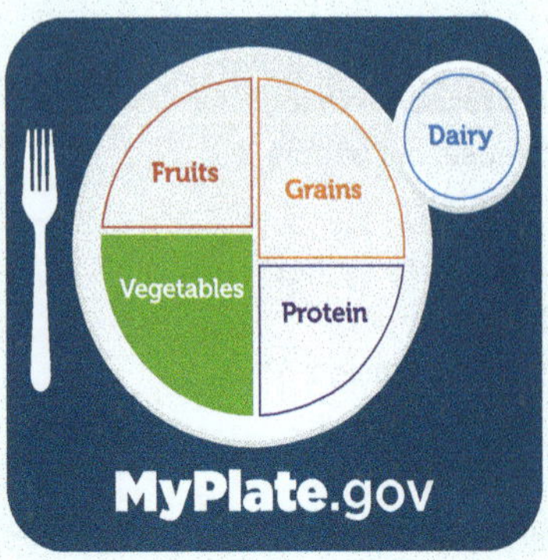

US Department of Agriculture or USDA

Approximate Percent of Calories from Macronutrients

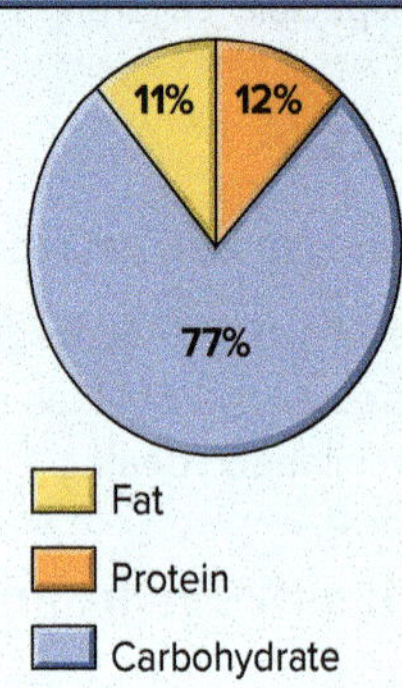

Percent of Daily Value/Serving

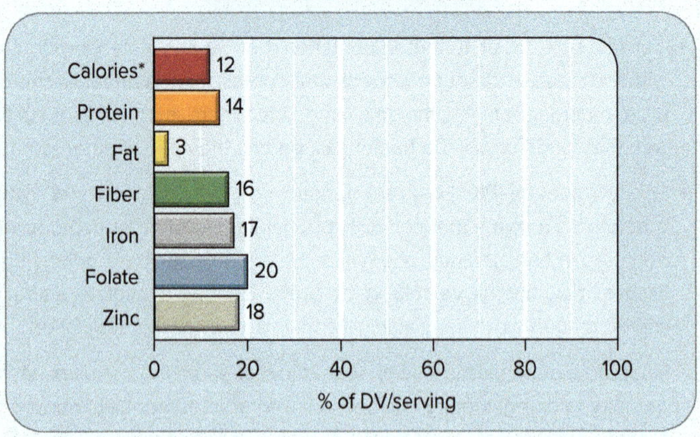

*2000 daily total kcal

Personal Dietary Analysis

1. Refer to the 3-day food log from the "Personal Dietary Analysis" feature in Chapter 3. Calculate your average protein intake by adding the grams of protein eaten each day, dividing the total by 3, and rounding the figure to the nearest whole number.

Sample Calculation:

Day 1 76 g

Day 2 55 g

Day 3 103 g

Total grams 234 g ÷ 3 days = 78 g of protein/day

Your Calculation:

Day 1 _____ g

Day 2 _____ g

Day 3 _____ g

Total grams _____ ÷ 3 days = _____ g/day

My average daily protein intake was _____ g.

2. The RDA for protein is based on body weight. Using the RDA of 0.8 g of protein/kg of body weight, calculate the amount of protein that you need to consume daily to meet the recommendation. To determine your body weight in kilograms, divide your weight (pounds) by 2.2, then multiply this number by 0.8 to obtain your RDA for protein. Then round the figure to the nearest whole number.

   My weight in pounds _____ ÷ 2.2 = _____ kg

   My weight in kg _____ × 0.8 = _____ g

   My RDA for protein = _____ g

   a. Did your average intake of protein meet or exceed your RDA level that was calculated in Step 1? _____ yes _____ no

   b. If your answer to 2a is yes, which foods contributed the most to your protein intake?

3. Review the log of your 3-day food intake. Calculate the average number of kilocalories that protein contributed to your diet each day during the 3-day period.

   a. Each gram of protein provides about 4 kcal; therefore, you must multiply the average number of grams of protein obtained in Step 1 by 4 kcal to obtain the average number of kcal from protein.

Sample Calculation:

78 g/day × 4 kcal/g = 312 kcal from protein

Your Calculation:

_____ g/day × 4 kcal/g = _____ average number of kcal from protein

Michael Lamotte/Cole Group/Photodisc/Getty Images

Bob Coyle/McGraw Hill

Lucy Stein/Image Source/Glow Images

4. Determine your average energy intake over the 3-day period by adding the kilocalories for each day and dividing the sum by 3, and round to the nearest whole number.

Sample Calculation:

Day 1 2500 kcal

Day 2 3200 kcal

Day 3 2750 kcal

Total kcal **8450 ÷ 3 days = 2817 kcal/day (average caloric intake)**

Your Calculation:

Day 1 ____ kcal

Day 2 ____ kcal

Day 3 ____ kcal

Total kcal ____ **÷ 3 days =** ____ **kcal/day (average)**

5. Determine the average percentage of energy that protein contributed to your diet by dividing the average kilocalories from protein obtained in Step 3 by the average total daily energy intake obtained in Step 4. Then round this figure to the nearest one-hundredth. Multiply this value by 100 (move the decimal point two places to the right, drop the decimal point, and add a percent symbol).

Sample Calculation:

312 kcal from protein ÷ 2817 kcal intake = 0.11 (rounded)

0.11 × 100 = 11%

Your Calculation:

____ kcal from protein ÷ ____ kcal intake = ____

____ × 100 = ____ %

6. Did your average intake of protein meet the recommendation of 10 to 35% of total calories? If your average protein intake was below 10%, list at least five foods you could eat that would boost your intake.

connect® Complete the Personal Dietary Analysis activity in Connect.

Critical Thinking

1. Have you used or are you currently using protein or amino acid supplements? _____ yes _____ no
   If you answered yes, explain why you use these supplements.
2. Are you a vegetarian? If so, describe your dietary practices (for example, vegan, lactoovovegetarian, or semivegetarian), and explain why you decided to become vegetarian. If you are not a vegetarian, explain why you would or would not consider this lifestyle.
3. Plan a day's meals and snacks for a healthy 132–pound (60 kg) adult lactoovovegetarian female who is not pregnant or breastfeeding. The menu should contain all essential amino acids but no animal foods other than eggs and foods from the dairy group. Your meal plan can range from 1800 to 2200 kcal, and it should follow the food group recommendations of the U.S. Department of Agriculture's MyPlate food guide (www.myplate.gov).
4. Using only plant foods, plan a day's meals and snacks for a healthy 154–pound (70 kg) adult vegan male. The menu should supply at least 2200 kcal and follow the food group recommendations of MyPlate except for the dairy group. Calcium–fortified soy milk may be used as a dairy substitute.

Ingram Publishing/Alamy Stock Photo

5. A recipe for bean salad has the following main ingredients:

   1 cup kidney beans
   1 cup green beans
   1 cup butter beans
   1 cup black beans
   1½ cups wine vinegar
   ⅓ cup canola oil
   ¼ cup chopped onion

   Explain why this recipe is not a complementary mixture of plant proteins. What plant foods could you add to the recipe to make it a complementary mixture?

Practice Test (Section numbers are listed below in red.)

Select the best answer.

1. (7.1) A protein
   a. is not a source of energy for the body.
   b. has fatty acids in its chemical structure.
   c. is composed of amino acids.
   d. is a complex inorganic molecule.
2. (7.1) Which of the following statements is false?
   a. Certain hormones are proteins.
   b. Nearly all enzymes are proteins.
   c. Proteins are part of triglycerides.
   d. The body uses amino acids to make antibodies.
3. (7.2) Which of the following foods generally provides the least amount of protein per serving?
   a. Fruits c. Nuts
   b. Milk d. Seeds
4. (7.2) Which of the following foods is not a source of complete protein?
   a. Peanut butter c. Fish
   b. Cheese d. Eggs
5. (7.3) In cells, _____ controls the assembly of amino acids into proteins.
   a. PEU c. IOM
   b. DNA d. ATP
6. (7.3) _____ is the process of removing nitrogen from an amino acid.
   a. Transamination c. Hydrogenation
   b. Denaturation d. Deamination
7. (7.3) Which of the following physical states is characterized by positive nitrogen balance?
   a. Starvation c. Puberty
   b. Illness d. Marasmus

8. (7.3) What is the RDA for protein of a healthy adult female who weighs 62 kg?
   a. 49.6 g
   b. 59.6 g
   c. 69.6 g
   d. 79.6 g
9. (7.4) There is strong scientific evidence linking red meat and processed red meat to which type of cancer?
   a. Colorectal
   b. Brain
   c. Lung
   d. Thyroid
10. (7.5) Which part of a food package can help you identify the quality of protein in a product?
    a. Grams of protein on the Nutrition Facts panel
    b. Net weight of the product
    c. Ingredient list
    d. %DV on the Nutrition Facts panel
11. (7.6) Which of the following foods is a source of complementary protein?
    a. Trail mix made with cashews, raisins, almonds, honey, and dried banana flakes
    b. Mashed chickpeas spread on pita bread
    c. Leafy greens salad topped with peanuts
    d. Whole–wheat bread with an apricot spread
12. (7.7) Which of the following foods is a source of high–quality protein that a vegan would eat?
    a. Quinoa
    b. Yogurt
    c. Brown rice
    d. Sweet potatoes
13. (7.8) A sign of marasmus is
    a. increased muscle growth.
    b. rapid weight gain.
    c. diabetes.
    d. wasting.
14. (7.8) Noemi is a registered dietitian nutritionist working in Texas with a client who has protein deficiency. The client is most likely
    a. a school–age child who is tall for his age.
    b. a 42–year–old female college math instructor who has a head cold.
    c. an athletic young male who is training for a triathlon.
    d. an older adult female who has memory loss.
15. (7.9) Nutritional genomics is the study of how genes
    a. vary according to a person's sex.
    b. interact with a person's diet and environment.
    c. determine what a person should eat.
    d. limit which medications and dietary supplements a person can safely take.
16. (7.9) People with celiac disease should
    a. take amino acid supplements.
    b. limit their protein intake to 20 g per day.
    c. avoid foods that contain gluten.
    d. eliminate protein from plant sources.
17. (7.10) Which of the following tips is a recommended way to reduce food costs?
    a. Buying food in bulk, if it can be used before spoiling
    b. Purchasing preportioned, packaged foods
    c. Shopping at convenience stores, if the stores are within 5 miles of your home
    d. Eating at fast–food outlets at least three times a week

Answers to Quiz Yourself

1. Animal foods such as meat and eggs are almost 100% protein. **False.** (Section 7.2)
2. Foods made from soybeans can be sources of high-quality protein. **True.** (Section 7.2)
3. Americans typically consume more protein from animal sources than from plant foods. **True.** (Section 7.6)
4. Registered dietitian nutritionists generally recommend that people with liver or kidney disease take amino acid supplements to increase their protein intake. **False.** (Section 7.8)
5. People can nourish their hair by using shampoo that contains protein. **False.** (Section 7.3)

Tatjana Baibakova/Shutterstock

Design Elements: (Stacked books): Siede Preis/Photodisc/Getty Images; (Nuts): C Squared Studios/Photodisc/Getty Images; (Recipe box, Raspberries, Blueberries): ©Wendy Schiff; (Pineapple): Stockbyte/Getty Images; (Radishes): karandaev/iStock/360/Getty Images; (MyPlate icon): US Department of Agriculture or USDA

CHAPTER 8

Vitamins

Quiz Yourself

Can vitamins give you more energy or reduce your chances of developing heart disease or cancer? Which cooking methods can increase the loss of vitamins from foods? Are vitamin deficiency disorders common in the United States? Test your knowledge of vitamins by taking the following quiz. The answers are at the end of the chapter.

1. Natural vitamins are better for you because they have more biological activity than synthetic vitamins. ______ T ______ F
2. Certain vitamins can be toxic if consumed in high doses. ______ T ______ F
3. Vitamin E is an antioxidant. ______ T ______ F
4. Vitamins are a source of "quick" energy. ______ T ______ F
5. According to scientific research, taking large doses of vitamin C daily prevents the common cold. ______ T ______ F

8.1 Vitamins: Basic Concepts

Learning Outcomes

1. Explain why all vitamins have been discovered.
2. Define *vitamin,* and explain how scientists determine whether a substance is a vitamin.
3. Classify vitamins according to whether they are fat soluble or water soluble.
4. Discuss ways to conserve the vitamin content of foods or increase vitamin bioavailability.
5. Explain the function of an antioxidant.
6. Explain the purpose of vitamin enrichment and fortification.

scurvy vitamin C–deficiency disease

For centuries, taking lengthy ocean voyages was a dangerous venture, not just because of the threat of severe storms and pillaging pirates but also because of a terrifying and deadly disease called **scurvy.** The first signs and symptoms of scurvy—fatigue and *petechiae* (*peh–tee'–key–eye*), pinpoint hemorrhages under the skin—occurred about 20 to 40 days after setting sail. As the disease progressed, the affected person's skin bruised easily; gums swelled, became spongy, and bled after being barely touched; teeth loosened and fell out. Wounds healed slowly or remained open sores. Not surprisingly, the person suffering from scurvy also became irritable and depressed. A particularly devastating sign of the disease was the opening up of old scars, exposing wounds that could become infected. Scurvy victims eventually died, generally from infections, brain hemorrhages, or heart complications.

In 1753, British physician James Lind published an article describing an experiment he had performed on 12 sailors suffering from scurvy. Lind divided the sick sailors into six pairs, and each pair received a different treatment. The six treatments were cider, vinegar, sulfuric acid, seawater, nutmeg, or oranges and lemons. Lind observed that the pair of sailors given the citrus fruit were the only ones to recover from scurvy. By today's standards, Lind's experiment was primitive, but as a result of his testing, Lind found the cure for scurvy: eating oranges and lemons. Eventually, food rations for British sailors included lemon juice to prevent the disease. The sailors earned the nickname "limeys" because at that time, people often referred to citrus fruits collectively as "limes."

Today, we know that scurvy results from a deficiency of vitamin C and that citrus fruits are among the richest dietary sources of the vitamin. Although Lind is often credited with having discovered the cure for scurvy, he did not suspect the disease resulted from the lack of something in the typical seafarer's diet. At that time, scientists were unaware that food contained vitamins. Lind thought scurvy was a digestive system disorder that could be treated with substances associated with warm climates.[1] In his experiment, Lind happened to administer citrus fruits to a pair of the sailors with scurvy because these fruits were associated with such climates.

Citrus fruits, including oranges, are rich sources of vitamin C.
Hortimages/Shutterstock

In 1911, Polish chemist Casimir Funk discovered a substance in an extract made from rice bran that he thought would cure the disease beriberi. Funk called the compound a "vitamine" (*vita* = necessary for life; *amine* = a type of nitrogen–containing substance) because of its chemical structure. The term *vitamine* was later modified to *vitamin* when scientists determined that there were several kinds of these substances in foods and not all were amines. By the end of the 20th century, scientists had added riboflavin, niacin, biotin, B–6, B–12, pantothenic acid, folate, ascorbic acid, A, D, E, and K to the list of vitamins. Humans also require choline, especially during *prenatal* (before birth) development. Like vitamin D, the body can make choline, but under certain conditions, the body does not synthesize enough to meet its needs. Choline is considered to be a *vitamin–like* essential nutrient.

It is unlikely that any vitamins still need to be discovered. Why? Babies grow and thrive on infant formulas, synthetic liquid diets containing vitamins and other nutrients known to be essential for health. Additionally, very ill people who cannot eat solid food can be kept alive for years on liquid synthetic feedings that contain all known nutrients, including vitamins. If a vitamin remained undiscovered, infants and people who are unable to consume solid foods would not be able to survive on formula diets. This chapter presents information about the 13 vitamins and choline, including physiological roles and major food sources.

What Is a Vitamin?

What is a vitamin? A **vitamin** is a complex organic compound that regulates certain metabolic processes in the body. A vitamin meets the following criteria:

- The body cannot synthesize the compound or make enough to maintain good health.
- The compound naturally occurs in commonly eaten foods.
- Signs and symptoms of a health problem (*deficiency disorder*) eventually occur when the substance is missing from the diet.
- Good health is restored if the deficiency disorder is treated early by supplying the missing substance.

Foods generally contain much smaller amounts of vitamins than of macronutrients. A slice of whole-wheat bread, for example, weighs 28 grams. Of that weight, only about 0.005% (1.48 mg) is composed of vitamins; carbohydrate, water, protein, fat, and minerals make up the bread's remaining weight. Furthermore, the body requires vitamins in milligram or microgram amounts, but it needs grams of macronutrients.

Amounts of certain vitamins may be expressed in *International Units* (*IUs*), but the use of IUs is being replaced by more precise measures. One microgram of vitamin D, for example, equals 40 IUs of the vitamin.

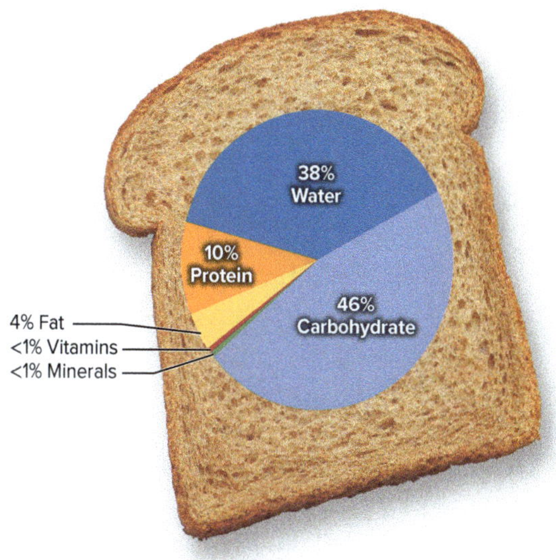

A slice of bread weighs about 1 ounce (28 g). Vitamins comprise only about 0.005% (1.48 mg) of the weight of the bread.
Jules Frazier/Photodisc/Getty Images

Roles of Vitamins

Vitamins play numerous roles in the body, and each of these micronutrients generally has more than one function (Fig. 8.1). Some vitamins, such as vitamin D, act as hormones; other vitamins, such as thiamin, riboflavin, and niacin, are parts of special molecules that enzymes need to function. In general, vitamins regulate a variety of body processes, including those involved in cell division and development, as well as the growth and maintenance of tissues.

Advertisements for vitamins often promote the notion that the micronutrients can "give" you energy. Vitamins, however, are not a source of energy because cells do not metabolize them for energy. Although the body does not use vitamins directly for energy, many vitamins participate in the chemical reactions that release energy from glucose, fatty acids, and amino acids. Figure 8.2 presents a simplified view of energy metabolism and indicates vitamins that are involved in the various steps of the process.

Most vitamins have more than one chemical form that functions in the body. For example, *retinol, retinal,* and *retinoic acid* are chemically related types of vitamin A that have roles in the body. Additionally, some vitamins have precursors (*provitamins* or *previtamins*) that do not function as vitamins until the body converts them into active forms. For example, the plant pigment beta-carotene is a provitamin for vitamin A, and vitamin D_3 is a previtamin for vitamin D.

vitamin complex organic molecule that regulates certain metabolic processes

oxidizing agent substance that removes electrons from atoms or molecules

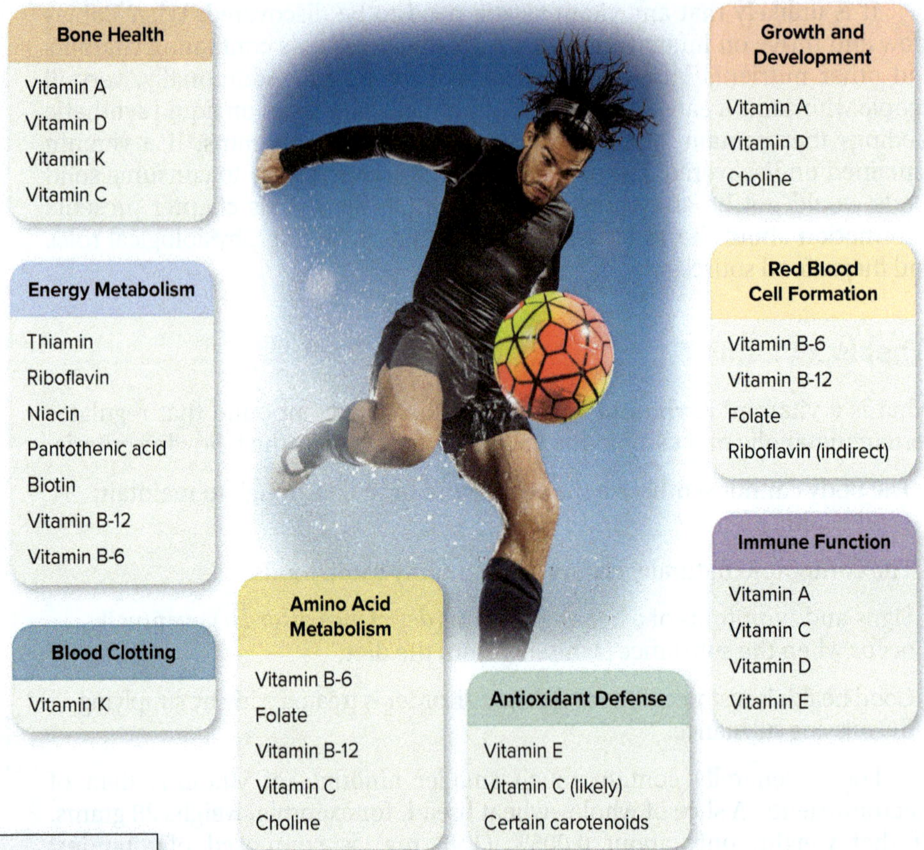

Figure 8.1 Functions of vitamins and related compounds. Groups of vitamins and related compounds (e.g., choline and certain carotenoids) work together to maintain good health.
Erik Isakson/Blend Images LLC

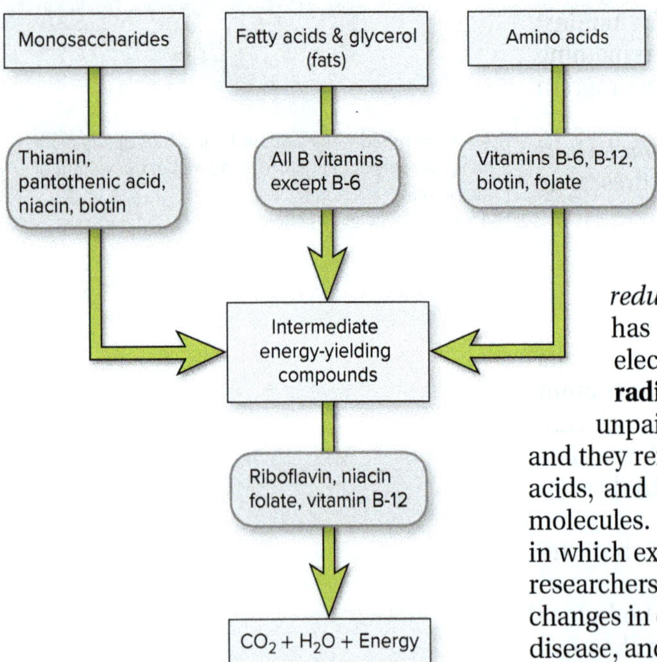

Figure 8.2 The metabolism of monosaccharides, fats, and amino acids for energy requires several vitamins.

radical substance with an unpaired electron

antioxidant substance that gives up electrons to radicals to protect cells

What Is an Antioxidant?

When many biochemical reactions take place, the compounds participating in the reactions lose or gain electrons. When an atom or molecule gains one or more electrons, it has been *reduced.* When an atom or molecule loses one or more electrons, it has been *oxidized.* An **oxidizing agent** is a substance that removes electrons from atoms or molecules. An oxidation reaction can form a **radical** (commonly referred to as a "free radical"), a substance with an unpaired electron. Radicals are highly reactive (chemically unstable), and they remove electrons from more stable molecules, such as proteins, fatty acids, and DNA (Fig. 8.3). As a result, radicals can damage or destroy these molecules. If the loss of electrons is uncontrolled, a chain reaction can occur in which excessive oxidation takes place and affects many cells. Many medical researchers suspect excess oxidation is responsible for promoting chemical changes in cells that ultimately lead to heart attack, stroke, cancer, Alzheimer's disease, and even the aging process.

Some radical formation in the body is necessary and provides some benefits.[2] Radicals, for example, stimulate normal cell growth and division. Additionally, white blood cells generate radicals as part of their activities that destroy infectious agents. Under normal conditions, cells regulate oxidation reactions by using antioxidants such as vitamin E. **Antioxidants** protect cells by giving up electrons to radicals. When the chemically unstable substance accepts an electron, it can form a more stable structure that does not pull electrons away from other compounds. By sacrificing electrons, antioxidants protect molecules such as polyunsaturated fatty acids in the membrane or DNA in the nucleus from being oxidized (Fig. 8.4).

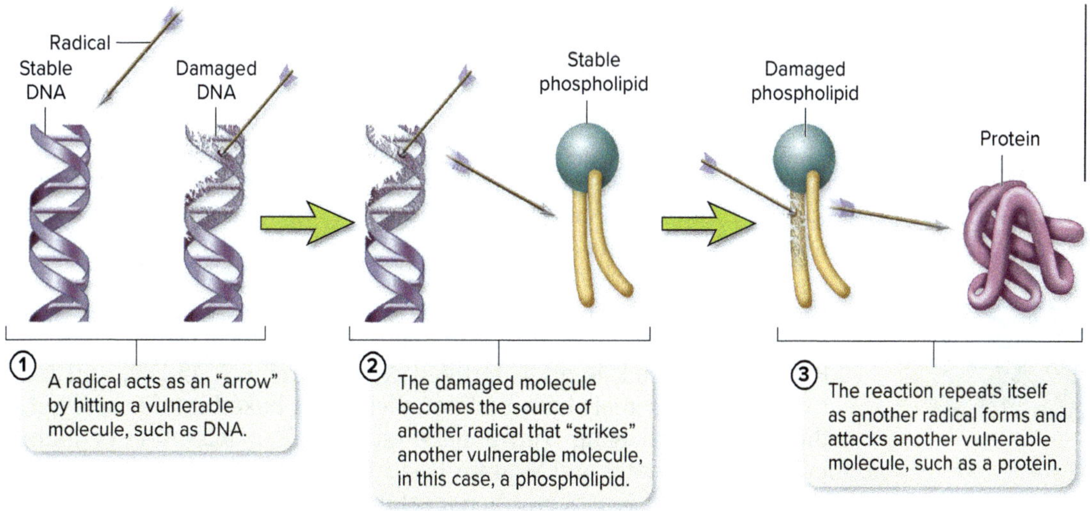

Figure 8.3 What is a radical? A radical is a highly reactive substance because it has an unpaired electron. Radicals remove electrons from more stable molecules, such as DNA in the cell's nucleus.

Sources of Vitamins

Plants, animals, fungi, and even bacteria supply natural forms of vitamins in our diets. In addition to foods, vitamin supplements are another source of these micronutrients. Although chemists can synthesize vitamins, certain types of bacteria and algae produce vitamins. These organisms can be grown in laboratory settings for the purpose of "harvesting" their vitamins to use in supplement production.

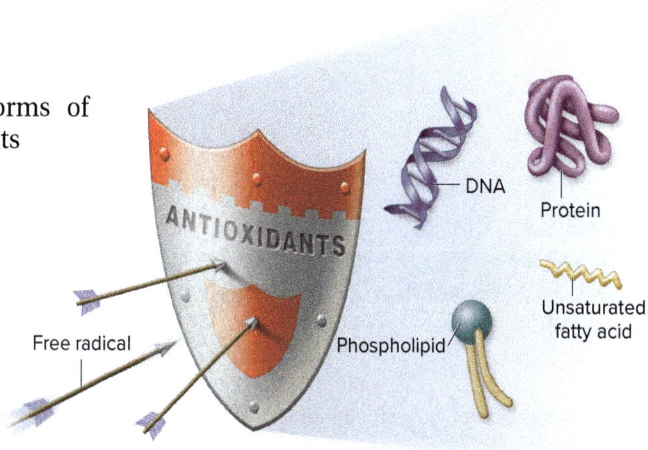

Figure 8.4 Antioxidant action. Antioxidants can protect the cell's plasma membrane and DNA from radicals.

Regardless of whether a particular vitamin is naturally in foods or synthesized in a laboratory, it generally has the same chemical structure and works equally well in the body—but there are exceptions. The natural form of vitamin E has more **biological activity,** that is, it produces more effects in the body, than synthetic vitamin E. On the other hand, *synthetic folic acid,* the type of folate that is added to flour and many ready–to–eat and cooked cereals, has almost twice the biological activity as the natural form of the vitamin.

Many adult Americans take multivitamin supplements that contain two or more vitamins and may also supply one or more minerals (*multivitamin/mineral supplement*). According to data collected in the 2017–2018 National Health and Examination Survey, 57% of adults reported regular use of a multivitamin/mineral supplement.[3] Such products are often marketed as "one–a–days" or for a particular target audience, such as males, people with diabetes, athletes, or older adults. However, recall from chapter 1 that there are no significant benefits for healthy people who take dietary supplements.

It is not necessary to consume 100% of every vitamin each day. If you are healthy and usually follow a nutritionally adequate diet, your cells should contain a supply of vitamins that can last for several days and possibly even years, depending on the vitamin. Furthermore, bacteria that reside in your lower intestinal tract produce certain vitamins, particularly biotin and vitamin K, and you can absorb these micronutrients to some extent. Additionally, your body can synthesize vitamin D and niacin under certain conditions.

Vitamin Enrichment and Fortification

In Chapter 3, we introduced the topic of micronutrient enrichment and fortification. In the United States, the federally regulated grain enrichment program specifies amounts of four B vitamins (thiamin, riboflavin, niacin, and folic acid) that manufacturers must add to their refined wheat flour and certain

biological activity describes vitamin's degree of potency or effects in the body

This refined wheat flour product (orzo) is enriched with vitamins thiamin, niacin, riboflavin, and folic acid and the mineral iron.
©Wendy Schiff

other milled grain products if they choose to participate in the program. Grain enrichment helps protect Americans from developing the deficiency diseases associated with the lack of these vitamins. Grain enrichment, however, does not replace vitamin E, vitamin B–6, and several other micronutrients that were lost from the grains during refinement. This is the major reason why registered dietitian nutritionists and other nutrition experts promote regular consumption of whole–grain products, such as whole–wheat bread and brown rice.

Fortification involves the addition during manufacturing of one or more vitamins (and/or other nutrients) to a wide array of commonly eaten foods. The vitamins that are added may or may not be in the food naturally. For example, milk is often fortified with vitamins A and D, and many ready–to–eat cereals have additional vitamins baked into or sprayed on them before packaging. Nutritional yeast is usually a rich source of thiamin, riboflavin, niacin, folate, and vitamin B–12 that adds a cheese–like flavor to foods. However, most Americans do not eat the vitamin–fortified product.

In the United States, vitamin fortification and enrichment of foods have improved vitamin intakes of Americans. However, some food manufacturers add vitamins to foods and beverages, particularly flavored, sugary drinks, that would otherwise be considered sources of empty calories. Many nutrition experts are concerned that by substituting such human–made products for more natural foods and beverages, Americans may consume excessive amounts of a few vitamins while reducing their intake of others.

TABLE 8.1 Classifying Vitamins

Fat-Soluble Vitamins
A (retinol)
D
E (alpha-tocopherol, other tocopherols)
K
Water-Soluble Vitamins
Thiamin (thiamine, B-1)
Riboflavin (B-2)
Niacin (B-3, nicotinamide, nicotinic acid)
B-6 (pyridoxine)
B-12 (cobalamin, cobalamine)
Biotin (H)
Pantothenic acid (B-5)
Folate (B-9, folic acid)
C (ascorbic acid)
Vitamin-Like
Choline

Classifying Vitamins

Vitamins A, D, E, and K are **fat–soluble vitamins.** These vitamins are in the lipid portions of foods and tend to associate with lipids in the body. Thiamin, riboflavin, niacin, vitamin B–6, pantothenic acid, folate, biotin, vitamin B–12 (collectively known as the B vitamins), and vitamin C are **water–soluble vitamins.** Water–soluble vitamins dissolve in the watery components of food and the body. (Choline is also water soluble.) Table 8.1 presents the vitamins and provides some other names that may be used to identify them.

Why is it important to know the difference between fat– and water–soluble vitamins? The body generally has more difficulty eliminating excess fat–soluble vitamins because these nutrients do not dissolve in watery substances such as urine. As a result, the body stores extra fat–soluble vitamins, primarily in the liver and in body fat. Over time, these vitamins can accumulate and cause toxicity, especially vitamins A and D. On the other hand, the body stores only limited amounts of most water–soluble vitamins; vitamin B–12 is an exception. Furthermore, kidneys can filter excesses of water–soluble vitamins from the bloodstream and eliminate them in urine. Thus, compared to fat–soluble vitamins there is a lower risk of developing toxicity from water–soluble vitamins.

Vitamin Absorption

The small intestine is the primary site of vitamin absorption. However, the intestine does not absorb 100% of the vitamins in food. Vitamin absorption tends to increase when the body's needs for the micronutrients are also higher than usual. The body's requirements for vitamins generally increase during periods of growth, such as infancy and adolescence, and during pregnancy and breastfeeding.

Fat–soluble vitamins are chemically similar to lipids, and the vitamins are in fatty portions of food. Thus, processes that normally occur during fat digestion facilitate the absorption of fat–soluble vitamins. For example, bile enhances lipid digestion, as well as fat–soluble vitamin absorption. In the small intestine, the presence of fat stimulates the secretion of a hormone that causes the gallbladder to release bile. Therefore, adding a small amount of fat

fat-soluble vitamins vitamins A, D, E, and K

water-soluble vitamins thiamin, riboflavin, niacin, vitamin B-6, pantothenic acid, folate, biotin, vitamin B-12, and vitamin C

to low–fat foods, such as tossing raw vegetables with some salad dressing, adding a pat of soft margarine to steamed carrots, or stir–frying green beans in peanut oil, can enhance the intestinal tract's ability to absorb the fat–soluble vitamins in these foods. To review lipid digestion, see Chapter 6.

Diseases or conditions that affect the GI tract can reduce vitamin absorption and result in deficiencies of these micronutrients. People with the inherited disease *cystic fibrosis* (*sis'–tik fie–bro'–sis*) are unable to digest fat properly because the disease causes blockages to form in ducts that convey pancreatic enzymes to the small intestine. As a result, cystic fibrosis reduces fat absorption, and people with the disease often develop deficiencies of fat–soluble vitamins. People who are unable to absorb vitamins may need to take large oral doses of vitamin supplements just to enable small amounts of the vitamins to be absorbed. In other cases, physicians inject vitamins into their patients' bodies, completely bypassing the need for the intestine to absorb the micronutrients.

Toss raw vegetables with an oil-based salad dressing to enhance absorption of fat-soluble vitamins in the intestinal tract.
©Wendy Schiff

Vitamin Deficiency and Toxicity Disorders

A diet that contains adequate amounts of a wide variety of foods, including minimally processed fruits, vegetables, and whole–grain breads and cereals, can help supply the vitamin needs of most healthy people. Vitamin deficiency disorders generally result from inadequate diets or conditions that increase the body's requirements for vitamins, such as reduced intestinal absorption or higher–than–normal excretion of the micronutrients. Today, severe vitamin deficiencies are uncommon in the United States, thanks in part to modern food preservation practices, food enrichment and fortification, and the year–round, widespread availability of fresh fruits and vegetables from other countries. Many Americans, however, consume considerably less than recommended amounts of certain vitamins, particularly vitamins E, D, and the vitamin–like compound choline.[4]

A few segments of the population have a high risk of vitamin deficiencies. These vulnerable people include people who have alcohol use disorder, older adults, and patients who are hospitalized for lengthy periods. Additionally, people who have anorexia nervosa, intestinal conditions that interfere with vitamin absorption, or rare metabolic defects that increase their vitamin requirements are more likely to develop vitamin deficiency disorders than people who do not have these conditions.

If your usual diet is nutritionally adequate but you occasionally have low intakes of vitamins, you are unlikely to develop vitamin deficiency diseases because your cells store these micronutrients to some extent. The likelihood of developing a deficiency disease increases when a person's diet consistently lacks the vitamin. When this happens, the person's body stores or tissue levels of the vitamin become depleted, and the signs and symptoms of the nutrient's deficiency disease begin to occur. A person, for example, can become vitamin C–deficient after about a month of consuming a diet that provides little or no vitamin C.[5]

If you are considering taking vitamin supplements or are taking them already, you need to recognize that more is not necessarily better. When cells are saturated with a vitamin, they contain all they need and cannot accept additional amounts of the micronutrient. When this situation occurs, continuing to take the vitamin can produce a toxicity disorder because exposure to the excess micronutrient or its by–products can damage cells.

Do you need to be concerned about developing a vitamin toxicity disorder? Probably not, unless you are taking excessive amounts (megadoses) of vitamin supplements or consuming large amounts of vitamin–fortified foods regularly. In their natural states, most commonly eaten foods do not contain toxic levels of vitamins. Taking a "one–a–day" type of multivitamin supplement regularly is unlikely to cause toxic effects in healthy adults because these products usually contain less than two times the Daily Values of each micronutrient component.

Did *YOU* Know?

A plant's vitamin content is largely determined by its genetic makeup; growing conditions, including soil composition and sunlight exposure; and maturity when harvested.

V J Matthew/Shutterstock

Food & Nutrition tips

Fresh vegetables naturally contain enzymes that, over time, contribute to the deterioration of the produce's quality, even when it is frozen. To retain desirable flavors, colors, and textures of most vegetables that will be frozen, you should *blanch* them before freezing.

Blanching involves immersing small batches of cut up fresh vegetables in boiling water for a short period. This application of heat denatures the plant enzymes that would otherwise cause the vegetables to deteriorate. To avoid overcooking, stop the blanching process by removing the produce from the hot water and plunging it immediately into a large container of very cold water. Before packing the food into freezer containers, drain as much water as possible from the vegetables.

©Wendy Schiff

Preserving the Vitamin Content of Foods

Regardless of whether you pick fruits and vegetables from your own garden or buy them from a farmers' market or supermarket, many kinds of produce, especially berries and leafy vegetables, are highly perishable. Therefore, these foods should be eaten as soon as they are harvested or purchased to ensure maximum vitamin retention. In many instances, unpackaged ("bulk") fresh fruits or vegetables that are sold in supermarkets do not have dates indicating when they should be used, and consumers have no way of knowing when the produce was harvested.

Fresh fruits and vegetables can lose substantial amounts of vitamins as a result of improper handling or lengthy storage conditions. Therefore, select fresh produce carefully when buying it in grocery stores. Avoid produce that is bruised, wilted, or shriveled, or shows signs of decay such as mold. If you are uncertain how to choose ripe fruits and vegetables, ask the person who manages the produce section of the store for advice.

Some vitamins, such as niacin and D, resist destruction by usual food storage conditions or preparation methods. Other vitamins—particularly vitamin C, thiamin, and folate—are easily destroyed or lost by improper food storage and cooking methods. Fresh produce is more likely to retain its natural vitamin content when stored at temperatures near freezing, in high humidity, and away from air. Therefore, you should keep most fresh fruits and vegetables in plastic packaging and chilled until you are ready to use them. Tomatoes, bananas, and garlic should be stored at room temperature. Although precut, packaged salad greens and other vegetables may be convenient to use, they are highly perishable and should be used soon after purchasing.

Exposure to excessive heat, alkaline substances, light, and air can destroy certain vitamins, especially vitamin C. To reduce such losses, trim, peel, and cut raw fruits and vegetables just before eating or serving them. The darker leaves of vegetable greens generally contain more vitamins than the inner, paler–color leaves or stems. Therefore, lightly trim away the outer leaves of lettuce and cabbage, and keep edible peels intact; just remove rotten or shriveled parts.

Water–soluble vitamins can leach out of food and dissolve in the cooking water, which is often discarded. By cooking vegetables in small amounts of water and reusing that water for soups or sauces, you are likely to consume those water–soluble nutrients. When preparing produce for cooking, cut the food into large pieces to reduce the amount of surface area that will be exposed to heat, water, and other conditions that can increase vitamin losses. Whenever possible, cook fruits and vegetables in their skins. If the skins are edible, eat them too because they are rich in fiber, vitamins, and minerals.

Quick cooking methods that involve little contact between produce and water, such as microwaving, steaming, and stir–frying, can conserve much of the vitamin content of the food. According to the Food and Drug Administration (FDA), microwave cooking does not reduce the nutrient content of foods any more than do conventional cooking methods.[6] Microwave cooking may help conserve more vitamins in food because the method cooks quickly and without the need to add much water.

To steam vegetables, place them in a steamer basket that fits inside a pot, add enough water to touch the bottom of the basket, cover the pot, and then heat the water until it boils (Fig. 8.5). As the steam gently

A farmers' market can be a source of locally grown fresh produce during the growing season.
Bill Tarpening/USDA

cooks the vegetables, add more water, if necessary. To stir–fry vegetables, heat a small amount of oil in a wok or pan that has deep sides, add small pieces of fresh vegetables, and stir the mixture, lightly coating the vegetables with oil. Stir–fried vegetables should be cooked until they are barely tender to retain their nutrients, as well as appealing textures, flavors, and colors. The "Recipe for Healthy Living" feature of this chapter has a recipe for stir–fried vegetables.

Are fresh fruits and raw vegetables better sources of vitamins than canned or frozen versions? Sometimes they are. During the canning process, the heating of food can cause the destruction of certain vitamins. However, produce that is frozen immediately after being harvested and then properly stored can be just as nutritious as fresh produce. Frozen fruits and vegetables are often economical alternatives to fresh produce, but they need to be cooked without thawing to conserve much of their vitamin content. The thawing process causes some of the water that was naturally in the produce to drip out, taking water–soluble vitamins with it. If this water is discarded, vitamins in the fluid are lost. The following "Food & Nutrition Tips" provide practical ways to preserve the vitamin content of food.

Figure 8.5 Conserving vitamins. Steaming vegetables can conserve much of the vitamin content of the produce.
©Wendy Schiff

Food & Nutrition tips

Ways to preserve the vitamin content of food include:

- Eat fresh fruits and vegetables along with their vitamin-rich edible peels or skins whenever possible.
- Cook fresh or frozen vegetables by microwaving, steaming, or stir-frying to minimize vitamin losses from cooking. Vegetables generally have high water content; therefore, add no water or just a small amount when microwaving vegetables.
- Do not overcook vegetables and minimize reheating because prolonged heating reduces vitamin content.
- Do not add margarine or butter to vegetables during cooking because fat-soluble vitamins and phytochemicals may enter the fat and be discarded when the fat is drained before serving. Fat in foods can enhance the body's absorption of fat-soluble vitamins; therefore, you can add some fat, such as olive oil or soft margarine, to vegetables after they are cooked.
- Store canned foods in a cool place. Canned foods can vary in the amount of nutrients they contain, largely because of differences in storage times and temperatures. If the can has been on the shelf for an extended period of time, the food's vitamin content may have deteriorated, and its taste and texture may be less than desirable.

Did YOU Know?

Have you ever wondered why a freshly peeled apple, eggplant, potato, or banana eventually turns brown? Damaged plant cells release an enzyme that results in the production of brown pigments when it is exposed to air. The action of this enzyme can be reduced by sprinkling salt or sugar on cut pieces of raw food; coating the pieces with an acidic solution, such as lemon juice; or covering them with airtight plastic wrap and chilling the food. Because heat destroys the enzyme, the unappetizing discoloring won't occur if you cook the fruit or vegetable immediately after peeling.

Photodisc/Getty Images

Concept CHECKPOINT

1. List at least three criteria used to designate a substance as a vitamin.
2. List three factors that distinguish vitamins from macronutrients.
3. Define the following terms: *provitamin, antioxidant,* and *radical.*
4. List the vitamins that must be added to refined grains to enrich them, according to the government's grain enrichment program.
5. Discuss at least five ways to preserve the vitamin content of fruits and vegetables during food preparation and storage.

See Appendix F for responses.

8.2 Fat-Soluble Vitamins

Learning Outcomes

1. List the four fat-soluble vitamins, and identify good dietary sources of each fat-soluble vitamin (or provitamin).
2. Discuss major functions of each fat-soluble vitamin.
3. Identify health problems associated with excesses and deficiencies of fat-soluble vitamins.

This section focuses on fat–soluble vitamins. Table 8.2 presents a summary of general information about all fat–soluble vitamins. Unless otherwise noted, RDA/AI values are for adults 19 to 50 years of age, excluding pregnant or breastfeeding females. Figure 8.6 indicates food groups of MyPlate that are good sources of fat–soluble vitamins.

TABLE 8.2 Summary of Fat-Soluble Vitamins

Vitamin	Major Functions in the Body	Adult RDA/AI (adult RDA = bold)	Major Dietary Sources	Major Deficiency Signs and Symptoms	Major Toxicity Signs and Symptoms
Vitamin A (preformed and provitamin A)	Normal vision and reproduction, cellular growth, and immune system function	**700–900 mcg RAE**	Preformed: liver, milk, fortified cereals Provitamin: yellow-orange, red, and dark green fruits and vegetables	Night blindness, xerophthalmia, poor growth, dry skin, reduced immune system functioning	**Adult Upper Level (UL) = 3000 mcg/day** Nausea and vomiting, headaches, bone pain and fractures, hair loss, liver damage, interference with vitamin K absorption
Vitamin D	Absorption of calcium and phosphorus, maintenance of normal blood calcium, calcification of bone, maintenance of immune function	**15–20 mcg**	Vitamin D–fortified milk, fortified cereals, fish liver oils, fatty fish	Rickets in children, osteomalacia in adults: soft bones, depressed growth, and reduced immune system functioning	**Adult UL = 100 mcg/day** Poor growth, calcium deposits in soft tissues
Vitamin E	Antioxidant, maintenance of nervous and immune system functions	**15 mg** (alpha-tocopherol)	Vegetable oils and products made from these oils, certain fruits and vegetables, nuts and seeds, fortified cereals	Loss of muscular coordination, nerve damage, reduced immune system function	**Adult UL = 1000 mg/day** Excessive bleeding as a result of interfering with vitamin K metabolism
Vitamin K	Production of active blood-clotting factors	**90–120 mcg**	Green leafy vegetables, canola and soybean oils, and products made from these oils	Excessive bleeding	**Adult UL = undetermined** Unknown

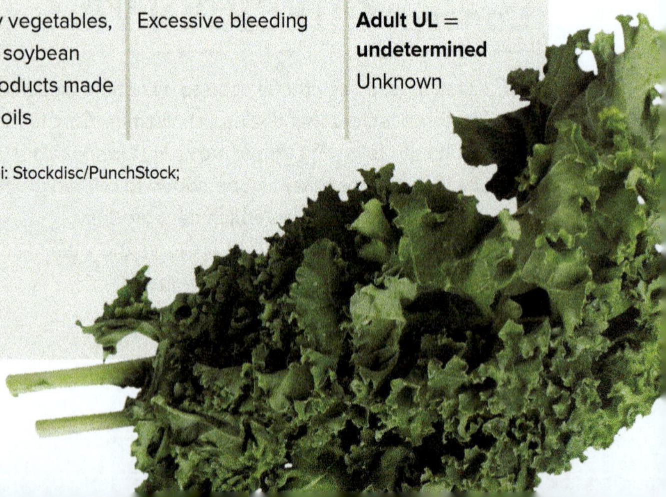

Oil: Jacques Cornell/McGraw Hill; Milk: Ken Karp/McGraw Hill; Broccoli: Stockdisc/PunchStock; Salmon: ©Wendy Schiff; Kale: Stockdisc/Getty Images

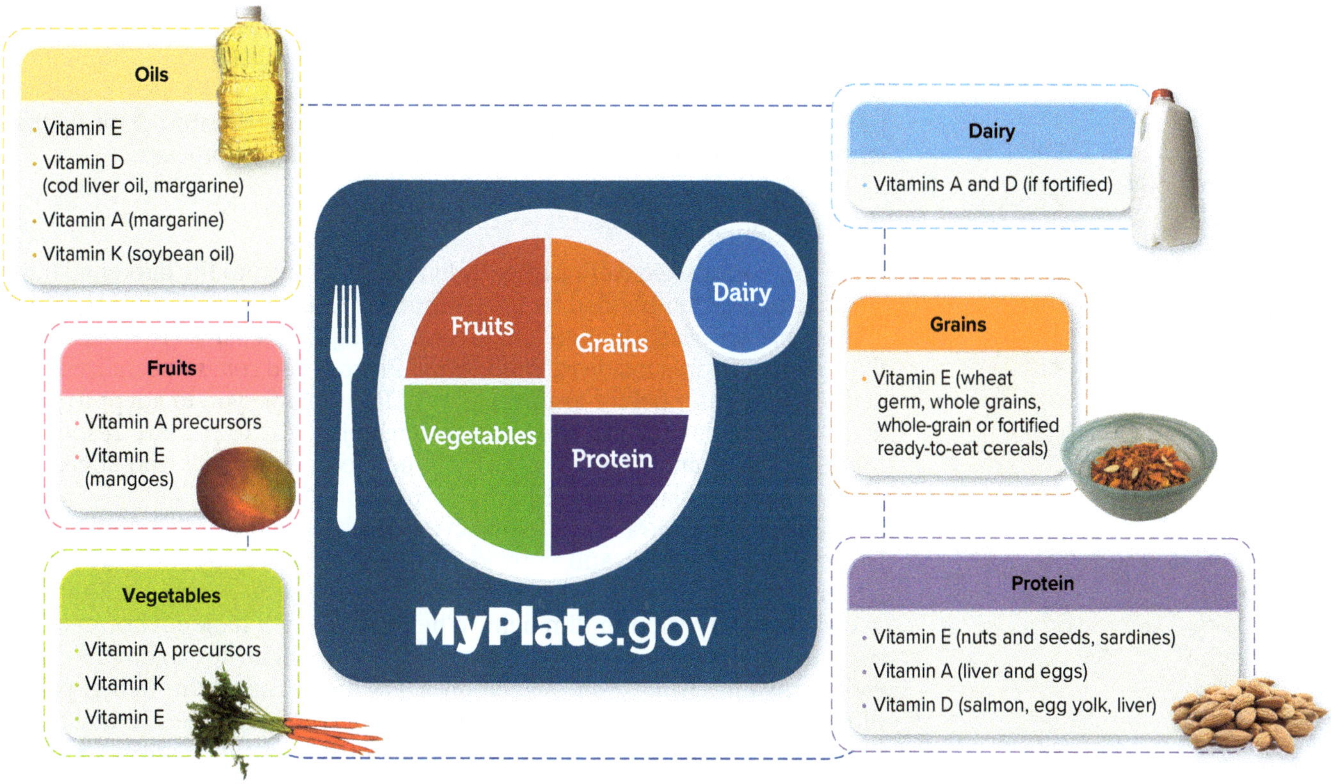

Figure 8.6 MyPlate and fat-soluble vitamins. This illustration highlights MyPlate food groups that are generally good sources of fat-soluble vitamins.
MyPlate food guide icon: U.S. Department of Agriculture; Oil: D. Hurst/Alamy Stock Photo; Mango, Carrots, Cereal: ©Wendy Schiff; Milk: Judith Collins/Alamy Stock Photo; Almonds: Ingram Publishing/SuperStock

Vitamin A

Do you associate vitamin A with eating carrots and good vision? It is true that vitamin A is involved in the visual process and carrots contain vitamin A precursors. However, vitamin A can multitask: It has numerous functions in the body. Furthermore, carrots are not the only source of vitamin A precursors; many fruits and vegetables are rich sources of these compounds.

Vitamin A is actually a family of compounds that includes retinol. **Retinol (preformed vitamin A)** is the most active form of the vitamin in the body. Although retinol and the other forms of vitamin A are only in animal foods, plants contain hundreds of yellow, orange, and red pigments called **carotenoids.** A few carotenoids are vitamin A precursors because the body can use them to make some retinol. Amounts of vitamin A in food are often reported as micrograms of retinol activity equivalents (RAE). One RAE is approximately 1 mcg of retinol.

All cells in the body need vitamin A to develop and function properly. Vitamin A participates in the processes of cell production, growth and development, function, and maintenance. For example, the vitamin is necessary for the production, maturation, and maintenance of **epithelial cells,** cells that form protective tissues that line the body, including skin and linings of the digestive, respiratory, and reproductive tracts. Certain epithelial cells secrete *mucus,* a sticky fluid that keeps the tissue moist and forms a barrier against many environmental pollutants and infectious agents. When the mucus–secreting epithelial cells do not have vitamin A, they deteriorate and no longer produce mucus. A lack of vitamin A can also reduce fertility because the vitamin is required for maintaining the epithelial cells that line the reproductive tracts of males and females.

Vitamin A plays a role in regulating the activity of the immune system. Recall from Chapter 4 that the immune system helps protect the body from infections. Thus, vitamin A–deficient people are at greater risk of infections than those with adequate levels of the vitamin in their bodies.

retinol (preformed vitamin A) most active form of vitamin A in the body

carotenoids yellow, orange, and red pigments in fruits and vegetables

epithelial cells cells that form protective tissues that line the body

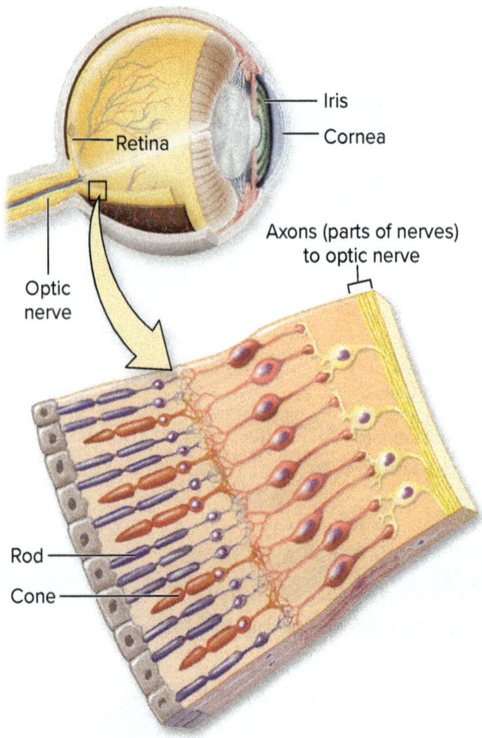

Figure 8.7 Vitamin A and vision. The retina, the light-sensitive area inside each eye, contains rods and cones, specialized nerve cells that are essential for vision.

rhodopsin vitamin A–containing protein that is needed for vision in dim light

Normal bone growth and development also require vitamin A. Although your bones do not appear to change their shape, they are constantly remodeled by processes that involve tearing down and rebuilding the tissues to meet the physical demands that you place on them each day. Vitamin A participates with other vitamins and minerals in the bone remodeling process.

What Is Night Blindness?

Have you ever walked from a brightly lit theater lobby into the darkened movie auditorium and felt blinded for a few seconds? This is a normal visual response to the sudden and dramatic reduction in light intensity. The *retina,* the light–sensitive area inside each eye, contains *rods* and *cones,* specialized nerve cells that are essential for vision (Fig. 8.7). Rods enable you to adapt to poorly lit environments and see objects as shades of black. Cones are responsible for color vision and function in well–lit environments. Rods and cones need vitamin A, particularly retinol, to function properly.

Figure 8.8 illustrates the visual cycle involving rods. Rods remove retinol from the bloodstream and convert it to retinal. In dark conditions, retinal binds to a protein called *opsin* to form **rhodopsin** (*visual purple*). Rhodopsin is necessary for vision in dim light. When you are in a dark environment and exposed to a small amount of light, such as moonlight, the light strikes your rods, activating rhodopsin. Activation alters the shape of the retinal portion of rhodopsin and bleaches the compound. This process causes retinal to split away, reforming opsin. The transformation of rhodopsin to opsin stimulates the rod to send a nervous impulse to the brain that signals the visual processing areas to interpret what is seen. After splitting away from opsin, retinal can bind to the protein, forming rhodopsin again.

As a result of the visual process shown in Figure 8.8, some retinal is destroyed. To replace the retinal, rods remove some retinol (vitamin A) from the bloodstream and convert it to retinal. *Night blindness,* the inability to see in dim light, occurs if retinol is unavailable. Night blindness is an early sign of vitamin A deficiency. Although cone cells also need vitamin A to function, the inability to see certain colors (color blindness) is not the result of a vitamin A deficiency.

Food & Nutrition tips

When preparing salads or other dishes, do not discard the edible dark green leaves of lettuce, cabbage, or broccoli. These darkly pigmented parts of the plants contribute more provitamin A carotenoids to your diet than the lighter-colored parts. Use them in salads or on sandwiches. In addition to carotenoids, many fruits and vegetables contain hundreds of other phytochemicals that may benefit your health. To obtain a variety of these compounds, include colorful vegetables and fruits in your meals and snacks.

C Squared Studios/Photodisc/Getty Images

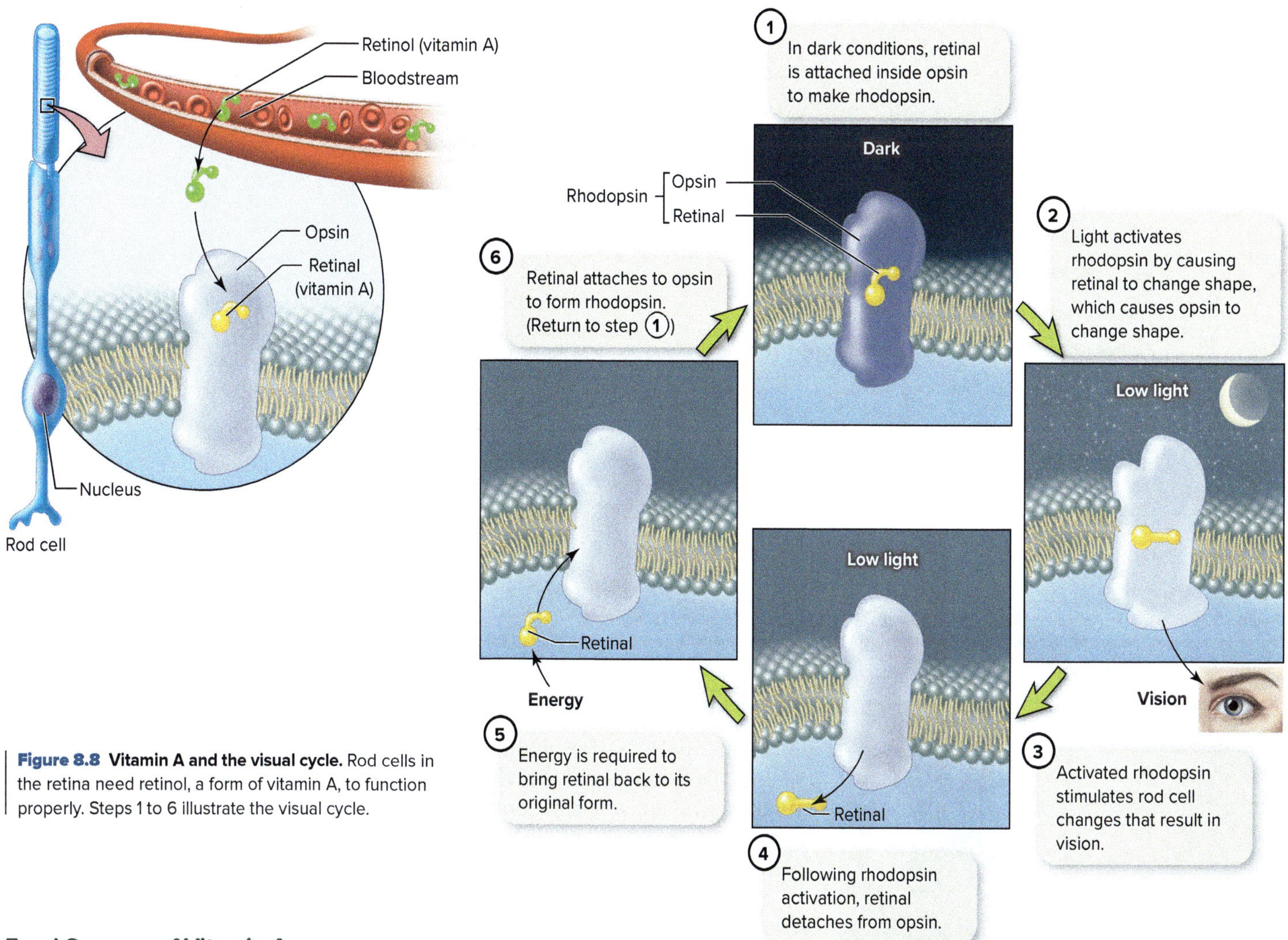

Figure 8.8 Vitamin A and the visual cycle. Rod cells in the retina need retinol, a form of vitamin A, to function properly. Steps 1 to 6 illustrate the visual cycle.

Food Sources of Vitamin A

Animal foods such as liver, butter, fish liver oils, and eggs are good sources of preformed vitamin A. Some foods are fortified with the vitamin during processing. Vitamin A–fortified milk, yogurt, margarine, and cereals are important sources of the nutrient for Americans. Carrots, spinach and other leafy greens, pumpkin, sweet potatoes, broccoli, mangoes, and cantaloupe are rich sources of **beta–carotene,** a provitamin carotenoid that the body can convert to vitamin A. However, the body obtains only 1 mcg of retinol from every 12 mcg of beta–carotene in a food.

beta-carotene provitamin carotenoid that the body can convert to vitamin A

In addition to beta–carotene, common carotenoids include lutein (*loo'–tee–en*), zeaxanthin (*zee–ah–zan'–thin*), and lycopene (*lie'–ko–peen*). Green, leafy vegetables, such as spinach and kale, have high concentrations of lutein and zeaxanthin. Tomato juice and other tomato products, including pizza sauce, contain considerable amounts of lycopene. Although lutein, zeaxanthin, and lycopene are carotenoids, the body does not convert them to vitamin A. Nevertheless, these plant pigments may function as beneficial antioxidants in the human body.

Fruit that is fully ripened contains higher concentrations of beta–carotene than less ripe fruit.[7] Darkly pigmented fruits and vegetables usually contain more beta–carotene and other provitamin A carotenoids than lightly colored produce. For example, carrots, sweet potatoes, mangoes, and peaches contain more beta–carotene than cauliflower, white potatoes, apples, and bananas. Dark green fruits and vegetables also contain carotenoids, but their green pigment *chlorophyll* contents hide the red, yellow, and orange pigments. Table 8.3 lists some foods that are sources of vitamin A and its precursors.

Did *YOU* Know?

Tomatoes are a good source of certain carotenoids, particularly beta-carotene and lycopene. Exposure to heat can break down plant cells, and as a result, the cells release carotenoids. Thus, cooking tomatoes can improve the intestinal tract's ability to absorb carotenoids in the food.

Ingram Publishing/Alamy Stock Photo

TABLE 8.3 Vitamin A Content of Selected Foods

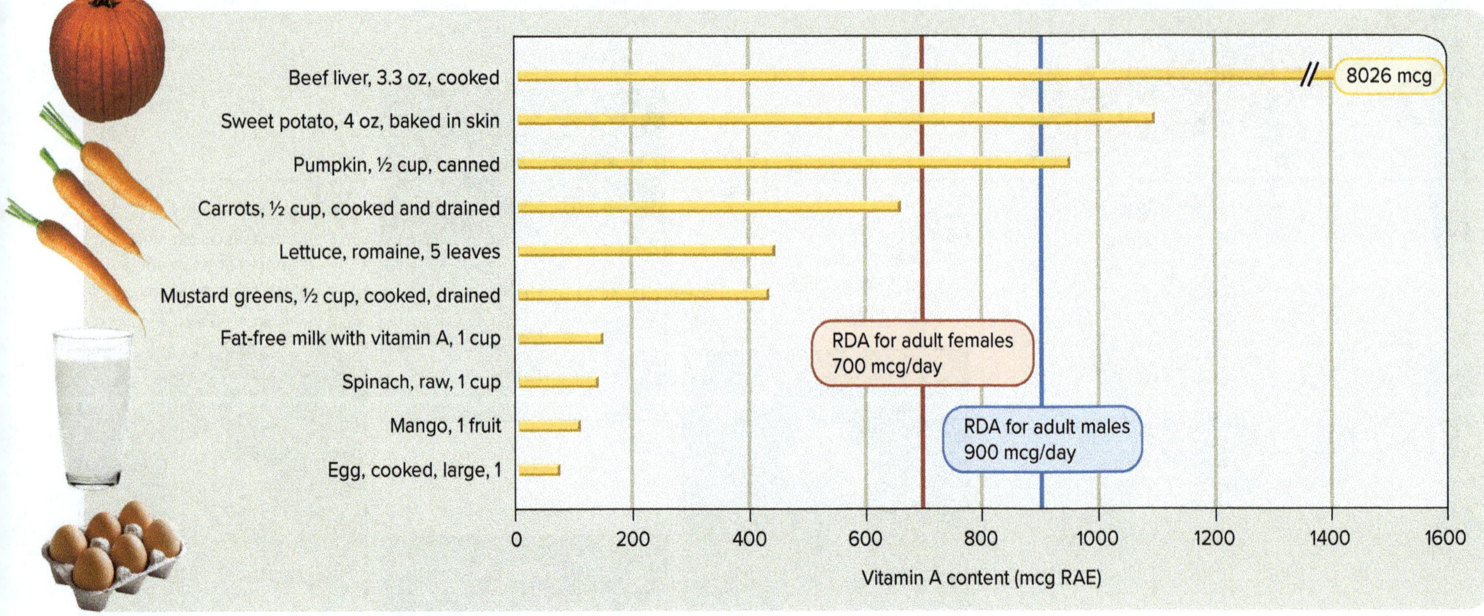

Source: Data from U.S. Department of Agriculture, Agricultural Research Service: *FoodData Central*, 2019.
Pumpkin: Ingram Publishing/Alamy Stock Photo; Carrots: Isabelle Rozenbaum/PhotoAlto/SuperStock; Glass of milk: Nipaporn Panyacharoen/Shutterstock; Eggs: muharrem ömer/iStockphoto

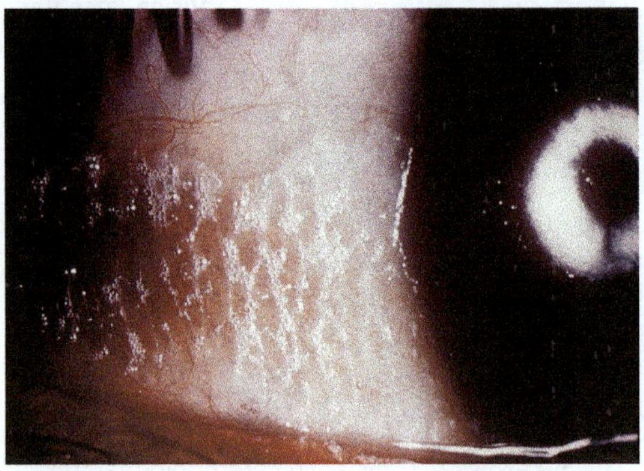

Figure 8.9 Early sign of xerophthalmia. Vitamin A deficiency can cause drying (the white foamy areas) of the surface of the eye. If untreated, this condition can lead to blindness.
Centers for Disease Control and Prevention

Dietary Adequacy

For adults, the RDA for vitamin A is 700 mcg RAE (females) to 900 mcg RAE (males). Deficiencies of the vitamin are rare in the United States.[8] Preschool children who do not eat enough vegetables, urban poor, older adults, and people with severe alcoholism, fat malabsorption, or liver diseases are at risk for vitamin A deficiency.

Vitamin A Deficiency Epithelial cells are among the first to become affected by a deficiency of vitamin A. In skin, vitamin A–deficient epithelial cells produce too much **keratin,** a tough protein found in hair, nails, and the outermost layers of skin. Keratin accumulates within the skin and makes the tissue rough and bumpy. Keratin also forms in tissues that do not normally contain the protein, such as the *cornea,* the clear covering over the iris of the eye (see Figure 8.7). The cornea enables light to enter the eye. The epithelial cells that line the inner eyelids secrete mucus that helps keep the cornea moist and clean. In a person suffering from chronic vitamin A deficiency, these cells accumulate keratin, and the white of the eye develops "foamy" areas (Fig. 8.9). Eventually, the epithelial cells harden and stop producing mucus. This condition is called **xerophthalmia** (*zir–op–thal′–me–a*) or "dry eye." Corneas affected by xerophthalmia can be damaged easily by dirt and bacteria. Unless a person with xerophthalmia receives vitamin A, the condition eventually leads to blindness.

Each year, thousands of children in developing nations, especially in Africa and Southeast Asia, become blind because of severe vitamin A deficiency.[9] Vitamin deficiency also reduces the effectiveness of the immune system, and many children suffering from vitamin A deficiency die from infections such as measles. In countries where vitamin A deficiency is widespread, public health efforts are being taken to reduce the prevalence of the condition. Such efforts include educating people about the need to eat regionally grown foods that are rich in provitamin A beta–carotene and giving vitamin A injections periodically to vulnerable populations. In some countries, governments encourage food manufacturers to fortify commonly eaten foods, such as sugar and margarine, with vitamin A.

keratin tough protein found in hair, nails, and the outermost layers of skin

xerophthalmia condition affecting the eyes that results from vitamin A deficiency

Vitamin–A deficient females are more likely to die during pregnancy than pregnant females who are not lacking the vitamin.[9] Furthermore, infants who are born to vitamin A–deficient females are more likely to die than babies born to healthy females. However, pregnant females should not take vitamin A supplements to prevent birth defects without consulting with their physicians. When taken during pregnancy, excess vitamin A is a **teratogen,** an agent that causes birth defects.

Did *YOU* Know?

According to the FDA, Golden Rice, which has been bioengineered to contain the vitamin A precursor beta-carotene, is safe to eat.[10] Golden Rice was originally developed to provide a rich source of beta-carotene for children in some of the poorest countries in the world, whose traditional diets contained rice but was often vitamin A deficient. Approximately 1½ cups cooked Golden Rice (about 50 grams dry) will meet almost 30% of the Estimated Average Requirement (EAR) of vitamin A for a preschool child.[11] Because Golden Rice is a bioengineered food, its promoters have had to overcome the general public's fears concerning the safety of such foods.

Vitamin A Toxicity

The UL for vitamin A intake is 3000 mcg/day for adults. Excessive consumption of vitamin A can damage the liver because the organ is the main site for vitamin A storage. Toxicity signs and symptoms include headache, nausea, vomiting, visual disturbances, hair loss, bone pain, and bone fractures.

Miscarriage and birth defects may result when excessive amounts of vitamin A are taken early in pregnancy. Females of childbearing age should limit their overall intake of vitamin A to about 100% of the Daily Value (20 mcg). In addition, females who may become pregnant or who are pregnant should restrict their intake of rich food sources of vitamin A, such as liver and fish liver oils.

People who eat too many foods rich in beta–carotene or take too many beta–carotene supplements can have high blood levels of beta–carotene called *carotenemia.* The excess beta–carotene in the blood accumulates in the skin, causing a yellow–orange discoloration of the skin called **carotenodermia** (*kar′–et–eh–no–der′–mee–ah*).[8] This condition occasionally develops in infants who eat a lot of baby foods that contain carrots, apricots, winter squash, or green beans. In most instances, carotenodermia is harmless. The skin's natural color eventually returns to normal when carotenoid–rich foods are reduced in the diet.

Physicians may prescribe medications derived from vitamin A, such as isotretinoin, to treat severe acne and other skin disorders. Although these medications are less toxic than natural vitamin A, ingesting excessive amounts can produce harmful symptoms. Furthermore, vitamin A derivatives can cause miscarriage or severe birth defects in the offspring of females who use them during pregnancy. Therefore, females of childbearing age should avoid pregnancy while using these medications.

Vitamin D

By the 1700s, some people in parts of northern Europe had learned that exposing children to sunlight or giving the youngsters fish liver oil could prevent or treat **rickets.** Children with rickets have bones that are soft and can become misshapen. Leg bones, for example, bow under the weight of carrying the upper part of the body (Fig. 8.10). Additionally, the affected child's joints, rib cage, and

teratogen agent that causes birth defects

carotenodermia yellow-orange discoloration of the skin that results from high blood levels of beta-carotene

rickets vitamin D deficiency disorder in children

Did *YOU* Know?

In the early 1900s, teams of explorers raced to discover and explore the North and South Poles. Starvation was a major risk of embarking on such expeditions. Some explorers learned survival techniques from the Inuits, an Eskimo population who inhabit the Arctic region of North America. The Inuits warned the explorers to avoid eating polar bear and seal livers because severe illness and death could result. Unfortunately, many Western explorers did not know that their sled dogs were also not safe for human consumption.

From 1911 to 1913, Douglas Mawson and Xavier Mertz were on an ill-fated expedition to explore an area near the South Pole. As the two men struggled to return to their winter base camp, they had to eat their sled dogs, particularly the animals' livers, to avoid starvation. Both men suffered terribly from the diet; Mertz did not survive. Over 50 years later, scientists determined that sled dogs can accumulate large amounts of vitamin A in their livers without showing ill effects. However, eating just a few ounces of sled dog liver can be toxic for humans.

Ingram Publishing/SuperStock

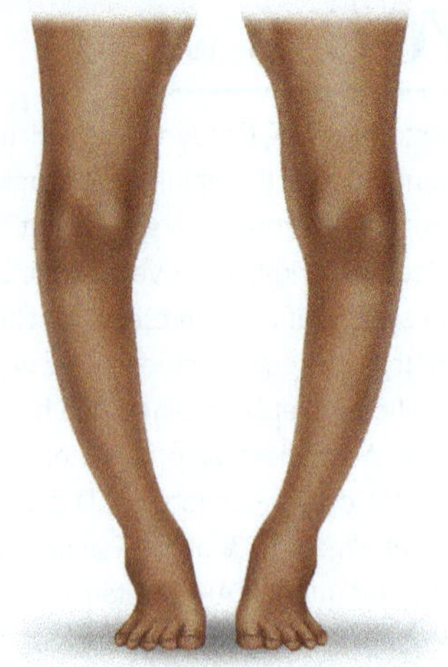

Figure 8.10 Rickets. A child suffering from rickets has soft bones that do not grow properly. The youngster's leg bones bow under the weight of carrying the upper part of the body.

hips (pelvis) become deformed, and the child may complain of muscle pain. In 1922, scientists discovered a fat–soluble factor in cod liver oil that was needed for proper bone health and preventing rickets. The scientists thought the factor that cured rickets was a vitamin, and they named the substance *vitamin D.* The fat–soluble factor in cod liver oil is still considered a vitamin because rickets can be prevented and treated by taking vitamin D supplements or eating vitamin D–rich foods. However, vitamin D is actually a hormone.

The body can make vitamin D after skin cells are exposed to the sun's ultraviolet radiation (*UVB,* in particular), which explains why the nutrient is often called the "sunshine vitamin." The radiation converts a substance in skin that is derived from cholesterol into a previtamin (*vitamin* D_3). Vitamin D_3 circulates to the liver, where the substance is converted into the inactive compound *25–OH vitamin D.* Eventually, the kidneys convert 25–OH vitamin D into the active hormone we call "vitamin D" (Fig. 8.11). Certain foods and dietary supplements are also sources of vitamin D_3. The kidneys convert the vitamin D_3 from these sources into the active form of the vitamin.

Why Is Vitamin D Necessary?

Vitamin D is necessary for the metabolism of the minerals calcium and phosphorus, and the production and maintenance of healthy bones (Fig. 8.12). Vitamin D stimulates small intestinal cells to absorb calcium and phosphorus from food. When the body lacks vitamin D, the intestine absorbs only 10 to 15% of the calcium in foods; when the body's vitamin D status is adequate, intestinal absorption of dietary calcium increases to 30 to 80%.[12] Vitamin D also stimulates bone cells to form *calcium phosphate,* the major mineral compound in bone. Without adequate vitamin D, bone cells cannot deposit enough calcium and phosphorus to produce strong bones.

When the level of calcium in blood drops below normal, a variety of complex physiological responses that involve vitamin D help raise the level to normal. Chapter 9 will discuss the regulation of blood calcium in more detail.

Vitamin D has other roles in the body, including regulating neuromuscular and immune function and reducing inflammation.[13] Vitamin D is also involved

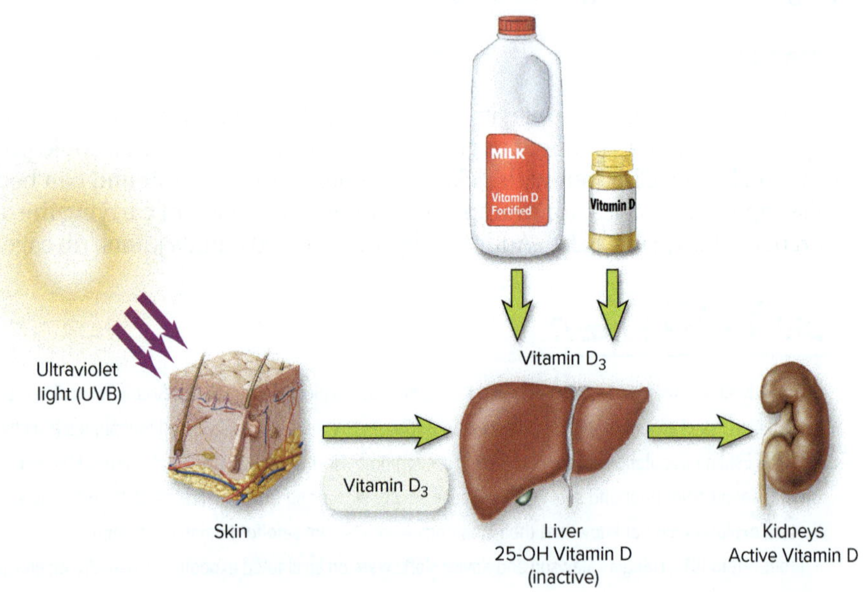

Figure 8.11 Vitamin D. After skin synthesizes vitamin D_3 from a cholesterol derivative, the previtamin undergoes changes in the liver and kidneys to become the active form of the vitamin. Forms of previtamin D that are in certain foods and dietary supplements must also undergo changes in the liver and kidneys to become active vitamin D.

REAL PEOPLE | REAL STORIES

Alicia Croker

Amanda Croker

Amanda Croker, a recent graduate of the University of Kansas, routinely tried to limit her sun exposure to avoid sunburn and the possibility of skin cancer in the future. "My skin is so light," Amanda says, "I even got a sunburn in Kansas in the middle of October because I was outside for a couple hours!" She was unaware that her efforts to protect her skin were also limiting her body's ability to make vitamin D, "the sunshine vitamin."

"I'm taking a prescription dose of vitamin D because my family doctor tested my blood and found out that I didn't have enough of the vitamin in my body," Amanda says. "She told me that people need sunlight to make vitamin D, but she said she wasn't surprised that I'm deficient, since I have red hair. I told her that I try to avoid being out in the sun for more than 30 minutes because I 'crisp' when I get exposed to sun. Anyway, I don't notice that much of a difference now that I'm on the medication." Although Amanda does not "feel" different now, her bones eventually may have become soft and weak if her doctor had not discovered that Amanda was vitamin D–deficient. Before the discovery and purification of vitamin D, children were given cod liver oil to prevent rickets, the vitamin's deficiency disease.

As a result of the treatment, Amanda's blood level of vitamin D became normal within a few weeks. Because few foods are good sources of vitamin D, she will need to take a daily supplement of the micronutrient, which is available without a prescription, for the rest of her life. By taking the supplement, she should be able to maintain a healthy blood level of the fat-soluble vitamin. Vitamin D can be toxic, so taking vitamin D supplements is not recommended unless people (like Amanda) have a deficiency of the nutrient.

in controlling cell growth, and as a result, the micronutrient may reduce the risk of certain cancers. However, more research is needed to clarify the vitamin's role in immune function and disease prevention.

Sources of Vitamin D

Fish liver oils and fatty fish, especially salmon, herring, and catfish, are among the few foods that naturally contain previtamin forms of vitamin D. Milk is routinely fortified with vitamin D, and some brands of ready–to–eat cereals, orange juice, nondairy milk alternatives, and margarine have the vitamin added to them as well. Mushrooms exposed to UV light can be a source of vitamin D. Mushrooms produce the vitamin through a process similar to human skin cells exposed to UV light.[14] Some grocery stores sell vitamin D–enhanced mushrooms. Read product ingredient lists and Nutrition Facts panels to identify products fortified or enhanced with vitamin D. Table 8.4 lists some food sources of vitamin D. Food composition tables often list the vitamin D content of foods in International Units (IU); 1 mcg of vitamin D equals 40 IU.

Vitamin D and Sunlight Vitamin D is not widespread in food; therefore, your body depends on sun exposure to synthesize the vitamin. The amount of time you need to spend in the sun to form adequate amounts of vitamin D depends primarily on your location, the time of day and year, and your age and skin color. If you live south of about the 33rd parallel in the United States and you are outdoors when sunlight is most intense (between 10 A.M. and 3 P.M.),

Did *YOU* Know?

Information provided by tanning parlors may include claims that the tanning process is safe. Nevertheless, the ultraviolet radiation emitted by tanning beds increases the risk of wrinkles and skin cancer. Therefore, the use of tanning beds should be avoided.

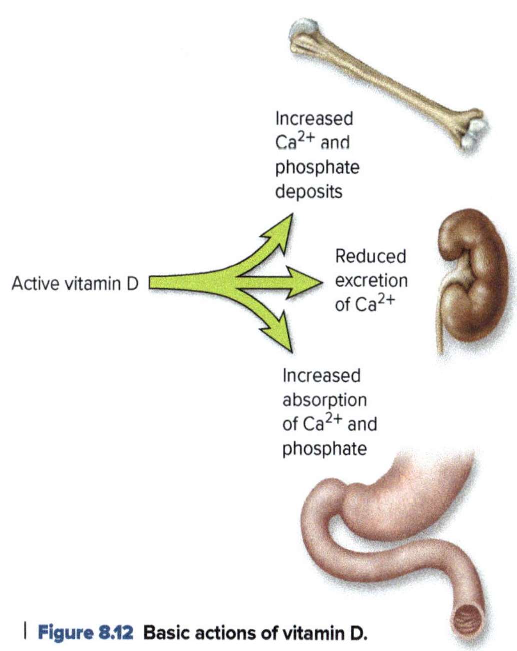

Figure 8.12 Basic actions of vitamin D.

TABLE 8.4 Vitamin D Content of Selected Foods

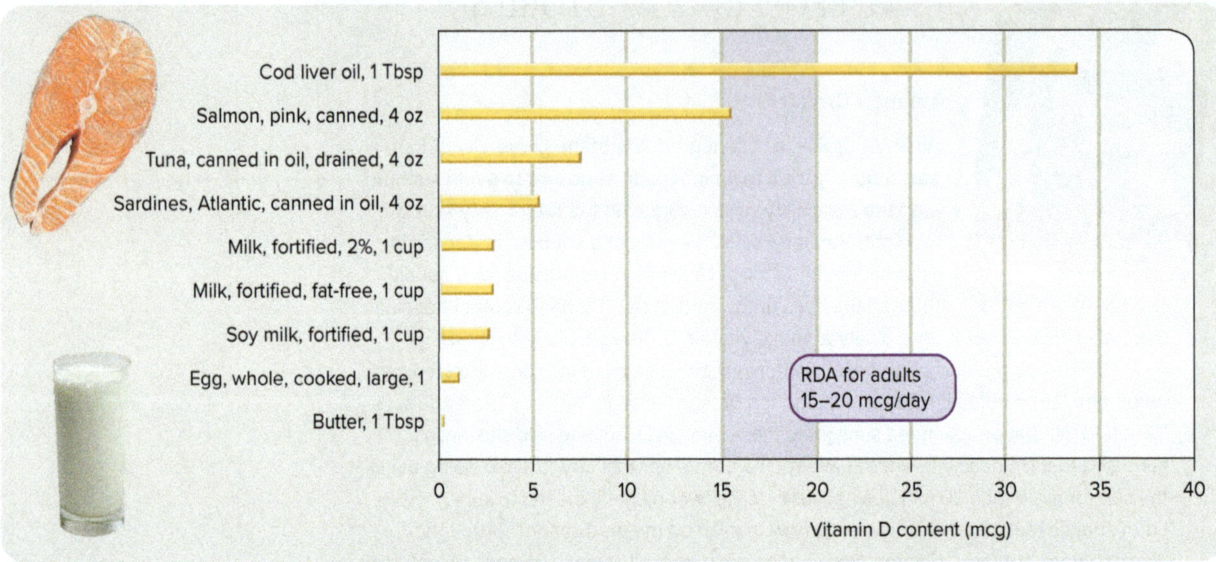

Source: Data from U.S. Department of Agriculture, Agricultural Research Service: *FoodData Central*, 2019.
Salmon: Julija Dmitrijeva/123RF; Milk: tysmith/iStock/Getty Images

Figure 8.13 Latitude and vitamin D status. If you live in North America south of the 33rd parallel and are outdoors when sunlight is most intense during the day, you are likely to obtain enough sun exposure to synthesize vitamin D most of the year.

you may obtain enough sun exposure to make vitamin D during most of the year.[13] In North America, the 33rd parallel extends from about Charleston, South Carolina, through Dallas, Texas, to San Diego, California (Fig. 8.13). The Earth's atmosphere blocks ultraviolet (UV) radiation. If you live north of the 33rd parallel, the angle of the winter sun is such that the sun's rays must pass through more of the atmosphere than at other times of the year (Fig. 8.14). As a result, your skin may not make sufficient amounts of previtamin D during the winter, and you may not have adequate vitamin D stored in your body to last until spring. Therefore, you may need to take a supplement that contains 100% of the adult Daily Value for vitamin D (see Appendix B), especially from November through February. Clouds, shade, window glass, and air pollution are environmental factors that can also limit the amount of UV radiation that reaches your skin.

Skin contains *melanin,* the brown pigment that can reduce the skin's ability to produce vitamin D from sunlight. Darker skin contains more melanin than

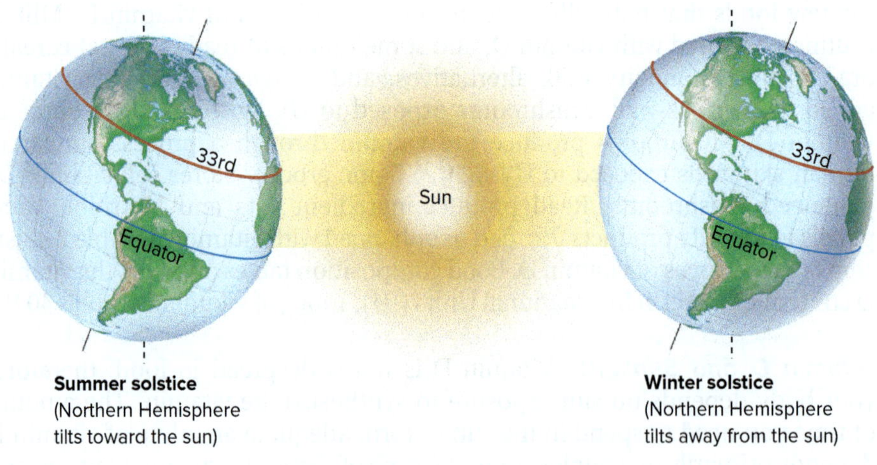

Figure 8.14 Seasonal variations in sunlight intensity. If you live north of the 33rd parallel, the angle of the winter sun is such that the sun's rays must pass through more of the atmosphere than at other times of the year. As a result, skin tends to form less previtamin vitamin D in the winter.

lighter skin. If you have dark skin, you may need to spend even more time in the sun to form adequate amounts of the vitamin. Otherwise, you should consider consuming foods that contain the vitamin or a vitamin D supplement regularly.

Although you may think tanned skin is attractive, tanning and severe sunburn increase the risk of developing wrinkles and skin cancer. UV radiation is a major risk factor for skin cancer, including melanoma, the most deadly form of skin cancer. Dermatologists often advise people to apply sunscreens consistently before going outdoors. Using a sunscreen, however, limits skin's ability to synthesize previtamin vitamin D_3. When properly applied, a sunscreen with a *sun protection factor* (*SPF*) of 8 or more blocks sunlight that is needed to form D_3.[15] To allow your body to make some active vitamin D, you can expose your skin to the sun for about 15 minutes *before* applying a commercial sunscreen. Concerns about the risk of skin cancer have made it difficult for scientists to develop guidelines for the amount of sunlight exposure that is necessary to optimize vitamin D synthesis.

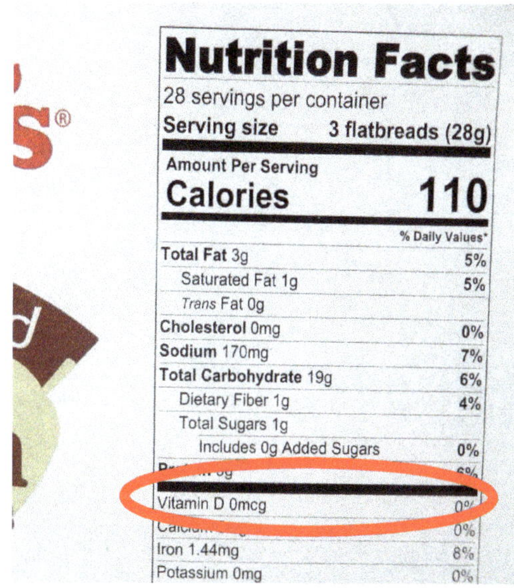

Vitamin D is included on food product Nutrition Facts panel.
©Wendy Schiff

Dietary Adequacy

For adults under 70 years of age, the RDA for vitamin D is 15 mcg/day. Many Americans who are 2 years of age and older do not consume enough vitamin D to meet the RDA.[4] According to the Dietary Guidelines, the vitamin is a "nutrient of public health concern." Younger people, however, generally have higher blood levels of the vitamin than older adults.

In the United States, a national effort to fortify milk with vitamin D began in the 1930s, and as a result, rickets became rare in this country. Although severe rickets is uncommon, public health officials are concerned about the recent increase in the number of cases reported among infants and toddlers. Breast milk contains insufficient amounts of vitamin D to prevent rickets. Infants who are most likely to develop rickets are breastfed, and they have dark skin, minimal sunlight exposure, and little or no vitamin D intake.[13] Exposing breastfed babies to sunlight reduces their risk of the disease, but as mentioned earlier, medical experts do not know how much sun exposure is necessary. Thus, breastfed infants should consume a supplement containing 10 mcg (400 IU) of vitamin D per day soon after birth. The adult form of rickets is called **osteomalacia** (*ahs'–tee–o–mah–lay'–she–a*). The bones of people with osteomalacia have normal amounts of *collagen,* the protein that provides structure for the skeleton, but they contain less–than–normal amounts of calcium. The bones are soft and weak, and break easily as a result. Muscle weakness is also a symptom of osteomalacia.

Adults who are confined indoors or almost fully covered during the day, such as for religious reasons, are at risk for osteomalacia (Fig. 8.15).[13] Osteomalacia is a risk for adults who have kidney, liver, or intestinal diseases because these conditions may reduce both vitamin D production and calcium absorption. Exposing skin to sunlight, eating vitamin D–rich foods, or taking vitamin D supplements can help adults avoid vitamin D deficiency and osteomalacia.

As a person ages, production of vitamin D_3 in skin declines and conversion of the previtamin to active vitamin D in kidneys also decreases. As a result of these age–related changes, older adults are more likely to develop vitamin D deficiency than younger persons. Older adults are also at risk for bone fractures. According to results of several studies, older adults who took vitamin D supplements that provided no more than 400 IU per day did not reduce their risk of hip fracture. When vitamin D supplementation increased to more than 400 IU of vitamin D per day, some studies found a lower risk of hip fracture. Results of other studies, however, indicated that supplementation with the higher amounts of vitamin D might *increase* risks of hip fracture and heart disease. Thus, more research is needed to understand the effects of vitamin D supplementation on risk of hip fractures and overall health.[13]

Figure 8.15 Clothing and vitamin D status. Adults who are almost fully covered during the day, such as for religious reasons, are at risk for osteomalacia.
John Lund/Blend Images LLC

osteomalacia adult rickets; condition characterized by poorly mineralized (soft) bones

Although *osteoporosis* is usually associated with inadequate calcium intakes, a long-term vitamin D deficiency contributes to the condition because calcium absorption is reduced.[13] The calcium section of Chapter 9 provides more information about osteoporosis.

Vitamin D Toxicity Recall that the body stores fat-soluble vitamins, including vitamin D, in body fat and the liver. Long-term ingestion of vitamin D supplements that supply 10,000 to 40,000 IU/day can accumulate in the body and produce toxicity.[13] When excess vitamin D is consumed, the small intestine absorbs too much calcium from foods, and the mineral is deposited in soft tissues, including the kidneys, heart, and blood vessels. The calcium deposits can interfere with cells' ability to function, which can cause cellular death. Other signs and symptoms of vitamin D toxicity include muscular weakness, loss of appetite, diarrhea, vomiting, and mental confusion. You do not have to be concerned about your body making toxic levels of vitamin D when exposed to sunlight because skin limits its production of the previtamin, vitamin D_3.[13] The Upper Level (UL) for vitamin D is 100 mcg/day (4000 IU/day).

Did *YOU* Know?

Rancidity results when fat in food, particularly the unsaturated fat, undergoes oxidation. Rancid fat makes the food smell and taste bad, and people usually refuse to eat it. To inhibit oxidation of fatty acids and increase a food's shelf life, manufacturers add antioxidants such as BHT and BHA to the food during production.

Vitamin E

Vitamin E has several forms, but only the *alpha-tocopherol (al′-fah toe-koff′-e-roll)* form is used by the body.[16] **Alpha-tocopherol** (*vitamin E*) is an antioxidant that protects polyunsaturated fatty acids in the phospholipids of cell membranes from being damaged by radicals (Fig. 8.16). Such oxidative damage may be associated with the development of atherosclerosis, the process that occurs within arteries and contributes to heart attack and stroke; cancer; and premature cellular aging and death. Other roles for vitamin E include maintaining immune system function.

alpha-tocopherol vitamin E

Food Sources of Vitamin E

Rich food sources of vitamin E include sunflower seeds, almonds, and plant oils, especially sunflower, safflower, canola, and olive oils. Products made from vitamin E-rich plant oils, such as margarines and salad dressings, also supply the micronutrient. Other important dietary sources of the vitamin include fish, whole grains, nuts, seeds, and certain vegetables. Meats, processed grain

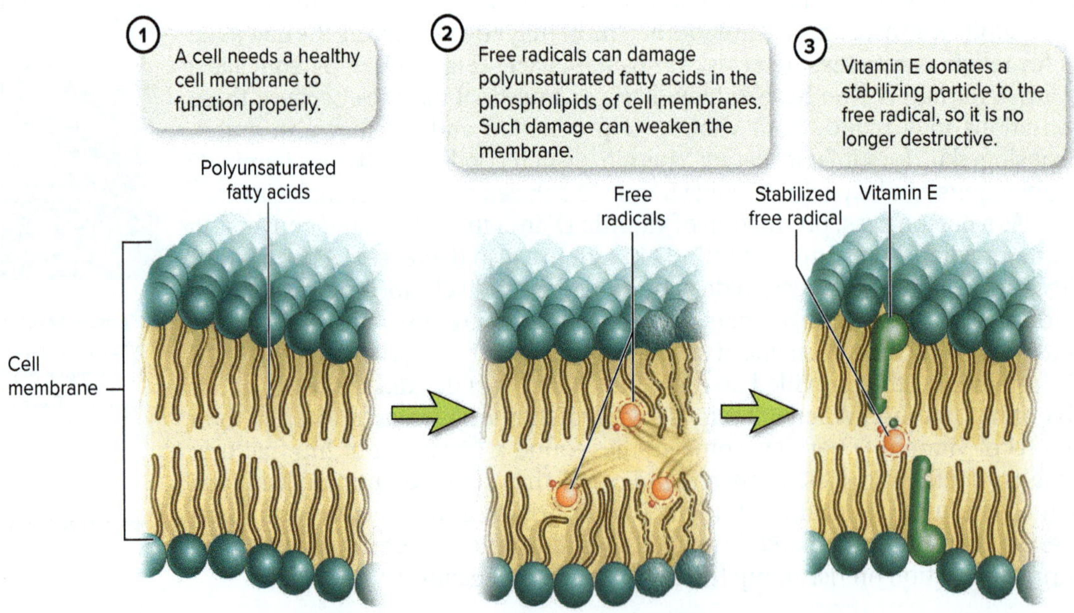

Figure 8.16 Vitamin E and the cell membrane. Vitamin E protects unsaturated fatty acids in cell membranes from being damaged (oxidized) by radicals.

products, and dairy products generally do not contain much vitamin E. Table 8.5 lists some common foods that are sources of the micronutrient.

Harvesting, processing, storage, and cooking methods influence the amount of vitamin E retained in food. During the milling process, most of the vitamin E that is in whole grains is lost, and it is not restored by enrichment. Furthermore, vitamin E is highly susceptible to destruction by exposure to oxygen, metals, and light, as well as high temperatures that occur when heating oil for deep–fat frying.

Dietary Adequacy

For adults, the RDA for vitamin E is 15 mg/day of alpha–tocopherol. Although many American adults do not consume recommended amounts of vitamin E, vitamin E deficiency is rare.[16] A healthy body stores the vitamin in body fat, skeletal muscle, and the liver.

Amounts of vitamin E may be reported as a number of IUs. One mg of alpha–tocopherol is about 1.5 IU of the natural form of the vitamin. Synthetic vitamin E is used to fortify foods and produce supplements that contain the vitamin. One mg of vitamin E is equal to 2.22 IU of the synthetic form of the vitamin.[16]

Vitamin E Deficiency People who have diseases that interfere with fat absorption may become deficient in vitamin E because dietary fat enhances intestinal absorption of the micronutrient. Vitamin E deficiency can reduce the functioning of the nervous and immune systems. Long–term vitamin E deficiency results in nerve damage, loss of neuromuscular control, and blindness.[16]

Infants who are born too early (premature infants) may have low vitamin E stores because the vitamin is transferred from mother to fetus late in pregnancy. If premature infants develop signs of vitamin E deficiency, they can be given special infant formulas and supplements that provide the vitamin.

Food & Nutrition tips

To help protect the vitamin E in vegetable oils from oxidative damage and avoid consuming rancid oils:

- Store unopened containers of oil in a cool, dark cabinet or pantry, away from heat (the stove) and light.
- Throw out any cooking oils that develop an "off" odor, such as a "fishy" smell, which may indicate the oil is rancid.
- Use cooking oil within 2–3 months after opening to decrease the risk of the oil turning rancid.

Maximilian Stock Ltd/Photodisc/Getty Images

TABLE 8.5 Vitamin E Content of Selected Foods

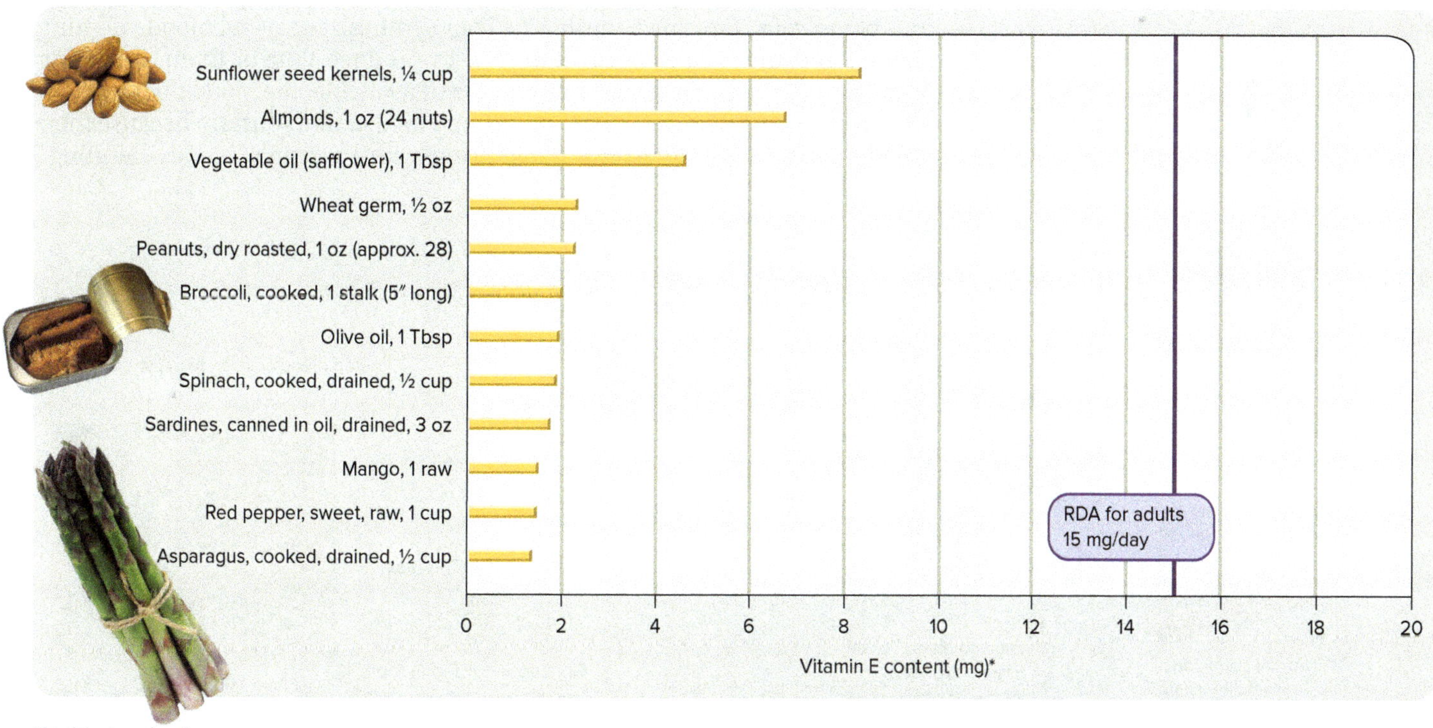

*As alpha-tocopherol.

Source: Data from U.S. Department of Agriculture, Agricultural Research Service: *FoodData Central*, 2019.
Amonds: onair/Shutterstock; Sardines: ©Wendy Schiff; Asparagus: Pixtal/age fotostock

Consuming dietary supplements that supply excessive amounts of vitamin E may interfere with vitamin K's role in blood clotting and lead to uncontrolled bleeding (hemorrhage).
Steve Wisbauer/Stockbyte/Getty Images

Vitamin E Toxicity For healthy adults, the UL for vitamin E is 1000 mg/day of alpha–tocopherol. Consuming amounts of vitamin E in foods has not been associated with any negative effects on health.[16] However, taking dietary supplements that supply excessive amounts of the vitamin may interfere with vitamin K's role in blood clotting and lead to uncontrolled bleeding (hemorrhage). Therefore, people who are taking medications that interfere with blood clotting ("blood thinners") should check with their physicians before using vitamin E supplements.

Vitamin K

In the early 1930s, Danish researcher Henrick Dam discovered a factor in alfalfa that played a role in blood clotting (coagulation). Dam named the factor vitamin "K" after *koagulation,* the Danish spelling of *coagulation.* Vitamin K is a family of compounds that includes *phylloquinone* (*fill–o–kwin′–own*) from plants and *menaquinones* (*men–eh–kwin′–owns*) in fermented foods, such as sauerkraut, cheddar cheese, natto, and miso. Natto and miso are made from fermented soybeans and are popular in Japan. Bacteria that normally live in the intestinal tract also make menaquinones, but the extent to which bacteria contribute to the body's vitamin K supply is unclear.[17]

The presence of dietary fat, bile, and pancreatic juice helps vitamin K be absorbed in the small intestine. Some vitamin K is stored in the liver, and some is incorporated in lipoproteins for transport in the bloodstream. The liver also breaks down vitamin K and eliminates the vitamin's by–products by adding them to bile.

Why Is Vitamin K Necessary?

Blood contains inactive clotting factors and cell fragments called *platelets* that are necessary for blood clotting to occur. When a blood vessel is cut, blood in the injured area undergoes a complex series of steps to form a clot that stops the bleeding (Fig. 8.17). A clot is composed of strands of the protein *fibrin* that trap blood cells, forming a mesh. The liver synthesizes several blood–clotting factors, and the organ needs vitamin K to produce four of them, including *prothrombin,* properly. When vitamin K is unavailable, the blood does not clot effectively. Some people take the prescribed medication warfarin because their blood clots too easily. Vitamin K can interfere with warfarin's "blood–thinning"

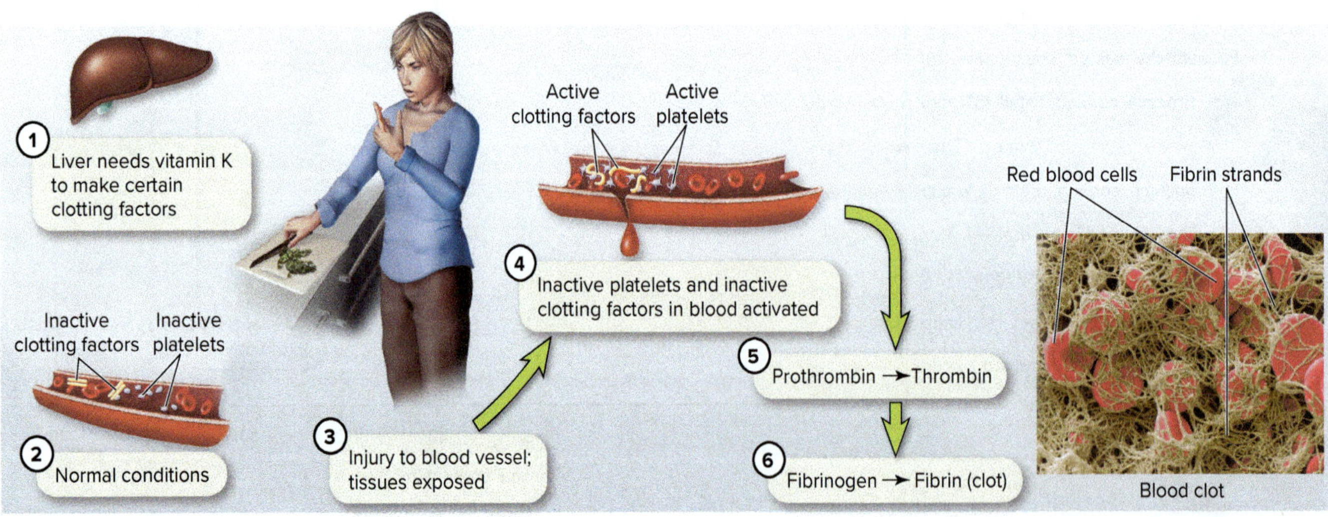

Figure 8.17 Vitamin K and blood clotting. When a blood vessel is cut, blood in the injured area undergoes a series of steps to form a clot that stops the bleeding.
Steve Gschmeissner/Science Photo Library/Getty Images

activity, so people who take the medication should not consume vitamin K supplements. Additionally, these patients should try to maintain consistent dietary intakes of the vitamin each day.

Vitamin K also helps bone–building cells produce *osteocalcin,* a protein needed for normal bone mineralization. Low blood levels of vitamin K are associated with the risk of *osteoporosis,* a condition characterized by bones with low mineral density. However, more research is needed to clarify the vitamin's role in bone health.

Brussels sprouts are an excellent source of vitamin K.
Pixtal/age fotostock

Food Sources of Vitamin K

Major food sources of vitamin K are green leafy vegetables such as kale, turnip greens, salad greens, cabbage, and spinach; broccoli; and green beans. Other reliable sources of the vitamin are soybean and canola oils, and products made from these oils, such as salad dressing. Table 8.6 lists some foods that are sources of the vitamin. The chemical structure of vitamin K is very stable and resists being destroyed by usual cooking methods.

Dietary Adequacy

No RDAs for vitamin K have been established, but AIs for the vitamin are 120 mcg/day for males and 90 mcg/day for females. The AIs can be met easily by eating a salad that contains 1 cup of leafy vegetables and 2 tablespoons of salad dressing made with soybean oil. Ingesting amounts of natural forms of vitamin K that exceed the AIs has not been reported to be harmful to humans.

Vitamin K: Deficiency Although the body stores very little vitamin K, deficiencies among adults rarely occur.[17] Vitamin K deficiency, however, can develop in people who have liver diseases or conditions that impair fat absorption, such as cystic fibrosis. Additionally, long–term antibiotic therapy can reduce the number of bacteria in the large intestine that synthesize vitamin K and, as a result, might contribute to a deficiency of the nutrient. The most reliable sign of vitamin K deficiency is an increase in the time it takes for blood to clot.

TABLE 8.6 Vitamin K Content of Selected Foods

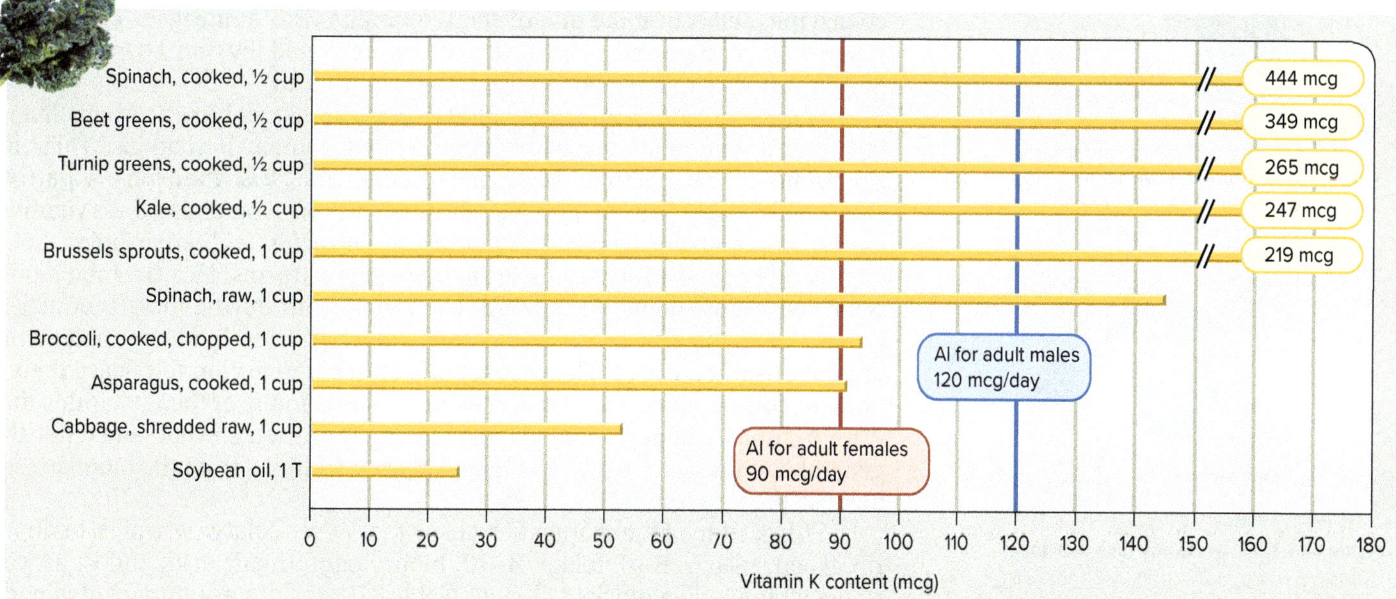

Source: Data from U.S. Department of Agriculture, Agricultural Research Service: *FoodData Central*, 2019.
Kale: ©Wendy Schiff

Babies are generally born with low vitamin K stores, and a deficiency of the vitamin can occur soon after birth. Vitamin K-deficient infants are at risk of serious bleeding because their bodies are unable to make certain blood–clotting factors. To prevent vitamin K deficiency from developing during infancy, newborns generally receive a single injection of vitamin K.

Concept CHECKPOINT

6. Prepare a table of fat-soluble vitamins. For each vitamin, indicate its major function in the body, major food sources, deficiency disorder (if it has a specific name), and major signs and symptoms of the deficiency disorder. If the vitamin is known to be toxic, also indicate major toxicity signs and symptoms. Check your table against the information provided in Table 8.2.

See Appendix F for responses.

8.3 Water-Soluble Vitamins

Learning Outcomes

1. Explain how coenzymes function in the body, and list three vitamins that function as coenzymes.
2. List the water-soluble vitamins, and identify good dietary sources of each vitamin (or its precursor).
3. Discuss major functions of each water-soluble vitamin.
4. Identify health problems associated with excesses and deficiencies of water-soluble vitamins.

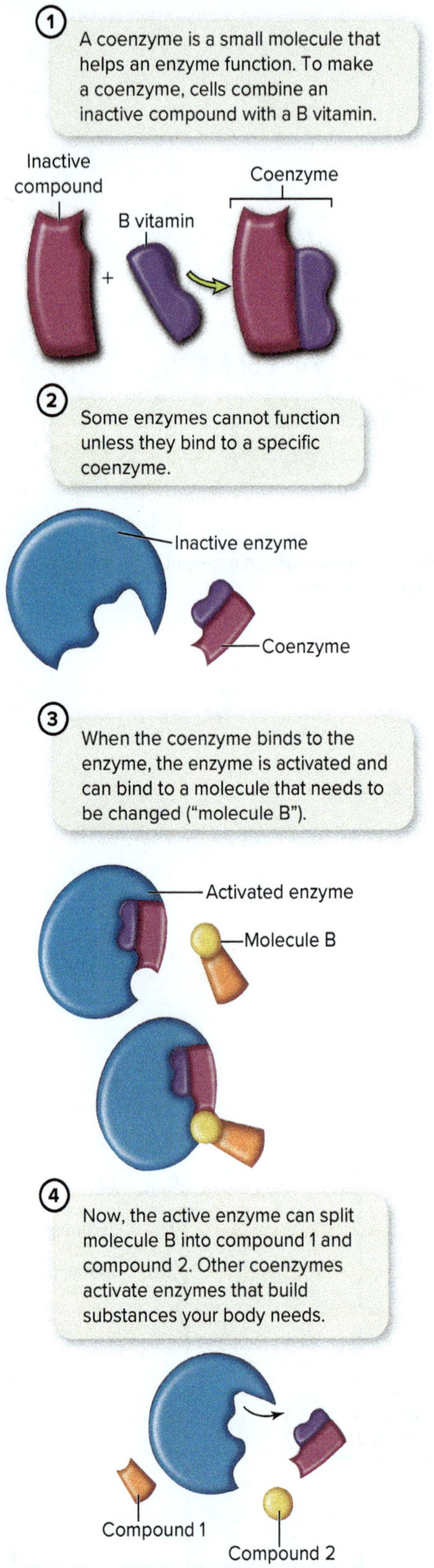

Figure 8.18 **Coenzymes and their function.**

coenzyme small, organic molecule that interacts with enzymes, enabling the enzymes to function

In the body, most water–soluble vitamins function as components of specific coenzymes. A **coenzyme** is a small, organic molecule that interacts with enzymes and, as a result, regulates a chemical reaction. To synthesize a coenzyme, cells combine one of the B vitamins with a nitrogen–containing, nonprotein compound (Fig. 8.18). When activated by the coenzyme, the enzyme enables the reaction to occur (see Figure 8.18).

Many of the chemical reactions involved in the metabolism of carbohydrates, fats, and amino acids involve coenzymes that contain B vitamins. Thiamin, riboflavin, niacin, vitamin B–6, and pantothenic acid function as part of coenzymes involved in energy metabolism. Coenzymes containing these vitamins are also necessary for synthesizing glucose, amino acids, and certain lipids.

Foods contain B vitamins in their coenzyme forms. Health–food stores often sell supplements that contain coenzymes, but buying these products is a waste of money. In the small intestine, coenzymes in food or supplements are not absorbed intact. The compounds undergo digestion to release their B vitamin components. The small intestine absorbs many of the B vitamins that were in foods and supplements, and the micronutrients eventually enter the general circulation. Cells then remove the free vitamins from the bloodstream and use them to rebuild coenzymes.

This section of the chapter focuses on the water–soluble vitamins: thiamin, riboflavin, niacin, B–6, folate, B–12, biotin, pantothenic acid, and C; as well as the vitamin–like nutrient choline. Table 8.7 provides a summary of general information about all the water–soluble vitamins and choline. Figure 8.19 indicates food groups that are good sources of water–soluble vitamins.

TABLE 8.7 Summary of Water-Soluble Vitamins and Choline

Vitamin	Major Functions in the Body	Adult RDA/AI (adult RDA = bold)	Major Dietary Sources	Major Deficiency Signs and Symptoms	Major Toxicity Signs and Symptoms
Thiamin	Part of coenzyme needed for carbohydrate metabolism and the metabolism of certain amino acids; may help neurotransmitter production	**1.1–1.2 mg**	Pork, wheat germ, enriched breads and cereals, nutritional yeast	Beriberi and Wernicke-Korsakoff syndrome: Weakness, abnormal nervous system functioning	None (Upper Level [UL] not determined)
Riboflavin	Part of coenzymes needed for carbohydrate, amino acid, and lipid metabolism	**1.1–1.3 mg**	Milk, yogurt, and other milk products; enriched breads and cereals; liver	Inflammation of the mouth and tongue, eye disorders	None (UL not determined)
Niacin	Part of coenzymes needed for energy metabolism	**14–16 mg**	Enriched breads and cereals, beef, liver, tuna, salmon, poultry, pork, mushrooms	Pellagra: Diarrhea Dermatitis Dementia Death	**Adult UL = 35 mg/day** Flushing of facial skin, itchy skin, nausea and vomiting, liver damage
Pantothenic acid	Part of the coenzyme that is needed for synthesizing fat and releasing energy from macronutrients	5 mg	Beef and chicken liver, sunflower seeds, mushrooms, yogurt, soy milk, fortified cereals	Rarely occurs	Unknown (UL not determined)
Biotin	Needed for synthesizing glucose and fatty acids	30 mcg	Liver, eggs, peanuts, salmon, pork, mushrooms, sunflower seeds	Rarely occurs: Skin rash, hair loss, seizures, and other neurological disorders; developmental delays in infants	Unknown (UL not determined)
Vitamin B-6	Part of coenzyme needed for amino acid metabolism, involved in neurotransmitter and hemoglobin synthesis	**1.3–1.7 mg**	Meat, fish, and poultry; potatoes, bananas, spinach, sweet red peppers, broccoli	Dermatitis, anemia, depression, confusion, and neurological disorders such as seizures	**Adult UL = 100 mg/day** Nerve destruction
Folate	Part of coenzyme needed for DNA synthesis and conversion of cysteine to methionine, preventing homocysteine accumulation	**400 mcg DFE**	Dark green, leafy vegetables; liver, legumes, asparagus, broccoli, orange juice, enriched breads and cereals (folic acid)	Megaloblastic anemia, diarrhea, neural tube defects in embryos	**Adult UL = 1000 mcg/day** May stimulate cancer cell growth
Vitamin B-12	Part of coenzymes needed for various cellular processes, including folate metabolism; maintenance of myelin sheaths	**2.4 mcg**	Animal foods, fortified cereals, fortified soy milk	Pernicious anemia: Megaloblastic anemia and nerve damage resulting in paralysis and death	None (UL not determined)
Ascorbic acid (vitamin C)	Needed for collagen synthesis and maintenance; antioxidant; synthesis of neurotransmitters and certain hormones; immune system functioning	**75–90 mg** (nonsmokers)	Peppers, citrus fruits, papaya, broccoli, cabbage, berries	Scurvy: Poor wound healing, pinpoint hemorrhages, bleeding gums, bruises, depression	**Adult UL = 2000 mg/day** Diarrhea and GI tract discomfort
Choline	Neurotransmitter and phospholipid synthesis; methionine metabolism	425–550 mg	Widely distributed in foods and human biosynthesis (production)	Liver damage	Fishy body odor and reduced blood pressure

Bowl of cereal, Mushrooms: ©Wendy Schiff; Orange: yurakp/123RF; Spinach: Florea Marius Catalin/E+/Getty Images

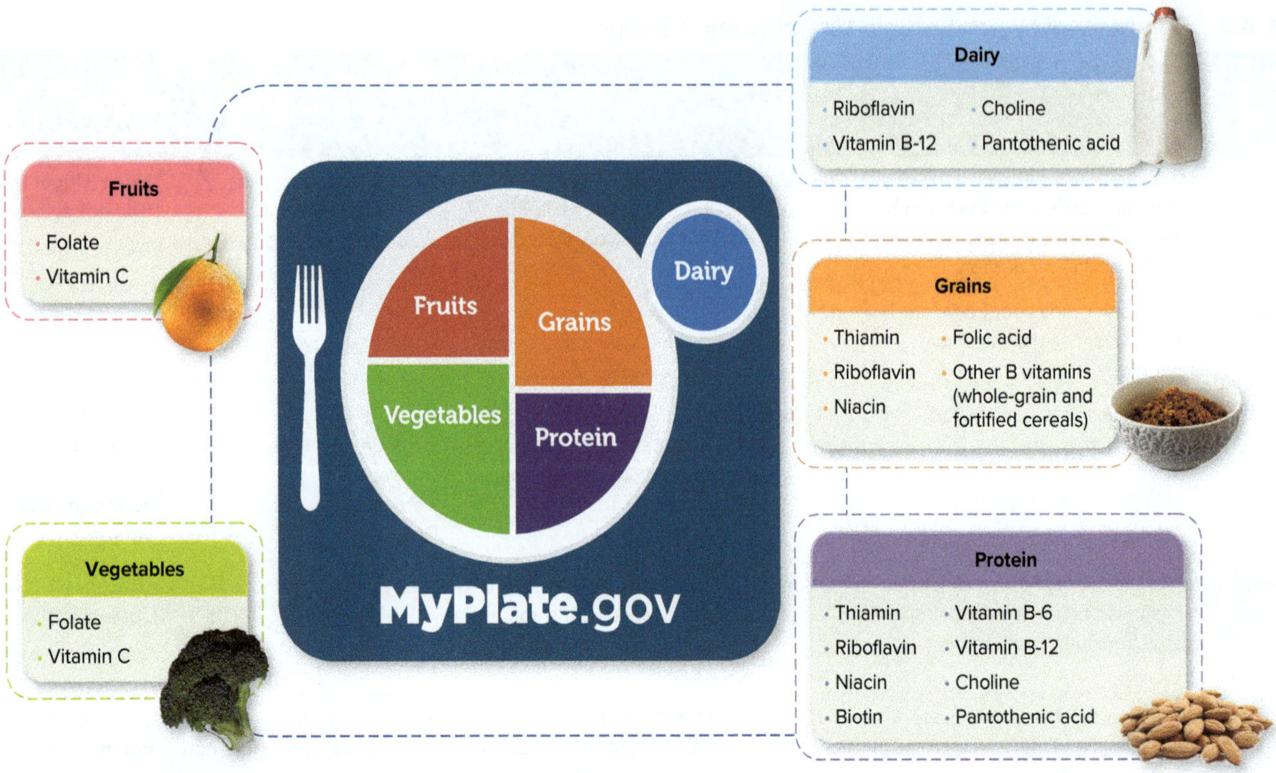

Figure 8.19 MyPlate: Water-soluble vitamins and choline. This illustration highlights MyPlate food groups that generally are good sources of water-soluble vitamins and choline.

MyPlate food guide icon: U.S. Department of Agriculture; Orange: yurakp/123RF; Broccoli, Bowl of cereal: ©Wendy Schiff; Milk: Judith Collins/Alamy Stock Photo; Almonds: Ingram Publishing/SuperStock

beriberi thiamin deficiency disease

Thiamin

In the body, thiamin is used to make a coenzyme that participates in chemical reactions involved in the release of energy from carbohydrates. Additionally, the thiamin–containing coenzyme plays a role in the metabolism of certain amino acids, and the coenzyme may be necessary for the synthesis of neurotransmitters. A neurotransmitter is a chemical produced by a nerve cell that enables the cell to communicate to other nerve cells.

Food Sources of Thiamin

Foods that contribute thiamin to American diets include whole–grain and enriched breads and cereals, pork, nuts, legumes, and orange juice. Table 8.8 lists some common foods that are good sources of thiamin. Overcooking and cooking food in alkaline solutions, such as water to which some baking soda has been added to brighten the color of green vegetables, can destroy thiamin.

Dietary Adequacy

The adult RDA for thiamin is 1.2 mg/day and 1.1 mg/day for males and females, respectively. There are no reports of toxicity from consuming high amounts of thiamin from food or supplements,[18] probably because the excess vitamin is readily excreted in urine. Thus, no UL has been established for thiamin.

The thiamin deficiency disease is called **beriberi.** People suffering from beriberi are very weak and have poor muscular coordination (Fig. 8.20). The severe lack of thiamin also negatively affects the functioning of the cardiovascular, digestive, and nervous systems.

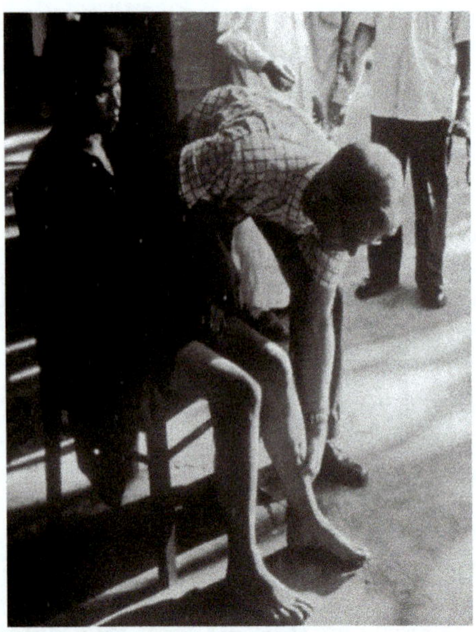

Figure 8.20 Beriberi. This woman has a form of beriberi that causes fluid to accumulate in the legs (edema).

Centers for Disease Control and Prevention

TABLE 8.8 Thiamin Content of Selected Foods

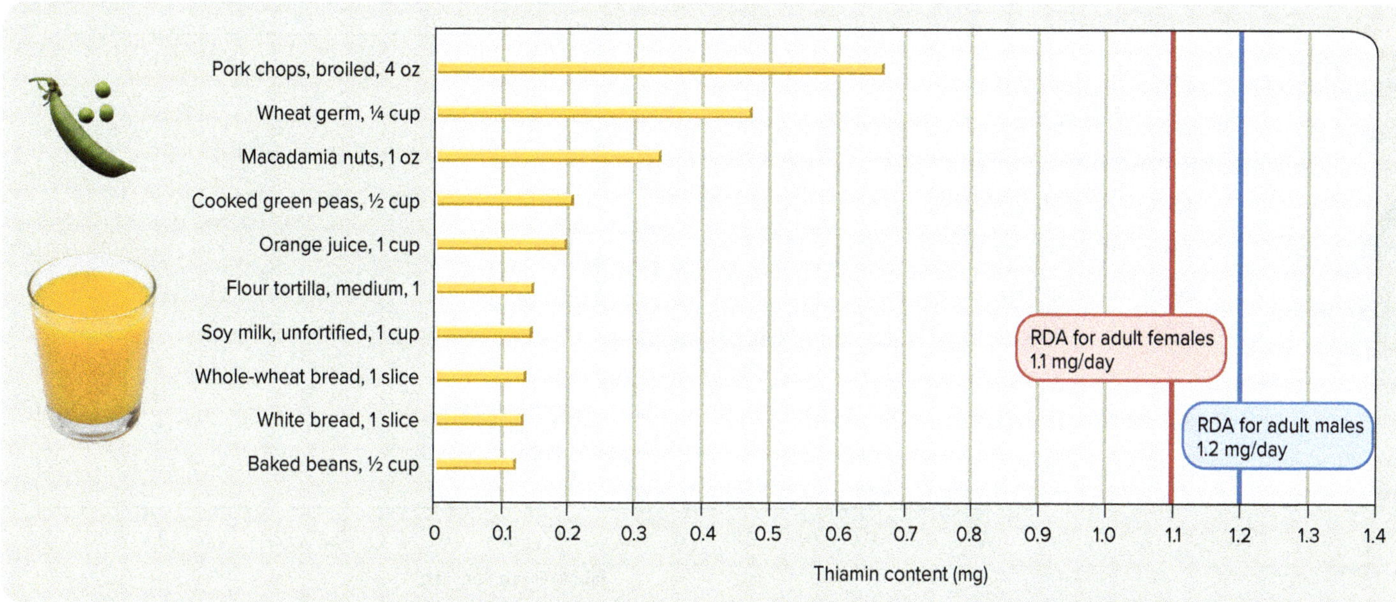

Source: Data from U.S. Department of Agriculture, Agricultural Research Service: *FoodData Central*, 2019.
Juice: Stockbyte/Getty Images; Peas: C Squared Studios/Photodisc/Getty Images

In the United States, the degenerative brain disorder associated with thiamin deficiency is called **Wernicke–Korsakoff syndrome** (*vear′–nih–key kor′–sah–koff*). Most cases of Wernicke–Korsakoff syndrome occur in people with alcohol use disorder (AUD) because alcohol reduces thiamin absorption and increases the vitamin's excretion. People with AUD also tend to have poor eating habits that contribute to deficiencies of thiamin and other vitamins. Signs of Wernicke–Korsakoff syndrome include abnormal eye movements, staggering gait, and distorted thought processes. Treatment involves avoiding alcohol and obtaining thiamin injections. Without prompt treatment, people with Wernicke–Korsakoff syndrome can become disabled permanently or die.

Wernicke-Korsakoff syndrome degenerative brain disorder resulting from thiamin deficiency that occurs primarily among people with alcohol use disorder

Riboflavin

Riboflavin is a component of two coenzymes that play key roles in metabolism of carbohydrates, lipids, and amino acids. Milk, yogurt, and other milk products; enriched cereals; and liver are among the best sources of riboflavin. Mushrooms, broccoli, asparagus, and spinach and other green leafy vegetables also contain substantial amounts of the vitamin. Table 8.9 lists these and other food sources of riboflavin. Riboflavin's chemical structure is fairly stable, but exposure to light causes the vitamin to break down rapidly. Therefore, riboflavin–rich foods, such as milk and milk products, should not be packaged or stored in clear glass containers.

Dietary Adequacy

The RDA for riboflavin is 1.1 mg/day for females and 1.3 mg/day for males. The average riboflavin intake of adult Americans is about 1.8 mg/day for females and 2.5 mg/day for males.[4] In the United States, riboflavin deficiency rarely occurs because many commonly eaten foods contain riboflavin (see Table 8.9). However, people who do not consume milk, milk products, or enriched breads and cereals may develop mild cases of riboflavin deficiency. A symptom of mild riboflavin deficiency is to become fatigued easily. Riboflavin is not well absorbed by the intestinal tract, so consuming large amounts of the vitamin does not appear to cause side effects.[19] Thus, no UL has been established for riboflavin.

Did *YOU* Know?

Riboflavin is naturally yellow. If you take a dietary supplement that contains high amounts of riboflavin, your kidneys will excrete the excess, and you may notice the bright yellow color of your urine.

TABLE 8.9 Riboflavin Content of Selected Foods

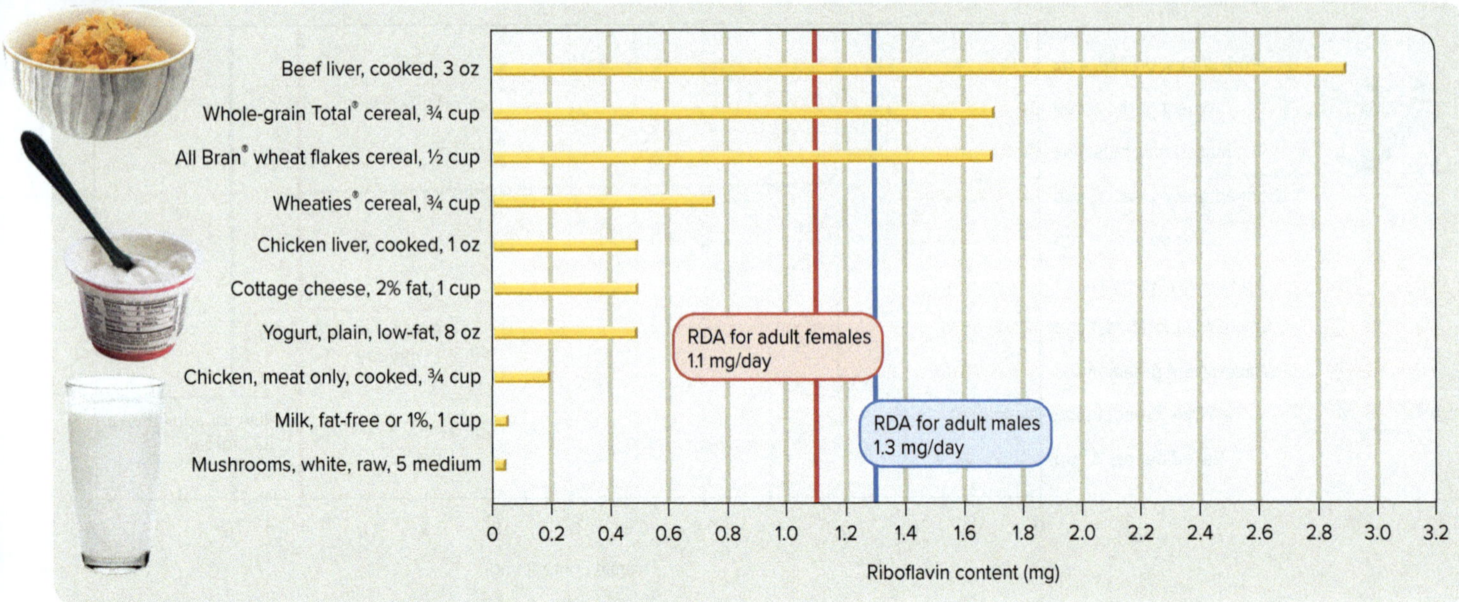

Source: Data from U.S. Department of Agriculture, Agricultural Research Service: *FoodData Central*, 2019.
Cereal: ©Wendy Schiff; Yogurt: Hutchings Photography/Digital Light Source/McGraw Hill; Milk: Nipaporn Panyacharoen/Shutterstock

Niacin

The body uses niacin to synthesize two coenzymes that participate in at least 200 reactions, including those involved in the release of energy from macronutrients. Major food sources of niacin include enriched cereals, beef liver, tuna, salmon, poultry, pork, and mushrooms (Table 8.10). The chemical structure of niacin is very heat stable, so food retains much of its niacin content during usual preparation and cooking methods.

TABLE 8.10 Niacin Content of Selected Foods

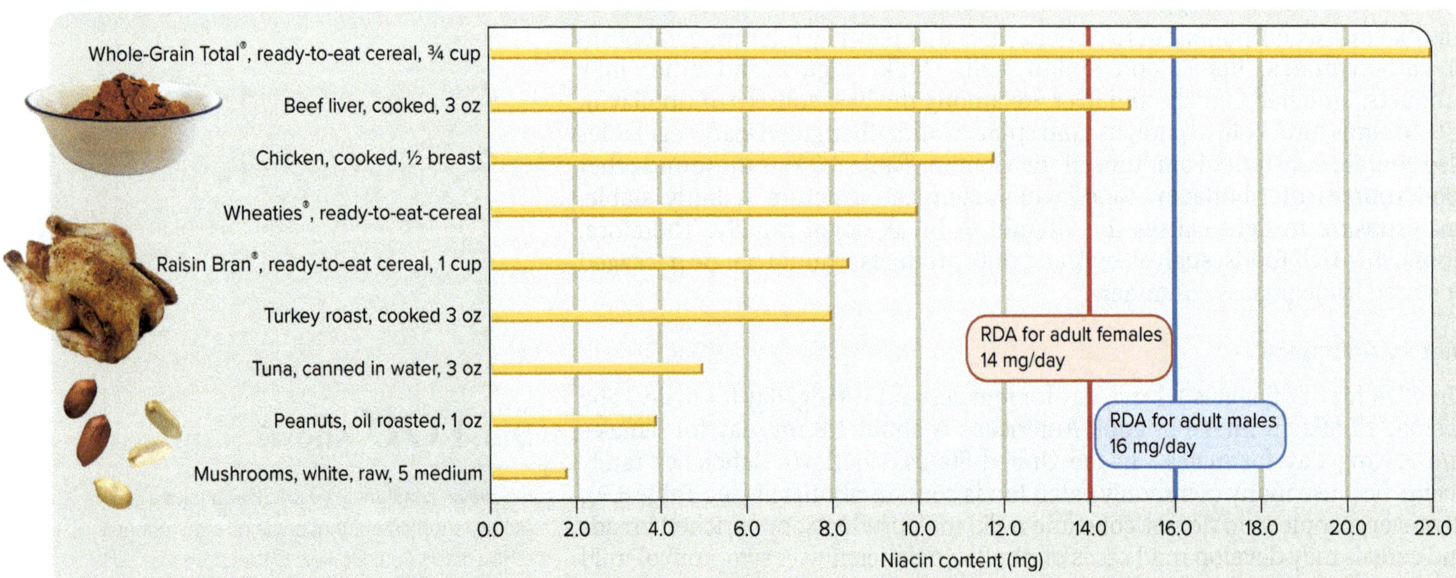

Source: Data from U.S. Department of Agriculture, Agricultural Research Service: *FoodData Central*, 2019.
Bowl of cereal: ©Wendy Schiff; Roast chicken: Louis Hiemstra/E+/Getty Images; Peanuts: C Squared Studios/Photodisc/Getty Images

When diets supply plenty of protein–rich foods, the human body can synthesize niacin from a precursor, the amino acid tryptophan. It takes about 60 mg of tryptophan to yield about 1 mg of niacin. For example, eggs and milk lack niacin, but they are rich sources of tryptophan that can be converted to the B vitamin.

The niacin content of corn is considerably higher than that of most other vegetables, but the B vitamin is tightly bound to a protein that resists digestion. Thus, people who eat corn as their staple food are prone to develop a niacin deficiency. The traditional Mexican diet is corn based, but niacin deficiency was not a widespread concern in Mexico when it was a major health concern in other parts of the world. Why? The Mexican practice of soaking corn kernels in limewater (an alkaline solution of water and calcium hydroxide) before using them to prepare tortillas helps free the niacin, enhancing its ability to be absorbed. In the United States, corn products such as hominy and grits are sources of niacin because they have been treated with limewater before cooking.

Figure 8.21 Pellagra. This person with pellagra shows one of the classic signs of the niacin deficiency disease: dermatitis.
Centers for Disease Control and Prevention

Dietary Adequacy

The adult RDA for niacin is 14 mg/day (females) and 16 mg/day (males). In the United States, people with alcoholism and those with rare disorders that disrupt tryptophan metabolism are generally the only groups at risk of niacin deficiency. Early signs and symptoms of mild niacin deficiency include poor appetite, weight loss, and weakness. If the affected person continues to consume a niacin–deficient diet, the condition worsens and **pellagra** (*peh–lah′–gra* or *peh–lay′–gra*) develops. The classic signs and symptoms of pellagra are dermatitis, diarrhea, dementia, and death—the "4 Ds of pellagra" (Fig. 8.21). In the early twentieth century, pellagra was widespread in the southeastern United States (see the "Did You Know?" box in Section 2.1 in Chapter 2). Today, the disease is rare in Western societies, but it still occurs among impoverished populations in developing countries, particularly in regions of Africa, India, and China.

The adult UL for niacin is 35 mg/day. There have been no reports indicating that the niacin naturally in foods can cause toxicity.[20] Physicians may prescribe megadoses of niacin supplements to treat elevated blood cholesterol levels. However, patients taking very high doses of the vitamin (more than 3 grams/day) may experience side effects such as gastrointestinal ulcers, loss of vision, and liver damage.[21] The "Vitamins as Medicines" section of this chapter discusses the use of niacin supplements to reduce cholesterol levels.

Did *YOU* Know?

In the early 1950s, some babies became unusually irritable and developed seizures after being fed a commercial infant formula. Scientists determined that the vitamin B-6 in the formula had been destroyed by excessive heating during the manufacturing process. Vitamin B-6 is needed for neurotransmitter synthesis, so the irritability and seizures may have resulted from a lack of neurotransmitters in the infants' brains. The babies were effectively treated with vitamin B-6.

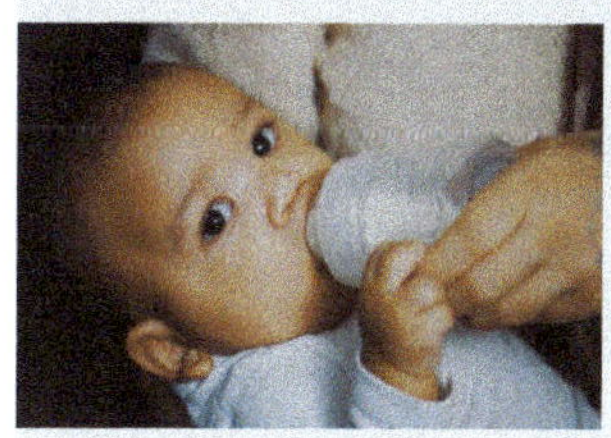

CameraCraft/Shutterstock

Vitamin B-6

The body requires vitamin B–6 to make a coenzyme needed for amino acid metabolism, including the conversion of the amino acid tryptophan to niacin and transamination reactions that form nonessential amino acids (see Figure 7.11). The coenzyme that contains vitamin B–6 also helps convert a toxic amino acid, homocysteine, to cysteine, a nonessential amino acid. During red blood cell (RBC) production, the coenzyme participates in the synthesis of heme. Heme is the iron–containing portion of **hemoglobin,** the protein in RBCs that transports oxygen. If vitamin B–6 is unavailable for heme synthesis, a type of anemia develops. **Anemia** is a disorder characterized by fewer–than–normal red blood cells and poor oxygen transport in blood. Vitamin B–6 is also involved in the synthesis of *neurotransmitters,* chemicals that nerves produce to transmit messages.

pellagra niacin deficiency disease

hemoglobin iron-containing protein in red blood cells that transports oxygen

anemia disorder characterized by too few red blood cells and poor oxygen transport in blood

Food Sources of Vitamin B-6

Liver, meat, fish, and poultry are among the best dietary sources of vitamin B–6. Additionally, potatoes, bananas, spinach, sweet red peppers, and broccoli

TABLE 8.11 Vitamin B-6 Content of Selected Foods

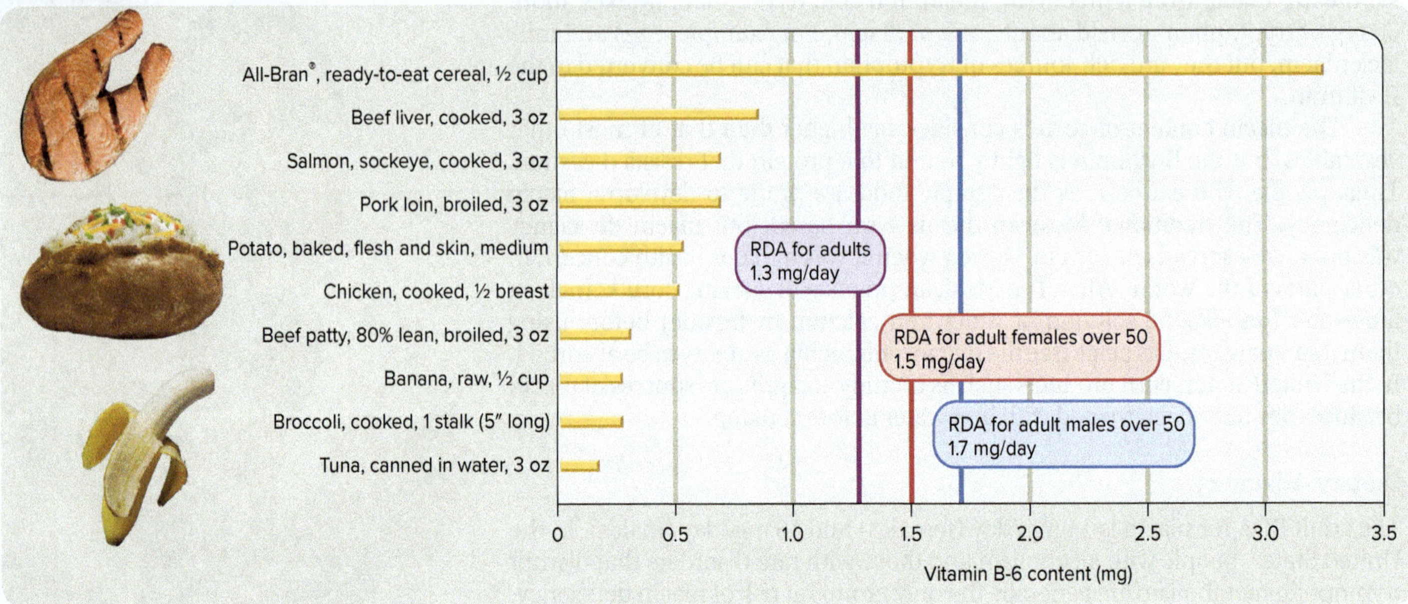

Source: Data from U.S. Department of Agriculture, Agricultural Research Service: *FoodData Central*, 2019.
Baked potato: DNY59/E+/Getty Images; Salmon: C Squared Studios/Photodisc/Getty Images; Banana: Stockdisc/Getty Images

are good sources of vitamin B–6. During the refining process, the vitamin B–6 that is naturally in grains is lost, and the nutrient is not added back to the grain products during enrichment. However, many ready–to–eat and cooked cereals have been fortified with the vitamin. Table 8.11 lists some foods that are major sources of vitamin B–6. During cooking, excessive heat can cause major losses of the vitamin.

Dietary Adequacy

The adult RDAs for vitamin B–6 range from 1.3 to 1.7 mg/day. In the United States, the average adult consumes more than the RDA of the vitamin.[4] Therefore, cases of vitamin B–6 deficiency are rare, but they can result from alcoholism or genetic conditions that affect vitamin B–6 metabolism. Signs and symptoms of vitamin B–6 deficiency include dermatitis, anemia, seizures, depression, and confusion.

The adult UL for vitamin B–6 is 100 mg/day. Unlike most B vitamins, megadoses of vitamin B–6 are toxic.[22] The "Vitamins as Medicines" section of this chapter provides information about vitamin B–6 toxicity.

Folate

"Folate" is the name for a group of related compounds that includes folate in foods and **folic acid.** Folic acid refers specifically to the synthetic form of the vitamin generally found in supplements and added to fortify foods.[23] In the body, cells convert all forms of folate to a group of folate–containing coenzymes collectively called **tetrahydrofolate** (*te′–tra–hi–drow–foe′–late*), or simply **THF.** THF accepts a single–carbon group, such as a methyl group (CH_3), from one compound and becomes 5–methyltetrahydrofolate (5–MTHF) as a result. 5–MTHF is the main form of folate in the bloodstream. 5–MTHF transfers its methyl group to another substance. As a result, THF participates in many chemical reactions involved in DNA synthesis and amino acid metabolism. As cells prepare to divide, they need THF to make DNA.

folic acid form of folate

tetrahydrofolate (THF) folate coenzyme

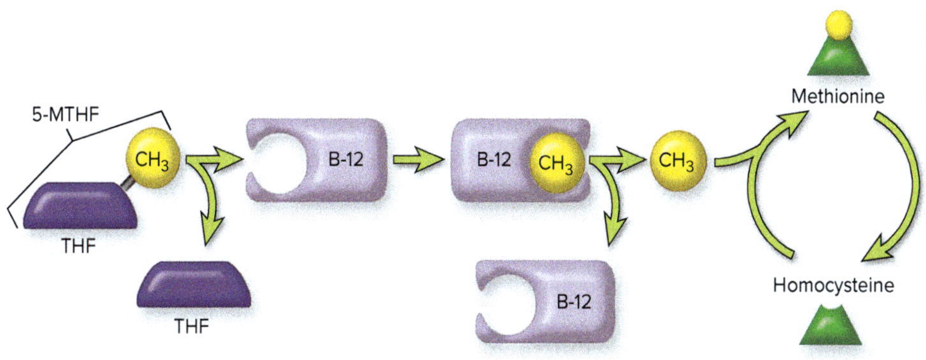

Figure 8.22 Folate and vitamin B-12: Working together. This diagram shows how folate in the form of methyl-THF works with vitamin B-12 to transfer a methyl group (CH_3).

Certain roles of folate and vitamin B–12 are interrelated. 5–MTHF can transfer a CH_3 group to vitamin B–12, which, in turn, transfers the CH_3 group to homocysteine, forming the essential amino acid methionine (Fig. 8.22). This process recycles methionine. When vitamin B–12 is unavailable, folate cannot be used, and a deficiency of the vitamin occurs, even though dietary intakes of folate are adequate.[24]

Food Sources of Folate

Green leafy vegetables, liver, legumes, asparagus, broccoli, and orange juice are good sources of naturally occurring folate. The synthetic folic acid that is used to fortify or enrich food is better absorbed than naturally occurring forms of folate.

The folate content of foods may be reported as micrograms of *dietary folate equivalents* (*DFE*s). DFE units account for differences in the body's ability to absorb folic acid and natural forms of folate. Table 8.12 lists the folate content of selected foods in micrograms of DFE.

Folate is extremely susceptible to destruction by heat, oxidation, and UV light. Food processing and preparation can destroy 50 to 90% of the folate in food. By eating fresh fruits and raw or lightly cooked vegetables, you are likely to obtain most of the foods' folate content.

TABLE 8.12 Folate Content of Selected Foods

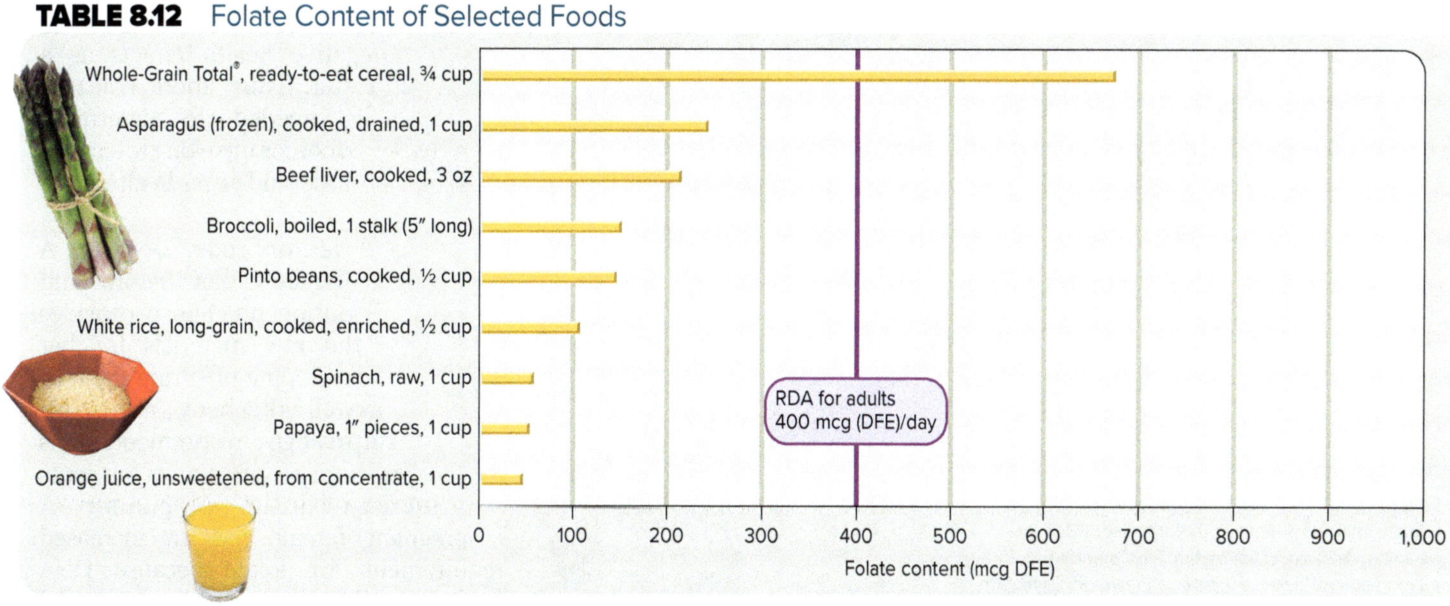

Source: Data from U.S. Department of Agriculture, Agricultural Research Service: *FoodData Central*, 2019.
Asparagus: Pixtal/age fotostock; Rice: ©Wendy Schiff; Juice: Stockbyte/Getty Images

Dietary Adequacy

The adult RDA for folate is 400 mcg (DFE)/day. The diets of most Americans provide adequate amounts of folate.[4] The risk of folate deficiency is highest in alcoholics, females of childbearing age, pregnant females, and people with disorders of the intestinal tract that reduce the vitamin's absorption.[23]

Folate deficiency usually results from nutritionally inadequate diets, but excess alcohol consumption and use of certain medications can negatively affect the body's ability to absorb and use folate, resulting in deficiencies of the vitamin. Initially, folate deficiency affects cells that rapidly divide, such as red blood cells. Mature RBCs do not have nuclei, and they live for only about 4 months. Thus, the body must replace old or worn–out RBCs constantly. To keep up with their rapid rate of cell division, the precursor cells that mature into RBCs must actively synthesize DNA. Without folate, RBC precursor cells that reside in bone marrow enlarge, but they cannot divide normally because they are unable to form new DNA. Bone marrow releases some of the abnormal RBCs into the bloodstream before they mature (Fig. 8.23). This condition, called megaloblastic (*mega* = large; *blast* = immature cell) anemia, is characterized by large, immature RBCs (*megaloblasts*) that still have nuclei and do not carry normal amounts of oxygen.

Because many of folate's metabolic roles are related to those of vitamin B–12, diets that lack either vitamin produce a number of identical deficiency signs and symptoms. For example, being deficient in folate or vitamin B–12 can cause megaloblastic anemia. Therefore, a person with this type of anemia needs further analysis of their blood to determine which vitamin is lacking.

Although the folate naturally in foods does not appear to be toxic, the UL for the synthetic form of the vitamin (folic acid) is 1000 mcg/day. The UL was established because taking folic acid supplements can cure not only the anemia that occurs in folate deficiency, but also the anemia that is a sign of vitamin B–12 deficiency. Folic acid supplementation, however, does not prevent the serious nervous system damage that accompanies the B–12 deficiency. Furthermore, some medical experts are concerned that taking folic acid supplements and consuming foods that contain the micronutrient may cause excess folic acid to accumulate in blood and produce negative health effects as a result. Thus, more research is needed to determine whether ingesting too much folic acid poses health risks.

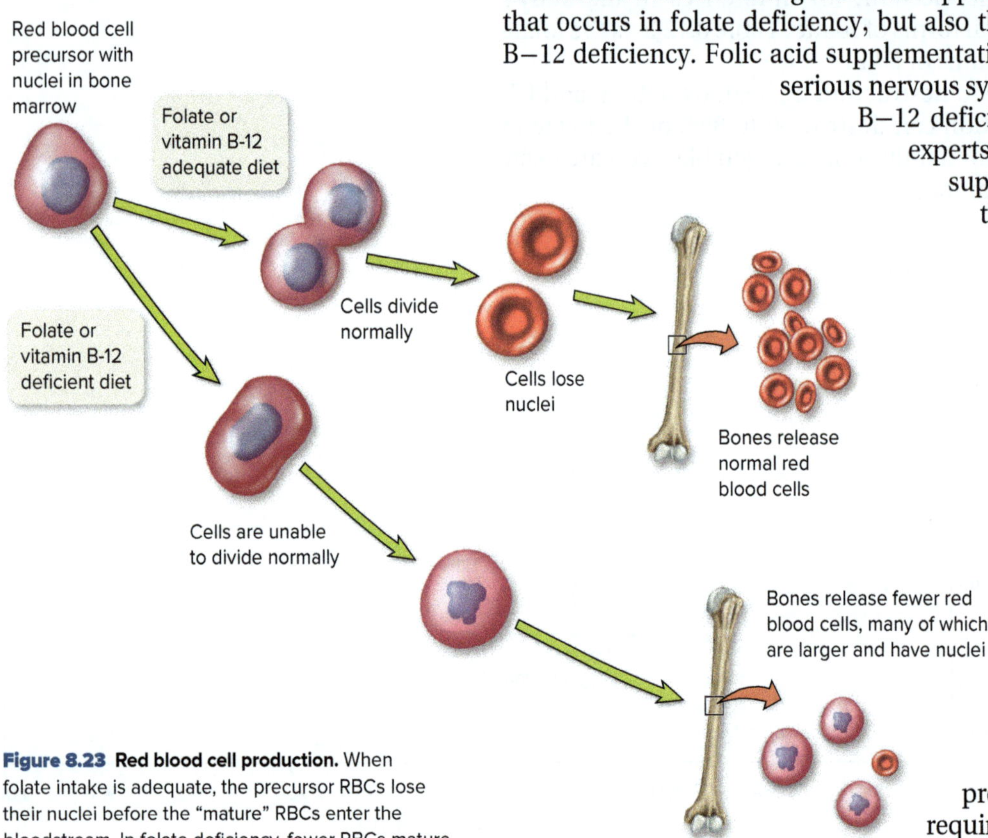

Figure 8.23 Red blood cell production. When folate intake is adequate, the precursor RBCs lose their nuclei before the "mature" RBCs enter the bloodstream. In folate deficiency, fewer RBCs mature. Many of the cells are larger and retain their nuclei. RBCs also need vitamin B-12 to divide normally.

Neural Tube Defects A female's diet before and during pregnancy provides the raw materials for her developing offspring's needs, as well as her body's needs. Not surprisingly, recommendations for intakes of many nutrients increase during pregnancy. A pregnant female has an increased requirement for folate because DNA synthesis and cell division take place at a rapid pace during embryonic development.

During the first few weeks after conception, the **neural tube** forms in the human embryo (Fig. 8.24a). This tube eventually develops into the brain and spinal cord. Pregnant females who suffer from folate deficiency have high risk of giving birth to infants with neural tube defects. The two most common neural tube defects are **spina bifida** (*spy′–na bif′–eh–dah*) and **anencephaly** (*an–en–sef′–ah–lee*). Spina bifida occurs when the embryo's spine does not form properly and the bones fail to enclose the spinal cord (Fig. 8.24b). Infants with severe spina bifida have a section of their spinal cord or a sac containing some spinal fluid bulging through an opening in their backs (Fig. 8.24c). Often, people with spina bifida are unable to use muscles in the lower part of their bodies, and as a result, they cannot walk independently. Infants born with anencephaly have much of their brain malformed or missing, and they usually die shortly after birth.

In 1992, officials with the U.S. Public Health Service recommended that all females capable of becoming pregnant consume 400 mcg of folic acid daily to prevent neural tube defects. According to medical experts, adequate folic acid intake before and during early pregnancy can prevent many cases of spinal tube defects.[23] For pregnant females, adequate folate status is critical early in pregnancy because the neural tube begins to form about 21 days after conception. This developmental milestone occurs when many females are not even aware they are pregnant.

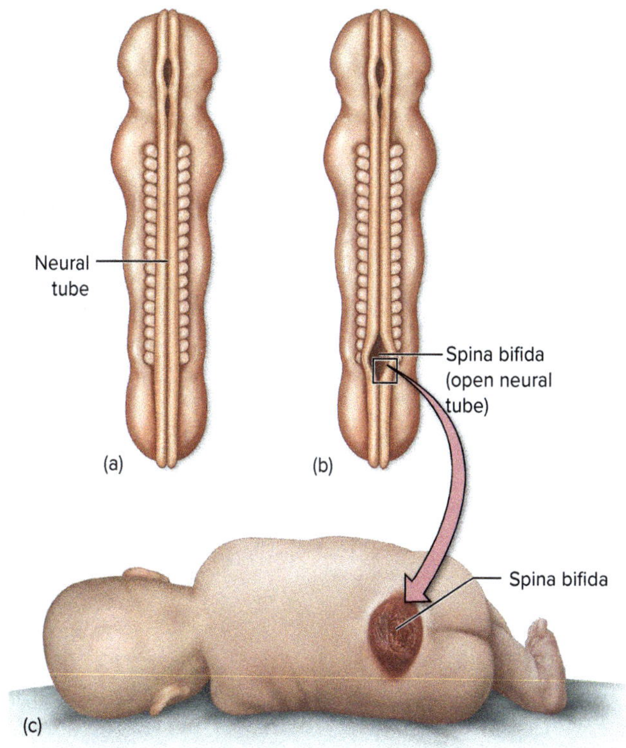

Figure 8.24 Neural tube. (a) Normal neural tube in an early embryo. (b) Spina bifida (abnormal neural tube) in an early embryo. (c) Infant with severe spina bifida.

Did *YOU* Know?

Recall from Figure 8.22 that cells bind a methyl group (CH_3) to folate (THF) to form MTHF. To convert MTHF to the biologically active form of folate (5-MTHF), cells need the enzyme *methylenetetrahydrofolate reductase (MTHFR)*. If a mutation occurs in the gene that codes for the synthesis of MTHFR, production of 5-MTHF is slowed or blocked. Therefore, fewer methyl groups are transferred to vitamin B-12, resulting in slower recycling of methionine and a buildup of homocysteine in the bloodstream.

Normally, people have very little homocysteine in their bloodstream. Recall from Chapter 6 that elevated blood homocysteine levels may increase the risk of cardiovascular disease (CVD). Because the gene that codes for the enzyme that helps convert folate to 5-MTHF is defective and 5-MTHF is needed for DNA production, females with the defective gene are at risk of giving birth to a baby with neural tube defects.

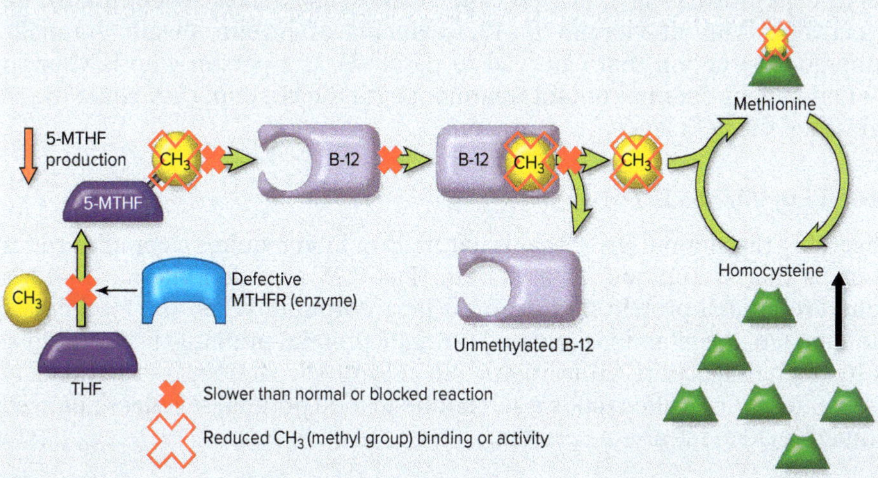

neural tube embryonic structure that eventually develops into the brain and spinal cord

spina bifida type of neural tube defect in which the spine does not form properly before birth, and it fails to enclose the spinal cord

anencephaly type of neural tube defect in which the brain does not form properly or is missing

Each year, about 3000 pregnancies in the United States are affected by anencephaly and spina bifida.[25] However, the prevalence of these neural tube defects has declined since enrichment of grains with folic acid began in 1998.

Because females of childbearing age may become pregnant at some point, they need to be aware of the association between the lack of folate and neural tube defects. In addition to including folate–rich foods in their diets, young females can prepare for pregnancy by taking a daily multivitamin supplement that contains 400 mcg of synthetic folic acid. Although folic acid intakes have increased among Americans since the United States began enriching grain products with the vitamin, only 34% of females who are of childbearing age report that they take a daily folic acid supplement.[25]

Depression Growing scientific research reports that people with low serum folate levels have an increased risk for depression, and many people diagnosed with depression have lower serum folate levels.[23,26] Scientists have studied folic acid and 5–MTHF supplementation as *adjuvant therapies,* combining the B–vitamin with traditional antidepressant medications. Supplementation with folic acid has not shown consistent results as adjuvant therapy for depression. However, researchers report that a high–dose 5–MTHF supplementation of 15,000 mcg per day, 15 times the UL for folate (1000 mcg), combined with traditional antidepressant medications, does reduce symptoms of depression for some people.[23] Thus, additional research is needed to understand the association between folate status and depression. People should always seek the help of a medical professional to diagnose and manage symptoms of depression and only take high–dose folate supplementation under medical supervision.

Vitamin B-12

Cells require vitamin B–12 to make coenzymes that participate in a variety of cellular processes, including the transfer of CH_3 groups in the metabolism of folate (see Figure 8.22). Vitamin B–12 is also needed to convert folate to coenzyme forms that are needed for metabolic reactions, including DNA synthesis. Vitamin B–12 also participates in homocysteine metabolism.

There is one vital function of vitamin B–12 that does not involve folate: maintaining the *myelin sheaths* that wrap around parts of certain nerve cells, insulating them. Myelin enables the nerves to communicate effectively. Without vitamin B–12, segments of myelin sheath gradually undergo destruction that can lead to paralysis. If a person who is vitamin B–12 deficient does not obtain treatment with the vitamin, they could die as a result of the deficiency.

Absorbing Vitamin B-12

Absorbing the vitamin B–12 that is naturally in food requires a complex series of steps that are unique for a vitamin (Fig. 8.25). Natural vitamin B–12 is bound to animal protein that prevents its absorption. When the food enters the stomach, the vitamin is released from the protein, primarily by the actions of hydrochloric acid (HCl) in gastric juice. Synthetic vitamin B–12 in dietary supplements or fortified foods is not bound to food protein, so it does not need stomach acid to release the protein.

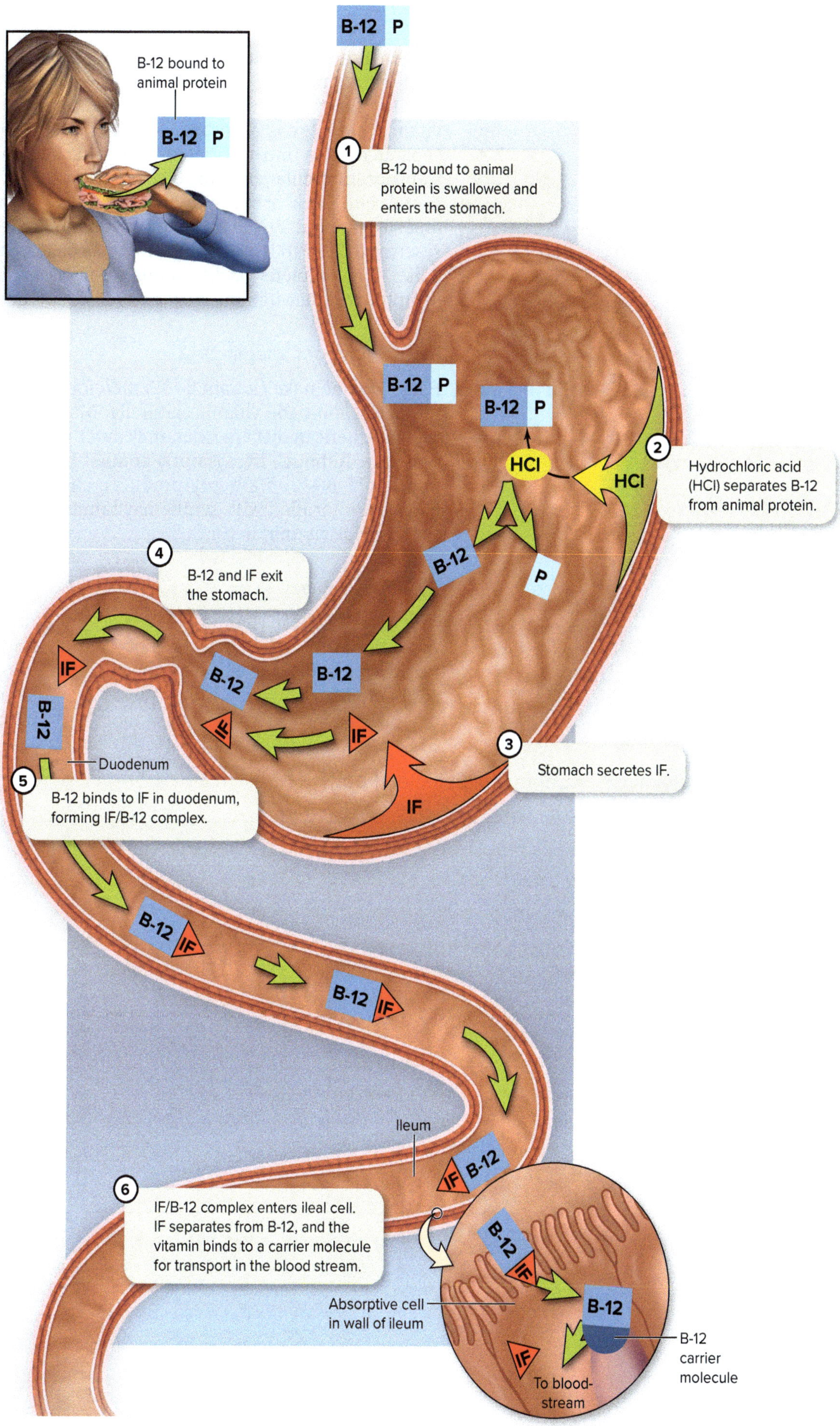

Figure 8.25 Natural vitamin B-12 absorption. Absorbing natural vitamin B-12 from food requires a complex series of steps.

Animal foods are naturally rich sources of vitamin B-12.
nehophoto/Shutterstock

In the small intestine, vitamin B–12 binds to **intrinsic factor (IF),** a compound produced by certain stomach cells. Eventually, the vitamin B–12/intrinsic factor complex reaches the ileum of the small intestine, where the vitamin complex is absorbed. Within the absorptive cells, vitamin B–12 is separated from intrinsic factor and attached to transport molecules. The transport molecules enter the bloodstream and travel to the liver via the hepatic portal vein. The liver removes vitamin B–12 from many of the carrier molecules and stores about 50% of the vitamin. A healthy liver has enough vitamin B–12 reserves to last 5 to 10 years.[27] Therefore, a healthy person who decides to follow a diet that completely lacks vitamin B–12 is not likely to experience signs and symptoms of the vitamin's deficiency disorder for as long as 10 years.

Food Sources of Vitamin B-12

Plants do not make vitamin B–12; therefore, we rely almost entirely on animal foods to supply the vitamin naturally. Major sources of vitamin B–12 in the typical American diet are meat, milk and milk products, poultry, fish, shellfish, and eggs. Although liver is not a popular food, it is one of the richest sources of vitamin B–12. Many soy products, such as soy milk, and ready–to–eat cereals are fortified with synthetic vitamin B–12. Table 8.13 lists some foods that provide vitamin B–12.

Dietary Adequacy

intrinsic factor (IF) substance produced in the stomach that facilitates intestinal absorption of vitamin B-12

The adult RDA for vitamin B–12 is 2.4 mcg/day. Most Americans consume more than the RDA.[4] No UL has been established for vitamin B–12 because no adverse effects have been observed with excess intake from food or dietary supplements.

TABLE 8.13 Vitamin B-12 Content of Selected Foods

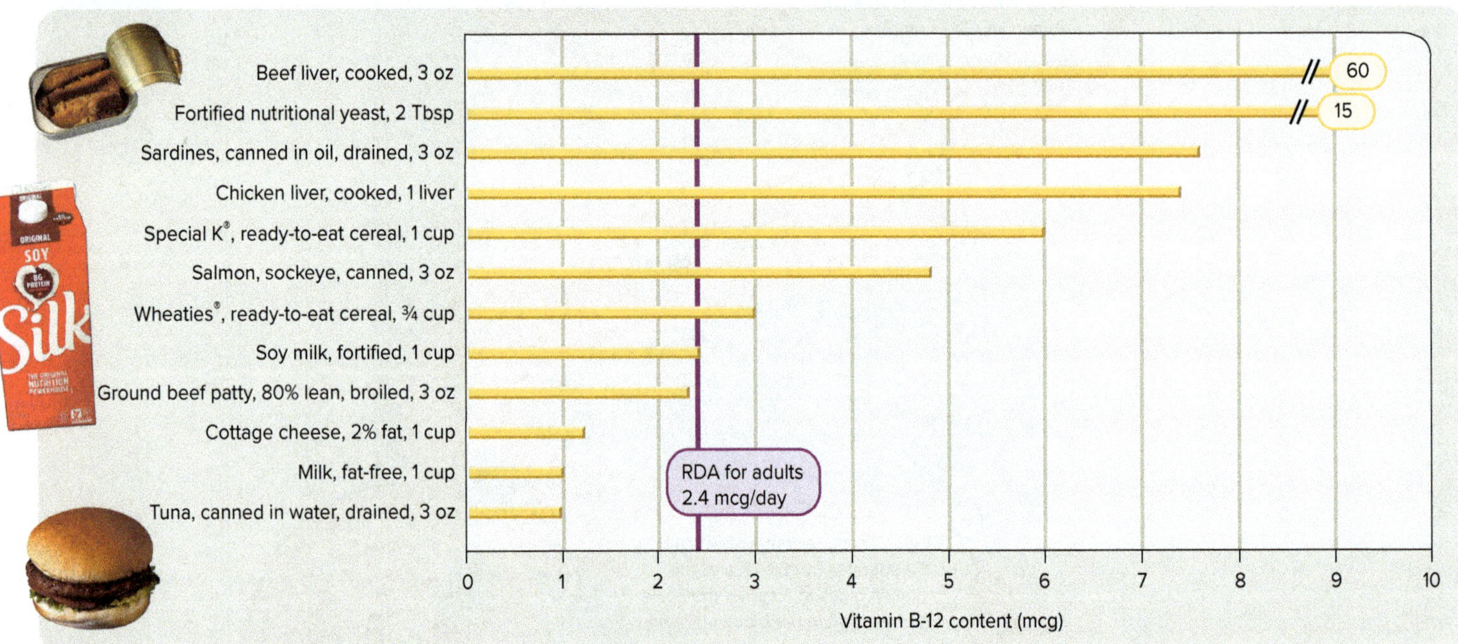

Source: Data from U.S. Department of Agriculture, Agricultural Research Service: *FoodData Central*, 2019.
Sardines, Milk: ©Wendy Schiff; Hamburger: Burke/Triolo Productions/Brand X Pictures/Getty Images

pernicious anemia condition caused by the lack of intrinsic factor and characterized by vitamin B-12 deficiency, nerve damage, and megaloblastic RBCs

Vitamin B-12 Deficiency As many as 15% of Americans have vitamin B–12 deficiency.[28] Vitamin B–12 deficiency is characterized by nerve damage and megaloblastic RBCs. Other common signs and symptoms of the deficiency include muscle weakness, sore mouth, smooth and shiny tongue, memory loss, confusion, difficulty walking and maintaining balance, and numbness and tingling sensations, particularly in the lower extremities. Researchers recently observed an increased risk for depression in people with a vitamin B–12 deficiency, especially in females and older adults.[29,30] Most cases of vitamin B–12 deficiency result from problems that interfere with intestinal absorption of the vitamin (cobalamin) in foods (*food–cobalamin malabsorption*) and not from inadequate intakes.[24]

As people age, HCl production in their stomachs declines. Thus, many older adults have food–cobalamin malabsorption because they are unable to release vitamin B–12 from animal protein.[24] People with this condition can develop vitamin B–12 deficiency despite consuming diets that contain the vitamin. In addition to advanced age, chronic alcoholism, gastric bypass surgeries for weight reduction, and certain medications, particularly those that reduce stomach acid secretion, can contribute to food–cobalamin malabsorption. People with this form of malabsorption produce intrinsic factor, so they can absorb synthetic forms of the vitamin in dietary supplements that contain B–12.

An estimated 2% of older adults have an autoimmune disorder that causes inflammation and destruction of the cells in the stomach that produce intrinsic factor.[24] (An autoimmune disorder results when a person's immune system attacks their own cells.) In these cases, diets may supply adequate amounts of vitamin B–12, but the lack of intrinsic factor prevents most of the micronutrient from being absorbed. Eventually, people who lack intrinsic factor develop a disease called **pernicious** ("deadly") **anemia.** As its name implies, pernicious anemia can lead to vitamin B–12 deficiency and death. Treatment involves bypassing the need for intrinsic factor and intestinal absorption, usually by providing routine vitamin B–12 injections. Furthermore, taking large doses of vitamin B–12 floods the intestinal tract with the vitamin and enables a small amount to be absorbed without the need for intrinsic factor.

In addition to advanced age, family history is a risk factor for pernicious anemia. Therefore, it is a good idea to have your blood tested for signs of vitamin B–12 deficiency as you grow older, especially if you have a close relative with pernicious anemia.

Vegans ("total" vegetarians) avoid eating animal products. Plant foods supply little vitamin B–12, so vegans need to be concerned about their intakes of the nutrient (see the "Vegetarianism" section of Chapter 7). People who eat few or no animal products should consume foods that have been fortified with vitamin B–12, such as fortified soy milk, cereals, and nutritional yeast, or take supplements that supply the vitamin. Vegan females who breastfeed their babies need to be aware that their breast milk may contain inadequate amounts of vitamin B–12. Babies who consume only vitamin B–12–deficient breast milk are likely to develop megaloblastic anemia and serious nervous system problems, including diminished brain growth and spinal cord damage. The signs and symptoms of the deficiency are likely to occur during the first few months of life, particularly when the infants' mothers did not supplement their diets with vitamin B–12 during pregnancy. Providing vitamin B–12 to the deficient babies promptly can prevent permanent damage to the children's nervous systems.

Fortified nutritional yeast can be a source of vitamin B-12 for people who consume few or no animal products. The yeast is sprinkled on foods to provide a cheese-like flavor.
stephaniefrey/123RF

Did *YOU* Know?

Avidin is a protein in raw egg whites that binds to biotin, making the vitamin resist digestion. People who consume raw eggs regularly for long periods of time may develop signs of biotin deficiency. Cooking eggs destroys avidin, making biotin available for absorption.

Clement Mok/Photodisc/Getty Images

Pantothenic Acid

Pantothenic acid is a component of *coenzyme A,* which is critical for energy metabolism and fatty acid production in the body. Pantothenic acid is so widespread in foods that a nutritional deficiency is unlikely to occur among healthy people who eat varied diets. Rich sources of pantothenic acid include cereals that have been fortified with the vitamin, beef and chicken liver, sunflower seeds, mushrooms, peas, and soy milk.

Most Americans consume adequate amounts of pantothenic acid, and as a result, deficiencies of the vitamin are rare. When scientists experimentally cause pantothenic deficiency in human subjects, they note signs and symptoms that include headache, fatigue, impaired muscle coordination, and GI tract disturbances. People who abuse alcohol may develop pantothenic acid and other B–vitamin deficiencies, particularly if their diets are lacking other nutrients as well. At this time, scientists have not set an UL for the vitamin because there have been no reports of toxicity from taking high doses.

Biotin

Biotin participates in chemical reactions that add carbon dioxide to other compounds. By doing so, the vitamin promotes the synthesis of glucose and fatty acids and the breakdown of certain amino acids. Severe deficiencies of biotin rarely occur because intestinal bacteria produce some biotin and the vitamin is in a wide variety of foods. Liver, eggs, peanuts, salmon, pork, mushrooms, and sunflower seeds are good sources of biotin. The typical diet of American adults provides adequate amounts of the micronutrient.

Conditions that reduce intestinal bacteria's biotin production, such as surgical removal of a large part of the colon or taking antibiotics for many months, can contribute to a biotin deficiency. Signs and symptoms of biotin deficiency include skin rash, hair loss, seizures, and some other nervous system disorders. Because reports of biotin toxicity are lacking, nutrition scientists have not set an UL for the micronutrient.

MY DIVERSE PLATE

Guava

Guava is a tropical fruit that is available in many varieties. Although guavas can be eaten fresh, the fruit is often used to make sauces, jellies, and juices. Guavas are an excellent source of fiber and vitamin C. The white or pink flesh of a guava contains more vitamin C than an orange.

©Wendy Schiff

Vitamin C

Most animals do not need dietary sources of vitamin C **(ascorbic acid)** because they can synthesize all the vitamin they need. Humans and guinea pigs are among the few species that are unable to make vitamin C, and for these animals, the micronutrient is essential.

Vitamin C absorption occurs in the small intestine. As intakes of the vitamin increase, the amount absorbed decreases. The small intestine absorbs less of the micronutrient when intakes of the vitamin exceed 200 mg/day.[31] Additionally, the kidneys increase their excretion of the vitamin in response to high intakes. Therefore, taking megadoses of vitamin C may be wasteful because such high amounts of the vitamin are not well absorbed and excesses are eliminated in urine and feces.

Functions of Vitamin C

Vitamin C does not function as part of a coenzyme as do most of the B vitamins, but the vitamin facilitates certain chemical reactions. Vitamin C also acts as an antioxidant by donating electrons to other compounds. In the body, vitamin C has widespread physiological roles; the following sections describe some of the micronutrient's most well–understood functions.

ascorbic acid vitamin C

Collagen Synthesis Vitamin C participates in reactions that form and maintain collagen. **Collagen** is a fibrous protein that gives strength to *connective tissue.* Connective tissues, such as bone, cartilage, and tendons, connect and support other structures in the body. During collagen formation, vitamin C helps create numerous cross–connections between the amino acids in collagen that greatly strengthen the connective tissue (Fig. 8.26). If vitamin C is unavailable, the body forms weak connective tissue and is unable to maintain existing collagen. Some of the more obvious signs of scurvy, such as swollen gums that bleed easily, teeth that loosen and fall out of their sockets, skin that bruises easily, and poor wound healing, are primarily the result of weak connective tissue.

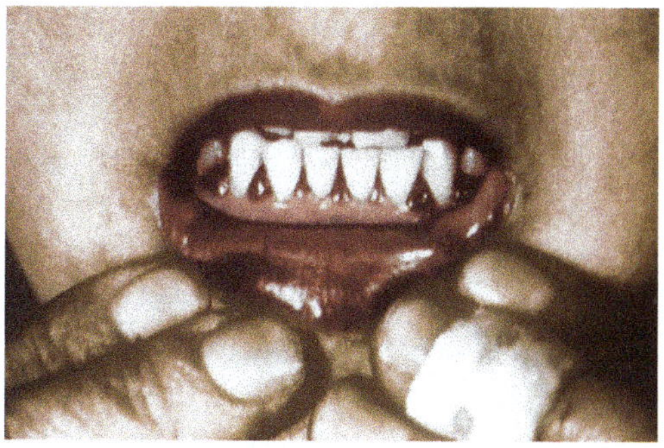

Swollen gums that bleed easily are a sign of scurvy.
Centers for Disease Control and Prevention

Antioxidant Activity Results of some experiments indicate that vitamin C can act as an antioxidant by donating electrons to radicals. Vitamin C also may donate electrons to another antioxidant—vitamin E. Thus, vitamin C recycles vitamin E so that it can regain its antioxidant function. Scientists, however, do not know the extent of vitamin C's antioxidant abilities in the human body. Taking excessive amounts of vitamin C may be harmful because in high doses, the vitamin has **prooxidant** effects. A prooxidant promotes radical production.

Other Roles of Vitamin C in the Body Vitamin C plays a role in the body's immune function, and the vitamin is necessary for the synthesis of bile and certain neurotransmitters. Vitamin C is also involved in the production of various hormones, including the "stress hormone" cortisol; adolsterone, a hormone involved in blood pressure regulation; and thyroxin, the thyroid hormone that regulates energy metabolism.

Food Sources of Vitamin C

Plant foods are the best dietary sources of vitamin C. Peppers, citrus fruit, papaya, broccoli, and berries contain relatively high amounts of the micronutrient (Table 8.14). Potatoes and vitamin C–fortified fruit drinks and cereals also supply vitamin C. Most animal foods are not sources of the micronutrient.

collagen fibrous protein that gives strength to connective tissue

prooxidant substance that promotes free radical production

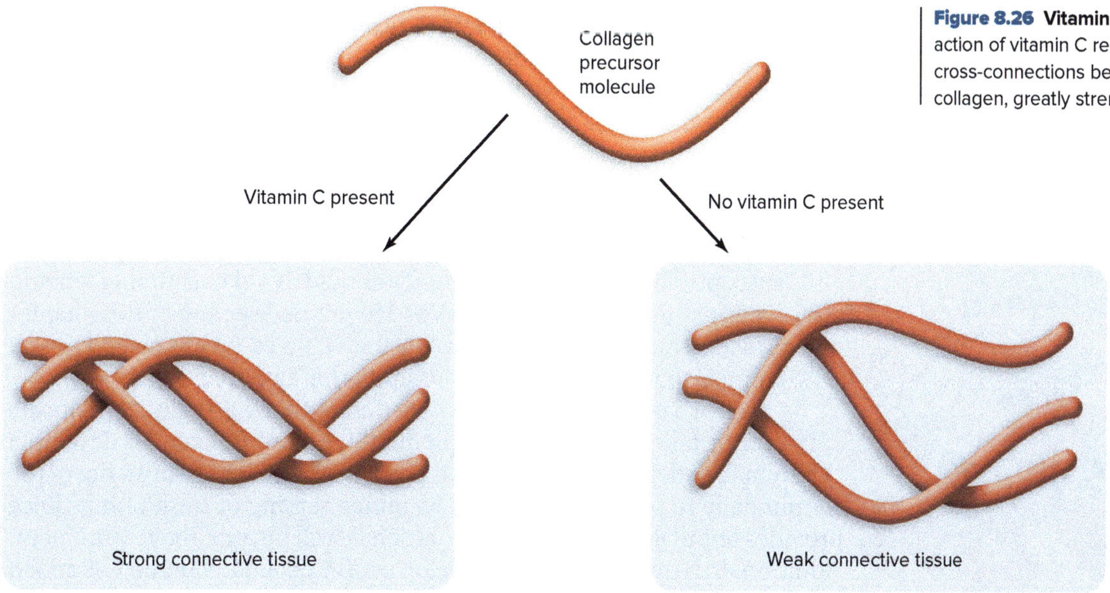

Figure 8.26 Vitamin C and collagen formation. The action of vitamin C results in the formation of numerous cross-connections between the amino acids in collagen, greatly strengthening the connective tissue.

TABLE 8.14 Vitamin C Content of Selected Foods

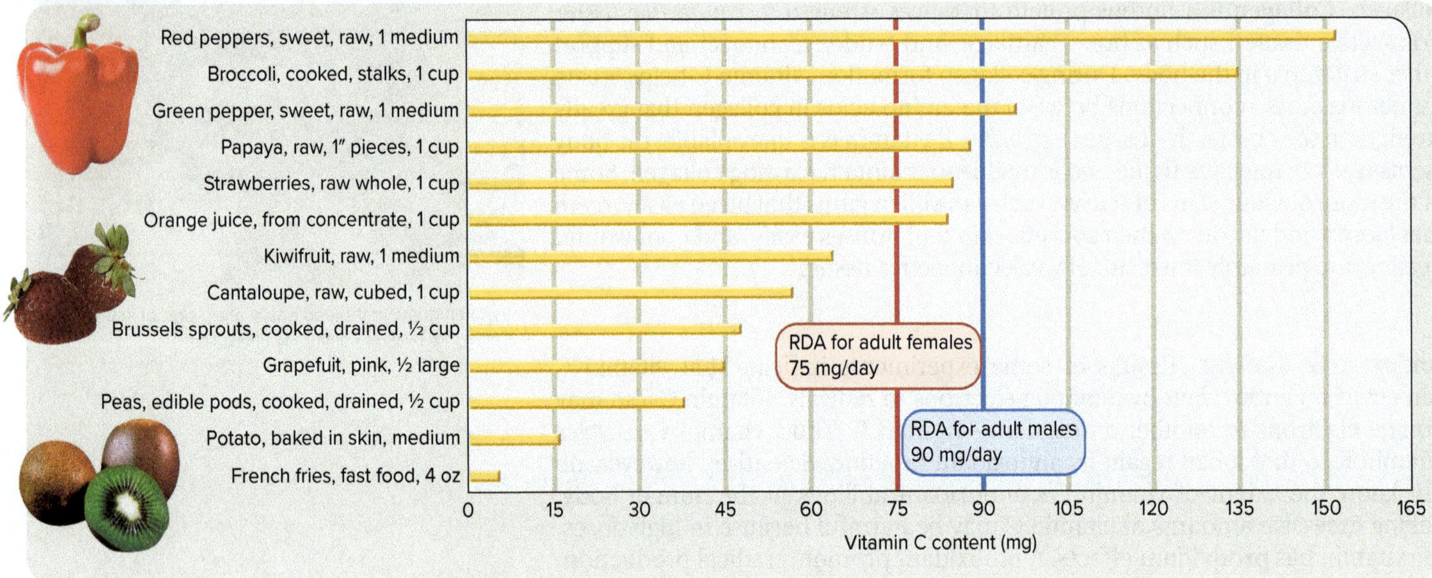

Source: Data from U.S. Department of Agriculture, Agricultural Research Service: *FoodData Central*, 2019.
Pepper: Manfred Zobrist/Hemera/Getty Images; Strawberries: Burke/Triolo Productions/Artville/Getty Images; Kiwifruit: Ingram Publishing/Alamy Stock Photo

Vitamin C is very unstable in the presence of heat, oxygen, light, alkaline conditions, and the minerals iron and copper. Storing vitamin C–rich foods in cool conditions, such as in the refrigerator, helps preserve the micronutrient. Because vitamin C is easily lost during cooking, eat raw fruits and vegetables whenever possible.

Did *YOU* Know?

People who are exposed to cigarette smoke exhaled by smokers ("passive smokers") need to consume more than the RDA for vitamin C, but no specific amount has been recommended for these persons.[5]

Dietary Adequacy

The adult RDA for vitamin C is 75 and 90 mg/day for females and males, respectively. Cigarette smokers need to add an extra 35 mg/day to their RDA because cigarette smoke contains free radicals and exposure to the smoke increases radical formation in their lungs. By including vitamin C–rich fruits and vegetables in their diets, healthy people can obtain adequate amounts of vitamin C.

The adult UL for vitamin C is 2000 mg/day. When people exceed this amount of the vitamin, gastrointestinal upsets, including diarrhea, often occur.[2] Recall that taking megadoses of vitamin C supplements is wasteful because high amounts of the vitamin are not well absorbed and the excesses are eliminated in urine and feces. Furthermore, when cells are saturated with vitamin C, the excess vitamin and *oxalate,* a by–product of breaking down vitamin C, circulate in the bloodstream. The kidneys filter and eliminate these unnecessary substances in urine. Excess oxalate excretion may raise the risk of kidney stones.[5] Therefore, people who are susceptible to develop such stones should avoid consuming vitamin C supplements. The "Vitamins as Medicines" section of this chapter provides more information about the value of taking megadoses of vitamin C to prevent or treat colds and other disorders.

©Wendy Schiff

Even if you do not regularly eat foods that naturally contain the vitamin or take a vitamin C supplement, you are unlikely to develop a vitamin C deficiency because most people require less than 10 mg of the vitamin daily to prevent scurvy. An 8–ounce serving of fresh orange juice provides about 80 mg of vitamin C, which is well beyond the requirement. Additionally, Americans rarely develop scurvy because vitamin C is added to many processed foods, including fruit and sports drinks, ready–to–eat cereals, and nutrition or power bars. If you would still like to take a vitamin C supplement daily, consider choosing a product that provides up to 200 mg of the vitamin in each tablet.

Choline

The body needs the vitamin–like nutrient choline to produce phospholipids in cell membranes and acetylcholine, a neurotransmitter associated with learning and memory, muscle control, and many other nervous system functions. Additionally, the liver uses choline to synthesize a part of a phospholipid (see Chapter 6). Choline is widely distributed in foods. Liver, wheat germ, eggs, beef, and pork are among the richest sources of the nutrient. The source of choline in egg yolk is the phospholipid lecithin (see Chapter 6).

The adult AI for choline is 425 mg/day (females) and 550 mg/day (males). On average, adult Americans consume about 332 mg of choline per person per day.[4] Although the amount of choline that Americans consume is often less than the AI for the micronutrient, the body can make choline from serine, an amino acid. However, the body cannot produce enough choline to meet the amount that is required, especially during some life stages, including pregnancy and lactation (see Chapter 13). Vegans may develop choline deficiency because they do not eat animal sources of food.[32] Choline deficiency can cause liver damage. Excessive choline intake is associated with a fishy body odor and reduced blood pressure.

Concept CHECKPOINT

7. Explain how most B vitamins function.
8. Prepare a table of water-soluble vitamins and choline. For each of these micronutrients, indicate its major function in the body, major food sources, deficiency disorder (if it has a specific name), and major signs and symptoms of the deficiency disorder. If the micronutrient is known to be toxic, also indicate major toxicity signs and symptoms. Check your table against the information provided in Table 8.7.

See Appendix F for responses.

8.4 Vitamins as Medicines

Learning Outcome

1. Evaluate the use of vitamin supplements with respect to their potential health benefits and hazards.

Vitamin supplements are effective for treating people with specific vitamin deficiency diseases, metabolic defects that increase vitamin requirements, and a few other medical conditions. However, scientific evidence generally does not support claims that megadoses of vitamins can prevent or treat everything from gray hair to lung cancer.

How did vitamins earn the reputation of being cure–alls? The fact that a very small amount of a vitamin can prevent or cure the vitamin's deficiency disease provides the foundation for beliefs that these micronutrients are useful for preventing or treating serious chronic diseases. Furthermore, many people think that vitamins are helpful and safe in any amount. When ingested in high doses, however, vitamins and related compounds such as carotenoids often

Did YOU Know?

Intravenous (IV) medical treatments are fluids and medications administered directly into a patient's bloodstream through a vein. Health care professionals usually provide IV treatments in hospitals or medical clinics. Intravenous nutrition therapies, also known as "IV vitamin infusions," are a popular health trend promoted by some celebrities. Regardless of health status, anyone can purchase vitamin infusions called "vitamin drips" as a feature of pop-up clinics and house-call-type nutritional services. Promoters of vitamin drips often claim that the nutrient infusions can improve the condition of skin and hair as well as help relieve hangovers, dehydration from the flu or "overexertion," food poisoning, and jet lag. These kinds of IV infusions are often costly and not medically necessary. Furthermore, people receiving the infusions are at risk of a skin infection where the needle is inserted or serious blood infection if sterile techniques are not used during needle insertion. Scientific research concerning the safety and effectiveness of the personal use of unprescribed IV vitamin and/or mineral infusions is limited.

It is best to consume fluids, especially water, by mouth for proper hydration and eat a balanced diet to provide the vitamins and other nutrients needed by the body. If people are too sick to consume fluids orally and are dehydrated, they should seek professional care at a medical facility.

have druglike effects in the body. In some cases, these physiological responses are beneficial, but in other instances, vitamin excesses cause unpleasant and even dangerous side effects. Therefore, people should be just as cautious about using megadoses of vitamins as they need to be when taking medications. This section of Chapter 8 focuses on some current scientific evidence regarding the usefulness and safety of using large doses of vitamins and related compounds as medications.

Vitamin A and Carotenoids as Medicine?

Findings of observational studies suggest an association between eating diets rich in fruits and vegetables and lower risk of certain cancers, heart disease, and *age-related macular degeneration* (see the following section, "Dietary Supplements and AMD"). Such diets provide plenty of beta-carotene and other antioxidant carotenoids. Scientists have conducted numerous large-scale studies to determine whether carotenoid supplements provide health benefits. The following sections summarize research findings regarding the health effects of taking these supplements.

Cancer

Results of clinical studies have not provided support for taking vitamin A or beta-carotene supplements to reduce the risk of cancer.[8] In two major studies, smokers who took beta-carotene supplements were *more* likely to die of lung cancer than smokers who took placebos.[8] Nevertheless, more research is needed to determine whether other carotenoids, such as lycopene and *beta-cryptoxanthin* (*bay′-ta krip′-toe-zan′-thin*), protect against various forms of cancer.

Dietary Supplements and AMD

In the United States, age-related macular degeneration (AMD) is one of the leading causes of blindness among older adults.[33] The disease is associated with changes in the *macula* the region within the eye that provides the most detailed central vision (Fig. 8.27). When the macula is damaged, objects appear to be distorted, as in the grid shown in Figure 8.28 (right side). Major risk factors for AMD are genetics, smoking, and advanced age, but diet also plays a role in the development of the condition.

The macula contains the carotenoids lutein and zeaxanthin. In the Age-Related Eye Disease Study (AREDS), patients who already had AMD took dietary supplements that contained relatively high amounts of vitamins C and E, beta-carotene, and the minerals copper and zinc.[8] The combination of these substances slowed the progression of vision loss associated with advanced macular degeneration. In a follow-up study, scientists determined that giving beta-carotene supplements to patients with AMD was not helpful. However, the patients who took lutein and zeaxanthin supplements experienced a delay in the progression of AMD. At this point, there is no scientific evidence that taking antioxidant supplements *prevents* healthy people from developing AMD.

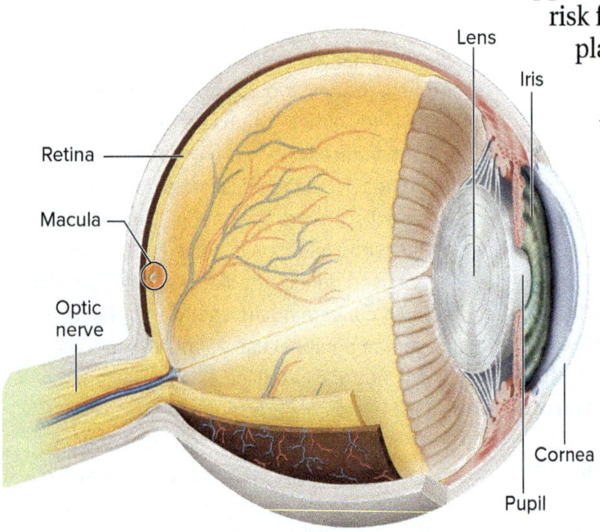

Figure 8.27 The macula. The macula is the region within the eye that provides the most detailed central vision.

Vitamin E as Medicine?

Major studies fail to show that high intakes of vitamin E consistently reduce the risk of CVD, cancer, and mild cognitive impairment, a condition characterized by declining thought-processing abilities that can progress to

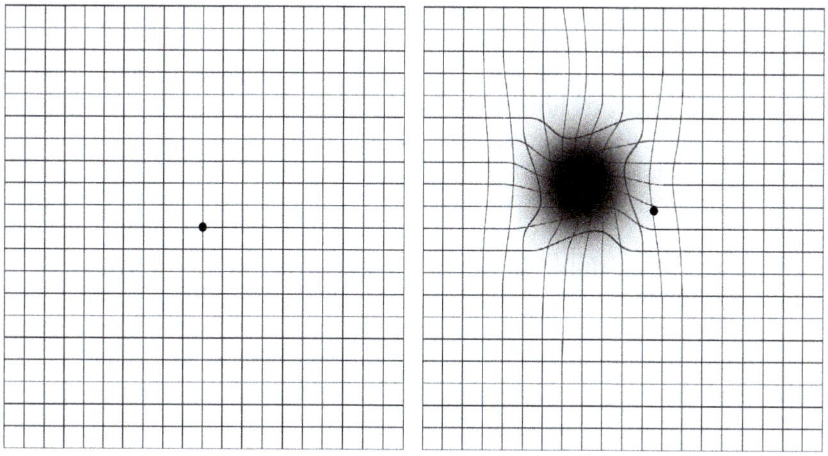

Figure 8.28 Visual effect of macular degeneration. When shown a grid composed of straight lines, people with normal vision see the grid on the left. However, people with age-related macular degeneration would see distorted gridlines, as shown in the right illustration.

Alzheimer's disease.[16] Furthermore, use of vitamin E supplements has been associated with an increased risk of prostate cancer (males) and hemorrhagic stroke. More clinical research is needed to determine whether combinations of vitamin E and other antioxidant nutrients can prevent or slow the development of chronic diseases, such as CVD, cancer, and Alzheimer's disease.

Did scientists *prove* vitamin E supplements are useless or "bad"? No. Vitamin E has several forms, and it is possible that the alpha–tocopherol form of vitamin E that is typically in supplements does not have any beneficial druglike effects on the body. The dosage of vitamin E and the length of time the supplements are administered during clinical tests may provide other reasons why research findings do not always support epidemiological observations. To detect positive effects on health, scientists may need to test higher doses of the vitamin than have been studied in the past. Additionally, studies that last longer may be necessary; the health benefits of vitamin E supplementation may take several years to become evident.

Niacin as Medicine?

In Chapter 6, we discussed the association between elevated blood levels of LDL cholesterol and increased risk of CVD, the number one killer of Americans. Heart disease and stroke are the major forms of CVD. Physicians may prescribe megadoses of niacin alone or in combination with statins to reduce elevated LDL cholesterol levels.[20] High doses of niacin can be toxic, so people should consult with qualified medical practitioners before using the vitamin as a medicine.

Vitamin B-6 as Medicine?

Popular sources of nutrition information often recommend large doses of vitamin B–6 to treat *premenstrual syndrome* (*PMS*), a condition that many females experience a few days before their menstrual period begins. Symptoms of PMS include bloating, fatigue, headaches, anxiety and irritability. However, studies investigating the value of vitamin B–6 supplementation for relieving PMS provide mixed results.[22] Scientists continue to study the effectiveness of using high doses of vitamin B–6 to treat PMS.

Unlike most other B vitamins, vitamin B–6 is toxic in high doses, causing severe sensory nerve damage when taken in doses that exceed the UL for extended periods. Signs and symptoms of vitamin B–6 toxicity include walking difficulties and numbness of the hands and feet. The nerve damage usually resolves when affected people stop ingesting megadoses of vitamin B–6.

Did *YOU* Know?

Consumers often search the Internet for advice concerning ways to prevent or treat health conditions, particularly by using vitamins. Some people also search for methods of hiding illegal behaviors, such as illicit drug use. In one reported case, a 17-year-old male followed advice that he obtained on the Internet concerning a nonnutritional use of niacin supplements. According to the Internet sources, high doses of niacin interfere with the reliability of urine testing to detect recent marijuana use. The teenager took megadoses of niacin prior to the urine testing and, as a result, experienced abnormal liver function that required hospitalization.[34] The young man was fortunate to recover without the need for a liver transplant.

Folic Acid, B-6, and B-12 as Medicine?

Deficiencies of folic acid, vitamin B–6, and vitamin B–12 are associated with elevated blood levels of homocysteine. High homocysteine levels may be a biochemical marker (an indicator in the body) or a risk factor for cardiovascular disease. However, results of research do not provide evidence that lowering homocysteine levels by taking these B vitamins reduces the risk of heart disease or stroke.[22]

Some scientists think that poor vitamin B–6 status contributes to the cognitive decline that often occurs with aging. Cognitive functions include the ability to remember, understand, and use information. According to the results of a few reliable studies, there is no evidence that short–term use of vitamin B–6, by itself or along with folic acid and/or vitamin B–12 can prevent or slow cognitive decline among older adults.[22]

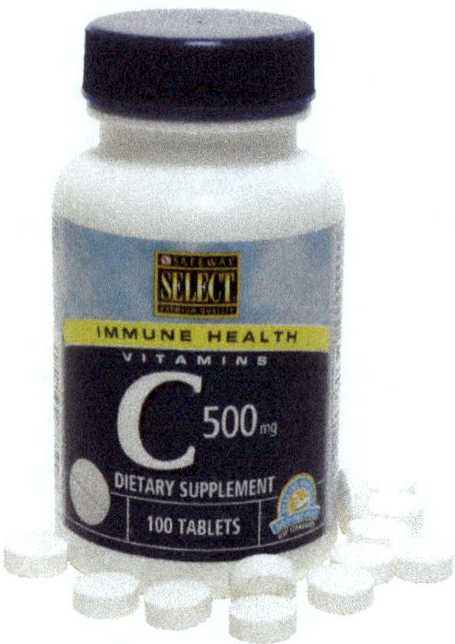

Routine vitamin C supplementation does not prevent the common cold in the general population.
John Flournoy/McGraw Hill

Vitamin C as Medicine?

In 1970, Nobel Prize–winning American chemist Dr. Linus Pauling (1901–1994) published *Vitamin C and the Common Cold.* This bestselling book established the popular belief that megadoses of vitamin C could prevent colds. As a result of Pauling's claim, many Americans take megadoses of the micronutrient when they notice the first cold symptoms.

In the years that followed the publication of *Vitamin C and the Common Cold,* Pauling became more convinced of vitamin C's health benefits. He claimed that large doses of vitamin C could battle a variety of diseases, including influenza, cancer, and CVD. He even believed the vitamin could slow the aging process. Despite Pauling's impressive credentials in chemistry, many conventional nutrition scientists have approached his ideas about vitamin C's health benefits with caution and skepticism.

Can taking vitamin C protect you against infection by cold viruses? The evidence collected from several scientific studies indicates that routine vitamin C supplementation (200 mg or more of the vitamin daily) does not prevent colds in the general population.[5] However, taking such large doses of the vitamin may reduce the duration of cold symptoms by a day or so. Additionally, vitamin C may reduce the severity of cold symptoms because the micronutrient acts like an antihistamine when taken in very large doses.

Military members living in highly populated training centers and athletes that frequently travel for sports competitions are at a higher risk of exposure to the cold virus. These populations also undergo daily, high–intensity training that also increases their risk of virus infection.[35] Recent scientific research reports these populations may reduce their risk of developing a cold by taking 1000 mg of vitamin C per day for no more than 30 days.[35] Taking high doses of vitamin C for longer periods of time may result in prooxidant effects and radical production (see the Vitamin C section in this chapter).

Beyond the Common Cold

When LDL cholesterol is oxidized, it is more likely to contribute to atherosclerosis than nonoxidized LDL cholesterol (see Chapter 6). As an antioxidant, vitamin C may reduce the oxidation of LDL cholesterol, lowering the risk of CVD. Studies to determine whether vitamin C can help prevent CVD have not provided consistent evidence that the micronutrient reduces the risk of CVD or dying from the condition.[5] Atherosclerosis takes years and probably decades to result in heart attack, strokes, and other forms of CVD. Thus, more research is needed to determine whether long–term vitamin C supplementation can reduce the risk of this leading killer of Americans.

Food & Nutrition tips

Vitamin C supplements that contain "ester" forms of the vitamin do not provide any advantage over "regular" vitamin C supplements. The ester forms of the supplements are usually more expensive, too.

According to findings of epidemiological studies, people who consume diets containing high amounts of vitamin C–rich fruits and vegetables have lower risk of cancer than people who do not eat much of these foods. Nevertheless, results of studies generally do not support the usefulness of taking vitamin C supplements to prevent cancer.[5] There is some encouraging scientific evidence that intravenous (IV) administration of megadoses of vitamin C may be useful in treating cancer.[5] The "Nutrition Matters" section in this chapter provides more information concerning the role of diet in cancer development and prevention.

It is important to note that complex chronic diseases such as CVD, cognitive decline, and cancer generally do not have simple causes. Inherited factors as well as lifestyle practices other than diet—smoking, obesity, and lack of physical activity—contribute to the development of these health conditions. Therefore, it is unlikely that people will be able to prevent or treat cancer simply by taking vitamin C (or any other supplement).

Did YOU Know?

In 2020, Americans spent over $55 billion on dietary supplements, which included vitamin supplements.[36]

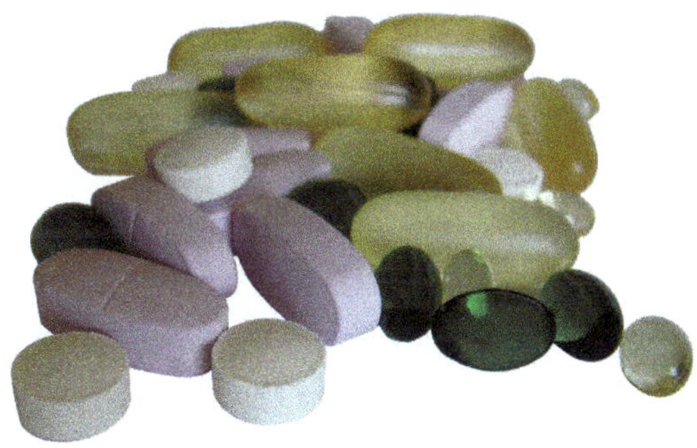

©Wendy Schiff

Some Final Thoughts

According to a report issued by the U.S. Preventive Services Task Force, there is a lack of credible scientific evidence for or against using dietary supplements, including those with vitamins, to prevent major chronic diseases.[37] However, there is strong evidence to advise against the use of beta–carotene supplements, especially among smokers.[37] Although the health benefits of vitamin supplementation are often uncertain, there is more consistent evidence that a diet high in fruits, vegetables, and legumes has important benefits. It is possible that other constituents besides vitamins may account for the benefits of such diets.

Consuming a wide variety of vitamins, antioxidants, and phytochemicals in their natural states and concentrations (in foods) may be the most effective way to lower your risk of CVD, cancer, and many other serious chronic diseases. Why? These substances probably work together to enhance health, and isolating them from their natural sources or synthesizing and concentrating them into supplements may reduce their usefulness and increase their risks.[38] Therefore, nutrition experts recommend that adults eat a variety of fruits and vegetables each day rather than take antioxidant or phytochemical supplements.

Nutrition experts tend to emphasize the importance of combining a nutritious diet with regular exercise for achieving and maintaining good health. However, an overwhelming amount of scientific evidence links tobacco use to several forms of cancer, heart disease, AMD, and other serious chronic conditions. If you smoke tobacco, quitting may have a greater beneficial impact on your long–term health than improving your diet or increasing your physical activity level while still continuing tobacco use.

Concept CHECKPOINT

9. Dorothy is 85 years of age. She has excellent vision, but she takes megadoses of beta-carotene because she thinks taking the phytochemical helps prevent macular degeneration. Based on the information in Section 8.4, what would you tell Dorothy about her beta-carotene use?

10. Explain why people should be careful about taking megadoses of vitamin supplements.

11. What is a serious side effect that can occur from taking megadoses of niacin?

12. Explain why people should avoid taking high doses of vitamin B-6.

13. A friend of yours takes 1000 mg of vitamin C daily because they thinks the vitamin prevents colds. After reading Section 8.4, what would you tell your friend about their vitamin C use?

See Appendix F for responses.

8.5 Nutrition Matters: Diet and Cancer

Learning Outcomes

1. Define key terms, including *metastasize, carcinogen,* and *malignant tumor.*
2. Identify common cancer sites and causes of cancer deaths for Americans.
3. Identify lifestyle practices associated with increased risk of certain cancers.
4. Identify foods that may increase risk of cancer.
5. Discuss some practical steps that people can take to reduce their risk of cancer.

For many people, no diagnosis evokes more fear than that of *cancer.* Although modern medical technologies have enabled physicians to make great progress in treating many types of cancer successfully, the disease is still the second–leading cause of death in the United States. However, the risk of dying from cancer has decreased by 32% over the past 30 years.[39] According to the American Cancer Society, one in five Americans dies of cancer.[40]

In 2022, the three leading sites of new cancer cases (excluding common types of skin cancer) for American males were prostate, lung, and colon and rectal (colorectal) cancers.[41] The three leading sites of new cancer cases in American females were breast, lung, and colorectal cancers. As shown in Figure 8.29, lung cancer is responsible for most cancer deaths. There is good news: A significant number of cancer deaths are preventable.

This section focuses on the roles of diet and exercise in the development, progression, and possible prevention of certain cancers. For more information about cancer, including specific forms of cancer, visit the American Cancer Society's website at www.cancer.org or the National Cancer Institute's website at www.cancer.gov.

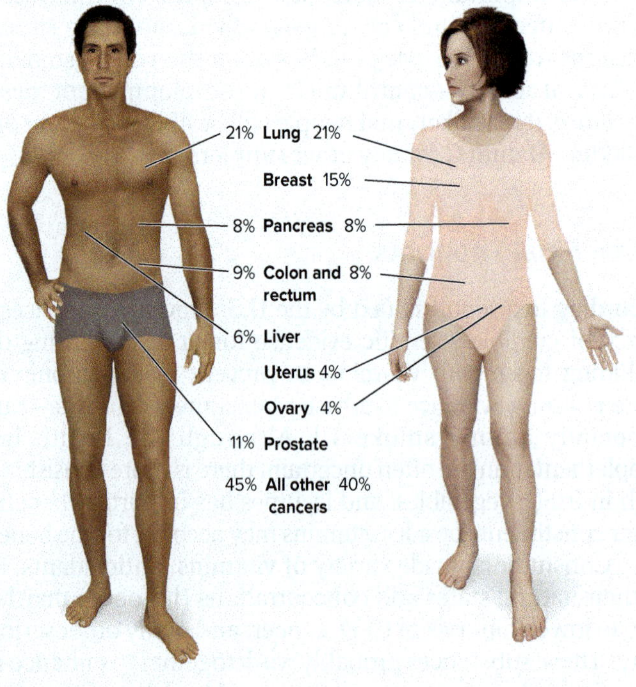

Figure 8.29 Leading cancer sites and deaths. This figure indicates leading sites of cancer deaths and percentages of cancer deaths, according to 2022 estimates.
Source: American Cancer Society, *Cancer facts & figures 2022.* www.cancer.org. Accessed: May 10, 2023.

What Is Cancer?

Cancer is the term for a group of chronic diseases characterized by cells that have undergone damage to certain genes (*mutations*). Genes are portions of DNA that code for the production of specific proteins (see Chapter 7). DNA dictates cellular growth, division, and eventual death. The rate at which healthy cells develop, grow, divide, and die occurs in a controlled fashion. As a normal body cell develops and matures, it acquires specialized functions. Examples include the ability of muscle cells to contract and that of certain stomach cells to secrete hydrochloric acid.

Cancerous (*malignant*) cells are literally "out of control": They divide repeatedly and frequently, and they do not die. If genes that regulate cellular growth, division, and death mutate, abnormal cell development, rapid cell growth, and unchecked cell division can result. When a cell becomes malignant, it does not perform the specialized functions of the cells from which it was derived. A cancerous liver cell, for example, does not remove toxins from the bloodstream or store nutrients properly.

As a result of their rapid growth rate, many types of cancer cells form masses, called **malignant tumors.** Malignant cells often break away from the tumor. These cells can move to and invade other parts of the body. When cancer spreads to other tissues, the disease has **metastasized** (*meh–tass'–tah–sized*).

malignant tumor mass of cancerous cells

metastasized spread to other tissues

Somecellsmultiplyexcessivelyandform*benign*(*bih–nine'*) masses or tumors as a result. A benign tumor is not cancerous because the tumor's cells do not destroy nearby tissues or metastasize. Benign tumors are usually harmless. In some cases, however, a benign tumor grows large enough to interfere with the functioning of healthy structures, such as a blood vessel or brain tissue. When this occurs, the tumor needs to be treated to reduce its size or remove it.

A **carcinogen** (*car–sin'–o–jin*) is an environmental factor, such as radiation, tobacco smoke, or a virus, that triggers cancer. Carcinogens can irritate tissues, causing an inflammatory response by the body. Over time, the inflammation may result in cancer development. Some carcinogens damage DNA, causing mutations to genes that control certain cell behaviors. Over time, repeated exposures to various carcinogens take their toll on cells, making malignant cells more likely to develop. In fact, it is not unusual for cancerous cells to develop during a person's lifetime. However, a healthy immune system identifies the malignant cells and destroys them before they multiply uncontrollably. Nutrient deficiencies, the aging process, and environmental insults, such as exposure to radiation and certain chemicals, reduce the effectiveness of the immune system. As a result, cancer cells are able to divide unchecked and eventually metastasize. Figure 8.30 illustrates major steps in the development and behavior of most cancerous cells.

Because of their rapid growth and frequent cell divisions, malignant cells require more nutrients than normal cells. To supply enough nutrients to meet their needs, cancerous tumors stimulate the body to form blood vessels that divert blood away from healthy cells and into the tumor (see Figure 8.30). Much of the severe body wasting that usually accompanies advanced cases of cancer happens because healthy tissues are unable to obtain adequate supplies of nutrients, and they die.

Regular health checkups and various screening methods can detect cancer in its early stages, when many forms can be effectively treated. Screening methods include using a special scope to view the inside of the lower gastrointestinal tract to detect colorectal cancer (*colonoscopy*). Conventional medical treatments for cancer typically involve surgical removal of cancerous tissue; chemotherapy, medications that are toxic to cancer cells or limit their growth; and radiation that kills cancer cells. Early diagnosis of cancer is very important. Once cancer metastasizes, it is far more difficult to treat.

What Causes Cells to Become Cancerous?

Why do many smokers develop lung cancer and some not develop it? Why do some females have breast cancer and others remain free of the disease? Medical researchers have discovered several risk factors for many forms of the disease, including the following:

- Aging (80% of cancers occur in people over 55 years of age)
- Having a family history of cancer
- Using tobacco
- Being exposed to some forms of radiation
- Being exposed to certain environmental substances, such as irritants
- Having certain viral and bacterial infections
- Having elevated levels of certain hormones
- Consuming alcohol and certain foods
- Being physically inactive and having excess body fat[41,42]

People can lower their risk of cancer by avoiding known carcinogens, especially tobacco smoke. In the United States, approximately one of every five cancer deaths is associated with smoking tobacco.[41] Smoking is responsible for many cases of lung cancer, but mouth, larynx ("voicebox"), esophageal, stomach, pancreatic, kidney, and bladder cancers are also associated with tobacco use.

Cancer causation is a complex process. Therefore, it may be impossible to pinpoint a single cause of a patient's cancer. Individuals differ widely in their genetic makeups, lifestyle practices, environmental exposures, and nutritional states. Some forms of the disease are likely to be inherited, but poor diet, alcohol consumption, physical inactivity, and excess body fat are responsible for 18% of the cancers that occur in the United States.[41] The following sections examine relationships between dietary factors and the risk of cancer.

The Role of Diet in Cancer Development

Rates of specific cancers vary widely among different populations. Males and females living in Asian countries, for example, have lower risks of prostate and breast cancer, respectively, than males and females living in Western countries. After Asian males and females migrate to Western countries, their risks of prostate and breast cancer increase and eventually become similar to the risks of their adopted country's native population. Such findings provide epidemiological evidence of associations between cancer and certain lifestyle choices.

According to results of observational and experimental studies, certain substances in foods and beverages promote cancer development. Alcohol, for example, is a carcinogen. People who consume alcoholic drinks daily have higher–than–average risks of cancers of the mouth,

carcinogen factor that triggers cancer

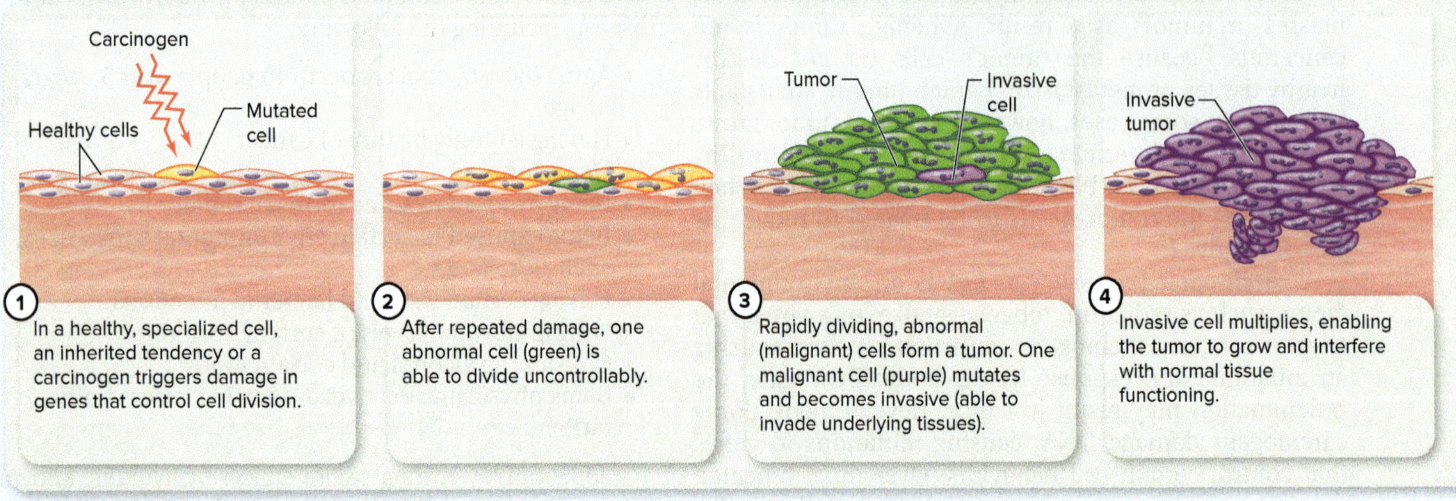

Figure 8.30 **Cancer development and progression.**

throat, esophagus, larynx, colorectum, liver, and breast.[41] The risk rises as the amount of alcohol consumed increases and when cigarette smoking is combined with drinking. To reduce your risk of cancer, avoid alcoholic beverages or drink in moderation: no more than one standard drink per day for females and no more than two standard drinks per day for males (see the "Nutrition Matters" section of Chapter 6).

Certain molds that can grow on nuts or grains produce a chemical called *aflatoxin* (*ah′–fla–tox′–in*). Consuming foods that are contaminated with these molds increases the risk of liver cancer. Although liver cancer is not common among Americans, populations that consume peanuts and grains that have been stored improperly have high rates of this type of cancer. Chapter 12 will discuss toxic molds and ways to protect the food supply from them and other food–borne health threats.

Diets that contain large amounts of processed meat and/or red meat (defined as beef, pork, and lamb) are associated with increased risk of colorectal and prostate cancer.[42] Processed meat ("deli" meat) has sodium nitrate and sodium nitrite added to it during production. Under certain conditions, these chemicals form *nitrosamines* (*ni–tros′–ah–menes*), which are carcinogens. Eating high amounts of smoked or salt–preserved (*pickled*) foods, such as salted fish, may increase the risk of stomach cancer.[43]

The high temperatures used to fry, grill, or broil meat may cause the formation of a group of substances in the food called *heterocyclic* (*het′–eh–ro–si′–klic*) amines. Results of animal studies indicate that these amines are carcinogens. However, it is not known whether heterocyclic amines are human carcinogens.[44] Nevertheless, you may want to consider limiting your intake of grilled meats and avoiding charred parts of charcoal–grilled meats. Also, covering the grate with aluminum foil, poking a few holes in the foil, and placing the meat on top of it will reduce charring of the meat.

Excess Body Fat

Diets that provide a surplus of calories result in a person becoming overweight or obese, conditions characterized by excess body fat (see Chapter 10). People who are obese have higher risks of cancers of the colon, rectum, breast (in postmenopausal females), uterus, kidney, and esophagus than people who have healthy amounts of body fat.[45] The reasons for the relationship between excess body fat and cancer development are unclear. However, people who are overweight and obese often have elevated insulin, estrogen, and testosterone levels. Excesses of these and possibly other hormones may stimulate the production of cell factors that promote malignant cell development and growth. In some cases, cancer triggers are easier to identify. For example, overweight and obesity are associated with the increased risk of esophageal cancer. Excess abdominal fat contributes to the reflux of stomach acid into the esophagus by restricting the ability of the stomach to expand during meals. As a result, the gastroesophageal sphincter is unable to keep the stomach's contents from moving into the esophagus. (See the Chapter 4 "Nutrition Matters" section.) Chronic exposure to stomach acid irritates the lining of the esophagus, making cancerous cells more likely to develop in the affected area.

Table 8.15 indicates dietary factors that may be associated with increased risk of various cancers. Currently, there is no scientific evidence that consuming artificial sweeteners in typical amounts, such as aspartame; coffee; fluoridated water; genetically modified foods; or foods preserved by radiation (*irradiated foods*)

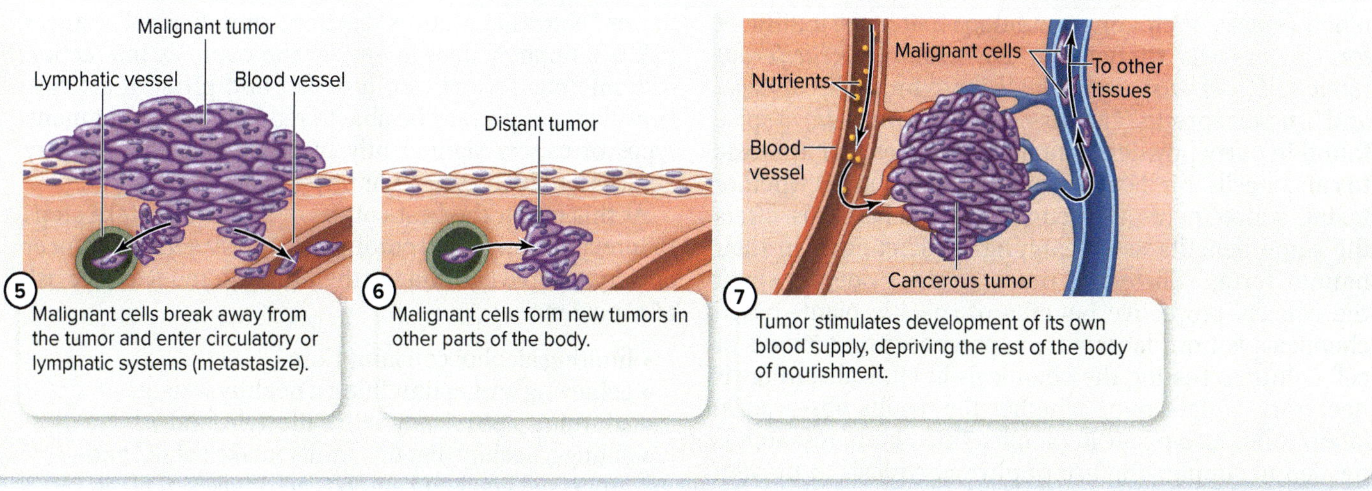

Ingram Publishing/Alamy Stock Photo

causes cancer.[46,47] However, physical activity and some dietary practices may *reduce* the risk of cancer.

Nutrition and Physical Activity Cancer Prevention

Consuming a nutritious diet that contains adequate fluids and a variety of foods, especially plant foods, may prevent or delay the development of certain cancers. Drinking enough water and other nonalcoholic fluids may reduce the risk of bladder cancer. Adequate fluid intake encourages frequent urination, which reduces the time that carcinogens remain in contact with the walls of the bladder.[48]

High intakes of fiber–rich foods, including fruits and vegetables, may reduce the risk of cancer.[46] Fruits and vegetables are rich sources of fiber and vitamin C, as well as carotenoids and other phytochemicals that may have antioxidant activity in the body. Antioxidants protect DNA from free radical damage, preventing potentially cancer–causing mutations from occurring. Although research is ongoing, results of scientific studies generally do not provide evidence that taking dietary supplements that contain antioxidants reduces the risk of cancer.[46] Some medical experts are concerned that taking antioxidant supplements may *increase* the risk of lung and prostate cancers, especially among smokers and males with family histories of prostate cancer.

Many plant foods naturally contain chemicals that reduce carcinogens' ability to damage DNA. Other phytochemicals inhibit tumor growth by being toxic to

TABLE 8.15 Possible Diet-Related Carcinogens

Possibly Carcinogenic	Common Dietary Sources	Possible or Known Action
Aflatoxin	Moldy nuts and grains	Can cause liver cancer
Arsenic	Natural contaminant in drinking water	May increase risk of bladder cancer
Alcohol 	Alcoholic beverages	Unclear but may be related to a toxic byproduct of alcohol metabolism
Excess calorie intake	All macronutrients	Excess calorie intake is linked to obesity, which can increase certain hormone levels
Heterocyclic amines	Foods exposed to high temperature, especially grilled meat that has charred areas	Formation of heterocyclic amines that alter DNA of colon cells
Sodium nitrate and nitrite	Processed meat, such as deli meats, hot dogs, and bacon	Formation of nitrosamines

©Wendy Schiff

cancer cells, interfering with their ability to form supportive blood vessels, or creating conditions that make it difficult for cancer cells to flourish. For example, *cruciferous* (*crew–siff′–er–us*) vegetables, such as broccoli, cabbage, and Brussels sprouts, and *turmeric* (*tur′–mur–ik*), a spice found in curry powder, contain substances that are toxic to cancer cells. Medical researchers do not know whether taking supplements that contain phytochemicals offers the same benefits as consuming the chemicals in their natural forms. Currently, much of the evidence about the cancer–protective benefits of specific plant–based chemicals is from laboratory experiments on animals or cell cultures. Testing the chemicals in clinical settings is necessary to determine whether the results observed in laboratories also occur in people. Thus, more research is needed to support the use of phytochemicals to prevent or treat cancer.

Reducing the Risk

Although medical researchers are studying the cancer–fighting potential of various chemicals naturally in food, no major breakthroughs have occurred. Avoiding obesity by establishing healthy eating and physical activity habits, especially early in life, may reduce the risk of cancer. According to results of epidemiological studies, people who exercise regularly have lower risks of certain cancers, including colon and breast cancer. Regular physical activity may reduce the risk by helping to control weight, improving immune system functioning, and enhancing the body's regulation of various hormones. Following the American Cancer Society's recommendations does not guarantee zero risk of cancer. As mentioned earlier in this "Nutrition Matters" section, an individual's cancer risk is ultimately determined by the complex interaction of numerous factors, not just diet and physical activity. However, people may be able to reduce their risk of many types of cancer significantly by improving their dietary choices and increasing their physical activity.

At this point, the best course of action is to take steps to prevent cancer. According to the American Cancer Society, people can change their diets to reduce their risk of cancer by:

- limiting alcohol consumption;
- achieving and maintaining a healthy weight;
- adopting a physically active lifestyle; and
- eating a healthy diet that limits intakes of red and processed meat and emphasizes intakes of plant foods, including fruits, vegetables, and whole grains.[46]

Concept CHECKPOINT

14. Which form of cancer is the leading cause of cancer deaths in the United States?

15. Margaret's breast cancer metastasized to her lungs. Define *metastasize.*

16. List three dietary factors that may contribute to cancer.

17. List three steps people can take to reduce their risk of cancer.

See Appendix F for responses.

Chapter References

See Appendix G.

Summary

8.1 Vitamins: Basic Concepts

- Vitamins are organic compounds that the body cannot synthesize or make enough of to maintain good health; that naturally occur in commonly eaten foods; that cause deficiency disease when they are missing from diets; and that restore good health when added back to the diet.

- Vitamins play numerous roles in the body, and each vitamin generally has more than one function. In general, vitamins regulate a variety of body processes, including those involved in cell division and development as well as the growth and maintenance of tissues. Vitamins are not a source of energy. Some vitamins have precursors that do not function as vitamins until the body converts them into active forms.
- An oxidation reaction can form a radical. Radicals damage or destroy molecules by removing electrons from them. Cells normally regulate oxidation reactions by using antioxidants such as vitamin E.
- Vitamins A, D, E, and K are fat-soluble vitamins; thiamin, riboflavin, niacin, vitamin B-6, pantothenic acid, folate, biotin, vitamin B-12, and vitamin C are water-soluble vitamins. The body stores extra fat-soluble vitamins. Over time, these vitamins can accumulate and cause toxicity. Water-soluble vitamins are generally not as toxic as fat-soluble vitamins.
- The enrichment program for grains specifies amounts of thiamin, riboflavin, niacin, and folic acid that manufacturers that participate in the program must add to their refined wheat flour and certain other milled grain products. Fortification involves the addition of one or more vitamins (and/or other nutrients) to a wide array of commonly eaten foods during manufacturing. The vitamins that are added may or may not be in the food naturally.
- Vitamin deficiency disorders generally result from inadequate diets or conditions that increase the body's requirements for vitamins, such as reduced intestinal absorption or higher-than-normal excretion of the micronutrients. Severe vitamin deficiencies are uncommon in the United States. The chances of developing a vitamin deficiency disease increase when the diet consistently lacks the micronutrient and levels of the nutrient in the body become depleted.
- Most commonly eaten foods do not contain toxic levels of vitamins. Vitamin toxicity is most likely to occur in people who take megadoses of vitamin supplements or consume large amounts of vitamin-fortified foods regularly.

Nataliia K/Shutterstock

8.2 Fat-Soluble Vitamins

- Vitamin A is involved in vision, immune function, and cell development. The vitamin A precursor, beta-carotene, functions as an antioxidant. Dietary sources of preformed vitamin A include liver and fish liver oils; provitamin A carotenoids are especially plentiful in dark green and orange fruits and vegetables. Excess vitamin A can be quite toxic and cause birth defects when taken during pregnancy.
- Vitamin D is both a hormone and a vitamin. Exposure to sunlight enables human skin to synthesize a precursor of the vitamin from a cholesterol-like substance. The body can convert this substance to the active form of the vitamin. A few foods, including fatty fish and fortified milk, are dietary sources of the vitamin. Vitamin D helps regulate the level of blood calcium by increasing calcium absorption from the intestine. Infants and children who do not obtain enough vitamin D may develop rickets, and adults with inadequate amounts of the vitamin in their bodies may develop osteomalacia. Older people and breastfed infants often need a supplemental source of the vitamin. Excess intakes of vitamin D can cause the body to deposit calcium in soft tissues.
- Vitamin E functions primarily as an antioxidant. By donating electrons to electron-seeking compounds, vitamin E neutralizes them. Plant oils and products made from these oils are generally rich sources of vitamin E.
- Vitamin K is necessary for the liver to make certain blood clotting factors and is needed for bone mineralization. Green leafy vegetables and soybean and canola oils are good dietary sources of vitamin K. The bacteria in the colon also synthesize vitamin K, which is used by the body. Vitamin K deficiency can develop in people with liver diseases or conditions impairing fat absorption. Long-term antibiotic therapy reduces the number of intestinal bacteria and may contribute to a vitamin K deficiency. Infants are born with low vitamin K stores. To prevent vitamin K deficiency from developing, newborns generally receive a single vitamin K injection.

Tim UR/Shutterstock

8.3 Water-Soluble Vitamins

- Thiamin, riboflavin, and niacin play key roles as part of coenzymes in energy-yielding reactions. These water-soluble vitamins help cells metabolize carbohydrates, fats, and proteins. Enriched grain products are common sources of all three of the vitamins. Beriberi is the severe thiamin deficiency disease; pellagra results from a severe lack of niacin. Excess intakes of niacin can cause toxicity.
- Vitamin B-6 is involved in protein metabolism, especially in synthesizing nonessential amino acids. The vitamin also participates in the synthesis of neurotransmitters and the metabolism of homocysteine. Healthy people can obtain enough vitamin B-6 by eating a varied diet that contains animal foods and rich plant sources of the micronutrient. High doses of vitamin B-6 should be avoided because the vitamin can cause nervous system damage.
- Folate plays important roles in DNA synthesis and homocysteine metabolism. Rich food sources of folate are leafy vegetables, organ meats, and orange juice. Enriched grains have folic acid added to them. Signs of folate deficiency include megaloblastic anemia. Pregnancy increases the body's needs for folate; a deficiency in early pregnancy can result in neural tube defects in offspring. Females of childbearing age can meet the RDA for the vitamin by taking dietary supplements that contain folic acid.
- The body needs vitamin B-12 to metabolize folate and homocysteine, and maintain the insulation surrounding nerves. Although vitamin B-12 does not occur naturally in plant foods, the vitamin is in animal foods and products that have been fortified with the micronutrient. Vitamin B-12 deficiency is more common in older adults. As people age, they produce less stomach acid, which can result in B-12 deficiency even though adequate amounts of the micronutrient are consumed in foods. Loss of intrinsic factor production in the stomach also causes vitamin B-12 deficiency and a condition called pernicious anemia.
- The body uses vitamin C to synthesize and maintain collagen, a major protein in connective tissue. Vitamin C also functions as an antioxidant. Fresh fruits and vegetables, especially citrus fruits, are generally good sources of the micronutrient. Because vitamin C is readily lost in cooking, diets should emphasize fresh or lightly cooked fruits and vegetables. Smoking increases the body's requirement for vitamin C. Scurvy is the vitamin C deficiency disease. Excess vitamin C may cause diarrhea and increase the risk of kidney stones in some people.

Pixtal/SuperStock

8.4 Vitamins as Medicines

- Although the health benefits of vitamin supplementation remain uncertain, there is consistent epidemiological evidence that a diet high in fruits, vegetables, and legumes may lower the risk of CVD, cancer, and other serious chronic disease. Consuming a wide variety of vitamins, antioxidants, and phytochemicals in their natural states and concentrations (in foods) may be the most effective way to achieve good health. Therefore, healthy people should consume a variety of fruits, vegetables, and other plant foods daily rather than take vitamin or antioxidant supplements.

8.5 Nutrition Matters: Diet and Cancer

- Cancer occurs when genes that regulate cellular growth, division, and death mutate, resulting in abnormal cell development, rapid cell growth, and unchecked cell division. Maintaining a healthy body weight, exercising regularly, consuming diets that supply plenty of fruits and vegetables, limiting intakes of red and/or processed meat, and avoiding tobacco smoke and excess alcohol may reduce the risk of cancer.

Recipe for Healthy Living

©Wendy Schiff

Stir-Fried Vegetable Medley

"Eat your vegetables!" Does this order bring back some not-so-fond dinnertime memories? If you are like many Americans, cooked vegetables were not high on your list of favorite foods when you were a child. Corn on the cob was fun to eat, but you may have picked at your peas and carrots, and refused to eat broccoli unless it was smothered in a gooey cheese sauce. You probably drew a "line in the sand" when offered Brussels sprouts. Now that you have read Chapter 8, you should appreciate the nutritional contribution that a wide variety of colorful vegetables can make to your diet. Your parents told the truth when they said, "Vegetables are good for you."

This vegetable medley can be the foundation of a vegetarian meal or a colorful accompaniment to main dishes. You don't need to use all the vegetables in the recipe; feel free to improvise, especially when certain fresh vegetables are in season and plentiful. If you use fresh vegetables, wash them in cool water and drain on paper towels before cutting them into bite-size pieces. If you use frozen vegetables, add them directly to the hot oil, without thawing them first. Be careful because ice from the vegetables can cause hot oil to spatter.

Your goal is to heat the vegetables in a small amount of hot oil only until they are still colorful and crisp—about 4 to 8 minutes. Add the quicker-cooking vegetables (e.g., mushrooms) last, so they don't overcook. Use a slotted spoon to drain oil from vegetables and serve immediately.

This recipe (without rice or noodles) makes about four ½-cup servings. Each serving supplies approximately 114 kcal, 3 g protein, 7 g fat, 3 g fiber, 46 mg vitamin C, 44 mcg folate, and a variety of phytochemicals from the various vegetables.

INGREDIENTS:

- 1 Tbsp low-sodium soy sauce
- 2 Tbsp water
- 1 Tbsp brown sugar
- ⅛ tsp ground black pepper
- 2 Tbsp vegetable oil
- 1 large garlic clove, minced
- ½ cup onions, sliced
- 1 cup asparagus, cut into 2″-long pieces
- ½ cup broccoli, small pieces
- ½ cup carrots, sliced into "coins"
- ½ medium green pepper, cut in 2″ strips, about ¼″ wide
- ½ medium sweet red pepper, cut in 2″ strips, about ¼″ wide
- 6 button mushrooms, crosscut into quarters

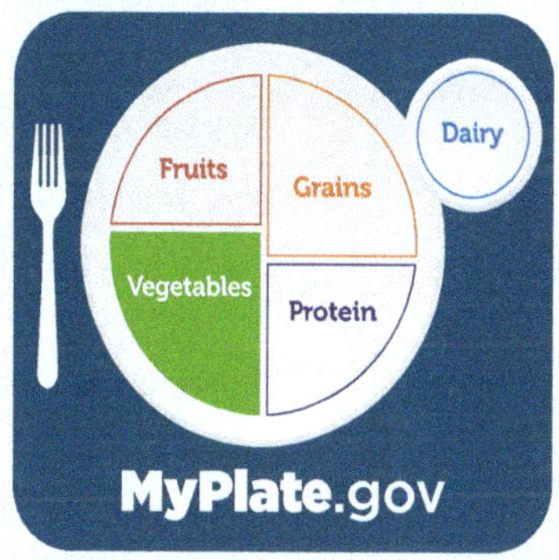

US Department of Agriculture or USDA

Approximate Percent of Calories from Macronutrients

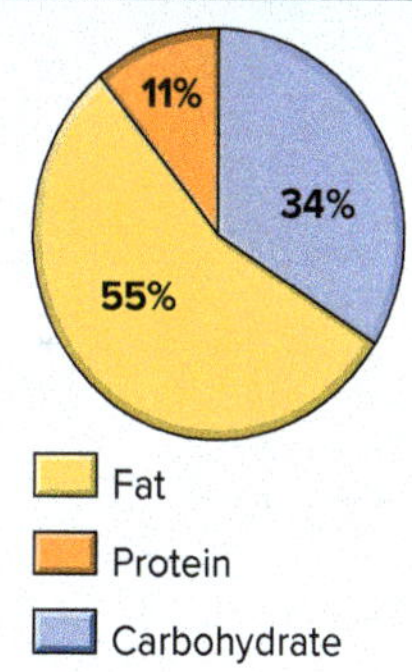

PREPARATION STEPS:

1. In a small dish, prepare mixture of soy sauce, water, and brown sugar. Set aside.
2. Heat oil in a wok or large deep frying pan over medium-high heat. Add garlic and onions. Stir.
3. When onions and garlic are translucent, add asparagus, broccoli, carrots, peppers, and mushrooms. Stir mixture constantly to coat vegetables with oil.
4. Add soy sauce, water, and brown sugar mixture.
5. Add zucchini. Sprinkle black pepper over vegetables.
6. Do not overcook. If the mixture seems too dry, add another tablespoon of water.
7. Serve immediately over cooked rice or noodles.

Percent of Daily Value/Serving

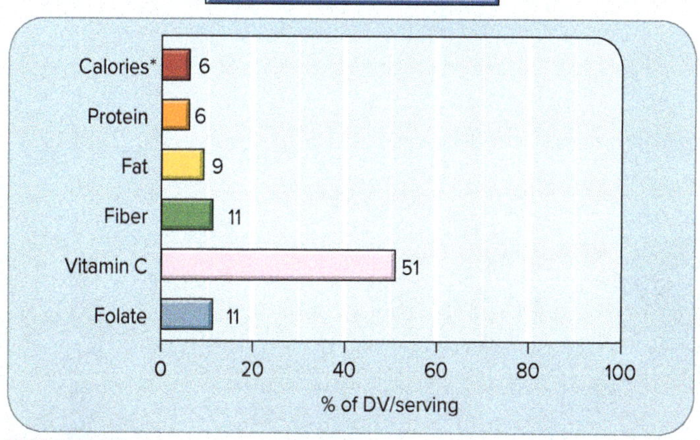

*2000 daily total kcal

Personal Dietary Analysis

Using the DRIs

1. Refer to your 1- or 3-day food log from the "Personal Dietary Analysis" feature in Chapter 3.
   - **a.** Find the RDA values for vitamins under your life stage/sex group category in the DRI tables (see Appendix H). Write those values under the "My RDA" column in the table of this analysis activity.
   - **b.** Review your personal dietary assessment. Find your 3-day average intakes of vitamins A, E, C, D, folate, B-12, thiamin, riboflavin, and niacin. Write those values under the "My Average Intake" column of the table.
   - **c.** Calculate the percentage of the RDA you consumed for each vitamin by dividing your intake by the RDA amount and multiplying the figure you obtain by 100. For example, if your average intake of vitamin C were 100 mg/day, and your RDA for the vitamin were 75 mg/day, you would divide 100 mg by 75 mg to obtain 1.25. To multiply this figure by 100, simply move the decimal point two places to the right, and replace the decimal point with a percentage sign (125%). Thus, your average daily intake of vitamin C was 125% of the RDA. Place the percentages for each vitamin under the "% of My RDA" column.
   - **d.** Under the ">, <, or =" column, indicate whether your average daily intake was greater than (>), less than (<), or equal to (=) the RDA.
2. Use the information you calculated in the first part of this activity to answer the following questions:
   - **a.** Which of your average vitamin intakes equaled or exceeded the RDA value?
   - **b.** Which of your average vitamin intakes was below the RDA value?
   - **c.** What foods would you eat to increase your intake of the vitamins that were less than the RDA levels? (Review sources of certain vitamins in Chapter 8.)
   - **d.** Turn in your completed table and answers to your instructor.

Greg Kuchik/Photodisc/Getty Images

Personal Dietary Analysis: Vitamins

Vitamin	My RDA	My Average Intake	% of My RDA	>, <, or =
A				
E				
C				
D				
Folate				
B-12				
Thiamin				
Riboflavin				
Niacin				

Complete the Personal Dietary Analysis activity in Connect.

Critical Thinking

1. Choose a vitamin and type the nutrient's name in the search box of an Internet browser. Locate three sites that sell products containing the vitamin. Review the pages of each site, making notes about claims, prices, and the kinds of links provided. Then write a three– to five–page report that describes and compares the information you found at the sites. In your paper, discuss any claims made on behalf of these products that you consider false or misleading. Include the URLs for the sites mentioned in your report.

2. Choose three different websites that sell vitamin supplements. Choose three different products and compare prices for the same product at the different websites. Then compare prices of the vitamins from these sites with the prices you would pay at a local supermarket, discount department store, or drugstore. Make sure to compare products with the same chemical composition (single vitamin, for example) and that have the same amount of the vitamin in each dose. Create an infographic or write a one–page report about your findings.

3. While watching a video on social media, you hear a so–called nutrition expert make claims for a substance that they claim to be a vitamin. What questions would you ask the expert to ascertain whether the substance truly is a vitamin?

4. One of your friends takes megadoses of vitamin A, C, B–6, and E because he thinks they help him stay healthy. What would you tell him about taking such large doses of these vitamins?

5. According to MyPlate, a person who needs 2000 kcal/day should consume 2 cups of fruit and 2.5 cups of vegetables daily. Plan a day's meals and snacks that provide these amounts of fruits and vegetables. In your plan, incorporate foods you like to eat. For information concerning how much of a certain form of fruit or type of vegetable equals 1 cup, visit www.myplate.gov and click on the "Fruits" and then the "Vegetables" tabs.

Practice Test (Section numbers are listed below in red.)

Select the best answer.

1. (8.1) Megadoses of vitamins
   a. are safe to take if the vitamins are water soluble.
   b. prevent certain chronic diseases.
   c. are available naturally from a wide variety of foods.
   d. may be toxic.

2. (8.1) Vitamins
   a. are metabolized to yield energy.
   b. occur in gram amounts in foods.
   c. are organic molecules.
   d. are macronutrients.

3. (8.1) People who are unable to absorb fat are likely to develop a _____ deficiency.
   a. vitamin A
   b. folate
   c. vitamin B–12
   d. riboflavin

4. (8.1) Enriched grain products have specific amounts of _____ added during processing.
   a. vitamin C
   b. vitamin A
   c. vitamin B–12
   d. thiamin

5. (8.1) The vitamin content of a vegetable can be affected by
   a. soil composition.
   b. the vegetable's maturity when harvested.
   c. sunlight exposure.
   d. All of the above are correct.

C Squared Studios/Photodisc/Getty Images

6. (8.2) Which of the following foods is *not* a rich source of provitamin A?
   a. Beef
   b. Carrots
   c. Squash
   d. Sweet potato

7. (8.2) Barbara has problems seeing when driving at night. A review of her diet reveals low fruit and vegetable consumption. Based on this information, a _______________ deficiency could be contributing to Barbara's condition.
   a. thiamin
   b. riboflavin
   c. vitamin D
   d. vitamin A

8. (8.2) Children who lack vitamin D can develop
   a. pellagra.
   b. rickets.
   c. beriberi.
   d. scurvy.

9. (8.2) During pregnancy, excess _____ intake is known to be teratogenic.
   a. vitamin A
   b. biotin
   c. vitamin K
   d. folate

10. (8.2) Vitamin K can be produced by
    a. skin exposure to ultraviolet radiation.
    b. hydrolysis of seawater.
    c. intestinal bacteria.
    d. conversion of lactic acid into lactate.

11. (8.2) Sunflower seeds, almonds, and plant oils are rich sources of this vitamin.
    a. Thiamin
    b. Folate
    c. Vitamin C
    d. Vitamin E

12. (8.3) Lack of vitamin _____ causes scurvy.
    a. A
    b. C
    c. D
    d. K

13. (8.3) Diets that lack niacin can lead to
    a. rickets.
    b. beriberi.
    c. pellagra.
    d. pernicious anemia.

14. (8.3) To reduce the likelihood of giving birth to babies with neural tube defects, females of childbearing age should obtain adequate

a. folate.
b. biotin.
c. cellulose.
d. niacin.

15. (8.3) Major food sources of vitamin B–12 include

a. enriched grain products.
b. meat and milk products.
c. fruit and vegetables.
d. nuts and seeds.

16. (8.3) Which of the following statements is *false*?

a. Intrinsic factor is needed for vitamin B–12 absorption.
b. Vitamin B–12 deficiency is common among older adults.
c. Patients with pernicious anemia are treated with high doses of folic acid.
d. If untreated, pernicious anemia can be deadly.

17. (8.3) Which of the following foods is a rich source of vitamin C?

a. Whole milk
b. Egg white
c. Hamburger patty
d. Green pepper

18. (8.3) Dark green vegetables such as spinach, asparagus, and broccoli are rich sources of which vitamins?

a. Vitamin K and folate
b. Thiamin and vitamin D
c. Riboflavin and niacin
d. Vitamins B–6 and B–12

19. (8.4) Lina is taking high–dose vitamin B–6 that exceeds the nutrient's UL to treat her premenstrual syndrome (PMS). She is at risk of experiencing the toxicity symptom of

a. excessive bleeding.
b. flushing of the skin.
c. decreased vision.
d. permanent nerve damage.

20. (8.5) Which of the following practices may reduce your risk of cancer?

a. Eating fruits and vegetables
b. Eating grilled meats regularly
c. Smoking no more than 10 cigarettes daily
d. Consuming two to three standard alcoholic drinks daily

Answers to Quiz Yourself

1. Natural vitamins are better for you because they have more biological activity than synthetic vitamins. **False.** (Section 8.1)
2. Certain vitamins can be toxic if consumed in high doses. **True.** (Section 8.1)
3. Vitamin E is an antioxidant. **True.** (Section 8.2)
4. Vitamins are a source of "quick" energy. **False.** (Section 8.1)
5. According to scientific research, taking large doses of vitamin C daily prevents the common cold. **False.** (Section 8.4)

Design Elements: (Stacked books): Siede Preis/Photodisc/Getty Images; (Nuts): C Squared Studios/Photodisc/Getty Images; (Recipe box, Raspberries, Blueberries): ©Wendy Schiff; (Pineapple): Stockbyte/Getty Images; (Radishes): karandaev/iStock/360/Getty Images; (MyPlate icon): US Department of Agriculture or USDA

CHAPTER 9

Water and Minerals

Michael Turner/123RF

Quiz Yourself

Can water be toxic when consumed in excess? What is osteoporosis? Which foods are rich sources of iron? Test your knowledge of water and minerals by taking the following quiz. The answers are at the end of the chapter.

1. Registered dietitian nutritionists recommend consuming 8 glasses of water a day for optimal health. ______ T ______ F

2. Ounce per ounce, cottage cheese contains more calcium than plain yogurt. ______ T ______ F

3. Potassium, sodium, and chloride ions are involved in fluid balance. ______ T ______ F

4. Selenium is an essential mineral. ______ T ______ F

5. In general, plants are good sources of iron because the plant pigment chlorophyll contains iron. ______ T ______ F

9.1 Introducing Water and Minerals

Learning Outcomes

1. Explain what can happen when dehydration occurs.
2. Classify mineral nutrients as major, trace, or possible essential minerals.
3. Explain the difference between a major mineral and a trace mineral.

In late August 1997, a 19-year-old student wrestler who weighed 233 pounds attended a university in North Carolina. The young man was anxious to lose enough weight to qualify for the 195-pound weight class when the collegiate wrestling season began later that semester. By November 6, the wrestler had lost 23 pounds, but the first tournament was to be held in 2 days and he was still 15 pounds too heavy. To lose ("cut") the extra weight, he engaged in an intense, almost nonstop training session. In a 12-hour period that spanned November 6 and 7, the wrestler restricted his water and food intake severely. Additionally, he wore a special suit made from a rubberized material that was *vapor impermeable* and he even covered it with a cotton warm-up outfit. By wearing this combination of clothing, the young man perspired profusely. Sweating can cause the body to lose a significant amount of "water weight"; however, sweating can also cause **dehydration** (body water depletion), especially when a person restricts their fluid intake. Severe dehydration is a life-threatening condition that requires urgent medical care.

dehydration body water depletion

hyperthermia very high body temperature

major minerals essential mineral elements required in amounts of 100 mg or more per day

trace minerals essential mineral elements required in amounts that are less than 100 mg per day

At 3 P.M. on November 6, the college wrestler began exercising vigorously in a hot environment. By 11:30 P.M., he had lost 9 pounds. After resting for about 2 hours, the young man resumed his exercise regimen in a desperate effort to lose the remaining 6 pounds. Around 2:45 A.M. on November 7, he had to discontinue exercising: He was extremely fatigued and unable to communicate. An hour later, he stopped breathing and his heart ceased beating. Attempts to revive him were unsuccessful. **Hyperthermia** (very high body temperature) is likely to have contributed to his death.[1]

The rapid-weight-loss practices that resulted in the death of the young wrestler were prohibited by National Collegiate Athletic Association (NCAA) regulations. However, many wrestlers were determined to lose enough weight prior to a meet so they could gain a competitive edge by wrestling in weight categories that were lower than their preseason weights. Today, NCAA rules prohibit wrestlers from losing weight rapidly by any means, including restricting food and fluid intakes excessively. Failure to follow the NCAA's wrestling guidelines can result in suspension from competition.

We often take water for granted, but this simple molecule is highly essential. You can survive for weeks, even months, if your diet lacks carbohydrates, lipids, proteins, and vitamins. But if you do not have any water, your life will end within a week or two.

Many of water's functions involve certain minerals. About 15 mineral elements have known functions in the body and are necessary for human health. The body requires these particular micronutrients in milligram or microgram amounts. The essential minerals are classified into two groups: **major minerals** and **trace minerals** (Table 9.1). If we require 100 mg or more of a mineral per day, the mineral is classified as a major mineral; otherwise, the micronutrient is a trace mineral. The body also contains very small amounts of other minerals, such as nickel and arsenic. The essential nature of this particular group of minerals has not been fully determined, so we will refer to them as "possible essential minerals" (see Table 9.1).

Melica/Shutterstock

TABLE 9.1 Minerals with Known or Possible Roles in the Body*

Major Mineral	Trace Mineral	Possible Essential Mineral
Calcium (Ca)	Chromium (Cr)	Arsenic (As)
Chloride (Cl)	Copper (Cu)	Boron (B)
Magnesium (Mg)	Fluoride (F)**	Lithium (Li)
Phosphorus (P)	Iodine (I)	Nickel (Ni)
Potassium (K)	Iron (Fe)	Silicon (Si)
Sodium (Na)	Manganese (Mn)	Vanadium (V)
Sulfur (S)	Molybdenum (Mo)	
	Selenium (Se)	
	Zinc (Zn)	

*Chemical symbol is shown in parentheses next to mineral's name.

**Although fluoride is not essential, the mineral plays an important role in strengthening teeth and bones.

The mineral nutrients are key components of body structures and play vital roles in metabolism, water balance, muscle movement, and various physiological processes. Mineral deficiencies can cause serious health problems and, in severe cases, death. In this chapter, we will focus on water first and then discuss mineral nutrients, some of which are of major concern for public health officials in the United States and other parts of the world.

Concept CHECKPOINT

1. What can happen to a person when dehydration occurs?
2. What is the difference between a major mineral, trace mineral, and possible essential mineral?
3. List at least five major minerals and five trace minerals.
4. List at least two possible essential minerals.

See Appendix F for responses.

9.2 Water

Learning Outcomes

1. Discuss the functions of water in the body, as well as typical sources of intake and loss.
2. Identify foods that have high and low water contents.
3. Discuss how the body maintains its fluid balance.
4. Explain why dehydration and water intoxication can be life-threatening conditions.

Compared to other nutrients, water is so unusual, it is in a class by itself. Water is a simple compound: A molecule of water is composed of two hydrogen atoms and one oxygen atom (H_2O). Water does not need to be digested, and it is easily absorbed by the intestinal tract.

Depending on a person's age, sex, and body composition, 50 to 75% of their body is water weight. Lean muscle tissue contains more water (about 73%) than fat tissue (about 20%). On average, young adult males have more lean tissue than young females. Approximately 60% of an average young male's body weight is water; the average young adult female's body has more fat and, therefore, slightly less water than an average young male's body. A person's percentage of body weight that is water declines from birth to old age.

Water is a major solvent that dissolves many body substances, including glucose. Water often participates directly in chemical reactions, such as those involved in digesting food. Water's other physiological roles include transporting substances, removing waste products, lubricating tissues, and regulating body temperature and acid–base balance (proper blood pH). Furthermore, water is a major component of body fluids, such as blood, saliva, sweat, tears, mucus, and the fluid in joints. Table 9.2 lists roles of water in the body. Although water has numerous functions in the body, this unique nutrient does not provide energy.

TABLE 9.2 Functions of Water in the Body

Water:
• Is a solvent
• Is a major component of blood, saliva, sweat, tears, mucus, joint fluid
• Removes wastes
• Helps transport substances
• Lubricates tissues
• Regulates body temperature
• Helps digest foods
• Participates in many chemical reactions
• Helps maintain proper blood pH

Membrane Transport

One reason why water is such a vital nutrient is its role in helping cells obtain materials from their environment and eliminate wastes. To understand some of water's functions in the body, it is necessary to understand how water-soluble substances can become distributed when they are in water. In some instances, substances move by **simple diffusion,** that is, they move from where they are highly concentrated to where they are less concentrated (Fig. 9.1). For example, simple diffusion occurs when a sugar cube is dropped into a cup of hot tea. As the cube dissolves, the concentration of sugar molecules around the cube becomes higher than the concentration of sugar molecules near the surface of the tea. However, the sugar molecules spontaneously diffuse throughout the tea, and eventually, the concentration of sugar molecules and water molecules becomes evenly distributed within the beverage.

Did *YOU* Know?

The layer of fat that is under the skin (*subcutaneous fat*) interferes with the transfer of heat to the skin. This explains why people who have excess body fat and are in warm conditions are more uncomfortable and perspire more than people who have healthy body weights.

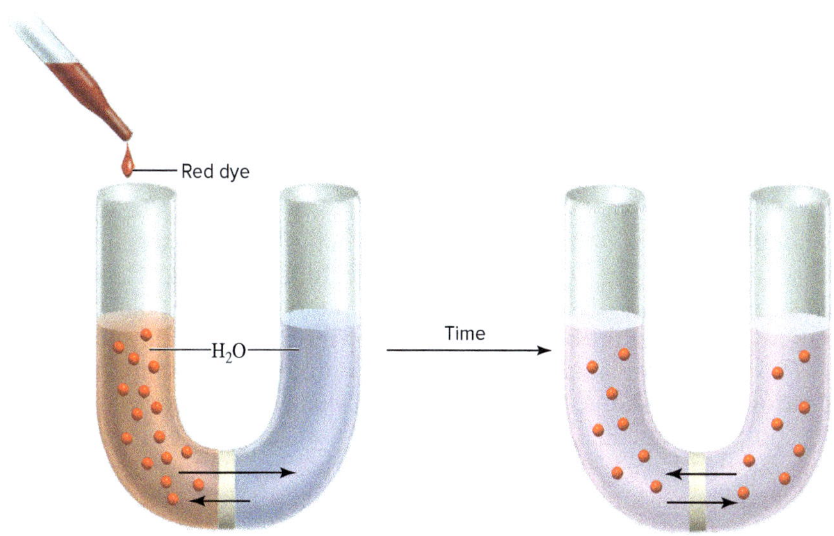

Figure 9.1 Simple diffusion. Simple diffusion can occur when there is a greater concentration of a substance in one region than in another. In this example, the dye is more concentrated on the left side of the container than on the right side. Eventually, the dye molecules diffuse (move) from where they are more concentrated to where they are less concentrated. The diffusion stops when the dye molecules are equally distributed throughout the container.

simple diffusion movement of substances from a region of higher to lower concentration

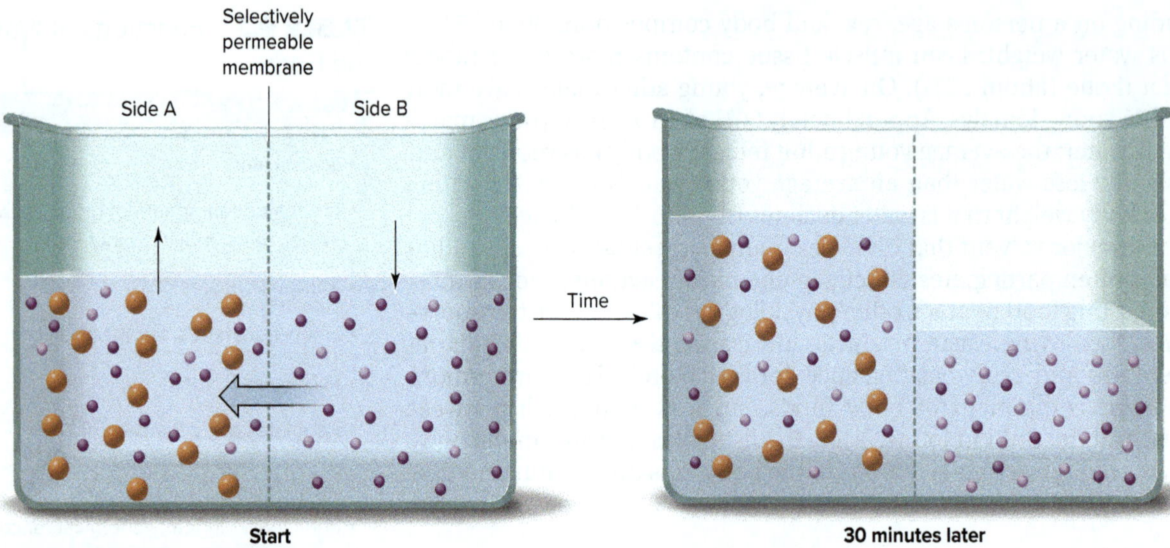

Figure 9.2 Osmosis. Osmosis is the diffusion of a solvent, such as water, through a selectively permeable membrane. Water moves from a compartment that is less concentrated to a compartment that has more material dissolved in it.

selectively permeable membrane barrier that allows the passage of certain substances and prevents the movement of other substances

osmosis diffusion of a solvent, such as water, through a selectively permeable membrane

extracellular fluid watery fluid that surrounds cells or is in liquid portion of blood

intracellular fluid watery fluid that is inside cells

hydration water status

Simple diffusion also occurs when there is a greater concentration of substances on one side of a *selectively permeable membrane* than on the other. A **selectively permeable membrane** is a barrier that allows the passage of certain substances and prevents the movement of other substances through it. **Osmosis** is the diffusion of a solvent, such as water, through a selectively permeable membrane, such as the plasma membrane of a human cell. The concentration of substances dissolved in the water, such as glucose, influences osmosis. Water moves from a region that has less material dissolved in it (dilute) to a region that has more material dissolved in it (Fig. 9.2). The diffusion stops when the concentrations of the material on either side of the plasma membrane are equal.

To survive, a human cell carefully controls the passage of substances through its plasma membrane. Some materials, such as oxygen and carbon dioxide, easily pass through the cell's plasma membrane by simple diffusion. Oxygen diffuses into cells; carbon dioxide (a cellular waste product) diffuses out of cells. Large molecules, such as proteins, may be unable to pass through the cell's membrane unless there are special carrier molecules to help them enter the cell.

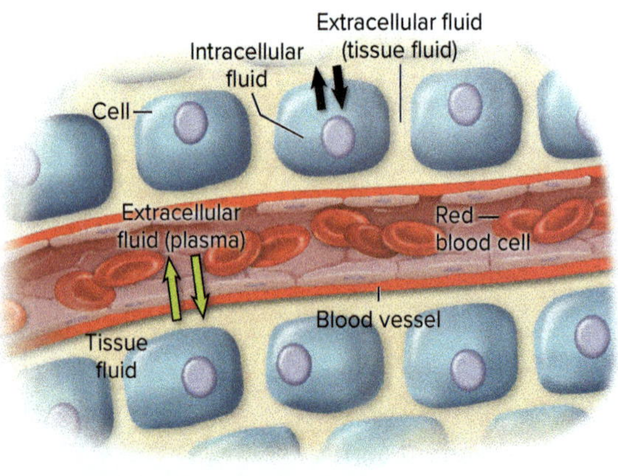

Figure 9.3 Fluid compartments in the body. Intracellular fluid is inside cells; extracellular fluid surrounds cells (tissue fluid) or is the watery fluid in blood (*plasma*). Water is exchanged between plasma and tissue fluid (green arrows) as well as between tissue and intracellular fluids (black arrows).

Body Fluid

The body has two major fluid compartments in the body: intracellular fluid and extracellular fluid (Fig. 9.3). **Intracellular fluid** is inside cells. **Extracellular fluid** surrounds cells (tissue fluid) or is the watery portion of blood (*plasma*). About two-thirds of the body's water is in the intracellular fluid compartment.

The body maintains the balance of compartmental fluids and proper **hydration,** water status, primarily by controlling ion concentration in each fluid compartment. Recall from Chapter 4 that ions are elements or small molecules that have electrical charges. Ions that conduct electricity, such as sodium and potassium ions, are called *electrolytes*. Table 9.3 lists several electrolytes that function in the body.

Maintenance of intracellular fluid volume depends to a large extent on the intracellular concentration of potassium and phosphate ions (Fig. 9.4). On the other hand, maintenance of extracellular fluid volume depends primarily on the extracellular concentration of sodium and chloride ions (see Figure 9.4). Changes in the normal concentrations of these ions can cause water to shift out

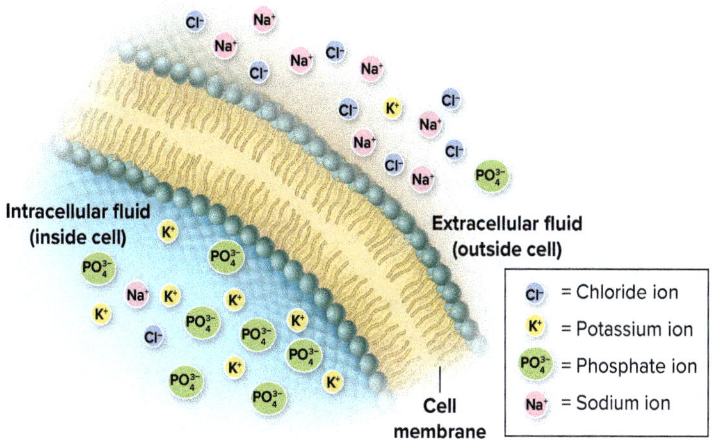

Figure 9.4 Major intracellular and extracellular ions involved in fluid balance.

TABLE 9.3 Common Electrolytes (Mineral Ions)

- Sodium (Na^+)
- Calcium (Ca^{2+})
- Potassium (K^+)
- Magnesium (Mg^{2+})
- Chloride (Cl^-)
- Phosphate (PO_4^{3-})
- Sulfate (SO_4^{2-})

of one compartment and move into the other. For example, if extracellular fluid has fewer than normal sodium ions, water moves from the extracellular compartment into cells. When this occurs, the cells swell and can burst (Fig. 9.5a). On the other hand, if extracellular fluid has an excess of sodium ions, water moves out of cells. As a result, the cells shrink and die because they lack enough intracellular fluid to function (Fig. 9.5b). Recall from Chapter 7 that edema occurs when an excessive amount of fluid is in the space surrounding cells (see Figure 7.1). To function normally, the body must maintain intracellular and extracellular fluid volumes within certain limits.

Sources of Water

How much water should a person drink to be healthy? Many people believe consuming 8 glasses of water a day is necessary for good health. Contrary to this popular belief, there is no "rule of thumb" recommendation that specifies how many glasses of water to consume each day.[2] Factors such as environmental temperatures, health conditions, physical activities, and dietary choices influence individual water requirements. Thus, total water intakes vary widely. **Total water intake** refers primarily to water ingested by consuming beverages and foods.

total water intake water ingested by consuming beverages and foods

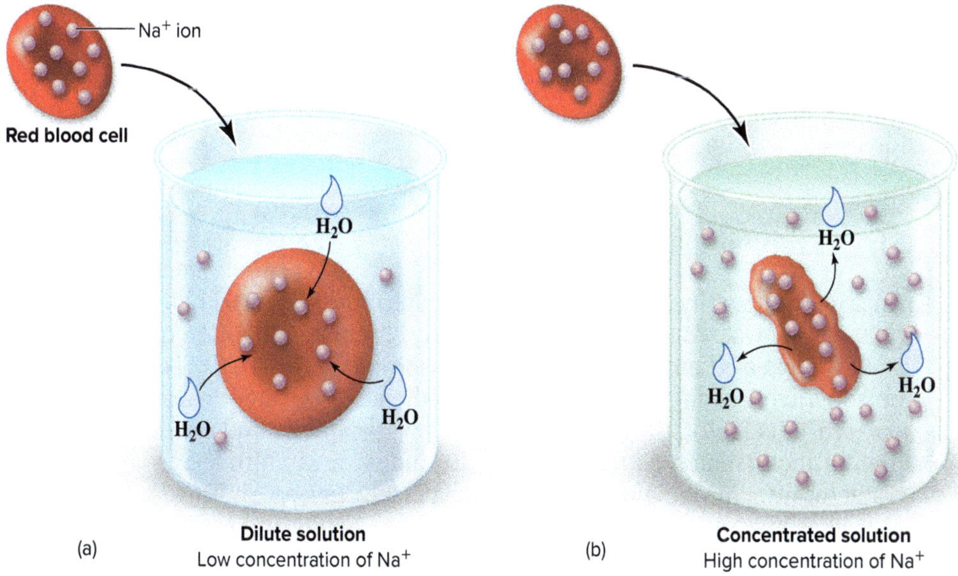

Figure 9.5 Maintaining proper hydration. Cells, such as red blood cells, need to maintain their fluid balance. The cell in beaker (a) is placed in a solution that contains few sodium ions (Na^+) in relation to the concentration of Na^+ in the cell. As a result, water moves from the solution and into the red blood cell. The cell can swell and burst. The solution in beaker (b) has an excess of sodium ions (Na^+) compared to the concentration of Na^+ inside the cell. Water moves out of the red blood cell placed in beaker (b). As a result, the cell shrinks and can die.

TABLE 9.4 How Much Water Is in That Food/ Beverage?

Food	Water % by Weight
Lettuce	95
Tomato	95
Watermelon	91
Milk, 1% fat	90
Apple, with skin	86
Avocado (Florida)	79
Potato, white, baked with skin	75
Banana	75
Chicken, white meat, roasted	65
Ground beef, 80% lean, broiled	56
Bread, whole wheat	39
Margarine, stick	16
Crackers, saltines	5
Vegetable oil	0

Source: Data from U.S. Department of Agriculture, Agricultural Research Service: *FoodData Central*, 2019.
Lettuce: lynx/iconotec.com/Glow Images; Tomato D. Hurst/Alamy Stock Photo

The Adequate Intake (AI) for *total* water intake is approximately 11 cups (2.7 L) for young females and approximately 15.5 cups (3.7 L) for young males.[3] These amounts do not need to be consumed in the form of water. Other sources of water include fruit juice, milk, soup, coffee, tea, soft drinks, and flavored bottled water. Most solid foods also contain some water. Fruits and vegetables appear to be solid, but they generally contain 60 to 95% water weight. Table 9.4 lists some commonly consumed foods and their water content by weight. About 80% of our total water intake is from water and other beverages; food supplies the remaining amount of our water intake.[3]

In addition to the water in beverages and foods, a considerable amount of water enters the digestive tract daily through secretions from the mouth, stomach, intestine, pancreas, and gallbladder. The intestinal tract absorbs most of this water, too. Each day, only about 0.4 to 0.8 cup (approximately 100 to 200 ml) of the water that enters the digestive tract is not absorbed. The body eventually eliminates the unabsorbed water in feces.

metabolic water water formed by cells as a metabolic by-product

Cells also form some water as a by–product of metabolism ("metabolic water"). **Metabolic water** also contributes to the body's water intake and fluid balance.

The Essential Balancing Act

An average healthy adult consumes and produces approximately 2.6 quarts (2500 ml) of water daily (Fig. 9.6).[4] The body eliminates about 2.6 quarts of water in urine, exhaled air, feces, and perspiration (see Figure 9.6). Thus, a healthy person's average daily water intake equals their average daily losses (output).

Various factors influence a person's water intake and output. Environ–mental factors such as temperature, humidity, and altitude can affect body water losses. Physiological conditions, especially fever, vomiting, and diarrhea, as well as lifestyle practices, including exercise habits and sodium and alcohol intakes, can also alter the body's fluid balance.

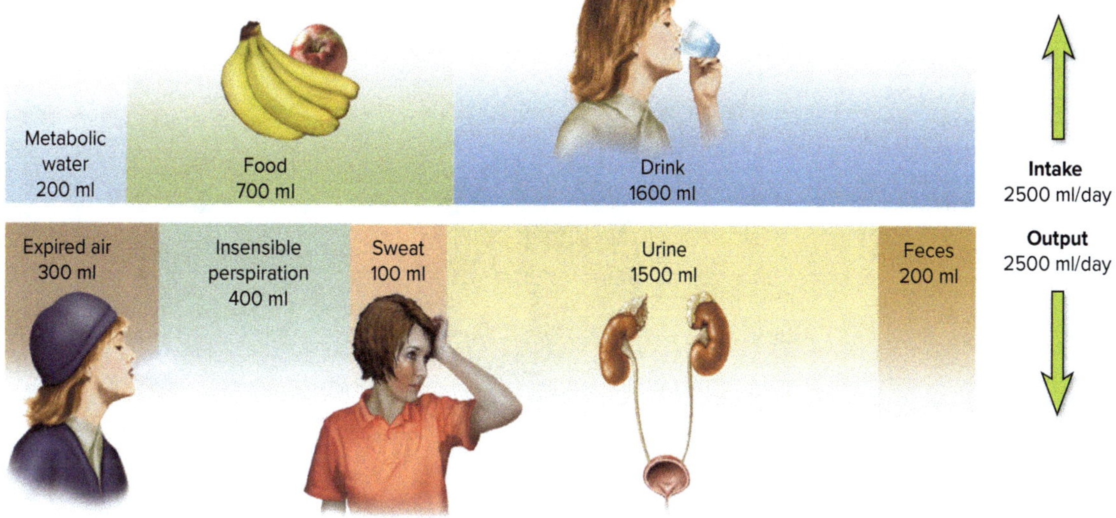

Figure 9.6 Daily water balance. An average healthy adult consumes and produces approximately 2500 ml of water and eliminates about 2500 ml of water daily.

Perspiration is body water that is secreted by sweat glands in skin. When perspiration reaches the skin's surface, it evaporates into the air. This process helps cool the body and maintain its normal temperature. *Insensible perspiration* is body water that diffuses through the layers of skin or is exhaled from the lungs instead of being secreted by sweat glands.[4] People are usually unaware that their bodies are constantly losing water in this manner; hence, the term *insensible* perspiration.

Kidneys and Hydration

The kidneys are the major regulator of the body's water content and ion concentrations. In a healthy person, the kidneys maintain proper hydration by filtering excess ions and water from blood as it flows through the kidney's tissues. Water is the main component of urine. If you drink more watery fluids than your body needs, your kidneys excrete the excess water in urine.

Kidneys also remove drugs and metabolic waste products, such as urea, from the bloodstream. Sometimes, minerals and waste products settle out of urine and collect into crystals. If the crystals enlarge and form a hard mass, the object is called a kidney stone (Fig. 9.7). Kidney stones often contain the mineral calcium.[5] As a kidney stone moves out of the kidney and enters the tube leading to the bladder, it may cause considerable pain and bloody urine until it passes out of the body. Dehydration increases the likelihood of forming kidney stones.

What Is a Diuretic? Caffeine is a **diuretic,** a substance that increases urine production. Coffee, tea, energy drinks, and soft drinks often contain caffeine or caffeine-related compounds. However, the water consumed in caffeinated beverages is not completely lost in urine, so drinking these fluids may still contribute to meeting your water needs.[3]

Water Conservation

As mentioned in the beginning of this chapter, body water depletion is called dehydration. Dehydration can be a life-threatening condition. When you are hot and perspiring heavily, your kidneys try to conserve as much water as possible to avoid dehydration. **Antidiuretic hormone (ADH)** and **aldosterone** (*al-dahs'-te-rown*) are two hormones that participate in the body's efforts

Did *YOU* Know?

Your weight may increase by a few pounds after you eat salty foods and then consume beverages. The weight gain is due to a temporary increase in body water volume. If you resume your usual intake of sodium, your kidneys will eliminate the excess sodium and water in urine within a day. As your body regains its normal fluid balance, your weight also returns to normal.

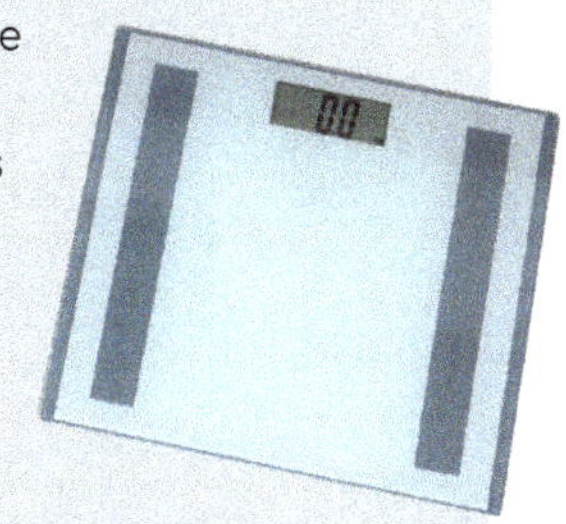

Weerapat Kiatdumrong/123RF

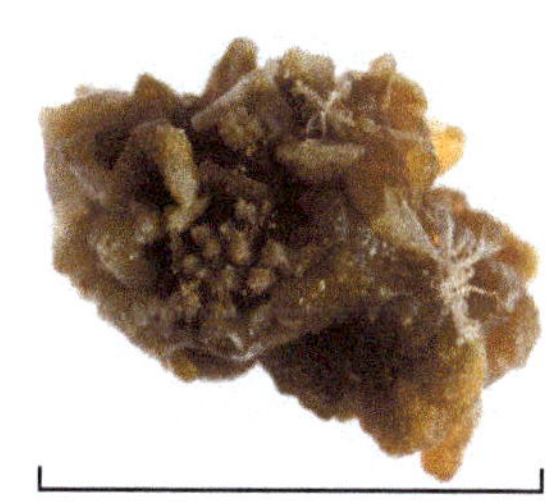

Figure 9.7 Kidney stones. Kidney stones can be quite painful when they move from the kidneys to the bladder and are eliminated in urine.
Jonathan Kirn/Stockbyte/Getty Images

diuretic substance that increases urine production

antidiuretic hormone (ADH) hormone that participates in water conservation

aldosterone hormone that participates in sodium and water conservation

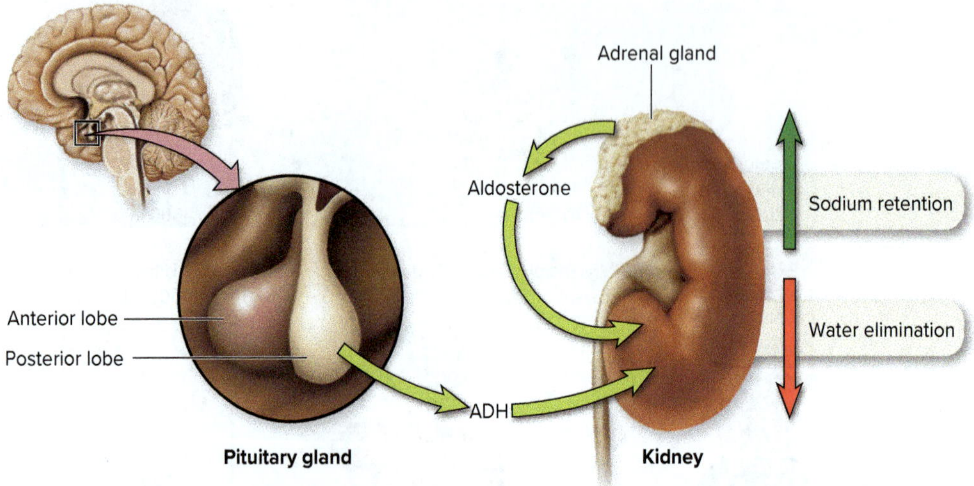

Figure 9.8 Effects of antidiuretic hormone and adolsterone on kidneys. In response to dehydration, the posterior pituitary gland in the brain secretes antidiuretic hormone, which signals the kidneys to conserve water. Additionally, the adrenal glands secrete the hormone aldosterone. Aldosterone reduces urinary excretion of sodium. When the kidneys retain sodium, they return the mineral to the general circulation. As a result of the actions of these two hormones, body water and sodium are conserved.

to maintain fluid balance. In response to dehydration, the posterior pituitary gland in the brain releases antidiuretic hormone. Antidiuretic hormone stimulates the kidneys to conserve water. Additionally, the adrenal glands secrete aldosterone. Aldosterone signals kidneys to reduce the elimination of sodium in urine, and as a result, the kidneys return the mineral and water to the general circulation. The diagram in Figure 9.8 summarizes the effects of antidiuretic hormone and aldosterone on the kidneys.

The simplest way to determine if you are consuming enough water is to notice the amount of urine you eliminate. When your fluid intake is adequate, your kidneys will produce enough urine to maintain fluid balance. If you consume more fluid than needed, your kidneys will eliminate the excess, and you will produce plenty of urine. On the other hand, if you limit your fluid intake or have high fluid losses such as in sweat, you will produce small amounts of urine.

In addition to urine volume, the color of urine may be a useful indicator of hydration status. Straw-colored (light yellow) urine can indicate adequate hydration, whereas dark-colored urine may be a sign of dehydration. However, the color of urine is not always a reliable guide for judging a person's hydration status.[6] It is important to recognize that having urinary tract infections or ingesting certain medications, foods, and dietary supplements, especially those containing the B-vitamin riboflavin, can alter urine's color.

Alcohol is a diuretic. Normally, antidiuretic hormone signals the kidneys to conserve water (see Figure 9.8). Alcohol, however, inhibits ADH secretion from the pituitary gland in the brain, enabling the kidneys to eliminate more urine than normal. Alcohol consumption actually results in urinary water losses that are greater than the volume of fluid consumed. Therefore, alcohol contributes to dehydration.

Did *YOU* Know?

Scientists do not know what causes a hangover, the headache, tiredness, thirst, and overall discomfort that occurs a few hours after drinking too much alcohol. Dehydration, the body's immune response, and congeners may be responsible for the unpleasant, delayed side effects of excess alcohol consumption.[7] Congeners are substances in alcoholic drinks that contribute to the taste and color of the beverages. (Beer and vodka have lower congener contents than red wine and whiskey.) Alcoholic drinks with high congener contents tend to produce more severe hangovers than drinks with lower contents of these substances.

©Wendy Schiff

Dehydration

Despite the body's mechanisms to balance its water content, some fluid is constantly being lost, primarily via the skin and lungs. If a person does not consume enough fluids to replace that water, dehydration can occur. Rapid weight loss is a sign of dehydration. Every 16 ounces (about 0.5 L) of water that the body loses represents a pound of body weight. If you lose 1 to 2% of your usual body weight in fluids, you will feel fatigued and thirsty. If you weigh 150 pounds, for example, and your weight drops 3 pounds after exercising in hot conditions, you have lost 2% of your body weight, primarily as water weight.

As the loss of body water approaches 4% of body weight, muscles lose considerable amounts of strength and endurance. By the time body weight is reduced by 7 to 10% as a result of body fluid losses, severe weakness results. At a 20% reduction of body weight, coma and death are likely.

Thirst is the primary regulator of fluid intake.[4] The thirst response alerts you to the need to replenish water that was lost by sweating and other means. The majority of healthy people meet their AI for water by letting thirst be their guide.[3] Thirst stimulates people to drink fluids *before* severe dehydration occurs. However, people who are dehydrated and older than 60 years of age do not sense thirst as accurately as younger adults.[8] Furthermore, older adults may be more susceptible to developing dehydration than younger persons because as kidneys age, they become less able to conserve water when fluid intakes are low. Therefore, it may be necessary to remind older adults to drink more watery fluids, especially when they are physically active or in warm conditions. Nevertheless, healthy older adults are generally able to maintain adequate hydration.[8]

People who are sick, especially children with fever, vomiting, diarrhea, and increased perspiration, may need to be given special solutions of water and electrolytes to prevent dehydration. Athletes and other people who work or exercise outdoors, especially in hot conditions, also need to stay properly hydrated to avoid dehydration and heat-related illnesses such as heat exhaustion. Chapter 11 provides information about heat-related illnesses.

People who work or exercise outdoors, especially in hot conditions, need to stay hydrated to avoid dehydration and heat-related illnesses.
Sculpies/Shutterstock

Did *YOU* Know?

If you reside in the southeastern United States, along the Gulf of Mexico, or on an island in the Caribbean Sea, or if you plan to visit these places, you need to be prepared for tropical cyclones, especially hurricanes. According to the U.S. National Oceanic and Atmospheric Administration (NOAA), the annual hurricane season for these regions begins on June 1 and lasts until November 30.

U.S. Army

Over the past few years, major hurricanes have killed thousands of people and devastated parts of Louisiana and several Caribbean islands, including Haiti, Puerto Rico, the Bahamas, and the Dominican Republic. For the storms' survivors, having access to drinkable (potable) water was crucial. People can live for a few days in such conditions before dehydration contributes to their deaths. This photo shows a Haitian woman receiving bottled water from a U.S. military serviceman during hurricane relief efforts in October 2016.

Can Too Much Water Be Toxic?

There is no Upper Level (UL) for water. **Water intoxication,** however, can occur when an excessive amount of water is consumed in a short time period or the kidneys have difficulty filtering water from blood. The excess water dilutes the normal sodium concentration of blood, causing hyponatremia (*hypo* = low; *natremia* = sodium in the bloodstream). As a result of the water imbalance, too much water moves into cells, including brain cells. Signs and symptoms of water intoxication may include drowsiness, nausea and vomiting, confusion, inability to coordinate muscular movements, and weight gain.[9] If the condition is not detected early and treated effectively, coma and death can result.

water intoxication condition that occurs when too much water is consumed in a short time period or the kidneys have difficulty filtering water from blood

Healthy people rarely drink enough water to become intoxicated. However, water intoxication can develop in people with disorders that interfere with the kidney's ability to excrete water normally. Endurance athletes such

Endurance athletes such as cyclists who consume large amounts of plain water in an effort to keep hydrated during competition may be at risk of water intoxication.
Lothar Schulz/fStop/Getty Images

as cyclists, triathletes, and marathon runners who consume large amounts of plain water in an effort to keep hydrated during competition may be at risk of water intoxication. Chapter 11 discusses the importance of proper hydration for athletes.

Concept CHECKPOINT

5. List at least five different functions of water in the body.
6. Define *osmosis*.
7. Which ions are found primarily in extracellular fluid? Which ions are found primarily in intracellular fluid?
8. Discuss ways the body obtains and loses water.
9. What can happen to cells if the body is unable to regulate its water balance?
10. How do antidiuretic hormone and aldosterone help maintain fluid balance in the body?
11. How much water do healthy young males and females need to consume daily (AI values)?
12. What is a diuretic? Identify two diuretics commonly consumed by Americans.
13. List at least three signs and symptoms of dehydration.
14. List at least three signs and symptoms of water intoxication.

See Appendix F for responses.

9.3 Minerals: Basic Concepts

Learning Outcomes

1. Identify general functions of minerals in the body.
2. Discuss factors that influence the body's ability to absorb and use minerals.
3. Describe factors that can affect retention of minerals during food preparation.

Minerals, such as iron and calcium, are a group of elements in Earth's rocks, soils, and natural water sources. Plants, animals, and other living things cannot synthesize minerals. Plants obtain the minerals they need from soil or fertilizer; animals generally obtain minerals when they consume plants and other animals or substances that contain these elements. About 15 mineral elements have known functions in the body and are necessary for human health.

Several minerals, including lead and mercury, are often found in the human body, but they are environmental contaminants that have no known functions. The body can eliminate most minerals in urine. However, exposure to excessive amounts of minerals can cause toxicity.

Unlike vitamins, minerals are indestructible. Because minerals cannot be destroyed, heating a food or exposing it to most other environmental conditions will not affect the food's mineral content. However, minerals are water soluble, and they can leach out of a food and into cooking water. By using the cooking water to make soups or sauces, you can obtain minerals from the food that would otherwise be discarded.

Cheese is a good source of calcium, phosphorus, selenium, zinc, and sodium.
Ingram Publishing/SuperStock

Why Are Minerals Necessary?

Essential minerals have diverse roles in the body (Fig. 9.9). Some minerals form inorganic structural components of tissues, such as calcium and phosphorus in

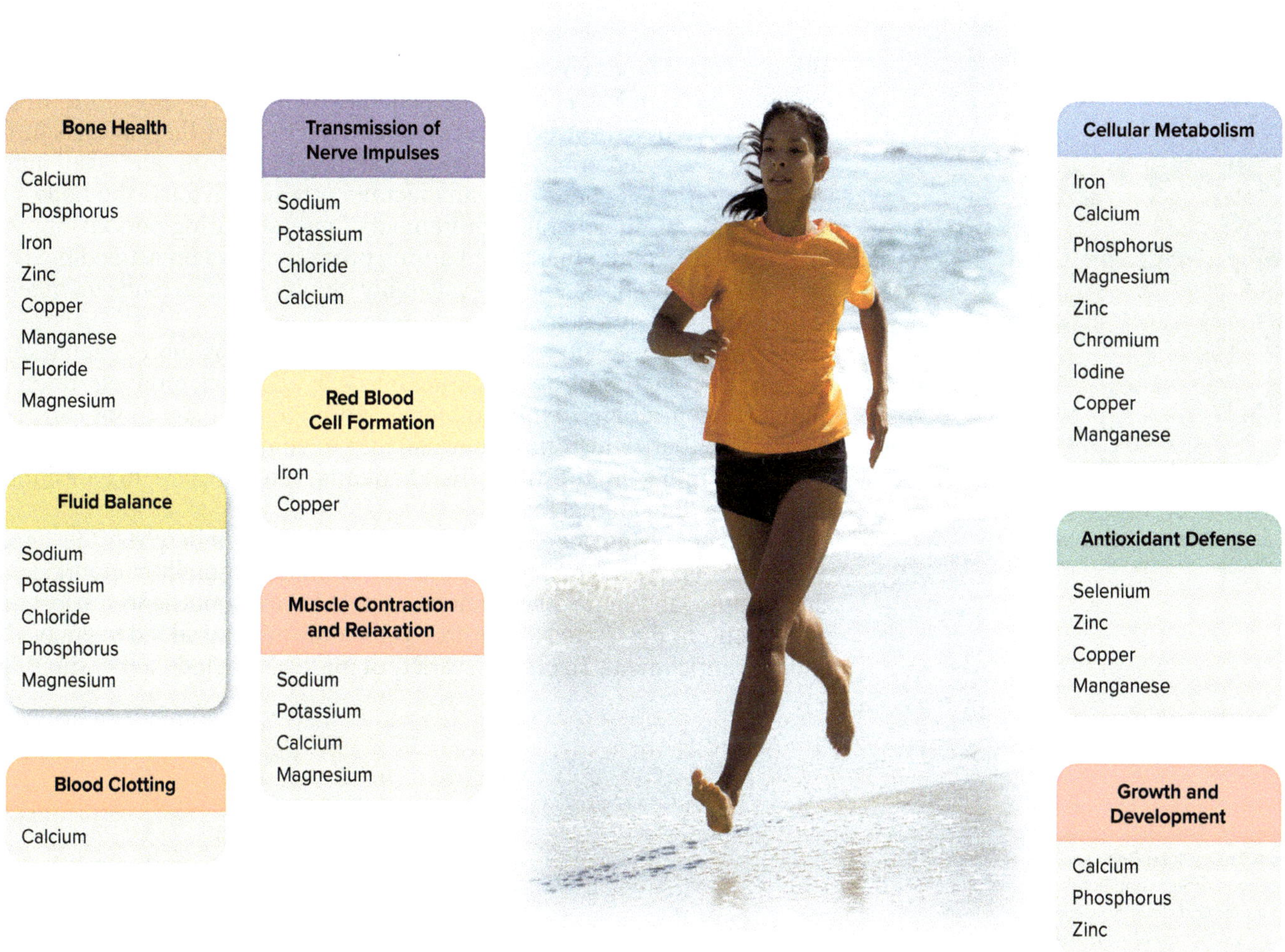

Figure 9.9 Minerals and their functions. Groups of minerals work together to maintain good health.
Stephen Simpson/Photodisc/Getty Images

bones and teeth. Minerals may also function as inorganic ions, substances that have negative or positive charges (see Chapter 4). For example, calcium ions (Ca^{2+}) participate in blood clotting and sodium ions (Na^+) help maintain fluid balance. Phosphate ions participate in acid–base balance. Some ions, such as magnesium (Mg^{2+}) and copper (Cu^{2+}), are cofactors. A **cofactor** is a metallic ion or small molecule that activates certain chemical reactions. Many minerals are components of various enzymes, hormones, or other organic molecules, such as cobalt in vitamin B–12, iron in hemoglobin, and sulfur in the amino acids, methionine and cysteine. Although cells cannot metabolize minerals for energy, certain minerals are involved in chemical reactions that release energy from macronutrients.

cofactor metallic ion or small molecule that activates certain chemical reactions

In some instances, the digestive tract absorbs more minerals than the body needs, but the excess is excreted, primarily in urine or feces. In other instances, the body stores the extra minerals in the liver, bones, or other tissues. Toxicity signs and symptoms occur when minerals accumulate in the body to such an extent that they interfere with the functioning of cells. Under normal conditions, the human body does not store large quantities of most minerals, and it loses small amounts of these essential elements every day. Therefore, people should choose their diets carefully so their bodies can maintain an adequate supply of minerals.

The bioavailability of minerals tends to increase during periods of growth, including infancy.
Darlene Schueller/McGraw Hill

Sources of Minerals

Although most foods contain small amounts of minerals, Figure 9.10 presents food groups from MyPlate that are generally rich sources of various minerals. The digestive tract, however, does not absorb 100% of the minerals in foods or dietary supplements. The body's ability to absorb and use minerals (*bioavailability*) depends on many factors. A major factor is the body's need for the mineral. In general, requirements increase during periods of growth, such as infancy and puberty, and during pregnancy and breastfeeding. During these critical life stages, the bioavailability of minerals also tends to increase to help meet the body's demand.

Compared to plant foods, animal foods tend to be more reliable sources of minerals, such as iron and calcium. Why? Animal products often have higher concentrations of these minerals. Additionally, plant foods can contain substances that reduce the bioavailability of minerals, particularly calcium, zinc, and iron. On the other hand, plants supply more magnesium and manganese than animal foods.

In general, the more processing a plant food undergoes, the lower its natural mineral content. Cereal grains, for example, naturally contain selenium, zinc, copper, and some other minerals, but these micronutrients are lost during refinement. In the United States, iron is the only mineral added to grains if they undergo enrichment. To obtain a variety of minerals, include some whole-grain

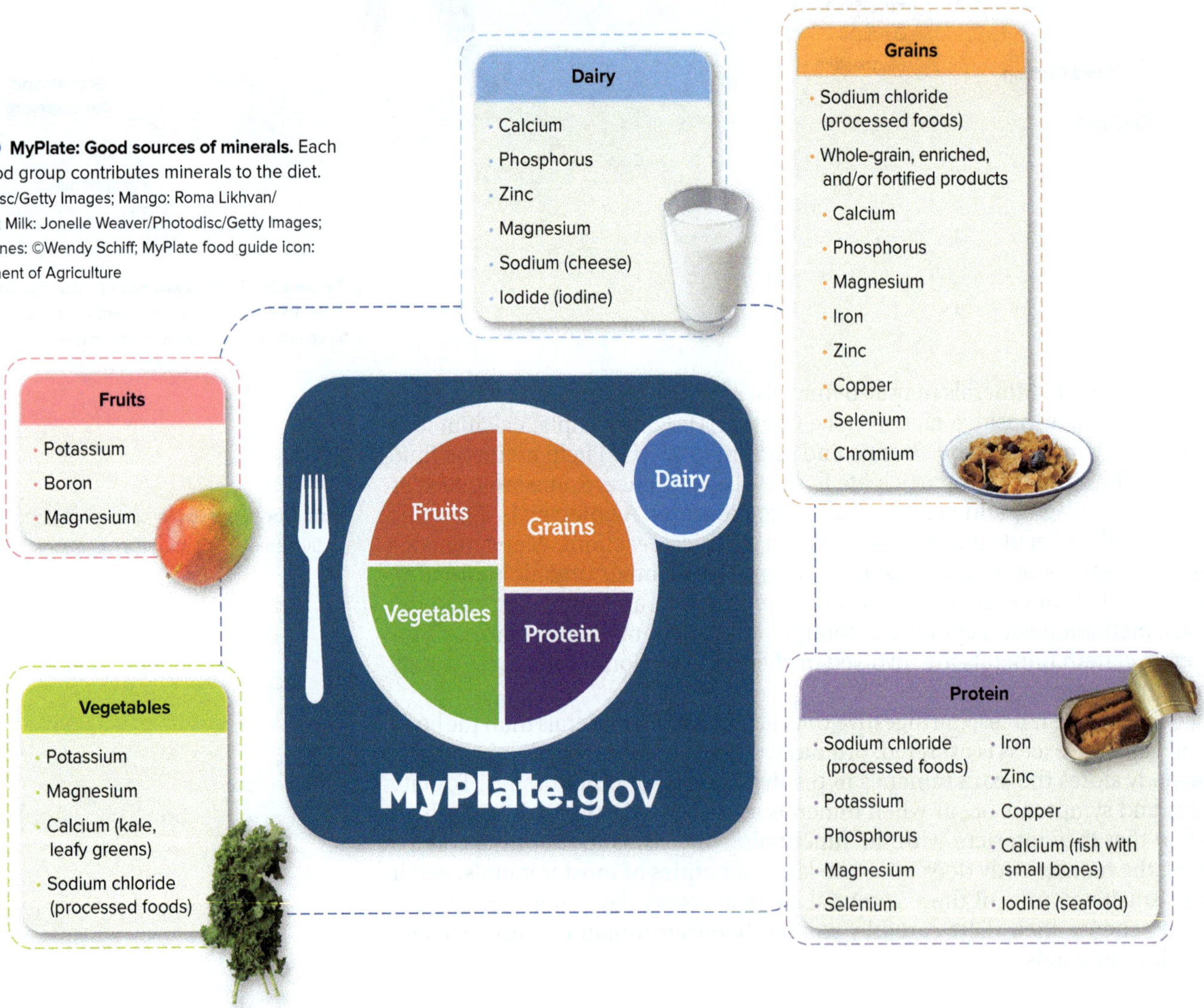

Figure 9.10 MyPlate: Good sources of minerals. Each MyPlate food group contributes minerals to the diet.
Kale: Stockdisc/Getty Images; Mango: Roma Likhvan/Shutterstock; Milk: Jonelle Weaver/Photodisc/Getty Images; Cereal, Sardines: ©Wendy Schiff; MyPlate food guide icon: U.S. Department of Agriculture

products in your diet each day. By following the recommendations of MyPlate (see Chapter 3) and eating a variety of plant and animal foods, you are likely to obtain adequate amounts of all essential minerals.

Other Sources of Minerals

The tap water in your community may be a source of minerals that you may have overlooked. Fluoride is often added to public water supplies. Although fluoride is not essential for life, the mineral strengthens bones and teeth when consumed in adequate amounts. "Hard" water naturally contains a variety of minerals, including calcium, magnesium, sulfur, iron, and zinc. Water with high mineral content often tastes and smells unpleasant. Many people drink bottled water as a substitute for tap water because they think bottled water tastes better and it is safer. To learn more about bottled water, read the "Nutrition Matters" section at the end of this chapter.

Dietary supplements are another source of minerals. A daily multiple vitamin and mineral supplement is generally safe for healthy people because a dose of this type of supplement does not provide high amounts of minerals. However, recall from Chapter 1, there are no significant benefits for healthy people who take dietary supplements. People need to be careful when taking dietary supplements that contain individual minerals, such as iron or selenium. Many minerals have a narrow range of safe intake; therefore, it is easy to consume a toxic amount, especially by taking supplements that contain only a particular mineral (Fig. 9.11). Additionally, an excess of one mineral can interfere with the absorption or metabolism of other minerals. For example, the presence of a large amount of zinc in the intestinal tract decreases copper absorption. Single–mineral supplements are usually unnecessary unless they are prescribed to treat a specific medical condition, such as iron deficiency.

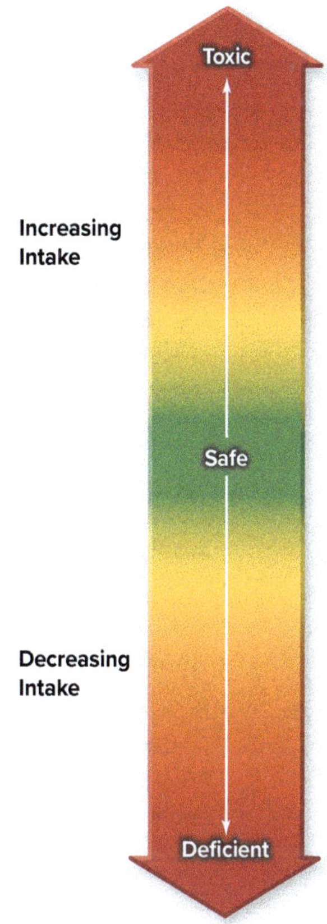

Figure 9.11 Mineral intake. Many minerals have a narrow range of safe intake. As a result, it is relatively easy to consume a toxic amount, especially by taking supplements that only contain a particular mineral.

Concept CHECKPOINT

15. List at least three different functions of minerals in the body, and provide an example of a mineral that performs each function.

16. Explain how foods that are naturally good sources of minerals can become poor sources of those minerals by the time you eat them.

17. Discuss factors that influence mineral absorption in the digestive tract.

18. In the United States, which mineral is added to grain products as part of the grain enrichment program?

See Appendix F for responses.

9.4 Major Minerals

Learning Outcomes

1. List key functions and good food sources of the major mineral nutrients.
2. Discuss deficiency and toxicity disorders associated with the major minerals.
3. Discuss hypertension and osteoporosis, including risk factors.
4. Discuss ways to improve intakes of major minerals without relying on dietary supplements.

This section focuses on the major minerals. Table 9.5 summarizes nutrition–related information about these micronutrients. Unless otherwise noted, RDA/AI values are for adults, excluding pregnant or breastfeeding females. For more information about Dietary Reference Intake values for minerals, see Appendix H.

TABLE 9.5 Summary of Major Minerals

Mineral	Major Functions in the Body	Adult RDA/AI (adult RDA = bold)	Major Dietary Sources	Major Deficiency Signs and Symptoms	Major Toxicity Signs and Symptoms
Calcium (Ca)	• Structural component of bones and teeth • Blood clotting • Transmission of nerve impulses • Muscle contraction • Regulation of metabolism	**1000–1200** mg	Milk and milk products, canned fish, tofu made with calcium sulfate, leafy vegetables, calcium-fortified foods such as orange juice	• Increased risk of osteoporosis • May increase risk of hypertension	Upper Level (UL) = 2.0–2.5 g/day • Intakes > 2.5 g/day may cause kidney stones and interfere with absorption of other minerals.
Sodium (Na)	• Maintenance of proper fluid balance • Transmission of nerve impulses • Muscle contraction • Transport of certain substances into cells	1500 mg (19–50 years of age)	Luncheon meats; processed and canned foods; pretzels, chips, and other snack foods; condiments; sauces; table salt	• Muscle cramps	UL = 2300 mg/day • Contributes to hypertension in susceptible individuals • Increases urinary calcium losses
Potassium (K)	• Maintenance of proper fluid balance • Transmission of nerve impulses	4700 mg	Fruits, vegetables, milk, meat, legumes, whole grains	• Irregular heartbeat • Muscle cramps	No UL has been determined. • Slowing of heart rate that can result in death
Magnesium (Mg)	• Bone strengthening • Serves as a cofactor for certain enzymes • Maintains normal heart and nerve functioning	Males: **400–420 mg** Females: **310–320 mg**	Wheat bran, green vegetables, nuts, chocolate, legumes	• Muscle weakness and pain • Poor heart function	UL = 350 mg/day (medication-related) • Diarrhea
Phosphorus (P)	• Structural component of bones and teeth • Maintenance of acid-base balance • Component of DNA, phospholipids, and other organic compounds	**700 mg**	Dairy products, processed foods, soft drinks, fish, baked goods, meat	• Poor appetite, anemia, muscle weakness, bone pain	UL = 4 g/day • Poor bone mineralization
Chloride (Cl)	• Maintenance of proper fluid balance • Production of stomach acid • Maintenance of acid-base balance	2300 mg (19–50 years of age)	Processed foods, salty snacks, table salt	• Seizures (observed in infants)	UL = 3600 mg/day • Hypertension (because of the association with sodium in sodium chloride [table salt])
Sulfur (S)	• Component of organic compounds such as certain amino acids and vitamins	None	Protein-rich foods	• None reported	• Unlikely from dietary sources

Cereal: ©Wendy Schiff; Vegetables: Ingram Publishing/Alamy Stock Photo; Pasta: HandmadePictures/iStock/Getty Images; Peanut butter: Burke Triolo Productions/Brand X Pictures/Getty Images

Calcium (Ca)

Calcium is the most plentiful mineral element in the human body. All cells need calcium, but more than 99% of the body's calcium is in an inorganic compound that forms the structural component of bones and teeth. The remaining calcium is in muscle tissue and extracellular fluid.

Why Is Calcium Necessary?

Although the body needs calcium to form bones and teeth, the mineral is vital to all cells. Calcium is involved in muscle contraction, blood clot formation, nerve impulse transmission, and cell metabolism. Additionally, calcium may play important roles in maintaining healthy blood pressure and functioning of the immune system.

Bone Development and Maintenance Although your bones do not appear to change shape, they are being remodeled continually in response to the physical stresses placed on them. The remodeling process involves breaking down bone where there is little stress and building bone where there is more stress. Bones have specialized cells called osteoclasts and osteoblasts. Simply put, **osteoclasts** tear down bone tissue where it is not needed; **osteoblasts** build bone tissue where it is needed. For example, if you begin to play tennis regularly and hold the racket in your right hand, osteoblasts in the bones of your right arm build bone tissue where it is needed to help support the muscular activity. As a result, the bones in that arm become denser than the bones in your left arm. Bones that are denser have greater bone mass, so they are stronger and less likely to fracture than less dense bones. Figure 9.12 shows x–rays of bone tissue. By just looking at the photos, can you tell which bone is denser and has greater mass?

Maintaining Normal Blood Calcium Levels The body has complex hormonal systems to maintain calcium homeostasis. The thyroid and parathyroid glands help regulate blood calcium levels (Fig. 9.13). In response to falling blood cal–cium levels, the parathyroid glands secrete **parathyroid hormone (PTH),** which signals osteoclasts to tear down bone tissue. This process releases calcium from bones so the mineral can enter the bloodstream. Recall from Chapter 8 that PTH also works with vitamin D to increase intestinal calcium absorption and reduce calcium excretion in urine (see Figure 8.12).

When the level of calcium in blood is too high, the thyroid gland secretes the hormone *calcitonin*. Calcitonin signals osteoclasts to stop tearing down bone tissue. High calcium levels in the blood also inhibit PTH secretion and the activation of vitamin D, thus decreasing calcium absorption and increasing calcium excretion in the urine. All these physiological responses help maintain your blood calcium level within the normal range.

Sources of Calcium

Table 9.6 includes some foods that are among the richest sources of calcium. Fluid milk, yogurt, and cheese provide most of the calcium in American diets.[11] Moreover, the calcium in milk products is well absorbed and used by the body. Not all products made from milk are rich sources of calcium. Cottage cheese, for example, does not supply as much calcium as the milk from which it is made because the milk loses as much as half of its calcium content when it is processed to make cottage cheese. Although butter, sour cream, and cream cheese are made from whole milk, people generally do not eat enough of these high–fat foods to contribute much calcium to their diets.

Did *YOU* Know?

Bone remodeling occurs throughout life, and as a result of this process, most of your skeleton is replaced about every 10 years.[10]

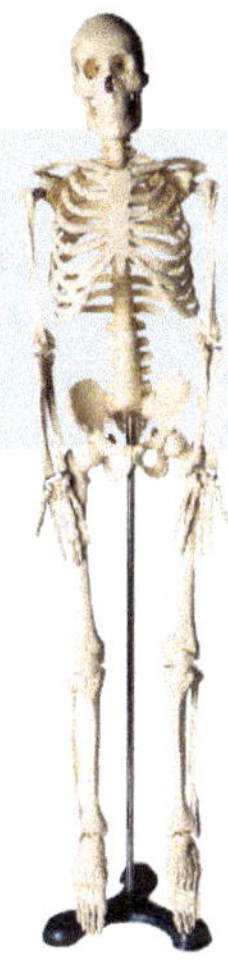

Comstock/Alamy Stock Photo

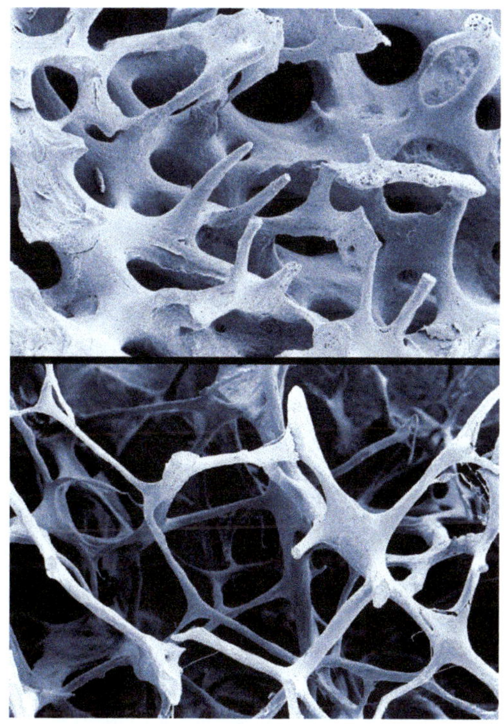

Figure 9.12 Bone tissue. Normal spine tissue is on the top; the spine tissue on the bottom is from a person with osteoporosis.
David Gregory & Debbie Marshall/Wellcome Images/Science Source

osteoclasts bone cells that tear down bone tissue

osteoblasts bone cells that add bone to where the tissue is needed

parathyroid hormone (PTH) hormone secreted by parathyroid glands when blood calcium levels are too low

Figure 9.13 **Maintaining blood calcium levels.**

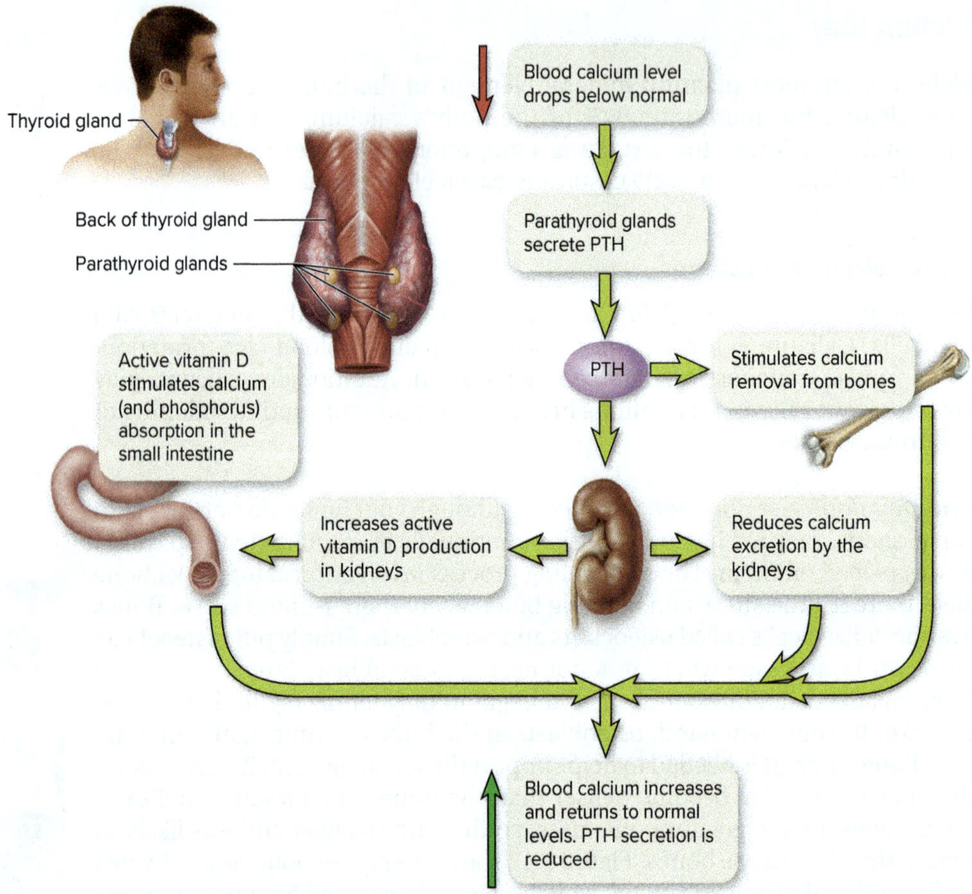

TABLE 9.6 Calcium Content of Selected Foods

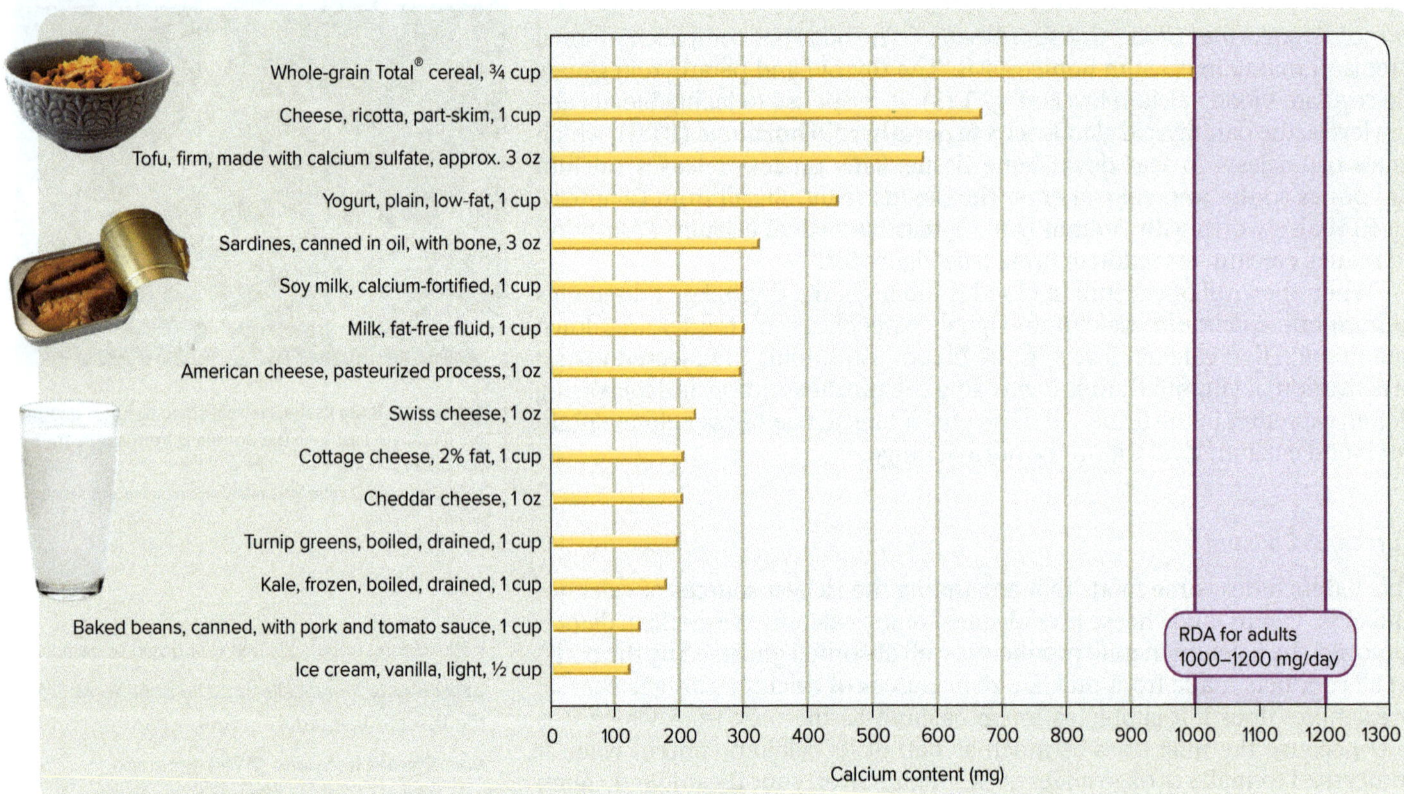

Source: Data from U.S. Department of Agriculture, Agricultural Research Service: *FoodData Central*, 2019.
Cereal, Sardines: ©Wendy Schiff; Milk: Nipaporn Panyacharoen/Shutterstock

Figure 9.14 Calcium bioavailability of various foods. To obtain approximately the same amount of calcium that is in 8 fluid ounces of fat-free milk, you would need to consume the amounts of the foods shown.
©Wendy Schiff

Certain foods from plants contain calcium, but the foods also contain *phytic (fi′–tik) acid* or *oxalic (awk–sal′–ik) acid,* naturally occurring substances that interfere with calcium absorption. Phytic acid is a compound in whole grains and in certain seeds and beans. Spinach, collard greens, and sweet potatoes have high amounts of oxalic acid that is also called oxalate. In fact, rhubarb leaves are toxic because they contain such high amounts of the chemical.

Good plant sources of calcium include broccoli and leafy greens, especially kale, collard, turnip, bok choy, and mustard greens. Nevertheless, the calcium in plant foods containing oxalic acid and/or phytic acid is not as bioavailable as the calcium in milk and milk products. For example, 1 cup of fat–free milk supplies almost 300 mg of calcium, and about 30% of the calcium in milk is bioavailable.[12] Although a cup of boiled and drained spinach, rich in oxalic acid, contains 116 mg of calcium, only about 5% of that amount is bioavailable.[12] (The oxalic acid in spinach reduces the availability of the plant's calcium content.) Figure 9.14 shows approximately how much cooked kale, broccoli, and spinach a person would need to eat to obtain about the same amount of bioavailable calcium that is in 1 cup of fat–free milk.

Calcium is added to a variety of foods, including fortified orange juice, margarine, cereals, and breakfast bars. Beverages used as milk alternatives made from nuts, cereals, and other plants such as soy, almond, cashew, oat, rice, and coconut may be fortified with calcium. Be sure to check the Nutrition Facts panel to see how much calcium the beverage provides. Another source of calcium is soybean curd (tofu) that is made with calcium sulfate. Figure 9.15 indicates food groups from MyPlate that are good sources of calcium.

Calcium Supplements Many adults find it difficult to consume enough milk products and other calcium–rich foods to achieve adequate intakes of the mineral. Thus, taking calcium supplements or antacids that contain calcium has become a common practice, especially among older adults. If you choose to take a calcium supplement, consider products that include vitamin D because the vitamin enhances calcium absorption. Taking only 500 mg of calcium at a time and ingesting the supplement with meals will also improve the mineral's absorption.

Results of some studies found that taking supplements containing high amounts of calcium (more than 1000 mg/day) increased the risk of dying from cardiovascular disease (CVD). Results of other studies, however, did not find that calcium supplementation posed such risk.[11] More research is needed to determine whether calcium supplementation poses serious risks to health.

MY DIVERSE PLATE

Bok Choy (Pak Choi or Chinese Cabbage)

image100/SuperStock

Bok choy is a cruciferous vegetable that is typically used in Asian cookery. Cruciferous vegetables contain phytochemicals that may have cancer-fighting activity. The leaves and stalks are steamed or sautéed with other vegetables. Bok choy is a good source of minerals, especially iron, calcium, and potassium; the vegetable also contains high amounts of vitamin C.

Did *YOU* Know?

Antacids generally contain calcium carbonate, which makes them an inexpensive way of obtaining some calcium. For example, an antacid pill that contains 750 mg of calcium carbonate provides 300 mg of elemental calcium. People, however, should not rely on antacids as a major source of calcium. Long-term ingestion of calcium from antacids may contribute to kidney stones.[5]

A few brands of antacids contain magnesium, as well as calcium carbonate. If taken in large doses over a long period, these particular antacids can be toxic.[13] Bone pain is a symptom of toxicity from taking antacids containing calcium carbonate and magnesium. To avoid health problems associated with calcium-containing antacids, consumers should follow dosage recommendations on the labels of the products.

©Wendy Schiff

Figure 9.15 MyPlate: Good food sources of calcium. Broccoli: Burke Triolo Productions/Artville/Getty Images; Orange juice: Stockbyte/Getty Images; Swiss cheese: Comstock/Jupiter Images/Getty Images; Cereals: Robert Santos/BlueMoon Stock/Getty Images; Almonds: onair/Shutterstock; MyPlate food guide icon: U.S. Department of Agriculture

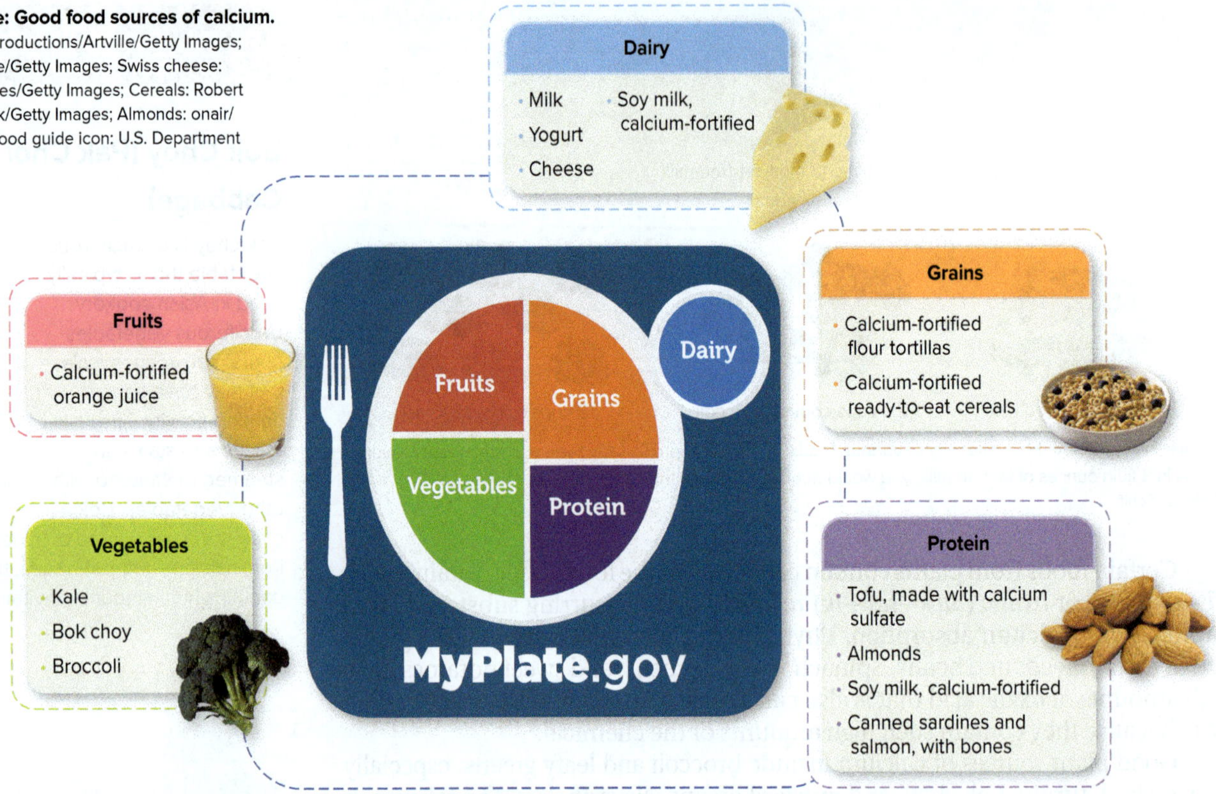

Food & Nutrition tips

Each of the following foods contains about the same amount of calcium that is in 1 cup of fat-free milk (approximately 300 mg):

- 1.25 cups low-fat (2% milk) cottage cheese
- 1 slice (1 oz) processed cheese (American cheese)
- ⅔ cup plain, low-fat yogurt
- 1.5 oz natural cheese (e.g., cheddar or Swiss)
- 1 cup calcium-fortified soy milk

Therefore, always check with your physician before taking any dietary supplement, and if you do use a calcium supplement, do not exceed the Upper Level. Dietary supplements, including those that contain calcium, may have a "seal of approval" displayed on the label. As mentioned in Chapter 2, such seals of approval are not a guarantee that products are safe or effective.

Dietary Adequacy

For adults 19–50 years old, the RDA for calcium is 1000 mg/day. In the United States, average calcium intakes were 1084 mg/day for males and 857 mg/day for females in 2017–2018.[13] Thus, females tended to consume less than the RDA for calcium. Total vegetarians (vegans) and people who are lactose-intolerant are at risk of calcium deficiency because they often avoid consuming milk and milk products, the most reliable dietary sources of calcium. According to the Dietary Guidelines, calcium is a "nutrient of public health concern" (see Chapter 3).

Healthy adults absorb about 30% of the calcium in foods, but this percentage varies, depending on the type of food.[5] During stages of life when the body needs extra calcium—such as infancy and pregnancy—absorption can be as high as 60%. Vitamin D enhances calcium absorption.

Older people, especially females, do not absorb calcium as well as younger people. Thus, the RDA for calcium is 1200 mg/day for females over 50 years of age, and adults over 70 years of age. In addition to advanced age, other factors that reduce calcium absorption include vitamin D deficiency and diarrhea.

Calcium Toxicity The Upper Level (UL) for calcium is 2000 to 2500 mg/day. Normally, the small intestine prevents too much calcium from being absorbed. However, taking too many calcium-containing antacids or supplements, or drinking too much vitamin D–fortified milk, can result in excessive calcium absorption and hypercalcemia. **Hypercalcemia** (*hyper* = excess; *calcemia* = calcium in the blood) is a condition characterized by a higher–than–normal concentration of calcium in blood. Signs and symptoms of hypercalcemia include bone pain, muscle weakness, and fatigue, and people with hypercalcemia can develop kidney stones. Treatment for hypercalcemia may include avoiding vitamin D and calcium supplements to reduce calcium absorption.

What Is Osteoporosis?

Osteoporosis is a chronic disease characterized by loss of bone mass and reduced bone structure (see Figure 9.12). People with osteoporosis have weak bones that are susceptible to fractures. In the United States, osteoporosis is a major public health problem. More than 10 million Americans have osteoporosis, and another 43 million are at risk of the disease because they have loss of bone mass.[14] Most people with osteoporosis are older adult females.

In the United States, half of females and up to one–fourth of males who are over 50 years of age will have an osteoporosis–related fracture at some point.[15] Many people do not realize their bones are becoming weaker and they have osteoporosis until they experience a fracture. People with osteoporosis may break a bone by falling, or they may experience spontaneous fractures in which the fragile bone shatters for no apparent reason. Osteoporosis–related fractures often involve the spine, hip, wrist, or ankle bones. In severe cases, bones in the upper spine fracture and then heal in an abnormally curved position, giving the obvious appearance associated with osteoporosis (Fig. 9.16).

Fractures, especially hip fractures, can be devastating events, especially for older adults. One in five older Americans who experience a broken hip dies of complications related to the fracture within 1 year of the injury.[15]

What Causes Osteoporosis? Several factors, including smoking and lack of physical activity, contribute to bone loss and osteoporosis (Table 9.7). Note that some of the factors cannot be modified, but other factors can be changed to prevent or delay the development of the disease.

TABLE 9.7 Risk Factors for Osteoporosis

Factors People Cannot Change:
• Being female
• Growing older
• Having white or Asian ancestry (confers the highest risk)
• Having a family history of osteoporosis
• Having a small, thin-boned body frame
Factors People Can Change:
• Having low estrogen levels in females, low testosterone levels in males
• Having untreated anorexia nervosa
• Following long-term diets that contain inadequate amounts of calcium and vitamin D
• Being physically inactive
• Smoking cigarettes
• Consuming excess alcohol

Source: Modified from www.niams.nih.gov/Health_Info/Bone/Osteoporosis/overview.asp

Did *YOU* Know?

Canned fish with edible soft bones, such as sardines and salmon, are good calcium sources. When making salmon patties using canned fish, mash up the bones along with the salmon. The bones are so soft, you won't notice them in the cooked product.

©Wendy Schiff

Figure 9.16 Osteoporosis. Severe curvature of the upper spine is a sign of osteoporosis.
Sally and Richard Greenhill/Alamy Stock Photo

hypercalcemia condition characterized by higher-than-normal concentration of calcium in blood

osteoporosis chronic disease characterized by loss of bone mass and reduced bone structure

Growing older; being female, especially a female with Asian ancestry; drinking too much alcohol; and having a small, thin-boned body frame are risk factors for osteoporosis.
Purestock/Superstock

Around 30 years of age, most healthy young males and females have acquired all of their adult bone mass.[16] Regardless of one's sex, loss of bone tissue begins in mid-adulthood. In males, bone loss is slow and steady beginning around age 30. In females, however, the rate of bone loss increases significantly after *menopause,* that is, after menstrual cycles have ceased. At this time of life, females have the highest risk of osteoporosis. Why? The sex hormones **testosterone** and **estrogen** are needed for normal bone development and maintenance. In females of childbearing age, ovaries are the primary source of estrogen. After menopause, a female's ovaries no longer produce estrogen, and as a result, her rate of bone loss exceeds the rate of bone replacement. Because estrogen is so important to maintaining strong bones, young adult females should see a physician if they have signs of estrogen deficiency, such as irregular menstrual cycles.

A simple way to monitor bone mass is by tracking height. Losing an inch or more of adult height may be the first sign that a person has experienced fractures of the spine due to osteoporosis.[17] If osteoporosis is suspected, a person can undergo special painless x-ray testing to determine the extent of bone loss. Individuals with a family history of osteoporosis, males who have low testosterone levels, and females who are postmenopausal should ask their physician if testing to determine bone mineral density is necessary. Treatment for osteoporosis may include certain medications, calcium and vitamin D supplements, and regular physical activity.

Reducing the Risk of Osteoporosis Efforts to reduce the risk of osteoporosis should begin early in life. Proper diet and regular exercise are especially important from early childhood through late adolescence because the body actively builds bone during these life stages. By following the recommendations of MyPlate, most people can obtain adequate amounts of calcium from foods. Exposing skin to sunlight can stimulate the body's ability to form vitamin D, but some people will need to take calcium and vitamin D supplements.

Performing weight-bearing activities increases bone mass because contracting muscles keep tension (physical stress) on bones. Table 9.8 lists examples of weight-bearing and non-weight-bearing activities. Regular physical activity that

testosterone hormone needed for normal bone development and maintenance

estrogen hormone that plays a role in normal bone development and maintenance

Food & Nutrition tips

The following suggestions can add more calcium to your diet:

- Read the Nutrition Facts panel on food packages and choose foods with at least 20% of the Daily Value for calcium per serving whenever possible.
- Sprinkle grated low-fat cheeses on top of salads, bean or pasta dishes, and cooked vegetables.
- For a snack, melt a slice of low-fat cheese on half a whole-wheat bagel, whole-wheat crackers, or a slice of rye bread.
- If a recipe calls for water, substitute fat-free milk for water when appropriate. For example, use fat-free milk when making cooked oatmeal or pancake batter.
- Add ¼ cup nonfat milk powder to 1 pound of raw ground meat when preparing hamburgers, meatballs, or meatloaf.
- Make homemade smoothies by blending plain low-fat Greek yogurt with fresh or frozen fruit (see "Recipe for Healthy Living" at the end of Chapter 4).

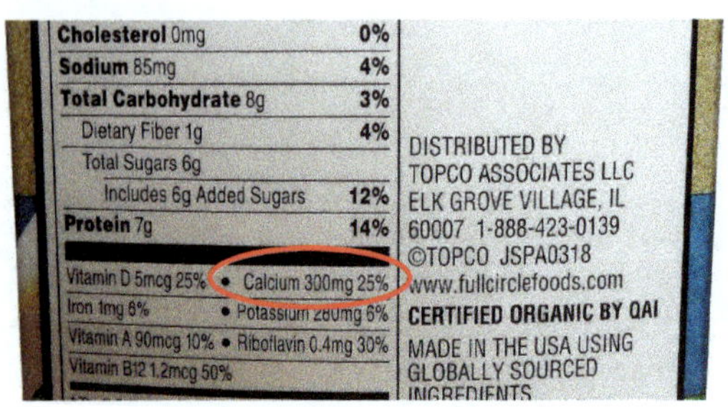

A serving of this soy milk provides 25% of the Daily Value for calcium, which is 1300 mg.
©Wendy Schiff

TABLE 9.8 Examples of Weight-Bearing and Non-Weight-Bearing Activities

Weight-bearing	Non-weight-bearing
Low-impact aerobics	Lying in bed
Basketball	Swimming
Running or jogging	Water aerobics
Walking or hiking	Cycling
Jumping rope	Traveling in reduced-gravity situations (e.g., space flight)
Dancing	
Stair climbing	
Strength training with weights	
Tennis and other racket sports	

These older adults are performing tai chi, a weight-bearing activity that can improve balance and may reduce the likelihood of falling.
Floresco Productions/OJO Images/age fotostock

includes weight–bearing muscular movements provides numerous benefits to health, such as improving balance and reducing the likelihood of falling.

Everyone needs to be concerned about their risk of osteoporosis and focus on maximizing bone mass while they are young. The interactive osteoporosis risk test at the International Osteoporosis Foundation's website (https://riskcheck.osteoporosis.foundation//) can help you determine whether you or someone you know is at risk for the disease.

Sodium (Na)

Salt is the primary source of sodium in American diets. The chemical commonly called "table salt" or simply "salt" is actually *sodium chloride,* a compound composed of two minerals, sodium and chloride. A teaspoon of table salt supplies 2325 mg of sodium. (Unless otherwise noted, we will refer to sodium chloride simply as "salt" or "table salt.") The human digestive tract absorbs almost all of the sodium that is in foods and beverages.

Why Is Sodium Necessary?

As mentioned in the "Water" section of this chapter, sodium plays a major role in maintaining normal fluid balance. The mineral is also necessary for the transmission of impulses by nerves, for transporting small substances such as glucose and amino acids into cells, and for functioning of muscles.

Sources of Sodium

Most uncooked vegetables, raw meats, and grain products are naturally low in sodium. Thus, most of the sodium Americans consume is from foods available from restaurants and the salt that is added to food during processing (including curing, pickling, and canning).[18] As a food additive, salt enhances flavors and can prevent the growth of microorganisms responsible for food spoilage. Other food additives that contain sodium include sodium nitrate, sodium citrate, and *monosodium glutamate* (*MSG*), a seasoning that is often added to foods served in Chinese restaurants.

In the United States, the leading sources of dietary sodium include breads and rolls, pizza, sandwiches, deli meats, soups, burritos, tacos, and salty snacks.[19] Table 9.9 lists some selected foods and indicates how much sodium is in a serving of each food. To rate your sodium intake, take the Sodium Intake Assessment that is in this section of the chapter.

A ½ cup serving of these black beans provides 6% of the Daily Value for sodium, which is 2300 mg.
©Wendy Schiff

TABLE 9.9 Sodium Content of Selected Foods/Food Additives

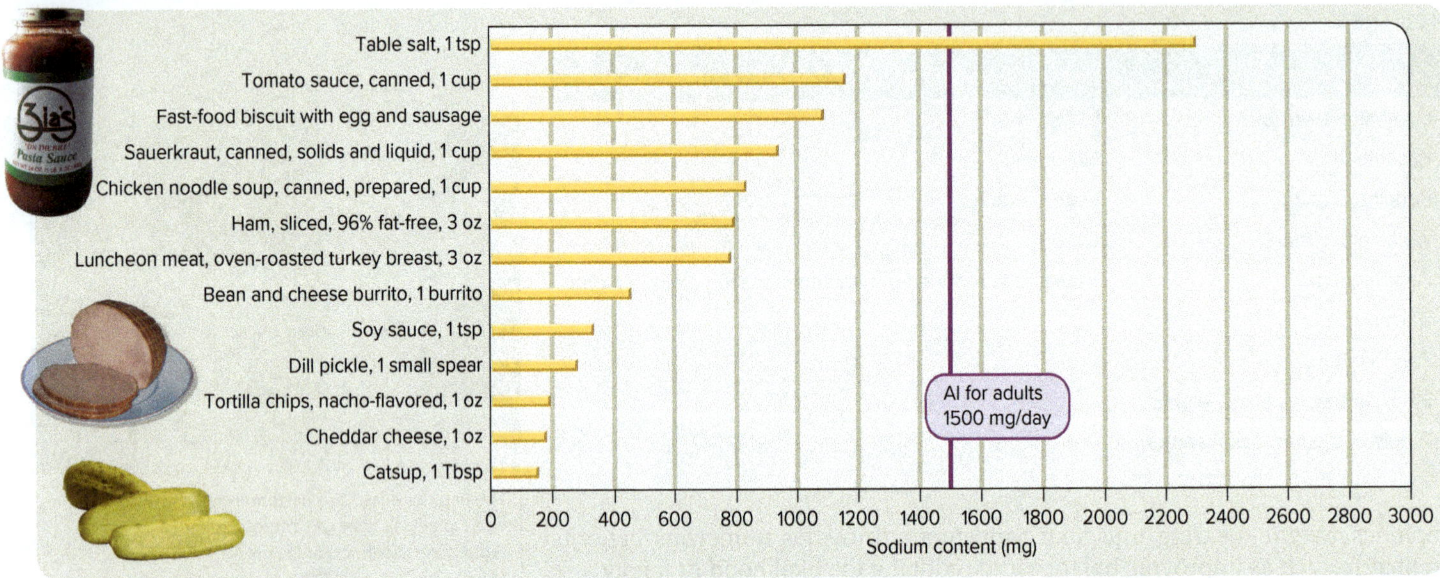

Source: Data from U.S. Department of Agriculture, Agricultural Research Service: *FoodData Central,* 2019.
Pasta sauce: ©Wendy Schiff; Ham: McGraw Hill; Sliced pickle: Binh Thanh Bui/Shutterstock

Did *YOU* Know?

In October 2021, the FDA released voluntary guidance for the food industry to reduce the amount of salt added to processed foods, including packaged and prepared food from restaurants. The agency provided sodium reduction targets for 16 different food categories to attempt to reduce the sodium intake of the average American from about 3400 mg to 3000 mg per day by spring 2024.[20] To meet these voluntary sodium targets, food manufacturers and restaurants must modify their food formulations and recipes to include less added salt and still appeal to consumers. In addition, modified food formulations should maintain the nutritional quality of the food. For example, after sodium is reduced in a food product, increasing the added sugars in a product to improve the flavor would reduce the food's nutritional quality and not be beneficial to Americans.

Some consumers may be concerned about food tasting "bland" if it contains less added salt. However, a person who consumes lower-sodium foods may find the foods bland at first, but their taste buds quickly adapt, and they can prefer foods lower in sodium. Some people report high-sodium foods they used to enjoy taste too salty!

Dietary Adequacy

Humans require only about 180 mg of sodium per day, but the AI for adults is 1500 mg/day. The AI for sodium does not apply for people who perspire heavily, such as marathon runners, or people who work in extremely hot conditions.[20] Sweat contains small amounts of sodium, chloride, and some other minerals. People who perspire extensively can lose large amounts of these minerals in their sweat.

Sodium Deficiency In the United States, diets generally supply far more sodium than the AI amount, and as a result, an average person is unlikely to become sodium deficient. A healthy body is able to regulate its sodium concentration effectively, but sodium depletion can occur in certain situations. A person who loses more than 2 to 3% of body weight as a result of excessive sweating is at risk of sodium depletion. In most cases, drinking fluids and simply eating some salty foods or adding salt to foods is usually effective for restoring the body's sodium content. However, endurance athletes may need to consume sports drinks during competition to avoid dehydration and sodium depletion. Salt tablets are generally not recommended for sodium replacement. Sodium depletion also can result from diarrhea or vomiting, especially in infants. In these cases, it is necessary to obtain medical care promptly to replace the lost fluids and electrolytes because infants can develop dehydration rapidly.

Sodium Toxicity The adult UL for sodium is 2300 mg/day. Recall from Chapter 3, the Institute of Medicine established the new DRI category Chronic Disease Risk Reduction Intake (CDRR) for sodium. According to the CDRR for sodium, healthy people 14 years and older should lower or maintain their sodium intake to below 2300 mg/day to reduce the risk of chronic disease.[20] High salt (sodium chloride) intakes are associated with increased risk of chronic disease including hypertension, heart disease, and stroke.[18] On average, adult Americans consumed 3531 mg of sodium per person per day in 2017–2018, which is 1.5 times more than the UL and CDRR for sodium.[14]

Hypertension

Hypertension, a condition characterized by persistently elevated blood pressure, is a serious public health problem in the United States. Compared to people with normal blood pressure, hypertensive individuals have greater risk of CVD, especially heart disease and stroke, as well as kidney failure and damage to other organs. Approximately one—half of adult Americans have hypertension.[21] Children can also develop hypertension. In the United States, one in nine children has chronically elevated blood pressure levels.[18] Hypertension is often called the "silent killer" because high blood pressure generally does not cause symptoms until the affected person's organs and blood vessels have been damaged.

The best way to detect hypertension is to have regular blood pressure screenings. When you have your blood pressure determined, two measurements are actually taken. The first measurement is the **systolic pressure,** which is the maximum blood pressure within an artery. This value occurs when the ventricles, the heart's pumping chambers, contract. The second measurement is the **diastolic pressure,** which measures the pressure in an artery when the ventricles relax between contractions. The systolic value is always higher than the diastolic value. For adults, healthy blood pressure readings are less than 120/80 millimeters of mercury (mm Hg). After having your blood pressure measured, ask the clinician for your systolic and diastolic readings and keep a record of the values.

A person who is under physical or emotional stress can expect their blood pressure to rise temporarily. However, persistent systolic blood pressure readings of 120 mm Hg or higher and diastolic readings of 80 mm Hg or higher are abnormal. If a person's blood pressure persists at systolic values that are greater than or equal to (≥) 130 mm Hg or diastolic values that are ≥ 80 mm Hg, they have a form of hypertension. Table 9.10 lists the American Heart Association's categories for blood pressure levels in adults.

High salt (sodium chloride) intakes are associated with increased risk of hypertension.
Marko Horvat/123RF

systolic pressure maximum blood pressure within an artery that occurs when the ventricles contract

diastolic pressure pressure in an artery that occurs when the ventricles relax between contractions

What Causes Hypertension? Most cases of hypertension do not have simple causes, but advanced age, African–American ancestry, obesity, physical inactivity, smoking cigarettes, and excess alcohol and sodium intakes are among the major risk factors for the condition (Table 9.11). Blood pressure usually increases as a person ages, which may be due to plaque that builds up in arteries

Sodium Intake Assessment

For each question, place a check in the column that best describes your sodium intake habits. How often do you . . .	Rarely	Occasionally	Often	Daily
1. Eat cured or processed meats ("deli" meats), such as bacon, sausage, or hot dogs?				
2. Eat canned or frozen vegetables with sauce?				
3. Eat commercially prepared meals, main dishes, or canned or dehydrated soups?				
4. Eat processed cheeses, such as cheese spreads?				
5. Eat salted nuts, popcorn, pretzels, corn chips, or potato chips?				
6. Add salt to cooking water for vegetables, rice, or pasta?				
7. Add salt, seasoning mixes, salad dressings, or condiments—such as soy sauce, steak sauce, pickles, and catsup—to foods during preparation or at the table?				
8. Salt your food before tasting it?				
9. Ignore reading the Nutrition Facts panel for sodium content when buying foods?				
10. Choose menu items that are salty or with sauces when dining out?				

Scoring: The more checks you put in the "often" or "daily" columns, the higher your dietary sodium intake is.

Source: Adapted from USDA, *Home and Garden Bulletin,* No. 232–6, April 1986.

TABLE 9.10 Categories for Blood Pressure Levels in Adults (Ages 18 Years and Older)

	Blood Pressure Level (mm Hg)		
Category	**Systolic**		**Diastolic**
Normal	< 120	**and**	< 80
Elevated	120–129	**and**	< 80
Hypertension Stage 1	130–139	**OR**	80–89
Hypertension Stage 2	≥ 140	**OR**	≥ 90
Hypertensive Crisis (consult doctor immediately)	> 180	**and/or**	> 120

Source: American Heart Association, *Understanding blood pressure readings*. www.heart.org/en/health-topics/high-blood-pressure/understanding-blood-pressure-readings. Accessed: May 10, 2023.

TABLE 9.11 Major Risk Factors for Hypertension

• Family history
• Advanced age
• African-American ancestry
• Obesity
• Physical inactivity
• Consuming excess sodium
• Cigarette smoking
• Consuming excess alcohol
• Type 2 diabetes

(atherosclerosis) and interferes with the normal functioning of the blood vessels. Healthy arteries are flexible tubes that expand with each heartbeat and recoil in between beats. Atherosclerotic arteries are less flexible and cannot expand as much as healthy arteries. As a result, the heart must work harder to pump blood through the stiff arteries, and blood pressure becomes chronically elevated.

Obesity, a condition characterized by excessive amounts of body fat, is a major risk factor for hypertension. Physical inactivity is another leading risk factor related to hypertension. Obesity and physical inactivity are modifiable risk factors. By exercising and losing some excess fat, obese people who have hypertension may experience reductions in their blood pressure.

Excessive alcohol intake increases the risk of hypertension. To reduce their chances of developing high blood pressure, people should avoid alcohol or limit their consumption to two or fewer drinks/day (males) and only one drink/day (females and older adults). Other important risk factors for hypertension include having diabetes and using tobacco.

Buy seasonings that do not have added salt.
©Wendy Schiff

Food & Nutrition tips

To reduce your sodium intake, consider taking these actions:

- Read the Nutrition Facts panels before purchasing packaged foods to determine sodium contents of the items.
- Prepare homemade meals and snacks as much as possible so you have control over your salt intake.
- Use herbs and spices that do not contain added salt while preparing foods.
- Add small amounts of vinegar, lemon juice, or plain yogurt to salads, fish, meats, or vegetables just before eating instead of salt. The acids enhance the flavors of these foods.
- Taste your food *before* salting it. Adjust to eating foods with less salt in them.
- Do not keep a salt shaker on your table.
- When ordering items in restaurants, request that no salt be added to your food while it is being prepared.

Pixtal/AGE Fotostock

As mentioned earlier, a high-sodium diet is associated with increased risk of hypertension.[18] Many medical researchers think some people are genetically "sodium sensitive." A person who is sodium sensitive is more likely to develop hypertension as a result of consuming a high-sodium diet than an individual who lacks this sensitivity.

As mentioned earlier, the CDRR for sodium recommends that healthy people should lower or maintain their sodium intake below 2300 mg/day to reduce the risk of chronic disease, including hypertension.[20] One teaspoon of salt contains about 2300 mg of sodium. Members of high-risk populations, particularly people whose blood pressure is within the "elevated" category of blood pressure guidelines (see Table 9.10), may need to limit their sodium intake to 1500 mg per day. People who are being treated for hypertension should check with their physician for advice concerning an acceptable sodium intake.

If you want to lower your sodium intake, try gradually reducing your use of salt and consumption of processed foods. By doing so, you will eventually become accustomed to the taste of less salty food. To replace salt as a seasoning, try using garlic, citrus juice, and herbs and spices to enhance the taste of foods. Furthermore, avoid buying seasonings with added salt, such as "garlic salt" or "onion salt," and check the ingredient list to purchase seasonings without added salt (garlic *powder* or onion *powder*) instead.

Information about a packaged food's sodium content is a mandatory component of the Nutrition Facts panel. If you take the time to read labels, you can find foods that have little or no salt added to them during processing. For example, an ounce of salted peanuts provides 230 mg of sodium; the same amount of unsalted peanuts has only 2 mg of sodium. The "Food & Nutrition Tips" feature in this section of the chapter provides some suggestions for reducing your salt (sodium) intake.

Even if your blood pressure is normal now, it is important to have regular blood pressure checks as you grow older because the risk of hypertension increases with age. Young adults, however, are not immune to hypertension. Justin Steinbruegge, the college student featured in "Real People, Real Stories" in this section, was 18 when he found out he had hypertension. When was the last time you had your blood pressure measured? What were the systolic and diastolic values?

Treatment for hypertension usually includes taking certain medications, following dietary modifications, and making some other lifestyle changes (Table 9.12). The *Dietary Approaches to Stop Hypertension* (*DASH*) diet is low in

TABLE 9.12 Practical Steps to Reduce Your Risk of Hypertension

• Follow the dietary recommendations of MyPlate concerning fruit, vegetable, and low-fat milk intakes. Furthermore, consider using fresh fruit and unsalted nuts to replace some empty calories in your daily meals and snacks. (See Chapter 3 for more information about empty calories.)
• Reduce your consumption of salty foods and have your blood pressure checked regularly.
• Attain and maintain a healthy body weight.
• Incorporate more physical activity into your daily schedule. For example, walk more often and use steps instead of elevators or escalators.
• If you drink alcohol, consume alcoholic beverages in moderation—no more than two drinks/day for males and one drink/day for females and older adults.
• Avoid using tobacco products.

Nutrition & Health *Equity*

Black Americans have a higher prevalence of hypertension (HTN) than all other ethnic groups in the United States. In 2018, about 57% of Blacks had high blood pressure and were 30% more likely to die of cardiovascular disease (CVD) than non-Hispanic white people with HTN.[22] Being of African-American ancestry is a risk factor for developing HTN, and social determinants of health (SDOH) (see Chapter 1), including lack of access to healthy foods, living in areas with fewer health care providers, and having jobs that make attending medical appointments more difficult, also contribute to this inequity.

Black adults diagnosed with high blood pressure and prescribed medication to manage their condition may not take their blood pressure medication as prescribed because historical racism often contributes to Black people not trusting the medical advice of a white health care provider.[23] Other Black people may not take medications due to focusing on a healthy diet, exercise, and religious faith, including hope of being cured of hypertension instead of taking medications in addition to lifestyle efforts. However, researchers report health education programs provided in Black churches and barbershops that use community health care workers to promote healthy lifestyle behaviors have been successful in lowering blood pressure levels.[24] These programs focus on consuming a low fat, low-sodium diet rich in fruits and vegetables; increasing physical activity; and strategies for meal planning, stress management, medication adherence, and structured goal setting. Establishing more community-based health programs to serve Black communities may be one way to help reduce the rates of uncontrolled blood pressure in this population.

©Wendy Schiff

REAL PEOPLE | REAL STORIES

©Wendy Schiff

Justin Steinbruegge

In the spring of his senior year of high school, Justin Steinbruegge was surprised to learn that he had hypertension. "It was a huge shock. I had no signs or symptoms of hypertension. I didn't even know what hypertension was," says Justin. To manage his high blood pressure, he took medication prescribed by his physician, but he had difficulty controlling his sodium intake. "Eating pizzas and salty snacks seemed to keep my blood pressure on a never-ending roller coaster ride," he says.

"When I was 23, I enrolled in a police academy. Four months prior to entering the academy, I started an exercise program. By the time classes began, I thought I was in fantastic shape," says Justin, "but I was training too much, and my diet was horrible—loaded with too many processed foods and not enough fresh fruits, vegetables, and meats. I was ignorant about the importance of proper nutrition to good health."

After finishing at the top of his class at the police academy, Justin focused on improving his health. "I developed a healthy combination of diet and exercise that I continue to follow," he says. "I avoid eating out . . . I eat mainly fresh meats, cheeses, nuts, fruits, and vegetables. I drink only water—no juice, soda, or alcohol. I work out 6 days a week, and I include lifting free weights, rowing, light running, and yoga in my exercise routine. My journey to achieving good health has become a rewarding experience. Now, I don't need to take blood pressure medicine."

sodium, total fat, and saturated fat, and high in fruits, vegetables, whole grains, and fat–free or low–fat dairy products. Research indicates that people can lower their blood pressure and reduce their risk of CVD by following the DASH diet, losing excess body fat, and increasing their physical activity level.[25] To obtain more information about this diet and some low–sodium recipes, visit the National Heart, Lung, and Blood Institute's website: www.nhlbi.nih.gov/health–topics/dash–eating–plan.

Nutrition Facts	
8 servings per container	
Serving size	**2/3 cup (55g)**
Amount per serving	
Calories	**230**
	% Daily Value*
Total Fat 8g	**10%**
Saturated Fat 1g	**5%**
Trans Fat 0g	
Cholesterol 0mg	**0%**
Sodium 160mg	**7%**
Total Carbohydrate 37g	**13%**
Dietary Fiber 4g	**14%**
Total Sugars 12g	
Includes 10g Added Sugars	**20%**
Protein 3g	
Vit. D 2mcg 10%	Calcium 260mg 20%
Iron 8mg 45%	Potas. 235mg 6%

* The % Daily Value (DV) tells you how much a nutrient in a serving of food contributes to a daily diet. 2,000 calories a day is used for general nutrition advice.

A serving of this food provides about 6% of the Daily Value for potassium, which is 4700 mg.
Food and Drug Administration

Potassium (K)

Potassium is the primary positively charged ion in the intracellular fluid. In fact, most of the body's potassium is in cells. Like sodium, potassium plays a key role in maintaining proper fluid balance. Unlike sodium, potassium is associated with lower, rather than higher, blood pressure values. Potassium is also necessary for transmitting nerve impulses, contracting muscles, and maintaining normal kidney function. Potassium–rich diets, such as the DASH diet, may lower blood pressure.[25] A natural way to counteract high sodium intakes is to consume foods naturally rich in potassium and low in sodium, such as fruits and vegetables.

Sources of Potassium

Overall, fresh fruits and vegetables and fruit and vegetable juices are good dietary sources of potassium. Milk, whole grains, dried beans, and meats are also major contributors of potassium to American diets. Table 9.13 lists foods that are among the richest sources of this mineral. Figure 9.17 indicates food groups that are naturally good sources of potassium. Consumers can read the new Nutrition Facts on food packages to obtain information about products' potassium content.

Dietary Adequacy

The AI for potassium for males is 3400 mg per day, and for females, it is 2600 mg per day.[20] On average, adult males consume 2937 mg per day and adult females consume 2324 mg per day.[14] Because Americans tend to consume less than the AI amount of potassium, the Dietary Guidelines include the mineral as a "nutrient of public health concern." People can raise their potassium intakes by following the DASH diet and/or increasing their consumption of foods rich in potassium (see Table 9.13).

The body is unable to conserve potassium as well as sodium; therefore, the risk of potassium deficiency is greater than that of sodium deficiency. Individuals suffering from excessive sweating, vomiting, diarrhea, or kidney diseases that increase potassium excretion are at risk for potassium depletion. Symptoms of the condition generally include loss of appetite, muscle cramps, confusion, and constipation.

Although there is no UL for potassium, taking potassium supplements can upset the gastrointestinal tract. Moreover, if a person's kidneys are not able to eliminate the excess potassium, the mineral accumulates in the blood and can cause the heart to stop beating. To avoid toxicity, do not take potassium supplements unless you are under a physician's care.

Did *YOU* Know?

Salt substitutes often contain a type of salt called potassium chloride, or potassium chloride salt. People who have severe kidney diseases may accumulate toxic levels of potassium in their blood. Therefore, kidney disease patients should consult their physicians before using salt substitutes made with potassium chloride. Fruits and vegetables are recommended sources of potassium instead of potassium chloride.

TABLE 9.13 Potassium Content of Selected Foods

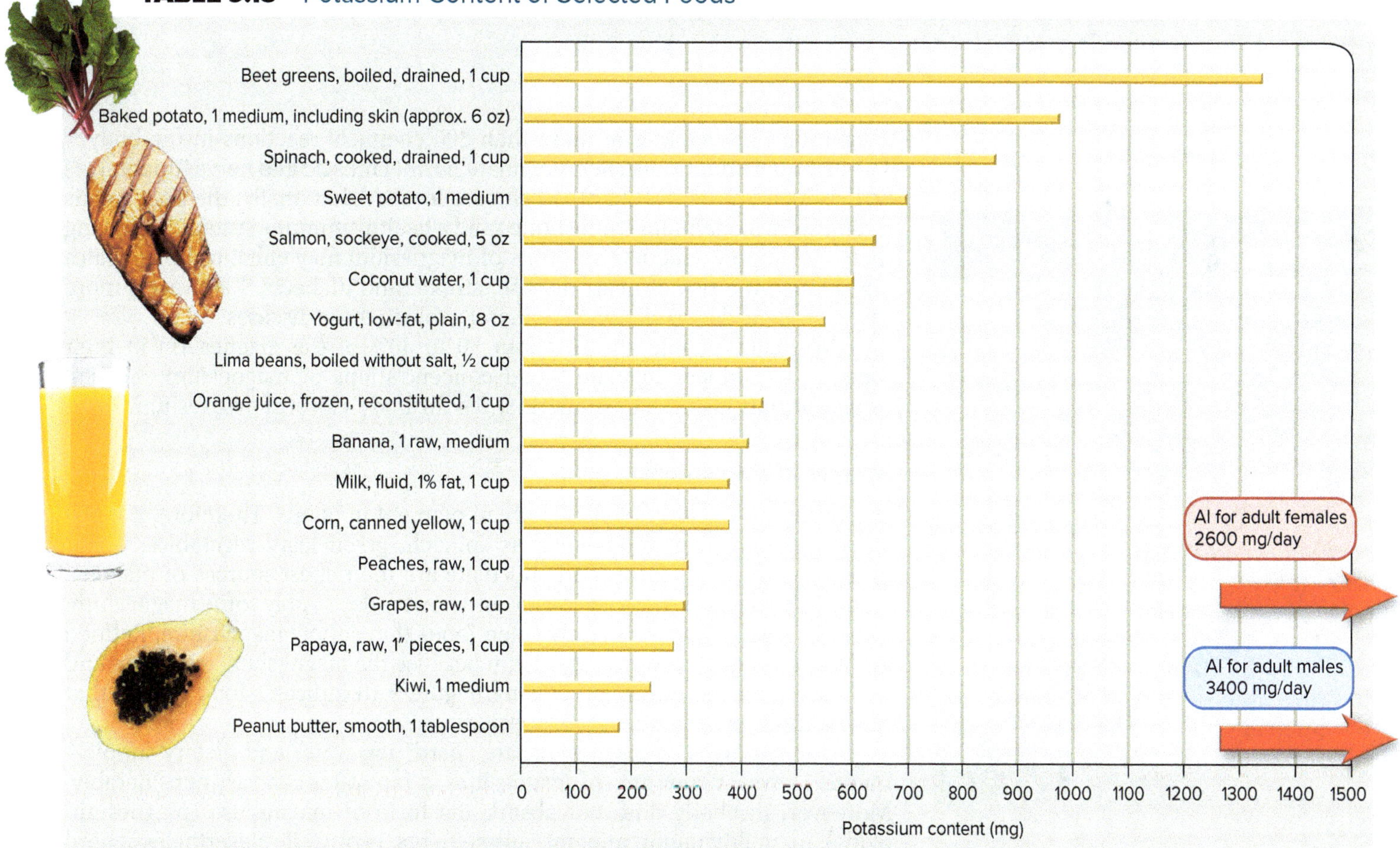

Source: Data from U.S. Department of Agriculture, Agricultural Research Service: *FoodData Central*, 2019.

Beet greens: ©Wendy Schiff; Grilled salmon: Michael Lamotte/Cole Group/Photodisc/Getty Images; Orange juice: Sergei Vinogradov/seralexvi/123RF; Papaya: Marilyn Barbone/Shutterstock

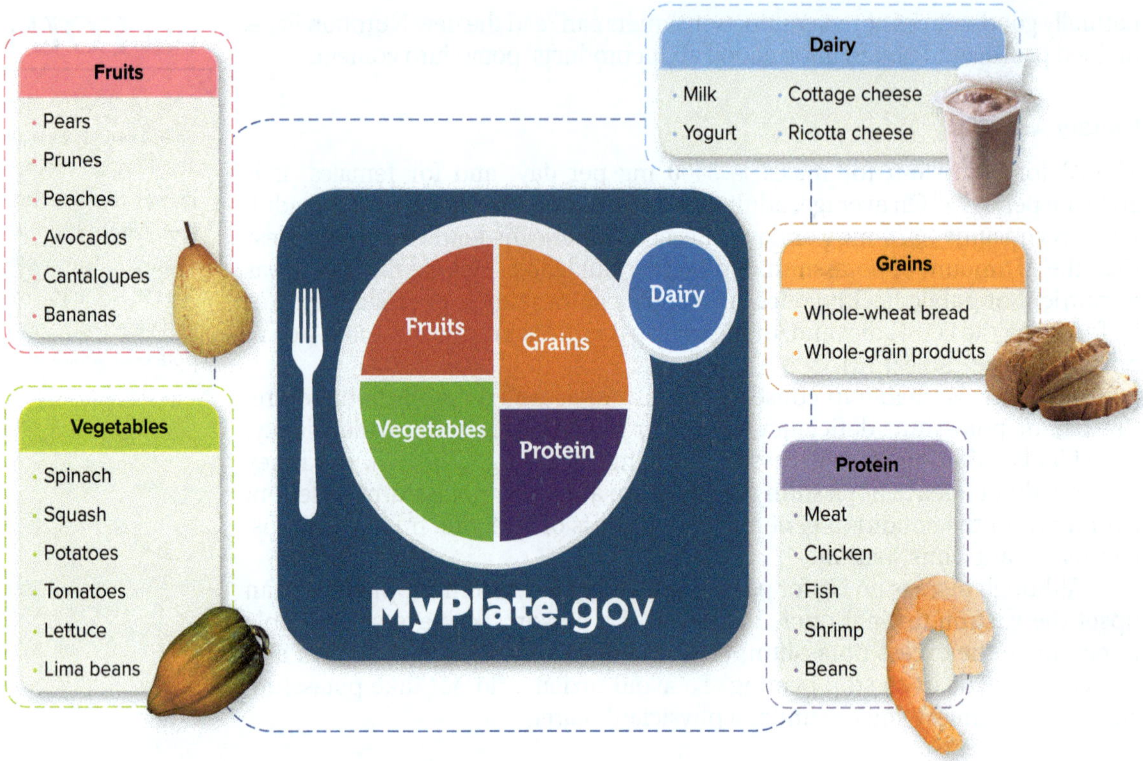

Figure 9.17 MyPlate: Good food sources of potassium. Acorn Squash: Burke/Triolo Productions/Artville/Getty Images; Pear: Stockbyte/Getty Images; Yogurt: Ingram Publishing/SuperStock; Bread: Fullerene/iStock/Getty Images; Shrimp: Ingram Publishing/Alamy Stock Photo; MyPlate food guide icon: U.S. Department of Agriculture

Magnesium (Mg)

Magnesium participates in more than 300 chemical reactions in the body.[26] The essential mineral also helps regulate normal muscle and nerve function, as well as blood pressure and blood glucose levels. Additionally, the body needs magnesium to maintain strong bones and a healthy immune system. Following diets that supply adequate amounts of magnesium may enhance bone health and reduce the risk of heart disease, stroke, and diabetes.[26] However, more research is needed to clarify the mineral's role in these diseases.

Normally, people absorb about 30 to 40% of the magnesium in their diets.[26] The kidneys regulate blood concentrations of magnesium and can reduce urinary losses of the mineral when the body's level of magnesium is low.

Tap water can be a source of magnesium and other mineral nutrients.
Burke/Triolo Productions/Photolibrary/Getty Images

Sources of Magnesium

Magnesium is in chlorophyll, the green pigment in plants. Therefore, it is not surprising that plant foods such as spinach, green leafy vegetables, whole grains, beans, nuts, seeds, and chocolate are the richest sources of magnesium. Animal products, such as milk and meats, also supply some magnesium. Table 9.14 lists some commonly eaten foods that supply magnesium. Refined grains are generally low in magnesium because the magnesium-rich bran and germ are removed during processing. Figure 9.18 indicates food groups that are naturally good sources of magnesium.

Other sources of magnesium are "hard" tap water and dietary supplements. However, amounts of magnesium in tap water can vary considerably. Moreover, the body does not absorb the form of magnesium (magnesium oxide) in multivitamin/mineral supplements very well. Nevertheless, hard water and magnesium supplements can still contribute to meeting a person's magnesium needs.

TABLE 9.14 Magnesium Content of Selected Foods

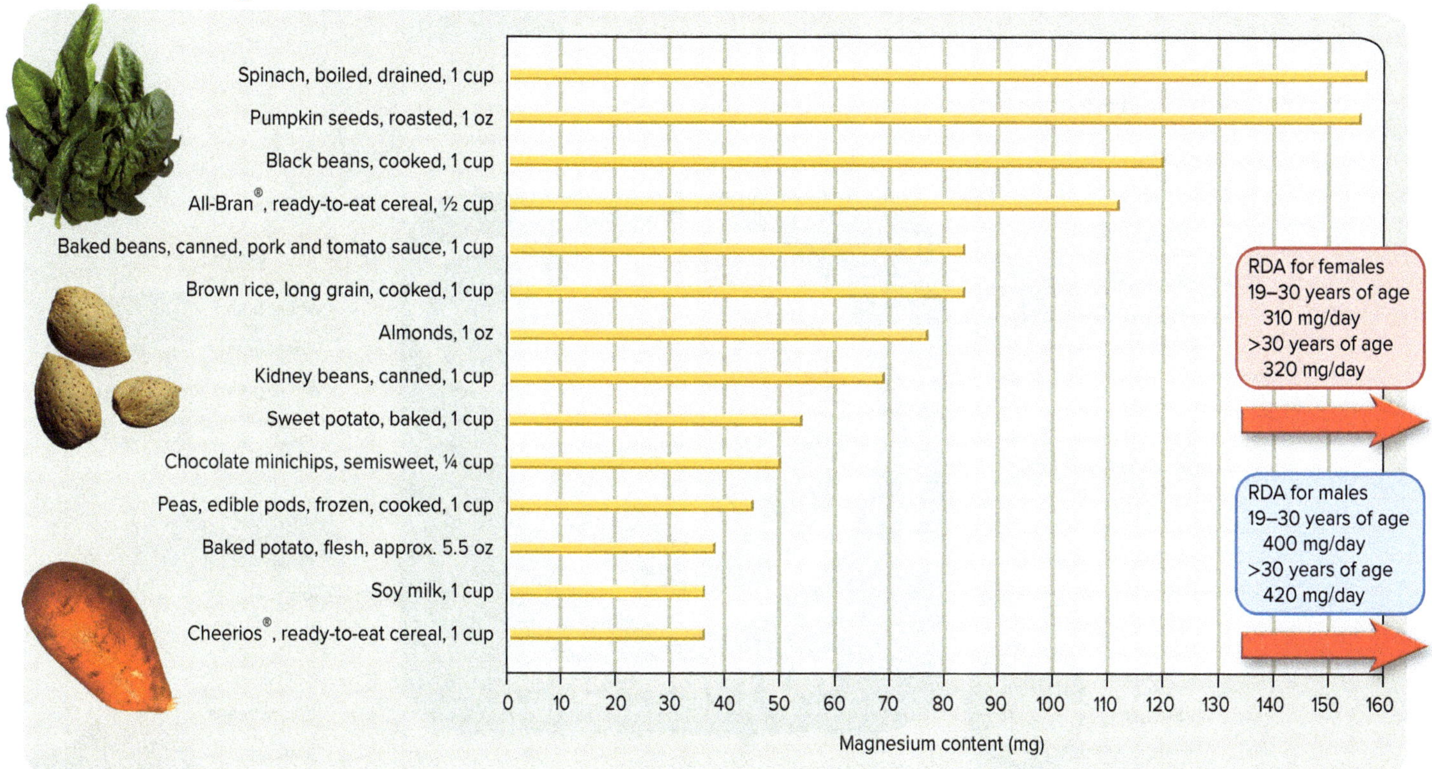

Source: Data from U.S. Department of Agriculture, Agricultural Research Service: *FoodData Central, 2019.*
Sweet potato, Spinach: ©Wendy Schiff; Almonds: C Squared Studios/Photodisc/Getty Images

Dietary Adequacy

Adult RDAs for magnesium range from 310 to 320 mg/day for females, and 400 to 420 mg/day for males. Many people in the United States do not consume recommended amounts of magnesium.[14] Despite this, cases of magnesium deficiency rarely occur among healthy members of the population.[26] Nevertheless, alcoholics, people with poorly controlled diabetes, or people who use certain medications (diuretics) that increase urinary excretion of magnesium have high risk of magnesium deficiency. Older adults are also at risk of magnesium deficiency because their bodies absorb less of the mineral and urinary losses increase with advancing age.

In humans, mild magnesium deficiency can cause tiredness, weakness, loss of appetite, nausea, and vomiting. Signs and symptoms of severe magnesium deficiency often include abnormal heartbeat rhythm, inability to relax muscles, personality changes, and seizures. Chronic magnesium deficiency may increase the risk of osteoporosis because the deficiency lowers the level of calcium in blood.

Magnesium Toxicity Magnesium toxicity rarely occurs from eating too much magnesium-rich food.[26] Toxicity is more likely to occur from ingesting excessive magnesium from laxatives, antacids, or dietary supplements that contain the mineral. Thus, the UL for the micronutrient (350 mg/day) is for magnesium-containing medications and not food sources. A person who consumes too much magnesium often develops diarrhea.

Patients suffering from kidney failure and older adults have high risk of magnesium toxicity because their kidneys do not excrete the mineral as effectively as the kidneys of younger, healthier individuals. In cases of kidney failure, the high concentration of magnesium in blood causes weakness, nausea, slowed breathing, very low blood pressure, and death.

Did *YOU* Know?

People who experience migraine headaches have lower serum and tissue magnesium levels than those who do not. Scientists have also discovered magnesium deficiency may increase a person's risk of developing migraines.[27] The evidence-based guidelines by the American Academy of Neurology and the American Headache Society state that magnesium therapy is "probably effective" for migraine prevention.[26] However, the typical dose of magnesium used for migraine prevention exceeds the UL; therefore, you should discuss using magnesium for headache prevention and treatment with your health care provider to avoid magnesium toxicity.

Blend Images/Image Source

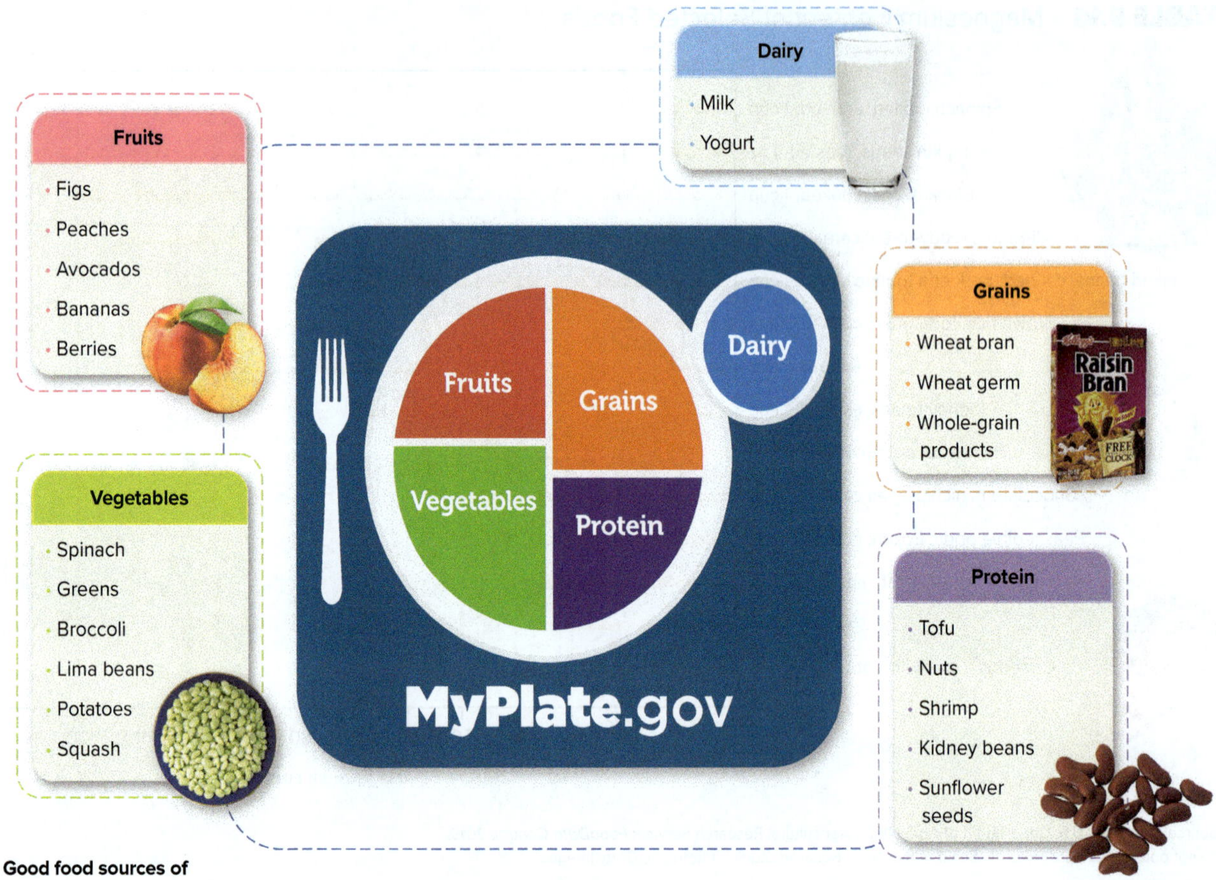

Figure 9.18 **MyPlate: Good food sources of magnesium.**
Bowl of beans: Brand Z Food/Alamy Stock Photo; Peach: Tim UR/Shutterstock; Milk: Nipaporn Panyacharoen/Shutterstock; Raisin Bran: Jill Braaten/McGraw Hill; Kidney beans: ©Wendy Schiff; MyPlate food guide icon: U.S. Department of Agriculture

Chloride (Cl^-)

Chloride is the primary negatively charged ion found in extracellular fluid. Chloride is essential for maintaining proper fluid and acid–base balance. The micronutrient is needed for the transmission of nerve impulses and production of hydrochloric acid (HCl) in the stomach. For most Americans, the major source of chloride is table salt (sodium chloride) that is in processed food. Seaweed, rye, tomatoes, celery, and olives are rich food sources of the mineral.

Chloride Deficiency and Toxicity

The AI for chloride is 2300 mg. A chloride deficiency disorder is unlikely because of high amounts of sodium chloride in the typical American diet. However, deficiencies may be seen in cases of extreme vomiting, diarrhea, and/or sweating, as well as with diuretic use. Fatigue and loss of appetite are common symptoms of chloride deficiency.

The UL for chloride is 3600 mg/day. Elevated blood chloride levels may result from excessive sodium chloride intake. As in cases of excess sodium ingestion, consuming too much chloride can contribute to hypertension.

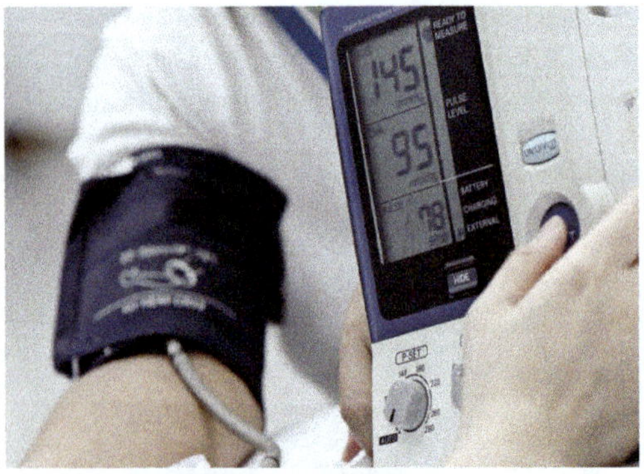

Consuming too much chloride can contribute to hypertension.
anucha maneechote/Shutterstock

Sulfur (S)

Sulfur is a component of several organic compounds, including the amino acids methionine and cysteine, and the vitamins biotin and thiamin. Sulfur is essential for the activity of many enzymes and as part of antioxidant molecules. Protein–rich foods are among the best food sources of sulfur. Some foods and beverages, including many wines, are also sources of sulfur (*sulfites*). In the United States, cases of sulfur deficiency are rare but can occur in people who have severe protein deficiency. Sulfur toxicity is unlikely from dietary sources. Thus, there is no AI, RDA, or UL established for sulfur.

Concept CHECKPOINT

19. What is osteoporosis, and why is it a major public health concern in the United States?

20. Identify at least four major risk factors for osteoporosis. Which risk factors can be modified to reduce the risk of osteoporosis?

21. Lisa's blood pressure is 115/68 mm Hg, which is in the ______ range. What is hypertension? What are major risk factors for hypertension?

22. What is the DASH diet? Aside from making dietary modifications, what other lifestyle changes can people with hypertension make to lower their blood pressure?

23. Prepare a table for the major minerals that includes information about each mineral's major roles in the body, primary food sources, and signs and symptoms of the mineral's deficiency, as well as toxicity disorders. Check your table against the information provided in Table 9.5.

See Appendix F for responses.

9.5 Trace Minerals

Learning Outcomes

1. List key functions and good food sources of trace minerals.
2. Discuss deficiency and toxicity disorders associated with the trace minerals.

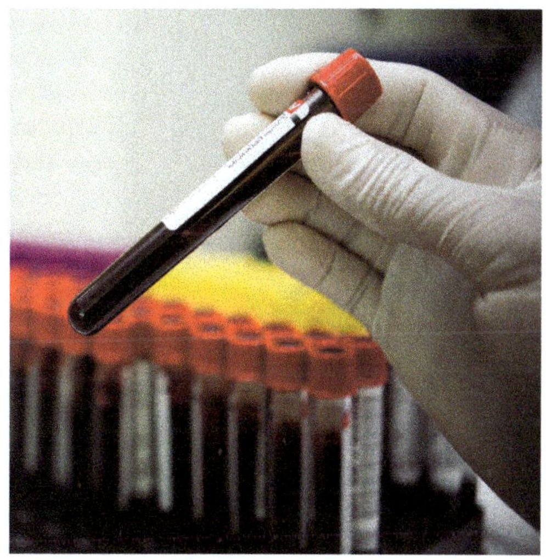

The iron-containing molecule hemoglobin is responsible for the red color of oxygenated blood.
Aomboon/Shutterstock

Although the body requires trace minerals in very small amounts, obtaining adequate amounts of these important nutrients can be difficult. Iron, for example, is one of Earth's most plentiful metals, but the total amount of iron in the human body is quite small, averaging only about 0.006% of a person's body weight.[4] Table 9.15 summarizes nutrition–related information about iron and other trace minerals. Unless otherwise noted, RDA/AI values are for adults, excluding pregnant or breastfeeding females.

Iron (Fe)

Do you think of "iron" when you think of "strength"? Associating iron with strength makes sense because muscular strength and endurance are reduced when the body lacks iron. Iron is a component of hemoglobin and *myoglobin* (*my'–o–glow–bin*). **Hemoglobin** is the iron–containing protein in red blood cells that transports oxygen to tissues and some carbon dioxide away from tissues. Hemoglobin is also responsible for the red color of oxygenated blood. **Myoglobin** is the iron–containing protein in muscle cells that controls oxygen uptake from red blood cells. Oxygen is critical for energy metabolism. Cells also contain iron in **cytochromes** (*sigh'–toe–crowms*), a group of proteins that are necessary for certain chemical reactions involved in the release of energy from macronutrients. If the body does not have enough iron to make hemoglobin, myoglobin, and the cytochromes, cells cannot obtain the energy they need to

hemoglobin iron-containing protein in red blood cells that transports oxygen

myoglobin iron-containing protein in muscle cells that controls oxygen uptake from red blood cells

cytochromes group of proteins involved in the release of energy from macronutrients

TABLE 9.15 Summary of Trace Minerals

Mineral	Major Functions in the Body	Adult RDA/AI (adult RDA = bold)	Major Dietary Sources	Major Deficiency Signs and Symptoms	Major Toxicity Signs and Symptoms
Iron (Fe)	• Component of hemoglobin and myoglobin that carries oxygen • Energy generation • Immune system function	Females: **18 mg** Males: **8 mg**	Meat and other animal foods, except milk; whole-grain and enriched breads and cereals; fortified cereals	• Fatigue upon exertion • Small, pale red blood cells • Low hemoglobin levels • Poor immune system function • Growth and developmental retardation in infants	UL = 45 mg/day • Intestinal upset • Organ damage • Death
Zinc (Zn)	• Component of numerous enzymes	Females: **8 mg** Males: **11 mg**	Seafood, meat, whole grains	• Skin rash • Diarrhea • Depressed sense of taste and smell • Hair loss • Poor growth and physical development	UL = 40 mg/day • Intestinal upset • Depressed immune system function • Supplement use can reduce copper absorption.
Copper (Cu)	• Promotion of iron metabolism • Component of antioxidant enzymes • Component of enzymes involved in connective tissue synthesis	**0.9 mg**	Liver, cocoa, legumes, whole grains, shellfish	• Anemia • Reduced immune system function • Poor growth and development	UL = 10 mg/day • Vomiting • Abnormal nervous system function • Liver damage
Selenium (Se)	• Component of an antioxidant system	**55 mcg**	Meat, eggs, fish, seafood, whole grains	• Muscle pain and weakness • Form of heart disease	UL = 400 mcg/day • Nausea • Vomiting • Hair loss • Weakness • Liver damage
Iodine (I)	• Component of thyroid hormones	**150 mcg**	Iodized salt, saltwater fish, dairy products	• Goiter • Cretinism (intellectual impairment and poor growth in infants of females who were iodine deficient during pregnancy)	UL = 1100 mcg/day • Reduced thyroid gland function
Fluoride (F)	• Increases resistance of tooth enamel to cavity formation • Stimulates bone formation	Females: **3 mg** Males: **4 mg**	Fluoridated water, tea, seaweed	• No true deficiency, but increased risk of tooth decay	UL = 10 mg/day • Stomach upset • Staining of teeth during development • Bone deterioration
Chromium (Cr)	• Enhances insulin action	Females: **20–25 mcg** Males: **30–35 mcg**	Egg yolks, whole grains, pork, nuts, mushrooms	• Blood glucose level remains elevated after meals.	• Unknown but currently under scientific investigation • May interact with certain medications
Manganese (Mn)	• Cofactor for certain enzymes, including some involved in carbohydrate metabolism	Females: **1.8 mg** Males: **2.3 mg**	Nuts, oats, beans, tea	• None in humans	UL = 11 mg/day • Abnormal nervous system function
Molybdenum (Mo)	• Component of certain coenzymes	**45 mcg**	Liver, peas, beans, cereal products, leafy vegetables, low-fat milk	• None in healthy humans	UL = 2000 mcg/day • Rarely occurs from usual dietary sources • Overdoses of dietary supplements containing molybdenum may cause joint pain; side, lower back, or stomach pain; and swelling of feet or lower legs.

Shrimp, mushroom, Spinach, Beans: ©Wendy Schiff

perform work. Thus, fatigue is a major symptom of iron deficiency. Iron also plays roles in immune system function and brain development.

Sources of Iron

Beef, fish, and poultry ("meat") contain more iron than most plant foods. Some of the iron in meat is present as hemoglobin and myoglobin. These forms of iron are collectively referred to as **heme iron.** The remaining iron in meat, as well as all the iron in vegetables, grains, and supplements, is **nonheme iron.**

heme iron form of iron in hemoglobin and myoglobin

nonheme iron form of iron that is primarily in vegetables, grains, meats, and supplements

The intestinal tract absorbs more of the heme iron than nonheme iron in foods.[28] Some plant foods, such as spinach, contain nonheme iron, but oxalic acid in spinach binds to the mineral, reducing its absorption. Other naturally occurring compounds that reduce iron absorption include phytic acid in whole grains, tannins in tea, and substances that are chemically related to tannins in coffee.

Meat is the major source of iron in the typical American diet. Other important sources of iron are fortified cereals and products made from enriched flour, such as breads and rolls. Dairy products are poor sources of iron. Figure 9.19 indicates food groups that are good sources of iron. Table 9.16 lists some foods that are among the richest sources of iron.

Regulating Iron Under normal conditions, the body regulates iron absorption and conservation. The digestive tract absorbs only 5 to 15% of the iron in foods. However, the intestinal tract can absorb more iron when the body's need for the trace mineral increases. Despite iron enrichment, only about 5% of the iron added to grain products is absorbed.[28] The liver, the body's main site for iron storage, incorporates the trace mineral into the protein ferritin (*fer′–ih–tin*) until it is needed.

After red blood cells die, the body breaks them down and conserves most of the iron that was in hemoglobin. By doing so, the body can recycle the trace mineral to make hemoglobin for new red blood cells. Nonetheless, some iron is lost each day via the GI tract, urine, and skin. Any form of bleeding, including menstruation, also contributes to iron losses. Replacing the iron is essential to good health.

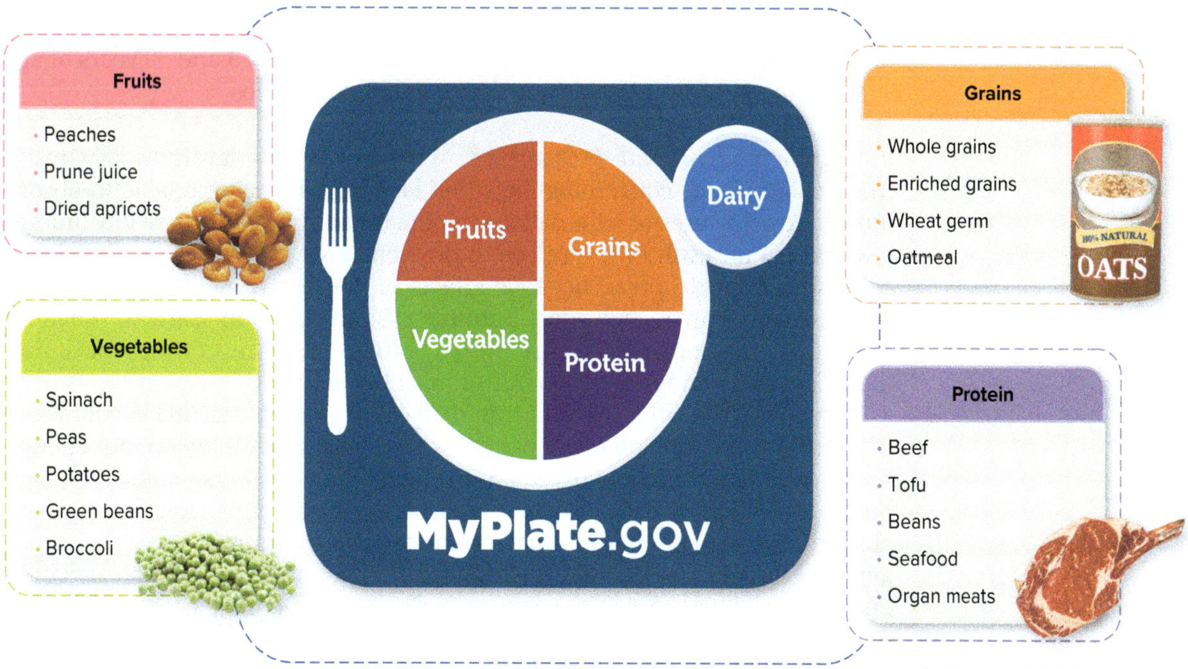

Figure 9.19 MyPlate: Good food sources of iron.
Peas: Ingram Publishing/SuperStock; Apricots: lynx/iconotec/Glowimages; Oatmeal: Jacques Cornell/McGraw Hill; Steak: Aoldman/iStock/Getty Images; MyPlate food guide icon: U.S. Department of Agriculture

TABLE 9.16 Iron Content of Selected Foods

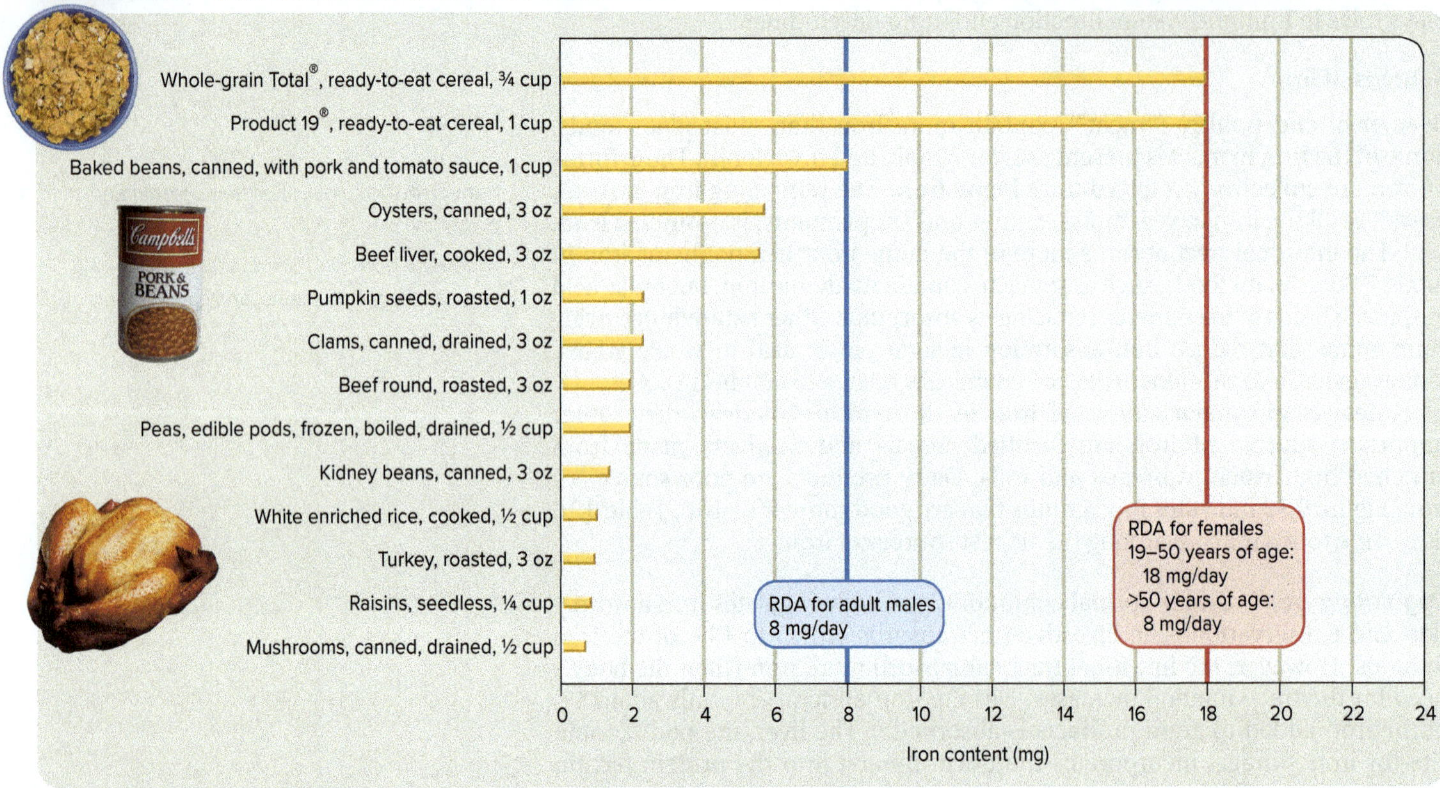

Source: Data from U.S. Department of Agriculture, Agricultural Research Service: *FoodData Central*, 2019.
Cereal: Joe Belanger/iStock/Getty Images; Beans: ©Wendy Schiff; Turkey: Ernie Friedlander/Cole Group/Photodisc/Getty Images

Dietary Adequacy

For adult males, the RDA for iron is 8 mg per day; for adult females between 19 and 50 years of age, the RDA for iron is 18 mg/day. The average daily intake for American males is 16.2 mg, whereas the average daily intake for American females is about 12.3 mg/day.[14] Thus, females between 19 and 50 years of age are more likely to have inadequate iron intakes than males.

Iron Deficiency–Related Disorders Low blood iron levels usually result from losing blood, consuming diets that lack iron, or being unable to absorb dietary iron. In cases of **iron deficiency,** the body's iron stores are low but not low enough to result in severe health problems. Nevertheless, iron deficiency can still have widespread negative effects on the body, including interfering with normal growth, behavior, immune system function, cardiac function, and energy metabolism.[29] In the United States and the rest of the world, iron deficiency is the most common nutritional deficiency.

Iron deficiency can lead to iron deficiency anemia. Anemia occurs when oxygen transport in blood is impaired, generally because there are not enough red blood cells to carry the oxygen or the red blood cells do not contain enough hemoglobin (Fig. 9.20). If oxygen is lacking, cells cannot release considerable amounts of energy from macronutrients, so symptoms of iron deficiency anemia include lack of energy and difficulty concentrating on mental activities. Furthermore, the heart of a person suffering from anemia has to work harder to circulate oxygen–poor blood throughout the body. Over time, anemia can cause rapid or irregular heartbeat, chest pain, an enlarged heart, and even heart failure. Table 9.17 lists common signs and symptoms of iron deficiency anemia.

There are many different kinds of anemia, but iron deficiency anemia is the most common form. According to the World Health Organization (WHO),

iron deficiency condition characterized by low body stores of iron

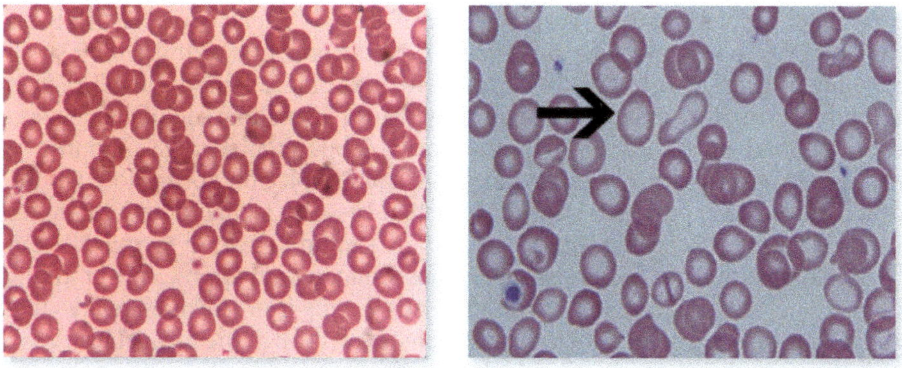

(a) (b)

Figure 9.20 Iron-deficient red blood cells. (a) Normal red blood cells. (b) Iron-deficient red blood cells. Arrow points to one of the deficient cells.

Normal RBCs: plenoy m/Shutterstock; Iron deficient RBCs: Dennis K. Burns, MD,Travis G. Brown, MD,Walter L. Kemp, MD/ McGraw Hill

about 30% of females of childbearing age in the world suffer from anemia, and many cases of the condition are due to iron deficiency.[30] An estimated 16 million American children and adolescents have iron deficiency anemia.[31]

Substantial blood loss is a common cause of iron deficiency anemia. Such losses of blood often result from serious intestinal diseases, severe physical injuries, and excessive menstrual bleeding. Diseases that reduce red blood cell formation or increase red blood cell destruction also cause anemia. It is important to note that some types of anemia are the result of genetic defects and not dietary deficiencies. Blood testing can determine which kind of anemia a person has developed, so the condition can be treated properly.

Females of childbearing age generally lose some iron during menstruation. Females with heavy menstrual blood losses are especially prone to iron deficiency anemia. Although pregnant females do not have to contend with menstrual blood losses, they still need to be concerned about their iron intake. According to the Dietary Guidelines, iron is a "nutrient of public health concern" for females who are pregnant or able to become pregnant. The "Food & Nutrition Tips" feature in this section suggests practical ways to increase dietary sources of iron.

Iron deficiency anemia is especially harmful for pregnant females, as well as children. During pregnancy, a female's need for iron increases as her blood supply expands and new tissues are added to both her body and that of her fetus. Pregnant females who suffer from iron deficiency anemia have higher risk of dying during pregnancy than healthy pregnant females. Anemic pregnant females are also more likely to give birth to premature or low–birth–weight infants. Premature babies are born before the 37th week of pregnancy; normally, pregnancies last about 40 weeks from the date of the mother's last menstrual period. A low–birth–weight baby weighs less than 5½ pounds at birth. Compared to healthy newborns, premature or low–birth–weight infants are more likely to die during their first year of life. Iron supplementation can prevent pregnant females from developing iron deficiency and iron deficiency anemia before they give birth.

Because of their rapid growth rates, infants and toddlers have higher needs for iron than older children. Furthermore, iron appears to be necessary for normal nervous system functioning, including brain development. Iron–deficient infants can experience delays in the development of normal motor and mental functions.

Consuming too much milk may play a role in the development of iron deficiency in children. Milk is a poor source of iron. Thus, children who drink excessive amounts of milk may not have the appetite to eat foods that are more reliable iron sources. The calcium that is in milk may interfere with iron absorption when the beverage is consumed with foods that contain the mineral.[33] To reduce the risk of iron deficiency, children should be encouraged to eat more iron–rich foods, such as meat, products made from soybeans, and iron–fortified cereals.

TABLE 9.17 Signs and Symptoms of Iron Deficiency Anemia

• Pale skin
• Fatigue and weakness
• Irritability
• Difficulty concentrating and thinking
• Brittle nails
• Headache

Source: U.S. Library of Medicine (Medline), National Institutes of Health, MedlinePlus: *Iron deficiency anemia*. 2022. https://medlineplus.gov/ency/article/000584.htm. Accessed: May 10, 2023.

Did *YOU* Know?

Donating 1 pint (approximately 0.5 L) of blood represents a loss of 200 to 250 mg of iron. Even when a person consumes adequate amounts of iron, their body generally needs several weeks to replace the iron that was in the donated blood. According to the American Red Cross, healthy people can give whole blood every 56 days.[32] Iron deficiency can result from donating blood too often.[33]

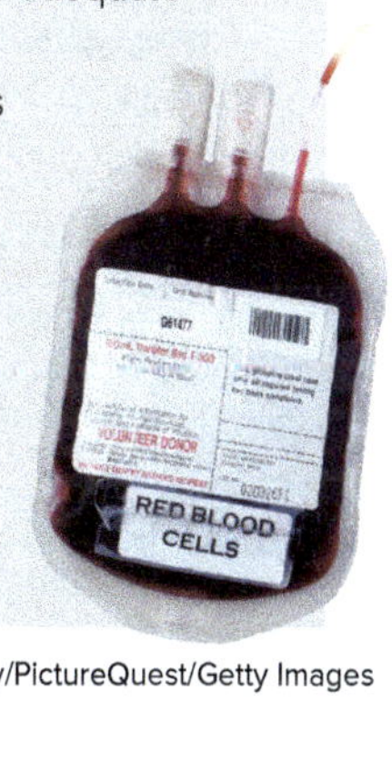

Liquidlibrary/PictureQuest/Getty Images

Did *YOU* Know?

Cooking utensils may be a source of dietary iron. When foods are cooked in cast-iron cookware, some iron migrates from the cookware and enters the food. Acidic ingredients such as tomato sauce absorb the most iron during cooking. Replacing the heavy iron cookware with lighter stainless steel and aluminum pots and pans can reduce the amount of iron in diets.

C Squared Studios/Photodisc/Getty Images

Total vegetarians have a higher risk of iron deficiency–related disorders than people who eat meat because meat provides heme iron. Combining a small amount of meat with plant foods improves the bioavailability of the plant's nonheme iron. Vegetarians, however, may reject recipes that include any meat, especially red meats. Recall that some plant foods contain high amounts of oxalic acid and phytic acid, substances that can inhibit iron absorption. On the other hand, vegetarian diets usually are rich in vitamin C, a factor that increases nonheme iron absorption. Thus, vegetarians should consume vitamin C–rich foods along with plant foods, especially those that contain appreciable amounts of iron, such as spinach, lentils, and soybeans. Eating iron–fortified ready–to–eat cereals can also be helpful for vegetarians, even though the form of iron used to fortify cereals is not as well absorbed as heme iron. Finally, vegetarians can take a multivitamin/mineral supplement that contains iron and vitamin C to ensure their iron and other mineral intakes are adequate.

Treatment for iron deficiency anemia generally includes iron supplements and the addition of iron–rich foods to the diet. It is also important to find and treat factors that may be causing the anemia, such as intestinal bleeding.

Iron Toxicity The UL for iron is 45 mg/day. Although not having enough iron in the body interferes with normal growth, development, and functioning, ingesting too much iron poses the risk of toxicity. Between 1983 and 2000, 43 American children died as a result of accidentally taking too many iron–containing dietary supplements.[33] Early signs of acute iron poisoning include vomiting and diarrhea that may progress to coma and death.

In 1997, the U.S. Food and Drug Administration (FDA) required a warning statement on the packaging of iron supplements to reduce the number of iron poisoning cases among young children. The agency also required unit–dose packaging of oral iron supplements that contained 30 mg of iron or more per dose. The individually wrapped supplements were designed to make it difficult for young children to ingest large quantities of the supplements at a time. After these rules were instituted, only one child died from iron poisoning between 1998 and 2002.[33] In 2003, however, a federal court ruled that the FDA did not have the legal authority to require special unit–dose packaging of iron supplements. The agency, however, could continue to require warning statements on supplement labels.

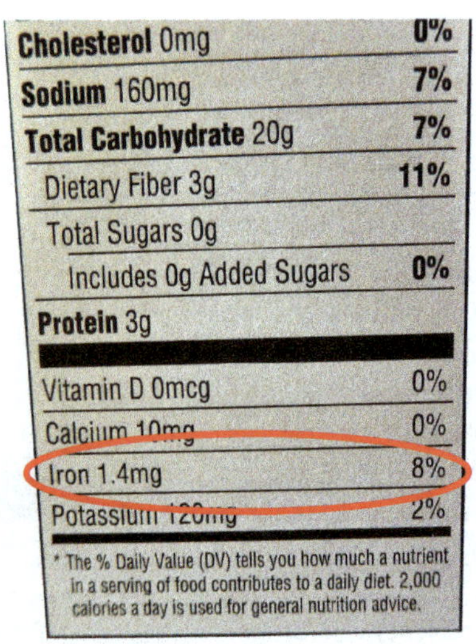

The Nutrition Facts panel indicates the %DV of iron in a serving of the food.
©Wendy Schiff

Food & Nutrition tips

The following suggestions can add more iron to your diet:

- Read the Nutrition Facts panel for information about a packaged food's iron content; choose foods with a Daily Value of 20% or greater whenever possible.
- Cook foods in a cast iron skillet.
- Eat lean meat, poultry, or fish with plant sources of iron.
- Add orange segments or chopped tomatoes to spinach salads or cooked spinach.
- Add chopped onions and green peppers to peas or beans.
- Serve sweet potatoes with fresh orange segments or dried apricots.
- Add raspberries, strawberries, raisins, or dried apricots to cereal.
- Drink orange juice when eating peanut butter or soy nut butter sandwiches.
- Consume watermelon, dried plums, dried apricots, or raisins for snacks.

Iron Overload: Hereditary Hemochromatosis *Iron overload* is a condition characterized by excess iron in the body. Iron overload occurs when toxic amounts of iron supplements are ingested, but the condition also results from certain genetic diseases. **Hereditary hemochromatosis** (*he′–mo–crow′–ma–toe–sis*) is the most common type of iron overload disease.[34] People who have hereditary hemochromatosis (HH) absorb too much iron. The body has no way to eliminate the excess iron, so the mineral accumulates in tissues and can cause joint pain, abnormal bronze skin color, diabetes, and damage to the liver, heart, adrenal glands, and pancreas.

hereditary hemochromatosis most common type of iron overload disease

HH most often affects people who have northern European ancestors. In the United States, about 5 of every 1000 non–Hispanic white Americans are susceptible to developing the disease.[34] Males are more likely to be diagnosed with HH than females. Additionally, males tend to develop health problems from the excess iron at a younger age than females with the condition.

Common signs and symptoms of HH include joint pain, fatigue, weight loss, abdominal pain, loss of sex drive, and abnormal skin color (gray or bronze). Even though people with HH begin accumulating iron early in life, they often do not report any signs and symptoms of the disease until they are over 40 years of age. Testing is available to determine the presence of the genes that are responsible for the disease.

Many people who have HH experience vague symptoms or no symptoms at all. If the disease is not detected early and treated effectively, the organ damage resulting from the condition can be deadly. Treatment usually includes visiting a clinic periodically to have blood removed. This process stimulates the tissues that produce red blood cells to use storage iron for hemoglobin production. People with HH should avoid taking dietary supplements that contain iron as well as vitamin C supplements (vitamin C enhances iron absorption).

MY DIVERSE PLATE

Although the kiwano (*Cucumis metuliferusis*) is native to southern Africa, this odd-looking relative of the cucumber is grown in California. The green pulpy seeds can be scooped out of the rind and eaten as a snack or added to salads. A single kiwano is a good source of protein (almost 4 g), iron (2.3 mg), and vitamin C (11 mg). The pulp, however, has a slimy texture and not much flavor. In Africa, kiwano is generally considered to be an edible weed that is not eaten unless other foods are unavailable.

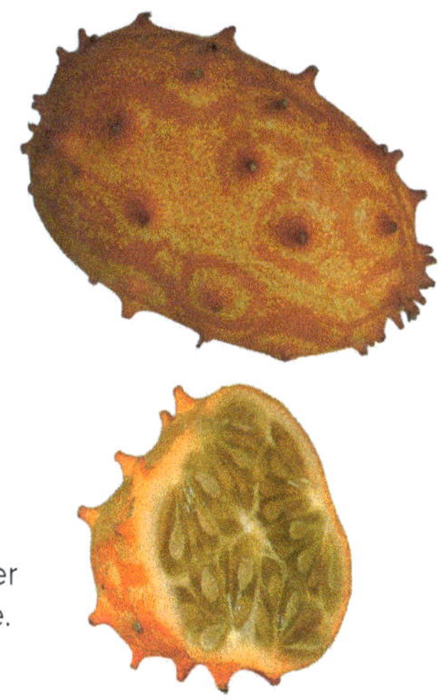

©Wendy Schiff

Zinc (Zn)

In 1958, physician Ananda Prasad was working in Iran when he examined a 21–year–old man with dwarfism, intellectual disability, iron deficiency anemia, and underdeveloped sexual organs.[35] Prasad noted that the young man ate unleavened ("flat") bread almost exclusively. After examining other patients in Iran and Egypt with similar health problems and dietary practices, Prasad hypothesized that diet was responsible for the condition. Eventually, medical researchers determined that Prasad's patients had severe zinc deficiencies. After these patients were given zinc supplements, they began to grow and develop normally. Prasad later determined that girls also experienced stunted growth and delays in sexual maturation as a result of zinc deficiency.[36]

In the regions where the males who were zinc deficient lived, the typical diet was comprised primarily of unleavened whole–wheat bread and little animal protein. Unleavened whole–wheat bread is naturally high in phytic acid and fiber, substances that decrease zinc bioavailability. In places where people use yeast to leaven (raise) bread dough, severe zinc deficiency is less likely to occur. Yeast reduces the binding effects of phytic acid and fiber, making zinc more bioavailable. Consuming zinc–rich sources of animal protein, such as meat and milk, also reduces the likelihood of zinc deficiency.

Other factors that influence the bioavailability of zinc include the body's need for the mineral and the presence of large amounts of certain other metals. During times when a healthy body needs zinc, the small intestine absorbs more. However, the presence of excess copper or iron in the small intestine interferes with zinc absorption. Thus, iron supplements should be taken between meals instead of with them.

Did *YOU* Know?

Results of some studies indicate that zinc, in the form of lozenges or syrup, reduces the duration of signs and symptoms of the common cold by 2 days.[37] However, more research is needed to determine the type of zinc and how much to consume before general recommendations can be established for ingesting zinc-containing products to treat colds.

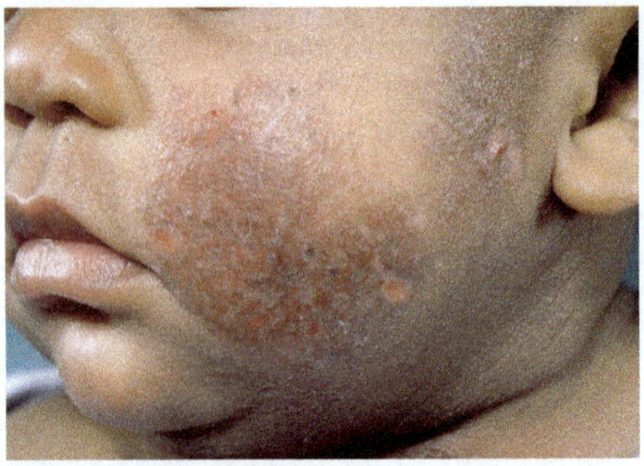

Figure 9.21 Sign of zinc deficiency in a baby. Zinc is necessary for healthy skin and normal physical growth. This child has a type of dermatitis that is a sign of zinc deficiency.

Medical-on-Line/Alamy stock photo

Why Is Zinc Necessary?

Zinc is a component of about a hundred enzymes.[37] Zinc is necessary for wound healing, the sense of taste and smell, DNA synthesis, healthy skin, and proper functioning of the immune system (Fig. 9.21). Zinc is also essential for growth and development during pregnancy, childhood, and adolescence.

Sources of Zinc

Zinc is widespread in foods (Table 9.18). Red meat and poultry products supply most of the zinc in the typical American diet. Figure 9.22 indicates MyPlate food groups that have foods that are good sources of zinc.

Dietary Adequacy

Adult RDAs for zinc are 8 mg/day (women) and 11 mg/day (men). In the United States, the average adult consumes adequate amounts of zinc.[14] Thus, zinc deficiency is not a widespread problem in the United States. However, alcoholics have high risk of zinc deficiency because alcohol reduces zinc absorption and increases excretion of the mineral in urine. Making matters worse, many people who suffer from alcoholism do not consume nutritious diets. Vegetarians need more zinc than people who eat meat because the GI tract does not absorb zinc from plant foods as well as from animal foods.[37] Older adults are at risk of zinc deficiency because they often consume foods that do not provide adequate amounts of the trace mineral.

Although breast milk contains zinc, the milk does not supply enough of the trace mineral for infants who are older than 6 months of age. To increase the likelihood that their diets contain enough zinc, breastfed babies who are between 6 and 12 months of age need to consume foods that contain the trace mineral, such as zinc–fortified infant cereal.

TABLE 9.18 Zinc Content of Selected Foods

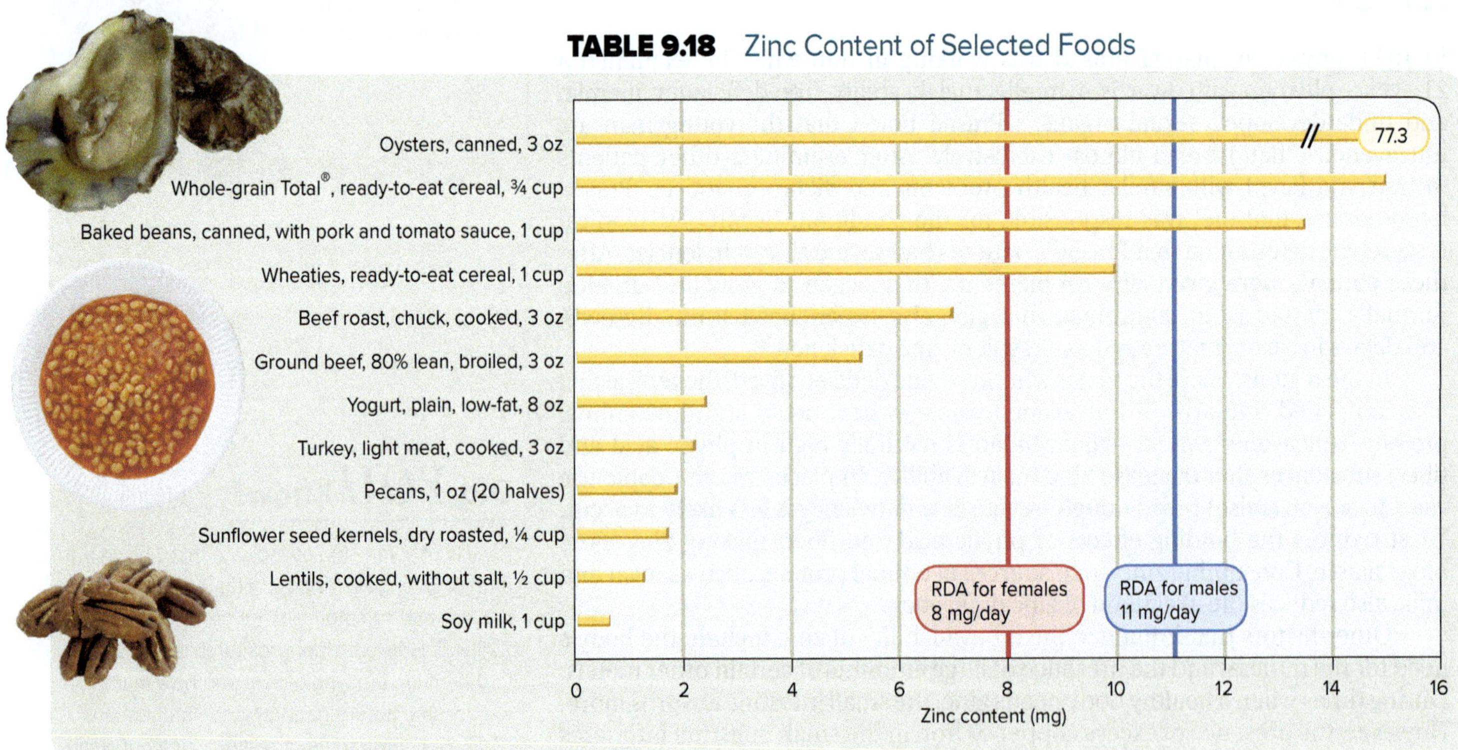

Source: Data from U.S. Department of Agriculture, Agricultural Research Service: *FoodData Central,* 2019.

Oysters: ©Wendy Schiff; Bowl of beans: Jasper White/Image Source; Pecans: Foodcollection

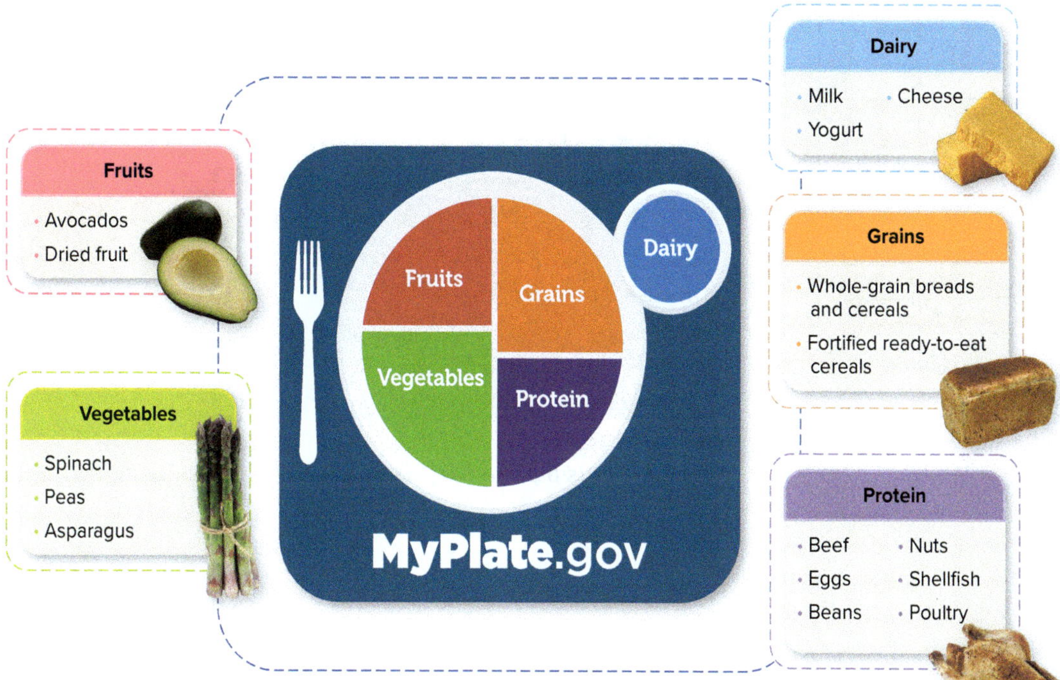

Figure 9.22 MyPlate: Good food sources of zinc.
Asparagus: Pixtal/age fotostock; Avocado: Ingram Publishing/SuperStock; Cheese: Brent Hofacker/Shutterstock; Bread: Everyday Images/Alamy Stock Photo; Roast chicken: Louis Hiemstra/E+/Getty Images; MyPlate food guide icon: U.S. Department of Agriculture

In children and adolescents, zinc deficiency can cause growth retardation and delayed sexual maturation. Adult males who are zinc deficient may experience sexual dysfunction, particularly the inability to attain an erection. Other signs of zinc deficiency include loss of appetite, diarrhea, hair loss, dermatitis, poor wound healing, impaired sense of taste, and mental slowness.[37]

Zinc Toxicity The UL for zinc is 40 mg/day. Zinc intakes that exceed the UL can reduce beneficial HDL cholesterol levels in blood. Ingesting more than 150 mg of zinc per day can also result in diarrhea, cramps, nausea, vomiting, and depressed immune system function.[37] Additionally, megadoses of zinc may interfere with copper absorption. Therefore, people should avoid high intakes of zinc, unless they are under a physician's supervision.

Copper (Cu)

The body uses copper to make several enzymes that act as antioxidants. The mineral is also involved in iron metabolism, immune function, and collagen production. A wide variety of foods contain copper, but the foods listed in Table 9.19 are among the richest sources of the micronutrient.

The adult RDA for copper is 0.9 mg/day, and Americans typically consume more than this amount.[14] Consuming high amounts of iron or zinc can interfere with the intestinal tract's ability to absorb copper, so taking these supplements may cause copper deficiency. People with chronic digestive system diseases may also develop copper deficiency because their bodies' ability to absorb the mineral is depressed. A deficiency of copper can lead to anemia, impaired functioning of the immune and nervous systems, abnormal skin pigmentation, and poor growth (children). Copper toxicity rarely occurs.

Did *YOU* Know?

Vegetarians typically avoid eating meat and often include lots of whole grains and beans and other legumes (sometimes called "pulses") in their diets. However, the phytic acid in seeds can bind to zinc, reducing the mineral's absorption.[37] The process of sprouting reduces a seed's phytic acid content. To increase the bioavailability of the zinc in beans and other seeds, soak the raw seeds (mung beans or "wheat berries," for example) in water for several hours. Then remove the seeds from the water and place them on a plate. Cook the seeds as soon as they begin to sprout.

Jupiterimages/PHOTOS.com/Getty Images

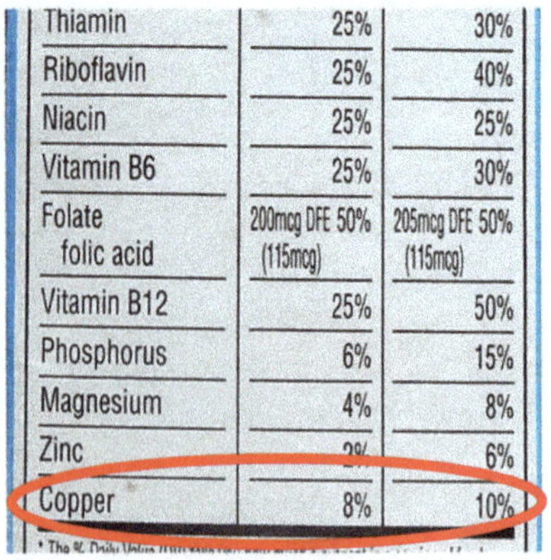

This Nutrition Facts panel indicates the %DVs of copper in a serving of plain cereal (middle column) and a serving of the cereal with some added milk (right column).
©Wendy Schiff

TABLE 9.19 Copper Content of Selected Foods

Food	mg
Beef liver, 3 oz, cooked	12.40
Crab meat, ½ cup, canned	1.10
Cashews, 1 oz, dry roasted	0.63
Mushrooms, ½ cup, white, cooked, drained	0.40
Soybeans, ½ cup, mature, boiled, drained	0.35
Chocolate, ¼ cup, semisweet	0.30
Baked beans, ½ cup, canned, with pork and tomato sauce	0.27
Sunflower seeds, 1 oz, dry roasted	0.26

RDA adults = 0.9 mg/day

Source: Data from U.S. Department of Agriculture, Agricultural Research Service: *FoodData Central*, 2019.
Pixfiction/Shutterstock

Iodine (I)

During World War I, physicians noted that males drafted into the U.S. military from the Great Lakes region were far more likely to have goiter (*goy′–ter*) than men from some other areas of the country. Goiter is an enlargement of the thyroid gland that is not the result of cancer (Fig. 9.23). Goiters often occur among populations living in areas that have iodine–depleted soil. In general, these regions are inland and far from an ocean. If people in these communities limit their diets to locally produced foods, they might not have enough iodine in their diets.

Iodine (I_2) is poisonous, but most ingested iodine loses an electron to become the iodide ion (I^-) in the digestive tract.[38] Iodide is the form of iodine that the body uses. Most of the iodide in an adult's body is located in the thyroid gland. Under normal conditions, the kidneys filter and eliminate excess iodide from blood.

thyroid hormone hormone that regulates the body's metabolic rate

From 1917 to 1922, researchers in Ohio conducted an experiment on a group of school–age females enrolled in 5th to 8th grades in which one group of the children received doses of iodine, whereas the other group (the control group) did not receive the trace mineral. The results of the study indicated iodine was nearly 100% effective in preventing goiter in the healthy children. Moreover, the majority of the young females who already had goiters when they received the iodine experienced a reduction in the size of their thyroid glands by the end of the study. In 1924, iodine was added to table salt in the United States, and as a result, cases of goiter caused by iodine deficiency rarely occur in this country. Today, use of iodized salt is the major method of preventing iodine deficiencies in developed nations, but inadequate iodine intake and goiters are still common in central Asia and central Africa.

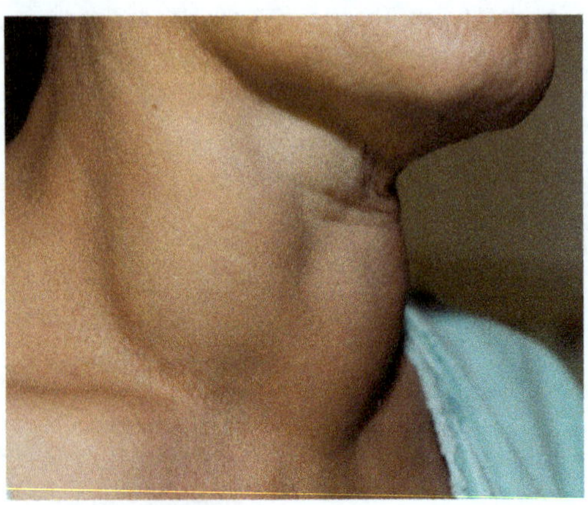

Figure 9.23 Goiter.
Karan Bunjean/Shutterstock

Why Is Iodine Necessary?

People require iodine for normal thyroid function and for the production of two thyroid hormones, collectively referred to as **thyroid hormone.** Thyroid hormone controls the rate of cell metabolism, that is, the rate at which cells obtain energy. The thyroid gland traps iodide from the bloodstream and accu–mulates the element for thyroid hormone synthesis. If a person's iodine intake is too low, the pituitary gland in the brain releases a hormone that causes the thyroid gland to enlarge. As a result, the larger thyroid gland tries to remove as much iodide as possible from the bloodstream. It is important to note that an enlarged thyroid gland can also be a sign of some diseases and conditions that are not related to iodine intake.

Sources of Iodine

Major sources of iodine include saltwater fish; seafood; seaweed; some plants, especially the leaves of plants grown near oceans; and iodized salt. A half teaspoon of iodine–fortified salt supplies the adult RDA for iodine. Iodine fortification of salt is voluntary in the United States, so not all salt has the trace mineral added to it, including some brands of Himalayan, Celtic, Kosher, and sea salts. Dairy products including milk are a good source of iodine if the cattle feed contains supplemental iodine. Some dairy manu–facturers sterilize the equipment used to process milk with iodine–based sanitizers, which contributes to the iodine content in milk. Other dietary sources of iodine include food additives that contain the mineral, such as certain dough conditioners and food dyes. Table 9.20 lists some foods that are good sources of iodine.

Dietary Adequacy

The adult RDA for iodine is 150 mcg/day. Most Americans have adequate iodine intakes.[38] As many Americans, particularly older adults, try to reduce their risk of hypertension by using less salt, iodine intakes may decline to marginal or inadequate levels.

Between 2007–2010, Swedish researchers studied the effect of the "Paleolithic" diet on iodine levels of 49 obese females.[39] About half of the females consumed the Paleolithic diet, which is also known as the "Paleo Diet" (PD). During the study, the PD group consumed lean meats, fish and seafood, fruits, vegetables and root vegetables, eggs, and nuts. Foods excluded from the PD included dairy products, cereals, beans, refined fats

Did *YOU* Know?

Sea salt forms when the water in naturally occurring saltwater evaporates, leaving mineral deposits, including sodium chloride. Table salt is made by processing natural forms of sodium chloride to remove other minerals. In the United States, manufacturers often add iodine to table salt to make "iodized salt." The sea salt that is usually sold in supermarkets is not a good source of iodine because it has not been fortified with the mineral. It is wise to avoid excess sodium from all sources, including sea salt.

©Wendy Schiff

TABLE 9.20 Iodine Content of Selected Foods

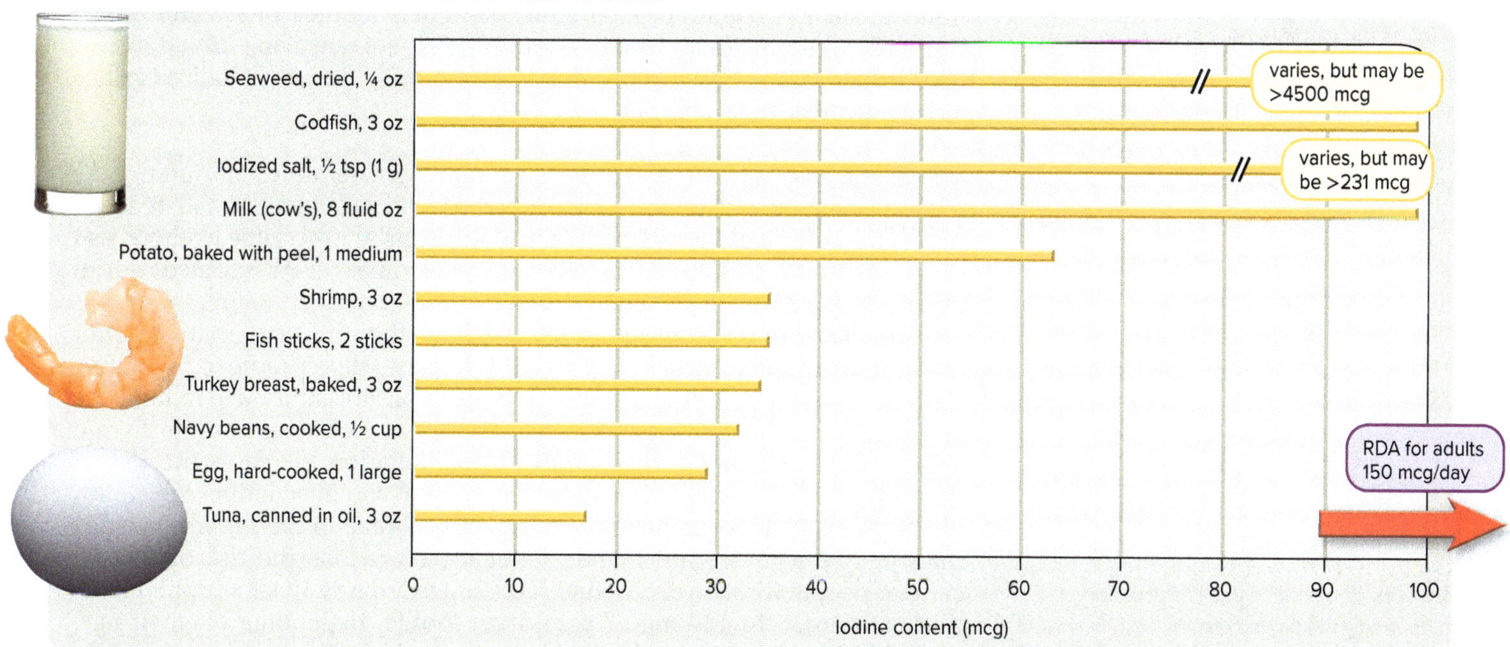

Source: Data from Linus Pauling Institute, Oregon State University. 2015. http://lpi.oregonstate.edu/infocenter/minerals/iodine/. Accessed: May 10, 2023.

Milk: Ken Karp/McGraw Hill; Shrimp: Ingram Publishing/Alamy Stock Photo; Egg: Siede Preis/Photodisc/Getty Images

Did *YOU* Know?

Raw vegetables, particularly turnips, cabbage, brussels sprouts, cauliflower, and broccoli, contain **goitrogens**. These compounds inhibit iodine metabolism by the thyroid gland and, as a result, reduce thyroid hormone production. Unless people eat large amounts of raw vegetables that contain goitrogens or they are iodine deficient, they do not need to be concerned about eating these foods.[38] Furthermore, cooking vegetables destroys goitrogens.

Steven P. Lynch

and sugars, added salt, bakery products, and sugar–sweetened soft drinks. The other half of subjects followed an unrestricted diet (control group). At the beginning of the study, the iodine status of all subjects was assessed by measuring urinary iodine levels; results indicated that subjects had normal iodine levels. The female's iodine levels were remeasured 6 months later, and the test revealed that the PD group's urinary iodine levels dropped below normal levels. Thus, these females had developed iodine deficiency. The researchers who conducted the study noted that the urinary iodine levels of the PD group did increase 2 years later, but the scientist contributed this to lack of adherence to the PD restrictions.

The researchers concluded the PD can cause iodine deficiency. This deficiency is likely to occur when iodized salt is not included in the diet, and supplemental iodine is not provided.[39] Dairy products and iodized salt were two iodine–rich foods eliminated in the PD group.

Iodine Deficiency In cases of iodine deficiency, the thyroid gland produces insufficient amounts of thyroid hormone and goiter develops. As a result of the lack of thyroid hormone, iodine–deficient people generally have low metabolic rates and elevated blood cholesterol levels. Other signs and symptoms of iodine deficiency include fatigue, difficulty concentrating on mental tasks, weight gain, intolerance of cold temperatures, constipation, and dry skin.

Throughout the world, millions of people are at risk of iodine deficiency. Pregnant females who are iodine deficient have high risk of stillbirths (giving birth to a dead infant) or low–birth–weight babies. During fetal life, thyroid hormone is crucial for normal brain development. Thus, infants of iodine–deficient females are likely to be born with a condition called **cretinism** (*kre'–tin–ih–zim*). Babies with cretinism have permanent brain damage, reduced intellectual functioning, and growth retardation. Worldwide, iodine deficiency is the most common cause of preventable intellectual disability.[38] Pregnant females can reduce the risk of giving birth to infants with cretinism by consuming adequate amounts of iodine throughout pregnancy.

Iodine deficiency is a serious threat to health in places where soils are iodine deficient and commonly eaten foods are not fortified with the trace mineral, such as regions of Latin America, India, Southeast Asia, and Africa. Currently, international health organizations are engaging in efforts to eliminate iodine deficiency, primarily by promoting the use of iodized salt or iodine–fortified vegetable oils.

Iodine Toxicity The UL for iodine is 1.1 mg/day. Over time, consuming very high amounts of iodine can cause thyroid gland enlargement and reduced production of thyroid hormone. These side effects are the same as those that occur when diets are deficient in iodine. Excess iodine is also associated with an increased risk of a form of thyroid cancer.[38]

Nuclear power plant accidents, such as the one that occurred after a major earthquake struck Japan in 2011, may release radioactive iodine (iodine–131) into the environment. When people are exposed to radioactive iodine, the element can enter their bodies and be picked up by their thyroid glands. Exposure to a high amount of the radiation increases the risk of thyroid cancer, especially in children.[38] People who have iodine deficiency are more likely to accumulate the radioactive iodine in their thyroid glands than people who are not iodine deficient. To reduce the risk of thyroid cancer, people who are exposed to radioactive iodine can take supplements that contain 16 to 130 mg of potassium iodide, depending on their age, until the risk of extreme radiation exposure ends.[38] The thyroid gland picks up the potassium iodide, which helps block the thyroid gland's uptake of the radioactive form of the element.

goitrogens compounds that inhibit iodide metabolism by the thyroid gland

cretinism condition affecting infants of females who were iodine deficient during pregnancy

Fluoride (F)

Although fluoride is not an essential nutrient, the mineral strengthens bones and teeth when ingested in small amounts. In some parts of the United States, well water contains high amounts of the trace mineral. To ensure that people living in other sections of the country obtain healthful amounts of fluoride, the mineral is often added to public water supplies, toothpastes, and dental rinses. Tea naturally contains fluoride.

Long–term consumption of too much fluoride can cause *fluorosis.* In cases of skeletal fluorosis, excess fluoride builds up in bones, causing joint stiffness and bone pain. Skeletal fluorosis rarely occurs in the United States. Dental fluorosis often affects young children who drink well water that contains high amounts of fluoride, but it also can occur in children who routinely swallow toothpaste and dental rinses. Children with severe fluorosis develop permanently stained teeth that do not resist decay as do healthy teeth (Fig. 9.24). Parents should teach their children to place a pea–sized amount of toothpaste on their brush, rinse with water, and spit out the excess fluid.

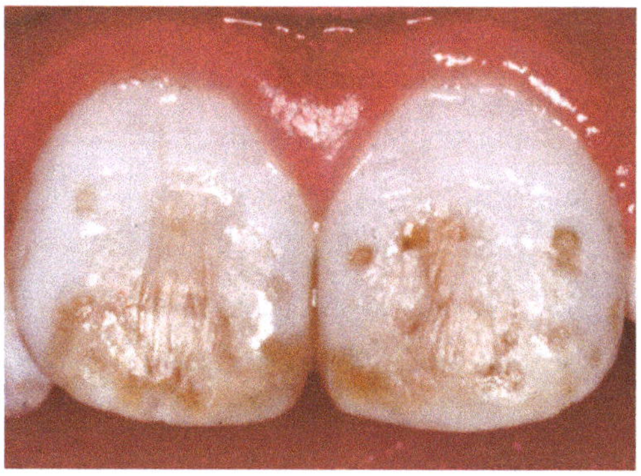

Figure 9.24 **Dental fluorosis.**
Centers for Disease Control and Prevention

Selenium (Se)

Selenium is widespread in Earth's crust, but soils can vary widely in their content of the essential trace mineral. Areas of the western United States, including parts of Colorado and South Dakota, have unusually high concentrations of selenium in soil. Certain types of plants that grow in these places accumulate toxic levels of the mineral. Livestock that graze on the selenium–rich plants often ingest poisonous amounts of the trace mineral. In horses and cattle, selenium toxicity can cause hair and weight loss, malformed hooves that can separate from the animals' feet, muscle weakness and loss of muscular function (paralysis), and death.

Why Is Selenium Necessary?

In the body, selenium functions as a component of several proteins referred to as selenoproteins (*sell'–in–oh–pro'–teens*). Many selenoproteins are anti–oxidants. Other selenoproteins are necessary for the normal functioning of the immune system and thyroid gland.

Sources of Selenium

Although the selenium content of foods varies, nuts, whole–grain products, seafood, and meats are generally rich sources of the trace mineral (Table 9.21). Because Brazil nuts can have very high selenium contents, people should not eat these nuts regularly.

Dietary Adequacy

Most Americans' diets meet the RDA for selenium. In the United States, selenium deficiency is uncommon, but the condition may occur in people who have serious digestive tract conditions that interfere with the mineral's absorption. Selenium deficiency reduces thyroid gland activity and can lead to goiter. The deficiency also depresses immune system function and may contribute to the development of heart disease and cancer. In parts of China where the soil lacks selenium and the population consumes locally produced foods, diets typically contain inadequate amounts of selenium. Certain types of cancer and a form of heart disease are common in these areas of China. There is not enough scientific evidence, however, to support the use of selenium supplements to prevent or treat cancer or CVD.[40]

Brazil nuts have high selenium contents.
©Wendy Schiff

Selenium Toxicity The UL for selenium is 400 mcg/day. In the United States, selenium toxicity (selenosis) (*sell'–in–o'–sis*) is rare. Chronic selenosis, however,

TABLE 9.21 Selenium Content of Selected Foods

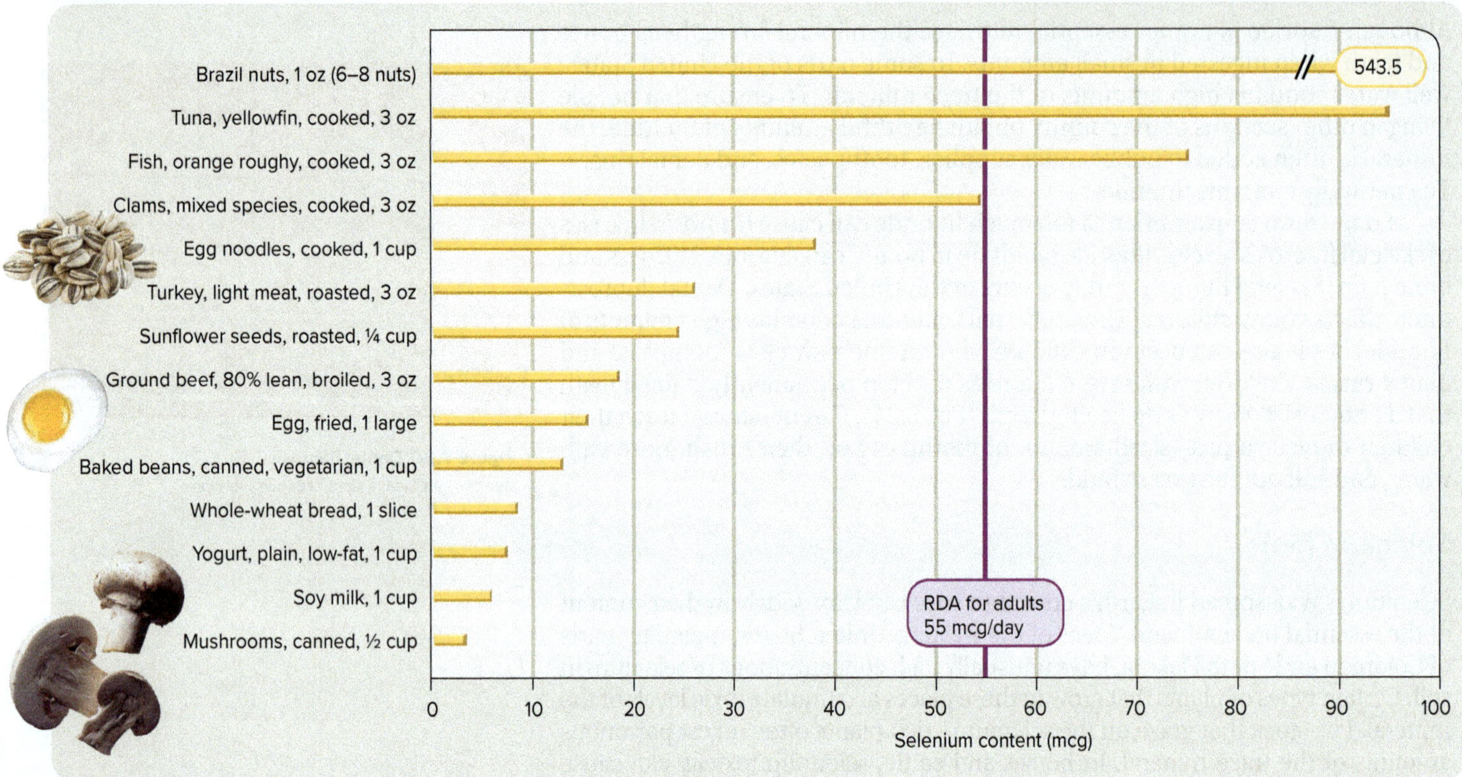

Source: Data from U.S. Department of Agriculture, Agricultural Research Service: *FoodData Central,* 2019.
Sunflower seeds: Jacques Cornell/McGraw Hill; Fried egg: Burke/Triolo Productions/Getty Images; Mushrooms: C Squared Studios/Getty Images

can occur from drinking well water that naturally contains too much selenium. Selenium toxicity can develop by taking megadoses of dietary supplements. In humans, signs and symptoms of chronic selenosis include brittle fingernails, loss of hair and nails, garlicky body odor, nausea, vomiting, and fatigue.

Chromium (Cr)

The importance of chromium as an essential trace mineral in human diets has been recognized only for about the past 40 years. The results of scientific studies suggest that chromium plays an important role in maintaining proper carbo–hydrate and lipid metabolism. The concentration of chromium in the body is generally low because the digestive tract absorbs only about 0.4 to 2.5% of the chromium in foods; the remainder is excreted in the feces.[41] Vitamin C and the B–vitamin niacin enhance the intestinal tract's ability to absorb chromium.

Why Is Chromium Necessary?

Most cells require the hormone insulin to receive glucose from the bloodstream. Chromium may enhance insulin's action on cell membranes and, in a way, help to "hold the door open" for glucose's entry into the cells. Can people who have diabetes experience better blood glucose regulation by taking chromium supplements? A review of scientific research indicated mixed results when people took chromium supplements to improve their blood glucose levels. In some of the studies, chromium supplements were helpful for people with diabetes, but the results of other studies indicated no such benefits.

Sources of Chromium

Although chromium is widely distributed in foods, most foods contain less than 2 mcg of the mineral per serving.[41] Information regarding the chromium content of various foods is difficult to find because most reliable food composition tables do not include this trace mineral. In general, meat, whole-grain products, yeast, fruits, and vegetables are good sources of chromium. Like selenium, the amount of chromium in plant foods reflects the chromium content of soils where crops are grown.

Whole grains, fruits, and vegetables are good sources of chromium.
Ken Karp/McGraw Hill

Dietary Adequacy

The adult AIs for chromium are 25 mcg/day for young females and 35 mcg/day for young males. Well-balanced diets typically contain these amounts of chromium. On average, American adults consume diets that meet or exceed their AIs for chromium.[41]

Chromium Deficiency Signs of chromium deficiency are impaired glucose tolerance and elevated blood cholesterol and triglyceride levels. The mechanism by which chromium influences cholesterol metabolism is not known but may involve enzymes that control the body's cholesterol production. Cases of chromium deficiency have been reported in people maintained on special-formula diets that did not contain chromium, as well as in severely malnourished children.

Chromium Toxicity The form of chromium that is naturally in foods has not been shown to produce toxicity, so no UL has been set for the trace mineral. The long-term safety of taking various chromium supplements is unknown.[41] Therefore, taking supplemental chromium may be risky.

Manganese (Mn)

Manganese is an important component and activator of many enzymes in the body. These manganese-dependent enzymes play a role in wound healing, metabolism, bone and cartilage formation, and antioxidant function.[42] Some experts suspect that the trace mineral may play a role in preventing osteoporosis, diabetes, and epilepsy. Food sources of manganese include leafy vegetables, whole grains, nuts, and tea. Small amounts of manganese are in natural sources of water.

Manganese deficiency rarely occurs in humans. Results of animal studies indicate that manganese deficiency interferes with reproduction, impairs growth, alters carbohydrate and lipid metabolism, and causes skeletal abnormalities.[42] Although a small amount of manganese is essential, too much is highly toxic. Overconsumption of manganese, which can occur from drinking water that is contaminated with the mineral, causes a nervous system disorder that is similar to Parkinson's disease. Cases of manganese toxicity from food consumption have not been reported.[42]

Molybdenum (Mo)

Molybdenum is a cofactor that is needed for the functioning of four enzymes:

- Aldehyde oxidase
- Mitochondrial amidoxime
- Sulfite oxidase
- Xanthine oxidase

Legumes, grains, and nuts are excellent sources of molybdenum. There have been no reports of molybdenum deficiency in healthy persons, and excesses of the trace mineral generally have low toxicity.[43]

Concept CHECKPOINT

24. What are the signs and symptoms of iron deficiency anemia? Which members of the population are most at risk of iron deficiency?

25. Identify at least three signs or symptoms of hemochromatosis. How is the condition treated?

26. What is a goiter? What is cretinism? How can cretinism be prevented?

27. Prepare a table for trace minerals that includes information about each trace mineral's major role or roles in the body, food sources, and signs and symptoms of the mineral's deficiency as well as toxicity disorders. Check your table against the information provided in Table 9.15.

See Appendix F for responses.

9.6 Minerals with Possible Physiological Roles

Learning Outcome

1 Discuss the six minerals, including arsenic, that may have roles in the body.

A few minerals, including nickel, arsenic, and silicon, are found in small amounts in the body, but their roles in the body are unclear (Table 9.22). At present, this group of minerals is not classified as essential nutrients. Although there are reports of severe illness and deaths resulting from environmental exposure to high amounts of these minerals, foods generally do not contain toxic amounts of them. Rice, however, can contain high levels of arsenic.

Did *YOU* Know?

Arsenic is naturally in water, soil, and air. Plants, especially rice, take up arsenic from the environment and store the metal. Inorganic arsenic is quite toxic, and rice is a leading source of inorganic arsenic in diets.[44] The FDA tests foods, including rice, for arsenic. In 2020, the agency issued a guidance document that sets a limit of 100 parts per billion for the amount of inorganic arsenic that is allowed in infant cereals that contain rice.

TABLE 9.22 Summary of Six Possible Essential Minerals

Mineral	Possible Functions	Suggested Daily Human Intakes	Dietary Sources
Arsenic (As)	Methionine metabolism, growth, and reproduction (animal studies)	No UL has been established; however, inorganic arsenic is highly toxic. Long-term exposure to arsenic may contribute to heart disease and certain cancers.	Fish, rice, fruits, vegetables
Boron (B)	Steroid hormone metabolism	1–13 mg (UL: 20 mg/day)	Fruit, leafy vegetables, peanuts, beans, wine
Lithium (Li)	Reproduction; maintenance of appropriate mood	1000 mcg (No UL has been established)	Water supply, grains, vegetables
Nickel (Ni)	Amino acid and fatty acid metabolism	25–35 mcg (UL: 1 mg/day)	Chocolate, nuts, beans, whole grains
Silicon (Si)	Connective tissue, including bone formation	25–30 mg (Insufficient data to determine UL)	Root vegetables, whole grains
Vanadium (V)	Glucose metabolism; tooth and bone mineralization	10 mcg (UL: 1.8 mg/day)	Shellfish, mushrooms, black pepper, parsley, dill

Sources: Data from Food and Nutrition Board, Institute of Medicine: *Dietary Reference Intakes for Vitamin A, Vitamin K, Arsenic, Boron, Chromium, Copper, Iodine, Iron, Manganese, Molybdenum, Nickel, Silicon, Vanadium, and Zinc* and *Standing Committee on the Scientific Evaluation of Dietary Reference Intakes*, National Academy Press, 2006; Szklarska D: Is Lithium a Micronutrient? From Biological Activity and Epidemiological Observation to Food Fortification. *Biological Trace Element Research* 189(1): 18, 2019.

Concept CHECKPOINT

28. Identify at least four minerals that are classified as possible essential minerals.

29. Prepare a table for arsenic, boron, lithium, nickel, silicon, and vanadium that includes information about each mineral's possible function and major food sources. Check your table against the information provided in Table 9.22.

See Appendix F for responses.

9.7 Nutrition Matters: Bottled Water versus Tap Water

Learning Outcomes

1. Explain the roles of the FDA and EPA in regulating water safety in the United States.
2. Discuss the pros and cons of drinking bottled water.
3. Discuss the significance of the Safe Drinking Water Act.
4. Explain health risks associated with BPA.

Today, preparing for a class often involves bringing into the classroom a notebook, something to write with, and a bottle of cold water. Even many college instructors keep a bottle of water nearby as they lecture. Not long ago, most Americans got their water from a faucet ("tap water"). Now, millions of Americans are turning away from the tap and choosing to drink bottled water instead. If you prefer drinking bottled water over tap water, you are part of a growing trend. According to the International Bottled Water Association (IBWA), each American consumed, on average, about 45 gallons of bottled water in 2020, which is an increase of 17 gallons from 2010.[45]

Why do so many Americans drink bottled water when tap water is much less costly? Among adults, taste, convenience, and health concerns were major reasons they chose bottled water over other beverages. For most Americans, however, bottled water is usually unnecessary and expensive, as it is often very similar to tap water. Consumers need to be aware that the water in some bottled water products actually comes from a municipal water supply. However, when public water supplies are disrupted by hurricanes, tornadoes, or earthquakes, drinking bottled water may be a consumer's only option. What is bottled water? Is it safe to drink?

According to FDA guidelines, bottled water manufacturers are responsible for producing safe products.
Mikhail Klyoshev/Shutterstock

The Environmental Protection Agency (EPA) regulates the sanitation of public water supplies in the United States. Another federal agency, the FDA, regulates bottled water products that are marketed for interstate commerce. Neither agency "certifies" bottled water.[46]

The FDA requires bottled water producers to:

- process, bottle, hold, and transport bottled water under sanitary conditions;
- protect water sources from bacteria, chemicals, and other contaminants;
- use quality control processes to ensure the bacteriological and chemical safety of the water; and
- sample and test both the source of water and the final product for contaminants.[47]

Bottled water is sealed in containers and has no added ingredients other than a substance that prevents the growth of microbes, such as bacteria. Bottled water may have fluoride added, but amounts must meet FDA guidelines. If bottled water manufacturers add flavorings or other ingredients to their products, the name of the product must indicate the added ingredients; "Bottled Water with Cherry Flavor," for example. These drinks are often called "flavored water beverages." Flavored waters may simply contain additives that make the beverage taste good, but a growing number of flavored waters also have added nutrients other than sugars, such as sodium, potassium, and amino acids. The beverage's label must identify the additives in the list of ingredients.

Safety standards for bottled water are similar to those established by the EPA for tap water. According to FDA guidelines, bottled water manufacturers are responsible for producing safe products. Production procedures for bottled water must follow manufacturing regulations established and enforced by the FDA. Additionally, the FDA inspects water bottling facilities regularly.

Because of FDA regulations and oversight, consumers can be assured that their supply of bottled water is safe. Additionally, most Americans can trust the safety of their tap water because the vast majority of municipal water systems in the United States are regulated by the *Safe Drinking Water Act (SDWA)*. As a result of this law, most tap water undergoes a thorough purification process and is constantly tested for safety. If such testing indicates that the water supply may pose a threat to public health, consumers are warned through media, and a "boil order"—requirement to boil water for 10 minutes to kill harmful microorganisms—may be issued.

Although bottled water is safe to drink, the plastic used to contain it may have toxic effects on health. *Bisphenol (biss′–feen–ol) A,* which is also called *BPA,* is a chemical used to make polycarbonate plastics and epoxy resins. Polycarbonate plastics are in many consumer products that come in contact with foods and beverages, including some water bottles and older baby bottles. The epoxy resin that contains BPA is used to coat the inside of certain food cans. The coating prevents the can's metal from coming in contact with and being damaged by the food (Fig. 9.25). However, BPA can leach from polycarbonate plastic containers or epoxy resin–coated cans and enter the food or beverage stored within them.

Scientists are studying BPA to determine the extent to which the chemical can affect human health. Concern over the safety of BPA has encouraged some manufacturers to discontinue using the chemical in their products. Plastic containers that are marked with recycle codes 1, 2, 4, 5, and 6 most likely do not contain BPA. Some plastics that are marked with recycle codes 3 or 7 may be made with BPA (Fig. 9.26).[48] To reduce exposure to BPA, consider taking these actions:

Figure 9.25 BPA and epoxy resins. The epoxy resin that is used to coat the inside of cans contains BPA.
©Wendy Schiff

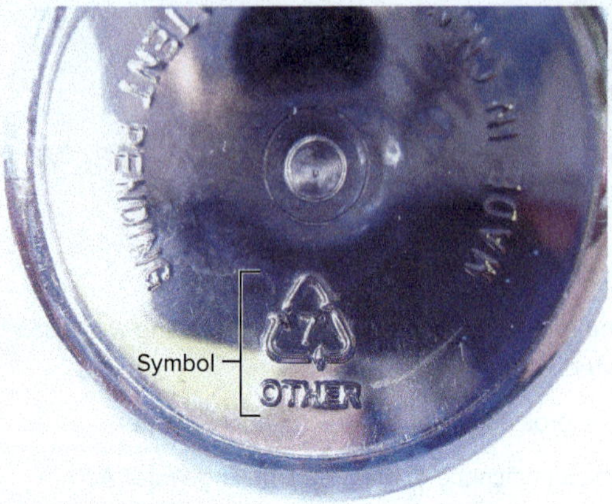

Figure 9.26 Where's the BPA? This symbol indicates that the plastic used to make this water bottle may contain BPA.
©Wendy Schiff

- Avoid heating foods in polycarbonate plastic containers.
- Avoid plastic containers that have the symbol shown in Figure 9.26.
- Do not wash polycarbonate containers in the dishwasher or with harsh detergents.
- Reduce intake of foods in cans lined with epoxy resins.
- Cook or store foods and beverages in glass, porcelain, or stainless steel containers.
- Avoid using polycarbonate dishware, cups, or eating utensils for serving foods (especially hot foods or liquids).
- Use baby bottles and drinking cups that are BPA free; avoid using old plastic baby bottles and toys that may contain BPA.

Concept CHECKPOINT

30. What is the significance of the Safe Drinking Water Act?
31. List three ways to reduce your exposure to BPA.

See Appendix F for responses.

Chapter References

See Appendix G.

Summary

9.1 Introducing Water and Minerals

- Dehydration can be deadly. A mineral nutrient can be classified as a major, trace, or possible essential mineral.

9.2 Water

- Water is a simple compound that does not undergo digestion. In the body, water is a major solvent that often participates directly in chemical reactions. Water's other physiological roles include transporting substances, removing waste products, lubricating tissues, and regulating body temperature and acid-base balance. Water does not provide energy for the body.
- The body maintains a balance between intracellular and extracellular fluids primarily by controlling concentrations of ions in each fluid compartment. If the normal concentrations of these ions change too much, water shifts out of a compartment, and cells shrink or swell as a result.
- Total water intake includes water from beverages and foods. Most of the water that enters the digestive tract is absorbed. Metabolic water is another source of water for the body.
- The body loses water in urine, perspiration, exhaled air, feces, and insensible perspiration. A healthy person's average daily total water intake equals their average output. Environmental factors, physiological conditions, and lifestyle practices can alter the body's fluid balance. The kidneys are the major regulator of the body's water content and mineral ion concentrations.
- The AI for total water intake is 2.7 L for young females and 3.7 L for young males. These amounts do not need to be consumed in the form of fluids because most solid foods and metabolic water contribute some water to the body. Thirst is the primary regulator of fluid intake.

ginasanders/123RF

9.3 Minerals: Basic Concepts

- Some minerals function as inorganic ions, cofactors, or structural components of tissues; other minerals are components of various enzymes, hormones, or other organic molecules. Cells cannot metabolize minerals for energy. The body, however, needs certain minerals to catalyze specific chemical reactions that release energy from macronutrients. Excessive amounts of minerals in the body can disrupt normal cell functioning, causing toxicity. Most foods contain small amounts of minerals. When the body's needs for minerals increase, the bioavailability of minerals generally increases to meet the demand.

9.4 Major Minerals

- Calcium is a major structural component of bones and teeth, and the mineral is necessary for blood clotting, muscle contraction, nerve transmission, and cell metabolism. Calcium absorption depends on vitamin D. Milk and milk products are rich calcium sources. Although people, especially females, are at risk of developing osteoporosis as they age, various lifestyle modifications may help reduce this risk.

Groceries: Burke/Triolo Productions/Brand X Pictures/Getty Images; Apple: lynx/Iconotec/Glow Images

- Sodium, the major positively charged ion found outside cells, is vital for maintaining fluid balance and transmitting nerve impulses. The typical American diet provides high amounts of sodium, primarily from processed and canned foods. Diets high in sodium are associated with increased risk of hypertension.
- Hypertensive individuals have greater risk of CVD, kidney failure, and damage to other organs than people with normal blood pressures. Advanced age, African-American ancestry, obesity, physical inactivity, cigarette smoking, and excess alcohol and sodium intakes are major risk factors for hypertension. Treatment for hypertension usually includes following dietary modifications and making some other lifestyle changes.
- Potassium, the major positively charged ion found inside cells, has functions that are similar to those of sodium. Potassium–rich diets may lower blood pressure. Plant foods, meat, and milk are good sources of potassium.
- Magnesium is a cofactor for numerous chemical reactions and is needed for nerve and heart function. Plant foods are good sources of magnesium.

9.5 Trace Minerals

- Iron is a critical component of hemoglobin, myoglobin, and cytochromes. Hemoglobin in red blood cells transports oxygen from the lungs to the tissues. Iron deficiency can result in iron deficiency anemia. Iron absorption depends on the body's need for the mineral and the form of iron in food. Heme iron is better absorbed than nonheme iron.
- Females of childbearing age have higher needs for iron than males because of menstrual blood loss. Throughout the world, iron deficiency–related disorders are common. People who have hereditary hemochromatosis develop iron toxicity because they absorb too much iron.
- Zinc functions as a cofactor that activates many enzymes. Zinc is involved in growth and development, antioxidant activity, immune function, and taste. Zinc deficiency can result in growth failure and decreased immune function. Iodine is needed to make thyroid hormone. When the diet lacks iodine, the thyroid gland enlarges, forming a goiter. Cretinism can occur in infants born to females who were iodine deficient during pregnancy. Iodine deficiency is rare in the United States because table salt is often fortified with iodine.
- Selenium functions as a component of selenoproteins, many of which are antioxidants. Other selenoproteins are involved in the normal functioning of the immune system and thyroid gland. Meat, fish, nuts, whole grains, and seeds are good sources of selenium. Chromium enhances the action of insulin. Chromium is found in meats and whole grains. More research is needed to determine the effects of taking chromium supplements.

9.6 Minerals with Possible Physiological Roles

- A few minerals, including nickel and silicon, are found in very small amounts in the body, but their roles in the body are unclear. At present, these particular minerals are not classified as essential nutrients. Inorganic arsenic, however, is very toxic.

9.7 Nutrition Matters: Bottled Water versus Tap Water

- In the United States, bottled water is usually unnecessary and expensive, and it is often derived from a municipal water supply. The EPA regulates the sanitation of public water supplies in the United States. The FDA regulates bottled water products that are marketed for interstate commerce. The Safe Drinking Water Act (SDWA) helps ensure the safety of tap water. Although more research is needed to clarify the role of BPA in health, it is wise to avoid products that contain the chemical.

©Wendy Schiff

Recipe for Healthy Living

Egg Salad

Egg salad is an easy-to-make meal or snack. This recipe makes about three ⅓-cup servings. Each serving supplies approximately 125 kcal, 8.5 g protein, 8.6 g fat, 250 mg cholesterol, 1.2 mg iron, and 191 mg sodium.

Jonell Weaver/Photodisc/Getty Images

INGREDIENTS:

- 2 Tbsp peeled, finely chopped yellow onion
- 4 hard-cooked large eggs*
- 2 Tbsp reduced-fat mayonnaise
- 1 tsp pickle relish
- dash of black pepper

* To hard-cook eggs:

1. Place the eggs in a small saucepan and cover the eggs completely with about 2 inches of water.
2. Heat water and eggs to a boil, then cover the saucepan with its lid, and remove it from the heat.
3. Allow the eggs to remain in the hot water for approximately 20 to 25 minutes (large eggs).
4. Remove eggs from the saucepan and cool by immersing in cold water.

PREPARATION STEPS:

1. Place chopped onion in small mixing bowl.
2. Peel hard-cooked eggs and discard peels. Chop eggs on a cutting board.
3. Add chopped eggs to onions.
4. Add reduced-fat mayonnaise and pickle relish to the egg and onion mixture. Blend together. Mixture should be moist.
5. Serve as a spread on whole-wheat crackers or on rye bread.
6. Egg salad is perishable, so cover any leftover salad and store in the refrigerator for no longer than a day.

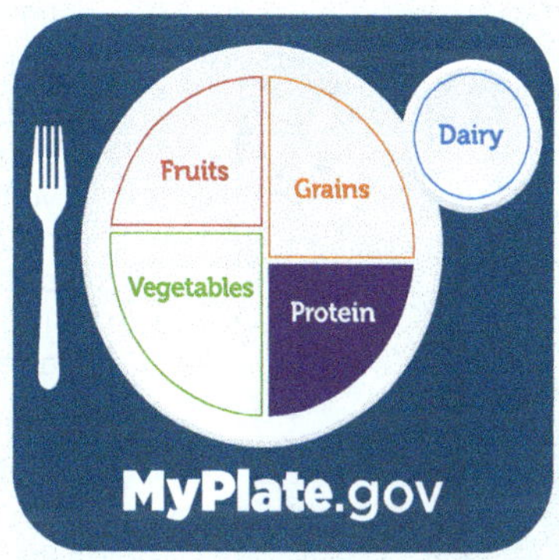

US Department of Agriculture or USDA

Approximate Percent of Calories from Macronutrients

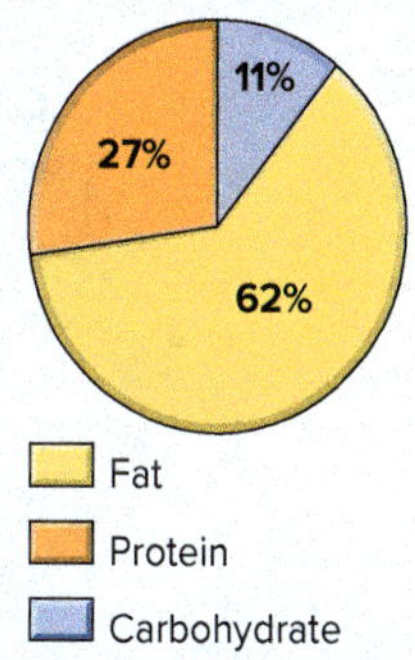

Percent of Daily Value/Serving

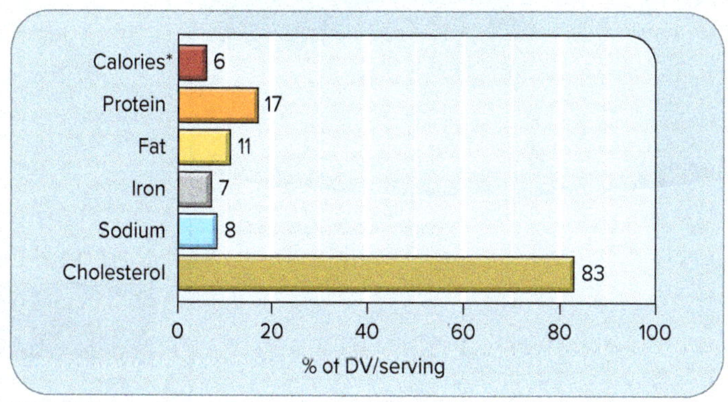

*2000 daily total kcal

Ingram Publishing/SuperStock

Personal Dietary Analysis

Using the DRIs

1. Refer to your 3-day food log from the "Personal Dietary Analysis" feature in Chapter 3.

   a. Find the RDA/AI values for minerals under your life stage/sex group category in the DRI tables (see Appendix H). Write those values under the "My RDA/AI" column in the table that accompanies this analysis activity.

   b. Review your personal dietary assessment. Find your 3-day average intakes of iron, calcium, zinc, sodium, potassium, and magnesium. Write those values under the "My Average Intake" column of the table.

   c. Calculate the percentage of the RDA/AI you consumed for each mineral by dividing your intake by the RDA/AI amount and multiplying the figure you obtain by 100. For example, if your average intake of iron was 9 mg/day, and your RDA for the mineral is 18 mg/day, you would divide 9 mg by 18 mg to obtain 0.50. To multiply this figure by 100, simply move the decimal point two places to the right, and replace the decimal point with a percentage sign (50%). Thus, your average daily intake of iron was 50% of the RDA. Place the percentages for each mineral under the "% of My RDA/AI" column.

   d. Under the ">, <, or =" column, indicate whether your average daily intake was greater than (>), less than (<), or equal to (=) the RDA/AI.

2. Use the information you calculated in the first part of this activity to answer the following questions:

   a. Which of your average mineral intakes equaled or exceeded the RDA/AI?

   b. Which of your average mineral intakes was below the RDA/AI?

   c. What foods would you eat to increase your intake of the minerals that were less than the RDA/AI levels? (Review sources of the minerals in this chapter.)

   d. Turn in your completed table and answers to your instructor.

Personal Dietary Analysis: Minerals

Mineral	My RDA/AI	My Average Intake	% of My RDA/AI	>, <, or =
Iron				
Calcium				
Zinc				
Sodium				
Potassium				
Magnesium				

Complete the Personal Dietary Analysis activity in Connect.

Critical Thinking

1. Before the advent of refrigeration, salting meat was a common way of preventing microbes from spoiling the food. Explain why salting was effective as a means of food preservation.
2. A friend of yours refuses to drink tap water because she thinks it is contaminated. She drinks only bottled water or well water. If she asked you to explain why you drink tap water, what would you tell her?
3. A group of food manufacturers is considering fortifying some of their products with iron, chromium, boron, and iodine. Explain why you think they should or should not fortify the foods with each of these minerals.
4. Ben is a total vegetarian (vegan). What advice would you give to him concerning his need for calcium, iron, potassium, magnesium, and zinc?

C Squared Studios/Photodisc/Getty Images

5. Consider your family history and lifestyle to determine whether you are at risk of osteoporosis. If you are at risk, what steps can you take at this point in your life to reduce your chances of developing this disease?
6. Consider your family history and lifestyle to determine whether you are at risk of hypertension. If you are at risk, what steps can you take at this point in your life to reduce your chances of developing this disease?
7. On social media, a person claiming to be a doctor recommends taking megadoses of zinc, iron, and selenium supplements to enhance muscular strength and endurance. Discuss why you would or would not follow this person's advice.

Practice Test (Section numbers are listed below in red.)

Select the best answer.

1. (9.1) Which of the following statements is false?
   a. Lean tissue contains more water than fat tissue.
   b. Water is a major solvent.
   c. Generally, young females have more body water than young males.
   d. Water does not provide energy.
2. (9.2) If the extracellular fluid has an excess of sodium ions,
   a. sodium ions move into cells.
   b. intracellular water moves to the outside of cells.
   c. phosphate and calcium ions are eliminated in feces.
   d. blood levels of arsenic and oxalate increase.
3. (9.2) Which of the following foods has the lowest percentage of water?
   a. Tomatoes
   b. Oranges
   c. Whole-grain bread
   d. Vegetable oil
4. (9.3) In the United States, refined grains are often enriched with
   a. iron
   b. selenium
   c. potassium
   d. iodine
5. (9.4) Which of the following foods is not a good source of calcium?
   a. Butter
   b. American cheese
   c. Canned sardines
   d. Kale
6. (9.4) Henry is concerned about his risk of osteoporosis. Which of the following characteristics is a modifiable risk factor for this chronic condition?
   a. Family history
   b. Racial/ethnic background
   c. Physical activity level
   d. Age

7. **(9.4)** The primary source of sodium in the typical American diet is
   a. bottled water.
   b. unprocessed food.
   c. fruit.
   d. salt.
8. **(9.4)** Which of the following populations has the highest risk of hypertension?
   a. People with African–American ancestry
   b. Young, physically active Asian males
   c. Hispanic females who do not drink alcohol
   d. Young adults who consume high amounts of fruit
9. **(9.4)** This morning, Jonas's blood pressure was 128/82 mm Hg, which is in the _____ range.
   a. normal
   b. elevated
   c. Stage 1 hypertension
   d. Stage 2 hypertension
10. **(9.5)** Sources of heme iron include
    a. fortified grain products.
    b. beef.
    c. spinach.
    d. cast–iron cookware.
11. **(9.5)** Worldwide, the most common nutrient deficiency disorder is _____ deficiency.
    a. iodine
    b. cobalt
    c. iron
    d. calcium
12. **(9.5)** Which of the following statements is false?
    a. Iodine is necessary for normal thyroid function.
    b. In the United States, milk is usually fortified with iodine.
    c. Having too much or too little iodine in the diet can cause the thyroid gland to enlarge.
    d. Saltwater fish and other seafood are sources of iodine.
13. **(9.6)** Which of the following statements is true?
    a. Safety standards for bottled water are similar to those for tap water.
    b. Most brands of water bottled in the United States contain unsafe amounts of arsenic and lead.
    c. The EPA inspects and certifies water bottling facilities at least three times a year.
    d. The majority of Americans should drink bottled water because water from municipal systems is unsafe.

Thaiview/Shutterstock

Answers to **Quiz Yourself**

1. Registered dietitian nutritionists recommend 8 glasses of water a day for optimal health. **False.** (Section 9.2)
2. Ounce per ounce, cottage cheese contains more calcium than plain yogurt. **False.** (Section 9.4)
3. Potassium, sodium, and chloride ions are involved in fluid balance. **True.** (Section 9.2)
4. Selenium is an essential mineral. **True.** (Section 9.5)
5. In general, plants are good sources of iron because the plant pigment chlorophyll contains iron. **False.** (Section 9.4)

Design Elements: (Stacked books): Siede Preis/Photodisc/Getty Images; (Nuts): C Squared Studios/Photodisc/Getty Images; (Recipe box, Raspberries, Blueberries): ©Wendy Schiff; (Pineapple): Stockbyte/Getty Images; (Radishes): karandaev/iStock/360/Getty Images; (MyPlate icon): US Department of Agriculture or USDA

CHAPTER 10

Energy Balance and Weight Control

Comstock Images/Getty Images

Quiz Yourself

What is the difference between overweight and obesity? Why do most people gain body fat as they age? How can you tell whether a person is following a fad diet or a diet that is likely to be safe and effective? Are there any medications or dietary supplements that help people lose weight? Test your knowledge of energy balance and weight management concepts by taking the following quiz. The answers are at the end of the chapter.

1. You can determine whether you have an unhealthy amount of body fat simply by measuring your waistline. ______ T ______ F
2. The safest way to lose a lot of weight and keep it off is to follow a low-carbohydrate, high-fat diet, such as the ketogenic diet. ______ T ______ F
3. As people age, their muscle cells turn into fat cells. ______ T ______ F
4. When a person consumes more protein than needed, the excess can be converted to fat and stored in fat cells. ______ T ______ F
5. Cellulite is a unique type of fat that can be eliminated by taking certain dietary supplements. ______ T ______ F

McGraw Hill connect

A wealth of proven resources are available on Connect® including SmartBook®, NutritionCalc Plus, and many other dynamic learning tools. Ask your instructor about Connect!

10.1 The Obesity Epidemic

Learning Outcomes

1. Explain the difference between overweight and obesity.
2. Discuss how the prevalence of obesity in the United States changed between 1988–1994 and 2017–March 2020.

In the United States, overweight and obesity are widespread nutritional problems that have reached epidemic proportions (Fig. 10.1). According to the National Heart, Lung, and Blood Institute, **overweight** refers to having extra body weight that is contributed by bone, muscle, body fat, and/or body water.[1] A professional basketball player, for example, may be "overweight" because of their muscular body build—not because they have excess body fat. **Obesity** is a condition characterized by excessive and unhealthy amounts of body fat. Unless noted otherwise, we will use the term *overweight* to describe people who have too much body fat but are not obese.

overweight having extra weight from bone, muscle, body fat, and/or body water

obesity condition characterized by excessive and unhealthy amounts of body fat

It is important to recognize that some body fat is essential for good health. However, people who are overweight or obese ("overfat") have a greater risk for serious chronic health conditions and diseases than people who have healthy body weights.

Prevalence of Overweight and Obesity

To determine whether a person's weight is healthy, overweight, or obese, medical experts generally use the **body mass index (BMI).** BMI is a numerical value that reflects body weight adjusted for height (Table 10.1). People whose

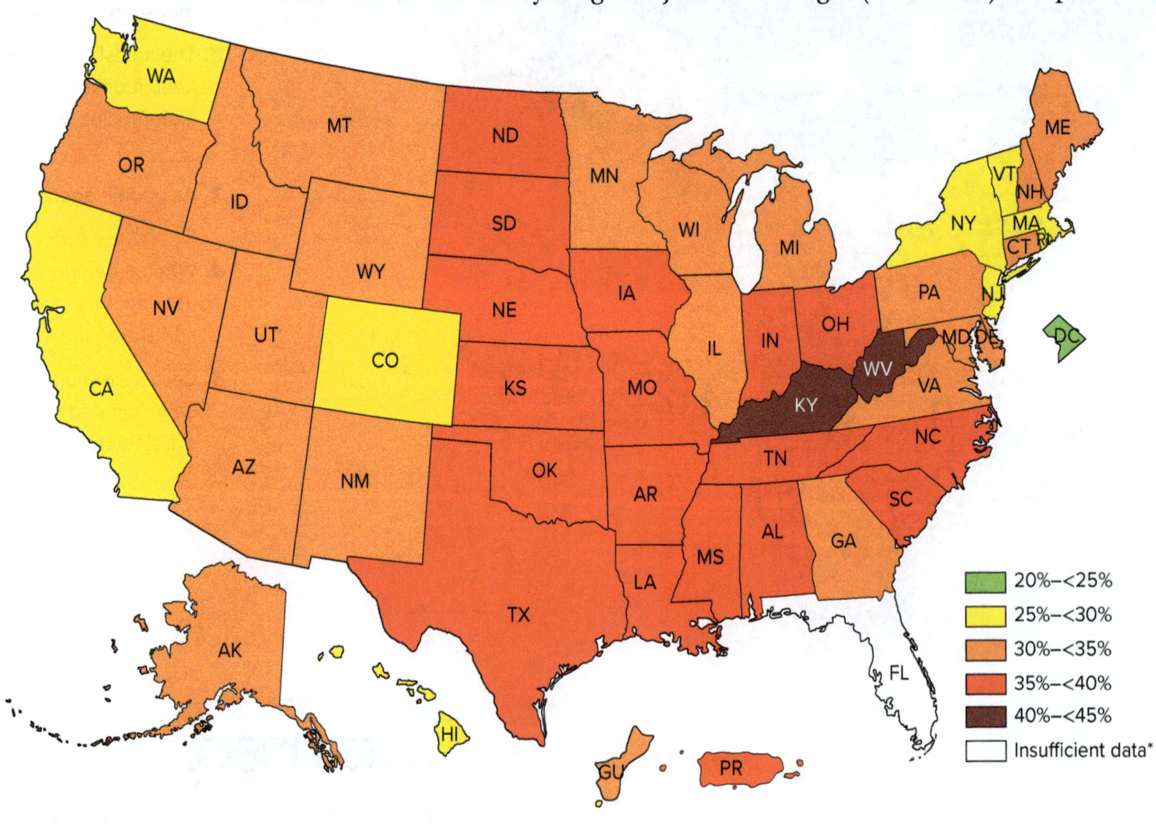

Figure 10.1 Obesity map of the United States (states and territories), 2021.
Source: Adapted from Centers for Disease Control and Prevention. Accessed: May 17, 2023. www.cdc.gov/obesity/data/prevalence-maps.html.

TABLE 10.1 Adult Weight Status Categories (BMI)

BMI	Weight Status
Below 18.5	Underweight
18.5–24.9	Healthy
25.0–29.9	Overweight
30.0–39.9	Obese
40 and above	Extremely obese

Source: National Heart, Lung, and Blood Institute: "The practical guide: Identification, evaluation, and treatment of overweight and obesity in adults," *NIH Publication* 00-4084, 2000. www.nhlbi.nih.gov/files/docs/guidelines/prctgd_c.pdf. Accessed August 7, 2023.

TABLE 10.2 Prevalence of Overweight, Obesity, and Extreme Obesity Among Adults: United States

Weight Status	1988–1994	2015–2016	2017–March 2020
Overweight	33.1%	31.6%	30.7%
Obese	22.9%	39.6%	41.9%
Extremely obese	2.8%	7.7%	9.2%

Source of data: Center for Disease Control and Prevention, National Center for Health Statistics: *Obesity and Overweight. 2023.* www.cdc.gov/nchs/fastats/obesity-overweight.htm. Accessed August 7, 2023.

BMI is within the overweight or obese categories are at a higher risk of chronic health problems associated with excess body fat. According to data collected from a national survey, almost 42% of American adults over the age of 20 were obese in 2017–March 2020.[2] As shown in Table 10.2, the percent of adults who are obese increased by 82% between 1988–1994 and 2019–March 2020. People who are classified as being obese can be further categorized into a subgroup of individuals who are "extremely" or "severely" obese." In 2019–March 2020, over 9% of adult Americans were classified as being extremely obese. This percentage tripled since 1988–1994.

The prevalence of obesity among American children and adolescents has also risen sharply over the past 25 years.[3] In 2017–2020, approximately 19.3% of children aged 2 through 19 were obese.[4] Public health experts are very concerned about the high prevalence of obesity among children because these youngsters are more likely to mature into adults who have obesity than are children who are not obese. See Chapter 13 for a discussion of childhood obesity.

Overweight and obesity rates are also rising rapidly throughout the world ("globesity"). In 2016, the World Health Organization (WHO) estimated that worldwide, 1.9 *billion* adults were overweight and 650 million adults were obese.[5] Throughout the world, overweight and obesity are responsible for more deaths than underweight.[5]

According to national target goals established in *Healthy People 2030*, the percentage of adults who are obese should decline to 36% by 2030.[6] Furthermore, the percentage of children aged 2 through 19 years who are obese should decline to 15.5% by 2030.

Regardless of whether a person wants to maintain, lose, or gain weight, the basic principles of energy balance apply. By reading this chapter, you will learn about energy balance, body composition, health consequences linked to having too much or too little body fat, and factors that contribute to unwanted weight gain. This chapter also provides practical tips for helping you achieve or maintain a healthy body weight through sensible eating and physical activity practices. Concern about body size can result in unhealthy eating practices.

body mass index (BMI) numerical value that reflects body weight adjusted for height

MY DIVERSE PLATE

In the United States, the prevalence of adult obesity varies among populations that make up the country's major racial/ethnic groups. Compared to non-Hispanic white adults, obesity is more common among non-Hispanic Black adults (49.9%) and Hispanic-American adults (45.6%).[2] The prevalence of obesity is 41.4% among non-Hispanic white adults. The prevalence of obesity is lowest among non-Hispanic Asian adults (16.1%), compared with all other racial and Hispanic-origin groups.

Public health experts think social determinants of health (SDOH) contribute to the differences in obesity in these racial/ethnic groups, such as lower high school graduation and employment rates, higher levels of food insecurity, lack of access to healthy foods, and decreased access to quality health care.

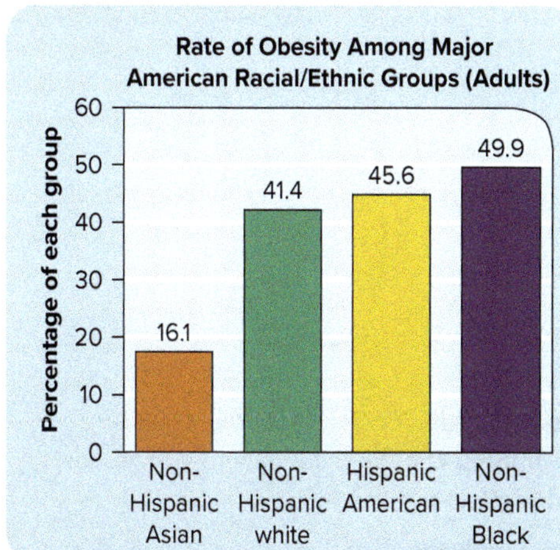

Concept CHECKPOINT

1. Calvin is overweight, but he has a healthy percentage of body fat. Explain why this situation can occur.
2. What percentage of adult Americans were obese in 2017–March 2020?

See Appendix F for responses.

10.2 Body Composition

Learning Outcomes

1. List the components of the two major body compartments.
2. Discuss the difference between subcutaneous fat and visceral fat.
3. Describe the various ways body fat is measured and the pros and cons of each method.
4. Explain why it is important to have healthy amounts of body fat.

The body is composed of two major compartments: **fat–free mass** (lean tissues) and **total body fat.** (Some scientists who study body composition divide the body into four compartments: body water, bone mineral, fat–free mass, and total body fat.) Fat–free mass is composed of body water, mineral–rich tissues such as bones and teeth, and protein–rich tissues, including muscles and organs. Total body fat includes "essential fat" and *adipose tissue.* Essential fat is in cell membranes, certain bones, and nervous tissue. Essential fat is vital for survival. Adipose tissue contains white adipose cells (fat cells) that are specialized for storing energy in the form of triglycerides (see Figure 6.13). People who are overweight and obese have excessive amounts of fat stored in their adipose tissue. Furthermore, people who are obese often store triglycerides in other tissues, particularly the liver and muscles. Such fat deposits are harmful because they interfere with normal cell function and contribute to inflammation.

Did *YOU* Know?

In addition to white fat cells, the body has deposits of beige fat and brown fat cells.[7] As a result of brown fat and beige fat cell activity, the body wastes energy as heat (thermogenesis). Thus, such cells can help regulate body temperature when a person is exposed to chilly conditions.

Adipose Tissue

When food is plentiful, adipose cells remove excess fat from the bloodstream for storage. As the amount of fat stored in adipose cells increases, the size of each cell expands, and the body gains weight. The body develops more fat cells when overeating occurs in adulthood, especially in cases of extreme obesity.[8] Once fat cells form, scientists think the cells remain until they die or are surgically removed.

When the body needs energy, adipose cells release fat for other cells to use as fuel. As each adipose cell loses some fat, it becomes smaller. If adipose cells continue to release fat, the body eventually loses weight as a result. In addition to storing and releasing fat, adipose cells secrete hormones and other substances, some of which have roles in regulating food intake, glucose metabolism, and immune responses.

Subcutaneous Fat and Visceral Fat

Subcutaneous (*sub* = under; *cutaneous* [*qu–tay′–nee–us*] = skin) tissue holds skin in place over underlying tissues such as muscles. Subcutaneous tissue also contains adipose cells. When subcutaneous tissue has more adipose cells than other kinds of cells, it is referred to as *subcutaneous fat.* Subcutaneous fat helps insulate the body against cold temperatures and protects muscles and bones from bumps and bruises. Subcutaneous fat is unevenly distributed. This layer of fat is thicker in certain regions of male and female bodies, especially in the abdominal area, thighs, and buttocks (Fig. 10.2).

fat-free mass body compartment that is composed of body water, mineral-rich tissues, and protein-rich tissues

total body fat essential fat and adipose tissue

In addition to subcutaneous fat, the body has *visceral* (*viss′–eh–rol*) fat. Visceral fat also contains adipose cells, but this type of body fat is in the *omentum,* a structure that is under the abdominal muscles and hangs over

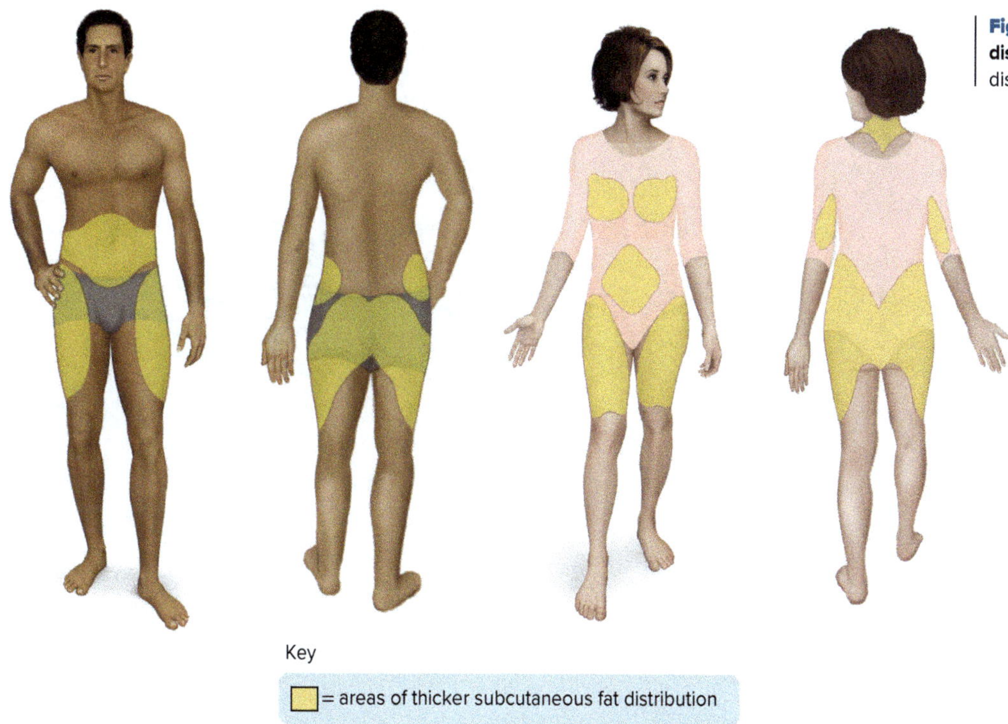

Figure 10.2 Uneven subcutaneous fat distribution. Subcutaneous fat is unevenly distributed.

the intestines (Fig. 10.3). Although there are some racial differences, females generally have more subcutaneous fat than males, whereas males tend to have more visceral fat than females.[9] Excessive amounts of visceral fat and/or subcutaneous abdominal fat result in what is commonly called a "beer belly" or the "middle–age spread."

Visceral fat makes substances that cause inflammation in the body. These inflammatory factors may increase risks of type 2 diabetes and cardiovascular disease (CVD). Section 10.5 discusses an easy way to determine whether you have too much visceral fat.

Did *YOU* Know?

Cellulite, lumpy-appearing skin on thighs and buttocks of most females, is not a unique type of fat or the result of a disease. Scientists have no clear understanding of why cellulite occurs, but it may simply be subcutaneous fat that is loosely held in place by irregular bands of connective tissue. At this point, there is no treatment that effectively smooths the skin's dimpled appearance.[10]

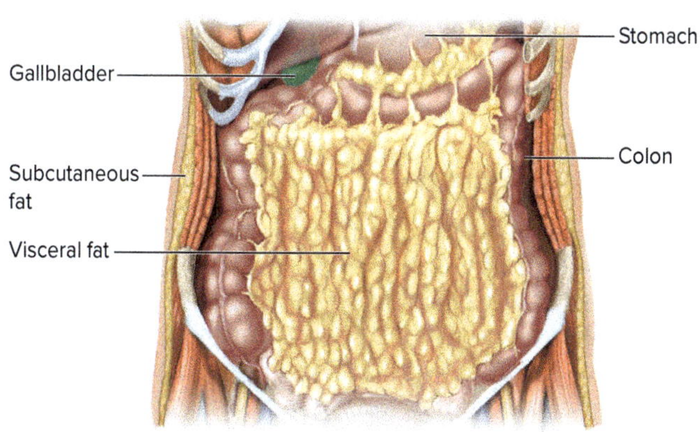

Figure 10.3 What is visceral fat? Visceral fat is in the omentum that is under abdominal muscles and hangs over the intestines.

Figure 10.4 Underwater weighing.
David Madison/Getty Images

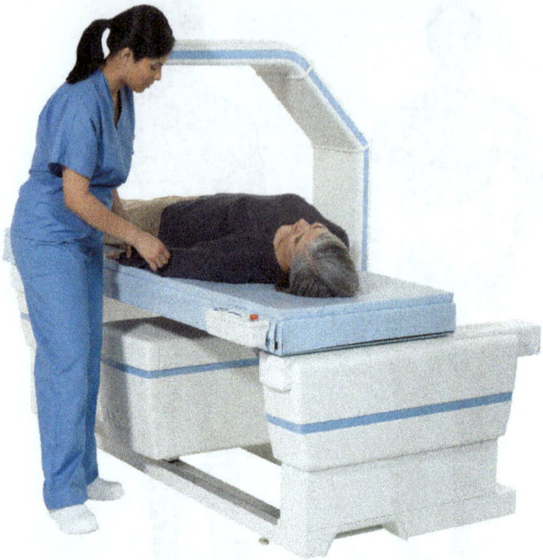

Figure 10.5 Dual-energy x-ray absorptiometry (DXA).
Photo Courtesy of Hologic, Inc.

Measuring Body Fat

When you weigh yourself on a scale, you cannot determine whether your weight is healthy. Why? It is healthier to have more fat-free tissue than fat tissue, but the scale does not distinguish between these two major components of your body. Information concerning your body's composition, especially its *percentage* of fat, can help you predict your risk of obesity-related diseases.

There is no direct way to measure a living person's percentage of body fat. The following sections describe some indirect methods of assessing body composition, including their advantages and disadvantages.

Underwater Weighing

Underwater weighing involves comparing a person's weight "on land" to their weight when completely submerged in a tank of water (Fig. 10.4). Lean tissue is denser than water; fat tissue is not as dense as water. Thus, a person who has more body fat will weigh less when under water than a person who has more lean tissue. The underwater weighing method can be an accurate way of assessing body composition. However, the method is not a convenient, easy, inexpensive, or practical way to estimate body fat because it requires special testing facilities.

underwater weighing technique of estimating body composition that involves comparing weight on land to weight when completely submerged in a tank of water

dual-energy x-ray absorptiometry (DXA) technique of estimating body composition that involves scanning the body with multiple low-energy x-rays

Dual-Energy X-Ray Absorptiometry

Dual-energy x-ray absorptiometry (DXA) involves the use of multiple low-energy x-rays to scan the entire body. The method provides a detailed "picture" of internal structures, including fat deposits (Fig. 10.5). During the scanning process, the equipment emits a dose of radiation that is lower than that used for a chest x-ray. Although DXA is a highly accurate way to estimate body fat content, the equipment is very expensive and not widely available outside of clinical settings.

Air Displacement

The **air displacement** method assesses the volume of a person's body, which can be used to calculate their body composition. After being weighed on a very precise scale, the subject sits in the chamber of a device called the BOD POD (Fig. 10.6). This device measures the volume of air in the chamber with the person in it and compares the value with the volume of air that was in the chamber when it was unoccupied. The person's volume is the volume of air that was displaced after the subject entered the chamber. Air displacement measurements provide highly accurate estimates of body fat content, but the measuring device is expensive and not practical for most consumers to use.

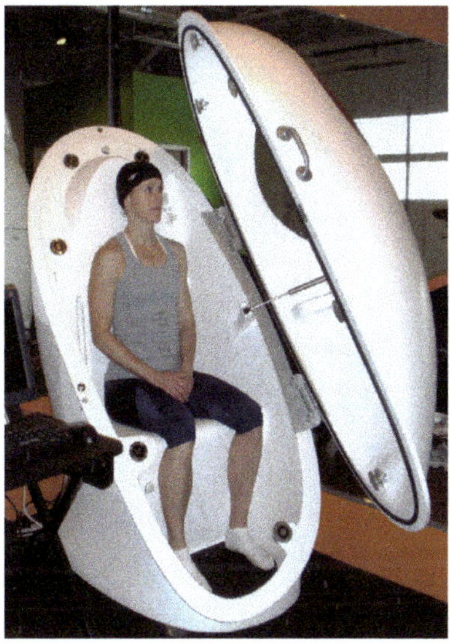

Figure 10.6 BOD POD.
Mary Dean Coleman-Kelly

Bioelectrical Impedance

Bioelectrical impedance is a quick way to estimate body fat content. This method is based on the principle that water and electrolytes conduct electricity. Body fat resists the flow of electricity because fat tissue contains less water and electrolytes than lean tissue. The bioelectrical impedance device sends a painless, low-energy electrical current via wires connected to electrodes placed on the subject's skin (Fig. 10.7). Within a few minutes, the device converts information about the body's electrical resistance into an estimate of total body fat. The method can be accurate, as long as the subject's hydration status is normal. Dehydration results in an overestimation of total body fat. Consumers can purchase a bioelectrical impedance device that resembles a bathroom scale, but scientific data about the machine's accuracy are lacking.

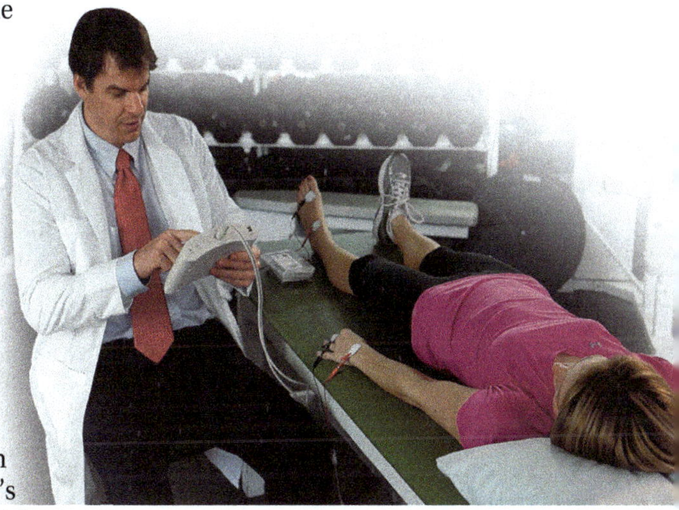

Figure 10.7 Bioelectrical impedance.
RJL Systems, Inc.

Skinfold Thickness

A common technique for estimating total body fat involves taking **skinfold thickness measurements** at multiple body sites, such as over the triceps muscle of the arm (Fig. 10.8). The width of a skinfold indicates the depth of the subcutaneous fat at that site. To perform the measurements, a trained person pinches a section of the subject's skin, gently pulls it away from underlying muscle tissue, and uses special calipers to measure the thickness of the fat. After taking the measurements, the values are incorporated into a mathematical formula that provides an estimate of the subject's amount of body fat.

Skinfold thickness measurements are relatively easy and inexpensive to perform, but the method's accuracy largely depends on the skill of the person performing the measurements. Also, the technique may underestimate total body fat when used on overfat subjects. However, by combining data collected from skinfold, waist and hip circumference, and body frame (skeletal joint) measurements, researchers can obtain more reliable estimates of an individual's total body fat.

air displacement method of estimating body composition by determining body volume

bioelectrical impedance technique of estimating body composition in which a device measures the conduction of a weak electrical current through the body

skinfold thickness measurements technique of estimating body composition in which calipers are used to measure the width of skinfolds at multiple body sites

How Much Body Fat Is Too Much?

Some body fat is essential for good health, but too much adipose tissue, especially visceral fat, can interfere with the body's ability to function normally.

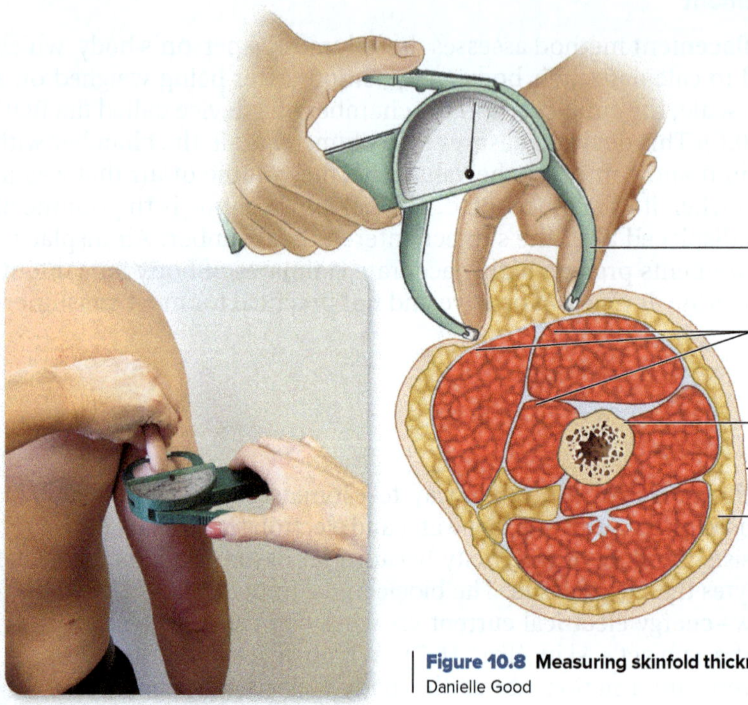

Figure 10.8 Measuring skinfold thickness.
Danielle Good

TABLE 10.3 Adult Body Weight Classification by Percentage of Body Fat

	Body Fat (%)	
Classification	**Males**	**Females**
Healthy	13–21%	23–31%
Overweight	22–25%	32–37%
Obese	26–31%	38–42%
Extremely obese	32% or more	43% or more

Source: Adapted from Food and Nutrition Board, National Institute of Health: *Dietary Reference Intakes for energy, carbohydrate, fiber, fat, fatty acids, cholesterol, protein, and amino acids (macronutrients)*. Table 5.5, page 126. Washington, DC: National Academies Press, 2005.

Percentages of body fat can be used to develop weight classifications for adults. According to one such classification system, a male is overweight when his body is 22 to 25% fat; a female is overweight when her body is 32 to 37% fat (Table 10.3).[11] It is important to note that the average healthy young female has more body fat than the average healthy young male because she needs the extra fat for hormonal and reproductive purposes.

Adults tend to gain adipose tissue as they age, but for older adults, some additional fat does not necessarily contribute to serious health problems. The extra fat may provide some health benefits, such as providing an energy reserve for a very ill person who cannot eat. Furthermore, the extra padding of fat may protect a person from being injured by falling.

Concept CHECKPOINT

3. Which tissues comprise total body fat and which comprise fat-free mass?
4. Why is it necessary to have some body fat?
5. List three roles for subcutaneous fat and visceral fat.
6. Describe three different methods of measuring body fat, including drawbacks of each method.
7. A young male's body is 23% fat. According to information in Table 10.3, is he overweight, obese, or healthy?

See Appendix F for responses.

10.3 Energy for Living

Learning Outcomes

1. Define *basal metabolism.*
2. Use the "rule of thumb" method to estimate a person's daily basal metabolic energy needs.
3. Discuss the three major ways the body uses energy.
4. Describe factors that influence the basal metabolic rate.

Regardless of what you are doing—eating, watching television, studying, exercising, even sleeping—your body needs a constant supply of energy to function. **Energy** is defined as the capacity to perform work. There are several forms of energy, but heat, mechanical, chemical, and even electrical energy occur in living things. The total amount of energy is *constant;* that is, the amount remains the same because energy cannot be created or destroyed. Nevertheless, the various forms of energy can be stored, released, moved, or transformed from one kind to another. The following sections discuss how human cells obtain and use energy.

Did *YOU* Know?

Advertisements for vitamin or mineral supplements often claim that the supplements can "boost" energy levels. Certain vitamins and minerals are needed to regulate energy metabolism, and deficiencies of these nutrients can reduce the body's ability to obtain energy from macronutrients. However, the body does not obtain energy from vitamins or minerals. Therefore, taking vitamin or mineral supplements will not increase the energy level of a person who is adequately nourished.

Stockbyte/Getty Images

Energy Intake

Just as a car engine uses a mixture of gasoline, ethanol, and oxygen to run properly, your body uses a mixture of *biological fuels* and oxygen to do its work. For humans, biological fuels are foods and beverages that contain macronutrients (**energy intake**). For some individuals, nonnutrient alcohol (ethanol) also provides energy. Under normal conditions, our cells metabolize primarily glucose and fatty acids, but small amounts of amino acids are also used for energy.

Cells release the energy stored in biological fuels by breaking bonds within the compounds' molecules. The energy that is released can be captured and stored in special compounds, such as **adenosine triphosphate (ATP),** until it is needed. Cells obtain only about 40% of the energy that was in macronutrients by forming ATP. Cells release the remaining energy as heat. Figure 10.9 summarizes events that result in macronutrient storage or breakdown for energy. The diagram in Appendix C summarizes the complex chemical pathways that most cells use to generate ATP from the metabolism of glucose, fat, and amino acids.

Energy Output

Energy output (*energy expenditure*) refers to the energy (calories) cells use to carry out their activities. For example, muscle cells need energy to contract, liver cells use energy to convert toxic compounds to safer substances, and intestinal cells need energy to absorb certain nutrients. The following sections discuss the major ways the body uses food energy.

Basal and Resting Metabolism

Metabolism refers to all chemical changes, or reactions, that constantly occur in living cells. Anabolic reactions require energy to occur; catabolic reactions release energy. **Basal metabolism** is the minimal number of calories the body uses for vital physiological activities after fasting and resting for 12 hours.

energy capacity to perform work

energy intake calories from foods and beverages that contain macronutrients and alcohol

adenosine triphosphate (ATP) molecule that stores energy

energy output calories cells use to carry out their activities

basal metabolism minimal number of kilocalories the body uses to support vital activities after fasting and resting for 12 hours

Figure 10.9 What happens to macronutrients? (a) After being absorbed, macronutrients are used for energy or stored as glycogen or fat. (b) When energy is needed to fuel cells, macronutrients are released from storage and metabolized to synthesize ATP.

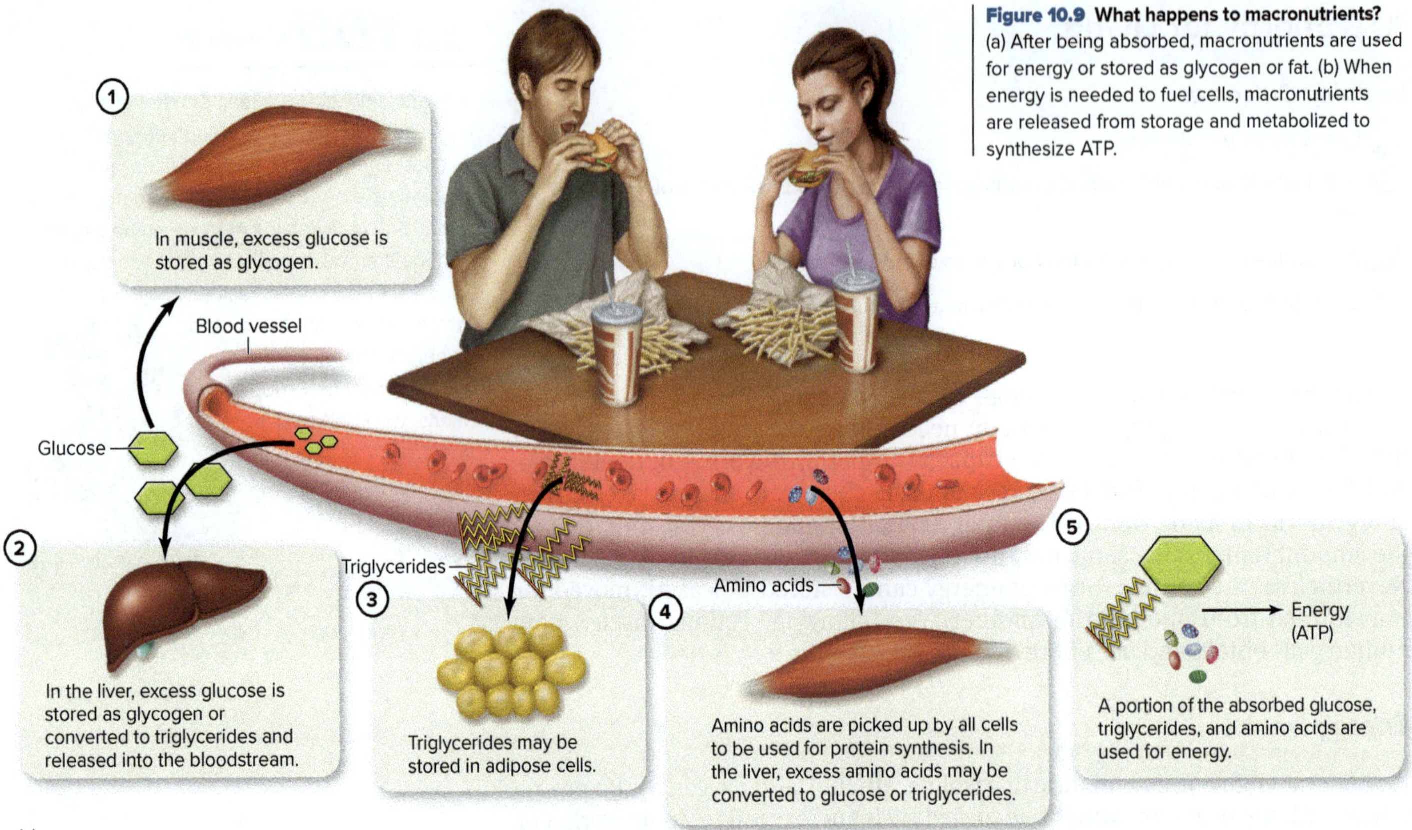

(a)

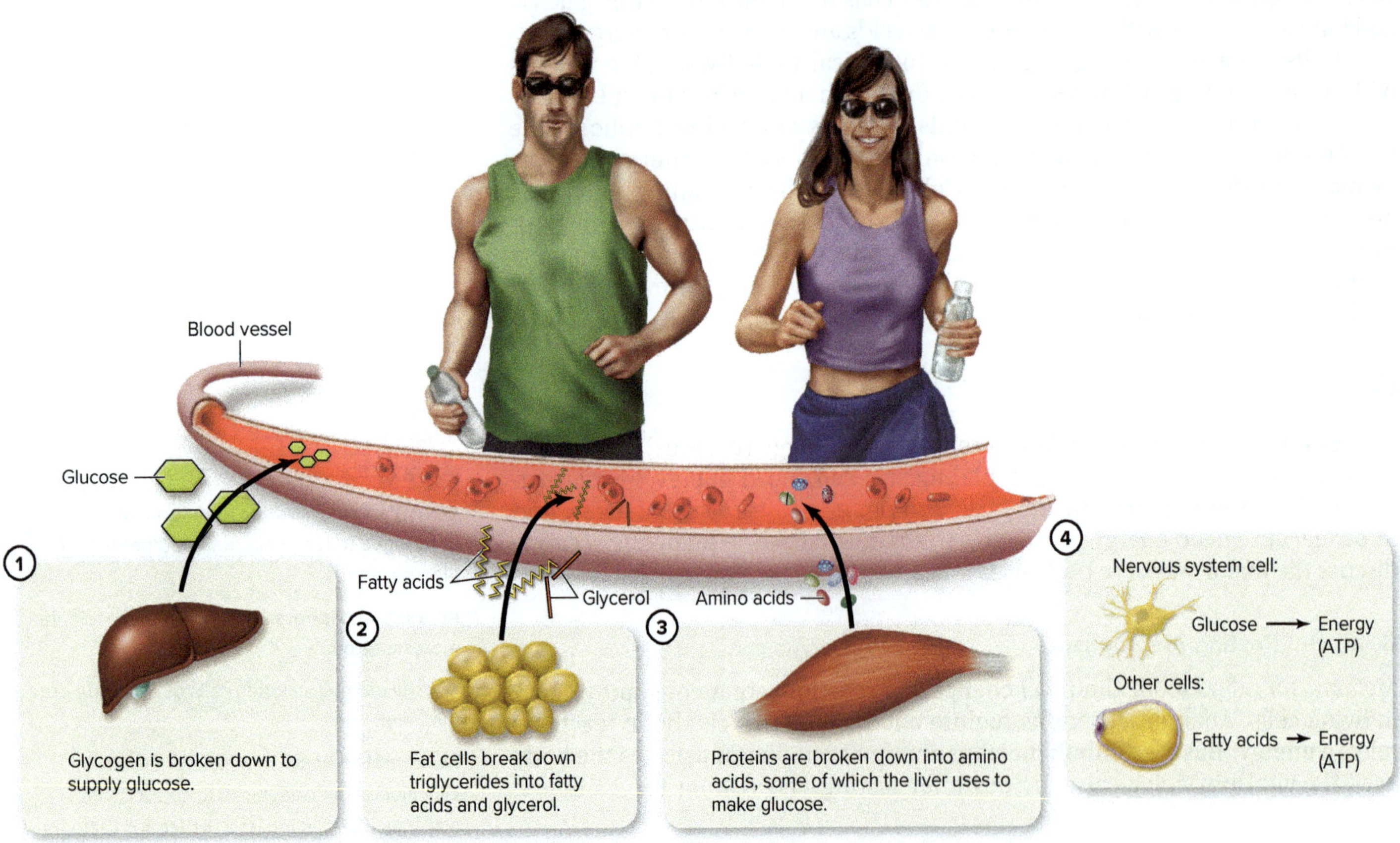

(b)

Basal metabolic processes include breathing, circulating blood, and maintaining constant liver, brain, and kidney functions. Basal metabolism does not encompass energy needed for skeletal muscle movement (physical activity), digestion of food, and absorption and processing of nutrients. Generally, basal metabolism accounts for most of the body's total energy use.[12]

The **resting metabolic rate (RMR)** refers to the body's rate of energy use a few hours after resting and eating. A person's RMR is slightly higher than their BMR (basal metabolic rate). Although there is a difference between the BMR and the RMR, researchers often use the terms interchangeably in their publications.

Thyroid hormone, secreted by the thyroid gland, regulates metabolism (see Figure 9.13). A person who has an overactive thyroid gland produces too much thyroid hormone. As a result, this person has a higher–than–normal metabolic rate. Signs and symptoms of excess thyroid hormone production (*hyperthyroidism*) include feeling warm, sweaty, nervous, and restless; having rapid heart rate and chronic diarrhea; and losing weight despite having an increased appetite. You might think that hyperthyroidism is the key to treating unwanted weight gain, but the condition can have serious side effects, such as elevated blood pressure and heart failure. Treatment can include medication or surgery to reduce thyroid gland activity. A person who suffers from *hypothyroidism* does not produce enough thyroid hormone, and as a result, they have a lower–than–normal metabolic rate. This individual typically complains of feeling cold, lacking energy and interest in usual activities, being constipated, and gaining weight easily. Treatment for a hypoactive thyroid gland generally includes medication that contains a form of thyroid hormone.

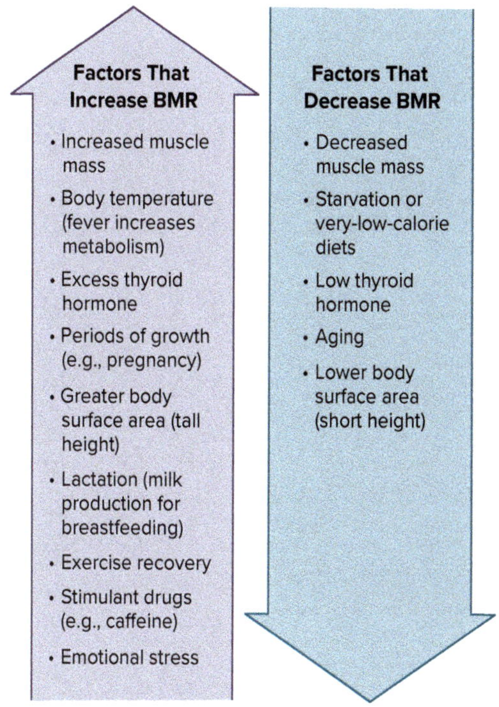

Figure 10.10 Factors that influence the metabolic rate.

Factors That Influence the Metabolic Rate In addition to thyroid hormone, numerous factors can increase or decrease basal metabolic rates (Fig. 10.10). Thus, metabolic rates vary among individuals. Factors that influence basal metabolism include:

- *Body composition.* Lean body mass is the major factor that influences the metabolic rate.[11] Muscle tissue, a component of lean body mass, is more metabolically active than fat tissue. In general, a person who has more muscle mass will have a higher metabolic rate than someone with less muscle tissue.
- *Sex.* Males generally have higher metabolic rates than females because they tend to have more lean body mass.
- *Body surface area.* A tall, slender person who weighs 150 pounds has a higher metabolic rate than a shorter person who also weighs 150 pounds. Why? The body constantly loses energy in the form of heat that moves to the skin's surface and then into the environment. Because the taller person's body has more surface area than the shorter person's body, the taller individual has to generate more heat energy to replace that which is lost.
- *Age.* Basal metabolism declines as one grows older, primarily due to the loss of fat–free tissues such as muscle. After 20 years of age, a female's BMR declines about 2% and a male's BMR about 3% per decade.[13] Therefore, the average adult needs about 150 fewer daily kilocalories per decade as they age. As many adults grow older, they think their muscles have "turned into" fat. A muscle cell, however, cannot transform itself into a fat cell. During the aging process, lean tissue mass shrinks, as cells from muscle, bone, and organs die and are not replaced. Fat cells, however, can continue to develop throughout life, especially when a person overeats consistently. When adipose tissue expands in size, it can fill in spaces formerly occupied by muscle

resting metabolic rate (RMR) body's rate of energy use a few hours after resting and eating

As people grow older, their calorie needs decline.
icsnaps/Shutterstock

and organ tissues. Regular exercise helps build and preserve lean body mass, and to some extent, people can maintain a higher metabolic rate by being physically active as they grow older.

- *Calorie intake.* Calorie intake also affects the metabolic rate. The body conserves energy use when calorie intakes are very low or lacking altogether. To enhance the rate of weight loss, a person who is overfat should reduce caloric intake while maintaining a normal metabolic rate. Because very-low-calorie diets reduce the metabolic rate, such diets are not generally recommended for weight loss.

As listed in Figure 10.10, the following factors increase the metabolic rate:

- *Fever.*
- *Stimulant drugs* (caffeine, for example).
- *Pregnancy.*
- *Milk production in a female who has given birth.*
- *Recovery after exercise* (metabolic after-effects of exercise).
- *Emotional stress.*

Calculating Metabolic Energy Needs The basal metabolic rate is fairly constant from day to day.[14] Thus, a young adult can estimate their daily metabolic rate by following a "rule of thumb" formula:

$$\text{Formula for males} = 1.0 \text{ kcal/kg/hr}$$

$$\text{Formula for females} = 0.9 \text{ kcal/kg/hr}$$

To estimate the number of calories you need for your basal metabolism, first convert your weight in pounds to kilograms by dividing your weight by 2.2. (A kilogram is approximately 2.2 pounds.)

_____ lb ÷ 2.2 lb = _____ kg

Then, depending on your sex, use one of the following formulas:

_____ kg × 0.9 (females) = _____ kcal/hr

_____ kg × 1.0 (males) = _____ kcal/hr

Finally, use this hourly basal metabolic rate to estimate your basal metabolic rate for an entire day by multiplying the hourly value by 24.

_____ kcal/hr × 24 hr = _____ kcal/day

These calculations only provide a rough estimate of your daily metabolic rate. To estimate your Estimated Energy Requirement (EER) for a 24-hour period, you need to add kilocalories used for physical and other activities to your BMR figure.

Physical Activity

Physical activity, voluntary skeletal muscle movement, increases energy expenditure above basal energy needs. The number of kilocalories expended for a particular physical activity depends largely on the type of activity, how long it is performed (duration), the degree of effort (intensity) used while performing the activity, and the weight of the person. A heavy person expends more kilocalories when performing the same activity, for the same duration, and at the same intensity than a lighter person. Why? The muscles of the heavier person must work harder to move the larger body.

To estimate your energy needs for physical activity, visit www.acefitness.org/education-and-resources/lifestyle/tools-calculators/physical-activity-calorie-counter/. Table 10.4 lists various physical activities and the approximate number of kilocalories an individual who weighs 150 pounds expends while performing each activity for 30 minutes. For example, a 150-pound person who walks for 30 minutes (3.5 mph) burns approximately 130 kcal during the walk.

The number of calories you need for physical activity can vary widely, depending on how active you are each day. Because you can control the type, intensity, and duration of your physical activities, you can manipulate your energy output to increase, decrease, or maintain your weight.

nonexercise activity thermogenesis (NEAT) involuntary skeletal muscular activities such as fidgeting

thermic effect of food (TEF) energy used to digest foods and beverages as well as absorb and further process the macronutrients

Nonexercise Activity Thermogenesis **Nonexercise activity thermogenesis** (*thermo* = heat; *genesis* = production) or **NEAT** refers to *involuntary* skeletal muscle activity, that is, physical activity that a person does not consciously control. NEAT activities include shivering, fidgeting, maintaining muscle tone, and maintaining body posture when not lying down. Some people may expend as much as 700 kcal daily as NEAT.[15] It is possible that some individuals resist weight gain from overeating because they have higher-than-average energy expenditures for NEAT. Nevertheless, the contribution of NEAT to overall calorie needs is fairly small for most people.

Thermic Effect of Food

The body needs a relatively small amount of energy to digest foods and beverages, as well as to absorb and further process the macronutrients. The energy used for these tasks, generally 5 to 10% of total caloric intake, is referred to as the **thermic effect of food (TEF).** For example, if your energy intake is 3000 kcal/day, TEF would account for 150 to 300 kcal.

TABLE 10.4 Approximate Caloric Output/30 minutes of Activity (150-pound Person)

Physical Activity	Approximate kcal/30 min
Cycling (15 mph)	408
Swimming (vigorous)	339
Running/jogging, steady pace (6 mph)	339
Martial arts	339
Cycling (12–13 mph)	270
Touch football	270
Tennis (singles)	270
Hiking	204
Walking (4 mph)	168
Walking (3.5 mph)	129
Dancing (casual)	102
Weight training	102
Housework	102
Sitting and sewing	33

Source: American Council on Exercise: Tools & calculators: Physical activity calorie counter. ND. www.acefitness.org/education-and-resources/lifestyle/tools-calculators/physical-activity-calorie-counter. Accessed: August 7, 2023.

Matthew Leete/Photodisc/Getty Images

Putting It All Together

To estimate your daily energy expenditure, you could add the kilocalories you burned for basal metabolism, physical activity, and TEF in a day. However, another method is to use one of the formulas published by the Food and Nutrition Board of the Institute of Medicine, which is in the Personal Dietary Analysis activity at the end of this chapter.[11]

Did *YOU* Know?

Results of scientific studies suggest that watching television is associated with obesity among children and adults.[16,17] Television watching is a sedentary activity that may include viewing food advertisements and eating foods that are high in empty calories. Health experts recommend that people limit their nonwork-related screen time (including activities involving computers and video games) to no more than 2 hours/day.[18]

David Moyer/McGraw Hill

Concept CHECKPOINT

8. Using the "rule of thumb" formula, estimate the daily basal metabolic energy needs of a female who weighs 185 pounds.
9. List the three major ways the body uses energy, and identify the one that is most easily altered.
10. For most people, which form of energy expenditure uses the most energy on a daily basis?
11. Discuss at least five factors that influence basal metabolic rate.

See Appendix F for responses.

10.4 Energy Balance

Learning Outcomes

1. Explain the concept of energy balance.
2. Compare states of positive and negative energy balance.

Understanding the concept of *energy balance* is critical to understanding why most people gain, lose, or maintain weight. Your body is in a state of **energy equilibrium** and "balanced" when your calorie intake from food and beverages equals your calorie output for basal metabolism, physical activity, and TEF (Fig. 10.11). By maintaining a balanced energy state, your weight will remain relatively stable over time.

energy equilibrium calorie intake equals calorie output

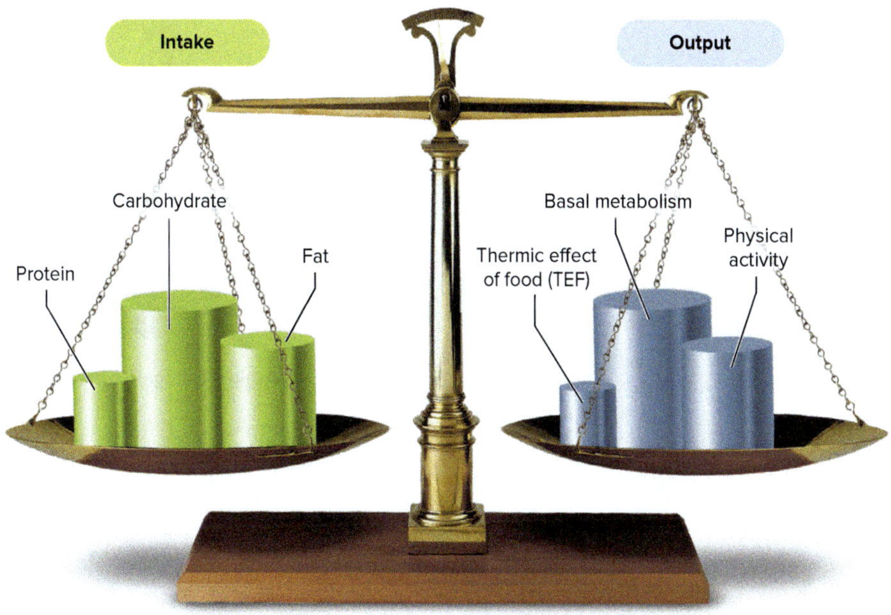

Figure 10.11 Energy balance. Your energy state is in equilibrium and "balanced" when your calorie intake from food and beverages equals your calorie output for basal metabolism, physical activity, and TEF.

If your calorie intake is lower than your calorie output, you are in **negative energy balance.** In this state, your body needs more calories to carry out its activities than your diet is supplying. Therefore, your body metabolizes stored fat for energy. Weight loss results from being in a negative energy state. Over time, you will notice your clothes have become baggy as your adipose tissue shrinks.

If your calorie intake from macronutrients (and alcohol) is greater than your calorie output, you are in a state of **positive energy balance.** In this state, your body stores excess dietary fat in adipose cells. Additionally, the body converts surplus dietary carbohydrate, protein, and alcohol to fat and stores that fat in adipose cells. Weight gain results from being in a positive energy state, and eventually, you will notice that your clothes seem to have shrunk.

Positive energy balance is necessary for pregnant females because extra calories are needed to add new tissues that support the pregnancy. Positive energy balance also occurs during periods of growth, such as during fetal development, infancy, childhood, and adolescence. Over time, however, even a small positive energy balance can cause anyone's weight to increase, regardless of their age. Maintenance of energy balance—matching calorie intake to calorie output over the long term—is critical for controlling body weight (Fig. 10.12).

Positive energy balance occurs during pregnancy.
JGI/Tom Grill/Blend Images LLC

Did *YOU* Know?

Have you heard of the "freshman 15," the popular belief that college students gain 15 pounds during their freshman year? Results of scientific studies confirm that freshmen are likely to gain weight, but the increase is much less than 15 pounds—7.5 pounds on average.[19] Factors that contribute to the weight gain include feeling "stressed out," eating too much unhealthy food, drinking too much alcohol, and being less physically active than before entering college.

negative energy balance calorie intake is less than calorie output

positive energy balance calorie intake is greater than calorie output

Figure 10.12 The body's possible energy states. This figure presents three possible energy states.

Concept CHECKPOINT

12. What happens to a person's body weight when they are in a state of positive energy balance?

13. When is it desirable for a person to be in a positive energy balance state?

14. When is a person in a negative energy balance state?

See Appendix F for responses.

10.5 Classifying Overweight and Obesity

Learning Outcomes

1. Use the formula to calculate BMI, and classify a person's BMI as underweight, healthy, overweight, obese, or extremely obese.
2. Discuss serious health problems that are associated with having too much body fat and being obese in particular.
3. Compare "apple" to "pear" body shapes in terms of fat distribution and effects on health.
4. Explain how to measure waist circumference and the usefulness of this information in regard to a person's health.

At one time, people referred to height/weight tables to determine whether their body weights were "ideal" or "desirable." Today, medical experts use the BMI to judge whether an adult's weight is healthy. As mentioned in the beginning of the chapter, BMI is a numerical value that reflects body weight adjusted for height. People whose BMI is within the overweight or obese categories are at a higher risk of chronic health problems associated with excess body fat.[11]

Recall that Table 10.1 presented adult weight classifications based on BMI ranges. Healthy BMIs range from 18.5 to 24.9. *Overweight* adults have BMIs that range from 25.0 to 29.9; *obese* adults have BMIs that range from 30.0 to 39.9. People whose BMIs are 40 or higher are classified as *extremely obese.* In 2015–2016, the average BMI for adult American males and females was 29.1 and 29.6, respectively.[20]

Muscle is denser than fat; therefore, many muscular people may have BMIs in the overweight range, yet have healthy percentages of body fat. For example, a muscular person with a BMI of 25.0 is more likely to be healthy than a sedentary (physically inactive) person who also has a BMI of 25.0. The BMI may underestimate body fat in people who have lost muscle tissue as a result of aging or illness. Therefore, BMIs should not be applied to highly muscular individuals, people who are elderly, or chronically ill persons. However, sedentary people with BMIs of 30 or higher generally have excess body fat.

Did *YOU* Know?

You can use the following formula to calculate your BMI:

$$\frac{\text{weight (lb)}}{[\text{height (in)}]^2} \times 703$$

For example, a person who weighs 140 pounds and is 5-feet, 3-inches (63 inches) tall has a BMI of approximately 24.8.

Calculation: $[140 \div (63)^2] \times 703$

$(140 \div 3969) \times 703 =$

$0.03527 \times 703 = 24.8$

Excess Body Fat: Effects on Health

People with BMIs greater than 25 have increased risks of type 2 diabetes, hypertension, and CVD. Many types of cancer, including cancers of the gallbladder, pancreas, cervix, uterus, breast (postmenopausal females), colon, rectum, and kidney, are more common in people with obesity. Furthermore, patients with obesity have high risk of experiencing serious complications during and after surgery. Such patients often require more anesthesia, and their incisions are more likely to become infected than surgical patients whose weights are within the healthy range. People who are extremely obese are far more likely to develop the serious chronic diseases associated with excess body fat and to die prematurely than people who have BMIs that range from 18.5 to 24.9.[21]

People who are overweight or obese are more likely to develop osteoarthritis, a painful chronic condition that affects joints and interferes with the person's ability to move. People with obesity typically have difficulty carrying out routine daily activities, especially those that require walking, carrying, kneeling, and stooping. Additionally, people with obesity are more likely to suffer from chronic heartburn, as well as *sleep apnea,* a condition that causes breathing to stop periodically during sleep.

People with obesity are more likely to have fertility problems than people with healthy BMIs. *Polycystic ovary disease* often affects females with obesity

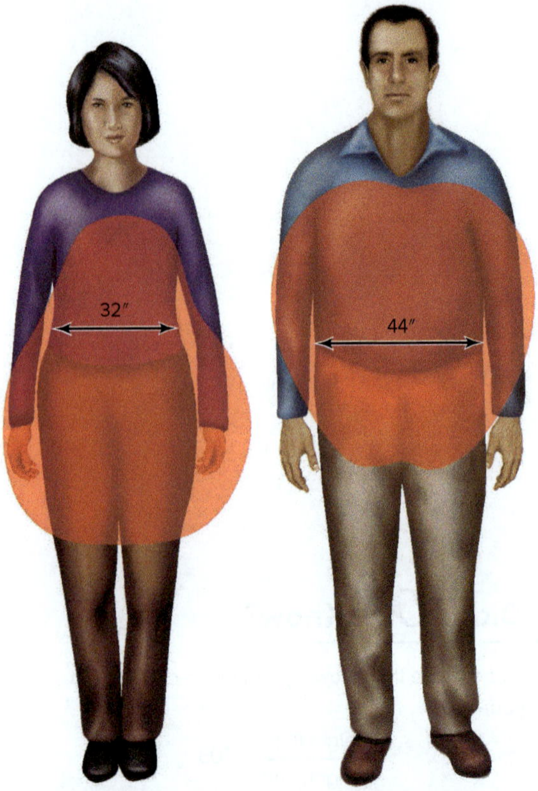

(a) Lower-body fat distribution (pear shape)

(b) Central-body fat distribution (apple shape)

Figure 10.13 Body fat distribution: Typical differences. Females tend to store extra subcutaneous fat below the waist (a). Males tend to store extra visceral fat in the central region of their bodies (b).

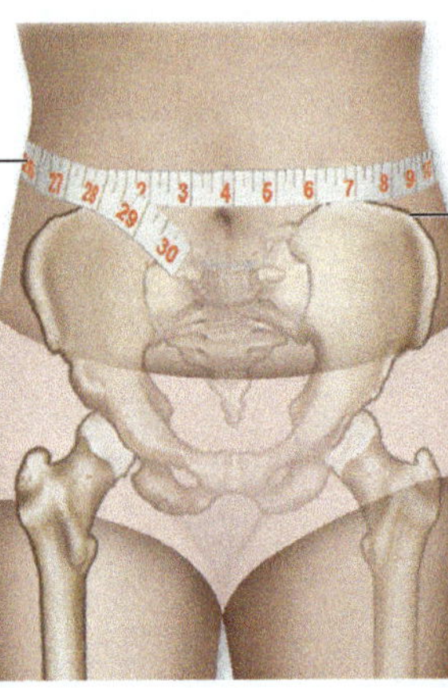

Figure 10.14 Measuring waist circumference.

central-body obesity condition characterized by excessive visceral fat

and can reduce their chances of becoming pregnant. During pregnancy, females who are obese have high risk of *gestational diabetes* and a form of hypertension that can be deadly (see Chapter 5). Additionally, pregnant females with obesity are at greater risk for stillbirths or giving birth to babies with birth defects than are pregnant females who have lower BMIs.

A person's self-esteem and mental health can be negatively affected by their "weight problem." Many Americans admire slim and muscular body builds over overfat body shapes. Thus, the general public may have *weight bias:* negative attitudes, beliefs, and stereotypes about a person because of their weight. For example, some people believe obesity is a condition that results from a lack of willpower and the inability to "push oneself away from the table"; and people that are obese are lazy, stupid, and sloppy.[22] Therefore, people that are obese must deal with the stigma (negative attitude) of obesity that others may have toward them.[22] The stigma of obesity can result in discriminatory practices that limit chances for career opportunities and access to adequate health care for people who are obese. Depression, higher anxiety levels, lower self-esteem, social isolation, increased amounts of stress, and abuse of drugs and alcohol are observed in the obese population.[22] Table 10.5 lists major health problems associated with excessive body fat.

Body Fat Distribution: Effects on Health

The results of medical research suggest that the distribution of excess body fat has a closer association with obesity-related diseases than does the percentage of total body fat. Females tend to store extra subcutaneous fat below the waist, primarily in the buttocks and thighs (Fig. 10.13a). Having this particular pattern of fat distribution (a "pear shape") adds stress to hip and knee joints that must carry the extra weight, but the pattern is not associated with increased risk of more serious chronic diseases such as type 2 diabetes.

Males tend to store extra visceral fat in the abdominal or central region of their bodies.[23] **Central-body obesity** is characterized by a large amount of visceral fat that spreads beyond buttocks and thighs. A person with central-body obesity is sometimes described as having an "apple" body shape (see Figure 10.13b). People with central-body obesity have higher risks of serious chronic diseases, especially CVD and type 2 diabetes ("diabesity"), than people who have waists that do not extend beyond their hips.

A quick and easy method to determine your risk of obesity-related disorders is to measure your *waist circumference,* which is an indicator of central-body obesity. Figure 10.14 shows a recommended placement of the

TABLE 10.5 Health Problems Associated with Excess Body Fat

Overweight and Obesity Increase the Risk of:	
• Cardiovascular disease	• Chronic low back pain
• Hypertension	• Loss of mobility
• Type 2 diabetes	• Fatty liver disease (not alcohol related)
• Metabolic syndrome	• Erectile dysfunction in males (impotence)
• Polycystic ovary syndrome in females	• Low-grade inflammation
• Infertility	• Gastroesophageal reflux disorder
• Elevated blood lipid levels	• Psychological depression
• Gallstones	• Certain cancers
• Sleep apnea	• Skin ulcers
• Osteoarthritis	• Asthma

Source: Garvey WT and others: American Association of Clinical Endocrinologists and American College of Endocrinology Comprehensive Clinical Practice Guidelines for Medical Care of Patients with Obesity. *Endocrine Practice* 22 Suppl 3:1-203, 2016.

tape measure. Note the positioning of the tape at the top of the hip bones and not necessarily at the narrowest point. It is also important to use a measuring tape that does not stretch and to hold the tape around the waist, parallel to the floor. Central–body obesity is defined by a waist circumference of greater than 40 inches in males and greater than 35 inches in females.[24]

Concept CHECKPOINT

15. A young female's BMI is 31.2. According to this information, is this female likely to be healthy, overweight, obese, or extremely obese?

16. List at least six different serious health problems that are associated with having too much body fat and being obese in particular.

17. What are "weight bias" and the "stigma" of obesity?

18. Which of the two major types of body fat distribution is more likely to pose serious health risks? Which chronic diseases are more likely to develop in people who have this pattern of excess fat deposition?

19. What is a quick and easy way to determine whether a person's body fat distribution is likely to result in serious health problems, such as type 2 diabetes?

See Appendix F for responses.

10.6 What Causes Overweight and Obesity?

Learning Outcomes

1. Identify factors that can influence the development of excess body fat.
2. Discuss factors that influence hunger and satiety.
3. Discuss the roles of hormones and peptides in regulating hunger.
4. Explain what is meant by having "thrifty genes."
5. Discuss how the set-point theory can explain why people often regain the weight they lose.

Although an excess intake of calories in relation to calorie output causes weight gain, there is no simple *cause* of obesity. To lose weight, a person needs to create a negative energy state by eating fewer calories, expending more calories than the amount consumed, or taking both actions. For many people, however, it is not easy to alter calorie input and output. Numerous factors, including physiological, environmental, behavioral, and psychological forces, influence a person's calorie intake and expenditure.

Physiological Factors

From a physiological standpoint, eating behavior is complex and largely involves interactions among the nervous, endocrine, and digestive systems as well as fat tissue. *Hunger* and *satiety* are key sensations that regulate eating behavior. **Hunger** is an uncomfortable feeling that drives a person to consume food. Thus, hunger is the need to eat. **Satiety** is the sense that enough food or beverages have been consumed to satisfy hunger.

Physical sensations influence eating behavior. As time between meals increases, the stomach signals "it's time to eat" by contracting, causing hunger pangs. As the contractions become stronger, the person usually eats or drinks something to relieve the discomfort. The size of the stomach influences satiety.

Nutrition & Health *Equity*

Body composition, including lean body mass and total body fat, can vary among racial and ethnic populations. These body composition differences may contribute to body mass index (BMI) not correctly assessing the body weight and disease risk of these populations. For example, Black people often have higher muscle mass and less visceral fat, the type of fat associated with chronic disease, compared to non-Hispanic white people.[25] However, developing type 2 diabetes is associated with a BMI of 28, classified as overweight, in Black people and a higher BMI of 30, classified as obese, in white people. Scientists also report that people of Asian ancestry may develop diabetes, heart disease, and high blood pressure at lower BMIs than the overweight BMI range of 25 to 29.9. This is due to the population's smaller body structure and higher distribution of visceral fat.[26] Therefore, the World Health Organization recommends a BMI of <23 as a healthy body weight for people of Asian ancestry.[27]

It is important to remember that BMI is only one part of a person's overall assessment of health. Other tools such as body fat distribution and waist circumference should also be used to assess a person's body weight and disease risk.

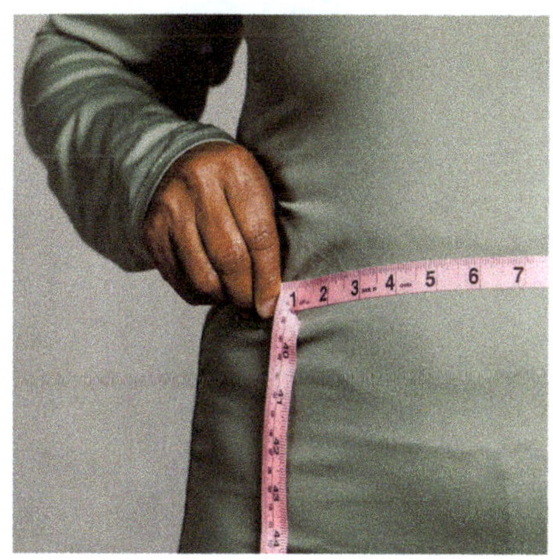

Rawpixel.com/Shutterstock

hunger uncomfortable feeling that drives a person to consume food

satiety sense that enough food or beverages have been consumed to satisfy hunger

During meals, the stomach stretches as it fills. The sensation that the stomach has reached its capacity can make a person stop eating. Nevertheless, many people who are overfat do not recognize the sensation of stomach fullness, and as a result, they may eat even when they should not be hungry.

An area of the *hypothalamus,* a structure in the brain, controls hunger and satiety. Scientists think this region functions as a "hunger/satiety center." The stomach, intestines, and fat tissue produce certain proteins, such as hormones and *peptide YY,* that stimulate nerve cells involved in the regulation of hunger and satiety. **Ghrelin** (*grel′–in*), a hormone secreted mainly by the stomach, stimulates eating behavior. Some scientists think that reducing ghrelin production or activity is the key to helping people lose or maintain their weight. The small and large intestines release peptide YY, a protein that signals the stomach to reduce ghrelin secretion. The small intestine also releases *cholecystokinin* (*CCK*), the hormone that stimulates the gallbladder to contract and the pancreas to release digestive enzymes (see Chapter 6). Additionally, CCK stimulates the brain and other nervous tissue, suppressing appetite as a result.

ghrelin protein that stimulates eating behavior

leptin hormone that reduces hunger and inhibits fat storage in the body

Fat cells secrete **leptin,** a hormone that reduces hunger and inhibits fat storage in the body. A person's blood leptin level is directly proportional to their amount of body fat.[28] The brain obtains information about the status of body fat stores by monitoring the level of leptin in blood. When researchers administer leptin to genetically engineered mice that cannot synthesize the hormone, the rodents lose weight. Studies involving humans, however, generally find that obese people produce high amounts of leptin, but their bodies resist the hormone's hunger–suppressing action.[28] Figure 10.15 illustrates the complex effects of ghrelin, leptin, CCK, and peptide YY on hunger and satiety.

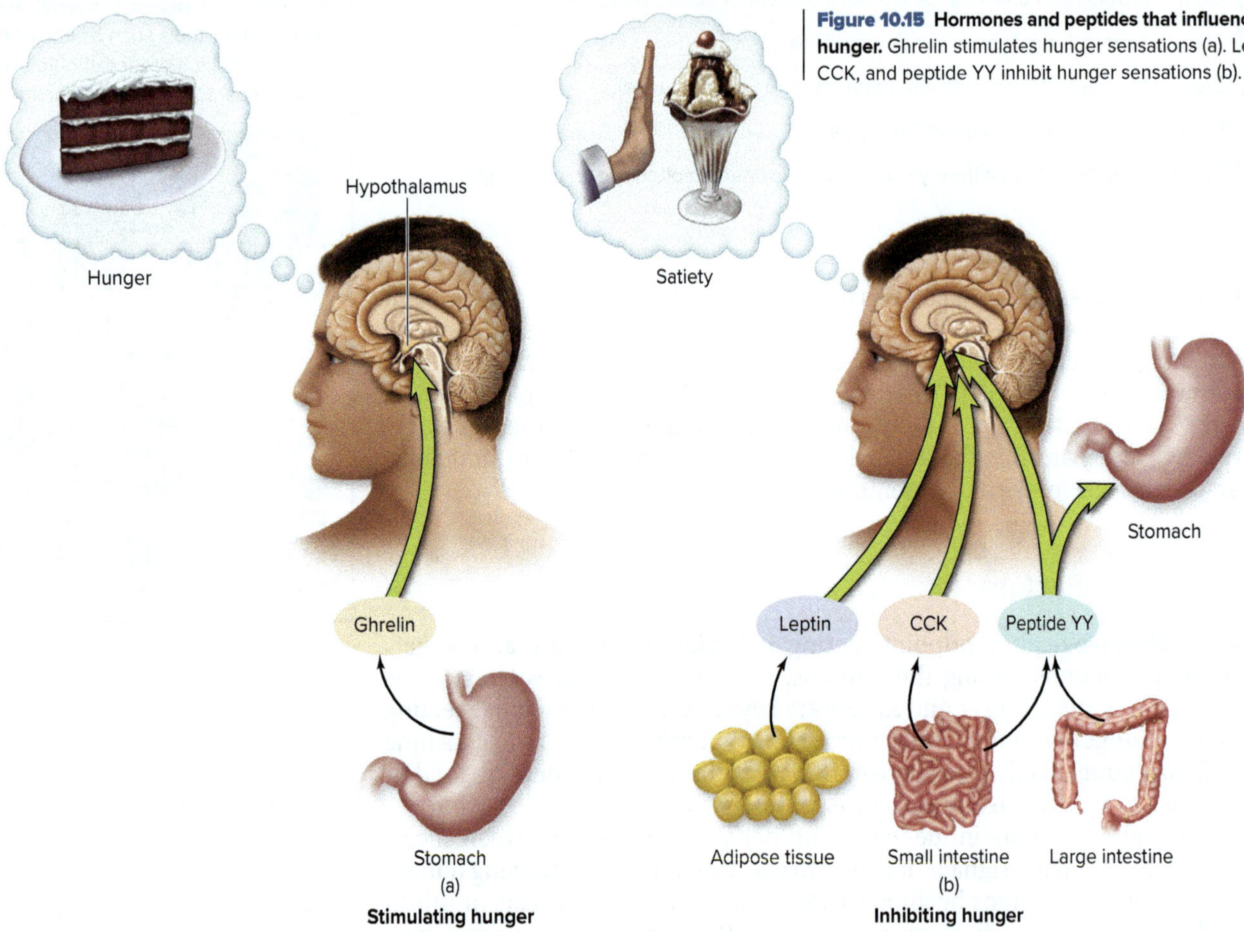

Figure 10.15 Hormones and peptides that influence hunger. Ghrelin stimulates hunger sensations (a). Leptin, CCK, and peptide YY inhibit hunger sensations (b).

Gut Microbiota

In Chapter 4, we discussed the importance of having a healthy diversity and balance of gut microbiota. Such microbes normally reside in the human intestinal tract, primarily the large intestine. In healthy people, gut microbiota perform a variety of useful functions, including synthesis of vitamin K and several B vitamins, energy metabolism (*fermentation*), and protection against infectious intestinal bacteria. Each person has a unique collection of microbes in their large intestine, and having a very diverse population of microbes in one's gut is generally associated with good health.[29]

A person's diet is one of many factors that influence the microbial diversity of their intestinal tract. Recall from Chapter 5 that dietary fiber is not digested by human enzymes. Certain kinds of gut microbiota ferment undigested carbohydrates, particularly dietary fiber, for energy and produce short–chain fatty acids (SCFAs) as metabolic by–products. Butyrate (*byu'–ter–ate*), propionate (*pro'–pee–eh–nate*), and acetate (*ah'–seh–tate*) are the major SCFAs that result from this fermentation. The cells that line the colon use most of the butyrate as a source of energy; the other fatty acids are absorbed and metabolized by the liver for energy. Any SCFAs that remain in the colon are excreted in feces.

Intestinal Dysbiosis The typical Western diet is low in fiber and supplies high amounts of fat, animal protein, and added sugars and other food additives. This diet is associated with *intestinal dysbiosis,* an unhealthy imbalance of various kinds of gut microbiota.[29] Recall from Chapter 4 that intestinal (gut) dysbiosis is thought to contribute to inflammation in the intestines that results in inflammatory bowel disease (IBD), which includes ulcerative colitis and Crohn's disease. Furthermore, the dysbiosis may also be a factor in the development of diseases that affect the entire body (*systemic*) and are characterized by chronic inflammation and impaired energy metabolism, such as type 2 diabetes and obesity.

According to the results of a scientific analysis of several human studies, subjects with obesity had higher amounts of SCFAs in their feces than people who had healthy body weights.[30] The person who is obese may have higher–than–normal amounts of SCFAs in their feces because excessive amounts of the fatty acids were produced by the unhealthy balance of microbes in the person's colon. At this point, however, scientists do not have a clear understanding of how intestinal bacteria and their production of SCFAs are associated with the development of obesity in humans.[30,31]

Does having obesity influence the kinds of microbes that populate the affected person's large intestine, or does having certain kinds of gut microbes lead to the development of obesity? This is basically a question of which comes first—obesity or gut dysbiosis. It would be quite a scientific breakthrough to show that having an imbalance of certain kinds of gut microbes con–tributes to obesity (causation). Medical researchers could develop dietary, pharmacological, or other kinds of treatments for obesity that modify the populations of gut microbiota to be similar to the gut microbiota of people who have healthy body weights. Thus, more research involving human subjects is needed to clarify whether there is a direct relationship between gut dysbiosis and obesity risk.

Food Composition Factors

Dietary factors, particularly amounts of fat and certain carbohydrates in diets, can influence body fat production, as well as appetite. Fatty foods are more energy dense than foods that contain more carbohydrate, protein, and water than fat. Thus, high–fat diets are associated with excess calorie intakes and rising obesity rates. Recall from Chapter 5 that foods high in refined carbohydrates, especially those with added sugars, are sources of empty calories. Such foods do not satisfy hunger as well as those that contain more protein or fat. Diets that supply high amounts of added sugars are associated with increased risk of gaining weight, having excess body weight, and being obese.[32]

Ultra-processed Foods In Chapters 1 and 3, we defined several food–related terms, including "refined," "processed," and "ultra–processed." Most Americans consume a wide variety of refined, processed, and ultra–processed foods each day. Some medical researchers are concerned about the possible unhealthy effects that may occur as a result of eating diets that contain high amounts of ultra–processed foods.

Many researchers use the NOVA system to classify processed foods. According to the NOVA system, an "ultra–processed diet" contains a lot of energy–dense foods, which are high in unhealthy fats, refined starches, added sugars, and salt.[33] In general, such diets are poor sources of fiber, vitamins, and minerals, and contain added colors, flavors, emulsifiers, and preservatives.[33] Ultra–processed foods generally contain one or more ingredients that are not used or are rarely used by typical consumers when preparing food in their kitchens. A "minimally processed diet" does not contain added fats and sugars, salt, preservatives, or other additives. Furthermore, minimally processed diets supply more fiber and micronutrients than ultra–processed diets.[33]

A diet rich in ultra-processed foods, such as this fruit-flavored, sugar-sweetened cereal, is associated with weight gain.
Elena Veselova/Shutterstock

Recently, researchers investigated the effects of ultra–processed and minimally processed diets on the calorie intake of 20 adults.[34] The study's participants lived in a special research facility (inpatient) during the duration of the study, so their food intake and physical activity could be controlled, observed, and measured. Subjects ate an ultra–processed diet for 2 weeks, before switching to a minimally processed diet for 2 weeks. Both diets contained the same number of calories, macronutrient distribution, and amounts of fiber, sugar, and sodium, but participants could eat as much or as little food as they desired (ad libitum). When subjects followed the ultra–processed diet, they ate faster and consumed higher amounts of carbohydrates and fats than when they followed the minimally processed diet. Furthermore, when subjects consumed the ultra–processed diet, they ate an average of 500 more kilocalories per day, resulting in a 2–pound weight gain, by the end of the 2–week period.[34] However, subjects consumed fewer calories and lost 2 pounds during the 2 weeks that they consumed the minimally processed diet. Thus, the researchers concluded that reducing consumption of ultra–processed foods may be an effective strategy for weight loss, and obesity prevention and treatment.[34] More research is needed to determine the effects of ultra–processed foods on health.

Fructose Some medical researchers think the consumption of the simple sugar fructose is associated with the current obesity epidemic. Fructose is often added to ultra–processed foods and beverages to sweeten them. High fructose intakes stimulate fat synthesis in the liver, which contributes to the development of nonalcoholic fatty liver disease.[35] For a more extensive review of the role that carbohydrates may play in weight gain, see the "Do Carbohydrates Cause Weight Gain?" section of Chapter 5. Information about nonalcoholic fatty liver disease is also included in Chapter 5.

Genetic Factors

Genetics play a major role in the development of obesity. Most physical characteristics are inherited, including metabolic rate, hormone production, body frame size, and pattern of fat distribution. All of these characteristics affect body weight.

Some rats and mice are genetically predisposed to become obese because they have inherited genes for "thrifty metabolisms." Rodents with thrifty metabolisms have bodies that are more efficient at storing excess energy as fat than rodents that do not have such metabolisms. It is possible that humans who gain weight easily have genes that code for thrifty metabolisms as well. In ancient times, food was often scarce, and people had to eat as much as they could when food was available. During times when food was plentiful, individuals who had thrifty metabolisms stored more of the excess energy from food as body fat than persons who did not have such efficient metabolisms. The people who lacked thrifty metabolisms wasted the excess food energy as body heat. As a result, the energy–"thrifty" people were more likely to survive periods of starvation than the other persons. In many modern societies, however, high–calorie food is available 24 hours a day and starvation is unlikely. As a result, having thrifty metabolisms is no longer beneficial because depositing excess body fat often results in serious health problems.

If you gain weight easily, you may have inherited a thrifty metabolism. To prevent becoming overfat, you need to be physically active and make careful food choices. On the other hand, you probably do not have a thrifty metabolism if you can eat a lot of food and have difficulty gaining weight.

Medical researchers have identified several genes that contribute to human fatness. Genes also control hormones that regulate growth and metabolism, as well as ghrelin and leptin production. Certain mice, for example, become obese because they lack genes for synthesizing leptin (Fig. 10.16). Researchers are interested in developing medications that regulate the influence of certain genes over metabolism and hormone production. If research indicates that such medications are safe and effective, they could help people manage their weight over the long term.

What's the Set-Point Theory? The majority of people who intentionally lose weight regain the weight over time. According to the **set–point theory,** the body's fat content (and, therefore, body weight) is genetically predetermined. The set point acts like a home thermostat, except that it regulates body weight instead of temperature. For example, a person infected with an intestinal virus tends to lose weight because they have no interest in eating for a few days. During and after recovery, the person generally regains the lost weight. This observation provides support for the set–point theory. Results of studies involving mice support the set–point theory.[36]

When calorie intakes are reduced, blood thyroid hormone levels decline, depressing the normal basal metabolic rate. Additionally, the caloric cost of performing weight–bearing activities decreases when a person loses weight. As a result, an activity that required 100 kcal before weight loss may burn only 80 kcal after weight loss. Furthermore, weight loss appears to make the body become more efficient at storing calories from macronutrients as fat. When a person gains weight and stays at that weight for a while, their body tends to

Did *YOU* Know?

Prader-Willi syndrome is a rare genetic condition that results from the lack of genes in a particular section of a chromosome. People with this condition have skeletal deformities, delayed motor development, decreased intellectual functioning, food cravings, and insatiable appetites. If their access to food is not restricted, children with Prader-Willi syndrome eat constantly and become obese. There is no cure for Prader-Willi syndrome.

Figure 10.16 Genetic obesity in mice. Certain genetically engineered mice (the mouse on the right) become obese because they lack genes for synthesizing leptin.

set-point theory scientific notion that body fat content is genetically predetermined

establish a new and higher set point. According to the theory, all these changes protect the body from losing weight and explain why weight loss is so difficult to achieve and maintain.

Opponents of the set–point theory argue that weight does not remain constant throughout adulthood: The average person gains weight slowly, at least until old age. Thus, body weight may result more from lifestyle practices and environmental influences than predetermined biological controls such as a set point.

Environmental Influences

appetite desire to eat appealing food

Consider your eating behavior during a typical holiday meal that includes a variety of attractive, tasty foods. After eating the meal, your hunger should be satisfied, but as soon as pie or cake is placed on the table, do you "find some room" in your stomach for some dessert? If you often eat when you are not hungry or when you see a favorite food, then you are probably aware of the effect your environment can have on your appetite. **Appetite** is the desire to eat appealing food.

Food advertising is an aspect of the environment that has a powerful influence on your food choices. To entice you to buy their products, food manufacturers usually appeal to your senses, emphasizing the appearance and taste of food, in particular. Recently, a television ad for a fast–food chain promoted a hamburger that has three beef patties, three slices of cheese, and six bacon slices topped with mayonnaise. According to information at the fast–food company's website, this burger provides 800 kcal. The site, however, does not tell you that 800 kcal is more than one–third of a day's calorie needs for an average person! How do you respond when you see food advertisements on television? Do the ads make you hungry or eager to try a new food product?

Our environment also affects whether we choose to be sedentary or physically active. In our homes, we rely on a variety of "energy–saving" devices such as dishwashing machines, TV remote controls, and garage door openers to work for us. Outside our homes, we use cars, elevators, escalators, and other motorized devices, instead of our feet, to move us from place to place. With the help of machines, our lives are considerably easier; however, we are consuming more calories than we did in the past.

How does the presence of an appealing food affect your appetite?
Edward ONeil Photography Inc./iStock/Getty Images

Americans eat, on average, about 300 more kilocalories per day than they consumed 20 years ago.[37] The increase in energy intake may help explain why the prevalence of obesity has increased among Americans over the past 20 years. According to the principles of energy balance, excess energy intake in relation to energy output results in weight gain.

Many Americans have "desk jobs" that require little muscular movement. When we have some leisure time, we often spend it performing tasks that involve sitting—watching television, playing computer games, or chatting with people on the Internet. Only 24% of American adults meet recommendations for leisure–time physical activity.[38] According to experts with the Centers for Disease Control and Prevention (CDC), healthy adults should perform at least 150 minutes of moderate–intensity physical activity (brisk walking, for example) per week. Adults who include exercise in their weight–loss plan may need to engage in at least 60 minutes of moderate–intensity activity each day to achieve long–term maintenance of lower body weight.

If you live on or near a university or college campus, your environment probably provides ample opportunities for engaging in exercise and sports, such as tennis courts, swimming pools, and weight–training rooms. It is relatively easy to be physically active while you are in college, *if you choose to be.* After you graduate, consider what you will do to maintain a healthy level of physical activity each day, especially if you have a sedentary job. How likely are you to use the staircase in a building when you see the elevator? Will you keep a pair of comfortable shoes at work so you can walk for at least 20 minutes during lunch?

Did *YOU* Know?

Regular exercise, such as jogging, can shrink abdominal fat, but it is not possible to "spot-reduce" by exercising a fatty body part intensely.[12] The energy needed to fuel muscle activity comes from fatty deposits within muscle tissue and the rest of the body. Exercise, however, can improve muscle tone, so that fat tissue appears less flabby.

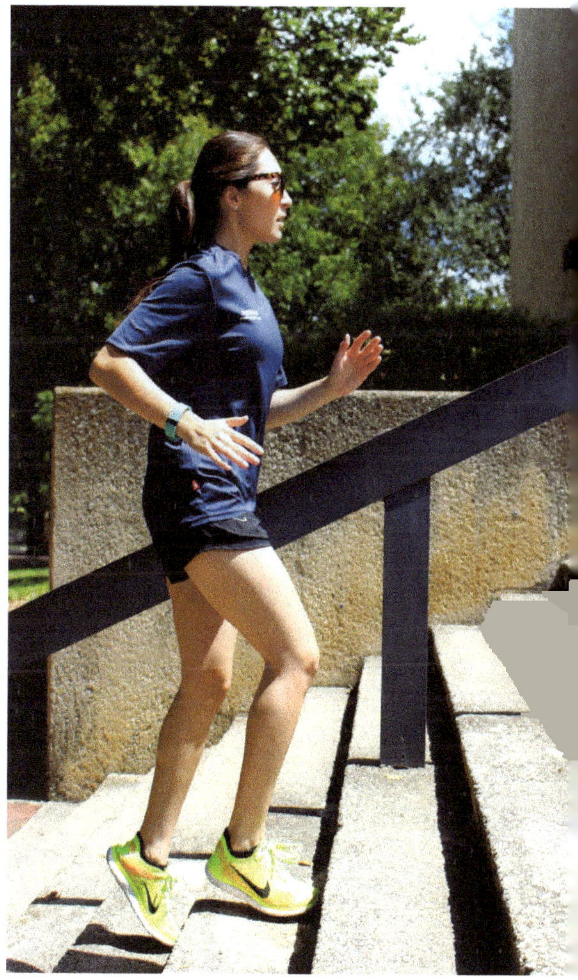

Elise Westmoreland/©Wendy Schiff

Genes and Environment: Interactions

It is difficult to determine the extent to which an obese person's genetic makeup or environment contributes to their excess body weight. Children are more likely to become overweight or obese if their biological parents were overfat prior to becoming pregnant.[39] This finding appears to support the hypothesis that obesity is an inherited trait. Nevertheless, genes do not control every–thing about our health, including our weight. Environmental and other factors can modify the expression of genes. For example, children whose mothers are obese may have inherited genes that increase risk for obesity. However, these children may avoid becoming obese if they adopt a physically active lifestyle and do not overeat. On the other hand, these children may become obese if their parents have poor eating habits and sedentary lifestyles and the young–sters follow their parents' practices.

Other Factors That Influence Weight

Psychological factors such as mood and self–esteem influence eating behav–iors and body weight. Many people eat not because they are hungry but because they are bored, anxious, angry, or depressed. Obesity may increase the likelihood of depression, especially in American non–Hispanic white females.[40] Researchers, however, cannot easily determine whether being obese causes depression or being depressed causes obesity.

Among some segments of American society, the ideal female figure is slim but curvy, and the ideal male physique is trim and muscular. As a result, societal pressures inspire many young females to idealize under–weight. Consider the body shapes of many fashion models, professional

Did *YOU* Know?

Infants born to females who smoked during pregnancy are more likely to become obese as adults than babies born to nonsmokers.[41]

Monkey Business Images/Getty Images

Societal pressures inspire many young females to idealize underweight people, especially thin female models and celebrities.
Iryna Kalchenko/123RF

ballerinas, and successful young actresses. These young females have so little subcutaneous fat that some of their bones protrude under their skin. In their relentless efforts to pursue such unrealistic body shapes, many young females adopt unhealthy and potentially life-threatening eating practices. The "Nutrition Matters" section of this chapter focuses on eating disorders, including *anorexia nervosa.*

Concept CHECKPOINT

20. What is hunger? What is satiety?
21. Discuss the roles of leptin and ghrelin in regulating hunger.
22. Under what conditions would having "thrifty genes" benefit a person?
23. What is the set-point theory?
24. What is the difference between hunger and appetite?
25. Describe how the environment influences a person's food intake and physical activity level.
26. Provide at least three examples of ways that physiological, psychological, and environmental factors can influence eating behavior.

See Appendix F for responses.

10.7 Weight Loss and Its Maintenance

Learning Outcomes

1. Identify four key elements that are important for weight loss and maintenance.
2. Describe features of reliable weight-loss plans or programs.
3. Discuss practical steps that people who are overfat can take to reduce their body fat.
4. Explain what members of the National Weight Control Registry do to maintain their reduced body weights.

Before you embark on an effort to lose (or gain) weight, an important first step is determining whether it is even necessary to change your weight. The need for changing your weight should be based on your overall health and family history of weight-related diseases. If you are dissatisfied with your body weight and shape, consider the following questions. Are you physically healthy at your present weight? If your BMI is within the healthy range, why do you think you need to lose or gain weight?

Using a BMI calculator (www.nhlbi.nih.gov/health/educational/lose_wt/BMI/bmicalc.htm) or consulting a BMI table such as the one in Figure 10.17 that indicates the range of healthy weights for a particular height (BMIs of 18.5 up to 24.9) can be helpful. According to this chart, an adult who is 5-feet, 5-inches tall has a healthy weight range of about 111 to 149 pounds.

To determine your BMI range using the table in Figure 10.17, locate your height in the left-most column with your left index finger, then locate your weight along the bottom line of the graph with your right index finger. Read across the row with your left finger and up from the bottom with your right finger, until your fingers meet. Note the color of the BMI range

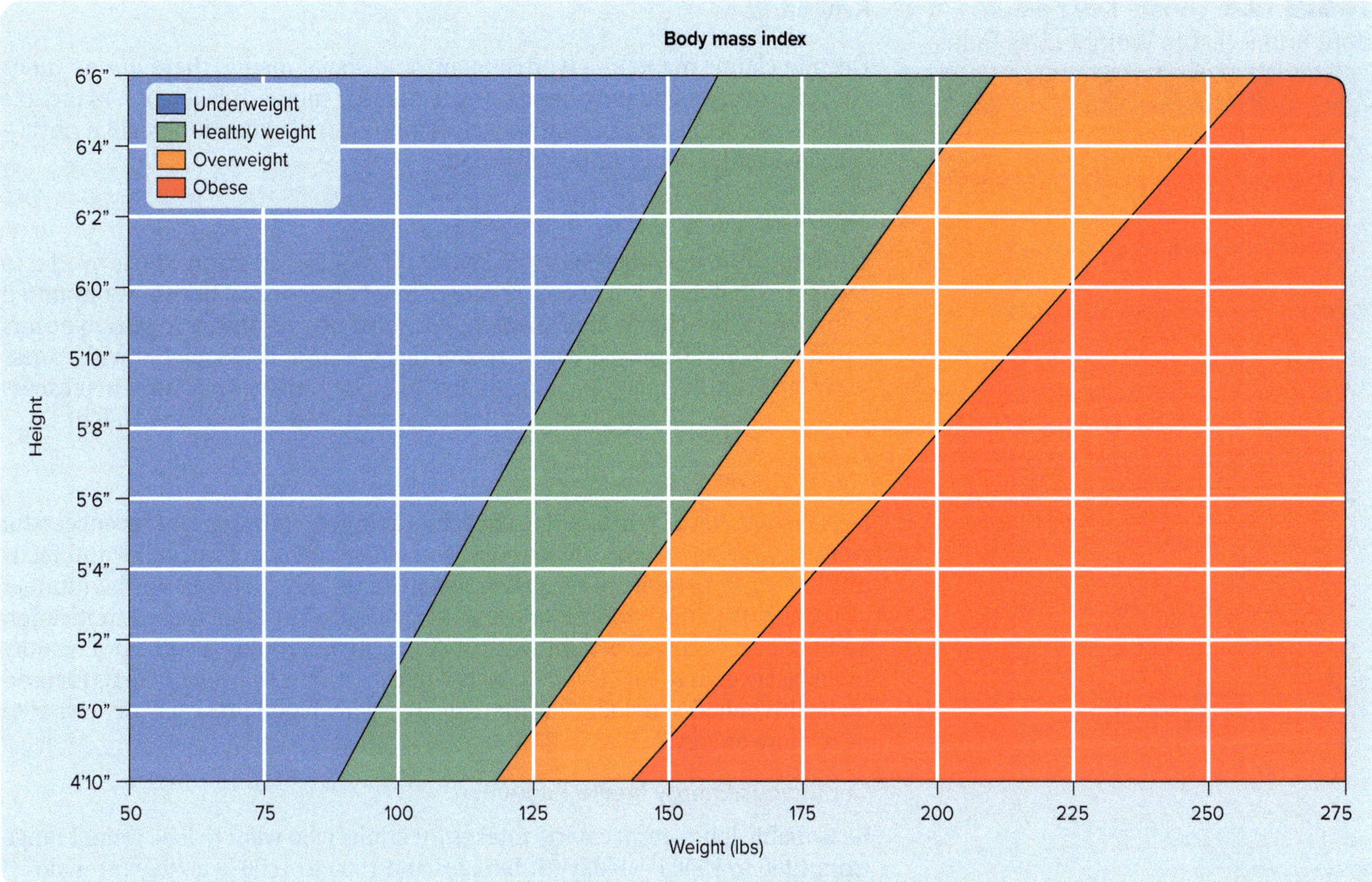

Figure 10.17 Adult BMI chart. Are you at a healthy weight?

where the two fingers meet and compare it to the color key in the upper left-hand corner of the figure. According to this graph, is your BMI in the healthy range?

The next step, if necessary, is setting a reasonable and realistic goal weight. It is important to note that a person who is overweight or obese does not have to shed a lot of weight to reduce risk factors associated with CVD, stroke, and type 2 diabetes. Just losing 5 to 10% of excess weight can increase beneficial high-density lipoprotein levels (HDL cholesterol), reduce elevated blood pressure and triglyceride levels, and improve glucose tolerance.[24]

Did *YOU* Know?

Only 10% of people who are obese in the United States seek medical treatment, including clinically proven lifestyle changes, prescription medications, and surgical procedures, to treat their obesity. Unfortunately, 82% of people who are overweight or obese believe that their excess body fat is their own fault and losing weight is solely their responsibility.[42] This incorrect belief, *weight bias* (see section 10.3), contributes to people who are obese not visiting a medical professional for treatment to manage their weight.

Scientists need to research ways to effectively reduce weight bias in individuals. However, education programs can bring awareness of the issue to the general public and health care professionals.

Features of Medically Sound Weight-Loss Plans

Table 10.6 presents some important features of reliable weight-loss plans. To lose excess pounds safely, the diet plan needs to be individualized as well as nutritionally adequate.[43] Such plans should be safe and effective, as well as flexible enough to meet the person's nutritional, psychological, and social needs. A medically sound weight-loss diet should emphasize a wide variety of low-calorie, readily available nutritious foods and be adaptable to the person's food likes and dislikes. Furthermore, reliable weight-loss plans should provide suggestions for altering environments that foster overeating and sedentary behaviors. Before beginning any weight-loss diet, people who are overfat should obtain their physician's approval, especially if they have serious health conditions, such as type 2 diabetes.

TABLE 10.6 Some Key Features of Safe and Reliable Weight-Loss Plans

A safe and reliable weight-loss plan:
• Is safe and effective.
• Meets nutritional, psychological, and social needs.
• Incorporates a variety of common foods from all food groups.
• Fosters slow but steady weight loss.
• Does not require costly devices or diet books.
• Accommodates family and restaurant meals, parties and special occasions, and food likes.
• Does not make the person feel deprived.
• Emphasizes readily available nutritious foods.
• Promotes changing habits that lead to overeating.
• Encourages regular physical activity.
• Provides suggestions for obtaining social support.
• Can be followed for a lifetime.

Kudryashova Alla/Shutterstock

Key Factors

Despite claims made in advertisements and social media, there are no quick cures for overweight and obesity. Key factors for successful weight loss include motivation, setting goals, continuous energy (calorie) reduction, regular physical activity, and behavior modification.

Motivation

The motivation to lose weight and keep it off requires a person who is overfat to recognize that there is a need to change their behavior and become committed to making those changes permanent. For some people, this recognition occurs when they are diagnosed with a health disorder that is associated with excess body fat. For some people, being unable to enjoy activities because their excess body weight restricts their movement triggers the decision to lose weight.

Setting Goals

Setting long-term, reasonable, and achievable goals is important for successful weight loss. People who create goals associated with healthy habits and focus on the health benefits of weight loss instead of only focusing on the number of pounds they need to lose or their "goal weight," have increased motivation and adherence to the weight loss process[44]; for example, "I will eat a healthy breakfast everyday" or "I want to walk 3 miles without stopping." Goals focused on health benefits include "I want to decrease my blood pressure" or "I want to have more energy."

Continuous Energy Intake Reduction

Reasonable long-term calorie intakes for adults who want to lose weight range from 1200 to 1500 kcal/day for females and 1500 to 1800 kcal/day for males.[24] At these calorie intake levels, careful food choices are necessary to obtain nutritionally adequate diets. Certain diets severely limit food intake and provide fewer than 800 kcal/day. Such *very-low-calorie diets* are not recommended without close medical supervision.[24] People who follow very-low-calorie diets may lose a lot of weight rapidly, but they tend to regain the weight quickly after they "go off" the diet and return to their former eating habits.

The healthy way to lower intake of total calories is to reduce consumption of added sugars, solid fats, and alcohol. People can also achieve negative energy states by reducing total calorie intake *and* increasing physical activity.[24]

No particular food has a "metabolic advantage" by promoting greater calorie burning by the body. A dieter's goal should be reducing total calorie intake while obtaining all essential nutrients by eating a variety of nutrient-dense foods.

Preparing nutritionally adequate but calorie-reduced meals and snacks can be easier when dietary experts have already done the calculations for consumers. For example, MyPlate provides patterns of food choices for 12 different calorie levels (1000 to 3200 kcal/day) that incorporate all food groups (www.myplate.gov/resources/MyPlatePlan). Chapter 3 provides more information about MyPlate. For more general information about weight control, obesity, and nutrition, visit the Weight-control Information Network (WIN) at www.niddk.nih.gov/. Other reliable sites for weight control information include the Academy of Nutrition and Dietetics at www.eatright.org.

Eventually, a person who intentionally loses weight by reducing energy intake and increasing physical activity reaches a weight *plateau.* When this occurs, the individual is in energy balance. To continue losing weight, the person must reduce their calorie intake or increase physical activity beyond the present levels. For someone who is consuming only 1000 to 1200 kcal/day, cutting calories even further can lower the metabolic rate, hindering weight loss.

A pound of body fat contains about 3500 kcal. This fact is often used to estimate how changes in caloric intake relate to weight changes: You should lose or gain approximately one pound of weight in a year for every 10 kcal/day reduction or increase of your caloric intake (10 kcal/day × 365 days = 3650 kcal). However, estimating the number of pounds that can be lost or gained in a period of time is not a simple calculation. Factors such as a person's age, weight, height, sex, and physical activity level contribute to their rate of weight change over time. According to a new formula, cutting your caloric intake by 10 kcal/day would enable you to lose only about half a pound in one year, and it would take over 3 years for you to lose a pound of weight, after which no further weight would be lost.[45] This formula helps explain why many people who reduce their usual calorie intakes do not lose as much weight as they expect over time and eventually their weight plateaus. To estimate your caloric intake to achieve weight change by using this formula, visit www.niddk.nih.gov/bwp.

Moderate-intensity activities, such as brisk walking, are recommended for people who want to lose or manage their weight.
Prasit photo/Getty Images

Regular Physical Activity

It is difficult to burn much energy without being physically active. By increasing their physical activity level and burning more calories, people do not have to limit their food intake as excessively as they would by relying on calorie reduction alone to lose weight. You do not need to jog for 10 miles to reap the benefits of engaging in regular physical activity. Moderate-intensity activities, such as brisk walking, are recommended for people who want to lose or manage their weight. Chapter 11 discusses the healthful benefits of a physically active lifestyle.

Behavior Modification

Controlling calorie intake and increasing physical activity are easier to accomplish if people who are overfat analyze their faulty behaviors, identify eating *cues* and "problem" habits, and develop ways to change the behaviors. Eating cues are usually environmental factors that stimulate eating behavior, such as seeing an ad for a fast-food restaurant or smelling freshly baked brownies when walking past a bakery. Identifying such cues can enable people to recognize effects of the signals and avoid inappropriate ones, whenever possible. By analyzing their food-related behaviors and habits, people who are overfat can often determine in which circumstances they tend to overeat.

Some people benefit from practicing *intuitive eating,* such as being aware of their physiological signs of hunger and satiety. Intuitive eating focuses on only eating when you are hungry and stopping eating when you are full. Scientists report that people who practice intuitive eating have fewer unhealthy behaviors such as binge eating (see section 10.11), reductions in BMI, and long-term weight maintenance.[44]

Although the process may seem slow, people are more likely to change ingrained habits by focusing on changing one behavior at a time. For example, many people snack on energy-dense foods and drinks while watching television. To change this habit, a person could decide to eat only at the kitchen table and avoid all food and beverage consumption while sitting in front of a TV.

For individuals who want to lose (or gain) weight, keeping records of food intake and physical activities can increase a person's self-awareness of their behaviors and can be helpful for estimating daily calorie input and output. However, people who are overfat often underestimate their calorie intake[48] and overestimate their energy output.[49] Therefore, people need to record information about food choices and physical activity habits accurately.

Tips for Modifying Food- and Exercise-Related Behaviors The following suggestions may help a person lose excess weight, as well as maintain a lower, healthier body weight.

Did *YOU* Know?

"Activity trackers" (also known as "fitness trackers") are available in smartphone apps and wearable devices similar to wrist watches. Such trackers are used to monitor specific aspects of fitness and health, including the number of steps taken while engaging in physical activities. Adults who use activity trackers often set a goal of walking 10,000 steps/day, which is about 5 miles for most people.[46] Because healthy adults generally walk 4000 to 18,000 steps a day, 10,000 steps a day is a reasonable activity goal.[47] However, people who are not physically active may find it difficult to walk 10,000 steps/day. Recently, researchers at Harvard Medical School determined that older females who walked 4400 to 7,500 steps a day lowered their risk of premature death by 41%.[46] Thus, people who increase their physical activity by walking at least 4400 steps/day may lower their risk of premature death.

Rawpixel.com/Shutterstock

Read Nutrition Facts panels to compare calorie, saturated fat, and fiber contents of packaged foods.
Andrew Resek/McGraw Hill

Using a small scale for weighing food can be helpful for limiting portion sizes.
©Wendy Schiff

- **Planning Menus**
  1. Plan meals and snacks to cover 3 or more days, then use the plan to prepare grocery lists. People who plan and prepare foods in advance have less stress around mealtime and improved diet quality![50]
  2. When menu planning, include sources of protein, unsaturated fat, and complex carbohydrates in meals and snacks.
  3. Avoid labeling certain foods as "off limits." Depriving yourself of such items can result in bingeing on the "forbidden" food. Learning to analyze why you have difficulty controlling your intake of these foods and developing strategies to learn how to reduce your intake of them can be very helpful.

- **Grocery Shopping**
  1. To reduce the likelihood of making impulsive food choices, shop for food *after* eating.
  2. Shop from a grocery list. If a food is not on your list, ask yourself if you really need to buy it and if it will hinder or help your weight-loss efforts.
  3. Read food labels to compare calorie, fat, and fiber contents per serving.

- **Food Preparation**
  1. Reduce the use of solid fat in cooking; bake, broil, or roast meats instead of frying them.
  2. Add less solid fat to foods such as cooked vegetables before serving or eating them.
  3. If you sample foods while preparing them, consider the amounts you ate and at mealtimes reduce your portion sizes accordingly.
  4. Prepare only enough food to provide one limited-size portion for yourself. Using measuring cups and a small scale for weighing food can be helpful.
  5. Take your usual-size portion and return one-third to one-half of it to the serving dish or container.
  6. Remove serving dishes from the table. Keeping foods or their containers in sight can encourage overeating.

- **Eating Behavior**
  1. Practice intuitive eating. Eat only when you are hungry and stop when you are full.
  2. Keep nutrient-dense, low-calorie snack foods, such as fresh fruits and vegetables, on hand.
  3. Eat meals and snacks at scheduled times; do not skip meals, especially breakfast.
  4. Eat all food in a "dining" area; avoid eating while engaged in other activities, such as reading a book, using your smartphone, or watching television.
  5. Leave some food on your plate.
  6. Become a "defensive eater." Practice ways to refuse food graciously or request smaller portions.

- **Holidays and Parties**
  1. Beforehand, think about what you will eat and drink while attending the event. Practice polite ways to decline food.
  2. Consider limiting your food intake before the special occasion to avoid consuming too many calories for the day.
  3. Eat a low-calorie snack about an hour before the occasion.
  4. Drink fewer alcoholic beverages. Replace alcoholic beverages with ice water or diet soft drinks.

- **Restaurants**
  1. Avoid fried menu items or those made with butter, gravy, or cream sauce.
  2. Choose pasta with red sauce instead of white sauce.
  3. Request salad dressing "on the side" so you can control the amount.
  4. Think "small." Order an entrée and share it with another person.
  5. Do not be a member of the "clean plate club." Ask your server for a carryout container when they bring your order to the table. Then divide the food in half and place one–half in the container. Be sure to refrigerate the leftovers within 2 hours after the meal and eat them for a meal the following day.
  6. If you choose a dessert item, share it with others.
  7. Avoid eating regularly at fast–food outlets.
  8. When at fast–food outlets, make substitutions, such as a salad instead of a fried fish sandwich, a regular hamburger instead of a specialty burger, or a roasted chicken sandwich instead of a breaded and fried chicken sandwich. Order a diet soft drink or water instead of a regular soft drink.

- **Physical Activity**
  1. Choose physical activities that you enjoy and can do without the need for expensive equipment.
  2. Increase the time you spend walking each day. Keep walking shoes where you can see them.
  3. Take breaks and move when sitting for long periods. For example, stand up and move for 3 minutes, twice per hour. Take a short walk, climb some stairs, or do some jumping jacks or squats.
  4. Take stairs instead of elevators or escalators whenever possible.
  5. Park your car farther from your destination and walk, if you feel it is safe to do so.
  6. Perform jumping jacks, push ups, sit ups, or lift handheld weights while watching television.
  7. Adopt moderate–intensity activities for your leisure time. For example, join a co–ed volleyball club, join a hiking or walking group, or take a ballroom dancing class.

- **Self–Monitoring**
  1. Track your food and physical activity habits using a special notebook, or the notepad application on your smartphone. There are also various smartphone apps designed to track these habits such as SparkPeople®.
  2. Using your notebook or smartphone, note the time and place of eating, as well as the type and amount of food eaten. Also record who was present and your mood when you ate meals and snacks.
  3. Use the information you record to identify your food–related problem areas, such as eating when bored or depressed.
  4. When tracking physical activity, record the form of exercise you performed and the number of minutes you spent engaging in that activity each day. Try to achieve at least 150 minutes of moderate–intensity activities each week.
  5. Measure your waistline weekly and keep a record of the measurements.
  6. Weigh yourself at least once a week, preferably at the same time and without clothing. However, do not rely only on your weight as an indication of your progress. Regular exercise often increases muscle mass that can result in weight gain or failure to lose weight. However, adding muscle mass is healthier than maintaining too much body fat.
  7. Ask a friend, family member, or medical professional to hold you accountable to your weight and health goals. Make sure you choose someone who

Food & Nutrition tips

- When you are hungry, drink some water or fat-free milk or eat a banana, an apple, a few whole dates, or a handful of raisins.
- Skipping meals can contribute to fatigue. Eating a nutritionally balanced breakfast and small between-meal snacks can provide the carbohydrates needed to maintain blood glucose levels.
- Include foods that supply some protein, healthy fat, and fiber in each meal to provide a sense of fullness and satiety.

Bananas: Ivan Danik/iStock/Getty Images; Apples: Photodisc/Getty Images; Raisins: Jacques Cornell/McGraw Hill

will not be critical but will support and encourage you. Scientists report accountability is more effective than only using self–monitoring to track your goal progress.[51]

- **Rewards for New Behaviors**
  1. Plan nonfood motivators or rewards for specific behaviors such as consistency in healthy eating habits or physical activity, or meeting a weight–loss goal. For example, "I'll spend an afternoon doing my favorite hobby once I walk 30 minutes a day, 5 days a week, consistently for 2 weeks," or "I'll buy those earbuds I've wanted when I lose 5 pounds."
  2. Encourage family and friends to provide praise and encouragement for your efforts to manage your weight. Let family, friends, and associates know that you are trying to lose weight and you would appreciate their help and support. Thank them when you receive their praise and support.

- **Changing Negative Thought Patterns**
  1. Do not get discouraged by occasional setbacks; relapses can be expected when changing behaviors. For example, instead of thinking, "I can't lose weight. I'm a failure," say to yourself, "OK, so I lost control and had too much to eat at the wedding. That's to be expected. I just need to get back on track. I'll pull in the reins on my eating for the next day and exercise more."
  2. Think positively about progress. "I didn't lose any weight this week, but I didn't gain any either. I must be losing fat and getting trimmer; those slacks fit better than they did 2 months ago."
  3. Counter negative thoughts with positive statements. "Next time, I'll eat two cookies instead of four."

JGI/Blend Images LLC

Community-Based Weight-Loss Programs

Many communities offer a variety of weight–loss programs. Registered dietitian nutritionists (RDNs) often conduct weight–loss classes at hospitals or universities. If a physician prescribes dietary counseling by an RDN, the patient's health insurance may cover the cost of such treatment.

Some commercial weight–loss programs, such as *Weight Watchers,* have diet plans that have been developed by RDNs. Members who have lost weight while following the plan often conduct local meetings. Before joining any weight–loss program, consumers should obtain answers to the following questions:

- How much does the program cost? Do I pay when I attend meetings? Do I need to sign a contract? If so, for what length of time?
- Do I have to buy special foods or dietary supplements?
- Is nutrition counseling provided? Do the persons providing nutrition counseling and information have degrees in nutrition and dietetics from accredited colleges or universities? How much contact will I have with a counselor?
- Was the diet plan developed by RDNs? Does the plan emphasize the importance of making lifestyle changes, including ways to increase physical activity?
- Does the program's advertising include questionable weight–loss claims and deceptive testimonials?

REAL PEOPLE | REAL STORIES

Valerie Haapala

Jan Haapala

Early in 2005, 24-year-old Jan Haapala did not feel well. He slept poorly at night, and during the day, extreme fatigue seemed to be his constant companion. Frequent indigestion added to his general discomfort. Concerned about her husband's health, Jan's wife, Valerie, urged him to see a doctor and have a physical exam. That trip to the doctor's office would prove to be a major turning point in Jan's life.

During the physical exam, Jan was shocked to learn he had hypertension. The physician prescribed medication to reduce his dangerously elevated blood pressure.

At the time of his doctor's appointment, Jan weighed 335 pounds and his BMI was over 45, which is in the "extremely obese" range. If Jan could not reduce his blood pressure and maintain it at a healthy level, he was very likely to die prematurely. Taking medication can reduce elevated blood pressure and keep it under control, but losing excess fat often cures hypertension, making the medication unnecessary.

Jan was aware that his lifestyle contributed to his extreme obesity. He typically ate two doughnuts and a 20-ounce serving of a sugar-sweetened soft drink for breakfast. Lunches and dinners were "big" meals, often composed of fast foods, especially double cheeseburgers and large portions of french fries and soft drinks. Candy bars were among his favorite snacks. "I knew those foods were unhealthy, but that didn't change what I ate once I sat down to eat." In addition to his poor food choices, Jan was physically inactive.

After returning home from the doctor's office, Jan made up his mind to improve his health. He recalls, "I didn't want to take medicine to stay alive." Determined to lose weight, Jan heeded his wife's suggestion to join Weight Watchers, one of the oldest and most reliable commercial group weight-loss programs in the country.

Although some people can lose weight on their own, others benefit from participating in group weight-loss programs that offer sensible and safe nutritional guidance, including tips for controlling the size of food portions and increasing physical activity. Group membership also provides emotional support from other members, a factor that is often critical for weight-loss success and maintenance.

By joining Weight Watchers, Jan acquired the information he needed to change his lifestyle to a healthier one. He also developed the motivation to follow a personalized diet and exercise plan that was very different from his past eating and physical activity patterns. In addition to choosing less fatty and sugary foods, he started a walking regimen that eventually became a running regimen. The pounds seemed to melt away from his body, and not long after he began to lose weight, he was able to discontinue taking the medication for hypertension.

When Jan joined the weight-loss program, his waistline measured 46 inches and he wore shirts with a 21-inch neck size. By the time he reached his goal weight of 210 pounds, his waistline was a trim 35 inches and his neck size measured only 15½ inches. Although it took 3 years, Jan had lost 125 pounds! The gradual but steady weight loss was an indication that he was making the kinds of behavioral and lifestyle changes that would be permanent.

Today, Jan is "feeling great" and running races from 5Ks to half-marathons. He has even completed his first marathon. His advice to people who are overfat and worried about what strangers will think while they exercise: "Don't be embarrassed about doing something in public that's good for you. People won't judge you negatively for trying to become healthier." Clearly, Jan is a role model and inspiration for anyone who is overweight or obese and wants to get their weight under control.

Did YOU Know?

Many people eat as a source of comfort and a way to cope with their negative emotions, such as stress, anxiety, anger, and boredom, which can contribute to excess calorie consumption and weight gain. Scientists report that people tend to seek high-calorie, high-fat foods during periods of stress, and when people are stressed, their bodies store more fat than when they are relaxed.[52] Some people emotionally eat because eating causes the release of dopamine, a neurotransmitter that makes them feel good. Thus, a person will eat to "feel better." Eating food also may be used as a distraction instead of dealing with an uncomfortable situation.

People who eat emotionally should find healthier ways to cope with their emotions, such as engaging in replacement activities instead of eating, such as going for a walk or taking a bath to relieve stress, or practicing *mindfulness* including guided meditation and yoga.[52] Most universities have free wellness programs for students that address mental health and emotional eating. A mental health professional or a registered dietitian is trained to help a person who struggles with emotional eating.

mheim3011/Getty Images

ilze79/123RF

Individuals who lose excess weight and maintain the weight loss tend to eat breakfast daily.
©Wendy Schiff

Successful Losers: How Do They Manage Their Weight?

The National Weight Control Registry tracks a group of over 10,000 adult Americans, mostly females, who have lost at least 30 pounds and maintained the weight loss for at least 1 year.[53] Information about members' nutrition–and exercise–related practices provides some insights into lifestyle practices that foster losing excess weight and maintaining the lower weight. The majority of registry members:[53]

- eat low–calorie, low–fat diets.
- eat breakfast every day.
- weigh themselves at least once a week.
- exercise, on average, for 60 minutes each day.
- limit television watching to less than 10 hours per week.

Concept CHECKPOINT

27. What are four key elements that are important for weight loss and maintenance?
28. List at least six features of reliable weight-loss plans or programs.
29. What questions would it be wise for consumers to have answered before they join a weight-loss group or plan?
30. List at least three steps that members of the National Weight Control Registry often take to maintain their reduced body weights.

See Appendix F for responses.

10.8 Medical Treatments for Obesity

Learning Outcomes

1. Discuss medications and surgeries that are options for treating obesity.
2. Explain the pros and cons of bariatric surgical procedures.

Patients with obesity are often unsatisfied with the amount of weight they lose while following fad as well as conventional diets. The frustration of repeated dieting leads some persons with obesity to turn to physicians for prescription medication and surgical procedures for managing their weight.

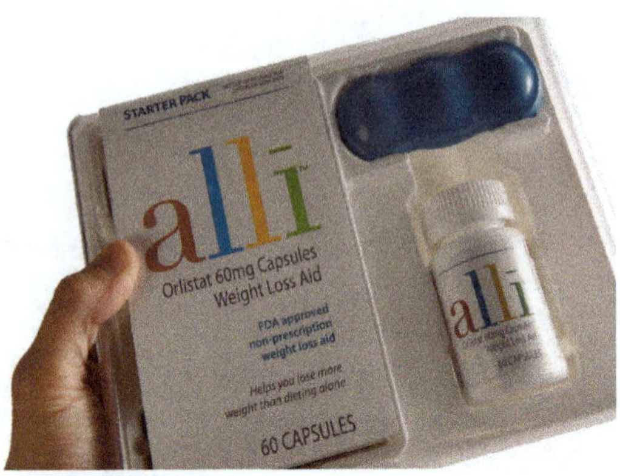

Alli® is a nonprescription form of orlistat available as a weight-loss aid for adults.
Jill Braaten/McGraw Hill

Weight-Loss Medication

As of January 2020, the U.S. Food and Drug Administration (FDA) has approved a few weight–loss medications that can be prescribed for long–term use, including Contrave®, Saxenda®, Qsymia®, Wegovy®, and Xenical® (orlistat). Some of these medications suppress or regulate appetite, but orli–stat reduces fat digestion by about 30%. The undigested fat is eliminated in the feces and can cause an oily, unpleasant discharge. The fat carries fat–soluble vitamins along with it, and these micronutrients are eliminated in feces as well. Therefore, patients using orlistat often need to take a vitamin supplement that contains fat–soluble vitamins. Females who are pregnant or may become pregnant should consult their physicians before using these medications. Although prescription medication can aid weight–loss efforts, their use does not replace the need to reduce calorie intake and increase physical activity.

Bariatric Surgical Procedures

Bariatric (*bar–ee–a'–tric*) **medicine** is the medical specialty that focuses on the treatment of obesity. Bariatric surgery can be an effective method of treating people who are extremely obese (have BMIs greater than or equal to 40) or have BMIs of greater than or equal to 35 and serious health problems because of their obesity.[24] Such surgical procedures drastically reduce the size of the stomach of a person with obesity, markedly limiting their food intake. Patients who have bariatric surgery may lose significant amounts of their excess weight. Most patients, however, regain some weight within 10 years after the surgery.[54] Aside from helping people with obesity lose considerable amounts of weight and maintain the loss, bariatric surgery can produce dramatic health benefits. Patients often achieve normal blood pressure, glucose levels, and triglyceride levels after surgery. Furthermore, overall death rates are lower for people with extreme obesity who lose weight after undergoing bariatric surgery. Such surgeries are relatively safe: Fewer than 1% of patients die as a result of a bariatric surgical procedure.[55]

In the United States, two common surgical approaches to treating obesity are the *Roux–en–Y* (*ru–en–why'*) *gastric bypass* and *gastric banding procedures.* Both of these operations can be performed *laparoscopically,* that is, by using several small incisions that allow surgeons to insert instruments and a video camera into the abdomen. Laparoscopic bariatric surgical procedures reduce recovery time and the risk of infections.

During the Roux–en–Y gastric bypass operation, the surgeon staples across the upper part of the stomach to create a small pouch. This procedure reduces the patient's stomach capacity to about 1.5 oz, which is approximately the volume of one egg. (Normally, the stomach's capacity is about 32 oz.) Additionally, the surgeon cuts the small intestine and attaches the lower end of it to the newly formed stomach pouch (Fig. 10.18a). The "bypassed" section of the intestine does not receive food, so digestion and absorption are reduced as a result of the surgery.

Complications often associated with gastric bypass surgery include intestinal blockage and bleeding, leaks along the staple site, blood clot formation, and wound infections. After surgery, gastric bypass patients can develop micronutrient deficiencies and some bone loss. However, patients can reduce their risk of nutrient deficiencies by taking vitamin and mineral supplements.

Sleeve gastrectomy is another form of bariatric surgery that reduces the stomach's size. During this procedure, the surgeon staples the stomach to form a banana–shaped pouch that holds about 2 to 5 ounces of food. The surgery does not involve bypassing a section of the small intestine, so nutrient absorption is not reduced (Fig. 10.18b). People who have undergone sleeve gastrectomy can eat only small portions of solid foods, which contributes to rapid weight loss. These patients also lose weight because they are not as hungry as they were before having the surgery. Why? The portion of the stomach that secretes ghrelin is removed during surgery, and as a result, hunger sensations are reduced. Sleeve gastrectomy surgery is irreversible because the unused portion of the stomach is removed.

By performing the *adjustable gastric banding* procedure, the bariatric surgeon creates the small stomach pouch with an adjustable band instead of fixed surgical staples (Fig. 10.19). By adjusting the tightness of the band, the surgeon determines the size of the stomach. Laparoscopic gastric banding

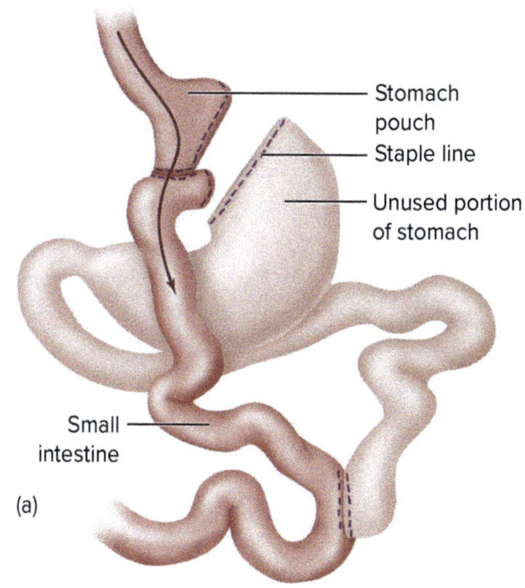

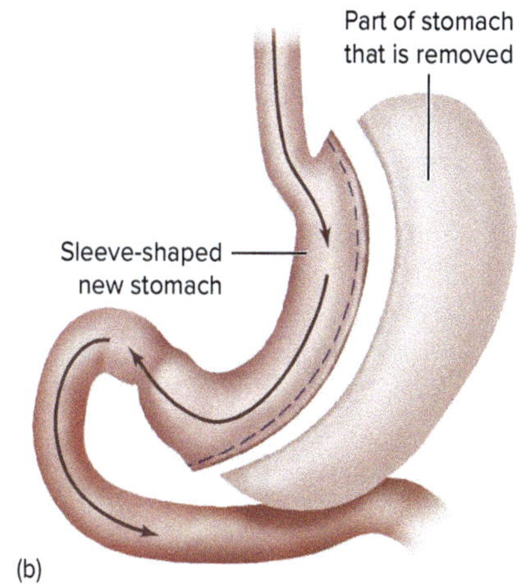

Figure 10.18 Common bariatric surgical procedures. (a) The Roux-en-Y gastric bypass procedure. (b) The sleeve gastrectomy procedure.

bariatric medicine medical specialty that focuses on the treatment of obesity

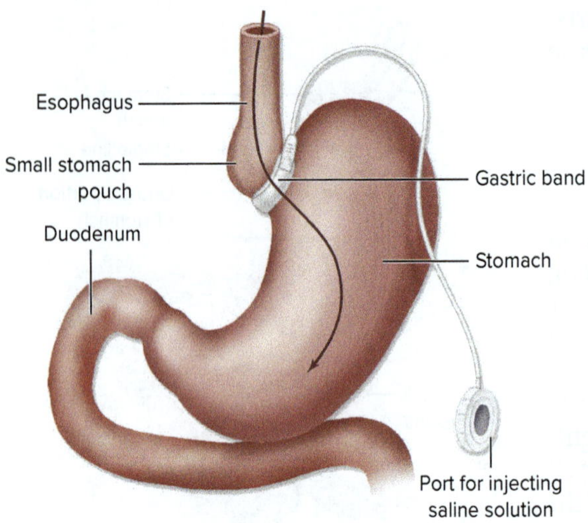

Figure 10.19 Adjustable gastric banding.

procedure is easier to perform and safer than the other types of bariatric surgery.[56] The procedure is also reversible. However, people with obesity who undergo the procedure lose weight more slowly and they lose less weight than those who undergo gastric bypass or sleeve gastrectomy procedures.

Bariatric surgeries result in weight loss partly because the stomach pouch fills quickly with food and patients experience satiety sooner than prior to surgery. Moreover, overeating causes discomfort or vomiting. Thus, people who undergo such surgical procedures must make major lifestyle changes, such as learning to plan and consume frequent, small meals. Engaging in regular exercise and avoiding "soft calories" (energy-dense, high-calorie foods such as milk shakes and ice cream) are also important for maximizing weight loss and maintaining the lower body weight over time.

People who participate in an intensive diet and lifestyle management program after surgery have more success losing weight and maintaining their weight loss over time when compared to bariatric surgery patients who do not have such treatment.[57] Nevertheless, some people require weight-loss medications along with dietary and other lifestyle changes to maintain weight loss after surgery.

FDA-Approved Devices

The FDA has approved a few unique weight-loss devices for certain adults with obesity. The Maestro Rechargeable System is a rechargeable electrical pulse generator that is surgically implanted under the skin in the patient's abdomen, along with its electrodes and wire leads. The device delivers intermittent electrical pulses to the major nerve that sends messages from the brain to the stomach. This particular nerve is involved in regulating stomach emptying after meals. The nerve also signals the brain to sense that the stomach is empty or full. The patient is able to recharge the pulse generator with a special external charger. According to results of a clinical study, the device helped adult patients with obesity lose weight, and most of the patients maintained their weight loss after surgery. Furthermore, the pulse generator was safe to use during the 2-year study.[58] Ongoing research will provide more information about the device's long-term safety and effectiveness.

As mentioned earlier in this chapter, people who are overfat often eat too much food because they do not sense stomach fullness as well as people who have BMIs in the healthy range. To help simulate the feeling of a full stomach, medical researchers developed intragastric (within the stomach) "balloons" as a temporary weight-loss aid for people who do not qualify for or do not want to undergo bariatric surgery.[59] Intragastric balloons do not require the need for surgery as do gastric bypass and banding procedures. The balloons help people lose weight because they effectively delay stomach emptying, which results in early satiety and reduced food intake.[60]

After positioning the balloon in the stomach, a physician fills the balloon with a salt and water (*saline*) solution. The saline expands the balloon to about the size of a grapefruit, so it takes up approximately one-third of the stomach's capacity.[61] The balloon remains in the patient's stomach for 6 months, at which time it is deflated and removed through the mouth. Results of several studies indicate intragastric balloons help patients achieve considerable weight loss after using the device for 6 months.[62] Patients, however, often regain much of that weight within the following 6 months.

To reduce the risk of serious complications or death, only physicians who have been properly trained should insert or remove intragastric balloons.[63] Furthermore, intragastric balloons should not be used in people who have had bariatric surgery or chronic digestive tract disorders.

Concept CHECKPOINT

31. Explain how certain prescribed medications can aid weight-loss efforts.

32. Describe common types of bariatric surgery in the United States. Why is bariatric surgery effective? Compared to gastric stapling procedures, what is a major advantage of gastric banding?

See Appendix F for responses.

10.9 Unreliable Weight-Loss Methods

Learning Outcome

1 Identify characteristics of weight-reduction methods that are likely to be unsafe and unreliable.

There are numerous products and services available for adults who want to lose weight. Some people who are overweight or obese join commercial weight–loss programs that have good track records for encouraging successful weight loss and maintenance. Other people who are overfat obtain individual dietary counseling from an RDN to help them lose weight. By recognizing that it took months and probably years to gain the excess fat, individuals may be more likely to accept advice and diets that result in slow but steady weight loss. However, many people who are overweight and obese seek quick fixes to lose weight, such as *fad* diets and dietary supplements promoted to "burn" or "melt" fat fast.

Did *YOU* Know?

According to a 2016 survey, 42% of American adults are trying to lose excess body fat.[64] To lose weight, the majority of people reported the following strategies:

- limiting portion sizes;
- performing different kinds of exercise;
- drinking more water;
- eating more fruits and vegetables;
- reducing sweets, junk food, and fast-food intake;
- reducing carbohydrate, fat, and calorie intake and skipping meals.

It is interesting to note that only 1 to 5% of people reported fad diets, liquid diets, "diet foods," or over-the-counter weight-loss products as weight-loss tools.

Fad Diets

A **fad** is a trendy practice that has widespread appeal among a population. After a period, however, people lose interest in the practice, and it becomes no longer fashionable. People often lose weight while following fad diets; however, they usually regain much of the weight that was lost while on the diet when they resume their prior eating and other lifestyle habits. Too often, people think of a weight–loss diet as a temporary change in their eating habits. Achieving a healthy body weight and maintaining that weight require making lifestyle changes that a person adopts for the rest of their life.

Fad diets often rely on *gimmicks.* A gimmick is a novel feature that makes the diet seem to be unique and more likely to work than other diets. Some fad diets use the gimmick of emphasizing one food or food group while excluding almost all others. The cabbage soup and grapefruit diets are examples of fad diets that promote eating single foods. People who are overfat may lose some weight while following eating plans that restrict food variety, but the weight loss occurs because the diet is low in calories, not because cabbage, grapefruit, or other "special food" contains compounds that cause rapid weight loss. Following such diets for a few days often results in boredom and monotony. Eventually, people who are trying to lose weight abandon such restrictive menu plans because they just cannot face another cup of cabbage soup or bowl of grapefruit. Weight regain occurs when the people return to their former eating habits—the habits that contributed to their original overweight or obese conditions.

Fad diets will come and go. However, if you examine any diet plan carefully, you can determine whether it is probably a fad. A typical fad diet:

- offers a quick fix; that is, the diet promotes rapid weight loss without calorie restriction and increased physical activity.
- is nutritionally imbalanced. The diet limits food selections from a few food groups, eliminates an entire food group, and may dictate specific rituals, such as eating only fruit for breakfast, eating only certain food combinations, or limiting the time when you can eat.

fad trendy practice that has widespread appeal for a period, then becomes no longer fashionable

- requires buying a book or various gimmicks, such as expensive dietary supplements, weight–loss patches, or cellulite–reducing creams.
- uses outlandish and unscientific claims to support its usefulness. For example, *The Beverly Hills Diet* book promoted the notion that people become fat because food "gets stuck" in their bodies, rots, and produces toxins. Another author of a fad diet book claimed people can lose weight by following diet plans based on their blood types. These notions and recommendations are not supported by scientific evidence.
- relies on testimonials from famous people or connects the diet to trendy places, such as Beverly Hills, California, and South Beach, Florida.
- does not emphasize the need to change eating habits and physical activity patterns.

Low-Carbohydrate Approaches

Fad weight–loss diets that limit carbohydrate intakes are high in protein and fat. A weight–reduction diet is low carbohydrate if it eliminates or severely restricts the intake of carbohydrate–rich foods such as breads, cereals, fruits, vegetables, and sweets. The lack of variety often leads to boredom with the selection of foods, and as a result, people lose weight because they tend to eat less.

Low–carbohydrate diets usually produce rapid weight loss initially, primarily because the body loses water. Why? The body produces less glycogen when carbohydrate intake is low and uses much of its stored glycogen to supply glucose for energy. Tissues maintain about 3 grams of water with each gram of glycogen, so a reduction in body glycogen content results in the need for less water to store with it. The kidneys eliminate the excess water in urine. Furthermore, a very–low–carbohydrate intake causes the liver to produce glucose, mostly from certain amino acids supplied by the body's tissue proteins. Protein tissue also contains a lot of water. When protein–rich tissues are dismantled and their amino acids used for energy, the water that was stored with the proteins also ends up in urine.

Not all forms of carbohydrate are unhealthy. Foods that are rich sources of dietary fiber, such as whole grains, beans, and whole fruits and vegetables, are considered to be sources of "healthy carbs." The human digestive tract digests such foods more slowly than foods that have more refined carbohydrates, which may promote satiety. Eating a low–carbohydrate diet may produce rapid weight loss, but choosing foods that are low in refined carbohydrates and rich sources of less–refined carbohydrates may help individuals achieve and maintain a healthy body weight.

An analysis of nearly 50 studies indicated that over the long term, people lose similar amounts of weight regardless of whether they follow a low–carbohydrate or a low–fat diet.[65] Low–carbohydrate diets, however, can result in more favorable high–density lipoprotein ("good cholesterol") and serum triglyceride levels.[66] Recall from Chapter 6 that increased high–density lipoprotein levels and decreased serum triglyceride levels are associated with lower risk of CVD.

Fad diets that eliminate an entire food group, such as fruits, are likely to be nutritionally imbalanced.
Pixtal/age fotostock

Very-Low-Fat Approaches

Very–low–fat diets are actually very–high–carbohydrate diets. These diets supply approximately 5 to 10% of calories from fat and generally result in rapid weight loss when followed consistently. The most notable are the Pritikin Diet and Dr. Dean Ornish's "Eat More, Weigh Less" diet plans. Very–low–fat diets are not harmful for healthy adults, but they are difficult to follow for the long term. Fat contributes to the flavor and texture of foods. Extremely low–fat diets are not tasty, and they eliminate many foods that are usually high on peoples' favorite foods lists, such as ice cream and meat. Although grains, fruits, and vegetables are nutrient–dense foods, eating them repeatedly and without fat can cause "diet boredom."

Did *YOU* Know?

In Section 10.7, we discussed the importance of continuous energy reduction as a key factor for losing weight, particularly excess body fat. Registered dietitian nutritionists generally recommend weight-loss plans that reduce calorie intake by a specified number of kilocalories per day. However, *intermittent fasting* (*IF*) is gaining popularity as an alternative weight-loss method that also limits calorie intake. There are several different forms of IF, and they typically feature a cyclic pattern of consuming solid foods as desired and then avoiding most or all foods (fasting) over a consistent period. One form of IF, for example, requires fasting every other day; another type of IF restricts food intake to certain hours of the day, such as between 8 A.M. and 4 P.M.

Sergii Gnatiuk/123RF

Medical researchers have conducted numerous studies using free-living human subjects who were overweight or obese and did not have diabetes to determine the effectiveness of IF and its possible health benefits beyond weight loss. In many instances, the findings of these studies were inconsistent. Some results indicated that people lost weight while following IF because they reduced their caloric intake, but other findings did not show IF was superior to a control diet (a "regular" diet or a continuous calorie-restricted diet) for losing weight. In one study that examined the effectiveness of IF for losing weight, almost 40% of the subjects in the IF group did not complete the study.[67] This "dropout" rate was higher than the dropout rates of the calorie-restricted and the control groups. A high dropout rate for a weight-loss plan may indicate that the plan is too difficult for people to follow for long periods.

An evaluation of results found in many studies (a *meta-analysis*) showed IF is not better for losing weight than a control diet.[68] Nevertheless, IF appeared to help people maintain lean body mass, lower BMI, reduce insulin resistance, and improve blood glucose control. Such physiological changes are beneficial for individuals who have too much body fat. People should consult with a physician to determine whether they are in good health and can safely use IF.

Dietary Supplements for Weight Loss

Many people who are overweight or obese are attracted to dietary supplements for weight loss because they believe promoters' claims that their products are "magic bullets" for shedding unwanted weight quickly and effortlessly. Although several different types of weight-loss supplements are available, these products generally have not been scientifically tested in humans for safety and effectiveness.

In 2004, the FDA banned the sale of most dietary supplements that contained the natural stimulant ephedra (*eh-feh'-dra*), or ephedra-related compounds, after the agency received reports of serious side effects and even deaths resulting from use of these products. Table 10.7 presents science-based findings about some popular weight-loss supplements, including their potential usefulness and safety concerns. At this point, medical experts do not recommend any dietary supplement for weight loss.[69]

At present, registered dietitian nutritionists do not recommend any dietary supplement for weight loss.
Don Farrall/Photodisc/Getty Images

Analyzing Advertising Hype for Weight-Loss Supplements

Maybe you have heard or read remarkable claims for weight-loss products such as "Lose weight while you sleep," "Lose 30 pounds in just 30 days," and "Eat anything you want and still lose weight." In general, advertising claims that a product promises quick and easy weight loss are too good to be true.

A Federal Trade Commission (FTC) study identified several features, including typical claims, for popular weight-loss products and services. The agency's report also provided reasons why you should be skeptical of such features when they are used in magazine ads, in television infomercials, or on Internet websites. According to the FTC, you should be wary of claims that the product or service:

- *Causes rapid and extreme weight loss.* Ads commonly use outrageous claims such as "Lose up to 18 pounds in one week!" to attract consumers. The use of the modifier "up to" means that the person using the product could lose considerably less than 18 pounds a week.
- *Requires no need to change dietary patterns or physical activity.* Principles of energy balance do not support claims such as "Lose weight without dieting or

TABLE 10.7 Summary of Selected Weight-Loss Supplements

Supplement	Usefulness	Side Effects/Safety Concerns (Usual Doses)
African mango (*Irvingia gabonensis*)	Possible modest effect on body weight	Headache, intestinal gas, and sleep disturbances
Beta-glucans	Not effective	Intestinal gas
Bitter orange	May increase metabolic rate and suppress appetite Mixed results concerning weight loss	Some negative effects reported, including chest pain and increased blood pressure and heart rate
Capsaicin and other capsaicinoids	Not effective	Gastrointestinal distress, increased insulin levels, and decreased HDL cholesterol
Carnitine	Possible modest effect on body weight	Nausea, vomiting, diarrhea, and abdominal discomfort
Chitosan	Little effect on weight loss	May cause gastrointestinal discomfort including nausea, constipation, and intestinal gas
Chromium	May enhance weight loss to a small extent	Headache, diarrhea, and dizziness
Conjugated linoleic acid	Has minimal effect on reducing body weight	Appears to be safe but can cause gastrointestinal upset
Ephedrine (ma huang, ephedra, ephedra sinica, sida cordifolia, pinellia, and ephedrine with caffeine)	Ephedrine-containing products promote short-term weight loss but also increase risk of serious side effects. Ephedra-containing dietary supplements are banned in the United States.	May cause rapid heart rate, elevated blood pressure, dizziness, sweating, headache, and sleep disturbances Linked to heart attacks, strokes, and deaths
***Garcinia cambogia* (hydroxycitric acid, HCA)**	Overall evidence does not suggest usefulness.	May cause headache, nausea, and gastrointestinal discomfort
Glucomannan	Little or no effect on body weight	Tablet forms can block esophagus. Diarrhea, intestinal gas, and abdominal discomfort
Green tea and green tea extracts	May enhance weight loss to a small extent	Contains caffeine Concentrated extracts linked to severe liver damage
Guar gum	Not effective	May cause diarrhea and intestinal gas
Hoodia	Not effective	Can increase blood pressure and heart rate
Pyruvate	Possible but minimal effect on body weight	None reported, but long-term safety is unknown
Raspberry ketone	Lack of scientific support for effectiveness	Unknown
Spirulina (blue-green algae)	Not effective	Unknown
Yohimbe	Not effective	Serious negative health effects including hypertension and rapid heart rate

Source: National Institutes of Health, Office of Dietary Supplements: *Dietary supplements for weight loss.* Updated 2022. https://ods.od.nih.gov/factsheets/WeightLoss-HealthProfessional/. Accessed January 2023.

strenuous exercise" and "Eat as much as you want—the more you eat, the more you'll lose." Regular exercise and moderate energy intake are necessary for weight loss and long–term maintenance.

everydayplus/123RF

- *Results in permanent weight loss.* Claims such as "Discover the secret to permanent weight loss" and "Lose weight and keep it off" often appear in ads. These claims target consumers who have lost weight but gained it back and are wary of weight–loss products. Long–term weight loss is difficult to achieve without calorie reduction and regular exercise, and claims that permanent weight loss can result simply from using a product are questionable.[70]
- *Is scientifically proven or doctor endorsed.* Some ads claim their product or service has been "clinically tested," "scientifically proven," or "physician

recommended." Scientific testing of the product or service supposedly occurred at "respected" or "leading" medical centers or universities. However, most ads do not provide information about testing sites or journals where the results were published. Such information is critical for assessing the reliability of claims. Endorsements by "doctors" or medical professionals can be misleading. A doctor could be someone with a bogus doctorate degree (PhD) or a PhD in a nonscientific field. Moreover, consumers need to be aware that the "professionals" pictured or featured in the ads may be models or fictional characters.

- *Includes a money–back guarantee.* Consumers should recognize that a product does not necessarily work just because it is guaranteed. The FTC frequently sues companies that fail to return money to dissatisfied consumers, as their ads guaranteed.
- *Is safe or natural.* Ads may include safety–related claims, such as "proven 100% safe" or "safe, immediate weight loss." Additionally, the term *natural* often accompanies safety claims, implying that "natural" weight–loss products are safer than prescribed weight–loss treatments. Despite such assurances, weight–loss supplement manufacturers usually have little scientific evidence to support safety claims, particularly concerning long–term use of their products. Furthermore, "natural" does not indicate safety. Mushrooms are natural, but many species of the fungi are highly toxic and even deadly.
- *Is supported by satisfied customers.* Ads for weight–loss products or services typically feature testimonials from satisfied users. The assumption is that if the product worked for the person providing the testimonial, it should work for anyone. According to the FTC, testimonials generally provide little reliable information about what consumers can expect from using the product.
- *Displays before–and–after photos.* Many ads use photos of "satisfied" customers to support claims that their weight–loss products are effective. In the typical "before" photo, the subject has poor posture, no smile, unkempt hair, and unfashionable, unflattering clothing. In the "after" photo, the person stands with their shoulders held back and abdomen tucked in. Additionally, the subject is usually smiling and appears more attractive than in the "before" photo. If you read carefully, you may find disclaimers in small print, such as "results not typical," at the bottom of the "after" photo.

Did *YOU* Know?

In 2017, the developers and promoters of a nationally marketed weight-loss program and its accompanying diet products settled a $2 million lawsuit that had been brought against them by the FTC. To meet the terms of the settlement, the promoters agreed to stop selling their weight-loss products and program by using what the FTC considered unfair and deceptive marketing practices.[71] According to the FTC's lawsuit, the weight-loss program's promoters claimed their new "technology" would enable people to lose 20 to 40 pounds in 40 days and keep the weight off without the need to follow a calorie-restricted diet. The FTC indicated that such claims were false or not supported by scientific evidence. For more information about this weight-loss program and its lawsuit and settlement, visit www.ftc.gov/news-events/news/press-releases/2017/04/marketers-nutrimost-ultimate-fat-loss-system-settle-ftc-charges.

U.S. Federal Trade Communication

Concept CHECKPOINT

33. What is a "fad" diet? List at least four typical features of fad diets.

34. Why do fad diets and dietary supplements promoted for weight loss appeal to people who are overfat?

35. Identify at least four popular dietary supplements that are promoted for weight loss, and indicate whether each supplement is safe and effective.

36. Discuss at least three features or claims that are commonly used in ads for weight-loss products or services.

See Appendix F for responses.

10.10 Gaining Weight

Learning Outcomes

1. Explain why some people are underweight.
2. Discuss ways to gain weight safely and sensibly.

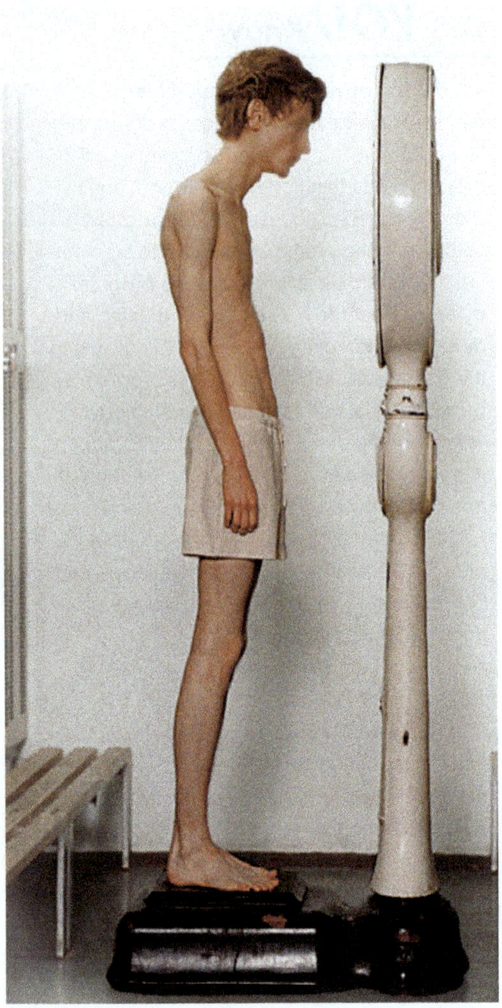

Many underweight individuals are interested in gaining weight, especially muscle mass.
Julia Christe/fStop/Getty Images

In 2017–2018, about 1.6% of Americans who were 20 to 39 years of age were underweight.[72] An **underweight** individual has a BMI that is less than 18.5. Factors that contribute to underweight include genetics, lifestyle practices, serious chronic diseases, and psychological disturbances.

It is often difficult to pinpoint a cause of underweight; multiple factors contribute to having a lower-than-average body weight. Individuals who inherit higher resting metabolic rates, tall body frames, or both may find it difficult to gain weight. Excessive physical activity can result in low body weight. Compared to sedentary adults, the bodies of rapidly growing, physically active children and adolescents have higher energy needs. If these children do not consume enough energy, they can lose weight. Chronic diseases such as cancer, tuberculosis, AIDS, and inflammatory bowel disease often result in severe weight loss that is difficult to treat. Some people who suffer from depression fail to eat enough food to support their energy needs, and they lose weight as a result.

If a person's BMI was within the healthy range before excessive weight loss occurred, an evaluation by a physician may be necessary to determine the cause or causes of the loss, especially when the underweight person has not tried to shed pounds. A thorough medical examination can rule out possible reasons for unintentional weight loss, such as hormonal imbalances, depression, cancer, and infectious or digestive tract diseases.

Many individuals who are underweight want to gain weight, especially muscle mass. For a person who is underweight, gaining weight can be just as challenging as losing weight is for a person who is overfat. To gain weight, adults who are underweight can gradually increase their consumption of calorie-dense foods, especially those high in healthy fats. Fatty fish, such as salmon; olives; avocados; seeds; low-fat cheeses; nuts and nut butters; and granola made with dried fruit, seeds, and nuts are high-calorie nutritious food choices with low saturated-fat content. Additionally, people who are underweight can replace beverages such as soft drinks with more nutritious calorie sources, such as 100% fruit juices, smoothies, and milk shakes made with peanut butter and reduced-fat ice cream. Following a regular meal and snack schedule also aids in weight gain and maintenance.

Sometimes people who are underweight are too busy to eat, and as a result, their caloric intakes are too low to support weight gain. If physical activity habits contribute to their inability to increase their weight, people who are underweight can find ways to be less active. If their weight remains low, people who are underweight can add muscle mass through a *resistance-training* (weight-lifting) program, but they must increase their calorie intake to support the additional exercise. Otherwise, gaining muscle tissue is not likely to occur.

underweight describes person with a BMI of less than 18.5

In many instances, healthy people who are underweight may have to accept their body builds. Furthermore, they can realize the health benefits of being lean and the sheer enjoyment of being able to eat a variety of foods without gaining weight. For the typical slim person, the passage of time is usually all that is necessary for weight gain to occur, as the aging process is often accompanied by increasing body fat.

Concept CHECKPOINT

37. List at least three health conditions that are often associated with being underweight.

38. Discuss at least three measures a person who is underweight can take to gain lean mass safely.

See Appendix F for responses.

10.11 Nutrition Matters: Eating Disorders and Disordered Eating

Learning Outcomes

1. Explain the difference between having an eating disorder and practicing disordered eating.
2. List the major types of eating disorders and their signs.
3. Identify risk factors for eating disorders, disordered eating practices, and the athlete triad.
4. Describe the health consequences of eating disorders and professional forms of treatment for these conditions.

Many American adolescents and young adults are dissatisfied with their body shape and weight. Among some segments of society, the ideal female figure is slim but curvy, and the ideal male physique is lean and muscular. The media constantly bombard consumers with images of the "ideal" body. Television shows and movies often portray thin females or muscular males as happy and successful. As a result, societal pressures inspire many young people to idealize underweight. In their efforts to pursue such unrealistic slender body shapes, young people may adopt unhealthy and potentially life-threatening eating practices. This "Nutrition Matters" section focuses on eating disorders and disordered eating behaviors.

Eating disorders are psychological disturbances that lead to abnormal physiological changes and dangerous health complications. An eating disorder is not the same as disordered eating. **Disordered eating** is chaotic and abnormal food-related practices, such as skipping meals, limiting food choices, following fad diets, and bingeing on food. Disordered eating behaviors are temporary, and they often occur when a person is under a lot of stress or wants to lose weight quickly. When a person adopts disordered eating behaviors as a lifestyle, the practices can become harmful and lead to serious eating disorders.

Major Types of Eating Disorders

According to guidelines published in the fifth edition of the American Psychiatric Association's (APA's) *Diagnostic and Statistical Manual of Mental Disorders* (*DSM*-5), physicians can diagnose three main types of eating disorders:[73]

- Anorexia nervosa (AN);
- Bulimia nervosa (BN); and
- Binge-eating disorder (BED).

Eating Disorders: Risk Factors

There are several known risk factors for eating disorders. These factors include being female, having a poor self-image, having low self-esteem and a perfectionist personality, being in a dysfunctional family environment, and being teased about weight. Table 10.8 lists these and several other risk factors.[74,75]

A person's age is a major factor in predicting their risk of a particular eating disorder. BED is more likely to occur in adults, whereas AN and BN tend to begin during adolescence.[76] Results of studies of adolescent females conducted over a long period of time indicated that 13% developed an eating disorder.[77] Factors that contribute to an adolescent's vulnerability to develop AN or BN generally include:

- increases in body weight and size, as well as changes in body shape that normally occur as a teenager develops into an adult;
- greater dependence on peers for their emotional support and social approval; and
- greater freedom from parental control over food choices and eating practices.

TABLE 10.8 Risk Factors for Eating Disorders

• Being female
• Being an adolescent (AN and BN)
• Having a history of frequent dieting
• Having a first-degree relative with an eating disorder
• Placing a high degree of importance on having an "ideal" body shape
• Being dissatisfied with one's body shape
• Having a poor self-image and low self-esteem
• Having a perfectionist personality
• Being from a dysfunctional family
• Being teased or bullied about one's weight
• Being in an occupation or sport that emphasizes a lean body build

eating disorders psychological disturbances that lead to certain physiological changes and serious health complications

disordered eating chaotic and abnormal food-related practices

Causes of Eating Disorders

The causes of eating disorders are complex and include genetic, biological, social, psychological, and environmental factors. The following sections discuss several of these factors.

Genetic and Biological Factors

People who have a first-degree relative (a sister or mother, for example) with AN are 10 times more likely to have an eating disorder than people who do not have any relatives with AN.[78] Specific genes that modify the brain's ability to regulate moods may have a role in development of eating disorders.[79] More research, however, is needed to clarify how genes affect eating behavior.

Psychological and Personality Factors

Mood and anxiety disorders, as well as substance abuse, often accompany eating disorders. It is not clear whether eating disorders cause psychological problems or are the result of emotional disturbances. Furthermore, people with eating disorders typically have a distorted view of their body weight or shape, and they often place a great deal of importance on "ideal" body weight, shape, or specific body parts. Having a poor self-image and being dissatisfied with one's body shape are risk factors for developing an eating disorder.

Other Factors

National origin, income level, occupation, and lifestyle choices contribute to the development of eating disorders in people who are vulnerable to developing them. Eating disorders are more prevalent in industrialized, high-income countries such as the United States, Australia, Canada, Japan, and many European nations.[80] Individuals who participate in activities or occupations that encourage thinness such as ballet, acting, and fashion modeling have an increased risk for developing an eating disorder, especially AN and BN (Fig. 10.20). Athletes, especially those participating in endurance running, swimming, and gymnastics, are also at risk. The "Athlete Triad" section of this "Nutrition Matters" provides more information about eating disorders among athletes.

Figure 10.20 Risky occupation. People in occupations that encourage thinness, such as ballet, have an increased risk of developing an eating disorder. Nanette_Grebe/iStockphoto/Getty Images

Adolescents who are at risk of an eating disorder, such as AN, may develop the condition in response to a stressful life event (a "trigger"). Moving away from home to attend college is an example of a trigger. Additionally, the manner in which a family interacts and functions can increase the risk of developing an eating disorder, especially among adolescents. Teenagers with eating disorders tend to think their parents are more controlling and less affectionate than the parents of healthy youths.[81] In general, individuals with eating disorders describe having a poor emotional connection to their family.

Parents can influence whether their children develop an eating disorder because they play a role in establishing healthy attitudes about body weight and shape, as well as healthy eating practices. Adolescents may model unhealthy behaviors of their parents, especially when the teens' mothers are overly focused on their own weight.[82] Furthermore, parents who tease their children about their body weight may foster the development of disordered eating behaviors.[83] On a positive note, parents and other caregivers can play an important role in preventing eating disorders among their children by modeling healthy eating behaviors and expressing a positive attitude toward their own bodies.

The following sections provide more specific information about AN, BN, and BED. We also discuss other conditions that are characterized by unhealthy eating and exercise patterns, and we describe measures to prevent and treat these conditions.

Anorexia Nervosa

It is not unusual for *anorexia,* loss of appetite, to occur when a person is ill. The anorexia generally resolves as the individual recovers. **Anorexia nervosa (AN)** is a severe psychological disturbance characterized by self-imposed starvation that results in malnutrition and very low body weight (*emaciation*). A BMI of 18.5 is the lowest limit of what is generally considered a normal or healthy body weight for adults.

anorexia nervosa (AN) severe psychological disturbance characterized by self-imposed starvation

According to estimates, about 0.5% of American females have AN.[84] Although females comprise three times more AN cases than males,[76] males often experience a delay in diagnosis of AN because people assume males are not at risk for eating disorders.[85]

Signs and Symptoms

People with AN have distorted body images and severely restrict their food intake, which leads to rapid weight loss and maintenance of an unhealthy low body weight. Even when they are emaciated, the patients deny they are too thin, and they are overly concerned or fearful about becoming fat. The fear of weight gain or becoming overweight compels people with AN to act in ways that hinder weight gain. These behaviors can include frequent food restriction (dieting), fasting, or exercising excessively.

Some forms of AN involve bingeing and purging behaviors. **Purging** includes activities that limit calorie intake (self-induced vomiting after eating, for example) or increase calorie output by performing more physical activity than is necessary for optimal health. Although AN does not typically involve purging behaviors, when fasting or food restriction becomes difficult to maintain, someone with AN may try to compensate for eating by self-induced vomiting, misuse of laxatives or diuretics, or excessive exercise.[84]

Because people with AN are so thin, they lack adequate amounts of subcutaneous fat to insulate their bodies and have trouble maintaining their body heat as a result. Therefore, individuals with the disorder often wear layers of clothing to keep warm and hide their emaciated appearance. People with AN also have lanugo (*la-new'-go*), which is dense, fine white hair that grows on the face and other parts of the body. Lanugo may help retain body heat. Table 10.9 lists lanugo and some other common signs of AN.

AN has the highest mortality rate of any psychological disorder.[84] As many as 6% of people with AN eventually die as a result of the disorder.[84] Primary causes of death include suicide, heart failure, and alcoholism.[86]

TABLE 10.9 Common Health Consequences of Anorexia Nervosa

• Severe constipation
• Slow heart rate
• Hypotension (low blood pressure)
• Loss of normal menstrual cycles (females)
• Low bone mineral density
• Lanugo (delicate, dense, white hair on the skin)
• Low thyroid hormone and decreased metabolic rate
• Muscle wasting and weakness

Bulimia Nervosa

Bulimia nervosa (BN) is a severe psychological condition characterized by repeated episodes of binge eating followed by purging, such as self-induced vomiting, to prevent weight gain. **Binge eating** involves consuming an amount of food that is much larger than what a normal person would eat in a brief period of time, such as 2 hours. People with BN feel they have no control over their eating behavior during binge episodes. After a binge, people with BN attempt to prevent weight gain with inappropriate purging methods, such as self-induced vomiting, abuse of laxatives and/or diuretics, fasting, or excessive exercise. As with AN, a major underlying personal characteristic of BN is having a distorted body image.

BN usually begins in adolescence or young adulthood. An estimated 2 to 3% of American females have BN.[84] Females are more likely to have BN than males.

Signs and Symptoms

Unlike people who have anorexia nervosa, individuals with bulimia nervosa are often difficult to identify by their appearance because they tend to have BMIs in the normal or overweight range (18.5 to less than 30). However, people with BN are often ashamed of their food-related behaviors and attempt to conceal their binge-purge practices from others.

People with BN frequently induce vomiting by thrusting fingers deep into their mouths and, as a result, scrape their knuckles. Thus, characteristic signs of bulimia nervosa are bite marks and scars on the knuckles. Dentists often identify people who practice bulimic behaviors because the acid in vomit erodes the enamel on the surfaces of teeth, especially the backs of teeth. Eroded teeth develop many dental caries (cavities) and may become chipped and ragged in appearance.

Persons with bulimia nervosa often have low self-esteem and feel guilty or depressed after a binge. Bingeing usually involves eating "forbidden" foods such as cakes, cookies, ice cream, and other high-fat, high-carbohydrate foods during binges. In cases of BN, different triggers may bring on a binge, including stressful situations, extensive dieting, and negative feelings about body weight, body shape, or food.

purging activities that limit calorie intake or increase calorie output

bulimia nervosa (BN) severe psychological condition characterized by repeated episodes of binge eating followed by unhealthy behaviors to prevent weight gain

binge eating consuming a much larger amount of food than is normally eaten in a brief period

The bingeing activities can help the person with BN feel better in the short term; but guilt, emotional distress, and depression persist in the long term and worsen the disorder. The compelling need to binge and purge eventually becomes a preoccupation for people with BN, and as a result, they become less involved in social activities and more isolated.

Some people show behavioral characteristics of both anorexia nervosa and bulimia nervosa because the illnesses can overlap. About half of the females diagnosed as having anorexia nervosa eventually develop signs or symptoms of bulimia.

Health Risks

Self-induced vomiting and diuretic abuse can cause fluid and electrolyte imbalances that can be life threatening. Frequent vomiting also causes tears, ruptures, or bleeding of the esophagus and stomach. Someone who abuses laxatives may become dependent on their use in order to stimulate bowel movements. As many as 4% of people with BN die as a result of the disorder.[80] Nearly one in four of these deaths is due to suicide.

Several psychological disorders are associated with BN. Compared to people who do not have an eating disorder, people with BN are more likely to have low self-esteem and bipolar, depressive, and personality disorders. Some people with BN also abuse alcohol or other drugs.

Binge-Eating Disorder

Binge-eating disorder (BED) is an eating disorder that features recurrent episodes of overeating that are not followed by purging behaviors.[73] As mentioned earlier, binge eating involves consuming a much larger amount of food than most people would eat under similar circumstances in a brief period of time, such as within 2 hours.

In the United States, BED affects more people than the other eating disorders.[75] According to one report, 3.5% of adult American females and 2% of adult American males have been diagnosed with binge-eating disorder.[80]

TABLE 10.10 Common Signs and Symptoms of Binge-Eating Disorder[76]

A person with BED:
• Is extremely distressed over the binge-eating behavior.
• Eats large amounts of food when not hungry.
• Eats more rapidly than normal and eats until uncomfortably full.
• Eats alone because of feeling embarrassed by how much is consumed.
• Feels depressed, guilty, or disgusted with themselves after the binge.
• Frequently restricts food intake to lose weight.

People with BED can be normal weight, overweight, and obese. Table 10.10 lists typical signs and symptoms of binge-eating disorder.

People with BED often have other psychological illnesses, such as bipolar, depressive, anxiety, and substance use disorders. Furthermore, stressful events and feelings of loneliness, anxiety, depression, anger, isolation, and frustration can trigger a food binge. During binges, a person with BED typically isolates themselves and consumes large quantities of calorie-dense foods, such as ice cream, cookies, sweets, and potato chips. While bingeing, the person may feel better, but after the episode of overeating, the individual usually feels depressed, ashamed, guilty, and disgusted with their abnormal eating behavior.

As with the other eating disorders, BED usually begins during adolescence and young adulthood. However, people with BED are usually older than people who have AN or BN.[76]

Other Unhealthy Eating Behaviors

The *DSM-5* also describes several "other specified" feeding and eating behaviors, which include less severe versions of the three main eating disorders, as well as a condition called *night eating syndrome.* The following sections describe night eating syndrome and some popularized unhealthy eating practices.

Night Eating Syndrome

People with **night eating syndrome (NES)** experience episodic food binges that are not followed by purging. These binges take place after the evening meal or when the person wakens from sleep during the night. People with NES are aware of their actions and can recall their overnight eating practices the next day. Approximately 1 to 2% of the general population may have NES.[87]

Diabulimia

Insulin treatment often causes weight gain in people who are recently diagnosed with type 1 diabetes. Accepting the weight can be difficult for a person who has a poor body image. However, some people with the chronic disease discover they can manipulate their body weight by skipping insulin injections or using less insulin than prescribed.

binge-eating disorder (BED) eating disorder featuring recurrent episodes of binge eating that are not followed by purging behaviors

night eating syndrome (NES) episodic food binges that are not followed by purging; binges take place after the evening meal or when the person wakens from sleep during the night

This abusive practice, which affects an estimated 11 to 42% of young females with type 1 diabetes, is often referred to as **"diabulimia."**[88] The signs and symptoms of diabulimia include those that are associated with poorly controlled type 1 diabetes (see Table 5.9 in Chapter 5). Complications of long-term diabulimia are very serious and can include blindness, kidney failure, hypertension, heart attack, permanent heart damage, stroke, and nerve damage.[89]

Muscle Dysmorphia

Muscle dysmorphia is an unhealthy preoccupation with the body being too small or not muscular enough. This condition occurs almost exclusively in males, and it is common among body builders. More than 50% of males with muscle dysmorphia have used anabolic steroids to help gain weight and increase muscle size.[90] People with muscle dysmorphia exercise obsessively and may lift weights for hours every day. They can become so obsessed with exercise that they continue to train despite injury and they allow exercise to take priority over work, school, and friends. Individuals who have muscle dysmorphia are more likely to be dissatisfied with their lives, abuse drugs, and commit suicide when compared to people who do not have the condition.[91]

The Athlete Triad

Young males and females participating in competitive sports or sports and activities that benefit from a low body weight, such as gymnastics, dance, and distance running, are at risk of developing one or more components of the **athlete triad** (Fig. 10.21). This condition is characterized by three interrelated components:

- **Low energy availability** (negative energy balance);
- Abnormal reproductive system function; and
- Reduced bone mineral density that may lead to osteoporosis.

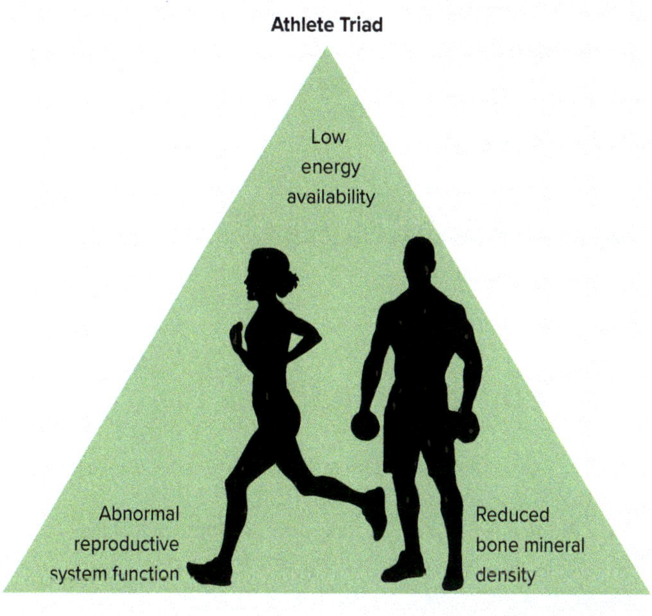

Figure 10.21 The athlete triad. People participating in competitive sports or sports and activities that benefit from a low body weight are at risk of developing one or more components of the athlete triad.

Female Athlete Triad

As many as 16% of female athletes have all three components of the triad.[92] Female athletes may restrict their calorie intake to improve athletic performance, lose weight, or maintain a lean physique. Other female athletes may lack the appetite, nutrition knowledge, or time to consume enough energy to meet their high needs and maintain a healthy BMI, especially during periods of intense physical training.

Females of childbearing age need adequate amounts of energy to maintain normal menstrual cycles. Low energy availability is associated with a deficiency of the "female hormone" estrogen. The lack of estrogen causes abnormal reproductive system function including menstrual disturbances, such as longer-than-normal intervals between menstrual periods and absence of menstrual periods. Estrogen is needed to build and maintain bone mineral mass. Female athletes who have low energy availability and no menstrual periods can have low bone mineral density and an increased risk for bone fractures.

When the lack of estrogen occurs during adolescence, a critical time for bone growth, the female athlete's bone mineral density may not improve even after she recovers from the triad.[92] Premature osteoporosis is a serious and long-term consequence of the athlete triad.

Male Athlete Triad

Male athletes may develop the "male athlete triad," a condition that is similar to the female athlete triad. Athletes with the male athlete triad excessively limit their calorie intake so they have low energy availability. This practice results in abnormal male reproductive function, including lower-than-normal blood levels of testosterone (the major reproductive hormone in males) and reduced bone

diabulimia popular term used to describe people with type 1 diabetes who skip insulin injections or use less insulin than prescribed to control their body weight

muscle dysmorphia unhealthy preoccupation with the body being too small or not muscular enough

athlete triad condition that is characterized by low energy availability, abnormal reproductive function, and reduced bone mineral density in athletes

low energy availability state of negative energy balance

mineral density.[93] Male athletes at risk of developing triad symptoms include athletes who participate in competitive sports that benefit from a lower body weight, such as distance runners and cyclists, and athletes who have to "make weight"to compete in a lower weight class, such as wrestlers and boxers.[93]

Normal testosterone levels are necessary for males to build and maintain bone mineral mass. Thus, male athletes with low testosterone levels may have reduced bone mineral density, which increases their risk of bone fractures. Although more research is needed to determine the long-term effects of the male athlete triad, it is important to recognize that some male athletes are at risk of developing this condition. For more information about the athlete triad, visit the website for the Female and Male Athlete Triad Coalition at www.femaleandmaleathletetriad.org/athletes/what-is-the-triad/.

Preventing Eating Disorders

Early identification of people with risk factors for eating disorders, such as disordered eating patterns and distorted body images, can help prevent the development of eating disorders. Prevention efforts generally focus on fostering the individual's body acceptance, improving their self-confidence, and encouraging the person to challenge the idea that being thin is ideal.

Treating Eating Disorders

If someone you know has an eating disorder, it is important to encourage the person to seek help as soon as possible to avoid serious health consequences. In academic environments, professional help is often available at student health centers and student guidance or counseling facilities on campus.

It may be difficult to talk with someone who you suspect has an eating disorder. When approaching the individual to discuss your concerns, you can follow these recommendations:[94]

- Meet with the person privately and without distractions.
- Share your feelings and concerns in a caring and supportive manner.
- Describe occasions when you observed behaviors that were signs of an eating disorder.
- Avoid using harsh, judgmental language and instead explain how you feel. For example, instead of stating, "You've lost too much weight. You look like you're starving!" say, "I feel very worried about how much weight you've lost."
- Do not suggest simple solutions, such as: "If you'd just have meals with the family again, you'd eat and feel better."
- Suggest that they discuss your concerns with an expert who treats eating disorders, such as a psychotherapist or RDN. Offer to help find such an expert, make an appointment, or accompany them on the first visit.
- Avoid conflict. If the person denies that they have an eating disorder, repeat your concerns and indicate that you will be available if the person decides to get help in the future.
- End the conversation by expressing your desire to be supportive. For example, remind the person that you want them to be happy and healthy.

A multidisciplinary group of experts in nutrition, mental health, and medicine is necessary for optimal treatment of eating disorders.[76] Treatment should include psychotherapy while also addressing critical nutritional needs and other medical conditions. RDNs play a critical role in assessing and treating eating disorders, including providing nutritional counseling. RDNs can help stabilize healthy eating patterns and improve patients' nutrient status by providing appropriate medical nutrition therapy. Furthermore, experts in nutrition can play an important role in helping patients avoid relapse once recovery is underway.

Treatment for eating disorders can occur in several settings. Inpatient hospitalization or residential care in a facility that specializes in the treatment of the disorders may be required when the patient's medical complications become life threatening. However, eating disorder treatment most often occurs in outpatient care settings. Outpatient care may involve individual, group, or family counseling. Treatment of eating disorders may also include prescription medications. Participation in eating disorder support groups may help prevent relapse.

Cognitive behavioral therapy (CBT) is a major treatment approach for BN and BED.[95] CBT is a general term used to describe psychological therapy approaches that address unhealthy emotions and behaviors. The therapy teaches people healthy coping strategies to use when under stress instead of fasting, bingeing, purging, or excessive exercise.

cognitive behavioral therapy (CBT) psychological therapy approaches that address unhealthy emotions and behaviors

Concept CHECKPOINT

39. Keira usually skips breakfast. Explain whether her behavior is a sign of an eating disorder.
40. List the major types of eating disorders and their signs.
41. List at least four risk factors for eating disorders.
42. What are three serious health consequences of anorexia nervosa?

See Appendix F for responses.

Chapter References

See Appendix G.

Summary

10.1 The Obesity Epidemic

- The prevalence of overweight and obesity has reached epidemic proportions in the United States and throughout the world. Over 40% of American adults were obese in 2017–March 2020; about 9% of the adults were extremely obese. The prevalence of obesity among Americans has risen sharply over the past 20 years.

10.2 Body Composition

- The body is composed of two major compartments: total body fat and fat-free mass. People who are overfat have excessive adipose tissue. Percentage of body fat is associated with risks of obesity-related diseases. Medical researchers use several methods to estimate body composition.

10.3 Energy for Living

- Biological fuels are foods and beverages that contain macronutrients and the nonnutrient alcohol. Under normal conditions, human cells metabolize primarily glucose and fatty acids, but small amounts of amino acids are also used for energy. Cells release the energy stored in biological fuels by breaking bonds within the molecules. Cells obtain only about 40% of the energy that was in macronutrients by forming ATP. Cells release the remaining energy as heat.
- Basal or resting metabolism, physical activity, and TEF account for total energy use by the body. Metabolism accounts for the largest share of an average person's daily energy needs. Various factors including thyroid hormone, body composition, gender, and age influence the metabolic rate. Physical activity is energy use by skeletal muscle movement. NEAT is energy use for involuntary skeletal muscle activities, such as shivering. TEF is the increase in energy needs that occurs during digestion, absorption, and processing of nutrients in food.

10.4 Energy Balance

- Energy balance is a state in which a person's calorie intake from food and beverages equals their calorie output for metabolism, physical activity, and TEF. Negative energy balance occurs when calorie output is greater than calorie intake, resulting in weight loss. Positive energy balance occurs when calorie intake is greater than calorie output, resulting in weight gain.

10.5 Classifying Overweight and Obesity

- In addition to percentage of body fat, medical experts use BMI to determine whether one's weight is healthy. BMIs of 18.5 to 24.9 are healthy; BMIs of 25.0 to 29.9 are in the overweight range. Persons with BMIs of 30.0 or more are obese.
- People with BMIs greater than 25 have increased risks of CVD, hypertension, type 2 diabetes, and certain cancers. Compared to people with healthy BMIs, obese people are more likely to die prematurely from all causes. Body fat distribution is associated with obesity-related diseases. Excessive central-body fat (apple shape) is associated with

MedicalRF.com

C Squared Studios/Photodisc/Getty Images

increased risks of CVD, type 2 diabetes, and hypertension. Waist circumferences that exceed 40 inches (males) or 35 inches (females) are linked to obesity-related diseases.

10.6 What Causes Overweight and Obesity?

- There is no simple cause for obesity. From a physiological standpoint, eating behavior is complex and largely involves interactions among the nervous, endocrine, and digestive systems, as well as fat tissue. Dietary and inherited factors also influence body weight. Additionally, people who are overfat may have inherited genes for "thrifty metabolisms." According to the set-point theory, the body's fat content is genetically predetermined. The set point may protect the body from losing weight and explain why weight loss is so difficult to achieve and maintain.
- Environmental factors can have a powerful influence on food choices and appetite. The environment also influences people's patterns of physical activity.

10.7 Weight Loss and Its Maintenance

- A reliable weight-loss plan is safe and effective, and the plan should meet the person's nutritional, psychological, and social needs. Successful weight loss and long-term weight maintenance involve motivation, calorie reduction, regular physical activity, and behavior modification.
- A pound of body fat contains about 3500 kcal. If energy output exceeds calorie intake, a person can expect to lose weight. A healthy way to lower total calorie intake is to reduce consumption of added sugars, added fats, and alcohol. A goal of a person who wants to lose weight should be reducing total calorie intake while obtaining all essential nutrients.

10.8 Medical Treatments for Obesity

- At present, the FDA has approved a few medications for weight loss. Orlistat can improve weight loss to a small degree, when it is combined with a plan that includes calorie restriction and regular exercise. Bariatric surgeries reduce the stomach volume of people with extreme obesity.

10.9 Unreliable Weight-Loss Methods

- People can lose weight while following fad diets, but they usually regain much of the weight that was lost when they resume their prior eating and other lifestyle habits. Achieving a healthy body weight and maintaining that weight require making lifestyle changes that a person adopts for the rest of their life. No dietary supplement is recommended for weight loss.

10.10 Gaining Weight

- Underweight can be caused by various factors, such as genetics, excessive physical activity, and certain diseases. To gain weight, a person who is underweight generally needs to increase portion sizes, eat more calorie-dense foods, and reduce physical activity, if excessive.

10.11 Nutrition Matters: Eating Disorders and Disordered Eating

- Disordered eating, involving temporary chaotic and abnormal food-related practices, often occurs when a person is under a lot of stress or wants to lose weight. When a person adopts disordered eating behaviors as a lifestyle, the practices can become harmful and difficult-to-treat eating disorders. Eating disorders are psychological disturbances that lead to behavioral changes and serious health complications. Anorexia nervosa, bulimia nervosa, and binge-eating disorder are the major types of eating disorders. Treatment of eating disorders is complex, involving more than just dietary counseling.

Wavebreakmedia Ltd/Getty Images

Recipe for Healthy Living

Did You Make That Dip?!

©Wendy Schiff

When it's time to entertain your friends, you can make a fiesta dip that will disappear quickly and provide plenty of compliments, too. Fresh cilantro and dried cumin give recipes a distinctive "Mexican" flavor. Bean dip recipes often include canned refried beans and cheeses that are high in fat. This lower-fat version uses canned pinto beans and reduced-fat cheese. You can also substitute commercially prepared salsa for the homemade version, but you may find that making salsa with fresh ingredients is worth the extra effort. If you like a hotter salsa, use jalapeño peppers instead of mild green chili peppers. When preparing hot peppers, be careful to avoid touching your eyes or the inside of your nose until after you have thoroughly washed your hands with soap and water. The peppers contain *capsaicin,* the highly irritating chemical used in pepper spray that causes intense burning when in contact with mucous membranes.

This recipe makes approximately eight ¼-cup servings. Each serving (without tortilla) supplies approximately 130 kcal, 8 g protein, 4 g fat, 48 mcg folate, 338 mg potassium, 1.2 mg iron, and 5 g fiber.

SALSA LAYER INGREDIENTS:

1 large, fresh ripe tomato
¼ medium onion
1 clove garlic, finely minced
4-oz can of mild green chilies (peeled and diced)
3 Tbsp fresh cilantro leaves

GUACAMOLE LAYER INGREDIENTS:

1 ripe, black-skinned avocado (slightly soft)
juice of ½ lime
¼ tsp dried cumin (a spice)
⅛ tsp ground black pepper

OTHER LAYERS:

14-oz to 16-oz can pinto beans (no added salt)
1 cup shredded fat-reduced cheddar cheese
⅓ cup fat-free plain yogurt (optional)

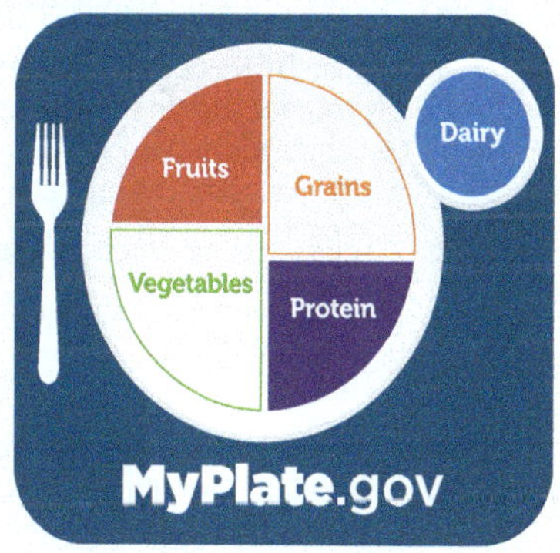

US Department of Agriculture or USDA

PREPARATION STEPS:

1. Set aside an 8″ dinner plate.
2. Wash cilantro in cool water and shake off excess water. Wash tomato, avocado, and lime in cool water. Set cilantro and fruits aside on paper towels.
3. Use a sharp knife to *mince* (finely chop) onion, garlic, and cilantro on a cutting board. Place in a small bowl.
4. Add chilies to the onion, garlic, and cilantro mixture. Stir gently until well mixed.
5. *Dice* (cut into small pieces) tomato and add to minced ingredients. You've made *salsa.*
6. To make guacamole, cut the avocado in half lengthwise and remove the round seed. Using a large spoon, scoop out the soft green flesh and place it in a small bowl. Discard the skin. Dice the avocado into small pieces and sprinkle with lime juice. Mash avocado and lime mixture. Add cumin and black pepper to avocado mixture, stir, and set aside.
7. Drain juice from beans and rinse them with cool water. Mash beans with the back of a large spoon; bean mixture will be lumpy.
8. Spread mashed beans evenly on the plate, leaving about ¾" from the edge of the plate free of beans.

9. Cover beans with individual layers of salsa and guacamole, and top with shredded cheese. If desired, place yogurt in the center as a low-fat substitute for sour cream and top with the cheese.
10. Loosely cover layered dip with clear plastic wrap and refrigerate.
11. To serve, spoon dip on pieces of soft whole-wheat tortillas.

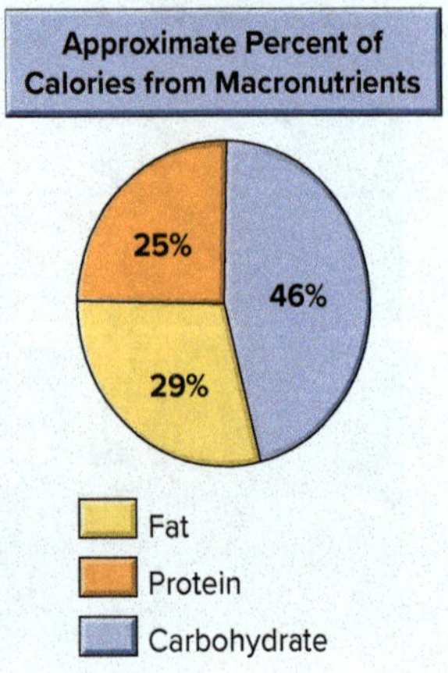

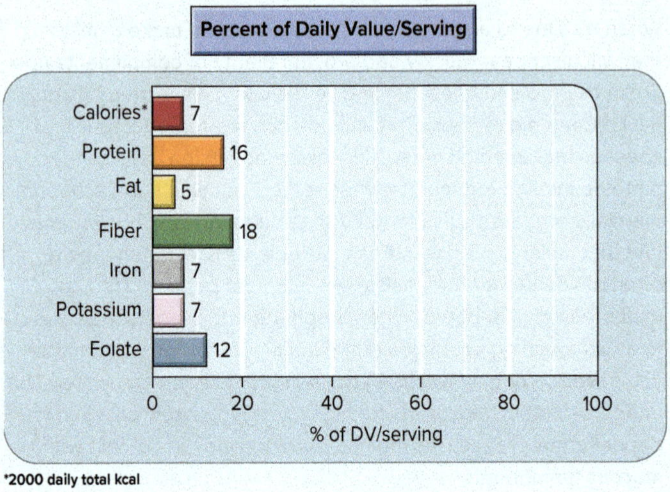

Personal Dietary Analysis

1. Consider your eating behaviors.
   a. Explain how your emotional state influences your eating behaviors. For example, do you eat certain foods when you feel happy, "stressed out," or depressed? Which foods do you eat under these circumstances? Describe a situation in which you ate in response to an emotional state instead of being hungry.
   b. Discuss the influence that food advertising has on your eating practices.
   c. Besides emotional state and food advertising, what other factors influence your eating habits?
2. Estimate your energy requirement by using one of the following formulas; choose the appropriate formula for your sex.

Males 19 Years and Older:

Estimated Energy Requirement (EER) $= 662 - (9.53 \times \text{AGE}) + \text{PA} \times (15.91 \times \text{WT} + 539.6 \times \text{HT})$

Females 19 Years and Older:

Estimated Energy Requirement (EER) $= 354 - (6.91 \times \text{AGE}) + \text{PA} \times (9.36 \times \text{WT} + 726 \times \text{HT})$

The variables in the formulas are as follows:

AGE = age in years

PA = physical activity estimate (see Table 10.11)

WT = weight in kg (lb ÷ 2.2)

HT = height in meters (inches ÷ 39.4)

Barbara Dudzinska/iStockphoto/Getty Images

TABLE 10.11 Physical Activity Level Estimates

Activity Level	PA (Males)	PA (Females)
Sedentary (no exercise)	1.00	1.00
Low activity (for example, walking the equivalent of 2 miles/day at 3–4 mph)	1.11	1.12
Active (for example, walking the equivalent of 7 miles/day at 3–4 mph)	1.25	1.27
Very active (for example, walking the equivalent of 17 miles/day at 3–4 mph)	1.48	1.45

Source of data: Food and Nutrition Board, National Institute of Medicine: *Dietary Reference Intakes for energy, carbohydrate, fiber, fat, fatty acids, cholesterol, protein, and amino acids (macronutrients).* Washington, DC: National Academies Press, 2005.

McGraw Hill connect® Complete the Personal Dietary Analysis activity in Connect.

Critical Thinking

1. Kim and Kevin have similar swimming skills. While in a swimming pool, Kim floats easily when she extends her arms and legs in the water. When Kevin extends his arms and legs and tries to float in the pool, he sinks. Why is Kim able to float more easily than Kevin?

2. An advertisement for a weight–loss supplement claims that the mixture of herbs in the product increases the metabolic rate by 150%. Explain why you would or would not recommend this product to someone who wants to lose weight.

Wavebreak Media Ltd/123RF

3. Explain why it is usually difficult to pinpoint a *cause* of obesity.

4. Why are most people who lose weight unable to maintain the lower body weight over time?

5. Determine your BMI using Table 10.7. If your BMI is outside the healthy range, discuss your reasons for being interested or not interested in losing weight. If you want to lose or gain weight, what lifestyle changes will you make? If your BMI is within the healthy range, discuss steps you can take to maintain a healthy body weight as you grow older.

6. People must first be aware of their weight biases to help stop weight stigma in society. Complete the weight bias quiz at www.stopweightbias.com/quiz/. How do you feel about your quiz results? Do you hold negative opinions about people with excess weight you were unaware of? Identify any personal assumptions about body weight and health that may be contributing to any biases. Can you recall specific examples of weight bias in health care? Education? Media?

Practice Test (Section numbers are listed below in red.)

Select the best answer.

1. (10.1) Body mass index (BMI) is
   a. a standard used to calculate a person's body fat percentage.
   b. based on body weight adjusted for height.
   c. gradually being replaced by more reliable height/weight tables.
   d. a chart used to predict one's caloric intake.
2. (10.2) _____ fat deposits are in the omentum, which is under the abdominal muscles.
   a. Subcutaneous
   b. Visceral
   c. Cellulite
   d. Bulimic
3. (10.2) _____ relies on the principle that lean tissue is denser than water.
   a. Bioelectrical impedance
   b. Direct calorimetry
   c. Underwater weighing
   d. Air displacement
4. (10.3) Basal metabolism includes energy needs for
   a. breathing and circulating blood.
   b. performing physical activity.
   c. digesting food.
   d. absorbing nutrients.
5. (10.4) Negative energy balance occurs when
   a. the body needs more calories than the diet supplies.
   b. fat storage in the body increases.
   c. energy intake is higher than energy output.
   d. the thermic effect of food equals NEAT.
6. (10.5) Kyran's BMI is 27. According to this information, he is likely
   a. healthy.
   b. overweight.
   c. obese.
   d. extremely obese.
7. (10.6) _____ is a hormone that reduces hunger and inhibits fat storage in the body.
   a. Coumadin
   b. Leptin
   c. Ghrelin
   d. Dexadrin
8. (10.7) Members of the National Weight Control Registry tend to
   a. skip breakfast regularly.
   b. follow low–carbohydrate/high–protein diets.
   c. exercise two to three times per week.
   d. eat meals regularly, including breakfast.
9. (10.8) Which bariatric surgery reduces the stom–ach size into a banana shape pouch that holds 2 to 5 ounces of food?
   a. Adjustable gastric band
   b. Roux–en–Y gastric bypass
   c. Intragastric balloon
   d. Sleeve gastrectomy
10. (10.9) Which of the following is a characteristic of fad diets?
    a. They provide a long term plan for weight loss and maintenance.
    b. It is nutritionally adequate.
    c. They do not require special foods or dietary supplements.
    d. It uses unscientific claims to support their usefulness.
11. (10.10) Which health condition is associated with being underweight?
    a. Sleep apnea
    b. Inflammatory bowel disease
    c. Gallstones
    d. Hypertension
12. (10.11) Joseph's BMI is 23. When he is under a lot of stress, he eats a large amount of cake, cookies, and ice cream. Soon after eating these foods, he goes to a bathroom and makes himself vomit. Based on this information, Joseph probably has
    a. diarexia psychosis.
    b. anorexia nervosa.
    c. cystic fibrosis.
    d. bulimia nervosa.

Answers to Quiz Yourself

1. You can determine whether you have an unhealthy amount of body fat simply by measuring your waistline. **True.** (Section 10.5)
2. The safest way to lose a lot of weight and keep it off is to follow a low-carbohydrate, high-fat diet, such as the ketogenic diet. **False.** (Section 10.9)
3. As people age, their muscle cells turn into fat cells. **False.** (Section 10.3)
4. When a person consumes more protein than needed, the excess can be converted to fat and stored in fat cells. **True.** (Section 10.4)
5. Cellulite is a unique type of fat that can be eliminated by taking certain dietary supplements. **False.** (Section 10.2)

CHAPTER

11

Nutrition for Physically Active Lifestyles

Quiz Yourself

What is adenosine triphosphate (ATP)? How much protein is needed for optimal muscular development? Are there any dietary supplements that can improve muscle strength and endurance safely? To test your nutrition and fitness knowledge, take the following quiz. The answers are at the end of the chapter.

1. People who exercise regularly can reduce their risk of type 2 diabetes. ______ T ______ F
2. Sports drinks are not useful for fluid replacement. ______ T ______ F
3. Protein is the body's preferred fuel for muscular activity. ______ T ______ F
4. Heatstroke is a serious illness that requires immediate professional medical treatment. ______ T ______ F
5. While at rest, skeletal muscles metabolize more glucose than fat for energy. ______ T ______ F

Paul Bradbury/Tony Talle/Caia Images/Alamy Stock Photo

11.1 Introduction to Nutrition for Fitness and Sport

Learning Outcomes

1. Explain the difference between physical activity and exercise.
2. Define *physical fitness.*
3. Discuss recommendations concerning the duration and intensity of physical activity to achieve good health.

| Shutterstock

Bicycling for weeks through quaint villages and over steep mountains in France; swimming for miles in the chilly English Channel while being buffeted by waves; lifting metal disks that weigh more than the weight lifter—the extent to which some people push their bodies is truly amazing. Millions of Americans admire competitive athletes for their physical accomplishments and enjoy watching them perform. Many Americans, however, lead *sedentary* lives; that is, their daily activities do not require much muscular exertion. In fact, Americans are less physically active than they were over 20 years ago. In 1998, 40% of Americans who were 18 years of age or older met the recommended amounts of moderate or vigorous physical activity. In 2018, only 25% of Americans met the physical activity recommendation.[1] Even when Americans have the time, many choose not to be physically active.

The human body is designed for **physical activity,** movement that results from skeletal muscle contraction. Most of the physical activities you perform each day are *unstructured,* for example, shopping for groceries or doing household tasks. **Exercise** refers to physical activities that are usually planned and structured for a particular purpose, such as having fun or increasing muscle mass. Both forms of physical activity can benefit your health.

Physical fitness may be defined as the ability to perform moderate–to vigorous–intensity activities for a reasonable amount of time, without becoming excessively fatigued. A *physically fit* person has the strength, endurance, flexibility, and balance to meet the physical demands of daily living, exercise, and sports. Proper nutrition is essential for optimal physical fitness and sports performance.

Regardless of whether you aspire to be a world–class athlete or simply want to be healthier, regular exercise should be a part of your daily routine. Physically inactive people do not have to perform high–intensity structured workouts daily to improve their health. In 2018, the U.S. Department of Health and Human Services issued the *2018 Physical Activity Guidelines for Americans,* 2nd edition (https://health.gov). According to these evidence–based guidelines, healthy adults under 65 years of age should perform:

- moderate–intensity physical activity for at least 150 minutes a week, or
- vigorous–intensity (high–intensity) physical activity for at least 75 minutes a week, and
- 8 to 10 strengthening exercises (8 to 12 repetitions of each exercise) that focus on major muscle groups at least twice a week.[2]

By reading this chapter, you will learn about the benefits of a physically active lifestyle, different cellular energy systems, and dietary practices that are appropriate for athletes and other physically active people. This chapter also provides practical tips for planning an exercise routine that you can follow for a lifetime.

physical activity movement resulting from contraction of skeletal muscles

exercise physical activities that are usually planned and structured for a purpose

physical fitness ability to perform moderate- to vigorous-intensity activities for a reasonable amount of time, without becoming excessively fatigued

Concept CHECKPOINT

1. What is the difference between exercise and physical activity?
2. How does a physically fit person differ from a physically unfit person?
3. What is the minimum amount of time adults should spend exercising each week?

See Appendix F for responses.

11.2 Benefits of Regular Exercise

Learning Outcomes

1. Describe the health benefits of performing exercise on a regular basis.
2. Describe the physical activity pyramid, and explain how it can be used as a guide for developing a personal exercise plan.
3. Calculate a person's target heart rate range.

People who exercise regularly can help reduce their risks of serious chronic conditions, including cardiovascular disease, type 2 diabetes, hypertension, obesity, osteoporosis, and certain cancers.[3] Maintaining a physically active lifestyle is one of the most important steps that people can take to improve their health. By following the *2018 Physical Activity Guidelines for Americans,* 2nd edition, Americans can reduce the risk of chronic diseases that contribute to millions of premature deaths each year.[4]

Determining the Intensity of Physical Activity

As Figure 11.1 illustrates, you can gain physical, as well as psychological, benefits by performing moderate–intensity physical activity regularly. Furthermore, you may achieve even greater health benefits by increasing the duration, frequency, and intensity of your exercise routine.

Intensity refers to the level of exertion (physical effort) used to perform an activity. Duration and type of physical activity, as well as body weight, influence the intensity of skeletal muscle movement. Thus, activities such as walking and bicycling can be classified as either moderate– or vigorous–intensity physical activity, depending on the rate at which the activities are performed, as well as the weight of the person performing them. Visit www.cdc.gov/physicalactivity/basics/measuring/index.html for examples of physical activities that are generally classified as moderate or vigorous intensity.

Some physical activities, such as wheelchair rugby, can be classified as vigorous intensity.
Thomas Busk/Cultura Creative/Alamy Stock Photo

There are a few ways to determine the intensity of exercise. One way is to judge your level of exertion based on physical signs, such as breathing rate and sweat production. While exercising at the moderate–intensity level, you should not only be aware of your muscular effort, but also be able to chat with an exercise partner comfortably.

A popular method of estimating the intensity of exercise is to use a percentage of your *age–related maximum heart rate.* To calculate your age–related maximum heart rate, subtract your age from 220. For example, a 27–year–old person would use 193 beats/minute (220 minus 27) for their age–related maximum heart rate. Your **target heart rate zone** is the range of heart rate that reflects the intensity of your exertion during physical activity. According to the Centers for Disease Control and Prevention (CDC) and American College of Sports Medicine (ACSM), for moderate–intensity physical activity, your target heart rate zone should be 64 to 76% of your age–related maximum heart rate.[5] To

target heart rate zone heart rate range that reflects intensity of physical exertion

Figure 11.1 Benefits of being physically active. Regular physical activity improves health in several ways.
Image Source/Getty Images

determine your moderate–intensity "zone," take your age–related maximum heart rate and multiply this figure by 0.64 and 0.76. For example, the age–related maximum heart rate of a 20–year–old person is 200 beats per minute (220 minus 20). Multiply 200 beats per minute (bpm) by 0.64 to calculate the 64% value and multiply 200 bpm by 0.76 to obtain the 76% level. This person's target heart rate zone for moderate–intensity activities is 128 to 152 bpm.

Moderate–intensity physical activities expend 3.5 to 7.0 kcal/min.[6] To "burn" (*oxidize* or *metabolize*) more energy and give your heart a more vigorous workout, you can engage in physical activities that expend more than 7 kcal/min. Such physical activities usually require considerable muscular effort and result in significant increases in breathing rate and perspiration. Examples of vigorous physical activities include jogging, running, aerobic dancing, swimming laps, and bicycling uphill.

To exercise vigorously, your target heart rate should be 77 to 93% of your age–related maximum heart rate.[5] To calculate this range, follow the same formula that you used to determine the range for moderate–intensity activity, except that you need to change "0.64 and 0.76" to "0.77 and 0.93." A 20–year–old person, for example, would have an estimated maximum age–related heart rate of 200 bpm, and the 77 to 93% levels would range from 154 to 186 bpm. If you have any serious health problems, ask your physician to help you determine your target zone.

You can easily measure your heart rate (pulse) by finding the *radial artery* in your wrist or the *carotid artery* in your neck. The radial artery is preferred for pulse measurement because heavy pressure on the carotid artery during measurement results in a slower heart rate.[7] To measure the radial heart rate, locate the radial artery by gently placing your index and middle fingers on the underside of your wrist by the thumb. Count your pulse for 10 seconds, and then multiply that number by 6 to determine your heart rate for 1 minute. Your heart rate begins to decline as soon as you stop exercising, so you need to practice taking your pulse while still working out. By wearing an activity tracking device on your wrist, however, you can easily check your heart rate without having to stop, find your radial artery, and count your pulse (Fig. 11.2).

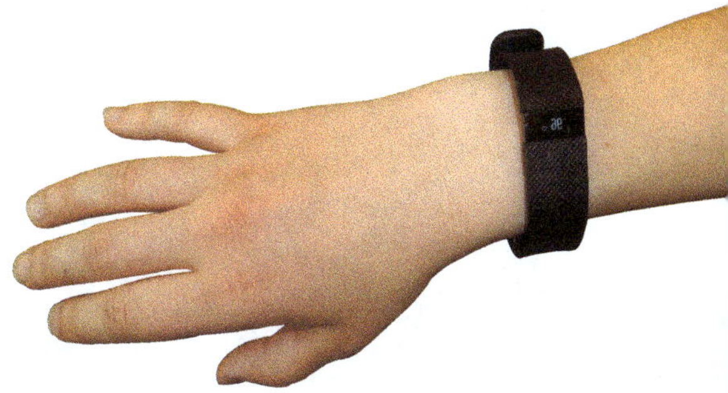

Figure 11.2 Using an activity tracking device to monitor one's pulse.
©Wendy Schiff

Physical Activity Pyramid

The physical activity pyramid shown in Figure 11.3 provides practical suggestions for increasing the intensity of various routine activities. For example, you will exert more physical effort and expend more energy if you use stairs instead of elevators or jog instead of walk the dog.

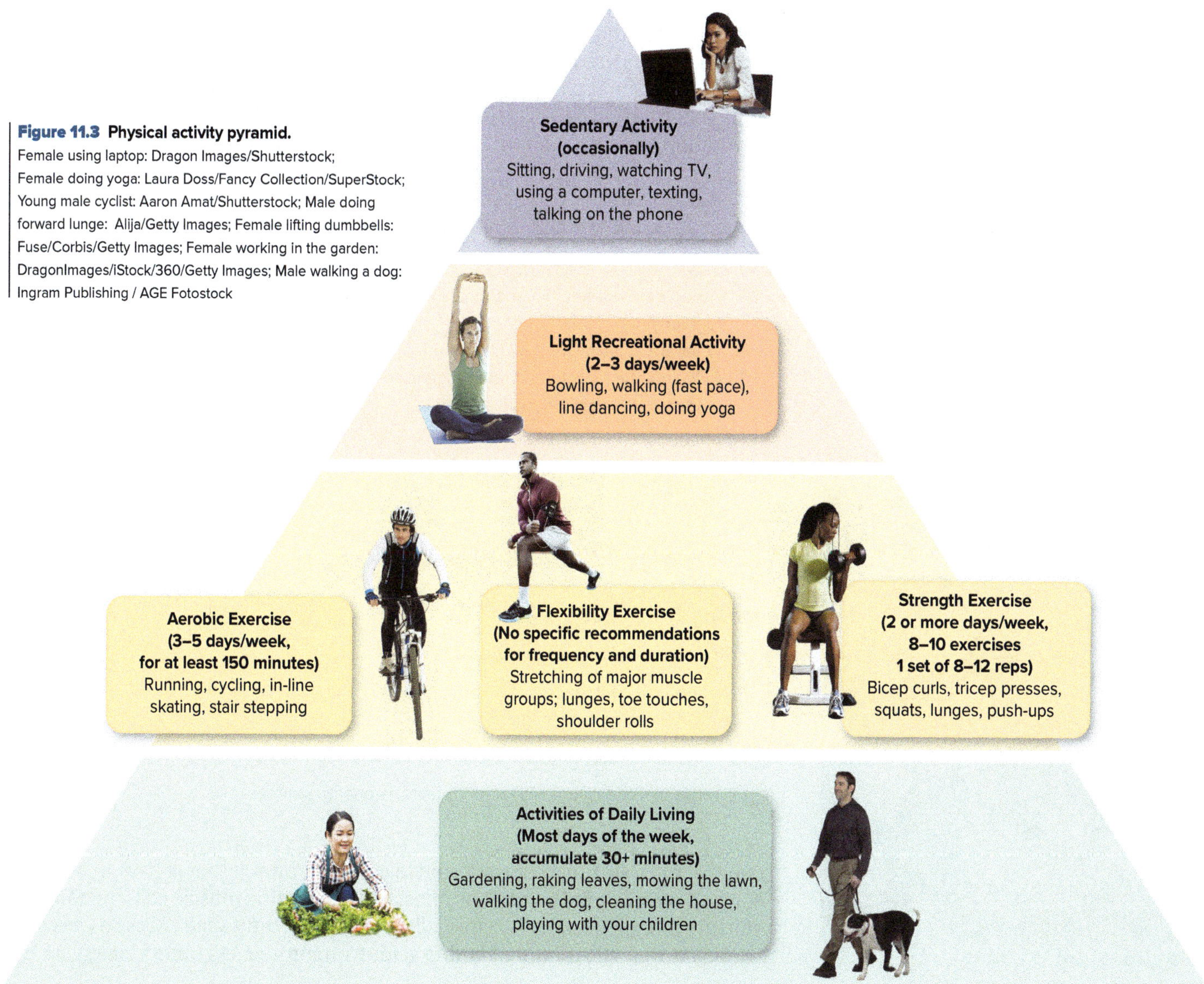

Figure 11.3 Physical activity pyramid.
Female using laptop: Dragon Images/Shutterstock; Female doing yoga: Laura Doss/Fancy Collection/SuperStock; Young male cyclist: Aaron Amat/Shutterstock; Male doing forward lunge: Alija/Getty Images; Female lifting dumbbells: Fuse/Corbis/Getty Images; Female working in the garden: DragonImages/iStock/360/Getty Images; Male walking a dog: Ingram Publishing / AGE Fotostock

Swimming is an aerobic exercise.
Fran Polito/Moment/Getty Images

aerobic exercise physical activities that involve sustained, rhythmic contractions of large muscle groups

Low-intensity, unstructured physical activities that involve usual daily living activities, such as routine household chores, form the foundation of the physical activity pyramid. The next level of the activity pyramid recommends adding *aerobic* exercise to foundation activities at least three times a week. **Aerobic exercise** involves sustained, rhythmic contractions of large muscle groups in the legs and arms. Such activities raise your heart rate, giving your heart a more effective workout. Running, jogging, rapid walking, and swimming are aerobic activities. Additionally, the second level of the pyramid recommends performing strength (resistance) exercises two or more times a week to increase muscle mass and strength. Resistance exercises, such as weight lifting, can also increase bone mass.

The second level of the activity pyramid also includes flexibility exercises, such as stretching, because they can improve the range of motion of joints. Having a high degree of flexibility is important for everyone, so exercises that develop and maintain flexibility are an appropriate feature of a basic physical activity program. Scientists, however, have not determined whether such exercises reduce the risk of injury.[4] As indicated in Figure 11.3, there are currently no specific recommendations for the number of days per week and the duration (hours or minutes) that a person should perform flexibility exercises.

The third level of the pyramid encourages regularly performing recreational activities that are physical, such as yoga and line dancing. The top of the pyramid depicts activities that expend little energy, such as watching television or using a personal computer. The pyramid recommends spending little time being sedentary. The "Nutrition Matters" section in this chapter provides information concerning how to design a more formal physical fitness plan.

Concept CHECKPOINT

4. List at least five health benefits of performing moderate-intensity exercise regularly.
5. What are at least two benefits of performing resistance exercise regularly?
6. Calculate the target heart rate range for a 24-year-old person performing moderate-intensity physical activity.

See Appendix F for responses.

Paolo Bona/Shutterstock

11.3 Energy for Muscular Work

Learning Outcomes

1. Explain how cells make and use ATP.
2. Discuss how each energy system supplies ATP for muscles and under what conditions a particular energy system functions.
3. Explain how a person can improve their aerobic capacity.
4. Discuss the body's use of macronutrients as energy sources.

To move, muscles must contract, and to contract, muscles must have a source of energy. Under normal conditions, most cells, including muscle cells, metabolize a mixture of biological fuels, especially glucose and fatty acids. Muscle cells also metabolize a small amount of amino acids from proteins to obtain energy.

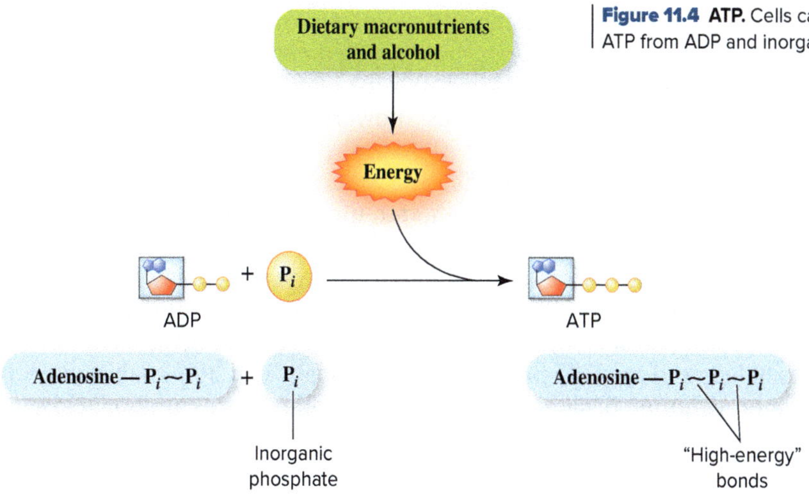

Figure 11.4 ATP. Cells capture and store energy by forming ATP from ADP and inorganic phosphate (P_i).

Energy Metabolism

Cells obtain energy by means of a complex series of chemical reactions that progressively break down (*catabolize*) macronutrients and alcohol to release the energy that is stored within them. Cells lose much of this energy as heat, but they capture some of the energy in *high–energy compounds* such as adenosine triphosphate (ATP). ATP forms when an *inorganic phosphate* group (P_i) bonds with **adenosine diphosphate (ADP)** and traps energy in the process (Fig. 11.4). Note in Figure 11.4 that ADP has two inorganic phosphate groups and ATP has three inorganic phosphate groups.

Recall from Chapter 5 that glucose is the primary energy source for human cells. The simple sugar can be catabolized when free oxygen (O_2) is unavailable (**anaerobic**) or available (**aerobic**). Catabolic processes involve *oxidation,* the removal of electrons from compounds to create new compounds. During **glycolysis** (*glyco* = carbohydrate [particularly sugar]; *lysis* = breakdown), the first stage of glucose oxidation, glucose is degraded to form two **pyruvate** molecules under anaerobic conditions (Fig. 11.5a). Glycolysis produces a small amount of ATP.

adenosine diphosphate (ADP) high-energy compound; by-product of ATP use

anaerobic conditions that lack free oxygen

aerobic conditions that require free oxygen

glycolysis first stage of glucose oxidation

pyruvate compound that results from anaerobic breakdown of glucose

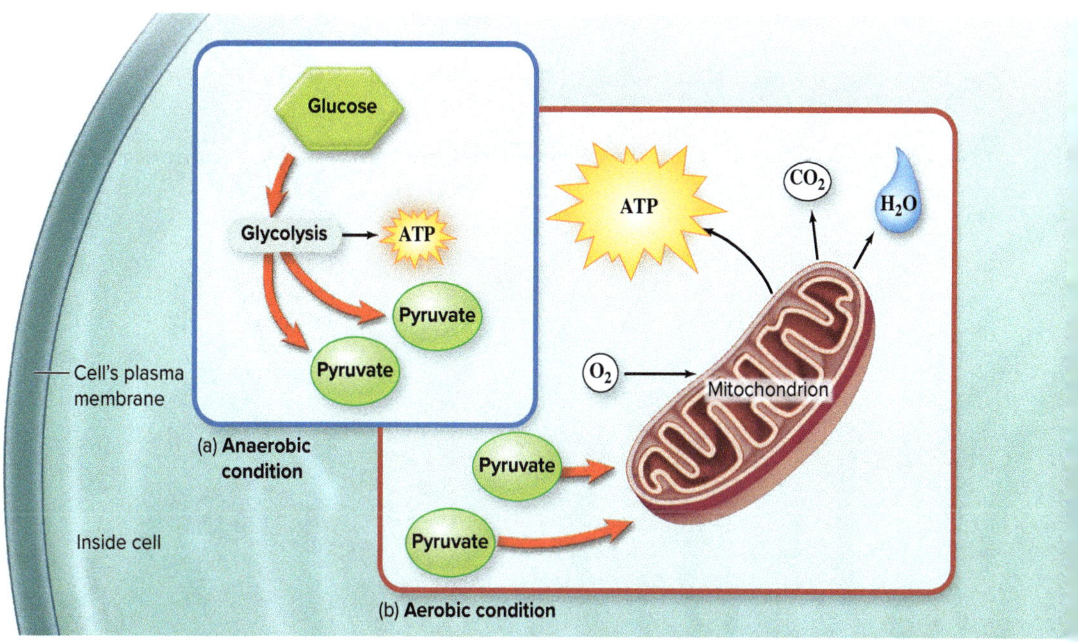

Figure 11.5 Obtaining energy. (a) During glycolysis, the first stage of glucose oxidation, cells degrade glucose into two pyruvate molecules under anaerobic conditions. (b) If oxygen is available, each pyruvate molecule undergoes further degradation in mitochondria, forming CO_2, H_2O, and ATP, at certain points.

Did YOU Know?

Human cells can convert certain amino acids into glucose (*gluconeogenesis*), but the cells are unable to make glucose from fatty acids.

mitochondria cellular structures that release energy from macronutrients

If oxygen is available, each pyruvate molecule undergoes further oxidation in a stepwise series of chemical pathways called *aerobic respiration.* Pyruvate moves from the fluid within cells (cytoplasm) into **mitochondria** (Fig. 11.5b). Mitochondria are often referred to as "powerhouses" because much of the energy stored in glucose or other biological fuels is released within these cellular structures. In mitochondria, pyruvate molecules undergo complete degradation, and as a result, cells generate more ATP than during glycolysis. Oxygen is a key player in this phase of the process because the element bonds to hydrogen atoms that were released from pyruvate molecules, forming water (H_2O) (see Figure 11.5b). When cells completely oxidize glucose to release energy, the end products are ATP, carbon dioxide (CO_2), and H_2O. Most of the CO_2 is exhaled, and the H_2O produced metabolically can help maintain proper body water volume. Besides glucose, triglycerides (fat), amino acids, and alcohol are also sources of ATP. Figure 11.6 summarizes the basic pathways that dietary protein, carbohydrate, and fat follow during energy metabolism.

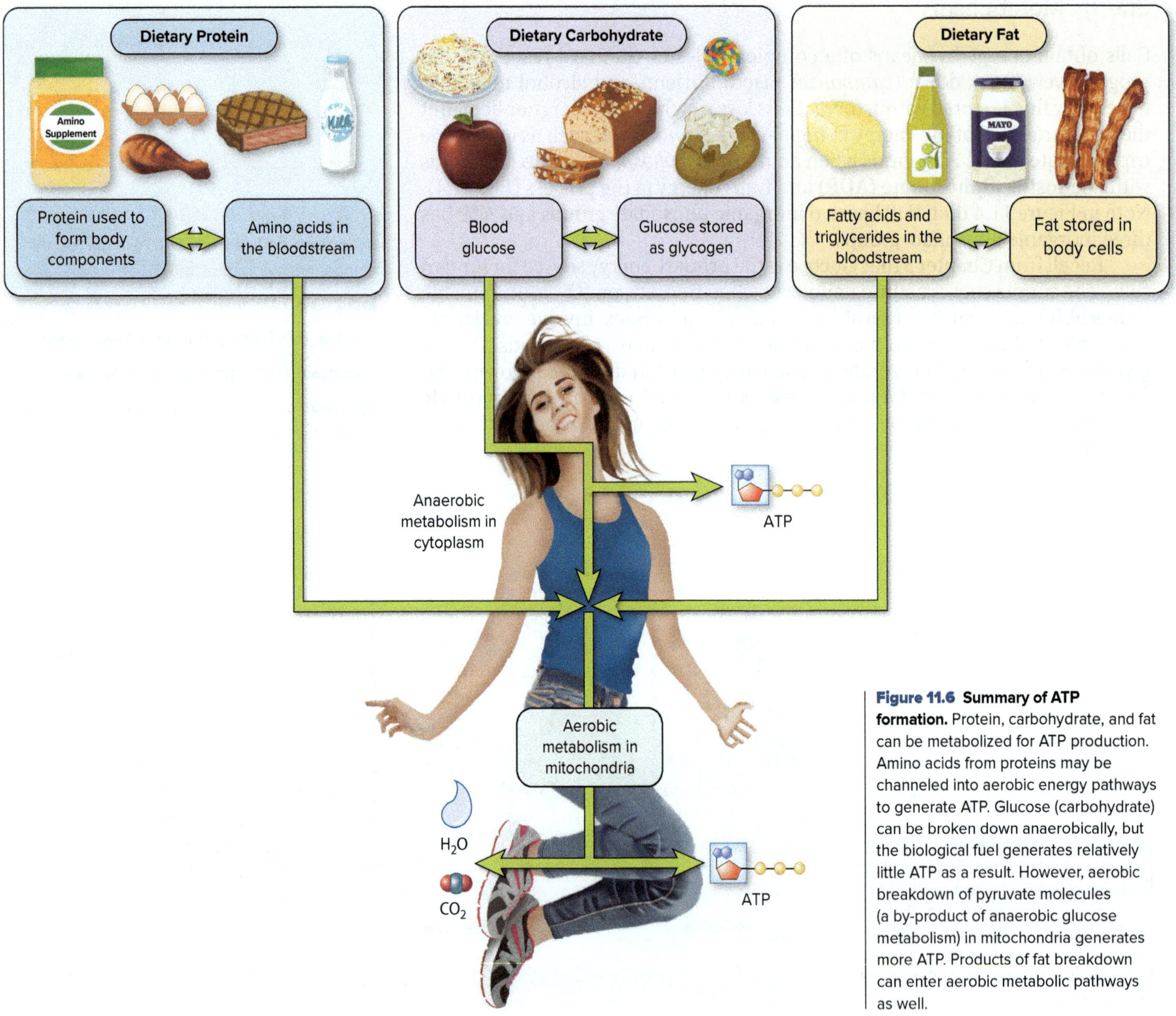

Figure 11.6 Summary of ATP formation. Protein, carbohydrate, and fat can be metabolized for ATP production. Amino acids from proteins may be channeled into aerobic energy pathways to generate ATP. Glucose (carbohydrate) can be broken down anaerobically, but the biological fuel generates relatively little ATP as a result. However, aerobic breakdown of pyruvate molecules (a by-product of anaerobic glucose metabolism) in mitochondria generates more ATP. Products of fat breakdown can enter aerobic metabolic pathways as well.

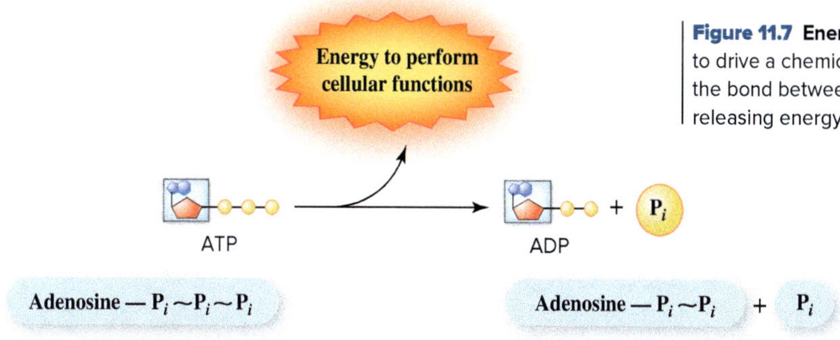

Figure 11.7 Energy from ATP. When a cell needs energy to drive a chemical reaction, it uses an enzyme to break the bond between the last two phosphate groups of ATP, releasing energy and reforming ADP and P_i.

For more detailed illustrations of the metabolic pathways that biological fuels undergo, see Appendix C.

How Do Cells Use ATP?

ATP is the primary source of direct energy for all cells. ATP is often referred to as "energy currency" because it functions like money. Just as you save money until it is needed to make a purchase, your cells save energy in ATP until it is needed to power cellular work.

When a cell needs some energy to drive a chemical reaction, it uses an enzyme to break the bond between the last two phosphate groups of ATP (Fig. 11.7). This process releases energy (*ATP–energy*) and reforms ADP and P_i. Thus, cells can recycle their supplies of ADP and P_i. Cells do not store much ATP, so they must constantly replace their supply of the high–energy compound by recycling ADP and P_i.

Energy Systems for Exercising Muscles

Gram for gram, fat supplies more energy than carbohydrate. Fatty acids, however, are not a very useful fuel for intense, brief exercise, such as a 100–meter sprint. Why? A fatty acid molecule has fewer oxygen atoms in relation to carbon atoms than a glucose molecule. Thus, cells need more oxygen to metabolize a fatty acid molecule than to burn a glucose molecule. During a brief bout of intense exercise, the heart and lungs do not have enough time to deliver much oxygen to muscles. Under these conditions, glucose is a major source of energy. For physical activities that last longer and are less intense, muscles can use more fat for energy because the lungs are able to supply them with enough oxygen.

Muscle cells rely on three major systems to obtain energy: the phosphocreatine *(PCr)–ATP, lactic acid,* and *oxygen systems.* The PCr–ATP and lactic acid systems do not need oxygen to produce ATP. Thus, these systems metabolize glucose under anaerobic conditions, such as when a person holds their breath while sprinting or lifting a heavy load. As the duration of the activity increases, muscle cells need to form considerably more ATP. To meet this demand, muscle cells depend heavily on the oxygen system to metabolize glucose and fat.

During a brief bout of intense anaerobic exercise, such as track-and-field shot put, glucose is a major source of energy for working muscles.
Robert Daly/OJO Images/age fotostock

Figure 11.8 PCr. Muscle cells break down PCr into creatine (Cr) and inorganic phosphate P_i, releasing energy to form ATP from ADP and P_i (a). When the intense activity stops and there is no need to maintain high levels of ATP, an inorganic phosphate group bonds with creatine to recycle PCr (b).

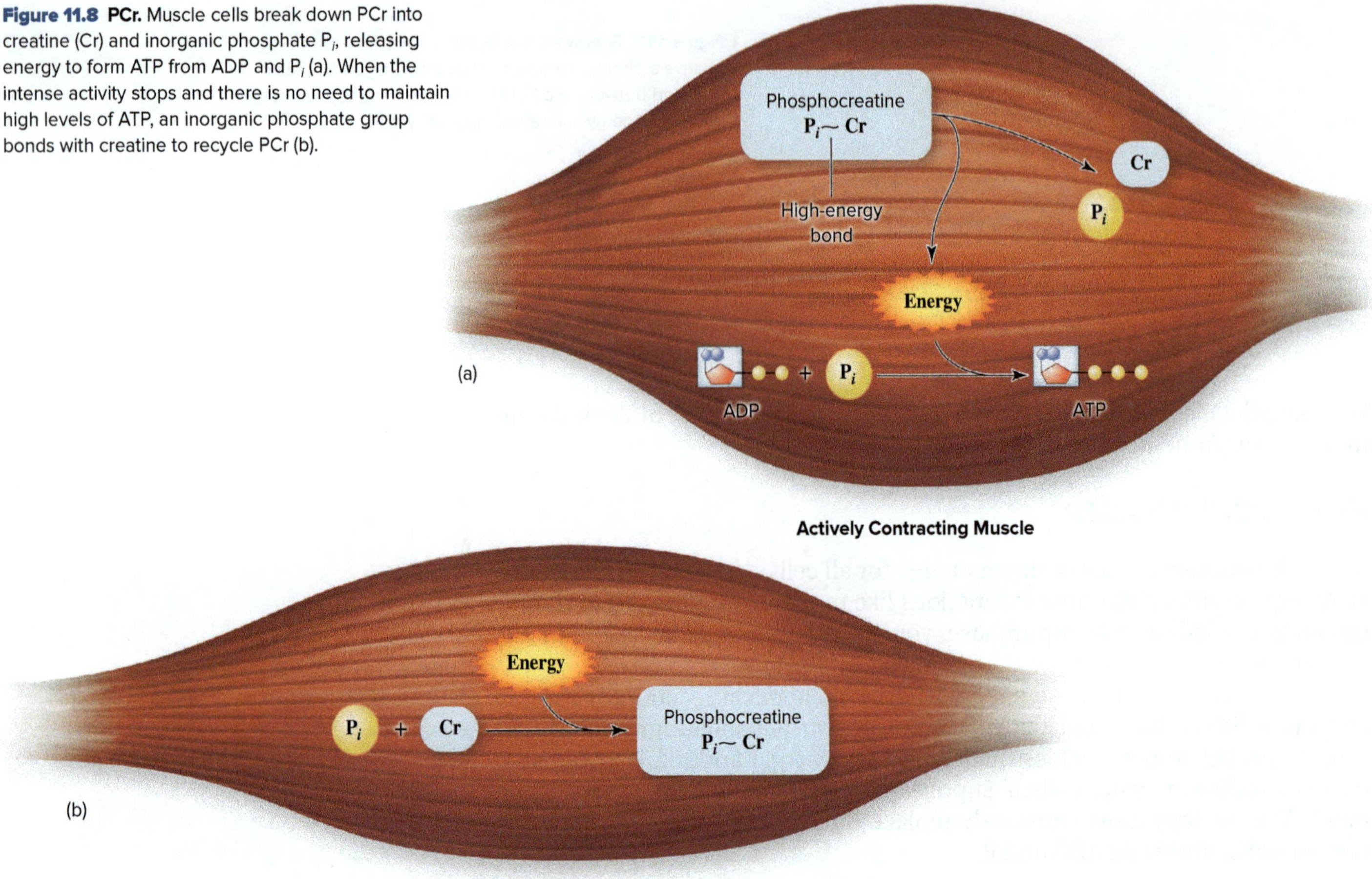

The three energy-releasing systems do not function independently of each other during intense physical exertion; each contributes ATP-energy to power intense muscular activity. The following sections provide more information about these major energy systems.

PCr-ATP Energy System

phosphocreatine (PCr) high-energy compound used to reform ATP under anaerobic conditions

A resting muscle cell contains only a small amount of ATP that can be used immediately. Muscle cells have another type of high-energy compound—**phosphocreatine (PCr)**—that enables the cells to produce more ATP quickly under anaerobic conditions. To make the ATP, cells break down PCr into *creatine* and inorganic phosphate (P_i), releasing energy to form ATP from ADP and P_i (Fig. 11.8a). Cells do not use PCr directly to power their activities; the compound provides the energy to resupply ATP.

By breaking down PCr to form ATP, muscle cells can instantly meet the energy demands of the swiftest and most powerful muscle movements, such as jumping, lifting, throwing, and sprinting.[8] However, during intense events, the PCr-ATP system can provide enough energy to muscle cells for only about 10 seconds.[8] When the intense activity stops and there is no need to maintain high levels of ATP, an inorganic phosphate group bonds with creatine to recycle PCr (see Figure 11.8b). Muscle cells, however, do not make or store much PCr.

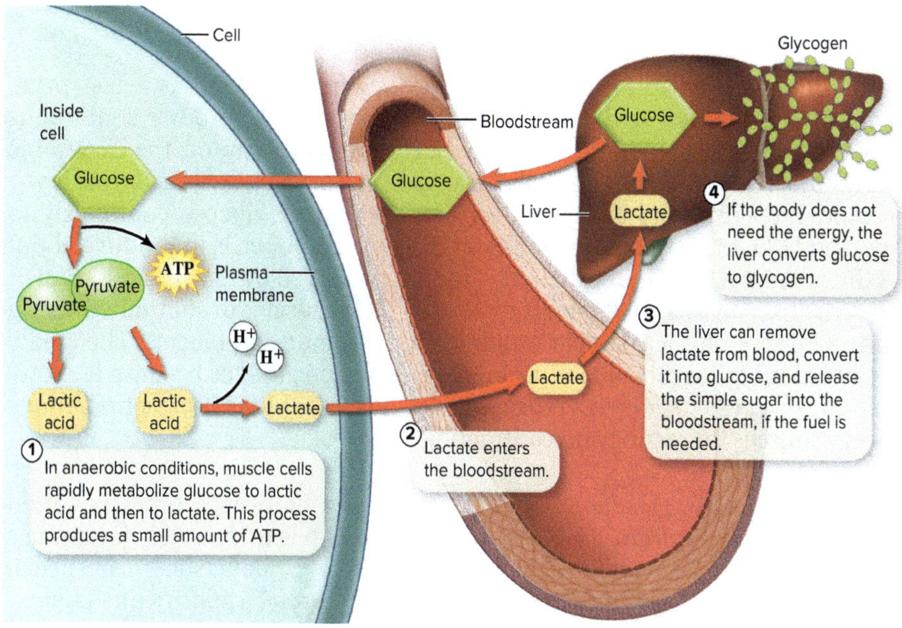

Figure 11.9 Cori Cycle.

Creatine Supplementation

The use of dietary supplements that contain creatine, such as *creatine monohydrate,* to improve physical performance is popular among athletes. Do these supplements work? Scientific evidence does support the use of oral creatine supplements to improve athletic performance.[9] Creatine supplementation increases the amount of creatine stored in muscles, which increases the production of muscle PCr. Having high levels of PCr in muscle tissue helps resupply the ATP needed for muscle activity. Creatine supplements are most beneficial for athletes who perform short-term, high-intensity exercises that require high levels of ATP.[9] Table 11.8 summarizes science-based findings regarding creatine supplementation.

In anaerobic physical activities such as sprinting, muscle cells metabolize glucose to pyruvate and then convert pyruvate to lactate. The liver then converts lactate into glucose to continue to fuel active muscles.
Aflo Co., Ltd./Alamy

Lactic Acid Energy System

When physical activity lasts longer than about 10 seconds, the PCr–ATP energy system cannot keep up with the demand for energy, and muscle cells must metabolize glucose to generate more ATP. The immediate source of glucose for working muscles is glycogen that is stored in muscles. The liver also helps supply glucose for muscles by degrading glycogen and releasing glucose molecules into the bloodstream.

In anaerobic conditions, muscle cells metabolize glucose to pyruvate and then convert pyruvate to **lactic acid.** This process is called the *Cori Cycle* (Fig. 11.9.1). The degradation of glucose to lactic acid produces a small amount of ATP—only enough to sustain maximum physical exertion for 10 to 180 seconds (Fig. 11.9.1).[8] Lactic acid accumulates in muscles and converts to a related substance, *lactate.* Although certain muscle cells can use lactate as a fuel, some of the compound enters the bloodstream (Fig. 11.9.2). The liver removes lactate from blood and can convert the compound into glucose (Fig. 11.9.3). The liver may then release the glucose into the bloodstream to help meet muscles' demand for fuel or use the simple sugar to make glycogen (Fig. 11.9.4).

Hydrogen ions (H^+) form as a result of the conversion of lactic acid to lactate. The accumulation of H^+ in muscle tissue contributes to muscle acidity, a condition that can lead to muscle fatigue and declining physical performance.

lactic acid compound formed from pyruvate during anaerobic metabolism

Even well-trained athletes experience muscle fatigue as the time they spend performing intense muscular exertion increases.
Torwai Suebsri/123RF

Oxygen (Aerobic) Energy System

You would not be able to enjoy activities such as walking at a fast pace, swimming laps, playing a game of soccer or basketball, or other continuous types of physical activity if your muscles depended only on the anaerobic energy systems. When muscle cells have plenty of oxygen, such as during low– to moderate–intensity exercise, they can metabolize glucose completely to CO_2 and H_2O. In fact, the availability of oxygen enables cells to produce about 18 times more ATP–energy than the amount produced by anaerobic systems.

The oxygen (aerobic) energy system enables muscles to continue working during intense physical activities that last for 2 minutes or longer.[8] The ability to obtain energy aerobically is useful for endurance athletes because it allows their muscle cells to contract repeatedly for hours. Table 11.1 presents various energy sources for resting and contracting muscles.

Aerobic Capacity The ability of your heart and lungs (sometimes referred to as the *cardiorespiratory system*) to deliver oxygen to muscles determines your capacity for intense aerobic physical activity. Scientists can use special equipment to estimate maximal oxygen intake (*aerobic capacity* or *VO_2max*) during vigorous physical exertion. A simple way to determine if you are nearing your aerobic capacity is to engage in vigorous exercise and note when your breathing rate increases to the point that you cannot carry on a conversation.

You can increase your aerobic exercise capacity by engaging in an endurance training program that gradually increases the intensity level of activities. Such training increases the number of mitochondria in your muscle cells, and thus your ability to generate ATP rapidly. However, even highly trained athletes experience muscle fatigue after increasing the time they usually spend performing intense muscular exertion.

Fat or Carbohydrate for Fueling Exercise?

The intensity of a physical activity largely influences the relative amounts of fatty acids and glucose that muscles metabolize for energy. Glucose supplies

TABLE 11.1 Energy Sources for Muscles*

Source/System	When in Use	Examples of Activities
ATP	At all times	All types
Phosphocreatine (PCr)	All exercise initially; short bursts of exercise thereafter	Shot put, high jump, bench press
Carbohydrate		
Anaerobic	High-intensity exercise, especially lasting 10–180 seconds	100-meter dash (sprint)
Aerobic	Exercise lasting 2 minutes–3 hours or more; the higher the intensity of exercise, the greater the use	Basketball, swimming, jogging
Fat	At rest	Sitting
	Exercise lasting more than a few minutes; low- to moderate-intensity physical activities	30-minute brisk walk
Protein	Low amounts during all exercise; slightly more during endurance exercise, especially when carbohydrate fuel is lacking	Long-distance running

* Note that at any given time, more than one energy system is operating.

Source: Adapted from Smith AM and others: *Wardlaw's Contemporary Nutrition.* 12th ed. New York: McGraw Hill, 2022.

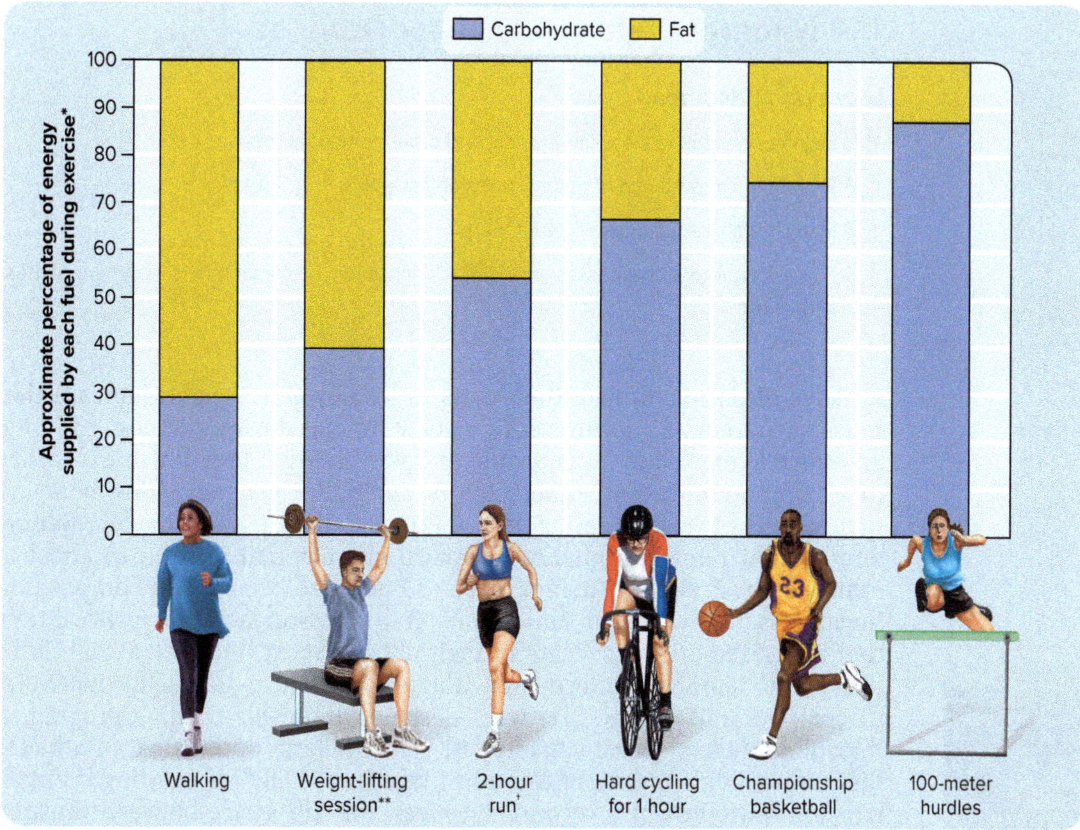

*Protein supplies a minor percentage of energy.
**Fat use generally is higher because much of the time spent weight lifting is for rest periods.
†The values shown are for a runner consuming carbohydrate during the run; more fat and less carbohydrate would be used if carbohydrate was not consumed.

Figure 11.10 Rough estimates of energy use during exercise. The intensity of an exercise largely influences the relative amounts of macronutrients that muscles metabolize for energy.

only about 40% of the energy needed to sustain a person who is resting or engaged in very light to light activities, such as watching TV, typing, and walking. Fat is the primary fuel muscles use while resting or engaged in low– to moderate–intensity physical activities.[10] During high–intensity exercise, the rate of fat oxidation decreases while that of glucose oxidation increases. The chart in Figure 11.10 illustrates rough estimates of carbohydrate and fat metabolism during six forms of exercise.

An individual's level of training influences the ratio of glucose to fatty acids that their muscles use during exercise. Trained endurance athletes tend to oxidize more fat when exercising at the same intensity than untrained persons.[11] As a result, muscle cells of trained athletes "spare" glycogen; that is, they conserve their supply of glucose. By sparing their glycogen supplies, athletes can enhance their capacity to exercise longer.

Concept CHECKPOINT

7. How is glycogen used during exercise?
8. How is ATP formed? How do cells use ATP?
9. Explain how each energy system supplies ATP for muscles. Which energy systems operate under anaerobic conditions?
10. How can you improve your aerobic capacity?
11. When is fat a major source of energy for muscles?

See Appendix F for responses.

Did *YOU* Know?

As people grow older, their aerobic capacities decline with each passing decade. By being physically active, however, even older adults can maintain a higher degree of aerobic capacity than their sedentary counterparts. It is never too late to begin a training program to improve physical fitness. If you have existing health problems, you should have a complete medical checkup and obtain your physician's OK before beginning a moderate-intensity fitness program. Males older than 40 years and females older than 50 years who plan a vigorous-intensity program should also consult their physician for help designing a safe, effective program.

11.4 General Dietary Advice for Athletes

Learning Outcomes

1. Estimate an athlete's energy, protein, and other nutrient needs.
2. Explain the importance of optimal nutrition for athletes.
3. Identify high-carbohydrate/low-fat foods.
4. Discuss heat-related illnesses, including identification, prevention, and treatment of these conditions.

Athletes often manipulate their diets to lose or gain weight, increase their muscular strength, and prevent or delay fatigue during exercise. Although an athlete's diet plays a major role in determining if they finish first or last in a competitive event, other factors, including genetic endowment and mental and physical training, also influence athletic performance. **Genetic endowment** refers to inherited physical characteristics that can affect an athlete's physical performance, such as body size, shape, and composition. Regardless of how well an athlete eats, if this person lacks the physical traits that are necessary for success in their chosen sport, the athlete will find it difficult to compete effectively. Athletes must also be highly motivated to compete and engage in a well–designed intensive training program to maximize their physical capabilities. Nevertheless, optimizing an athlete's diet may provide a competitive advantage, especially for sporting events in which hundredths of a second can mean the difference between finishing first and finishing second.

Athletes and coaches often believe misinformation concerning the value of dietary supplements, certain foods, and fad diets for optimizing physical health and performance. Such beliefs can lead to diet–related practices that are useless and a waste of money. In some cases, these practices are harmful or even deadly.

Sports nutrition focuses on applying nutrition principles and research findings to improving athletic performance. This section provides specific dietary recommendations that are appropriate for athletes and other physically active people. If you would like additional information on sports nutrition, contact a registered dietitian nutritionist. Other reliable sources of sports nutrition include websites of the American College of Sports Medicine (www.acsm.org) and *The American Journal of Sports Medicine.* If you are interested in studying sports nutrition, check with your academic advisor to determine whether your college or university offers sports nutrition courses.

Optimizing an athlete's diet may provide a competitive advantage.
Nicholas Piccillo/Shutterstock

genetic endowment inherited physical characteristics that can affect physical performance

Energy for Athletic Performance

Compared to nonathletes, athletes generally need more energy to support their physically active lifestyles. Athletes who do not consume enough food energy can lose muscle mass and bone density, experience fatigue and menstrual problems (females), and be at risk of injury.[8] Thus, low energy intakes can hinder an athlete's chances of performing well.

Male athletes who train or compete aerobically for more than 90 minutes daily need at least 50 kcal/kg/day; their female counterparts need at least 45 to 50 kcal/kg/day.[12] Thus, athletes may require 3000 kcal/day or more to support their energy needs and maintain their weight. Table 11.2 presents a sample daily menu that is nutritionally adequate, supplies approximately 5000 kcal/day, and provides ample amounts of carbohydrate.

How can athletes tell if they are consuming enough energy? One way is to have them keep accurate food records and use the information to estimate their daily calorie intakes. Athletes can also monitor their body weights and have their skinfold thicknesses measured regularly. If their weights were within the healthy BMI range before training and they start to lose weight during training, the individuals should consume more food until they regain their pretraining weights. Consuming an additional 500 to 700 kcal/day, especially by eating calorie– and nutrient–dense foods such as nuts and dried fruit, is a healthy way for anyone to boost their calorie intake. Athletes who gain too much body fat can increase their energy output by spending more time in training. Athletes who are overfat can also reduce their food intake by about 200 to 500 kcal/day until they are in the healthy BMI range. In general, a good way to reduce energy intake is to limit portions of fatty foods. Chapter 3 provides general information to help you plan nutritious menus.

For most physically active people, fat should supply 20 to 35% of energy, which is the Acceptable Macronutrient Distribution Range (AMDR) for fat that is recommended for the general population.[8] High–fat diets, such as the ketogenic diet, are used by some athletes. Scientific studies evaluating high–fat diets in athletes show conflicting results. Some studies report that a high–fat diet improves athletic performance, while other studies show no improvement in athletic performance. More research is needed to determine if athletes can benefit from consuming high–fat diets.[13,14] Furthermore, very–low–fat diets (<20% of total energy from fat) are not beneficial for athletic performance.[8]

Focusing on Carbohydrate Intake

Recommended diets for athletes should supply adequate amounts of energy from carbohydrates. To maintain adequate muscle glycogen, athletes who perform moderate– to vigorous–intensity activity for 1 to 3 hours daily should consume 6 to 10 g of carbohydrate per kilogram of body weight each day.[8] Glycogen depletion is a major cause of fatigue during endurance exercise. By consuming several servings of grains, starchy vegetables, and fruits daily, an athlete can obtain enough carbohydrate to maintain adequate liver and muscle glycogen stores.

To calculate your recommended range of carbohydrate intake for endur–ance activities, multiply your weight in kilograms by 6 and then by 10. For example, a 145–pound (66 kg) female athlete (endurance) should consume between 396 and 660 g of carbohydrate each day. If she requires 3000 kcal/day to maintain her weight and physical activity level and she consumes 60% of her energy from carbohydrates, she will obtain about 450 g of carbohydrate, which is within the recommended range.

It is important to keep in mind that there is no "one size fits all" diet plan that specifies amounts of carbohydrate–rich foods for pre–event, event, or post–event meals and snacks. Diets for athletes should be individualized and based on factors such as the athlete's sex, body size and weight, sport and training level, and exposure to environmental conditions, as well as personal experiences and food preferences.[8] Furthermore, athletes should test any dietary strategies during practices or trials several days or weeks before a competitive event.

Pre-event Meals and Snacks

About 1 to 4 hours before competing in events lasting longer than 1 hour, athletes should consider eating a meal that supplies 1 to 4 g of carbohydrate per kg of body weight because such meals can improve performance.[8] Table 11.3 provides some menu ideas for high–carbohydrate, low–fat pre–event

TABLE 11.2 Sample Daily 5000 kcal Menu

5000 kcal 62% Carbohydrate 21% Fat 17% Protein
Breakfast
Fat-free milk, 1 cup
Cheerios™ cereal, 2 cups
Bran muffins, 2 medium
Orange, 1
Snack
Low-fat plain yogurt, 1 cup
Chopped dates, 1 cup
Lunch
Chicken enchilada, with beans and cheese, 2
Romaine lettuce, 2 cups
Garbanzo beans, 1 cup
Shredded carrots, ¾ cup
Chopped celery, ½ cup
Seasoned croutons, ½ cup
French dressing, 2 Tbsp
Whole-wheat bread, 2 slices
Avocado, mashed, 2 Tbsp
Snack
Banana, 1
Bagel, 1
Cream cheese, 1 Tbsp
Milk, 1%, 1 cup
Dinner
Lean beef, pot roast, 6 oz
Mashed sweet potatoes, with milk, 2 cups
Olive oil, 2 tsp
Spinach egg noodles, 1½ cups cooked
Grated parmesan cheese, ¼ cup
Green beans, 1 cup
Oatmeal-raisin cookies, 3, large
Milk, 1%, 1 cup
Snack
Whole-wheat bagel
Peanut butter, 1 Tbsp
Raisins, ½ cup
Cranberry juice, 1½ cups

Athletes who eat a high-carbohydrate, low-fat pre-event meal, such as pasta with vegetables, can improve their athletic performance.
Nesavinov/Shutterstock

TABLE 11.3 High-Carbohydrate, Low-Fat Pre-Event Meals

Meal A	Meal B	Meal C
Instant oatmeal, plain, prepared with water, 1½ cup Cinnamon, ½ tsp Honey, 1 Tbsp Fat-free or soy milk, 8 oz	Pasta salad, 1½ cups French bread, 4 oz Olive oil, 2 tsp	Cornflakes, ready-to-eat cereal, 1½ cup Fat-free milk, 8 oz Banana, medium
Pear, sliced, 1 cup	Apple juice, 8 oz	Pineapple juice, 8 oz
Orange juice, 8 oz Raisin bread, toasted 2 slices Peanut butter, 1 Tbsp Fruit spread, 2 Tbsp	Frozen yogurt, 1 cup	English muffin, toasted Avocado, mashed, 1 Tbsp Mango, chopped, ¼ cup Nutty energy bar*

*See this chapter's "Recipe for Healthy Living."

meals that supply approximately 200 g of carbohydrate. Eating high–fat foods such as sausage, bacon, sauces, and gravies is not recommended because they may contribute to intestinal discomfort while competing.[8] Although some nutritionists advise athletes to exclude high–fiber foods from pre–event meals, there is a lack of scientific evidence to support this recommendation. Nevertheless, athletes can have different responses to eating high–fiber diets prior to events; individuals who react negatively may find it necessary to avoid eating high–fiber foods until after competing.

High–carbohydrate, low–fat food choices for pre–event meals or snacks include cereal with fat–free milk, bagels, dried fruit, pretzels and a sports drink, cooked oatmeal with fruit, pasta, baked potato topped with yogurt, and toasted bread with jelly. Table 11.4 presents commonly eaten foods that are high in carbohydrate and relatively low in fat. Food eaten about an hour before competing should be blended or liquid to promote rapid stomach emptying. Examples of such foods are low–fat smoothies or liquid meal–replacement formulas.

TABLE 11.4 Energy and Macronutrient Contents of Selected Foods

Food and Amount	kcal	Carbohydrate (g)	Protein (g)	Fat (g)
Spaghetti with sauce, meatless, 1 cup	293	47.47	8.93	7.14
Egg noodles, cooked, 1 cup	221	40.02	7.22	3.30
Macaroni, plain, cooked, 1 cup	209	40.7	7.66	1.23
Rice, instant, white, cooked, 1 cup	204	44.22	3.60	0.83
Pretzels, 1 cup	173	36.18	4.52	1.32
Bagel, ½, regular	164	32.00	6.70	1.00
Baked sweet potato, with peel, 1 medium	138	30.09	3.00	0.22
Orange juice, unsweetened, 1 cup	117	27.40	1.69	0.37
Corn, canned, drained, 1 cup	110	23.52	3.79	2.0
Banana, 1 med. (approx. 7″ long)	105	26.95	1.29	0.39
Mango, chunks, 1 cup	100	24.80	1.35	0.50
Pita, white, 1 small	77	15.60	2.55	0.33

Source: U.S. Department of Agriculture, Agricultural Research Service, *FoodData Central*, 2019.

©Wendy Schiff

What Is Carbohydrate (Glycogen) Loading? A healthy person stores about 6 g of glycogen per kilogram of body weight. Therefore, an individual who weighs 165 pounds (75 kg) stores about 450 g of glycogen. This amount of glycogen supplies 1800 kcal, which is enough energy to enable the person to bicycle at 13 mph for about 4 hours and 15 minutes. Endurance athletes who have more muscle glycogen at the start of an event may be able to exercise longer than those who do not have as much muscle glycogen. **Carbohydrate (glycogen) loading** involves manipulating dietary and physical activity patterns 3 to 7 days before an event to increase muscle glycogen stores well above the normal range. The practice helps delay fatigue in athletes participating in events lasting more than 90 minutes.[8]

Evgenia Sh/Shutterstock

About 3 g of water are incorporated into muscle tissue along with each gram of glycogen. Thus, carbohydrate loading adds water to muscles. Although this fluid aids in maintaining proper hydration status, some individuals experience muscle stiffness and unwanted weight gain as a result of carbohydrate loading. Table 11.5 presents the classic 7–day method of carbohydrate loading and two modified versions. Athletes who would like to determine whether a carbohydrate–loading regimen helps their performance should try one of the regimens during training to experience its effects. Refer to the references below Table 11.5 for more information on each carbohydrate loading method. Rather than promote carbohydrate loading, many nutrition and human performance experts simply recommend that athletes routinely follow a high–carbohydrate diet and consume certain forms of carbohydrate during prolonged exercise.

carbohydrate (glycogen) loading practice of manipulating physical activity and dietary patterns to increase muscle glycogen stores

TABLE 11.5 Carbohydrate Loading: Comparing 7-Day Methods

Day	Classic Method	Modified Method 1	Modified Method 2
1	**Exercise:** Depletion **Diet:** Low carbohydrate	**Exercise:** Depletion **Diet:** High carbohydrate	**Exercise:** Tapering (reduced physical activity) **Diet:** Moderate carbohydrate
2	**Exercise:** Tapering **Diet:** Low carbohydrate	**Exercise:** Tapering followed by supplemental carbohydrate feeding **Diet:** High carbohydrate	**Exercise:** Tapering **Diet:** Moderate carbohydrate
3	**Exercise:** Tapering **Diet:** Low carbohydrate	**Exercise:** Tapering followed by supplemental carbohydrate feeding **Diet:** High carbohydrate	**Exercise:** Tapering **Diet:** Moderate carbohydrate
4	**Exercise:** Tapering **Diet:** Low carbohydrate	**Exercise:** Tapering followed by supplemental carbohydrate feeding **Diet:** Moderate carbohydrate	**Exercise:** Tapering **Diet:** Moderate carbohydrate
5	**Exercise:** Tapering **Diet:** High carbohydrate	**Exercise:** Tapering followed by supplemental carbohydrate feeding **Diet:** Moderate carbohydrate	**Exercise:** Tapering **Diet:** High carbohydrate
6	**Exercise:** Tapering or none (rest) **Diet:** High carbohydrate	**Exercise:** Tapering followed by supplemental carbohydrate feeding **Diet:** Moderate carbohydrate	**Exercise:** Tapering or rest **Diet:** High carbohydrate
7	**Exercise:** Tapering or rest **Diet:** High carbohydrate	**Exercise:** Tapering followed by supplemental carbohydrate feeding **Diet:** Moderate carbohydrate	**Exercise:** Tapering or rest **Diet:** High carbohydrate
8	Competitive event	Competitive event	Competitive event

Sources of data: Goforth HW Jr: Effects of depletion exercise and light training on muscle glycogen supercompensation in men. *American Journal of Physiology, Endocrinology, & Metabolism* 285(6): E1304, 2003; Kreider RB and others: ISSN exercise & sport review update: Research & recommendations. *Journal of International Society of Sports Nutrition* 15:38, 2018; Williams MH and others: *Nutrition for Health, Fitness, & Sport.* 12th ed. New York: McGraw Hill, 2020.

Consuming Carbohydrate During Events

When athletes exercise vigorously for longer than 60 minutes, their glycogen supplies become depleted. At this point, athletes report they have "hit the wall"; that is, they feel unable to maintain a competitive pace. While performing prolonged physical activity, athletes can delay reaching "the wall" by consuming 30 to 60 g of carbohydrate per hour of activity.[8]

Sports drinks are a convenient way to obtain a source of glucose during lengthy and vigorous physical activities. Commercially available sports drinks are usually sweetened with nutritive sweeteners, such as sucrose, glucose, fructose, or maltodextrin. Such beverages typically provide 15 to 27 g of carbohydrate per 12-ounce serving. Foods or drinks that are concentrated sources of fructose are not recommended because large amounts of this particular simple sugar may cause gastrointestinal upset. In addition to supplying carbohydrate, sports drinks contain water and electrolytes, such as sodium, that can benefit athletes during prolonged physical effort. Sports gels are also good sources of simple carbohydrate, but they generally supply very little fluid. Therefore, it is important for athletes who consume these products to drink enough water to maintain proper hydration during endurance events.

During a post-event meal, starchy foods, such as this risotto made with rice and mushrooms, can be served to boost an athlete's carbohydrate consumption.
Nickola_Che/Shutterstock

Consuming Carbohydrate During Exercise Recovery

After completing exhaustive physical activity, trained athletes can replenish nearly all of their glycogen stores within a few days, provided they rest and eat a high-carbohydrate diet. During a post-event meal, starchy foods such as whole-grain bread, baked or mashed potatoes, rice, and pasta can be served to boost athletes' carbohydrate consumption.

During the exercise recovery period that occurs after an endurance event, athletes need to consume 6 to 12 g of carbohydrate/kg of body weight to replenish their muscle glycogen stores.[8] To restore their supply of muscle glycogen quickly after an event, athletes can consume sports drinks, candy, sugar-sweetened soft drinks, and fruit or fruit juices. However, if an athlete's dietary carbohydrate intake is inadequate, consuming beverages that contain carbohydrates and protein, such as chocolate milk, can increase glycogen levels faster than carbohydrate alone.[15,16]

Did *YOU* Know?

Milk contains two major kinds of protein: casein and whey. *Whey proteins* can be isolated (separated) from the liquid that forms during cheese production. Whey protein is a source of essential amino acids, vitamins, and minerals.

Most dietary supplements that are made from whey are powders that contain *whey protein isolates.* The powder can be mixed with water, milk, or juice to form a drink; sprinkled on cereal; or incorporated into a smoothie. For healthy young adults engaged in heavy resistance training, whey protein supplementation can enhance exercise recovery and increase muscle protein production.[18,19]

Be sure to review the Nutrition Facts panel and ingredient list of protein supplements because they may contain added sugars and other added ingredients.

©Wendy Schiff

What About Protein?

The adult RDA for protein is 0.8 g/kg of body weight.[17] During prolonged physical activity, muscles lose some protein because they metabolize certain amino acids for energy. Furthermore, extra amino acids are needed to repair and increase muscle tissue after exercise (recovery).[15] According to a recent review, athletes who engage in moderate-intensity exercise generally need 1.4 to 2.0 g of protein/kg of body weight/day.[13] Athletes who perform vigorous intensity or resistance exercise may benefit from consuming more than 2.0 g/kg of body weight daily. Recommended sources of protein include milk and other dairy products, lean meats, and dietary supplements that contain whey, casein, soy, and egg proteins.

Consuming extra protein does not appear to be harmful for healthy physically active people. Eating excessive amounts of protein, however, can be dangerous for people who cannot metabolize amino acids or ammonia properly (see Figure 7.12).

Under normal conditions, carbohydrate and fat are the primary fuels for cellular activity. Protein is not a major biological fuel. To spare protein so the nutrient can be used for muscle tissue growth and repair during recovery, it is very important for physically active people to consume adequate amounts of carbohydrate and fat.

Raising the Bar?

"Lasting energy," "fast fuel," "optimal energy"—such claims are used to promote so–called energy bars, gels, and drinks. Many people, whether physically active or sedentary, consume energy bars as a snack or before and/or after physical activity because they are convenient. However, some energy or sports bars contain processed ingredients, such as soy, pea, and milk proteins that are fortified with vitamins, minerals, and fiber. Sugars such as brown rice syrup, corn syrup, or honey are added to hold these ingredients together. *High–protein* energy bars may appeal to athletes who think proteins are a source of energy for physical activity. However, proteins are not a major biological fuel, and they are not "quick energy" sources because the liver must process amino acids before they can be used for energy.

Proteins in high-protein energy bars are not "quick energy" sources. The liver must process amino acids before they can be used for energy.
©Wendy Schiff

Energy and sports bars made with minimally processed ingredients such as dried fruits and nuts are more natural sources of energy, nutrients, and fiber. Granola bars and fresh or dried fruits are less expensive than energy or sports bars. You can make your own "energy" bars by following the recipe in the "Recipe for Healthy Living" feature. If you prefer to eat commercial energy bars, check the products' Nutrition Facts panels to determine amounts of carbohydrates, added sugars, and other nutrients that are in a serving. By eating several energy bars daily, you may ingest high amounts of iron, vitamin A, and other micronutrients. Therefore, consider energy bars as occasional snacks and not as meal replacements. Table 11.6 presents approximate energy and macronutrient contents per typical serving of various popular energy bars and gels.

Energy drinks usually contain sugars and a lot of **caffeine,** a stimulant drug that is naturally in coffee and tea. When consumed in moderate amounts, caffeine can increase alertness and decrease fatigue. Some energy drinks also contain the amino acid *taurine* and herbal substances such as ginseng. Although the body uses taurine, the compound is not required by humans. At present, there is no scientific evidence that ginseng enhances the stimulating effects of caffeine.[20] The "Ergogenic Aids: Separating Fact from Fiction" section of this chapter provides more information about caffeine, ginseng, and some other substances that are promoted for enhancing physical performance.

caffeine naturally occurring stimulant drug

TABLE 11.6 Popular Energy Bars and Gels: Energy and Nutrient Contents/Serving

Product	Energy kcal	Total Carbohydrates g	Added Sugars g	Fiber g	Protein g	Fat g
PowerBar® Protein Plus (chocolate peanut butter), 1 bar	210	24	11	4	20	6
PowerBar® PowerGel (vanilla), 1 packet	100	26	26	0	0	0
Luna Bar® (lemon zest), 1 bar	190	28	9	3	8	6
Clif Bar® (chocolate chip), 1 bar	250	43	17	5	10	6
Clif Shot® (vanilla gel), 1 packet	100	25	25	0	0	0
Balance Bar® (yogurt honey peanut), 1 bar	200	22	17	1	15	6
KIND® (Dark Chocolate Cherry Cashew), 1 bar	170	22	6	4	4	10
Larabar™ (apple pie), 1 bar	200	25	6	4	4	9

Source: Clif Bar & Company, www.clifbar.com/; Balance Bar Food Company, www.balance.com/; Power Bar, www.powerbar.com/ and www.shop.powerbar.com/; KIND, www.kindsnacks.com/; Larabar, www.larabar.com/. Accessed: January 10, 2023.

TABLE 11.7 Heat-Related Illnesses: Signs and Symptoms

Heat Cramps
• Painful muscle cramps, usually in legs or abdomen • Very heavy sweating • Tiredness • Thirst
Heat Exhaustion
• Dark urine • Cool, moist skin • Weakness • Light-headedness or dizziness • Headache • Nausea and vomiting
Heatstroke
• Dry, hot, red skin • High fever (over 104°F) • Rapid, shallow breathing • Confusion, irrational behavior • Seizures, loss of consciousness

Source: National Library of Medicine: MedlinePlus: *Heat emergencies.* Updated November 2021. https://medlineplus.gov/ency/article/000056.htm. Accessed: January 10, 2023.

Focusing on Fluids

The Adequate Intake (AI) for total water intake is approximately 11 cups (2.7 L) and 15.5 cups (3.7 L) for young females and males, respectively. Athletes generally require more water than nonathletes to keep their bodies cool during muscular activity. Many factors, however, influence a person's hydration status. Among athletes in particular, differences in sports, fitness levels, and environmental conditions affect fluid needs. For example, sweating can cause runners to lose 3 to 8 cups (750 to 2000 ml) of water per hour.[21] Even a small degree of dehydration can lead to declines in an athlete's endurance, strength, and overall performance. Moreover, body temperature rises when dehydration occurs, increasing the risk of heat–related illness.

Heat-Related Illness

As the environmental temperature and humidity increase, the evaporation of sweat from skin slows, and the body has difficulty cooling itself by perspiring. Ineffective sweating contributes to fatigue, makes the heart work harder, and raises the risk of *heat–related illness.* Table 11.7 presents three major types of heat–related illness and their common signs and symptoms.

Heat cramps are the earliest symptoms of heat illness. These painful muscle *spasms* (involuntary contractions) can affect any muscle, but they usually occur in the abdominal or calf muscles. Other early signs and symptoms of heat illness include heavy sweating, tiredness, and thirst.[22] If heat cramps are not treated effectively, the person develops heat exhaustion.

Heat exhaustion can occur after heavy exercise in warm conditions, especially when fluid and salt intakes have been inadequate. A person suffering from heat exhaustion experiences weakness, nausea, and vomiting; complains of having a headache and feeling dizzy; and has dark urine and cool, moist skin.

To treat heat cramps or heat exhaustion, move the victim to a cool or shady place and have the person lie on their back with legs elevated 12 inches. To relieve leg cramps, massage the affected muscles until they relax. Cool the person by fanning, spraying with cool water, or giving a cool sponge bath. It is also very important to have the victim drink cool water or a sports drink, unless the person is vomiting or has lost consciousness. People with heat exhaustion should be monitored closely because the condition can rapidly develop into *heatstroke.*

Heatstroke is the most dangerous form of heat–related illness. Signs of heatstroke include dry, red, hot skin; elevated body temperature (often more than 104°F, measured rectally); rapid, shallow breathing; irrational and confused behavior; and loss of consciousness (coma).[22] Heatstroke is a medical emergency that needs to be treated by trained medical staff. If you suspect that a person has heatstroke, summon emergency medical assistance immediately (dial 911) and move the victim to a cool environment. While waiting for professional medical care to arrive, give the patient cool water to drink (if they are conscious and not vomiting) and spray the person's skin with cool water.

Replenishing Fluids

To reduce the risk of heat illnesses, athletes and other physically active people should avoid exercising under extremely hot, humid conditions and should replace fluid losses that occur during prolonged exertion. Athletes participating in lower–intensity activities in cooler weather, lasting 1 to ½ hours, may meet water needs by drinking fluids in response to becoming thirsty. However, athletes participating in higher–intensity activities lasting 1½ to 2 hours in warm or hot environments may benefit from a pre–established fluid consumption plan focused on minimizing fluid losses.[23] Athletes following these drinking plans must be monitored to ensure fluids are not overconsumed, leading to

heat cramps heat-related illness characterized by painful muscle contractions

heat exhaustion heat-related illness that can occur after intense exercise

heatstroke most dangerous form of heat-related illness

water intoxication (*hyponatremia*). Nevertheless, athletes should avoid losing more than 2% of their body weight during exercise.

To estimate the amount of fluids needed to replace water loss during exercise (*rehydration*), athletes can weigh themselves prior to exercising and then calculate 2% of their body weight (0.02 × weight). After working out, athletes should weigh themselves again. If the difference between preexercise and postexercise body weights is more than 2%, fluid replacement is necessary during such activities. For example, if you weigh 150 pounds, 2% of that weight is 3 pounds (0.02 × 150). Therefore, you should drink enough watery fluids to avoid losing 3 or more pounds when you train or compete. In general, you can replace each pound that you lose during exercise by drinking 1¼ to 1½ pints (2½ to 3 cups) of water.[8] For example, if your usual weight is 150 pounds and you weigh 149 pounds immediately after exercising, you have lost 1 pound of body water during the activity. The next time you exercise under similar conditions, you can consume 3 cups of water during the activity to maintain your body's water balance. Why do you need to drink more fluids than you have lost in sweat? After exercise, rehydration may stimulate the kidneys to produce more urine than normal; thus, you need to drink the extra watery fluids to achieve proper hydration. As noted in Chapter 9, alcoholic beverages have a diuretic effect on the body and are not recommended for rehydration.

Sports drinks provide some nutritional benefits beyond those of plain water.

Do I Need a Sports Drink? Sports drinks provide some nutritional benefits beyond those of plain water. These beverages usually contain simple carbo-hydrates, a source of energy that can enhance performance during endurance activities. Recommended products contain about 21 g of carbohydrate per 12-ounce serving, or about 6% carbohydrate by weight. Drinks with sugar contents above 10%, such as soft drinks or fruit juices, are not recommended because they may cause intestinal discomfort. Sodium and other electrolytes in sports beverages help maintain blood volume, enhance the absorption of water and carbohydrate from the intestinal tract, and stimulate thirst.

Should you drink water or a sports drink during competition? The International Society of Sports Nutrition (ISSN) recommends athletes consume sports drinks during high-intensity activities lasting more than 1 hour. Sport drinks that provide 30 to 60 g of carbohydrates per hour of activity are recommended for these athletes.[13] Although electrolytes are lost in sweat, the quantities lost in shorter periods can be easily replaced by consuming foods and beverages, such as water or fruit juice, after the event.

It is possible to drink too much water and develop water intoxication (see Chapter 9). Endurance athletes, especially poorly trained individuals, may compete at relatively low exercise intensities for prolonged periods. Under these conditions, the athletes do not sweat as much; therefore, they do not need to replace as much water as better-trained athletes who exercise at higher intensities. Regardless of training level, athletes who drink too much water can dilute the level of sodium in their blood and develop serious and even deadly side effects. Although sports drinks generally contain sodium, these beverages are mostly water; therefore, consuming excessive amounts of sports drinks can contribute to fluid overload. To avoid water intoxication, athletes should drink water according to their thirst or follow a closely monitored fluid consumption plan. If an athlete gains weight while exercising, they may be retaining too much fluid.

Food & Nutrition tips

Compared to commercially available sports drinks, dilute fruit juices with added sugar and salt are less expensive sources of water, sodium, and simple sugars. You can prepare your own "sports" drink by adding ¼ cup of orange juice, ¼ cup of sugar, and ⅛ tsp of table salt to 30 oz of water or club soda. Pour the beverage into a quart pitcher, stir, cover, and refrigerate until needed.

Antioxidant Vitamins

During aerobic physical activities, skeletal muscles use more oxygen and generate more radicals than resting muscle tissue.[24] Exercise can produce

Intense exercise may stimulate the body's natural antioxidant defense system.
i love images/Alamy Stock Photo

a temporary imbalance between radical ("free radical") generation and the ability of antioxidants to counteract them. Scientific evidence suggests that such *oxidative stress* may contribute to muscle fatigue and damage. Nevertheless, results of studies that examined whether antioxidant vitamin supplements enhanced athletic performance generally concluded that performance was not improved, unless there was a preexisting deficiency of those particular vitamins.

Findings of some scientific studies indicate that radicals generated during intense exercise may stimulate the body's natural antioxidant defense system. Thus, the oxidative stress produced during exercise might have benefits, and blocking this process by taking antioxidant vitamin supplements may not be desirable.[8]

Athletes should be cautious about taking dietary supplements that contain antioxidants based on anecdotes or advertising claims because there is not enough scientific evidence concerning the long-term effects of using these products.[8] Rather than experiment on themselves with antioxidant supplements, athletes should follow diets that contain foods naturally rich in antioxidants, such as fruits, vegetables, whole-grain breads and cereals, and vegetable oils.

Iron

The body needs iron to produce red blood cells, transport oxygen, and obtain energy. Thus, iron deficiency can negatively affect athletic performance. Young female athletes are likely to develop iron deficiency because of their menstrual blood losses: Bleeding is a cause of iron deficiency. Athletes who follow low-calorie or vegetarian (especially vegan) diets are also at risk of iron deficiency because their food choices may be low in iron.

In the early phase of their training, endurance athletes often develop *sports anemia,* a temporary condition that results from an increase in the liquid portion of blood (*plasma*), rather than iron deficiency. This particular type of anemia does not appear to have negative effects on physical performance. Nevertheless, it can be difficult to differentiate between sports anemia and true iron deficiency anemia. It is a good idea for athletes, especially females, to have their iron status checked at the beginning of a training season and at least once during midseason. If an athlete is iron deficient, a physician needs to determine the cause and prescribe treatment.

Calcium

Athletes, especially those who are total vegetarians or restrict their consumption of dairy products to lose weight, can have marginal or low calcium intakes. This practice may result in weak bones that fracture easily, as well as in osteoporosis later in life. Additionally, female athletes who have irregular or no menstrual cycles may be deficient in the hormone estrogen. Although weight-bearing exercise, such as jogging, improves bone density, estrogen is also necessary for maintaining healthy bones. Female athletes who develop menstrual cycle abnormalities should consult a physician to determine the cause. Decreasing the amount of training or gaining weight may restore a regular menstrual pattern. For information about the *female athlete triad,* see the Chapter 10 "Nutrition Matters" section; Chapter 9 provides information about osteoporosis.

Concept CHECKPOINT

12. What are some practical ways to assess whether an athlete's energy intake is adequate?
13. Why should athletes be concerned about their carbohydrate intakes before, during, and after prolonged intense physical activity?
14. Identify at least five high-carbohydrate/low-fat foods.
15. Explain why athletes may need to consume more than the RDA for protein.
16. Why should athletes be concerned about their fluid status?
17. What are major signs and symptoms of heat cramps, heat exhaustion, and heatstroke?
18. When is consuming a sports drink a better choice than plain water for rehydration?
19. Explain why you would or would not recommend that an athlete take antioxidant supplements.
20. Explain why iron deficiency can impair an athlete's physical performance.
21. For young female athletes, what is the significance for bone health of having irregular or no menstrual cycles?

See Appendix F for responses.

11.5 Ergogenic Aids: Separating Fact from Fiction

Learning Outcomes

1. Discuss the pros and cons of using amino acid supplements and other ergogenic aids for athletic performance.
2. Identify three ergogenic aids that have been banned by at least one athletic association.
3. Identify rich sources of caffeine, and discuss the effects that the substance can have on the body.

Athletes often use **ergogenic aids**—foods, devices, dietary supplements, and even drugs ("doping")—to improve their physical performance. Bee pollen, dried adrenal glands from cattle, seaweed, freeze–dried liver flakes, and ginseng are among the dietary supplements that athletes consume as they hope to gain the competitive edge over their rivals. However, no reliable scientific evidence supports the effectiveness of most dietary supplements purported to have ergogenic effects.[8] Nevertheless, many athletes firmly believe in the value of the performance enhancing aids that they use. In many instances, the perceived benefits are more likely to result from the placebo effect than from actual physiological changes (see Chapter 2).

A few dietary substances and practices can enhance physical performance. These ergogenic aids include sufficient water and electrolytes, carbohydrates, and a balanced and varied diet consistent with MyPlate recommendations. For athletes, meeting carbohydrate and fluid needs—along with overall nutrient needs—is the most important ergogenic aid.

Athletes should be skeptical of claims made for any substance until its ergogenic effects and long–term safety have been determined by researchers who are not associated with the supplement industry. Rather than searching for a "magic bullet" to enhance their performance, athletes should concentrate their efforts on improving their dietary habits, training routines, and sports techniques. Nutrient supplements should be used for specific dietary deficiencies, such as preventing iron deficiency or boosting calcium intakes.

ergogenic aids foods, devices, dietary supplements, or drugs used to improve physical performance

Table 11.8 summarizes science-based findings regarding caffeine, creatine, and some other dietary supplements and ergogenic aids that are popular among athletes. It is important to note that certain dietary supplements are known to be unsafe, and the use of others is restricted or banned by major athletic organizations. For example, in 2004, the U.S. Food and Drug

TABLE 11.8 Evaluation of Some Popular Ergogenic Supplements/Aids

Substance	Claim	Current Evidence-Based Findings Concerning Claims	Side Effects
Beta-alanine	Enhances training capacity and improves performance during brief bouts of intense exercise	May be beneficial	May cause tingling sensations
Beta-hydroxy-beta-methylbutyrate (HMB)	Decreases protein metabolism, increasing muscle mass	May increase muscle mass, but evidence is weak	None reported, but results of long-term use are unknown
Branched-chain amino acids (BCAA)	Provide energy for muscles	May enhance muscle recovery after intense physical activity	None reported
Caffeine (guarana)	Enhances fat metabolism Delays depletion of muscle glycogen Increases alertness	Consuming 3–6 mg of caffeine/kg of body weight about 1 hour before exercise can benefit certain athletes.	High doses can cause nervousness, shakiness, and sleep disturbances; very high doses can be deadly. Intakes that cause the kidneys to eliminate more than 15 mcg of caffeine per ml of urine are banned by the National Collegiate Athletic Association.
Calcium pyruvate	Increases endurance and promotes weight loss	Not effective	Gastrointestinal upset May be harmful
Chromium	Increases lean mass	No benefit	Toxic intakes are above 400 mcg/day.
Creatine	Enhances muscular endurance and strength Increases lean muscle mass	Increases lean and total body mass, improves performance for intense activities that are not endurance activities, increases strength Such positive effects are not consistent among athletes. Consuming up to 30 g/day for 5 years is safe for healthy adults.	May cause weight gain and gastrointestinal upset
Energy drinks	Increase alertness and focus Enhance anaerobic and endurance performance Increase fat metabolism	Results vary based on the composition of the energy drink. Some energy drinks may enhance the performance of athletes and promote a small amount of fat loss.	See "Caffeine." Elevated heart rate and blood pressure Increased likelihood of engaging in risky behaviors
Ginseng	Combats fatigue and improves stamina	Does not improve performance but may decrease muscle damage during exercise	May cause gastrointestinal upset, itching, sleeplessness, and, in rare cases, allergic reactions and liver damage
Nitrate	Improves exercise endurance and performance	Benefits are not evident in elite athletes.	Nitrate concentrates, such as beetroot juice, may cause intestinal upset and red urine.
Sodium bicarbonate (baking soda)	Reduces lactic acid accumulation	May enhance performance, particularly in exercises that rely primarily on the lactic acid energy system (e.g., 400-meter sprint)	May cause nausea, vomiting, and diarrhea

Source: Position of the Academy of Nutrition and Dietetics, Dietitians of Canada, and the American College of Sports Medicine: Nutrition and athletic performance. *Journal of the Academy of Nutrition and Dietetics* 116(3):501, 2016; Kreider RB and others: ISSN exercise & sport nutrition review update: Research & recommendations. *Journal of the International Society of Sports Nutrition* 15(38), 2018; NIH, MedlinePlus: *American ginseng*. 2021. Accessed: January 10, 2023. https://medlineplus.gov /druginfo/natural/967.html.

Terry Wild Studio/McGraw Hill

Administration (FDA) banned the use of ephedrine–containing supplements. Ephedrine (ephedra, ma huang) increases central nervous system activity, but the drug can cause serious, even deadly side effects. The following section discusses the ergogenic properties of caffeine.

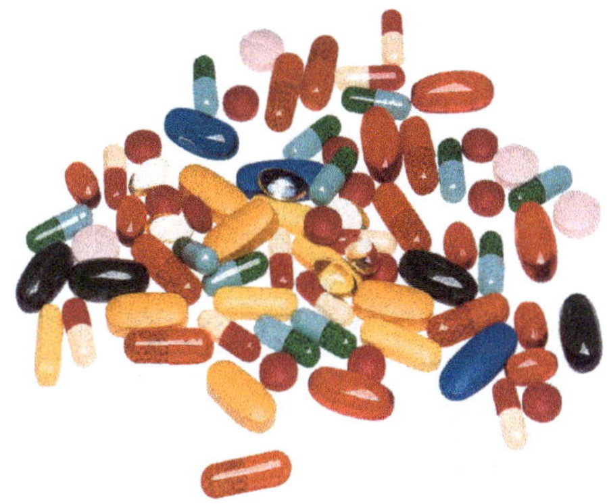

Certain dietary supplements are unsafe, and their use by athletes is restricted or banned by major athletic organizations.
Comstock/Alamy Stock Photo

Did *YOU* Know?

The human body makes DHEA, which certain cells convert to estrogen and testosterone, the "sex hormones." Testosterone stimulates muscle cells to remove amino acids from the bloodstream, which enables the cells to synthesize protein. As a result of this anabolic effect, muscular growth occurs. Anabolic steroids are synthetic substances that are chemically similar to testosterone. In the United States, a physician's prescription is needed to legally obtain anabolic steroids. When abused, such substances can cause serious side effects. Anabolic steroids, for example, can cause bloody liver cysts; increased risk of cardiovascular disease, hypertension, and reproductive problems; and aggressive behavior ("roid rage").

The body also synthesizes growth hormone (GH). As its name implies, GH stimulates muscle and bone growth, and it helps regulate metabolism. Long-term use of GH may increase the risk of cancer and have other potentially harmful side effects, such as elevated blood glucose levels. The FDA has few approved uses for medications that contain GH, including the treatment of GH deficiency in children.[25] Without access to the hormone, deficient children do not grow properly and, as adults, have heights that are much shorter than average. The National Collegiate Athletic Association (NCAA), however, bans the use of GH, as well as anabolic steroids, by athletes.[26]

Caffeine

Worldwide, caffeine is the most widely used ergogenic aid. Caffeine raises the level of fatty acids in blood, and as a result, exercising muscles can use more fat for energy. Caffeine also enhances the ability of skeletal and heart muscles to contract and increases mental alertness. Although consuming even small amounts of caffeine may help endurance athletes, the NCAA bans the use of caffeine and limits the amount of caffeine that athletes can have in their bodies during competition.[26] Athletes who have more than 15 micrograms of caffeine per milliliter (mcg/ml) in their urine can be banned.[8] Nevertheless, the World Anti–Doping Agency (WADA) does not prohibit the use of caffeine.[27]

People who are not regular caffeine consumers may experience shakiness, rapid heart rate, sleep disturbances, diarrhea, and frequent urination after ingesting relatively high amounts of the stimulant drug. Caffeine is addictive; discontinuing its use results in *withdrawal,* temporary unpleasant side effects, especially headache.

It is important to understand that the quick "energy" boost provided by "energy drinks" is a result of caffeine or other stimulants contained in the products. Nevertheless, energy drinks also provide calories if they are sweetened with added sugars. Table 11.9 compares the caffeine and calorie contents of popular beverages, including an energy drink (e.g., Red Bull®). "Stay awake" pills and chewing gums are concentrated sources of caffeine that can be purchased without a prescription. In September 2015, the FDA warned consumers to avoid "powdered" caffeine. At that time, the agency was aware of two deaths that were attributed to use of such products.[28]

Coffee is a rich source of caffeine.
Holly Curry/McGraw Hill

Tea: John A. Rizzo/Photodisc/Getty Images; Coffee beans: lynx/iconotec.com/Glow Images

TABLE 11.9 Caffeine Content of Selected Beverages

Beverage and Amount	Approximate Energy (kcal)	Approximate Caffeine (mg)
Chocolate drink, milk based, 8 oz	120	2
Caffé Latte Starbucks, 16 oz*	190	150
Coffee, brewed from grounds, unsweetened, 8 oz	2	95
Coffee, instant plain, prepared, 8 oz	5	62
Cola, with added sugar and caffeine, 12-oz can	155	33
5-hour ENERGY shot®, original, 1.9-oz bottle**	4	200
Monster® energy drink, 8-oz can	121	81
Mountain Dew®, 12-oz can***	170	54
Red Bull®, 8.4-oz can	112	75
Tea, black, brewed, unsweetened, 8 oz	2	47
Tea, sugar-sweetened, ready-to-drink, 12-oz can	183	37

Source: U.S. Department of Agriculture: Agricultural Research Service. *FoodData Central,* 2019.

*Starbucks Beverages: *Nutrition information.* www.starbucks.com. Accessed: January 10, 2023.

**5-Hour Energy: *Product Facts.* https://5hourenergy.com. Accessed: January 10, 2023.

***Pepsico: Facts about your favorite beverages. www.pepsicobeveragefacts.com. Accessed: January 10, 2023.

Concept CHECKPOINT

22. Why do many athletes use ergogenic aids?

23. Identify three ergogenic aids that have been banned by the FDA or an athletic association.

24. Identify at least three beverages that are rich sources of caffeine.

25. Discuss the ergogenic effects that caffeine can have on the body.

See Appendix F for responses.

11.6 Nutrition Matters: Developing a Personal Fitness Plan

Learning Outcomes

1. Identify the key components of an aerobic workout regimen.
2. Design a personal fitness program that suits your interests and lifestyle.

When developing your personal physical fitness plan, first consider your fitness goals. For example, if you want to lose weight, how much do you want to lose and how many weeks will it take to lose that amount? Do you want to focus more on strengthening your muscles or on improving your aerobic capacity? Then determine when you can work out and whether you'll need to join a fitness facility, such as a gym, or purchase special equipment, such as handheld weights. For a comprehensive fitness program, make sure to include aerobic, resistance, and flexibililty activities in your weekly exercise regimen. The following fitness plan has three stages: initial, improvement, and maintenance phases.

Initiation

The first 3 to 6 weeks of your new exercise program is the *initiation* stage. Start by incorporating short periods of physical activity into your daily routine. For example, you can walk more often, take the stairs instead of the elevator, and do more housework, gardening, or other activities that cause you to "huff and puff" a bit. Furthermore, you can strive to reduce the time that you spend in sedentary activities.

The goal is to accumulate a total of 150 minutes of moderate–intensity types of activity each week. If necessary, the time that you spend engaging in the activity can occur in several short intervals, each lasting at least 10 minutes. If you do not have 150 minutes to spend on exercising, try increasing the intensity of the activities during shorter bouts of exercise to obtain some health benefits.

Improvement and Maintenance

The next 5 or 6 months of the program is the *improvement* stage, in which you increase the intensity and duration of exercises. When you begin the improvement phase, exercise at an intensity that is near the lower end of your target heart rate zone. As you progress and become more physically fit, you can increase the intensity by exercising at a higher heart rate.

By the end of the improvement stage, you may notice that you have reached your goals, and you do not seem to be making further gains in your fitness. This plateau marks the beginning of your *maintenance* stage. At this point, you can evaluate your personal fitness plan, and if you would like to make new goals, this is the time to develop them. If you are satisfied with your fitness level, continue with your present program. Discontinuing exercise gradually results in *detraining,* declining physical fitness.

FatCamera/E+/Getty Images

Components of an Aerobic Workout Regimen

Ideally, you should establish a regular time for exercising that fits into your daily routine. To be effective, your aerobic workout program needs to include the following components:

1. **Warm–up** Warming up muscles can increase your joints' range of motion (flexibility) and may decrease your risk of injury. Stretching for 5 to 10 minutes is a good way to warm up. Start with smaller muscle groups such as the arms and progressively work toward stretching larger muscle groups in the legs and abdomen. Hold your position in the stretch for 15 seconds and do not bounce. If stretching causes pain, stop immediately. "No pain, no gain" is not true: Pain is an indication of injury. Another way to warm up is to perform 5 to 10 minutes of the anticipated activity but at a low intensity. For example, if you walk for fitness, warm up by walking at a slower pace.
2. **Aerobic workout** To obtain substantial health benefits, you should engage in some form of aerobic activity regularly. A comprehensive aerobic workout emphasizes the *type, duration, frequency, intensity,* and *progression* of exercise.
   - **Type:** The kinds of exercise you choose should increase your heart and breathing rates and involve rhythmic movements of large muscle groups in the legs. Examples include brisk walking, running, swimming, and cycling. If you swim, add some *weight–bearing activities,* such as walking, to your comprehensive fitness plan. Weight–bearing exercises place stress on your bones, increasing their strength.
   - **Duration:** Duration is the amount of time spent in an exercise session. A session should generally last at least 20 to 30 minutes, depending on intensity, not including time spent warming up and cool–ing down. Ideally, the exercise session should be continuous (without stopping), but multiple 10–minute bouts of moderate to intense activity with rest periods in between are also acceptable.
   - **Frequency:** The frequency of exercise describes the number of times that the activity is performed, generally on a weekly basis. To derive significant health benefits, the frequency of aerobic exercise should be at least five times per week. By exercis–ing daily, you can enjoy even greater benefits.
   - **Intensity:** Health benefits can occur when you achieve at least a moderate level of intensity during exercise.
   - **Progression:** Progression, the final component of a comprehensive fitness plan, refers to the gradual increase in the frequency, intensity, and duration of exercise that occurs over a period.

3. **Cool–down:** To cool down, you can repeat the same stretches you performed during warming up. Stretch for 5 to 10 minutes. Cooling down may prevent injury and reduce muscle soreness.

What About Strength (Resistance) Training?

Strength training, such as weight lifting, is an important part of a comprehensive physical fitness plan. Strength training should be done at least 2 days per week. To start, warm up by performing flexibility exercises for 5 to 10 minutes. Then perform a group of 8 to 10 exercises that strengthen major muscle groups of the upper body and lower body. Cool down for 5 to 10 minutes at the end of each session.

Fitness centers have machines that provide resistance for various muscle groups. For resistance training outside of gymnasiums or fitness clubs, you can purchase simple elastic exercise cords designed to increase muscular strength (Fig. 11.11). For increasing upper arm strength, a set of inexpensive handheld weights can be kept in a convenient location for performing resistance exercise regularly, such as near the TV. The weights should allow you to perform at least one set of 8 to 15 repetitions. When you can do more than 15 repetitions with relative ease, consider increasing the weight slightly.

Mixing It Up

To make your exercise routine more enjoyable, include several types of physical activity, called *cross-training,* in your weekly regimen. For example, jogging one day might be followed by swimming the next day. Cross–training not only keeps you from becoming bored with your workouts, but also strengthens different muscle groups in your body and reduces your risk of injury. Additionally, invite a friend or relative to be your exercise partner. Having an exercise partner may provide additional motivation and encouragement to exercise regularly.

Figure 11.11 Increasing muscle strength. This individual is using a simple rubber exercise cord to increase muscular strength.
©Wendy Schiff

Some people who are overfat do not experience significant weight loss while following an exercise regimen. However, they still benefit from regular physical activity. Initially, exercise programs for people who are obese should emphasize non–weight–bearing activities, such as swimming, water aerobics, and bicycling. As people who are obese lose weight and become more fit, they can add weight–bearing activities to their plans.

Whatever physical activities you choose to include in your fitness program, they should be enjoyable and easy to incorporate into your routine. You can apply the dietary principles of variety, balance, and moderation to your exercise routine:

- **Variety:** Perform several different activities to exercise different muscle groups.
- **Balance:** For overall fitness, balance your exercise regimen by including activities that build cardiovascular endurance, muscular strength, and flexibility.
- **Moderation:** Focus on exercising to keep fit without overdoing it and injuring yourself. You do not need to work out vigorously every day to become healthier.

To learn more about the health benefits of physical fitness, determine your level of fitness, or develop a personal fitness program, access the following websites:

www.cdc.gov/physicalactivity/
www.hhs.gov/fitness/index.html.

Concept CHECKPOINT

26. List three key components of an aerobic workout regimen.

27. Develop a personal fitness regimen that you can follow for the rest of your life.

See Appendix F for responses.

Chapter References

See Appendix G.

Summary

11.1 Introduction to Nutrition for Fitness and Sport

- Many Americans lead sedentary lives. People can manage their weight more effectively and reduce their risk of developing the major causes of death and disability by engaging in at least 150 minutes of moderate-intensity physical activity weekly. Furthermore, most people can achieve even greater health benefits by increasing the duration, frequency, and intensity of their physical activities.

11.2 Benefits of Regular Exercise

- Regular physical activity reduces the risk of dying prematurely and developing heart disease, diabetes, and high blood pressure. Additionally, regular exercise builds and maintains healthy bones, muscles, and joints; helps reduce blood pressure in people who have high blood pressure; reduces the risk of developing certain cancers; aids weight-control efforts; helps older adults become stronger and reduces their risk of falls; reduces risk of sleep disorders, depression, and anxiety; and promotes psychological well-being. Millions of Americans suffer from chronic illnesses that can be prevented or improved by exercising more often.

Vasily Pindyurin/Westend61 GmbH/Alamy Stock Photo

11.3 Energy for Muscular Work

- ATP is the major form of energy used by cells. Phosphocreatine (PCr) can rapidly reform ATP from its breakdown product ADP, but PCr supplies are limited. To generate ATP, muscle cells can metabolize carbohydrate, fat, and protein. In muscle cells, glucose molecules are broken down through a series of steps to yield lactic acid (in anaerobic conditions) or CO_2 plus H_2O (in aerobic conditions).
- The proportions of macronutrients used for energy largely depend on the intensity of the physical activity. Fat is a key aerobic fuel for muscle cells, especially at low-intensity exercise. At rest and during light activity, muscles burn primarily fat for energy needs. In comparison, little protein generally is used to fuel muscles. During brief bouts of intense physical effort, the cardiorespiratory system is unable to deliver adequate oxygen to muscles. Under such anaerobic conditions, muscle cells metabolize glucose rather than fat for energy, but their ability to sustain the release of energy for intense activity is limited. Muscle cells can obtain far more energy when in aerobic conditions.

11.4 General Dietary Advice for Athletes

- Sports nutrition focuses on applying nutrition principles and research findings to improving athletic performance. Athletes often manipulate their diets to enhance their body contours, increase their muscular strength, and prevent or delay fatigue during exercise. Although diet is important, genetic endowment, mental training, and physical training are crucial factors that influence athletic performance.
- A high-carbohydrate diet can be beneficial for athletes, and carbohydrate-rich foods should form the foundation of pre-event meals. Many athletes need more protein than the RDA.
- Physically active people need to be concerned about their fluid intakes. Fluid replacement should be based on thirst and loss of body weight while exercising. It is important to recognize signs and symptoms of heat illness. Consuming a source of electrolytes, such as a sports drink, can be helpful. It is important to avoid overconsumption of water because of the risk of water intoxication.

11.5 Ergogenic Aids: Separating Fact from Fiction

- Athletes often ingest certain substances, including herbal products, because they believe these substances have ergogenic effects. Scientific evidence, however, does not suggest that most of these substances are effective. Furthermore, long-term safety of many ergogenic aids has not been determined, and use of some substances is restricted or banned by various athletic organizations.

11.6 Nutrition Matters: Developing a Personal Fitness Plan

- The three components of an aerobic workout are warming up, performing the moderate- or vigorous-intensity activity, and cooling down.

Recipe for Healthy Living

©Wendy Schiff

Nutty Energy Bars

You can use the following recipe to make your own "energy bars." This recipe makes eight 4-by-2-inch bars. Each bar supplies approximately 281 kcal, 8 g protein, 13 g fat, 33 g digestible carbohydrates, 3 g fiber, and 1.4 mg iron.

INGREDIENTS:

- Oil cooking spray
- 1 egg, large
- 1½ cups instant oats, dry
- ¼ cup almonds, slivered
- ¼ cup enriched wheat flour
- ½ cup smooth peanut butter
- 1 Tbsp canola oil
- ½ cup honey
- ½ tsp ground cinnamon
- 1 tsp vanilla

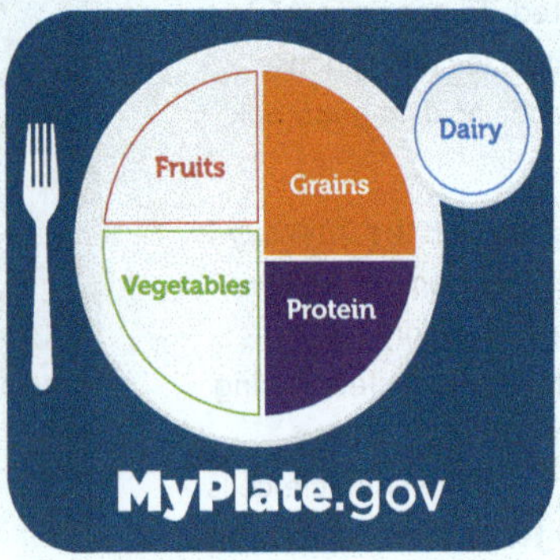

US Department of Agriculture or USDA

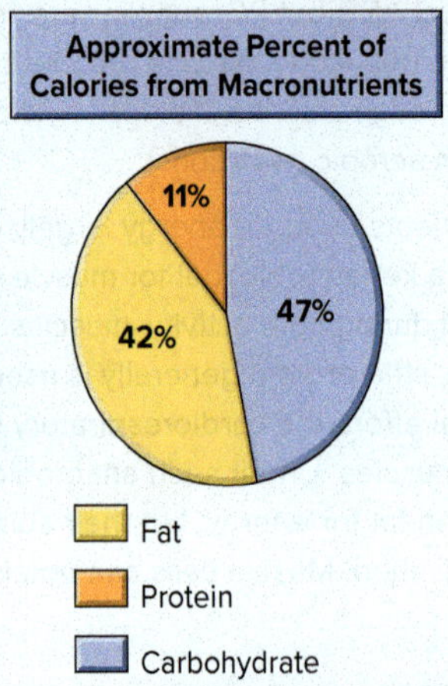

PREPARATION STEPS:

1. Preheat oven to 350°. Spray "nonstick" oil on the inside bottom and sides of an 8″ × 8″ pan.
2. Crack open the egg, drop the egg's contents into a medium-size bowl, and discard the shell. Using a fork, beat the egg until its yolk is completely mixed with the egg white.
3. Add remaining ingredients to the egg and stir until well blended. Mixture will be thick, like cookie dough. Using a large spoon, press dough into the bottom of the pan, covering the inside of the pan evenly.
4. Bake for 12 to 14 minutes. Cool completely before cutting into eight 4-by-2-inch bars.

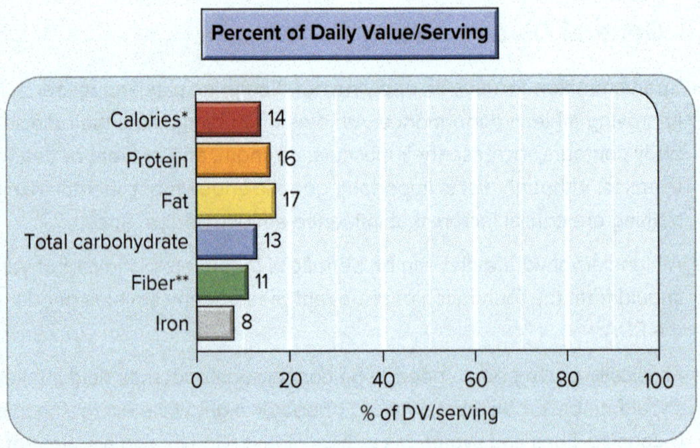

*2000 daily total kcal
**Component of total carbohydrates

Critical Thinking

1. Analyze your weekly physical activity habits. Does your participation in various physical activities meet the minimum recommendations? If not, which physical activities are you willing to include in your weekly routine to improve your fitness level?
2. Calculate your target heart rate zone for moderate–intensity as well as vigorous–intensity activities.
3. Why do human cells rely far more on glucose and fat for energy than on protein?
4. What advice would you give a runner concerning fluid intake before and during a marathon?
5. One of your friends is a student athlete who uses ginseng, nitrate, and caffeine to enhance their physical performance. What would you tell your friend concerning the pros and cons of taking these substances?
6. Sunny is an athlete who weighs 143 pounds. According to the information in Section 11.4, calculate the recommended amount of protein (g) she should consume daily.

Practice Test (Section numbers are listed below in red.)

Select the best answer.

1. (11.1) Saki is physically fit. She has
   a. an increased risk of osteoporosis.
   b. the strength, endurance, and flexibility to meet the demands of daily living.
   c. a greater need for vitamins and minerals than other women.
   d. no need to exercise regularly.
2. (11.2) A _____ physical activity generally requires a high degree of exertion.
   a. vigorous
   b. basic
   c. moderate
   d. precise
3. (11.2) Aerobic activities
   a. enable muscles to use less oxygen than normal.
   b. do not require voluntary muscular contractions.
   c. force muscle cells to use more vitamins and minerals for energy.
   d. involve sustained, rhythmic contractions of certain large skeletal muscles.
4. (11.2) Which of the following statements is true?
   a. Resistance exercises do not help build bone mass.
   b. Sedentary activities do not require much energy to perform.
   c. Anaerobic energy systems need large quantities of oxygen to produce ATP.
   d. All of the above are correct.
5. (11.2) Which of the following target heart rate ranges would be appropriate for a 26–year–old person who wants to perform moderate–intensity exercise? (The answers are approximations of beats per minute.)
   a. 124 to 147
   b. 149 to 180
   c. 97 to 136
   d. 107 to 165
6. (11.3) During glycolysis, the body
   a. converts two fatty acid molecules into one glucose molecule.
   b. synthesizes one amino acid molecule from one carbon dioxide molecule and two oxygen molecules.
   c. breaks down one glucose molecule to form two pyruvate molecules.
   d. metabolizes fatty acids into carbon monoxide and peroxide.
7. (11.3) Adebola is studying quietly. Under these conditions, her muscles are using primarily _____ for energy.
   a. fat
   b. glucose
   c. amino acids
   d. ketone bodies
8. (11.3) Human cells release much of the energy stored in carbohydrates, fats, and amino acids as
   a. electricity.
   b. phosphocreatine.
   c. ATP.
   d. heat.

9. **(11.3)** To obtain energy under aerobic conditions, cells need

a. ribose.
b. oxygen.
c. methionine.
d. alanine.

10. **(11.3)** _____ is an immediate and direct source of energy for cells.

a. LDP
b. Glycogen
c. ATP
d. Phospholipid

11. **(11.4)** Carbohydrate loading

a. provides a competitive edge for award–winning sprinters, bodybuilders, and weight lifters.
b. involves manipulating dietary patterns and physical activities prior to an endurance event.
c. often results in short–term weight loss and positive energy balance.
d. is generally recommended for long–term weight control for athletes.

12. **(11.4)** Which of the following foods is high carbohydrate and low fat?

a. Roast chicken
b. Swiss cheese
c. Peanut butter
d. Toast spread with strawberry jam

13. **(11.4)** Last August, Damien experienced heavy sweat–ing, painful leg cramps, and fatigue while exercising outdoors. Based on this information, he probably had

a. heatstroke.
b. heat cramps.
c. hypothermia.
d. hypothyroidism.

14. **(11.5)** Caffeine

a. is the most widely used ergogenic aid in the world.
b. reduces the level of fatty acids in blood.
c. is nonaddictive.
d. decreases mental alertness.

15. **(11.6)** In a comprehensive aerobic workout, __________ is how to gradually increase the intensity, number of days, and the amount of time spent on exercise.

a. progression
b. duration
c. moderation
d. frequency

Answers to Quiz Yourself

1. People who exercise regularly can reduce their risk of type 2 diabetes. **True.** (Section 11.2)
2. Sports drinks are not useful for fluid replacement. **False.** (Section 11.4)
3. Protein is the body's preferred fuel for muscular activity. **False.** (Section 11.4)
4. Heatstroke is a serious illness that requires immediate professional medical treatment. **True.** (Section 11.4)
5. While at rest, skeletal muscles metabolize more glucose than fat for energy. **False.** (Section 11.3)

Design Elements: (Stacked books): Siede Preis/Photodisc/Getty Images; (Nuts): C Squared Studios/Photodisc/Getty Images; (Recipe box, Raspberries, Blueberries): ©Wendy Schiff; (Pineapple): Stockbyte/Getty Images; (Radishes): karandaev/iStock/360/Getty Images; (MyPlate icon): US Department of Agriculture or USDA

CHAPTER 12

Food Safety Concerns

Pixtal/age fotostock

Quiz Yourself

Which foods are most likely to be responsible for food-borne illness? How can you reduce the risk of contracting one of these illnesses? Which government agencies monitor the safety of the U.S. food supply? By reading this chapter, you will learn answers to these questions. Test your knowledge of food safety by taking the following quiz. The answers are at the end of the chapter.

1. Aflatoxins are the most common sources of food-borne illness in the United States. ______ T ______ F
2. In case of emergency, it is not safe to drink the water that is in an undamaged water heater. ______ T ______ F
3. Certain fungi, such as button mushrooms, are safe to eat. ______ T ______ F
4. The Environmental Protection Agency (EPA) regulates the proper use of pesticides in the United States. ______ T ______ F
5. The best way to tell if a food is safe to eat is to smell it. ______ T ______ F

Food can become contaminated with pathogens or their toxins during production, processing, and preparation.
shironosov/iStock/Getty Images Plus

12.1 Introduction to Food Safety

Learning Outcomes

1. Identify the most common cause of food-borne disease outbreaks in the United States.
2. Explain the difference between a food-borne infection and a food-borne intoxication.

Most Americans look forward to taking vacations, and for many people, this involves boarding a cruise ship to visit tropical, Caribbean island destinations such as Cozumel, Grenada, or Aruba. The attractions of a cruise vacation include enjoyable activities for people of all ages, as well as dining opportunities that showcase plenty of delicious food. Every year, however, some people's dream vacation becomes a cruising nightmare, as hundreds of guests and crew members become miserably ill with stomach pain, diarrhea, and vomiting as a result of norovirus infection. In the United States, norovirus is the most common cause of **food-borne illness** outbreaks.[1]

People who are infected with norovirus will have the agent of infection in their vomit and feces. If people with the infection do not wash their hands thoroughly after vomiting or having bowel movements, they can spread the infection to others by handling food or ice cubes, shaking hands, or touching objects that other people touch. The infection spreads rapidly among people who are living in close quarters, such as on a cruise ship. To control the spread of the infection, sick passengers should be isolated from healthy people, and bathrooms, guest rooms, and any surface that may have been touched by ill persons should be disinfected with a strong bleach solution.[2]

People develop food-borne illnesses when they consume foods or beverages that contain certain microbes, natural or human-made toxins (poisons), or other contaminants. Disease-causing agents, such as norovirus, are referred to as **pathogens.** When pathogens are in food, they can make the item unsafe to eat. Many kinds of food-borne pathogens infect the digestive tract, inflaming the tissues and causing an "upset stomach" within a few hours after being ingested. A few types of food-borne pathogens multiply in the human intestinal tract, enter the bloodstream, and cause general illness when they invade other tissues. Such illnesses are called **food-borne infections.** Other pathogens do not sicken humans directly, but these microbes contaminate food and secrete poisons. When the contaminated food is eaten, the toxins irritate the intestinal tract and cause a type of food-borne illness called **food intoxication** (or food poisoning).

Illnesses spread through the ingestion of food or water are common, debilitating, and sometimes life-threatening diseases for millions of people around the world. There are more than 250 kinds of food-borne illnesses.[3] Each year, an estimated 48 million Americans become sick from various food-borne illnesses.[4] Of those people who contract such ailments, 128,000 require hospitalization and 3000 die. Promoting food safety efforts to prevent these diseases is a major goal of U.S. public health officials.

This chapter focuses primarily on food-borne illness, including common sources of these infections in the United States. Additionally, this chapter presents safe food-handling practices.

food-borne illness disorder that is caused by consuming disease-causing agents in food or water

pathogens disease-causing agents

food-borne infection illness that results when a pathogen in food inflames the intestinal tract or other body tissues

food intoxication illness that results when poisons produced by certain microbes contaminate food and irritate the intestinal tract

Concept CHECKPOINT

1. Which agent of infection is responsible for most cases of food-borne illness in the United States?
2. Is the illness caused by norovirus the result of a food-borne infection or food intoxication?

See Appendix F for responses.

12.2 Protecting Our Food

Learning Outcomes

1. Identify government agencies that are responsible for ensuring the safety of the food supply in the United States.
2. Discuss the roles of the FDA, FSIS, and EPA.

The United States has one of the safest food supplies in the world, primarily the result of a team effort conducted by cooperating federal, state, and local agencies that regulate and monitor the production and distribution of food. The Food and Drug Administration (FDA) of the U.S. Department of Health and Human Services and the U.S. Department of Agriculture (USDA) are the key federal agencies that protect consumers by regulating the country's food industry. Other team members include the Environmental Protection Agency (EPA), the Centers for Disease Control and Prevention (CDC), the Federal Trade Commission (FTC), and state and local governments.

To help protect our food supply, the FDA performs many important tasks, such as regulating nearly all domestic and imported food sold in interstate commerce and enforcing federal food safety laws. Additionally, the FDA establishes standards for safe food manufacturing practices, such as Hazard Analysis and Critical Control Point (HACCP) programs. HACCP is a science-based, systematic approach to preventing food-borne illness by predicting which hazards are most likely to occur in a food production facility. When a hazard is identified, food manufacturers can then take appropriate measures to prevent the illness. If necessary, FDA officials can take certain enforcement actions, such as requesting that a food manufacturer recall an unsafe item so that it is removed from store shelves. Another important function of the FDA is educating the general public about safe food-handling practices.

Although the FDA oversees the safety of most foods, the USDA's Food Safety and Inspection Service (FSIS) enforces food safety laws for domestic and imported meat and poultry products. FSIS staff inspect beef, poultry, and other food animals for diseases before and after slaughter, and the agency also ensures that meat and poultry processing plants meet federal standards. Additionally, FSIS staff collect and analyze food samples to check for the presence of microbial and other unwanted and potentially harmful material in foods. If a food hazard is identified, FSIS officials can ask meat and poultry processors to recall their unsafe products. Additionally, food safety experts with FSIS conduct programs to educate people about proper food-handling practices. For more information, visit the FSIS website at www.fsis.usda.gov/.

The EPA oversees the quality of our drinking water. EPA staff establish safe drinking water standards and assist state officials in their efforts to monitor water quality. Furthermore, EPA staff regulate toxic substances and wastes to prevent their entry into foods and the environment. In the United States, water-borne disease-causing microbes are not major public health threats because most of the population has access to water that has been treated to make it safe to drink (potable).

State and local officials work with the FDA and other federal agency staff to implement national food safety standards for foods produced and sold within their state's borders. Local health departments, for example, are

Food and Drug Administration (FDA)

The FDA oversees the safety of most foods.

In many communities, restaurants must display their rating for sanitation where customers can easily see it. Local health departments can close restaurants that do not receive high enough ratings until food safety hazards have been corrected.
©Wendy Schiff

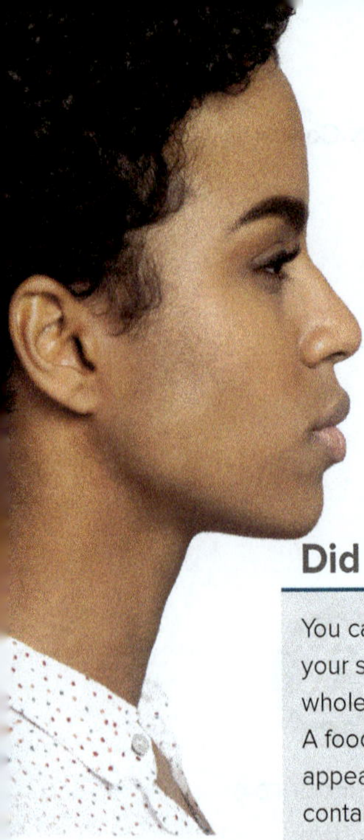

fizkes/Shutterstock

Did *YOU* Know?

You cannot always rely on your senses to judge the wholesomeness of a food. A food can taste, smell, and appear safe to eat, but it may contain pathogens and/or their toxic by-products.

responsible for inspecting restaurants, grocery stores, dairy farms, and local food processing companies. In many communities, restaurants are required to post their sanitation rating where customers can easily see it. Local health departments can close restaurants that do not receive high enough ratings and prevent them from reopening until food safety hazards have been corrected.

After you obtain foods and bring them into your home, it becomes your responsibility to reduce the risk of food–borne illness by handling the items properly. However, if you suspect that something you consumed made you or a family member very sick, you should contact your physician for treatment. The physician may decide to report the case of food–borne illness to local public health officials so they can investigate and determine the source of your infection.

Concept CHECKPOINT

3. Discuss the roles of the FDA, FSIS, and EPA in protecting the U.S. food supply.

4. How do local health departments protect consumers from food-borne illness?

See Appendix F for responses.

Wine production relies on the use of certain kinds of microbes.
lynx/ iconotec.com/Glow Images

12.3 Pathogens in Food

Learning Outcomes

1. Discuss ways that pathogens can contaminate human foods.
2. Explain how cross-contamination of food occurs.
3. Discuss the purpose of pasteurization.
4. Discuss conditions that favor the survival and multiplication of food-borne pathogens.
5. Identify foods that are high risk for supporting pathogens.

For thousands of years, people have used certain microbes to produce a variety of foods, including hard cheeses, raised breads, and alcoholic beverages. When microorganisms metabolize nutrients in food, they often secrete substances that alter the color, texture, taste, and other characteristics of the food in beneficial and desirable ways. Other kinds of microbes grow and multiply in food, but their metabolic byproducts spoil the food, making it unfit for human consumption.

Some microorganisms are parasites. A **parasite** needs to live on or within a host organism to survive. Parasites often obtain their nourishment from the host's tissues, so the host's health suffers as a result of a parasitic infection.

A **contaminated food** (or beverage) is no longer wholesome—pure or safe for human consumption. Contamination generally occurs when something that may or may not be harmful enters food or beverages unintentionally. Such contaminants include pathogens, insect parts, residues of compounds used to kill insects that destroy food crops, and metal fragments from food

parasite organism that lives in or on another organism, often deriving nourishment from its host

contaminated food item that is impure or unsafe for human consumption

processing equipment. The following section discusses how pathogens can contaminate our food.

To reduce your risk of food-borne illness, keep flies, cockroaches, and other vermin away from your food.
Simon Murrell/Dynamic Graphics Group/IT Stock Free/Alamy Stock Photo

Sources of Pathogens

The pathogens that cause food–borne illness can live practically anywhere—in air, water, soil, and sewage and on various surfaces. Our skin, nasal passages, and large intestines have vast colonies of various kinds of microbes, some of which can be pathogenic. Animals, including cats, dogs, reptiles, cattle, and poultry, can also harbor harmful microbes on and in their bodies, especially in their intestinal tracts. Because pathogens are found throughout your environment, there are numerous ways that the microbes can contaminate your food.

Common Routes for Transmitting Pathogens

One common route for transmitting harmful microbes involves *vermin,* animals that often live around sewage or garbage, such as flies, cockroaches, mice, and rats. When vermin land on or crawl across filth, they pick up pathogens on their feet. When the vermin come into contact with food, they can transfer the pathogens to humans. To reduce your risk of food–borne illness, keep flies, cockroaches, and other vermin away from your food.

Poor personal hygiene practices frequently transfer microbes to food. People can contaminate their hands with pathogens when they come in contact with feces, such as while using the toilet or changing a baby's soiled diaper. Furthermore, animals harbor pathogens in their feces, as well as on their skin and fur. If children prepare or eat foods after stroking animals at petting zoos or playing with pets, they can transmit these microbes to themselves or others. Thus, it is important for people to wash their hands before preparing or eating foods. Children (and many adults) often need to be reminded to wash their hands.

The intestinal tracts of turtles and other animals are sources of pathogens.
Comstock Images/Alamy Stock Photo

Improper food handling frequently results in food–borne illness. A common practice is failing to wash cutting boards and food preparation utensils after they come in contact with raw meat or poultry. The contaminated boards and utensils are then used to prepare other foods. As a result of this practice, **cross–contamination** is likely to occur because the pathogens in one food are transferred to another food, contaminating it. If that food is eaten raw, such as carrots in a salad, it carries a high risk of food–borne illness. Failing to cook foods properly can also increase the likelihood of food–borne illness. **Pasteurization** is a special heating process used by many commercial food producers to destroy pathogens. In the United States, for example, most juices and milk have been pasteurized before they are marketed. The "Food Preservation" section of this chapter discusses other ways of preserving foods.

High-Risk Foods

Not all foods are likely to harbor pathogens. To survive and multiply, most microbes need warmth, moisture, and a source of nutrients, and some also require oxygen. In general, high–risk foods are warm, moist, and protein–rich, and they have a neutral or slightly acidic pH. Many of the foods we eat every day, such as meats, eggs, milk, and products made from milk, fit this description. Table 12.1 presents some high–risk foods and the primary food–borne pathogens they may contain.

cross-contamination unintentional transfer of pathogens from one food to another

pasteurization process that kills pathogens in foods and beverages as well as many microbes responsible for spoilage

Ingram Publishing

©Wendy Schiff

lynx/iconotec.com/Glow Images

TABLE 12.1 Some High-Risk Foods and Their Primary Pathogens

Raw/Undercooked Food	Typical Menu Item	Common Pathogens
Beef	Rare hamburger	*Salmonella* species *E. coli* O157:H7
Pork	Undercooked pork (rarely in the United States)	*Trichinella*
Poultry	Chicken, turkey, duck	*Salmonella* species *Campylobacter jejuni*
Eggs	Salad dressings, sauces made with raw eggs, custards, cream-filled desserts, toppings made with raw eggs	*Salmonella enteritidis*
Fresh produce	Tomatoes, peppers, cantaloupes, lettuce, spinach, sprouts	*Salmonella* species *E. coli* O157:H7
Raw fish/finfish	Sushi; lightly cooked fish	*Anisakis*
Shellfish	Oysters	*Vibrio vulnificus*
	Clams	*Vibrio* species Hepatitis A virus Norovirus
Milk and milk products	Unpasteurized milk, some soft cheeses such as Camembert and Brie	*Listeria monocytogenes* *Salmonella* species *Campylobacter jejuni* *E. coli* O157:H7 *Brucella* species

Source of data: Food and Drug Administration: *Bad bug book,* 2nd ed. 2012. www.fda.gov/food/foodborne-pathogens/bad-bug-book-second-edition. Accessed: January 15, 2023.

Concept CHECKPOINT

5. Discuss at least three ways pathogens can contaminate human foods.

6. With regard to food preparation, what is cross-contamination?

7. What is pasteurization?

8. List at least three conditions that favor the survival and multiplication of food-borne pathogens.

9. Identify at least four foods that are high risk for supporting pathogens.

See Appendix F for responses.

12.4 Food-Borne Illness

Learning Outcomes

1. Describe typical signs and symptoms of food-borne illness.
2. Explain when a person who has a food-borne illness should seek professional medical help.
3. Explain why it is incorrect to call a food-borne illness the "flu" or "stomach flu."

Signs and symptoms of food-borne (and water-borne) illnesses generally involve the digestive tract and include nausea, vomiting, diarrhea, and intestinal cramps. However, most pathogens have an *incubation period,* a length of time in which they grow and multiply in food or the digestive tract before they can cause illness. Thus, if you develop signs and symptoms of a food-borne illness, you might have difficulty identifying the source of the infection or intoxication. Was the vomiting and diarrhea that you experienced at 3 A.M. the result of eating soft-cooked eggs for breakfast 18 hours earlier or the sliced deli chicken you ate for lunch 2 days ago?

Various factors influence whether an individual becomes ill after consuming a food or beverage that has been contaminated with a pathogen or toxin. The number of pathogenic microbes in a food or the amount of toxin it contains can contribute to the risk and severity of a food-borne illness. Furthermore, individuals vary in their vulnerability to many food-borne pathogens. In general, high-risk groups are pregnant females, very young children, older adults, and people who suffer from serious chronic illnesses or weakened immune systems.

In most cases, otherwise healthy individuals who suffer from common types of food-borne illness recover completely and without professional medical care within a few days.[5] However, many people mistakenly report that they have the "stomach *flu*" when they actually are suffering from a food- or water-borne illness. The "flu," or *influenza,* and COVID-19 are infectious diseases caused by specific viruses that invade the respiratory tract. Influenza and COVID-19 are characterized by coughing, fever, weakness, and body aches. Intestinal cramps, diarrhea, and vomiting are *not* typical signs and symptoms of influenza, but they can be a sign of COVID-19. On the other hand, food-borne illness primarily affects the digestive system, not the respiratory system; thus, coughing is not a typical sign of food-borne illness. As a result, it can be difficult to determine what is causing your diarrhea and intestinal cramps.

You should consult a physician for proper testing, diagnosis, and treatment an intestinal disorder is accompanied by one or more of the signs and symptoms listed in Table 12.2.[6] Hospitalization may be required if vomiting, diarrhea, and other signs of illness are severe.

TABLE 12.2 Serious Signs and Symptoms of Food-Borne Illness: When to See a Physician (Adults)

• Decreased urination and other signs of dehydration
• Prolonged vomiting
• Diarrhea for more than 3 days
• Severe intestinal or rectal pain
• Fever exceeding 102°F
• Stools containing blood or pus
• Trouble breathing, muscular weakness, mental confusion, or other symptoms or signs of nervous system impairment

Concept CHECKPOINT

10. Identify at least three typical signs and symptoms of food-borne illness.
11. When should a person suffering from a food-borne illness seek professional medical help?
12. Discuss the differences between a food-borne illness, influenza, and COVID-19.

See Appendix F for responses.

12.5 Common Food-Borne Pathogens

Learning Outcomes

1. Identify common pathogens that are sources of food-borne illness in the United States.
2. Discuss typical signs, symptoms, and incubation periods of food-borne illnesses that are caused by each major pathogen.
3. Identify foods that are often sources of pathogens or their toxins.

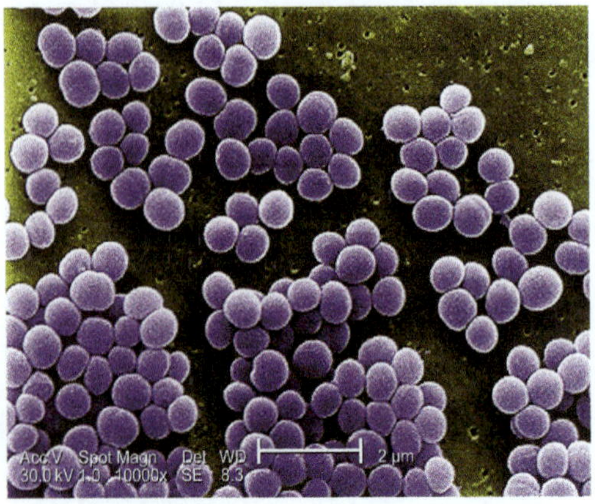

Improperly cooked or handled beef, poultry, eggs, and foods made with eggs are common sources of *Staphylococcus aureus*.
Janice Haney Carr/CDC

The major kinds of pathogens are bacteria, viruses, protozoa, and fungi. In the United States, *bacteria* and *viruses* are responsible for most cases of food–borne illness. The following sections take a closer look at some of the major pathogens that can cause food–borne illness.

Bacteria

Bacteria are single–cell microorganisms that do not have the complex array of organelles that plant and animal cells contain. Some bacteria can live without oxygen, such as in canned or vacuum–packed foods. Other types of bacteria transform into inactive resistant forms called *spores* when living conditions are less than ideal. If the environment becomes more hospitable, the spores revert to the active bacterial state.

Many kinds of bacteria are pathogens that cause food–borne illness, including forms of *Campylobacter, Clostridium, Escherichia, Listeria, Salmonella,* and *Staphylococcus* (*kam'–pih–low–bak'–ter, klah–strih'–dee–um, esh–eh–rik'–ee–ah, lih–stear'–ee–ah, sal–meh–nel'–lah, and staff–eh–lo–cah'–kes*). Some types of pathogenic bacteria do not cause infections when they are consumed in food, but these microbes produce toxins that cause food intoxication. Table 12.3 summarizes some general information about common bacterial sources of food–borne illness in the United States.

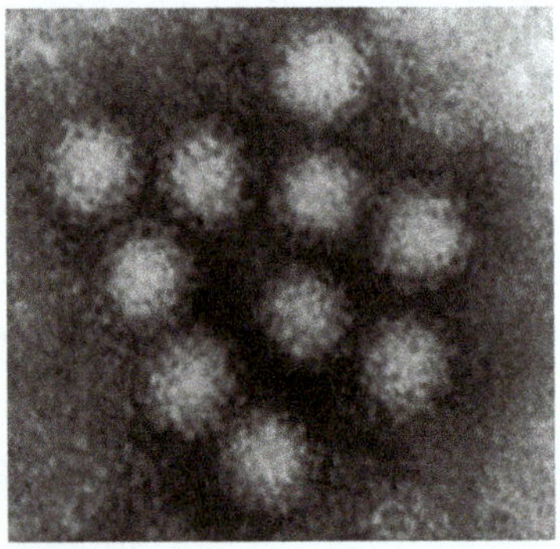

Figure 12.1 Norovirus.
F.P. Williams/U.S. Environmental Protection Agency

Viruses

Viruses are another common source of food–borne infection (Table 12.4). A **virus** is simply a piece of genetic material coated with protein (Fig. 12.1). Viruses must invade a living cell to produce more viruses. Unlike certain bacteria, viruses do not secrete toxins, and therefore, they do not cause food intoxication. Contaminated food or water, however, can transmit viruses to humans and cause food infection.

Vaccines that protect against hepatitis A virus and rotaviral infections are available. As of January 2021, a vaccine to prevent norovirus infections was in development.

bacteria simple single-celled microorganisms

virus microbe consisting of a piece of genetic material coated with protein

fungi simple organisms that live on dead or decaying organic matter

Fungi

Fungi such as molds, yeast, and mushrooms are simple life forms that live on dead or decaying organic matter. Certain fungi, such as button mushrooms and the mold in blue cheese, are beneficial and edible. Other fungi are responsible for spoiling foods, such as bread molds, or causing respiratory problems or allergic reactions in sensitive people. A serious concern is

TABLE 12.3 Common Sources of Food-Borne Illness: Bacteria

Bacterium	High-Risk Foods	Approx. Time of Onset	Typical Signs and Symptoms
***Bacillus cereus* (toxin)**	Starchy foods (rice, potatoes, puddings, some soups)	6–15 hours	Abdominal cramps, watery diarrhea, nausea
Campylobacter jejuni	Raw, undercooked poultry; raw milk; contaminated water	2–5 days	Diarrhea (sometimes bloody), abdominal cramping, vomiting, fever
***Clostridium botulinum* (toxin)**	Vacuum-packed foods, improperly canned foods, low-acid foods Honey may contain spores.	18–36 hours	Vomiting, diarrhea, blurry or double vision, difficulty swallowing, muscular weakness Can be fatal
***Clostridium perfringens* (toxin)**	Cooked meat, poultry, casseroles, gravies	8–16 hours	Watery diarrhea (may be bloody), mild abdominal cramps
***Escherichia coli* O157:H7 (toxin)**	Raw ground beef, raw seed sprouts, raw leafy greens, fresh fruit, raw milk, unpasteurized juices, foods contaminated with feces	3–4 days	Intestinal cramps, bloody diarrhea, kidney failure Can be fatal
Listeria monocytogenes	Raw meat and poultry, raw milk, fresh soft cheese made from raw milk, liver pâté, smoked seafood, deli meats, hot dogs, raw vegetables	Few hours to 3 days; 3 months for invasive disease	Fever, muscular aches, vomiting, diarrhea In pregnant women, the infection can lead to stillbirth (birth of a dead fetus) or premature birth. Can be fatal
***Salmonella* species**	Meat, poultry, seafood, and eggs; raw seed sprouts; raw vegetables; unpasteurized juice	6 hours to 3 days	Nausea, vomiting, fever, headache, abdominal cramps, diarrhea Can be fatal.
***Shigella* species**	Unclean water; raw vegetables, herbs, and other foods that were contaminated with feces as a result of poor food-handling and personal hygiene practices	8–50 hours	Abdominal cramps, fever, diarrhea that may contain blood and mucus Can be fatal in infants, older adults, and people with a weak immune system
***Staphylococcus aureus* (toxin)**	Meat and meat products, poultry, and eggs; foods made with eggs, such as potato, egg, macaroni, and tuna salads; custards, and cream-filled pastries	1–7 hours	Diarrhea, nausea, vomiting, abdominal cramps
Vibrio vulnificus	Raw oysters, clams, crabs, and other raw or undercooked seafood; unclean water	12 hours to 3 weeks	Fever, vomiting, diarrhea, abdominal cramps Can be fatal
Yersinia enterocolitica	Raw vegetables, undercooked pork, contaminated water, unpasteurized milk	1 day to 2 weeks	Diarrhea, high fever, abdominal pain, vomiting (in some cases)

Source of data: Food and Drug Administration: *Bad bug book*, 2nd ed. 2012. www.fda.gov/food/foodborne-pathogens/bad-bug-book-second-edition. Accessed: January 15, 2023.

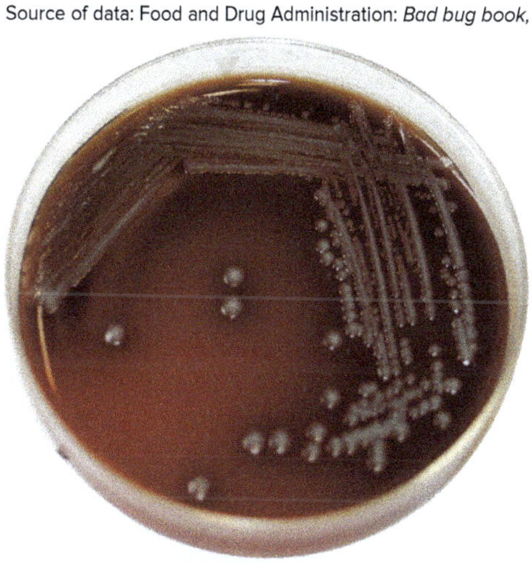

Scientists can often "grow" (culture) bacteria under laboratory conditions and identify them.
Megan Mathias and J. Todd Parker/Centers for Disease Control and Prevention

TABLE 12.4 Common Sources of Food-Borne Illness: Viruses

Virus	High-Risk Foods	Approx. Time of Onset	Typical Signs and Symptoms
Norovirus	Food, water, or surfaces that have been contaminated with infected feces Virus can be spread person to person	12 hours to 2 days	"Explosive" vomiting; watery, nonbloody diarrhea; abdominal cramps; nausea; low-grade fever (occasionally)
Rotavirus	Food, water, or objects that have been contaminated with infected feces	2 days	Fever, abdominal pain, vomiting, watery diarrhea
Hepatitis A (HAV)	Food or water that has been contaminated with HAV from feces	2 to 4 weeks	Fever, loss of appetite, nausea, vomiting, diarrhea, muscle aches, general weakness May have signs of liver inflammation, such as jaundice, liver enlargement, and dark-colored urine

Source of data: Food and Drug Administration: *Bad bug book,* 2nd ed. 2012. www.fda.gov/food/foodborne-pathogens/bad-bug-book-second-edition. Accessed: January 15, 2023.

Figure 12.2 Deadly mushrooms. Severe illness and even death can occur if people pick and eat toxic mushrooms, such as this *Amanita* mushroom.
Jorgen Bausager/Folio Images/Getty Images

the toxicity of several varieties of wild mushrooms. Cases of severe illness and death have been reported as a result of people picking and eating toxic wild mushrooms after mistaking them for edible varieties (Fig. 12.2). Nevertheless, fungi are not a major source of food-borne illness in the United States.

Certain molds produce *aflatoxins,* substances that can cause severe illness, particularly liver damage, and even death when consumed. Tree nuts, peanuts, rice, and corn that are stored under warm, humid conditions can become sources of aflatoxins. In some regions of the world, especially Africa and Southeast Asia, people often eat foods that are contaminated with aflatoxin-producing molds. Rates of liver cancer are high in these places. Thus, medical researchers think there is an association between exposure to aflatoxins and development of liver cancer. No outbreaks of food-borne intoxication caused by aflatoxins have been reported among the U.S. population.[5]

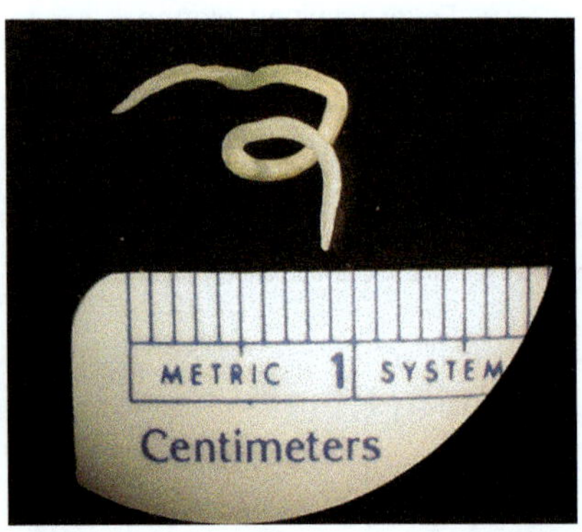

Figure 12.3 ***Anisakis.***
DPDx/Centers for Disease Control and Prevention

protozoa single-celled microorganisms that have complex cell structures

Protozoa and Parasitic Worms

Giardia (*jee-ar'-dee-ah*) and *Cryptosporidium* (*krip-toe-spor-ih'-dee-im*) are **protozoa** (*pro-toe-zoe'-ah*), single-celled microorganisms that have a more complex cell structure than bacteria. Table 12.5 lists some protozoa that often cause food-borne and water-borne illnesses in the United States. The "Nutrition Matters" section discusses traveler's diarrhea, which can result from a protozoal infection.

Certain types of worms such as *Trichinella* (*trick'-ah-nell'-ah*) and *Anisakis* (*ah-neh-say'-kes*) are parasites that can contaminate food or water (Fig. 12.3). A parasite is a life form that obtains its nourishment from another living thing (the host). Most Americans who become infected with parasitic worms that are found in the United States recover when they receive proper treatment. Nevertheless, some infected persons suffer long-term health problems and even die as a result of the illnesses. Table 12.5 provides information about the kinds of parasitic worms that can infect people.

TABLE 12.5 Common Sources of Food-Borne Illness: Protozoa and Parasitic Worms

Pathogen	High-Risk Foods	Approx. Time of Onset	Typical Signs and Symptoms
Protozoa			
Cryptosporidium	Foods prepared by people whose hands were contaminated with infected feces; fresh juices, produce, and milk	7–10 days	Profuse, watery diarrhea; abdominal pain; fever; nausea; vomiting
Giardia	Consumption of contaminated water, including water from lakes, streams, swimming pools Travelers to certain countries, hikers, and people who swim in or camp by lakes and streams.	1–2 weeks	Diarrhea, abdominal pain, weight loss
Toxoplasma	Raw or partially cooked wild game (especially bear and wild pig), clams, oysters Accidentally ingesting infected cat feces	5–23 days	Fever, headache, muscle aches, rash Pregnant females should avoid contact with cat feces because the parasite can infect their unborn offspring, causing eye or brain damage and even fetal death.
Parasitic Worms			
Trichinella	Raw or undercooked infected meat, especially pork, bear, seal, and walrus meat	1–4 weeks	Nausea; vomiting; diarrhea; fatigue; fever; abdominal discomfort; facial swelling, especially around the eyes; muscle pain Death can occur in severe cases.
Anisakis	Raw or undercooked infected seafood	1 day to 2 weeks	Abdominal pain, nausea, vomiting, diarrhea

Source of data: Food and Drug Administration: *Bad bug book,* 2nd ed. 2012. www.fda.gov/food/foodborne-pathogens/bad-bug-book-second-edition. Accessed: January 15, 2023.

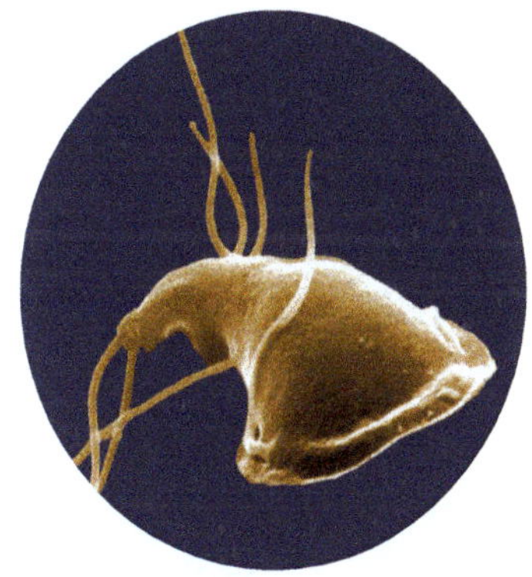

Giardia
Dr. Stan Erlandsen and Dr. Dennis Feely/CDC

Concept CHECKPOINT

13. Identify at least three kinds of bacteria that are common sources of food-borne illness in the United States.

14. Identify three viruses that are common sources of food-borne illness in the United States.

15. What are aflatoxins?

16. Identify a parasitic worm and a parasitic protozoan that can cause food- or water-borne illness in the United States.

See Appendix F for responses.

12.6 Preventing Food-Borne Illness

Learning Outcomes

1. Explain how to reduce the risk of food-borne illness when purchasing, preparing, cooking, and storing foods and beverages.
2. Explain why ground meat and poultry often are sources of food-borne illness.
3. Identify the temperature range that encourages rapid multiplication of pathogens.
4. Discuss food safety educators' four simple actions for reducing the risk of food-borne illness.

While you cannot control the safety of foods that are prepared in restaurants or other places outside your home, you can greatly reduce your risk of food–borne illness by following some important rules, most of which require changing risky food selection, preparation, and storage practices.

Before purchasing eggs, open the carton and check the eggs. Do not buy cartons that have any cracked eggs. ©Wendy Schiff

Purchasing Food

The following tips can help reduce your risk of food–borne illness:

- When shopping in a supermarket, select frozen foods and highly perishable foods, such as meat, poultry, or fish, last.
- Check "best by" dates on packaged perishable foods. Choose products with the latest dates.
- Do not buy food in damaged containers; avoid containers that leak, bulge, or are severely dented, or jars that are cracked or have loose or bulging lids.
- Open egg cartons and examine eggs; avoid cartons that have cracked eggs.
- Purchase only pasteurized milk and fresh juices, as well as cheese made with pasteurized milk (check the label).
- Request that butchers, "deli" counter workers, and bakery staff wear new sanitary gloves when handling meat, fish, poultry, cheese, "deli" foods, and baked goods for purchase.
- Pack meat, fish, and poultry in separate plastic bags, so their drippings do not contaminate each other and your other groceries.
- Place fresh fruits and vegetables in plastic bags or cloth bags that are only used for these foods to prevent contamination from the grocery cart.
- After shopping for food, take groceries home immediately. Refrigerate or freeze meat, fish, egg, and dairy products promptly.
- Store whole eggs in their cartons, even if your refrigerator has a place for storing eggs. Egg cartons are designed to keep eggs fresh longer than a refrigerator's egg compartment.
- Do not store reusable grocery bags in your vehicle. During warmer weather, the warm temperatures inside a parked vehicle promote growth of bacteria.
- Wash reusable bags after every use to prevent contamination of your groceries when you shop.

Setting the Stage for Food Preparation

Contaminated hands and food preparation surfaces spread pathogens. The following tips can help reduce the risk of food–borne illness:

- Wash hands thoroughly with very warm, soapy water for at least 20 seconds before and after touching food. Be sure to rub the soap over all surfaces of your hands, including under your fingernails. The friction from scrubbing your hands lifts the dirt and microbes from your skin, and the water rinses the microbes off your hands. If soap and clean water for handwashing are not available, use a hand sanitizing gel that contains at least 60% alcohol, which kills most bacteria on skin, and thus your hands.[7] For the gel to be most effective, you should follow the instructions provided on the hand sanitizer product's label.
- Before preparing food, clean food preparation surfaces, including kitchen counters, cutting boards, dishes, knives, and other food preparation equip– ment, with hot, soapy water. You can destroy most pathogens when you clean and *sanitize* food preparation surfaces with a solution made by adding a table– spoon of bleach to 1 gallon of water.
- The FDA recommends cutting boards with unmarred surfaces made of easy– to–clean, nonporous materials, such as plastic, marble, or glass. If you prefer to use wooden cutting boards, make sure they are made of a nonabsorbent hard– wood, such as oak or maple, and have no obvious seams or cracks.
- Replace cutting boards when they become streaked with cuts because these grooves can be difficult to clean thoroughly and may harbor bacteria.
- If possible, have a cutting board reserved for meats, fish, and poultry; have another cutting board for fruits and vegetables; and have a third board for breads. Clean all cutting boards in the dishwasher or with hot, soapy water. You can also sanitize cutting boards with a dilute bleach solution.
- Sanitize food preparation surfaces and equipment that have come in con– tact with raw meat, fish, poultry, and eggs as soon as possible to destroy pathogens that may be present. In addition, sanitize damp kitchen sponges by heating them in a microwave oven for 60 seconds and machine–wash and –dry kitchen towels frequently.

Preparing Food

The following tips can help reduce your risk of food–borne illness:

- Do not use foods from containers that leak, bulge, or are severely dented or from jars that are cracked or have loose or bulging lids.
- Do not use foods from containers that have damaged safety seals because the food they contain may have been contaminated.
- Do not taste or use food that spurts liquid or has a bad odor when the can is opened.
- Read product labels to determine whether foods need to be refrigerated after their packages are opened.
- Before preparing fresh produce, carefully wash the foods under running water to remove dirt and bacteria clinging to the surface. Bacteria can be sticky, so scrub the peel with a vegetable brush if it is to be eaten. Even if you plan to remove the skin or peel, wash the produce before you cut it.
- Do not rinse prewashed bagged lettuce or leafy greens. Rinsing the greens again increases the risk of contamination.

Did *YOU* Know?

Most alcohol gel hand sanitizers do not kill *all* viruses that cause food-borne illness, especially norovirus, which is the most common cause of outbreaks of food-borne illness in the United States (see Section 12.1). Alcohol concentrations of at least 60% found in some gel hand sanitizers are effective at killing bacteria on the skin because the alcohol easily penetrates the cell membrane of bacteria, thus destroying the microbes on a person's hands. However, norovirus is difficult to kill because it has a protective protein coat called a *capsid* that alcohol does not easily penetrate.

Workers who handle and prepare food in restaurants and other food establishments are ***required*** to clean their hands with soap and warm water[8] because handwashing is more effective than alcohol sanitizers at removing microorganisms, including norovirus, from hands.[2] However, people who handle and prepare food may use a hand sanitizer in addition to washing their hands. Nonetheless, sanitizers are not allowed as a substitute for hand washing.

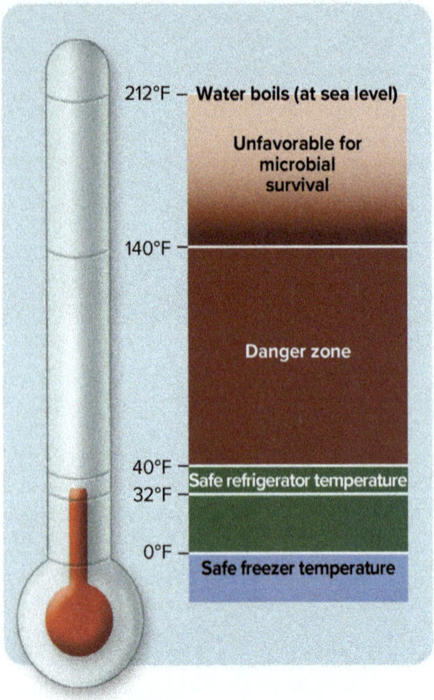

Figure 12.4 Temperature guide for food safety. According to HHS and the USDA, most pathogenic microbes grow well when the temperature of risky food is between 40°F and 140°F—the "danger zone."

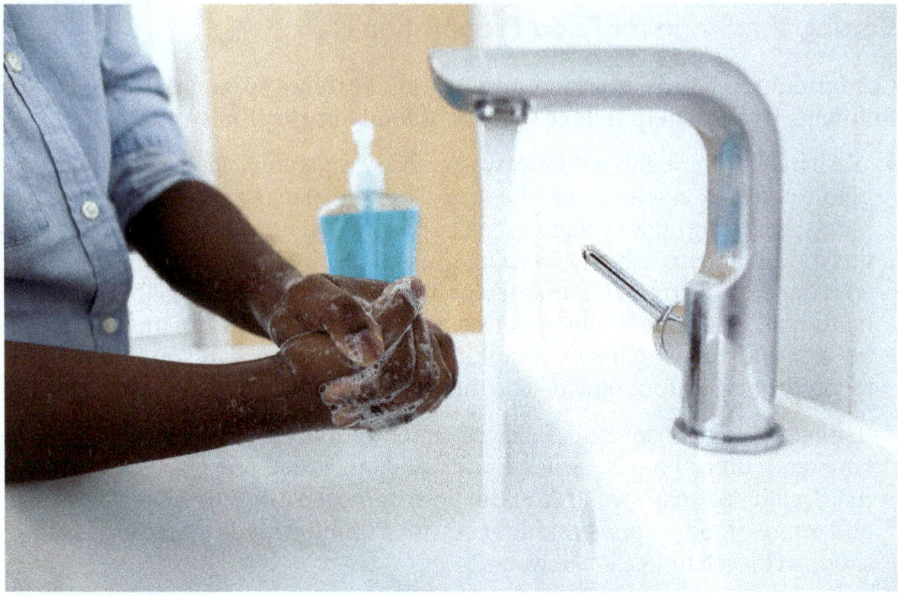

Daisy Daisy/Shutterstock

- Avoid eating moldy foods. Mold does not grow well on low-moisture foods. Thus, small amounts of mold on hard cheeses and on firm fruits and vegetables can be removed by cutting away the mold along with at least 1 inch of food that surrounds the moldy area.[9]
- **When in doubt, throw the food out.**

Maintaining the Proper Temperature of Foods

According to the U.S. Department of Health and Human Services (HHS) and USDA, most microbes grow well when the temperature of a high-risk food is between 40°F and 140°F—the "danger zone" (Fig. 12.4).[10] Cooking foods to the proper temperature destroys food-borne viruses and bacteria, such as norovirus and *E. coli* O157:H7. To be safe, a product must be cooked to an internal temperature that is high enough to destroy harmful pathogens and bacterial toxins. Using a meat thermometer is a reliable way to ensure that meat, poultry, thick pieces of fish, and egg-containing dishes have reached the proper internal temperature without overcooking. Table 12.6 indicates the USDA's recommended minimum internal temperatures for cooking these foods.

Meat thermometers must be used properly. In general, the thermometer should be placed in the thickest part of the muscle tissue, away from bone, fat, or gristle (Fig. 12.5). If the thermometer is inserted incorrectly or placed in the

TABLE 12.6 Safe Minimum Internal Temperatures

Food	Safe Minimum Internal Temperature (°F)
Beef steaks and roasts	145
Fish	145
Fresh pork	145
Ground beef, pork, lamb, and veal	160
Egg dishes	160
Poultry products	165

Source: U.S. Department of Health and Human Services: *Safe minimum cooking temperatures charts*. 2022. https://www.foodsafety.gov/food-safety-charts/safe-minimum-internal-temperatures. Accessed: August 15, 2023.

Figure 12.5 Use of a meat thermometer.
©Wendy Schiff

wrong area, the reading may not accurately reflect the internal temperature of the product.

Microwave cooking can result in uneven heating that does not destroy microbes in the cool spots. While cooking a food in a microwave oven, keep the dish covered, and stop the oven occasionally to stir the food. Stirring the food reduces uneven heating. Cook the food until it reaches 165°F. Microwave cooking is not recommended for stuffed foods because during cooking, the temperature of the stuffing may not be high enough to eliminate pathogens.[11]

Chilling food slows the growth of microbes in the items, but some bacteria can grow even at proper refrigeration temperatures. Freezing does not destroy bacteria or inactivate viruses in food; the process just halts the microbes' ability to multiply. As frozen food thaws, the bacteria and viruses resume their activities and can cause illness.

A major challenge for food handlers involves keeping large amounts of high-risk foods at safe temperatures when they are served from a single container. Food that is near the sides of the container may stay hotter or colder than food that is near or in the center of the container. As shown in Figure 12.6, chilled foods should be kept covered and served from a shallow container filled with ice. Hot foods should be kept covered and served from shallow, heated pans. The best simple advice to follow: "Keep hot foods hot and cold foods cold."

Figure 12.6 **Keep cold foods cold.**
©Wendy Schiff

The following tips can help reduce your risk of food-borne illness:

- Always thaw high-risk foods in the refrigerator, under cold running water, or in a microwave oven.
- Cook foods immediately after thawing.
- Marinate food in the refrigerator, and if marinating meat, fish, or poultry, discard the marinade.
- Do not remove cold foods from the refrigerator or hot foods from the stove until it is time to serve them.

Raw Fish

Eating raw fish, such as sushi, can be safe for most healthy people if the fish is very fresh before being commercially frozen and then thawed. While frozen, the fish must maintain an internal temperature of 10°F for 7 days. The freezing step is important because very cold temperatures can kill parasites that are often in fish tissucs. If you choose to eat uncooked fish, purchase it from reputable

Eating raw fish, such as some forms of sushi, can be safe for most healthy people if the fish is very fresh before being commercially frozen and then thawed.
Tom Craig/Alamy Stock Photo

Did *YOU* Know?

Have you heard about the "5-second rule"? Supposedly, food that drops on the floor will not pick up microbes if it is picked up within 5 seconds. A study determined that this is a food-related myth. As soon as food touches a contaminated surface such as a floor, some microbes adhere to it.[12]

Adam Gault/OJO Images/age fotostock

Raw ground meats are often contaminated with dangerous microbes because the grinding process mixes the pathogens throughout the meat.
Food Safety and Inspection Service/U.S. Department of Agriculture

establishments that have high standards for quality and sanitation. Nevertheless, it is prudent to not eat any raw animal products, including fish.

Ground Meats, Poultry, and Fish

Ground meats, poultry, and fish are highly perishable and must be thoroughly cooked to avoid being a source of food–borne illness. The interior portion of an intact piece of raw animal flesh is free of bacteria because the tissues are not exposed to air. However, ground meats, fish, and poultry products are often contaminated with microbes. Prior to being ground up, the surface of a chunk of meat, fish, or poultry may contain relatively harmless concentrations of pathogens. The grinding process, however, mixes the pathogens throughout the meat. At the same time, grinding the meat greatly increases its surface area, exposing more of the protein–rich tissues to microbes in air. Furthermore, the meat grinder can be a source of pathogens and spread them to the food product, especially if the machine was not properly cleaned after its last use. The particles of food that remain in the grinder can provide food for pathogenic microorganisms. Therefore, surfaces that touch ground meats should be cleaned carefully.

The following tips can help reduce your risk of food–borne illness:

- Cook beef, poultry, pork, thick pieces of fish, and egg–containing dishes thoroughly, using a meat thermometer to check for doneness.
- Cook eggs until the yolk and white have solidified and no “runniness” remains.
- Bake stuffing separately from poultry, or wash the poultry cavity thoroughly and stuff the bird immediately before cooking. Make sure the temperature of the stuffing reaches 165°F. After cooking, transfer the stuffing to a clean bowl for serving or storage.
- Serve meat, poultry, and fish on a clean plate. Never use the same plate that held the raw product. For example, when grilling hamburgers, do not put cooked items on the plate that was used to carry the raw meat to the grill.

Did *YOU* Know?

Food processing methods, including cooking, canning, and freezing, do not make toxic mushrooms safe to eat. To avoid being poisoned by such mushrooms, simply do not eat them.[5]

Toxic Jack O' Lantern mushrooms (*Omphalotus olearius*)
AwakenedEye/Getty Images

Storing Food

After food is cooked, careless food handling continues to set the stage for the growth of pathogens. Food–borne pathogens thrive at “room temperature,” temperatures that are between 60°F and 110°F. A common practice is to let hot or cold foods remain on the table at room temperature for a few hours. Although you may not feel like clearing the table after eating, it is a good idea to cover leftovers and refrigerate or freeze them as soon as you have finished eating, or within 2 hours. If environmental temperatures are above 90°F, refrigerate the leftovers within 1 hour.[10] There is no need to let hot foods cool before chilling or freezing them.

To reduce your risk of food–borne illness, follow these food storage tips:

- Check your refrigerator's temperature regularly to make sure it stays below 41°F. Keep the refrigerator as cold as possible without freezing milk and lettuce.
- Cook ground meats and poultry soon after purchasing. If this is not possible, freeze the ground items.
- Note that raw fish, shellfish, and poultry are highly perishable. It is best to cook these foods or freeze them the day they are purchased.

- Use refrigerated ground meat and patties within 1 to 2 days, and use frozen meat and patties within 3 to 4 months after purchasing them. Table 12.7 presents recommended time limits for refrigeration and freezer storage of foods. Foods that are stored in the freezer for longer than recommended periods are safe to eat after being thawed and cooked properly, but they often develop unappealing flavors.[13] Note that ground meats are more perishable than intact cuts of meat.
- Use refrigerated leftovers within 4 days.

TABLE 12.7 Cold Storage Time Limits for Perishable Foods

Product	Storage Period in Refrigerator (40°F)	Storage Period in Freezer (0°F)
Fresh meat		
Ground meat	1–2 days	3–4 months
Steaks and roasts	3–5 days	4–12 months
Fresh pork		
Chops	3–5 days	4–12 months
Ground	1–2 days	3–4 months
Cured meats		
Luncheon meat, deli sliced or open package	3–5 days	1–2 months
Unopened package	2 weeks	1–2 months
Fresh fish	1–3 days	—
Fresh chicken or turkey (whole)	1–2 days	12 months
Parts	1–2 days	9 months
Eggs		
Fresh, in shell	3–5 weeks	—
Hard-cooked	1 week	—

Source: U.S. Department of Health and Human Services, FoodSafety.gov: *Cold food storage chart.* 2021. www.foodsafety.gov/food-safety-charts/cold-food-storage-charts. Accessed: January 15, 2023.

CoolPhotoGirl/Shutterstock

People who work in food service and handle food directly should wear gloves.
lightfieldstudios/123RF

Food & Nutrition tips

Be alert for signs of unsafe food-handling practices when you eat away from home:

- Make sure tabletops, dishes, and eating utensils are clean.
- Make sure servers who handle food directly are wearing gloves.
- Check pudding, cream-filled or custard desserts, and salad bar foods to make sure they are chilled and kept on ice.
- Make sure hot foods are served hot or served from a heated food bar.
- If your serving of meat, poultry, or fish does not appear to be thoroughly cooked, ask the waiter to return the item to the kitchen to be heated again.

Figure 12.7 Storing risky foods in the refrigerator. Are these foods stored properly?
Source: U.S. Department of Agriculture, Food Safety and Inspection Service.
Be Safe Food Campaign/USDA

Cross–contamination is a threat not only during food preparation, but can also become a problem during food storage. Therefore, keep all foods, including leftovers, covered while they are in the refrigerator. This practice can prevent drippings from foods that are often contaminated, such as raw chicken, from tainting other foods. Furthermore, store raw meats, fish, poultry, and shellfish on lower shelves of the refrigerator, so they are separated from foods that are to be eaten raw. Examine Figure 12.7. Can you find any examples of improperly stored refrigerated foods?

Food safety educators condensed these rules into four simple actions as a part of their Check Your Steps program:

1. **CLEAN. Wash hands and surfaces often.**
2. **SEPARATE. Do not cross–contaminate.**
3. **COOK. Cook to proper temperatures.**
4. **CHILL. Refrigerate promptly.**

You can find reliable information about food–borne illness at www.foodsafety.gov. For general food safety questions, you can visit "Ask USDA" (https://ask.usda.gov/s/) and request food safety–related information.

Concept CHECKPOINT

17. List at least four rules for reducing the risk of food-borne illness when you purchase foods and beverages.
18. List at least five rules for reducing the risk of food-borne illness when you prepare and cook foods.
19. List at least three rules for reducing the risk of food-borne illness when you store cooked foods.
20. Explain why ground meat and poultry often are sources of food-borne illness.
21. What preparation step can be taken to make fish safer to eat raw?
22. What temperature range encourages rapid multiplication of pathogens?
23. What are the simple steps that food safety educators promote for reducing the risk of food-borne illness?

See Appendix F for responses.

12.7 Food Preservation

Learning Outcomes

1. Discuss food preservation methods, including how each method extends the shelf life of foods.
2. Explain how to prepare home-canned, low-acid foods that are safe to eat.

If nothing is done to preserve a fresh food, the item soon undergoes various chemical changes that eventually result in spoilage. Preserving food extends its shelf life. **Shelf life** refers to the period of time that a food can be stored before it spoils. Heating is one of the oldest ways to preserve foods. Heat can kill or deactivate pathogens, and the process also destroys naturally occurring enzymes in foods that can contribute to food spoilage.

Fermentation is an ancient method of food preservation that is still used to produce a variety of foods, including yogurt, wine, pickles, and sauerkraut. The fermentation process involves adding certain bacteria or yeast to food. These microbes use sugars in the food to make acids and alcohol, chemicals that hinder the growth of other types of bacteria and yeast that can spoil food.

For centuries, people preserved meats, fruits, and other foods that had high water contents by adding salt or sugar to them. To grow, bacteria need plenty of water; yeasts and molds can grow when less water is available. Adding sugar or salt to foods draws water out of cells, including bacteria, fungi, protozoa, and worms. As a result, these pathogens are less likely to survive in sugary or salty foods. Drying reduces a food's water content. Dried fruits such as raisins, for example, have a longer shelf life than grapes, their natural counterparts.

Today, we can add pasteurization, refrigeration, freezing, canning, irradiation, additives, and *aseptic* processing to the list of food preservation techniques (Table 12.8). Aseptic processing involves sterilizing a food and its package separately, before the food enters the package. The **sterilization** process destroys all microorganisms and viruses. After undergoing aseptic packaging, boxes of sterile foods and beverages, such as milk or juices, can remain free of microbial growth for several years while sitting on store or pantry shelves. However, once the containers of these products are opened, the foods or beverages have the same shelf life as their counterparts that have not undergone aseptic processing.

Fluid milk that has undergone aseptic processing does not need to be refrigerated, as long as the package remains sealed.
©Wendy Schiff

Home-Canned Foods

When food is canned, commercial food production methods require heating the food to certain temperatures for specified times. Thus, unless the can or jar has been damaged, properly processed canned foods should be free of pathogens. Certain home-canned foods, however, may contain the microorganism *Clostridium botulinum (C. botulinum)* that causes *botulism* or its toxin. The home-canning process may kill *C. botulinum* bacteria in the food, but their spores or toxin may remain. That is why home-canned, *low-acid* foods such as beans and corn should be boiled for 10 minutes before eating. The botulinum toxin is

MY DIVERSE PLATE

Kimchi (*kim-chee'*) is a traditional Korean food made with vegetables, such as cabbage, cucumbers, or Korean radishes. Kimchi is served with meals, but some people also eat it as a snack. A very popular kimchi uses bunches of leaves from fresh napa cabbages. The leaves are first carefully washed and then salted. The salt kills harmful bacteria and draws water out of the leaves so they become soft and limp. Next, the leaves are rinsed to remove the salt, and then each leaf is completely smeared with a paste-like mixture that is made with hot red chili pepper flakes, garlic, ginger, brown sugar, fish or shrimp sauce, and chopped Korean radishes, green onions, and *buchu* (Asian chives). The seasoned cabbage leaves are folded and stuffed into a sealed container, and the mixture is allowed to ferment for 1 to 5 days.

©Wendy Schiff

shelf life period of time that a food can be stored before it spoils

fermentation process used to preserve or produce a variety of foods, including pickles and wine

sterilization process that destroys all microorganisms and viruses

TABLE 12.8 Summary of Food Preservation Methods

Method	Means of Effectiveness	Examples of Foods
Heating (cooking, pasteurization, aseptic processing)	Kills or deactivates spoilage and pathogenic microbes, destroys enzymes that result in food spoilage	Most foods
Adding salt/sugar	Binds water, decreasing the amount available for microbes	Ham, bacon, pickled foods, jams, jellies
Smoking	Kills spoilage microbes, destroys enzymes that result in food spoilage Smoking is a method of heating. The process involves salting food before smoking and refrigerating or freezing the food after smoking.	Meats, fish
Curing	Retards the growth of *C. botulinum* and stabilizes the flavor of the food Additives such as sodium nitrate and nitrite are used to cure meat, fish, or poultry.	Processed meat, smoked fish
Chilling/freezing	Slows molecular movement, retarding microbial and enzymatic activity	Most foods
Drying (dehydration)	Removes much of the moisture in food that microbes need to survive	Fruit, herbs, meat jerkies, seeds
Fermenting	Produces acids and alcohol that interfere with the survival of unwanted microbes	Alcoholic beverages, yogurt, cheeses, soy sauce
Canning	Kills spoilage microbes, destroys enzymes in food that result in spoilage, removes oxygen that certain microbes need to survive	Meat, fish, poultry, fruits, vegetables, milk
Irradiating	Destroys most pathogens, delays sprouting (potatoes)	Spices, raw meat and poultry, fresh fruits and vegetables

Source: Complete Guide to Home Canning, 2015 revision. www.uga.edu/nchfp/publications/publications_usda.html. Accessed: June 9, 2023.

Dried fruit: ©Wendy Schiff ; Pickles: Kevin Sanchez/Cole Group/Photodisc/Getty Images; Spam: Elite Images/McGraw Hill

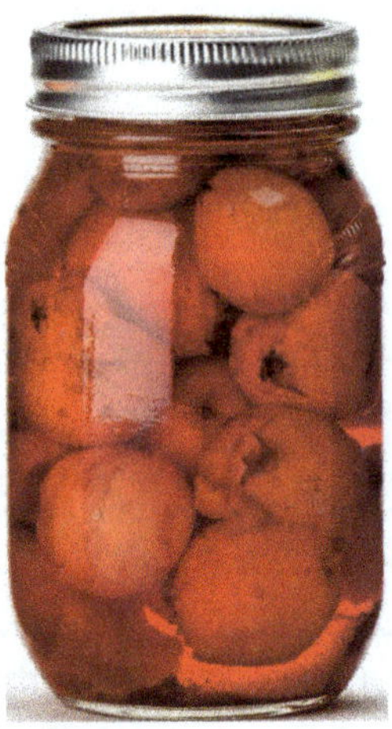

Home-canned, low-acid foods should be boiled for 10 minutes before eating them.
Robert George Young/Photographer's Choice/Getty Images

highly poisonous; never taste a home–canned, low–acid food before boiling it. Foods made with vinegar, fruit, or citrus juices are usually high–acid foods, and as a result, such items are not likely to be sources of *C. botulinum*. For more information about proper home canning of foods, obtain a copy of the U.S. Department of Agriculture's *Complete Guide to Home Canning, 2015 revision* (https://nchfp.uga.edu/publications/publications_usda.html) or contact a food and nutrition specialist at your community's university extension office.

Irradiation

The process of food irradiation preserves food by using a high amount of energy to kill pathogens such as *Salmonella* and *E. coli* O157:H7, as well as microbes that contribute to food spoilage (Fig. 12.8). The processes used to irradiate foods do not make the items radioactive. The energy passes through the food, as in microwave cooking, and no radioactive material is left behind. The energy is strong enough to destroy the genetic material, as well as cell membranes or cell walls of insects and microbes. As a result, irradiation is a highly effective way of destroying insects and microorganisms that may be in foods. It is important to recognize that even when foods, especially meats, have been irradiated, once their packaging has been opened, the foods can still become contaminated. Table 12.9 lists foods that the FDA allows to be irradiated (as of 2022).[14]

Irradiation extends the shelf life of spices, dry vegetable seasonings, meats, seeds, shell eggs, and fresh fruits and vegetables. Except for dried seasonings, packages that contain irradiated foods must be labeled with the international food irradiation symbol, the Radura, and include a statement indicating the product has been treated by irradiation (Fig. 12.9). Produce that is not packed for sale ("bulk foods") must display the Radura label on each item or on the bin from which consumers choose the produce.

According to medical experts with the World Health Organization (WHO), USDA, and CDC, irradiated foods are safe to eat.[14] Furthermore, irradiation causes few or no nutritive losses. Despite such assurances, many Americans are skeptical about the safety of irradiated foods, and they avoid purchasing such products.

Figure 12.8 **Comparing nonirradiated peaches and irradiated peaches.**
Ted Foxx/Alamy Stock Photo

TABLE 12.9 Foods Approved for Irradiation in the United States (2022)

- Beef, pork, and poultry
- Shrimp, lobsters, and crabs
- Fresh fruits and vegetables
- Seeds for sprouting (e.g., alfalfa sprouts)
- Shell eggs
- Spices and seasonings

Concept CHECKPOINT

24. Define *shelf life.*
25. Identify four methods of food preservation and explain how each method extends the shelf life of foods.
26. What preparation step can be taken to make home-canned, low-acid foods safe to eat?
27. List three foods that may undergo irradiation to extend their shelf life.

See Appendix F for responses.

Figure 12.9 **Radura symbol.**
tatadonets/123RF

12.8 Preparing for Disasters

Learning Outcomes

1. Explain actions people in households can take to have safe food and water available after a disaster.
2. Discuss ways to safely prepare food when a power outage occurs.

On August 29, 2021, Hurricane Ida moved onto land in Lousiana with high winds, heavy rains, and severe flooding. As a result of the storm, over 1,000,000 utility customers in the affected region lost power.[15] What steps can you take to have enough water and food available to survive hurricanes, earthquakes, or other serious emergency situations?

A supply of clean water and wholesome food is necessary for surviving disasters such as hurricanes and earthquakes. This chapter's "Nutrition Matters"

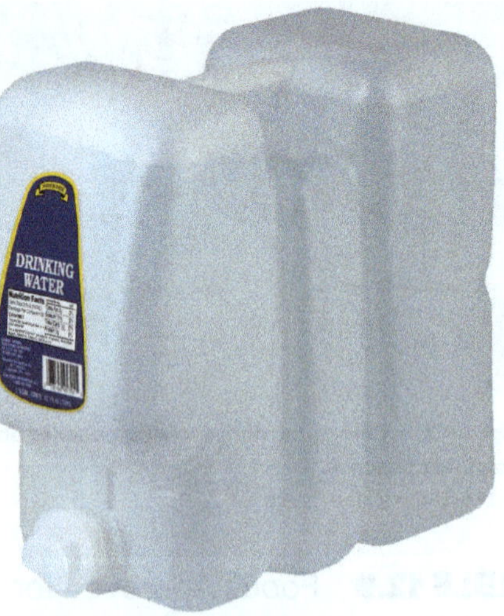

Store 1 gallon of water per person per day for emergency situations. This 2½ gallon jug will provide water for 1 person for 2½ days.
Sven Kahns/McGraw Hill

section provides instructions for sanitizing water. The following recommendations may help sustain you and your loved ones for several days:

- Store at least 1 gallon of water per person per day. Ideally, you should have at least a 3- to 5-day supply of drinking water, or at least 5 gallons of water for each person in a household. Children and breastfeeding females may need more than 1 gallon of water per day. Also, more water may be necessary for people living in warm climates. Furthermore, store extra water for food preparation, personal hygiene, dishwashing, and pets.
- Water should be maintained in a cool place and in sturdy plastic bottles with tight-fitting lids.
- Change stored water every 6 months.
- Drink only bottled, boiled, or treated water until you are certain the public water supply is safe.
- If you have time to prepare, fill a bathtub with water to use if it becomes necessary. The water, however, will need to be sanitized before being consumed.
- If you drink bottled water, make sure the seal has not been broken.

If your emergency water supply is inadequate, you can consume melted ice cubes from the freezer, canned fruit juices, and water drained from an undamaged water heater.[16] Water stored in the tank of the toilet (not the bowl) is also fit to drink. Water in swimming pools and spas can be used for personal hygiene needs but not for drinking. Never drink water from car radiators, home heating systems, or water beds.[16] Alcoholic beverages contribute to dehydration and therefore should be avoided.

TABLE 12.10 Foods to Store for Emergency Situations

- Canned and aseptic pouches of meats, fish, beans, fruits, and vegetables
- Aseptic individual serving "boxes" of fruit juices, cow's milk, or fortified plant beverages
- Unopened boxes of cereal, low-salt crackers, and trail mix
- Prewrapped granola bars, protein bars, or energy bars
- Peanut or other nut butters
- Dried fruit, nuts, and meat jerky

Emergency Food Supply

A disaster can easily disrupt your access to safe food; therefore, you should store at least a 3-day supply of food for emergency use. Choose foods that have a long storage life, require no refrigeration, and can be eaten without cooking, such as canned meats, beans, fruits, and vegetables. Table 12.10 lists foods that can be included in your emergency food supply. Also store a manual can opener, paper plates, and eating utensils.

If stored under proper conditions, unopened canned, aseptic pouches, and boxed foods will remain fresh for about 2 years. Before storing food, use a permanent marking pen to write the date on the package. You should use and replace foods before they lose their freshness or reach their expiration dates. An ideal food storage location is a cool, dry, dark place. Do not store foods near gasoline, oil, paints, or petroleum-based solvents because some food products absorb their odors. You can protect foods from rodents and insects by storing them in airtight containers or plastic storage bins.

If you have no electricity, consume perishable food in your refrigerator or freezer before using your emergency food supply. However, discard cooked foods after they have been at room temperature for 2 hours. Do not eat food that appears spoiled or is from cans that are leaking or bulging.

To prepare meals safely after a disaster, you will need to store:

- a camp stove or charcoal grill.
- fuel for cooking, such as charcoal. **Never** cook food on a camp stove or charcoal grill indoors.[18] The fumes contain *carbon monoxide,* an odorless gas that can be deadly when it accumulates in a closed space.

Did *YOU* Know?

During an electrical power outage, you can keep refrigerated food safely for only about 4 hours—if you do not open the refrigerator door. A full freezer can maintain its cold temperature for about 48 hours (24 hours if the freezer is half full), if the freezer door remains closed.[17]

- matches.
- cooking and eating utensils.
- paper plates, cups, and towels.

For more information about emergency preparedness, visit www.foodsafety.gov/keep-food-safe/food-safety-in-disaster-or-emergency.

Never cook food on a charcoal grill indoors or in enclosed areas because of the high risk of inhaling deadly carbon monoxide fumes.
Conny Fridh/Johner RF/Getty Images

Concept CHECKPOINT

28. Develop an emergency food and water supply plan for your home. Identify at least five foods that are appropriate for an emergency food supply.

29. In case public water supplies are disrupted, identify at least three sources of drinking water in homes that are safe for humans. Identify at least two sources of water that are unsafe to drink.

See Appendix F for responses.

12.9 Food Additives

Learning Outcomes

1. Explain the difference between direct and indirect food additives, and provide examples of each type of additive.
2. Discuss the impact of the GRAS list and the Delaney Clause on the use of food additives.
3. Identify unintentional food additives.
4. Discuss pesticides, including the pros and cons of their use.
5. Explain the usefulness of integrated pest management.

A **food additive** is any substance that becomes incorporated into food during production, packaging, transportation, or storage. Food manufacturers incorporate *direct* or *intentional additives* into their products for various reasons. Such additives may make food easier to process, more nutritious, able to stay fresh longer, or better tasting. Most additives are added to influence a food's sensory characteristics, including taste or color.

Many direct food additives help maintain the safety of foods by limiting the growth of bacteria that cause food-borne illness. Other additives protect against the action of enzymes that can lead to undesirable changes in the food's color and taste. Such unwanted chemical changes occur when enzymes that are naturally in certain foods are exposed to the oxygen in air. *Antioxidant additives,* including vitamins E and C and a variety of *sulfites,* can prevent oxygen from reacting with these enzymes. **Color additives** are dyes, pigments, or other substances that provide color to foods, drugs, or cosmetics, such as beta-carotene in margarine and FD&C (Food, Drug, & Cosmetic) Red No. 40 in cherry-flavored cough syrup. Table 12.11 lists common types of direct food additives (including color additives), their

food additive any substance that becomes incorporated into food during production, packaging, transport, or storage

color additives dyes, pigments, or other substances that provide color to food

TABLE 12.11 Common Types of Direct Food Additives

Types of Additive	Functions	Typical Products	Examples of Specific Ingredients
Preservatives	Prevent food spoilage	Jellies, beverages, baked goods, processed meat, cereals, snack foods	Ascorbic acid, citric acid, sodium benzoate, calcium propionate, BHA, BHT, sodium nitrite
Sweeteners	Add sweetness	Processed foods, beverages, baked goods, sugar substitutes	Sucrose, glucose, mannitol, corn syrup, aspartame, sucralose
Flavors and spices	Add specific flavors	Puddings, pie fillings, gelatins, cake mixes, candies, soft drinks	Natural flavorings, artificial flavorings, spices
Nutrients	Replace nutrients lost during processing, boost levels of nutrients naturally in food	Flour, cereals, margarine, fruit beverages, energy bars	Thiamin hydrochloride, riboflavin, niacin, niacinamide, folic acid, beta-carotene
Emulsifiers	Keep oily and watery ingredients from separating	Salad dressings, peanut butter, chocolate, frozen desserts	Soy lecithin, mono- and diglycerides, polysorbates
Stabilizers, thickeners, binders	Provide uniform texture and improve "mouth feel"	Frozen desserts, puddings, sauces	Gelatin, pectin, guar gum, carrageenan, whey
Color additives	Enhance natural colors, provide color to colorless and "fun" foods	Processed foods, including candies, snack foods, margarine, cheese, soft drinks, gelatin	FD&C Blue No. 1, FD&C Red No. 40, beta-carotene, caramel color, fruit and vegetable juices.

Source: U.S. Food and Drug Administration: *Food: Types of food ingredients.* 2022. https://www.fda.gov/food/food-additives-and-gras-ingredients-information-consumers/types-food-ingredients. Accessed: January 15, 2023.

Examples of foods that contain color and flavor additives.
©Wendy Schiff

uses, examples of products that contain the additives, and names of specific additives.

Indirect additives, such as compounds from a food's wrapper or container, can enter food as it is packaged, transported, or stored. Indirect additives, however, have no purpose. The FDA and certain international organizations regulate all food additives to ensure that processed foods and their packaging are safe.[19]

Food Safety Legislation: Food Additives

By the 1950s, hundreds of ingredients were being added to foods during processing. Many of these substances had long histories of being safe; others were deemed safe after undergoing scientific testing. In 1958, the U.S. Congress enacted the **Food Additives Amendment.** According to this amendment, an ingredient that had been in use prior to 1958 was **Generally Recognized as Safe (GRAS)** when qualified experts generally agreed that the substance was safe for its intended use. The Food Additives Amendment excluded GRAS substances from being defined as food additives.[19] Thus, modern food manufacturers can include substances on the GRAS list as ingredients without testing them for safety or getting prior approval from the FDA.

As a result of the 1958 Food Additives Amendment, the manufacturer of a *new* food additive (one developed after 1958) must provide evidence of the substance's safety to the FDA before the additive can be used.[19] When evaluating the safety of a newly developed food additive, FDA experts consider the chemical composition and characteristics of the substance, the amount of the

Food Additives Amendment U.S. legislation that requires evidence that a new food additive is safe before it can be marketed for use

Generally Recognized as Safe (GRAS) ingredients considered to be safe

substance that Americans would typically ingest, and the additive's effects on the body. If the additive is safe in amounts that people are likely to consume, FDA experts establish a level of the substance that can be added to foods.

According to the **Delaney Clause** of the Food Additives Amendment, food manufacturers cannot add a new compound that causes cancer at *any* level of intake. Thus, if an additive causes cancer, even though very high doses may be necessary to cause the disease, no amount of the additive is considered to be safe, and none is allowed in food. Evidence for cancer risk could come from either laboratory animal or human studies. The FDA allows very few exceptions to this clause.

The FDA cannot ban **unintentional food additives**—various industrial chemicals, pesticide residues, and mold toxins—from foods, even though some of these contaminants may be carcinogenic. The Food Quality Protection Act of 1996 established the safety standard of "a reasonable certainty of no harm" for pesticide residues in foods.[20] As a result of this act, the "no risk" provision of the Delaney Clause does not apply to pesticide residues. However, the Delaney Clause remains in effect for food additives. The following section discusses chemical contaminants and other unintentional food additives.

Did *YOU* Know?

According to the results of numerous scientific studies, food additives do not cause hyperactivity or learning disabilities in children. However, a small percentage of people may be allergic to FD&C Yellow No. 5, so this color additive must be listed on the package label if it is used as an ingredient.[19]

Delaney Clause component of the 1958 Food Additives Amendment that prevents manufacturers from adding carcinogenic compounds to food

unintentional food additives substances that are accidentally in foods

Other Substances in Foods

Some substances can accidentally enter food during processing. Although such contaminants can blend into a food, they are not food additives. Common biological and physical food contaminants are insect parts, rodent feces or urine, dust and dirt, and bits of metal or glass from machinery used to process food. Although some of these substances may not be harmful to health, most people find it unappealing to have filth and other unintentional ingredients in their foods.

According to the Federal Food, Drug, and Cosmetic Act, adulterated food contains objectionable and unsanitary material, and it cannot be distributed. However, the FDA permits very small amounts of unavoidable, naturally occurring substances such as dirt and insect parts in foods because they are not harmful when consumed in minute amounts. The FDA established guidelines concerning amounts (*action levels*) of certain materials that are permitted in specific foods, such as mold and rodent hairs in paprika or insect eggs in canned orange juice. According to the FDA, ". . . it is economically impractical to grow, harvest, or process raw products that are totally free of non–hazardous, naturally occurring, unavoidable defects. Products harmful to consumers are subject to regulatory action whether or not they exceed the action levels."[21]

Herbert Spichtinger/Image Source

Chemical contaminants also enter foods unintentionally. Toxic metals, such as lead, cadmium, and mercury, are naturally in our environment, and these elements may also be in our food. Poisonous human–made compounds such as *benzene* and *polychlorinated biphenols* are in the environment as well. Toxic metals or poisonous compounds resulting from human manufacturing practices can pollute sources of water used by consumers (well water, for example). Americans who drink water from municipal supplies can be assured that the water is analyzed regularly to determine its concentrations of toxic substances. However, people who rely on privately owned wells should have the water tested routinely.

What Are Pesticides?

A **pesticide** is any substance that people use to control or kill unwanted insects, weeds, rodents, fungi, or other harmful organisms. There are several different kinds of pesticides. **Insecticides** control or kill insects; **rodenticides** kill mice

pesticide substance that people use to kill or control unwanted insects, weeds, or other harmful organisms

insecticides substances used to control or kill insects

rodenticides substances used to kill mice and rats

Did YOU Know?

The Environmental Working Group (EWG) is a nonprofit organization that has scientists who use data from government sources to protect American consumers. Each year, the EWG analyzes results of government testing to develop a list that ranks conventionally grown produce according to amounts of pesticide residues. The list can serve as a consumer's guide for making decisions regarding which produce to purchase. In 2023, strawberries were at the top of the list because they had the highest pesticide residue levels of the produce that was tested. Avocados were among the produce that contained the lowest level of pesticide residues. To access the produce list, visit www.ewg.org/foodnews/full-list.php.

Pixtal/age fotostock

Figure 12.10 Helpful insect. The spined soldier bug (*left*) makes a meal of a Mexican bean beetle larva. Bean beetle larvae are devastating pests of snap beans and soybeans.
U.S. Department of Agriculture (USDA)

herbicides substances used to destroy weeds

fungicides substances used to limit the spread of fungi

tolerances maximum amounts of pesticide residues that can be in or on each treated food crop

and rats; **herbicides** destroy weeds; and **fungicides** limit the spread of fungi, such as mold and mildew.

Pesticide Residue Tolerances The use of pesticides in modern farming practices has helped increase crop yields, reduce food costs, and protect the quality of many agricultural products. However, many pesticides leave small amounts (*pesticide residues*) in or on treated crops, including fruits, vegetables, and grains, even when they are applied correctly. Concentrations of pesticide residues often decrease as food crops are washed, stored, processed, and prepared. Nevertheless, some of these substances may remain in fresh produce, such as apples or peaches, as well as in processed foods, such as canned applesauce or peaches.

The EPA regulates the proper use of pesticides. The agency can limit the amount of a pesticide that is applied on crops, restrict the frequency or location of the pesticide's application, or require the substance be used only by specially trained, certified persons. The EPA also sets pesticide **tolerances,** which are the maximum amounts of pesticide residues that can be in or on each treated food crop.

Nonchemical Methods of Pest Management Although the EPA focuses on chemical methods of managing pests, the agency also promotes nonchemical pest management techniques that may be safer for humans and the environment. Integrated pest management (IPM) involves using a variety of methods for controlling pests while limiting damage to the environment. IPM methods include growing pest-resistant crops, using predatory wasps to control crop-destroying insects, and trapping adult insect pests before they can reproduce. Biologically based pesticides, such as sex hormones (*pheromones*) that attract pesky insects to predators or traps and viruses that infect insects and weeds, are becoming increasingly popular among farmers. Such methods are often safer for humans than traditional chemical pesticides (Fig. 12.10).

Fruits and vegetables grown without use of most pesticides are available and may bear an "organic" label (see Chapter 3 for information about organic foods). These products generally are more expensive than those grown using pesticides, and they are not necessarily safer or more nutritious than conventionally produced foods.

How Safe Are Pesticides? Pesticides used in agriculture have both beneficial and unwanted effects. Pesticides help protect the food supply and make food crops available at reasonable cost. Nevertheless, pesticides have the potential to harm humans, animals, or the environment because they are designed to kill or otherwise negatively affect organisms. If a pesticide is applied improperly to cropland, it may remain in the soil, be taken up by plant roots, decompose to other compounds, or enter groundwater and waterways. Winds may carry pesticides in air and dust to distant locations. Each path can be a route to the human food chain (Fig. 12.11).

The potential harmful effects of a pesticide in food depend on the particular chemical and how effectively the body can eliminate it, its concentration in the food, how much and how often it is eaten, and the consumer's vulnerability to the substance. Tolerable amounts of pesticide residues on or in foods are extremely small. It is possible that regular exposure to small amounts of these chemicals may enable the substances to accumulate in the body and produce toxicity or initiate cancer.[22] However, according to the American Cancer Society, there is not enough evidence that these small pesticide exposures from foods cause cancer.[23] The overall health benefits of consuming fruits and vegetables is greater than the potential risks from small pesticide exposures.[23]

Figure 12.11 Pesticide pathways. If a pesticide is applied improperly to cropland, it may remain in the soil, be taken up by plant roots, decompose to other compounds, or enter groundwater and waterways. Winds may carry pesticides in air and dust to distant locations. Each path can be a route to the human food chain.

According to American Cancer Society, the overall health benefits of consuming fruits and vegetables is greater than the potential risks from small pesticide exposures in these foods. Alexander Raths/Shutterstock

Health experts have studied rates of cancers among people who have close contact with pesticides, such as farmers and pesticide applicators. People who are exposed to certain pesticides have a greater risk of developing specific cancers, including non–Hodgkin lymphoma and lung, liver, prostate, and breast cancers, than people who have not been exposed to the chemicals.[23] More research is needed to explain how exposure to pesticides influences cancer risk. Thus, environmental health experts continue to monitor the effects of pesticides on humans.

Concept CHECKPOINT

30. What is the difference between direct and indirect food additives?
31. What is the GRAS list?
32. Explain the role of the Delaney Clause in protecting the U.S. food supply.
33. What is an unintentional food additive? Provide at least three examples of such additives.
34. What is a pesticide? Provide at least three examples of types of pesticides.
35. Define *integrated pest management,* and provide an example of this method.
36. Which U.S. agency regulates the use of pesticides?
37. What is a pesticide tolerance?
38. Explain how pesticides can enter the human food chain.

See Appendix F for responses.

12.10 Nutrition Matters: Avoiding "The Revenge"

Learning Outcomes

1. Discuss ways to prevent traveler's diarrhea.
2. Identify nations where visitors are most likely to develop traveler's diarrhea.

Crawford A. Wilson III/Moment/Flickr/Getty Images

If you like traveling to foreign countries and enjoy sampling exotic cuisines in these places, one thing you do not need to encounter is *travelers' diarrhea (TD)*. Also referred to as "Montezuma's Revenge" and "Tut's Tummy," TD can ruin your vacation. By taking some precautions, however, your digestive system can stay healthy while you travel abroad.

TD results from consuming food or water contaminated with pathogens. The illness is characterized by the abrupt onset of abdominal cramps and loose or watery bowel movements. Additional symptoms may include nausea, vomiting, intestinal bloating, and fever. TD generally lasts 3 to 7 days without treatment, but signs and symptoms may persist in a small percentage of infected travelers.

Unless precautions are taken, TD is likely to occur during or shortly after traveling in regions where sanitary water supplies are not always available, people who prepare food have less than ideal personal hygiene practices, and untreated human feces are used for fertilizer. Health experts estimate that bacterial pathogens are responsible for 80 to 90% of cases of TD.[24] Protozoa such as *Giardia* account for about 10% of TD cases. When travelers suffer from persistent symptoms, the illness is likely to be the result of a protozoal rather than a bacterial infection.

Thirty to seventy percent of travelers to international places will develop TD, depending on their destinations and the season in which they travel.[24] Low-risk countries include the United States, Canada, Australia, and Japan, and countries in Northern and Western Europe. Intermediate-risk countries include those in Eastern Europe, South Africa, and some of the Caribbean islands. High-risk areas generally have high-density populations, widespread pollution, and inadequate water treatment systems. These regions include Mexico and other Central American countries, and most of Asia, the Middle East, Africa, and South America. Before traveling to a foreign country, consider reviewing travelers' health information that is provided by the CDC at wwwnc.cdc.gov/travel/notices.

Reducing Your Risk of TD

Experts at the CDC recommend several approaches for reducing your risk of TD. These include following instructions regarding food and beverage selection, avoiding contact with contaminated waterways, sanitizing drinking water, and using medications that may prevent TD.

If you do not have access to soap and clean water, carry a small container of an alcohol-based (at least 60% alcohol) hand cleaner with you. Before you eat, wash your hands with the hand cleaner. When you visit places where risk of contracting TD is moderate to high, avoid consuming food or beverages purchased from street vendors. Before you eat meat or other high-risk foods, make sure they are fully cooked and served hot. Avoid raw foods that have been washed in water, such as fresh fruits, raw vegetables, and salads. Do not eat fresh fruit without peeling it first. Avoid water, ice, and beverages diluted with water or ice, such as reconstituted fruit juices and iced drinks. Do not drink fresh fluid milk or milk products that have not been pasteurized. Safe beverages include those that are bottled and sealed; bottled beer and wine may also be safe to drink. If boiled water is used to make tea and coffee, these beverages may be safe to consume. Use bottled water to wash hands, brush teeth, and take medication. Figure 12.12 shows foods and beverages that are generally safe to eat and menu items that should be avoided because they are too risky to eat.

TD can be spread when pathogens from human or animal feces are in sources of water, including swimming pools, lakes, rivers, and other bodies of water. Accidentally swallowing even small amounts of contaminated water

Figure 12.12 What's generally safe to eat?
Source: wwwnc.cdc.gov/travel/page/food-water-safety. Accessed: January 21, 2023

can cause illness. Swimming pools that contain chlorinated water may be safe places to swim, if the disinfectant and pH levels are properly maintained. However, some pathogens are resistant to levels of chlorine commonly used to disinfect swimming pools. Thus, you should also avoid swallowing chlorinated swimming pool water.

How Can I Sanitize Drinking Water?

Boiling is the most reliable method to kill most of the pathogens that water may contain.[25] Water should be boiled for 1 minute, covered, and allowed to cool to room temperature without adding ice.

Chemical Disinfection

Disinfection methods destroy large numbers of microorganisms, reducing the likelihood of infection. Chemical disinfection with chlorine (in household bleach) or iodine is an alternative method of sanitizing water when boiling water is not practical. Iodine and chlorine are available in tablets and drops. To reduce the risk of TD, follow the manufacturer's instructions for sanitizing water with these chemicals.

Filtering Water

Most portable water filters can remove protozoa from drinking water, but they may not remove bacteria or viruses. To destroy bacteria and viruses by chemical means, you should disinfect the water with iodine or chlorine after using water filters. Reverse osmosis filters remove bacteria, as well as viruses from water.[25]

Filters collect organisms from water, so wash your hands with sanitized water after handling used filters. For more information about water filters and other water disinfection methods, visit wwwnc.cdc.gov/travel/page/water-disinfection.

Preventive Medications

Before you leave the United States to enter a high-risk region, see your physician for their recommendations for

Boiling is the most reliable way to sanitize water.
Maya Kruchankova/Shutterstock

preventing TD. You can also take medications that contain bismuth subsalicylate (BSS), the active ingredient in Pepto–Bismol, to reduce the risk of TD. Side effects of BSS commonly include nausea, constipation, and blackening of the tongue and bowel movements. BSS is not recommended for children under 12 years of age, and the medication should be avoided by people who are allergic to aspirin or have certain chronic conditions. Therefore, check with your physician before taking BBS along when you travel to other countries.

Treating TD

Because pathogenic bacteria are often the cause of TD, antibiotics are the primary method of treating the condition. In the United States, physicians often prescribe antibiotics for their patients to carry with them when they travel out of the country. When used in combination with antibiotics, *antimotility* agents can be helpful. These medications, such as loperamide, the active ingredient in Imodium, slow the muscular activity of the digestive tract and can provide relief from the diarrhea associated with TD. Before you leave the United States, consider asking your physician for recommendations for treating TD should you become infected while in a foreign country.

Dehydration can be a serious complication of TD. To prevent dehydration, it is important for people suffering from TD to replace fluids and electrolytes lost by vomiting and diarrhea. However, travelers should remember to use only beverages that are in sealed containers, including carbonated soft drinks. If fluid loss is severe, it is best to obtain professional medical treatment.

Concept CHECKPOINT

39. List three ways to reduce the risk of traveler's diarrhea.

40. List three foods that are generally safe to eat while traveling in Africa, India, and Central America, according to Figure 12.12.

See Appendix F for responses.

Chapter References

See Appendix G.

Summary

12.1 Introduction to Food Safety

- In the United States, millions of people become ill from various food-borne illnesses each year. Food-borne illness occurs when microscopic pathogens or their toxic by-products enter food (or beverages) and are consumed. Many kinds of pathogens infect the digestive tract; other types of food-borne pathogens do not sicken humans directly, but these microbes secrete toxins into food. When the food is eaten, the toxins irritate the intestinal tract and cause food intoxication.

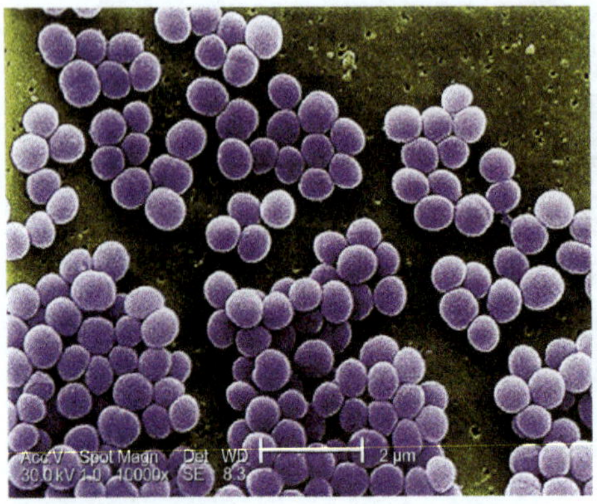

Janice Haney Carr/CDC

12.2 Protecting Our Food

- The United States has one of the safest food supplies in the world, primarily the result of a team effort conducted by cooperating federal, state, and local agencies that regulate and monitor the production and distribution of food. The FDA and the USDA are the key federal agencies that protect consumers by regulating the country's food industry.

12.3 Pathogens in Food

- When certain microorganisms metabolize nutrients in food, they often secrete substances that alter the color, texture, taste, and other characteristics of the food in beneficial and desirable ways. When pathogens are in food, they can make the item unsafe to eat. Food contaminants include pathogens, insect parts, residues of compounds used to kill insects that destroy food crops, and metal fragments from food processing equipment.
- The microbes that cause food-borne illness can live practically anywhere. Common routes for transmitting harmful microbes to food involve vermin, poor personal hygiene practices, and improper food preparation and storage practices.
- To grow, most microbes need warmth, moisture, and a source of nutrients, and some microorganisms also need oxygen. In general, high-risk foods are warm, moist, and protein-rich, and they have a neutral or slightly acidic pH. Such foods include meat, poultry, milk and milk products, and eggs.

John A. Rizzo/Photodisc/Getty Images

12.4 Food-Borne Illness

- Many factors influence whether an individual becomes ill after eating food or drinking a beverage that has been contaminated with a pathogen. The number of pathogens in a food or the amount of toxin it contains can contribute to the risk and severity of a food-borne illness. In general, high-risk groups are pregnant females, very young children, older adults, and persons who suffer from serious chronic illnesses or weakened immune systems.
- Signs and symptoms of food- or water-borne illnesses primarily involve the digestive tract and include nausea, vomiting, diarrhea, and intestinal cramps. Medical treatment should be obtained when an intestinal disorder is accompanied by fever, bloody bowel movements, prolonged vomiting and diarrhea, and/or dehydration.

12.5 Common Food-Borne Pathogens

- The major kinds of pathogens are bacteria, viruses, protozoa, and fungi. In the United States, bacteria and viruses are responsible for most cases of food-borne illness.

12.6 Preventing Food-Borne Illness

- People can reduce their risk of food-borne illness by following safe food-handling practices including proper food selection, preparation, and storage.
- Cooking foods is an effective way to destroy pathogens; however, careless handling of cooked food sets the stage for the growth of pathogens.
- A simple rule to follow is: "Keep hot foods hot and cold foods cold." Other simple rules include: Wash hands and surfaces often, do not cross-contaminate, heat foods to proper temperatures, and chill foods promptly.

12.7 Food Preservation

- Modern methods of food preservation include pasteurization, refrigeration, freezing, canning, aseptic processing, use of preservative additives, and irradiation.

12.8 Preparing for Disasters

- It is important to have a supply of clean water and wholesome food on hand to survive disasters.

12.9 Food Additives

- Direct food additives can make food easier to process, more nutritious, able to stay fresh longer, better tasting, or more attractive in appearance. An indirect food additive is a substance that becomes incorporated into food during production, packaging, transport, or storage.
- The 1958 Food Additives Amendment established the GRAS list and requires the manufacturer of a new food additive to provide evidence of the substance's safety to the FDA before the additive can be marketed for use. Because of the Delaney Clause of the Food Additives Amendment, food manufacturers cannot add a new compound that causes cancer at *any* level of exposure.
- Unintentional food additives are substances that accidentally are in foods. A pesticide is any substance that people use to kill or control unwanted insects, weeds, rodents, or other organisms. The EPA regulates the proper use of pesticides and sets pesticide tolerances. IPM methods control agricultural pests while limiting damage to the environment.
- Agricultural pesticides have both beneficial and unwanted effects. Pesticides help protect the food supply and make food crops available at reasonable cost. Nevertheless, these substances may harm humans, animals, or the environment. Environmental health experts continually monitor the effects of pesticides on humans.

12.10 Nutrition Matters: Avoiding "The Revenge"

- Traveler's diarrhea (TD) results from consuming food or water contaminated with pathogens, especially bacteria. TD is likely to occur during or shortly after travel in regions where sanitary water supplies are lacking, people who prepare food have less than ideal personal hygiene practices, and untreated human feces are used for fertilizer.
- To reduce the risk of TD, follow instructions regarding safe food and beverage selection, avoid contact with contaminated waterways, sanitize drinking water, and consult with your physician concerning the use of medications that may prevent the condition.

Recipe for Healthy Living

Very Easy Chicken Salad

Chicken that is properly prepared, cooked, and stored is a safe, nutritious, economical food. A good way to use up day-old cooked chicken is to make chicken salad. This recipe is unusual because it includes seedless grapes and sunflower seeds. (You can substitute day-old cooked turkey for the chicken and chopped apple or dried apricots for the grapes.)

The chicken salad recipe makes approximately three 1-cup servings. A serving of this salad supplies approximately 252 kcal, 17 g protein, 16 g fat, 2.1 g fiber, and 184 mg sodium.

National Cancer Institute

INGREDIENTS:

1¼ cups roasted, chopped, and chilled chicken meat
1 medium stalk celery, washed and chopped
½ cup raw onion, skinned and finely chopped
½ cup washed seedless red or green grapes, halved
½ cup unsalted roasted sunflower seeds
⅛ tsp curry powder
⅛ tsp ground black pepper
¼ cup low-fat mayonnaise-type salad dressing

PREPARATION STEPS:

1. Place chopped, cooked chicken in a large bowl.
2. Add the rest of the ingredients to the chicken and gently mix until well blended.
3. Serve on a bed of washed leaf lettuce or as a sandwich filling.

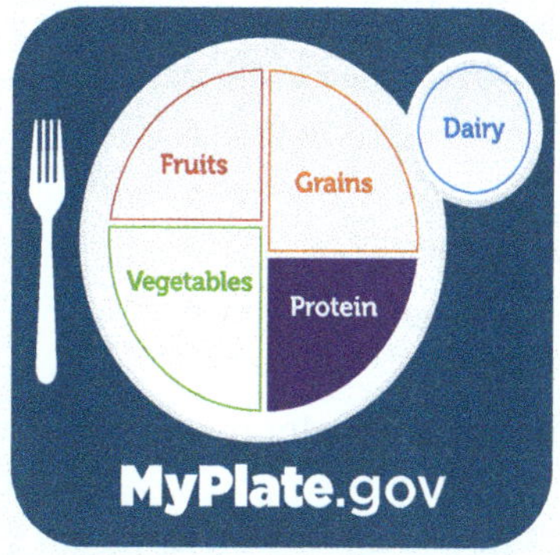

US Department of Agriculture or USDA

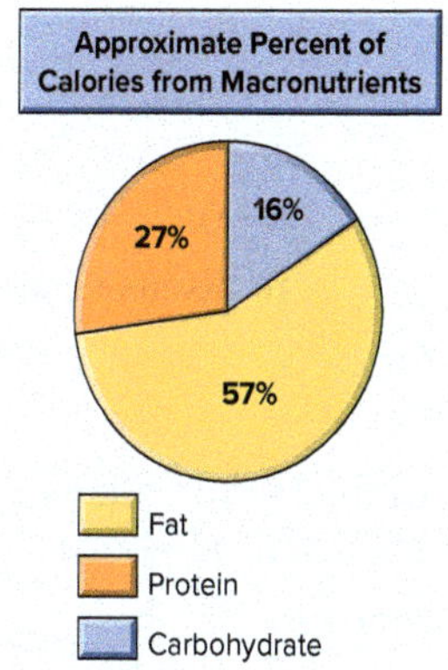

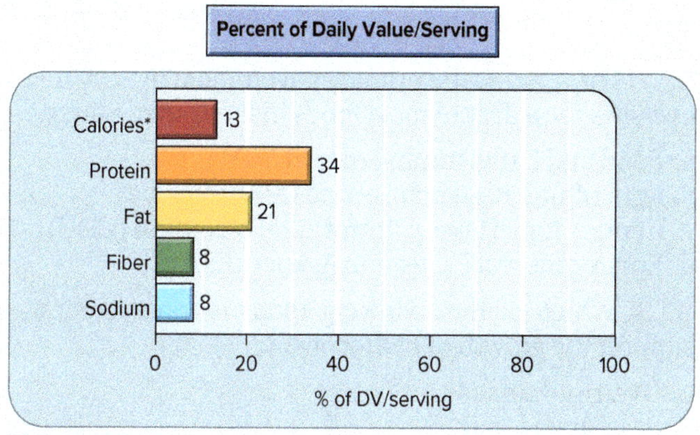

*2000 daily total kcal

Critical Thinking

1. Consider your usual food preparation practices. After having read this chapter, list any food preparation practices you have that are unsafe.
2. Consider your usual food storage practices. After having read this chapter, list any of your food storage practices that are unsafe.
3. For lunch on Monday, you visited a restaurant and ordered a hamburger, French fries, baked beans, and a sealed container of pasteurized orange juice. Two days later, you developed nausea, vomiting, fever, chills, headache, abdominal cramps, and diarrhea. You suspect Monday's lunch made you sick. Considering the foods that you ate for lunch on Monday, which one was the most likely source of your infection? Why? Which food-borne pathogen was the most likely suspect for causing the intestinal disorder? In the future, what steps can you take to reduce the likelihood that the food will make you sick again?
4. One of your friends plans to vacation in South America next week. What practical advice can you provide to help them avoid traveler's diarrhea?
5. Develop an infographic or pamphlet to educate consumers about food-borne illness, including ways to reduce the risk of such illnesses.

Comstock Images/Getty Images

Practice Test (Section numbers are listed below in red.)

Select the best answer.

1. (12.1) ____ are disease–causing microbes.
   a. Pathogens
   b. Toxins
   c. Teratogens
   d. Oxidants
2. (12.1) In the United States, ____ is responsible for most outbreaks of food–borne illness.
   a. *norovirus*
   b. *Clostridium botulinum*
   c. stomach flu
   d. *E. coli*
3. (12.2) The ____ is the primary government agency that oversees the safety of most foods in the United States.
   a. Food and Drug Administration (FDA)
   b. Agricultural Research Service (ARS)
   c. Centers for Disease Control and Prevention (CDC)
   d. Environmental Protection Agency (EPA)
4. (12.3) Which of the following foods is most likely to support the growth of pathogens?
   a. Overripe bananas
   b. Pasteurized milk
   c. Raw ground meat
   d. Commercially canned tomato soup
5. (12.3) ____ is the commercial heating process that destroys harmful bacteria in milk and fruit juices.
   a. Sporulation
   b. Sedimentation
   c. Detoxification
   d. Pasteurization
6. (12.4) Food–borne illnesses are usually characterized by
   a. flulike signs and symptoms.
   b. coughing, sneezing, and respiratory inflammation.
   c. megaloblastic anemia and nervous system defects.
   d. abdominal cramps, diarrhea, and vomiting.
7. (12.5) In the United States, common sources of food–borne illness include all of the following, except
   a. *norovirus.*
   b. *Staphylococcus aureus.*
   c. fungi.
   d. *Salmonella.*
8. (12.6) Which of the following practices can help reduce the growth of food–borne pathogens?
   a. Washing hands before preparing food
   b. Keeping cold foods cold and hot foods hot
   c. Cooking foods to proper internal temperatures
   d. All of the above are correct.
9. (12.7) Irradiation of food is
   a. an untested technology.
   b. not recommended because the process increases nutrient losses.
   c. considered to be a safe process by the FDA.
   d. a widely accepted method of food preservation in the United States.
10. (12.7) Which of the following food preservation processes effectively destroys microbes?
    a. Freezing
    b. Sterilization
    c. Adding salt
    d. Chilling
11. (12.8) How many gallon(s) of water per person per day should an emergency food supply include?
    a. One
    b. Two
    c. Three
    d. Four
12. (12.9) Which of the following substances are *not* direct food additives?
    a. Sulfites
    b. Enrichment nutrients
    c. Pesticide residues
    d. Emulsifiers
13. (12.10) When you travel to countries outside the United States, you can reduce your risk of "traveler's diarrhea" by
    a. eating whole fresh fruits and vegetables.
    b. avoiding water that has not been bottled and sealed.
    c. purchasing foods from street vendors.
    d. consuming ice with beverages.

Answers to Quiz Yourself

1. Aflatoxins are the most common sources of food-borne illness in the United States. **False.** (Section 12.1)
2. In case of emergency, it is not safe to drink the water that is in an undamaged water heater. **False.** (Section 12.8)
3. Certain fungi, such as button mushrooms, are safe to eat. **True.** (Section 12.5)
4. The Environmental Protection Agency (EPA) regulates the proper use of pesticides in the United States. **True.** (Section 12.9)
5. The best way to tell if a food is safe to eat is to smell it. **False.** (Section 12.3)

Design Elements: (Stacked books): Siede Preis/Photodisc/Getty Images; (Nuts): C Squared Studios/Photodisc/Getty Images; (Recipe box, Raspberries, Blueberries): ©Wendy Schiff; (Pineapple): Stockbyte/Getty Images; (Radishes): karandaev/iStock/360/Getty Images; (MyPlate icon): US Department of Agriculture or USDA

CHAPTER 13

Nutrition for a Lifetime

Karena Rigby

Quiz Yourself

What steps can young females take to prepare for pregnancy? Compared to breast milk, do infant formulas provide the same health benefits for infants? When is the best time to begin feeding solid foods to infants? Which nutrients are most likely to be deficient in diets of older adults? While reading this chapter, you will learn answers to these questions. Test your knowledge of nutrition during various life stages by taking the following quiz. The answers are at the end of the chapter.

1. During pregnancy, a mother-to-be should double her food intake because she's "eating for two." ______ T ______ F
2. Infant formulas provide the same health benefits to infants as breast milk. ______ T ______ F
3. Caregivers should add solid foods to an infant's diet within the first month after the baby is born. ______ T ______ F
4. Over the past few decades, the prevalence of obesity has increased among American children. ______ T ______ F
5. Compared to younger persons, older adults have lower risks of nutritional deficiencies. ______ T ______ F

13.1 Introduction to Nutrition for a Lifetime

Learning Outcomes:

1. Identify the life stages.
2. Explain why it is important to learn about nutrition concerns for pregnancy, lactation, childhood, and the older adult years.

Rachel Weill/UpperCut Images/age fotostock

The chapter opening portrait features a young female during her eighth month of pregnancy, looking forward to a bright future. Thanks to technological and medical advances that occurred during the 21st century, her baby has an excellent chance of enjoying a long, healthy life.

If you are a female, are you pregnant? Do you already have children? If you have children, were they breastfed or formula–fed? Do you live with and help care for an elderly parent or grandparent? These may seem to be personal questions, but many undergraduate college students do not fit the stereotype of being 18 to 22 years of age, having no children, and residing away from home.

Most of the nutrition recommendations presented in this textbook apply to people who are "adults"—loosely defined as the period (*life stage*) when a person is 19 to 70 years of age. This chapter focuses on the differing nutrition needs and health concerns of people who are in specific life stages. These particular life stages are the **prenatal period** or pregnancy, the time between conception and birth; **lactation** (milk production for breastfeeding); infancy; childhood; adolescence; and the older adult period that generally spans from about 70 years of age until death.

Why is it necessary to learn some basic information about nutrition–related concerns during various life stages? If you do not have children, you may become a parent in the future. If your parents and grandparents are relatively young and vigorous now, you can expect them to experience declining physical functioning as they grow older. Finally, you need to recognize that most of these changes are normal and will affect you as well.

Today in the United States, many of the leading causes of death are chronic diseases, such as heart disease and cancer. Long–term health–related practices, including dietary and physical activity habits, often contribute to the development of these diseases. Whether you enjoy overall good health or suffer from one or more disabling physical ailments as you grow older depends not only on your lifestyle choices but also on several other factors, including your heredity, relationships, environment, income, education level, and access to health care. Achieving and maintaining a healthy body weight are critical to enjoying good health throughout life. This chapter also includes information about obesity during childhood and adolescence.

prenatal period time between conception and birth; pregnancy

lactation milk production

Concept CHECKPOINT

1. Matthias is 79 years of age. He is in the ______ life stage.
2. For females, which life stage involves milk production for breastfeeding?

See Appendix F for responses.

13.2 Pregnancy

Learning Outcomes

1. Discuss the importance of having preconception health care and prenatal care, including adopting healthy lifestyles.
2. Explain the difference between an embryo and a fetus.
3. Discuss the roles of the placenta and uterus during pregnancy.
4. List major physiological changes that occur to females during pregnancy, and identify typical nutrition-related discomforts associated with this stage of life.
5. Discuss the importance of a nutritionally adequate diet and an appropriate weight gain during pregnancy.
6. Explain why gestational diabetes, preeclampsia, and eclampsia are serious health concerns for pregnant females and their unborn offspring.

"It's positive!" Each day, the results of pregnancy testing are a source of excitement, relief, or concern for thousands of females. Today, 45% of pregnancies are unplanned in the United States.[1] Therefore, all sexually active females with male partners of childbearing age should be aware of their likelihood of becoming pregnant. Regardless of whether a pregnancy is planned or not, dietary practices before and during pregnancy play a major role in the course of the pregnancy and the primary outcome—a healthy infant.

Preconception Health Care

Conception occurs the moment a male's sperm cell enters a female's egg cell. The *preconception* period occurs before a female of childbearing age becomes pregnant (conceives) and a male who is sexually mature fathers a child. Health professionals who provide optimal preconception health care typically advise a female who may become pregnant to make healthy lifestyle choices. Such decisions can improve her fertility and likelihood of having a normal pregnancy and healthy baby. A father–to–be, however, also needs to be aware that his lifestyle choices can negatively affect his body's ability to produce enough healthy sperm to be fertile. Obesity, for example, may result in erectile dysfunction (the inability to have or maintain an erection [males]) and sperm defects, both of which reduce fertility.[2,3] Excess alcohol consumption, as well as the use of nicotine–containing products, certain medications, and illegal drugs, can interfere in the production of healthy sperm and reduce sexual functioning. Thus, males should consider obtaining preconception counseling so they can make healthy lifestyle decisions well before they become fathers.

When planning a pregnancy, both females and males should consider adopting healthy lifestyles that include:

- Consuming a nutritionally adequate diet,
- Achieving and maintaining a healthy body mass index (BMI),
- Avoiding harmful drugs, and
- Exercising regularly.

The Prenatal Period

The prenatal period (*gestation*) encompasses the time from conception until the birth of a *full–term* infant, about 38 to 42 weeks later. During the first

conception moment when a sperm enters an egg (fertilization)

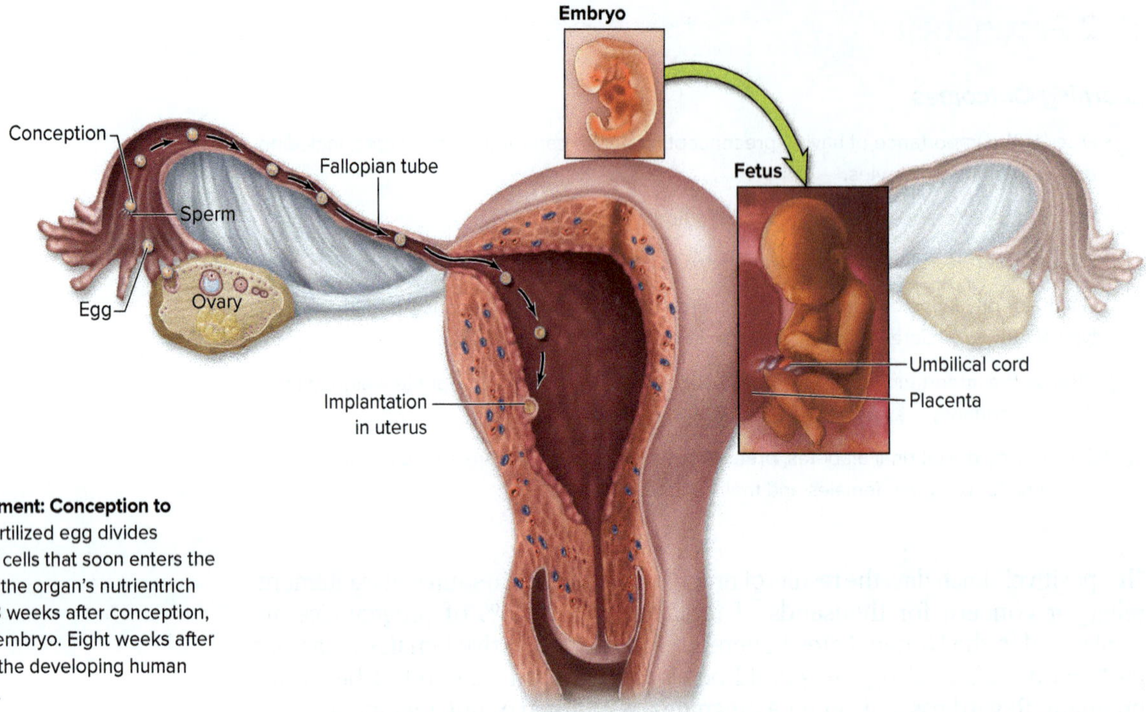

Figure 13.1 Prenatal development: Conception to fetus. After conception, the fertilized egg divides repeatedly, forming a mass of cells that soon enters the uterus and implants itself into the organ's nutrientrich lining. From 14 days through 8 weeks after conception, the mass of cells is called an embryo. Eight weeks after conception and until its birth, the developing human being is referred to as a fetus.

2 weeks after conception, the fertilized egg divides repeatedly, forming a mass of cells that enters the female's **uterus,** the female reproductive organ that protects the developing organism. The mass of cells buries itself into the nutrient–rich lining of the uterus (implantation) and continues to develop. For the next 6 weeks, the rapidly dividing mass of cells, called an **embryo,** increases in size and forms organs. Eight weeks after conception, the developing human being is referred to as a **fetus** (Fig. 13.1).

The prenatal period is often divided into three stages or *trimesters.* During the first trimester, the embryo/fetus develops most of its organs, and by the end of this period, the fetus can move. The first trimester is a critical stage in human development because nutrient deficiencies or excesses and exposure to toxic compounds, such as alcohol, are most likely to have devastating effects on the embryo/fetus. However, many females who are in their first trimester do not realize they are pregnant.

As the second trimester begins, the fetus is still very tiny, about 2½ to 3 inches in length, and weighs only about an ounce. However, the fetus is beginning to look more like a human infant: It has fully formed arms, hands, fingers, legs, feet, and toes. The fetus's organs continue to grow and mature in their ability to function. As the fetus moves around, its mother becomes increasingly aware of its presence within her body.

By the beginning of the third trimester, the fetus is approximately 10 inches long (from crown to rump) and weighs about 2 pounds. During this trimester, the fetus will nearly double in length and multiply its weight by three to four times. Thus, a healthy fetus usually weighs about 6 to 8 pounds and is 19 to 21 inches long by the time it is full–term and ready to be born.

Throughout the prenatal period, the embryo/fetus depends entirely on its mother for survival. During most of the pregnancy, the expectant mother nourishes her embryo/fetus through the **placenta,** the organ of pregnancy that connects the uterus to the embryo/fetus via the *umbilical cord* (see Figure 13.1). The role of the placenta is to transfer nutrients and oxygen from the mother's

uterus female reproductive organ that protects the developing organism during pregnancy

embryo human organism from 14 days to 8 weeks after conception

fetus human organism from 8 weeks after conception until birth

placenta organ of pregnancy that connects the uterus to the embryo/fetus via the umbilical cord

bloodstream to the embryo/fetus. Additionally, the placenta transfers wastes from the embryo/fetus to the mother's bloodstream, so her body can eliminate them. Unfortunately, the placenta does not filter many microbes and toxic substances, such as alcohol and nicotine, from the mother's blood. Thus, agents of infection and harmful chemicals can pass through the placenta, enter the embryo/fetus, and cause disease, birth defects, or embryonic/fetal death.

A fetus generally needs to spend at least 37 weeks developing within the uterus for its organs to be physiologically mature enough to survive after birth without the need for special care. A fetus's weight partly depends on the supply of nutrients that it receives through the placenta.[4] If the placenta fails to grow properly, the developing fetus is likely to be born too soon and be lighter than average at birth.

Birth Weight

Birth weight is a major factor that determines whether a baby is healthy and survives their first year of life. **Low–birth–weight (LBW)** infants generally weigh less than 5½ pounds at birth. LBW is the second leading cause of death among American infants (birth defects are the leading cause).[5] In 2020, about 8% of infants born in the United States were low birth weight.[6] Pregnant females who are under 20 years of age or 40 years of age or older are more likely to give birth to LBW infants than females in other age groups. Additionally, females who smoke during pregnancy are at risk to have LBW babies.[7] Low birth weight is often associated with premature or preterm births. In 2021, 10.5% of births in the United States were preterm; that is, they occurred before the 37th week of pregnancy.[8]

Early preterm infants are born before 34 weeks of gestation. In 2021, 2.81% of babies born in the United States were early preterm infants.[6] Such infants are more likely to have serious health problems or die soon after birth than babies delivered after 34 weeks of pregnancy. An early preterm infant who is born after about 26 weeks of pregnancy may survive if cared for in a hospital nursery for high–risk newborns (Fig. 13.2). However, the tiny infant's body will not have stores of fat and certain minerals that normally accumulate during the last month of pregnancy. Additionally, early preterm babies are likely to have conditions that complicate their medical care and food intake, such as breathing difficulties and weak sucking and swallowing abilities.

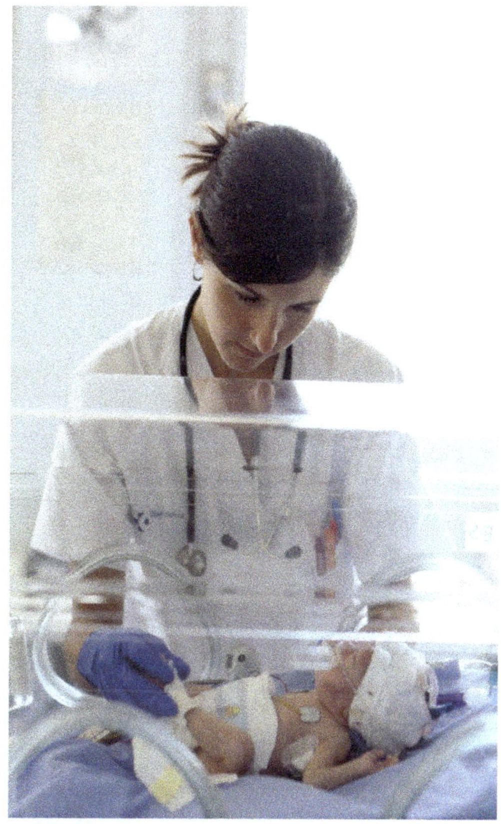

Figure 13.2 **Early preterm infant.**
Javier Larrea/age fotostock/Alamy Stock Photo

Nutrition-Related Signs of Pregnancy

During pregnancy, a female's body undergoes major physiological changes, such as increased blood volume, breast size, and levels of several hormones. These adaptations enable her body to nourish and maintain the developing embryo/fetus, as well as produce milk for her infant after its birth. However, some of the physical changes cause discomfort for the pregnant female.

In the first trimester, most females experience physical signs that they are pregnant, such as enlarged breasts and "morning sickness." Other common nutrition–related signs, as well as complaints, of pregnancy include extreme tiredness, swollen feet, constipation, and heartburn. In most cases, such discomforts do not create serious complications, and they resolve within a few months.

Breast Changes

During pregnancy, hormones signal the breasts to increase in size in preparation for lactation. The mother's pituitary gland in the brain produces **prolactin,**

low-birth-weight (LBW) describes baby generally weighing less than 5½ pounds at birth

prolactin hormone that stimulates milk production after delivery

a hormone that stimulates the development of milk–producing tissue in the breasts. However, a pregnant female's breasts do not form milk because high levels of progesterone, a hormone that helps maintain pregnancy, inhibit milk production.[9] After birth, the level of progesterone drops rapidly, essentially removing the "brakes" from the breasts' ability to produce milk.

Morning Sickness

A common sign of pregnancy is **morning sickness,** nausea that is sometimes accompanied by vomiting. The name "morning sickness" is misleading because the queasy feeling can occur at any time of the day. The cause of this unpleasant condition is unclear, but it may be the result of the pregnant female's body adapting to higher levels of female hormones. Additionally, emotional stress and certain foods can contribute to nausea. The condition generally begins early in the first trimester, and most females are no longer affected by the 16th week of pregnancy.[10] However, some females experience nausea and vomiting occasionally throughout their pregnancies.

To help control mild morning sickness, pregnant females can avoid odors and foods, such as fried or greasy foods, that trigger nausea. Some females find that eating crackers and drinking some water helps reduce the likelihood of feeling nauseated, especially before they get out of bed in the morning. Furthermore, eating smaller but more frequent meals and nutritious snacks rich in protein can be helpful. Products that contain ginger, such as ginger tea, can also help relieve the nausea of morning sickness.[10]

If the nausea and vomiting occur more than three times a day or the vomit looks like "coffee grounds," the pregnant female should contact her physician for treatment.[10] Morning sickness that persists beyond the fourth month of pregnancy should also be brought to the attention of a physician. During pregnancy, excessive vomiting is harmful because it can lead to dehydration, weight loss, and electrolyte imbalances.

Did *YOU* Know?

Early in pregnancy, 50–80% of pregnant females experience nausea, and about 50% experience vomiting. "Morning sickness" and its symptoms of nausea and vomiting are often minimized by pregnant females and medical professionals, resulting in undertreatment.[11]

Medical professionals often recommend acupressure wristbands and ginger-containing foods such as ginger candies, ginger ale, and ginger tea to manage the symptoms of mild cases of nausea and vomiting.[11]

A pregnant female may not seek medical treatment for severe nausea and vomiting (vomiting more than 3 times per day) because she is concerned about the safety of medications for her fetus.[11] However, there are safe and effective prescription medications that pregnant females can take to manage symptoms and improve their quality of life.

Nonetheless, if severe nausea and vomiting are not managed, symptoms can progress and be difficult to control. Therefore, females should seek early medical treatment for nausea and vomiting to minimize complications, including dehydration and hospitalization.

Antonio Guillem/Shutterstock

morning sickness nausea and vomiting associated with pregnancy

Fatigue in Pregnancy

Early in pregnancy, the mother's blood volume expands to approximately 150% of normal. The number of red blood cells, however, increases by only 20 to 30%, and this change occurs more gradually. As a result, the pregnant female develops *physiological anemia,* a condition characterized by a lower concentration of red blood cells in the bloodstream. This form of anemia is a normal response to pregnancy, rather than the result of inadequate nutrient intake. Nevertheless, physiological anemia may be responsible for the extreme tiredness experienced by pregnant females during their first trimester. As their red blood cell numbers increase, expectant mothers report having more energy, especially during the second trimester. By the third trimester, however, most pregnant females are easily fatigued again, possibly because carrying a rapidly growing fetus is physically demanding.

What Is Edema?

High levels of certain hormones can cause various tissues to retain fluid (mostly water) during pregnancy. Although the extra fluid causes some minor swelling (edema), especially in the hands and feet, the condition is normal. In most cases, mild edema does not require treatment such as restricting salt intake or taking diuretics. Edema, however, can be a sign of trouble if hypertension and the appearance of protein in the urine accompany the swelling. The "Hypertension in Pregnancy" section of this chapter discusses various forms of hypertension that can occur during pregnancy.

Digestive Tract Discomforts

During pregnancy, certain hormones produced by the placenta relax muscles of the digestive tract. As a result, intestinal movements slow down and digested material takes longer to pass through the tract, increasing the likelihood of constipation (see the "Fiber and Health" section of Chapter 5). To help prevent constipation, pregnant females should consume adequate amounts of fiber and fluids. During pregnancy, the Adequate Intake (AI) for fiber is 28 g/day, and the AI for total water is 3 L/day.[12] If constipation persists after making these dietary changes, the pregnant female should discuss this concern with her physician.

Heartburn is another common complaint of pregnant females. As the fetus grows, the uterus pushes upward in the mother's abdominal cavity and applies pressure on her stomach. When this occurs, stomach acid can enter the esophagus, causing heartburn (see the "Nutrition Matters" section of Chapter 4). To help avoid heartburn, the pregnant female can consume smaller meals, avoid lying down after eating, eat less fatty foods, and learn to identify and avoid foods that seem to contribute to heartburn. If heartburn continues to be bothersome, the female should consult her physician and discuss other ways to treat the condition.

Having an adequate fiber intake can help prevent constipation during pregnancy.
Vladimir Pcholkin/Getty Images

Pregnancy: General Dietary Recommendations

As mentioned earlier in this section, females of childbearing age should take steps to ensure good health before becoming pregnant. For example, females can analyze the nutritional adequacy of their diets and choose to eat foods that correct any marginal or deficient intakes. Prior to pregnancy, sedentary females can begin an exercise regimen, females who are overweight or obese can lose some excess weight, and females who smoke can join smoking cessation programs. The time to remedy faulty lifestyle practices and increase chances of having a healthy pregnancy and baby is long before pregnancy occurs.

During pregnancy, the mother–to–be should follow a diet that meets her own nutritional needs, as well as those of her developing offspring. Depending on the trimester, an expectant female's requirements for energy (calories), protein, and many other nutrients are greater than her needs prior to pregnancy. Nevertheless, a pregnant female does not need to double her usual food intake just because she is "eating for two." Table 13.1 compares Recommended Dietary Allowances (RDAs) for energy and selected nutrients that apply to healthy 25–year–old nonpregnant and pregnant females.

TABLE 13.1 Comparing Selected DRIs: 25-Year-Old Nonpregnant and Pregnant Females

Energy/ Nutrient*	Nonpregnant	Pregnant
Kilocalories	Estimated Energy Requirement (EER)	First trimester = EER + 0 Second trimester = EER + 340 Third trimester = EER + 452
Protein	46 g/day	71 g/day
Vitamin C	75 mg/day	85 mg/day
Thiamin	1.1 mg/day	1.4 mg/day
Niacin	14 mg/day	18 mg/day
Folate	400 mcg/day	600 mcg/day
Vitamin D	15 mcg/day	15 mcg/day
Calcium	1000 mg/day	1000 mg/day
Iron	18 mg/day	27 mg/day
Iodine	150 mcg/day	220 mcg/day

* RDA
Source: Data from Institute of Medicine: *Dietary Reference Intakes.*

Energy Needs

In the first trimester, a pregnant female's daily energy requirement (*Estimated Energy Requirement* or *EER*) is essentially the same as a nonpregnant female's because the embryo/fetus is quite small (see Table 13.1). However, the fetus grows rapidly during the second and third trimesters, and the pregnant female requires more energy and nutrients to support its growth, as well as her own body's needs. During the second trimester, the expectant mother should consume approximately 340 more kilocalories per day than her prepregnancy EER. Throughout the third trimester, she should add about 450 kcal per day to her prepregnancy EER (see Table 13.1). If a female is physically active during her pregnancy, she may need to increase her kilocalorie intake by even more than these levels. Why? As the pregnant female gains weight, her muscles require more energy to move her body. The average expectant mother, however, reduces her physical activity level during the third trimester, conserving energy.

Liver, beans, many kinds of vegetables, and fortified cereals are sources of folate.
©Wendy Schiff

TABLE 13.2 Sample Menu for a 25-Year-Old Pregnant Female (Second Trimester)

Breakfast
¾ cup cooked oatmeal, made with ½ cup fat-free milk and sprinkled with ¼ cup raisins
½ cup calcium-fortified orange juice
Mid-morning Snack
1 cup apple slices
2 Tbsp peanut butter
Lunch
Cheese sandwich
2 slices whole-grain bread
1 slice American cheese
2 slices tomato
¼ cup leaf lettuce
2 tsp mayonnaise-type low-calorie salad dressing
1 cup strawberries
1 cup fat-free milk
½ cup reduced-sugar or no-sugar-added Greek yogurt
Mid-afternoon Snack
⅓ cup low-sodium, whole-grain crackers
⅓ cup hummus
½ cup raw carrots
½ cup fat-free milk
Dinner
Broiled salmon filet, 5 oz
1¼ cups mixed salad greens
1 Tbsp Italian dressing
½ cup enriched white rice
1 cup steamed broccoli
2 small dinner rolls
1 Tbsp olive oil spread
Evening Snack
1 cup frozen yogurt

Folate and Iron Needs

A pregnant female's requirements for folate and iron are 50% higher than those of a nonpregnant female. It is important for females to enter pregnancy with adequate folate status because embryos need the vitamin to support rapid cell division. As discussed in Chapter 8, pregnant females who are folate deficient have high risk of giving birth to infants with neural tube defects, such as spina bifida (see Figure 8.24). To obtain adequate folate, females of childbearing age as well as pregnant females should include rich food sources of folate in their diets, such as liver, ready-to-eat cereals that have been fortified with folic acid, asparagus, broccoli, dried beans, and green leafy vegetables. Additionally, females of childbearing age should take a multivitamin/mineral supplement that supplies at least 400 mcg of folic acid, a form of folate.

As the pregnant female's blood volume expands, her need for iron increases because her body must make more hemoglobin for the extra red blood cells. Additionally, the female's body transfers iron to the fetus to build its stores of the mineral. If females fail to meet their iron needs during pregnancy, their iron stores can be severely depleted, and they can develop iron deficiency anemia. Pregnant females who have anemia are at risk of giving birth prematurely and having low-birth-weight infants.[12] As mentioned in Chapter 3, *The Dietary Guidelines for Americans* identifies iron as a nutrient of public health concern for pregnant females.

Even when their diets include good sources of iron such as red meats, liver, and enriched cereals, pregnant females often need a supplemental source of iron. Thus, most physicians recommend special prenatal multivitamin/mineral supplements that contain iron for their pregnant patients.

Menu Planning for Pregnant Females

Rather than view pregnancy as a time to splurge by eating foods that are high in empty calories, the mother-to-be should obtain the extra energy from nutrient-dense foods. For example, drinking an additional cup of fat-free milk, eating a bowl of an enriched whole-grain cereal, and taking a prenatal supplement each day can supply extra kilocalories, as well as protein, fiber, and micronutrients. Table 13.2 presents a day's meals and snacks for a sedentary 25-year-old female who is in her second trimester of pregnancy. Her prepregnancy EER was 2000 kcal, so her sample menu is based on MyPlate recommendations for 2400 kcal, enough to cover her increased EER during this trimester.

Is Fish Safe to Eat During Pregnancy? Fish and shellfish (e.g., clams, shrimp, and crabs) are excellent sources of many minerals, omega-3 fatty acids, and high-quality protein. However, most fish and shellfish contain small amounts of methylmercury, a compound that contains the toxic mineral mercury.[13] Certain kinds of fish, however, contain higher levels of methylmercury than others. Therefore, pregnant and breastfeeding females should avoid eating swordfish, shark, king mackerel, tilefish from the Gulf of Mexico, marlin, orange roughy, and bigeye tuna.[13] When a pregnant female eats these foods, the methylmercury in them can eventually reach the developing fetus and damage its nervous system. The Dietary Guidelines section of Chapter 3 provides recommendations for safe levels of fish consumption by pregnant females.

What About Cravings?

The stereotype of a pregnant female who craves pickles and ice cream is not simply a myth. Cravings are common during this stage of life. However, ask pregnant females to identify the foods they crave, and you are likely to get a variety of responses. The causes of cravings are unknown, but they may be responses to the hormonal changes associated with pregnancy or to the emotional state of the mother–to–be. In other instances, specific food cravings may simply reflect the pregnant female's family traditions. Unless food cravings contribute to excess weight gain, they are generally harmless.

Some females develop **pica,** the practice of eating nonfood items such as laundry starch, chalk, cigarette ashes, and soil. Some studies have linked pica with iron and zinc deficiency, but it is not clear if pica is the result or the cause of such deficiencies. Pregnant females should refrain from practicing pica, especially eating clay or soil. Soil may contain substances that interfere with the absorption of minerals in the intestinal tract. Furthermore, eating soil can be harmful because the dirt may be contaminated with toxic substances, such as lead and pesticides, and pathogenic microbes.

pica craving nonfood items

Weight Gain During Pregnancy

Nearly all pregnant females experience weight gain. In fact, gaining an appropriate amount of weight is crucial during pregnancy. How much weight a female should gain depends on her prepregnancy weight (Table 13.3). Females who were underweight prior to pregnancy should gain 28 to 40 pounds; females whose prepregnancy weights were within the healthy range can expect to gain 25 to 35 pounds.[14] Females who were overweight before they became pregnant should gain 15 to 25 pounds, and females who were obese should gain 11 to 20 pounds during pregnancy. The recommendations are higher for females who are pregnant with more than one fetus. For example, a healthy female who is carrying twins may gain as much as 54 pounds during pregnancy.[14]

In 2015, 48% of females in the United States who were pregnant with one fetus gained more than the recommended amount of weight during pregnancy.[14] Females who gain excess weight during pregnancy are likely to retain the extra pounds long after their babies are born. Furthermore, expectant mothers who gain excessive amounts of weight are more likely to give birth to high–birth–weight (HBW) babies, which may be referred to as large–for–gestational–age infants. An HBW baby weighs more than 8 pounds 13 ounces (4000 g) at birth. When compared to newborns with healthy weights, HBW infants have higher risk of being injured during the birth process. Furthermore, HBW infants may be more likely to develop obesity, diabetes, and hypertension at some point in their lives.[15] About 9% of newborns are HBW babies.[15]

Females who are underweight and do not gain enough weight during pregnancy are at risk of having preterm or low–birth–weight infants. Underweight pregnant females should try to reach healthy weights by the end of the first trimester and then meet the recommended weight–gain goals. Females who are obese have a greater risk of developing hypertension, as well as type 2 diabetes, during pregnancy. Obese pregnant females are also at risk of giving birth to HBW babies. However, females should not try to lose weight while they are pregnant because calorie restriction may harm the fetus.

TABLE 13.3 Recommended Weight Gain During Pregnancy*

Weight Classification Before Pregnancy (BMI)	Range of Weight Gain During Pregnancy (pounds)
Underweight (< 18.5)	28–40
Healthy weight (18.5–24.9)	25–35
Overweight (25.0–29.9)	15–25
Obese (30.0 or more)	11–20

*Pregnant with one fetus

Andersen Ross/Photodisc/Getty Images

Accounting for the Weight Gain

It is important to understand that much of the weight a female gains during a healthy pregnancy is not body fat. By the end of a full–term pregnancy, the average fetus weighs about 7½ pounds, and the placenta and *amniotic fluid* that surrounds the fetus account for about 3½ pounds. The remaining weight is composed of tissues and fluids the mother's body gains during pregnancy (*maternal weight* gain). Table 13.4 indicates the typical distribution of weight that is gained during this life stage.

TABLE 13.4 Distribution of Weight Gain During Pregnancy*

Tissue	Approximate Pounds
Maternal	
Blood	4
Breasts	2
Uterus	2
Fat, protein, and retained fluid	11
Fetus	7.5
Placenta	1.5
Amniotic fluid	2.0

* Values are for a 30-pound total weight gain.
Source: March of Dimes, *Weight gain during pregnancy.* January 2022. www.marchofdimes.org/pregnancy/weight-gain-during-pregnancy.aspx. Accessed January 29, 2023.

Rate of Weight Gain

For an expectant mother, the rate of weight gain is important, as well as the amount of weight gained. Most pregnant females add up to 5 pounds of weight during the first trimester.[16] Throughout the rest of their pregnancies, healthy females typically gain at a faster rate, 3 to 4 pounds *each* month. Figure 13.3 charts the course of weight gain in a healthy pregnancy. The rate of gain reaches a steady pace of about 1 pound/week during the second and third trimesters. During the second and third trimesters, recommended weight gains are higher for females who are underweight and lower for females who are overweight or obese.

The Importance of Prenatal Care

Ideally, females of childbearing age should plan for pregnancy and receive dietary advice before becoming pregnant. If this is not possible, *prenatal care* should begin early in pregnancy because many medical problems that may occur during this life stage can be diagnosed and treated before the health of the mother or her fetus is threatened. **Prenatal care** is specialized to meet the health care needs of pregnant females. Routine prenatal health care includes measuring and monitoring the pregnant female's weight, blood pressure, blood glucose level, and uterine growth. The prenatal health care provider may also discuss various concerns with the expectant mother, such as morning sickness, safe types of physical activity, what to expect during the birth process, and basic infant care skills. Additionally, the health care provider can advise the pregnant female to make appropriate lifestyle choices, such as avoiding the use of tobacco, alcohol, and illegal substances. Females who receive adequate prenatal care are more likely to have fewer complications (health problems) and healthier infants than females who do not receive such care.[17]

During pregnancy, it is important for a female to decide whether she will breastfeed her baby. Pregnant females who want to breastfeed their babies should inform their physicians and learn as much as they can about breastfeeding early in their pregnancy.

prenatal care specialized health care for pregnant females

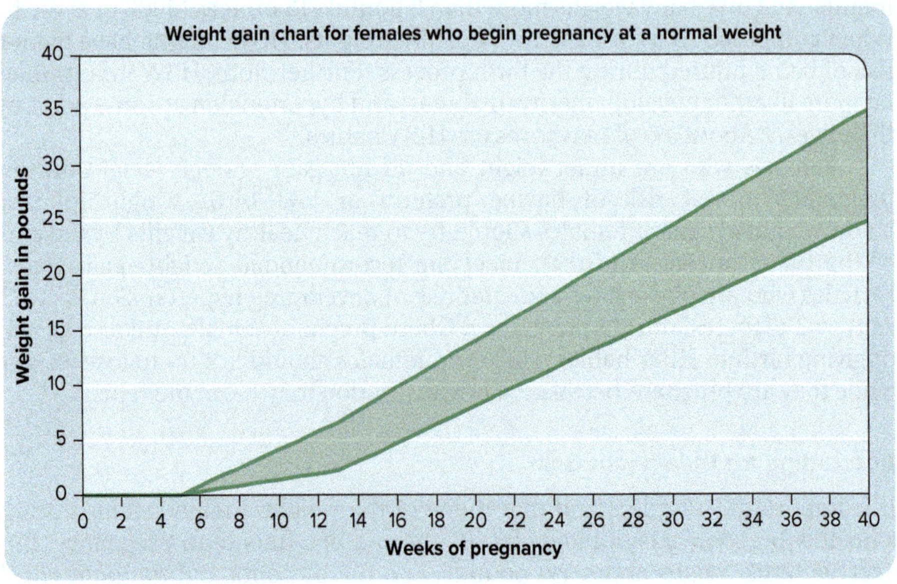

Figure 13.3 Rate of weight gain: Healthy pregnancy. This chart illustrates the rate of maternal weight gain that typically occurs during a healthy pregnancy (single fetus).
Source: www.cdc.gov/reproductivehealth/pdfs/maternal-infant-health/pregnancy-weight-gain/tracker/single/Normal_Weight_Tracker_508Tagged.pdf

Gestational Diabetes

According to estimates, as many as 10% of pregnant females in the United States develop type 2 diabetes during pregnancy (*gestational diabetes*).[18] When a female has gestational diabetes, her fetus receives too much glucose and converts the excess into fat. Thus, females with this form of diabetes often give birth to HBW babies. After birth, these infants often have difficulty controlling their own blood glucose levels and are at risk of becoming overweight as children.

Gestational diabetes can be detected during routine prenatal care. For more information about diabetes during pregnancy, see Chapter 5.

gestational hypertension type of hypertension that can develop during pregnancy

Hypertension in Pregnancy

Some females who had normal blood pressure values before they became pregnant develop hypertension, after the 20th week of pregnancy. This condition is called **gestational hypertension.** This form of hypertension is often mild and resolves within a few months after the baby is delivered.[19] However, some females with gestational hypertension develop a more severe form of the condition, called *preeclampsia* (*pre–ee–klamp′–see–ah*). Preeclampsia is characterized by a sudden, dramatic increase in weight that is due to edema, particularly of the hands and face; severe hypertension; persistent headache; vision problems; trouble breathing; and protein in urine (Fig. 13.4).

In the United States, about 4% of pregnant females develop preeclampsia.[20] Pregnant females who have high risk of preeclampsia are those who are over 40 years of age, are obese, have a history of diabetes or hypertension, and are carrying more than one fetus.[20] If a female suffering from preeclampsia develops seizures, her condition is called *eclampsia* (*ee–klamp′–see–ah*). Worldwide, preeclampsia and eclampsia are major causes of death among pregnant females.

At present, the most effective treatment for preeclampsia and eclampsia is delivering the fetus, but infants born before the 24th week of pregnancy are unlikely to survive. If the fetus is older than 24 weeks, its mother may be hospitalized for treatment. This practice helps physicians closely monitor the

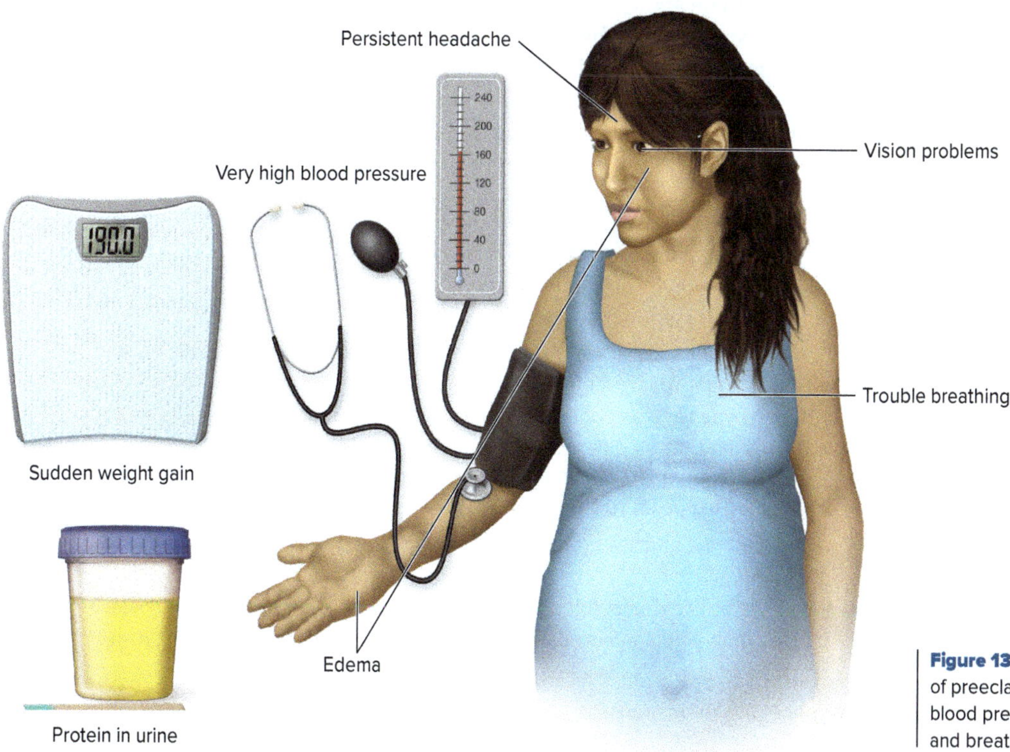

Figure 13.4 Preeclampsia. Major signs and symptoms of preeclampsia include sudden weight gain, very high blood pressure, edema, persistent headache, vision and breathing problems, and protein in urine.

Did *YOU* Know?

Caffeine is a stimulant drug that can pass through the placenta and enter the embryo/fetus. The drug may depress the flow of blood in the placenta, which may be harmful to the developing offspring. At present, there is no clear scientific evidence that a pregnant female's consumption of caffeinated beverages contributes to *miscarriage*. A miscarriage is the death of an embryo or fetus that occurs before the 20th week of gestation. Nevertheless, pregnant females should avoid drinking concentrated sources of caffeine, especially "energy shots," and limit their caffeine consumption to 200 mg/day.[21] This is approximately the amount of caffeine in 1.5 cups of regular brewed coffee.

Klattistock/iStock/Getty Images

mother's condition and enables the fetus to mature until it has a better chance of surviving after a premature birth.

Drug Use

Exposure to alcohol and tobacco is harmful to the embryo/fetus. Females who drink alcohol during pregnancy are at risk of having a child with a fetal alcohol spectrum disorder such as FAS (see Figure 6.31). Scientists do not know if there is a "safe" amount of alcohol that pregnant females can consume; therefore, females of childbearing age who are sexually active or pregnant should avoid alcoholic beverages. The "Nutrition Matters" section of Chapter 6 provides more information about fetal alcohol spectrum disorder.

Compared to pregnant females who do not smoke cigarettes, expectant mothers who smoke have higher risk of giving birth too early and having LBW babies. Furthermore, expectant mothers who smoke cigarettes may increase the risk of having babies with birth defects or that die of *sudden infant death syndrome (SIDS)*. SIDS is the sudden, unexplained death of an infant younger than 1 year of age and the leading cause of death for babies between 1 month and 1 year of age.[22]

The use of illegal drugs, herbal supplements, and medications during pregnancy can also harm the embryo/fetus. Ideally, the time to quit abusing illegal drugs is before pregnancy. Pregnant females should consult their physicians before using herbal supplements or taking any drugs, even over–the–counter medications.

What About Physical Activity?

Females can derive many benefits from being physically active during pregnancy, including enhanced muscle tone and strength, reduced edema, and improved mood and sleep. Most pregnant females can continue their prepregnancy exercise regimens, especially those that included low– or moderate–intensity and flexibility activities. Recommended activities generally include walking, cycling, swimming, prenatal yoga, or light aerobics. Pregnancy is not the time to begin an intense fitness regimen or perform high–risk physical activities. Activities that are risky and should be avoided include hot yoga; downhill skiing; contact sports such as volleyball, soccer, and basketball; and horseback riding.[23] Pregnant females should discuss their physical activity practices and needs with their physicians. Some expectant females, such as those experiencing hypertension or premature labor contractions, may need to restrict their physical activity.

Most pregnant females can continue their prepregnancy exercise regimens, especially those that included low- or moderate-intensity and flexibility activities.
wong sze yuen/Shutterstock

Concept CHECKPOINT

3. When is an embryo referred to as a fetus?
4. What is the role of the placenta?
5. What is a major factor that determines whether a newborn baby is healthy and survives its first year of life?
6. Why can hypertension during pregnancy be dangerous?
7. Identify at least three different nutrition-related signs of pregnancy.
8. How much weight should a female at a healthy weight gain during pregnancy? How much weight should she gain if she was underweight before becoming pregnant? How much weight should she gain if she was overweight or obese before pregnancy?
9. Why is having adequate folate and iron status important for pregnant females?
10. Discuss the harmful effects that a pregnant female's alcohol consumption and cigarette smoking can have on her embryo/fetus.

See Appendix F for responses.

13.3 Infant Nutrition

Learning Outcomes

1. Discuss the physiological processes of lactation and breastfeeding.
2. Discuss the nutrient and energy needs of a lactating female.
3. Describe benefits that females derive from breastfeeding and infants derive from consuming their mother's milk.
4. Compare the energy and nutrient contents of human milk, cow's milk, and infant formulas.
5. Describe signs that an infant is ready to eat solid foods, and identify appropriate foods for infants.
6. Discuss a healthy infant's rate of growth.

Rapid physical growth characterizes infancy, the life stage that extends from birth to about 2 years of age. During the first 4 to 6 months of life, a healthy baby doubles its birth weight, and by 1 year of age, an infant's birth weight has tripled. Additionally, an infant's length increases by 50% during its first year of life. Thus, if a baby girl weighs 7 pounds and is 20 inches long at birth, you would expect her to weigh 21 pounds and be 30 inches long by her first birthday (Fig. 13.5).

Compared to older children, an infant needs more energy and nutrients per pound of body weight to support its rapid growth. If an infant's diet lacks adequate energy and nutrients, the baby's growth may slow or even stop. The following sections take a closer look at infant nutrition, including breastfeed–ing and other infant feeding practices.

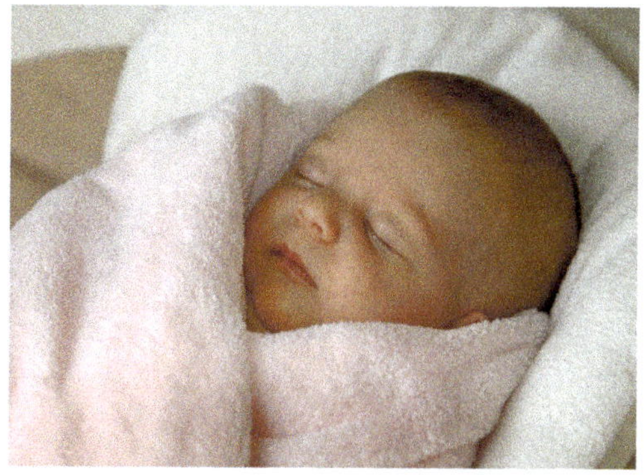

(a)

(b)

Figure 13.5 Growth rates during infancy. A healthy newborn baby (a) grows rapidly during its first year. By 1 year of age, an infant's birth weight has tripled (b). Additionally, an infant's length increases by 50% during its first year of life.

Newborn: Travis VanDenBerg/Alamy Stock Photo; Birthday baby: Elyse Lewin/Brand X Pictures/Getty Images

Breast Milk Is Best Milk

Two hundred years ago, if a new mother was unable to breastfeed her baby, the child faced certain death—unless a female who was producing breast milk could be located to suckle (*nurse*) the infant. Today, a new mother can choose to nurse her baby or feed the child an **infant formula,** a synthetic food that simulates human milk. Although both foods provide adequate nutrition for young babies, breastfeeding provides benefits beyond nutrition for the new mother as well as her infant.

Human milk is uniquely formulated to meet the nutrient needs of a newborn baby. During the first couple of days after giving birth, the new mother's breasts produce **colostrum** (*co–loss'–trum*), a yellowish fluid that does not look like milk. By the end of the first week of lactation, colostrum has undergone a transition to *mature milk*. If you compare the appearance of mature human milk to cow's milk, you will notice that breast milk is more watery than cow's milk and may have a slightly bluish color.

Colostrum is a very important first food for babies because the fluid contains antibodies and immune system cells that may be absorbed by the infant's immature digestive tract. Colostrum also contains prebiotics that encourage the growth of beneficial bacteria in the infant's GI tract (see Chapter 4). Biologically active substances in colostrum help an infant's body fight infections and hasten the maturation of the baby's immune and digestive systems.[24] Thus, breastfed infants have lower risks of allergies and gastrointestinal, respiratory, and ear infections than formula–fed infants.[25] Furthermore, breastfed babies may be less likely to develop childhood asthma, leukemia, obesity, sudden infant death syndrome, and type 1 diabetes than infants who are not breastfed.

infant formula synthetic food that simulates human milk

colostrum initial form of breast milk that contains anti-infective properties

Infants who are breastfed have a lower risk of sudden infant death syndrome (SIDS) than infants who do not consume breast milk.
Jiang Jin/Purestock/SuperStock

Human milk is a rich source of lipids, including cholesterol, and fatty acids such as linoleic acid, *arachidonic acid (AA),* and *docosahexaenoic acid (DHA).* An infant's nervous system, especially the brain and eyes, depends on AA and DHA, and the vitamin–like nutrient choline, for proper development. Furthermore, the fat in breast milk helps supply the energy needed to maintain the infant's overall growth.

The practice of breastfeeding also provides some important advantages for parents, particularly the new mother. Breastfeeding is more convenient and economical than using infant formula. Human milk is readily available; there is no need to purchase cans of infant formula and have them on hand. As milk leaves the breast, it is always fresh, free of bacteria, and ready to feed without mixing, bottling, or warming. Because human milk production requires a considerable amount of energy, lactating females can lose the extra body fat gained during pregnancy faster than mothers who use infant formula. Additionally, females who breastfeed their babies have lower risks of breast cancer (before menopause) and ovarian cancer than females who do not breastfeed. Some of these benefits depend on whether a female breastfeeds exclusively, that is, provides no other foods, and the number of months the mother nurses her infant. Table 13.5 lists these and several other advantages of breastfeeding.

The Milk Production Process: Lactation

When an infant suckles, nerves in the mother's nipple signal her brain to release prolactin and **oxytocin** (*ox–e–tose'–in*) into her bloodstream. Prolactin stimulates specialized cells in breasts to form milk. These cells carry out the lactation process by synthesizing some nutrients and removing others from the mother's bloodstream and adding them to her milk. Oxytocin plays a different role in establishing successful lactation. This hormone signals breast tissue to "let down" milk. The *let–down reflex* enables milk to travel in several tubes (ducts) to the nipple area. A reflex is a physical response that is automatic and not under conscious control. When let–down occurs, the infant removes the milk by continued sucking (Fig. 13.6). Shortly before the flow of milk begins, the lactating female often feels a tingling sensation in her nipples, a signal that let–down is occurring.

Did *YOU* Know?

When a new mother breastfeeds her newborn immediately after delivery, oxytocin signals her uterus to contract, reducing the risk of excessive uterine bleeding.

TABLE 13.5 Advantages of Breastfeeding

Advantages for Infants
Human milk • Is free of bacteria as it leaves the breast. • Supplies antibodies and immune cells. • Is easily digested. • Reduces risk of food allergies, especially to proteins in infant formulas and cow's milk. • Changes in composition over time to meet the changing needs of a growing infant. • Contains zinc, iron, and other minerals in highly absorbable forms. • Decreases risks of ear, intestinal, and respiratory infections. • May reduce the risk of asthma, obesity, and type 1 diabetes in childhood.
Advantages for New Mothers
Breastfeeding • Reduces uterine bleeding after delivery. • Promotes shrinkage of the uterus to its prepregnancy size. • Decreases the risk of breast cancer (before menopause) and ovarian cancer. • May promote maternal weight loss. • May enhance bonding with the infant. • Is less expensive and more convenient than feeding infant formula.

oxytocin hormone that elicits the "let-down" response and causes the uterus to contract

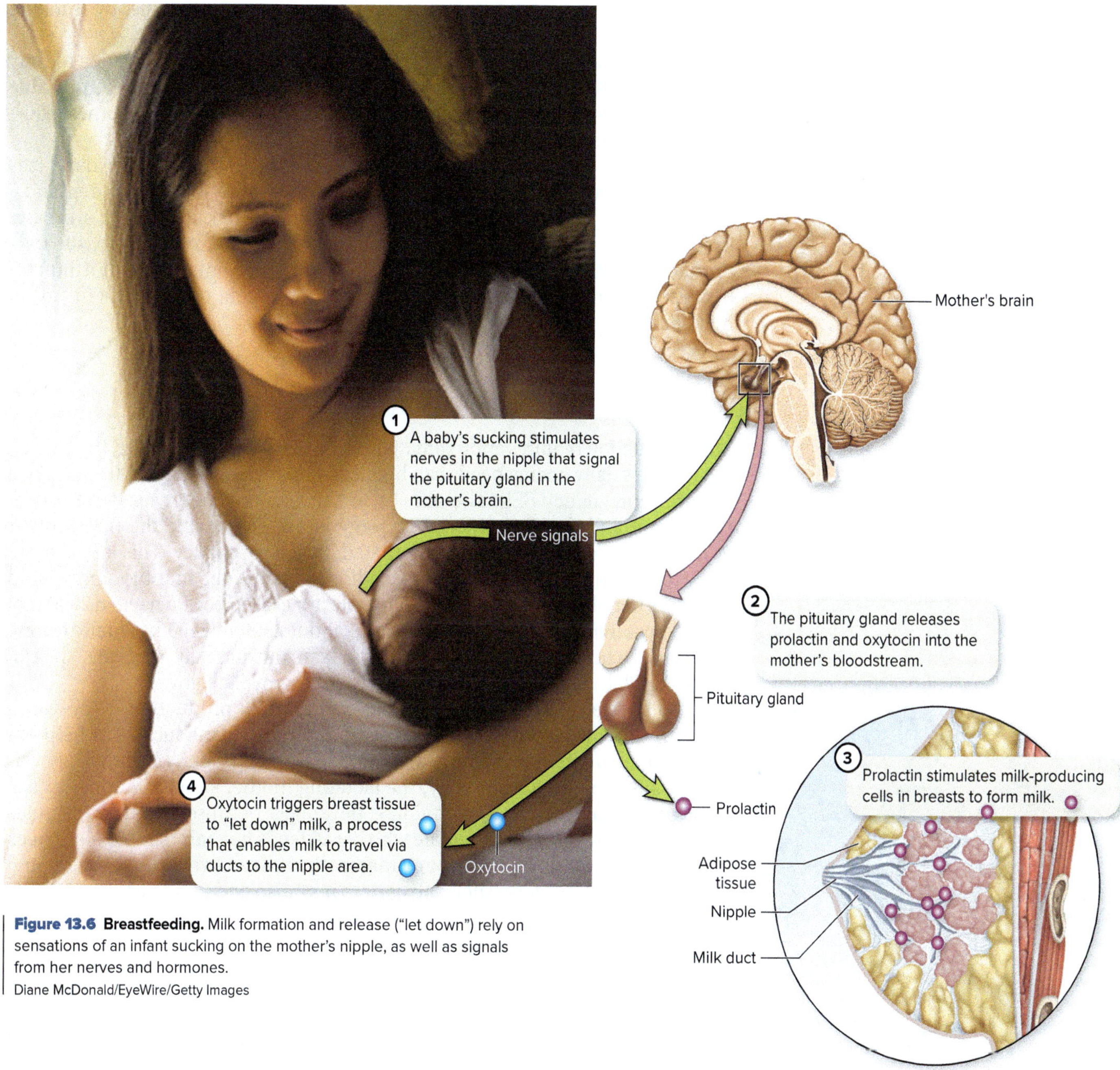

Figure 13.6 Breastfeeding. Milk formation and release ("let down") rely on sensations of an infant sucking on the mother's nipple, as well as signals from her nerves and hormones.
Diane McDonald/EyeWire/Getty Images

Embarrassment, emotional stress and tension, pain, and fatigue can easily block the let–down reflex. For example, if a lactating mother is tense or upset, let–down does not occur, and her infant will not be able to obtain milk when it suckles. When this happens, the hungry infant becomes frustrated and angry, and the mother may respond by becoming even more tense and upset, setting up a vicious cycle. At this point, new mothers often give up breastfeeding, reporting that they tried to suckle their babies but were unable to "produce" milk.

Lactating females need to be aware of the connection between their emotional state and failure to let down. To smooth the path to successful lactation, it helps if new mothers are in a comfortable, relaxed environment when they breastfeed their babies. When lactation and breastfeeding are well established, the let–down response often occurs without the need for suckling.

Did *YOU* Know?

The size of a female's breasts does not influence her ability to breastfeed her infant. However, certain surgical procedures used to enlarge or decrease breast size can disrupt the nerves and milk-producing tissue in the breasts.[26] Females who had surgery to alter their breasts may be able to produce milk after giving birth, but their infants' growth rates should be monitored to make sure the babies are obtaining enough milk.

For example, the mother's let–down reflex may be triggered just by thinking about nursing her infant or hearing it cry.

Breastfeeding is a skill, and like other skills, it takes some practice to fully master. Thus, it may take a few weeks for the new mother to feel comfortable with the process. By persevering, mother and baby are likely to become a successful breastfeeding team.

Typically, a lactating female produces over 3 cups of milk per day.[9] It is important to recognize that milk production relies on "supply and demand." The more the infant suckles (demand), the more milk its mother's breasts produce (supply). However, if milk is not fully removed from the breasts, milk production soon ceases. This is likely to occur when infants are not hungry because they have been given baby food and formula to supplement breast milk feedings.

Dietary Planning for Lactating Females

Milk production requires approximately 800 kcal every day. However, the lactating female's daily energy needs can be met by adding only about 300 to 400 kcal to her prepregnancy EER. The difference between the energy needed for milk production and the recommended energy intake can enable the new mother to lose the extra body fat she accumulated during pregnancy. This loss is more likely to occur if she continues breastfeeding her baby for 6 months or more and increases her physical activity level. A female, for example, who needed 2000 kcal before becoming pregnant would require about 2300 to 2400 kcal daily during lactation. To help plan meals and snacks that are nutritionally adequate, she can follow MyPlate recommendations at www.choosemyplate.gov.

No special foods are necessary to sustain milk production. However, a lactating female should drink fluids every time her infant suckles to help her maintain adequate milk volume and keep her body properly hydrated. For as long as she breastfeeds her baby, the lactating mother should limit her intake of alcohol– and caffeine–containing beverages because her body secretes these drugs into her milk. A female who breastfeeds her baby should also check with her physician before using any medications, even over–the–counter and herbal products, because such substances may also end up in her breast milk.

Figure 13.7 Vitamin D and infants. Exposing an infant's skin to some direct sunlight enables the baby's body to form vitamin D.
BananaStock/age fotostock

Is Breast Milk a Complete Food?

Registered dietitian nutritionists (RDNs) and pediatricians generally recommend that new mothers breastfeed their healthy infants exclusively for about 6 months.[25,27] It is not necessary to supplement healthy young infants' diets with other fluids, such as water, juice, and infant formula. After an infant reaches 4 to 6 months of age, the breastfed infant should also be given some appropriate solid foods. Breastfeeding may be combined with infant foods until the child's first birthday. However, there is no reason why children cannot be breastfed for longer periods. Throughout the world, many mothers continue to nurse their babies well past the babies' first birthdays, but in the United States, this practice is uncommon.

Although breast milk is highly nutritious, it is not a complete food for all infants. Human milk may contain inadequate amounts of vitamins D and B–12 and the minerals iron and fluoride. The American Academy of Pediatrics (AAP) recommends all breastfed infants be given a supplement that supplies 400 IU of vitamin D per day until they are consuming that amount of the vitamin from food or infant formula. Exposing the infant to some sun can also help meet part of the child's vitamin D needs (Fig. 13.7). If a lactating female is a total vegetarian and does not consume a source of vitamin B–12, she should consult her physician concerning the need for vitamin B–12 supplementation. When breastfed infants are about 6 months old, they should also be consuming some zinc– and iron–containing solid foods because the amount of zinc and iron in their mother's milk may no longer meet their needs. Furthermore, a fluoride supplement may be necessary for breastfed babies. Before giving any dietary supplements to their baby, parents or caregivers should discuss their infant's nutritional needs with the child's physician.

Did *YOU* Know?

A newborn has a tiny stomach. Typically, a healthy newborn will want to nurse every 1 to 3 hours. Within a few weeks, the baby's stomach's capacity increases, which enables the child to have longer intervals between feedings and fewer feedings per day.

Quitting Too Soon

Nearly all healthy females are physically capable of breastfeeding their infants. In 2019, 83% of American females started breastfeeding their babies soon after birth.[28] Within 2 days of birth, 19% of the breastfed babies were also consuming formula. By the time the infants were 6 months old, almost 56% continued to be breastfed. By their first birthday, approximately 36% of the babies were still consuming their mother's milk.

Females who breastfeed their newborns often stop the practice within 6 months. There are many reasons why females discontinue nursing their infants too soon. New mothers often quit because they lack information about and support for breastfeeding their babies. Some females discontinue breastfeeding because of uncertainty over how much milk their babies are consuming. Baby bottles are marked to indicate ounces, so a mother who bottle–feeds her infant can easily measure the amount of formula consumed. A lactating mother, however, has to observe her baby for cues indicating the child is full. When a breastfeeding baby is no longer interested in nursing and stops, its mother has to assume the infant is satisfied with the feeding. A well–nourished breastfed infant will gain weight normally and generally have six or more wet diapers, as well as one or two bowel movements that have soft stools per day. Parents or caregivers who are concerned about their infants' food intake or nutritional status should consult their physician immediately.

Caregivers can use a bottle to feed a baby breast milk that has been properly stored.
Ariel Skelley/DigitalVision/Getty Images

Many new mothers discontinue breastfeeding before their babies are 6 months old because they need to return to work and have caregivers feed their babies. Although lactating females can learn to express milk from their breasts and preserve it for later feedings, many workplaces do not have comfortable, private facilities for females to express milk and then store it safely.

To enhance the likelihood that a nursing mother continues to breastfeed, it is helpful to enlist the support of a female relative or friend who has successfully breastfed her children. Furthermore, the female's partner needs to understand and appreciate the function of the human breast as a source of nearly perfect nourishment for infants. New mothers are unlikely to begin and continue nursing their babies without their partners' support. La Leche League is an international organization dedicated to providing education and support for breastfeeding females (1–877–4–LALECHE or www.llli.org). Also, hospitals may employ lactation consultants or specialists. Lactation consultants are often nurses or registered dietitian nutritionists who are trained to provide information and advice about breastfeeding. For more information about breastfeeding, visit www.cdc.gov/breastfeeding/.

Did *YOU* Know?

Concentrated or ready-to-use infant formulas are highly perishable. Once a container of liquid infant formula has been opened, the product can remain in the refrigerator for only 1 day. Any formula that remains in the bottle after a feeding should be discarded because it will be contaminated by bacteria and enzymes in the infant's saliva.

Infant Formula Feeding

Not every female wants or is able to breastfeed her baby. Infant formulas are a safe and nutritionally adequate source of nutrients for babies who are not breastfed. To produce artificial milk for babies, infant formula manufacturers

Food & Nutrition tips

©Wendy Schiff

- Do not heat infant formula or human milk in a microwave oven. The heat can destroy immune factors in human milk and create hot spots that can scald an infant's tongue.
- If an infant does not finish a bottle of freshly expressed human milk during a feeding, the baby should be offered the remaining milk to drink within 2 hours. After that feeding, any milk that is still in the bottle should be discarded.

alter cow's milk to improve its digestibility and nutrient content. Infant formulas generally contain heat-treated proteins from cow's milk, lactose and/or sucrose, and vegetable oil. Infant formulas generally lack cholesterol, but some of these products have prebiotics and the fatty acids DHA and AA added to them. Vitamins and minerals are added to the product, and infant formulas generally contain higher levels of iron and calcium than human milk. Table 13.6 compares the nutrient and energy contents of human milk, fresh cow's milk, and a couple of cow's milk-based and soy-based infant formulas. Note the differences between human milk and the formulas; formulas are not identical to human milk. Furthermore, formula manufacturers have been unable to duplicate human antibodies and other unique immune system factors that are in breast milk.

An interesting feature of human milk is that its fat content changes during each feeding, which usually lasts about 20 minutes. In the beginning of the session, the mother's milk is lower in fat, but as her infant continues to suckle, the fat content of her milk gradually increases.[29] The higher fat content of the "hind milk" may make the baby feel satisfied and, as a result, discontinue feeding. Infant formulas, however, have uniform composition; that is, they do not change their fat content during a feeding session. Thus, the mother or infant caregiver is more likely to control the amount of formula the baby consumes, possibly leading to overfeeding. Nevertheless, the overall energy content of human milk is about the same as that of infant formulas (22 kcal per ounce).

Experts with the AAP recommend that caregivers provide an iron-fortified infant formula for babies who are not breastfed. Not all infant formulas contain iron, so it is important to read the product's label before purchasing it. Formula-fed babies may also need a source of fluoride, but caregivers should check with their infants' physicians before providing a supplement containing the mineral. For babies who are allergic to infant formulas made from cow's milk proteins, similar products made with soy or other proteins are available.

TABLE 13.6 Comparing Approximate Compositions of Human Milk, Cow's Milk, and Iron-Fortified Infant Formulas (per Ounce)

Milk or Iron-Fortified Formula	Energy (kcal/oz)	Protein (g/oz)	Carbohydrate (g/oz)	Fat (g/oz)	Cholesterol (mg/oz)	Iron (mg/oz)	Calcium (mg/oz)
Human milk, mature	22.0	0.32	2.12	1.35	4.31	0.01	9.86
Cow's milk, whole	18.6	0.96	1.46	0.99	3.05	0.01	34.5
Cow's milk, fat-free	11.3	1.09	1.54	0.07	0.61	0.02	39.5
Cow's milk protein–based formulas (ready to feed)							
Similac Advance®	20.0	0.41	2.06	1.12	0.61	0.36	16.1
Enfamil Neuropro Infant®	20.0	0.42	2.25	1.06	0.00	0.36	15.6
Soy protein–based formulas (ready to feed)							
ProSobee®	20.0	0.50	2.12	1.06	0.00	0.36	21.7
Similac Isomil Advance®	20.0	0.49	2.04	1.10	0.00	0.36	21.0

©Wendy Schiff

Source: Data from U.S. Department of Agriculture, Agricultural Research Service: *FoodData Central, 2019.* https://fdc.nal.usda.gov/. Accessed February 4, 2023.

What About Cow's Milk?

Why not feed fresh fluid cow's milk to an infant? Cow's milk is too high in calcium, sodium, potassium, and protein and does not contain enough vitamin E, iron, and essential fatty acids to meet an infant's nutrient needs.[30] In addition, infants have more difficulty digesting the fat and **casein** (*kay'–seen*), the major protein in cow's milk, than the fat and major proteins in human milk. Cow's milk can also contribute to intestinal bleeding and iron deficiency.[31] Thus, cow's milk should not be fed to infants until they are 1 year of age.[32]

Whole milk is a major source of saturated fat in young children's diets. According to the AAP, reduced–fat forms of milk can be given to children who are between 1 and 2 years of age, but only as part of a diet that supplies 30% of total calories from fat.[32] To reduce the risk of developing cardiovascular disease (CVD) later in life, children over age 2 years should consume low–fat or fat–free dairy products.

Commercially prepared baby foods are available in a variety of forms and packaging.
©Wendy Schiff

Allergies Allergies are immune system responses to the presence of foreign proteins in the body. Allergies to proteins in foods, especially cow's milk proteins, often begin in infancy and may persist through childhood. Signs and symptoms of food allergies typically include the following:

- Vomiting, diarrhea, intestinal gas and pain, bloating, or constipation.
- Skin rash called eczema (*ek'–zuh–mah*).
- Runny nose and breathing difficulties, such as asthma.

Compared to breastfed infants, formula–fed babies have a greater risk of cow's milk allergy and eczema.[33] When a female has a personal or family history of food allergies, she may be able to prevent her children from developing such allergies if she breastfeeds her babies exclusively for 4 months.

Did *YOU* Know?

Many parents think adding solid foods to infants' diets helps babies sleep through the night. Actually, this developmental milestone generally occurs around 3 to 4 months of age, regardless of what infants are eating.

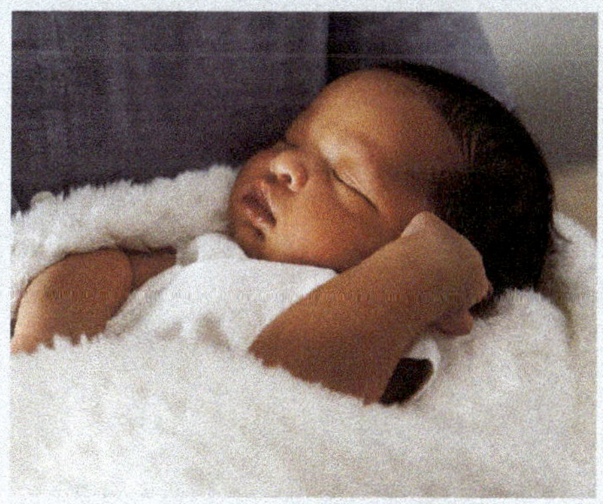

Monkey Business Images/Shutterstock

Introducing Solid Foods

Before about 6 months of age, babies' nutritional needs can generally be met with human milk or infant formula. Solid ("complementary" or "baby") foods should not be introduced to infants until they are about 4 to 6 months of age.[33] At this age, many infants need the additional calories supplied by solid foods. Breastfed babies may also need a dietary source of zinc and iron because their stores of the mineral are usually exhausted about 6 months after birth. Nevertheless, caregivers should continue to provide human milk or iron–fortified infant formula as the foundation of the baby's diet for the first year.

Many new parents are anxious to start feeding their young infants solid food. However, babies are not physically mature enough to consume solids before they are 4 to 6 months of age. For example, a baby's kidney functions are quite limited until the child is about 4 to 6 weeks of age. Additionally, an infant's digestive tract cannot readily digest starch before the child is about 3 months old.

Infants are born with the **extrusion reflex,** an involuntary response that occurs when a solid or semisolid object is placed in an infant's mouth. As a result of this reflex, a young baby thrusts its tongue forward, pushing the object out of its mouth. Thus, trying to feed the infant solid foods with a spoon is a messy, frustrating process, as the child automatically pushes the food out of its mouth. Liquid foods, such as breast milk or infant formula, do not elicit the extrusion reflex, so the baby swallows fluids.

As the infant reaches 6 months of age, the extrusion reflex disappears, and the child develops the physiological abilities to digest a wider range of foods and metabolize the nutrients they contain. Moreover, a 6–month–old infant can usually sit up with back support and coordinate muscular control

casein major protein in cow's milk

extrusion reflex involuntary response in which a young infant thrusts its tongue forward when a solid or semisolid object is placed in its mouth

weaning gradual process of shifting from breastfeeding or bottle-feeding to drinking from a cup and eating solid foods

over their mouth and neck movements. These signs indicate the baby is ready physically to eat pureed or soft foods from a spoon, is less likely to choke on such foods, and can turn their head away from food when full. **Weaning** is the gradual process of shifting an infant from breastfeeding or bottle–feeding to drinking from a cup and eating solid foods.

Typical first foods should be rich in iron and zinc, including soft, cooked, fortified cereal and ground meat or chicken mixed with breast milk or formula. Vegetables should be cooked and pureed or mashed, such as green beans, squash, or commercially prepared baby vegetables. Fresh, peeled, mashed fruit, such as ripe bananas or avocados, are great first foods. The second Food & Nutrition Tips feature in this section provides some basic advice for feeding infants.

Each new food should be introduced into a baby's diet one at a time. If the child consumes the food for about 3 to 5 days and does not show signs of an allergic response, such as diarrhea, vomiting, eczema (skin rash), or runny nose, the caregiver can consider that food safe for the child to eat. The caregiver should continue to introduce a single new food into the infant's diet over a period of several days and watch the baby for signs of allergic reactions.

In the past, highly allergenic foods, such as peanut butter, fish, and eggs, were not introduced into a baby's diet until the child was at least 1 year of age. However, results of current studies indicate that *delaying* such foods may *increase* the child's risk of food allergies. According to experts with the National Institutes of Health (NIH), infants who have severe eczema, egg allergy, or both conditions are at high risk of developing peanut allergy. Such high–risk infants should be fed a small amount of a peanut–containing food (such as a mixture of smooth peanut butter and water) when they are 4 to 6 months of age.[34] This practice has been shown to prevent the infants from developing peanut allergy later in childhood. Caregivers should always check with the infant's physician for advice on when to start feeding a new food to the child.

Figure 13.8 Learning to drink from a "sippy" cup.
Amanda Ice

At about 6 to 8 months of age, the baby's first set of teeth, the "primary teeth," begin to appear. These teeth are important for proper nutrition because they help the child bite and chew food. By 8 to 12 months of age, most infants can feed themselves, using their hands to grab food or their thumb and index finger to pick up items such as crackers, strips of cooked meat or poultry, and cooked string beans. Babies can also hold a bottle and practice drinking from a special cup ("sippy cup") that has a lid with a spout (Fig. 13.8). Babies need to practice self–feeding skills, even if it means playing with food and creating messes. By about 10 months of age, many infants are mastering self–feeding. In most cases, a healthy infant can be offered foods that the rest of the family consumes, as long

mae_chaba/Shutterstock

Food & Nutrition tips

When feeding solid foods to an infant:

- Use a baby-sized spoon—a small spoon with a broad handle.
- Seat the child in an upright position to ease swallowing.
- Add some breast milk or infant formula to the food, and place a small dab of the semisolid food on the spoon's tip. Gently place the spoon on the infant's tongue and tilt it so the food slides onto the tongue. If the infant spits it out, the baby is not ready to eat solid foods, so do not continue with the feeding.
- When the infant accepts and swallows the food, expect the child to take only two or three bites during these early feeding sessions.

as the items are part of a healthy diet (i.e., nutrient–dense and low in sodium with no added sugars), age appropriate, and in small amounts.

Babies need to practice self-feeding skills, even if it means playing with food and creating messes.
Andrew Olney/OJO Images/age fotostock

What Not to Feed an Infant

By the end of the first year, an infant should be consuming many different nutrient–dense foods—grain products, meats, fruits, and vegetables—along with breast milk or infant formula. Introducing a baby to various foods helps the child learn about different tastes, odors, and textures. However, certain foods and beverages are not appropriate for infants. **Avoid feeding an infant the following things:**

- **Honey.** This product may contain spores of *Clostridium botulinum* that can produce a potentially fatal toxin in infants (children who are under 1 year of age [see Chapter 5]).
- **Excessive infant formula or human milk.** Depending on their age, most infants need less than 30 ounces of human milk or infant formula daily. A child who drinks too much milk may not eat enough solid foods that contain nutrients lacking in milk.
- **Semisolid baby food in a baby bottle that has the nipple opening enlarged.** This practice contributes to overfeeding and does not help the child learn self–feeding skills.
- **Candy, flavored gelatin water, or soft drinks.** These items provide few micronutrients and too much added sugar. According to the Dietary Guidelines, children should not consume added sugars before they are 2 years of age.
- **Small pieces of hard or coarse foods.** Pieces of food that are difficult to chew, whole nuts, grapes, chunks of cooked meat, raw carrot "coins," and popcorn can cause choking. Caregivers should supervise meals to keep young children from stuffing too much food in their mouths.
- **Fruit juice.** Babies should not be given fruit juices before they are 1 year of age. Fruit juice lacks the fiber content of whole or mashed fruit, and children may consume the beverages instead of more nutritious foods. Also, the fructose and sorbitol (a sugar alcohol) in pear, apple, and prune juice may cause diarrhea, gas, and abdominal pain.
- **Unpasteurized (raw) milk, yogurt, cheese, or juice.** These beverages and foods may be contaminated with bacteria or viruses.
- **Goat's milk.** Goat's milk is low in iron, folate, and vitamins C and D.
- **Plant–based milk alternatives.** These beverages, typically made from oat, rice, coconut, or almond, should not be offered to infants the first year of life. They do not provide the same calories or nutrients as human milk or infant formula and, thus, do not support proper growth and development.[35]

Baby Bottle Caries

At bedtime, many caregivers place infants in their cribs with a baby bottle containing formula, juice, or a sugar–sweetened beverage. This practice is not recommended because the sleepy infant sucks slowly, allowing the carbohydrate–containing fluid to bathe the child's teeth and provide a source of nutrients for bacteria that stick to teeth. These bacteria produce acids that dissolve tooth enamel, causing cavities to form in the teeth (*dental caries*). Dentists often refer to this condition as "baby bottle caries." To reduce the risk of baby bottle caries, infants should be given only water in their bedtime bottles.

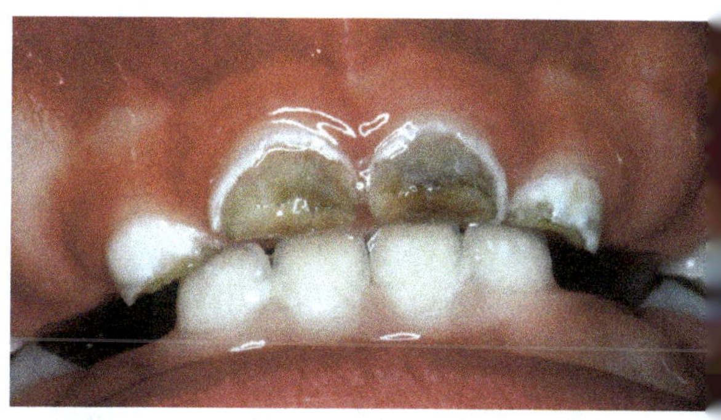

Baby bottle caries.
Ted Croll/Science Source

Elizaveta Vladimerovna Koifman/123RF

Concept CHECKPOINT

11. What is the "let-down reflex"?
12. How does lactation affect a new mother's energy needs?
13. What is colostrum, and why is it a valuable first food for breastfed babies?
14. RDNs recommend that healthy infants be breastfed exclusively during their first _____ months of life.
15. List at least five benefits that infants derive from breastfeeding.
16. Identify at least three benefits that women derive from breastfeeding their babies.
17. Compare the energy, macronutrient, and calcium contents of an ounce of human milk with those of an ounce of cow's milk.
18. Identify at least three physiological indications that an infant is ready to eat solid foods.
19. How much should a healthy infant's weight increase during its first year of life?

See Appendix F for responses.

13.4 The Preschool Years

Learning Outcomes

1. Identify some major nutrition-related health concerns facing American preschool children.
2. Summarize practical suggestions for encouraging healthy eating habits among preschool children.
3. Describe steps caregivers can take to improve the nutritional status of preschool children.

Childhood can be divided into the preschool period (2 to 5 years of age), the school–age period (6 to 11 years of age), and adolescence (12 to 19 years of age). In this section, we focus on some major nutrition concerns of preschool children. The rapid growth rate that characterizes the first 12 months of life tapers off quickly during the preschool years and proceeds at a slow but steady rate until the end of childhood. If an average infant's growth rate did not slow down, they might weigh about 190 pounds and be about 5 feet, 7 inches tall by 3 years of age! However, the average 3–year–old weighs about 32 pounds and is about 3 feet in height.

As the growth rate slows after infancy, preschoolers' appetites decrease because they do not need as much food. Parents and other caregivers must recognize that a 3–year–old child should not be expected to eat as eagerly as they did as an infant. Furthermore, children do not have the stomach capacity to eat adult–size portions of foods.

Figure 13.9 Fun food. A slice of whole-grain bread topped with nut butter and fruit is the basis of a healthy meal for a preschool-age child.
©Wendy Schiff

When planning meals and snacks for children, caregivers should emphasize nutrient–dense foods, such as nuts and nut butters, lean meats, low–fat dairy products, whole–grain breads and cereals, fruits, and vegetables (Fig. 13.9). Although many ready–to–eat cereals are sweetened with sugar, it is not necessary to eliminate such foods. Caregivers, however, should read product labels and choose whole–grain varieties with less added sugar. Additionally, it is important to monitor children's intake of sweets because sugary items can crowd out more nutritious foods from their diets. The *Dietary Guidelines for Americans* recommend children not consume foods sweetened with low– and no–calorie sweeteners, also called high–intensity sweeteners (see Chapter 4), for the first 2 years of life. During these early preschool years, food preferences are being developed, and a child fed foods sweetened with high–intensity sweeteners may build a preference for highly sweetened foods.[35]

MyPlate plan provides a day's food group selections for healthy preschoolers that can be adjusted for their age, sex, and physical activity level (www.myplate.gov/myplate–plan). Three–year–old males, for example, generally need 1200 to 1400 kcal/day, depending on their physical activity level. Three–year–old females generally need 1000 to 1400 kcal/day, depending on their level of physical activity.

Snacks

Snacking is not necessarily a bad habit, especially if snacks are nutrient dense and fit into the child's overall diet. Nutritious snacks can be offered at midmorning or midafternoon, when the child is likely to become hungry between meals. A healthy 4– or 5–year–old child can safely eat raw vegetables without fear of choking. Nutritious dips, such as the yogurt dip in this chapter's "Recipe for Healthy Living," may make raw vegetables more appealing to children. Table 13.7 lists some nutritious snacks that children tend to like.

Fostering Positive Eating Behaviors

Parents often refer to their preschool children as "picky eaters" because the youngsters do not eat everything offered to them. Furthermore, it is not unusual for children to have "food jags," periods in which they refuse to eat a food that they liked in the past or want to eat only a particular food, such as peanut butter and jelly sandwiches or cereal and milk. Picky eating and food jags may be expressions of a child's growing need for independence. Caregivers should avoid nagging, forcing, and bribing children to eat. Instead, caregivers can offer the children a variety of healthy foods each day and allow the youngsters to choose which items and how much to eat.

Many children, especially preschool children, resist eating new foods. The temperature, appearance, texture, and taste of a food influence whether children will sample it. For example, young children often reject lumpy or hot–temperature foods. Sometimes children object to having foods mixed together such as in stews and casseroles, even though they like the ingredients when served separately. It is not unusual for young children to dislike vegetables that have strong flavors or odors, such as broccoli, onions, and asparagus. The idea

Hummus with vegetable sticks is a nutritious snack.
©Wendy Schiff

TABLE 13.7 Nutritious Snacks

- Nut butters spread on graham crackers
- Fruit smoothies (see Chapter 4 "Recipe for Healthy Living")
- Fruit salad (or cut-up fruit)
- Mini-pizzas (half an English muffin, topped with tomato sauce and mozzarella cheese, and heated in toaster oven or microwave oven)
- Plain, low-fat yogurt topped with granola or fresh fruit
- Peanuts, cashews, or sunflower seeds
- Cheese melted on whole-wheat crackers
- Dried fruit
- Trail mix
- Reduced-sugar or no-added-sugar ready-to-eat cereal
- Vegetable sticks dipped in hummus

MY DIVERSE PLATE

A recent study compared the snacking practices of over 3000 American and Chinese children who were between 4 and 13 years of age.[36] Nearly all of the American children who participated in the study consumed snacks, including fruit, flavored-water beverages, and candy. For the American children, snacks contributed about 25% of their daily calorie intake. The young Chinese subjects of the study were less likely to snack than the American children. Furthermore, the Chinese children consumed only 10% of their daily energy intake from snacks. Fruit was the leading calorie-containing snack of the Chinese children, whereas candy was the leading calorie-containing snack of the American youngsters. Regardless of one's age, consuming too many sugary snacks can contribute to an unhealthy amount of body fat.

Hieng Ling Tie/age fotostock

Eating vegetables may become more appealing when children help grow, select, or prepare fresh produce.
Ariel Skelley/Blend Images/Image Source

of eating vegetables may become more appealing when children help grow, select, or prepare fresh produce. Nevertheless, it is important to recognize that everyone, including a child, is entitled to dislike certain foods.

The social atmosphere can make mealtimes enjoyable or unbearable, which in turn influences a child's desire to eat. Mealtimes should be happy, social occasions to enjoy healthful foods with parents or other caregivers. The kitchen table should not become a battleground in which adults use threats or bribes to force children to eat unfamiliar foods. For example, avoid telling a child, "You'll just sit there until you clean your plate" or "You can have a cupcake if you eat your peas." When a child refuses to eat, have the youngster remain at the table for a while. If the child continues to be disinterested in eating, remove the food and wait until the next scheduled meal or snack.

Healthy children are not in danger of starving if they skip a meal. A child who is not hungry at mealtimes may have eaten a snack before the meal at a friend's home. If a child's lack of appetite persists, caretakers should consult the child's physician to rule out illness.

Food & Nutrition tips

- Foods with bright colors, crisp textures, and sweet or mild flavors usually appeal to children. When planning meals, consider including foods with these attractive characteristics.
- As a parent or caregiver, encourage children to try at least one bite of each new food provided at mealtimes. It can take 8 to 10 offerings of a food before a child will accept it.[35] Offering children the same food in various forms (raw or cooked) or prepared in different ways can increase food acceptance.
- To stimulate your child's appetite, try serving food on a small colorful plate that is designed to appeal to young children.
- Keep in mind that you are your children's role model for food choices and physical activity habits. If you eat a variety of nutrient-dense foods and are physically active, your children are likely to eat such foods and be active as well.

Common Food-Related Concerns

Nutrition–related problems that often affect preschool children are iron deficiency, dental caries, food allergies, and obesity. Section 13.7 focuses on obesity during childhood and adolescence.

Iron Deficiency

Iron deficiency can lead to decreased physical stamina, learning ability, and resistance to infection. The best way to prevent iron deficiency in children is to provide foods that are good sources of iron, such as lean meat and enriched breads and cereals. Milk and other dairy products are poor sources of iron, so caregivers may need to limit daily servings of foods from this food group. Healthy preschool children should consume 2 to 2.5 cups/day of fat–free or low–fat milk or equivalent dairy products, which include milk substitutes such as soy beverages. Chapter 9 provides more information about iron deficiency.

Dental Caries

Many preschool children have had one or more dental caries by the time they enter school. If dental caries are not treated, jaw pain, gum infection, and tooth loss can occur. The following tips can help reduce the risk of dental caries in children:

- Brush teeth with a pea–sized amount of fluoride–containing toothpaste twice daily.
- Provide routine pediatric dental care and fluoridated drinking water.
- Any carbohydrate can contribute to dental caries, so reduce the number of snacks the child eats. Also, provide the child with more nutrient–dense snacks, such as raw fruits and vegetables or pieces of cheese.

Concept CHECKPOINT

20. What is a "food jag"?
21. Discuss effects that iron deficiency can have on children.
22. List at least three steps caregivers can take to reduce the risk of dental caries in children.

See Appendix F for responses.

13.5 School-Age Children

Learning Outcomes

1. Compare the typical eating patterns of school-age children with those of preschool children.
2. Identify some major nutrition-related health concerns facing American school-age children.
3. Summarize practical suggestions for encouraging healthy eating habits among school-age children.

Many school–age children adopt diets that are nutritionally inadequate.[37] Compared to preschoolers, older children often skip breakfast. Furthermore, school–age children typically consume more foods away from home and more fried items and sugar–sweetened beverages than younger children. Diets of school–age children tend to provide excessive amounts of solid fat, added sugars, and sodium.[37] Excessive intakes of fat and sugar contribute to the development of obesity, and high intake of sodium may contribute to hypertension among children.

Children who routinely eat breakfast are more likely to have healthier body weights than children who skip this meal.
Ariel Skelley/Blend Images LLC

School–age children often do not eat recommended amounts of fruits and vegetables, and the youngsters typically consume less than recommended amounts of dietary fiber. Low fiber intake contributes to constipation. In children, constipation is defined as having dry, lumpy, or hard stools that are difficult to eliminate or having fewer than two bowel movements per week.[38] Iron and calcium supplements, as well as certain medications, including narcotics, contribute to constipation among children. In otherwise healthy children, adding fiber to the diet and giving enough water to prevent dehydration may help avoid constipation. Caregivers should not give laxatives to a child unless under the direction of the child's physician.

Caregivers can help improve children's diets by encouraging youngsters to eat breakfast regularly. Children who routinely eat breakfast are more likely to have better diets and healthier body weights than children who skip this meal.[39] Breakfast menus do not need to feature traditional fare such as bacon, eggs, waffles, or pancakes. For example, leftovers from the previous night's dinner can be eaten for breakfast. Convenient "fast breakfasts" that school–age children can prepare quickly include ready–to–eat cereal with milk and fruit, cottage cheese and fruit, a peanut butter and jelly sandwich, or yogurt topped with trail mix or pieces of fresh fruit.

To help with menu planning for children, MyPlate Plan provides a site that enables caregivers to find the recommended number of daily servings of grains, protein foods, fruits, vegetables, and dairy items for children. The food group recommendations are based on the child's age (2 years of age or older), sex, weight, height, and physical activity level. To access this site, visit www.myplate.gov/myplate–plan.

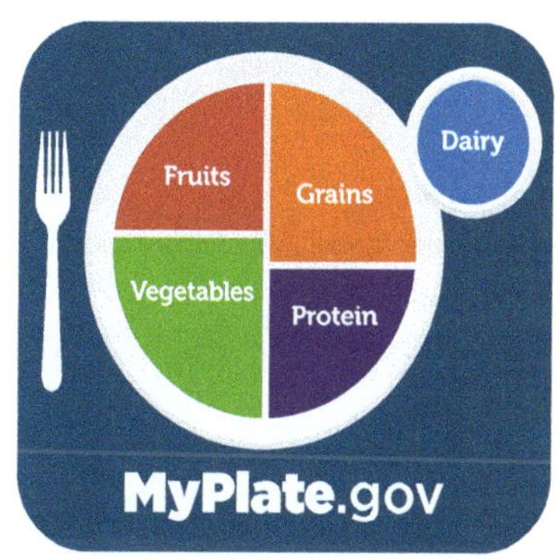

U.S. Department of Agriculture

Parents and other caregivers need to be concerned about foods that are available at school and obtain answers to the following questions: What kinds of foods are offered for school breakfasts and lunches? Does the school have vending machines accessible to youngsters? If so, what kinds of foods and beverages are sold from these machines? Do vending machines offer competitively priced, nutrient–dense foods? What can be done to ensure that the machines and the school's cafeteria provide nutritious foods that children will eat? Furthermore, what efforts are being made to teach children about proper nutrition while they are in school?

Parents and other caregivers need to be concerned about their children's food choices, especially while they are at school.
Stuart Pearce/Pixtal/age fotostock

Food & Nutrition tips

The following tips can help you improve your child's diet:

- Guide your family's food choices instead of dictating what they eat.
- Eat meals together as a family as often as possible.
- If necessary, reduce the amount of fat, especially saturated and trans fat, in your family's diet.
- Do not place your child on a restrictive diet unless the diet is recommended by the child's physician.
- Avoid using food as a reward or punishment.
- Encourage the child to drink water instead of sugar-sweetened beverages.
- Keep healthy snacks (such as fat-free or low-fat milk, fresh fruit, and vegetables) on hand.
- Serve at least five servings of fruits and vegetables each day.
- Discourage eating meals or snacks while watching TV or playing electronic games.
- Encourage the child to eat a nutrient-dense breakfast daily.

Concept CHECKPOINT

23. Discuss how young children's eating patterns often change when they enter school.

24. List at least three tips for improving diets of school-age children.

See Appendix F for responses.

13.6 Adolescence

Learning Outcomes

1. Discuss puberty and how it affects an adolescent's energy and nutrient needs.
2. Discuss some major nutrition-related health concerns facing American adolescents.

adolescence life stage in which a child matures physically into an adult

Adolescence is the life stage in which a child matures physically into an adult. During adolescence, the reproductive organs increase in size and begin functioning properly. Furthermore, individuals typically attain their full height by the end of adolescence. During this life stage, youth also develop emotionally, intellectually, and socially as they prepare for their adult roles.

Healthy adolescents learn to function independently of their adult caregivers. Thus, youths face a variety of lifestyle choices, including decisions regarding eating and physical activity habits. Such decisions often set the stage for the quality of their health in adulthood. For many teens, however, pressure to conform to fads and be influenced by other adolescents ("peer pressure") negatively affects their diets and overall health.

Puberty signals the end of childhood. Most males begin puberty when they are between 9 and 14 years of age; most females begin puberty between 8 and 13 years of age.[40] Puberty is a period characterized by dramatic physical changes, including increases in height and weight, known as the adolescent "growth spurt."

Most females begin their growth spurt between 10 and 13 years of age. Males begin their growth spurt later than females—generally when they are between 12 and 15 years of age. Females usually begin menstruating during their growth spurt. A female's skeletal growth is almost complete about 2 years after her first menstrual period, whereas males typically continue to gain stature until they are in their early twenties. The timing of puberty and growth spurts can vary widely, primarily due to genetic, environmental, and nutritional factors. Figure 13.10 shows a group of adolescents who are about the same age but are at different stages of physical maturity.

The amount of energy and nutrients required during adolescence reflects the individual's stage of growth and physical activity level. Early in adolescence, when a female is at the peak of her growth spurt, she typically consumes more energy than a male of the same age, especially if she is active in sports. By the time females' growth rate slows, males' growth spurt begins. As a result, adolescent males' food intake generally surpasses that of adolescent females. If maturing adolescents choose to eat nutritious foods and maintain a high level of physical activity, they can take advantage of their increased hunger and gain lean body mass without gaining excess body fat.

Did *YOU* Know?

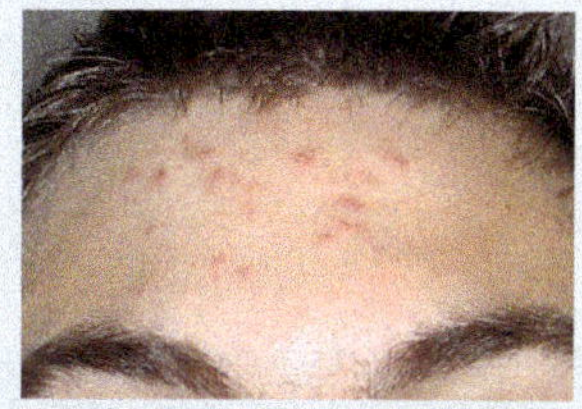

PhD.Neal R. Chamberlin/McGraw Hill

Many teenagers are plagued by *acne*—pimples, blackheads, and reddening of the skin—that often occur on the face, upper back, and chest. The cause of acne is unknown, but hormonal, emotional, and genetic factors may contribute to the condition's development. In the past, dermatologists (physicians who treat skin disorders) did not think diet played much of a role in acne. According to a few studies, diets that limit fluid milk consumption and/or have a low glycemic index may reduce acne breakouts.[41] See Section 5.8 of Chapter 5 for information about the glycemic index.

Nutrition-Related Concerns of Adolescents

People generally establish their future eating habits and physical activity practices when they are teenagers. According to results of a nationwide survey, many American youth are not following healthy diets or getting enough exercise. In 2019, about 31% of the high school students did not consume milk during the week before they were questioned about their eating habits.[42] More females than males did not consume milk. About 15% of high school students drank one or more servings of a carbonated, sugar–sweetened soft drink each day of the week before the survey. About 8% of the teenagers did not eat vegetables other than fried potatoes, potato chips, or French fries during the previous week. Almost 67% of the students skipped breakfast each day during the same period of time. According to findings of this survey, only about 56% of the high school students met the recommended levels of physical activity on 5 or more of the 7 days before the survey.

Adolescents who are physically inactive and whose diets rely heavily on unhealthy foods purchased at fast–food restaurants, convenience stores, or vending machines may be setting the stage for the development

Figure 13.10 Different rates of physical maturity. Although these adolescents are about the same age, they are in different stages of physical maturity.
Cathy Yeulet/123RF

of obesity, type 2 diabetes, heart disease, and other serious chronic diseases later in life. Obesity, eating disorders, and low iron and calcium intakes are major nutrition-related concerns of adolescents. Youth, especially teenage females, are at risk of developing disordered eating practices and eating disorders. You can learn more about eating disorders by reading the "Nutrition Matters" section of Chapter 10. In this chapter, Section 13.7 discusses adolescent obesity.

Adolescents should consume 3 cups/day of fat-free or low-fat milk or equivalent dairy products.
Fabrice Lerouge/Onoky/Corbis

Iron and Calcium Intakes

Adolescent males may become iron deficient during their growth spurt because their iron intakes do not keep up with their bodies' needs for the mineral. Adolescent females are also at risk of iron deficiency, especially if their diets lack iron-rich foods and they have heavy menstrual blood losses. Iron deficiency leads to increased fatigue and decreased ability to concentrate and learn. Teenagers need to understand why iron is important for good health and incorporate reliable food sources of the mineral in their diets (see Chapter 9).

Over the last 20 years, many adolescents have switched from drinking milk to drinking soft drinks. RDNs and other nutrition experts are concerned that these adolescents have inadequate calcium intakes because of this practice. Inadequate calcium intake during adolescence is associated with decreased bone mass and increased likelihood of bone fractures later in life. To encourage youth to consume recommended amounts of calcium, parents or other caregivers should explain the importance of the mineral to bone health and provide calcium-rich foods and beverages during meals and snacks. Adolescents should consume 3 cups/day of fat-free or low-fat milk or equivalent dairy products. Furthermore, teenagers need to be aware that physical activity can strengthen bones, whereas smoking cigarettes is a risk factor for *osteoporosis,* a condition characterized by loss of bone density. To learn more about this condition, see the "Major Minerals" section of Chapter 9.

Vegetarianism

Some teenagers adopt vegetarian diets as a way of defining their identity and asserting independence from their caregivers. Although vegetarian diets can be healthy alternatives to the typical American diet, some youth use vegetarianism to mask disordered eating behaviors (see Chapter 10's "Nutrition Matters" section).[43] When planning their diets, teenage vegans need to include foods that supply adequate amounts of calcium, iron, and zinc, and vitamins D and B-12. For information to help plan well-balanced, nutritionally adequate diets, teens can use the recommendations of the MyPlate food guide that are appropriate for their age, sex, and physical activity level. Chapter 7 discusses vegetarianism in more detail.

Concept CHECKPOINT

25. At what age does puberty usually occur in males? At what age does puberty generally occur in females?

26. Why are intakes of iron and calcium important during adolescence?

See Appendix F for responses.

13.7 Childhood Obesity

Learning Outcomes

1. Provide definitions for *overweight, obesity,* and *extreme obesity* in childhood and adolescence.
2. Identify health consequences of childhood obesity.
3. Discuss factors that contribute to childhood obesity.
4. Discuss strategies for preventing and treating childhood obesity.

Over the past few decades, U.S. public health officials became concerned about the increasing prevalence of obesity among children and adolescents ("childhood obesity"). Over 19% of American children who are between the ages of 2 and 19 years are obese.[44] Between 1988–1994 and 2017–March 2020, the prevalence of obesity among American children who were between 12 and 19 had increased dramatically (Fig. 13.11).

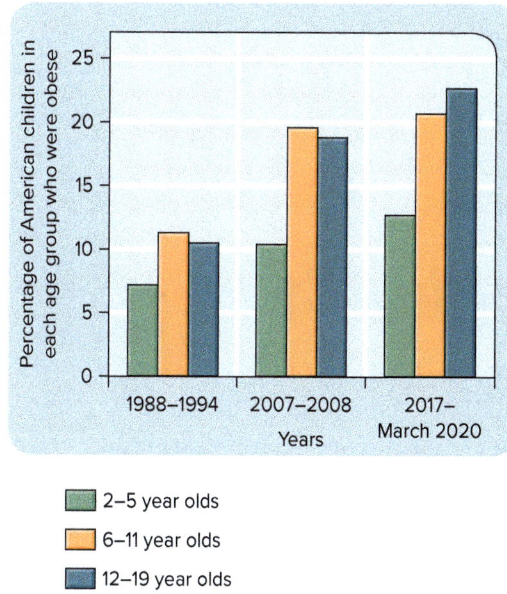

Figure 13.11 Prevalence of obesity among American children aged 2 to 19 years, for selected years 1988–1994 through 2017–March 2020 (prepandemic COVID-19).

Source: www.cdc.gov/nchs/data/hestat/obesity_child_07_08/obesity_child_07_08.htm, www.stacks.cdc.gov/view/cdc/106273. Accessed February 11, 2023.

Defining Obesity in Children

Health care professionals use BMI–for–age charts that are available from the CDC to determine children's and adolescents' weight status. The BMI for children is calculated in the same way as for adults, but BMIs for children are plotted on sex–specific growth charts (see Appendix E). Children and adolescents whose BMI–for–age is greater than or equal to (≥) the 85th percentile and less than (<) the 95th percentile are overweight, and youngsters who are at or above the 95th percentile are obese (Table 13.8). There is no generally accepted definition for extreme obesity among children. However, a BMI at or above 120% of the 95th percentile has been used to identify children who have extreme obesity.[45]

TABLE 13.8 Weight Status Classifications: Children and Adolescents (Ages 2 to 19)

Age	BMI-for-Age Percentile
Desirable weight	> 5th to < 85th
Overweight	≥ 85th to < 95th
Obese	≥ 95th

Health Problems Associated with Childhood Obesity

Compared to children who have healthy body weights, children and adolescents who are obese are more likely to have elevated blood pressure, cholesterol, and glucose levels.[46] These chronic conditions are risk factors for CVD. Obesity also affects children's psychological and social health. Children who have obesity, for example, are more likely to experience weight–based bullying and teasing than children who are not obese.[46] Table 13.9 lists serious health problems that children who are obese are more likely to have than children who have healthy body weights.

Many children and adolescents who are obese do not "grow out" of their excess body fat. Such children are more likely than children who have healthy weights to be obese as adults.[46]

TABLE 13.9 Serious Health Problems Associated with Childhood Obesity

- Impaired glucose tolerance, insulin resistance, and type 2 diabetes
- Elevated blood lipids
- Breathing problems, including sleep apnea and asthma
- Musculoskeletal problems, including joint discomfort
- Fatty liver disease, gallstones, and gastroesophageal reflux (heartburn)
- Social and psychological problems, such as discrimination and poor self-esteem

Source: Centers for Disease Control and Prevention, Overweight & obesity: *Obesity consequences*. Updated July 2022. www.cdc.gov/obesity/basics/consequences.html. Accessed February 11, 2023.

Childhood Obesity: Contributing Factors

There is no single cause of excess body fat in children, but researchers have identified multiple factors that are associated with the development of the condition. As in cases of adult obesity, the development of childhood obesity results from complex interactions among multiple factors (see Chapter 10). The following sections examine the roles of genetic, biological, and environmental factors that contribute to childhood obesity.

Genetic and Other Biological Factors

Scientific evidence supports the role of inherited (genetic) and other biological factors in the development of childhood obesity, especially in cases of obesity that begin early in life.[47] Some of the biological factors that contribute to obesity include:

- **Having parents who are overfat.** Children who have parents that are obese have a greater risk of obesity than children whose parents have healthy weights. If both parents are obese, the child has 10 times the risk of obesity as a child who has only one obese parent.
- **Having a mother who was overfat during pregnancy.** A mother's weight during pregnancy is a more important predictor of childhood obesity than the father's body weight. When compared to children whose mothers have a healthy body weight during pregnancy, babies born to mothers who are overweight are nearly three times as likely to be overweight. Furthermore, females who are obese are more likely than females who have healthy body weights to have large babies, and such infants have a high risk of childhood obesity.
- **Having a mother who gained too much weight and/or had diabetes during pregnancy.** Females who gain too much weight during pregnancy or have diabetes are more likely to give birth to high–birth–weight babies. Such children are also at risk of developing excess body fat in childhood.
- **Having a mother who smoked during pregnancy.** Although the reasons are unclear, females who smoke during pregnancy set the stage for the future development of overweight and obesity in their babies.
- **Being undernourished during prenatal development.** Poor nutrition (undernutrition) during pregnancy contributes to delivery of a low–birth–weight infant. Such infants are more likely to develop hypertension, CVD, and type 2 diabetes later in life, which are chronic diseases associated with obesity. At this point, scientists have been unable to explain why prenatal undernutrition contributes to hypertension, CVD, and type 2 diabetes in adulthood.

Being physically active can help children who are overweight or obese lose excess body fat.
Ingram Publishing/SuperStock

Biological factors, however, are not entirely responsible for the current obesity epidemic. In many cases, it is difficult to determine whether genes play a more important role than environmental factors in the development of childhood obesity. Is a child obese because the youngster inherited genes from their parents that "program" for obesity? Or is a child obese because their caregivers provide an excess of empty calories and do not encourage the child to be physically active?

Impact of the Child's Environment

In the United States, many children are exposed to an environment that encourages overeating and consumption of foods that supply a lot of empty calories. Additionally, the environment often does not provide children with opportunities to participate in enough physical activity. Table 13.10 lists these and other environmental factors that contribute to childhood obesity.

TABLE 13.10 Environmental Factors That Contribute to Childhood Obesity

- Easy access to foods and drinks that are high in empty calories at or near schools
- Limited access to healthy and affordable foods, particularly in areas that have many convenience stores and fast-food restaurants
- Advertising of unhealthy foods that targets youth
- Lack of set periods for daily physical activity in schools and safe places to be active in many communities
- Large portion sizes of foods sold in some restaurants and grocery stores
- Excess exposure to digital media, including television. Such exposure contributes to childhood obesity because the sedentary activities can reduce the time children spend being physically active.
- Personal and family stress, which may lead to eating as a way to cope with emotions
- Sleeping less than the recommended hours of sleep per age

Source: Hampl SE and others: Clinical practice guideline for the evaluation and treatment of children and adolescents with obesity. *Pediatrics* 151(2): e2022060640, 2023.

Preventing Childhood Obesity

Recommendations that may reduce the likelihood of obesity among children include:

- New mothers of healthy infants should breastfeed their babies exclusively for about the first 6 months of life and continue breastfeeding along with introducing appropriate foods until the children are old enough to drink cow's milk safely.
- Child care providers and schools should provide opportunities for children to be physically active throughout the day.
- Caregivers should limit children's exposure to video games, smartphones, tablets, computers, and TV ("screen time"). For specific recommendations concerning such limits, visit www.healthychildren.org/ and search for "Screen time."
- Community leaders can support ways to increase physical activity for children and adults. Such efforts may include creating safe parks, playgrounds, and bicycling and walking trails as well as organizing and promoting supervised recreational sports activities.

To help reduce the prevalence of childhood obesity, community leaders can organize and promote supervised recreational sports activities for children. FatCamera/E+/Getty Images

Treating Childhood Obesity

The American Academy of Pediatrics (AAP) recommends the overall treatment plan for obesity should focus on the child, but it should also include the entire family.[48] A healthy lifestyle, including adoption of a healthy, nutrient–dense diet, should be adopted by all family members, which includes everyone eating the same foods. Caregivers should not place a child or youth on a weight reduction diet without consulting the child's physician.

The treatment goal for managing childhood obesity is maintaining weight rather than losing weight.[47] As the child who is obese grows taller, their BMI may fall into the healthy range. To ensure the child's growth rate is not slowed, caregivers need to balance the calories children consume with the calories they use for physical activity and need for growth. Encouraging more physical activity, especially for sedentary children, is also recommended.

For adolescents who are severely obese, treatment approaches that go beyond dietary changes and increased physical activity are often necessary. Such interventions may include prescription medication and weight loss surgery (see Chapter 10).

Bariatric Surgery for Youth

Bariatric (weight loss) surgery can improve the health of adolescents with extreme obesity.[49] According to the AAP obesity treatment guidelines, bariatric surgery in the United States is fairly safe for adolescents.[48] However, not all adolescents that are severely obese should have bariatric surgery. Nevertheless, medical researchers are studying youth who have undergone bariatric surgery to determine whether the procedures have negative long–term effects.

To qualify for bariatric surgery, an adolescent should have:

- severe obesity (BMI of 35 or more);
- attained their adult height (skeletal maturity), which is generally at 13 years of age for females and 15 years of age for males; and
- developed serious chronic conditions that are associated with obesity, such as hypertension, type 2 diabetes, or sleep apnea, which may improve after surgery.[49]

Regardless of one's age, bariatric surgery is a drastic measure to lose excess body weight and involves some risks. Thus, caregivers and health care providers should carefully determine whether an adolescent who is extremely obese is a suitable candidate for bariatric surgery. For example, is the adolescent emotionally ready to handle the surgery and make the necessary lifestyle changes to achieve good health and well–being after the procedure?

Concept CHECKPOINT

27. Define *overweight* and *obese* for children as defined by the Centers for Disease Control and Prevention.
28. List at least three health consequences of childhood obesity.
29. List at least three factors that contribute to the development of childhood obesity.
30. What information is used to determine whether an adolescent who is obese qualifies for bariatric surgery?

See Appendix F for responses.

13.8 Nutrition for Older Adults

Learning Outcomes

1. Explain the difference between life expectancy and life span.
2. List the characteristics of a person who experiences *healthy aging*.
3. Identify physiological changes that are associated with the normal aging process.
4. Discuss nutrient needs for older adults.
5. Identify nutrients that are often lacking in diets of older adults.
6. Discuss steps caregivers can take to improve nutrient intakes of older persons.

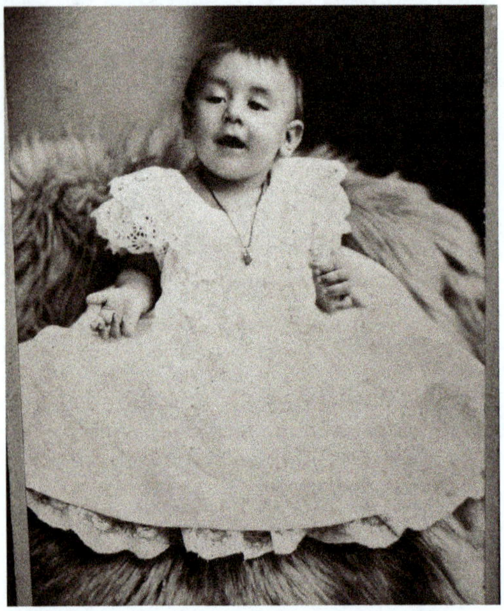

George John Blum of St. Louis, Missouri, circa 1900. In 1900, the life expectancy of a baby born in the United States was only 47 years.
©Wendy Schiff

In 1900, the life expectancy of a baby born in the United States was only 47 years. **Life expectancy** is the length of time a person born in a specific year, such as 1900, can expect to live. One hundred years ago, the top three leading causes of death for Americans were pneumonia, influenza, and other infectious diseases. By 2021, life expectancy in the United States rose to almost 77 years.[50] In 1950, an American who was 65 years of age could expect to live for about 14 more years. In 2019, a 65–year–old American could expect to live for 19.5 more years. Major factors that have contributed to increased life expectancy during the past century include improved diets, housing conditions, and public sanitation, as well as advances in medicine.

According to the U.S. Census Bureau, about 15% of the U.S. population were 65 years of age or older in 2016 (Fig. 13.12).[51] By 2060, government experts estimate that about 23% of Americans will be in this age group. Americans who are the "oldest old," 85 years of age or older, comprise one of the fastest–growing

life expectancy length of time an average person born in a specific year can expect to live

segments of the U.S. population. In 2016, almost 2% of Americans were in that age group. By 2060, almost 5% of Americans will be 85 years of age or older.[52]

Healthy Aging

Although more Americans are living longer than their ancestors, they are not necessarily living well. Chronic diseases are among the leading causes of death in the United States (see Figure 1.2). These diseases are associated with lifestyles that include smoking, eating a poor diet, and being physically inactive.

Growing old is a normal and natural process. Your body ages, regardless of dietary and other health–related practices you follow. People who experience *healthy aging* avoid developing serious chronic diseases, such as cancer and cardiovascular disease. Additionally, such persons enjoy good quality of life, maintaining both adequate thinking abilities and positive psychological health as they grow older.[53]

Scientists have found a strong genetic component to human longevity. If you have ancestors who are very old or lived to be 90 years of age or more, you may have inherited "longevity genes." Your focus, however, should not be simply on living longer but on living longer *and* healthier.

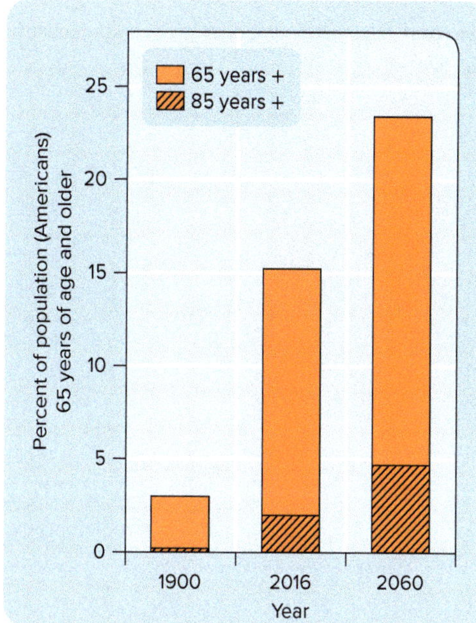

Figure 13.12 **Older adult population of the United States: 1900, 2016, and projection for 2060.** www.census.gov/data/tables/2017/demo/popproj/2017-summary-tables.html; Administration for Community Living: *2017 Profile of older Americans*. 2018. https://acl.gov/. Accessed March 14, 2023.

The Aging Process

The aging process begins at conception and is characterized by numerous predictable physical changes. By the time you are about 70 years of age, you will have reached the final life stage, *older adulthood.* What causes people to age is unclear. Scientists who study the aging process have learned that cell structure and function inevitably decline with time, leading to many of the physiological changes shown in Table 13.11. Eventually, most cells lose the ability to regenerate their internal parts, and they die. As more and more cells in an organ die, the organ loses its functional capacity, and as a result, other organs fail and body systems are adversely affected. When this happens, the person soon dies. **Senescence** (*seh–ness'–enz*) refers to declining organ functioning and increased vulnerability to disease that occur after a person reaches physical maturity.

senescence declining organ functioning and increased vulnerability to disease that occur after reaching physical maturity

TABLE 13.11 Aging: Normal Physiological Changes

Body System	Changes
Digestive	Reduced saliva, gastric acid, and intrinsic factor secretion; increased heartburn and constipation
Skin, hair, and nails (integument)	Graying hair; drier skin and hair; skin loses elasticity and forms wrinkles; skin bruises easily
Musculoskeletal	Bone-forming cells become less active, resulting in bone loss that can lead to tooth loss and bones that fracture easily; fractures heal more slowly; joints become stiff and painful; muscle mass declines, resulting in loss of strength and stamina
Nervous	Decreased brain weight, reduced production of neurotransmitters, delayed transmission of nervous impulses, loss of short-term memory, and reduced sensory abilities (e.g., vision, hearing, smell, and taste)
Lymphatic/Immune	Reduced functioning resulting in increased vulnerability to cancer and infections
Circulatory	Hardening of the arteries, reduced cardiac output, increased risk of blood clots
Endocrine	Decreased production of reproductive, growth, and thyroid hormones
Respiratory	Reduced lung capacity, increased vulnerability to respiratory infections
Urinary	Increased loss of functional kidney cells, resulting in decreased blood filtration rate; loss of bladder control
Reproductive	Males: Decreased male hormone production and sperm count Females: Declining female hormone production, cessation of menstrual cycles, and loss of fertility

life span maximum number of years an organism can live

The Human Life Span

Life span refers to the maximum number of years an organism such as a human can live. To date, a Frenchwoman, Jeanne Calment, had the longest documented human life span—122 years. Some scientists think the human life span can be lengthened considerably just by making certain dietary changes. This chapter's "Nutrition Matters" section takes a closer look at the role of diet and longevity.

Common Nutrition-Related Concerns

Compared to younger persons, older adults have greater risk of nutritional deficiencies because of physiological changes associated with the normal aging process. An older adult's food intake tends to decrease as their metabolic rate and physical activity level decline. Despite having lower energy needs, older adults need the same or even higher amounts of vitamins and minerals. Meeting micronutrient needs while eating less food can be difficult for older adults to accomplish. Other factors that can influence an older person's nutritional status include illnesses, medications, low income, and lack of social support.

Critical nutrients for older adults are protein, omega–3 fats, fiber, calcium, magnesium, potassium, and the vitamins folate, B–12, B–6, D, E, and K.[54] Older adults tend to consume more–than–recommended amounts of sodium, saturated fat, and added sugars. Their diets, however, generally do not provide enough fiber, calcium, magnesium, potassium, zinc, folate, choline, and vitamins A, B–6, C, D, and E.[53,55] Table 13.12 shows results of a study that focused on the food choices of older American adults. Many American males and females ages 71 years and older do not meet the recommendations concerning amounts of foods from MyPlate food groups.[55] Intakes of vegetables, dairy foods, and fruits are low compared to intakes of grains and protein foods. The following sections discuss some of the major health concerns that often affect the nutritional status of older adults. Some of these conditions have been discussed in previous chapters.

Changes in Body Weight

As the human body ages, its need for energy decreases. In senescence, muscle mass declines as some muscle cells shrink or die (*sarcopenia*). The loss of muscle mass leads to a decrease in muscular strength and basal metabolism. The aging body typically loses lean tissue and gains some fat tissue. Increased body fat results from overeating and lack of physical activity, but even athletic males and lean females usually gain some central body fat after they are 50 years of age. Being overfat may protect the older adult from being injured in falls, but having too much body fat

TABLE 13.12 Percentages of American Adults Ages 71 Years and Older Who Follow USDA Dietary Recommendations

Food Group	Older Adult Males Who Meet the Recommendations (%)	Older Adult Females Who Meet the Recommendations (%)
Fruit (cups)	14	21
Vegetables (cups)	18	15
Grains (oz equivalents)	36	40
Dairy (cups)	6	1
Protein (oz equivalents)	52	30

Source: Adapted from Steinfeldt LC and others: Meeting dietary guidelines recommendations: Older adults: What we eat in America, NHANES 2013–2016. *FSRG Dietary Data Brief, no. 28.* 2020.

increases the risk of type 2 diabetes, hypertension, CVD, and osteoarthritis. For more information about overweight and obesity, see Chapter 10.

Older people may develop *sarcopenic obesity*—the loss of skeletal muscle mass and strength combined with an excessive increase in body fat. In general, people with sarcopenic obesity are more likely to experience mobility problems and other physical limitations than people who do not have the condition.[56] Regardless of one's age, eating too many calories and being physically inactive contribute to the development of excess body fat. To treat sarcopenic obesity, an age–appropriate, individualized program of resistance exercises that includes a nutrient–dense, calorie–controlled diet may be useful.

Older adults, especially those who are over 75 years of age, may unintentionally lose weight. Several factors can contribute to weight loss among older adults (Table 13.13). Older adults may eat less because they have lost the ability to taste and smell food. Loss of teeth and difficulty swallowing (*dysphagia*) can also result in decreased food consumption. Declines in normal *cognitive* functioning (thinking abilities) resulting from conditions such as Alzheimer's disease or reduced blood flow to the brain can contribute to poor nutritional status. People who lack normal cognitive functioning may be unable to make decisions regarding planning nutritious meals, as well as to shop for and prepare food. Conditions that interfere with mobility and flexibility, such as arthritis and osteoporosis, can also interfere with food shopping and preparation activities. Social and economic factors often play a role in reduced food intake. Many older people live alone and on fixed incomes, circumstances that are associated with depression and inability to afford adequate amounts of nutritious food. In many instances, very old people refuse to eat and, as a result, lose considerable amounts of weight, a situation that hastens their death.

For older adults who find that food no longer tastes "good," adding more spices may improve the taste of food and, as a result, stimulate weak appetites. Efforts to make mealtimes social events, such as inviting friends to share potluck meals together, can enhance older adults' mental outlooks and spark their interest in eating. Older adults can increase or maintain their weight by consuming energy–dense snacks between meals, such as cheese, milkshakes, nuts, or oatmeal cookies. If weight loss becomes significant, a physician should be consulted to determine the cause.

TABLE 13.13 Reduced Food Intake Among Older Adults: Contributing Factors

- Reduced ability to taste and smell food
- Difficulty swallowing
- Loss of teeth
- Loss of normal cognitive function
- Lack of income
- Depression
- Reduced mobility and flexibility

Physical Inactivity

Many of the undesirable physical changes we associate with growing old are the result of a lifetime of physical inactivity. Regardless of a person's age, a physically active lifestyle increases muscle strength and mobility, improves balance, slows bone loss, and boosts emotional well–being. Most older adults can benefit from performing aerobic and strength–training activities regularly. Before embarking on a program to increase physical fitness, however, sedentary older adults should consult their physicians concerning appropriate activities. The "Real People, Real Stories" feature in this section is about Carol Neumann, a 91–year–old who works out regularly and is physically fit.

Digestive System Problems

Constipation is a major complaint of older adults. By increasing their intakes of fiber–rich foods, such as whole–grain products and vegetables, older adults may be able to have more regular bowel movements (see the "Fiber" section of Chapter 5). Dehydration contributes to constipation, so older people should make sure their fluid intake is adequate.

Did *YOU* Know?

You are never too old to gain some benefits from exercise. Any form of physical activity, from doing household chores to performing yoga or tai chi, can help extend an older adult's longevity.

Donna Day/Big Cheese Photo/SuperStock

Ruined steps at Machu Picchu
Rodrigo Torres/Glow Images

REAL PEOPLE | REAL STORIES

©Wendy Schiff

Carol Neumann

Carol Neumann is a petite 91-year-old grandmother of three who embodies good health and physical fitness. Two days a week, she walks for a total of four miles on a gym's indoor track. Two other days of the week, she participates in the gym's fitness class for older adults, which lasts for an hour each session. Her physical activity habits also include shopping for groceries, preparing meals, and cleaning her home. Carol is so healthy, the only medication she takes is an over-the-counter vitamin D supplement. "I don't know why I'm healthy, but I think a combination of things have contributed, including 'good genes.' My two siblings are also in their 90s."

Carol's history of maintaining a healthy body weight and being physically active offers another possible reason for her good health. "I've always been slim and physically active. When I was a teenager, I had to walk back and forth to school. I grew up in the upper peninsula of Michigan, where winters were long and cold. During cold weather, I'd ice skate with my friends for at least two hours, three or four nights a week. In summers, I liked to swim in lakes and hike on trails.... When I was in college, I didn't have a car, so I walked places.

In addition to having a physically active lifestyle as a teenager, Carol recalls eating healthy foods, especially vegetables and fruits. "I think my diet was good, too. I ate vegetables—fresh and canned. Mother grew vegetables in her garden, and she canned what we didn't eat when it was fresh.... I used to pick wild blueberries, so I ate lots of them. Mother canned blueberries, too."

After Carol graduated from the University of Michigan, she became a middle-school teacher. She got married, and she quit teaching to raise two children. During that time, she remained physically active; she tended to her garden and did yardwork, played golf (without using a golf cart), and went cross-country skiing in the winter. After her children graduated and moved away from home, Carol and her husband enjoyed traveling to all 50 states and more than 50 countries. She and her husband climbed Machu Picchu in Peru when they were in their early 70s. "When we visited Egypt, I rode on a camel!" she exclaimed.

In her later years, Carol kept her mind active, too. She did a variety of volunteer work for her church and community, including helping elementary and middle-school children in her school district improve their reading skills. Today, she likes to read.

When asked what advice she would give young adults to help them live long and healthy lives, Carol said, "Be thankful! Face adversity with courage and grace, have a positive attitude, and do some good in this world."

As a person ages, their stomach secretes less hydrochloric acid and intrinsic factor. These changes can contribute to poor absorption of vitamin B–12 and the development of vitamin B–12 deficiency and pernicious anemia. Older adults may be able to meet their vitamin B–12 needs by eating foods fortified with the micronutrient or taking vitamin B–12 supplements. Some older adults, however, must take injections of the B vitamin to prevent pernicious anemia (see Chapter 8). Older persons are also at risk of iron deficiency because reduced stomach acid production may hinder iron absorption. Furthermore, many older adults take aspirin regularly, and this practice can cause intestinal bleeding that can lead to iron deficiency anemia. Intestinal ulcers and cancer can also cause blood loss from the digestive tract. The discovery of blood in bowel movements needs to be reported to a physician—regardless of one's age.

Many older adults take one or more prescription drugs daily. Although such medications can improve health and quality of life of older adults, some drugs interfere with the body's absorption and/or use of certain nutrients. Additionally, older adults often take one or more dietary supplements regularly. Certain dietary supplements, including herbal products, can reduce or amplify the effects of prescribed medications. Therefore, older adults should notify their physicians about their use of all dietary supplements. A few foods also interfere with prescribed drugs. Grapefruit juice, for example, can alter the potency of certain medications that are used to lower blood pressure or cholesterol.

Many older adults take more than one prescription medication daily. Some drugs interfere with the body's absorption and/or use of certain nutrients.
Jose Luis Pelaez Inc/Blend Images LLC

Tooth Loss In the 1950s, surveys of Americans indicated that the majority of older adults had lost all their natural teeth. Since then, the percentage of older adults who retain all or most of their teeth has increased in the United States. Adults should have 28 natural teeth. In 2019, about 17% of Americans who were 65 years of age or older had lost all of their teeth.[57] Tooth loss is related to long-term poor dental hygiene, cigarette smoking, and poor dietary practices. By following recommended dental hygiene practices, obtaining regular dental care, and avoiding tobacco use, you can greatly increase your chances of keeping most of your teeth as you age.

Excessive tooth loss can lead to faulty eating habits. People who lack teeth often avoid crisp or chewy foods, such as fresh fruits, vegetables, whole-grain cereals, and meat. According to results of a study involving over 300 older adults, subjects who had 10 or fewer teeth consumed considerably less than the recommended amounts of fruits, vegetables, meat and beans, and oils when compared to older adults who had more than 10 teeth.[58] Although dentures that replace natural teeth can enable some people to chew normally, many older adults do not like to wear them because they can be uncomfortable. When a person has difficulty chewing food, serving soft foods such as ground meats, cooked vegetables, pureed fruits, and puddings can stimulate the individual's appetite.

Many older adults do not like to wear dentures because they can be uncomfortable.
Stockbyte/PunchStock/Getty Images

Depression in Older Adults

In 2018, 13% of females and 9% of males aged 55 years and over reported having symptoms of depression.[59] Situations that contribute to depression among the older adult population include coping with chronic illness or loss of mobility, and isolation and loneliness as family members and friends move away or die. If the depressed person loses interest in cooking and eating, weight loss and nutrient deficiencies are likely to occur. In many instances, depression can be managed with medication, but social support and psychological counseling may be necessary as well. Without proper treatment, depressed persons, especially males, are at risk of alcoholism and suicide. Nevertheless, suicide is not a leading cause of death among older adults.[60]

Dietary Planning in Older Adulthood

MyPlate can provide the basis for planning nutritionally adequate meals and snacks for healthy older adults. However, amounts of foods recommended in these diet plans may not provide enough vitamin D and calcium for older adults. By regularly consuming fortified and/or enriched foods, older adults can increase their intakes of these micronutrients. In many instances, older adults can also benefit from taking a daily multiple vitamin/mineral supplement.

Friends, relatives, and health care personnel should be alert for indications of poor nutrient intakes among older people, especially those who are at risk and live in nursing homes or other long-term care facilities. For example, family members can make sure the older adult's nutrient needs are met by visiting

Many communities in the United States offer the Meals on Wheels program in which volunteers deliver nutritious meals prepared at a senior center to qualified, homebound older adults.
Ken Hammond/USDA

the person's residence during mealtimes, observing the foods that are offered, and, if necessary, helping the older adult eat. Additionally, monitoring the older person's weight can indicate whether long–term food intake has been adequate. Older adults who live at home may need help planning nutritionally adequate diets. In these instances, registered dietitian nutritionists can be consulted to provide personalized dietary advice.

Community Nutrition Services for Older Adults

The Chapter 1 "Nutrition Matters" section provides information about popular community–based nutrition services for American older adults. You can obtain information regarding locally available nutrition services for older people from medical clinics, private practitioners, hospitals, and health maintenance organizations in your area. To learn more about nutrition–related programs for older adults, visit the following websites: National Institute on Aging, www.nia.nih.gov/; National Council on Aging, www.ncoa.org/economic–security/benefits/food–and–nutrition/food–assistance–benefits/; and Administration for Community Living, www.acl.gov/.

Food & Nutrition tips

The following suggestions can help caregivers improve nutrient intakes of older adults:

- Emphasize nutrient-dense foods when planning daily menus.
- Try new foods, seasonings, and ways of preparing foods.
- Have easy-to-prepare, nutrient-dense foods on hand for times when the older person is too tired to cook large meals.
- Serve meals in well-lit or sunny areas, and plan appealing meals by using foods with different flavors, colors, shapes, textures, and smells.
- Plan occasions for the older adult to share cooking responsibilities and eat meals with friends or relatives.
- Encourage the older person to eat at a senior center whenever possible. Investigate community resources for helping the older adult obtain groceries, cook, or manage other daily care needs.
- If biting and chewing are difficult for an older adult, chop, grind, or blend tough or crisp foods.
- Prepare extra amounts of soup, stew, or casserole so leftovers can be frozen for future meals.

Concept CHECKPOINT

31. What is the difference between *life expectancy* and *life span?*

32. Identify at least five physiological changes that are associated with the normal aging process.

33. Explain why nutrient needs for older adults are often higher than those for younger persons.

34. List at least four nutrients that are often lacking in diets of older adults.

35. Suggest at least three ways caregivers can improve nutrient intakes of older persons.

See Appendix F for responses.

13.9 Nutrition Matters: In Search of the Fountain

Learning Outcomes

1. Discuss the role of telomere length in longevity.
2. Explain the effects of calorie restriction on the aging process.
3. Discuss actions that young people can take that may help them live longer, healthier lives.

In 1513, Spanish explorer Juan Ponce de León sailed to the southeastern coastal region of North America and discovered an area he named "Land of Flowers" (Florida). Ponce de León was on a mission to find gold, but he was also eager to locate a natural spring that supposedly had magical powers. According to Native Americans in the area, elderly people who drank the spring's water regained their youthful looks and vigor. Unfortunately, Ponce de León never found gold or the mythical "fountain of youth" in Florida. Nevertheless, many older adults still seek ways to combat aging, especially by taking certain hormones and dietary supplements promoted as having age-defying properties. Older Americans spend considerable amounts of money on such products, some of which may be harmful in the long run. Promoters of antiaging formulas or therapies claim their treatments can stop, and even reverse, the process of aging. Is there any reliable scientific evidence to support claims that you can take something to stay young longer?

Claims that a nutritional fountain of youth exists are simply not true. Currently, there is a lack of reliable scientific evidence that supports the use of antiaging therapies that include taking dietary supplements.[61] Scientists have determined that consuming a nutritious diet and obtaining adequate amounts of physical activity can promote good health as one ages.[62]

Juan Ponce de León.
Library of Congress, Prints and photographs Division [LC-USZ62-3106]

Biology of Aging

Biogerontologists are scientists who study the biology of aging. Biogerontologists have proposed several theories to explain why aging occurs and why some people live longer than others. According to one of these theories, longevity results from cells' ability to maintain and repair the damage done by a lifetime of exposure to the environment and the effects of everyday wear and tear. Some multicellular organisms are able to live longer when they can improve their abilities to repair damage to their DNA, reduce the toxicity of free radicals, and replace nonfunctioning cells. Antioxidants reduce the toxic effects of free radicals (see Chapter 8), and organisms produce a variety of antioxidants to control free radical production.

Calorie Restriction and Aging

Another area of biogerontological research that shows some promise is the use of *calorie restriction (CR)* to extend longevity. CR is not the same as fasting. Although CR provides fewer calories than an organism's usual diet, the dietary pattern supplies all essential nutrients. According to some research findings, CR can increase life spans of various organisms, including some kinds of rodents, flies, and snails.[62] Results of other studies, however, have not been so encouraging because the calorie-restricted animals had either the same or shorter life spans when compared to controls.

Even if scientists provide evidence that any form of CR adds some years to the human life span, would you be interested in reducing your food intake considerably to extend your life? How enjoyable would your life be if you ate only 900 kcal daily? What is important to you: how long you can live or how *well* you live?

Telomeres and Aging

Telomere length (*TL*) is associated with longevity. Telomeres are at the tips of chromosomes. Recall that a cell's nucleus contains DNA, which codes for the production of proteins. Before a cell divides, the cell makes a copy of its DNA. This process enables the cell's two "daughter" cells to contain the same DNA as the parent cell. Each time cell division occurs, a tiny bit of the telomeres is lost and the telomeres' length shortens as a result.

Although TL differs among humans, older people tend to have shorter telomeres than younger persons. TL can also reflect health status. Long telomeres are associated with having a healthy body weight, following a healthy diet, avoiding tobacco use, limiting alcohol consumption, and being physically active.[63] People with short telomeres have a higher risk of chronic diseases that are associated with aging, such as cancer and CVD. Thus, TL is a *biomarker* for aging and risk of certain chronic diseases.[53] A biomarker is a component of body fluids or cells that is measurable and reflects the body's health status.

Scientific observations concerning the length of telomeres and aging have led to investigations to determine factors that influence TL. The presence of inflammation and *oxidative stress* (an excess of radicals in the body) shortens TL. We discussed the role of inflammation in the disease process, particularly CVD, in Chapter 6. As mentioned earlier, we highlighted the importance of antioxidants in controlling excess radical formation in Chapter 8. Because plants are often rich sources of anti–inflammatory factors and antioxidants, some scientists think plant–based diets hold the key to longevity.[53] Nevertheless, more scientific studies that investigate the long–term effects of various dietary patterns on TL are needed to determine the extent to which plant–based diets promote longevity.

Some Keys to Healthy Aging

The science of biogerontology is still in its beginning stages; researchers have much to learn about the aging process and aging biomarkers before they can develop safe ways to enhance longevity. We already know that you can reduce your risk of dying prematurely from chronic diseases such as heart disease, hypertension, type 2 diabetes, and many forms of cancer by adopting healthy lifestyles. Rather than wait until the fountain of youth becomes a reality, you can take charge of your health now by consuming a nutritionally adequate diet, obtaining regular moderate– to vigorous–intensity physical activity, maintaining a healthy weight, avoiding tobacco products, limiting your alcohol consumption, and having regular physical checkups.

Concept CHECKPOINT

36. According to scientific evidence, which dietary modification may extend the life expectancies of certain organisms?

37. List three steps you can take that may extend your life expectancy.

See Appendix F for responses.

Chapter References

See Appendix G.

Summary

13.1 Introduction to Nutrition for a Lifetime

- The stages of life are the prenatal period or pregnancy, lactation, infancy, childhood, adolescence, and the older adult period. In a healthy person, the physical changes that occur during these stages are normal. Most of the leading causes of death among Americans are chronic diseases, such as heart disease and cancer. Long-term health-related practices, including dietary and physical activity habits, often contribute to the development of these diseases. Having good health throughout life depends not only on lifestyle choices, but also on several other factors, including heredity, relationships, environment, income, education level, and access to health care. Furthermore, achieving and maintaining a healthy body weight are critical to enjoying good health as one grows older.

Alamy Stock Photo

13.2 Pregnancy

- Embryonic/fetal life is characterized by rapid rates of cell division, resulting in a dramatic increase in cell numbers. The first trimester of pregnancy is a critical stage in human development because inadequate or excessive nutrient intakes as well as exposure to toxic compounds can have devastating effects on the embryo/fetus during this period. The placenta is the organ of pregnancy that transfers nutrients and oxygen from the mother's bloodstream to her embryo/fetus. The placenta also transfers wastes from the embryo/fetus to the mother's bloodstream so her body can eliminate them.
- Females of childbearing age should take steps to ensure they are in good health prior to becoming pregnant. During pregnancy, the female's body undergoes various physiological changes. These changes enable her body to nourish and maintain the developing fetus, as well as to produce milk for her infant after its birth. A pregnant female should follow a diet that meets her own nutritional needs, as well as those of her developing offspring. The mother-to-be can use MyPlate to develop nutritionally adequate daily menus, but she may also need to take a prenatal vitamin/mineral supplement.
- Females whose prepregnancy weights were within the healthy range can expect to gain 25 to 35 pounds during pregnancy. Females who gain excess weight during pregnancy may retain the extra weight long after delivery.
- Monitoring weight gain is an important aspect of prenatal care. Underweight females who do not gain enough weight during pregnancy are at risk of having preterm or low-birth-weight infants. females who are obese have a greater risk of developing hypertension and type 2 diabetes during pregnancy.

13.3 Infant Nutrition

- Breast milk is the best first food for infants. Compared to babies who are not breastfed, breastfed infants have lower risks of allergies and of gastrointestinal, respiratory tract, and ear infections. Females who breastfeed their babies can also derive some important benefits from the practice, such as losing extra fat gained during pregnancy.
- When an infant suckles, nerves in the nipple signal the mother's brain to release prolactin and oxytocin into her bloodstream. Oxytocin is necessary for the let-down reflex and causes the uterus to contract. Milk production relies on "supply and demand." If the breasts are not emptied fully, milk production soon ceases. To support milk production, the lactating mother needs about 400 to 500 extra kilocalories daily.
- Growth is very rapid during infancy. Health care practitioners can assess growth in infants and children by measuring body weight, height (or length), and head circumference over time. For about the first 6 months, a healthy infant's nutrient needs can be met by human milk or iron-fortified infant formula. Breastfed babies need vitamin D and possibly iron and fluoride supplements.

hannamariah/123RF

- Most infants do not need solid foods before 4 to 6 months of age. The first solid food offered to babies should be an iron-containing food such as an iron-fortified infant cereal, meat, or poultry. New foods should be added one at a time, and the child should be observed for allergy signs and symptoms. Whole cow's milk should not be fed to babies until they are 1 year of age.

13.4 The Preschool Years

- It is normal for a preschooler's appetite to decline as the child's growth rate tapers off. Additionally, it is not unusual for preschool children to be "picky eaters" or embark on "food jags." Caregivers should avoid nagging, forcing, and bribing children to eat but instead offer a variety of healthy food choices and allow the child to choose what and how much to eat. Nutrition-related problems that often affect preschool children are iron deficiency, dental caries, obesity, and food allergies.

13.5 School-Age Children

- School-age children often skip breakfast, and they tend to consume more foods away from home, larger portions of food, and more fried foods and sweetened beverages. Children who eat breakfast are more likely to have better diets and healthier body weights than children who skip this meal. Diets of many school-age children fail to supply recommended amounts of calcium and potassium while providing too much sodium.

13.6 Adolescence

- A child matures physically into an adult during adolescence. For many teens, pressure to conform to fads and be influenced by other adolescents negatively affects their diets and overall health. Obesity, eating disorders, and low iron and calcium intakes are major nutrition-related concerns of adolescents.

13.7 Childhood Obesity

- Public health experts are very concerned about the increasing prevalence of obesity among children in the United States. Children who are overfat have higher risks of elevated blood pressure, cholesterol, and glucose levels than children whose weights are within the healthy range. Children who are overfat may also have higher risk of hypertension, heart disease, and type 2 diabetes later in life. Such children are also more likely to have low self-esteem and become obese as adults.

13.8 Nutrition for Older Adults

- The aging process is characterized by numerous predictable physical changes. Senescence refers to declining organ functioning and increased vulnerability to disease that occurs after a person reaches physical maturity. Although genetics play a role in determining longevity, lifestyle practices and environmental conditions influence a person's rate of aging. People may be able to live longer and healthier by making responsible healthy lifestyle decisions while they are still young.
- Factors that can influence an older person's nutritional status include illnesses, tooth loss, medications, low income, depression, and lack of social support. Diets of older adults often provide inadequate amounts of fiber, calcium, magnesium, potassium, zinc, folate, choline, and vitamins A, B-6, C, D, and E. Friends, relatives, and health care personnel should be alert for indications of poor nutrient intakes among older people, especially those who live in nursing homes or other long-term care facilities.

Shutterstock / Olena Yakobchuk

13.9 Nutrition Matters: In Search of the Fountain

- Genetic and lifestyle factors influence longevity. Some biogerontologists think preserving telomere length, as well as restricting calorie intake, promotes living longer. Manipulating dietary patterns to control inflammation and oxidative damage may hold the keys to longevity, but more long-term research is needed. At present, there is a lack of reliable scientific evidence to support the use of antiaging therapies that include taking dietary suppplements.

Recipe for Healthy Living

Vegetable Dip

You can make low-fat versions of fruit or vegetable dips by replacing sour cream or mayonnaise in recipes with plain, low-fat yogurt. This vegetable dip recipe makes approximately five ¼-cup servings of dip. Each serving of dip provides approximately 48 kcal, 3 g protein, 2 g fat, 90 mg calcium, 135 mg sodium, and 126 mg potassium.

©Wendy Schiff

INGREDIENTS:

1 cup plain, low-fat yogurt
3 Tbsp calorie-reduced ranch dressing
¼ tsp curry powder (optional)
Cleaned raw vegetables, cut into bite-sized pieces (enough for five ½-cup servings)

PREPARATION STEPS:

1. In a small bowl, combine yogurt with the dressing and stir until well blended.
2. Refrigerate until ready to serve.
3. Serve chilled in a bowl that is surrounded with fresh pieces of raw vegetables.
4. Discard any remaining dip.

Approximate Percent of Calories from Macronutrients

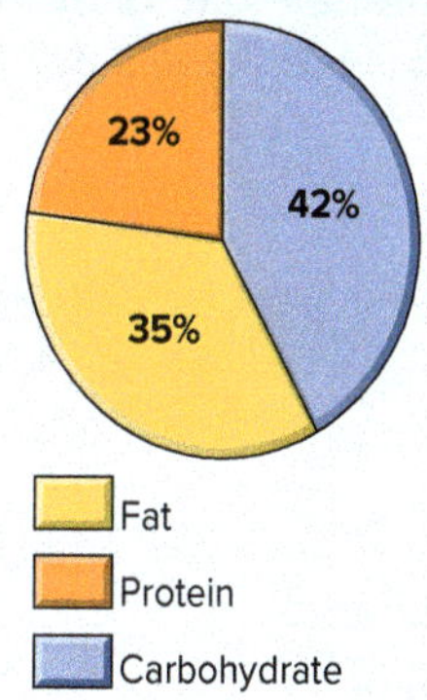

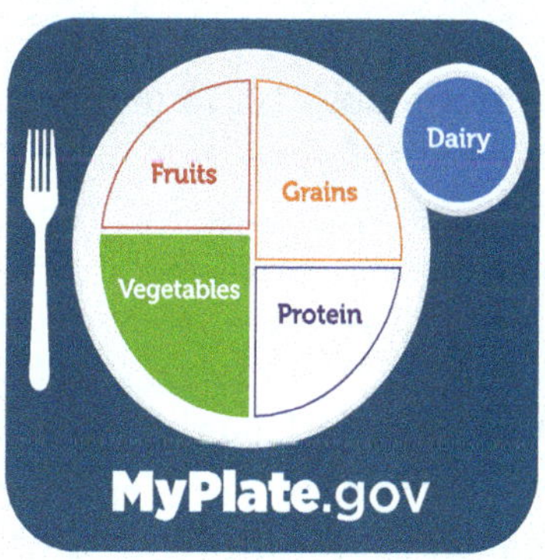

US Department of Agriculture or USDA

Percent of Daily Value/Serving

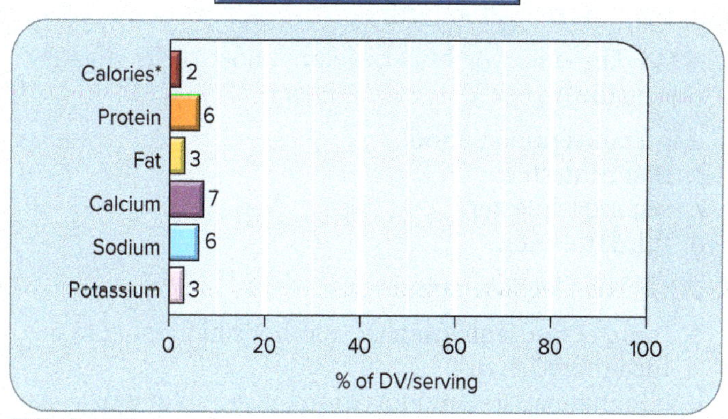

*2000 daily total kcal

Critical Thinking

1. One of your friends just found out that she is pregnant. Although her BMI is within the healthy range, she is concerned about gaining too much weight during pregnancy. What advice would you provide concerning the need to gain some weight during this life stage? If your friend's prepregnancy weight was 125 pounds, how much weight would be appropriate for her to gain during pregnancy?
2. Your pregnant friend wants your advice concerning whether she should breastfeed or formula-feed her baby. After reading Chapter 13, what information would you provide to help your friend decide to breastfeed?
3. Makena is a healthy 2-month-old baby. Makena's mother, Ada, wants to replace Makena's iron-fortified infant formula with the same fresh fluid 2% milk that she drinks. What advice would you give to Ada concerning the appropriateness of making such a decision?
4. Omar is 8 years old, and his weight is in the overweight range. His caregivers are also overweight, but they seem to be concerned about Omar's excess body weight. What advice would you provide his caregivers to help Omar achieve a healthy body weight?
5. Are your parents, grandparents, and great-grandparents still alive? If any of your ancestors died before they were 60 years of age, can you identify their causes of death and factors that contributed to their deaths? What lifestyle changes can you make now that can help you achieve a longer, healthier lifetime?

Practice Test (Section numbers are listed below in red.)

Select the best answer.

1. (13.1) The prenatal period refers to
   a. the time before conception.
   b. pregnancy.
   c. birth of an infant.
   d. the time right after birth.
2. (13.2) The embryo/fetus develops most of its organs during the
   a. preconception period.
   b. first trimester.
   c. second trimester.
   d. third trimester.
3. (13.2) The placenta cannot
   a. transfer nutrients from the mother's bloodstream to the embryo/fetus.
   b. eliminate waste products from the embryo/fetus.
   c. prevent all toxic substances from reaching the embryo/fetus.
   d. transfer oxygen from the mother's bloodstream to the embryo/fetus.
4. (13.2) During the first trimester, a pregnant female's daily energy requirement is _____ her daily energy needs before she became pregnant.
   a. 300 kcal lower than
   b. about the same as
   c. 300 kcal higher than
   d. 500 kcal higher than
5. (13.2) Females who are obese and pregnant with one fetus should gain _____ pounds during pregnancy.
   a. 11–20
   b. 20–25
   c. 25–35
   d. 35–50
6. (13.2) Preeclampsia is a form of _____ that can develop during pregnancy.
   a. hypertension
   b. diabetes
   c. cancer
   d. anemia
7. (13.3) Which of the following statements is true?
   a. A female's energy needs are higher during the first trimester than at any other time in pregnancy.
   b. Using infant formula to bottle-feed a baby is more convenient and less expensive than breastfeeding a baby.
   c. Oxytocin is necessary for the "let-down" reflex to occur.
   d. The American Pediatric Association recommends feeding fresh whole milk to infants when they are 6 months of age.

8. (13.3) A healthy infant who weighs 6.8 pounds at birth can be expected to weigh about _____ pounds by her first birthday.

a. 12.2
b. 13.6
c. 20.4
d. 24.6

9. (13.3) Breastfed infants are _____ than babies who are fed infant formula.

a. more likely to have diarrhea
b. less likely to have cystic fibrosis
c. more likely to have respiratory infections
d. less likely to have ear infections

10. (13.4) Preschool children should brush their teeth with fluoride–containing toothpaste to prevent

a. constipation.
b. iron deficiency.
c. food jags.
d. dental caries.

11. (13.5) School age children typically

a. drink adequate amounts of water.
b. consume recommended amounts of fruit.
c. skip breakfast.
d. consume diets rich in fiber.

12. (13.6) Which habit directly contributes to adolescents being at risk for decreased bone mass?

a. Consuming fruit juice instead of iron–rich foods
b. Participating in school sports daily
c. Replacing milk with sugar–sweetened beverages
d. Consuming high–fat diets

13. (13.7) Which of the following factors is associated with increased risk of obesity during childhood?

a. Having a family history of obesity
b. Eating 3 to 5 servings of fresh fruit daily
c. Being a full–term infant
d. Taking dietary supplements that contain zinc and vitamin C

14. (13.8) Which of the following characteristics is not associated with the normal aging process?

a. Increased risk of blood clots
b. Increased brain weight
c. Reduced saliva production
d. Reduced immunity to infectious diseases

15. (13.8) The highest number of years that a human being can expect to live is the human

a. gerontological scale.
b. life expectancy.
c. life span.
d. age achievement.

16. (13.8) Older adults often do not consume adequate amounts of

a. simple carbohydrates.
b. sodium.
c. vitamin K.
d. zinc.

17. (13.9) is associated with longevity.

a. Telomere length
b. Cognitive dysfunction
c. Gray hair
d. Vitamin supplementation

Answers to Quiz Yourself

1. During pregnancy, a mother-to-be should double her food intake because she's "eating for two." **False.** (Section 13.2)
2. Infant formulas provide the same health benefits to infants as breast milk. **False.** (Section 13.3)
3. Caregivers should add solid foods to an infant's diet within the first month after the baby is born. **False.** (Section 13.3)
4. Over the past few decades, the prevalence of obesity has increased among American children. **True.** (Section 13.7)
5. Compared to younger persons, older adults have lower risks of nutritional deficiencies. **False.** (Section 13.8)

Appendix A

English-Metric Conversions and Metric-to-Household Units

Ablestock/Hermera Technologies/Alamy Stock Photo

English-Metric Conversions

Length

English (USA)	Metric
inch (in)	= 2.54 cm, 25.4 mm
foot (ft)	= 0.30 m, 30.48 cm
yard (yd)	= 0.91 m, 91.4 cm

Metric	English (USA)
millimeter (mm)	= 0.039 in (thickness of a dime)
centimeter (cm)	= 0.39 in
meter (m)	= 3.28 ft, 39.4 in
kilometer (km)	= 0.62 mi, 1091 yd, 3273 ft

Weight

English (USA)	Metric
ounce (oz)	= 28.35 g
pound (lb)	= 453.60 g, 0.45 kg

Metric	English (USA)
milligram (mg)	= 0.000035 oz
gram (g)	= 0.04 oz ($^1/_{28}$ of an oz)
kilogram (kg)	= 35.27 oz, 2.20 lb

Volume

English (USA)	Metric
teaspoon (tsp)	= 5 ml
tablespoon (Tbsp)	= 15 ml
fluid ounce	= 0.03 liter (30 ml)
cup (c)	= 237 ml
pint (pt)	= 0.47 liter
quart (qt)	= 0.95 liter
gallon (gal)	= 3.79 liters

Metric	English (USA)
milliliter (ml)	= 0.03 oz
liter	= 1.06 qt
liter (l or L)	= 0.27 gal

Fahrenheit-Celsius Conversion Scale

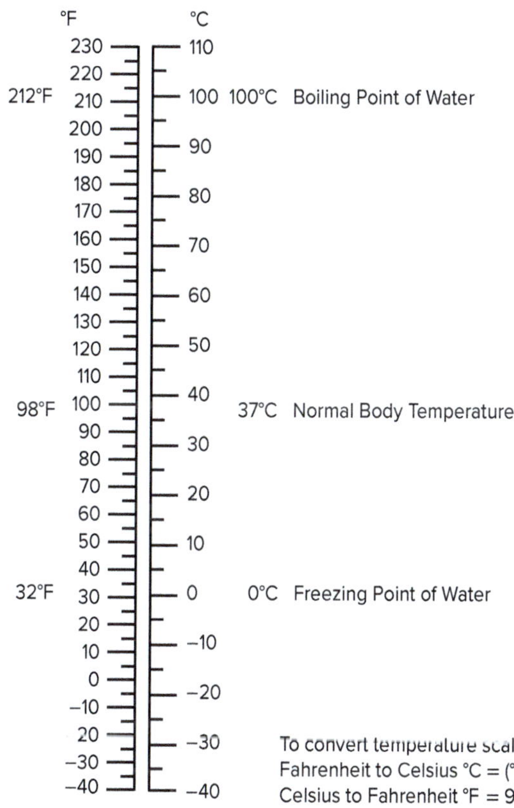

To convert temperature scales:
Fahrenheit to Celsius °C = (°F − 32) × 5/9
Celsius to Fahrenheit °F = 9/5(°C) + 32

Household Units

3 teaspoons	= 1 tablespoon
4 tablespoons	= ¼ cup
5⅓ tablespoons	= ⅓ cup
8 tablespoons	= ½ cup
10⅔ tablespoons	= ⅔ cup
16 tablespoons	= 1 cup
1 tablespoon	= ½ fluid ounce
1 cup	= 8 fluid ounces
1 cup	= ½ pint
2 cups	= 1 pint
4 cups	= 1 quart
2 pints	= 1 quart
4 quarts	= 1 gallon

Metric and Other Common Units

Unit/Abbreviation	Other Equivalent Measure
milligram/mg	$^1/_{1000}$ of a gram
microgram/µg/mcg	$^1/_{1,000,000}$ of a gram
deciliter/dl	$^1/_{10}$ of a liter (about ½ cup)
milliliter/ml	$^1/_{1000}$ of a liter (5 ml is about 1 tsp)

Appendix B

Daily Values Table

Dietary Constituent	Unit of Measure	Daily Values for People Over 4 Years of Age
Total fat	g	78
Saturated fatty acids	"	20
Protein	"	50
Cholesterol	mg	300
Total carbohydrate	g	275
Added sugars	"	50
Fiber	"	28
Vitamin A	µg (RAEs)	900
Vitamin D	µg	20
Vitamin E	mg	15
Vitamin K	µg	120
Vitamin C	mg	90
Folate	µg	400
Thiamin	mg	1.2
Riboflavin	"	1.3
Niacin	"	16
Vitamin B-6	"	1.7
Vitamin B-12	µg	2.4
Biotin	"	30
Pantothenic acid	mg	5
Choline	"	550
Calcium	"	1300
Phosphorus	"	1250
Iodine	µg	150
Iron	mg	18
Magnesium	"	420
Copper	"	0.9
Zinc	"	11
Sodium	"	2300
Potassium	"	4700
Chloride	"	2300
Manganese	"	2.3
Selenium	µg	55
Chromium	"	35
Molybdenum	"	45

Abbreviations: g = gram, mg = milligram, µg = microgram, RAEs = retinol activity equivalents

©Wendy Schiff

Appendix C

Energy Metabolism

Figure C.1 Glycolysis, step by step. This metabolic pathway begins with glucose and ends with pyruvate. Text in boxes explains the reactions.

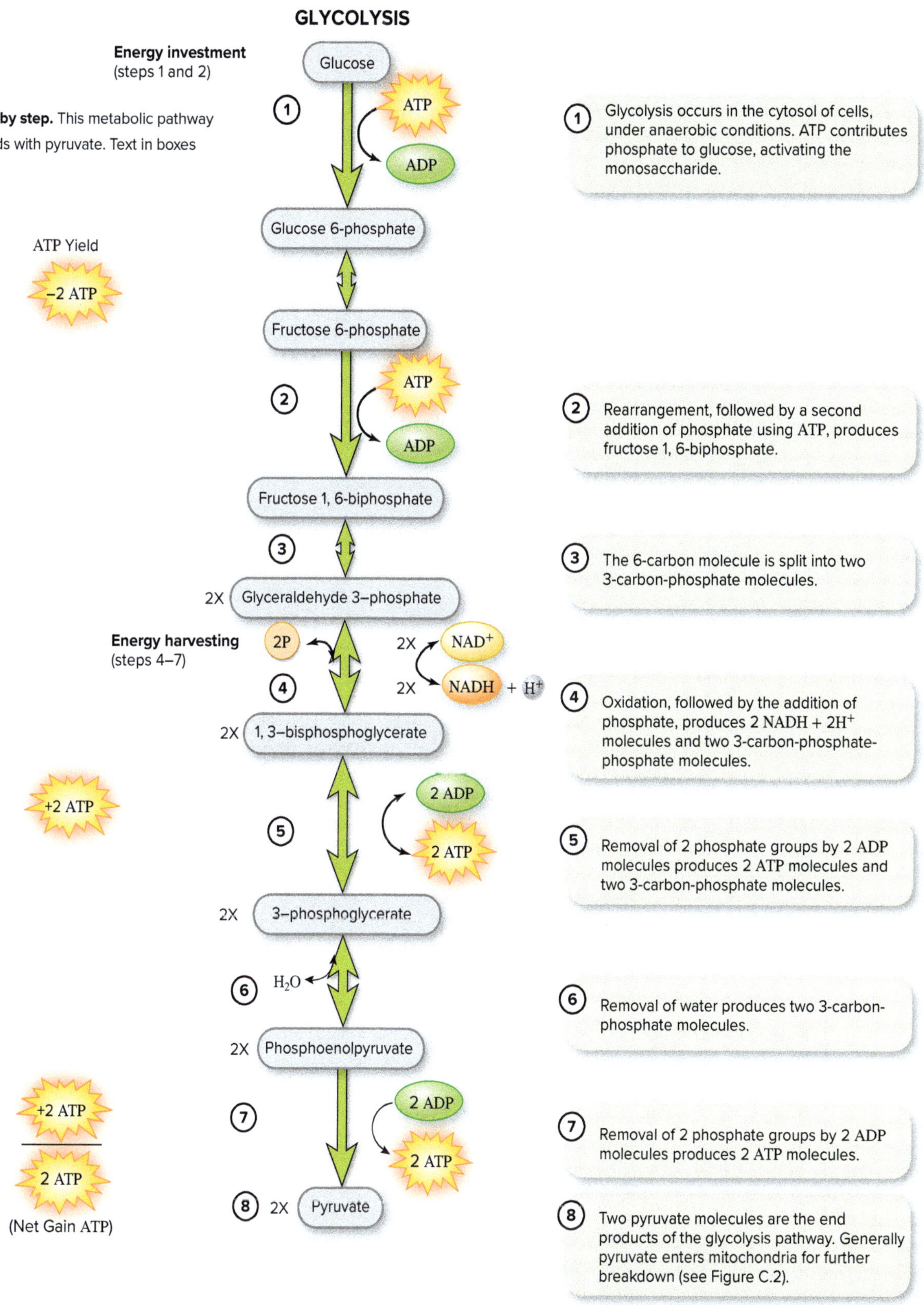

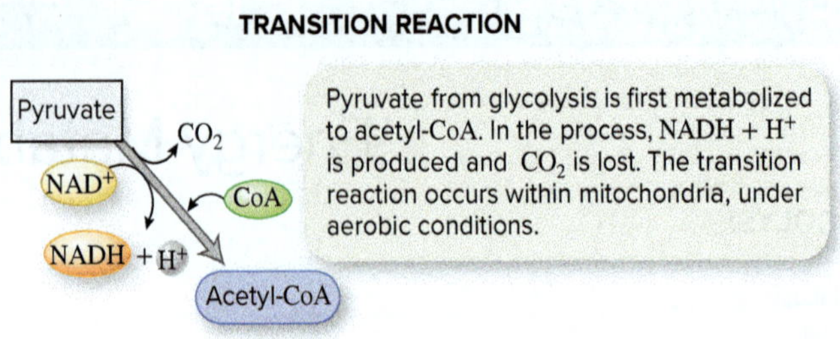

Figure C.2 The transition reaction.

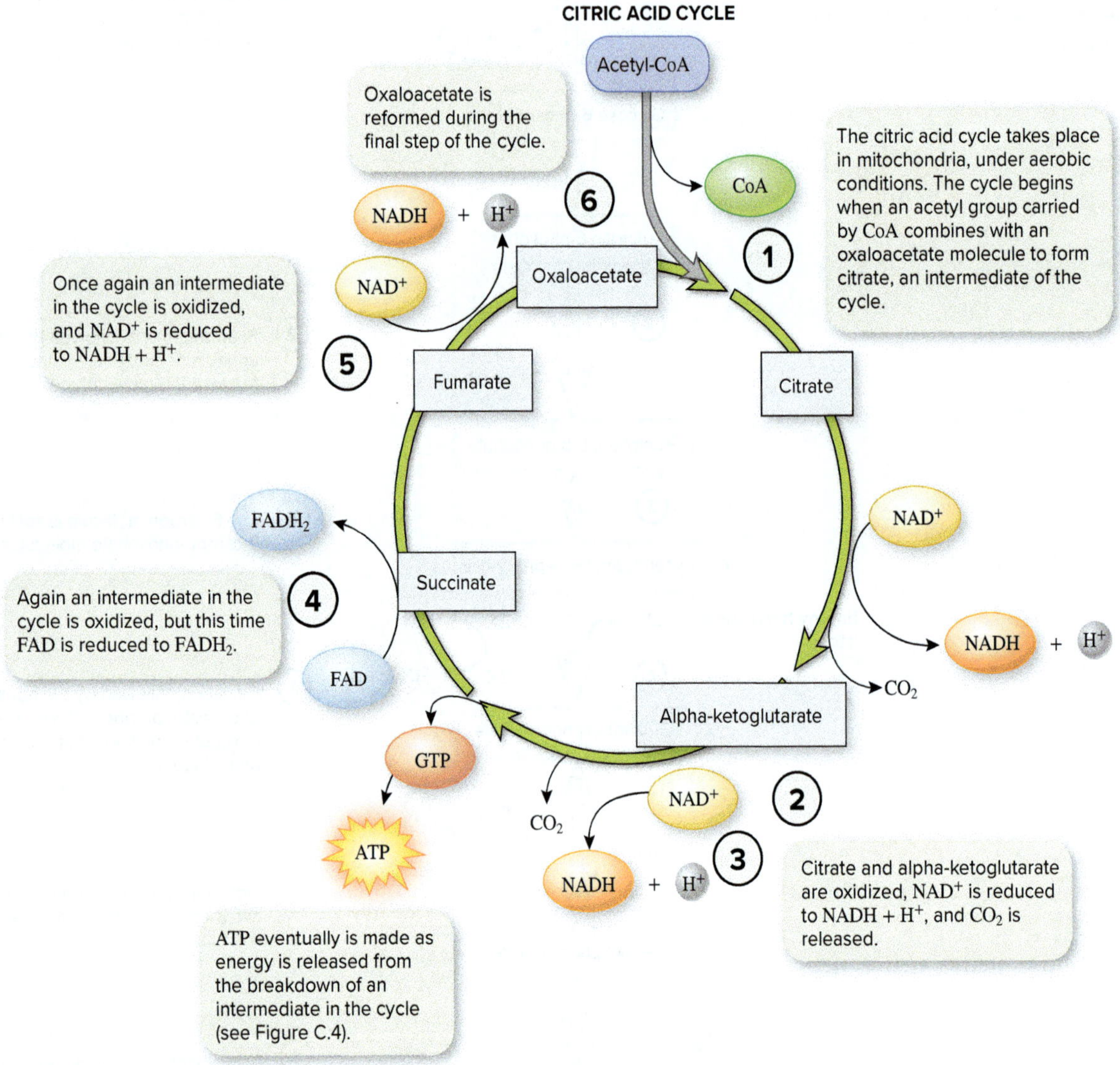

Figure C.3 The transition reaction and the citric acid cycle. The net result of one turn of the citric acid cycle (steps 1–6) is the oxidation of an acetyl group to two molecules of CO_2 and the formation of three molecules of $NADH + H^+$ and one molecule of $FADH_2$. One GTP molecule also results, which eventually forms ATP. The citric acid cycle turns twice per glucose molecule. Note that oxygen does not participate in any of the steps in the citric acid cycle. It instead participates in the electron transport chain.

Figure C.4 Simplified depiction of electron transfer in energy metabolism. High-energy compounds, such as glucose, give up electrons and hydrogen ions to NAD^+ and FAD. The $NADH + H^+$ and $FADH_2$ that are formed transfer these electrons and hydrogen ions to oxygen to form water (H_2O). This transfer process, which is called the electron transport chain, occurs in mitochondria. The energy yielded by the entire process generates ATP from ADP and P_i.

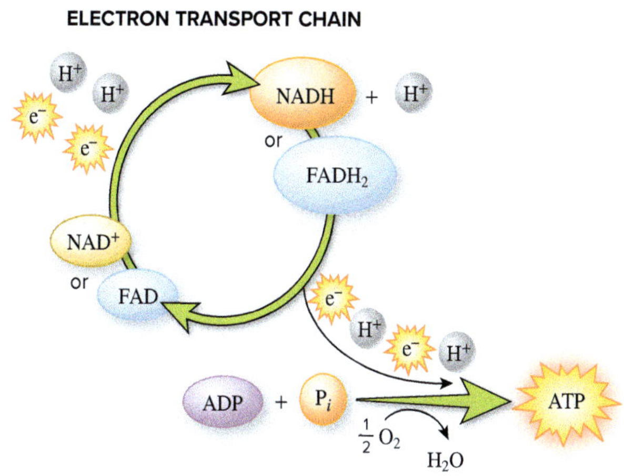

Appendix D

Ham: McGraw Hill;
Dairy products: D. Hurst/Alamy Stock Photo; Egg: Ken Karp/McGraw Hill

Amino Acids

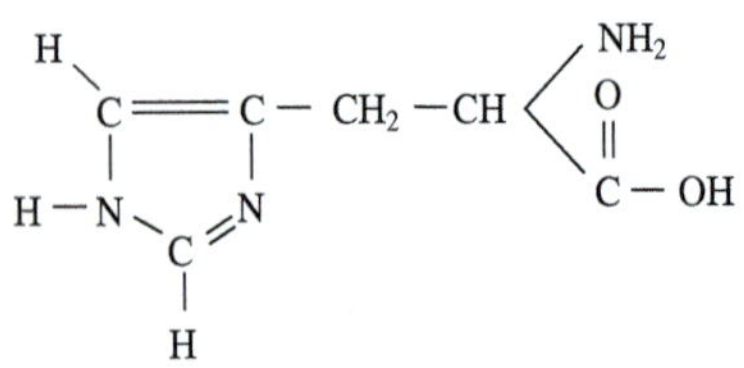

Histidine (His)
(essential)

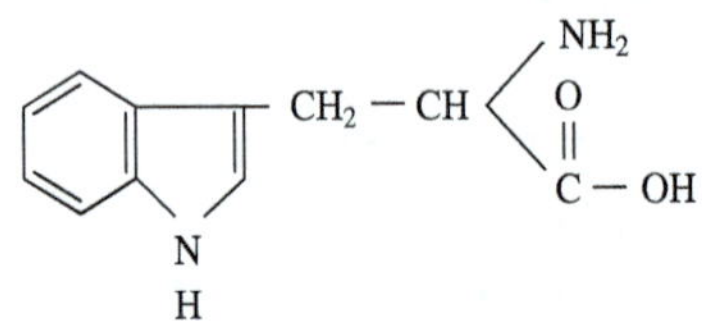

Tryptophan (Trp)
(essential)

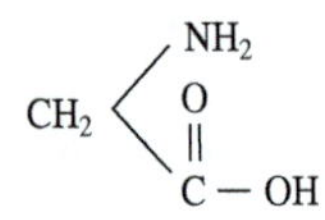

Glycine (Gly)

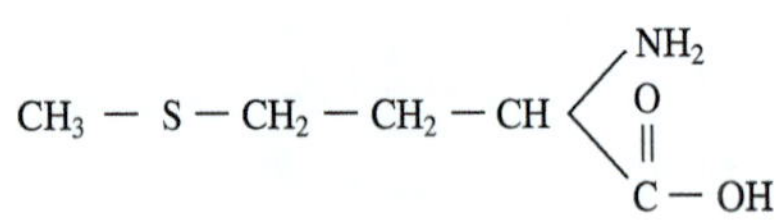

Methionine (Met)
(essential)

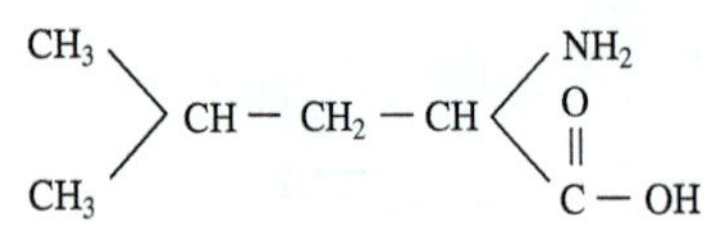

Leucine (Leu)
(essential)

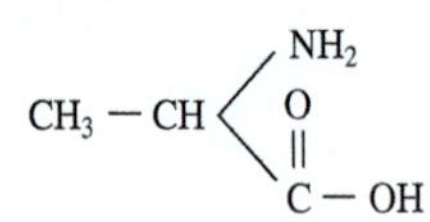

Alanine (Ala)

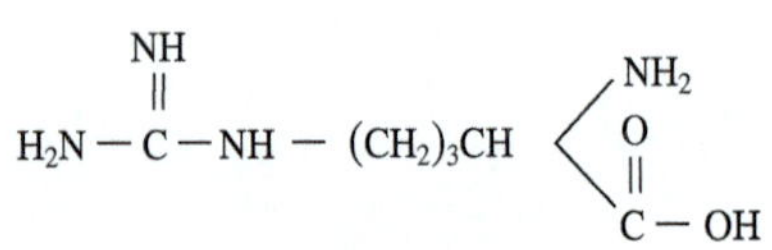

Arginine (Arg)
(essential in infancy)

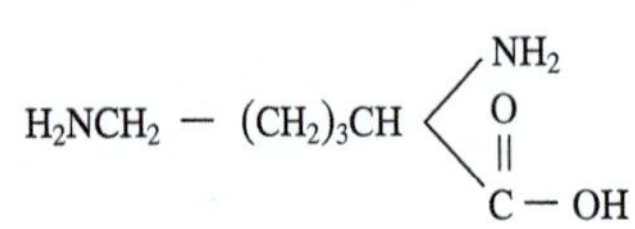

Lysine (Lys)
(essential)

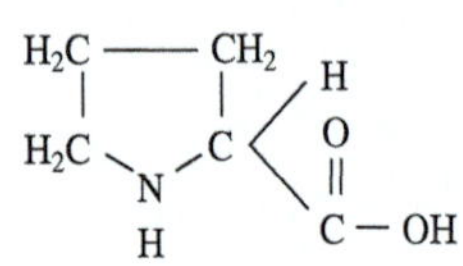

Proline (Pro)

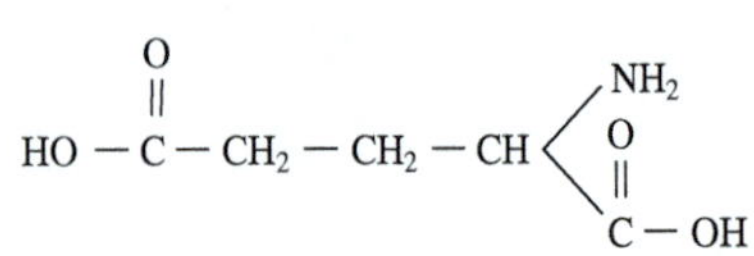

Glutamic Acid (Glu)

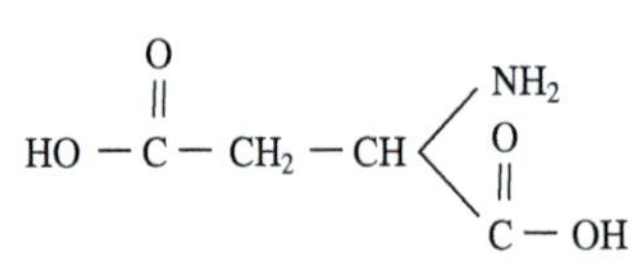

Aspartic Acid (Asp)

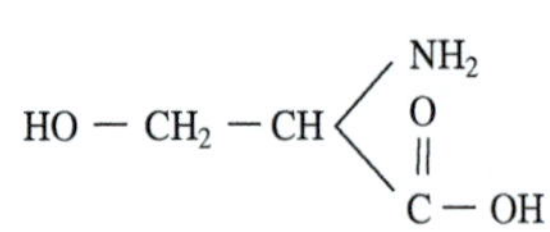

Serine (Ser)

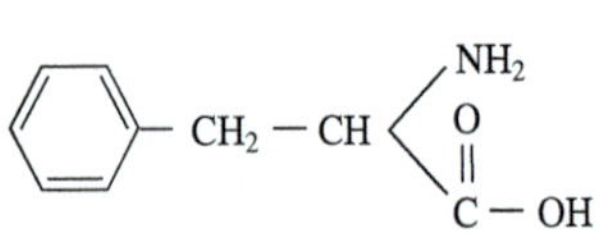

Phenylalanine (Phe)
(essential)

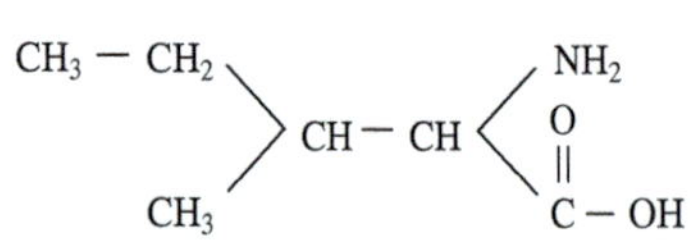

Isoleucine (Ile)
(essential)

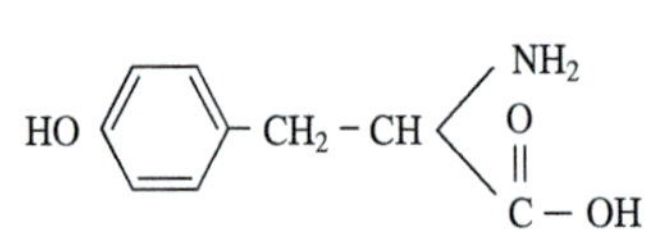

Tyrosine (Tyr)

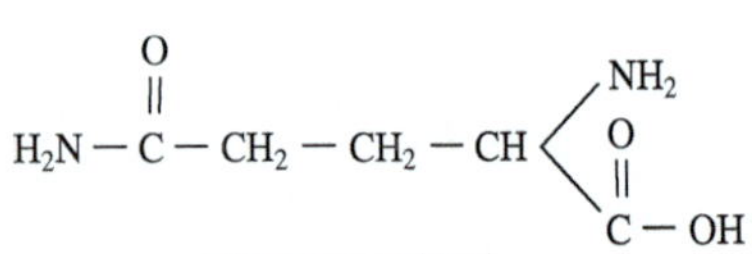

Glutamine (Gln)

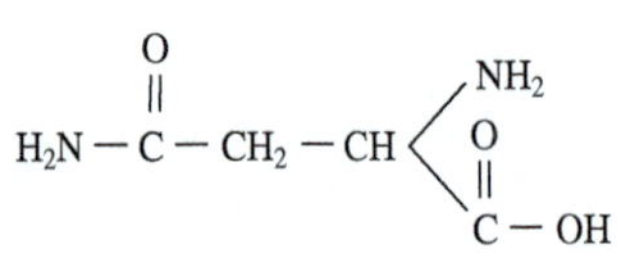

Asparagine (Asn)

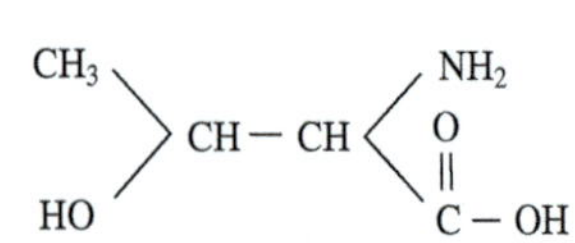

Threonine (Thr)
(essential)

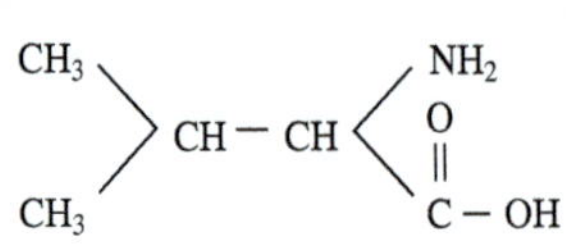

Valine (Val)
(essential)

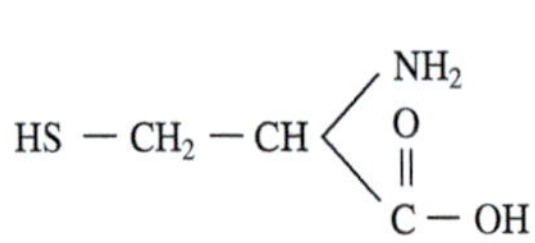

Cysteine (Cys)

Body Mass Index-for-Age Percentiles

2 to 20 years: Boys
Body mass index-for-age percentiles

NAME ____________________

RECORD # ____________

Date	Age	Weight	Stature	BMI*	Comments

***To Calculate BMI:** Weight (kg) ÷ Stature (cm) ÷ Stature (cm) x 10,000
or Weight (lb) ÷ Stature (in) ÷ Stature (in) x 703

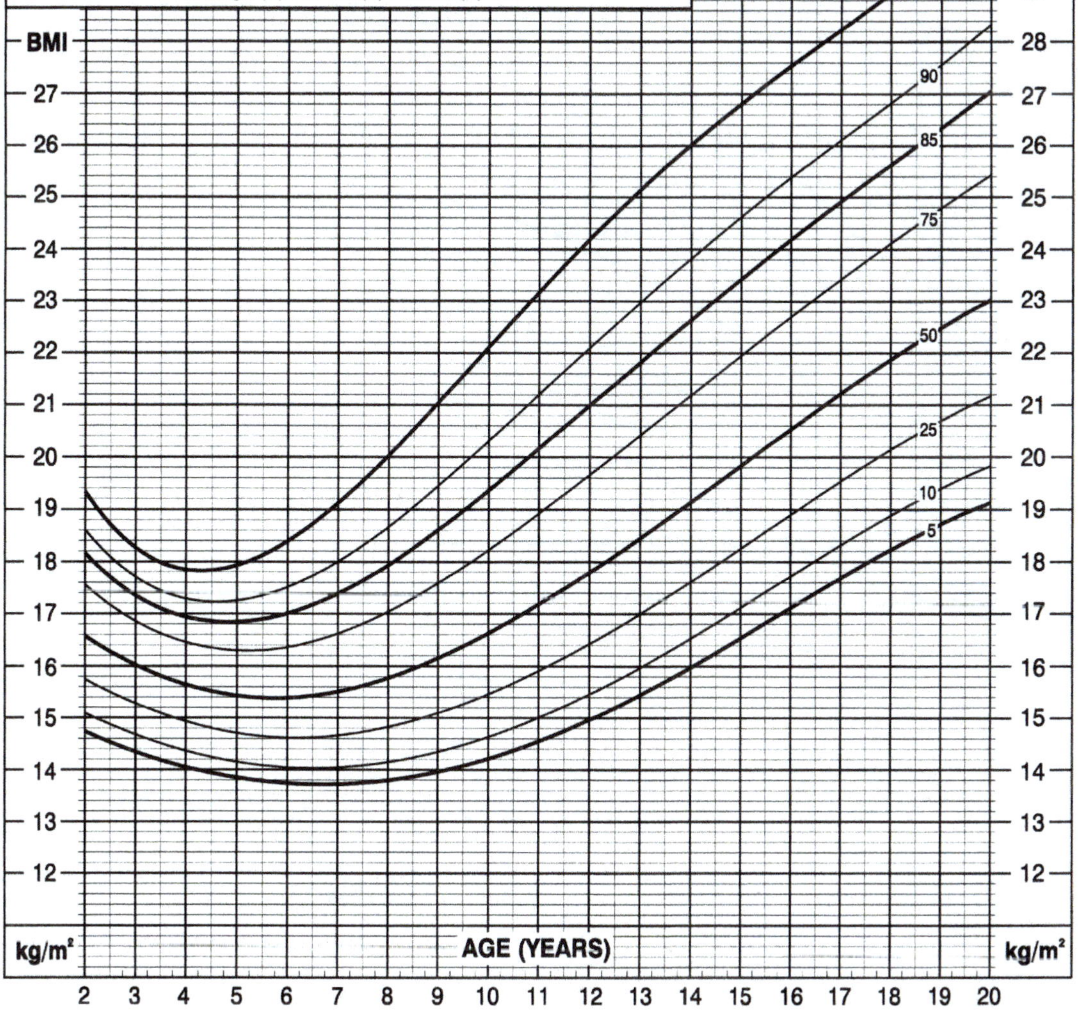

Published May 30, 2000 (modified 10/16/00).
SOURCE: Developed by the National Center for Health Statistics in collaboration with the National Center for Chronic Disease Prevention and Health Promotion (2000).

CDC

2 to 20 years: Girls
Body mass index-for-age percentiles

NAME ________________

RECORD # __________

Date	Age	Weight	Stature	BMI*	Comments

***To Calculate BMI:** Weight (kg) ÷ Stature (cm) ÷ Stature (cm) x 10,000
or Weight (lb) ÷ Stature (in) ÷ Stature (in) x 703

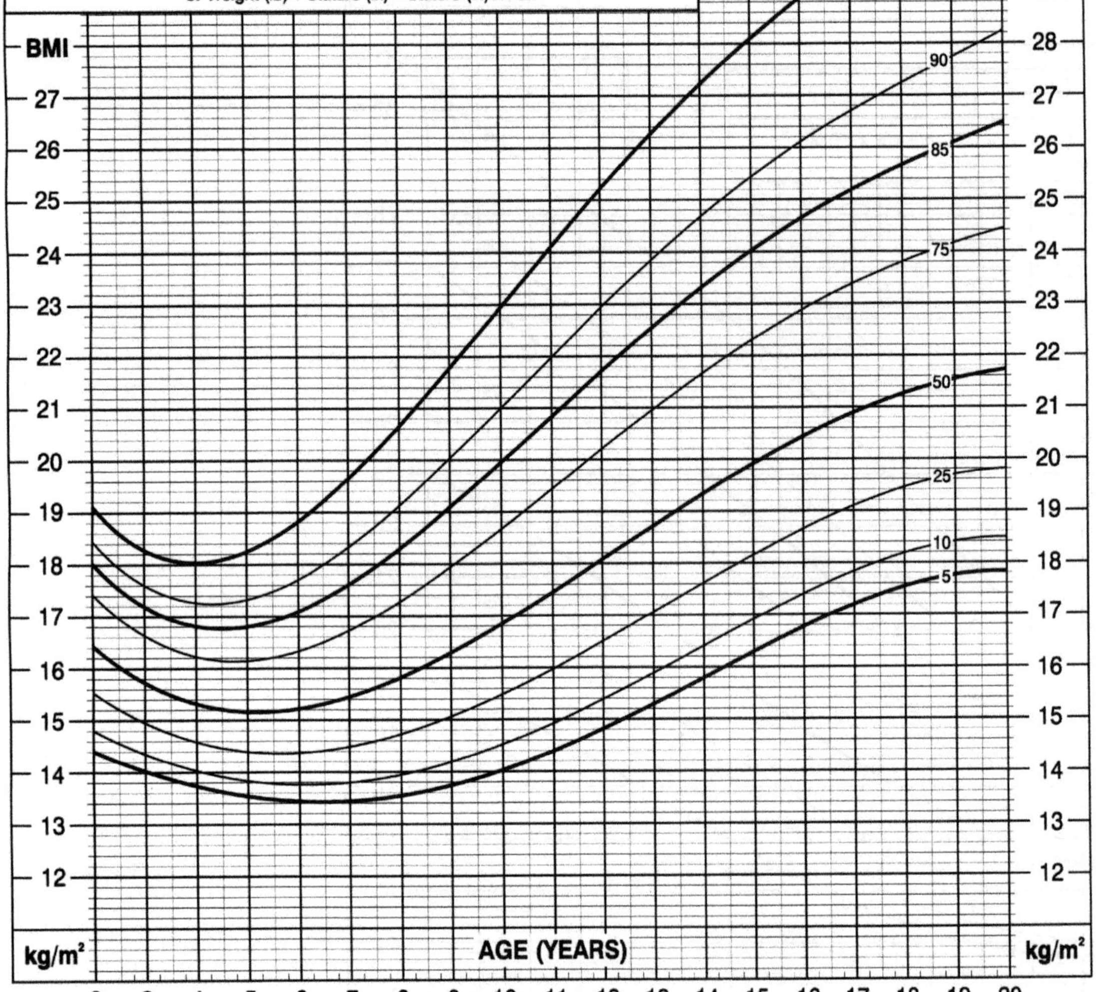

Published May 30, 2000 (modified 10/16/00).
SOURCE: Developed by the National Center for Health Statistics in collaboration with the National Center for Chronic Disease Prevention and Health Promotion (2000).

CDC

Appendix F

Answers to Concept Checkpoint Questions and Practice Test Questions

Concept Checkpoint Answers

Chapter 1

1. Heart disease, some types of cancer, stroke, and type 2 diabetes are among the 10 leading causes of death that are diet related.
2. Answers will vary but should include factors shown in Figure 1.1.
3. The six major classes of nutrients are carbohydrates, fats and other lipids, proteins, vitamins, minerals, and water.
4. The key features that indicate a substance is an essential nutrient include the following: A deficiency disease results when the substance is missing from the diet; the deficiency disease is corrected when the missing substance is added to the diet; and scientists can identify the nutrient's specific roles in the body and explain why abnormalities occurred when the substance was missing from the diet.
5. A phytochemical is a substance made by plants. According to the DSHEA, a dietary supplement is a product (excluding tobacco) that contains a vitamin, a mineral, an herb or other plant product, an amino acid, or a dietary substance that supplements the diet by increasing total intake.
6. A risk factor is a personal characteristic that increases chances of developing a chronic disease.
7. Answers will vary but should include lifestyle factors such as diet; alcohol, tobacco, and other drug use; and physical activity patterns.
8. Americans eat less red meat, milk, and eggs, and more fish and shellfish, cheese, and chicken and turkey, than in 1970. The diet supplies more peanuts and tree nuts, yogurt, and grains, but refined grain foods make up the majority of the grain products. Americans are eating additional servings of fruits and vegetables, but not enough to meet recommended amounts. Over the past 50 years, Americans' intake of added sugars has increased.
9. The main nutrition-related goal of *Healthy People 2030* is to improve health by promoting healthy eating and making nutritious foods available.
10. Health professionals generally use liters, grams, and meters to report volume, weight, and length.
11. A 198-pound person weighs 90 kg. To obtain the answer, divide 198 by 2.2 (90 kg).
12. To estimate the number of kilocalories in the slice of bread, add the number of kilocalories contributed by starch and sugar (carbohydrates) ($11 \text{ g} \times 4 \text{ kcal} = 44 \text{ kcal}$), fat ($3 \text{ g} \times 9 \text{ kcal} = 27 \text{ kcal}$), and protein ($3 \text{ g} \times 4 \text{ kcal} = 12 \text{ kcal}$). The answer is 83 kcal/slice.
13. Carbohydrates, fats, and proteins are macronutrients; vitamins and minerals are micronutrients.
14. Answers will vary but should be taken from the key nutrition-related concepts listed in Table 1.6.
15. A food that contains a lot of empty calories supplies excessive amounts of solid fat, added sugars, and/or alcohol. A nutrient-dense food contains more key beneficial nutrients (such as protein, fiber, calcium, and iron) in relation to its total calories per serving.
16. The physiological dose of a nutrient is the amount that is within the range of safe intake and enables the body to function optimally. A megadose is generally defined as an amount of a vitamin or mineral that is much greater than the recommended amount of the nutrient. Megadose amounts of certain nutrients can be toxic.
17. A person with excess body fat can be undernourished if their diet lacks essential nutrients, especially micronutrients.
18. Unfavorable environmental conditions that contribute to poor health in low-income countries include lack of access to clean water for drinking and preparing foods. Disruptive political conditions, including wars and civil unrest, contribute to poverty and food shortages.
19. Undernutrition increases the risk of death among pregnant females and children. Children who survive undernutrition often have stunted and delayed physical growth, blindness, and impaired intellectual development.
20. WIC is the Women, Infants, and Children feeding program, which is federally subsidized. WIC provides access to specific nutritious foods to improve the diets of low-income infants, children under the age of 5 years, and pregnant or breastfeeding women who are at nutritional risk.
21. Conventional farming methods require considerable amounts of water and pesticides that can harm the environment. Sustainable agriculture involves farming methods that provide food without depleting natural resources or harming the environment.
22. A bioengineered food results when humans change an organism's DNA and the products of such efforts are eaten.

Chapter 2

1. After performing the research study and collecting study data, researchers will next analyze their study data and draw conclusions based on the study results.
2. A control group is used to compare the results of having a treatment against those of not having a treatment.
3. A prospective study follows a large group of healthy people over time to determine whether those with a certain characteristic develop a disease and those without that characteristic remain disease free. A retrospective study follows a group of people who already suffer from a disease and compares them to a group of people with similar characteristics who do not have the disease.
4. A placebo is a fake treatment, such as a sham pill, injection, or medical procedure. Providing placebos to members of the control group enables scientists to compare the extent of the treatment's response with that of the placebo.
5. A direct or positive correlation in a research study is the relationship between two variables that change or move together in the same direction. An inverse or negative correlation is the relationship between two variables that change in opposite directions.
6. A peer–reviewed article is one that has been critically analyzed by a group of "peers." If peers agree that a study was well conducted, its results are fairly represented, and the research is of interest to the journal's readers, these scientists are likely to recommend that the journal's editors publish the article.
7. Conflicting research findings often result from differences in the ways various studies are designed and results analyzed. For example, the numbers, ages, and physical conditions of subjects; the type and length of the study; the amount of the treatment provided; and the statistical tests used to analyze results typically vary among studies. Other factors, such as genetic and lifestyle differences among human subjects, can influence the results of nutrition research.
8. A testimonial is a personal endorsement of a product, and anecdotes are reports or personal experiences. These two sources of nutrition information are not based on scientific facts and, therefore, are not reliable and contribute to misinformation.
9. Answers will vary but should be clues that indicate a source of nutrition information is unreliable.
10. Answers will vary but should include points presented in Table 2.1.
11. Although some states regulate and license people who call themselves nutritionists, consumers cannot always rely on those who refer to themselves as a "nutritionist" because there are no standard legal definitions for this descriptor. Registered dietitians and registered dietitian nutritionists are college–trained professionals who have extensive knowledge of foods, nutrition, and dietetics, the application of nutrition and food information to treat many health–related conditions. The titles "registered dietitian" and "registered dietitian nutritionist" are legally protected.
12. Consumers may find reliable nutrition experts, including registered dietitians, among faculty members in universities or colleges. Other ways to find registered dietitians include contacting the local dietetic association, calling the dietary department of a local hospital, or visiting the Academy of Nutrition and Dietetics website (www.eatright.org) or the Dietitians of Canada website (www.dietitians.ca).
13. Answers may vary but can include chiropractic manipu–lations, yoga, meditation, and use of herbal remedies and nonvitamin, nonmineral products.
14. According to the FDA, a drug is a substance, natural or human–made, that alters body functions. Although many dietary supplements act as drugs in the body, the FDA regulates dietary supplements as foods, not medications. Therefore, consumers do not need to obtain a physician's prescription to purchase a dietary supplement.
15. Answers will vary but can include the dietary supplements listed in Table 2.2. Responses should not include vitamins or minerals.

Chapter 3

1. To establish an RDA for a nutrient, nutrition scientists first determine its estimated average requirement. Then scientists add a "margin of safety" amount to the EAR that allows for individual variations in nutrient needs and helps maintain tissue stores.
2. RDAs are standards for recommending daily intakes of several nutrients. RDAs meet the nutrient needs of nearly all healthy individuals (about 98%) in a particular life stage/sex group. In some instances, scientists set AIs for certain nutrients because they do not have enough information to establish RDAs for them. An AI represents a group's average daily intake that appears to be adequate for health.
3. The EER is more specific than AMDRs or CDRRs because it takes into account the person's physical activity level, height, and weight, as well as sex and life stage. Furthermore, the EER estimates a person's energy requirements, while the AMDRs are macronutrient intake ranges that are nutri–tionally adequate and may reduce the risk of diet–related chronic diseases and the CDRRs are nutrient intake levels that reduce chronic disease risk within a healthy population.
4. Dietitians refer to DRIs as standards for planning nutritious diets for groups of people and evaluating the nutritional adequacy of a population's diet. Pharmaceutical companies refer to DRIs when developing formulas that replace breast milk or regular foods. For nutrition labeling purposes, the FDA uses RDAs to develop Daily Values (DVs).
5. Answers will vary but should include grains, such as products made from wheat, rice, and oats. Pasta, noodles, and flour tortillas are grains because wheat flour is their

main ingredient. Corn is a type of grain; therefore, cornmeal and popcorn are usually grouped with grains.

6. Enrichment is the replacement of some nutrients that were lost during processing. Fortification is the addition of nutrients to food to boost its nutritional content.
7. Foods generally classified as dairy include forms of fluid milk, yogurt, hard cheese, cottage cheese, ice cream, pudding, and frozen yogurt.
8. Dry beans, eggs, and nuts can substitute for meats because they are protein–rich foods.
9. According to the USDA, one–half cup of dried fruit is nutritionally equivalent to 1 cup of fresh fruit. Therefore, 1 cup of dried apricots is nutritionally equivalent to 2 cups of fresh apricots.
10. The nutritional content of foods within each food group often varies widely. Therefore, dietary guides generally recommend that people choose a variety of foods from each food group when planning daily meals and snacks to ensure nutrient needs are met.
11. Solid fats are usually hard at room temperature, such as beef fat, butter, lard, and shortening. Cream and coconut oil are also considered solid fats.
12. A minimally processed food has undergone simple kinds of processing, such as peeling, canning, or freezing. An ultra–processed food contains one or more additives that are intended to improve its shelf life or other desirable characteristics.
13. Answers will vary but should include the guidelines listed in Table 3.3.
14. A healthy person should limit their sodium intake to less than 2300 mg/day.
15. Answers will vary but should include the recommendations listed in the "Follow a Healthy Dietary Pattern at Every Life Stage" section.
16. Replace solid fats with oils.
17. Your diet should have less than 10% of total calories from saturated fat.
18. Limit added sugars intake to less than 10% of total calories.
19. At least one–half of your grains should be whole grains.
20. Pregnant females should not drink alcohol.
21. Calcium, vitamin D, potassium, and dietary fiber are nutrients that Americans tend to consume in limited amounts (Table 3.5).
22. Answers are fruit, dairy, or vegetables.
23. Answers may vary but should include comparing dietary intakes against recommended amounts of foods and beverages within one's diet pattern. Furthermore, a response can involve using the interactive menu planning features at the www.myplate.gov website to monitor food and nutrient intakes.
24. Answers may vary but should be either grains or meat.
25. Food Lists for diabetes and weight management focus on the energy, protein, carbohydrate, and fat content of foods. Unlike the MyPlate food groupings, classification of foods into lists is based on energy and macronutrient contribution.
26. Answers may vary but should discuss how %DVs are percentages instead of specific quantities, such as grams or milligrams of nutrients. Also, fresh or raw foods generally do not have labels indicating %DVs. Therefore, people may underestimate their intakes of nutrients if they rely on nutrition labels for the information.
27. Information provided on food labels and dietary supplements can be used to compare ingredients, as well as energy and nutrient contributions of similar products.
28. A health claim describes the relationship between a food, food ingredient, or dietary supplement and the reduced risk of a nutrition–related condition. Examples of health claims can vary, but students can use the ones listed in Table 3.8. A structure/function claim describes the role a nutrient or dietary supplement plays in maintaining a structure, such as "calcium strengthens bone," or in promoting a normal function, such as "fiber aids bowel regularity." Examples of nutrient claims can vary, but students can use the ones listed in Table 3.9.
29. The FDA does not permit manufacturers to market a dietary supplement product as a treatment or cure for a disease, or to relieve signs or symptoms of a disease. If FDA officials question the safety of a dietary supplement or the truthfulness of claims that appear on supplement labels, manufacturers are responsible for providing the agency with evidence that their products are safe and the claims on labels are honest and not misleading. The FDA requires dietary supplement manufacturers to evaluate the purity, quality, strength, and composition of their products before marketing them.
30. The organic milk would be from cattle that were fed 95 to 100% organic feed and not given antibiotics or hormones.
31. The USDA permits food processors to use the organic seal on packaging when the food contains at least 95% organic ingredients.
32. The Nutrition Facts panels on food labels, nutrient analysis software programs, and government websites can be reliable sources of information about the energy and nutrient contents of foods and beverages.
33. Traditional ethnic diets such as the Latin American, Mediterranean, African Heritage, and Asian diets are based on dishes containing small amounts of animal foods and larger amounts of plant foods, including locally grown fruits and vegetables, whole grains, legumes, nuts, and seeds. Scientific research shows these traditional eating patterns are associated with a lower incidence of chronic disease.
34. Answers will vary but should include information from Table 3.12.

Chapter 4

1. Electrons are negatively charged particles that surround the nucleus; protons are positively charged particles in the nucleus of an atom. An element is a substance that cannot be broken down into distinctive components under usual conditions. A chemical bond is an attraction that forms when atoms interact. A molecule results from the formation of chemical bonds. A compound is a molecule that contains two or more different elements. A solution is an evenly distributed mixture of two compounds. The solvent is the primary compound of a solution, and the solute is the substance dissolved in the solvent.
2. An ion is an atom or group of atoms that has an electrical charge because it has gained or lost one or more electrons.
3. Acids are substances that lose H^+ when dissolved in water; bases are substances that remove and accept H^+ when dissolved in water.
4. The pH of a watery solution refers to its hydrogen ion concentration.
5. A chemical reaction is a process that changes the arrangement of atoms in molecules.
6. Enzymes are proteins that initiate or facilitate (catalyze) chemical reactions.
7. Enzymes are sensitive to environmental conditions, including pH, temperature, and the presence of certain vitamins and minerals.
8. A cell is the smallest living functional unit in an organism. An organelle is a structure within the cell that has specific functions. DNA is a molecule that provides coded instructions for synthesizing proteins. A tissue is composed of cells usually joined together that have similar characteristics and functions.
9. Homeostasis is an internal chemical and physical environment that supports life and good health.
10. Answers will vary but should include information from Table 4.2.
11. Answers will vary but should relate to Figure 4.33 and begin with the mouth and end with the anus.
12. Examples of mechanical digestion include the biting and grinding actions of teeth and the muscular contractions of the GI tract. Chemical digestion includes actions of enzymes and HCl.
13. Choking can occur while eating if the epiglottis does not function properly and food enters the larynx and windpipe (trachea).
14. The lower esophageal sphincter keeps stomach contents from backing up into the esophagus.
15. The three sections of the small intestine are the duodenum, jejunum, and ileum. Most nutrient digestion and absorption take place in the upper portion of the small intestine, primarily in the jejunum.
16. Removal of the pancreas would reduce nutrient digestion because the pancreas produces digestive enzymes.
17. Some gut microbes can cause inflammation, especially if their numbers are not kept in control by populations of other, more beneficial gut microbes.
18. These conditions are types of inflammatory bowel disease (IBD) that may be the result of a faulty immune response by the body. IBDs are chronic and painful, and can cause anemia and weight loss due to malabsorption. Having an IBD increases the patient's risk of developing colon cancer.
19. Possible causes of constipation include low dietary fiber intake, low water intake, changes in usual routine, anxiety or depression, and ignoring usual urge to defecate.
20. Vomiting and diarrhea can cause dehydration, which can be dangerous.
21. Answers will vary but should include the information in Table 4.3.
22. Answers will vary but should include avoiding smoking, heavy consumption of alcohol, and use of NSAIDs.
23. Common signs and symptoms of irritable bowel syndrome are intestinal cramps, diarrhea, constipation, or alternating episodes of both diarrhea and constipation. Loose stools are often accompanied with mucus, and after bowel movements, the affected person feels as though elimination of stools was incomplete.
24. Answers will vary but should include the signs and symptoms listed in Table 4.4.

Chapter 5

1. Plants use carbon dioxide and water to make carbohydrates.
2. Plants use sunlight to obtain energy they need to live.
3. A human cell needs glucose to obtain energy.
4. The three most important dietary monosaccharides are glucose, fructose, and galactose.
5. Glucose is blood sugar; sucrose is table sugar; lactose is milk sugar; and maltose is malt sugar. Maltose comprises two glucose molecules; glucose and galactose comprise lactose; and glucose and fructose comprise sucrose.
6. A nutritive sweetener provides energy, whereas a nonnutritive sweetener does not provide energy in the amounts typically used to sweeten foods.
7. The parents should give their child the beverage sweetened with sucralose because the sweetener does not provide phenylalanine.
8. Starch and glycogen are both storage forms of polysaccharides, but starch is in plant foods and glycogen is primarily in liver and muscle tissue.
9. Most forms of dietary fiber are nondigestible carbohydrates. Oat bran and oatmeal, beans, apples, carrots, oranges and other citrus fruits, and psyllium seeds are rich sources of

soluble fiber; whole–grain products, including brown rice, contain high amounts of insoluble fiber.

10. A minor amount of starch is digested in the mouth by the action of salivary amylase. Starch digestion stops soon after the food enters the acid environment of the stomach. In the small intestine, pancreatic amylase breaks down remaining starch molecules into maltose molecules. The enzyme maltase splits each maltose molecule into two glucose molecules. The glucose molecules are absorbed by small intestinal cells; they enter the bloodstream and travel to the liver via the portal vein. In the small intestine, sucrase splits the sucrose molecules in the fruit jam spread into glucose and fructose molecules. These monosaccharides are absorbed by small intestinal cells and eventually enter the hepatic portal vein and travel to the liver. The fiber in the roti bread was not digested. The fiber moves through his intestinal tract and may be fermented by bacteria in his large intestine or contribute to his feces.

11. Insulin helps regulate blood glucose levels because the hormone enables glucose to enter most cells. The hormone enhances energy storage by promoting fat, glycogen, and protein production. Glucagon opposes insulin's effects by promoting the breakdown of glycogen. This process releases glucose into the bloodstream and, as a result, boosts the blood glucose level back to normal. Glucagon also stimulates liver and kidney cells to produce glucose from certain amino acids, the basic molecules that make up proteins. Furthermore, glucagon stimulates the break–down of triglyceride into glycerol and fatty acids. As a result, glycerol and fatty acids rapidly enter into the bloodstream.

12. A ketone body is a chemical that results from the incomplete breakdown of fat. Cells form ketone bodies when they must use greater–than–normal amounts of fat for energy. Under these conditions, there is not enough glucose available for cells to metabolize the fat efficiently, and excessive ketone bodies form as a result.

13. Sugar–sweetened beverages, sugary snacks, and sweets are the main sources of added sugars in Americans' diets.

14. Fruit juice is a better choice because juice is a natural product that contains nutrients and phytochemicals in fruit. "Orange–ade" is not a nutritionally equivalent substitute for orange juice because it may contain only 10% fruit juice.

15. To estimate the grams of starch in the serving of the cereal, subtract the amount of starch and total sugars in a serving (20 grams) from the amount of total carbohy–drates (44 grams), and the remainder is the amount of starch (24 grams).

16. On a Nutrition Facts panel, a serving of food that supplies 5%DV contains a low amount of the nutrient. A serving of food that supplies 20%DV or more contains a high amount of the nutrient. The food product that provides 8%DV for added sugars would be the better choice because it contains less added sugars. Dietary guidelines recommend people limit the amount of added sugars in their diet because these foods may replace other nutrient–dense foods or provide extra calories that can lead to weight gain.

17. Answers will vary but should be the signs and symptoms listed in Table 5.9.

18. To reduce her risk of developing type 2 diabetes, Lata should avoid being overweight, be physically active, and eat a healthy diet.

19. Answers may vary but should include signs presented in Table 5.11.

20. Lactose intolerance is a condition characterized by the inability to digest lactose completely.

21. Answers will vary but should include increasing one's intake of plant foods, especially whole–grain breads and cereals.

22. Eating soluble fiber can reduce the risk of cardiovascular disease by reducing blood cholesterol levels. The liver recycles the cholesterol from bile that is absorbed by the intestinal tract to make new bile. Soluble fiber interferes with the absorption of bile salts in the intestinal tract. As a result, the liver must use other sources of cholesterol, such as blood cholesterol, to produce bile. The fiber in food attracts water and swells in the digestive tract, forming a large soft mass that applies pressure to the inner muscular walls of the large intestine, stimulating the muscles to push the residue quickly through the tract. As a result, people who eat adequate amounts of fiber are less likely to strain while having bowel movements than people whose diets lack fiber.

23. Of this group of foods, raw apple has the lowest glycemic index.

Chapter 6

1. Rose's excess body fat can provide a source of energy for her, if she is ill and cannot eat. Also, the fat provides some insulation against cold temperatures and may protect her bones from breaking if she falls.

2. The cake is likely to taste better and have a more desirable texture because the fat in the milk contributes to flavor and "mouth feel."

3. Major lipids include fatty acids, triglycerides, phospholipids, and sterols, particularly cholesterol.

4. A saturated fatty acid has each carbon within the fatty acid chain completely filled with hydrogen atoms. An unsaturated fatty acid has two neighboring carbons within the chain that are missing two hydrogen atoms. Mono–unsaturated fatty acids have one double bond within the carbon chain; polyunsaturated fatty acids have two or more double bonds within the carbon chain.

5. Answers will vary but may include the fats and oils in Table 6.2.

6. Alpha–linolenic acid and linoleic acid are essential fatty acids Signs of essential fatty acid deficiency include poor growth in infants, scaly skin, poor wound healing, and hair loss.

7. Phospholipids have a hydrophylic region that contains phosphorus and a hydrophobic region with two fatty acids. A triglyceride has three fatty acids attached to a glycerol "backbone."
8. An omega–3 fatty acid has the first double bond in the polyunsaturated fatty acid's carbon chain appearing at the third carbon, when you start counting carbons at the omega end of the molecule. Alpha–linolenic acid, eicosapen–taenoic acid (EPA), and docosahexaenoic acid (DHA) are omega–3 fatty acids.
9. Egg yolk or whole egg would keep the oil and milk emulsified.
10. Cholesterol is found only in animal foods. Egg yolk, liver, meat, poultry, whole milk, cheese, and ice cream are rich sources of cholesterol. Cholesterol is a component of every cell membrane; cells use cholesterol to synthesize vitamin D, bile, and steroid hormones.
11. Answers will vary but should follow the steps outlined in Figures 6.10 to Figure 6.12, and the summary of lipid digestion and absorption presented in Figure 6.13.
12. Plants contain substances that interfere with enterohepatic circulation. This interference can reduce blood cholesterol levels because the liver must remove cholesterol from the bloodstream to make new bile salts.
13. The body needs energy to survive. The triglycerides stored in fat cells can be used for energy. Thus, a person with excessive body fat has more energy stored in their body than a slim person.
14. On average, fat contributes almost 38% of the typical American diet's total energy intake.
15. The AMDR for fat is 20 to 35% of total calories.
16. According to the Dietary Guidelines, people should consume less than 10% of their total calories from saturated fatty acids.
17. The serving of chips provides 9 g of fat. Each gram of fat supplies 9 kcal. Thus, fat contributes 81 of the 150 kcal in a serving of the chips (9 g × 9 kcal/g).
18. Atherosclerosis is a long–term disease process in which plaques build up inside arterial walls. Arterial plaques are deposits of cholesterol and other material under the lining of arteries. A thrombus is a fixed bunch of clots that remains in place. An embolus is a thrombus or part of a plaque that breaks free and travels through the bloodstream. Atherosclerosis can occur when something in the blood–stream irritates the lining of an artery. The body's immune system responds by producing inflammation within the artery. Inflammation can stimulate healing, but the process can also trigger certain cells within the arterial wall to form plaque. Plaque interferes with or blocks circulation in the affected area. Plaque also roughens the normally smooth surface that lines the artery. The rough lining slows blood flow in the area and makes clots more likely to form. If a plaque ruptures, repairing the damage also involves clot formation, and such blood clots can be life threatening.
19. Answers will vary but should include king mackerel, marlin, swordfish, or shark.
20. Answers will vary but should relate to information about risk factors presented in Table 6.4.
21. According to this information, Bernard has a low risk of CVD because his HDL cholesterol level is high.
22. The liver releases high–sensitivity C–reactive protein in response to infection and inflammation. Elevated CRP may be a marker for atherosclerosis, which means it may be an early warning sign for the condition. A lipoprotein profile can provide information about blood levels of HDL, LDL, and triglycerides.
23. Answers will vary but should include reducing intakes of meats; foods made from whole milk; and crackers, cakes, cookies, pies, and other high–fat processed foods. Increasing intakes of most nuts, fatty fish, and sources of polyunsaturated oils can increase intakes of unsaturated fats.
24. A statin is a prescription drug that reduces elevated blood lipid levels. Statins interfere with the liver's metabolism of cholesterol, effectively reducing LDL cholesterol and/or triglyceride levels as a result.
25. Answers will vary but should include points in Figure 6.29.
26. BAC is "blood alcohol concentration." As of January 2021, the legal limit for BAC was 0.8 for people who are 21 years of age or older.
27. Dontel is a heavy drinker.

Chapter 7

1. The amino acid is the chemical unit that makes up a protein.
2. Answers may vary but may include cell development and maintenance, and the production of enzymes, antibodies, certain hormones, structural and contractile components, and blood–clotting factors. Additionally, proteins transport nutrients and oxygen in the bloodstream, help maintain acid–base and fluid balance, and can be used for energy.
3. The three groups that make up an amino acid are the amino or nitrogen–containing group, R group, and acid group.
4. A carbon skeleton is the part of an amino acid that remains after the amino group is removed.
5. Twenty amino acids are needed to make human proteins. Nine of these amino acids are essential.
6. A high–quality or complete protein contains all essential amino acids in amounts that will support protein deposition in muscles and other tissues or support a young child's growth. A low–quality protein lacks one or more essential amino acids.
7. Most animal proteins are sources of high–quality proteins. Quinoa, amaranth, buckwheat, and processed soy proteins

are also good sources of high–quality protein. Other plant foods, particularly fruits and vegetables, are sources of low–quality proteins.

8. Tryptophan, lysine, threonine, methionine, and cysteine are often the limiting amino acids in foods.
9. Answers will vary but should include key steps noted in Figure 7.6.
10. A protein undergoes denaturation when it is exposed to various conditions that alter the macronutrient's natural folded and coiled shape. Deamination is the removal of the nitrogen–containing group from an amino acid. Transamination is the transfer of an amino group from an amino acid to a receiving compound such as a carbon skeleton.
11. Negative balance occurs during starvation, serious illnesses, and severe injuries. Positive balance occurs during periods of rapid growth, such as pregnancy, infancy, and puberty, and when people are recovering from illness or injury. Performing weight–training activities also leads to nitrogen retention.
12. To calculate the answer, convert the female's weight to kilograms by dividing 143 pounds by 2.2 (65 kg). Multiply her weight in kilograms (65) by 0.8 gram of protein/kg to calculate her RDA for protein (52 g).
13. Answers will vary but should follow the path illustrated in Figure 7.10.
14. The AMDR for adult protein intake is 10 to 35% of energy from protein.
15. Adult Americans' intake of unprocessed poultry products has increased since 1999, and the population's intake of unprocessed red meat has declined since 1999. Intakes of fish and shellfish have remained the same since 1999.
16. Consumption of red meat and processed meat is associ–ated with increased risk of chronic diseases, especially heart disease and hypertension.
17. Answers will vary based on individual food intakes.
18. To calculate the answer, divide the amount of protein/serving (5 g) by the Daily Value for protein, which is 50 g (5/50 = .10). Then, convert .10 to a percentage by moving the decimal point two spaces to the right and replacing it with a percent symbol. The answer is 10%.
19. High–quality sources of proteins contain all the essential amino acids. In Figure 7.15b, quinoa and chicken breast are high–quality proteins that support growth and development. Low–quality proteins lack one or more of the essential amino acids or contain inadequate amounts of these nutrients. In Figure 7.15b, the chickpeas, bell peppers, spinach, and carrots are low–quality protein sources.
20. Foods that substitute for meat include fish, poultry, eggs, soy products, or protein–rich dairy products such as cheese and yogurt. Extending high–quality protein refers to preparing dishes that incorporate small amounts of animal proteins with much larger amounts of plant proteins. Spaghetti with meat sauce is an example of a dish that extends high–quality protein in meat with more low–quality proteins in pasta.
21. The recipe does not provide a complementary mixture of proteins because fruit is not a good source of plant proteins to combine with peanuts.
22. Answers will vary but should include quinoa, amaranth, and buckwheat or legumes to complement or improve the cereal protein.
23. Semivegetarians consume small amounts of meat, fish, or poultry, whereas other vegetarians do not eat animal flesh.
24. Vegan diets may not contain enough energy, high–quality protein, omega–3 fatty acids, vitamins B–12 and D, and minerals zinc, iron, and calcium to meet a person's nutri–tional needs.
25. Children have higher protein and energy needs per pound of body weight than adults have. Because plant foods add bulk to the diet, vegan children are more likely to eat far less food than adult vegans because they become full sooner during meals. Thus, very young vegans may be unable to eat enough plant foods to meet their protein and energy needs.
26. People suffering from chronic diseases are at risk of protein undernutrition. People with low incomes, especially older adults, are also at risk of protein deficiency.
27. Children with PEM do not have enough protein and energy to grow properly and fight infections.
28. According to this information, the child probably suffers from severe protein–energy malnutrition because of the parental neglect and presence of muscular wasting.
29. Common signs and symptoms of food allergy are hives, swollen or itchy lips, skin flushing, eczema, diffi–culty swallowing, wheezing and difficulty breathing, and abdominal pain, vomiting, and diarrhea. Common signs and symptoms of celiac disease are abdominal bloating, chronic diarrhea, weight loss, and poor growth in children.
30. They must provide their children with a special diet that is very low in phenylalanine.
31. Information about a patient's genetic makeup could enable physicians to prescribe more effective treatments, includ–ing special diets.
32. Answers will vary but may include the points in the "Food & Nutrition Tips" feature in this section.
33. Answers may vary, depending on the person's living situation. The 3.5–pound bag of frozen peaches cost $7.99; the fresh peaches cost $2.49/lb, so 3.5 pounds of the peaches would cost about $8.72. So, people can save about $0.73 by purchasing the frozen peaches. However, they need to have access to a large freezer if they intend to store the fruit. Buy–ing smaller quantities of the fresh fruit may be a better option for people who do not have a way of storing the frozen fruit.

Chapter 8

1. Vitamins are complex organic compounds that regulate metabolic reactions. The body cannot synthesize most vitamins; vitamins naturally occur in commonly eaten foods; signs and symptoms of deficiency disease eventually occur when a vitamin is missing from the diet; if the deficiency disorder is treated early, good health is restored by supplying the missing vitamin.
2. Foods generally contain much smaller amounts of vitamins than of macronutrients; the body requires vitamins in much smaller amounts than macronutrients; and the body does not use vitamins directly for energy.
3. A provitamin is a vitamin precursor—a substance that does not function as a vitamin until the body converts it into an active form. An antioxidant is a substance that protects other compounds from oxidizing agents. A radical is a substance with an unpaired electron; radicals are highly reactive.
4. The federally regulated grain enrichment program specifies amounts of thiamin, riboflavin, niacin, and folic acid that manufacturers must add to their refined flour and other milled grain products.
5. Answers will vary but should relate to information concerning food preparation, such as cooking methods, and food storage practices.
6. Answers will vary, but the table should include vitamins A, D, E, and K and conform to information provided in Table 8.2.
7. Most water–soluble vitamins function as components of specific coenzymes. A coenzyme is a small, organic molecule that interacts with enzymes and, as a result, regulates a chemical reaction. To synthesize a coenzyme, cells combine one of the B vitamins that functions as a coenzyme with a nitrogen–containing, nonprotein compound. When activated by the coenzyme, the enzyme enables the reaction to occur.
8. Answers will vary, but the table should include all B vitamins, vitamin C, and choline, and conform to information provided in Table 8.7.
9. There is no scientific evidence that taking beta–carotene prevents healthy people from developing macular degeneration.
10. Taking megadoses of vitamins can produce druglike effects in the body, including unpleasant and even dangerous side effects.
11. Liver damage is one of most serious side effects of taking megadoses of niacin (see Table 8.7).
12. High doses of vitamin B–6 have been associated with sensory nerve damage. The damage is usually reversible—after affected people stop taking excessive amounts of the vitamin.
13. Routine vitamin C supplementation does not protect against infection by common cold viruses, but the practice may shorten the duration of cold symptoms. However, members of the military and athletes may benefit from short–term use of 1000 mg of vitamin C to reduce their risk of virus infection, especially if living in highly populated training centers or when frequently traveling.
14. Lung cancer is the leading cause of cancer deaths in the United States.
15. *Metastasize* means "spread," so Margaret's breast cancer has spread to her lungs.
16. Answers will vary but should include intakes of alcohol; processed and red meats; fried, grilled, and broiled meats; salted fish; and moldy nuts (see Table 8.15).
17. Answers will vary but should include not smoking, limiting alcohol consumption, and maintaining a healthy body weight.

Chapter 9

1. Dehydration occurs when a person loses significant body fluid, resulting in body water depletion. Severe dehydration is life–threatening and can cause a very high body temperature (hyperthermia) and death if not medically treated.
2. A mineral nutrient is classified as a major mineral if humans require 100 mg or more per day or a trace mineral if humans require less than 100 mg/day. A mineral is a possible essential mineral if the body has very small amounts of the mineral and its function or functions are unclear.
3. List should include at least five of the minerals in the major minerals column and five of the minerals in the trace minerals column of Table 9.1.
4. List should include at least two of the minerals listed in the possible essential minerals column of Table 9.1.
5. Water is a major solvent, and the nutrient often participates directly in chemical reactions. Water's other physiological roles include transporting substances, removing waste products, lubricating tissues, and regulating body temperature and acid–base balance. Additionally, water is a major component of blood, saliva, sweat, tears, mucus, and the fluid in joints.
6. Osmosis is the diffusion of a solvent, usually water, through a selectively permeable membrane.
7. Extracellular fluid contains higher concentrations of sodium and chloride ions than intracellular fluid. Intracellular fluid contains higher concentrations of potassium and phosphate ions than extracellular fluid.
8. The body obtains water by consuming various foods and liquids, such as fruit juice, milk, soup, coffee, tea, soft drinks, tap water, and plain or flavored bottled water. Cells also form some water as a metabolic by–product of metabolism. The body loses water in urine, perspiration, exhaled air, feces, vomit, and insensible perspiration.

9. If the body cannot regulate its fluid balance, cells can shrink, swell, and even die.
10. In response to dehydration, the posterior pituitary gland releases antidiuretic hormone, which stimulates kidneys to conserve water. Additionally, the adrenal glands secrete aldosterone, a hormone that signals kidneys to reduce the elimination of sodium in urine. As a result, the kidneys return the mineral to the general circulation. Because water follows sodium, it is conserved as well.
11. The AI for total water intake is approximately 11 cups (2.7 L) for young females and approximately 15.5 cups (3.7 L) for young males.
12. A diuretic is a substance that increases urine production. Alcohol and caffeine have diuretic effects on the body.
13. Signs and symptoms of dehydration include weight loss, fatigue, thirst, loss of muscular strength and endurance, severe weakness, and coma.
14. Signs and symptoms of water intoxication may include drowsiness, nausea and vomiting, confusion, inability to coordinate muscular movements, and weight gain. If not treated early, overhydration can result in death.
15. Answers may vary but may include inorganic structural components of tissues, such as calcium and phosphorus in bones and teeth. Minerals may also function as inorganic ions, such as calcium ions (Ca^{++}) that participate in blood clotting and sodium ions (Na^{+}) that help maintain fluid balance. Some ions, such as magnesium (Mg^{++}) and copper (Cu^{++}), are cofactors for chemical reactions. Many minerals are components of various enzymes, hormones, or other organic molecules, such as cobalt in vitamin B–12, iron in hemoglobin, and sulfur in the amino acids methio–nine and cysteine.
16. Refining foods can result in mineral losses. Minerals are water soluble and can be lost by leaching into watery cooking fluids.
17. The bioavailability of minerals depends largely on the body's need for the micronutrient. Additionally, minerals from animal foods are often better absorbed than minerals in plant foods.
18. The enrichment program for grains requires the addition of iron to grain products.
19. Osteoporosis is a chronic disease characterized by low bone mass and reduced bone structure. People with osteoporosis have weak bones that are susceptible to fractures that can be costly to treat and result in disability and death, particularly among older adults.
20. Answers may vary but should relate to risk factors listed in Table 9.7.
21. Lisa's blood pressure is within the normal range. Hypertension is characterized by persistent systolic values $\geq$140 mm Hg and diastolic values $\geq$90 mm Hg. Risk factors for hypertension include advanced age, African–American ancestry, obesity, physical inactivity, cigarette smoking, and excess alcohol and sodium intakes. Table 9.11 presents major risk factors for hypertension.
22. The Dietary Approaches to Stop Hypertension (DASH) diet is low in sodium, total fat, and saturated fat and high in fruits, vegetables, and low–fat dairy products. Other lifestyle changes that can reduce blood pressure include losing excess body fat and increasing physical activity.
23. Answers should conform to information in Table 9.5.
24. Iron deficiency is a condition characterized by low blood iron levels that are not low enough to result in severe health problems. Iron deficiency anemia occurs when oxygen transport in blood is impaired, generally because there are not enough red blood cells to carry the oxygen or because red blood cells do not contain enough hemoglobin. Signs and symptoms of iron deficiency anemia include conditions listed in Table 9.17. People who are likely to develop iron deficiency include those who have lost substantial amounts of blood and young females who exclude meat and enriched bread and cereal products from their diets. Additionally, others at risk of iron deficiency are females of childbearing age (who generally lose some iron during menstruation), pregnant females, and infants born to females who are iron deficient.
25. Hemochromatosis is an inherited disorder in which the body absorbs too much iron. Signs and symptoms include joint pain, fatigue, lack of energy, abdominal pain, loss of sex drive, abnormal bronze skin color, and heart problems. The condition is treated by avoiding iron supplements and periodically having blood removed.
26. Goiter is an enlarged thyroid gland. Cretinism occurs in infants born to females who were iodine deficient during pregnancy. The condition is characterized by permanent brain damage, reduced intellectual functioning, and growth retardation. Cretinism can be prevented by including a source of iodine in diets of pregnant females.
27. Answers will vary but should conform to information presented in Table 9.15.
28. Answers will vary but should include arsenic, boron, lithium, nickel, silicon, or vanadium.
29. Answers will vary but should conform to information provided in Table 9.22.
30. In the United States, most tap water undergoes a thorough purification process and is constantly tested for safety.
31. Answers will vary but should include avoiding containers that have the symbol that is shown in Figure 9.26.

Chapter 10

1. Calvin can be "overweight" if he has more muscle and less body fat than other people who are his height.
2. In 2017–2020, more than 42% of adult Americans were obese.

3. Total body fat includes adipose cells and fat in cell membranes, nervous tissue, and certain bones. Fat–free mass is composed of body water; mineral–rich tissues such as bones and teeth; and protein–rich tissues, including muscles and organs.

4. Body fat stores energy. Additionally, fat cells secrete numerous proteins, some of which have roles in regulating food intake, glucose metabolism, and immune responses.

5. Subcutaneous fat helps insulate the body against cold temperatures and protects muscles and bones from bumps and bruises. Subcutaneous fat also provides body contours.

6. Answers will vary but may include "underwater weighing" and measuring dual–energy x–ray absorptiometry, air displacement, bioelectrical impedance, and skinfold thicknesses. For many of the techniques, drawbacks include cost and size of equipment needed to measure body fat.

7. He is overweight.

8. To estimate daily basal metabolic needs of a 185–pound female, first convert the female's weight to kilograms by dividing 185 pounds by 2.2 (approximately 84 kg). Then multiply 84 kg by 0.9 kcal to determine the amount of energy expended for metabolism in an hour (approximately 76 kcal/hr). Multiply 76 kcal/hour by 24 hours to obtain the amount of energy expended for basal metabolic needs in a day (1824 kcal).

9. The three major ways the body uses energy are basal metabolism, physical activity, and TEF. Energy needs for physical activity can be easily altered.

10. For most people, basal metabolism uses the most energy each day.

11. Answers will vary but may include body composition, thyroid hormone level, sex, body surface area, age, calorie intake, fever, stimulants, pregnancy, milk production after childbirth, and recovery after exercise.

12. A person gains weight when in positive energy balance.

13. Positive energy balance is desirable for pregnant females and during periods of growth, such as during fetal development, infancy, childhood, and puberty.

14. Weight loss is characterized by negative energy balance.

15. This female is likely to be obese.

16. Answers will vary but should include conditions listed in Table 10.5.

17. Weight bias is the negative attitudes, beliefs, and stereo–types about a person because of their weight. The stigma of obesity is the negative attitude that many people have toward persons who are obese, such as characterizing these people as lazy, stupid, and sloppy.

18. Central–body obesity ("apple shape") is more likely to pose serious health risks than having excess fat below the waist, or a pear–shaped fat distribution. Central–body fat distribution is associated with increased risks of type 2 diabetes and CVD.

19. Measuring a person's waist circumference is a quick and easy way to determine whether body fat distribution is likely to result in serious chronic health problems.

20. Hunger is an uncomfortable feeling that drives a person to consume food. Satiety is the sense that enough food or beverages have been consumed to satisfy hunger.

21. Adipose cells secrete leptin, a hormone that reduces hunger and inhibits fat storage in the body. The stomach secretes ghrelin, which stimulates eating behavior.

22. Thrifty genes could benefit people during periods of food deprivation, such as famine.

23. According to the set–point theory, the body's fat content (and, therefore, body weight) is genetically predetermined. The set point acts like a home thermostat, except that it regulates body weight instead of temperature.

24. Hunger is a drive to eat food, whereas appetite is a desire to eat appealing food.

25. Answers will vary but should include food marketing and sensory appeal, such as appearance, taste, and odor. Our work and home environments influence whether we are sedentary or physically active by providing opportunities to be active.

26. Answers will vary but should include genetics, occupations that do not provide enough physical activity, and psycho–logical factors, such as mood and self–esteem.

27. Four key elements necessary for weight loss and mainte–nance are motivation, calorie reduction, regular physical activity, and behavior modification.

28. Answers will vary but should include points listed in Table 10.6.

29. Before joining a weight–loss group or plan, consumers should find out how much the program costs and how often they will need to pay for the program. Consumers should ask whether they need to sign a contract and the length of the contract. Do special foods or supplements need to be purchased? If nutrition counseling is provided, what are the credentials of the persons providing nutrition counseling and information? Furthermore, does the plan emphasize the importance of making lifestyle changes? Does the program's advertising include questionable weight–loss claims and deceptive testimonials?

30. Answers will vary but should include these: eat a low–calorie diet; eat regular meals, including breakfast almost every day; weigh themselves at least once a week; exercise for an average of 60 minutes daily; and limit television watching.

31. Weight–loss medications may suppress appetite or interfere with fat digestion and absorption.

32. Roux–en–Y gastric bypass and gastric banding proce–dures are common forms of bariatric surgery performed in the United States. Bariatric surgeries are effective because they reduce the size of the stomach. As a result, patients can consume only small portions of food at a time. Gastric banding is easier to perform, safer than other types of bariatric surgery, and relatively easy to reverse.

33. A fad diet is a trendy dietary practice that has widespread appeal among a population; after a period, people lose interest in the practice and it becomes no longer fashionable. Fad diets typically promote rapid weight loss without calorie restriction and increased physical activity; limit food selections to be from only a few food groups and dictate specific diet rituals; require buying a book or various gimmicks; use outlandish and unscientific claims to support their usefulness; rely on testimonials from famous people; connect the diet to trendy places; and do not emphasize the need to change eating habits and physical activity patterns.

34. Many people who are overweight or obese are attracted to fad diets and dietary supplements for weight loss because they believe promoters' claims that the diets or products are "magic bullets" for shedding unwanted weight quickly and effortlessly.

35. Answers will vary but should include supplements presented in Table 10.7.

36. Answers will vary but should include features or claims that the product or service causes rapid and extreme weight loss; requires no need to change dietary patterns or physical activity; results in permanent weight loss; is scientifically proven or doctor endorsed; includes a money–back guarantee; is safe; is supported by satisfied customers; displays before–and–after photos; and provides disclaimers in small print, such as "Results not typical" at the bottom of the ad or with an "after" photo.

37. Chronic diseases such as cancer, tuberculosis, AIDS, inflammatory bowel disease, depression, and hormonal imbalances can result in severe weight loss that is difficult to treat.

38. To gain weight, adults who are underweight can gradually increase their consumption of calorie–dense foods, especially those high in healthy fats. Eating regular meals and snacks also aids in weight gain and maintenance. If excess physical activity habits contribute to the inability to gain weight, people who are underweight can find ways to be less active. Furthermore, individuals who are underweight can add muscle mass by following resistance–training programs regularly, but their calorie intake must support extra energy demands of those programs.

39. Skipping breakfast regularly is a sign of disordered eating, not an eating disorder.

40. The three major types of eating disorders are anorexia nervosa (AN), bulimia nervosa (BN), and binge–eating disorder (BED). Major signs of AN are severely restricted food intake and maintenance of an unhealthy low body weight. People with BN are difficult to recognize, but characteristic signs of the condition are bite marks and scars on the knuckles that result from self–induced vomiting. The major sign of BED is consuming a much larger amount of food than most people would eat under similar circumstances in a brief period of time.

41. Answers will vary but should include the risk factors shown in Table 10.8.

42. Answers will vary but should include the health consequences listed in Table 10.9, as well as death.

Chapter 11

1. Physical activity is any movement that results from skeletal muscle contraction. Exercise refers to physical activities that are usually planned and structured for a particular purpose.

2. A physically fit person can perform moderate– to vigorous–intensity activities, for a reasonable amount of time, without becoming excessively fatigued.

3. Healthy adults under 65 years of age should engage in at least 150 minutes of moderate—intensity physical activity a week or 75 minutes of vigorous–intensity physical activity a week. Additionally, adults should include 8 to 10 strengthening exercises at least twice a week.

4. Answers will vary but should include benefits indicated in Figure 11.1.

5. Performing resistance exercises regularly can increase muscle mass and strength. Resistance exercises can also increase bone mass.

6. To calculate the target heart rate range of a 24–year–old person performing moderate–intensity activity, subtract the person's age (24) from 220 to obtain 196, the age–related maximum heart rate. Moderate–intensity activity raises the heart rate to 64 to 76% of the age; therefore, multiply 196 × .64 to obtain the lower limit of the range (125 bpm), and 196 × .76 to obtain the upper limit of the range (approximately 149 bpm). Thus, the heart rate range for this individual is 125 to 149 bpm.

7. Muscle cells break down glycogen to provide a source of glucose during exercise.

8. ATP is formed as a result of glycolysis and aerobic respiration. Some ATP is also formed by the breakdown of triglycerides, amino acids, and alcohol. Muscle cells also use phosphocreatine to produce ATP quickly under anaerobic conditions.

9. Muscle cells rely on the PCr–ATP, lactic acid, and oxygen systems to obtain energy. The PCr–ATP system forms ATP by breaking down PCr into creatine and P_i. The energy that is released helps synthesize ATP from ADP and P_i. In anaerobic conditions, muscle cells metabolize glucose to pyruvate and then convert pyruvate to lactic acid. The degradation of glucose to lactic acid produces a small amount of ATP. Lactic acid accumulates in muscles and converts to a related substance, lactate. Although certain muscle cells can use lactate as a fuel, some of the compound enters the bloodstream. The liver removes lactate from blood and can convert the compound into glucose. The oxygen system metabolizes glucose completely to CO_2 and H_2O, resulting in the production of far more ATP–energy than the amount produced by anaerobic systems. The PCr–ATP and lactic acid systems do not need oxygen to produce ATP.

10. Individuals can increase their aerobic exercise capacity by engaging in endurance–training programs that gradually increase the intensity level of activities. Such training improves muscle cells' ability to generate ATP rapidly.
11. Fat is a major energy source for muscles while one is resting or engaged in low– to moderate–intensity physical activities.
12. Practical ways to assess whether an athlete's energy intake is adequate include having the athlete keep accurate food records and use the information to estimate their daily calorie intakes. An athlete can also monitor their body weight and have skinfold thicknesses measured regularly.
13. Carbohydrate intake contributes to glycogen stores; glycogen depletion is a major cause of fatigue during endurance exercise. To maintain adequate muscle glycogen, athletes should consume 6 to 10 g of carbohydrates per kilogram of body weight daily. By consuming several servings of grains, starchy vegetables, and fruits daily, an athlete can obtain enough carbohydrates to maintain adequate liver and muscle glycogen stores.
14. Answers will vary but should include grains, starchy vegetables, and fruits. Also, see Table 11.4 for high–carbohydrate, low–fat foods.
15. Compared to nonathletes, athletes may need more than the RDA for protein because they need to have amino acids available to repair muscle tissue damaged during intense physical activity and to build muscle tissue.
16. Dehydration and overhydration detract from optimal physical performance and can cause serious health problems.
17. Answers will vary but should include signs and symptoms listed in Table 11.7.
18. The International Society of Sports Nutrition (ISSN) recommends athletes consume sports drinks during high–intensity activities lasting more than one hour. Although electrolytes are lost in sweat, the quantities lost in shorter periods can be easily replaced by consuming foods and beverages. For example, drinking water or fruit juice can replace fluid losses after shorter periods of exercise.
19. The practice of taking antioxidant supplements may not be helpful and could be harmful. Radicals generated during intense exercise may stimulate the body's natural antioxidant defense system. Thus, the oxidative stress produced during exercise might have benefits, and blocking this process by taking antioxidant vitamin supplements may not be desirable.
20. Iron deficiency can negatively affect athletic performance because the body needs iron to produce red blood cells, transport oxygen, and obtain energy.
21. Female athletes who have irregular or no menstrual cycles may lack adequate amounts of the hormone estrogen. Although weight–bearing exercise, such as jogging, improves bone density, estrogen is necessary for maintaining healthy bones. Female athletes who develop menstrual cycle abnormalities should consult a physician to determine the cause.
22. Athletes often use ergogenic aids to improve their physical performance because they hope to gain the competitive edge over their sports rivals.
23. Answers will vary but should include ephedra–containing supplements, caffeine (high intakes), anabolic steroids, and growth hormone.
24. Coffee, energy drinks, caffeinated soft drinks, and tea (see Table 11.9).
25. Caffeine raises the level of fatty acids in blood, and as a result, exercising muscles can use more fat for energy. Caffeine also enhances the ability of skeletal and heart muscles to contract and increases mental alertness.
26. The three key components of an aerobic workout regimen are warm–up, workout, and cool–down.
27. Answers will vary, but the fitness program should include the principles of exercise variety, balance, and moderation. Furthermore, the person should include activities that are enjoyable and easy to incorporate into their routine.

Chapter 12

1. Norovirus is responsible for most cases of food–borne illness in the United States.
2. The illness caused by norovirus is usually the result of a food–borne infection.
3. The FDA regulates nearly all domestic and imported food sold in interstate commerce and enforces federal food safety laws. Additionally, the FDA establishes standards for safe food manufacturing practices, such as Hazard Analysis and Critical Control Point programs. FDA officials can take certain enforcement actions, such as requesting that a food manufacturer recall an unsafe item, so that it is removed from store shelves. The FDA also educates the general public about safe food–handling practices. The Food Safety and Inspection Service (FSIS) enforces food safety laws for domestic and imported meat and poultry products. Additionally, FSIS staff collect and analyze food samples to check for the presence of microbial and other unwanted and potentially harmful material in foods. FSIS officials can ask meat and poultry processors to recall unsafe products. FSIS also educates people about proper food–handling practices. The EPA oversees drinking water quality and assists state officials in their efforts to

monitor water quality. Furthermore, EPA staff regulate toxic substances and wastes to prevent their entry into foods and the environment.

4. Local health departments are responsible for inspecting restaurants, grocery stores, dairy farms, and local food processing companies.
5. Three common ways harmful microbes are transmitted to food are by vermin, poor personal hygiene practices, and improper food handling.
6. Cross–contamination occurs when pathogens in one food are transferred to another food, contaminating it.
7. Pasteurization is a special heating process used by many commercial food producers to kill pathogens.
8. To survive and multiply, most microbes need warmth, moisture, and a source of nutrients, and some also require oxygen.
9. In general, high–risk foods are warm, moist, and protein rich, and they have a neutral or slightly acidic pH (see Table 12.1).
10. Signs and symptoms of food–borne illnesses generally involve the digestive tract and include nausea, vomiting, diarrhea, and intestinal cramps.
11. People who suffer from a food–borne illness should consult a physician when an intestinal disorder is accompanied by one or more of the signs that are listed in Table 12.2.
12. Food–borne illness generally causes gastrointestinal tract symptoms, whereas influenza and COVID–19 are primarily respiratory tract infections.
13. Answers will vary but should include bacteria listed in Table 12.3.
14. Answers will vary but should include viruses listed in Table 12.4.
15. Aflatoxins are poisonous substances produced by certain molds.
16. Answers will vary but should include a protozoan and a parasitic worm listed in Table 12.5.
17. Answers will vary but should include points under the "Purchasing Food" section of Chapter 12.
18. Answers will vary but should include points under the "Setting the Stage for Food Preparation" and "Preparing Food" sections of Chapter 12.
19. Answers will vary but should include points under the "Storing and Reheating Food" section of Chapter 12.
20. Ground meats and poultry are often sources of food–borne illness because grinding increases the surface area of the protein–rich food, exposing it to contaminants.
21. Raw fish should be frozen for recommended amounts of time before being eaten.
22. According to the HHS and USDA, the "danger zone" for rapid pathogen multiplication is between 40°F and 140°F.
23. The Check Your Steps program recommends four simpli–fied actions: clean, separate, cook, and chill.
24. Shelf life is the period of time that a food can be stored before it spoils.
25. Answers will vary but should include methods highlighted in Table 12.8.
26. Boiling home–canned, low–acid foods for 10 minutes can make them safe to eat.
27. Answers may vary but can include meats, fresh fruits and vegetables, spices, dry vegetable mixes, seeds, and shell eggs.
28. Answers will vary but should include points made under the "Preparing for Disasters" section of Chapter 12. Table 12.10 lists foods that can be stored safely for emergency situations.
29. Safe sources of drinking water include bottled water, an undamaged water heater, toilet tanks (not bowls), melted water from ice cubes, and fruit juices. Water from pools, spas, radiators, heating systems, and water beds is unsafe to drink.
30. A food additive is a substance added to food that influences the product's characteristics. Direct food additives may maintain the safety of foods, prevent undesirable changes, or improve taste or color of foods. Indirect additives, such as compounds from a food's wrapper or container, can enter food as it is packaged, transported, or stored. Indirect additives have no purpose.
31. Food ingredients that had been in use prior to 1958 were listed as Generally Recognized as Safe (GRAS) when qualified experts generally agreed that the substance was safe for its intended use. Food manufacturers can include items on the GRAS list as ingredients without testing them for safety or getting prior approval from the FDA.
32. According to the Delaney Clause, food manufacturers cannot add a new compound that causes cancer at any level of intake. Thus, if an additive causes cancer, even though very high doses may be necessary to cause the disease, no amount of the additive is considered to be safe, and none is allowed in food. Evidence for cancer risk could come from either laboratory animal or human studies. The FDA allows very few exceptions to this clause.
33. An unintentional food additive is a substance that acci–dentally enters food during processing. Common biological and physical food contaminants are insect parts, rodent feces or urine, dust and dirt, and bits of metal or glass from machinery used to process food.
34. A pesticide is any substance that people use to control or kill unwanted insects, weeds, rodents, fungi, or other organisms. Types of pesticides include insecticides, roden–ticides, herbicides, and fungicides.
35. Integrated pest management (IPM) involves using a variety of methods for controlling pests while limiting dam–age to the environment. IPM methods include grow–ing pest–resistant crops, using predatory wasps to control

crop–destroying insects, trapping adult insect pests before they can reproduce, and using biologically based pesticides.

36. The EPA regulates the use of pesticides.

37. A pesticide tolerance is the maximum amount of pesticide residue that can be in or on each treated food crop. A pesticide tolerance includes a margin of safety, so the maximum pesticide residue that is allowed to be in or on a food is much lower than amounts that can cause negative health effects.

38. Once a pesticide is applied to cropland, it may remain in the soil, be taken up by plant roots, decompose to other compounds, or enter groundwater and waterways. Winds may carry pesticides in air and dust to distant locations, where the toxic material can land on food crops.

39. Answers will vary but should include using hand sanitizers, drinking only commercially bottled water, boiling water before drinking it, eating foods that are fully cooked and served hot, and not eating locally provided fresh fruits and vegetables that have been washed in unsanitary water.

40. Answers will vary but should include the foods shown in the safe–to–eat category in Figure 12.12.

Chapter 13

1. Matthias is an older adult.

2. For females, lactation is the life stage that involves milk production for breastfeeding.

3. Eight weeks after conception, the developing human being is referred to as a fetus.

4. The role of the placenta is to transfer nutrients and oxygen from the mother's bloodstream to the embryo/fetus. Additionally, the placenta transfers wastes from the embryo/fetus to the mother's bloodstream, so her body can eliminate them.

5. Birth weight is a major factor that determines whether a baby is healthy and will survive their first year of life.

6. Hypertension during pregnancy can be dangerous because it increases the risk of maternal death and premature delivery.

7. Answers will vary but should include enlarged breasts, "morning sickness," tiredness, swollen feet, constipation, and heartburn.

8. Females whose prepregnancy weights were within the healthy range can expect to gain 25 to 35 pounds. Females who are underweight should gain 28 to 40 pounds; Females who are overweight should gain 15 to 25 pounds during pregnancy. Females who are obese should gain 11 to 20 pounds during pregnancy. The recommendations are higher for females who are pregnant with more than one fetus.

9. Pregnant females who are folate deficient have a high risk of giving birth to infants with neural tube defects, such as spina bifida. Pregnant females who are iron deficient can develop iron deficiency anemia, a condition that increases the risk of giving birth prematurely and having low–birth–weight (LBW) infants.

10. Females who drink alcohol during pregnancy are at risk of having a child with a fetal alcohol spectrum disorder. Compared to pregnant females who do not smoke cigarettes, expectant mothers who smoke have higher risk of giving birth too early and having LBW babies. Furthermore, expectant mothers who smoke cigarettes may increase the risk of having babies with birth defects or that die of sudden infant death syndrome.

11. In response to an infant sucking on its mother's nipple, the brain releases oxytocin, a hormone that signals breast tissue to "let down" milk. The let–down reflex enables milk to travel in several tubes to the nipple area. When let–down occurs, the infant removes the milk by continued sucking.

12. Lactation increases a new mother's energy needs by about 300 to 400 kcal more than her prepregnancy energy needs.

13. Colostrum is the first secretion of the breasts that occurs after delivery. Colostrum is a very important first food for babies because the fluid contains antibodies and immune system cells that can be absorbed by the infant's immature digestive tract. Colostrum also contains substances that encourage the growth of beneficial bacteria in the baby's intestinal tract that reduce the risk of infection.

14. New mothers should exclusively breastfeed their healthy babies for about the first 6 months of life.

15. Answers will vary but should include points in Table 13.5.

16. Answers will vary but should include points in Table 13.5.

17. Compared to whole cow's milk, human milk provides about the same amount of energy; less protein, sodium, potas–sium, and calcium; and more carbohydrate, essential fatty acids, vitamin E, and iron.

18. Signs indicating a baby is ready physically to eat solid foods include loss of the extrusion reflex; physiological maturity to swallow, digest, and metabolize a wider range of foods; and ability to sit up with back support and coordinate muscular control over mouth and neck movements.

19. A healthy infant should triple its birthweight during the first year of life.

20. A food jag is a period in which the child refuses to eat a food that was liked in the past or in which the child wants to eat only a particular food.

21. In children, iron deficiency can lead to decreased physical stamina, learning ability, and resistance to infection.

22. Answers will vary but should include brushing teeth with a peasized amount of fluoride–containing toothpaste twice daily; providing routine pediatric dental care; offering fluoridated drinking water; not eating carbohydrate–rich snacks, especially between meals; and offering sugarless chewing gum.

23. Compared to preschoolers, older children often skip breakfast, and they typically consume more foods away from home, larger portions of food, and more fried foods and sweetened beverages. School–age children also tend to consume less milk and fewer fruits and vegetables, except fried potatoes.

24. Answers will vary but should include encouraging children to eat breakfast and other tips from the "Food & Nutrition Tips" feature in the "School–Age Children" section of Chapter 13.

25. Males usually begin puberty when they are between 9 and 14 years of age. Most females begin puberty between 8 and 13 years of age.

26. Iron deficiency leads to increased fatigue and decreased ability to concentrate and learn. Inadequate calcium intake during adolescence is associated with decreased bone mass and increased likelihood of bone fractures later in life.

27. Answers should provide the information in Table 13.8. Children and adolescents whose BMI–for–age is greater than or equal to the 85th percentile and less than the 95th percentile are overweight, and youngsters who are at or above the 95th percentile are obese.

28. Answers will vary but should include elevated blood pressure, cholesterol, and glucose levels and the factors listed in Table 13.9.

29. Answers will vary but should include spending too much time on sedentary activities; consuming too many empty calories; and having a family history of obesity, high birth weight, and family members who are obese.

30. To qualify for bariatric surgery, an adolescent should have severe obesity, attained their adult height, and developed serious chronic conditions that are associated with obesity.

31. Life expectancy is the length of time a person born in a specific year, such as 1900, can expect to live. Life span refers to the maximum number of years a human can live.

32. Answers will vary but should include points listed in Table 13.11.

33. Compared to younger persons, older adults have greater risk of nutritional deficiencies because of physiological changes associated with the normal aging process. Other factors that can influence an older person's nutritional status include illnesses, medications, low income, and lack of social support.

34. Diets of older adults often provide inadequate amounts of fiber, magnesium, potassium, calcium, zinc, folate, choline, and vitamins A, B–6, C, D, and E.

35. Caregivers can improve nutrient intakes of older persons by adding more spices to foods, making mealtimes social events, and serving energy–dense snacks between meals, such as cheese, milkshakes, nuts, or oatmeal cookies. If an older adult has difficulty chewing, serving soft items such as ground meats, cooked vegetables, pureed fruits, and puddings can increase food intake. The "Food & Nutrition Tips" feature in the "Nutrition for Older Adults" section of Chapter 13 presents more suggestions for improving nutrient intakes of older adults.

36. Under certain conditions, calorie restriction extends the life expectancies of organisms that include fruit flies and rodents.

37. Answers may vary but should include adopting healthy life–styles by consuming a nutritionally adequate diet, obtaining regular moderate– to vigorous–intensity physical activity, maintaining a healthy weight, avoiding tobacco products, limiting alcohol consumption, and having regular physical checkups.

Answers to Practice Test Questions

Chapter 1

1. d	**2.** a	**3.** c	**4.** a	**5.** b
6. d	**7.** d	**8.** c	**9.** a	**10.** c
11. a	**12.** c	**13.** d	**14.** a	**15.** b
16. c				

Chapter 2

1. d	**2.** b	**3.** a	**4.** d	**5.** c
6. d	**7.** a	**8.** d	**9.** b	**10.** b
11. b	**12.** a	**13.** a		

Chapter 3

1. b	**2.** c	**3.** c	**4.** a	**5.** d
6. c	**7.** d	**8.** a	**9.** b	**10.** b
11. c	**12.** b	**13.** a	**14.** c	**15.** d
16. a	**17.** c	**18.** c	**19.** b	

Chapter 4

1. c	**2.** a	**3.** b	**4.** b	**5.** b
6. a	**7.** d	**8.** b	**9.** c	**10.** b
11. a	**12.** d	**13.** b	**14.** d	**15.** a
16. b	**17.** a			

Chapter 5

1. c	**2.** a	**3.** b	**4.** b	**5.** c
6. b	**7.** b	**8.** b	**9.** b	**10.** d
11. c	**12.** b	**13.** a	**14.** c	**15.** b

Chapter 6

1. a	**2.** b	**3.** c	**4.** a	**5.** b
6. d	**7.** b	**8.** c	**9.** b	**10.** c
11. b	**12.** c	**13.** c	**14.** b	**15.** d
16. b	**17.** a	**18.** d		

Chapter 7

1. c	**2.** c	**3.** a	**4.** a	**5.** b
6. d	**7.** c	**8.** a	**9.** a	**10.** c
11. b	**12.** a	**13.** d	**14.** c	**15.** b
16. c	**17.** a			

Chapter 8

1. d	**2.** c	**3.** a	**4.** d	**5.** d
6. a	**7.** d	**8.** b	**9.** a	**10.** c
11. d	**12.** b	**13.** c	**14.** a	**15.** b
16. c	**17.** d	**18.** a	**19.** d	**20.** a

Chapter 9

1. c	**2.** b	**3.** d	**4.** a	**5.** a
6. c	**7.** d	**8.** a	**9.** c	**10.** b
11. c	**12.** b	**13.** a		

Chapter 10

1. b	**2.** b	**3.** c	**4.** a	**5.** a
6. a	**7.** b	**8.** d	**9.** b	**10.** d
11. b	**12.** d			

Chapter 11

1. b	**2.** a	**3.** d	**4.** b	**5.** a
6. c	**7.** a	**8.** d	**9.** b	**10.** c
11. b	**12.** d	**13.** b	**14.** a	**15.** a

Chapter 12

1. a	**2.** a	**3.** a	**4.** c	**5.** d
6. d	**7.** c	**8.** d	**9.** c	**10.** b
11. a	**12.** c	**13.** b		

Chapter 13

1. b	**2.** b	**3.** c	**4.** b	**5.** a
6. a	**7.** c	**8.** c	**9.** d	**10.** d
11. c	**12.** c	**13.** a	**14.** b	**15.** c
16. d	**17.** a			

Burke Triolo Productions/Artville/Getty Images

Appendix G

References

Siede Preis/Photodisc/Getty Images

Chapter 1

1. Centers for Disease Control and Prevention, Division of Nutrition, Physical Activity, and Obesity: *Health Equity. 2021.* www.cdc.gov/nccdphp/dnpao/health–equity/index.html. Accessed May 10, 2023.
2. Sotos–Prieto M and others: Association of changes in diet quality with total and cause–specific mortality. *New England Journal of Medicine* 377(2): 143, 2017.
3. U.S. Office of Dietary Supplements: *Dietary Supplement Health and Education Act of 1994. Public Law 103–417, 103rd Congress.* Approved 1994. https://ods.od.nih.gov/About/DSHEA_Wording.aspx. Accessed May 10, 2023.
4. Xu J and others: Mortality in the United States, 2021. *NCHS Data Brief,* no. 456. Hyattsville, MD: National Center for Health Statistics. 2021.
5. American Cancer Society: *Cancer facts & figures 2022.* 2022. www.cancer.org/research/cancer–facts–statistics/all–cancer–facts–figures/cancer–facts–figures–2022.html. Accessed May 10, 2023.
6. American Cancer Society: *American Cancer Society Guideline for Diet and Physical Activity for Cancer Prevention.* Revised: June 2020. www.cancer.org/healthy/eat–healthy–get–active/acs–guidelines–nutrition–physical–activity–cancer–prevention/guidelines.html. Accessed May 10, 2023.
7. U.S. Department of Agriculture, Economic Research Service: *Food availability (per capita) data system: Loss–Adjusted Food Availability.* Last updated January 2020. www.ers.usda.gov/data–products/food–availability–per–capita–data–system/. Accessed May 10, 2023.
8. U.S. Department of Agriculture, Human Nutrition Information Service: Food and nutrient intakes by individuals in the United States, 1 day, 1987–88. *Nationwide Food Consumption Survey 1987–88,* NFCS Rep. No. 87–1–1. 1993. www.ars.usda.gov/ARSUserFiles/80400530/pdf/8788/nfcs8788_rep_87–i–1.pdf. Accessed May 10, 2023.
9. U.S. Department of Agriculture, Agricultural Research Service: *Nutrient intakes from food and beverages: Mean amounts consumed per individual, by gender and age, What We Eat in America, NHANES 2017–March 2020.* www.ars.usda.gov/ARSUserFiles/80400530/pdf/1720/Table_5_EIN_GEN_1720.pdf. Accessed March 25, 2023.
10. Congressional Research Service: U.S. Health Care Coverage and Spending, 2021. https://sgp.fas.org/crs/misc/IF10830.pdf. Accessed May 10, 2023.
11. U.S. Department of Health and Human Services, Assistant Secretary for Planning and Evaluation (ASPE): *Trends in the U.S. Uninsured Population, 2010–2020.* https://aspe.hhs.gov/reports/trends–us–uninsured–population–2010–2020. Accessed May 10, 2023.
12. Office of Disease Prevention and Health Promotion: *Healthy People 2030.* 2020. https://health.gov/healthypeople/about/healthy–people–2030–framework. Accessed May 10, 2023.
13. U.S. Department of Agriculture, Agricultural Research Service: *Solid fats and added sugars in foods.* 2015. www.ars.usda.gov/news–events/news/research–news/2015/solid–fats–and–added–sugars–in–foods/. Accessed May 10, 2023.
14. U.S. Department of Health and Human Services, U.S. Department of Agriculture: *Dietary Guidelines for Americans, 2020–2025.* 2020. www.dietaryguidelines.gov. Accessed May 10, 2023.
15. National Institutes of Health, National Center for Complementary and Integrative Health: *Vitamins and minerals.* 2018. https://nccih.nih.gov/health/vitamins. Accessed May 10, 2023.
16. McManus KD (contributor): 10 superfoods to boost a healthy diet. 2018. *Harvard Health Blog,* Harvard Medical School, Harvard Health Publishing. www.health.harvard.edu/blog/10–superfoods–to–boost–a–healthy–diet–2018082914463. Accessed May 10, 2023.
17. Position of the Academy of Nutrition and Dietetics: Functional foods. *Journal of the Academy of Nutrition and Dietetics* 113(8): 1096, 2013.
18. World Health Organization: *Obesity and overweight.* 2021. www.who.int/en/news–room/fact–sheets/detail/obesity–and–overweight. Accessed May 10, 2023.
19. United Nations: *World population prospects 2022.* https://population.un.org/wpp/Publications/. Accessed May 10, 2023.
20. Webb P and others: Hunger and malnutrition in the 21st century. *BMJ* 361: k2238, 2018.
21. World Health Organization: *Impact of COVID–19 on people's livelihoods, their health and our food systems.* 2020.

www.who.int/news/item/13–10–2020–impact–of–covid–19–on–people's–livelihoods–their–health–and–our–food–systems. Accessed March 25, 2023.

22. World Health Organization: *Drinking–water.* 2022. www.who.int/news–room/fact–sheets/. Accessed May 10, 2023.

23. World Health Organization: *Levels and trends in child malnutrition, Key findings of the 2021 edition.* www.who.int/publications–detail–redirect/9789240025257. Accessed May 10, 2023.

24. World Health Organization: *Malnutrition.* 2021. www.who.int/news–room/fact–sheets/detail/malnutrition. Accessed May 10, 2023.

25. Gebre B and others: Determinants of malnutrition among pregnant and lactating women under humanitarian setting in Ethiopia. *BMC Nutrition* 4(11), 2018. https://bmcnutr.biomedcentral.com/track/pdf/10.1186/s40795–018–0222–2.

26. World Health Organization: *Infant and young child feeding.* 2021. www.who.int/en/news–room/fact–sheets/detail/infant–and–young–child–feeding. Accessed May 10, 2023.

27. World Health Organization: *Micronutrient deficiencies: Vitamin A deficiency.* ND. www.who.int/data/nutrition/nlis/info/vitamin–a–deficiency. Accessed May 10, 2023.

28. U.S. Census Bureau: *Income and poverty in the United States: 2020.* www.census.gov/library/publications/2020/demo/p60–270.html. Accessed May 10, 2023.

29. U.S. Department of Health and Human Services: *Poverty guidelines for 2022.* January 2022. https://aspe.hhs.gov/topics/poverty–economic–mobility/poverty–guidelines. Accessed May 10, 2023.

30. Economic Research Report No. (ERR–275): A report summary: *Household food security in the United States in 2020.* 2021. www.ers.usda.gov/webdocs/publications/102076/err–298_summary.pdf?v=7704.9. Accessed May 10, 2023.

31. United Nations Children's Fund: *Supply annual report 2020.* 2021. www.unicef.org/supply/media/8246/file/Supply–Annual–Report–2020.pdf. Accessed May 10, 2023.

32. Popkin BM and others: Dynamics of the double burden of malnutrition and the changing nutrition reality. *Lancet* 395(10217): 65–74, 2020.

33. Prentice AM and others: The double burden of malnutrition in countries passing through the economic transition. *Annals of Nutrition and Metabolism* 72 Suppl 3:47–54, 2018.

34. World Food Program: *Double Burden to Double Duty: Policy Implications of Double–Duty Actions to Address the Full Spectrum of Malnutrition.* 2021. www.wfp.org/publications/double–burden–double–duty–policy–implications–double–duty–actions–address–full. Accessed May 10, 2023.

35. Food and Drug Administration, Food: *Consumer info about food from genetically engineered plants.* 2020. www.fda.gov/food/food–new–plant–varieties/consumer–info–about–food–genetically–engineered–plants. Accessed May 10, 2023.

36. International Rice Research Institute (IRRI): *Philippines becomes first country to approve nutrient–enriched "Golden Rice" for planting.* July 23, 2021. www.irri.org/news–and–events/news/philippines–becomes–first–country–approve–nutrient–enriched–golden–rice. Accessed May 10, 2023.

37. Food and Drug Administration, Animal & Veterinary: *AquAdvantage Salmon fact sheet.* 2022. www.fda.gov/animal–veterinary/aquadvantage–salmon/aquadvantage–salmon–fact–sheet. Accessed May 10, 2023.

38. U.S. Department of Agriculture: *Bioengineered Food Disclosure Standard.* www.ams.usda.gov/rules–regulations/be. Accessed May 10, 2023.

39. National Academy of Sciences: *Genetically engineered crops: Experiences and prospects.* 2016. The National Academies Press. www.nap.edu/catalog/23395/genetically–engineered–crops–experiences–and–prospects. Accessed May 10, 2023.

40. World Resources Report: *Creating a sustainable food future.* 2018. World Resources Institute. https://research.wri.org/sites/default/files/2019–07/creating–sustainable–food–future_2_5.pdf. Accessed May 10, 2023.

Chapter 2

1. Kraut A: *Dr. Joseph Goldberger & the war on pellagra.* ND. Office of NIH History. https://history.nih.gov/pages/viewpage.action?pageId=8883184. Accessed May 10, 2023.

2. Simoni RD and others: Copper as an essential nutrient and nicotinic acid as the anti–black tongue (pellagra) factor: The work of Conrad Arnold Elvehjem. *Journal of Biological Chemistry* 277(34): e22, 2002.

3. Darnall BD, Colloca L: Optimizing placebo and minimizing nocebo to reduce pain, catastrophizing, and opioid use: A review of the science and an evidence–informed clinical toolkit. *International Review of Neurobiology* 139: 129, 2018.

4. Centers for Disease Control and Prevention: *National Diabetes Statistics Report, 2020.* 2020. www.cdc.gov/diabetes/pdfs/data/statistics/national–diabetes–statistics–report.pdf. Accessed May 10, 2023.

5. Shin S and others: Sugar–sweetened beverage consumption in relation to obesity and metabolic syndrome among Korean adults: A cross–sectional study from the 2012–2016 Korean National Health and Nutrition Examination Survey (KNHANES). *Nutrients* 10(10), pii: E1467, 2018.

6. Yu G and others: A dietary pattern of higher fish, egg, milk, nut, vegetable and fruit, and lower salt intake correlates with the prevalence and control of hypertension. *American Journal of Hypertension* 31(6): 679, 2018.

7. Kearns CE and others: Sugar industry and coronary heart disease research: A historical analysis of internal industry documents. *JAMA Internal Medicine* 176(11): 1680, 2016.
8. Meyer–Rochow VB: Food taboos: their origins and purposes. *Journal of Ethnobiology and Ethnomedicine* 5: 18, 2009.
9. U.S. Department of Justice, Eastern District of Oklahoma: *Court orders Oklahoma company to stop selling colloidal silver as treatment for COVID–19.* May 15, 2020. www.justice.gov/usao–edok/pr/court–orders–oklahoma–company–stop–selling–colloidal–silver–treatment–covid–19. Accessed May 10, 2023.
10. Ferrara G and others: A focused review of smartphone diet–tracking apps: Usability, functionality, coherence with behavior change theory, and comparative validity of nutrient intake and energy estimates. *Journal of Medical Internet Research Mhealth Uhealth* 7(5): e9232, 2019.
11. Centers for Disease Control and Prevention, National Center for Health Statistics: *Dietary supplement use among adults: United States, 2017–2018.* NCHS Data Brief No. 399. www.cdc.gov/nchs/products/databriefs/db399.htm. Updated February 2021. Accessed May 10, 2023.
12. National Institutes of Health, National Center for Complementary and Integrative Health: *Using dietary supplements wisely.* January 2019. www.nccih.nih.gov/health/using–dietary–supplements–wisely. Accessed May 10, 2023.
13. National Institutes of Health, National Center for Complementary and Integrative Health: *Complementary, alternative, or integrative health: What's in a name?* 2021. https://nccih.nih.gov/health/integrative–health. Accessed May 10, 2023.
14. Levine GN and others: Meditation and cardiovascular risk reduction: A scientific statement from the American Heart Association. *Journal of the American Heart Association* 6(10), pii: e002218, 2017.
15. Lorenc A and others: Scoping review of systematic reviews of complementary medicine for musculoskeletal and mental health conditions. *BMJ Open* 8(10): e020222, 2018.
16. U.S. Food and Drug Administration: *Dietary supplements.* 2022. www.fda.gov/Food/DietarySupplements/default.htm. Accessed May 10, 2023.
17. National Institutes of Health, Office of Dietary Supplements: *What you need to know about dietary supplements.* 2020. https://ods.od.nih.gov/factsheets/WYNTK–Consumer/. Accessed May 10, 2023.
18. Statista, Pharmaceutical Products & Market: *Total U.S. dietary supplements market size from 2016 to 2024 (in billion U.S. dollars).* 2023. www.statista.com/statistics/828481/total–dietary–supplements–market–size–in–the–us/#statisticContainer. Accessed May 10, 2023.
19. Chen F and others: Association between dietary supplement use, nutrient intake, and mortality among US adults: A cohort study. *Annals of Internal Medicine,* April 9, 2019. doi:10.7326/M18–2478.
20. Kalt W and others. Recent research on the health benefits of blueberries and their anthocyanins. *Advances in Nutrition* 11(2): 224–236. 2020.
21. National Institutes of Health, MedlinePlus: *Blueberry.* 2019. https://medlineplus.gov/druginfo/natural/1013.html. Accessed May 10, 2023.

Chapter 3

1. Food Marketing Institute: *Supermarket facts.* 2021. www.fmi.org/our–research/supermarket–facts. Accessed April 11, 2023.
2. U.S. Department of Agriculture, Economic Research Service: *New products.* December 2021. www.ers.usda.gov/topics/food–markets–prices/processing–marketing/. Accessed May 10, 2023.
3. Institute of Medicine: *Dietary Reference Intakes for vitamin C, vitamin E, selenium, and carotenoids.* Washington, D.C.: National Academies Press, 2000.
4. Otten JJ and others (eds.): *Dietary Reference Intakes: The essential guide to nutrient requirements.* Institute of Medicine of the National Academies. Washington, D.C.: National Academies Press, 2006.
5. National Academies of Sciences, Engineering, and Medicine: *Guiding principles for developing Dietary Reference Intakes based on chronic disease.* Washington, D.C.: The National Academies Press, 2017.
6. U.S. Food and Drug Administration: *Draft guidance for industry and FDA staff: Whole grain label statements.* 2006. www.fda.gov/regulatory–information/search–fda–guidance–documents/draft–guidance–industry–and–fda–staff–whole–grain–label–statements. Accessed May 10, 2023.
7. U.S. Food and Drug Administration: *Draft guidance for industry: Labeling of plant–based milk alternatives and voluntary nutrient statements.* 2023. www.fda.gov/regulatory–information/search–fda–guidance–documents/draft–guidance–industry–labeling–plant–based–milk–alternatives–and–voluntary–nutrient–statements. Accessed April 11, 2023.
8. U.S. Department of Agriculture: *Fruit: Cup of fruit table.* 2020. www.myplate.gov/eat–healthy/fruits. Accessed May 10, 2023.
9. Monteiro CA and others: Ultra–processed foods: What they are and how to identify them. *Public Health Nutrition* 22(5): 936–941, 2019.
10. U.S. Department of Health and Human Services, U.S. Department of Agriculture: *Dietary Guidelines for Americans, 2020–2025.* 2020. www.dietaryguidelines.gov. Accessed May 10, 2023.
11. U.S. Department of Health and Human Services, U.S. Department of Agriculture: *Scientific report of the 2020 Dietary Guidelines advisory committee.* 2020. www.dietary

guidelines.gov/2020–advisory–committee–report. Accessed May 10, 2023.

12. U.S. Department of Agriculture: *Use oils.* 2020. www.myplate.gov/eat–healthy/more–key–topics. Accessed May 10, 2023.
13. U.S. Department of Health and Human Services, U.S. Department of Agriculture: *Top 10 Things You Need to Know About the Dietary Guidelines for Americans, 2020–2025.* www.dietaryguidelines.gov/resources/2020–2025–dietary–guidelines–online–materials/top–10–things–you–need–know–about–dietary. Accessed May 10, 2023.
14. U.S. Department of Agriculture, Economic Research Service: *Food availability (per capita) data system: Consumption compared to dietary recommendations.* Last updated January 2020. www.ers.usda.gov/data–products/food–availability–per–capita–data–system/interactive–charts–and–highlights/. Accessed May 10, 2023.
15. Campbell A: Are food exchanges for you? *Diabetes Self–Management,* September 2021. www.diabetesselfmanagement.com/healthy–living/nutrition–exercise/are–food–exchanges–for–you/. Accessed May 10, 2023.
16. Restrepo B, Minor T: New national menu labeling provides information consumers can use to help manage their calorie intake. *Amber Waves.* 2018. www.ers.usda.gov/amber–waves/2018/october/new–national–menu–labeling–provides–information–consumers–can–use–to–help–manage–their–calorie–intake/. Accessed May 10, 2023.
17. U.S. Government Publishing Office, Electronic Code of Federal Regulations: *Title 21: Food and drugs, part 101—food labeling.* Current as of June 6, 2022. www.ecfr.gov/current/title–21/chapter–I/subchapter–B/part–101. Accessed May 10, 2023.
18. U.S. Food and Drug Administration: *How to understand and use the Nutrition Facts label.* Last updated February 2022. www.fda.gov/food/new–nutrition–facts–label/how–understand–and–use–nutrition–facts–label. Accessed May 10, 2023.
19. U.S. Food and Drug Administration: *Use of the term healthy on food labeling.* Last updated October 27, 2022. www.fda.gov/food/food–labeling–nutrition/use–term–healthy–food–labeling. Accessed May 10, 2023.
20. World Resources Report: *Creating a sustainable food future.* 2018. World Resources Institute. https://research.wri.org/sites/default/files/2019–07/creating–sustainable–food–future_2_5.pdf. Accessed May 10, 2023.
21. U.S. Environmental Protection Agency: *Pesticides and food: Healthy, sensible food practices.* Last updated May 3, 2022. www.epa.gov/safepestcontrol/pesticides–and–food–healthy–sensible–food–practices. Accessed May 10, 2023.
22. The University of Maine: Cooperative Extension Publications. Bulletin #4336, *Best ways to wash fruits and vegetables.* 2022. https://extension.umaine.edu/publications/4336e/. Accessed May 10, 2023.
23. U.S. Department of Agriculture, Agricultural Marketing Service: *Agricultural Marketing Service's National Organic Program.* 2016. www.ams.usda.gov/publications/content/about–national–organic–program. Accessed May 10, 2023.
24. Organic Trade Association, Press Release: *2022 organic industry survey shows steady growth, stabilizing purchasing patterns.* June 2, 2022. https://ota.com/news/press–releases/22284. Accessed May 10, 2023.
25. Vigar V and others: A systematic review of organic versus conventional food consumption: Is there a measurable benefit on human health? *Nutrients* 12(1): 7, 2019.
26. Defagó MD and others: Dietary patterns and blood pressure in Southern Cone of Latin America. *Nutrition Metabolism and Cardiovascular Disease* 31(12): 3326–3334, 2021.
27. Reicks M and others: Impacts of A Taste of African Heritage: A culinary heritage cooking course. *Journal of Nutrition Education and Behavior* 54(5): 388–396, 2022.
28. Gavrilova NA, Gavrilov LA: Comments on dietary restriction, Okinawa diet and longevity. *Gerontology* 58: 221, 2012.
29. Takase H and others: Dietary sodium consumption predicts future blood pressure and incident hypertension in the Japanese normotensive general population. *Journal of the American Heart Association* 4(8): e001959, 2015.

Chapter 4

1. Sorokina M and others: A catalog of natural products occurring in watermelon—*Citrullus lanatus. Frontiers in Nutrition* 14(8): 729822, 2021.
2. Helander HF, Fändriks L: Surface area of the digestive tract—Revisited. *Scandinavian Journal of Gastroenterology* 49(6): 681, 2014.
3. Costanzo A and others: Fat taste sensitivity is associated with short–term and habitual fat intake. *Nutrients* 9(7): 781, 2017.
4. Feng P and others: Taste bud homeostasis in health, disease, and aging. *Chemical Senses* 39: 3, 2014.
5. Boscolo–Rizzo P and others: High prevalence of long–term olfactory, gustatory, and chemesthesis dysfunction in post–COVID–19 patients: A matched case–control study with one–year follow–up using a comprehensive psychophysical evaluation. *Rhinology* 59(6): 517–527, 2021.
6. Coelho DH and others: Decreasing incidence of chemosensory changes by COVID–19 variant. *Otolaryngology–Head and Neck Surgery* 168(4): 704–706, 2023.
7. Nesti L and others: Impact of nutrient type and sequence on glucose tolerance: Physiological insights and therapeutic implications. *Frontiers of Endocrinology* 10: 144, 2019.
8. Vanputte CL and others: *Seeley's essentials of anatomy & physiology 11e.* New York: McGraw Hill Publishing, 2022.

9. Mark EB and others: Assessment of colorectal length using the electromagnetic capsule tracking system: A comparative validation study in healthy subjects. *Colorectal Disease* 19(9): O350, 2017.
10. Khan A and others: Etiological aspects of intragastric bezoars and its associations to the gastric function implications: A case report and a literature review. *Medicine (Baltimore)* 97(27): e11320, 2018.
11. Bowen R: Gastrointestinal transit: How long does it take? ND. *VIVO Pathophysiology.* www.vivo.colostate.edu/hbooks/pathphys/digestion/basics/transit.html. Accessed May 10, 2023.
12. Sender R and others: Revised estimates for the number of human and bacteria cells in the body. *PLoS Biology* 14(8): e1002533, 2016.
13. Rinninella E and others: What is the healthy gut microbiota composition? A changing ecosystem across age, environment, diet, and diseases. *Microorganisms* 7(1): 14, 2019.
14. National Institutes of Health, National Institute of Diabetes and Digestive and Kidney Diseases: *Appendicitis.* 2014. www.niddk.nih.gov/health–information/digestive–diseases/appendicitis/definition–facts. Accessed May 10, 2023.
15. Girard–Madoux MJH and others: The immunological functions of the appendix: An example of redundancy? *Seminars in Immunology* 36(April): 31, 2018.
16. Zinöcker M, Lindseth I: The Western diet–microbiome–host interaction and its role in metabolic disease. *Nutrients* 10(3): 365, 2018.
17. Leustean AM and others: Implications of the intestinal microbiota in diagnosing the progression of diabetes and the presence of cardiovascular complications. *Journal of Diabetes Research* 12: 5205126, 2018.
18. Hui W and others: Fecal microbiota transplantation for treatment of recurrent *C. difficile* infection: An updated randomized controlled trial meta–analysis. *PLoS ONE* 14(1): e0210016, 2019.
19. National Center for Complementary and Integrative Health: *Probiotics: What you need to know.* August 2019. www.nccih.nih.gov/health/probiotics–what–you–need–to–know. Accessed May 10, 2023.
20. Centers for Disease Control and Prevention: *Inflammatory bowel disease: Data and statistics: Inflammatory bowel disease prevalence in the United States.* Last reviewed April 2022. www.cdc.gov/ibd/data–and–statistics/prevalence.html. Accessed May 10, 2023.
21. Centers for Disease Control and Prevention; Disease or Condition of the Week: *Inflammatory bowel disease (IBD).* 2020. www.cdc.gov/dotw/ibd/index.html. Accessed May 10, 2023.
22. Narula N and others: Association of ultra–processed food intake with risk of inflammatory bowel disease: Prospective cohort study. *British Medical Journal* 374: n1554, 2021.
23. National Institutes of Health, National Institute of Diabetes and Digestive and Kidney Diseases: *Ulcerative colitis: Definitions and Facts.* September 2020. www.niddk.nih.gov/health–information/digestive–diseases/ulcerative–colitis/definition–facts. Accessed May 10, 2023.
24. National Institutes of Health, National Institute of Diabetes and Digestive and Kidney Diseases: *Definition & facts for Crohn's disease.* 2017. www.niddk.nih.gov/health–information/digestive–diseases/crohns–disease/definition–facts. Accessed May 10, 2023.
25. National Institutes of Health, National Institute of Diabetes and Digestive and Kidney Diseases: *Definition & facts for constipation: What is constipation?* 2018. www.niddk.nih.gov/health–information/digestive–diseases/constipation/symptoms–causes. Accessed May 10, 2023.
26. National Institutes of Health, U.S. National Library of Medicine, Medline Plus: *Nausea and vomiting—Adults.* 2017. https://medlineplus.gov/ency/article/003117.htm. Accessed May 10, 2023.
27. National Institutes of Health, National Institute of Diabetes and Digestive and Kidney Diseases: *Eating, diet, & nutrition for GER & GERD.* 2020. www.niddk.nih.gov/health–information/digestive–diseases/acid–reflux–ger–gerd–adults/eating–diet–nutrition. Accessed May 10, 2023.
28. National Institutes of Health, National Institute of Diabetes and Digestive and Kidney Diseases: *Peptic ulcers (stomach and duodenal ulcers).* 2022. www.niddk.nih.gov/health–information/digestive–diseases/peptic–ulcers–stomach–ulcers. Accessed May 10, 2023.
29. American College of Gastroenterology: *Irritable bowel syndrome.* January 2022. https://gi.org/topics/irritable–bowel–syndrome/. Accessed May 10, 2023.
30. National Institutes of Health, National Institute of Diabetes and Digestive and Kidney Diseases: *Irritable bowel syndrome.* 2019. www.niddk.nih.gov/health–information/digestive–diseases/irritable–bowel–syndrome/definition–facts. Accessed May 10, 2023.
31. Werlang ME and others: Irritable bowel syndrome and dietary interventions. *Gastroenterology & Hepatology* 15(1): 2019.
32. American Cancer Society: *Colorectal cancer: Key statistics for colorectal cancer.* 2022. www.cancer.org/cancer/colon–rectal–cancer/about/key–statistics.html. Accessed May 10, 2023.
33. American Cancer Society: *Colorectal cancer: Colorectal cancer signs and symptoms.* 2020. www.cancer.org/cancer/colon–rectal–cancer/detection–diagnosis–staging/signs–and–symptoms.html. Accessed May 10, 2023.
34. American Cancer Society: *Colorectal cancer: Colorectal cancer risk factors.* 2020. www.cancer.org/cancer/colon–rectal–cancer/causes–risks–prevention/risk–factors.html. Accessed May 10, 2023.

35. American Cancer Society: *American Cancer Society guideline for colorectal cancer screening: For people at average risk.* 2020. www.cancer.org/cancer/types/colon–rectal–cancer/detection–diagnosis–staging/acs–recommendations.html. Accessed May 10, 2023.

Chapter 5

1. U.S. Department of Agriculture, Economic Research Service: *U.S. Consumption of caloric sweeteners.* Tables 51, 52, and 53. Last updated September 16, 2021. www.ers.usda.gov/data–products/sugar–and–sweeteners–yearbook–tables/. Accessed May 10, 2023.
2. U.S. Food and Drug Administration: *Guidance for industry: Ingredients declared as evaporated cane juice.* May 2016. www.fda.gov/regulatory–information/search–fda–guidance–documents/guidance–industry–ingredients–declared–evaporated–cane–juice. Accessed May 10, 2023.
3. Centers for Disease Control and Prevention: *Botulism.* Last updated August 2019. www.cdc.gov/botulism/index.html. Accessed May 10, 2023.
4. Sylvetsky AC, Rother KI. Nonnutritive sweeteners in weight management and chronic disease: A review. *Obesity* 26(4): 635–640, 2018.
5. U.S. Food and Drug Administration: *Additional information about high–intensity sweeteners permitted for use in food in the United States.* 2018. www.fda.gov/food/food–additives–petitions/additional–information–about–high–intensity–sweeteners–permitted–use–food–united–states. Accessed May 10, 2023.
6. Williams RA and others: Phenylketonuria: An inborn error of phenylalanine metabolism. *The Clinical Biochemists Reviews* 29(1): 31–41, 2008.
7. Arbind KC, Etheresia P. Revisiting the safety of aspartame. *Nutrition Reviews* 75(9): 718–730, 2017.
8. Academy of Nutrition and Dietetics: Position of the Academy of Nutrition and Dietetics: Use of nutritive and nonnutritive sweeteners. *Journal of the Academy of Nutrition and Dietetics* 112(5): 739, 2012.
9. Lita T and others: Failure of d–psicose absorbed in the small intestine to metabolize into energy and its low large intestinal fermentability in humans. *Metabolism* 59(2): 206, 2010.
10. U.S. Food and Drug Administration: *Guidance for industry: The declaration of allulose and calories from allulose on nutrition and supplement facts labels.* October 2020. www.fda.gov/regulatory–information/search–fda–guidance–documents/guidance–industry–declaration–allulose–and–calories–allulose–nutrition–and–supplement–facts–labels. Accessed May 10, 2023.
11. U.S. Food and Drug Administration: *Draft guidance for industry and FDA staff: Whole grain label statements.* 2006. www.fda.gov/regulatory–information/search–fda–guidance–documents/draft–guidance–industry–and–fda–staff–whole–grain–label–statements. Accessed May 10, 2023.
12. Slavin J, Carlson J: Carbohydrates. *Advances in Nutrition* 5(6): 760, 2014.
13. U.S. Department of Agriculture, Agricultural Research Service: *What we eat in America, NHANES 2017–2018: Nutrient intakes from food and beverages: Mean amounts consumed by individuals: By gender and age.* 2020. www.ars.usda.gov/ARSUserFiles/80400530/pdf/usual/Usual_Intake_gender_WWEIA_2015_2018.pdf. Accessed May 10, 2023.
14. Patel R, DuPont HL: New approaches for bacteriotherapy: Prebiotics, new–generation probiotics, and synbiotics. *Clinical Infectious Diseases* 60(Suppl 2): S108, 2015. Published online April 28, 2015.
15. Markowiak P, Śliżewska K: Effects of probiotics, prebiotics, and synbiotics on human health. *Nutrients* 9(9): 1021, 2017.
16. Otten JJ and others (eds.): *Dietary Reference Intakes: The essential guide to nutrient requirements.* Washington, D.C.: National Academies Press, 2006.
17. U.S. Department of Health and Human Services, U.S. Department of Agriculture: *Dietary guidelines for Americans, 2020–2025.* 2020. www.dietaryguidelines.gov. Accessed May 10, 2023.
18. U.S. Department of Agriculture, Agricultural Research Service. *FoodData Central, 2019.* https://fdc.nal.usda.gov/. Accessed May 10, 2023.
19. Hall KD, Guo J. Obesity energetics: Body weight regulation and the effects of diet composition. *Gastroenterology* 152(7): 1718–1727, 2017.
20. Hall KD and others. The energy balance model of obesity: Beyond calories in, calories out. *American Journal of Clinical Nutrition* 115(5): 1243–1254, 2020.
21. Long G and others: Comparison of dietary macronutrient patterns of 14 popular named dietary programmes for weight and cardiovascular risk factor reduction in adults: Systematic review and network meta–analysis of randomised trials. *British Medical Journal* 369: m696, 2020.
22. Saris WH: Sugars, energy metabolism, and body weight control. *American Journal of Clinical Nutrition* 78 (4 Suppl): 850S, 2003.
23. U.S. Department of Agriculture, Human Nutrition Information Service: *Nationwide Food Consumption Survey: Food and nutrient intakes by individuals in the United States, 1 day, 1987–88.* Report no. 87–I–1, 1993.
24. Bailey RL and others: Sources of added sugars in young children, adolescents, and adults with low and high intakes of added sugars. *Nutrients* 10(1): 102, 2018.
25. Augustin LSA and others: Glycemic index, glycemic load and glycemic response: An International Scientific Consensus Summit from the International Carbohydrate Quality

Consortium (ICQC). *Nutrition, Metabolism & Cardiovas–cular Diseases* 25: 795, 2015.

26. Malik VS, Hu FB. The role of sugar–sweetened beverages in the global epidemics of obesity and chronic diseases. *Nature Reviews Endocrinology* 18(4): 205–218, 2022.

27. Chhimwal J and others. Beverages and non–alcoholic fatty liver disease (NAFLD): Think before you drink. *Clinical Nutrition* 40(5): 2508–2519, 2021.

28. Alonso–Alonso M and others: Food reward system: Current perspectives and future research needs. *Nutrition Reviews* 73(5): 296, 2015.

29. Al–Dayyat HM and others: Non–alcoholic fatty liver disease and associated dietary and lifestyle risk factors. *Diabetes & Metabolic Syndrome: Clinical Research & Reviews* (12): 560–575, 2018.

30. National Institutes of Health, Institute of Diabetes and Digestive and Kidney Diseases: *Nonalcoholic fatty liver disease & NASH.* 2021. www.niddk.nih.gov/health–information/liver–disease/nafld–nash/definition–facts. Accessed May 10, 2023.

31. Centers for Disease Control and Prevention: *What is type I diabetes?* Updated 2022. www.cdc.gov/diabetes/basics/what–is–type–1–diabetes.html. Accessed May 10, 2023.

32. Murphy SL and others: Mortality in the United States, 2020. *NCHS Data Brief,* no. 427. Hyattsville, MD: National Center for Health Statistics. Accessed May 10, 2023.

33. Centers for Disease Control and Prevention: *2020 national diabetes statistics report.* Updated 2020. www.cdc.gov/diabetes/pdfs/data/statistics/national–diabetes–statistics–report.pdf. Accessed May 10, 2023.

34. Horta BL, de Lima NP: Breastfeeding and type 2 diabetes: Systematic review and meta–analysis. *Current Diabetes Reports* 19(1): 1, 2019.

35. Milken Institute: *Health literacy in the United States: Enhancing assessments and reducing disparities.* 2022. https://milkeninstitute.org/sites/default/files/2022–05/Health_Literacy_United_States_Final_Report.pdf. Accessed May 10, 2023.

36. Association of Diabetes Care and Education Specialist: Practice Paper: *Cultural competency, health literacy and numeracy.* 2019. www.diabeteseducator.org/practice/practice–tools/app–resources/cultural–competency–health–literacy–and–numeracy. Accessed May 10, 2023.

37. Brewster S and others: Non–attendance at diabetes outpatient appointments: A systematic review. *Diabetic Medicine* 37(9): 1427–1442, 2020.

38. Pojednic R and others: Physician nutrition advice and referrals to registered dietitians. *American Journal of Life–style Medicine,* May 2022.

39. National Institute of Diabetes and Digestive and Kidney Diseases (NIDDK): *Tests & diagnosis for gestational diabetes.* Last updated May 2017. www.niddk.nih.gov/health–information/diabetes/overview/what–is–diabetes/gestational/tests–diagnosis. Accessed May 10, 2023.

40. National Institute of Diabetes and Digestive and Kidney Diseases (NIDDK): *Gestational diabetes, after your baby is born.* Last updated 2017. www.niddk.nih.gov/health–information/diabetes/overview/what–is–diabetes/gestational/after–your–baby–is–born. Accessed May 10, 2023.

41. American Diabetes Association: *A1C.* ND. www.diabetes.org/a1c. Accessed May 10, 2023.

42. National Institutes of Health, Institute of Diabetes and Digestive and Kidney Diseases: *Diabetes prevention program (DPP).* ND. www.niddk.nih.gov/about–niddk/research–areas/diabetes/diabetes–prevention–program–dpp/Pages/default.aspx. Accessed May 10, 2023.

43. National Institute of Diabetes and Digestive and Kidney Diseases (NIDDK): *Diabetes diet, eating, & physical activity.* 2016. www.niddk.nih.gov/health–information/diabetes/overview/diet–eating–physical–activity. Accessed May 10, 2023.

44. Martín–Peláez S and others. Mediterranean diet effects on type 2 diabetes prevention, disease progression, and related mechanisms. A review. *Nutrients* 12(8): 2236, 2020.

45. National Institutes of Health, Institute of Diabetes and Diges–tive and Kidney Diseases: *Low blood glucose (hypoglycemia).* Updated 2021. www.niddk.nih.gov/health–information/diabetes/overview/preventing–problems/low–blood–glucose–hypoglycemia. Accessed May 10, 2023.

46. Earl J: Dog saves the life of a sleeping boy with type 1 diabetes. *CBS News,* March 7, 2016. www.cbsnews.com/news/dog–saves–the–life–of–sleeping–boy–with–type–1–diabetes/. Accessed May 10, 2023.

47. Gonder–Frederick LA and others: Variability of diabetes alert dog accuracy in a real–world setting. *Journal of Diabetes Science and Technology* 11(4): 714, 2017.

48. Neupane S and others: Exhaled breath isoprene rises during hypoglycemia in type 1 diabetes. *Diabetes Care* 39(7): e97, 2016.

49. Baidal DA and others: Correspondence: Bioengineering of an intraabdominal endocrine pancreas. *The New England Journal of Medicine* 376: 1887, 2017.

50. National Institutes of Health, National Heart, Lung, and Blood Institute: *Metabolic syndrome: Diagnosis.* Updated 2022. www.nhlbi.nih.gov/node/92324. Accessed May 10, 2023.

51. Saklayen MG: The global epidemic of the metabolic syndrome. *Current Hypertension Reports* 20: 12, 2018.

52. Bellini M and others: Low FODMAP diet: Evidence, doubts, and hopes. *Nutrients* 12(1): 148, 2020.

53. National Institutes of Health, Institute of Diabetes and Digestive and Kidney Diseases: *Lactose intolerance.* 2018.

www.niddk.nih.gov/health–information/health–topics/digestive–diseases/lactose–intolerance/Pages/facts.aspx. Accessed May 10, 2023.

54. Facioni MS and others: Nutritional management of lactose intolerance:Tthe importance of diet and food labelling. *Journal of Translational Medicine* 18(1): 260, 2020.
55. Centers for Disease Control and Prevention, Attention Deficit/Hyperactivity Disorder: *Facts about ADHD.* Last updated September 2021. www.cdc.gov/ncbddd/adhd/diagnosis.html. Accessed May 10, 2023.
56. Centers for Disease Control and Prevention, Attention Deficit/Hyperactivity Disorder: *Data & statistics.* Last updated October 2019. www.cdc.gov/ncbddd/adhd/data.html. Accessed May 10, 2023.
57. Rezapour M and others: Diverticular disease: An update on pathogenesis and management. *Gut Liver* 12(2): 125–132, 2018.
58. World Health Organization: *Cancer: Fact sheets.* Updated February 2022. www.who.int/mediacentre/factsheets/fs297/en/. Accessed May 10, 2023.
59. Centers for Disease Control and Prevention: *What can I do to reduce my risk of colorectal cancer.* Last updated February 2022. www.cdc.gov/cancer/colorectal/basic_info/prevention.htm. Accessed May 10, 2023.
60. Augustin LSA and others: Glycemic index, glycemic load and glycemic response: An International Scientific Consensus Summit from the International Carbohydrate Quality Consortium (ICQC). *Nutrition, Metabolism & Cardiovascular Diseases* 25(9): 795, 2015.

Chapter 6

1. Position of the Academy of Nutrition and Dietetics: Dietary fatty acids for healthy adults. *Journal of the Academy of Nutrition and Dietetics* 114(1): 136, 2014.
2. Bradbury J and others: Docosahexaenoic acid (DHA): An ancient nutrient for the modern human brain. *Nutrients* 3(5): 529, 2011.
3. Martin CR and others: Review of infant feeding: Key features of breast milk and infant formula. *Nutrients* 8(5): 279, 2016.
4. Otten JJ and others (eds.): *Dietary Reference Intakes: The essential guide to nutrient requirements.* Washington, D.C.: National Academies Press, 2006.
5. Remig V and others: Trans fat in America: A review of their use, consumption, health implications, and regulation. *Journal of the American Dietetic Association* 110(4): 585, 2010.
6. U.S. Food and Drug Administration: *Final determination regarding partially hydrogenated oils (removing trans fat).* Updated May 2018. www.fda.gov/food/food–additives–petitions/final–determination–regarding–partially–hydrogenated–oils–removing–trans–fat. Accessed May 10, 2023.
7. Mensick RP and others: The increasing use of interesterified lipids in the food supply and their effects on health parameters. *Advances in Nutrition* 7(4): 719, 2016.
8. U.S. Department of Health and Human Services: *Dietary Guidelines for Americans, 2020–2025.* 2020. www.dietaryguidelines.gov. Accessed May 10, 2023.
9. Oregon State University, Linus Pauling Institute: *Choline.* Updated 2015. https://lpi.oregonstate.edu/mic/other–nutrients/choline. Accessed May 10, 2023.
10. Bosner MS and others: Percent cholesterol absorption in normal women and men quantified with dual stable isotopic tracers and negative ion mass spectrometry. *Journal of Lipid Research* 40: 302, 1999.
11. U.S. Department of Agriculture, Agricultural Research Service: *What we eat in America, NHANES 2017–2018: Nutrient intakes from food and beverages: Mean amounts consumed by individuals: By gender and age.* 2020. www.ars.usda.gov/ARSUserFiles/80400530/pdf/usual/Usual_Intake_gender_WWEIA_2015_2018.pdf. Accessed May 10, 2023.
12. Murphy SL and others: Mortality in the United States, 2020. *NCHS Data Brief,* no. 427, figure 4 data. Hyattsville, MD: National Center for Health Statistics. Accessed May 10, 2023.
13. Hernaez A and others: Role of HDL function and LDL atherogenicity on cardiovascular risk: A comprehensive examination. *PLOSOne* 14(6): e0218533, 2019.
14. Jingyuan Li and others: Nicotine induces endothelial dysfunction and promotes atherosclerosis via GTPCH1. *Journal of Cellular and Molecular Medicine* 22(11): 5406, 2018.
15. Khatana C and others: Mechanistic insights into the oxidized low–density lipoprotein–induced atherosclerosis. *Oxidative Medicine and Cellular Longevity* 5: 5245308, 2020.
16. National Institutes of Health, National Heart, Lung, and Blood Institute: *Who is at risk for atherosclerosis?* Updated March 2022. www.nhlbi.nih.gov/health–topics/atherosclerosis. Accessed May 10, 2023.
17. Centers for Disease Control and Prevention, Million Hearts 2027 Initiative: *Estimated hypertension prevalence, treatment, and control among U.S. adults.* March 2021. https://millionhearts.hhs.gov/data–reports/hypertension–prevalence.html. Accessed May 10, 2023.
18. Quasim H and others: Impact of electronic cigarettes on the cardiovascular system. *Journal of the American Heart Association* 6(9): e006353, 2017.
19. American Heart Association: *Cardiovascular disease & diabetes.* 2021. www.heart.org/en/health–topics/diabetes/diabetes–complications–and–risks/cardiovascular–disease–diabetes. Accessed May 10, 2023.
20. Centers for Disease Control and Prevention, Smoking & Tobacco Use: *Health effects of cigarette smoking.* Last

updated January 2018. www.cdc.gov/tobacco/data_statistics/fact_sheets/health_effects/effects_cig_smoking/index.htm. Accessed May 10, 2023.

21. Centers for Disease Control and Prevention, Smoking & Tobacco Use: *Secondhand smoke (SHS) facts.* Updated January 2018. www.cdc.gov/tobacco/data_statistics/fact_sheets/secondhand_smoke/general_facts/index.htm. Accessed May 10, 2023.

22. American Heart Association: *Stress and heart health.* Updated June 21, 2021. www.heart.org/en/healthy–living/healthy–lifestyle/stress–management/stress–and–heart–health. Accessed May 10, 2023.

23. Centers for Disease Control and Prevention, National Center for Health Statistics: *Cholesterol.* Last updated September 6, 2022. www.cdc.gov/nchs/fastats/cholesterol.htm. Accessed May 10, 2023.

24. Mayo Clinic: *Cholesterol ratio or non–HDL cholesterol: Which is most important?* January 2022. www.mayoclinic.org/diseases–conditions/high–blood–cholesterol/expert–answers/cholesterol–ratio/faq–20058006. Accessed May 10, 2023.

25. American Heart Association: *How to get your cholesterol tested.* Last reviewed November 9, 2020. www.heart.org/en/health–topics/cholesterol/how–to–get–your–cholesterol–tested. Accessed May 10, 2023.

26. Grundy SM and others: 2018 AHA/ACC/AACVPR/AAPA/ABC/ACPM/ADA/AGS/APhA/ASPC/NLA/PCNA Guide–line on the Management of Blood Cholesterol: A report of the American College of Cardiology/American Heart Association Task Force on Clinical Practice Guidelines. *Circulation* 139(25): e1082–e1143, 2019.

27. Sun Y and others: Influence of cigarette smoking on burden and characteristics of coronary artery plaques in Chinese men. *Acta Cardiologica Sinica* 31(5): 398, 2015.

28. Fernandez ML, West KL: Mechanisms by which dietary fatty acids modulate plasma lipids. *Journal of Nutrition* 135(9): 2075, 2005.

29. Krandikar RS and others: Comparison between the effect of cow ghee and butter on memory and lipid profile of Wistar rats. *Journal of Clinical and Diagnostic Research* 10(9): FF11, 2016.

30. Hamley, S: The effect of replacing saturated fat with mostly n–6 polyunsaturated fat on coronary heart disease: A meta–analysis of randomized controlled trials. *Nutrition Journal* 16(1): 30, 2017.

31. Djuricic I, Calder PC: Beneficial outcomes of omega–6 and omega–3 polyunsaturated fatty acids on human health: An update for 2021. *Nutrients* 13(7): 2421, 2021.

32. National Institutes of Health, Office of Dietary Supplements: *Omega–3 fatty acids.* Updated July 18, 2022. https://ods.od.nih.gov/factsheets/Omega3FattyAcids–HealthProfessional/. Accessed May 10, 2023.

33. Carson JS and others: American Heart Association Nutri–tion Committee of the Council on Lifestyle and Cardio–metabolic Health; Council on Arteriosclerosis, Thrombosis and Vascular Biology; Council on Cardiovascular and Stroke Nursing; Council on Clinical Cardiology; Council on Peripheral Vascular Disease; and Stroke Council: Dietary cholesterol and cardiovascular risk: A science advisory from the American Heart Association. *Circulation* 141(3): e39–e53, 2020.

34. Lavelle F and others: Modern transference of domestic cooking skills. *Nutrients* 11(4): 870, 2019.

35. Mosko JE, Delach MJ: Cooking, creativity, and well–being: An integration of quantitative and qualitative methods. *The Journal of Creative Behavior* 55(2): 348–361, 2021.

36. Medline Plus: *How to lower cholesterol.* Updated 2019. https://medlineplus.gov/howtolowercholesterol.html. Accessed May 10, 2023.

37. Newman CB and others: Statin safety and associated adverse events: A scientific statement from the American Heart Association. *Arteriosclerosis, Thrombosis, and Vascular Biology* 39: e38, 2019.

38. National Institutes of Health, National Institute on Alcohol Abuse and Alcoholism: *Women and alcohol.* 2021. www.niaaa.nih.gov/publications/brochures–and–fact–sheets/women–and–alcohol. Accessed May 10, 2023.

39. Substance Abuse and Mental Health Services Administra–tion, Center for Behavioral Health Statistics and Quality: *Key substance use and mental health indicators in the United States: Results from the 2020 National Survey on Drug Use and Health.* 2020. www.samhsa.gov/data/report/2020–nsduh–annual–national–report. Accessed May 10, 2023.

40. Centers for Disease Control and Prevention, Alcohol and Public Health: *Frequently asked questions.* 2020. www.cdc.gov/alcohol/faqs.htm. Accessed May 10, 2023.

41. National Institutes of Health, National Institute on Alcohol Abuse and Alcoholism: *Understanding alcohol use disorder.* 2020. Updated April 2021. www.niaaa.nih.gov/publications/brochures–and–fact–sheets/understanding–alcohol–use–disorder. Accessed May 10, 2023.

42. Centers for Disease Control and Prevention, Alcohol and Public Health: *Fact sheets—Preventing excessive alcohol use.* 2022. www.cdc.gov/alcohol/fact–sheets/prevention.htm. Accessed May 10, 2023.

43. National Institutes of Health, National Cancer Institute: *Alcohol and cancer risk.* Last updated July 14, 2021. www.cancer.gov/about–cancer/causes–prevention/risk/alcohol/alcohol–fact–sheet. Accessed May 10, 2023.

44. American Heart Association: *Is drinking alcohol part of a healthy lifestyle?* Last reviewed December 2019. www.heart.org/en/healthy–living/healthy–eating/eat–smart/nutrition–basics/alcohol–and–heart–health. Accessed May 10, 2023.

45. Baj J and others: Magnesium, calcium, potassium, sodium, phosphorus, selenium, zinc, and chromium levels in alcohol use disorder: A review. *Journal of Clinical Medicine* 9(6): 1901, 2020.

46. Godos J and others: Alcohol consumption, bone mineral density, and risk of osteoporotic fractures: A dose–response meta–analysis. *International Journal of Environmental Research Public Health* 19(3): 1515, 2022.

Chapter 7

1. Shier D and others: *Hole's human anatomy & physiology.* 15th ed. New York: McGraw Hill Publishing, 2019.

2. Kurtz JA and others: Taurine in sports and exercise. *Journal of International Society of Sports Nutrition* 18(1): 39, 2021.

3. Reeds PJ: Dispensable and indispensable amino acids for humans. *Journal of Nutrition* 130: 1835S, 2000.

4. Institute of Medicine of the National Academies, Food and Nutrition Board: *Dietary Reference Intakes for energy, carbohydrate, fiber, fat, fatty acids, cholesterol, protein, and amino acids.* Washington, D.C.: National Academies Press, 2005.

5. Martínez–Villaluenga C and others: Pseudocereal grains: Nutritional value, health benefits and current applications for the development of gluten–free foods. *Food and Chemical Toxicology* 137: 111178, 2020.

6. Oelke EA and others: Quinoa. *Alternative field crops manual.* University of Wisconsin–Extension, University of Minnesota: Center for Alternative Plant & Animal Productions and the Minnesota Extension Service. ND. https://hort.purdue.edu/newcrop/afcm/quinoa.html. Accessed May 10, 2023.

7. Messina M and others: The health effects of soy: A reference guide for health professionals. *Frontiers in Nutrition* 9: 970364, 2022.

8. de Miranda RB and others: Effects of hydrolyzed collagen supplementation on skin aging: A systematic review and meta–analysis. *International Journal of Dermatology* 60(12): 1449–1461, 2021.

9. Woo T and others: Efficacy of oral collagen in joint pain—Osteoarthritis and rheumatoid arthritis. *Journal of Arthritis* 6: 233, 2017.

10. U.S. Department of Agriculture, Agricultural Research Service: *What we eat in America, NHANES 2017–2018: Nutrient intakes from food and beverages: Mean amounts consumed by individuals: By gender and age.* 2020. www.ars.usda.gov/ARSUserFiles/80400530/pdf/usual/Usual_Intake_gender_WWEIA_2015_2018.pdf. Accessed May 10, 2023.

11. Gardner CD and others: Maximizing the intersection of human health and the health of the environment with regard to the amount and type of protein produced and consumed in the United States. *Nutrition Reviews* 77(4): 197, 2019.

12. Zhong VW and others: Associations of processed meat, unprocessed red meat, poultry, or fish intake with incident cardiovascular disease and all–cause mortality. *Journal of the American Medical Association Internal Medicine* 180(4): 503–512, 2020.

13. World Cancer Research Fund/American Institute for Cancer Research: Continuous update project expert report 2018. Revised May 2018. www.dietandcancerreport.org. Accessed May 10, 2023.

14. Zeng L and others: Trends in processed meat, unprocessed red meat, poultry, and fish consumption in the United States. *Journal of the Academy of Nutrition and Dietetics* 119(7): 1085, 2019.

15. U.S. Department of Agriculture, MyPlate.gov: *What is MyPlate: Protein foods.* ND. www.myplate.gov/eat–healthy/protein–foods. Accessed May 10, 2023.

16. U.S. Department of Health and Human Services: *Scientific report of the 2020 Dietary Guidelines Advisory Committee:* Current dietary intakes through the life course, part D, chapter 1. U.S. Department of Agriculture, Agricultural Research Service, Washington, DC. 2020. www.dietaryguidelines.gov/2020–advisory–committee–report. Accessed May 10, 2023.

17. Gallup.com: *In U.S., 5% of Americans consider themselves vegetarians.* September 27, 2019. https://news.gallup.com/poll/267074/percentage–americans–vegetarian.aspx. Accessed May 10, 2023.

18. Bowman SA: A vegetarian–style dietary pattern is associated with lower energy, saturated fat, and sodium intakes; and higher whole grains, legumes, nuts, and soy intakes by adults: National Health and Nutrition Examination Surveys 2013–2016. *Nutrients* 12(9): 2668, 2020.

19. Neufingerl N and others: Nutrient intake and status in adults consuming plant–based diets compared to meat–eaters: A systematic review. *Nutrients* 14(1): 29, 2021.

20. Caspero A: *Building a healthy vegetarian diet: Myths and facts.* October 2021. www.eatright.org/health/wellness/vegetarian–and–plant–based/building–a–healthy–vegetarian–diet–myths. Accessed May 10, 2023.

21. Goraya AS and others: Neurology of nutritional vitamin B12 deficiency in infants: Case series from India and literature review. *Journal of Child Neurology* 30(13): 1831, 2015.

22. Forestell CA and others: Flexitarian diet and weight control: Healthy or risky eating behavior? *Frontiers in Nutrition* 5: 59, 2018.

23. Impossible Foods: Frequently asked questions: What is soy leghemoglobin? ND. https://faq.impossiblefoods.com/hc/en–us/articles/360019100553–What–is–soy–leghemoglobin–or–heme. Accessed May 10, 2023.

24. Shams–White MM and others: Dietary protein and bone health: A systematic review and meta–analysis from the

National Osteoporosis Foundation. *American Journal of Clinical Nutrition* 105(6): 1528, 2017.

25. Isanejad M and others: Association of protein intake with bone mineral density and bone mineral content among elderly women: The OSTPRE Fracture Prevention Study. *Journal of Nutrition Health and Aging* 21(6): 622, 2017.

26. National Institutes of Health, National Institute of Diabetes and Digestive and Kidney Diseases: *Eating, diet, & nutrition for kidney stones.* 2017. www.niddk.nih.gov/health–information/urologic–diseases/kidney–stones/eating–diet–nutrition. Accessed May 10, 2023.

27. Leidy HJ and others: The role of protein in weight loss and maintenance. *American Journal of Clinical Nutrition* 101(6): 1320S, 2015.

28. Van Baak MA and others: Dietary intake of protein from different sources and weight regain, changes in body composition and cardiometabolic risk factors after weight loss: The DIOGenes Study. *Nutrients* 9(12): 1326, 2017.

29. Malnutrition Quality Improvement Initiative: *Briefing: The value of quality malnutrition care, care burden of mal–nutrition in hospitalized adults.* ND. https://malnutritionquality.org/wp–content/uploads/briefing–the–value–of–quality–malnutrition–care.pdf. Accessed May 10, 2023.

30. White JV and others: Consensus statement: Academy of Nutrition and Dietetics and American Society for Parenteral and Enteral Nutrition: Characteristics recommended for the identification and documentation of adult malnutrition (undernutrition). *Journal of Parenteral and Enteral Nutri–tion* 36(3): 275, 2012.

31. Neighborhood Access Outreach to Health (NOAH): *Get well with food: How nutrition helps you recover.* ND. www.noahhelps.org/get–well–with–food–how–nutrition–helps–you–recover/. Accessed May 10, 2023.

32. Khan M and others: What is a registered dietitian nutritionist's role in addressing malnutrition? *Journal of the Academy of Nutrition and Dietetics* 118(9): 1804, 2018.

33. The World Health Organization: *Malnutrition fact sheet.* 2021. www.who.int/news–room/fact–sheets/detail/malnutrition. Accessed May 10, 2023.

34. Platts–Mills TAE and others: Diagnosis and management of patients with the α–gal syndrome. *Journal of Allergy and Clinical Immunology in Practice* 8(1): 15–23, 2020.

35. Greer FR and others, The American Academy of Pediatrics: Clinical report: The effects of early nutritional interven–tions on the development of atopic disease in infants and children: The role of maternal dietary restriction, breast–feeding, hydrolyzed formulas, and timing of introduction of allergenic complementary foods. *Pediatrics* 143(4): e20190281, 2019.

36. U.S. Food and Drug Administration: News Release: *FDA takes new steps regarding evaluating public health importance of additional food allergens.* April 2022. www.fda.gov/news–events/press–announcements/fda–takes–new–steps–regarding–evaluating–public–health–importance–additional–food–allergens. Accessed May 10, 2023.

37. American College of Allergy, Asthma, and Immunology: *Food allergy.* September 2022. https://acaai.org/allergies/types/food–allergy. Accessed May 10, 2023.

38. National Institutes of Health, National Institute of Diabetes and Digestive and Kidney Diseases: *Celiac disease: Defini–tion and facts for celiac disease.* ND. www.niddk.nih.gov/health–information/digestive–diseases/celiac–disease. Accessed May 10, 2023.

39. National Institutes of Health, National Library of Medicine, Genetics Home Reference: *Phenylketonuria.* Last updated October 2017. https://medlineplus.gov/genetics/condition/phenylketonuria/. Accessed May 10, 2023.

40. National Institutes of Health, MedlinePlus: *Phenylketonuria.* Last updated April 2019. https://medlineplus.gov/ency/article/001166.htm. Accessed May 10, 2023.

41. Prasad C and others: Introducing nutritional genomics teaching in undergraduate dietetic curricula. *Journal of Nutrigenetics and Nutrigenomics* 4(3): 165, 2011.

42. Kohlmeier M and others: Guide and position of the Inter–national Society of Nutrigenetics/Nutrigenomics on personalized nutrition: Part 2—Ethics, challenges and endeavors of precision nutrition. *Journal of Nutrigenetics and Nutrigenomics* 9: 28, 2016.

43. U.S. Department of Agriculture, Food Safety and Inspection Service: *Food product dating.* October 2019. www.fsis.usda.gov/wps/portal/fsis/topics/food–safety–education/get–answers/food–safety–fact–sheets/food–labeling/food–product–dating/food–product–dating. Accessed May 10, 2023.

Chapter 8

1. Bartholomew M: James Lind's *Treatise of the Scurvy* (1753). *Postgraduate Medicine* 78: 695, 2002.

2. Food and Nutrition Board: *Dietary Reference Intakes for vitamin C, vitamin E, selenium, and carotenoids.* Washington, D.C.: National Academies Press, 2000.

3. Centers for Disease Control and Prevention, National Center for Health Statistics: *Data brief 399: Dietary supplement use among adults: United States, 2017–2018.* 2021. www.cdc.gov/nchs/products/databriefs/db399.htm. Accessed May 10, 2023.

4. U.S. Department of Agriculture, Agricultural Research Service: *What we eat in America, NHANES 2017–2018: Nutrient intakes from food and beverages: Mean amounts consumed by individuals: By gender and age.* 2020. www.ars.usda.gov/ARSUserFiles/80400530/pdf/usual/Usual_Intake_gender_WWEIA_2015_2018.pdf. Accessed May 10, 2023.

5. National Institutes of Health, Office of Dietary Supplements: *Vitamin C.* March 2022. https://ods.od.nih.gov/factsheets/VitaminC–HealthProfessional/. Accessed May 10, 2023.
6. Food and Drug Administration: *Radiation–emitting products: Microwave oven radiation.* Updated 2017. www.fda.gov/radiation–emitting–products/resources–you–radiation–emitting–products/microwave–oven–radiation. Accessed May 10, 2023.
7. Carazo A and others: Vitamin A update: Forms, sources, kinetics, detection, function, deficiency, therapeutic use and toxicity. *Nutrients* 13(5): 1703, 2021.
8. National Institutes of Health, Office of Dietary Supplements: *Vitamin A: Fact sheet for health professionals.* February 2020. https://ods.od.nih.gov/factsheets/VitaminA–HealthProfessional/. Accessed May 10, 2023.
9. World Health Organization, Nutrition; Nutrition Landscape Information System (NLIS): *Vitamin A deficiency.* ND. www.who.int/data/nutrition/nlis/info/vitamin–a–deficiency. Accessed May 10, 2023.
10. Owens B: Golden rice is safe to eat, says FDA. *Nature Biotechnology* 36: 559, 2018.
11. International Rice Research Institute: *Golden rice communication tool.* 2021. www.irri.org/golden–rice–communication–toolkit. Accessed March 18, 2023.
12. Holick MF: Vitamin D deficiency. *New England Journal of Medicine* 357: 266, 2007.
13. National Institutes of Health, Office of Dietary Supplements: *Vitamin D.* Updated: August 2022. https://ods.od.nih.gov/factsheets/VitaminD–HealthProfessional/. Accessed May 10, 2023.
14. Cardwell G and others: A review of mushrooms as a potential source of dietary vitamin D. *Nutrients* 10(10): 1498, 2018.
15. Wacker M, Holick MF: Sunlight and vitamin D: Global perspective for health. *Dermato–Endocrinology* 5(1): 51, 2013.
16. National Institutes of Health, Office of Dietary Supplements: *Vitamin E.* 2021. https://ods.od.nih.gov/factsheets/VitaminE–HealthProfessional/. Accessed May 10, 2023.
17. National Institutes of Health, Office of Dietary Supplements: *Vitamin K.* 2021. https://ods.od.nih.gov/factsheets/VitaminK–HealthProfessional/. Accessed May 10, 2023.
18. National Institutes of Health, Office of Dietary Supplements: *Thiamin.* 2021. https://ods.od.nih.gov/factsheets/Thiamin–HealthProfessional. Accessed May 10, 2023.
19. National Institutes of Health, Office of Dietary Supplements: *Riboflavin.* 2022. https://ods.od.nih.gov/factsheets/Riboflavin–HealthProfessional/. Accessed May 10, 2023.
20. Oregon State University, Linus Pauling Institute, Micronutrient Information Center: *Niacin.* 2018. https://lpi.oregonstate.edu/mic/vitamins/niacin. Accessed May 10, 2023.
21. National Institutes of Health, Office of Dietary Supplements: *Niacin.* 2022. https://ods.od.nih.gov/factsheets/Niacin–HealthProfessional/. Accessed May 10, 2023.
22. National Institutes of Health, Office of Dietary Supplements: *Vitamin B6.* 2020. https://ods.od.nih.gov/factsheets/VitaminB6–HealthProfessional/. Accessed May 10, 2023.
23. National Institutes of Health, Office of Dietary Supplements: *Folate.* 2020. https://ods.od.nih.gov/factsheets/Folate–HealthProfessional/. Accessed May 10, 2023.
24. Oregon State University, Linus Pauling Institute, Micronutrient Information Center: *Vitamin B–12.* 2015. https://lpi.oregonstate.edu/mic/vitamins/vitamin–B12. Accessed May 10, 2023.
25. March of Dimes, News: *Fewer than half of U.S. women take recommended vitamins prior to pregnancy, according to March of Dimes new Prenatal Health & Nutrition Survey.* September 2017. www.marchofdimes.org/about/news/fewer–half–us–women–take–recommended–vitamins–prior–to–pregnancy–according–to–march. Accessed May 10, 2023.
26. Bender A and others: The association of folate and depression: A meta–analysis. *Journal of Psychiatric Research* 95: 9–18, 2018.
27. Dali–Youcef N, Andrès R: An update on cobalamin deficiency in adults. *Quarterly Journal of Medicine* 102(1): 17, 2008.
28. National Institutes of Health, Office of Dietary Supplements: *Vitamin B12.* Updated: March 9, 2022. https://ods.od.nih.gov/factsheets/VitaminB12–HealthProfessional/. Accessed May 10, 2023.
29. Wu Y and others: Associations of dietary vitamin B1, vitamin B2, vitamin B6, and vitamin B12 with the risk of depression: A systematic review and meta–analysis. *Nutrition Reviews* 80(3): 351–366, 2022.
30. Laird E and others: Low vitamin B12 but not folate is associated with incident depressive symptoms in community–dwelling older adults: A 4 year longitudinal study. *British Journal of Nutrition* 13: 1–22, 2021.
31. Oregon State University, Linus Pauling Institute, Micronutrient Information Center: *Vitamin C.* 2018. https://lpi.oregonstate.edu/mic/vitamins/vitamin–C. Accessed May 10, 2023.
32. Oregon State University, Linus Pauling Institute, Micronutrient Information Center: *Choline.* 2015. https://lpi.oregonstate.edu/mic/other–nutrients/choline. Accessed May 10, 2023.
33. Centers for Disease Control: *Learn about age–related macular degeneration.* 2020. www.cdc.gov/visionhealth/resources/features/macular–degeneration.html. Accessed May 10, 2023.
34. Ellsworth MA and others: Acute liver failure secondary to niacin toxicity. *Case Reports in Pediatrics* 2014: 692530, 2014.

35. Cerullo G and others: The long history of vitamin C: From prevention of the common cold to potential aid in the treatment of COVID–19. *Frontiers in Immunology* 11: 574029, 2020.

36. National Institutes of Health, Office of Dietary Supplements: *Multivitamin/mineral supplements.* 2022. https://ods.od.nih.gov/factsheets/MVMS–HealthProfessional/. Accessed May 10, 2023.

37. U.S. Preventive Services Task Force: *Final recommendation statement: Vitamin supplementation to prevent cancer and CVD: Counseling,* June 2022. www.uspreventiveservicestaskforce.org/uspstf/recommendation/vitamin–supplementation–to–prevent–cvd–and–cancer–preventive–medication. Accessed May 10, 2023.

38. Liu RH: Health benefits of fruits and vegetables are from additive and synergistic combinations of phytochemicals. *American Journal of Clinical Nutrition* 78: 517S, 2003.

39. American Cancer Society: *Risk of dying from cancer continues to drop at an accelerated pace.* January 2022. www.cancer.org/latest–news/facts–and–figures–2022.html. Accessed May 10, 2023.

40. American Cancer Society: *Lifetime risk of developing or dying from cancer.* May 2022. www.cancer.org/healthy/cancer–causes/general–info/lifetime–probability–of–developing–or–dying–from–cancer.html. Accessed May 10, 2023.

41. American Cancer Society: *Cancer facts & figures 2022.* www.cancer.org/research/cancer–facts–statistics/all–cancer–facts–figures/cancer–facts–figures–2022.html. Accessed May 10, 2023.

42. American Cancer Society: *What's wrong with hotdogs, hamburgers, and bacon?* 2018. www.cancer.org/latest–news/hot–dogs–hamburgers–bacon.html. Accessed May 10, 2023.

43. American Cancer Society: *Stomach cancer risk factors.* 2021. www.cancer.org/cancer/stomach–cancer/causes–risks–prevention/risk–factors.html. Accessed May 10, 2023.

44. National Institutes of Health, National Cancer Institute: *Chemicals in meat cooked at high temperatures and cancer risk.* 2017. www.cancer.gov/about–cancer/causes–prevention/risk/diet/cooked–meats–fact–sheet. Accessed May 10, 2023.

45. American Cancer Society: *Body weight and cancer risk.* 2020. www.cancer.org/healthy/cancer–causes/diet–physical–activity/body–weight–and–cancer–risk.html. Accessed May 10, 2023.

46. American Cancer Society: *American Cancer Society guideline for diet and physical activity.* 2020. www.cancer.org/healthy/eat–healthy–get–active/acs–guidelines–nutrition–physical–activity–cancer–prevention/guidelines.html. Accessed May 10, 2023.

47. American Cancer Society: *Water fluoridation and cancer risk.* 2022. www.cancer.org/healthy/cancer–causes/chemicals/water–fluoridation–and–cancer–risk.html. Accessed May 10, 2023.

48. American Cancer Society: *Bladder cancer risk factors.* 2019. www.cancer.org/cancer/bladder–cancer/causes–risks–prevention/risk–factors.html. Accessed May 10, 2023.

Chapter 9

1. Centers for Disease Control and Prevention: Hyperthermia and dehydration–related deaths associated with intentional rapid weight loss in three collegiate wrestlers—North Carolina, Wisconsin, and Michigan, November–December 1997. *Morbidity and Mortality Weekly Report* 47: 105, 1998. www.cdc.gov/mmwr/preview/mmwrhtml/00051388.htm. Accessed May 10, 2023.

2. Negoianu D, Goldfarb S: Just add water. *Journal of the American Society of Nephrology* 19(6): 1041, 2008.

3. Food and Nutrition Board, Institute of Medicine: *Dietary Reference Intakes for water, potassium, sodium, chloride, and sulfate.* Washington, D.C.: National Academies Press, 2004.

4. Saladin KS: *Anatomy & physiology.* 9th ed. New York: McGraw Hill Publishing, 2021.

5. Medline Plus: Kidney stones. Updated January 1, 2019. https://medlineplus.gov/genetics/condition/kidney–stones/#references. Accessed May 10, 2023.

6. Valtrin H: "Drink at least eight glasses of water a day." Really? Is there evidence for "8 × 8"? *American Journal of Physiological Regulation and Integrative Comparative Physiology* 283: R993, 2002.

7. Verster JC, Berthélemy O: Consumer satisfaction and efficacy of the hangover cure After–Effect©. *Advances in Preventive Medicine* 2012: 617942, 2012.

8. Denovan PB: Disturbances of thirst and fluid balance associated with aging. *Physiology & Behavior* 178: 28–34, 2017.

9. Hew–Butler T and others: Statement of the 3rd International Exercise–Associated Hyponatremia Consensus Development Conference, Carlsbad, California. *British Journal of Sports Medicine* 49: 1432, 2015.

10. U.S. Department of Health and Human Services: *Bone health and osteoporosis: A report of the Surgeon General.* Rockville, MD: U.S. Department of Health and Human Services, Office of the Surgeon General, 2004. www.ncbi.nlm.nih.gov/books/NBK45513/. Accessed May 10, 2023.

11. National Institutes of Health, Office of Dietary Supplements: *Calcium.* 2022. https://ods.od.nih.gov/factsheets/Calcium–HealthProfessional/. Accessed May 10, 2023.

12. Oregon State University, Linus Pauling Institute, Micronutrient Information Center: *Calcium.* 2017. https://lpi.oregonstate.edu/mic/minerals/calcium. Accessed May 10, 2023.

13. National Institutes of Health, Medline Plus: *Calcium carbonate with magnesium overdose.* 2021. https://medlineplus.gov/ency/article/002605.htm. Accessed May 10, 2023.

14. U.S. Department of Agriculture, Agricultural Research Service: *What we eat in America, NHANES 2017–2018: Nutrient intakes from food and beverages: Mean amounts consumed by individuals: By gender and age.* 2020. www.ars.usda.gov/ARSUserFiles/80400530/pdf/1718/Table_1_NIN_GEN_17.pdf. Accessed February 2, 2023.

15. National Osteoporosis Foundation: *What is osteoporosis and what causes it?* ND. www.nof.org/patients/what–is–osteoporosis/. Accessed May 10, 2023.

16. National Institute of Arthritis and Musculoskeletal and Skin Diseases: *Osteoporosis overview.* 2018. www.bones.nih.gov/health–info/bone/osteoporosis/overview. Accessed May 10, 2023.

17. National Osteoporosis Foundation: *Osteoporosis and your spine.* 2018. www.nof.org/patients/fracturesfall–prevention/exercisesafe–movement/osteoporosis–and–your–spine/. Accessed May 10, 2023.

18. Centers for Disease Control and Prevention: *Most Americans should consume less sodium.* 2019. www.cdc.gov/salt/index.htm. Accessed May 10, 2023.

19. Quader ZS and others: Sodium intake among persons aged ≥2 years—United States, NHANES 2013–2014. *Morbidity and Mortality Weekly Report* 66(12): 324, 2017.

20. National Academies of Science, Engineering, and Medicine: *Dietary Reference Intakes for sodium and potassium.* Washington, D.C.: National Academies Press, 2019.

21. Centers for Disease Control and Prevention: *High blood pressure: Facts about hypertension.* Last reviewed January 28, 2020. www.cdc.gov/bloodpressure/facts.htm. Accessed May 10, 2023.

22. U.S. Department of Health and Human Services, Office of Minority Health: *Heart disease and African Americans.* Last updated January 31, 2022. https://minorityhealth.hhs.gov/omh/browse.aspx?lvl=4&lvlid=19. Accessed May 10, 2023.

23. American Heart Association: *High blood pressure among Black people.* Last reviewed March 2022. www.heart.org/en/health–topics/high–blood–pressure/why–high–blood–pressure–is–a–silent–killer/high–blood–pressure–and–african–americans. Accessed May 10, 2023.

24. Sussman JB and others: Of barbershops and churches. *Circulation Cardiovascular Quality and Outcomes* 1(10): e005149, 2018.

25. National Heart, Lung, and Blood Institute: *DASH eating plan.* Updated December 29, 2021. www.nhlbi.nih.gov/health–topics/dash–eating–plan. Accessed May 10, 2023.

26. National Institutes of Health, Office of Dietary Supplements: *Magnesium.* 2022. https://ods.od.nih.gov/factsheets/Magnesium–HealthProfessional/. Accessed May 10, 2023.

27. Maier JA and others: Headaches and magnesium: Mechanisms, bioavailability, therapeutic efficacy and potential advantage of magnesium pidolate. *Nutrients* 12(9): 2660, 2020.

28. Food and Nutrition Board: *Dietary Reference Intakes for vitamin A, vitamin K, arsenic, boron, chromium, copper, iodine, iron, manganese, molybdenum, nickel, silicon, vanadium, and zinc.* Washington, D.C.: National Academies Press, 2000.

29. Ghosh K: Non–haematological effects of iron deficiency—A perspective. *Indian Journal of Medical Sciences* 60: 30, 2006.

30. World Health Organization: *Global anaemia estimates, 2021 Edition.* www.who.int/data/gho/data/themes/topics/anaemia_in_women_and_children. Accessed May 10, 2023.

31. Oregon State University, Linus Pauling Institute, Micronutrient Information Center: *Iron.* 2016. https://lpi.oregonstate.edu/mic/minerals/iron. Accessed May 10, 2023.

32. American Red Cross: *Eligibility requirements: Requirements by donation type.* ND. www.redcrossblood.org/donate–blood/how–to–donate/eligibility–requirements.html. Accessed May 10, 2023.

33. National Institutes of Health, Office of Dietary Supplements: *Iron.* 2022. https://ods.od.nih.gov/factsheets/Iron–HealthProfessional/ Accessed May 10, 2023.

34. National Institutes of Health, National Institute for Diabetes and Digestive and Kidney Diseases: *Hemochromatosis.* 2014. www.niddk.nih.gov/health–information/liver–disease/hemochromatosis. Accessed May 10, 2023.

35. Prasad A: Zinc deficiency. *British Medical Journal* 326: 409, 2003.

36. Prasad AS and others: Zinc deficiency in sickle cell disease. *Clinical Chemistry* 21: 582, 1975.

37. National Institutes of Health, Office of Dietary Supplements: *Dietary supplement fact sheet: Zinc.* 2022. https://ods.od.nih.gov/factsheets/Zinc–HealthProfessional/. Accessed May 10, 2023.

38. National Institutes of Health, Office of Dietary Supplements: *Dietary supplement fact sheet: Iodine.* 2022. https://ods.od.nih.gov/factsheets/Iodine–HealthProfessional/. Accessed May 10, 2023.

39. Manousou S and others: A paleolithic–type diet results in iodine deficiency: A 2–year randomized trial in postmenopausal obese women. *European Journal of Clinical Nutrition* 72(1): 124, 2018.

40. National Institutes of Health, Office of Dietary Supplements: *Dietary supplement fact sheet: Selenium.* 2021. https://ods.od.nih.gov/factsheets/Selenium–HealthProfessional/. Accessed May 10, 2023.

41. National Institutes of Health, Office of Dietary Supplements: *Dietary supplement fact sheet: Chromium.* 2022. https://ods.od.nih.gov/factsheets/Chromium–HealthProfessional/. Accessed May 10, 2023.

42. National Institutes of Health, Office of Dietary Supplements: *Dietary supplement fact sheet: Manganese.* 2021. https://ods.od.nih.gov/factsheets/Manganese–HealthProfessional. Accessed May 10, 2023.

43. National Institutes of Health, Office of Dietary Supplements: *Dietary supplement fact sheet: Molybdenum.* 2021. https://ods.od.nih.gov/factsheets/Molybdenum–HealthProfessional/. Accessed May 10, 2023.

44. Food and Drug Administration: *FDA issues final guidance for industry on action level for inorganic arsenic in infant rice cereals.* 2020. www.fda.gov/food/cfsan–constituent–updates/fda–issues–final–guidance–industry–action–level–inorganic–arsenic–infant–rice–cereals. Accessed May 10, 2023.

45. International Bottled Water Association: *Bottled water consumption shift.* 2021. https://bottledwater.org/bottled–water–consumption–shift/. Accessed May 10, 2023.

46. Environmental Protection Agency: *Water health series: Bottled water basics.* 2005. www.epa.gov/ground–water–and–drinking–water/bottled–water–fact–sheet. Accessed May 10, 2023.

47. Food and Drug Administration: *FDA regulates the safety of bottled water beverages including flavored water and nutrient–added water beverages.* 2018. www.fda.gov/food/buy–store–serve–safe–food/fda–regulates–safety–bottled–water–beverages–including–flavored–water–and–nutrient–added–water. Accessed May 10, 2023.

48. National Institutes of Health, National Institute of Environmental Health Sciences: *Bisphenol A (BPA).* 2021. www.niehs.nih.gov/health/topics/agents/sya–bpa/index.cfm. Accessed May 10, 2023.

Chapter 10

1. National Institutes of Health, National Heart, Lung, and Blood Institute: *Aim for a healthy weight: Why is a healthy weight important?* ND. www.nhlbi.nih.gov/health/heart–healthy–living/healthy–weight. Accessed May 10, 2023.

2. Stierman B and others: National Health and Nutrition Examination Survey 2017–March 2020: Prepandemic data files—Development of files and prevalence estimates for selected health outcomes, 2017–2020. 2021. *NCHS Data Brief,* no. 158. https://stacks.cdc.gov/view/cdc/106273. Accessed May 10, 2023.

3. Sanyaolu A and others: Childhood and adolescent obesity in the United States: A public health concern. *Global Pediatric Health* 6: 1, 2019.

4. Centers for Disease Control and Prevention: *Childhood obesity facts: Prevalence of childhood obesity in the United States, 2019–2020.* 2022. www.cdc.gov/obesity/data/childhood.html. Accessed May 10, 2023.

5. World Health Organization: *Obesity and overweight.* 2021. www.who.int/news–room/fact–sheets/detail/obesity–and–overweight. Accessed May 10, 2023.

6. U.S. Department of Health and Human Services: *Healthy People 2030: Nutrition and weight status.* https://health.gov/healthypeople/objectives–and–data/browse–objectives/overweight–and–obesity/reduce–proportion–adults–obesity–nws–03. Accessed May 10, 2023.

7. Hamza EH and others: Food ingredients involved in white–to–brown adipose tissue conversion and in calorie burning. *Frontiers in Physiology* 9: 1954, 2018.

8. Singh P and others: Effects of weight gain and weight loss on regional fat distribution. *American Journal of Clinical Nutrition* 96: 229, 2012.

9. Bosch TA and others: Identification of sex–specific thresholds for accumulation of visceral adipose tissue in adults. *Obesity (Silver Spring)* 23(2): 375, 2015.

10. Sadick N: Treatment for cellulite. *International Journal of Women's Dermatology* 5(1): 68, 2018.

11. National Institutes of Health, Food and Nutrition Board: *Dietary Reference Intakes for energy, carbohydrate, fiber, fat, fatty acids, cholesterol, protein, and amino acids.* Washington, D.C.: National Academies Press, 2005.

12. Rawson ES and others: *Williams' nutrition for health, fitness and sport.* 12th ed. New York: McGraw Hill, 2020.

13. Roberts SB, Dallal GE: Energy requirements and aging. *Public Health Nutrition* 8: 1028, 2005.

14. Haugen HA and others: Variability of measured resting metabolic rate. *American Journal of Clinical Nutrition* 78: 1141, 2005.

15. Garland T Jr. and others: The biological control of voluntary exercise, spontaneous physical activity and daily energy expenditure in relation to obesity: Human and rodent perspectives. *Journal of Experimental Biology* 214(2): 206, 2011.

16. Carson V and others: Systematic review of sedentary behavior and health indicators in school–aged children and youth: An update. *Applied Physiology, Nutrition, and Metabolism* 41(6 Suppl. 3): S240, 2016.

17. Tajir MJ and others: The association of television viewing in childhood with overweight and obesity throughout the life course. *American Journal of Epidemiology* 188(2): 282, 2019.

18. National Institutes of Health, National Heart, Lung, and Blood Institute: Reduce screen time. 2013. www.nhlbi.nih.gov/health/educational/wecan/reduce–screen–time/index.htm. Accessed May 10, 2023.

19. Vadeboncoeur C and others: A meta–analysis of weight gain in first year university students: Is freshman 15 a myth? *BMC Obesity* 2: 22, 2015.

20. Fryar CD and others: Mean body weight, height, waist circumference, and body mass index among adults: United States,1999–2000 through 2015–2016. *National Center for Health Statistics Report,* no. 122: 2018. www.cdc.gov/nchs/data/nhsr/nhsr122–508.pdf. Accessed May 10, 2023.

21. Kitahara CM and others: Association between class III obesity (BMI of 40–59 kg/m2) and mortality: A pooled analysis of 20 prospective studies. *PLoS Medicine* 11(7): e1001673, 2014.

22. Rubino F and others: Joint international consensus statement for ending stigma of obesity. *Nature Medicine* 26: 485–497, 2020.

23. Bidulescu A and others: Gender differences in the association of visceral and subcutaneous adiposity with adiponectin in African Americans: The Jackson Heart Study. *BMC Cardiovascular Disorders* 13: 9, 2013.

24. Jensen MD and others: 2013 AHA/ACC/TOS Guideline for the management of overweight and obesity in adults: A report of the American College of Cardiology/American Heart Association Task Force on Practice Guidelines and The Obesity Society. *Circulation* 129(25 Suppl. 2): S102, 2014.

25. Heymsfield SB and others: Why are there race/ethnic differences in adult body mass index–adiposity relationships? A quantitative critical review. *Obesity Reviews* 17(3): 262–275, 2016.

26. Caleyachetty R and others: Ethnicity–specific BMI cutoffs for obesity based on type 2 diabetes risk in England: A population–based cohort study. *The Lancet Diabetes and Endocrinology* 9(7): 419–426, 2021.

27. World Health Organization, Regional Office for the Western Pacific: The Asia–Pacific perspective: Redefining obesity and its treatment. 2000. https://apps.who.int/iris/handle/10665/206936. Accessed May 10, 2023.

28. Münzberg H, Morrison CD: Structure, production and signaling of leptin. *Metabolism* 64(1): 13, 2015. Published online September 28, 2014.

29. Rinninella E and others: Food components and dietary habits: Keys for a healthy gut microbiota composition. *Nutrients* 11(10), pii: E2393, 2019.

30. Kim KN and others: Short chain fatty acids and fecal microbiota abundance in humans with obesity: A systematic review and meta analysis. *Nutrients* 11(10): 2512, 2019.

31. Farup PG and others: Are nonnutritive sweeteners obesogenic? Associations between diet, faecal microbiota, and short–chain fatty acids in morbidly obese subjects. *Journal of Obesity* 2019: 4608315, 2019.

32. Steele EM and others: Ultra–processed foods and added sugars in the US diet: Evidence from a nationally representative cross–sectional study. *BMJ Open* 6(3): e009892, 2016.

33. Monteiro CA and others: Ultra–processed foods: What they are and how to identify them. *Public Health Nutrition* 22(5): 936, 2019.

34. Hall KD and others: Ultra–processed diets cause excess calorie intake and weight gain: An inpatient randomized controlled trial of ad libitum food intake. *Cell Metabolism* (30)1: 67, 2019.

35. Jensen T and others: Fructose and sugar: A major mediator of non–alcoholic fatty liver disease. *Journal of Hepatology* 68(5): 1063, 2018.

36. McNay DEG, Speakman JR: High fat diet causes rebound weight gain. *Molecular Metabolism* 2(2): 103, 2013. Published online November 2, 2012.

37. U.S. Department of Agriculture, Economic Research: *Food Availability (Per Capita) Data System: Percent of total daily per capita calories from the U.S. food availability, adjusted for spoilage and other waste.* Last updated January 9, 2020. www.ers.usda.gov/data–products/food–availability–per–capita–data–system/. Accessed May 10, 2023.

38. Elgaddal N and others: Physical Activity Among Adults Aged 18 and Over: United States, 2020. *NCHS Data Brief, no. 443.* Hyattsville, MD: National Center for Health Statistics. 2022.

39. Morandi A and others: Estimation of newborn risk for child or adolescent obesity: Lessons from longitudinal birth cohorts. *PLoS One* 7(11): e49919, 2012.

40. Hicken MT: Racial and ethnic differences in the association between obesity and depression in women. *Journal of Women's Health* 22(5): 445, 2013.

41. Moore BF and others: Fetal exposure to maternal active and secondhand smoking with offspring early–life growth in the Healthy Start study. *International Journal of Obesity (London)* 43(4): 652, 2019.

42. Kaplan LM and others: Perceptions of barriers to effective obesity care: Results from the National ACTION Study. *Obesity* 26(1): 61–69, 2018.

43. Raynor HA, Champagne CM: Position of the Academy of Nutrition and Dietetics: Interventions for the treatment of overweight and obesity in adults. *Journal of the Academy of Nutrition and Dietetics* 116(1): 129, 2016.

44. Kelley CP and others: Behavioral modification for the management of obesity. *Primary Care* 43(1):159–175, 2016.

45. Hall KD and others: Energy balance and its components: Implications for body weight regulation. *American Journal of Clinical Nutrition* 95(4): 989, 2012.

46. Lee I and others: Association of step volume and intensity with all–cause mortality in older women. *JAMA Internal Medicine* 179(8): 1105, 2019.

47. Tudor–Locke C and others: How many steps/day are enough? For adults. *The International Journal of Behavioral Nutrition and Physical Activity* 8: 79, 2011.

48. Schoeller DA and others: Self–report–based estimates of energy intake offer an inadequate basis for scientific conclusions. *American Journal of Clinical Nutrition* 97(6): 1413, 2013.

49. Dyrstad SM and others: Comparison of self–reported versus accelerometer–measured physical activity. *Medicine & Science in Sports & Exercise* 46(1): 99, 2014.

50. Ducrot P and others: Meal planning is associated with food variety, diet quality and body weight status in a large sample of French adults. *International Journal Behavior Nutrition and Physical Activity* 14(1): 12, 2017.

51. de Bruin WE and others: "Am I really hungry?" A qualitative exploration of patients' experience, adherence and behaviour change during hunger training: A pilot study. *BMJ Open* 9(12): e032248, 2019.

52. Tomiyama AJ: Stress and obesity. *Annual Reviews of Psychology* 70: 703–718, 2019.

53. National Weight Control Registry: ND. www.nwcr.ws/. Accessed May 10, 2023.

54. Bradley LE and others: Project HELP: A remotely delivered behavioral intervention for weight regain after bariatric surgery. *Obesity Surgery* 27(3): 586, 2017.

55. Kassir R and others: Complications of bariatric surgery: Presentation and emergency management. *International Journal of Surgery* 27: 77, 2016.

56. Neff KJ and others: Bariatric surgery: The challenges with candidate selection, individualizing treatment and clinical outcomes. *BMC Medicine* 11: 8, 2013. Published online January 10, 2013.

57. Srivastava G and others: A specialized medical management program to address post–operative weight regain in bariatric patients. *Obesity Surgery* 28(8): 2241, 2018.

58. Apovian CM and others: Two–year outcomes of vagal nerve blocking (vBloc) for the treatment of obesity in the ReCharge Trial. *Obesity Surgery* 27(1): 169, 2017.

59. Vyas D and others: Advances in endoscopic balloon therapy for weight loss and its limitations. *World Journal of Gastroenterology* 23(44): 7813, 2017.

60. Choi SJ, Choi HS: Various intragastric balloons under clinical investigation. *Clinical Endoscopy* 51(5): 407, 2018.

61. Apollo Endosurgery, Inc: *Orbera® managed weight loss system.* ND. www.orbera.com/. Accessed May 10, 2023.

62. Stavrou G and others: Life–threatening visceral complications after intragastric balloon insertion: Is the device, the patient or the doctor to blame? *Endoscopy International Open* 7(2): E122, 2019.

63. U.S. Food and Drug Administration: *Update: Potential risks with liquid–filled intragastric balloons—Letter to health care providers.* December 2019. www.fda.gov/medical–devices/letters–health–care–providers/update–potential–risks–liquid–filled–intragastric–balloons–letter–health–care–providers. Accessed May 10, 2023.

64. Han L and others: Trends in self–perceived weight status, weight loss attempts, and weight loss strategies among adults in the United States, 1999–2016. *JAMA Network Open* 2(11): e1915219, 2019.

65. Gardner CD and others: Effect of low–fat vs. low–carbohydrate diet on 12–month weight loss in overweight adults and the association with genotype pattern or insulin secretion. *Journal of the American Medical Association* 319(7): 667, 2018.

66. Bazzano LA and others: Effects of low–carbohydrate and low–fat diets: A randomized trial. *Annals of Internal Medicine* 161(5): 309, 2014.

67. Trepanowski JF and others: Effect of alternate–day fasting on weight loss, weight maintenance, and cardioprotection among metabolically healthy obese adults: A randomized clinical trial. *JAMA Internal Medicine* 117(7): 930, 2017.

68. Cho Y and others: The effectiveness of intermittent fasting to reduce body mass index and glucose metabolism: A systematic review and meta–analysis. *Journal of Clinical Medicine* 8(10): 1645, 2019.

69. National Institutes of Health, Office of Dietary Supplements: *Dietary supplements for weight loss.* 2019. https://ods.od.nih.gov/factsheets/WeightLoss–HealthProfessional/. Accessed May 10, 2023.

70. Federal Trade Commission, Bureau of Consumer Protection: *Claims in weight loss ads might be false.* 2019. consumer.ftc.gov/consumer–alerts/2019/08/claims–weight–loss–ads–might–be–false. Accessed May 10, 2023.

71. Federal Trade Commission: *FTC returns $1.95 million to consumers who bought NutriMost Ultimate Fat Loss System in the Pittsburgh area.* 2019. www.ftc.gov/news–events/news/press–releases/2019/02/ftc–returns–195–million–consumers–who–bought–nutrimost–ultimate–fat–loss–system–pittsburgh–area. Accessed May 10, 2023.

72. Fryar CD and others: *Prevalence of underweight among adults aged 20 and over: United States, 1960–1962 through 2017–2018.* www.cdc.gov/nchs/data/hestat/underweight–adult–17–18/underweight–adult.htm. Accessed May 10, 2023.

73. American Psychiatric Association: Feeding and eating disorders. In *Diagnostic and statistical manual of mental disorders.* 5th ed. 2013.

74. Eating Disorders Victoria: *Eating disorders explained.* www.eatingdisorders.org.au/eating–disorders–a–z/eating–disorders–explained/. Accessed May 10, 2023.

75. National Institutes of Health, National Institute of Mental Health: *Eating disorders.* 2016. www.nimh.nih.gov/health/topics/eating–disorders/index.shtml. Accessed May 10, 2023.

76. Ozier AD, Henry BW: Position of the American Dietetic Association: Nutrition intervention in the treatment of eating disorders. *Journal of the American Dietetic Association* 111(8): 1236, 2011.

77. Stice E and others: Prevalence, incidence, impairment, and course of the proposed DSM–5 eating disorder diagnoses in an 8–year prospective community study of young women. *Journal of Abnormal Psychology* 122(2): 455, 2013.

78. Pinheiro AP and others: The genetics of anorexia nervosa: Current findings and future perspectives. *International Journal of Child and Adolescent Health* 2(2): 153, 2009.

79. Baker JH and others: Genetics of anorexia nervosa. *Current Psychiatry Reports* 19(11): 84, 2017.

80. Smink FR and others: Epidemiology of eating disorders: Incidence, prevalence, and mortality rates. *Current Psychiatry Reports* 14: 406, 2012.

81. Dimitropoulos G and others: Inpatients with severe anorexia nervosa and their siblings: Non–shared experi–ences and family functioning. *European Eating Disorders Review* 21: 284, 2013.

82. Bould H and others: Do eating disorders in parents predict eating disorders in children? Evidence from a Swedish cohort. *Acta Psychiatrica Scandinavica* 132(1): 51, 2015.

83. Golden NH and others: Preventing obesity and eating disorders in adolescents. *Pediatrics* 138: 3, 2016.

84. Harrington BC and others: Initial evaluation, diagnosis, and treatment of anorexia nervosa and bulimia nervosa. *American Family Physician* 91(1): 46, 2015.

85. Mond JM and others: Prevalence and implications of eating disordered behavior in men. In Cohn L, R Lemberg (eds.), *Current findings on males with eating disorders* (pp. 195–215). Routledge/Taylor and Francis Group, 2014.

86. Ishigami A and others: Postmortem diagnosis of anorexia nervosa: An endocrinological and immunohistochemical approach. *Journal of Medical Investigation* 63(3–4): 305, 2016.

87. McCuen–Wurst C and others: Disordered eating and obesity: Associations between binge–eating disorder, night–eating syndrome, and weight–related comorbidities. *Annals of the New York Academy of Sciences* 1411(1): 96, 2018.

88. Gagnon C and others: Comorbid diabetes and eating disorders in adult patients: Assessment and considerations for treatment. *Diabetes Educator* 35: 537, 2012.

89. Matheiu J: What is diabulimia? *Journal of the American Dietetic Association* 108(5): 769, 2008.

90. Murray SB and others: Muscle dysmorphia and the DSM–V conundrum: Where does it belong? A review paper. *International Journal of Eating Disorders* 43: 483, 2010.

91. Tod D and others: Muscle dysmorphia: Current insights. *Psychology Research and Behavior Management* 9: 179, 2016.

92. Weiss Kelly AK and others: The female athlete triad. *Pediatrics* 138(2): e20160922, 2016.

93. De Souza MJ and others: The role of energy availability in reproductive function in the female athlete triad and extension of its effects to men: An initial working model of a similar syndrome in male athletes. *Sports Medicine* 49(Suppl. 2): 125, 2019.

94. National Eating Disorder Association: *How to help a loved one.* 2018. www.nationaleatingdisorders.org/learn/help/caregivers. Accessed May 10, 2023.

95. Linardon J and others: The efficacy of cognitive–behavioral therapy for eating disorders: A systematic review and meta–analysis. *Journal of Consulting and Clinical Psychology* 85(11): 1080, 2017.

Chapter 11

1. Whitfield GP and others: Participation in Leisure–Time Aerobic Physical Activity Among Adults, National Health Interview Survey, 1998–2018. *Journal of Physical Activity and Health* 18(S1): S25–S36, 2021.

2. Centers for Disease Control and Prevention: *Physical activity: How much physical activity do adults need?* Reviewed January 2020. www.cdc.gov/physicalactivity/basics/adults/. Accessed May 10, 2023.

3. Centers for Disease Control and Prevention: *Physical activity and health: The benefits of physical activity.* Reviewed January 2020. www.cdc.gov/physicalactivity/basics/pa–health/index.htm. Accessed May 10, 2023.

4. Department of Health and Human Services: *Physical activity guidelines for Americans, 2nd edition.* 2018. https://health.gov/our–work/nutrition–physical–activity/physical–activity–guidelines/current–guidelines. Accessed May 10, 2023.

5. Centers for Disease Control and Prevention: *Target heart rate and estimated maximum heart rate.* 2020. www.cdc.gov/physicalactivity/everyone/measuring/heartrate.html. Accessed May 10, 2023.

6. Centers for Disease Control and Prevention: *General physical activities defined by level of intensity.* ND. www.cdc.gov/nccdphp/dnpa/physical/pdf/pa_intensity_table_2_1.pdf. Accessed May 10, 2023.

7. Powers SK and others: *Exercise Physiology.* 11th ed. New York: McGraw Hill Publishing, 2021.

8. Position of the Academy of Nutrition and Dietetics, Dietitians of Canada, and the American College of Sports Medicine: Nutrition and athletic performance. *Journal of the Academy of Nutrition and Dietetics* 116(3): 501, 2016.

9. Kreider KB and others: International Society of Sports Nutrition position stand: Safety and efficacy of creatine supplementation in exercise, sport, and medicine. *Journal of the International Society of Sports Nutrition* 14: 18, 2017.

10. Williams MH and others: *Nutrition for Health, Fitness, & Sport.* 12th ed. New York: McGraw Hill, 2020.

11. Yeo WK and others: Fat adaptation in well–trained athletes: Effects on cell metabolism. *Applied Physiology, Nutrition, and Metabolism* 36: 12, 2011.

12. Economos CD and others: Nutritional practices of elite athletes: Practical recommendations. *Sports Medicine* 16: 381, 1993.
13. Kersick CM and others: ISSN exercise & sports nutrition review update: Research & recommendations. *Journal of the International Society of Sports Nutrition* 15: 38, 2018.
14. Rothschild JA and others: What should I eat before exercise? Pre–exercise nutrition and the response to endurance exercise: Current prospective and future directions. *Nutrients* 12(11): 3473, 2020.
15. Kerksick CM and others: International Society of Sports Nutrition position stand: Nutrient timing. *Journal of the International Society of Sports Nutrition* 14: 33, 2017.
16. Born KA and others: Chocolate milk versus carbohydrate supplements in adolescent athletes: A field based study. *Journal of the International Society of Sports Nutrition* 16: 6, 2019.
17. Otten JJ and others (eds.): *Dietary Reference Intakes: The essential guide to nutrient requirements.* Institute of Medicine of the National Academies. Washington, D.C.: National Academies Press, 2006.
18. West DW and others: Whey protein supplementation enhances whole body protein metabolism and perfor–mance recovery after resistance exercise: A double–blind crossover study. *Nutrients* 9: 735, 2017.
19. Jager R and others: International Society of Sports Nutrition position stand: Protein and exercise. *Journal of the Inter–national Society of Sports Nutrition* 14: 20, 2017.
20. National Institutes of Health, National Center for Complementary and Integrative Health: *Asian ginseng.* 2016. https://nccih.nih.gov/health/asianginseng/ataglance.htm. Accessed February 7, 2020.
21. Coyle EF: Fluid and fuel intake during exercise. *Journal of Sports Science* 22: 39, 2004.
22. National Library of Medicine, MedLinePlus: *Heat emergencies.* Updated January 2020. https://medlineplus.gov/ency/article/000056.htm. Accessed May 10, 2023.
23. Kenefick RW and others: Drinking strategies: Planned drinking versus drinking to thirst. *Sports Medicine* 48 (Suppl. 1): S31, 2018.
24. Pingitore A and others: Exercise and oxidative stress: Potential effects of antioxidant dietary strategies in sports. *Nutrition* 31(7–8): 916, 2015.
25. U.S. Food and Drug Administration: *Import alert 66–71.* December 2017. www.accessdata.fda.gov/cms_ia/importalert_204.html. Accessed May 10, 2023.
26. National Collegiate Athletic Association: *NCAA 2020–2021 banned substances.* www.ncaa.org/sport–science–institute/topics/2020–21–ncaa–banned–substances. Accessed May 10, 2023.
27. U.S. Anti–doping Agency: *Prohibited list.* January 2021. www.usada.org/athletes/substances/prohibited–list/. Accessed May 10, 2023.
28. U.S. Food and Drug Administration: *Dietary supplement products and ingredients: Pure and highly concentrated caffeine.* 2018. www.fda.gov/food/dietary–supplement–products–ingredients/fda–warns–consumers–about–pure–and–highly–concentrated–caffeine. Accessed May 10, 2023.

Chapter 12

1. Centers for Disease Control and Prevention: *Trends and outbreaks: Burden of norovirus illness in the U.S.* Updated February 2021. www.cdc.gov/norovirus/trends–outbreaks/burden–US.html. Accessed January 15, 2023.
2. Centers for Disease Control and Prevention: *Preventing norovirus.* Updated November 2022. www.cdc.gov/norovirus/about/prevention.html. Accessed January 15, 2023.
3. Centers for Disease Control and Prevention: *The Foodborne Disease Outbreak Surveillance System: Frequently asked questions.* Updated December 2019. www.cdc.gov/fdoss/faq.html. Accessed January 15, 2023.
4. Centers for Disease Control and Prevention: *Estimates of foodborne illness in the United States: Burden of foodborne illness: Overview.* Updated 2018. www.cdc.gov/foodborneburden/estimates–overview.html. Accessed January 15, 2023.
5. U.S. Food and Drug Administration: *Bad bug book: Food–borne pathogenic microorganisms and natural toxins handbook.* 2nd ed. 2012. www.fda.gov/food/foodborne–pathogens/bad–bug–book–second–edition. Accessed January 15, 2023.
6. Centers for Disease Control and Prevention: *Food poisoning.* Updated December 2022. www.cdc.gov/foodsafety/symptoms.html. Accessed January 15, 2023.
7. Centers for Disease Control and Prevention: *Handwashing: Clean hands save lives: Show me the science—When & how to use hand sanitizer in community settings.* 2019. www.cdc.gov/handwashing/show–me–the–science–hand–sanitizer.html. Accessed January 15, 2023.
8. U.S. Food and Drug Administration: *Food Code 2017.* 2017. www.fda.gov/food/fda–food–code/food–code–2017. Accessed January 15, 2023.
9. U.S. Department of Agriculture, Food Safety and Inspection Service: *Molds on food: Are they dangerous?* 2013. www.fsis.usda.gov/food–safety/safe–food–handling–and–preparation/food–safety–basics/molds–food–are–they–dangerous. Accessed January 15, 2023.
10. U.S. Department of Health and Human Services, Food Safety.gov: *4 Steps to food safety.* 2019. www.foodsafety.gov/keep–food–safe/4–steps–to–food–safety#cook. Accessed January 15, 2023.

11. U.S. Department of Agriculture, Food Safety and Inspection Service: *Cooking safely in the microwave oven.* 2013. www.fsis.usda.gov/food–safety/safe–food–handling–and–preparation/food–safety–basics/cooking–microwave–ovens. Accessed January 15, 2023.

12. Miranda RC, Schaffner DW: Longer contact times increase cross–contamination of *Enterobacter aerogenes* from surfaces to food. *Applied and Environmental Microbiology* 82(21): 6490, 2016.

13. U.S. Department of Agriculture, Food Safety and Inspection Service: *Freezing and food safety.* 2015. www.fsis.usda.gov/food–safety/safe–food–handling–and–preparation/food–safety–basics/freezing–and–food–safety. Accessed January 15, 2023.

14. Food and Drug Administration: *Food irradiation: What you need to know.* 2022. www.fda.gov/food/buy–store–serve–safe–food/food–irradiation–what–you–need–know. Accessed January 15, 2023.

15. *The New York Times*: *Hurricane Ida updates: In Ida's grip, Louisiana struggles to assess the damage.* August 30, 2021. Updated June 2022. www.nytimes.com/live/2021/08/30/us/hurricane–ida–updates. Accessed January 15, 2023.

16. Centers for Disease and Prevention, Water, Sanitation, & Hygiene (WASH)–related Emergencies & Outbreaks: *Finding other water sources in an emergency.* 2021. www.cdc.gov/healthywater/emergency/finding–other–sources.html. Accessed January 15, 2023.

17. U.S. Department of Agriculture, Food Safety and Inspection Service: *Keeping food safe during an emergency.* 2013. www.fsis.usda.gov/food–safety/safe–food–handling–and–preparation/food–safety–basics/keeping–food–safe–during. Accessed January 15, 2023.

18. University of Rochester Medical Center, Health Encyclopedia: *Avoid injury around barbeque grills.* ND. www.urmc.rochester.edu/encyclopedia/content.aspx?contenttypeid=1&contentid=1128. Accessed January 15, 2023.

19. U.S. Food and Drug Administration: *Food: Overview of food ingredients, additives & colors.* Current as of February 2018. www.fda.gov/food/food–ingredients–packaging/overview–food–ingredients–additives–colors. Accessed January 15, 2023.

20. U.S. Environmental Protection Agency: *Summary of the Food Quality Protection Act.* September 2022. www.epa.gov/laws–regulations/summary–food–quality–protection–act. Accessed January 15, 2023.

21. U.S. Food and Drug Administration: *Food defect levels handbook.* 2018. www.fda.gov/food/ingredients–additives–gras–packaging–guidance–documents–regulatory–information/food–defect–levels–handbook. Accessed January 15, 2023.

22. Centers for Disease Control and Prevention, U.S. Department of Health and Human Services: *National report on human exposure to environmental chemicals.* Updated March 2022. www.cdc.gov/exposurereport/. Accessed January 15, 2023.

23. American Cancer Society: *Cancer prevention & early detection facts & figures 2021–2022.* 2021. www.cancer.org/research/cancer–facts–statistics/cancer–prevention–early–detection.html. Accessed January 15, 2023.

24. Connor BA: *Chapter 2: Preparing international travelers.* 2019. Centers for Disease Control and Prevention. wwwnc.cdc.gov/travel/yellowbook/2020/preparing–international–travelers/travelers–diarrhea. Accessed January 21, 2023.

25. Centers for Disease Control and Prevention: *Water disinfection.* February 20, 2020. wwwnc.cdc.gov/travel/page/water–disinfection. Accessed January 21, 2023.

Chapter 13

1. Centers for Disease Control and Prevention: *Unintended pregnancy.* 2021. www.cdc.gov/reproductivehealth/contraception/unintendedpregnancy/. Accessed January 28, 2023.

2. Centers for Disease Control and Prevention: *Infertility.* 2022. www.cdc.gov/reproductivehealth/infertility/index.htm. Accessed January 28, 2023.

3. Moon KH and others: Obesity and erectile dysfunction: From bench to clinical implication. *World Journal of Men's Health* 37(2): 138, 2019.

4. Young MF, Ramakrishnan U: Maternal undernutrition before and during pregnancy and offspring health and development. *Annals of Nutrition and Metabolism* 76 (suppl 3): 41–53, 2020.

5. Centers for Disease Control and Prevention: *Infant mortality.* 2019. www.cdc.gov/reproductivehealth/maternalinfanthealth/infantmortality.htm. Accessed February 28, 2023.

6. Oysterman M and others: Births: Final data for 2020. *National Vital Statistics Report* 70(17), February 7, 2022. www.cdc.gov/nchs/data/nvsr/nvsr70/nvsr70–17.pdf. Accessed January 28, 2023.

7. U.S. Department of Health and Human Services: *The health consequences of smoking—50 years of progress: A report of the Surgeon General.* 2014. www.hhs.gov/surgeongeneral/reports–and–publications/tobacco/consequences–smoking–factsheet/index.html. Accessed January 28, 2023.

8. Centers for Disease Control and Prevention: *Preterm birth.* 2021. www.cdc.gov/reproductivehealth/maternalinfanthealth/pretermbirth.htm. Accessed January 28, 2023.

9. Lawrence RA, Lawrence RM: Physiology of Lactation. In *Breastfeeding.* 9th ed. Elsevier Inc. Philadelphia, PA, 2022.

10. National Library of Medicine, MedlinePlus: *Morning sickness.* 2021. https://medlineplus.gov/ency/article/003119.htm. Accessed January 28, 2023.

11. Committee on Practice Bulletins–Obstetrics. ACOG Practice Bulletin No. 189: Nausea and vomiting of pregnancy. *Obstetrics Gynecology* 131(1): e15–e30, 2018.

12. Otten JJ and others (eds.): *Dietary Reference Intakes: The essential guide to nutrient requirements.* Institute of Medicine of the National Academies. Washington, D.C.: National Academies Press, 2006.

13. Food and Drug Administration: *Advice about eating fish.* 2021. www.fda.gov/food/consumers/advice-about-eating-fish. Accessed January 29, 2023.

14. Centers for Disease Control and Prevention: *Maternal and infant health: Weight gain during pregnancy.* 2022. www.cdc.gov/reproductivehealth/maternalinfanthealth/pregnancy-weight-gain.htm#cdc. Accessed January 29, 2023.

15. Mayo Clinic Staff: *Diseases and conditions: Fetal macrosomia.* 2018. www.mayoclinic.org/diseases-conditions/fetal-macrosomia/symptoms-causes/syc-20372579. Accessed January 29, 2023.

16. Centers for Disease Control and Prevention: *Tracking your weight for women who begin pregnancy at a normal weight.* ND. www.cdc.gov/reproductivehealth/pdfs/maternal-infant-health/pregnancy-weight-gain/tracker/single/Normal_Weight_Tracker_508Tagged.pdf. Accessed January 29, 2023.

17. U.S. Department of Health and Human Services, National Institutes of Health: *What is prenatal care and why is it important?* 2017. www.nichd.nih.gov/health/topics/pregnancy/conditioninfo/prenatal-care. Accessed February 4, 2023.

18. Centers for Disease Control and Prevention: *Gestational diabetes.* 2022. www.cdc.gov/diabetes/basics/gestational.html. Accessed February 4, 2023.

19. National Heart, Lung, and Blood Institute, MedlinePlus: *High blood pressure in pregnancy.* 2019. https://medlineplus.gov/highbloodpressureinpregnancy.html. Accessed February 4, 2023.

20. Centers for Disease Control and Prevention: *High blood pressure: High blood pressure during pregnancy.* Last reviewed December 2022. www.cdc.gov/bloodpressure/pregnancy.htm. Accessed February 4, 2023.

21. Hinkle SN and others. Assessment of caffeine consumption and maternal cardiometabolic pregnancy complications. *JAMA Network Open* 4(11): e2133401, 2021.

22. National Institutes of Health: *Sudden infant death syndrome (SIDS).* 2017. www.nichd.nih.gov/health/topics/sids/Pages/default.aspx. Accessed February 4, 2023.

23. American College of Obstetricians and Gynecologists: *Frequently asked questions: Exercise during pregnancy.* 2022. www.acog.org/womens-health/faqs/exercise-during-pregnancy. Accessed February 4, 2023.

24. Le Doare K and others: Mother's milk: A purposeful contribution to the development of the infant microbiota and immunity. *Frontiers in Immunology* 9: 361, 2018.

25. Lessen R, Kavanagh K: Position of the Academy of Nutrition and Dietetics: Promoting and supporting breastfeeding. *Journal of the Academy of Nutrition and Dietetics* 115(3): 444, 2015.

26. Cheng F and others: Do breast implants influence breastfeeding? A meta-analysis of comparative studies. *Journal of Human Lactation* 34(3): 424–432, 2018.

27. Meek JY, Noble L: Section on Breastfeeding. Policy Statement: Breastfeeding and the use of human milk. *Pediatrics* 150(1): e2022057988, 2022.

28. Centers for Disease Control and Prevention*: Breastfeeding, Results: Breastfeeding rates. National Immunization Survey (NIS).* 2019. www.cdc.gov/breastfeeding/data/nis_data/results.html. Accessed February 4, 2023.

29. Yi DY, Kim SY: Human breast milk composition and function in human health: From nutritional components to microbiome and microRNAs. *Nutrients* 13(9): 3094, 2021.

30. National Institutes of Health, MedlinePlus: *Cow's milk—infants.* 2021. https://medlineplus.gov/ency/article/002448.htm. Accessed February 4, 2023.

31. Centers for Disease Control and Prevention: *Cow's milk and milk alternatives.* 2022. www.cdc.gov/nutrition/infantandtoddlernutrition/foods-and-drinks/cows-milk-and-milk-alternatives.html. Accessed January 4, 2023.

32. American Academy of Pediatrics: Expert panel on integrated guidelines for cardiovascular health and risk reduction in children and adolescents: Summary report. *Pediatrics* 128(S5): S213, 2011.

33. Hayes D, Klems S: *Reducing the risk of food allergies.* 2022. www.eatright.org/health/health-conditions/allergies-and-intolerances/reducing-the-risk-of-food-allergies. Accessed February 4, 2023.

34. National Institutes of Health, National Institute of Allergies and Infectious Diseases: *Guidelines for clinicians and patients for diagnosis and management of food allergy in the United States. 2017 addendum guidelines for the prevention of peanut allergy in the United States.* 2018. www.niaid.nih.gov/diseases-conditions/guidelines-clinicians-and-patients-food-allergy. Accessed February 4, 2023.

35. U.S. Department of Health and Human Services, U.S. Department of Agriculture: *Dietary guidelines for Americans, 2020–2025.* 2020. www.dietaryguidelines.gov. Accessed February 11, 2023.

36. Wang D and others: Snacking patterns in children: A comparison between Australia, China, Mexico, and the United States. *Nutrients* 10(2), pii: E198, 2018.

37. Liu J and others: Trends in diet quality among youth in the United States, 1999–2016. *Journal of the American Medical Association* 323(12): 1161–1174, 2020.

38. National Institutes of Health, National Institute of Diabetes and Digestive and Kidney Disease: *Constipation in children.* 2018. www.niddk.nih.gov/health-information/digestive

–diseases/constipation–children/definition–facts. Accessed March 14, 2023.

39. Wicherski J and others: Association between breakfast skipping and body weight—A systematic review and meta–analysis of observational longitudinal studies. *Nutrients* 13(1): 272, 2021.

40. National Institutes of Child Health and Human Develop–ment: *Puberty and precocious puberty.* 2021. www.nichd .nih.gov/health/topics/puberty. Accessed March 14, 2023.

41. American Academy of Dermatology: *Can the right diet get rid of acne?* ND. www.aad.org/diet. Accessed March 14, 2023.

42. Centers for Disease Control and Prevention: *Adolescent and Youth Risk Behavior Surveillance System (YRBSS).* 2018. www .cdc.gov/healthyyouth/yrbs/index.htm. Accessed March 14, 2023.

43. Position of the Academy of Nutrition and Dietetics: Veg–etarian diets. *Journal of the Academy of Nutrition and Dietetics* 116(12): 1970, 2016.

44. Stierman B and others: *National Health and Nutri–tion Examination Survey 2017–March 2020 prepandemic data files—Development of files and prevalence estimates for selected health outcomes.* 2021. https://stacks.cdc.gov /view/cdc/106273. Accessed February 11, 2023.

45. Centers for Disease Control and Prevention: *Overweight and obesity: Defining childhood weight status.* 2022. www.cdc .gov/obesity/basics/childhood–defining.html. Accessed February 11, 2023.

46. Centers for Disease Control and Prevention: *Obesity conse–quences.* 2022. www.cdc.gov/obesity/basics/consequences .html. Accessed February 11, 2023.

47. Dattilo AM and others: Need for early interventions in the prevention of pediatric overweight: A review and upcoming directions. *Journal of Obesity* 2012: 123023, 2012. Published online May 7, 2012.

48. Hampl SE and others: Clinical practice guideline for the evaluation and treatment of children and adolescents with obesity. *Pediatrics* 151(2): e2022060640, 2023.

49. National Institute of Diabetes and Digestive and Kidney Dis–eases, Weight management: *Bariatric surgery: Who is a good teen candidate for bariatric surgery?* 2020. www.niddk.nih .gov/health–information/weight–management/bariatric –surgery/potential–candidates. Accessed March 14, 2023.

50. Xu JQ and others: Mortality in the United States, 2021. *NCHS Data Brief, no. 456.* Hyattsville, MD: National Center for Health Statistics. 2022.

51. U.S. Census Bureau: *An aging nation: Projected number of children and older adults.* 2021. www.census.gov/library /visualizations/2018/comm/historic–first.html. Accessed March 14, 2023.

52. U.S. Census Bureau, Data: *2017 National population pro–jections tables,* Table 2. 2018. www.census.gov/programs –surveys/popproj/data/tables.html. Accessed March 14, 2023.

53. Crous–Bou M and others: Plant–rich dietary patterns, plant foods and nutrients, and telomere length. *Advances in Nutrition* 10: S296, 2019.

54. National Academies of Sciences, Engineering, and Medicine: *Meeting the dietary needs of older adults: Exploring the impact of the physical, social, and cultural environment: Workshop summary.* Washington, D.C.: The National Academies Press, 2016.

55. Roberts SB and others: Healthy Aging–Nutrition Matters: Start Early and Screen Often. *Advances in Nutrition* 12(4):1438–1448, 2021.

56. Batsis JA, Villareal DT: Sarcopenic obesity in older adults: Aetiology, epidemiology and treatment strategies. *National Review of Endocrinology* 14(9): 513, 2018.

57. Centers for Disease Control and Prevention, National Center for Health Statistics: *Table 35. Percentage of adults 65 years or older who have lost all their natural teeth.* www .cdc.gov/oralhealth/publications/OHSR2019–table–35 .html. Accessed March 16, 2023.

58. Savoca MR and others: Severe tooth loss in older adults as a key indicator of compromised diet quality. *Public Health Nutrition* 13(4): 466, 2010.

59. Federal Interagency Forum on Aging Related Statistics: *Older Americans 2020: Key indicators of well–being.* 2020. https:// agingstats.gov/docs/LatestReport/OA20_508_10142020 .pdf. Accessed March 16, 2023.

60. Centers for Disease Control and Prevention, Injury Pre–vention and Control: *Leading causes of death reports 1989–2020.* 2020. www.cdc.gov/injury/wisqars/leadingcauses .html. Accessed March 16, 2023.

61. Son DH and others: Recent advances in anti–aging medicine. *Korean Journal of Family Medicine* 40(5): 289, 2019.

62. National Institutes of Health, National Institute on Aging: *Calorie restriction and fasting diets: What do we know?* 2018. www.nia.nih.gov/health/calorie–restriction–and –fasting–diets–what–do–we–know. Accessed February 23, 2020.

63. Navarro Ibarra MJ and others: Diet, physical activity and telomere length in adults. *Nutrición Hospitalaria* 36(6): 1403, 2019.

Appendix H

Dietary Reference Intakes (DRIs): Tables

Dietary Reference Intakes (DRIs): Recommended Intakes for Individuals, Vitamins
Food and Nutrition Board, Institute of Medicine, National Academies

Life Stage Group	Vitamin A (μg/d)[a]	Vitamin C (mg/d)	Vitamin D (μg/d)[b,c]	Vitamin E (mg/d)[d]	Vitamin K (μg/d)	Thiamin (mg/d)	Riboflavin (mg/d)	Niacin (mg/d)[e]	Vitamin B-6 (mg/d)	Folate (μg/d)[f]	Vitamin B-12 (μg/d)	Pantothenic Acid (mg/d)	Biotin (μg/d)	Choline (mg/d)[g]
Infants														
0–6 mo	400*	40*	10	4*	2.0*	0.2*	0.3*	2*	0.1*	65*	0.4*	1.7*	5*	125*
7–12 mo	500*	50*	10	5*	2.5*	0.3*	0.4*	4*	0.3*	80*	0.5*	1.8*	6*	150*
Children														
1–3 y	300	15	15	6	30*	0.5	0.5	6	0.5	150	0.9	2*	8*	200*
4–8 y	400	25	15	7	55*	0.6	0.6	8	0.6	200	1.2	3*	12*	250*
Males														
9–13 y	600	45	15	11	60*	0.9	0.9	12	1.0	300	1.8	4*	20*	375*
14–18 y	900	75	15	15	75*	1.2	1.3	16	1.3	400	2.4	5*	25*	550*
19–30 y	900	90	15	15	120*	1.2	1.3	16	1.3	400	2.4	5*	30*	550*
31–50 y	900	90	15	15	120*	1.2	1.3	16	1.3	400	2.4	5*	30*	550*
51–70 y	900	90	15	15	120*	1.2	1.3	16	1.7	400	2.4[h]	5*	30*	550*
>70 y	900	90	20	15	120*	1.2	1.3	16	1.7	400	2.4[h]	5*	30*	550*
Females														
9–13 y	600	45	15	11	60*	0.9	0.9	12	1.0	300	1.8	4*	20*	375*
14–18 y	700	65	15	15	75*	1.0	1.0	14	1.2	400[i]	2.4	5*	25*	400*
19–30 y	700	75	15	15	90*	1.1	1.1	14	1.3	400[i]	2.4	5*	30*	425*
31–50 y	700	75	15	15	90*	1.1	1.1	14	1.3	400[i]	2.4	5*	30*	425*
51–70 y	700	75	15	15	90*	1.1	1.1	14	1.5	400	2.4[h]	5*	30*	425*
>70 y	700	75	20	15	90*	1.1	1.1	14	1.5	400	2.4[h]	5*	30*	425*
Pregnancy														
14–18 y	750	80	15	15	75*	1.4	1.4	18	1.9	600[j]	2.6	6*	30*	450*
19–30 y	770	85	15	15	90*	1.4	1.4	18	1.9	600[j]	2.6	6*	30*	450*
31–50 y	770	85	15	15	90*	1.4	1.4	18	1.9	600[j]	2.6	6*	30*	450*
Lactation														
14–18 y	1200	115	15	19	75*	1.4	1.6	17	2.0	500	2.8	7*	35*	550*
19–30 y	1300	120	15	19	90*	1.4	1.6	17	2.0	500	2.8	7*	35*	550*
31–50 y	1300	120	15	19	90*	1.4	1.6	17	2.0	500	2.8	7*	35*	550*

mg = milligram, μg = microgram

NOTE: This table (taken from the DRI reports; see **www.nap.edu**) presents Recommended Dietary Allowances (RDAs) in **bold type** and Adequate Intakes (AIs) in ordinary type followed by an asterisk (*). RDAs and AIs may both be used as goals for individual intake. RDAs are set to meet the needs of almost all (97–98%) individuals in a group. For healthy breastfed infants, the AI is the mean intake. The AI for other life stage and gender groups is believed to cover needs of all individuals in the group, but lack of data or uncertainty in the data prevents being able to specify with confidence the percentage of individuals covered by this intake.

[a]As retinol activity equivalents (RAEs). 1 RAE = 1 μg retinol, 12 μg β-carotene, 24 μg α-carotene, or 24 μg β-cryptoxanthin. To calculate RAEs from REs of provitamin A carotenoids in foods, divide the REs by 2. For preformed vitamin A in foods or supplements and for provitamin A carotenoids in supplements, 1 RE = 1 RAE.

[b]cholecalciferol. 1 μg cholecalciferol = 40 IU vitamin D.

[c]In the absence of adequate exposure to sunlight.

[d]As α-tocopherol. α-Tocopherol includes RRR-α-tocopherol, the only form of α-tocopherol that occurs naturally in foods, and the 2R-stereoisomeric forms of α-tocopherol (RRR-, RSR-, RRS-, and RSS-α-tocopherol) that occur in fortified foods and supplements. It does not include the 2S-stereoisomeric forms of α-tocopherol (SRR-, SSR-, SRS-, and SSS-α-tocopherol), also found in fortified foods and supplements.

[e]As niacin equivalents (NE). 1 mg of niacin = 60 mg of tryptophan; 0–6 months = preformed niacin (not NE).

[f]As dietary folate equivalents (DFE). 1 DFE = 1 μg food folate = 0.6 μg of folic acid from fortified food or as a supplement consumed with food = 0.5 μg of a supplement taken on an empty stomach.

[g]Although AIs have been set for choline, there are few data to assess whether a dietary supply of choline is needed at all stages of the life cycle, and it may be that the choline requirement can be met by endogenous synthesis at some of these stages.

[h]Because 10 to 30% of older people may malabsorb food-bound B-12, it is advisable for those older than 50 years to meet their RDA mainly by consuming foods fortified with B-12 or a supplement containing B-12.

[i]In view of evidence linking folate intake with neural tube defects in the fetus, it is recommended that all women capable of becoming pregnant consume 400 μg from supplements or fortified foods in addition to intake of food folate from a varied diet.

[j]It is assumed that females will continue consuming 400 μg from supplements or fortified food until their pregnancy is confirmed and they enter prenatal care, which ordinarily occurs after the end of the periconceptional period—the critical time for formation of the neural tube.

Dietary Reference Intakes (DRIs): Recommended Intakes for Individuals, Elements
Food and Nutrition Board, Institute of Medicine, National Academies

Life Stage Group	Calcium (mg/d)	Chromium (μg/d)	Copper (μg/d)	Fluoride (mg/d)	Iodine (μg/d)	Iron (mg/d)	Magnesium (mg/d)	Manganese (mg/d)	Molybdenum (μg/d)	Phosphorus (mg/d)	Selenium (μg/d)	Zinc (mg/d)
Infants												
0–6 mo	200*	0.2*	200*	0.01*	110*	0.27*	30*	0.003*	2*	100*	15*	2*
7–12 mo	260*	5.5*	220*	0.5*	130*	**11**	75*	0.6*	3*	275*	20*	**3**
Children												
1–3 y	**700**	11*	**340**	0.7*	**90**	**7**	**80**	1.2*	**17**	**460**	**20**	**3**
4–8 y	**1000**	15*	**440**	1*	**90**	**10**	**130**	1.5*	**22**	**500**	**30**	**5**
Males												
9–13 y	**1300**	25*	**700**	2*	**120**	**8**	**240**	1.9*	**34**	**1250**	**40**	**8**
14–18 y	**1300**	35*	**890**	3*	**150**	**11**	**410**	2.2*	**43**	**1250**	**55**	**11**
19–30 y	**1000**	35*	**900**	4*	**150**	**8**	**400**	2.3*	**45**	**700**	**55**	**11**
31–50 y	**1000**	35*	**900**	4*	**150**	**8**	**420**	2.3*	**45**	**700**	**55**	**11**
51–70 y	**1000**	30*	**900**	4*	**150**	**8**	**420**	2.3*	**45**	**700**	**55**	**11**
>70 y	**1200**	30*	**900**	4*	**150**	**8**	**420**	2.3*	**45**	**700**	**55**	**11**
Females												
9–13 y	**1300**	21*	**700**	2*	**120**	**8**	**240**	1.6*	**34**	**1250**	**40**	**8**
14–18 y	**1300**	24*	**890**	3*	**150**	**15**	**360**	1.6*	**43**	**1250**	**55**	**9**
19–30 y	**1000**	25*	**900**	3*	**150**	**18**	**310**	1.8*	**45**	**700**	**55**	**8**
31–50 y	**1000**	25*	**900**	3*	**150**	**18**	**320**	1.8*	**45**	**700**	**55**	**8**
51–70 y	**1200**	20*	**900**	3*	**150**	**8**	**320**	1.8*	**45**	**700**	**55**	**8**
>70 y	**1200**	20*	**900**	3*	**150**	**8**	**320**	1.8*	**45**	**700**	**55**	**8**
Pregnancy												
14–18 y	**1300**	29*	**1000**	3*	**220**	**27**	**400**	2.0*	**50**	**1250**	**60**	**12**
19–30 y	**1000**	30*	**1000**	3*	**220**	**27**	**350**	2.0*	**50**	**700**	**60**	**11**
31–50 y	**1000**	30*	**1000**	3*	**220**	**27**	**360**	2.0*	**50**	**700**	**60**	**11**
Lactation												
14–18 y	**1300**	44*	**1300**	3*	**290**	**10**	**360**	2.6*	**50**	**1250**	**70**	**13**
19–30 y	**1000**	45*	**1300**	3*	**290**	**9**	**310**	2.6*	**50**	**700**	**70**	**12**
31–50 y	**1000**	45*	**1300**	3*	**290**	**9**	**320**	2.6*	**50**	**700**	**70**	**12**

NOTE: This table presents Recommended Dietary Allowances (RDAs) in **bold type** and Adequate Intakes (AIs) in ordinary type followed by an asterisk (*). RDAs and AIs may both be used as goals for individual intake. RDAs are set to meet the needs of almost all (97–98%) individuals in a group. For healthy breastfed infants, the AI is the mean intake. The AI for other life stage and gender groups is believed to cover needs of all individuals in the group, but lack of data or uncertainty in the data prevents being able to specify with confidence the percentage of individuals covered by this intake.

Source: Dietary Reference Intakes for Calcium, Phosphorus, Magnesium, Vitamin D, and Fluoride (1997); *Dietary Reference Intakes for Thiamin, Riboflavin, Niacin, Vitamin B-6, Folate, Vitamin B-12, Pantothenic Acid, Biotin, and Choline* (1998); *Dietary Reference Intakes for Vitamin C, Vitamin E, Selenium, and Carotenoids* (2000); *Dietary Reference Intakes for Vitamin A, Vitamin K, Arsenic, Boron, Chromium, Copper, Iodine, Iron, Manganese, Molybdenum, Nickel, Silicon, Vanadium, and Zinc* (2001); and *Dietary Reference Intakes for Calcium and Vitamin D* (2011). These reports may be accessed via **www.nap.edu.**

Dietary Reference Intakes (DRIs): Recommended Intakes for Individuals, Macronutrients
Food and Nutrition Board, Institute of Medicine, National Academies

Life Stage Group	Carbohydrate (g/d)	Total Fiber (g/d)	Fat (g/d)	Linoleic Acid (g/d)	α-Linolenic Acid (g/d)	Protein[a] (g/d)
Infants						
0–6 mo	60*	ND	31*	4.4*	0.5*	9.1*
7–12 mo	95*	ND	30*	4.6*	0.5*	**11.0**
Children						
1–3 y	**130**	19*	ND[b]	7*	0.7*	**13**
4–8 y	**130**	25*	ND	10*	0.9*	**19**
Males						
9–13 y	**130**	31*	ND	12*	1.2*	**34**
14–18 y	**130**	38*	ND	16*	1.6*	**52**
19–30 y	**130**	38*	ND	17*	1.6*	**56**
31–50 y	**130**	38*	ND	17*	1.6*	**56**
51–70 y	**130**	30*	ND	14*	1.6*	**56**
>70 y	**130**	30*	ND	14*	1.6*	**56**
Females						
9–13 y	**130**	26*	ND	10*	1.0*	**34**
14–18 y	**130**	26*	ND	11*	1.1*	**46**
19–30 y	**130**	25*	ND	12*	1.1*	**46**
31–50 y	**130**	25*	ND	12*	1.1*	**46**
51–70 y	**130**	21*	ND	11*	1.1*	**46**
>70 y	**130**	21*	ND	11*	1.1*	**46**
Pregnancy						
14–18 y	**175**	28*	ND	13*	1.4*	**71**
19–30 y	**175**	28*	ND	13*	1.4*	**71**
31–50 y	**175**	28*	ND	13*	1.4*	**71**
Lactation						
14–18 y	**210**	29*	ND	13*	1.3*	**71**
19–30 y	**210**	29*	ND	13*	1.3*	**71**
31–50 y	**210**	29*	ND	13*	1.3*	**71**

NOTE: This table presents Recommended Dietary Allowances (RDAs) in **bold type** and Adequate Intakes (AIs) in ordinary type followed by an asterisk (*). RDAs and AIs may both be used as goals for individual intake. RDAs are set to meet the needs of almost all (97–98%) individuals in a group. For healthy breastfed infants, the AI is the mean intake. The AI for other life stage and gender groups is believed to cover needs of all individuals in the group, but lack of data or uncertainty in the data prevents being able to specify with confidence the percentage of individuals covered by this intake.

[a]Based on 0.8 g protein/kg body weight for reference body weight.

[b]ND = not determinable at this time.

Source: Dietary Reference Intakes for Energy, Carbohydrate, Fiber, Fat, Fatty Acids, Cholesterol, Protein, and Amino Acids (2002). This report may be accessed via **www.nap.edu.**

Dietary Reference Intakes (DRIs): Recommended Intakes for Individuals, Electrolytes and Water
Food and Nutrition Board, Health and Medicine Division, National Academies

Life Stage Group	Sodium (mg/d)	Sodium CDRR[a]	Potassium (mg/d)	Chloride (mg/d)	Water (l/d)
Infants					
0–6 mo	110*	ND[b]	400*	180*	0.7*
7–12 mo	370*	ND[b]	860*	570*	0.8*
Children					
1–3 y	800*	Reduce intakes if above 1,200 mg/day[c]	2000*	1500*	1.3*
4–8 y	1000*	Reduce intakes if above 1,500 mg/day[c]	2300*	1900*	1.7*
Males					
9–13 y	1200*	Reduce intakes if above 1,800 mg/day[c]	2500*	2300*	2.4*
14–18 y	1500*	Reduce intakes if above 2,300 mg/day[c]	3000*	2300*	3.3*
19–30 y	1500*	Reduce intakes if above 2,300 mg/day	3400*	2300*	3.7*
31–50 y	1500*	Reduce intakes if above 2,300 mg/day	3400*	2300*	3.7*
51–70 y	1500*	Reduce intakes if above 2,300 mg/day	3400*	2000*	3.7*
> 70 y	1500*	Reduce intakes if above 2,300 mg/day	3400*	1800*	3.7*
Females					
9–13 y	1200*	Reduce intakes if above 1,800 mg/day[c]	2300*	2300*	2.1*
14–18 y	1500*	Reduce intakes if above 2,300 mg/day[c]	2300*	2300*	2.3*
19–30 y	1500*	Reduce intakes if above 2,300 mg/day	2600*	2300*	2.7*
31–50 y	1500*	Reduce intakes if above 2,300 mg/day	2600*	2300*	2.7*
51–70 y	1500*	Reduce intakes if above 2,300 mg/day	2600*	2000*	2.7*
> 70 y	1500*	Reduce intakes if above 2,300 mg/day	2600*	1800*	2.7*
Pregnancy					
14–18 y	1500*	Reduce intakes if above 2,300 mg/day[c]	2600*	2300*	3.0*
19–30 y	1500*	Reduce intakes if above 2,300 mg/day	2900*	2300*	3.0*
31–50 y	1500*	Reduce intakes if above 2,300 mg/day	2900*	2300*	3.0*
Lactation					
14–18 y	1500*	Reduce intakes if above 2,300 mg/day[c]	2500*	2300*	3.8*
19–30 y	1500*	Reduce intakes if above 2,300 mg/day	2800*	2300*	3.8*
31–50 y	1500*	Reduce intakes if above 2,300 mg/day	2800*	2300*	3.8*

NOTE: The table is adapted from the *DRI reports*. See **www.nap.edu.** Adequate Intakes (AIs) are followed by an asterisk (*). These may be used as a goal for individual intake. For healthy breastfed infants, the AI is the average intake. The AI for other life stage and gender groups is believed to cover the needs of all individuals in the group, but lack of data prevent being able to specify with confidence the percentage of individuals covered by this intake; therefore, no Recommended Dietary Allowance (RDA) was set.

[a]CDRR = Chronic Disease Risk Reduction Intakes; [b]Not determined owing to insufficient strength of evidence for causality and intake-response; [c]Extrapolated from the adult CDRR based on sedentary Estimated Energy Requirements.

Source: National Academies of Sciences, Engineering, and Medicine. *Dietary Reference Intakes for Sodium and Potassium*. The National Academies Press, Washington, D.C., 2019; and *Dietary Reference Intakes for Water, Potassium, Sodium, Chloride, and Sulfate* (2005). These reports may be accessed via **www.nap.edu.**

Acceptable Macronutrient Distribution Ranges

	Range (percent of energy)		
Macronutrient	**Children, 1–3 y**	**Children, 4–18 y**	**Adults**
Fat	30–40	25–35	20–35
omega-6 polyunsaturated fats (linoleic acid)	5–10	5–10	5–10
omega-3 polyunsaturated fats[a] (α-linolenic acid)	0.6–1.2	0.6–1.2	0.6–1.2
Carbohydrate	45–65	45–65	45–65
Protein	5–20	10–30	10–35

[a]Approximately 10% of the total can come from longer-chain n-3 fatty acids.

Source: Dietary Reference Intakes for Energy, Carbohydrate, Fiber, Fat, Fatty Acids, Cholesterol, Protein, and Amino Acids (2002). The report may be accessed via **www.nap.edu.**

Dietary Reference Intakes (DRIs): Tolerable Upper Intake Levels (ULs[a]), Vitamins
Food and Nutrition Board, Institute of Medicine, National Academies

Life Stage Group	Vitamin A (μg/d)[b]	Vitamin C (mg/d)	Vitamin D (μg/d)	Vitamin E (mg/d)[c,d]	Vitamin K	Thiamin	Riboflavin	Niacin (mg/d)[d]	Vitamin B-6 (mg/d)	Folate (μg/d)[d]	Vitamin B-12	Pantothenic Acid	Biotin	Choline (g/d)	Carotenoids[e]
Infants															
0–6 mo	600	ND[f]	25	ND	ND	ND	ND	ND	ND	ND	ND	ND	ND	ND	ND
7–12 mo	600	ND	38	ND	ND	ND	ND	ND	ND	ND	ND	ND	ND	ND	ND
Children															
1–3 y	600	400	63	200	ND	ND	ND	10	30	300	ND	ND	ND	1.0	ND
4–8 y	900	650	75	300	ND	ND	ND	15	40	400	ND	ND	ND	1.0	ND
Males, Females															
9–13 y	1700	1200	100	600	ND	ND	ND	20	60	600	ND	ND	ND	2.0	ND
14–18 y	2800	1800	100	800	ND	ND	ND	30	80	800	ND	ND	ND	3.0	ND
19–70 y	3000	2000	100	1000	ND	ND	ND	35	100	1000	ND	ND	ND	3.5	ND
> 70 y	3000	2000	100	1000	ND	ND	ND	35	100	1000	ND	ND	ND	3.5	ND
Pregnancy															
14–18 y	2800	1800	100	800	ND	ND	ND	30	80	800	ND	ND	ND	3.0	ND
19–50 y	3000	2000	100	1000	ND	ND	ND	35	100	1000	ND	ND	ND	3.5	ND
Lactation															
14–18 y	2800	1800	100	800	ND	ND	ND	30	80	800	ND	ND	ND	3.0	ND
19–50 y	3000	2000	100	1000	ND	ND	ND	35	100	1000	ND	ND	ND	3.5	ND

[a]UL = The maximum level of daily nutrient intake likely to pose no risk of adverse effects. Unless otherwise specified, the UL represents total intake from food, water, and supplements. Due to lack of suitable data, ULs could not be established for vitamin K, thiamin, riboflavin, vitamin B-12, pantothenic acid, biotin, or carotenoids. In the absence of ULs, extra caution may be warranted in consuming levels above recommended intakes.

[b]As preformed vitamin A only.

[c]As α-tocopherol; applies to any form of supplemental α-tocopherol.

[d]The ULs for vitamin E, niacin, and folate apply to synthetic forms obtained from supplements, fortified foods, or a combination of the two.

[e]Beta-carotene supplements are advised only to serve as a provitamin A source for individuals at risk of vitamin A deficiency.

[f]ND = Not determinable due to lack of data of adverse effects in this age group and concern with regard to lack of ability to handle excess amounts. Source of intake should be from food only to prevent high levels of intake.

Source: Dietary Reference Intakes for Calcium and Vitamin D (2011); *Dietary Reference Intakes for Calcium, Phosphorus, Magnesium, Vitamin D, and Fluoride* (1997); *Dietary Reference Intakes for Thiamin, Riboflavin, Niacin, Vitamin B-6, Folate, Vitamin B-12, Pantothenic Acid, Biotin, and Chlorine* (1998); *Dietary Reference Intakes for Vitamin C, Vitamin E, Selenium, and Carotenoids* (2000); and *Dietary Reference Intakes for Vitamin A, Vitamin K, Arsenic, Boron, Chromium, Copper, Iodine, Iron, Manganese, Molybdenum, Nickel, Silicon, Vanadium, and Zinc* (2001). These reports may be accessed via **www.nap.edu.**

Dietary Reference Intakes (DRIs): Tolerable Upper Intake Levels (ULs[a]), Elements and Electrolytes[b]
Food and Nutrition Board, Institute of Medicine, National Academies

Life Stage Group	Boron (mg/d)	Calcium (g/d)	Copper (μg/d)	Fluoride (mg/d)	Iodine (μg/d)	Iron (mg/d)	Magnesium (mg/d)[c]	Manganese (mg/d)	Molybdenum (μg/d)	Nickel (mg/d)	Phosphorus (g/d)	Selenium (μg/d)	Vanadium (mg/d)[d]	Zinc (mg/d)	Sodium	Potassium	Chloride (mg/d)
Infants																	
0–6 mo	ND[e]	1	ND	0.7	ND	40	ND	ND	ND	ND	ND	45	ND	4	ND	ND	ND
7–12 mo	ND	1.5	ND	0.9	ND	40	ND	ND	ND	ND	ND	60	ND	5	ND	ND	ND
Children																	
1–3 y	3	2.5	1000	1.3	200	40	65	2	300	0.2	3	90	ND	7	ND	ND	2300
4–8 y	6	2.5	3000	2.2	300	40	110	3	600	0.3	3	150	ND	12	ND	ND	2900
Males, Females																	
9–13 y	11	3	5000	10	600	40	350	6	1100	0.6	4	280	ND	23	ND	ND	3400
14–18 y	17	3	8000	10	900	45	350	9	1700	1.0	4	400	ND	34	ND	ND	3600
19–70 y	20	2.5[f]	10,000	10	1100	45	350	11	2000	1.0	4	400	1.8	40	ND	ND	3600
>70 y	20	2	10,000	10	1100	45	350	11	2000	1.0	3	400	1.8	40	ND	ND	3600
Pregnancy																	
14–18 y	17	3	8000	10	900	45	350	9	1700	1.0	3.5	400	ND	34	ND	ND	3600
19–50 y	20	2.5	10,000	10	1100	45	350	11	2000	1.0	3.5	400	ND	40	ND	ND	3600
Lactation																	
14–18 y	17	3	8000	10	900	45	350	9	1700	1.0	4	400	ND	34	ND	ND	3600
19–50 y	20	2.5	10,000	10	1100	45	350	11	2000	1.0	4	400	ND	40	ND	ND	3600

[a]UL = The maximum level of daily nutrient intake that is likely to pose no risk of adverse effects. Unless otherwise specified, the UL represents total intake from food, water, and supplements. Due to lack of suitable data, ULs could not be established for arsenic, chromium, and silicon. In the absence of ULs, extra caution may be warranted in consuming levels above recommended intakes.

[b]Although silicon has not been shown to cause adverse effects in humans, there is no justification for adding silicon to supplements.

[c]The ULs for magnesium represent intake from a pharmacological agent only and do not include intake from food and water.

[d]Although vanadium in food has not been shown to cause adverse effects in humans, there is no justification for adding vanadium to food, and vanadium supplements should be used with caution. The UL is based on adverse effects in laboratory animals and this data could be used to set a UL for adults but not children and adolescents.

[e]ND = Not determinable due to lack of data of adverse effects in this age group and concern with regard to lack of ability to handle excess amounts.

[f]Upper Limit declines to 2 after age 50.

Source: National Academies of Sciences, Engineering, and Medicine. *Dietary Reference Intakes for Sodium and Potassium.* The National Academies Press, Washington, DC, 2019; *Dietary Reference Intakes for Calcium and Vitamin D* (2011); *Dietary Reference Intakes for Calcium, Phosphorus, Magnesium, Vitamin D, and Fluoride* (1997); *Dietary Reference Intakes for Thiamin, Riboflavin, Niacin, Vitamin B-6, Folate, Vitamin B-12, Pantothenic Acid, Biotin, and Choline* (1998); *Dietary Reference Intakes for Vitamin C, Vitamin E, Selenium, and Carotenoids* (2000); *Dietary Reference Intakes for Vitamin A, Vitamin K, Arsenic, Boron, Chromium, Copper, Iodine, Iron, Manganese, Molybdenum, Nickel, Silicon, Vanadium, and Zinc* (2001); *Dietary Reference Intakes for Water, Potassium, Sodium, Chloride, and Sulfate* (2004); and *Dietary Reference Intakes for Sodium and Potassium* (2019). These reports may be accessed via **www.nap.edu.**

Glossary

A

absorption process by which substances are taken up from the GI tract and enter the bloodstream or the lymph

Acceptable Macronutrient Distribution Ranges (AMDRs) macronutrient intake ranges that are nutritionally adequate and may reduce the risk of diet-related chronic diseases

acid-base balance maintaining the proper pH of body fluids

acid group acid portion of a compound

acids substances that donate hydrogen ions

added sugars sugars added to foods during processing or preparation

adenosine diphosphate (ADP) high-energy compound; by-product of ATP use

adenosine triphosphate (ATP) molecule that stores energy

Adequate Intakes (AIs) dietary recommendations for nutrients that scientists do not have enough information to set RDAs

adipose cells fat cells; specialized cells that store fat

adolescence life stage in which a child matures physically into an adult

aerobic conditions that require free oxygen

aerobic exercise physical activities that involve sustained, rhythmic contractions of large muscle groups

air displacement method of estimating body composition by determining body volume

aldosterone hormone that participates in sodium and water conservation

alpha-linolenic acid an essential fatty acid

alpha-tocopherol vitamin E

alternative health care health care practices that are not widely accepted and used by conventional medical practitioners

amino acids nitrogen-containing chemical units that comprise proteins

amino or **nitrogen-containing group** portion of an amino acid that contains nitrogen

anaerobic conditions that lack free oxygen

anatomy scientific study of cells and other body structures

anecdotes reports of personal experiences

anemia disorder characterized by too few red blood cells and poor oxygen transport in blood

anencephaly type of neural tube defect in which the brain does not form properly or is missing

anorexia nervosa (AN) severe psychological disturbance characterized by self-imposed starvation

antibodies infection-fighting proteins

antidiuretic hormone (ADH) hormone that participates in water conservation

antioxidant substance that gives up electrons to radicals to protect cells

appetite desire to eat appealing food

arterial plaque lipid-filled deposits that form under the lining of certain arteries

arteries vessels that carry blood away from the heart

ascorbic acid vitamin C

atherosclerosis long-term disease process in which plaques build up inside arterial walls

athlete triad condition that is characterized by low energy availability, abnormal reproductive function, and reduced bone mineral density in athletes

B

bacteria simple single-celled microorganisms

bariatric medicine medical specialty that focuses on the treatment of obesity

basal metabolism minimal number of kilocalories the body uses to support vital activities after fasting and resting for 12 hours

bases substances that accept hydrogen ions

beriberi thiamin deficiency disease

beta-carotene provitamin A carotenoid that the body can convert to vitamin A

bile substance that is produced by the liver to prepare fat and fat-soluble vitamins for digestion and absorption

bile salts components of bile that enhance lipid digestion and absorption

binge eating consuming a much larger amount of food than is normally eaten in a brief period

binge-eating disorder (BED) eating disorder featuring recurrent episodes of binge eating that are not followed by purging behaviors

bioavailability extent to which the digestive tract absorbs a nutrient and how well the body uses it

bioelectrical impedance technique of estimating body composition in which a device measures the conduction of a weak electrical current through the body

bioengineered food (BE food) food that contains genetic material that has been modified by certain scientific methods

bioengineering scientific techniques that alter an organism's hereditary material

biological activity describes vitamin's degree of potency or effects in the body

biotechnology using living things to manufacture new products

blood alcohol concentration (BAC) percentage of alcohol in the bloodstream

body mass index (BMI) numerical value that reflects body weight adjusted for height

buffer substance that can help maintain the normal pH of a solution

bulimia nervosa (BN) severe psychological condition characterized by repeated episodes of binge eating followed by unhealthy behaviors to prevent weight gain

C

caffeine naturally occurring stimulant drug

capillaries smallest blood vessels

carbohydrate (glycogen) loading practice of manipulating physical activity and dietary patterns to increase muscle glycogen stores

carbohydrates class of nutrients that includes glucose, a major source of energy for the body

carcinogen factor that triggers cancer

cardiovascular disease (CVD) group of diseases that affect the heart and blood vessels

carotenodermia yellow-orange discoloration of the skin that results from high blood levels of beta-carotene

carotenoids yellow, orange, and red pigments in fruits and vegetables

case-control study study that compares individuals who have a health condition with persons who have similar characteristics but are healthy

casein major protein in cow's milk

cell smallest living functional unit in an organism

central-body obesity condition characterized by excessive visceral fat

chemical bond attraction that holds atoms together

chemical reactions processes that change the atomic arrangements of molecules

chemistry study of the composition and characteristics of matter and changes that can occur to it

cholecystokinin (CCK) hormone that stimulates the gallbladder to release bile and pancreas to secrete digestive enzymes

cholesterol sterol in animal foods and precursor for steroid hormones, bile, and vitamin D

choline water-soluble compound in lecithin

chronic refers to diseases and many other medical conditions that are long term and often difficult to treat

Chronic Disease Risk Reduction Intakes (CDRRs) nutrient intake levels that reduce chronic disease risk within a healthy population

chylomicron particle formed by small intestinal cells that transports lipids in the bloodstream

chyme mixture of gastric juice and partially digested food

coenzyme small, organic molecule that interacts with enzymes, enabling the enzymes to function

cofactor metallic ion or small molecule that activates certain chemical reactions

cognitive behavioral therapy (CBT) psychological therapy approaches that address unhealthy emotions and behaviors

cohort study study that measures variables of a group of people over time

collagen fibrous protein that gives strength to connective tissue

color additives dyes, pigments, or other substances that provide color to food

colorectal cancer cancer that develops in the colon or the rectum

colostrum initial form of breast milk that contains anti-infective properties

complementary combinations mixing certain plant foods to provide all essential amino acids without adding animal protein

complex carbohydrates (polysaccharides) compounds composed of monosaccharides bonded together, forming large chainlike structures

compounds molecules that contain two or more different elements in specific proportions

conception moment when a sperm enters an egg (fertilization)

contaminated food item that is impure or unsafe for human consumption

control group group being studied that does not receive a treatment

conventional medical care health care practices that are widely accepted and used by mainstream medical practitioners

coronary artery disease (CAD) a major form of CVD; heart disease

correlation relationship between two variables

cretinism condition affecting infants of females who were iodine deficient during pregnancy

Crohn's disease common form of inflammatory bowel disease

cross-contamination unintentional transfer of pathogens from one food to another

cytochromes group of proteins involved in the release of energy from macronutrients

D

Daily Values (DVs) set of nutrient intake standards developed for labeling purposes

deamination removal of the nitrogen-containing group from an amino acid

deficiency disease state of health that occurs when a nutrient is missing from the diet

dehydration body water depletion

Delaney Clause component of the 1958 Food Additives Amendment that prevents manufacturers from adding carcinogenic compounds to food

denaturation altering a protein's natural shape and function by exposing it to conditions such as heat, acids, and physical agitation

diabetes mellitus (diabetes) group of serious chronic diseases characterized by abnormal glucose, fat, and protein metabolism

diabulimia popular term used to describe people with type I diabetes who skip insulin injections or use less insulin than prescribed to control their body weight

diastolic pressure pressure in an artery that occurs when the ventricles relax between contractions

diet usual pattern of food choices

dietary fiber (fiber) nondigestible plant material; most types are polysaccharides

Dietary Reference Intakes (DRIs) various energy and nutrient intake standards for Americans

dietary supplements nutrient preparations, certain hormones, and herbal products that are taken to increase total dietary intake

digestion process by which large food molecules are mechanically and chemically broken down

disaccharide simple sugar composed of two monosaccharides

disordered eating chaotic and abnormal food-related practices

diuretic substance that increases urine production

diverticula abnormal, small sacs that form in wall of colon

DNA molecule that provides coded instructions for making proteins

double-blind study experimental design in which neither the participants nor the researchers are aware of each participant's group assignment

dual-energy x-ray absorptiometry (DXA) technique of estimating body composition that involves scanning the body with multiple low-energy x-rays

duodenum first segment of the small intestine

E

eating disorders psychological disturbances that lead to certain physiological changes and serious health complications

edema accumulation of fluid in tissues

electrolytes ions of minerals that conduct electricity when they are dissolved in water

electrons small, negatively charged particles that surround the nucleus of an atom

element each type of atom; substance that cannot be separated into simpler substances by ordinary chemical or physical means

embolus thrombus or part of a plaque that breaks free and travels through the bloodstream

embryo human organism from 14 days to 8 weeks after conception

empty calories energy supplied by unhealthy solid fats, added sugars, and/or alcohol

emulsifier substance that helps water-soluble and water-insoluble compounds mix with each other

energy capacity to perform work

energy density energy value of a food in relation to the food's weight

energy equilibrium calorie intake equals calorie output

energy intake calories from foods and beverages that contain macronutrients and alcohol

energy output calories cells use to carry out their activities

enrichment replacement of some nutrients that were removed during processing

enterohepatic circulation process that recycles cholesterol in the body

enzyme molecule (usually a protein) that speeds up a particular chemical reaction

epidemiology study of the occurrence, distribution, and causes of health problems in populations

epiglottis flap of tissue that folds down over the larynx to keep food from entering the trachea during swallowing

epithelial cells cells that form protective tissues that line the body

ergogenic aids foods, devices, dietary supplements, or drugs used to improve physical performance
esophagus structure of the GI tract that connects the pharynx with the stomach
essential amino acids amino acids the body cannot make or make enough of to meet its needs
essential fatty acids lipids that must be supplied by the diet
essential nutrient nutrient that must be supplied by food
Estimated Average Requirement (EAR) amount of a nutrient that meets the needs of 50% of healthy people in a life stage/sex group
Estimated Energy Requirement (EER) average daily energy intake that meets the needs of a healthy person maintaining their weight
estrogen hormone that plays a role in normal bone development and maintenance
evidence based information that is based on results of high-quality scientific studies
exercise physical activities that are usually planned and structured for a purpose
extracellular fluid watery fluid that surrounds cells or is in liquid portion of blood
extrusion reflex involuntary response in which a young infant thrusts its tongue forward when a solid or semisolid object is placed in its mouth

F

fad trendy practice that has widespread appeal for a period, then becomes no longer fashionable
fat-free mass body compartment that is composed of body water, mineral-rich tissues, and protein-rich tissues
fat-soluble vitamins vitamins A, D, E, and K
fermentation process used to preserve or produce a variety of foods, including pickles and wine
fetus human organism from 8 weeks after conception until birth
folic acid form of folate
food additive any substance that becomes incorporated into food during production, packaging, transport, or storage
Food Additives Amendment U.S. legislation that requires evidence that a new food additive is safe before it can be marketed for use
Food and Nutrition Board (FNB) group of nutrition scientists who develop DRIs
food-borne illness disorder that is caused by consuming disease-causing agents in food or water
food-borne infection illness that results when a pathogen in food inflames the intestinal tract or other body tissues
food insecurity situation in which individuals or families are concerned about running out of food or not having enough money to buy more food
food intoxication illness that results when poisons produced by certain microbes contaminate food and irritate the intestinal tract
Food Lists For Diabetes and Weight Management method of classifying foods into lists based on macronutrient composition
fortification addition of any nutrient to food to boost its nutritional content
fructose monosaccharide in fruits, honey, and certain vegetables; "levulose" or "fruit sugar"
fungi simple organisms that live on dead or decaying organic matter
fungicides substances used to limit the spread of fungi

G

galactose monosaccharide that is a component of lactose
gastroesophageal reflux disease (GERD) condition that can result when acid reflux damages the wall of the lower esophagus and causes ulcers
gastrointestinal (GI) tract muscular tube that extends from the mouth to the anus
gene portion of DNA
Generally Recognized as Safe (GRAS) ingredients considered to be safe
genetic endowment inherited physical characteristics that can affect physical performance
gestational diabetes form of diabetes that pregnant females can develop
gestational hypertension type of hypertension that can develop during pregnancy
ghrelin protein that stimulates eating behavior
glucagon hormone that helps regulate blood glucose levels
glucose monosaccharide that is a primary fuel for muscles and other cells; "dextrose" or "blood sugar"
glycogen storage polysaccharide in animals
glycogenolysis glycogen breakdown
glycolysis first stage of glucose oxidation
goitrogens compounds that inhibit iodide metabolism by the thyroid gland
gut microbiota various microbes, mostly bacteria, that reside in the GI tract
gut microbiota transplantation (GMT) fecal transplants

H

H^+ chemical symbol for hydrogen ion
heat cramps heat-related illness characterized by painful muscle contractions
heat exhaustion heat-related illness that can occur after intense exercise
heatstroke most dangerous form of heat-related illness
heme iron form of iron in hemoglobin and myoglobin
hemoglobin iron-containing protein in red blood cells that transports oxygen
herbicides substances used to destroy weeds
hereditary hemochromatosis most common type of iron overload disease
high-density lipoprotein (HDL) lipoprotein that transports cholesterol away from tissues and to the liver, where it can be eliminated
high-fructose corn syrup (HFCS) caloric sweetener that is often added to food
high-intensity sweeteners group of nonnutritive sweeteners that are extremely sweet tasting compared to the same amount of sugar
high-quality (complete) protein protein that contains all essential amino acids in amounts that support the deposition of protein in tissues and the growth of a young person
high-sensitivity C-reactive protein (hs-CRP) protein produced primarily by the liver in response to inflammation
homeostasis maintenance of an internal chemical and physical environment that is critical for good health and survival
homocysteine amino acid that may play a role in the development of atherosclerosis
hunger uncomfortable feeling that drives a person to consume food
hydration water status
hydrocarbon chain chain of carbon atoms bonded to each other and to hydrogen atoms
hydrophilic part of a molecule that attracts water
hydrophobic part of a molecule that avoids water and attracts lipids
hypercalcemia condition characterized by higher-than-normal concentration of calcium in blood
hyperglycemia abnormally high blood glucose level
hypertension abnormally high blood pressure levels that persist
hyperthermia very high body temperature
hypoglycemia condition that occurs when the blood glucose level is abnormally low

I

ileum last segment of the small intestine
infant formula synthetic food that simulates human milk
inflammation sign of an immune system response to an agent of infection or a disease process
insecticides substances used to control or kill insects
insoluble fiber forms of dietary fiber that generally do not dissolve in water
insulin hormone that helps regulate blood glucose levels
integrative medical care medical care provided by health care professionals that coordinates conventional medical therapies with alternative health care approaches
intestinal dysbiosis condition characterized by the imbalance of the normal diversity of microbial populations in the GI tract
intracellular fluid watery fluid that is inside cells
intrinsic factor (IF) substance produced in the stomach that facilitates intestinal absorption of vitamin B-12
ion atom or group of atoms that has a positive or negative charge
iron deficiency condition characterized by low body stores of iron

J

jejunum middle segment of the small intestine

K

keratin tough protein found in hair, nails, and the outermost layers of skin
ketoacidosis condition that occurs when the body forms excessive ketone bodies
ketone bodies chemicals that result from incomplete fat breakdown
kilocalorie (kcal) or **Calorie** heat energy needed to raise the temperature of 1 liter of water 1° Celsius; measure of food energy
kwashiorkor form of undernutrition that results from consuming adequate energy and insufficient high-quality protein

L

lactase enzyme that splits lactose molecule
lactation milk production
lacteal lymph vessel in villus
lactic acid compound formed from pyruvate during anaerobic metabolism
lactoovovegetarian vegetarian who consumes milk products and eggs for animal protein
lactose disaccharide composed of a glucose and a galactose molecule; "milk sugar"
lactose intolerance inability to digest lactose properly
lactovegetarian vegetarian who consumes milk and milk products for animal protein
lecithin major phospholipid in food
legumes plants that produce pods with a single row of seeds
leptin hormone that reduces hunger and inhibits fat storage in the body
life expectancy length of time an average person born in a specific year can expect to live
life span maximum number of years an organism can live
lifestyle usual way of living, including dietary practices and physical activity habits
linoleic acid an essential fatty acid
lipases enzymes that break down lipids
lipids class of nutrients that generally do not dissolve in water
lipolysis fat breakdown
lipoprotein water-soluble structure that transports lipids through the bloodstream
lipoprotein lipase enzyme in capillary walls that breaks down triglycerides
low-birth-weight (LBW) describes baby generally weighing less than 5½ pounds at birth
low-density lipoprotein (LDL) lipoprotein that carries cholesterol into tissues
low energy availability state of negative energy balance
lower esophageal sphincter section of esophagus next to the stomach that controls the opening to the stomach
low-quality (incomplete) protein protein that lacks or has inadequate amounts of one or more of the essential amino acids
lumen open space within a structure such as the small intestine
lymph fluid in the lymphatic/immune system

M

macronutrients nutrients needed in gram amounts daily and that provide energy; carbohydrates, proteins, and fats
major minerals essential mineral elements required in amounts of 100 mg or more per day
malignant tumor mass of cancerous cells
malnutrition state of health that occurs when the body is improperly nourished
maltase enzyme that splits maltose molecule
maltose disaccharide composed of two glucose molecules; "malt sugar"
marasmus starvation
megadose amount of a vitamin or mineral that greatly exceeds the recommended amount
metabolic syndrome condition that increases risk of type 2 diabetes and CVD
metabolic water water formed by cells as a metabolic by-product
metabolism total of all chemical processes that take place in living cells
metastasized spread to other tissues
micelle lipid-rich particle that is surrounded by bile salts; transports lipids to absorptive cells
micronutrients vitamins and minerals
minerals elements that are found in the Earth's crust
mitochondria cellular structures that release energy from macronutrients
moderation obtaining adequate amounts of nutrients while balancing calorie intake with calorie expenditure
molecule matter that forms when two or more atoms interact and are held together by a chemical bond
monoglyceride single fatty acid attached to a glycerol backbone
monosaccharide simple sugar that is the basic chemical unit of carbohydrates
monounsaturated fatty acid (MUFA) fatty acid that has one double bond within the carbon chain
morning sickness nausea and vomiting associated with pregnancy
mucus slippery fluid made by certain cells in the body
multivitamin/mineral (MV/M) describes a dietary supplement that contains vitamins and minerals
muscle dysmorphia unhealthy preoccupation with the body being too small or not muscular enough
myocardial infarction heart attack
myoglobin iron-containing protein in muscle cells that controls oxygen uptake from red blood cells
MyPlate USDA's interactive Internet dietary and menu planning guide

N

negative energy balance calorie intake is less than calorie output
negative nitrogen balance state in which the body loses more nitrogen than it retains
neural tube embryonic structure that eventually develops into the brain and spinal cord
neutrons particles in the nucleus of an atom that carry no charge
night eating syndrome (NES) episodic food binges that are not followed by purging; binges take place after the evening meal or when the person wakens from sleep during the night
nitrogen balance (equilibrium) balancing nitrogen intake with nitrogen losses

nonalcoholic fatty liver disease (NAFLD) abnormal accumulation of fat in the liver that is not related to alcohol intake
nonessential amino acids group of amino acids that the body can make
nonexercise activity thermogenesis (NEAT) involuntary skeletal muscular activities such as fidgeting
nonheme iron form of iron that is primarily in vegetables, grains, meats, and supplements
nonnutritive sweeteners substances that sweeten foods while providing few or no kilocalories
nutrient dense describes a food or beverage that contains more key beneficial nutrients in relation to its total calories
nutrients chemicals necessary for proper body functioning
nutrigenetics study of how a person's genetic makeup affects the way their body responds to food
nutrigenomics study of how nutrients and other food components can affect one's genetic expression
nutrition scientific study of nutrients and how the body uses these substances
nutritional genomics science that investigates the complex interactions among gene functioning, dietary choices, and the environment
nutritive sweeteners substances that sweeten and contribute energy to foods

O

obesity condition characterized by excessive and unhealthy amounts of body fat
omega end first carbon of a fatty acid chain that has three hydrogen atoms attached to it
omega-3 fatty acid type of polyunsaturated fatty acid
omnivore organism that can digest and absorb nutrients from plants, animals, fungi, and bacteria
organ collection of tissues that function in a related fashion
organelles structures in cells that have specific functions
organic foods foods produced without the use of antibiotics, hormones, synthetic fertilizers and most synthetic pesticides, genetic improvements, or spoilage-killing radiation
organ system group of organs that work together for a similar purpose
osmosis diffusion of a solvent, such as water, through a selectively permeable membrane
osteoblasts bone cells that add bone to where the tissue is needed
osteoclasts bone cells that tear down bone tissue
osteomalacia adult rickets; condition characterized by poorly mineralized (soft) bones
osteoporosis chronic disease characterized by loss of bone mass and reduced bone structure
overweight having extra weight from bone, muscle, body fat, and/or body water
ovovegetarian vegetarian who eats eggs for animal protein
oxidizing agent substance that removes electrons from atoms or molecules
oxytocin hormone that elicits the "let-down" response and causes the uterus to contract

P

pancreatic amylase enzyme secreted by the pancreas that breaks down starch into maltose molecules
pancreatic lipase digestive enzyme that removes two fatty acids from a triglyceride
parasite organism that lives in or on another organism, often deriving nourishment from its host
parathyroid hormone (PTH) hormone secreted by parathyroid glands when blood calcium levels are too low
partial hydrogenation food manufacturing process that added hydrogen atoms to some unsaturated fatty acids, forming trans fats
pasteurization process that kills pathogens in foods and beverages as well as many microbes responsible for spoilage
pathogens disease-causing agents
peer review expert critical analysis of a research article before it is published
pellagra niacin deficiency disease
pepsin gastric enzyme that breaks down proteins into smaller polypeptides
peptide bond chemical attraction that connects two amino acids together
peptides small chains of amino acids
peristalsis type of muscular contraction of the gastrointestinal tract
pernicious anemia condition caused by the lack of intrinsic factor and characterized by vitamin B-12 deficiency, nerve damage, and megaloblastic RBCs
pescatarian vegetarian who eats seafood but no other animal proteins
pesticide substance that people use to kill or control unwanted insects, weeds, or other harmful organisms
pH measure of the acidity or alkalinity of a solution
phosphocreatine (PCr) high-energy compound used to reform ATP under anaerobic conditions
phospholipid type of lipid that is chemically similar to a triglyceride, except it contains phosphorus
physical activity movement resulting from contraction of skeletal muscles
physical fitness ability to perform moderate- to vigorous-intensity activities for a reasonable amount of time, without becoming excessively fatigued
physiological dose amount of a nutrient that is within the range of safe intake and enables the body to function optimally
physiology scientific study of the functioning of cells and other body structures
phytochemicals compounds made by plants that are not nutrients
pica craving nonfood items
placebo fake treatment, such as a sham pill, injection, or medical procedure
placebo effect positive response to a placebo
placenta organ of pregnancy that connects the uterus to the embryo/fetus via the umbilical cord
polypeptide protein composed of two or more amino acids
polyunsaturated fatty acid (PUFA) fatty acid that has two or more double bonds within the carbon chain
positive energy balance calorie intake is greater than calorie output
positive nitrogen balance state in which the body retains more nitrogen than it loses
prebiotics forms of dietary fiber that are poorly digested by humans but support the growth of beneficial gut microbes
pre-diabetes condition characterized by fasting blood glucose levels that are 100 to 125 mg/dl
prenatal care specialized health care for pregnant females
prenatal period time between conception and birth; pregnancy
probiotics products that contain healthful microbes; beneficial gut microbes
prolactin hormone that stimulates milk production after delivery
prooxidant substance that promotes free radical production
protein-energy malnutrition (PEM) condition that results from diets that provide inadequate amounts of protein and energy
proteins large, complex organic molecules made up of amino acids
protein turnover cellular process of breaking down proteins and recycling their amino acids
protons positively charged particles in the nucleus of an atom
protozoa single-celled microorganisms that have complex cell structures
pseudoscience presentation of information masquerading as factual and obtained by scientific methods

purging activities that limit calorie intake or increase calorie output

purines nitrogen-containing chemicals in genetic material

pyruvate compound that results from anaerobic breakdown of glucose

Q

quackery promotion of useless medical treatments

R

radical substance with an unpaired electron

randomized controlled trial (RCT) experimental study involving humans that usually randomly divides a group of people into treatment and control groups

Recommended Dietary Allowances (RDAs) daily nutrient recommendations that meet the needs of 97 to 98% of healthy people in a life stage/sex group

rectum lower section of the large intestine

registered dietitian (RD) or **registered dietitian nutritionist (RDN)** college-trained health care professional who has extensive knowledge of foods, nutrition, and dietetics

requirement smallest amount of a nutrient that maintains a defined level of nutritional health

resting metabolic rate (RMR) body's rate of energy use a few hours after resting and eating

retinol (preformed vitamin A) most active form of vitamin A in the body

R group part of amino acid that identifies the molecule as a particular amino acid

rhodopsin vitamin A–containing protein that is needed for vision in dim light

rickets vitamin D deficiency disorder in children

risk factor personal characteristic that increases a person's chances of developing a disease

rodenticides substances used to kill mice and rats

S

salivary amylase enzyme secreted by salivary glands that begins starch digestion

salt substance that forms when an acid and a base combine

satiety sense that enough food or beverages have been consumed to satisfy hunger

saturated fatty acid fatty acid that has each carbon atom within the chain filled with hydrogen atoms

scurvy vitamin C–deficiency disease

selectively permeable membrane barrier that allows the passage of certain substances and prevents the movement of other substances

semivegetarian (or **flexitarian)** vegetarian who eats animal products, except red meat

senescence declining organ functioning and increased vulnerability to disease that occur after reaching physical maturity

set-point theory scientific notion that body fat content is genetically predetermined

shelf life period of time that a food can be stored before it spoils

simple carbohydrates group of carbohydrates that includes sugars

simple diffusion movement of substances from a region of higher to lower concentration

skinfold thickness measurements technique of estimating body composition in which calipers are used to measure the width of skinfolds at multiple body sites

solid fats fats that are fairly hard at room temperature

solubility describes a substance's ability to dissolve and form a solution

soluble fiber forms of dietary fiber that dissolve or swell in water

solute lesser component of a solution that dissolves in solvent

solution evenly distributed mixture of two or more compounds

solvent primary component of a solution

spina bifida type of neural tube defect in which the spine does not form properly before birth, and it fails to enclose the spinal cord

starch storage polysaccharide in plants

sterilization process that destroys all microorganisms and viruses

sucrase enzyme that splits sucrose molecule

sucrose disaccharide composed of a glucose and a fructose molecule; "table sugar"

sustainable agriculture farming methods that do not deplete natural resources or harm the environment while meeting the demand for food

syndrome group of signs and symptoms that occur together and indicate a specific health problem

systolic pressure maximum blood pressure within an artery that occurs when the ventricles contract

T

target heart rate zone heart rate range that reflects intensity of physical exertion

teratogen agent that causes birth defects

testimonial personal endorsement of a product

testosterone hormone needed for normal bone development and maintenance

tetrahydrofolate (THF) folate coenzyme

thermic effect of food (TEF) energy used to digest foods and beverages as well as absorb and further process the macronutrients

thrombus fixed bunch of clots that remains in place

thyroid hormone hormone that regulates the body's metabolic rate

tissues masses of cells that have similar characteristics and functions

Tolerable Upper Intake Level (Upper Level or **UL)** highest average amount of a nutrient that is unlikely to be harmful when consumed daily

tolerances maximum amounts of pesticide residues that can be in or on each treated food crop

total body fat essential fat and adipose tissue

total water intake water ingested by consuming beverages and foods

trace minerals essential mineral elements required in amounts that are less than 100 mg per day

transamination transfer of the nitrogen-containing group from an unneeded amino acid to a carbon skeleton to form an amino acid

trans fatty acid unsaturated fatty acid that has a trans double bond

treatment group group being studied that receives a treatment

triglyceride lipid that has three fatty acids attached to a glycerol molecule

U

ulcerative colitis (UC) common form of inflammatory bowel disease

ultra-processed food food that contains numerous ingredients that were added to improve desirable characteristics

underwater weighing technique of estimating body composition that involves comparing weight on land to weight when completely submerged in a tank of water

underweight describes person with a BMI of less than 18.5

unintentional food additives substances that are accidentally in foods

unsaturated fatty acid fatty acid that is missing hydrogen atoms and has one or more double bonds within the carbon chain

urea waste product of amino acid metabolism

uterus female reproductive organ that protects the developing organism during pregnancy

V

variable personal characteristic or other factor that changes and can influence an outcome

vegan vegetarian who eats only plant foods

vegetarians people who eat plant-based diets

veins vessels that return blood to the heart

very-low-density lipoprotein (VLDL) lipoprotein that carries much of the triglycerides in the bloodstream

villi (singular, villus) tiny, fingerlike projections of the small intestinal lining that participate in digesting and absorbing food

virus microbe consisting of a piece of genetic material coated with protein

vitamin complex organic molecule that regulates certain metabolic processes

W

water intoxication condition that occurs when too much water is consumed in a short time period or the kidneys have difficulty filtering water from blood

water-soluble vitamins thiamin, riboflavin, niacin, vitamin B-6, pantothenic acid, folate, biotin, vitamin B-12, and vitamin C

weaning gradual process of shifting from breastfeeding or bottle-feeding to drinking from a cup and eating solid foods

Wernicke-Korsakoff syndrome degenerative brain disorder resulting from thiamin deficiency that occurs primarily among people with alcohol use disorder

X

xerophthalmia condition affecting the eyes that results from vitamin A deficiency

Index

Note: Page numbers followed by b indicates boxes; f indicates figures; t indicates tables.

E

F

J

K

L

M

Q

R

S